SI Base Units

Base Quantity	Name of Unit	Sy[illegible]
Length	meter	[illegible]
Mass	kilogram	[illegible]
Time	second	s
Electrical current	ampere	A
Temperature	kelvin	K
Amount of substance	mole	mol
Luminous intensity	candela	cd

Derived Units in the SI System

Physical Quantity	Name	Symbol	Units
Energy	joule	J	$kg\ m^2\ s^{-2}$
Force	newton	N	$kg\ m\ s^{-2}$
Power	watt	W	$kg\ m^2\ s^{-3}$
Electric charge	coulomb	C	A s
Electrical resistance	ohm	Ω	$kg\ m^2\ s^{-3}\ A^{-2}$
Electrical potential difference	volt	V	$kg\ m^2\ s^{-3}\ A^{-1}$
Electrical capacitance	farad	F	$kg^{-1}\ m^{-2}\ s^4\ A^2$
Frequency	hertz	Hz	s^{-1}

Prefixes Used with SI Units

Prefix	Symbol	Meaning
Peta-	P	10^{15} or 1,000,000,000,000,000
Tera-	T	10^{12} or 1,000,000,000,000
Giga-	G	10^{9} or 1,000,000,000
Mega-	M	10^{6} or 1,000,000
Kilo-	k	10^{3} or 1,000
Deci-	d	10^{-1} or 1/10
Centi-	c	10^{-2} or 1/100
Milli-	m	10^{-3} or 1/1,000
Micro-	μ	10^{-6} or 1/1,000,000
Nano-	n	10^{-9} or 1/1,000,000,000
Pico-	p	10^{-12} or 1/1,000,000,000,000
Femto-	f	10^{-15} or 1/1,000,000,000,000,000
Atto-	a	10^{-18} or 1/1,000,000,000,000,000,000

PHYSICAL CHEMISTRY
for the Chemical Sciences

PHYSICAL CHEMISTRY
for the Chemical Sciences

Raymond Chang
WILLIAMS COLLEGE

John W. Thoman, Jr.
WILLIAMS COLLEGE

University Science Books
www.uscibooks.com

University Science Books
www.uscibooks.com

Production Management: Jennifer Uhlich at Wilsted & Taylor
Manuscript Editing: John Murdzek
Design: Robert Ishi, with Yvonne Tsang at Wilsted & Taylor
Composition & Illustrations: Laurel Muller
Cover Design: Genette Itoko McGrew
Printing & Binding: Marquis Book Printing, Inc.

This book is printed on acid-free paper.

ISBN 978-1-891389-69-6 (hard cover)

ISBN 978-1-78262-087-7 (soft cover), only for distribution outside of North America and Mexico by the Royal Society of Chemistry.

Library of Congress Cataloging-in-Publication Data

Chang, Raymond.
Physical chemistry for the chemical sciences / Raymond Chang, John W. Thoman, Jr.
pages cm
Includes index.
ISBN 978-1-891389-69-6 (alk. paper)
1. Chemistry, Physical and theoretical—Textbooks. I. Thoman, John W., Jr., 1960– II. Title.
QD453.3.C43 2014
541—dc23 2013038473

Printed in Canada
10 9 8 7 6 5 4 3 2 1

About the cover art: Tunneling in the quantum harmonic oscillator. The red horizontal line represents the zero-point energy $\left(\frac{1}{2}h\nu\right)$ and the shaded region is the classically forbidden region in which $K_0 < 0$ (see Chapter 11).

Contents

Preface

Physical Chemistry for the Chemical Sciences is intended for use in a one-year introductory course in physical chemistry that is typically offered at the junior level (the third year in a college or university program). Students in the course will have taken general chemistry and introductory organic chemistry. In writing this book, our aim is to present the standard topics at the appropriate level with emphasis on readability and clarity. While mathematical treatment of many topics is necessary, we have provided a physical picture wherever possible for understanding the concepts. Only the basic skills of differential and integral calculus are required for working with the equations. The limited number of integral equations needed to solve the end-of-chapter problems may be readily accessed from handbooks of chemistry and physics or software such as Mathematica.

The 20 chapters of the text can be divided into three parts. Chapters 1–9 cover thermodynamics and related subjects. Quantum mechanics and molecular spectroscopy are treated in Chapters 10–14. The last part (Chapters 15–20) describes chemical kinetics, photochemistry, intermolecular forces, solids and liquids, and statistical thermodynamics. We have chosen a traditional ordering of topics, starting with thermodynamics because of the accessibility of the concrete examples and the closeness to everyday experience. For instructors who prefer the "atoms first" or molecular approach, the order can be readily switched between the first two parts without loss of continuity.

Within each chapter, we introduce topics, define terms, and provide relevant worked examples, pertinent applications, and experimental details. Many chapters include end-of-chapter appendices, which cover more detailed derivations, background, or explanation than the body of the chapter. Each chapter concludes with a summary of the most important equations introduced within the chapter, an extensive and accessible list of further readings, and many end-of-chapter problems. Answers to the even-numbered numerical problems may be found in the back of the book. The end-of-book appendices provide some review of relevant mathematical concepts, basic physics definitions relevant to chemistry, and thermodynamic data. A glossary enables the student to quickly check definitions. Inside of the front and back covers, we include tables of information that are generally useful throughout the book. The second color (red) enables the student to more easily interpret plots and elaborate diagrams and adds a pleasing look to the book.

An accompanying *Solutions Manual*, written by Helen O. Leung and Mark D. Marshall, provides complete solutions to all of the problems in the text. This supplement contains many useful ideas and insights into problem-solving techniques.

The lines drawn between traditional disciplines are continually being modified as new fields are being defined. This book provides a foundation for further study at the more advanced level in physical chemistry, as well as interdisciplinary subjects that include biophysical chemistry, materials science, and environmental chemistry fields such as atmospheric chemistry and biogeochemistry. We hope that you find our book useful when teaching or learning physical chemistry.

It is a pleasure to thank the following people who provided helpful comments and suggestions: Dieter Bingemann (Williams College), George Bodner (Purdue University), Taina Chao (SUNY Purchase), Nancy Counts Gerber (San Francisco State University), Donald Hirsh (The College of New Jersey), Raymond Kapral (University of Toronto), Sarah Larsen (University of Iowa), David Perry (University of Akron), Christopher Stromberg (Hood College), and Robert Topper (The Cooper Union).

We also thank Bruce Armbruster and Kathy Armbruster of University Science Books for their support and general assistance. We are fortunate to have Jennifer Uhlich of Wilsted & Taylor as our production manager. Her high professional standard and attention to detail greatly helped the task of transforming the manuscript into an attractive final product. We very much appreciate Laurel Muller for her artistic and technical skills in laying out the text and rendering many figures. Robert Ishi and Yvonne Tsang are responsible for the elegant design of the book. John Murdzek did a meticulous job of copyediting. Our final thanks go to Jane Ellis, who supervised the project and took care of all the details, big and small.

Raymond Chang
John W. Thoman, Jr.

CHAPTER 1

Introduction and Gas Laws

And it's hard, and it's hard, ain't it hard, good Lord.
— Woody Guthrie*

1.1 Nature of Physical Chemistry

Physical chemistry can be described as a set of characteristically quantitative approaches to the study of chemical problems. A physical chemist seeks to predict and/or explain chemical events using certain models and postulates.

Because problems encountered in physical chemistry are diversified and often complex, they require a number of different approaches. For example, in the study of thermodynamics and rates of chemical reactions, we employ a phenomenological, macroscopic approach. But a microscopic, molecular approach based on quantum mechanics is necessary to understand the kinetic behavior of molecules and reaction mechanisms. Ideally, we study all phenomena at the molecular level, because it is here that change occurs. In fact, our knowledge of atoms and molecules is neither extensive nor thorough enough to permit this type of investigation in all cases, and we sometimes have to settle for a good, semiquantitative understanding. It is useful to keep in mind the scope and limitations of a given approach.

1.2 Some Basic Definitions

Before we discuss the gas laws, it is useful to define a few basic terms that will be used throughout the book. We often speak of the *system* in reference to a particular part of the universe in which we are interested. Thus, a system could be a collection of helium molecules in a container, a NaCl solution, a tennis ball, or a Siamese cat. Having defined a system, we call the rest of the universe the *surroundings.* There are three types of systems. An *open system* is one that can exchange both mass and energy with its surroundings. A *closed system* is one that does not exchange mass with its surroundings but can exchange energy. An *isolated system* is one that can exchange neither mass nor energy with its surroundings (Figure 1.1). To completely define a system, we need to understand certain experimental variables, such as pressure, volume, temperature, and composition, which collectively describe the *state of the system.*

A system is separated from the surroundings by a definite boundary, such as walls or surfaces.

* "Hard, Ain't It Hard." Words and Music by Woody Guthrie. TRO-© Copyright 1952 Ludlow Music, Inc., New York, N.Y. Used by permission.

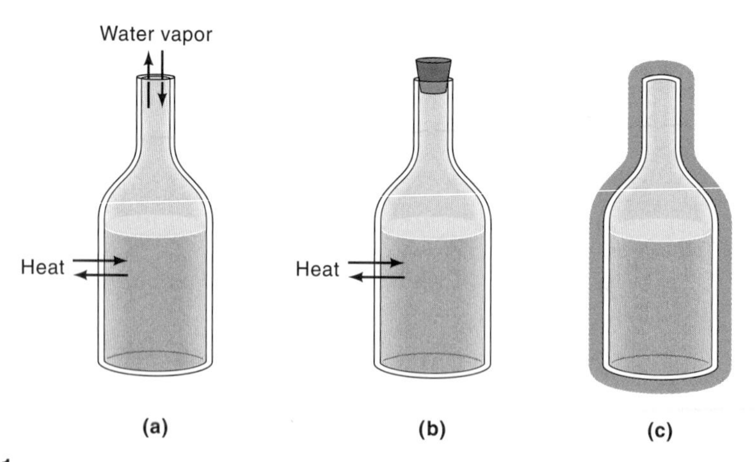

Figure 1.1
(a) An open system allows the exchange of both mass and energy; (b) a closed system allows the exchange of energy but not mass; and (c) an isolated system allows exchange of neither mass nor energy.

Most of the properties of matter may be divided into two classes: extensive properties and intensive properties. Consider, for example, two beakers containing the same amounts of water at the same temperature. If we combine these two systems by pouring the water from one beaker to the other, we find that the volume of the water is doubled and so is its mass. On the other hand, the temperature and the density of the water do not change. Properties whose values are directly proportional to the amount of the material present in the system are called *extensive properties*; those that do not depend on the amount are called *intensive properties*. Extensive properties include mass, area, volume, energy, and electrical charge. As already mentioned, temperature and density are both intensive properties, and so are pressure and electrical potential. Note that intensive properties are normally defined as ratios of two extensive properties, such as

$$\text{pressure} = \frac{\text{force}}{\text{area}}$$

$$\text{density} = \frac{\text{mass}}{\text{volume}}$$

1.3 An Operational Definition of Temperature

Temperature is a very important quantity in many branches of science, and not surprisingly, it can be defined in a number of different ways. Daily experience tells us that temperature is a measure of coldness and hotness, but for our purposes we need a more precise operational definition of temperature. Consider the following system of a container of gas A. The walls of the container are flexible so that its volume can expand and contract. This is a closed system that allows heat, but not mass, to flow into and out of the container. Initially, the pressure and volume are P_A and V_A, respectively. Now we bring the container in contact with a similar container of gas B at P_B and V_B. Heat exchange will take place until thermal equilibrium is reached. At equilibrium the pressure and volume of A and B will be altered to P'_A, V'_A and P'_B, V'_B. It is possible to remove container A temporarily, readjust its pressure and volume to P_A'' and V_A'', and still have A in thermal equilibrium with B at P_B' and V_B'. In fact, an infinite set of such values (P_A', V_A'), (P_A'', V_A''), (P_A''', V_A'''), ... can be obtained that will satisfy the equilibrium conditions. Figure 1.2 shows a plot of these points.

For all these states of A to be in thermal equilibrium with B, they must have the same value of a certain variable, which we call temperature. It follows from the discussion above that if two systems are in thermal equilibrium with a third system, then they must also be in thermal equilibrium with each other. This statement is generally known as the *zeroth law of thermodynamics*. The curve in Figure 1.2 is the locus of all the points that represent the states that can be in thermal equilibrium with system B. Such a curve is called an *isotherm*, or "same temperature." At another temperature, a different isotherm is obtained.

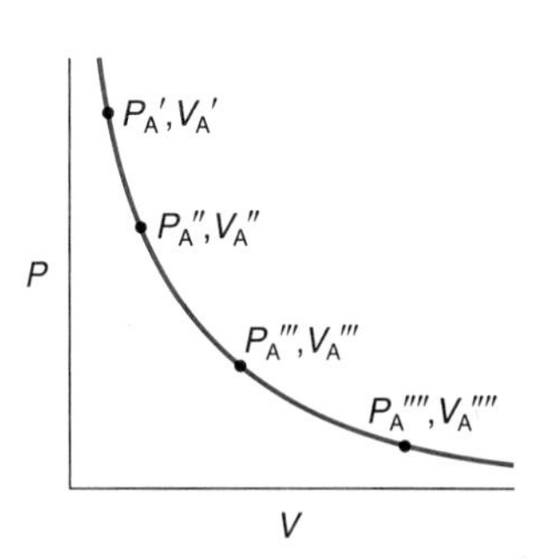

Figure 1.2
Plot of pressure versus volume at constant temperature for a given amount of a gas. Such a graph is called an isotherm.

1.4 Units

In this section we shall review the units chemists use for quantitative measurements.

For many years scientists recorded measurements in *metric units*, which are related decimally, that is, by powers of 10. In 1960, however, the General Conference of Weights and Measures, the international authority on units, proposed a revised metric system called the *International System of Units* (abbreviated SI). The advantage of the SI system is that many of its units are derivable from natural constants. For example, the SI system defines meter (m) as the length of the path traveled by light in vacuum during a time interval of 1/299,792,458 of a second. The unit of time, the second, is equivalent to 9,192,631,770 cycles of the radiation associated with a certain electronic transition of the cesium atom. In contrast, the fundamental unit of mass, the kilogram (kg), is defined in terms of an artifact, not in terms of a naturally occurring phenomenon. One kilogram is the mass of a platinum–iridium alloy cylinder kept by the International Bureau of Weights and Measures in Sevres, France.

In 2007, it was discovered that the alloy had mysteriously lost about 50 μg!

Table 1.1 gives the seven SI base units and Table 1.2 shows the prefixes used with SI units. Note that in SI units, temperature is given as K without the degree sign ° and the unit is plural—for example, 300 kelvins or 300 K. (More will be said of the Kelvin

Table 1.1
SI Base Units

Base quantity	Name of unit	Symbol
Length	meter	m
Mass	kilogram	kg
Time	second	s
Electrical current	ampere	A
Temperature	kelvin	K
Amount of substance	mole	mol
Luminous intensity	candela	cd

Table 1.2
Prefixes Used with SI and Metric Units

Prefix	Symbol	Meaning	Example
Tera-	T	1,000,000,000,000, or 10^{12}	1 terameter (Tm) = 1×10^{12} m
Giga-	G	1,000,000,000, or 10^9	1 gigameter (Gm) = 1×10^9 m
Mega-	M	1,000,000, or 10^6	1 megameter (Mm) = 1×10^6 m
Kilo-	k	1,000, or 10^3	1 kilometer (km) = 1×10^3 m
Deci-	d	1/10, or 10^{-1}	1 decimeter (dm) = 0.1 m
Centi-	c	1/100, or 10^{-2}	1 centimeter (cm) = 0.01 m
Milli-	m	1/1,000, or 10^{-3}	1 millimeter (mm) = 0.001 m
Micro-	μ	1/1,000,000, or 10^{-6}	1 micrometer (μm) = 1×10^{-6} m
Nano-	n	1/1,000,000,000, or 10^{-9}	1 nanometer (nm) = 1×10^{-9} m
Pico-	p	1/1,000,000,000,000, or 10^{-12}	1 picometer (pm) = 1×10^{-12} m

temperature scale in Section 1.4.) A number of physical quantities can be derived from the list in Table 1.1. We shall discuss only a few of them here.

Force

1 N is roughly equivalent to the force exerted by Earth's gravity on an apple.

The unit of force in the SI system is the *newton* (N), after the English physicist Sir Isaac Newton (1642–1726), defined as the force required to give a mass of 1 kg an acceleration of 1 m s^{-2}; that is,

$$1 \text{ N} = 1 \text{ kg m s}^{-2}$$

Pressure

Pressure is defined as

$$\text{pressure} = \frac{\text{force}}{\text{area}}$$

The SI unit of pressure is the *pascal* (Pa), after the French mathematician and physicist Blaise Pascal (1623–1662), where

$$1 \text{ Pa} = 1 \text{ N m}^{-2}$$

The following three relations are exact:

$$1 \text{ bar} = 1 \times 10^5 \text{ Pa} = 100 \text{ kPa}$$

$$1 \text{ atm} = 1.01325 \times 10^5 \text{ Pa} = 101.325 \text{ kPa}$$

$$1 \text{ atm} = 760 \text{ torr}$$

The torr is named after the Italian mathematician Evangelista Torricelli (1608–1674). The standard atmosphere (1 atm) is used to define the normal melting point and boiling point of substances and the bar is used to define standard states in physical chemistry. We shall use all of these units in this text.

Pressure is sometimes expressed in millimeters of mercury (mmHg), where 1 mmHg is the pressure exerted by a column of mercury 1 mm high when its density is 13.5951 g cm^{-3} and the acceleration due to gravity is 980.67 cm s^{-2}. The relation between mmHg and torr is

With currently accepted definitions and accuracy, 1 mmHg is 0.142 parts per million larger than 1 torr. In metrology, this is significant.

$$1 \text{ mmHg} = 1 \text{ torr}$$

One instrument that measures atmospheric pressure is the barometer. A simple barometer can be constructed by filling a long glass tube, closed at one end, with mercury, and then carefully inverting the tube in a dish of mercury, making sure that no air enters the tube. Some mercury will flow down into the dish, creating a vacuum at the top (Figure 1.3). The weight of the mercury column remaining in the tube is supported by atmospheric pressure acting on the surface of the mercury in the dish.

The device that is used to measure the pressure of gases other than the atmosphere is called a manometer. The principle of operation of a manometer is similar to that of a barometer. There are two types of manometers, shown in Figure 1.4. The closed-tube manometer is normally used to measure pressures lower than atmospheric pressure (Figure 1.4a). The open-tube manometer is more suited for measuring pressures equal to or greater than atmospheric pressure (Figure 1.4b).

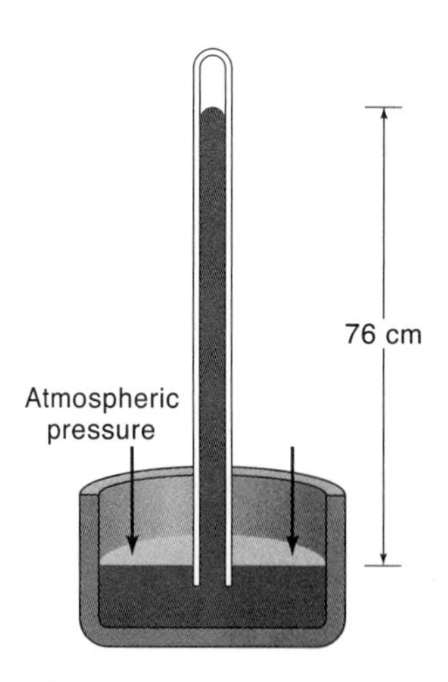

Figure 1.3
A barometer for measuring atmospheric pressure. Above the mercury in the tube is a vacuum. The column of mercury is supported by atmospheric pressure.

Energy

The SI unit of energy i· ... (J) [after the English physicist James Prescott Joule (1818–1889)]. Beca· ... ability to do work and work is force × distance, we have

1 N m

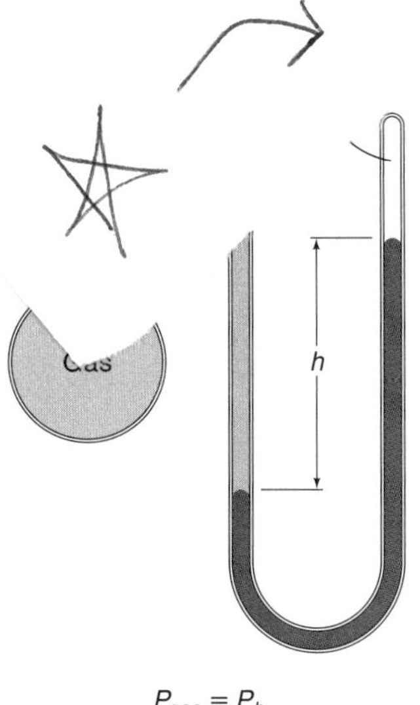

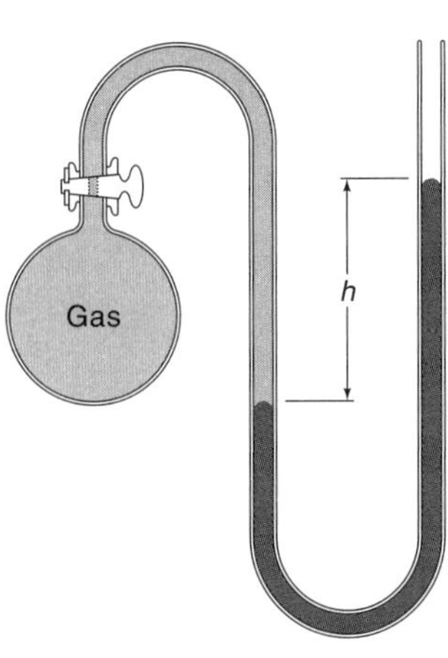

Figure 1.4
Two types of manometers used to measure gas pressures. (a) Gas pressure is less than atmospheric pressure. (b) Gas pressure is greater than atmospheric pressure.

Some chemists have continued to use the non-SI unit of energy, the calorie (cal), where

$$1 \text{ cal} = 4.184 \text{ J} \quad \text{(exactly)}$$

Most physical quantities have units and in general we can express such a quantity as

$$\text{physical quantity} = \text{numerical value} \times \text{unit}$$

For example, the speed of light (c) in vacuum is given by

$$c = 3.00 \times 10^8 \text{ m s}^{-1}$$

Thus, we can write

$$\frac{c}{\text{m s}^{-1}} = 3.00 \times 10^8$$

We shall use this convenient format in tables and figures.

Atomic Mass, Molecular Mass, and the Chemical Mole

By international agreement, an atom of the carbon-12 isotope, which has six protons and six neutrons, has a mass of exactly 12 atomic mass units (amu). One atomic mass unit is defined as a mass exactly equal to one-twelfth the mass of one carbon-12 atom. Experiments have shown that a hydrogen atom is only 8.400 percent as massive as the standard carbon-12 atom. Thus, the atomic mass of hydrogen must be $0.08400 \times 12 = 1.008$ amu. Similar experiments show that the atomic mass of oxygen is 16.00 amu and that of iron is 55.85 amu.

When you look up the atomic mass of carbon in a table such as the one on the inside front cover of this book, you will find it listed as 12.01 amu rather than 12.00 amu. The reason for the difference is that most naturally occurring elements (including carbon) have more than one isotope. This means that when we measure the atomic mass of an element, we must generally settle for the average mass of the naturally occurring mixture of isotopes. For example, the natural abundances of carbon-12 and carbon-13 are 98.90 percent and 1.10 percent, respectively. The atomic mass of carbon-13 has been determined to be 13.00335 amu. Thus, the average atomic mass of carbon can be calculated as follows:

$$\begin{aligned}\text{average atomic mass of carbon} &= (0.9890)(12 \text{ amu}) + (0.0110)(13.00335 \text{ amu}) \\ &= 12.01 \text{ amu}\end{aligned}$$

Because there are many more carbon-12 isotopes than carbon-13 isotopes, the average atomic mass is much closer to 12 amu than 13 amu. Such an average is called a *weighted average.*

If we know the atomic masses of the component atoms, then we can calculate the mass of a molecule. Thus, the molecular mass of H_2O is

$$2(1.008 \text{ amu}) + 16.00 \text{ amu} = 18.02 \text{ amu}$$

A *mole* (abbreviated mol) of any substance is the mass of that substance which contains as many atoms, molecules, ions, or any other entities as there are atoms in exactly 12 g of carbon-12. It has been determined experimentally that the number of atoms in one mole of carbon-12 is 6.0221415×10^{23}. This number is called *Avogadro's number*, after the Italian physicist and mathematician Amedeo Avogadro (1776–1856). Avogadro's number has no units, but dividing this number by mol gives us Avogadro's constant (N_A), where

$$N_A = 6.0221415 \times 10^{23} \text{ mol}^{-1}$$

For most purposes, N_A can be taken as 6.022×10^{23} mol^{-1}. The following examples indicate the number and kind of particles in one mole of any substance.

1. One mole of helium atoms contains 6.022×10^{23} He atoms.
2. One mole of water molecules contains 6.022×10^{23} H_2O molecules, or $2 \times (6.022 \times 10^{23})$ H atoms and 6.022×10^{23} O atoms.
3. One mole of NaCl contains 6.022×10^{23} NaCl units, or 6.022×10^{23} Na^+ ions and 6.022×10^{23} Cl^- ions.

The *molar mass* of a substance is the mass in grams or kilograms of 1 mole of the substance. Thus, the molar mass of atomic hydrogen is 1.008 g mol^{-1}, of molecular hydrogen is 2.016 g mol^{-1}, and of hemoglobin is 65,000 g mol^{-1}. In many calculations, molar masses are more conveniently expressed as kg mol^{-1}.

1.5 The Ideal Gas Law

Studying the behavior of gases has given rise to a number of chemical and physical theories. In many ways, the gaseous state is the easiest to investigate. We start by examining the properties of an ideal gas, which has the following characteristics: the molecules of an ideal gas possess no intrinsic volume and they neither attract nor repel one another. The *equation of state*, that is, the equation that relates the state variables of the (gaseous) system for an ideal gas is

$$PV = nRT \tag{1.1}$$

where n is the number of moles of the gas, T is the temperature in kelvins, and R is the gas constant, to be defined shortly. No ideal gas exists in nature, but under relatively high temperatures (≥ 25°C) and low pressures (≤ 10 atm) this equation roughly predicts the behavior of most gases.

The ideal gas equation is the accumulation of the work of the English chemist Robert Boyle (1627–1691) and the French physicists Jacques Charles (1746–1823) and Joseph Gay-Lussac (1778–1850). The gas laws associated with these scientists can be derived from Equation 1.1 under different conditions. For example, at constant temperature and amount of gas (n) we write

$$PV = \text{constant}$$

which is *Boyle's law.* At constant pressure and amount of gas, Equation 1.1 becomes

$$\frac{V}{T} = \text{constant}$$

The relation is called the *law of Charles and Gay-Lussac* or simply *Charles' law.* Charles' law takes the following form if the volume and amount of gas are kept constant:

$$\frac{P}{T} = \text{constant}$$

Another law, attributed to Avogadro, states that at constant pressure and temperature, equal volume of gases contain the same number of molecules. From Equation 1.1, we write

$$\frac{V}{n} = \text{constant}$$

The Kelvin Temperature Scale

As mentioned, the ideal gas equation holds only at low pressures. Therefore, in the limit of P approaching zero, Equation 1.1 can be rearranged to yield

$$T = \lim_{P \to 0} \frac{P\overline{V}}{R} \tag{1.2}$$

where $\overline{V}$ is called the *molar volume*, equal to V/n. Equation 1.2 defines the fundamental temperature scale, called the Kelvin scale, which is based on the ideal gas equation. Because P and $\overline{V}$ cannot take on negative values, the minimum value of T is zero.

The relationship between kelvins and degrees Celsius is obtained by studying the variation of the volume of a gas with temperature at constant pressure. At any given pressure, the plot of volume versus temperature yields a straight line. By extending the line to zero volume, we find the intercept on the temperature axis to be −273.15°C. At another pressure, we obtain a different straight line for the volume–temperature plot, but we get the *same* zero-volume temperature intercept at −273.15°C (Figure 1.5). (In practice, we can measure the volume of a gas over only a limited temperature range, because all real gases condense at low temperatures to form liquids.)

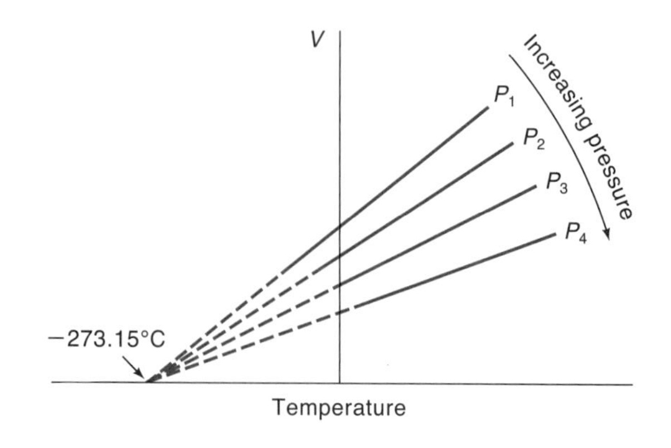

Figure 1.5
Plots of the volume of a given amount of gas versus temperature (t°C) at different pressures. All gases ultimately condense if they are cooled to low enough temperatures. When these lines are extrapolated, they all converge at the point representing zero volume and a temperature of −273.15°C.

In 1848 the Scottish mathematician and physicist William Thomson (Lord Kelvin, 1824–1907) realized the significance of this phenomenon. He identified −273.15°C as *absolute zero*, which is theoretically the lowest attainable temperature. Then he set up an absolute temperature scale, now called the Kelvin temperature scale, with absolute zero as the starting point. On the Kelvin scale, one kelvin (K) is equal in *magnitude* to one degree Celsius. The only difference between the absolute temperature scale and the Celsius scale is that the zero position is shifted. The relation between the two scales is

$$T/\text{K} = t/°\text{C} + 273.15 \tag{1.3}$$

Note that the Kelvin scale does not have the degree sign. Important points on the two scales match up as follows:

	Kelvin Scale	Celsius Scale
Absolute zero	0 K	−273.15°C
Freezing point of water	273.15 K	0°C
Boiling point of water	373.15 K	100°C

The normal boiling point and normal freezing point are measured at 1 atm.

In most cases we shall use 273 instead of 273.15 as the term relating K and °C. In this text we shall use T to denote absolute (Kelvin) temperature and t to indicate temperature on the Celsius scale. The Kelvin temperature scale has major theoretical significance; absolute temperatures must be used in gas law problems and thermodynamic calculations.

The Gas Constant R

The value of R can be obtained as follows. Experimentally it is found that 1 mole of an ideal gas occupies 22.414 L at 1 atm and 273.15 K (a condition known as *standard temperature and pressure*, or *STP*). Thus,

$$R = \frac{(1\ \text{atm})(22.414\ \text{L})}{(1\ \text{mol})(273.15\ \text{K})} = 0.08206\ \text{L atm K}^{-1}\ \text{mol}^{-1}$$

To express R in units of J K^{-1} mol^{-1}, we use the conversion factors

$$1\ \text{atm} = 1.01325 \times 10^5\ \text{Pa} \quad (1\ \text{Pa} = 1\ \text{N m}^{-2})$$

$$1\ \text{L} = 1 \times 10^{-3}\ \text{m}^3$$

and obtain

$$R = \frac{(1.01325 \times 10^5\ \text{N m}^{-2})(22.414 \times 10^{-3}\ \text{m}^3)}{(1\ \text{mol})(273.15\ \text{K})}$$

$$= 8.314\ \text{N m K}^{-1}\ \text{mol}^{-1}$$

$$= 8.314\ \text{J K}^{-1}\ \text{mol}^{-1} \qquad (1\ \text{J} = 1\ \text{N m})$$

From the two values of R we can write

$$0.08206 \text{ L atm K}^{-1} \text{ mol}^{-1} = 8.314 \text{ J K}^{-1} \text{ mol}^{-1}$$

or

$$1 \text{ L atm} = 101.3 \text{ J}$$

and

$$1 \text{ J} = 9.872 \times 10^{-3} \text{ L atm}$$

To express R in units of bar, we use the conversion factor 1 atm = 1.01325 bar and write

$$R = (0.08206 \text{ L atm K}^{-1} \text{ mol}^{-1})(1.01325 \text{ bar atm}^{-1})$$
$$= 0.08315 \text{ L bar K}^{-1} \text{ mol}^{-1}$$

EXAMPLE 1.1

Air entering the lungs ends up in tiny sacs called alveoli from which oxygen diffuses into the blood. The average radius of the alveoli is 0.0050 cm and the air inside contains 14 mole percent oxygen. Assuming that the pressure in the alveoli is 1.0 atm and the temperature is 37°C, calculate the number of oxygen molecules in one of the alveoli.

ANSWER

The volume of one alveoli, assumed spherical, is

$$V = \frac{4}{3}\pi r^3 = \frac{4}{3}\pi(0.0050 \text{ cm})^3$$
$$= 5.2 \times 10^{-7} \text{ cm}^3 = 5.2 \times 10^{-10} \text{ L} \qquad (1 \text{ L} = 10^3 \text{ cm}^3)$$

The number of moles of air in one alveoli is given by

$$n = \frac{PV}{RT} = \frac{(1.0 \text{ atm})(5.2 \times 10^{-10} \text{ L})}{(0.08206 \text{ L atm K}^{-1} \text{ mol}^{-1})(310 \text{ K})} = 2.0 \times 10^{-11} \text{ mol}$$

Because the air inside the alveoli is 14% oxygen, the number of oxygen molecules is

$$2.0 \times 10^{-11} \text{ mol air} \times \frac{14\% \text{ O}_2}{100\% \text{ air}} \times \frac{6.022 \times 10^{23} \text{ O}_2 \text{ molecules}}{1 \text{ mol O}_2}$$
$$= 1.7 \times 10^{12} \text{ O}_2 \text{ molecules}$$

1.6 Dalton's Law of Partial Pressures

So far we have discussed the pressure–volume–temperature behavior of a pure gas. Frequently, however, we work with mixtures of gases. For example, a chemist researching the depletion of ozone in the atmosphere must deal with several gaseous components. For a system containing two or more different gases, the total pressure (P_T) is the sum of the individual pressures that each gas would exert if it were alone and occupied the same volume. Thus,

$$P_T = P_1 + P_2 + \cdots = \sum_i P_i \tag{1.4}$$

where P_1, P_2, ... are the individual or *partial* pressures of components 1, 2, ..., and Σ is the summation sign. Equation 1.4 is known as *Dalton's law of partial pressures*, after the English chemist and school teacher John Dalton (1766–1844).

Consider a system containing two gases (1 and 2) at temperature T and volume V. The partial pressures of the gases are P_1 and P_2, respectively. From Equation 1.4,

$$P_1V = n_1RT \qquad \text{or} \qquad P_1 = \frac{n_1RT}{V}$$

$$P_2V = n_2RT \qquad \text{or} \qquad P_2 = \frac{n_2RT}{V}$$

where n_1 and n_2 are the numbers of moles of the two gases. According to Dalton's law,

$$\begin{aligned} P_T &= P_1 + P_2 \\ &= n_1\frac{RT}{V} + n_2\frac{RT}{V} \\ &= (n_1 + n_2)\frac{RT}{V} \end{aligned}$$

Dividing the partial pressures by the total pressure and rearranging, we get

$$P_1 = \frac{n_1}{n_1 + n_2}P_T = x_1P_T$$

and

$$P_2 = \frac{n_2}{n_1 + n_2}P_T = x_2P_T$$

where x_1 and x_2 are the mole fractions of gases 1 and 2, respectively. A mole fraction, defined as the ratio of the number of moles of one gas to the total number of moles of all gases present, is a dimensionless quantity. Furthermore, by definition, the sum of all the mole fractions in a mixture must be unity:

$$\sum_i x_i = 1 \tag{1.5}$$

In general, the partial pressure of the ith component (P_i) is related to the total pressure as follows:

$$P_i = x_i P_T \tag{1.6}$$

How are partial pressures determined? A manometer can measure only the total pressure of a gaseous mixture. To obtain partial pressures, we need to know the mole fractions of the components. The most direct method of measuring partial pressures is using a mass spectrometer. The relative intensities of the peaks in a mass spectrum are directly proportional to the amounts, and hence to the mole fractions, of the gases present.

The gas laws played a key role in the development of atomic theory, and there are many practical illustrations of the gas laws in everyday life. Here we shall briefly describe two examples that are particularly important to scuba divers. Seawater has a slightly higher density than fresh water—about 1.03 g mL^{-1} compared to 1.00 g mL^{-1}. The pressure exerted by a column of 33 ft (10 m) of seawater is equivalent to 1 atm pressure. What would happen if a diver were to rise to the surface rather quickly, holding his breath? If the ascent started at 40 ft under water, the decrease in pressure from this depth to the surface would be $(40 \text{ ft}/33 \text{ ft}) \times 1$ atm, or 1.2 atm. Assuming constant temperature, when the diver reached the surface, the volume of air trapped in his lungs would have increased by a factor of $(1 + 1.2)$ atm/1 atm, or 2.2 times! This sudden expansion of air could damage or rupture the membranes of his lungs, seriously injuring or killing the diver.

Dalton's law has a direct application to scuba diving. The partial pressure of oxygen in air is about 0.2 atm. Because oxygen is essential for our survival, it is sometimes hard to believe that it could be harmful to breathe more than our normal share. In fact, the toxicity of oxygen is well documented.* Physiologically, our bodies function best when the partial pressure of oxygen is 0.2 atm. For this reason, the composition of the air in a scuba tank is adjusted when the diver is submerged. For example, at a depth where the total pressure (hydrostatic plus atmospheric) is 4 atm, the oxygen content in the air supply should be reduced to 5 percent by volume to maintain the optimal partial pressure (0.05×4 atm $= 0.2$ atm). At a greater depth, the oxygen content must be even lower. Although nitrogen seems to be the obvious choice for mixing with oxygen in a scuba tank, because it is the major component of air, it is not the best choice. When the partial pressure of nitrogen exceeds 1 atm, a sufficient amount will dissolve in the blood to cause *nitrogen narcosis*. Symptoms of this condition, which resembles alcohol intoxication, include light-headedness and impaired judgment. Divers suffering from nitrogen narcosis have been known to do strange things, such as dancing on the sea floor and chasing sharks. For this reason, helium is usually employed to dilute oxygen in diving tanks. Helium, an inert gas, is much less soluble in blood than nitrogen, and it does not produce narcotic effects.

---

* At partial pressures above 2 atm, oxygen becomes toxic enough to produce convulsions and coma. Years ago, newborn infants placed in oxygen tents often developed *retrolental fibroplasia*, damage of the retinal tissues by excess oxygen. This damage usually resulted in partial or total blindness.

1.7 Real Gases

Under conditions when a gas behaves nonideally, we need to replace Equation 1.1 with a different equation of state for real gases. When a gas is being compressed, the molecules are brought closer to one another, and the gas will deviate appreciably from ideal behavior. One way to measure the deviation from ideality is to plot the compressibility factor (Z) of a gas versus pressure. Starting with Equation 1.1, we write

$$PV = nRT$$

and then

$$Z = \frac{P\overline{V}}{RT} \tag{1.7}$$

where $\overline{V}$ is the molar volume (L mol^{-1}) of the gas. For an ideal gas, $Z = 1$ for any value of P at a given T. However, the compressibility factors for real gases exhibit fairly divergent dependence on pressure (Figure 1.6). At low pressures, the compressibility factors of most gases are close to unity. In fact, in the limit of P approaching zero, we have $Z = 1$ for all gases. This finding is what we would expect, because all real gases behave ideally at low pressures. As pressure increases, some gases have $Z < 1$, which means that they are easier to compress than an ideal gas. Then, as pressure increases further, all gases have $Z > 1$. Over this region, the gases are harder to compress than an ideal gas. These behaviors are consistent with our understanding of intermolecular forces. In general, attractive forces are long-range forces, whereas repulsive forces operate only within a short range (more on this topic in Chapter 17). When molecules are far apart (e.g., at low pressures), the predominant intermolecular interaction is attraction. As the distance of separation between molecules decreases, the repulsive interaction among molecules becomes more significant.

Over the years, considerable effort has gone into modifying the ideal gas equation for real gases. Of the numerous such equations proposed, we shall consider three: the van der Waals equation, the Redlich–Kwong equation, and the virial equation of state.

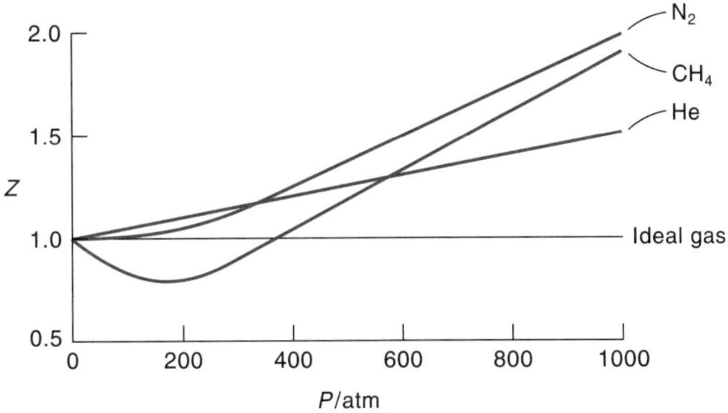

Figure 1.6
Plot of the compressibility factor versus pressure for real gases and an ideal gas at 273 K. Note that for an ideal gas $Z = 1$, no matter how great the pressure.

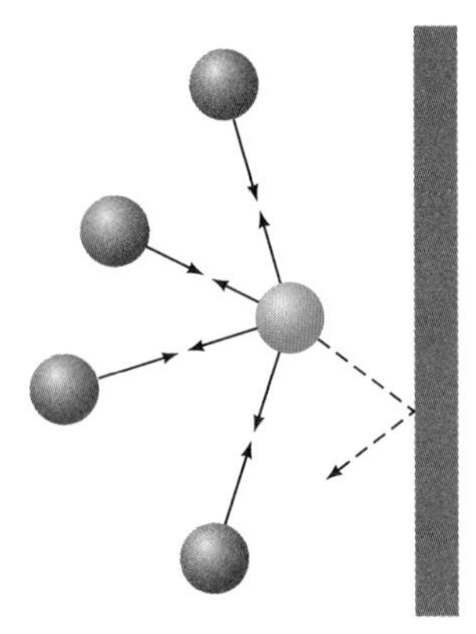

Figure 1.7
Effect of intermolecular forces on the pressure exerted by a gas. The speed of a molecule that is moving toward the container wall (red sphere) is reduced by the attractive forces exerted by its neighbors (gray spheres). Consequently, the impact this molecule makes with the wall is not as great as it would be if no intermolecular forces were present. In general, the measured gas pressure is always lower than the pressure the gas would exert if it behaved ideally.

The van der Waals Equation

The *van der Waals equation of state* [after the Dutch physicist Johannes Diderick van der Waals (1837–1923)] attempts to account for the finite volume of individual molecules in a nonideal gas and the attractive forces between them.

$$\left(P + \frac{an^2}{V^2}\right)(V - nb) = nRT \tag{1.8}$$

The pressure exerted by the individual molecules on the walls of the container depends on both the frequency of molecular collisions with the walls and the momentum imparted by the molecules to the walls. Both contributions are diminished by the attractive intermolecular forces (Figure 1.7). In each case, the reduction in pressure depends on the number of molecules present or the density of the gas, n/V, so that

$$\begin{aligned}\text{reduction in pressure}\\ \text{due to attractive forces}\end{aligned} \propto \left(\frac{n}{V}\right)\left(\frac{n}{V}\right)$$

$$= a\frac{n^2}{V^2}$$

where a is a proportionality constant.

Note that in Equation 1.8 P is the *experimentally measured* pressure of the gas and $(P + an^2/V^2)$ would be the pressure of the gas if there were no intermolecular forces present. Because an^2/V^2 must have units of pressure, a is expressed as atm $\text{L}^2\ \text{mol}^{-2}$. To allow for the finite volume of molecules, we replace V in the ideal gas equation with $(V - nb)$, where nb represents the total effective volume of n moles of the gas. Thus, nb must have the units of volume and b has the units L mol^{-1}. Both a and b are constants characteristic of the gas under study. Table 1.3 lists the values of

Recall that dividing the symbol by the unit gives us a pure number. Thus, for He we have $b = 0.0237\ \text{L mol}^{-1}$.

Table 1.3
van der Waals Constants and Boiling Points of Some Substances

Substance	a/atm L^2 mol^{-2}	b/L mol^{-1}	Boiling point/K
He	0.0341	0.0237	4.2
Ne	0.214	0.0174	27.2
Ar	1.34	0.0322	87.3
H_2	0.240	0.0264	20.3
N_2	1.35	0.0386	77.4
O_2	1.34	0.0312	90.2
CO	1.45	0.0395	83.2
CO_2	3.60	0.0427	195.2
CH_4	2.26	0.0430	109.2
C_2H_6	5.47	0.0651	184.5
H_2O	5.54	0.0305	373.15
NH_3	4.25	0.0379	239.8

a and b for a number of gases. The value of a is somehow related to the magnitude of attractive forces. Using the boiling point as a measure of the strength of intermolecular forces (the higher the boiling point, the stronger the intermolecular forces), we see that there is a rough correlation between the values of a and the boiling point of these substances. The quantity b is more difficult to interpret. Although b is proportional to the size of the molecule, the correlation is not always straightforward. For example, the value of b for helium is 0.0237 L mol^{-1} and that for neon is 0.0174 L mol^{-1}. Based on these values, we might expect that helium is larger than neon, which we know is not true. The values of a and b of a gas can be determined by several methods. The common practice is to apply the van der Waals equation to the gas in the critical state. We shall return to this point in Section 1.8.

The Redlich–Kwong Equation

The van der Waals equation is of historical importance—it was the first equation of state that provided a molecular interpretation of nonideal behavior. Of the many similar equations that have been proposed since van der Waals first presented his analysis, a particularly useful one is the Redlich–Kwong equation,

$$P = \frac{RT}{\overline{V} - B} - \frac{A}{\sqrt{T}(\overline{V})(\overline{V} + B)} \tag{1.9}$$

where A and B are constants. Like the van der Waals equation, the Redlich–Kwong equation also involves two constants for a gas. It yields more accurate results, however, over a wider range of temperature and pressure.

EXAMPLE 1.2

The molar volume of ethane (C_2H_6) at 350 K is 0.1379 L mol^{-1}. Calculate the pressure of the gas using (a) the ideal gas equation, (b) the van der Waals equation, and (c) the Redlich–Kwong equation, given that A = 96.89 L^2 atm mol^{-2} K$^{0.5}$ and B = 0.04515 L mol^{-1}.

ANSWER

(a) From Equation 1.1

$$P = \frac{RT}{\overline{V}}$$

$$= \frac{(0.08206 \text{ L atm K}^{-1} \text{ mol}^{-1})(350 \text{ K})}{0.1379 \text{ L mol}^{-1}} = 208.3 \text{ atm}$$

(b) Equation 1.8 can be rearranged to give

$$P = \frac{RT}{\overline{V} - b} - \frac{a}{\overline{V}^2}$$

Using the a and b values for ethane in Table 1.3, we write

$$P = \frac{(0.08206 \text{ L atm K}^{-1} \text{ mol}^{-1})(350 \text{ K})}{(0.1379 \text{ L mol}^{-1} - 0.0651 \text{ L mol}^{-1})} - \frac{(5.47 \text{ L}^2 \text{ atm mol}^{-2})}{(0.1379 \text{ L mol}^{-1})^2}$$

$$= 106.9 \text{ atm}$$

(c) From Equation 1.9

$$P = \frac{RT}{\overline{V} - B} - \frac{A}{\sqrt{T}(\overline{V})(\overline{V} + B)}$$

$$= \frac{(0.08206 \text{ L atm K}^{-1} \text{ mol}^{-1})(350 \text{ K})}{(0.1379 \text{ L mol}^{-1} - 0.04515 \text{ L mol}^{-1})} - \frac{96.89 \text{ L}^2 \text{ atm mol}^{-2} \text{ K}^{0.5}}{\sqrt{350 \text{ K}}(0.1379 \text{ L mol}^{-1})(0.1379 \text{ L mol}^{-1} + 0.04515 \text{ L mol}^{-1})}$$

$$= 104.5 \text{ atm}$$

COMMENT

The experimentally measured pressure is 98.69 atm. Thus, the Redlich–Kwong equation gives the closest value (the two values differ by 6 percent).

The Virial Equation of State

The word "virial" comes from Latin meaning "force." The cause of nonideal gas behavior is intermolecular forces.

Another way of representing gas nonideality is the *virial equation of state.* In this relationship, the compressibility factor is expressed as a series expansion in inverse powers of molar volume $\overline{V}$:

$$Z = 1 + \frac{B}{\overline{V}} + \frac{C}{\overline{V}^2} + \frac{D}{\overline{V}^3} + \cdots \tag{1.10}$$

where B, C, D, ... are called the second, third, fourth ... virial coefficients. The first virial coefficient is 1, which represents noninteracting molecules, as in an ideal gas. The second term describes the interaction between a pair of molecules, the third term describes the interaction among three molecules, and so on. For a given gas, they are evaluated from the P–V–T data of the gas by a curve-fitting procedure using a computer. For an ideal gas, the second and higher virial coefficients are zero and Equation 1.10 becomes Equation 1.1.

An alternate form of the virial equation is given by a series expansion of the compressibility factor in terms of the pressure P:

$$Z = 1 + B'P + C'P^2 + D'P^3 + \cdots \tag{1.11}$$

Because P and V are related, it is not surprising that relationships exist between B and B', C and C', and so on (see Problem 1.62). In each equation, the values of the coefficients decrease rapidly. For example, in Equation 1.11, the magnitude of the coefficients are such that $B' \gg C' \gg D'$ so that at pressures between zero and 10 atm, say, we need to include only the second term, provided the temperature is not very low:

$$Z = 1 + B'P \tag{1.12}$$

Equations 1.8 and 1.10 exemplify two rather different approaches. The van der Waals equation (and the Redlich–Kwong equation) accounts for the nonideality of gases by correcting for the finite molecular volume and intermolecular forces. Although these corrections do result in a definite improvement over the ideal gas equation, Equation 1.8 is still an approximate equation. The reason is that our present knowledge of intermolecular forces is insufficient to quantitatively explain macroscopic behavior. On the other hand, Equation 1.10 is accurate for real gases, but it does not provide us with any direct molecular interpretation. The nonideality of the gas is accounted for mathematically by a series expansion in which the coefficients B, C, ... can be determined experimentally. These coefficients do not have any physical meaning, although they can be related to intermolecular forces in an indirect way. Thus, our choice in this case is between an approximate equation that gives us some physical insight and an equation that describes the gas behavior accurately (if the coefficients are known), but tells us nothing about molecular behavior.

EXAMPLE 1.3

Calculate the molar volume of methane at 300 K and 100 atm, given that the second virial coefficient (B) of methane is -0.042 L mol^{-1}. Compare your result with that obtained using the ideal gas equation.

ANSWER

From Equation 1.10, neglecting terms containing C, D, ...,

$$\begin{aligned} Z &= 1 + \frac{B}{\overline{V}} \\ &= 1 + \frac{BP}{RT} \\ &= 1 + \frac{(-0.042 \text{ L mol}^{-1})(100 \text{ atm})}{(0.08206 \text{ L atm K}^{-1} \text{ mol}^{-1})(300 \text{ K})} \\ &= 1 - 0.17 = 0.83 \end{aligned}$$

$$\begin{aligned} \overline{V} &= \frac{ZRT}{P} \\ &= \frac{(0.83)(0.08206 \text{ L atm K}^{-1} \text{ mol}^{-1})(300 \text{ K})}{100 \text{ atm}} \\ &= 0.20 \text{ L mol}^{-1} \end{aligned}$$

For an ideal gas

$$\overline{V} = \frac{RT}{P}$$

$$= \frac{(0.08206 \text{ L atm K}^{-1} \text{ mol}^{-1})(300 \text{ K})}{100 \text{ atm}}$$

$$= 0.25 \text{ L mol}^{-1}$$

COMMENT

At 100 atm and 300 K, methane is more compressible than an ideal gas ($Z = 0.83$ compared to $Z = 1$) due to the attractive intermolecular forces between the CH_4 molecules.

1.8 Condensation of Gases and the Critical State

The condensation of gas to liquid is a familiar phenomenon. The first quantitative study of the pressure–volume relationship of this process was made in 1869 by the Irish chemist Thomas Andrews (1813–1885). He measured the volume of a given amount of carbon dioxide as a function of pressure at various temperatures and obtained a series of isotherms like those shown in Figure 1.8. At high temperatures the curves are roughly hyperbolic, indicating that the gas obeys Boyle's law. As the temperature is lowered, deviations become evident and a drastically different behavior is observed at T_4. Moving along the isotherm from right to left, we see that although the volume of the

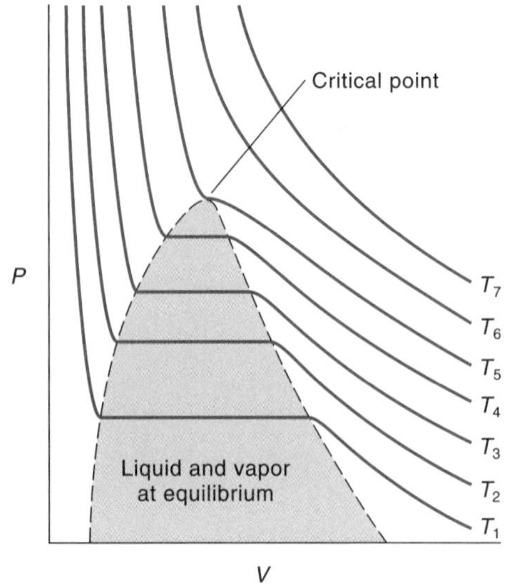

Figure 1.8
Isotherms of carbon dioxide at various temperatures (temperature increases from T_1 to T_7). As temperature increases, the horizontal line becomes shorter until it becomes a point at T_5, the critical point. Above this temperature carbon dioxide cannot be liquefied no matter how great the pressure.

gas decreases with pressure, the product PV is no longer a constant (because the curve is no longer a hyperbola). Increasing the pressure further, we reach a point that is the intersection between the isotherm and the dashed curve on the right. If we could observe this process, we would note the formation of liquid carbon dioxide at this pressure. With the pressure held constant, the volume continues to decrease (as more and more vapor is converted to liquid) until all the vapor has condensed. Beyond this point (the intersection between the horizontal line and the dashed curve on the left), the system is entirely liquid, and any further increase in pressure will result in only a very small decrease in volume, because liquids are much less compressible than gases. Figure 1.9 shows the liquifaction of carbon dioxide at T_1.

The pressure corresponding to the horizontal line (region in which vapor and liquid coexist) is called the *equilibrium vapor pressure* or simply the *vapor pressure* of the liquid at the temperature of the experiment. The length of the horizontal line decreases with increasing temperature. At a particular temperature (T_5 in Figure 1.8) the isotherm is tangent to the dashed curve and only one phase is present. The horizontal line is now a point known as the *critical point*. The corresponding temperature, pressure, and volume at this point are called the critical temperature (T_c), critical pressure (P_c), and critical volume (V_c). The critical temperature is the temperature above which no condensation can occur no matter how great the pressure. The critical constants of several gases are listed in Table 1.4. Note that the critical volume is usually expressed as a molar quantity, called the molar critical volume ($\overline{V}_c$), given by V_c/n, where n is the number of moles of the substance present.

The phenomenon of condensation and the existence of a critical temperature are direct consequences of the nonideal behavior of gases. After all, if molecules did not attract one another, then no condensation would occur, and if molecules had no volume, then we would be unable to observe liquids and solids. As mentioned earlier, the nature of molecular interaction is such that the force among molecules is attractive when they are relatively far apart, but as they get closer to one another (e.g., a liquid under pressure)

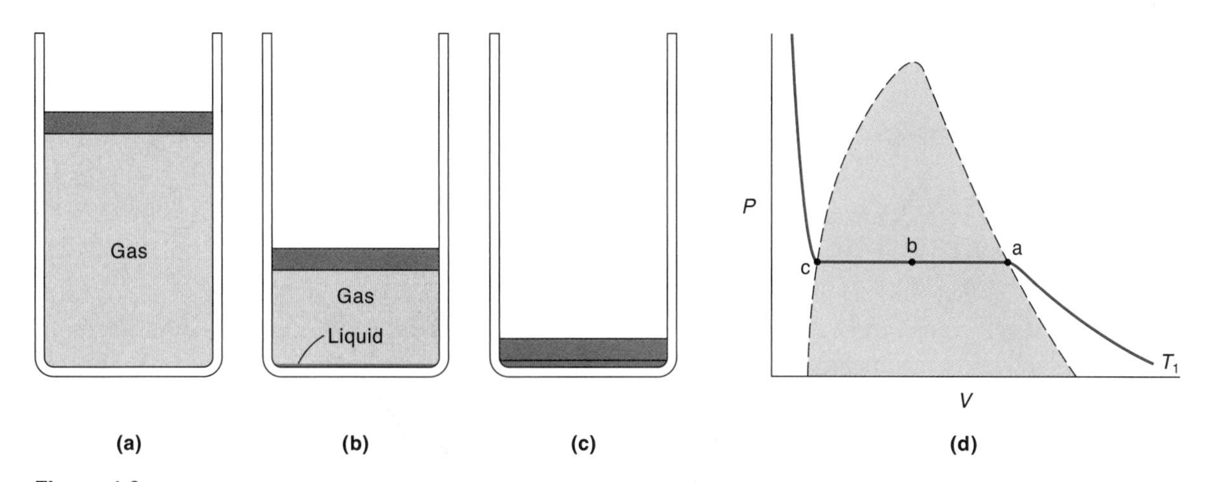

Figure 1.9
The liquefaction of carbon dioxide at T_1 (see Figure 1.8). At (a), the first drop of the liquid appears. From (b) to (c), the gas is gradually and completely converted to liquid at constant pressure. Beyond (c), the volume decreases only slightly with increasing pressure because liquids are highly incompressible. Part (d) shows an overall plot of the stages shown in (a), (b), and (c).

Table 1.4
Critical Constants of Some Substances

Substance	P_c/atm	$\overline{V}_c$/L mol^{-1}	T_c/K
He	2.25	0.0578	5.2
Ne	26.2	0.0417	44.4
Ar	49.3	0.0753	151.0
H_2	12.8	0.0650	32.9
N_2	33.6	0.0901	126.1
O_2	50.8	0.0764	154.6
CO	34.5	0.0931	132.9
CO_2	73.0	0.0957	304.2
CH_4	45.4	0.0990	190.2
C_2H_6	48.2	0.1480	305.4
H_2O	217.7	0.0560	647.6
NH_3	109.8	0.0724	405.3
SF_6	37.6	0.2052	318.7

this force becomes repulsive, because of electrostatic repulsions between nuclei and between electrons. In general, the attractive force reaches a maximum at a certain finite intermolecular distance. At temperatures below T_c, it is possible to compress the gas and bring the molecules within this attractive range, where condensation will occur. Above T_c, the kinetic energy of the gas molecules is such that they will always be able to break away from this attraction and no condensation can take place. The critical phenomenon of sulfur hexafluoride (SF_6) is shown in Figure 1.10.

An interesting relationship exists between the van der Waals constants a and b and the critical constants. Dividing Equation 1.8 throughout by n and rearranging, we obtain

$$\overline{V}^3 - \left(b + \frac{RT}{P}\right)\overline{V}^2 + \frac{a\overline{V}}{P} - \frac{ab}{P} = 0 \tag{1.13}$$

where $\overline{V}$ is the molar volume. This is a cubic equation and the solution yields three values of $\overline{V}$. At temperatures below T_c, $\overline{V}$ has three real roots; two of them correspond to the intersections of the horizontal line with the dashed curve in Figure 1.8, but the third root has no physical significance. Above T_c, $\overline{V}$ has one real root and two imaginary roots. At T_c, however, all three roots of $\overline{V}$ are real and identical; that is,

An imaginary root contains the $\sqrt{-1}$ term.

$$(\overline{V} - \overline{V}_c)^3 = 0$$

or

$$\overline{V}^3 - 3\overline{V}_c\overline{V}^2 + 3\overline{V}_c^2\overline{V} - \overline{V}_c^3 = 0 \tag{1.14}$$

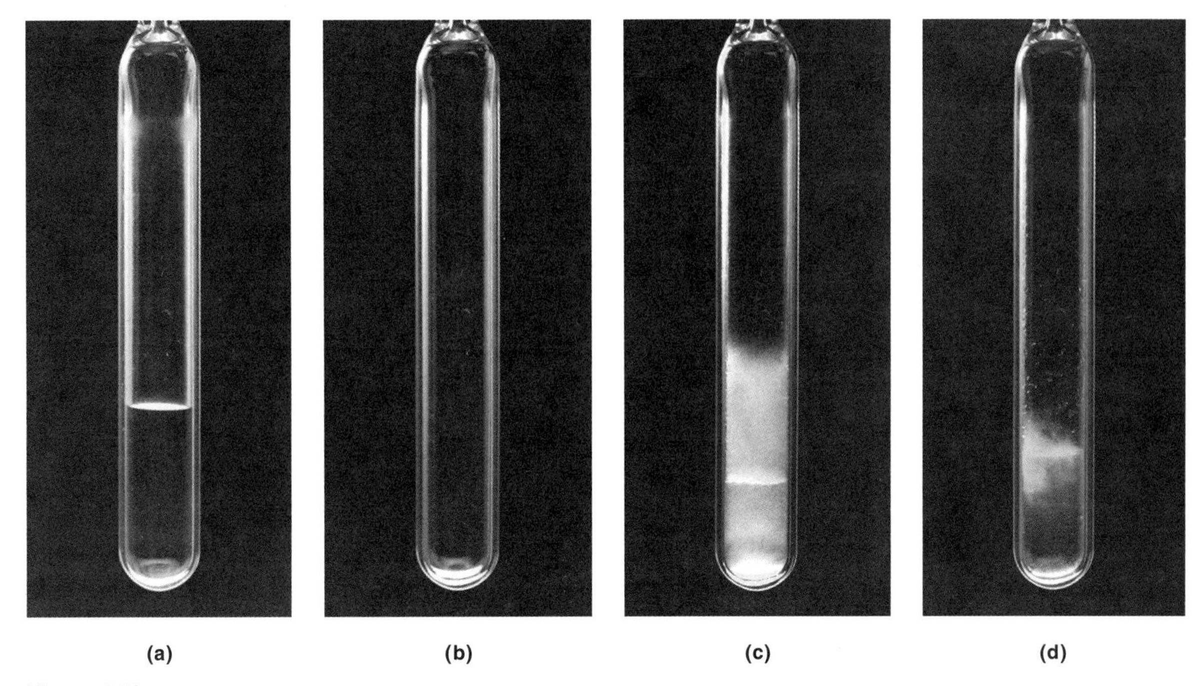

Figure 1.10
The critical phenomenon of sulfur hexafluoride (T_c = 45.5°C; P_c = 37.6 atm). (a) Below the critical temperature, a clear liquid phase is visible. (b) Above the critical temperature, the liquid phase disappears. (c) The substance is cooled just below its critical temperature. The fog is a phenomenon known as critical opalescence and is due to scattering of light by large density fluctuation in the critical fluid. (d) Finally, the liquid phase reappears.

Comparison of the coefficients for $\overline{V}^3$, $\overline{V}^2$, and $\overline{V}$ in Equations 1.13 and 1.14 gives the following:

For $\overline{V}^2$:

$$3\overline{V}_c = b + \frac{RT_c}{P_c} \tag{1.15}$$

For $\overline{V}$:

$$3\overline{V}_c^2 = \frac{a}{P_c} \tag{1.16}$$

Also,

$$\overline{V}_c^3 = \frac{ab}{P_c} \tag{1.17}$$

From Equations 1.15, 1.16, and 1.17, we obtain

$$a = 3P_c\overline{V}_c^2 \tag{1.18}$$

$$b = \frac{\overline{V}_c}{3} \tag{1.19}$$

$$R = \frac{8a}{27T_cb} \tag{1.20}$$

Therefore, if the critical constants of a substance are known, we can calculate both a and b. Actually, any two of the foregoing three equations can be used to obtain a and b. If the van der Waals equation were accurately obeyed in the critical region, the choice would be unimportant. However, this is not the case. The values of a and b depend significantly on whether we use the P_c–T_c, T_c–$\overline{V}_c$, or P_c–$\overline{V}_c$ combinations. It is customary to choose P_c and T_c because $\overline{V}_c$ is usually the least accurate of the critical constants. From Equation 1.19, we have

$$\overline{V}_c = 3b \tag{1.21}$$

which, when substituted into Equation 1.18, yields

$$a = 3P_c(3b)^2 = 27P_cb^2 \tag{1.22}$$

Substituting Equation 1.22 into Equation 1.20 gives

$$b = \frac{RT_c}{8P_c} \tag{1.23}$$

and

$$a = \frac{27R^2T_c^2}{64P_c} \tag{1.24}$$

Although Equations 1.23 and 1.24 should be considered approximate relationships because of the unreliability of the van der Waals equation in the critical region, they are frequently used to obtain the van der Waals constants. Thus, the values of a and b listed in Table 1.3 are mostly calculated from the critical constants (see Table 1.4) and Equations 1.23 and 1.24. The above method can also be used to relate the Redlich–Kwong constants (A and B) to the critical constants. The mathematical procedure is more involved, however.

In recent years there has been much interest in the practical applications of supercritical fluids (SCF), that is, the state of matter above the critical temperature. One of the most studied SCFs is carbon dioxide. Under appropriate conditions of temperature and pressure, SCF CO_2 can be used as a solvent for removing caffeine from raw coffee beans and cooking oil from potato chips to produce crisp, oil-free chips. It is also being used in environmental cleanups because it dissolves chlorinated hydrocarbons. SCFs of CO_2, NH_3, and certain hydrocarbons, such as hexane and heptane, are used in chromatography. SCF CO_2 has been shown to be an effective carrier medium for substances such as antibiotics and hormones, which are unstable at the high temperatures required for normal chromatographic separations.

1.9 The Law of Corresponding States

One of the remarkable consequences of Equation 1.8 was pointed out by van der Waals himself. First, we express Equation 1.8 in terms of molar volume $\overline{V}$:

$$\left(P + \frac{a}{\overline{V}^2}\right)(\overline{V} - b) = RT \tag{1.25}$$

If we divide the pressure, volume, and temperature of a gas by its critical constants, we obtain

$$P_R = \frac{P}{P_c} \qquad \overline{V}_R = \frac{\overline{V}}{\overline{V}_c} \qquad T_R = \frac{T}{T_c} \tag{1.26}$$

where P_R, $\overline{V}_R$, and T_R are called the reduced pressure, reduced volume, and reduced temperature, respectively. Expressed in terms of the reduced variables, Equation 1.25 becomes

$$\left(P_R P_c + \frac{a}{\overline{V}_R^2 \overline{V}_c^2}\right)(\overline{V}_R \overline{V}_c - b) = RT_R T_c \tag{1.27}$$

From Equations 1.15, 1.16, and 1.17, we can show that

$$P_c = \frac{a}{27b^2} \qquad \overline{V}_c = 3b \qquad T_c = \frac{8a}{27Rb} \tag{1.28}$$

Substituting the expressions for P_c, $\overline{V}_c$, and T_c in Equation 1.27, we obtain

$$\left(P_R + \frac{3}{\overline{V}_R^2}\right)(3\overline{V}_R - 1) = 8T_R \tag{1.29}$$

Note that Equation 1.29 does not contain the constants a and b, so it is applicable to *all* substances. Equation 1.29 is a mathematical statement of the *law of corresponding states*. In words, it says that all gases have the same properties if they are compared at the corresponding conditions of P_R, $\overline{V}_R$, and T_R. Put another way, if two gases have the same T_R and P_R values, then their $\overline{V}_R$ value must also be the same.

As an illustration, consider the situation in which 1 mole of nitrogen gas is kept at 6.58 atm and 189 K and 1 mole of carbon dioxide gas is kept at 14.3 atm and 456 K. Under these conditions, the gases are in corresponding states because they have the same reduced pressure ($P_R = 0.20$) and reduced temperature ($T_R = 1.5$). It follows, therefore, that they must also have the same reduced molar volume ($\overline{V}_R \sim 20$). Keep in mind that Equation 1.29 suffers from the same limitations as the original van der Waals equation (Equation 1.25).

Figure 1.11 gives a nice graphical demonstration of the law of corresponding states. For a collection of gases under different conditions, we can calculate the compressibility factor $Z(P\overline{V}/RT)$ from their P–V–T data. At a given reduced temperature T_R, we find that a plot of Z versus P_R of these gases roughly fall on the same curve, showing that the compressibility is the same for these gases at the same reduced pressure and temperature.* The law of corresponding states is particularly useful in engineering because it provides a fast way to obtain a large amount of information about gases under extreme conditions.

*Because $P = P_R P_c$, $\overline{V} = \overline{V}_R \overline{V}_c$, and $T = T_R T_c$, we have $Z = (P_R \overline{V}_R / T_R)(P_c \overline{V}_c / RT_c)$ or $(P_R \overline{V}_R / T_R) Z_c$. From Table 1.4 we see that the value of Z_c is roughly the same. Thus, at the same P_R and T_R, we expect the gases to have approximately the same $\overline{V}_R$ and hence the same Z value.

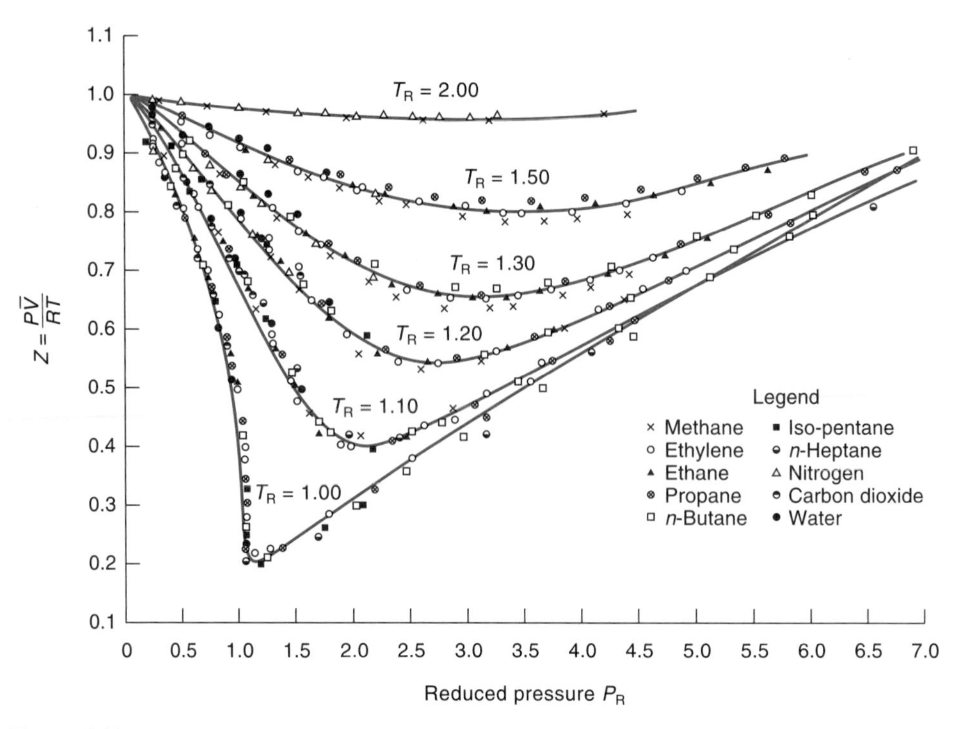

Figure 1.11
Compressibility factor (Z) of various gases as a function of reduced pressure and temperature. [*Source:* Gouq-Jen Su, *Ind. Chem.* **38**, 803 (1946).]

EXAMPLE 1.4

Use Figure 1.11 to estimate the molar volume of water at 504°C and 435 atm.

ANSWER

From Table 1.4, we find that $P_R = 2.0$ and $T_R = 1.2$ for water. According to Figure 1.11, $Z \approx 0.60$; therefore, the molar volume is

$$\overline{V} = \frac{RTZ}{P} = \frac{(0.08206 \text{ L atm K}^{-1} \text{ mol}^{-1})(777 \text{ K})(0.60)}{435 \text{ atm}}$$

$$= 0.088 \text{ L mol}^{-1}$$

From the ideal gas equation, we find $\overline{V} = 0.15 \text{ L mol}^{-1}$.

Key Equations

$PV = nRT$	(Ideal-gas equation)	(1.1)
$P_T = \sum_i P_i$	(Dalton's law of partial pressures)	(1.2)
$Z = \frac{PV}{nRT} = \frac{P\overline{V}}{RT}$	(Compressibility factor)	(1.7)
$\left(P + \frac{an^2}{V^2}\right)(V - nb) = nRT$	(van der Waals equation)	(1.8)
$Z = 1 + \frac{B}{\overline{V}} + \frac{C}{\overline{V}^2} + \frac{D}{\overline{V}^3} + \cdots$	(Virial equation)	(1.10)
$Z = 1 + B'P + C'P^2 + D'P^3 + \cdots$	(Virial equation)	(1.11)
$\left(P_R + \frac{3}{\overline{V}_R^2}\right)(3\overline{V}_R - 1) = 8T_R$	(Law of corresponding states)	(1.29)

Suggestions for Further Reading

STANDARD PHYSICAL CHEMISTRY TEXTS

Atkins, P. W., and J. de Paula, *Physical Chemistry*, 8th ed. W. H. Freeman, New York, 2006.

Laidler, K. J., J. H. Meiser, and B. C. Sanctuary, *Physical Chemistry*, 4th ed., Houghton Mifflin Company, Boston, 2003.

Levine, I. N., *Physical Chemistry*, 5th ed., McGraw-Hill, New York, 2009.

McQuarrie, D. A., and J. D. Simon, *Physical Chemistry*, University Science Books, Sausalito, CA, 1997.

Noggle, J. H., *Physical Chemistry*, 3rd ed., Harper Collins College Publishers, New York, 1996.

Silbey, R. J., R. A. Alberty, and M. G. Bawendi, *Physical Chemistry*, 4th ed., John Wiley & Sons, New York, 2004.

HISTORICAL DEVELOPMENT OF PHYSICAL CHEMISTRY

"One Hundred Years of Physical Chemistry," E. B. Wilson, Jr., *Am. Sci.* **74**, 70 (1986).

Laidler, K. J., *The World of Physical Chemistry*, Oxford University Press, New York, 1993.

Cobb, C., *Magick, Mayhem, and Mavericks: The Spirited History of Physical Chemistry*, Prometheus Books, Amherst, NY, 2002.

PHYSICAL PROPERTIES OF GASES

Tabor, D., *Gases, Liquids, and Solids*, 3rd ed., Cambridge University Press, New York, 1996.

Walton, A. J., *The Three Phases of Matter*, 2nd ed., Oxford University Press, New York, 1983.

ARTICLES

"The van der Waals Gas Equation," F. S. Swinbourne, *J. Chem. Educ.* **32**, 366 (1955).
"A Simple Model for van der Waals," S. S. Winter, *J. Chem. Educ.* **33**, 459 (1959).
"The Critical Temperature: A Necessary Consequence of Gas Nonideality," F. L. Pilar, *J. Chem. Educ.* **44**, 284 (1967).
"The Cabin Atmosphere in Manned Space Vehicles," W. H. Bowman and R. M. Lawrence, *J. Chem. Educ.* **48**, 152 (1971).
"Comparisons of Equations of State in Effectively Describing *PVT* Relations," J. B. Ott, J. R. Goales, and H. T. Hall, *J. Chem. Educ.* **48**, 515 (1971).
"Scuba Diving and the Gas Laws," E. D. Cooke, *J. Chem. Educ.* **50**, 425 (1973).
"The Invention of the Balloon and the Birth of Modern Chemistry," A. F. Scott, *Sci. Am.* January 1984.
"Derivation of the Ideal Gas Law," S. Levine, *J. Chem. Educ.* **62**, 399 (1985).
"Supercritical Fluids: Liquid, Gas, Both, or Neither? A Different Approach," E. F. Meyer and T. P. Meyer, *J. Chem. Educ.* **63**, 463 (1986).
"The Ideal Gas Law at the Center of the Sun," D. B. Clark, *J. Chem. Educ.* **66**, 826 (1989).
"The Many Faces of van der Waals's Equation of State," J. G. Eberhart, *J. Chem. Educ.* **66**, 906 (1989).
"Applying the Critical Conditions to Equations of State," J. G. Eberhart, *J. Chem. Educ.* **66**, 990 (1989).
"Does a One-Molecule Gas Obey Boyle's Law?" G. Rhodes, *J. Chem. Educ.* **69**, 16 (1992).
"A Lecture Demonstration of the Critical Phenomenon," R. Chang and J. F. Skinner, *J. Chem. Educ.* **69**, 158 (1992).
"Mountain Sickness," C. S. Houston, *Sci. Am.* October 1992.
"Equations of State," M. Ross in *Encyclopedia of Applied Physics*, G. L. Trigg, Ed., VCH Publishers, New York, 1993, Vol. 6, p. 291.
"Past, Present, and Possible Future Applications of Supercritical Fluid Extraction Technology," C. L. Phelps, N. G. Smart, and C. M. Wai, *J. Chem. Educ.* **73**, 1163 (1996).
"Interpretation of the Second Virial Coefficient," J. Wisniak, *J. Chem. Educ.* **76**, 671 (1999).
"Virial Coefficients Using Different Equations of State," C. B. Wakefield and C. Phillips, *J. Chem. Educ.* **77**, 1371 (2000).
"The Thermometer—From the Feeling to the Instrument," J. Wisniak, *Chem. Educator* [Online] **5**, 88 (2000) DOI 10.1007/s0089700037a.
"Updated Principle of Corresponding States," D. Ben-Amotz and A. D. Gift, *J. Chem. Educ.* **81**, 142 (2004).
"The Determination of Absolute Zero: An Accurate and Rapid Method," J. Gordon, L. Williams, J. James, and R. Bernard, *Chem. Educator* [Online] **13**, 351 (2008) DOI 10.1333/s00897082170a.
"How Heavy Is a Balloon? Using the Gas Laws," B. O. Johnson and H. Van Milligan, *J. Chem. Educ.* **86**, 224A (2009).
"An Interesting Algebraic Rearrangement of Semi-empirical Gaseous Equations of State. Partitioning of the Compressibility Factor into Attractive and Repulsive Parts," F. Watson, *Chem. Educator* [Online] **15**, 10 (2010) DOI 10.1007/s00897102210a.
"Measurement of the Compressibility Factor of Gases: A Physical Chemistry Laboratory Experiment," T. D. Varburg, A. J. Bendelsmith, and K. T. Kuwata, *J. Chem. Educ.* **88**, 1166 (2011).
"A Simple Mercury-Free Laboratory Apparatus To Study the Relationship between Pressure, Volume, and Temperature in a Gas," D. McGregor, W. V. Sweeney, and P. Mills, *J. Chem. Educ.* **89**, 509 (2012).

Problems

Ideal-Gas Laws

1.1 Classify each of the following properties as intensive or extensive: force, pressure (P), volume (V), temperature (T), mass, density, molar mass, molar volume ($\overline{V}$).

1.2 Some gases, such as NO_2 and NF_2, do not obey Boyle's law at any pressure. Explain.

1.3 An ideal gas originally at 0.85 atm and 66°C was allowed to expand until its final volume, pressure, and temperature were 94 mL, 0.60 atm, and 45°C, respectively. What was its initial volume?

1.4 Some ballpoint pens have a small hole in the main body of the pen. What is the purpose of this hole?

1.5 Starting with the ideal-gas equation, show how you can calculate the molar mass of a gas from a knowledge of its density.

1.6 At STP (standard temperature and pressure), 0.280 L of a gas weighs 0.400 g. Calculate the molar mass of the gas.

1.7 Ozone molecules in the stratosphere absorb much of the harmful radiation from the sun. Typically, the temperature and partial pressure of ozone in the stratosphere are 250 K and 1.0×10^{-3} atm, respectively. How many ozone molecules are present in 1.0 L of air under these conditions? Assume ideal-gas behavior.

1.8 Calculate the density of HBr in g L^{-1} at 733 mmHg and 46°C. Assume ideal-gas behavior.

1.9 Dissolving 3.00 g of an impure sample of $CaCO_3$ in an excess of HCl acid produced 0.656 L of CO_2 (measured at 20°C and 792 mmHg). Calculate the percent by mass of $CaCO_3$ in the sample.

1.10 The saturated vapor pressure of mercury is 0.0020 mmHg at 300 K and the density of air at 300 K is 1.18 g L^{-1}. **(a)** Calculate the concentration of mercury vapor in air in mol L^{-1}. **(b)** What is the number of parts per million (ppm) by mass of mercury in air?

1.11 A very flexible balloon with a volume of 1.2 L at 1.0 atm and 300 K is allowed to rise to the stratosphere, where the temperature and pressure are 250 K and 3.0×10^{-3} atm, respectively. What is the final volume of the balloon? Assume ideal-gas behavior.

1.12 Sodium bicarbonate ($NaHCO_3$) is called baking soda because when heated, it releases carbon dioxide gas, which causes cookies, doughnuts, and bread to rise during baking. **(a)** Calculate the volume (in liters) of CO_2 produced by heating 5.0 g of $NaHCO_3$ at 180°C and 1.3 atm. **(b)** Ammonium bicarbonate (NH_4HCO_3) has also been used as a leavening agent. Suggest one advantage and one disadvantage of using NH_4HCO_3 instead of $NaHCO_3$ for baking.

1.13 A common, non-SI unit for pressure is pounds per square inch (psi). Show that 1 atm = 14.7 psi. An automobile tire is inflated to 28.0 psi gauge pressure when cold, at 18°C. **(a)** What will the pressure be if the tire is heated to 32°C by driving the car? **(b)** What percent of the air in the tire would have to be let out to reduce the pressure to the original 28.0 psi? Assume that the volume of the tire remains constant with temperature. (A tire gauge measures not the pressure of the air inside but its excess over the external pressure, which is 14.7 psi.)

1.14 **(a)** What volume of air at 1.0 atm and 22°C is needed to fill a 0.98-L bicycle tire to a pressure of 5.0 atm at the same temperature? (Note that 5.0 atm is the gauge pressure, which is the difference between the pressure in the tire and atmospheric pressure. Initially, the gauge pressure in the tire was 0 atm.) **(b)** What is the total pressure in the tire when the gauge reads 5.0 atm? **(c)** The tire is pumped with a hand pump full of air at 1.0 atm; compressing the gas in the cylinder adds all the air in the pump to the air in the tire. If the volume of the pump is 33% of the tire's volume, what is the gauge pressure in the tire after 3 full strokes of the pump?

1.15 A student breaks a thermometer and spills most of the mercury (Hg) onto the floor of a laboratory that measures 15.2 m long, 6.6 m wide, and 2.4 m high. **(a)** Calculate the mass of mercury vapor (in grams) in the room at 20°C. **(b)** Does the concentration of mercury vapor exceed the air quality regulation of 0.050 mg Hg m^{-3} of air? **(c)** One way to treat small quantities of spilled mercury is to spray powdered sulfur over the metal. Suggest a physical and a chemical reason for this treatment. The vapor pressure of mercury at 20°C is 1.7×1.0^{-6} atm.

1.16 Nitrogen forms several gaseous oxides. One of them has a density of 1.27 g L^{-1} measured at 764 mmHg and 150°C. Write the formula of the compound.

1.17 Nitrogen dioxide (NO_2) cannot be obtained in a pure form in the gas phase because it exists as a mixture of NO_2 and N_2O_4. At 25°C and 0.98 atm, the density of this gas mixture is 2.7 g L^{-1}. What is the partial pressure of each gas?

1.18 An ultra-high-vacuum pump can reduce the pressure of air from 1.0 atm to 1.0×10^{-12} mmHg. Calculate the number of air molecules in a liter at this pressure and 298 K. Compare your results with the number of molecules in 1.0 L at 1.0 atm and 298 K. Assume ideal-gas behavior.

1.19 An air bubble with a radius of 1.5 cm at the bottom of a lake, where the temperature is 8.4°C and the pressure is 2.8 atm, rises to the surface, where the temperature is 25.0°C and the pressure is 1.0 atm. Calculate the radius of the bubble when it reaches the surface. Assume ideal-gas behavior. [*Hint:* The volume of a sphere is given by $(4/3)\pi r^3$, where r is the radius.]

1.20 The density of dry air at 1.00 atm and 34.4°C is 1.15 g L^{-1}. Calculate the composition of air (percent by mass) assuming that it contains only nitrogen and oxygen and behaves like an ideal gas. (*Hint:* First calculate the "molar mass" of air, then the mole fractions, and then the mass fractions of O_2 and N_2.)

1.21 A gas that evolved during the fermentation of glucose has a volume of 0.78 L when measured at 20.1°C and 1.0 atm. What was the volume of this gas at the fermentation temperature of 36.5°C? Assume ideal-gas behavior.

1.22 Two bulbs of volumes V_A and V_B are connected by a stopcock. The number of moles of gases in the bulbs are n_A and n_B, respectively, and initially the gases are at the same pressure, P, and temperature, T. Show that the final pressure of the system, after the stopcock has been opened, is equal to P. Assume ideal-gas behavior.

1.23 The composition of dry air at sea level is 78.03% N_2, 20.99% O_2, and 0.033% CO_2 by volume, **(a)** Calculate the average molar mass of this air sample. **(b)** Calculate the partial pressures of N_2, O_2, and CO_2 in atm. (At constant temperature and pressure, the volume of a gas is directly proportional to the number of moles of the gas.)

1.24 A mixture containing nitrogen and hydrogen weighs 3.50 g and occupies a volume of 7.46 L at 300 K and 1.00 atm. Calculate the mass percent of these two gases. Assume ideal-gas behavior.

1.25 The relative humidity in a closed room with a volume of 645.2 m^3 is 87.6% at 300 K, and the vapor pressure of water at 300 K is 0.0313 atm. Calculate the mass of water in the air. [*Hint:* The relative humidity is defined as $(P/P_s) \times 100\%$, where P and P_s are the partial pressure and saturated partial pressure of water, respectively.]

1.26 Death by suffocation in a sealed container is normally caused not by oxygen deficiency but by CO_2 poisoning, which occurs at about 7% CO_2 by volume. For what length of time would it be safe to be in a sealed room $10 \times 10 \times 20$ ft? [Source: "Eco-Chem," J. A. Campbell, *J. Chem. Educ.* **49**, 538 (1972).]

1.27 A flask contains a mixture of two ideal gases, A and B. Show graphically how the total pressure of the system depends on the amount of A present; that is, plot the total pressure versus the mole fraction of A. Do the same for B on the same graph. The total number of moles of A and B is constant.

1.28 A mixture of helium and neon gases is collected over water at 28.0°C and 745 mmHg. If the partial pressure of helium is 368 mmHg, what is the partial pressure of neon? (*Note:* Vapor pressure of water at 28°C is 28.3 mmHg.)

1.29 If the barometric pressure falls in one part of the world, it must rise somewhere else. Explain why.

1.30 A piece of sodium metal reacts completely with water as follows:

$$2Na(s) + 2H_2O(l) \longrightarrow 2NaOH(aq) + H_2(g)$$

The hydrogen gas generated is collected over water at 25.0°C. The volume of the gas is 246 mL measured at 1.00 atm. Calculate the number of grams of sodium used in the reaction. (*Note:* Vapor pressure of water at 25°C is 0.0313 atm.)

1.31 A sample of zinc metal reacts completely with an excess of hydrochloric acid:

$$Zn(s) + 2HCl(aq) \longrightarrow ZnCl_2(aq) + H_2(g)$$

The hydrogen gas produced is collected over water at 25.0°C. The volume of the gas is 7.80 L, and the pressure is 0.980 atm. Calculate the amount of zinc metal in grams consumed in the reaction. (*Note:* Vapor pressure of water at 25°C is 23.8 mmHg.)

1.32 Helium is mixed with oxygen gas for deep sea divers. Calculate the percent by volume of oxygen gas in the mixture if the diver has to submerge to a depth where the total pressure is 4.2 atm. The partial pressure of oxygen is maintained at 0.20 atm at this depth.

1.33 A sample of ammonia (NH_3) gas is completely decomposed to nitrogen and hydrogen gases over heated iron wool. If the total pressure is 866 mmHg, calculate the partial pressures of N_2 and H_2.

1.34 The partial pressure of carbon dioxide in air varies with the seasons. Would you expect the partial pressure in the Northern Hemisphere to be higher in the summer or winter? Explain.

1.35 A healthy adult exhales about 5.0×10^2 mL of a gaseous mixture with each breath. Calculate the number of molecules present in this volume at 37°C and 1.1 atm. List the major components of this gaseous mixture.

1.36 Describe how you would measure, by either chemical or physical means (other than mass spectrometry), the partial pressures of a mixture of gases: **(a)** CO_2 and H_2, **(b)** He and N_2.

1.37 The gas laws are vitally important to scuba divers. The pressure exerted by 33 ft of seawater is equivalent to 1 atm pressure. **(a)** A diver ascends quickly to the surface of the water from a depth of 36 ft without exhaling gas from his lungs. By what factor would the volume of his lungs increase by the time he reaches the surface? Assume that the temperature is constant. **(b)** The partial pressure of oxygen in air is about 0.20 atm. (Air is 20% oxygen by volume.) In deep-sea diving, the composition of air the diver breathes must be changed to maintain this partial pressure. What must the oxygen content (in percent by volume) be when the total pressure exerted on the diver is 4.0 atm?

1.38 A 1.00-L bulb and a 1.50-L bulb, connected by a stopcock, are filled, respectively, with argon at 0.75 atm and helium at 1.20 atm at the same temperature. Calculate the total pressure, the partial pressures of each gas, and the mole fraction of each gas after the stopcock has been opened. Assume ideal-gas behavior.

1.39 A mixture of helium and neon weighing 5.50 g occupies a volume of 6.80 L at 300 K and 1.00 atm. Calculate the composition of the mixture in mass percent.

Nonideal–Gas Behavior and the Critical State

1.40 Suggest two demonstrations to show that gases do not behave ideally.

1.41 Which of the following combinations of conditions most influences a gas to behave ideally: **(a)** low pressure and low temperature, **(b)** low pressure and high temperature, **(c)** high pressure and high temperature, and **(d)** high pressure and low temperature.

1.42 The van der Waals constants a and b for benzene are 18.00 atm L^2 mol^{-2} and 0.115 L mol^{-1}, respectively. Calculate the critical constants for benzene.

1.43 Using the data shown in Table 1.3, calculate the pressure exerted by 2.500 moles of carbon dioxide confined in a volume of 1.000 L at 450 K. Compare the pressure with that calculated assuming ideal behavior.

1.44 Without referring to a table, select from the following list the gas that has the largest value of b in the van der Waals equation: CH_4, O_2, H_2O, CCl_4, Ne.

1.45 Referring to Figure 1.6, we see that for He the plot has a positive slope even at low pressures. Explain this behavior.

1.46 At 300 K, the second virial coefficients (B) of N_2 and CH_4 are -4.2 cm^3 mol^{-1} and -15 cm^3 mol^{-1}, respectively. Which gas behaves more ideally at this temperature?

1.47 Calculate the molar volume of carbon dioxide at 400 K and 30 atm, given that the second virial coefficient (B) of CO_2 is -0.0605 L mol^{-1}. Compare your result with that obtained using the ideal-gas equation.

1.48 Consider the virial equation $Z = 1 + B'P + C'P^2$, which describes the behavior of a gas at a certain temperature. From the following plot of Z versus P, deduce the signs of B' and C' ($< 0, = 0, > 0$).

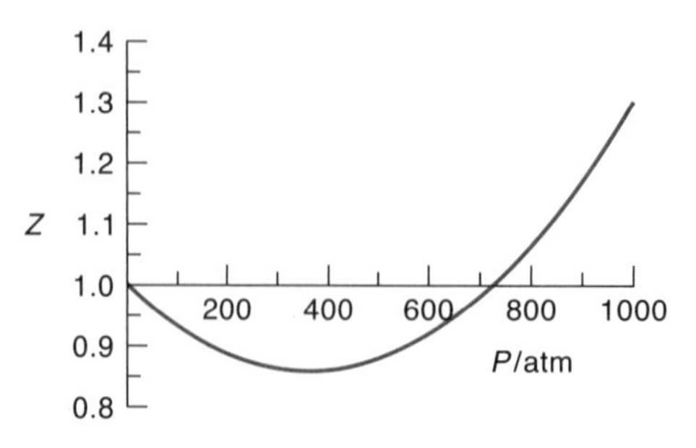

1.49 The critical temperature and critical pressure of naphthalene are 474.8 K and 40.6 atm, respectively. Calculate the van der Waals constants a and b for naphthalene.

1.50 Derive the van der Waals constants a and b in terms of the critical constants by recognizing that at the critical point, $(\partial P/\partial \overline{V})_T = 0$ and $(\partial^2 P/\partial \overline{V}^2)_T = 0$. (This problem requires a knowledge of partial differentiation.)

1.51 From the relationships among the van der Waals constants and the critical constants, show that $Z_c = P_c\overline{V}_c/RT_c = 0.375$, where Z_c is the compressibility factor at the critical point.

1.52 A CO_2 fire extinguisher is located on the outside of a building in Massachusetts. During the winter months, one can hear a sloshing sound when the extinguisher is gently shaken. In the summertime, there is often no sound when it is shaken. Explain. Assume that the extinguisher has no leaks and that it has not been used.

Additional Problems

1.53 A barometer with a cross-sectional area of 1.00 cm^2 at sea level measures a pressure of 76.0 cm of mercury. The pressure exerted by this column of mercury is equal to the pressure exerted by all the air on 1 cm^2 of Earth's surface. Given that the density of mercury is 13.6 g cm^{-3} and the average radius of Earth is 6371 km, calculate the total mass of Earth's atmosphere in kilograms. (*Hint:* The surface area of a sphere is $4\pi r^2$, where r is the radius of the sphere.)

1.54 It has been said that every breath we take, on average, contains molecules once exhaled by Wolfgang Amadeus Mozart (1756–1791). The following calculations demonstrate the validity of this statement. **(a)** Calculate the total number of molecules in the atmosphere. (*Hint:* Use the result from Problem 1.53 and 29.0 g mol^{-1} as the molar mass of air.) **(b)** Assuming the volume of every breath (inhaled or exhaled) is 500 mL, calculate the number of molecules exhaled in each breath at 37°C, which is the body temperature. **(c)** If Mozart's life span was exactly 35 years, how many molecules did he exhale in that period (given that an average person breathes 12 times per minute)? **(d)** Calculate the fraction of molecules in the atmosphere that were exhaled by Mozart. How many of Mozart's molecules do we inhale with each breath of air? Round your answer to one significant digit. **(e)** List three important assumptions in these calculations.

1.55 A stockroom supervisor measured the contents of a partially filled 25.0-gallon acetone drum on a day when the temperature was 18.0°C and the atmospheric pressure was 750 mmHg, and found that 15.4 gallons of the solvent remained. After tightly sealing the drum, an assistant dropped the drum while carrying it upstairs to the organic laboratory. The drum was dented and its internal volume was decreased to 20.4 gallons. What is the total pressure inside the drum after the accident? The vapor pressure of acetone at 18.0°C is 400 mmHg. (*Hint:* At the time the drum was sealed, the pressure inside the drum, which is equal to the sum of the pressures of air and acetone, was equal to the atmospheric pressure.)

1.56 A relation known as the barometric formula is useful for estimating the change in atmospheric pressure with altitude. **(a)** Starting with the knowledge that atmospheric pressure decreases with altitude, we have $dP = -\rho g dh$, where ρ is the density of air, g is the acceleration due to gravity (9.81 m s^{-2}), and P and h are the pressure and height, respectively. Assuming ideal-gas behavior and constant temperature, show that the pressure P at height h is related to the pressure at sea level P_0 ($h = 0$) by $P = P_0 e^{-g\mathcal{M}h/RT}$. (*Hint:* For an ideal gas, $\rho = P\mathcal{M}/RT$, where $\mathcal{M}$ is the molar mass.) **(b)** Calculate the atmospheric pressure at a height of 5.0 km, assuming the temperature is constant at 5.0°C, given that the average molar mass of air is 29.0 g mol^{-1}.

1.57 In terms of the hard-sphere gas model, molecules are assumed to possess finite volume, but there is no interaction among the molecules. **(a)** Compare the P–V isotherm for an ideal gas and that for a hard-sphere gas. **(b)** Let b be the effective volume of the gas. Write an equation of state for this gas. **(c)** From this equation, derive an expression for $Z = P\overline{V}/RT$ for the hard-sphere gas and make a plot of Z versus P for two values of T (T_1 and T_2, $T_2 > T_1$). Be sure to indicate the value of the intercepts on the Z axis. **(d)** Plot Z versus T for fixed P for an ideal gas and for the hard-sphere gas.

1.58 One way to gain a physical understanding of b in the van der Waals equation is to calculate the "excluded volume." Assume that the distance of closest approach between two similar spherical molecules is the sum of their radii ($2r$). **(a)** Calculate the volume around each molecule into which the center of another molecule cannot penetrate. **(b)** From your result in **(a)**, calculate the excluded volume for one mole of molecules, which is the constant b. How does this compare with the sum of the volumes of 1 mole of the same molecules?

1.59 You may have witnessed a demonstration in which a burning candle standing in water is covered by an upturned glass. The candle flame goes out and the water rises in the glass. The explanation usually given for this phenomenon is that the oxygen in the glass is consumed by combustion, leading to a decrease in volume and hence the rise in the water level. However, the loss of oxygen is only a minor consideration. **(a)** Using $C_{12}H_{26}$ as the formula for paraffin wax, write a balanced equation for the combustion. Based on the nature of the products, show that the predicted rise in water level due to the removal of oxygen is far less than the observed change. **(b)** Devise a chemical process that would allow you to measure the volume of oxygen in the trapped air. (*Hint:* Use steel wool.) **(c)** What is the main reason for the water rising in the glass after the flame is extinguished?

1.60 Express the van der Waals equation in the form of Equation 1.10. Derive relationships between the van der Waals constants (a and b) and the virial coefficients (B, C, and D) given that

$$\frac{1}{1-x} = 1 + x + x^2 + x^3 + \cdots \quad |x| < 1$$

1.61 The Boyle temperature is the temperature at which the virial coefficient B is zero. Therefore, a real gas behaves like an ideal gas at this temperature. **(a)** Give a physical interpretation of this behavior. **(b)** Using your result for B for the van der Waals equation in Problem 1.60, calculate the Boyle temperature for argon, given that $a = 1.345$ atm L^2 mol^{-2} and $b = 3.22 \times 10^{-2}$ L mol^{-1}.

1.62 From Equations 1.10 and 1.11, show that $B' = B/RT$ and $C' = (C - B^2)/(RT)^2$. (*Hint:* From Equation 1.10, first obtain expressions for P and P^2. Next, substitute these expressions into Equation 1.11.)

1.63 Estimate the distance (in Å) between molecules of water vapor at 100°C and 1.0 atm. Assume ideal-gas behavior. Repeat the calculation for liquid water at 100°C, given that the density of water at 100°C is 0.96 g cm^{-3}. Comment on your results. (The diameter of a H_2O molecule is approximately 3 Å. 1 Å = 10^{-8} cm.)

1.64 A mixture of methane (CH_4) and ethane (C_2H_6) is stored in a container at 294 torr. The gases are burned in air to form CO_2 and H_2O. If the pressure of CO_2 is 356 torr measured at the same temperature and volume as the original mixture, then calculate the mole fractions of the gases.

1.65 A 5.00-mole sample of NH_3 gas is kept in a 1.92-L container at 300 K. If the van der Waals equation is assumed to give the correct answer for the pressure of the gas, calculate the percent error made in using the ideal-gas equation to calculate the pressure.

1.66 A gaseous hydrocarbon in a container of volume 20.2 L at 350 K and 6.63 atm reacts with an excess of oxygen to form 205.1 g of CO_2 and 168.0 g of H_2O. What is the molecular formula of the hydrocarbon?

1.67 Which of the following has a greater mass: a sample of air of volume V at temperature T and pressure P or a sample of air plus water vapor having the same volume, temperature, and pressure?

1.68 **(a)** Show that the pressure P (in pascals) exerted by a fluid is given by $P = hdg$, where h is the column of the fluid in meters, d is the density in kg m^{-3}, and g is the acceleration due to gravity (9.81 m s^{-2}). **(b)** The volume of an air bubble that starts at the bottom of a lake at 5.24°C increases by a factor of 6 as it rises to the surface of water where the temperature is 18.73°C and the air pressure is 0.973 atm. The density of the lake water is 1.02 g cm^{-3}. Use the equation in **(a)** to determine the depth of the lake in meters.

1.69 Use the van der Waals constants in Table 1.3 to estimate the radius of argon in picometers.

1.70 A closed 7.8-L flask contains 1.0 g of water. At what temperature will half of the water be in the vapor phase? (*Hint*: Look up the vapor pressure of water in the inside back matter.)

1.71 The ideal gas equation (Equation 1.1) can be written in microscopic form as $PV = Nk_BT$ where N is the number of molecules and k_B is the Boltzmann constant, given by 1.381×10^{-23} J K^{-1}. Estimate the average distance (in nanometers) between air molecules at 273 K and 1.0 atm. [*Hint*: Assume that each air molecule is at the center of a sphere (with volume V/N) of diameter d.]

CHAPTER 2

Kinetic Theory of Gases

It is odd to think that there is a word for something, which, strictly speaking, does not exist, namely, "rest."

Max Born*

The study of gas laws exemplifies the phenomenological, macroscopic approach to physical chemistry. Equations describing the gas laws are relatively simple, and experimental data are readily accessible. Yet studying gas laws gives us no real physical insight into processes that occur at the molecular level. Although the van der Waals equation attempts to account for nonideal behavior in terms of intermolecular interactions, it does so in a rather vague manner. It does not answer such questions as: How is the pressure of a gas related to the motion of individual molecules, and why do gases expand when heated at constant pressure? The next logical step, then, is to try to explain the behavior of gases in terms of the dynamics of molecular motion. To interpret the properties of gas molecules in a more quantitative manner, we turn to the kinetic theory of gases.

2.1 The Model

Any time we try to develop a theory to account for experimental observations, we must first define our system. If we do not understand all the properties of a system, as is usually the case, we must make a number of assumptions. Our *model for the kinetic theory of gases* is based on the following assumptions:

1. A gas is made up of a great number of atoms or molecules, separated by distances that are large compared to their size.
2. The molecules have mass, but their volume is negligibly small.
3. The molecules are constantly in random motion.
4. Collisions among molecules and between molecules and the walls of the container are *elastic*; that is, kinetic energy may be transferred from one molecule to another, but it is not converted to other forms of energy.
5. There is no interaction, attractive or repulsive, between the molecules.

Assumptions 2 and 5 should be familiar from our discussion of ideal gases in Chapter 1. The difference between the ideal-gas laws and the kinetic theory of gases is that for the latter, we shall use the foregoing assumptions in an explicit manner to

* Born, M. *The Restless Universe*, 2nd ed., Dover., New York, 1951. Used by permission.

derive expressions for macroscopic properties, such as pressure and temperature, in terms of the motion of individual molecules.

2.2 Pressure of a Gas

Using the model for the kinetic theory of gases, we can derive an expression for the pressure of a gas in terms of its molecular properties. Consider an ideal gas made up of N molecules, each of mass m, confined in a cubic box of length l. At any instant, the molecular motion inside the container is completely random. Let us analyze the motion of a particular molecule with velocity v. Because velocity is a *vector* quantity—it has both magnitude and direction—v can be resolved into three mutually perpendicular components v_x, v_y, and v_z. These three components give the rates at which the molecule is moving along the x, y, and z directions, respectively; v is simply the resultant velocity (Figure 2.1).

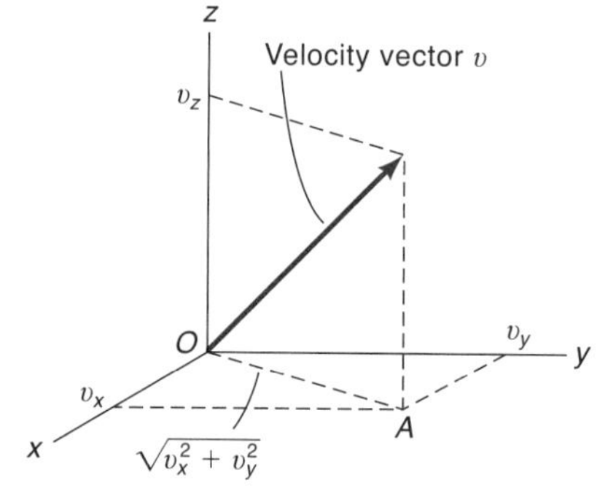

Figure 2.1
Velocity vector v and its components along the x, y, and z directions.

The projection of the velocity vector on the xy plane is $\overline{\text{OA}}$, which, according to Pythagoras' theorem, is given by

$$\overline{\text{OA}}^2 = v_x^2 + v_y^2$$

Similarly,

$$\begin{aligned} v^2 &= \overline{\text{OA}}^2 + v_z^2 \\ &= v_x^2 + v_y^2 + v_z^2 \end{aligned} \tag{2.1}$$

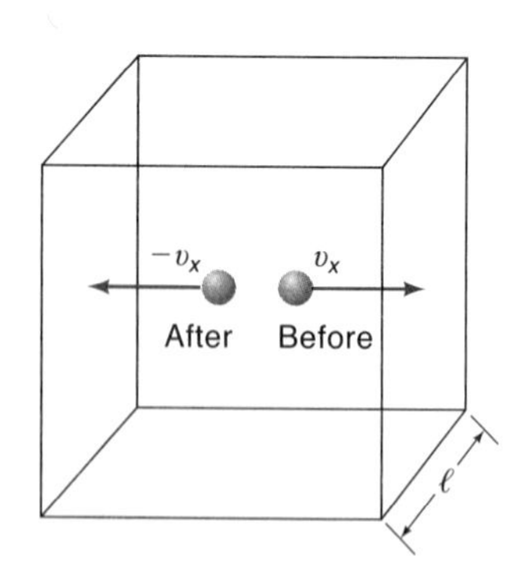

Figure 2.2
Change in velocity upon collision of a molecule moving with v_x with the wall of the container.

Let us for the moment consider the motion of a molecule only along the x direction. Figure 2.2 shows the changes that take place when the molecule collides with the wall of the container (the yz plane) with velocity component v_x. Because the collision is elastic, the velocity after collision is the same as before but opposite in direction. The momentum of the molecule is mv_x, where m is its mass, so that the *change* in momentum is given by

$$mv_x - m(-v_x) = 2mv_x$$

The sign of v_x is positive when the molecule moves from left to right and negative when it moves in the opposite direction. Immediately after the collision, the molecule will take time l/v_x to collide with the other wall, and in time $2l/v_x$ the molecule will strike the same wall again.* Thus, the frequency of collision between the molecule and a given wall (i.e., the number of collisions per unit time) is $v_x/2l$, and the change in momentum per unit time is $(2mv_x)(v_x/2l)$, or mv_x^2/l. According to Newton's second law of motion,

$$\begin{aligned} \text{force} &= \text{mass} \times \text{acceleration} \\ &= \text{mass} \times \text{distance} \times \text{time}^{-2} \\ &= \text{momentum time}^{-1} \end{aligned}$$

Therefore, the force (F) exerted by one molecule on one wall as a result of the collision is mv_x^2/l, and the total force due to N molecules is Nmv_x^2/l. Because pressure (P) is force/area and area (A) is l^2, we can now express the total pressure exerted on one wall as

$$\begin{aligned} P &= \frac{F}{A} \\ &= \frac{Nmv_x^2}{l(l^2)} = \frac{Nmv_x^2}{V} \end{aligned}$$

or

$$PV = Nmv_x^2 \tag{2.2}$$

where V is the volume of the cube (equal to l^3). When we are dealing with a large collection of molecules (e.g., when N is on the order of 6×10^{23}), there is a tremendous spread of molecular velocities. It is more appropriate, therefore, to replace v_x^2 in Equation 2.2 with the mean or average quantity, $\overline{v_x^2}$. Referring to Equation 2.1, we see that the relation between the averages of the squares of the velocity components and the average of the square of the velocity, $\overline{v^2}$, is still

$$\overline{v^2} = \overline{v_x^2} + \overline{v_y^2} + \overline{v_z^2}$$

The quantity $\overline{v^2}$ is called the *mean-square velocity*, defined as

$$\overline{v^2} = \frac{v_1^2 + v_2^2 + \cdots + v_N^2}{N} \tag{2.3}$$

When N is a large number, it is correct to assume that molecular motions along the x, y, and z directions are equally probable. This means that

$$\overline{v_x^2} = \overline{v_y^2} = \overline{v_z^2} = \frac{\overline{v^2}}{3}$$

and Equation 2.2 can now be written as

$$P = \frac{Nm\overline{v^2}}{3V}$$

* We assume that the molecule does not collide with other molecules along the way. A more rigorous treatment including molecular collisions gives the same result.

Translational motion is the movement of an object from one place to another.

Multiplying the top and bottom by 2 and recalling that the average kinetic energy of the molecule $\overline{E}_{trans}$ is given by $\frac{1}{2}m\overline{v^2}$ (where the subscript *trans* denotes translational motion), we obtain

$$P = \frac{2N}{3V}\left(\frac{1}{2}m\overline{v^2}\right) = \frac{2N}{3V}\overline{E}_{trans} \tag{2.4}$$

This is the pressure exerted by N molecules on one wall. The same result can be obtained regardless of the direction (x, y, or z) we describe for the molecular motion. We see that the pressure is directly proportional to the average kinetic energy or, more explicitly, to the mean-square velocity of the molecule. The physical meaning of this dependence is that the larger the velocity, the more frequent the collisions and the greater the change in momentum. Thus, these two independent terms give us the quantity $\overline{v^2}$ in the kinetic theory expression for the pressure.

2.3 Kinetic Energy and Temperature

Let us compare Equation 2.4 with the ideal-gas equation (Equation 1.1)

$$\begin{aligned} PV &= nRT \\ &= \frac{N}{N_A}RT \end{aligned}$$

or

$$P = \frac{NRT}{N_A V} \tag{2.5}$$

where N_A is the Avogadro constant. Combining the pressures in Equations 2.4 and 2.5, we get

$$\frac{2}{3}\frac{N}{V}\overline{E}_{trans} = \frac{N}{N_A}\frac{RT}{V}$$

or

$$\overline{E}_{trans} = \frac{3}{2}\frac{RT}{N_A} = \frac{3}{2}k_BT \tag{2.6}$$

where $R = k_B N_A$ and k_B is the Boltzmann constant, equal to $1.3806488 \times 10^{-23}$ J K^{-1} [after the Austrian physicist Ludwig Eduard Boltzmann (1844–1906)]. (In most calculations, we shall round k_B to 1.381×10^{-23} J K^{-1}.) We see that the mean kinetic energy of one molecule is proportional to absolute temperature.

The significance of Equation 2.6 is that it provides an explanation for the temperature of a gas in terms of molecular motion. For this reason, random molecular motion is sometimes referred to as *thermal motion*. It is important to keep in mind that the kinetic theory is a *statistical* treatment of our model; hence, it is meaningless to associate temperature with the kinetic energy of just a few molecules. Equation 2.6 also tells us that whenever two ideal gases are at the same temperature T, they must have the *same* average kinetic energy. The reason is that $\overline{E}_{\text{trans}}$ in Equation 2.6 is independent of molecular properties such as size or molar mass or amount of the gas present, as long as N is a large number.

It is easy to see that $\overline{v^2}$ would be a very difficult quantity to measure, if indeed it could be measured at all. To do so, we would need to measure each individual velocity, square it, and then take the average (see Equation 2.3). Fortunately, $\overline{v^2}$ can be obtained quite directly from other quantities. From Equation 2.6, we write

$$\frac{1}{2}m\overline{v^2} = \frac{3}{2}\frac{RT}{N_A} = \frac{3}{2}k_B T$$

so that

$$\overline{v^2} = \frac{3RT}{mN_A} = \frac{3k_B T}{m}$$

or

$$\sqrt{\overline{v^2}} = v_{\text{rms}} = \sqrt{\frac{3RT}{\mathcal{M}}} = \sqrt{\frac{3k_B T}{m}} \quad (\mathcal{M} = mN_A) \tag{2.7}$$

Remember that k_B is associated with the mass of a molecule and R is associated with the molar mass.

where v_{rms} is the *root-mean-square velocity** and m is the mass (in kg) of one molecule; $\mathcal{M}$ is the molar mass (in kg mol^{-1}). Note that v_{rms} is directly proportional to the square root of temperature and inversely proportional to the square root of molar mass of the molecule. Therefore, the heavier the molecule, the slower its motion.

2.4 The Maxwell Distribution Laws

The root-mean-square velocity gives us an average measure that is very useful in the study of a large number of molecules. When we are studying, say, one mole of a gas, it is impossible to know the velocity of each individual molecule for two reasons. First, the number of molecules is so huge that there is no way we can follow all their motions. Second, although molecular motion is a well-defined quantity, we cannot measure its velocity exactly. Therefore, rather than concerning ourselves with individual molecular velocities, we ask this question: For a given system at some known temperature, how many molecules are moving at velocities between v and $v + \Delta v$ at any moment? Or, how many molecules in a macroscopic gas sample have velocities, say, between 306.5 m s^{-1} and 306.6 m s^{-1} at any moment?

Because the total number of molecules is very large, there is a continuous spread, or *distribution*, of velocities as a result of collisions. We can therefore make the velocity range Δv smaller and smaller, and in the limit it becomes dv. This fact has great significance, because it enables us to replace the summation sign with the integral sign in calculating the number of molecules whose velocities fall between v and v +

* Because velocity is a vector quantity, the average molecular velocity, $\overline{v}$, must be zero; there are just as many molecules moving in the positive direction as there are in the negative direction. On the other hand, v_{rms} is a scalar quantity; that is, it has magnitude but no direction.

dv. Mathematically speaking, it is easier to integrate than to sum a large series. This distribution-of-velocities approach was first employed by the Scottish physicist James Clerk Maxwell (1831–1879) in 1860 and later refined by Boltzmann. They showed that for a system containing N ideal gas molecules at thermal equilibrium with its surroundings, the fraction of molecules dN/N moving at velocities between v_x and $v_x + dv_x$ along the x direction is given by

$$\begin{aligned}\frac{dN}{N} &= \left(\frac{m}{2\pi k_B T}\right)^{1/2} e^{-mv_x^2/2k_BT}\, dv_x \\ &= f(v_x)\, dv_x \end{aligned} \tag{2.8}$$

where m is the mass of the molecule, k_B the Boltzmann constant, and T the absolute temperature. The quantity $f(v_x)$, given by

$$f(v_x) = \left(\frac{m}{2\pi k_B T}\right)^{1/2} e^{-mv_x^2/2k_BT} \tag{2.9}$$

Velocity varies between $+\infty$ and $-\infty$.

is the *Maxwell velocity distribution function* in one direction (along the x axis). Figure 2.3 shows a plot of $f(v_x)$ versus v_x for nitrogen gas at three different temperatures. The fact that the maximum of $f(v_x)$ is at $v_x = 0$ does not necessarily mean that these molecules are standing still. Rather, it merely indicates that they are moving perpendicular to the x axis so that the projections of their velocities onto this axis are zero.

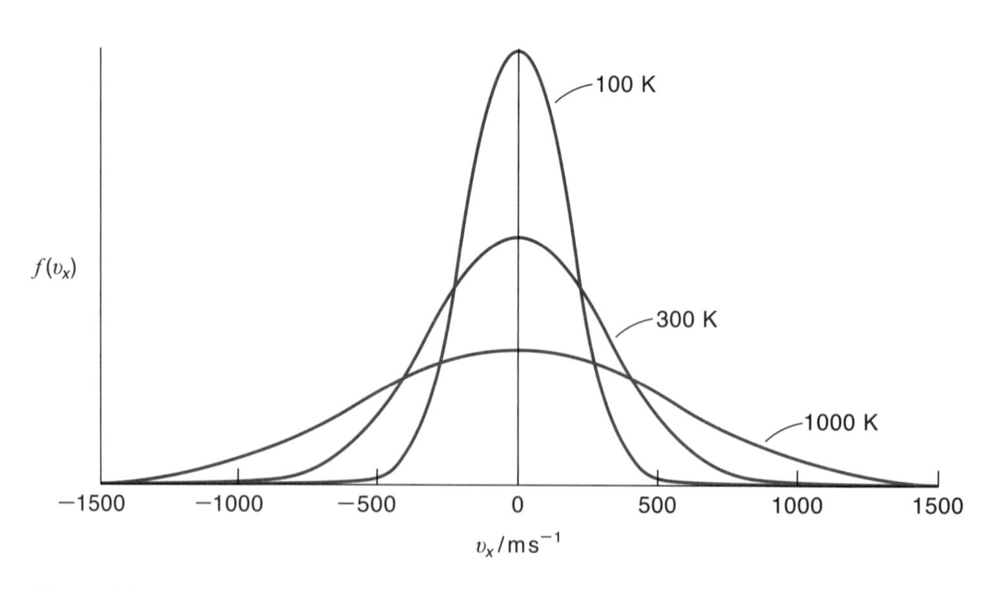

Figure 2.3
Velocity distribution along the x axis for the nitrogen molecule at three different temperatures. Each plot shows the fraction of molecules moving between v_x and $v_x + dv_x$ (and between $-v_x$ and $-v_x - dv_x$). The curves are symmetrically centered at $v_x = 0$.

As we mentioned earlier, velocity is a vector quantity. In many cases, we need to deal only with the speed of molecules (c), which is a scalar quantity; that is, it has magnitude but no directional properties. The fraction of molecules dN/N moving between speeds c and $c + dc$ is given by

$$\frac{dN}{N} = 4\pi c^2\left(\frac{m}{2\pi k_B T}\right)^{3/2} e^{-mc^2/2k_B T}\, dc$$
$$= f(c)dc \qquad (2.10)$$

where $f(c)$, the *Maxwell speed distribution function*, is given by

$$f(c) = 4\pi c^2\left(\frac{m}{2\pi k_B T}\right)^{3/2} e^{-mc^2/2k_B T} \qquad (2.11)$$

Speed varies between 0 and ∞.

Figure 2.4 shows the dependence of the speed distribution curve on temperature and molar mass. At any given temperature the general shape of the curve can be explained as follows. Initially, at small c values, the c^2 term in Equation 2.11 dominates, so $f(c)$ increases with increasing c. At larger values of c, the term $e^{-mc^2/2k_B T}$ becomes more important. These two opposing terms cause the curve to reach a maximum beyond which it decreases roughly exponentially with increasing c. The speed corresponding to the maximum value $f(c)$ is called the *most probable speed*, c_{mp}, because it is the speed of the largest number of molecules.

Figure 2.4a shows how the distribution curve is influenced by temperature. At low temperatures the distribution has a rather narrow range. As the temperature increases, the curve becomes flatter, meaning that there are now more fast-moving molecules. This temperature dependence of the distribution curve has important implications in chemical reaction rates. As we shall see in Chapter 15, in order to react, a molecule must possess a minimum amount of energy, called *activation energy*. At low temperatures the number of fast-moving molecules is small; hence, most reactions proceed at a slow rate. Raising the temperature increases the number of energetic molecules and causes an increase in the reaction rate. In Figure 2.4b we see that heavier gases have a narrower range of speed distribution than lighter gases at the same temperature. This is to be expected considering that heavier gases move slower, on the average, than the lighter gases.

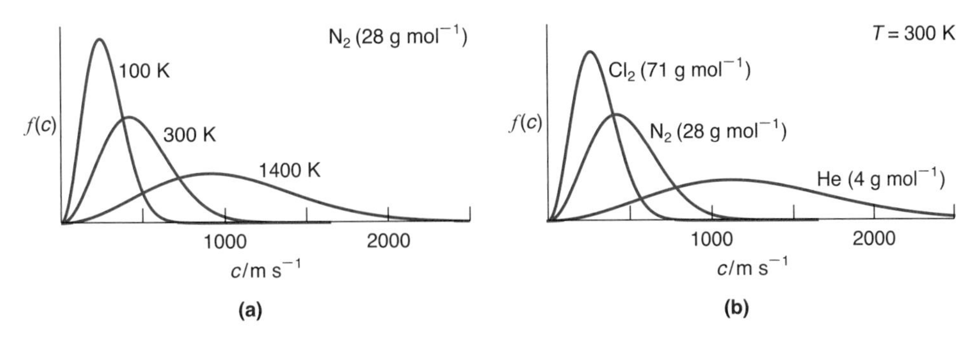

Figure 2.4
(a) Speed distribution for nitrogen gas as a function of temperature. At the higher temperatures, more molecules are moving at faster speeds. (b) Speed distribution of gases having different molar masses. At a given temperature, the lighter molecules are moving faster, on the average.

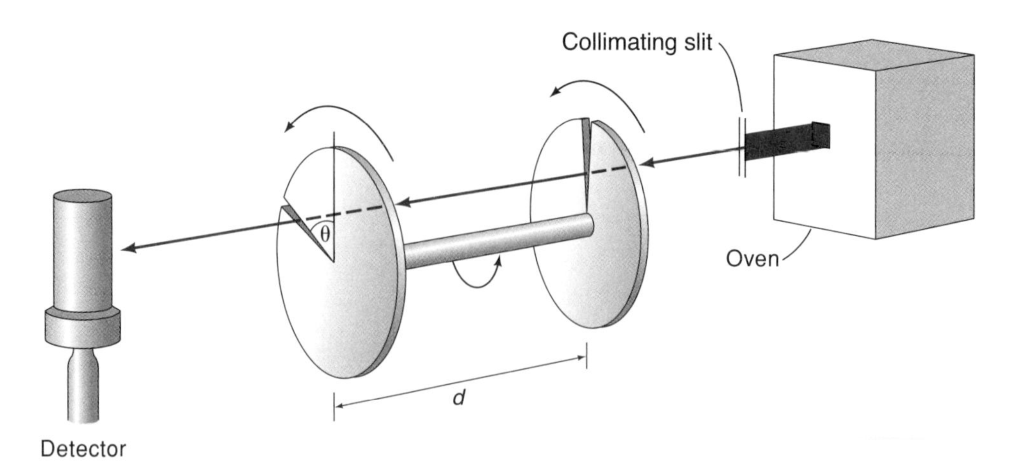

Figure 2.5
Atoms or molecules exiting from an oven at a known temperature are collimated into a well-defined beam. The beam encounters two rotating discs with wedge-shaped openings mounted on the same shaft. The two wedges are displaced from each other by an angle θ. Only a fraction of the molecules that pass through the first opening will pass through the second opening to reach the detector, which registers the number of molecules. Changing the rotational speed and d will allow molecules of different speeds to reach the detector. In this manner we can obtain a plot of the number of molecules versus their speeds, a plot known as the Maxwell speed distribution.

The Maxwell distribution has been verified experimentally by measurement of the number of molecules as a function of their speed. One version of the apparatus devised for this purpose consists of two discs with wedge-shaped openings mounted on a common axle (Figure 2.5). Molecules (or atoms) emitted from an oven at a certain temperature are collimated into a fine beam by the slits. With the axle turning, a molecule passing through the gap in the first disc will pass through the gap in the second disc only if the time required to travel between the discs is equal to an integral multiple of the time necessary for the discs to rotate from one gap to the next. A detector registers the number of molecules that pass through the second disc. Knowledge of the rotational speed enables the researcher to determine the speed of the molecules that pass through. A plot of the number of molecules versus speed gives a curve similar to the one shown in Figure 2.4.

Because kinetic energy is related to speed by $E = \frac{1}{2}mc^2$, it is not surprising that there is also an energy distribution function. A corresponding expression for the kinetic energy distribution is

$$\begin{aligned}\frac{dN}{N} &= 2\pi E^{1/2}\left(\frac{1}{\pi k_B T}\right)^{3/2} e^{-E/k_B T}\, dE \\ &= f(E)dE \end{aligned} \tag{2.12}$$

Kinetic energy varies between 0 and ∞.

where dN/N is the fraction of molecules that have kinetic energies between E and $E + dE$, and $f(E)$ is the energy distribution function. Figure 2.6 shows the energy distribution curves at 300 K and 1000 K. Keep in mind that these curves apply to *any* ideal gas. As mentioned earlier, two different gases at the same temperature have the same average kinetic energy and hence the same energy distribution curve.

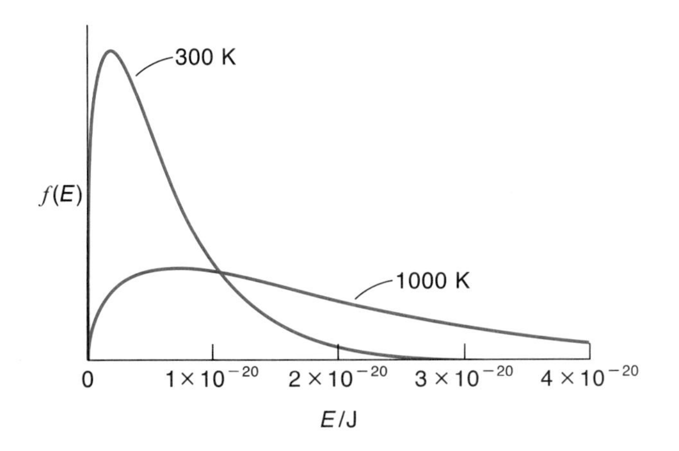

Figure 2.6
Energy distribution curves for an ideal gas at 300 K and 1000 K. Each plot shows the fraction of molecules having energies between E and $E + dE$.

The usefulness of the Maxwell speed distribution function is that it enables us to calculate average quantities. Consider the mean speed, $\overline{c}$, of a large collection of molecules at some temperature, T. The mean speed is calculated by multiplying each speed by the fraction of molecules that have that speed and then adding all the products together. Because the fraction of molecules with a speed in the range c to $c + dc$ is $f(c)dc$, the product of this fraction and the speed is $cf(c)dc$. Thus, the mean speed, $\overline{c}$, is obtained by evaluating the integral between the limits $c = 0$ and $c = \infty$:

$$\overline{c} = \int_0^\infty cf(c)dc \tag{2.13}$$

From Equation 2.11, we obtain

$$\overline{c} = 4\pi\left(\frac{m}{2\pi k_B T}\right)^{3/2}\int_0^\infty c^3 e^{-mc^2/2k_B T}\,dc$$

Looking up the standard integral in a handbook of chemistry and physics,* we find that

$$\int_0^\infty x^3 e^{-ax^2}dx = \frac{1}{2a^2}$$

where

$$a = \frac{m}{2k_B T}$$

Thus,

$$\overline{c} = 4\pi\left(\frac{m}{2\pi k_B T}\right)^{3/2} \times \frac{1}{2}\left(\frac{2k_B T}{m}\right)^2$$

$$= \sqrt{\frac{8k_B T}{\pi m}} = \sqrt{\frac{8RT}{\pi \mathcal{M}}} \tag{2.14}$$

* For example, see the *Handbook of Chemistry and Physics*, Haynes, W. M., Ed., 93rd ed., CRC Press, Boca Raton, FL, 2012.

The v_{rms} discussed earlier is the same as the root-mean-square speed, c_{rms},* which can also be calculated by first evaluating the integral:

$$\int_0^\infty c^2 f(c)\,dc$$

and then taking the square root of the result (see Problem 2.18).

To calculate the value of the most probable speed, c_{mp}, we differentiate the distribution function in Equation 2.11 with respect to c and set the result equal to zero. We then solve for c, which is the maximum value, that is, c_{mp} (see Problem 2.19). This procedure yields

$$c_{mp} = \sqrt{\frac{2k_B T}{m}} = \sqrt{\frac{2RT}{\mathcal{M}}} \tag{2.15}$$

---

* Because the square of the average velocity is a scalar quantity, it follows that $\overline{v^2} = \overline{c^2}$; hence, $v_{rms} = c_{rms}$.

EXAMPLE 2.1

Calculate the values of c_{mp}, $\overline{c}$, and c_{rms} for O_2 at 300 K.

ANSWER

The constants are:

$$R = 8.314 \text{ J K}^{-1} \text{ mol}^{-1}$$

$$\mathcal{M} = 0.03200 \text{ kg mol}^{-1}$$

The most probable speed is given by Equation 2.15

$$\begin{aligned} c_{mp} &= \sqrt{\frac{2 \times 8.314 \text{ J K}^{-1} \text{ mol}^{-1} \times 300 \text{ K}}{0.03200 \text{ kg mol}^{-1}}} \\ &= \sqrt{1.56 \times 10^5 \text{ J kg}^{-1}} \\ &= \sqrt{1.56 \times 10^5 \text{ m}^2 \text{ s}^{-2}} \\ &= 395 \text{ m s}^{-1} \end{aligned}$$

Similarly, we can show that

$$\overline{c} = \sqrt{\frac{8RT}{\pi\mathcal{M}}} = 446 \text{ m s}^{-1}$$

and

$$c_{\text{rms}} = \sqrt{\frac{3RT}{\mathcal{M}}} = 484 \text{ m s}^{-1}$$

COMMENT

The calculation shows, and indeed it is generally true, that $c_{\text{rms}} > \bar{c} > c_{\text{mp}}$. That c_{mp} is the smallest of the three speeds is due to the asymmetry of the curve (see Figure 2.4). The reason c_{rms} is greater than $\bar{c}$ is that the squaring process in Equation 2.3 is weighted toward larger values of c.

Finally, note that both the $\bar{c}$ and c_{rms} values of N_2 and O_2 are close to the speed of sound in air. Sound waves are pressure waves. The propagation of these waves is directly related to the movement of molecules and hence to their speeds.

2.5 Molecular Collisions and the Mean Free Path

Now that we have an explicit expression for the average speed, $\bar{c}$, we can use it to study some dynamic processes involving gases. We know that the speed of a molecule is not constant but changes frequently as a result of collisions. Therefore, the question we ask is: How often do molecules collide with one another? The collision frequency depends on the density of the gas and the molecular speed, and therefore on the temperature of the system. In the kinetic theory model, we assume each molecule to be a hard sphere of diameter d. A molecular collision is one in which the separation between the two spheres (measured from each center) is d.

Let us consider the motion of a particular molecule. A simple approach is to assume that at a given instant, all molecules except this one are standing still. In time t, this molecule moves a distance $\bar{c}t$ (where $\bar{c}$ is the average speed) and sweeps out a collision tube that has a cross-sectional area πd^2 (Figure 2.7). The volume of the cylinder is $(\pi d^2)(\bar{c}t)$. Any molecule whose center lies within this cylinder will collide with the

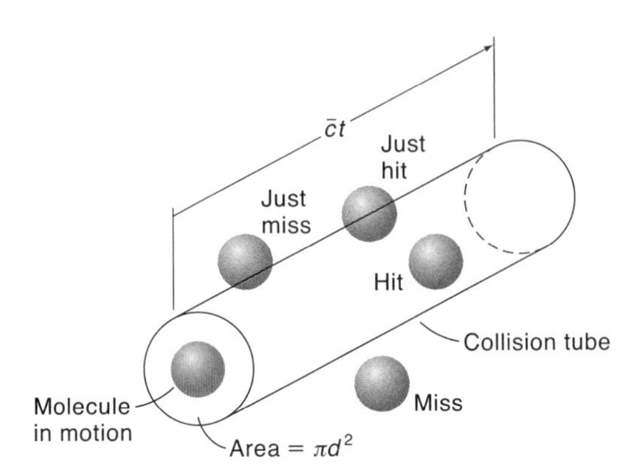

Figure 2.7
The collision cross section and the collision tube. Any molecule whose center lies within or touches the tube will collide with the moving molecule (red sphere).

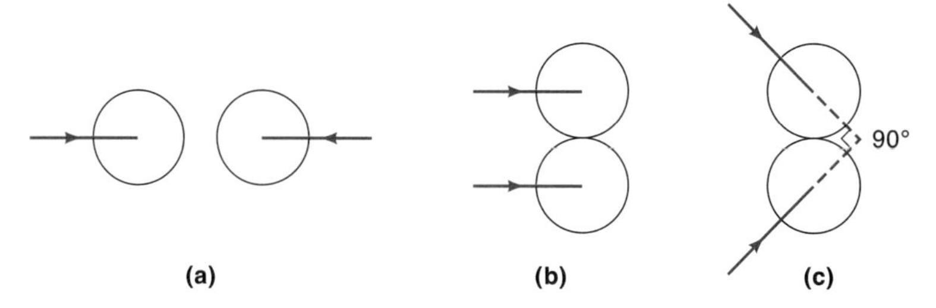

Figure 2.8
Three different approaches for two colliding molecules. The situations shown in (a) and (b) represent the two extreme cases, while that shown in (c) may be taken as the "average" case for molecular encounter.

moving molecule. If there are altogether N molecules in volume V, then the number density of the gas is N/V, the number of collisions in time t is $\pi d^2 \overline{c} t(N/V)$, and the number of collisions per unit time, or the *collision frequency*, Z_1, is $\pi d^2 \overline{c}(N/V)$. The expression for the collision frequency needs a correction. If we assume that the rest of the molecules are not frozen in position, we should replace $\overline{c}$ with the average *relative* speed. Figure 2.8 shows three different collisions for two molecules. The relative speed for the case shown in Figure 2.8c is $\sqrt{2}\overline{c}$, so that

$$Z_1 = \sqrt{2}\pi d^2 \overline{c}\left(\frac{N}{V}\right) \quad \text{collisions s}^{-1} \tag{2.16}$$

This is the number of collisions a *single* molecule makes in one second. Because there are N molecules in volume V and each makes Z_1 collisions per second, the total number of binary collisions, or collisions between two molecules, per unit volume per unit time, Z_{11}, is given by

$$\begin{aligned} Z_{11} &= \frac{1}{2} Z_1\left(\frac{N}{V}\right) \\ &= \frac{\sqrt{2}}{2}\pi d^2 \overline{c}\left(\frac{N}{V}\right)^2 \quad \text{collisions m}^{-3}\text{ s}^{-1} \end{aligned} \tag{2.17}$$

The factor $1/2$ is introduced in Equation 2.17 to ensure that we are counting each collision between two molecules only once. The probability of three or more molecules colliding at once is very small except at high pressures. Because the rate of a chemical reaction generally depends on how often reacting molecules come in contact with one another, Equation 2.17 is important in gas-phase chemical kinetics. We shall return to this equation in Chapter 15.

A quantity closely related to the collision frequency is the average distance traveled by a molecule between successive collisions. This distance, called the *mean free path*, λ (Figure 2.9), is defined as

Figure 2.9
The distances traveled by a molecule between successive collisions. The average of these distances is called the mean free path.

$$\lambda = (\text{average speed}) \times (\text{average time between collisions})$$

Because the average time between collisions is the reciprocal of the collision frequency, we have

$$\lambda = \frac{\overline{c}}{Z_1} = \frac{\overline{c}}{\sqrt{2}\pi d^2 \overline{c}(N/V)} = \frac{1}{\sqrt{2}\pi d^2 (N/V)} \tag{2.18}$$

Notice that the mean free path is inversely proportional to the number density of the gas (N/V). This behavior is reasonable because in a dense gas, a molecule makes more collisions per unit time and hence travels a shorter distance between successive collisions. The mean free path can also be expressed in terms of the gas pressure. Assuming ideal behavior,

$$P = \frac{nRT}{V}$$

$$= \frac{(N/N_A)RT}{V}$$

$$\frac{N}{V} = \frac{PN_A}{RT}$$

Equation 2.18 can now be written as

$$\lambda = \frac{RT}{\sqrt{2}\pi d^2 P N_A} \tag{2.19}$$

It might appear that λ is directly proportional to T and inversely proportional to P, but it is not. At constant volume and amount of gas, the effects of T and P cancel each other, and so λ depends only on the density of the gas.

EXAMPLE 2.2

The concentration of dry air at 1.00 atm and 298 K is about 2.5×10^{19} molecules cm^{-3}. Assuming that air contains only nitrogen molecules, calculate the collision frequency, the binary collision number, and the mean free path of nitrogen molecules under these conditions. The collision diameter of nitrogen is 3.75 Å. (1 Å = 10^{-8} cm.)

ANSWER

Our first step is to calculate the average speed of nitrogen. From Equation 2.14, we find $\overline{c} = 4.8 \times 10^2$ m s^{-1}. The collision frequency is given by

$$Z_1 = \sqrt{2}\pi(3.75 \times 10^{-8}\ \text{cm})^2(4.8 \times 10^4\ \text{cm s}^{-1})(2.5 \times 10^{19}\ \text{molecules cm}^{-3})$$

$$= 7.5 \times 10^9\ \text{collisions s}^{-1}$$

Note that we have replaced the unit "molecules" with "collisions" because, in the derivation of Z_1, every molecule in the collision volume represents a collision. The binary collision number is

$$Z_{11} = \frac{Z_1}{2}\left(\frac{N}{V}\right)$$

$$= \frac{(7.5 \times 10^9 \text{ collisions s}^{-1})}{2} \times 2.5 \times 10^{19} \text{ molecules cm}^{-3}$$

$$= 9.4 \times 10^{28} \text{ collisions cm}^{-3}\text{ s}^{-1}$$

Again, we converted molecules to collisions in calculating the total number of binary collisions. Finally, the mean free path is given by

$$\lambda = \frac{\bar{c}}{Z_1} = \frac{4.8 \times 10^4 \text{ cm s}^{-1}}{7.5 \times 10^9 \text{ collisions s}^{-1}}$$

$$= 6.4 \times 10^{-6} \text{ cm collision}^{-1}$$

$$= 640 \text{ Å collision}^{-1}$$

COMMENT

It is usually sufficient to express mean free path in terms of distance alone rather than distance per collision. Thus, in this example, the mean free path of nitrogen is 640 Å, or 6.4×10^{-6} cm.

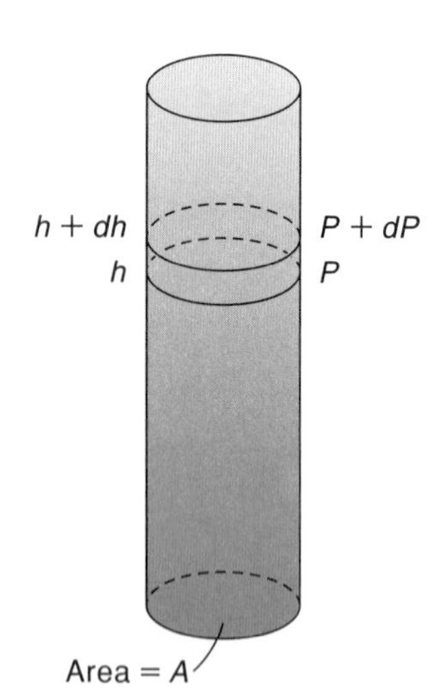

Figure 2.10
A column of air of area A extending from Earth's surface outward.

2.6 The Barometric Formula

In this section we shall consider the distribution of air molecules under the influence of Earth's gravitational field, and its effect on atmospheric pressure.

Unlike temperature, pressure varies with altitude in a fairly straightforward way. Gases in the atmosphere do not settle down on the surface under the influence of Earth's gravitational attraction because the translational kinetic motion of the molecules competes with the sedimentation forces. As a result, there is a density gradient of air molecules that decreases with increasing height. Consider a column of air of area A (Figure 2.10). The difference in pressure, dP, between height h and $h + dh$ is equal to the weight of a section of air of volume Adh. Because pressure = force/area, we write

$$dP = -\frac{\rho g A dh}{A} = -\rho g dh \tag{2.20}$$

where ρ is the density of air and g the gravitational constant (9.81 m s^{-2}) and the negative sign indicates that the pressure decreases with increasing height. Assuming ideal behavior, we can express P as

$$P = \frac{nRT}{V} = \frac{m}{V}\frac{RT}{\mathcal{M}} = \rho\frac{RT}{\mathcal{M}}$$

or

$$\rho = \frac{P\mathcal{M}}{RT} \tag{2.21}$$

where m is the mass of the molecules (assuming one type of species only) and $\mathcal{M}$ its molar mass. Substituting Equation 2.21 for ρ in Equation 2.20 yields

$$-\frac{dP}{P} = \frac{g\mathcal{M}}{RT}dh \tag{2.22}$$

If we assume the temperature to be constant, we can integrate Equation 2.22 between height zero (sea level, pressure = P_0) and a certain height h (pressure = P)

$$\int_{P_0}^{P}\frac{dP}{P} = -\frac{g\mathcal{M}}{RT}\int_0^h dh$$

$$\ln\frac{P}{P_0} = -\frac{g\mathcal{M}h}{RT}$$

or

$$P = P_0 e^{-g\mathcal{M}h/RT} \tag{2.23}$$

Equation 2.23, which is known as the *barometric formula*, can also be written as

$$P = P_0 e^{-gmh/k_BT} \tag{2.24}$$

where k_B is the Boltzmann constant. In our isothermal (constant-temperature) atmosphere model, the pressure decreases exponentially with height extending from the surface of Earth. The term k_BT/gm has the dimensions of length and represents the characteristic distance over which the pressure drops by a factor $1/e$ (i.e., when $h = k_BT/gm$). Either Equation 2.23 or 2.24 provides a rough estimate of pressure with altitude.

EXAMPLE 2.3

The partial pressure of oxygen is 0.20 atm at sea level at 25°C. Calculate its partial pressure at an altitude of 30 km (the stratosphere). Assume temperature to remain constant.

ANSWER

From Equation 2.23

$$P = 0.20\text{ atm} \times \exp\left[-\frac{(9.81\text{ m s}^{-2})(0.03200\text{ kg mol}^{-1})(30\times10^3\text{ m})}{(8.314\text{ J K}^{-1}\text{ mol}^{-1})(298\text{ K})}\right]$$

$$= 4.5\times10^{-3}\text{ atm}$$

COMMENT

Despite the fact that the temperature of the stratosphere is considerably below 25°C (about −23°C), the barometric formula provides a rough estimate of atmospheric pressure in that region.

2.7 Gas Viscosity

So far, we have mostly concentrated on the properties of gas molecules in the equilibrium state. Now let us consider how to measure the rate at which gases flow through a tube. At a given temperature and density, the rate will vary from gas to gas because each gas has a different viscosity, or resistance to flow.

A simple equation for gas viscosity can be derived from the kinetic theory as follows. The flow of molecules along a tube can be analyzed by dividing the gas into a set of laminas as shown in Figure 2.11. Laminas are layers with negligible thickness. The layer immediately adjacent to the surface of the tube's inner wall is stationary because of adhesion, but the velocity of the layers increases as we move away from the surface with the result that there is a velocity gradient along the z axis. Consider two layers separated by a distance λ, where λ is the mean free path. If v is the velocity of the slower-moving layer, then the velocity of the faster-moving layer is $v + \lambda(dv/dz)$, where (dv/dz) is the velocity gradient. In addition to the flow along the x axis, there is up-and-down motion along the z axis. When a molecule moves from a faster layer to a slower one, it transports additional momentum to the latter and tends to speed it up. The reverse is true when a molecule moves from a slower layer to a more quickly flowing one. The consequence is a drag or frictional force between these two layers that produces a viscosity effect. To maintain the velocity gradient, however, an external

A gradient is a measure of how a certain parameter changes with distance. Other examples are temperature gradient, concentration gradient, and electric field gradient.

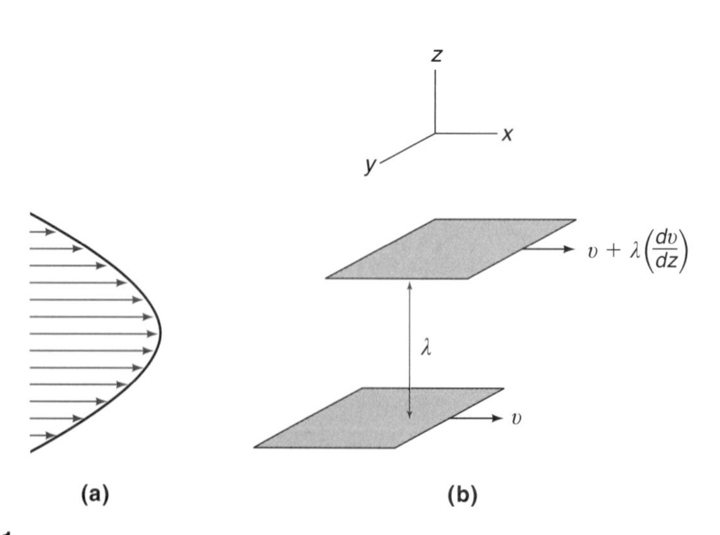

Figure 2.11
(a) Profile of the frontal motion of a gas in laminar flow along a tube. (b) Motion of two laminas separated by distance λ, the mean free path.

force, F, must be applied along x; this force is directly proportional to both the area of the layer, A, and the velocity gradient. Thus,

$$F \propto A\left(\frac{dv}{dz}\right)$$

$$= \eta A\left(\frac{dv}{dz}\right) \tag{2.25}$$

where η, the constant of proportionality, is known as the *coefficient of viscosity*, or simply the *viscosity*. Its units are newton second per meter squared (N s m^{-2}).

Next, we shall obtain an explicit expression for η. Consider a particular lamina at height h cm above the stationary plane (i.e., the wall of the tube). All molecules coming from a distance λ below height h will make their first collisions, and hence their first transfer of momentum, when they reach height h. If the velocity of the molecules moving in the plane is $h(dv/dz)$, then the velocity of molecules moving in a plane λ cm below this plane is $(h - \lambda)(dv/dz)$. The momentum transferred by each arriving molecule from the slower-moving plane to the faster one is $m(h - \lambda)(dv/dz)$. Similarly, for a molecule arriving from plane $(h + \lambda)$ cm above the wall, the momentum transferred is $m(h + \lambda)(dv/dz)$. Although a viscosity experiment measures the rate of flow of gas, within the gas sample molecular motion is random nevertheless. Therefore, we can assume, as an approximation, that one-third of the molecules are moving along each of the x, y and z directions. Consequently, at any instant along one particular axis, one-sixth of the molecules will be moving upward and one-sixth downward. The number of molecules going from one layer to the next per unit area per second is $(\frac{1}{6})(N/V)\bar{c}$, where $\bar{c}$ is the average speed, so that the rate of transfer of momentum by molecules moving upward per unit area is

$$\left(\frac{1}{6}\right)\left(\frac{N}{V}\right)\bar{c}m(h - \lambda)\left(\frac{dv}{dz}\right)$$

Similarly, for the downward movement, the rate of transfer of momentum per unit area is

$$\left(\frac{1}{6}\right)\left(\frac{N}{V}\right)\bar{c}m(h + \lambda)\left(\frac{dv}{dz}\right)$$

The difference between these two quantities, $(\frac{1}{3})(N/V)\bar{c}m\lambda(dv/dz)$, gives the *net* transfer of momentum per unit area per unit time, or F/A. Thus,

$$\frac{F}{A} = \frac{1}{3}\left(\frac{N}{V}\right)\bar{c}m\lambda\left(\frac{dv}{dz}\right)$$

From Equation 2.25

$$\frac{F}{A} = \eta\left(\frac{dv}{dz}\right)$$

Combining these two equations, we obtain

$$\eta = \frac{1}{3}\left(\frac{N}{V}\right)m\lambda\bar{c} \tag{2.26}$$

Substituting Equation 2.18 into Equation 2.26, we obtain

$$\eta = \frac{m\bar{c}}{3\sqrt{2}\pi d^2} \tag{2.27}$$

This is an interesting and rather unexpected result, for it tells us that viscosity is *independent* of density. As we can see from Equation 2.26, η seems to increase with increasing density of the gas (N/V). We must realize, however, that in a more dense gas, the number of molecular collisions is higher, and hence there is a smaller mean free path. These two opposing contributions to viscosity exactly cancel each other. Furthermore, from Equations 2.26 and 2.27, we see that the viscosity of a gas *increases* with the square root of temperature because the average speed, $\bar{c}$, is proportional to $\sqrt{T}$ (see Equation 2.14). This result is certainly contrary to what we know about liquids. Daily experience tells us, for example, that hot syrup pours more easily than cold syrup. We can understand this apparent contradiction by recognizing that the viscosity of gases arises as a result of transfer of momentum: at higher temperatures, the rate of transfer is greater so a larger force is required to maintain the motion of the layers of gas.

Experiments have generally confirmed the validity of Equations 2.26 and 2.27. Table 2.1 summarizes the viscosity and molecular collision diameter of a number of gases. Note that the collision diameter of a gas molecule can be calculated from Equation 2.27.

Table 2.1
Viscosity and Collision Diameter of Some Gases at 288 K[a]

Gas	$\eta/10^{-4}$ N s m^{-2}	Collision diameter/Å
Ar	0.2196	3.64
Kr	0.2431	4.16
Hg	0.4700[b]	4.26
H_2	0.0871	2.74
Air	0.1796	3.72
N_2	0.1734	3.75
O_2	0.2003	3.61
CH_4	0.1077	4.14
CO_2	0.1448	4.59
H_2O	0.0926	4.60
NH_3	0.0970	4.43

[a] From Kennard, E. H. *Kinetic Theory of Gases*, Copyright 1938 by McGraw-Hill. Used with permission of McGraw-Hill, New York.

[b] Measured at 492.6 K.

EXAMPLE 2.4

Calculate the viscosity of oxygen gas at 288 K.

ANSWER

Useful information:

$$\bar{c} = 437 \text{ m s}^{-1}$$

$$d = 3.61 \text{ Å} = 3.61 \times 10^{-10} \text{ m (from Table 2.1)}$$

$$m = 32.00 \text{ amu} \times 1.661 \times 10^{-27} \text{ kg amu}^{-1} = 5.315 \times 10^{-26} \text{ kg}$$

From Equation 2.27 we write

$$\eta = \frac{(5.315 \times 10^{-26} \text{ kg})(437 \text{ m s}^{-1})}{3\sqrt{2}\pi(3.61 \times 10^{-10} \text{ m})^2}$$

$$= 1.34 \times 10^{-5} \text{ kg m}^{-1} \text{ s}^{-1} = 1.34 \times 10^{-5} \text{ N s m}^{-2}$$

COMMENT

The calculated viscosity differs somewhat from that given in Table 2.1 because Equation 2.27 is only an approximate equation. A more rigorous derivation gives $\eta = m\bar{c}/2\sqrt{2}\pi d^2$. If we used this equation, then our answer would be $\eta = 2.00 \times 10^{-5}$ N s m^{-2}.

2.8 Graham's Laws of Diffusion and Effusion

Perhaps without thinking about it, we witness molecular motion on a daily basis. The scent of perfume and the shrinking of an inflated helium rubber balloon are examples of diffusion and effusion, respectively. We can apply the kinetic theory of gases to both processes.

The phenomenon of gas diffusion offers direct evidence of molecular motion. Were it not for diffusion, there would be no perfume industry, and skunks would be just another cute, furry species. Removing a partition separating two different gases in a container quickly leads to a complete mixing of molecules. These are spontaneous processes for which we shall discuss the thermodynamic basis in Chapter 4. During effusion, a gas travels from a high-pressure region to a low-pressure one through a pinhole or orifice (Figure 2.12). For effusion to occur, the mean free path of the molecules must be large compared with the diameter of the orifice. This ensures that a molecule is unlikely to collide with another molecule when it reaches the opening but will pass right through it. It follows then that the number of molecules passing through the orifice is equal to the number that would normally strike an area of wall equal to the area of the hole.

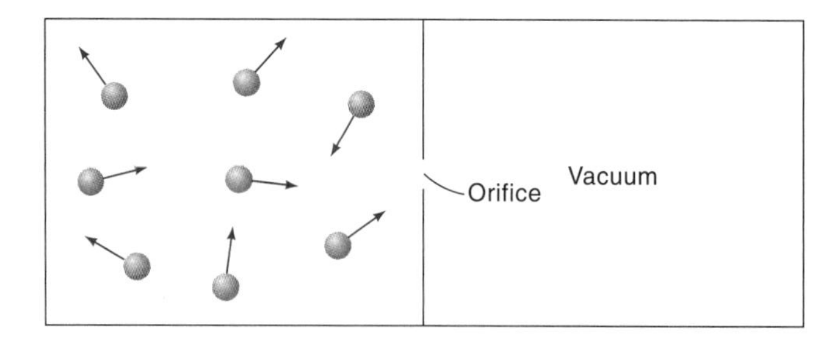

Figure 2.12
An effusion process in which molecules move through a small opening (orifice) into an evacuated region. The conditions for effusion are that the mean free path of the molecules is large compared to the size of the opening and the wall containing the opening is thin so that no molecular collisions occur during the exit. Also, the pressure in the right chamber must be low enough so as not to obstruct the molecular movement through the hole.

Although the basic molecular mechanisms for diffusion and effusion are quite different (the former involves *bulk* flow, whereas the latter involves *molecular* flow), these two phenomena obey laws of the same form. Both laws were discovered by the Scottish chemist Thomas Graham (1805–1869), the law of diffusion in 1831 and the law of effusion in 1864. These laws state that under the same conditions of temperature and pressure, the rates of diffusion (or effusion) of gases are inversely proportional to the square roots of their molar masses. Thus, for two gases 1 and 2, we have

$$\frac{r_1}{r_2} = \sqrt{\frac{\mathcal{M}_2}{\mathcal{M}_1}} \tag{2.28}$$

where r_1 and r_2 are the rates of diffusion (or effusion) of the two gases.

Practical illustrations of effusion are quite common. As mentioned above, a helium-filled balloon usually deflates much faster than an air-filled balloon. Although the conditions are not exact (because the holes in the stretched rubber are not small enough), this is essentially an effusion process, the rate of which is given by Equation 2.28. The pressure of the gas inside the balloon is greater than atmospheric pressure, and the stretched surface of the rubber has many tiny holes that allow the gas molecules to escape. Perhaps the best-known application of effusion is the separation of uranium isotopes ^{235}U and ^{238}U, whose natural abundances are approximately 0.72% and 99.28%, respectively. Only ^{235}U is a fissionable material. Although uranium is a solid, it can be converted readily to uranium hexafluoride (UF_6), which is easily vaporized above room temperature. Thus, these two isotopes can be separated from each other by effusion because $^{238}UF_6$ is heavier than $^{235}UF_6$ and therefore effuses more slowly.* We define a quantity, s, called the separation factor, such that

$$s = \frac{\text{rate of effusion of } ^{235}\text{UF}_6}{\text{rate of effusion of } ^{238}\text{UF}_6}$$

From Equation 2.28, we have

* Fluorine has only one stable isotope, and so the separation involves only two species, $^{235}UF_6$ and $^{238}UF_6$.

$$s = \sqrt{\frac{238 + (6 \times 19)}{235 + (6 \times 19)}} = 1.0043$$

Thus, in a single-stage effusion, the separation factor is very close to unity. After a second effusion process, however, the overall separation factor becomes $(1.0043)^2$, or 1.0086, a slight improvement. In general, then, the value of s for an n-stage process is $(1.0043)^n$. If n is a large number, say 2000, then it is indeed possible to obtain uranium with over 90 percent enrichment of the ^{235}U isotope.

In the study of effusion, it is important to know the rate at which molecules strike an area (e.g., the orifice). The number of collisions per unit area per unit time, Z_A, is related to the pressure and temperature of the gas by

$$Z_A = \frac{P}{(2\pi m k_B T)^{1/2}} \tag{2.29}$$

where P is the pressure or partial pressure of the gas (Appendix 2.1 on p. 63 gives the derivation). As the following example shows, this equation is useful for studying the rate of photosynthesis.

EXAMPLE 2.5

Calculate the mass in grams of CO_2 that collides every second with one side of a leaf that has an area of 0.020 m^2 at 25°C. Assume that the composition of CO_2 in air is 0.033% by volume and atmospheric pressure is 1.0 atm.

ANSWER

The relevant quantities are

$$P = \frac{0.033}{100} \times 1.00 \text{ atm} \times \frac{101325 \text{ Pa}}{1 \text{ atm}} = 33.4 \text{ Pa}$$

$$m = 44.01 \text{ amu} \times 1.661 \times 10^{-27} \text{ kg amu}^{-1} = 7.31 \times 10^{-26} \text{ kg}$$

From Equation 2.29 we write

$$Z_A = \frac{33.4 \text{ Pa}}{[2\pi(7.31 \times 10^{-26} \text{ kg})(1.381 \times 10^{-23} \text{ J K}^{-1})(298 \text{ K})]^{1/2}}$$

$$= 7.7 \times 10^{23} \text{ m}^{-2} \text{ s}^{-1}$$

The conversion factors are: 1 Pa = 1 kg m^{-1} s^{-2} and 1 J = 1 kg m^2 s^{-2}.

The number of CO_2 molecules colliding with the leaf per second is

$$7.7 \times 10^{23} \text{ m}^{-2} \text{ s}^{-1} \times 0.020 \text{ m}^2 = 1.5 \times 10^{22} \text{ s}^{-1}$$

Finally, the mass of CO_2 colliding with the leaf in one second is

$$1.5 \times 10^{22} \text{ molecules s}^{-1} \times 7.31 \times 10^{-23} \text{ g molecule}^{-1} = 1.1 \text{ g s}^{-1}$$

2.9 Equipartition of Energy

We saw in Section 2.3 that the average kinetic energy is $\frac{3}{2}k_BT$ for one molecule, or $\frac{3}{2}RT$ for 1 mole of a gas. How is this energy distributed? In this section, we consider a theorem for the distribution of energy within a molecule, a theorem based on classical mechanics and directly applicable to gas behavior.

According to the *equipartition of energy theorem*, the energy of a molecule is equally divided among all types of motion or *degrees of freedom*. For monatomic gases, each atom needs three coordinates (x, y, and z) to completely define its position. It follows that an atom has three translational degrees of freedom, each of which has energy $\frac{1}{2}k_BT$. Other types of motion, such as rotation and vibration, are also present in molecules, so there must be additional degrees of freedom for them. For a molecule containing N atoms, we would need a total of $3N$ coordinates to describe its motion. Three coordinates are required to describe the translational motion (e.g., the motion of the center of mass of the molecule), leaving $(3N - 3)$ coordinates for rotation and vibration. Three angles are needed to define the rotation of the molecule about the three mutually perpendicular axes through its center of mass, leaving $(3N - 6)$ degrees of freedom for vibration. If the molecule is linear, two angles will suffice. Because rotation of the molecule about the internuclear axis does not change the positions of the nuclei, this motion does not constitute a rotation.* Thus, a linear molecule such as HCl or CO_2 has $(3N - 5)$ degrees of freedom for vibration. Figure 2.13 shows the translational, rotational, and vibrational motions of a diatomic molecule.

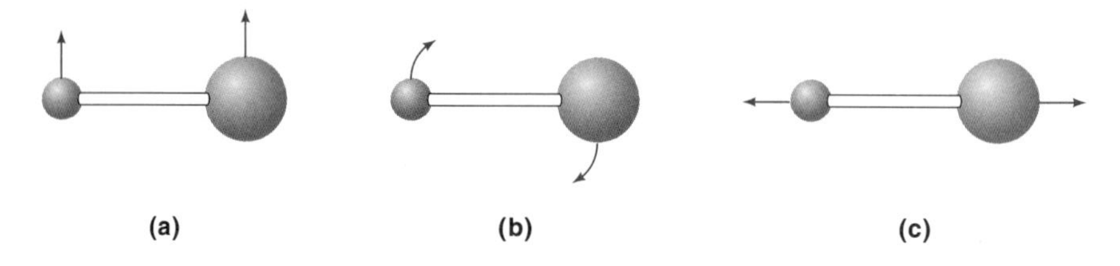

Figure 2.13
(a) Translational, (b) rotational, and (c) vibrational motion of a diatomic molecule, such as HCl.

Table 2.2 lists the equipartition of energy for different types of molecules. For 1 mole of a gas, each translational and rotational degree of freedom possesses energy $\frac{1}{2}RT$, whereas each vibrational degree of freedom possesses energy RT. The reason is that the vibrational energy contains two terms, one kinetic ($\frac{1}{2}RT$) and one potential ($\frac{1}{2}RT$). Consequently, the total energy contributions to 1 mole of a diatomic gas is

$$\overline{U} = \underset{\text{(translation)}}{\tfrac{3}{2}RT} + \underset{\text{(rotation)}}{RT} + \underset{\text{(vibration)}}{RT} = \tfrac{7}{2}RT \tag{2.30}$$

where $\overline{U}$ is the molar internal energy of the system.

* The same is true for the spinning motion of an atom about an axis through its center. Thus, an atom cannot rotate like a molecule and has no rotational degree of freedom.

Table 2.2
Equipartition of Energy for 1 Mole of Gas

Species	Translation	Rotation	Vibration
Atom	$\frac{3}{2}RT$	—	—
Linear molecule	$\frac{3}{2}RT$	RT	$(3N-5)RT$
Nonlinear molecule	$\frac{3}{2}RT$	$\frac{3}{2}RT$	$(3N-6)RT$

We can test the equipartition of energy theorem using heat-capacity measurements. The *specific heat* of a substance is the energy required to raise the temperature of 1 g of the substance by 1°C, or 1 K. It has the units joules per gram per kelvin (J g^{-1} K^{-1}). On the other hand, *heat capacity* (C) is the energy required to raise the temperature of a given quantity of a substance by 1 K. Heat capacity is given by the product of specific heat and mass of the substance in grams and is measured in joules per kelvin (J K^{-1}). Heat capacity is defined as

$$C = \frac{\Delta U}{\Delta T} \tag{2.31}$$

The Greek delta (Δ) means "change of" or "final − initial."

where ΔU is the amount of energy required to raise the temperature of the substance by ΔT.

As we shall see in more detail in Chapter 3, the heat capacity of a substance also depends on the manner in which heating takes place. For example, C can have quite different values, depending on whether we heat the gas at constant volume or at constant pressure. In the latter case, the gas will expand. Although both conditions are important in heat capacity measurements, we discuss only the heat capacity at constant volume (C_V) here. For an infinitesimal process at constant volume, Equation 2.31 can be written as

$$C_V = \left(\frac{\partial U}{\partial T}\right)_V \tag{2.32}$$

where ∂ denotes the partial derivative, and the subscript V reminds us that this is a constant-volume process.

For chemists, expressing heat capacity as a molar quantity is often more convenient. The *molar heat capacity*, $\overline{C}$, is given by C/n, where n is the number of moles of the gas present. Note that both specific heat and molar heat capacity are intensive properties, but heat capacity is an extensive property because it depends on the amount of the substance present.

For a monatomic gas that has molar translational energy $\frac{3}{2}RT$, the molar heat capacity is given by

$$\overline{C}_V = \left[\frac{\partial(\frac{3}{2}RT)}{\partial T}\right]_V = \frac{3}{2}R = 12.47 \text{ J K}^{-1} \text{ mol}^{-1}$$

Table 2.3
Predicted and Measured Constant-Volume Molar Heat Capacities of Gases at 298 K

Gas	$\overline{C}_V$/J K^{-1} mol^{-1} Predicted	$\overline{C}_V$/J K^{-1} mol^{-1} Measured
He	12.47	12.47
Ne	12.47	12.47
Ar	12.47	12.47
H_2	29.10	20.50
N_2	29.10	20.50
O_2	29.10	21.05
CO_2	54.06	28.82
H_2O	49.87	25.23
SO_2	49.87	31.51

For a diatomic molecule (see Equation 2.30), the molar heat capacity is

$$\overline{C}_V = \left[\frac{\partial(\frac{7}{2}RT)}{\partial T}\right]_V = \frac{7}{2}R = 29.10 \text{ J K}^{-1} \text{ mol}^{-1}$$

For polyatomic molecules, $\overline{C}_V$ can be obtained in a similar way.

Table 2.3 compares predicted values with experimentally measured molar heat capacities for several gases. The agreement is excellent for monatomic gases, but considerable discrepancies are found for molecules. To explain the discrepancies, we need to apply quantum mechanics. According to quantum mechanics, the electronic, vibrational, and rotational energies of a molecule are quantized (further discussed in Chapters 10 and 11). That is, different molecular energy levels are associated with each type of motion, as shown in Figure 2.14. Note that the spacing between successive electronic energy levels is much larger than that between the vibrational energy levels, which in turn is much larger than that between the rotational energy levels. The spacing between successive translational energy levels is so small that the levels practically merge into a continuum of energy. In fact, for most practical purposes, they can be treated as a continuum. Translational motion in a macroscopic environment, then, is treated as a classical rather than a quantum-mechanical phenomenon because its energy can vary continuously.

What do these energy levels have to do with heat capacities? When a system (e.g., a sample of gas molecules) absorbs heat from the surroundings, the energy is used to promote various kinds of motion. In this sense, the term *heat capacity* really means *energy capacity* because its value tells us the capacity of the system to store energy. Energy may be stored partly in vibrational motion; the molecules may be promoted to a higher vibrational energy level, or energy may be stored partly in electronic or rotational motion. In each case, the molecules are promoted to a higher energy level.

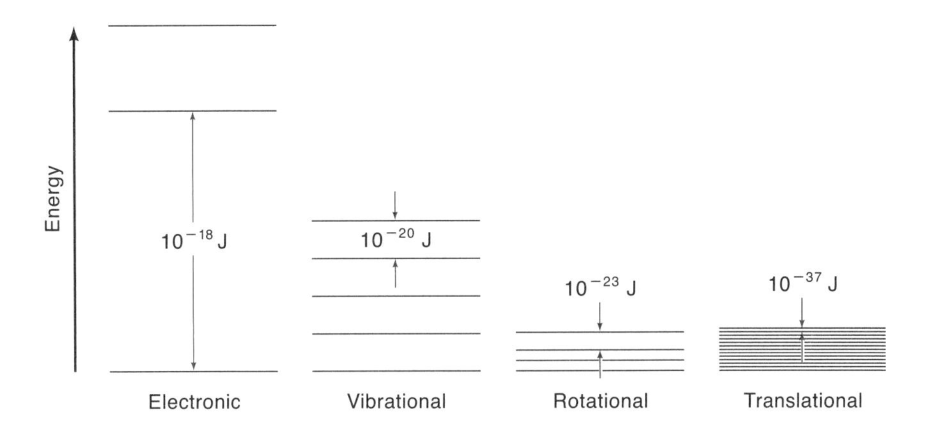

Figure 2.14
Approximate energy level spacings associated with translational, rotational, vibrational, and electronic motions.

Figure 2.14 suggests that it is much easier to excite a molecule to a higher rotational energy level than to a higher vibrational or electronic energy level, and this is indeed true. Quantitatively, the ratio of the populations (i.e., the number of molecules) in any two energy levels, E_2 and E_1, N_2/N_1, is given by the *Boltzmann distribution law*:

$$\frac{N_2}{N_1} = e^{-\Delta E/k_B T} \tag{2.33}$$

where $\Delta E = E_2 - E_1$ (we assume that $E_2 > E_1$), k_B is the Boltzmann constant, and T is the absolute temperature. Equation 2.33 tells us that for a system at thermal equilibrium at a finite temperature, $N_2/N_1 < 1$, which means that the number of molecules in the upper level is *always* less than that in the lower level (Figure 2.15).

We can make some simple estimates using Equation 2.33. For translational motion, the spacing between adjacent energy levels, ΔE, is approximately 10^{-37} J, so that $\Delta E/k_B T$ at 298 K is

$$\frac{10^{-37}\text{ J}}{(1.381 \times 10^{-23}\text{ J K}^{-1})(298\text{ K})} = 2.4 \times 10^{-17}$$

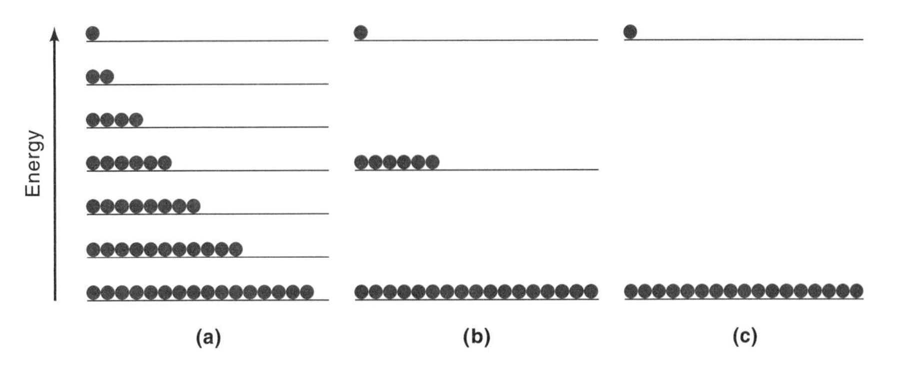

Figure 2.15
Qualitative illustration of the Boltzmann distribution law at some finite temperature T for three different types of energy levels. Note that if the energy spacing is large compared to k_BT, the molecules will crowd into the lowest energy level.

This number is so small that the exponential term on the right side of Equation 2.33 is essentially unity. Thus, the number of molecules in a higher energy level is the same as in the one below it. The physical meaning of this result is that the kinetic energy is not quantized, and a molecule can absorb any arbitrary amount of energy to increase its kinetic motion.

For rotational motion, we find that ΔE is also small compared with the $k_B T$ term; therefore, the ratio N_2/N_1 is close to (although smaller than) unity. This result means that the molecules are distributed fairly evenly among the rotational energy levels. The difference between rotational and translation motions is that only the energies of the former are quantized.

The situation is quite different for vibrational motion. Here, the spacing between levels is large (i.e., $\Delta E > k_B T$), so that the ratio N_2/N_1 is much smaller than 1. Thus, at 298 K, most of the molecules are in the lowest vibrational energy level, and only a small fraction are in the higher levels. Finally, because the spacing between electronic energy levels is so large, almost all the molecules are found in the lowest electronic energy level at room temperature.

From this discussion, we can draw several conclusions about heat capacities. We would expect that for molecules at room temperature, only the translational and rotational motions could contribute to the heat capacity. Take the O_2 molecules as an example. If we neglect both the vibrational and electronic contributions to heat capacity, the energy of the system becomes (see Equation 2.30)

$$\overline{U} = \tfrac{3}{2}RT + RT = \tfrac{5}{2}RT$$

and

$$\overline{C}_V = \tfrac{5}{2}R = 20.79 \text{ J K}^{-1} \text{ mol}^{-1}$$

This calculated value is indeed close to the measured value of 21.05 J K^{-1} mol^{-1} (see Table 2.3). The discrepancy means that the vibrational motion actually makes a small contribution to $\overline{C}_V$ at 298 K. Good agreement is also obtained for other diatomic and polyatomic molecules by including only translational and rotational motion. Furthermore, we predict that as temperature increases, more molecules will be promoted to higher vibrational levels. Therefore, at elevated temperatures, vibrational motions should begin to make an appreciable contribution to the heat capacity. This prediction is supported by experimental measurements. The molar heat capacities at constant volume of O_2 at several temperatures are as follows:

T/K	$\overline{C}_V$/J K^{-1} mol^{-1}
298	21.05
600	23.78
800	25.43
1000	26.56
1500	28.25
2000	29.47

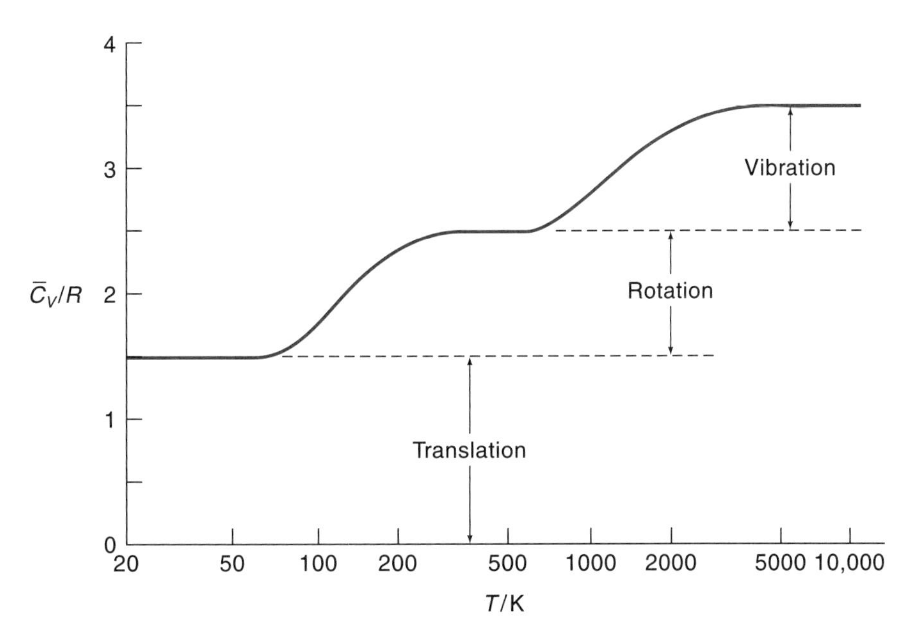

Figure 2.16
The molar heat capacity ($\overline{C}_V$) of an ideal diatomic molecule as a function of temperature. At low temperatures ($T < 100$ K) only translational motion contributes to $\overline{C}_V$. Vibrational motion contributes to $\overline{C}_V$ at temperatures above 1000 K. Note that the temperature scale is logarithmic and that the unit of $\overline{C}_V$ is R, the gas constant. We neglect the contribution due to electronic motion in this plot.

As temperature increases, the measured $\overline{C}_V$ for O_2 gets closer to the value calculated from the equipartition of energy theorem. The value at 1500 K is quite close to 29.10 J K^{-1} mol^{-1}. In fact, at 2000 K, the measured value is greater than the calculated value! The only way we can explain this behavior is to assume that electronic motion begins to contribute to the heat capacity at this temperature.

In summary, we should keep in mind that at room temperature, only translational motion and rotational motion contribute to heat capacity. At elevated temperatures, vibrational motion must also be taken into account. Only at very high temperatures does electronic motion play a role in determining the $\overline{C}_V$ values. Figure 2.16 shows the temperature variation of molar heat capacity of an ideal diatomic molecule, assuming no contribution to $\overline{C}_V$ from electronic motion.

Key Equations

$\bar{E}_{\text{trans}} = \frac{3}{2}\frac{RT}{N_A} = \frac{3}{2}k_BT$	(Average kinetic energy)	(2.6)
$v_{\text{rms}} = \sqrt{\frac{3k_BT}{m}} = \sqrt{\frac{3RT}{\mathcal{M}}}$	(Root-mean-square velocity)	(2.7)
$\frac{dN}{N} = \left(\frac{m}{2\pi k_BT}\right)^{1/2} e^{-mv_x^2/2k_BT}\,dv_x$	(Maxwell velocity distribution)	(2.8)
$\frac{dN}{N} = 4\pi c^2\left(\frac{m}{2\pi k_BT}\right)^{3/2} e^{-mc^2/2k_BT}\,dc$	(Maxwell speed distribution)	(2.10)
$\frac{dN}{N} = 2\pi E^{1/2}\left(\frac{1}{\pi k_BT}\right)^{3/2} e^{-E/k_BT}\,dE$	(Maxwell energy distribution)	(2.12)
$\bar{c} = \sqrt{\frac{8k_BT}{\pi m}} = \sqrt{\frac{8RT}{\pi\mathcal{M}}}$	(Average or mean speed)	(2.14)
$c_{\text{mp}} = \sqrt{\frac{2k_BT}{m}} = \sqrt{\frac{2RT}{\mathcal{M}}}$	(Most probable speed)	(2.15)
$Z_1 = \sqrt{2}\pi d^2\bar{c}\left(\frac{N}{V}\right)$	(Collision frequency)	(2.16)
$Z_{11} = \frac{\sqrt{2}}{2}\pi d^2\bar{c}\left(\frac{N}{V}\right)^2$	(Binary collision number)	(2.17)
$\lambda = \frac{1}{\sqrt{2}\pi d^2(N/V)}$	(Mean free path)	(2.18)
$P = P_0 e^{-gmh/k_BT}$	(Barometric formula)	(2.24)
$\eta = \frac{m\bar{c}}{3\sqrt{2}\pi d^2}$	(Gas viscosity)	(2.27)
$\frac{r_1}{r_2} = \sqrt{\frac{\mathcal{M}_2}{\mathcal{M}_1}}$	(Graham's laws of diffusion and effusion)	(2.28)
$C_V = \left(\frac{\partial U}{\partial T}\right)_V$	(Heat capacity at constant volume)	(2.32)
$\frac{N_2}{N_1} = e^{-\Delta E/k_BT}$	(Boltzmann distribution law)	(2.33)

APPENDIX 2.1

Derivation of Equation 2.29

Consider a molecule with velocity v_x moving toward a wall of area A along the x axis, which is perpendicular to the wall (Figure 2.17). It will impact the wall in time Δt if its distance from the wall is $v_x\Delta t$. If all the molecules in the container move with the same $|v_x|$, then those within the volume $Av_x\Delta t$ with positive x-component velocity will strike the wall in the interval Δt. (The $|\ |$ symbol means that we consider only the magnitude, not the sign, of v_x.) The total number of collisions in this interval is given by the product of the volume and the number density (N/V), that is, $Av_x\Delta t(N/V)$, where N is the total number of molecules and V is the volume of the container. There is a distribution in velocities, however, and we must sum the result over all the positive values of v_x according to Equation 2.8. The total number of collisions in time Δt should therefore be given by

$$\left(\frac{N}{V}\right)A\Delta t\int_0^\infty v_x f(v_x)dv_x$$

and the number of collisions per unit area per unit time, Z_A, is, from Equation 2.9,

$$Z_A = \left(\frac{N}{V}\right)\int_0^\infty v_x f(v_x)dv_x$$

$$= \left(\frac{N}{V}\right)\left(\frac{m}{2\pi k_B T}\right)^{1/2}\int_0^\infty v_x\, e^{-mv_x^2/2k_BT}\, dv_x \qquad (1)$$

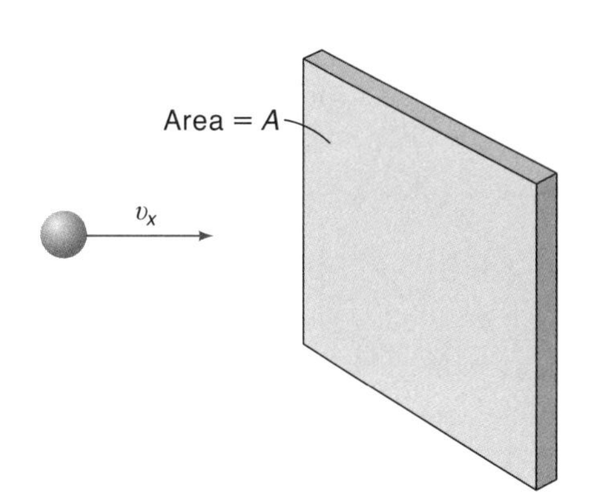

Figure 2.17
A molecule moving with velocity v_x in the positive direction (from left to right) toward the wall of area A. Only molecules within a distance $v_x\Delta t$ ($v_x > 0$) can reach the wall in a time interval Δt.

Evaluating the standard integral and rearranging, we obtain

$$Z_{\mathrm{A}} = \left(\frac{N}{V}\right)\left(\frac{k_{\mathrm{B}}T}{2\pi m}\right)^{1/2} \tag{2}$$

For an ideal gas,

$$\begin{aligned} PV &= nRT \\ &= \left(\frac{N}{N_{\mathrm{A}}}\right)RT \end{aligned} \tag{3}$$

so that

$$\begin{aligned} \left(\frac{N}{V}\right) &= \frac{PN_{\mathrm{A}}}{RT} \\ &= \frac{P}{k_{\mathrm{B}}T} \quad (R = k_{\mathrm{B}}N_{\mathrm{A}}) \end{aligned} \tag{4}$$

Substituting Equation 4 into 2, we get

$$\begin{aligned} Z_{\mathrm{A}} &= \frac{P}{k_{\mathrm{B}}T}\left(\frac{k_{\mathrm{B}}T}{2\pi m}\right)^{1/2} \\ &= \frac{P}{(2\pi m k_{\mathrm{B}}T)^{1/2}} \end{aligned} \tag{5}$$

which is Equation 2.29.

Suggestions for Further Reading

BOOKS

Hildebrand, J. H., *An Introduction to Molecular Kinetic Theory*, Chapman & Hall, London, 1963 (Van Nostrand Reinhold Company, New York).

Hirschfelder, J. O., C. F. Curtiss, and R. B. Bird, *The Molecular Theory of Gases and Liquids*, John Wiley & Sons, New York, 1954.

Tabor, D., *Gases, Liquids, and Solids*, 3rd ed., Cambridge University Press, New York, 1996.

Walton, A. J., *The Three Phases of Matter*, 2nd ed., Oxford University Press, New York, 1983.

ARTICLES

"Kinetic Energies of Gas Molecules," J. C. Aherne, *J. Chem. Educ.* **42**, 655 (1965).

"Kinetic Theory, Temperature, and Equilibrium," D. K. Carpenter, *J. Chem. Educ.* **43**, 332 (1966).

"Graham's Laws of Diffusion and Effusion," E. A. Mason and B. Kronstadt, *J. Chem. Educ.* **44**, 740 (1967).

"Heat Capacity and the Equipartition Theorem," J. B. Dence, *J. Chem. Educ.* **49**, 798 (1972).

"The Assumption of Elastic Collisions in Elementary Gas Kinetic Theory," B. Rice and C. J. G. Raw, *J. Chem. Educ.* **51**, 139 (1974).

"Velocity and Energy Distribution in Gases," B. A. Morrow and D. F. Tessier, *J. Chem. Educ.* **59**, 193 (1982).

"Temperature, Cool but Quick," S. M. Cohen, *J. Chem. Educ.* **63**, 1038 (1986).

"Applications of Maxwell-Boltzmann Distribution Diagrams," G. D. Peckham and I. J. McNaught, *J. Chem. Educ.* **69**, 554 (1992).

"Misuse of Graham's Laws," S. J. Hawkes, *J. Chem. Educ.* **70**, 836 (1993).

"Graham's Law and Perpetuation of Error," S. J. Hawkes, *J. Chem. Educ.* **74**, 1069 (1997).

"An Alternative Derivation of Gas Pressure Using the Kinetic Theory," F. Rioux, *Chem. Educator* [Online] **4**, 237 (2003) DOI 10.1333/s00897030704a.

Problems

Kinetic Theory of Gases

2.1 Apply the kinetic theory of gases to explain Boyle's law, Charles' law, and Dalton's law.

2.2 Is temperature a microscopic or macroscopic concept? Explain.

2.3 In applying the kinetic molecular theory to gases, we have assumed that the walls of the container are elastic for molecular collisions. Actually, whether these collisions are elastic or inelastic makes no difference as long as the walls are at the same temperature as the gas. Explain.

2.4 If 2.0×10^{23} argon (Ar) atoms strike 4.0 cm^2 of wall per second at a 90° angle to the wall when moving with a speed of 45,000 cm s^{-1}, what pressure (in atm) do they exert on the wall?

2.5 A square box contains He at 25°C. If the atoms are colliding with the walls perpendicularly (at 90°) at the rate of 4.0×10^{22} times per second, calculate the force and the pressure exerted on the wall given that the area of the wall is 100 cm^2 and the speed of the atoms is 600 m s^{-1}.

2.6 Calculate the average translational kinetic energy for a N_2 molecule and for 1 mole of N_2 at 20°C.

2.7 To what temperature must He atoms be cooled so that they have the same v_{rms} as O_2 at 25°C?

2.8 The c_{rms} of CH_4 is 846 m s^{-1}. What is the temperature of the gas?

2.9 Calculate the value of the c_{rms} of ozone molecules in the stratosphere, where the temperature is 250 K.

2.10 At what temperature will He atoms have the same c_{rms} value as N_2 molecules at 25°C? Solve this problem without calculating the value of c_{rms} for N_2.

Maxwell Speed Distribution

2.11 List the conditions used for deriving the Maxwell speed distribution.

2.12 Plot the speed distribution function for **(a)** He, O_2, and UF_6 at the same temperature, and **(b)** CO_2 at 300 K and 1000 K.

2.13 Account for the maximum in the Maxwell speed distribution curve (Figure 2.4) by plotting the following two curves on the same graph: (1) c^2 versus c and (2) $e^{-mc^2/2k_BT}$ versus c. Use neon (Ne) at 300 K for the plot in (2).

2.14 A N_2 molecule at 20°C is released at sea level to travel upward. Assuming that the temperature is constant and that the molecule does not collide with other molecules, how far would it travel (in meters) before coming to rest? Do the same calculation for a He atom. [*Hint:* To calculate the altitude, h, the molecule will travel, equate its kinetic energy with the potential energy, mgh, where m is the mass and g the acceleration due to gravity (9.81 m s^{-2}).]

2.15 The speeds of 12 particles (in cm s^{-1}) are 0.5, 1.5, 1.8, 1.8, 1.8, 1.8, 2.0, 2.5, 2.5, 3.0, 3.5, and 4.0. Find **(a)** the average speed, **(b)** the root-mean-square speed, and **(c)** the most probable speed of these particles. Explain your results.

2.16 At a certain temperature, the speeds of six gaseous molecules in a container are 2.0 m s^{-1}, 2.2 m s^{-1}, 2.6 m s^{-1}, 2.7 m s^{-1}, 3.3 m s^{-1}, and 3.5 m s^{-1}. Calculate the root-mean-square speed and the average speed of the molecules. These two average values are close to each other, but the root-mean-square value is always the larger of the two. Why?

2.17 The following diagram shows the Maxwell speed distribution curves for a certain ideal gas at two different temperatures (T_1 and T_2). Calculate the value of T_2.

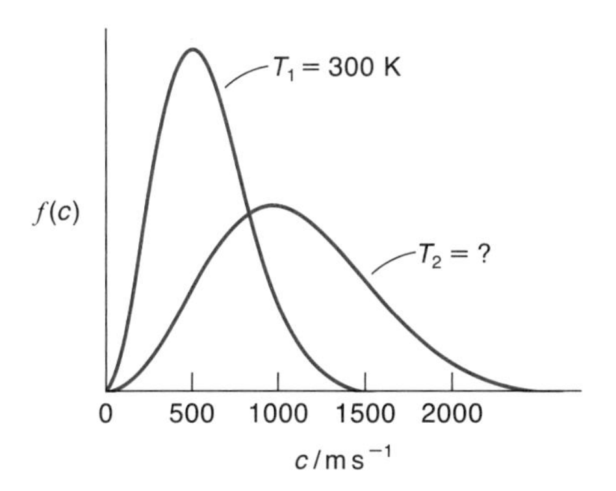

2.18 Following the procedure used in the chapter to find the value of $\overline{c}$, derive an expression for c_{rms}. (*Hint:* You need to consult the *Handbook of Chemistry and Physics* to evaluate definite integrals.)

2.19 Derive an expression for c_{mp}, following the procedure described in the chapter.

2.20 Calculate the values of c_{rms}, c_{mp}, and $\overline{c}$ for argon at 298 K.

2.21 Calculate the value of c_{mp} for C_2H_6 at 25°C. What is the ratio of the number of molecules with a speed of 989 m s^{-1} to the number of molecules with this value of c_{mp}?

2.22 Derive an expression for the most probable translational energy for an ideal gas. Compare your result with the average translational energy for the same gas.

2.23 Considering the magnitude of molecular speeds, explain why it takes so long (on the order of minutes) to detect the odor of ammonia when someone opens a bottle of concentrated ammonia at the other end of a laboratory bench.

2.24 How does the mean free path of a gas depend on **(a)** the temperature at constant volume, **(b)** the density, **(c)** the pressure at constant temperature, **(d)** the volume at constant temperature, and **(e)** the size of molecules?

2.25 A bag containing 20 marbles is being shaken vigorously. Calculate the mean free path of the marbles if the volume of the bag is 850 cm^3. The diameter of each marble is 1.0 cm.

2.26 Calculate the mean free path and the binary number of collisions per liter per second between HI molecules at 300 K and 1.00 atm. The collision diameter of the HI molecules may be taken to be 5.10 Å. Assume ideal-gas behavior.

2.27 Ultra-high vacuum experiments are routinely performed at a total pressure of 1.0×10^{-10} torr. Calculate the mean free path of N_2 molecules at 350 K under these conditions.

2.28 Suppose that helium atoms in a sealed container all start with the same speed, 2.74×10^4 cm s^{-1}. The atoms are then allowed to collide with one another until the Maxwell distribution is established. What is the temperature of the gas at equilibrium? Assume that there is no heat exchange between the gas and its surroundings.

2.29 Compare the collision number and the mean free path for air molecules at **(a)** sea level (T = 300 K and density = 1.2 g L^{-1}) and **(b)** in the stratosphere (T = 250 K and density = 5.0×10^{-3} g L^{-1}). The molar mass of air may be taken as 29.0 g, and the collision diameter as 3.72 Å.

2.30 Calculate the values of Z_1 and Z_{11} for mercury (Hg) vapor at 40°C, both at P = 1.0 atm and at P = 0.10 atm. How do these two quantities depend on pressure?

Gas Viscosity

2.31 Account for the difference between the dependence of viscosity on temperature for a liquid and a gas.

2.32 Calculate the values of the average speed and collision diameter for ethylene at 288 K. The viscosity of ethylene is 99.8×10^{-7} N s m^{-2} at the same temperature.

2.33 The viscosity of sulfur dioxide at 21.0°C and 1.0 atm pressure is 1.25×10^{-5} N s m^{-2}. Calculate the collision diameter of the SO_2 molecule and the mean free path at the given temperature and pressure.

Gas Diffusion and Effusion

2.34 Derive Equation 2.28 from Equation 2.14.

2.35 A flammable gas is generated in marsh lands and sewage by a certain anaerobic bacterium. A pure sample of this gas was found to effuse through an orifice in 12.6 min. Under identical conditions of temperature and pressure, oxygen takes 17.8 min to effuse through the same orifice. Calculate the molar mass of the gas, and suggest what this gas might be.

2.36 Nickel forms a gaseous compound of the formula $Ni(CO)_x$. What is the value of x given the fact that under the same conditions of temperature and pressure, methane (CH_4) effuses 3.3 times faster than the compound?

2.37 In 2.00 min, 29.7 mL of He effuse through a small hole. Under the same conditions of temperature and pressure, 10.0 mL of a mixture of CO and CO_2 effuse through the hole in the same amount of time. Calculate the percent composition by volume of the mixture.

2.38 Uranium-235 can be separated from uranium-238 by the effusion process involving UF_6. Assuming a 50:50 mixture at the start, what is the percentage of enrichment after a single stage of separation?

2.39 An equimolar mixture of H_2 and D_2 effuses through an orifice at a certain temperature. Calculate the composition (in mole fractions) of the gas that passes through the orifice. The molar mass of deuterium is 2.014 g mol^{-1}.

2.40 The rate (r_{eff}) at which molecules confined to a volume V effuse through an orifice of area A is given by $(\frac{1}{4})nN_A\bar{c}A/V$, where n is the number of moles of the gas. An automobile tire of volume 30.0 L and pressure 1,500 torr is punctured as it runs over a sharp nail. **(a)** Calculate the effusion rate if the diameter of the hole is 1.0 mm. **(b)** How long would it take to lose half of the air in the tire through effusion? Assume a constant effusion rate and constant volume. The molar mass of air is 29.0 g, and the temperature is 32.0°C.

Equipartition of Energy

2.41 Calculate the number of various degrees of freedom for the following molecules: **(a)** Xe, **(b)** HCl, **(c)** CS_2, **(d)** C_2H_2, **(e)** C_6H_6, and **(f)** a hemoglobin molecule containing 9272 atoms.

2.42 Explain the equipartition of energy theorem. Why does it fail for diatomic and polyatomic molecules?

2.43 A quantity of 500 joules of energy is delivered to one mole of each of the following gases at 298 K and the same fixed volume: Ar, CH_4, H_2. Which gas will have the highest rise in temperature?

2.44 Calculate the mean kinetic energy ($\overline{E}_{\text{trans}}$) in joules of the following molecules at 350 K: **(a)** He, **(b)** CO_2, and **(c)** UF_6. Explain your results.

2.45 A sample of neon gas is heated from 300 K to 390 K. Calculate the percent increase in its kinetic energy.

2.46 Calculate the value of $\overline{C}_V$ for H_2, CO_2, and SO_2, assuming that only translational and rotational motions contribute to the heat capacities. Compare your results with the values listed in Table 2.3. Explain the differences.

2.47 One mole of ammonia initially at 5°C is placed in contact with 3 moles of helium initially at 90°C. Given that $\overline{C}_V$ for ammonia is $3R$, if the process is carried out at constant total volume, what is the final temperature of the gases?

2.48 The typical energy differences between successive rotational, vibrational, and electronic energy levels are 5.0×10^{-22} J, 5.0×10^{-20} J, and 1.0×10^{-18} J, respectively. Calculate the ratios of the numbers of molecules in the two adjacent energy levels (higher to lower) in each case at 298 K.

2.49 The first excited electronic energy level of the helium atom is 3.13×10^{-18} J above the ground level. Estimate the temperature at which the electronic motion will begin to make a significant contribution to the heat capacity. That is, at what temperature will the ratio of the population of the first excited state to the ground state be 5.0%?

2.50 Consider 1 mole each of gaseous He and N_2 at the same temperature and pressure. State which gas (if any) has the greater value for: **(a)** $\overline{c}$, **(b)** c_{rms}, **(c)** $\overline{E}_{\text{trans}}$, **(d)** Z_1, **(e)** Z_{11}, **(f)** density, **(g)** mean free path, and **(h)** viscosity.

2.51 The root-mean-square velocity of a certain gaseous oxide is 493 m s^{-1} at 20°C. What is the molecular formula of the compound?

2.52 At 298 K, the $\overline{C}_V$ of SO_2 is greater than that of CO_2. At very high temperatures (>1000 K), the $\overline{C}_V$ of CO_2 is greater than that of SO_2. Explain.

2.53 Calculate the total translational kinetic energy of the air molecules in a spherical balloon of radius 43.0 cm at 24°C and 1.2 atm. Is this enough energy to heat 200 mL of water from 20°C to 90°C for a cup of tea? The density of water is 1.0 g cm^{-3}, and its specific heat is 4.184 J g^{-1} °C^{-1}.

Additional Problems

2.54 The following apparatus can be used to measure atomic and molecular speed. A beam of metal atoms is directed at a rotating cylinder in a vacuum. A small opening in the cylinder allows the atoms to strike a target area. Because the cylinder is rotating, atoms traveling at different speeds will strike the target at different positions. In time, a layer of the metal will deposit on the target area, and the variation in its thickness is found to correspond to Maxwell's speed distribution. In one experiment, it is found that at 850°C, some bismuth (Bi) atoms struck the target at a point 2.80 cm from the spot directly opposite the slit. The diameter of the cylinder is 15.0 cm, and it is rotating at 130 revolutions per second. **(a)** Calculate the speed (m s^{-1}) at which the target is moving. (*Hint:* The circumference of a circle is given by $2\pi r$, where r is the radius.) **(b)** Calculate the time (in seconds) it takes for the target to travel 2.80 cm. **(c)** Determine the speed of the Bi atoms. Compare your result in **(c)** with the c_{rms} value for Bi at 850°C. Comment on the difference.

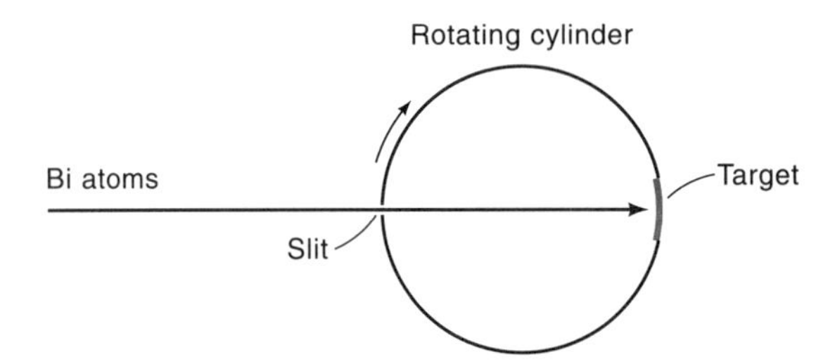

2.55 From your knowledge of heat capacity, explain why hot, humid air is more uncomfortable than hot, dry air and why cold, damp air is more uncomfortable than cold, dry air.

2.56 The escape velocity, v, from Earth's gravitational field is given by $(2GM/r)^{1/2}$, where G is the universal gravitational constant (6.67×10^{-11} m^3 kg^{-1} s^{-2}), M is the mass of Earth (6.0×10^{24} kg), and r is the distance from the center of Earth to the object, in meters. Compare the average speeds of He and N_2 molecules in the thermosphere (altitude about 100 km, $T = 250$ K). Which of the two molecules will have a greater tendency to escape? The radius of Earth is 6.4×10^6 m.

2.57 Suppose you are traveling in a space vehicle on a journey to the moon. The atmosphere in the vehicle consists of 20% oxygen and 80% helium by volume. Before takeoff, someone noticed a small leakage that, if left unchecked, would lead to a continual loss of gas at a rate of 0.050 atm day^{-1} through effusion. If the temperature in the space vehicle is maintained at 22°C and the volume of the vehicle is 1.5×10^4 L, calculate the amounts of helium and oxygen in grams that must be stored on a 10-day journey to allow for the leakage. (*Hint:* First calculate the quantity of gas lost each day, using $PV = nRT$. Note that the rate of effusion is proportional to the pressure of the gas. Assume that effusion does not affect the pressure or the mean free path of the gas in the space vehicle.)

2.58 Calculate the ratio of the number of O_3 molecules with a speed of 1300 m s^{-1} at 360 K to the number with that speed at 293 K.

2.59 Calculate the collision frequency for 1.0 mole of krypton (Kr) at equilibrium at 300 K and 1.0 atm pressure. Which of the following alterations increases the collision frequency more: **(a)** doubling the temperature at constant pressure or **(b)** doubling the pressure at constant temperature? (*Hint:* Use the collision diameter in Table 2.1.)

2.60 Apply your knowledge of the kinetic theory of gases to the following situations. **(a)** Two flasks of volumes V_1 and V_2 (where $V_2 > V_1$) contain the same number of helium atoms at the same temperature. **(i)** Compare the root-mean-square (rms) speeds and average kinetic energies of the helium (He) atoms in the flasks. **(ii)** Compare the frequency and the force with which the He atoms collide with the walls of their containers. **(b)** Equal numbers of He atoms are placed in two flasks of the same volume at temperatures T_1 and T_2 (where $T_2 > T_1$). **(i)** Compare the rms speeds of the atoms in the two flasks. **(ii)** Compare the frequency and the force with which the He atoms collide with the walls of their containers. **(c)** Equal numbers of He and neon (Ne) atoms are placed in two flasks of the same volume. The temperature of both gases is 74°C. Comment on the validity of the following statements: **(i)** The rms speed of He is equal to that of Ne. **(ii)** The average kinetic energies of the two gases are equal. **(iii)** The rms speed of each He atom is 1.47×10^3 m s^{-1}.

2.61 Compare the difference in barometer readings between the top floor of the Empire State Building (about 1225 ft) and sea level. Assume temperature to be at 20°C.

2.62 Use the kinetic theory of gases to explain why hot air rises.

2.63 Estimate the height of the atmosphere (measured from sea level) in meters that contains half of all the air molecules. Assume an average air temperature of 250 K.

2.64 Two ideal gases A and B are heated to different temperatures. If their pressures and densities are the same at these temperatures, show that their average speeds must also be the same.

2.65 At what temperature will He atoms have the same c_{rms} value as N_2 molecules at 25°C?

2.66 Identify the gas whose c_{rms} is 2.82 times that of HI at the same temperature.

2.67 Apply dimensional analysis to show that the term k_BT/mg (see p. 49) has the unit of length.

CHAPTER 3

The First Law of Thermodynamics

Some like it hot.

Thermodynamics is the science of heat and temperature and, in particular, of the laws governing the conversion of thermal energy into mechanical, electrical, or other forms of energy. It is a central branch of science that has important applications in chemistry, physics, biology, and engineering. What makes thermodynamics such a powerful tool? It is a completely logical discipline and can be applied without any sophisticated mathematical techniques. The immense practical value of thermodynamics lies in the fact that it systematizes the information obtained from experiments performed on systems and enables us to draw conclusions, without further experimentation, about other aspects of the same systems and about similar aspects of other systems. It enables us to predict whether a certain reaction will proceed and what the maximum yield might be.

Thermodynamics is a macroscopic science concerning such properties as pressure, temperature, and volume. Unlike quantum mechanics, thermodynamics is not based on a specific molecular model, and therefore it is unaffected by our changing concepts of atoms and molecules. Indeed, the major foundations of thermodynamics were laid long before detailed atomic theories became available. This fact is one of its major strengths. On the negative side, equations derived from laws of thermodynamics do not provide us with a molecular interpretation of complex phenomena. Furthermore, although thermodynamics helps us predict the direction and extent of chemical reactions, it tells us nothing about the *rate* of a process; that issue is addressed by chemical kinetics, the topic of Chapter 15.

This chapter introduces the first law of thermodynamics and discusses some examples of thermochemistry.

3.1 Work and Heat

In this section, we shall study two concepts that form the basis of the first law of thermodynamics: work and heat.

Work

In classical mechanics, *work* is defined as force times distance. In thermodynamics, work becomes a more subtle concept; it encompasses a broader range of processes, including surface work, electrical work, work of magnetization, and so on (Table 3.1).

Table 3.1
Different Types of Work

Type of work	Expression[a]	Meaning of symbols
Mechanical work	$f\,dx$	f: force; dx: distance traveled
Surface work	$\gamma\,dA$	γ: surface tension; dA: change in surface area
Electrical work	$E\,dQ$	E: potential difference; dQ: electric charge transferred
Gravitational work	$mg\,dh$	m: mass; g: acceleration due to gravity; dh: change in height
Expansion work	$P\,dV$	P: pressure; dV: change in volume

[a] The work done in each case corresponds to an infinitesimal process, as indicated by the d symbol.

Let us consider a particularly useful example of a system doing work—the expansion of a gas. A sample of a gas is placed in a cylinder fitted with a weightless and frictionless piston. We assume the temperature of the system is kept constant at T. The gas is allowed to expand from its initial state—P_1, V_1, T—to P_2, V_2, T as shown in Figure 3.1. We assume no atmospheric pressure is present, so the gas is expanding only against the weight of an object of mass m placed on the piston. The work done (w) in lifting the mass from the initial height, h_1, to the final height, h_2, is given by

$$w = -\text{force} \times \text{distance}$$

$$\begin{aligned} w &= -\text{mass} \times \text{acceleration} \times \text{distance} \\ &= -mg(h_2 - h_1) \\ &= -mg\Delta h \end{aligned} \tag{3.1}$$

where g is the acceleration (9.81 m s^{-2}) due to gravity and $\Delta h = h_2 - h_1$. Because m is in kilograms (kg) and h in meters (m), w has the unit of energy (J). The minus sign in Equation 3.1 has the following meaning: in an expansion process, $h_2 > h_1$ and w is negative. This notation follows the convention that when a system does work on its

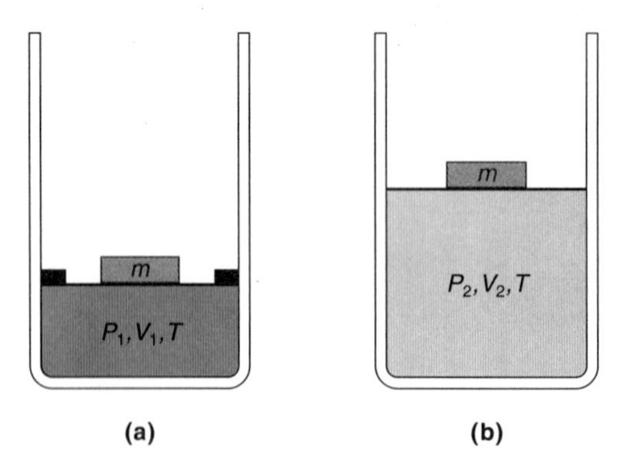

Figure 3.1
An isothermal expansion of a gas. (a) Initial state. (b) Final state.

surroundings, the work performed is a negative quantity. In a compression process, $h_2 < h_1$, so work is done on the system, and w is positive.

The external, opposing pressure, P_{ex}, acting on the gas is equal to force/area, so that

$$P_{ex} = \frac{mg}{A}$$

or

$$w = -P_{ex}A\Delta h = -P_{ex}(V_2 - V_1)$$
$$= -P_{ex}\Delta V \quad (3.2)$$

where A is the area of the piston, and the product $A\Delta h$ gives the change in volume. Equation 3.2 shows that the amount of work done during expansion depends on the value of P_{ex}. Depending on experimental conditions, the amount of work performed by a gas during expansion from V_1 to V_2 at T can vary considerably from one case to another. In one extreme, the gas is expanding against a vacuum (e.g., if the mass m is removed from the piston). Because $P_{ex} = 0$, the work done, $-P_{ex}\Delta V$, is also zero. A more common arrangement is to have some mass resting on the piston so that the gas is expanding against a *constant* external pressure. As we saw earlier, the amount of work performed by the gas in this case is $-P_{ex}\Delta V$, where $P_{ex} \neq 0$. Note that as the gas expands, the pressure of the gas, P_{in}, decreases constantly. For the gas to expand, however, we must have $P_{in} > P_{ex}$ at every stage of expansion. For example, if initially $P_{in} = 5$ atm and the gas is expanding against a constant external pressure of 1 atm ($P_{ex} = 1$ atm) at constant temperature T, then the piston will finally come to a halt when P_{in} decreases to exactly 1 atm.

Is it possible to have the gas perform a greater amount of work for the same increase in volume? Suppose we have an infinite number of identical weights exerting a total pressure of 5 atm on the piston. Because $P_{in} = P_{ex}$, the system is at mechanical equilibrium. Removing one weight will decrease the external pressure by an infinitesimal amount so that $P_{in} > P_{ex}$ and the gas will very slightly expand until P_{in} is again equal to P_{ex}. When the second weight is removed, the gas expands a bit further and so on until enough weights have been lifted from the piston to decrease the external pressure to 1 atm. At this point, we have completed the expansion process as before. How do we calculate the amount of work done in this case? At every stage of expansion (i.e., each time one weight is lifted), the infinitesimal amount of work done is given by $-P_{ex}dV$, where dV is the infinitesimal increase in volume. The total work done in expanding from V_1 to V_2 is therefore

$$w = -\int_{V_1}^{V_2} P_{ex}dV \quad (3.3)$$

Because P_{ex} is no longer a constant value, the integral cannot be evaluated in this form.* We note, however, that at every instant, P_{in} is only infinitesimally greater than P_{ex}; that is,

$$P_{in} - P_{ex} = dP$$

* If P_{ex} were constant, this integral would be $-P_{ex}(V_2 - V_1)$, or $-P_{ex}\Delta V$.

so that we can rewrite Equation 3.3 as

$$w = -\int_{V_1}^{V_2} (P_{\text{in}} - dP)dV$$

Realizing that $dPdV$ is a product of two infinitesimal quantities, we have $dPdV \approx 0$, and we can write

$$w = -\int_{V_1}^{V_2} P_{\text{in}}dV \tag{3.4}$$

Equation 3.4 is a more manageable form, because P_{in} is the pressure of the system (i.e., the gas), and we can express it in terms of a particular equation of state. For an ideal gas,

$$P_{\text{in}} = \frac{nRT}{V}$$

so that

$$w = -\int_{V_1}^{V_2} \frac{nRT}{V} dV$$

$$= -nRT \ln \frac{V_2}{V_1} = -nRT \ln \frac{P_1}{P_2} \tag{3.5}$$

because $P_1V_1 = P_2V_2$ (at constant n and T).

Equation 3.5 looks quite different from our earlier expression for work done ($-P_{\text{ex}}\Delta V$), and in fact it represents the *maximum* amount of work of expansion from V_1 to V_2. The reason for this result is not difficult to see. Because work in expansion is performed against external pressure, we can maximize the work done by adjusting the external pressure so that it is only infinitesimally smaller than the internal pressure at every stage, as described above. Under these conditions, expansion is a *reversible* process. By reversible, we mean that if we increase the external pressure by an infinitesimal amount, dP, we can bring the expansion to a stop. A further increase in P_{ex} by dP would actually result in compression. Thus, a reversible process is one in which the system is always infinitesimally close to equilibrium.

In chemical kinetics, *reversible* means the reaction can proceed in both directions.

A truly reversible process would take an infinite amount of time to complete, and therefore it can never really be done. We could set up a system so that the gas does expand very slowly and try to approach reversibility, but actually attaining it is impossible. In the laboratory, we must work with *real* processes that are always irreversible. The reason we are interested in a reversible process is that it enables us to calculate the maximum amount of work that could possibly be extracted from a process. This quantity is important in estimating the efficiency of chemical and biological processes, as we shall see in Chapter 4.

EXAMPLE 3.1

A quantity of 0.850 mole of an ideal gas initially at a pressure of 15.0 atm and 300 K is allowed to expand isothermally until its final pressure is 1.00 atm. Calculate the value of the work done if the expansion is carried out (a) against a vacuum, (b) against a constant external pressure of 1.00 atm, and (c) reversibly.

ANSWER

(a) Because $P_{ex} = 0$, $-P_{ex}\Delta V = 0$, so that no work is performed in this case.
(b) Here the external, opposing pressure is 1.00 atm, so work will be done in expansion. The initial and final volumes can be obtained from the ideal gas equation:

$$V_1 = \frac{nRT}{P_1} \qquad V_2 = \frac{nRT}{P_2}$$

Furthermore, the final pressure of the gas is equal to the external pressure, so $P_{ex} = P_2$. From Equation 3.2, we write

$$\begin{aligned} w &= -P_2(V_2 - V_1) \\ &= -nRTP_2\left(\frac{1}{P_2} - \frac{1}{P_1}\right) \\ &= -(0.850\text{ mol})(0.08206\text{ L atm K}^{-1}\text{ mol}^{-1})(300\text{ K})(1.00\text{ atm}) \\ &\quad\times\left(\frac{1}{1.00\text{ atm}} - \frac{1}{15.0\text{ atm}}\right) \\ &= -19.5\text{ L atm} \end{aligned}$$

To express the result in the more convenient unit of joules, we use the conversion factor 1 L atm = 101.3 J so that the work done is

$$\begin{aligned} w &= -19.5\text{ L atm} \times \frac{101.3\text{ J}}{1\text{ L atm}} \\ &= -1.98 \times 10^3\text{ J} \end{aligned}$$

(c) For an isothermal, reversible process, the work done is given by Equation 3.5:

$$\begin{aligned} w &= -nRT\ln\frac{V_2}{V_1} \\ &= -nRT\ln\frac{P_1}{P_2} \\ &= -(0.850\text{ mol})(8.314\text{ J K}^{-1}\text{ mol}^{-1})(300\text{ K})\ln\frac{15.0\text{ atm}}{1.00\text{ atm}} \\ &= -5.74 \times 10^3\text{ J} \end{aligned}$$

COMMENT

The negative signs in (b) and (c) indicate that work is done by the system on the surroundings. As expected, the reversible expansion does the maximum amount of work.

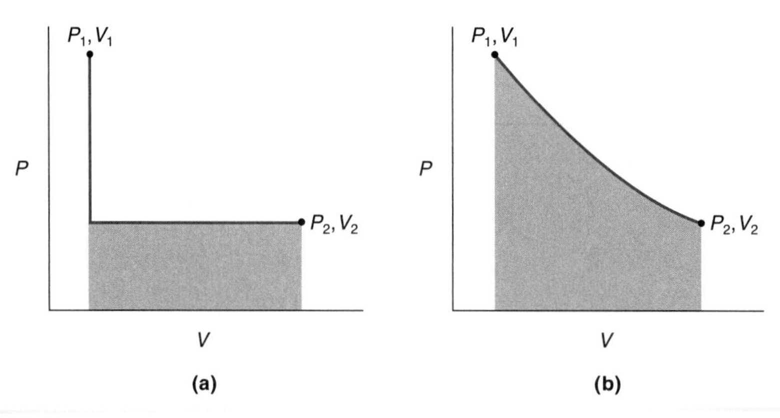

Figure 3.2
Isothermal gas expansion from P_1, V_1 to P_2, V_2. (a) An irreversible process. (b) A reversible process. In each case the shaded area represents the work done during expansion. The reversible process does the most work.

Figure 3.2 shows graphically the work done for cases (b) and (c) in Example 3.1. In an irreversible process (Figure 3.2a), the amount of work done is given by $P_2(V_2 - V_1)$, which is the area under the curve. For a reversible process, the amount of work is also given by the area under the curve (Figure 3.2b). Because the external pressure is no longer held constant, however, the area is considerably greater.

From the foregoing discussion, we can draw several conclusions about work. First, work should be thought of as a mode of energy transfer. Gas expands because there is a pressure difference. When the internal and external pressure are equalized, the word *work* is no longer applicable. Second, the amount of work done depends on how the process is carried out—that is, the *path* (e.g., reversible versus irreversible)—even though the initial and final states are the same in each case. Thus, work is not a *state function*, a property that is determined by the state of the system, and we cannot say that a system has, within itself, so much work or work content.

Some of the terms used in thermodynamics are explained on p. 1.

An important property of state functions is that when the state of a system is altered, a change in any state function depends only on the initial and final states of the system, not on how the change is accomplished. Let us assume that the change involves the expansion of a gas from an initial volume $V_1(2\ \text{L})$ to a final volume $V_2(4\ \text{L})$ at constant temperature. The change or the increase in volume is given by

$$\Delta V = V_2 - V_1$$
$$= 4\ \text{L} - 2\ \text{L} = 2\ \text{L}$$

The change can be brought about in many ways. We can let the gas expand directly from 2 L to 4 L as described above, or first allow it to expand to 6 L and then compress the volume down to 4 L, and so on. No matter how we carry out the process, the change in volume is always 2 L. Similarly, we can show that pressure and temperature, like volume, are state functions.

Heat

Heat is the transfer of energy between two bodies that are at different temperatures. Like work, heat appears only at the boundary of the system and is defined by a process. Energy is transferred from a hotter object to a colder one because there is a temperature difference. When the temperatures of the two objects are equal, the word *heat* is no longer applicable. Heat is not a property of a system and is not a state function. It is therefore path dependent. Suppose that we raise the temperature of 100.0 g of water initially at 20.0°C and 1 atm to 30.0°C and 1 atm. What is the heat transfer for this process? We do not know the answer because the process is not specified. One way to raise the temperature is to heat the water using a Bunsen burner or electrically using an immersion heater. The heat, q (transferred from the surroundings to the system), is given by

$$\begin{aligned} q &= ms\Delta T \\ &= (100.0 \text{ g})(4.184 \text{ J g}^{-1} \text{ K}^{-1})(10.0 \text{ K}) \\ &= 4184 \text{ J} \end{aligned}$$

where s is the specific heat of water. Alternatively, we can bring about the temperature increase by doing mechanical work on the system; for example, by stirring the water with a magnetic stirring bar until the desired temperature is reached as a result of friction. The heat transfer in this case is zero. Or we could first raise the temperature of water from 20.0°C to 25.0°C by direct heating and then stir the bar to bring it up to 30.0°C. In this case, q is somewhere between zero and 4184 J. Clearly, then, an infinite number of ways are available to increase the temperature of the system by the same amount, but the heat change always depends on the path of the process.

In conclusion, work and heat are not functions of state. They are measures of energy transfer, and changes in these quantities are path dependent. The conversion factor between the *thermochemical calorie* and the joule, which is the mechanical equivalent of heat, is

$$1 \text{ cal} = 4.184 \text{ J} \quad \text{exactly}$$

EXAMPLE 3.2

A 73-kg person drinks 500 g of milk, which has a "caloric" value of approximately 720 cal g^{-1}. If only 17% of the energy in milk is converted to mechanical work, how high (in meters) can the person climb based on this energy intake?

ANSWER

The energy intake for mechanical work is

$$\Delta E = 0.17 \times 500 \text{ g} \times \frac{720 \text{ cal}}{1 \text{ g}} \times \frac{4.184 \text{ J}}{1 \text{ cal}} = 2.6 \times 10^5 \text{ J}$$

Next, we equate this amount of energy to the potential energy in raising the person:

$$\Delta E = 2.6 \times 10^5 \text{ J} = mgh$$

where m the person's mass, g the acceleration due to gravity (9.81 m s^{-2}), and h the height in meters. Because 1 J = 1 kg m^2 s^{-2}, we write

$$h = \frac{\Delta E}{mg}$$

$$= \frac{2.6 \times 10^5 \text{ kg m}^2 \text{ s}^{-2}}{73 \text{ kg} \times 9.81 \text{ m s}^{-2}} = 3.6 \times 10^2 \text{ m}$$

3.2 The First Law of Thermodynamics

The *first law of thermodynamics* states that energy can be converted from one form to another but cannot be created or destroyed. Put another way, this law says that the total energy of the universe is a constant, or energy is conserved. In general, we can divide the energy of the universe into two parts:

$$E_{\text{univ}} = E_{\text{sys}} + E_{\text{surr}}$$

where the subscripts denote the universe, system, and surroundings, respectively. For any given process, the changes in energies are

$$\Delta E_{\text{univ}} = \Delta E_{\text{sys}} + \Delta E_{\text{surr}} = 0$$

or

$$\Delta E_{\text{sys}} = -\Delta E_{\text{surr}}$$

Thus, if one system undergoes an energy change, ΔE_{sys}, the rest of the universe, or the surroundings, must undergo a change in energy that is equal in magnitude but opposite in sign; energy gained in one place must have been lost somewhere else. Furthermore, because energy can be changed from one form to another, the energy lost by one system can be gained by another system in a different form. For example, the energy lost by burning oil in a power plant may ultimately turn up in our homes as electrical energy, heat, light, and so on.

In chemistry, we are normally interested in the energy changes associated with the system, not with the surroundings. We have seen that because heat and work are not state functions, it is meaningless to ask how much heat or work a system possesses. On the other hand, the internal energy of a system is a state function, because it depends only on the thermodynamic parameters of the state, such as temperature, pressure, and composition. Note that the adjective *internal* implies that other kinds of energy may be associated with the system. For example, the whole system may be in motion and therefore possess kinetic energy (K). The system may also possess potential energy (V). Thus, the total energy of the system, E_{total}, is given by

$$E_{\text{total}} = K + V + U$$

where U denotes the internal energy of the system. This internal energy consists of translational, rotational, vibrational, electronic, and nuclear energies, as well as intermolecular interactions. In most cases we shall consider, the system will be at rest, and external fields (e.g., electric or magnetic fields) will not be present. Thus, both K and V are zero and $E_{\text{total}} = U$. As mentioned earlier, thermodynamics is not based on a particular model; therefore, we have no need to know the exact nature of U. In fact, we normally have no way to calculate this quantity accurately. All we are interested in, as you will see below, are methods for measuring the *change* in U for a process. For simplicity, we shall frequently refer to internal energy as energy and write its change, ΔU, as

$$\Delta U = U_2 - U_1$$

where U_2 and U_1 are the internal energies of the system in the final and initial states, respectively.

Energy differs from both heat and work in that it always changes by the same amount in going from one state to another, regardless of the nature of the path. Mathematically, the first law of thermodynamics can be expressed as

$$\Delta U = q + w \tag{3.6}$$

We assume all work done is of the P–V type.

or, for an infinitesimal change,

$$dU = đq + đw \tag{3.7}$$

Equations 3.6 and 3.7 tell us that the change in the internal energy of a system in a given process is the sum of the heat exchange, q, between the system and its surroundings and the work done, w, on (or by) the system. The sign conventions for q and w are summarized in Table 3.2. Note that we have deliberately avoided the Δ sign for q and w, because this notation represents the difference between the final and initial

Table 3.2
Sign Conventions for Work and Heat

Process	Sign
Work done by the system on the surroundings	−
Work done on the system by the surroundings	+
Heat absorbed by the system from the surroundings (endothermic process)	+
Heat absorbed by the surroundings from the system (exothermic process)	−

states and is therefore not applicable to heat and work, which are not state functions. Similarly, although dU is an *exact differential* (see Appendix 3.1 on p. 116), that is, an integral of the type $\int_1^2 dU$ is independent of the path—the $đ$ notation reminds us that $đq$ and $đw$ are *inexact differentials* and therefore are path dependent. In this text, we shall use capital letters for thermodynamic quantities (such as U, P, T, and V) that are state functions, and lowercase letters for thermodynamic quantities (such as q and w) that are not.

A straightforward application of Equation 3.6 is an irreversible gas expansion against a constant pressure P, where $w = -P\Delta V$. The equation can now be written as

$$\Delta U = q - P\Delta V$$

Because $\Delta V > 0$ for gas expansions, the internal energy of the system decreases as a result of the gas doing work on the surroundings. If the gas absorbs heat from the surroundings, then $q > 0$ and the internal energy of the system increases. We see that the sign conventions in Table 3.2 for work and heat are consistent with the mathematical statement of the first law.

The first law of thermodynamics is a law of conservation of energy: its formulation is based on our vast experience in the study of relationships between different forms of energies. Conceptually, the first law is easy to comprehend, and it can be readily applied to any practical system. Consider, for example, the thermochemical changes in a constant-volume adiabatic bomb calorimeter (Figure 3.3). This device enables us to measure the heat of combustion of substances. It is a tightly sealed, heavy-walled, stainless steel container that is thermally isolated from its surroundings. (The word

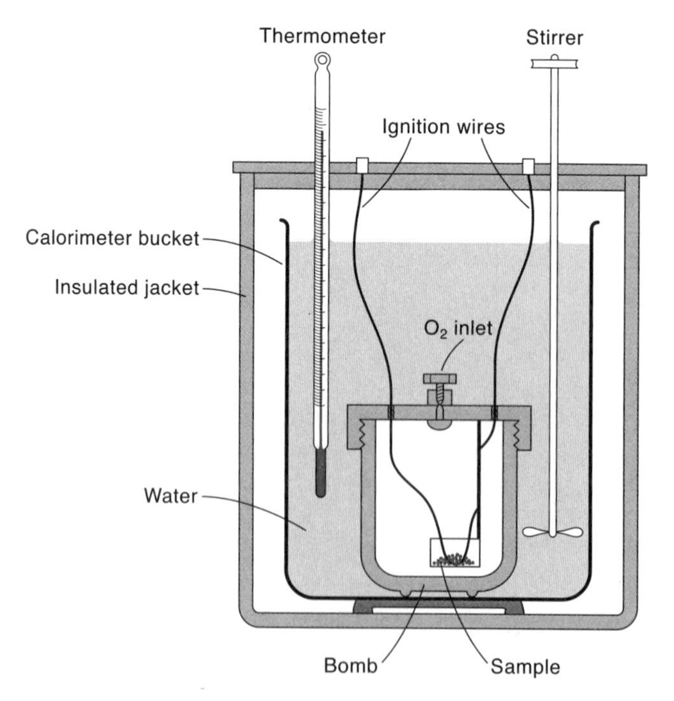

Figure 3.3
Schematic diagram of a constant-volume bomb calorimeter. The heat capacity of the calorimeter is the sum of the heat capacities of the bomb and the water bath.

adiabatic means no heat exchange with the surroundings.) The substance under investigation is placed inside the container, which is filled with oxygen at about 30 atm. The combustion is started by an electrical discharge through a pair of wires that are in contact with the substance. The heat released by the reaction can be measured by registering the rise in the temperature of the water filling the inner jacket of the calorimeter. Using the heat capacity of the calorimeter, we can calculate the change in energy as a result of combustion, given by

$$\begin{aligned}\Delta U &= q_V + w \\ &= q_V - P\Delta V \\ &= q_V \end{aligned} \tag{3.8}$$

In our arrangement, the volume is kept constant so that $P\Delta V = 0$ and $\Delta U = q_V$. We have used the subscript V on q to remind us of this condition. Equation 3.8 may seem strange at first: ΔU is equated to the heat released, which, as we said earlier, is not a state function. However, we have restricted ourselves to a particular process or path, that is, one that takes place at constant volume; hence q_V can have only one value for a given amount of the particular compound burned in the calorimeter.

3.3 Enthalpy

In the laboratory, most chemical and physical processes are carried out under constant pressure rather than constant volume conditions. In such cases, we write

$$\Delta U = q + w = q_P - P\Delta V$$

or

$$U_2 - U_1 = q_P - P(V_2 - V_1)$$

where the subscript P denotes the constant-pressure process. Rearrangement of the equation above gives

$$q_P = (U_2 + PV_2) - (U_1 + PV_1) \tag{3.9}$$

Earlier we discussed that q is not a state function, but U, P, and V are. Therefore, we define a function, called *enthalpy* (H), as follows:

$$H = U + PV \tag{3.10}$$

where U, P, and V are the internal energy, pressure, and volume of the system. All the terms in Equation 3.10 are functions of state; H has the units of energy. From Equation 3.10, we can write the change in H as

$$\Delta H = H_2 - H_1 = (U_2 + P_2V_2) - (U_1 + P_1V_1)$$

Setting $P_2 = P_1 = P$ for a constant-pressure process, we obtain, by comparison with Equation 3.9,

$$\Delta H = (U_2 + PV_2) - (U_1 + PV_1) = q_P$$

Again, we have restricted the change to a specific path—this time at constant pressure—so that the heat, q_P, can be equated directly to the change in the state function, H.

In general, when a system undergoes a change from state 1 to state 2, the change in enthalpy is given by

$$\begin{aligned}\Delta H &= \Delta U + \Delta(PV) \\ &= \Delta U + P\Delta V + V\Delta P + \Delta P\Delta V \qquad (3.11)\end{aligned}$$

This equation applies if neither pressure nor volume is kept constant. The last term, $\Delta P\Delta V$, is not negligible.* Recall that both P and V in Equation 3.11 refer to the system. If the change is carried out, say, at constant pressure, and if the pressure exerted by the system on the surroundings (P_{in}) is equal to the pressure exerted by the surroundings on the system (P_{ext}), that is,

$$P_{in} = P_{ext} = P$$

then we have $\Delta P = 0$, and Equation 3.11 now becomes

$$\Delta H = \Delta U + P\Delta V \qquad (3.12)$$

Similarly, for an infinitesimal change that occurs under constant-pressure conditions with the external pressure equal to the internal pressure,

$$dH = dU + PdV$$

A Comparison of ΔU and ΔH

What is the difference between ΔU and ΔH? Both terms represent the change in energy, but their values differ because the conditions are not the same. Consider the following situation. The heat evolved when two moles of sodium react with water,

$$2\text{Na}(s) + 2\text{H}_2\text{O}(l) \rightarrow 2\text{NaOH}(aq) + \text{H}_2(g)$$

is 367.5 kJ. Because the reaction takes place at constant pressure, $q_P = \Delta H = -367.5$ kJ. To calculate the change in internal energy, from Equation 3.12, we write

$$\Delta U = \Delta H - P\Delta V$$

* Note that for an infinitesimal change, we would write $dH = dU + PdV + VdP + dPdV$. Because $dPdV$ is the product of two infinitesimal quantities, we would ignore it and have $dH = dU + PdV + VdP$.

If we assume the temperature to be 25°C and ignore the small change in the volume of solution, we can show that the volume of 1 mole of H_2 generated at 1 atm is 24.5 L, so that $-P\Delta V = -24.5$ L atm or -2.5 kJ. Finally,

$$\Delta U = -367.5 \text{ kJ} - 2.5 \text{ kJ}$$

$$= -370.0 \text{ kJ}$$

This calculation shows that ΔU and ΔH are slightly different. The reason ΔH is smaller than ΔU in this case is that some of the internal energy released is used to do gas expansion work (the H_2 generated has to push the air back), so less heat is evolved. In general, the difference between ΔH and ΔU in reactions involving gases is $\Delta(PV)$ or $\Delta(nRT) = RT\Delta n$ (if T is constant), where Δn is the change in the number of moles of gases; that is,

$$\Delta n = n_{\text{products}} - n_{\text{reactants}}$$

For the above reaction, $\Delta n = 1$ mol. Thus, at $T = 298$ K, $RT\Delta n$ is approximately 2.5 kJ, which is a small but not negligible quantity in accurate work. On the other hand, for chemical reactions occurring in condensed phases (liquids and solids), ΔV is usually a small number (≤ 0.1 L per mole of reactant converted to product) so that $P\Delta V = 0.1$ L atm, or 10 J, which can be neglected in comparison with ΔU and ΔH. Thus, changes in enthalpy and internal energy in reactions not involving gases or in cases for which $\Delta n = 0$ are one and the same for all practical purposes.

EXAMPLE 3.3

A 0.5122-g sample of naphthalene ($C_{10}H_8$) was burned in a constant-volume bomb calorimeter. Consequently, the temperature of the water in the inner jacket (see Figure 3.3) rose from 20.17°C to 24.08°C. If the effective heat capacity (C_V) of the bomb calorimeter plus water is 5267.8 J K^{-1}, calculate ΔU and ΔH for the combustion of naphthalene in kJ mol^{-1}.

ANSWER

The reaction is

$$C_{10}H_8(s) + 12O_2(g) \rightarrow 10CO_2(g) + 4H_2O(l)$$

The amount of heat evolved is given by

$$C_V\Delta T = (5267.8 \text{ J K}^{-1})(3.91 \text{ K}) = 20.60 \text{ kJ}$$

From the molar mass of naphthalene (128.2 g), we write

$$q_V = \Delta U = -\frac{(20.60 \text{ kJ})(128.2 \text{ g mol}^{-1})}{0.5122 \text{ g}} = -5156 \text{ kJ mol}^{-1}$$

The negative sign indicates that the reaction is exothermic.

To calculate ΔH, we start with $\Delta H = \Delta U + \Delta(PV)$. When all reactants and products are in condensed phases, $\Delta(PV)$ is negligible in comparison with ΔH and ΔU. When gases are involved, $\Delta(PV)$ cannot be ignored. Assuming ideal-gas behavior, we have $\Delta(PV) = \Delta(nRT) = RT\Delta n$, where Δn is the change in the number of moles of gas in the reaction. Note that T here refers to the *initial* temperature because we are comparing reactants and products under the same conditions. For our reaction, $\Delta n = (10 - 12)$ mol $= -2$ mol so that

$$\begin{aligned}\Delta H &= \Delta U + RT\Delta n \\ &= -5156 \text{ kJ mol}^{-1} + \frac{(8.314 \text{ J K}^{-1} \text{ mol}^{-1})(293.32 \text{ K})(-2)}{1000 \text{ J/kJ}} \\ &= -5161 \text{ kJ mol}^{-1}\end{aligned}$$

COMMENT

(1) The difference between ΔU and ΔH is quite small for this reaction. The reason is that $\Delta(PV)$ (which in this case is equal to $RT\Delta n$) is small compared to ΔU or ΔH. Because we assumed ideal-gas behavior (we ignored the volume change of condensed phases), ΔU has the same value (-5156 kJ mol^{-1}) whether the process occurs at constant V or at constant P because the internal energy is independent of pressure or volume. Similarly, $\Delta H = -5161$ kJ mol^{-1}, whether the process is carried out at constant V or at constant P. The heat change q, however, is -5156 kJ mol^{-1} at constant V and -5161 kJ mol^{-1} at constant P, because it is path dependent. (2) In our calculation we ignored the heat capacities of the products (water and carbon dioxide) and the excess oxygen gas used in the combustion because the amounts of these substances are small compared to the bomb calorimeter itself. This omission does not introduce serious errors.

EXAMPLE 3.4

Compare the difference between ΔH and ΔU for the following physical changes: (a) 1 mol ice $\rightarrow$ 1 mol water at 273 K and 1 atm and (b) 1 mol water $\rightarrow$ 1 mol steam at 373 K and 1 atm. The molar volumes of ice and water at 273 K are 0.0196 L mol^{-1} and 0.0180 L mol^{-1}, respectively, and the molar volumes of water and steam at 373 K are 0.0188 L mol^{-1} and 30.61 L mol^{-1}, respectively.

ANSWER

Both cases are constant-pressure processes:

$$\Delta H = \Delta U + \Delta(PV) = \Delta U + P\Delta V$$

or

$$\Delta H - \Delta U = P\Delta V$$

(a) The change in molar volume when ice melts is

$$\begin{aligned}\Delta V &= \overline{V}(l) - \overline{V}(s)\\ &= (0.0180 - 0.0196)\ \text{L mol}^{-1}\\ &= -0.0016\ \text{L mol}^{-1}\end{aligned}$$

Hence,

$$\begin{aligned}P\Delta V &= (1\ \text{atm})(-0.0016\ \text{L mol}^{-1})\\ &= -0.0016\ \text{L atm mol}^{-1}\\ &= -0.16\ \text{J mol}^{-1}\end{aligned}$$

(b) The change in molar volume when water boils is

$$\begin{aligned}\Delta V &= \overline{V}(g) - \overline{V}(l)\\ &= (30.61 - 0.0188)\ \text{L mol}^{-1}\\ &= 30.59\ \text{L mol}^{-1}\end{aligned}$$

Hence,

$$\begin{aligned}P\Delta V &= (1\ \text{atm})(30.59\ \text{L mol}^{-1})\\ &= 30.59\ \text{L atm mol}^{-1}\\ &= 3100\ \text{J mol}^{-1}\end{aligned}$$

COMMENT

This example clearly shows that $(\Delta H - \Delta U)$ is negligibly small for condensed phases but can be quite appreciable if the process involves gases. Further, in (a), $\Delta U > \Delta H$; that is, the increase in the internal energy of the system is greater than the heat absorbed by the system because when ice melts, there is a decrease in volume. Consequently, work is done on the system by the surroundings. The opposite situation holds for (b), because in this case, steam is doing work on the surroundings.

For many physical processes (such as phase transitions) and chemical reactions (such as acid–base neutralization), ΔH can be measured in a constant-pressure calorimeter. To make a constant-pressure calorimeter, we can stack two plastic coffee cups together as shown in Figure 3.4. Because the whole apparatus is open to the atmosphere, the pressure is constant, and the heat change for the process is equal to the enthalpy change ($q_P = \Delta H$). Provided we know the heat capacity of the calorimeter, as well as the temperature change, we can determine the heat or enthalpy change. In this respect, the constant-pressure calorimeter is similar to the constant-volume calorimeter.

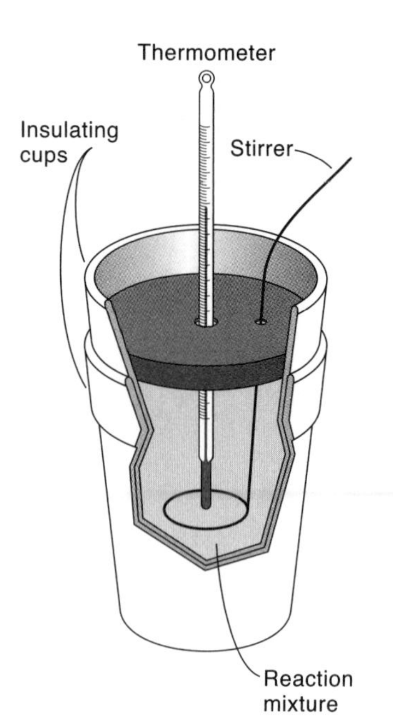

Figure 3.4
A constant-pressure calorimeter made of two plastic coffee cups. The outer cup helps to insulate the reacting mixture from the surroundings. Two solutions of known volume containing the reactants at the same temperature are carefully mixed in the calorimeter. The heat produced or absorbed by the reaction can be determined from the temperature change, the quantities of the solutions used, and the heat capacity of the calorimeter.

3.4 A Closer Look at Heat Capacities

The concept of heat capacity was first introduced in Chapter 2. So far in this chapter, we have examined the use of heat capacity in calorimetry, that is, the measurement of heat changes in chemical and physical processes. In this section, we shall examine the measurement of heat capacity more closely. In particular, we shall see that the heat capacity of a gas depends on the nature of a process; that is, whether it is constant pressure or constant volume.

For this discussion of heat capacity, we assume no phase change.

When heat is added to a substance, its temperature will rise. This fact we know well. But just how much the temperature will rise depends on (1) the amount of heat delivered, (2) the amount of the substance present, (3) the chemical nature and physical state of the substance, and (4) the conditions under which energy is added to the substance. The temperature rise (ΔT) for a given amount of a substance is related to the heat added (q) by the equation

$$q = C\Delta T$$

or

$$C = \frac{q}{\Delta T} \tag{3.13}$$

where C, a proportionality constant, is called the heat capacity. Because the increase in temperature depends on the amount of substance present, it is often convenient to speak of the heat capacity of 1 mole of a substance, or molar heat capacity, $\overline{C}$, where

$$\overline{C} = \frac{C}{n} = \frac{q}{n\Delta T} \tag{3.14}$$

where n is the number of moles of the substance present in a particular measurement. Note that C is an extensive property, but $\overline{C}$, like all other molar quantities, is intensive.

Heat capacity is a directly measurable quantity. Knowing the amount of the substance present, the heat added, and the temperature rise, we can readily calculate the value of $\overline{C}$ using Equation 3.14. It turns out, however, that the value we calculate also depends on how the heating process is carried out. Although many different conditions can be realized in practice, we shall consider only two important cases here: constant volume and constant pressure. We have already seen in Section 3.2 that for a constant-volume process, the heat absorbed by the system is equal to the increase in internal energy, that is, $\Delta U = q_V$. Hence, the heat capacity at constant volume, C_V, is given by

$$C_V = \frac{q_V}{\Delta T} = \frac{\Delta U}{\Delta T}$$

or, expressed in partial derivatives,

$$C_V = \left(\frac{\partial U}{\partial T}\right)_V \tag{3.15}$$

or

$$dU = C_V dT \tag{3.16}$$

Similarly, for a constant-pressure process we have $\Delta H = q_P$, so that the heat capacity at constant pressure is

$$C_P = \frac{q_P}{\Delta T} = \frac{\Delta H}{\Delta T}$$

or, expressed in partial derivatives,

$$C_P = \left(\frac{\partial H}{\partial T}\right)_P \tag{3.17}$$

or

$$dH = C_P dT \tag{3.18}$$

From the definitions of C_V and C_P, we can calculate the values of ΔU and ΔH for processes carried out under constant-volume or constant-pressure conditions. Integrating Equations 3.16 and 3.18 between temperatures T_1 and T_2, we obtain

$$\Delta U = \int_{T_1}^{T_2} C_V dT = C_V(T_2 - T_1) = C_V \Delta T = n\overline{C}_V \Delta T \tag{3.19}$$

$$\Delta H = \int_{T_1}^{T_2} C_P dT = C_P(T_2 - T_1) = C_P \Delta T = n\overline{C}_P \Delta T \tag{3.20}$$

where n is the number of moles of the substance present. We have assumed that both C_V and C_P are independent of temperature. This is not always true, however. At low temperatures ($\leq$ 300 K), only the translational and rotational motions described in Section 2.9 make a major contribution to heat capacity. At high temperatures, when transitions among vibrational energy levels become appreciable, the heat capacity will increase accordingly. Studies of the temperature dependence of heat capacity for many substances show that it can be represented by an equation $C_P = a + bT$, where a and b are constants for a given substance over a particular temperature range. Such an expression must be used in Equation 3.20 in accurate work. In most cases, however, if the temperature change in a process is small, say 50 K or less, we can often treat C_V and C_P as if they were independent of temperature.

In general, C_V and C_P for a given substance are not equal to each other. The reason is that work has to be done on the surroundings in a constant-pressure process, so that *more* heat is required to raise the temperature by a definite amount in a constant-pressure process than in a constant-volume process. It follows, therefore, that $C_P > C_V$. This is true mainly for gases. The volume of a liquid or solid does not change appreciably with temperature; consequently, the work done as it expands is quite small so that for most purposes, C_V and C_P are nearly the same for condensed phases.

Our next step is to see how C_P differs from C_V for an ideal gas. We start by writing

$$H = U + PV = U + nRT$$

For an infinitesimal change in temperature, dT, the change in enthalpy for a given amount of an ideal gas is

$$dH = dU + d(nRT)$$
$$= dU + nRdT$$

Substituting $dH = C_P dT$ and $dU = C_V dT$ into the above equation, we get

$$C_P dT = C_V dT + nRdT$$
$$C_P = C_V + nR$$
$$C_P - C_V = nR \qquad (3.21a)$$

or

$$\overline{C}_P - \overline{C}_V = R \qquad (3.21b)$$

Thus, for an ideal gas, the molar constant-pressure heat capacity is greater than the molar constant-volume heat capacity by R, the gas constant. Appendix B lists the $\overline{C}_P$ values of many substances.

EXAMPLE 3.5

Calculate the values of ΔU and ΔH for the heating of 55.40 g of xenon from 300 K to 400 K. Assume ideal-gas behavior and that the heat capacities at constant volume and constant pressure are independent of temperature.

ANSWER

Xenon is a monatomic gas. In Section 2.9 we saw that $\overline{C}_V = \frac{3}{2}R = 12.47\ \text{J K}^{-1}\ \text{mol}^{-1}$. Thus, from Equation 3.21b, we have $\overline{C}_P = \frac{3}{2}R + R = \frac{5}{2}R = 20.79\ \text{J K}^{-1}\ \text{mol}^{-1}$. The quantity 55.40 g of Xe corresponds to 0.4219 mole. From Equations 3.19 and 3.20,

$$\begin{aligned}\Delta U &= n\overline{C}_V\Delta T \\ &= (0.4219\ \text{mol})(12.47\ \text{J K}^{-1}\ \text{mol}^{-1})(400 - 300)\text{K} \\ &= 526\ \text{J}\end{aligned}$$

$$\begin{aligned}\Delta H &= n\overline{C}_P\Delta T \\ &= (0.4219\ \text{mol})(20.79\ \text{J K}^{-1}\ \text{mol}^{-1})(400 - 300)\text{K} \\ &= 877\ \text{J}\end{aligned}$$

EXAMPLE 3.6

The molar heat capacity of oxygen at constant pressure is given by $(25.7 + 0.0130T/\text{K})\ \text{J K}^{-1}\ \text{mol}^{-1}$ over a certain temperature range. Calculate the enthalpy change when 1.46 moles of O_2 are heated from 298 K to 367 K.

ANSWER

From Equation 3.20,

$$\begin{aligned}\Delta H &= \int_{T_1}^{T_2} n\overline{C}_P dT = \int_{298\ \text{K}}^{367\ \text{K}} (1.46\ \text{mol})(25.7 + 0.0130T/\text{K})\ \text{J K}^{-1}\ \text{mol}^{-1}\, dT \\ &= (1.46\ \text{mol})\left[25.7T + \frac{0.130T^2}{2\ \text{K}}\right]_{298\ \text{K}}^{367\ \text{K}} \text{J K}^{-1}\ \text{mol}^{-1} \\ &= 3.02 \times 10^3\ \text{J}\end{aligned}$$

3.5 Gas Expansion

The expansion of an ideal gas is a simple process to which we can apply what we have learned so far about the first law of thermodynamics. Although ideal-gas expansion does not have much chemical significance, it enables us to use some of the equations derived in previous sections to calculate changes in thermodynamic quantities. We consider two special cases: isothermal expansion and adiabatic expansion.

Isothermal Expansion

An *isothermal* process is one in which the temperature is held constant. The work done in isothermal reversible and irreversible expansions was discussed in some detail in Section 3.1 and will not be repeated here. Instead, we shall look at changes in heat, internal energy, and enthalpy.

Because temperature does not change in an isothermal process, the change in internal energy is zero; that is, $\Delta U = 0$. This follows from the fact that ideal-gas molecules neither attract nor repel each other. Consequently, their total energy is independent of the distance of separation between molecules and therefore the volume. Mathematically, this relationship is expressed as

$$\left(\frac{\partial U}{\partial V}\right)_T = 0$$

This partial derivative tells us that the change in internal energy of the system with respect to volume is zero at constant temperature. For example, for one mole of a monatomic ideal gas, $U = \frac{3}{2}RT$ and $[\partial(\frac{3}{2}RT)/\partial V]_T = 0$. From Equation 3.6,

$$\Delta U = q + w = 0$$

or

$$q = -w$$

Thus, in isothermal expansion, the heat absorbed by the gas is equal to the work done by the ideal gas on its surroundings. Referring to Example 3.1, we see that the heat absorbed by an ideal gas when it expands from a pressure of 15 atm to 1 atm at 300 K is zero in (a), 1980 J in (b), and 5740 J in (c). Because maximum work is performed in a reversible process, it is not surprising that the heat absorbed is greatest for (c).

Finally, we also want to calculate the enthalpy change for such an isothermal process. Starting from

$$\Delta H = \Delta U + \Delta(PV)$$

we see that $\Delta U = 0$, as mentioned above, and because PV is constant at constant T and n (Boyle's law), $\Delta(PV) = 0$, so that $\Delta H = 0$. Alternatively, we could write $\Delta(PV) = \Delta(nRT)$. Because the temperature is unchanged and no chemical reaction occurs, both n and T are constant and $\Delta(nRT) = 0$; hence $\Delta H = 0$.

Adiabatic Expansion

Referring to Figure 3.1, suppose we now isolate the cylinder thermally from its surroundings so that there is no heat exchange during the expansion. This means that $q = 0$ and the process is *adiabatic*. Consequently, there will be a temperature drop and T will no longer be a constant. Let's consider two cases here.

Reversible Adiabatic Expansion. Suppose that the expansion is reversible. The two questions we ask are: What is the P–V relationship for the initial and final states, and how much work is done in the expansion?

For an infinitesimal adiabatic expansion, the first law takes the form

$$\begin{aligned} dU &= đq + đw \\ &= đw = -PdV = -\frac{nRT}{V}\,dV \end{aligned}$$

or

$$\frac{dU}{nT} = -R\frac{dV}{V}$$

Note that $đq = 0$ and we have replaced the external, opposing pressure with the internal pressure of the gas because it is a reversible process. Substituting $dU = C_V dT$ into the above equation, we obtain

$$\frac{C_V dT}{nT} = \overline{C}_V \frac{dT}{T} = -R\frac{dV}{V} \tag{3.22}$$

Integration of Equation 3.22 between the initial and final states gives

$$\int_{T_1}^{T_2} \overline{C}_V \frac{dT}{T} = -R\int_{V_1}^{V_2} \frac{dV}{V}$$

Assuming $\overline{C}_V$ to be independent of temperature, we get

$$\overline{C}_V \ln\frac{T_2}{T_1} = R \ln\frac{V_1}{V_2}$$

Because $\overline{C}_P - \overline{C}_V = R$ for an ideal gas, we write

$$\overline{C}_V \ln\frac{T_2}{T_1} = (\overline{C}_P - \overline{C}_V) \ln\frac{V_1}{V_2}$$

Dividing by $\overline{C}_V$ on both sides, we get

$$\begin{aligned} \ln\frac{T_2}{T_1} &= \left(\frac{\overline{C}_P}{\overline{C}_V} - 1\right)\ln\frac{V_1}{V_2} \\ &= (\gamma - 1)\ln\frac{V_1}{V_2} \\ &= \ln\left(\frac{V_1}{V_2}\right)^{\gamma-1} \end{aligned}$$

where γ is called the *heat capacity ratio*, given by

$$\gamma = \frac{\overline{C}_P}{\overline{C}_V} \tag{3.23}$$

We assume that the equipartition of energy theorem holds here.

For a monatomic gas, $\overline{C}_V = \frac{3}{2}R$ (see Section 2.8) and $\overline{C}_P = \frac{5}{2}R$, so that $\gamma = \frac{5}{3}$, or 1.67. For diatomic molecules, we have $\overline{C}_V = \frac{5}{2}R$ and $\overline{C}_P = \frac{7}{2}R$, so that $\gamma = \frac{7}{5}$, or 1.4. Finally, we arrive at the following useful results:

$$\left(\frac{V_1}{V_2}\right)^{\gamma-1} = \frac{T_2}{T_1} = \frac{P_2V_2}{P_1V_1} \qquad \left(\frac{P_1V_1}{T_1} = \frac{P_2V_2}{T_2}\right)$$

or

$$\left(\frac{V_1}{V_2}\right)^{\gamma} = \frac{P_2}{P_1}$$

Thus, for an adiabatic process, the P–V relationship becomes

$$P_1V_1^{\gamma} = P_2V_2^{\gamma} \tag{3.24}$$

Recall the conditions under which this equation was derived: it applies to an ideal gas, and it applies to a reversible adiabatic change. Equation 3.24 differs from Boyle's law ($P_1V_1 = P_2V_2$) in the exponent γ, because temperature is *not* kept constant in an adiabatic expansion.

The work done in an adiabatic process is given by

$$w = \int_1^2 dU = \Delta U = \int_{T_1}^{T_2} C_V dT$$

$$= C_V(T_2 - T_1)$$

$$= n\overline{C}_V(T_2 - T_1) \tag{3.25}$$

where $T_2 < T_1$ and $w < 0$, because the gas expands. We assume $\overline{C}_V$ to be temperature independent.

Appearance of the quantity $\overline{C}_V$ in Equation 3.25 may seem strange because the volume is not held constant. However, adiabatic expansion (from P_1, V_1, T_1 to P_2, V_2, T_2) can be thought of as a two-step process, as shown in Figure 3.5. First, the gas is isothermally expanded from P_1, V_1 to P_2', V_2 at T_1. Because temperature is constant, $\Delta U = 0$. Next, the gas is cooled at constant volume from T_1 to T_2 and its pressure drops from P_2' to P_2. In this case, $\Delta U = n\overline{C}_V(T_2 - T_1)$, which is Equation 3.25. The fact that U is a state function enables us to analyze the process by employing a different path.

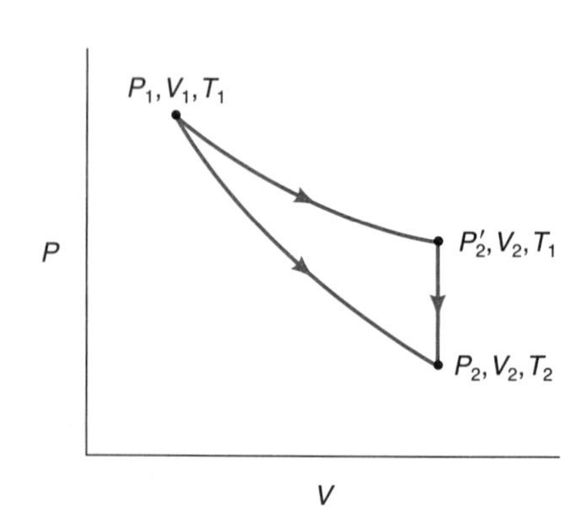

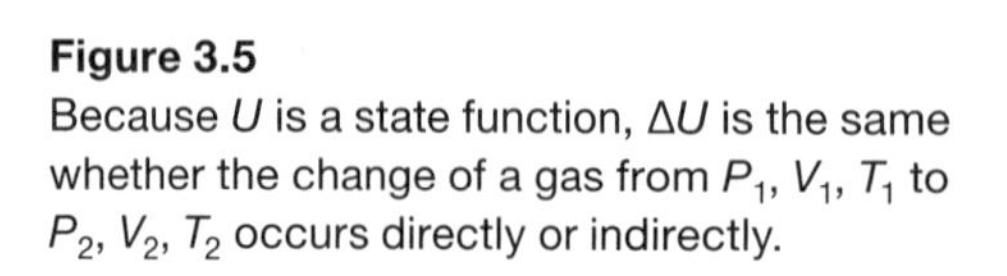

Figure 3.5
Because U is a state function, ΔU is the same whether the change of a gas from P_1, V_1, T_1 to P_2, V_2, T_2 occurs directly or indirectly.

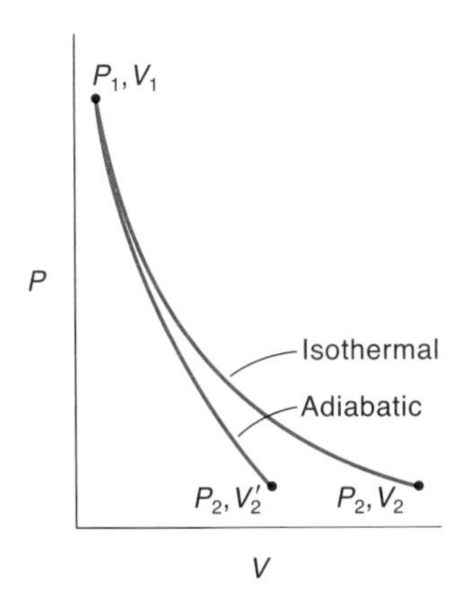

Figure 3.6
P–V plots of an adiabatic, reversible and an isothermal, reversible expansion of an ideal gas. In each case, the work done in expansion is represented by the area under the curve. Note that the curve is steeper for the adiabatic expansion because γ is greater than one. Consequently, the area is smaller than the isothermal case.

Examples 3.1 and 3.7 show that less work is performed in a reversible, adiabatic expansion than in a reversible, isothermal expansion. In the latter case, heat is absorbed from the surroundings to make up for the work done by the gas, but this does not occur in an adiabatic process, so the temperature drops. Plots of reversible isothermal and reversible adiabatic expansions are shown in Figure 3.6.

This comparison shows that not all reversible processes do the same amount of work.

EXAMPLE 3.7

A quantity of 0.850 mole of a monatomic ideal gas initially at a pressure of 15.0 atm and 300 K is allowed to expand until its final pressure is 1.00 atm (see Example 3.1). Calculate the work done if the expansion is carried out adiabatically and reversibly.

ANSWER

Our first task is to calculate the final temperature, T_2. This is done in three steps. First, we need to evaluate V_1, given by $V_1 = nRT_1/P_1$.

$$V_1 = \frac{(0.850 \text{ mol})(0.08206 \text{ L atm K}^{-1} \text{ mol}^{-1})(300 \text{ K})}{15.0 \text{ atm}}$$

$$= 1.40 \text{ L}$$

Next, we calculate the value of V_2 using the following relation:

$$P_1 V_1^\gamma = P_2 V_2^\gamma$$

$$V_2 = \left(\frac{P_1}{P_2}\right)^{1/\gamma} V_1 = \left(\frac{15.0}{1.00}\right)^{3/5} (1.40 \text{ L}) = 7.11 \text{ L}$$

Finally, we have $P_2 V_2 = nRT_2$, or

$$T_2 = \frac{P_2 V_2}{nR} = \frac{(1.00 \text{ atm})(7.11 \text{ L})}{(0.850 \text{ mol})(0.08206 \text{ L atm K}^{-1} \text{ mol}^{-1})}$$

$$= 102 \text{ K}$$

Hence,

$$\Delta U = w = n\overline{C}_V(T_2 - T_1)$$
$$= (0.850 \text{ mol})(12.47 \text{ J K}^{-1} \text{ mol}^{-1})(102 - 300) \text{ K}$$
$$= -2.10 \times 10^3 \text{ J}$$

Irreversible Adiabatic Expansion. Finally, consider what happens in an irreversible adiabatic expansion. Suppose we start with an ideal gas at P_1, V_1, and T_1 and P_2 is the constant external pressure. The final volume and temperature of the gas are V_2 and T_2. Again, $q = 0$ so that

$$\Delta U = n\overline{C}_V(T_2 - T_1) = w = -P_2(V_2 - V_1) \tag{3.26}$$

Furthermore, from the ideal-gas equation we write

$$V_1 = \frac{nRT_1}{P_1} \quad \text{and} \quad V_2 = \frac{nRT_2}{P_2}$$

Substituting the expressions for V_1 and V_2 in Equation 3.26, we obtain

$$n\overline{C}_V(T_2 - T_1) = -P_2\left(\frac{nRT_2}{P_2} - \frac{nRT_1}{P_1}\right)$$

Thus, knowing the initial conditions and P_2, we can solve for T_2 and hence the work done (see Problem 3.34).

The decrease in temperature, or the cooling effect that occurs in an adiabatic expansion, has some interesting practical consequences. A familiar example is the formation of fog when the caps of carbonated soft drinks or corks of champagne bottles are removed. Initially, the bottles are pressurized with carbon dioxide and air, and the space above the liquid is saturated with water vapor. When the cap is removed, the gases inside rush out. The process takes place so rapidly that the expansion of the gases can be compared to adiabatic expansion. As a result, the temperature drops and water vapor condenses to form the observed fog.

3.6 The Joule–Thomson Effect

As mentioned in the last section, the internal energy of an ideal gas does not depend on its volume—it is a function only of its temperature. Therefore, if an ideal gas expands but does no work; that is, if it expands against a vacuum, then there is no change in its temperature. Consequently, its internal energy remains unchanged and we have $(\partial U/\partial V)_T = 0$. What about a real gas? In 1845 the English physicist James Prescott Joule (1831–1879) carried out an experiment to answer this question

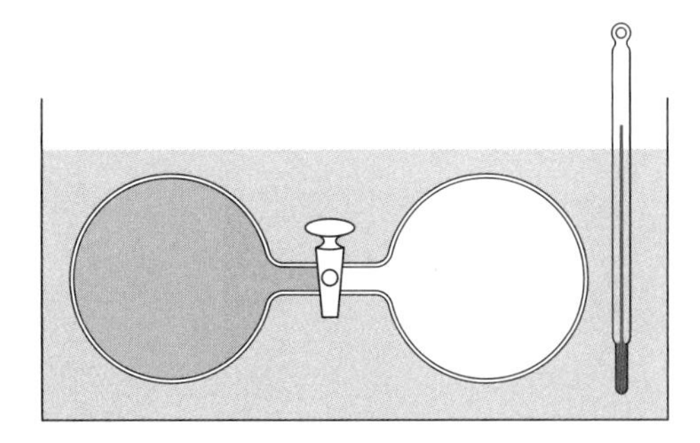

Figure 3.7
The Joule expansion experiment. As the gas expands into the vacuum, the change in temperature is registered by the thermometer.

(Figure 3.7). By opening the stopcock, a gas was allowed to expand into a vacuum. Because there was no opposing pressure, no work was done. Joule was unable to detect a change of the water bath's temperature, which meant that there was no heat exchange between the gas and the water bath. Therefore, he concluded that, like an ideal gas, the internal energy of a real gas is also independent of its volume at constant temperature.

It turned out that Joule's experiment was unreliable because the heat capacity of the water bath was much greater than that of the gas, so any change in the gas's temperature would be difficult to detect. In 1853 William Thomson (Lord Kelvin) collaborated with Joule in a similar experiment that allowed far more accurate results to be obtained. As shown in Figure 3.8, a gas initially at P_1, V_1, T_1 is expanded (by pushing the piston from left to right) through a porous barrier against a pressure $P_2 (P_2 < P_1)$. Because the rate of gas transport is slow, the pressures are held constant at P_1 and P_2. In addition, the apparatus is thermally isolated from its surroundings, so this is an adiabatic process; that is, $q = 0$.

In the initial experiment, Joule supposedly borrowed his wife's silk scarf to use as the porous barrier.

The work done in this process consists of two parts. The work done on the gas on the left, w_L, is given by

$$\begin{aligned} w_{\mathrm{L}} &= -P_1(0 - V_1) \\ &= P_1 V_1 \end{aligned}$$

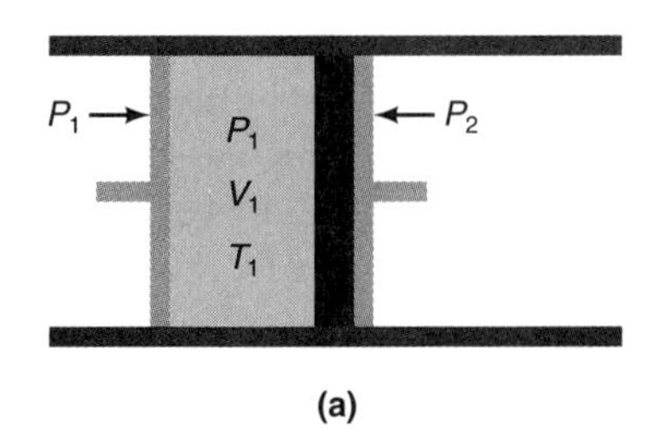

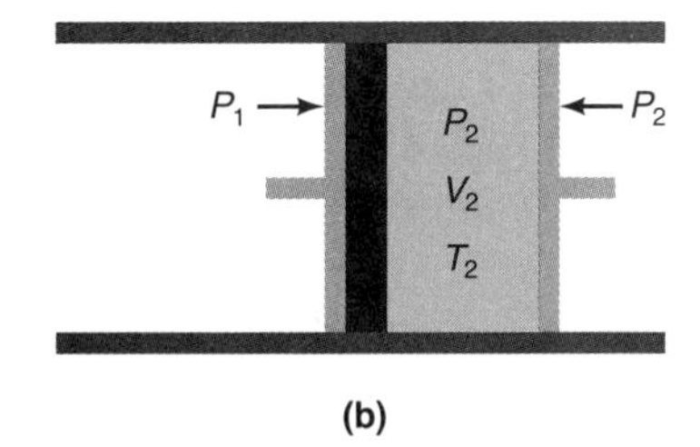

Figure 3.8
Apparatus for studying the Joule–Thomson effect. Initially the gas on the left of the porous barrier (shown in red) is at P_1, V_1, T_1 and the piston on the right is in contact with the barrier. As the piston on the left is slowly pushed toward the right, the gas moves through the barrier and expands against P_2. Eventually, the gas reaches the final state at P_2, V_2, T_2. The apparatus is thermally isolated from its surroundings so $q = 0$.

because the final volume of the gas is zero. (Note that this is an irreversible process, so the work done is $-P\Delta V$.) After the gas passes through the barrier, it now does work against a constant pressure P_2, so the work done by the gas on the right side is

$$\begin{aligned} w_R &= -P_2(V_2 - 0) \\ &= -P_2V_2 \end{aligned}$$

Thus, the net work done is

$$w = P_1V_1 - P_2V_2$$

and the change in internal energy is

$$\Delta U = U_2 - U_1 = q + w = w = P_1V_1 - P_2V_2$$

Rearranging, we obtain

$$U_2 + P_2V_2 = U_1 + P_1V_1 \tag{3.27}$$

From the definition of enthalpy ($H = U + PV$), we write

$$H_2 = H_1$$

That is, the gas expansion takes place at constant enthalpy, or $\Delta H = 0$.

The Joule–Thomson experiment showed there was a temperature drop as a result of gas expansion. We define the *Joule–Thomson coefficient*, μ_{JT}(in units K atm^{-1}), as

μ_{JT} is zero for an ideal gas.

$$\mu_{JT} = \left(\frac{\partial T}{\partial P}\right)_H \tag{3.28}$$

which is a measure of the change in temperature with respect to pressure at constant enthalpy. For a finite change, the coefficient takes the form

$$\mu_{JT} = \left(\frac{\Delta T}{\Delta P}\right)_H \tag{3.29}$$

As Equation 3.28 shows, μ_{JT} can be obtained from the slope of a plot of T versus P. To construct such a curve shown in Figure 3.9a, we start with some initial P_i and T_i (point a) and choose a value of P_f (less than P_i) as the opposing pressure, resulting in T_f (point b). We repeat the experiment by using the same P_i and T_i but a different opposing pressure P_f and measure the final temperature T_f (point c). Continuing with this procedure, we obtain points d, e, ... and then join these points with a smooth curve. Because all the states (represented by the points) have the same enthalpy, such a plot yields an *isenthalpic* (same enthalpy) curve. To the left of point d the slope of the curve is positive. A positive μ_{JT} corresponds to a cooling effect as the gas expands (temperature decreases with decreasing pressure; that is, both ΔT and ΔP are negative). The opposite happens to the right of point d. Here the slope is negative and a negative μ_{JT} means that the gas actually warms up on expansion. In terms of intermolecular interactions, we can understand both the cooling and heating phenomenon.

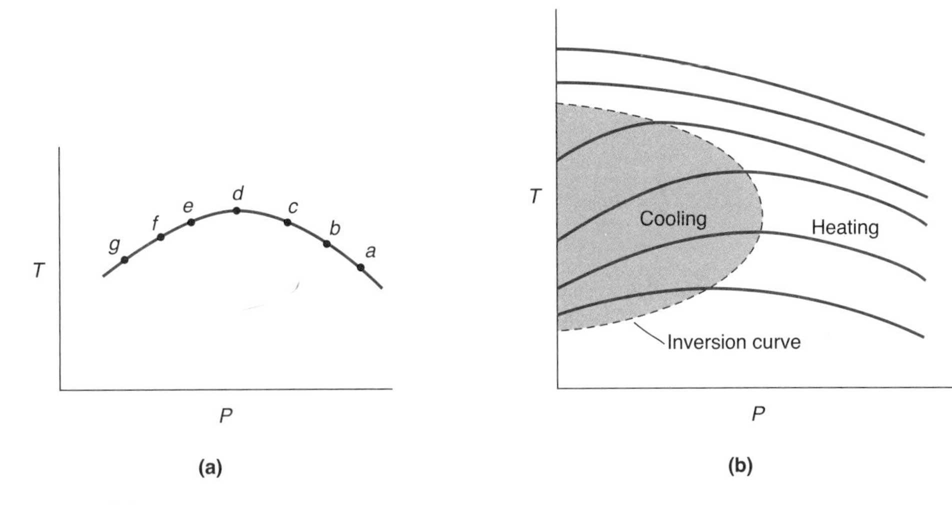

Figure 3.9
(a) An isenthalpic curve obtained from a series of Joule–Thomson experiments (see text). The slope of the line at any point *P* and *T* is equal to the Joule–Thomson coefficient, (b) *T* versus *P* plots of isenthalpic curves. The boundary line is the locus of the maxima of the isenthalpic curves where $\mu_{JT} = 0$. [Adapted from Kirkwood, J. G. and I. Oppenheim, *Chemical Thermodynamics,* McGraw-Hill Book Company, New York (1961)]

The temperature falls as the gas passes through the porous barrier because some of the gas molecules' kinetic energy is used to break away from the attractive forces. Consequently, the gas cools. Under different temperature and pressure conditions, intermolecular forces are less important than molecular size, so repulsion (due to the proximity of molecules) outweighs attraction. On expansion, the potential energy of the molecules drops, resulting in a release of heat to the surroundings. At point *d*, which is the maximum of the curve, the slope is zero and there is no cooling or heating effect. The temperature at which μ_{JT} is zero is called the *inversion temperature.*

To obtain another isenthalpic curve, we start with a different T_i with P_i being kept the same as before. P_f is again varied and the corresponding T_f's are measured. Repeating the process generates a series of curves shown in Figure 3.9b. The boundary line is the locus of the maxima of the individual isenthalpic curves. Inside the boundary (shaded area) μ_{JT} is positive, meaning that any combination of *T* and *P* that fall within this region will result in the cooling of the gas in a Joule–Thomson expansion. The opposite holds for *T* and *P* combinations outside the boundary.

The Joule–Thomson effect has important practical applications in the liquefaction of gases. According to Figure 3.9b, for the gas expansion to give rise to cooling, its initial temperature must be below the inversion temperature. With the exceptions of helium and hydrogen, most gases have maximum inversion temperatures above room temperature. In the industrial Linde process shown in Figure 3.10, a gas such as nitrogen (maximum inversion temperature = 348°C) at room temperature is first compressed and then subjected to an adiabatic expansion. The cooled gas is circulated back to the compressor and lowers the temperature of the exiting gas before it undergoes another expansion. After many repetitions of the compression−expansion cycle, the gas will reach −196°C, the boiling point of nitrogen, and begin to liquefy. Because the maximum inversion temperature of hydrogen is −68°C, its temperature must first be lowered by liquid nitrogen before it is liquefied by the Linde process.

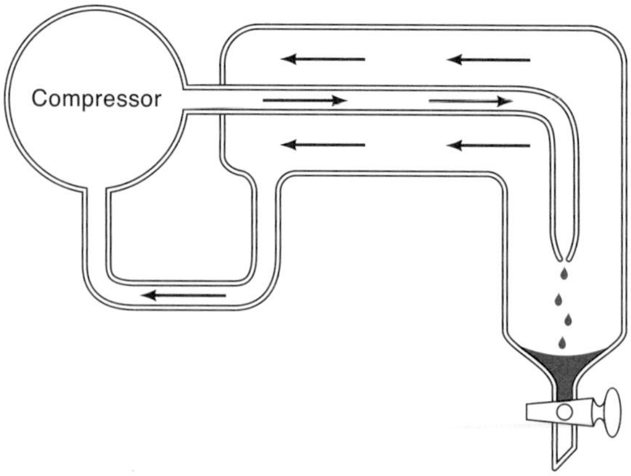

Compressor

Figure 3.10
The Linde process for the liquefaction of gases.

EXAMPLE 3.8

If the Joule–Thomson coefficient of argon is 0.32 K atm^{-1}, estimate the final temperature of Ar at 30 atm and 50°C that is forced through a small orifice to a final pressure of 1 atm.

ANSWER

Using the approximate form of the Joule–Thomson coefficient (Equation 3.29),

$$\mu_{JT} = \left(\frac{\Delta T}{\Delta P}\right)_H$$

we have $\Delta P = (1 - 30)$ atm or -29 atm. Thus,

$$0.32 \text{ K atm}^{-1} = \frac{\Delta T}{-29 \text{ atm}}$$

or $\Delta T = -9$ K and the final temperature is (50 − 9)°C or 41°C.

3.7 Thermochemistry

In this section we shall apply the first law of thermodynamics to thermochemistry, the study of energy changes in chemical reactions.

Standard Enthalpy of Formation

Chemical reactions almost always involve changes in energy. The *enthalpy of reaction* can be defined as the heat absorbed or released in the transformation of reactants at some temperature and pressure to products at the same temperature and pressure. For a constant-pressure process, q_P is equal to the enthalpy change of the reaction,

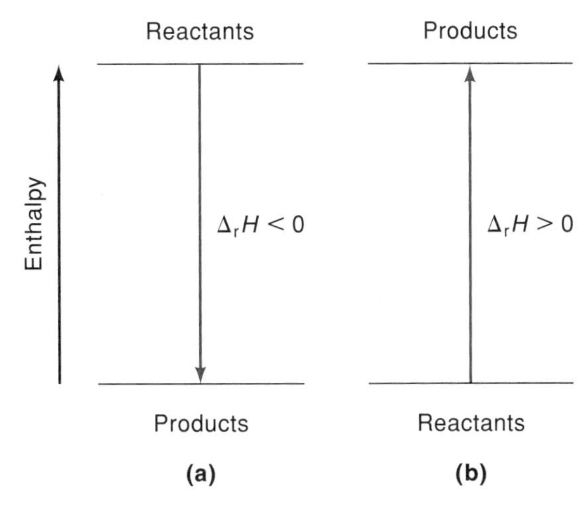

Figure 3.11
Enthalpy changes for (a) an exothermic reaction and (b) an endothermic reaction.

$\Delta_r H$, where the subscript "r" denotes reaction. An *exothermic reaction* is a process that gives off heat to its surroundings and for which $\Delta_r H$ is negative; for an *endothermic reaction* $\Delta_r H$ is positive because the process absorbs heat from the surroundings (Figure 3.11).

Consider the following reaction:

$$\text{C(graphite)} + \text{O}_2(g) \rightarrow \text{CO}_2(g)$$

When 1 mole of graphite is burned in an excess of oxygen at 1 bar and 298 K to form 1 mole of carbon dioxide at the same temperature and pressure, 393.5 kJ of heat is given off.* The enthalpy change for this process is called the *standard enthalpy of reaction*, denoted by $\Delta_r H°$. It has the units of kJ (mol reaction)$^{-1}$. One mole of reaction is when the appropriate numbers of moles of substances (as specified by the stoichiometric coefficients) on the left-hand side are converted to the substances on the right-hand side of the equation. For simplicity, we express $\Delta_r H°$ for the combustion of graphite as -393.5 kJ mol^{-1}, which is defined as the enthalpy change when the reactants in their standard states are converted to products in their standard states. The standard state is defined as follows: For a pure solid or liquid, it is the state at a pressure $P = 1$ bar (see Section 1.2) and some temperature T. For a pure gas, the standard state refers to the hypothetical ideal gas at a pressure of 1 bar and some temperature of interest. The symbol for a standard state is a "circle" superscript.

The standard state is defined only in terms of pressure (1 bar).

In general, the standard enthalpy change of a chemical reaction can be thought of as the total enthalpy of the products minus the total enthalpy of the reactants:

$$\Delta_r H° = \Sigma \nu \overline{H}° \text{ (products)} - \Sigma \nu \overline{H}° \text{ (reactants)}$$

* The temperature during combustion is much higher than 298 K, but we are measuring the total heat change from reactants at 1 bar and 298 K to product at 1 bar and 298 K. Therefore, the heat given off when the product cools to 298 K is part of the enthalpy of reaction.

where $\overline{H}°$ is the standard molar enthalpy and v is the stoichiometric coefficient. The units of $\overline{H}°$ are kJ mol^{-1} and v is a number without units. For the hypothetical reaction

$$aA + bB \rightarrow cC + dD$$

where a, b, c, and d are the respective stoichiometric coefficients of the chemical species A, B, C, and D, the standard enthalpy of reaction is given by

$$\Delta_r H° = c\overline{H}°(C) + d\overline{H}°(D) - a\overline{H}°(A) - b\overline{H}°(B)$$

However, there is no way of measuring the *absolute* values of molar enthalpy of any substance. Only values *relative* to an arbitrary reference can be determined. This problem is similar to the one geographers face in expressing the elevations of specific mountains or valleys. By common agreement all geographic heights and depths are expressed relative to sea level, an arbitrary reference with a defined elevation of "zero" meters. Similarly, chemists have agreed on an arbitrary reference point for enthalpy.

The "sea level" reference point for all enthalpy expressions is called the *standard molar enthalpy of formation* ($\Delta_f\overline{H}°$), where the subscript "f" denotes formation. The standard molar enthalpy of formation is the enthalpy change when one mole of a compound is formed from its constituent elements at 1 bar and 298 K. (Keep in mind that the standard state is defined only in terms of pressure. We shall always use $\Delta_f\overline{H}°$ values at 298 K because most of the thermodynamic data are collected at this temperature.) Referring to the above hypothetical reaction, we can now write the standard enthalpy change as

$$\Delta_r H° = c\Delta_f\overline{H}°(C) + d\Delta_f\overline{H}°(D) - a\Delta_f\overline{H}°(A) - b\Delta_f\overline{H}°(B) \quad (3.30)$$

In general, we write

When the stoichiometric coefficient is one, it is not shown in Equation 3.31.

$$\Delta_r H° = \Sigma v\Delta_f\overline{H}°(\text{products}) - \Sigma v\Delta_f\overline{H}°(\text{reactants}) \quad (3.31)$$

Applying Equation 3.31 to the combustion of graphite shown earlier, we express the standard enthalpy of reaction as

$$\Delta_r H° = \Delta_f\overline{H}°(CO_2) - \Delta_f\overline{H}°(\text{graphite}) - \Delta_f\overline{H}°(O_2) = -393.5 \text{ kJ mol}^{-1}$$

Allotropes are two or more forms of the same element that differ in their physical and chemical properties.

By convention, we arbitrarily assign a value of zero to $\Delta_f\overline{H}°$ for elements in their most stable allotropic forms at 298 K. Thus,

$$\Delta_f\overline{H}°(O_2) = 0$$

$$\Delta_f\overline{H}°(\text{graphite}) = 0$$

because O_2 and graphite are the stable allotropic forms of oxygen and carbon at 1 bar and this temperature. On the other hand, neither ozone nor diamond is the most stable allotropic form at 1 bar and 298 K, and we have $\Delta_f\overline{H}°(O_3) = 142.7\ \text{kJ mol}^{-1}$ and $\Delta_f\overline{H}°(\text{diamond}) = 1.90\ \text{kJ mol}^{-1}$. The standard enthalpy for the combustion of graphite can now be written as

$$\Delta_r H° = \Delta_f\overline{H}°(CO_2) = -393.5\ \text{kJ mol}^{-1}$$

Thus, the standard molar enthalpy of formation for CO_2 is equal to the standard enthalpy of reaction.

There is no mystery in the assignment of a zero value to $\Delta_f\overline{H}°$ for the elements. As mentioned above, we cannot determine the absolute value of the enthalpy of a substance. Only values *relative* to an arbitrary reference can be given. In thermodynamics we are primarily interested in the changes of H. Although any arbitrarily assigned value of $\Delta_f\overline{H}°$ for an element would do, zero makes calculations simpler. The importance of the standard molar enthalpies of formation is that once we know their values, we can calculate the standard enthalpies of reaction. The $\Delta_f\overline{H}°$ values are obtained by either the direct method or the indirect method, described below.

The Direct Method. This method of measuring $\Delta_f\overline{H}°$ works for compounds that can be readily synthesized from their elements. The formation of CO_2 from graphite and O_2 is such an example. Other compounds that can be directly synthesized from their elements include SF_6, P_4O_{10}, and CS_2. The equations representing their syntheses are

$$\text{S(rhombic)} + 3F_2(g) \rightarrow SF_6(g) \qquad \Delta_r H° = -1209\ \text{kJ mol}^{-1}$$

$$4\text{P(white)} + 5O_2(g) \rightarrow P_4O_{10}(s) \qquad \Delta_r H° = -2984.0\ \text{kJ mol}^{-1}$$

$$\text{C(graphite)} + 2\text{S(rhombic)} \rightarrow CS_2(l) \qquad \Delta_r H° = 87.86\ \text{kJ mol}^{-1}$$

Note that S(rhombic) and P(white) are the most stable allotropes of sulfur and phosphorus, respectively, at 1 bar and 298 K, so their $\Delta_f\overline{H}°$ values are zero. As is the case for CO_2, the standard enthalpy of reaction ($\Delta_r H°$) for the three reactions shown is equal to $\Delta_f\overline{H}°$ for the product compound in each case.

The Indirect Method. Most compounds cannot be synthesized directly from their elements. In some cases, the reaction proceeds too slowly or not at all, or side reactions produce compounds other than the desired product. In these cases, the value of $\Delta_f\overline{H}°$ can be determined by an indirect approach, which is based on Hess's law. *Hess's law* (after the Swiss chemist Germain Henri Hess, 1802–1850) can be stated as follows: When reactants are converted to products, the change in enthalpy is the same whether the reaction takes place in one step or in a series of steps. In other words, if we can break down a reaction into a series of reactions for which the value of $\Delta_r H°$ can be measured, we can calculate the $\Delta_r H°$ value for the overall reaction. The logic of Hess's law is that because enthalpy is a state function, its change is path independent.

A simple analogy for Hess's law is as follows. Suppose you go from the first floor to the sixth floor of a building by elevator. The gain in your gravitational potential energy (which corresponds to the enthalpy change for the overall process) is the same whether you go directly there or stop at each floor on your way up (breaking the reaction into a series of steps).

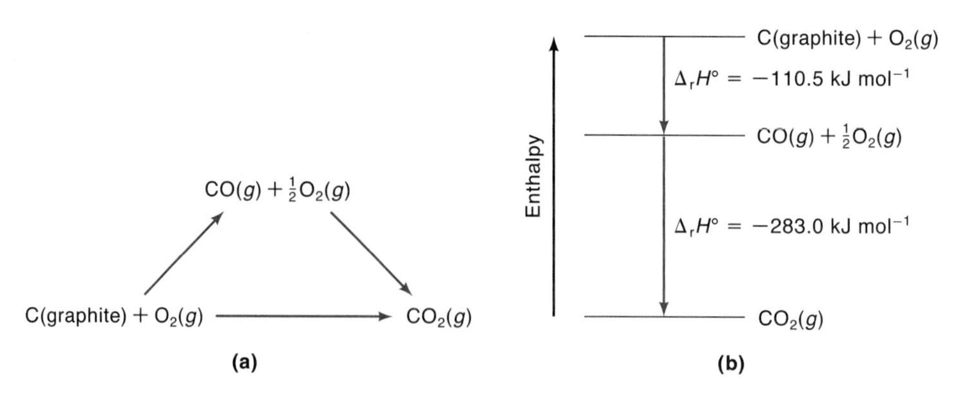

Figure 3.12
(a) The formation of CO_2 from graphite and O_2 can be broken into two steps. (b) The enthalpy change for the overall reaction is equal to the sum of the enthalpy changes for the two steps.

Let us apply Hess's law to find the value of $\Delta_f\overline{H}°$ for carbon monoxide. We might represent the synthesis of CO from its elements as

$$\text{C(graphite)} + \tfrac{1}{2}\text{O}_2(g) \rightarrow \text{CO}(g)$$

We cannot burn graphite in oxygen without also forming some CO_2, however, so this approach will not work. To circumvent this difficulty, we can carry out the following two separate reactions, which do go to completion:

(1) $\text{C(graphite)} + \text{O}_2(g) \rightarrow \text{CO}_2(g) \qquad \Delta_r H° = -393.5 \text{ kJ mol}^{-1}$

(2) $\text{CO}(g) + \tfrac{1}{2}\text{O}_2(g) \rightarrow \text{CO}_2(g) \qquad \Delta_r H° = -283.0 \text{ kJ mol}^{-1}$

First, we reverse equation 2 to get

When we reverse an equation, $\Delta_r H°$ changes sign.

(3) $\text{CO}_2(g) \rightarrow \text{CO}(g) + \tfrac{1}{2}\text{O}_2(g) \qquad \Delta_r H° = +283.0 \text{ kJ mol}^{-1}$

Because chemical equations can be added and subtracted just like algebraic equations, we carry out the operation 1 + 3 and obtain

(4) $\text{C(graphite)} + \tfrac{1}{2}\text{O}_2(g) \rightarrow \text{CO}(g) \qquad \Delta_r H° = -110.5 \text{ kJ mol}^{-1}$

Thus, $\Delta_f\overline{H}°(\text{CO}) = -110.5 \text{ kJ mol}^{-1}$. Looking back, we see that the overall reaction is the formation of CO_2 (reaction 1), which can be broken down into two parts (reactions 2 and 4). Figure 3.12 shows the overall scheme of our procedure.

The general rule in applying Hess's law is that we should arrange a series of chemical equations (corresponding to a series of steps) in such a way that, when added together, all species cancel except for the reactants and products that appear in the overall reaction. This means that we want the elements on the left and the compound of interest on the right of the arrow. To achieve this goal, we often need to multiply some or all of the equations representing the individual steps by the appropriate coefficients. The following is another example showing this approach.

EXAMPLE 3.9

Calculate the standard molar enthalpy of formation of acetylene (C_2H_2) from its elements:

$$2C(graphite) + H_2(g) \rightarrow C_2H_2(g)$$

The equations for combustion and the corresponding enthalpy changes are:

$$(1)\quad C(graphite) + O_2(g) \rightarrow CO_2(g) \qquad \Delta_r H° = -393.5 \text{ kJ mol}^{-1}$$

$$(2)\quad H_2(g) + \tfrac{1}{2}O_2(g) \rightarrow H_2O(l) \qquad \Delta_r H° = -285.8 \text{ kJ mol}^{-1}$$

$$(3)\quad 2C_2H_2(g) + 5O_2(g) \rightarrow 4CO_2(g) + 2H_2O(l) \qquad \Delta_r H° = -2598.8 \text{ kJ mol}^{-1}$$

ANSWER

Because we want to obtain one equation containing C and H_2 as reactants and C_2H_2 as product, we need to eliminate O_2, CO_2, and H_2O from equations 1, 2, and 3. We note that equation 3 contains 5 moles of O_2, 4 moles of CO_2, and 2 moles of H_2O. First, we reverse equation 3 to get C_2H_2 on the product side:

$$(4)\quad 4CO_2(g) + 2H_2O(l) \rightarrow 2C_2H_2(g) + 5O_2(g) \qquad \Delta_r H° = +2598.8 \text{ kJ mol}^{-1}$$

Next, we multiply equation 1 by 4 and equation 2 by 2 and carry out the operation 4(eqn. 1) + 2(eqn. 2) + (eqn. 4):

$$4C(graphite) + 4O_2(g) \rightarrow 4CO_2(g) \qquad \Delta_r H° = -1574.0 \text{ kJ mol}^{-1}$$

$$+\quad 2H_2(g) + O_2(g) \rightarrow 2H_2O(l) \qquad \Delta_r H° = -571.6 \text{ kJ mol}^{-1}$$

$$+\quad 4CO_2(g) + 2H_2O(l) \rightarrow 2C_2H_2(g) + 5O_2(g) \qquad \Delta_r H° = +2598.8 \text{ kJ mol}^{-1}$$

$$4C(graphite) + 2H_2(g) \rightarrow 2C_2H_2(g) \qquad \Delta_r H° = +453.2 \text{ kJ mol}^{-1}$$

$$\text{or}\quad 2C(graphite) + H_2(g) \rightarrow C_2H_2(g) \qquad \Delta_r H° = +226.6 \text{ kJ mol}^{-1}$$

Because the above equation represents the synthesis of C_2H_2 from its elements, we have $\Delta_f \overline{H}°(C_2H_2) = \Delta_r H° = +226.6 \text{ kJ mol}^{-1}$. (When the equation is divided by 2, the value of $\Delta_r H°$ is halved.)

Table 3.3 lists the $\Delta_f \overline{H}°$ values for a number of common inorganic and organic compounds. (A more extensive listing is given in Appendix B.) Note that compounds differ not only in the magnitude of $\Delta_f \overline{H}°$ but in the sign as well. Water and other compounds that have negative $\Delta_f \overline{H}°$ values lie "downhill" on the enthalpy scale relative to their constituent elements (Figure 3.13). These compounds tend to be more stable than those that have positive $\Delta_f \overline{H}°$ values. The reason is that energy has to be supplied to the former to decompose them into the elements, while the latter decompose with the evolution of heat.

Table 3.3
Standard Molar Enthalpies of Formation at 298 K and 1 Bar for Some Inorganic and Organic Substances

Substance	$\Delta_f\overline{H}^\circ$/kJ mol^{-1}	Substance	$\Delta_f\overline{H}^\circ$/kJ mol^{-1}
C(graphite)	0	$CH_4(g)$	−74.85
C(diamond)	1.90	$C_2H_6(g)$	−84.7
$CO(g)$	−110.5	$C_3H_8(g)$	−103.8
$CO_2(g)$	−393.5	$C_2H_2(g)$	226.6
$HF(g)$	−273.3	$C_2H_4(g)$	52.3
$HCl(g)$	−92.3	$C_6H_6(l)$	49.04
$HBr(g)$	−36.4	$CH_3OH(l)$	−238.7
$HI(g)$	26.48	$C_2H_5OH(l)$	−277.0
$H_2O(g)$	−241.8	$CH_3CHO(l)$	−192.3
$H_2O(l)$	−285.8	$HCOOH(l)$	−424.7
$NH_3(g)$	−46.3	$CH_3COOH(l)$	−484.2
$NO(g)$	90.4	$C_6H_{12}O_6(s)$	−1274.5
$NO_2(g)$	33.9	$C_{12}H_{22}O_{11}(s)$	−2221.7
$N_2O_4(g)$	9.7		
$N_2O(g)$	81.56		
$O_3(g)$	142.7		
$SO_2(g)$	−296.1		
$SO_3(g)$	−395.2		

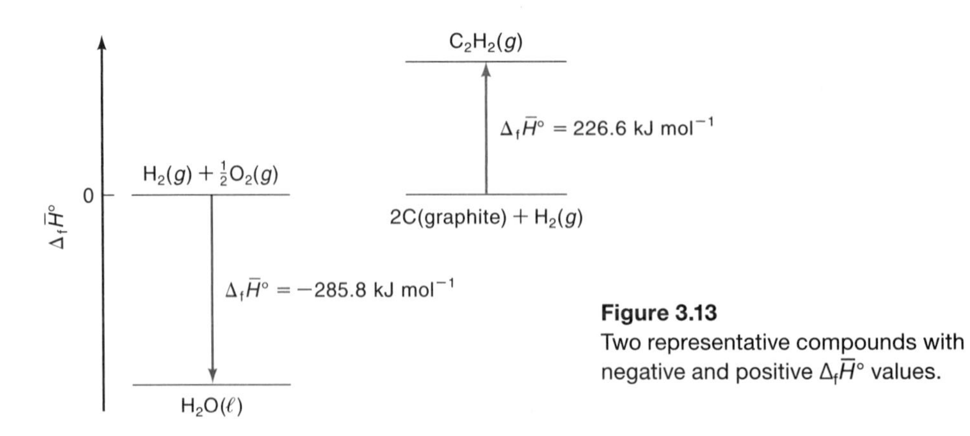

Figure 3.13
Two representative compounds with negative and positive $\Delta_f\overline{H}^\circ$ values.

EXAMPLE 3.10

Metabolism is the stepwise breakdown of the food we eat to provide energy for growth and function. A general overall equation for this complex process represents the degradation of glucose ($C_6H_{12}O_6$) to CO_2 and H_2O:

$$C_6H_{12}O_6(s) + 6O_2(g) \rightarrow 6CO_2(g) + 6H_2O(l)$$

Calculate the standard enthalpy of the reaction at 298 K.

ANSWER

From Equation 3.31

$$\Delta_r H^\circ = [6\Delta_f \overline{H}^\circ(CO_2) + 6\Delta_f \overline{H}^\circ(H_2O)] - [\Delta_f \overline{H}^\circ(C_6H_{12}O_6) + 6\Delta_f \overline{H}^\circ(O_2)]$$

Using the $\Delta_f \overline{H}^\circ$ values from Table 3.3, we write

$$\Delta_r H^\circ = 6(-393.5 \text{ kJ mol}^{-1}) + 6(-285.8 \text{ kJ mol}^{-1}) - (-1274.5 \text{ kJ mol}^{-1}) - 6(0 \text{ kJ mol}^{-1})$$

$$= -2801.3 \text{ kJ mol}^{-1}$$

COMMENT

(1) The enthalpy of reaction in this case is also called the enthalpy of combustion. The same quantity of heat is evolved whether we burn 1 mole of glucose in air or let the metabolic process break it down. (2) In tables of $\Delta_f \overline{H}^\circ$ values, you need to find not only the correct compound but also the proper physical state. For example, the $\Delta_f \overline{H}^\circ$ value of liquid water at 298 K and 1 bar is -285.8 kJ mol^{-1}, whereas that of water vapor is -241.8 kJ mol^{-1}. The difference, 44.0 kJ mol^{-1}, is the molar enthalpy of vaporization at 298 K:

$$H_2O(l) \rightarrow H_2O(g) \qquad \Delta_{vap}\overline{H}^\circ = 44.0 \text{ kJ mol}^{-1}$$

(3) The metabolic process is extremely complex, involving many steps. Because H is a state function, however, we can calculate the value of $\Delta_r H^\circ$ simply from the $\Delta_f \overline{H}^\circ$ values of the reactants and final products.

Dependence of Enthalpy of Reaction on Temperature

Suppose you have measured the standard enthalpy of a reaction at a certain temperature, say 298 K, and want to know its value at 350 K. One way to find out is to repeat the measurement at the higher temperature. Fortunately, we also can obtain the desired quantity from tabulated thermodynamic data without doing another experiment. For any reaction, the change in enthalpy at a particular temperature is

$$\Delta_r H = \Sigma H(\text{products}) - \Sigma H(\text{reactants})$$

To see how the enthalpy of reaction ($\Delta_r H$) itself changes with temperature, we differentiate this equation with respect to temperature at constant pressure as follows

$$\begin{aligned}\left(\frac{\partial \Delta_r H}{\partial T}\right)_P &= \left(\frac{\partial \Sigma H(\text{products})}{\partial T}\right)_P - \left(\frac{\partial \Sigma H(\text{reactants})}{\partial T}\right)_P \\ &= \Sigma C_P(\text{products}) - \Sigma C_P(\text{reactants}) \\ &= \Delta C_P \end{aligned} \tag{3.32}$$

because $(\partial H/\partial T)_P = C_P$. Integration of Equation 3.32 gives

$$\int_1^2 d\Delta_r H = \Delta_r H_2 - \Delta_r H_1 = \int_{T_1}^{T_2} \Delta C_P dT = \Delta C_P(T_2 - T_1) \tag{3.33}$$

where $\Delta_r H_1$ and $\Delta_r H_2$ are the enthalpies of reaction at T_1 and T_2, respectively. Equation 3.33 is known as *Kirchhoff's law* (after the German physicist Gustav Robert Kirchhoff, 1824–1887). This law says that the difference between the enthalpies of a reaction at two different temperatures is just the difference in the enthalpies of heating the products and reactants from T_1 to T_2 (Figure 3.14). Note that in deriving this equation, we have assumed the C_P values are all independent of temperature. Otherwise, they must be expressed as a function of T in the integration as mentioned in Section 3.4.

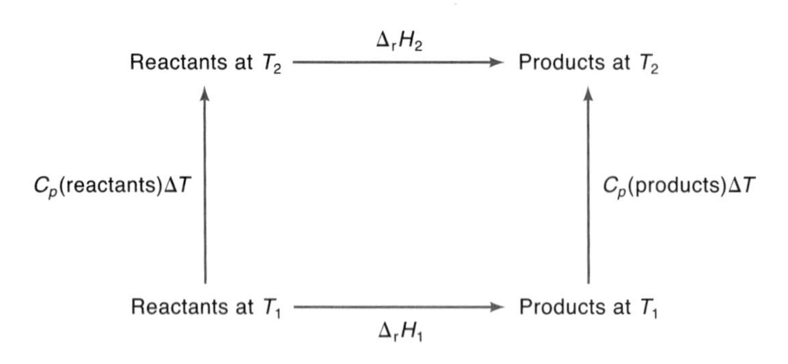

Figure 3.14
Schematic diagram showing Kirchhoff's law (Equation 3.33). The enthalpy change is $\Delta_r H_2 = \Delta_r H_1 + \Delta C_P(T_2 - T_1)$, where ΔC_P is the difference in heat capacities between products and reactants.

EXAMPLE 3.11

The standard enthalpy change for the reaction

$$3O_2(g) \rightarrow 2O_3(g)$$

is given by $\Delta_r H° = 285.4$ kJ mol^{-1} at 298 K and 1 bar. Calculate $\Delta_r H°$ at 380 K. Assume that the $\overline{C}_P$ values are all independent of temperature.

ANSWER

In Appendix B, we find the molar heat capacities at constant pressure for O_2 and O_3 to be 29.4 J K^{-1} mol^{-1} and 38.2 J K^{-1} mol^{-1}, respectively. From Equation 3.33,

$$\Delta_r H^\circ_{380} - \Delta_r H^\circ_{298} = \Delta C_P(T_2 - T_1)$$

$$= \frac{[(2)38.2 - (3)29.4]\text{J K}^{-1}\text{ mol}^{-1}}{(1000 \text{ J/kJ})} \times (380 - 298) \text{ K}$$

$$= -0.97 \text{ kJ mol}^{-1}$$

$$\Delta_r H^\circ_{380} = \Delta_r H^\circ_{298} + (-0.97 \text{ kJ mol}^{-1})$$

$$= (285.4 - 0.97) \text{ kJ mol}^{-1}$$

$$= 284.4 \text{ kJ mol}^{-1}$$

COMMENT

Note that the value of $\Delta_r H^\circ_{380}$ is not appreciably different from that of $\Delta_r H^\circ_{298}$. For gas-phase reactions, the increase in enthalpy from T_1 to T_2 for products tends to cancel that of reactants.

3.8 Bond Energies and Bond Enthalpies

Because chemical reactions involve the breaking and making of chemical bonds in the reactant and product molecules, a proper understanding of the thermochemical nature of reactions clearly requires a detailed knowledge of bond energies. Bond energy is the energy required to break a bond between two atoms. Consider the dissociation of 1 mole of H_2 molecules at 298 K and 1 bar:

$$H_2(g) \rightarrow 2H(g) \qquad \Delta_r H^\circ = 436.4 \text{ kJ mol}^{-1}$$

Assigning the energy of the H—H bond a value of 436.4 kJ mol^{-1} might be tempting, but the situation is more complicated. What is measured is actually the bond enthalpy of H_2, not its bond energy. To understand the difference between these two quantities, consider first what we mean by bond energy.

Figure 3.15 shows the *potential-energy curve* of the H_2 molecule. Let us start by asking how the molecule is formed. At first, the two hydrogen atoms are far apart and exert no influence on each other. As the distance of separation r is shortened, both Coulombic attraction (between electrons and nuclei) and Coulombic repulsion (between electron and electron and nucleus and nucleus) begin to affect each atom. Because attraction outweighs repulsion, the potential energy of the system decreases with decreasing distance of separation. This process continues until the net attraction force reaches a maximum, leading to the formation of a hydrogen molecule. Further shortening of the distance increases the repulsion, and the potential rises steeply. The reference state (zero potential energy) corresponds to the case of two infinitely separated H atoms. Potential energy is a negative quantity for the bound state (i.e., H_2), and energy in the form of heat is given off as a result of the bond formation.

The most important features in Figure 3.15 are the minimum point on the potential-energy curve for the bonding state, which represents the most stable state for the

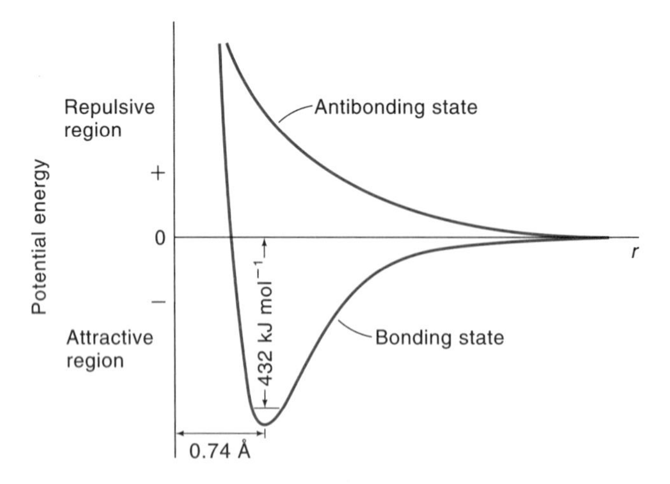

Figure 3.15
Potential-energy curve for the H_2 molecule. The short horizontal line represents the lowest vibrational energy level of the molecule. The intercepts of this line with the curve shows the maximum and minimum bond lengths during a vibration.

molecule, and the corresponding distance of separation, called the *equilibrium distance*. However, molecules are constantly executing vibrational motions that persist even at absolute zero. Furthermore, the energies associated with vibration, like the energies of an electron in an atom, are quantized. The lowest vibrational energy is not zero but equal to $\frac{1}{2}h\nu$, called the *zero-point energy*, where ν is the fundamental frequency of vibration of H_2. Consequently, the two hydrogen atoms cannot be held rigidly in the molecule, as is the case when a molecule is situated at the minimum point. Instead, the lowest vibrational state for H_2 is represented by the horizontal line. The intercepts between this line and the potential-energy curve represent the two extreme bond lengths during the course of a vibration. We can still speak of equilibrium distance in this case, although it is more like the average of the two extreme bond lengths. The bond energy of H_2 is the vertical distance from lowest vibrational energy level to the reference state of zero potential energy.

The measured enthalpy change (436.4 kJ mol^{-1}) should not be identified with the bond energy of H_2 for two reasons. First, upon dissociation, the number of moles of gas doubles, and hence gas-expansion work is done on the surroundings. The enthalpy change (ΔH) is not equal to the internal energy change (ΔU), which is the bond energy, but is related to it at constant pressure by the equation

$$\Delta H = \Delta U + P\Delta V$$

Second, the hydrogen molecules have vibrational, rotational, and translational energy before they dissociate, whereas the hydrogen atoms have only translational energy. Thus, the total kinetic energy of the reactant differs from that of the product. Although these kinetic energies are not relevant to the bond energy, their difference is unavoidably incorporated in the $\Delta_r H°$ value. Thus, despite the fact that bond energy has a firmer theoretical basis, for practical reasons we shall use bond enthalpies to help us study energy changes of chemical reactions.

Bond Enthalpy and Bond Dissociation Enthalpy

With respect to diatomic molecules such as H_2 and the following examples,

$$N_2(g) \rightarrow 2N(g) \qquad \Delta_r H° = 941.4 \text{ kJ mol}^{-1}$$

$$HCl(g) \rightarrow H(g) + Cl(g) \qquad \Delta_r H° = 430.9 \text{ kJ mol}^{-1}$$

bond enthalpy has a special significance because there is only one bond in each molecule, so the enthalpy change can be assigned unequivocally to that bond. For this reason, we shall use the term *bond dissociation enthalpy* for diatomic molecules. Polyatomic molecules are not so straightforward. Measurements show that the energy needed to break the first O–H bond in H_2O, for example, is different from that needed to break the second O–H bond:

$$H_2O(g) \rightarrow H(g) + OH(g) \qquad \Delta_r H° = 502 \text{ kJ mol}^{-1}$$

$$OH(g) \rightarrow H(g) + O(g) \qquad \Delta_r H° = 427 \text{ kJ mol}^{-1}$$

Table 3.4
Average Bond Enthalpies/kJ mol[-1]

Bond	Bond enthalpy[a]	Bond	Bond enthalpy
H−H	436.4	C−S	255
H−N	393	C=S	477
H−O	460	N−N	393
H−S	368	N=N	418
H−P	326	N≡N	941.4
H−F	568.2	N−O	176
H−Cl	430.9	N−P	209
H−Br	366.1	O−O	142
H−I	298.3	O=O	498.8
C−H	414	O−P	502
C−C	347	O=S	469
C=C	619	P−P	197
C≡C	812	P=P	490
C−N	276	S−S	268
C=N	615	S=S	351
C≡N	891	F−F	158.8
C−O	351	Cl−Cl	242.7
C=O[b]	724	Br−Br	192.5
C−P	264	I−I	151.0

[a] Bond enthalpies for diatomic molecules have more significant figures than those for polyatomic molecules because they are directly measurable quantities and are not averaged over many compounds as for polyatomic molecules.

[b] The C=O bond enthalpy in CO_2 is 799 kJ mol^{-1}.

In each case, an O–H bond is broken, but the first step is more endothermic than the second. The difference between the two $\Delta_r H^\circ$ values suggests that the second O–H bond itself undergoes change, because the chemical environment has been altered. If we were to study the O–H breaking process in other compounds, such as H_2O_2, CH_3OH, and so on, we would find still other $\Delta_r H^\circ$ values. Thus, for polyatomic molecules, we can speak only of the *average* bond enthalpy of a particular bond. For example, we can measure the bond enthalpy of the O–H bond in 10 different polyatomic

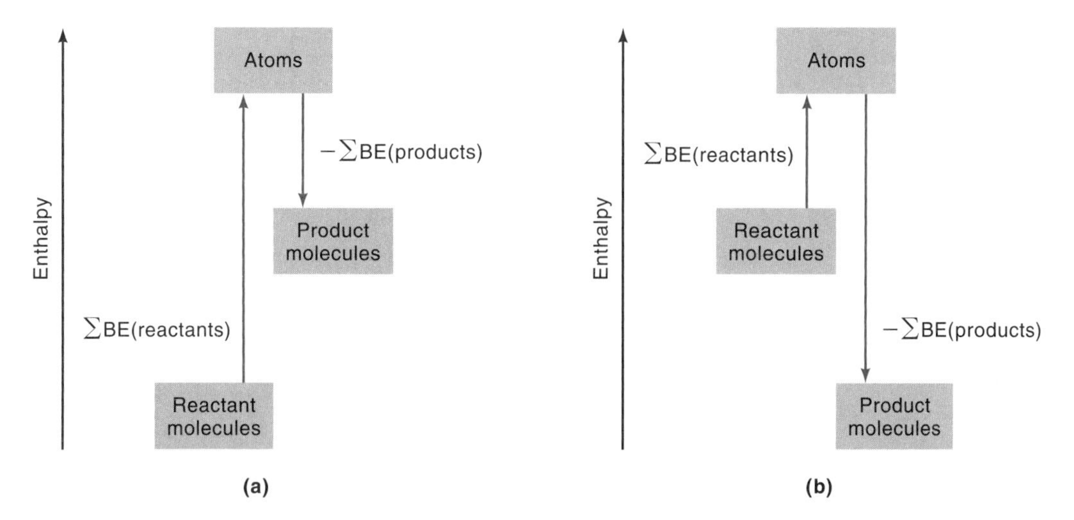

Figure 3.16
Bond enthalpy changes in (a) an endothermic reaction and (b) an exothermic reaction.

molecules and obtain the average O–H bond enthalpy by dividing the sum of the bond enthalpies by 10. When we use the term *bond enthalpy*, then, it is understood that we are referring to an average quantity, whereas *bond dissociation enthalpy* means a precisely measured value. Table 3.4 lists the bond enthalpies of a number of common chemical bonds. As you can see, triple bonds are stronger than double bonds, which, in turn, are stronger than single bonds.

The usefulness of bond enthalpies is that they enable us to estimate $\Delta_r H^\circ$ values when precise thermochemical data (i.e., $\Delta_f H^\circ$ values) are not available. Because energy is required to break chemical bonds and chemical bond formation is accompanied by a release of heat, we can estimate $\Delta_r H^\circ$ values by counting the total number of bonds broken and formed in the reaction and recording all the corresponding energy changes. The enthalpy of reaction in the *gas phase* is given by

$$\Delta_r H^\circ = \Sigma\text{BE(reactants)} - \Sigma\text{BE(products)}$$
$$= \text{total energy input} - \text{total energy released} \qquad (3.34)$$

Note that unlike all other thermodynamic equations, here Δ means "initial state − final state."

where BE stands for average bond enthalpy. As written, Equation 3.34 takes care of the sign convention for $\Delta_r H^\circ$. If the total energy input is greater than the total energy released, the $\Delta_r H^\circ$ value is positive and the reaction is endothermic. Conversely, if more energy is released than absorbed, the $\Delta_r H^\circ$ value is negative and the reaction is exothermic (Figure 3.16). If reactants and products are all diatomic molecules, then Equation 3.34 will yield accurate results because the bond dissociation enthalpies of diatomic molecules are accurately known. If some or all of the reactants and products are polyatomic molecules, Equation 3.34 will yield only approximate results because the bond enthalpies for calculation will be average values.

EXAMPLE 3.12

Estimate the enthalpy of combustion for methane

$$CH_4(g) + 2O_2(g) \rightarrow CO_2(g) + 2H_2O(g)$$

at 298 K and 1 bar using the bond enthalpies in Table 3.4. Compare your result with that calculated from the enthalpies of formation of products and reactants.

ANSWER

The first step is to count the number of bonds broken and the number of bonds formed. This is best done by creating a table:

Type of bonds broken	*Number of bonds broken*	*Bond enthalpy* / $kJ\ mol^{-1}$	*Enthalpy change* / $kJ\ mol^{-1}$
C–H	4	414	1656
O=O	1	498.8	997.6

Type of bonds formed	*Number of bonds formed*	*Bond enthalpy* / $kJ\ mol^{-1}$	*Enthalpy change* / $kJ\ mol^{-1}$
C=O	2	799	1598
O–H	4	460	1840

From Equation 3.34

$$\Delta_r H^\circ = [(1656\ kJ\ mol^{-1} + 997.6\ kJ\ mol^{-1}) - (1598\ kJ\ mol^{-1} + 1840\ kJ\ mol^{-1})]$$

$$= -784.4\ kJ\ mol^{-1}$$

To calculate the $\Delta_r H^\circ$ value using Equation 3.31, we obtain the $\Delta_f H^\circ$ values from Table 3.3 and write

$$\Delta_r H^\circ = [\Delta_f H^\circ(CO_2) + 2\Delta_f H^\circ(H_2O)] - [\Delta_f H^\circ(CH_4) + 2\Delta_f H^\circ(O_2)]$$

$$= [-393.5\ kJ\ mol^{-1} + 2(-241.8\ kJ\ mol^{-1})] - [(-74.85\ kJ\ mol^{-1}) + 2(0)]$$

$$= -802.3\ kJ\ mol^{-1}$$

COMMENT

The agreement between the estimated $\Delta_r H^\circ$ value using bond enthalpies and the actual $\Delta_r H^\circ$ value is fairly good in this case. In general, the more exothermic (or endothermic) the reaction, the better the agreement. If the actual $\Delta_r H^\circ$ value is a small positive or negative quantity, then the value obtained from bond enthalpies becomes unreliable. Such values may even give the wrong sign for the reaction enthalpy.

Key Equations

$w = -P_{ex}\Delta V$	(Work of irreversible gas expansion)	(3.2)
$w = -nRT\ln\dfrac{V_2}{V_1} = -nRT\ln\dfrac{P_1}{P_2}$	(Work of isothermal, reversible ideal gas expansion)	(3.5)
$\Delta U = q + w$	(First law of thermodynamics)	(3.6)
$dU = đq + đw$	(First law of thermodynamics)	(3.7)
$H = U + PV$	(Definition of enthalpy)	(3.10)
$C_V = \left(\dfrac{\partial U}{\partial T}\right)_V$	(Heat capacity at constant volume)	(3.15)
$C_P = \left(\dfrac{\partial H}{\partial T}\right)_P$	(Heat capacity at constant pressure)	(3.17)
$\Delta U = n\overline{C}_V\Delta T$	(Change in internal energy)	(3.19)
$\Delta H = n\overline{C}_P\Delta T$	(Change in enthalpy)	(3.20)
$C_P - C_V = nR$	(Difference in heat capacities of an ideal gas)	(3.21)
$P_1V_1^{\gamma} = P_2V_2^{\gamma}$	(Adiabatic, reversible expansion of an ideal gas)	(3.24)
$\mu_{JT} = \left(\dfrac{\partial T}{\partial P}\right)_H$	(Joule–Thomson coefficient)	(3.28)
$\Delta_r\overline{H}° = \Sigma v\Delta_f\overline{H}°(\text{products}) - \Sigma v\Delta_f\overline{H}°(\text{reactants})$	(Standard enthalpy of reaction)	(3.31)
$\Delta_rH_2 - \Delta_rH_1 = \Delta C_P(T_2 - T_1)$	(Kirchhoff's law)	(3.33)
$\Delta_rH° = \Sigma\text{BE(reactants)} - \Sigma\text{BE(products)}$	(Standard enthalpy of reaction)	(3.34)

APPENDIX 3.1

Exact and Inexact Differentials

Let us consider a function z of two variables x and y:

$$z = f(x, y)$$

If there is an infinitesimal change in x at constant y, the corresponding change in z is dz, given by $dz = (\partial z/\partial x)_y dx$. Similarly, for an infinitesimal change dy while keeping x constant, we have $dz = (\partial z/\partial y)_x dy$. Now if both x and y undergo infinitesimal changes, the change in z is the sum of the changes due to dx and dy:

$$dz = \left(\frac{\partial z}{\partial x}\right)_y dx + \left(\frac{\partial z}{\partial y}\right)_x dy \tag{1}$$

Here, dz is called the *total differential* because it is expressed in terms of both dx and dy.

Consider the following example of a total differential. The pressure of one mole of a gas is a function of volume and temperature:

$$P = f(V, T)$$

We can write the total differential, dP, as

$$dP = \left(\frac{\partial P}{\partial V}\right)_T dV + \left(\frac{\partial P}{\partial T}\right)_V dT \tag{2}$$

For one mole of a van der Waals gas, we first rearrange Equation 1.8 to show P as a function of V and T

$$P = \frac{RT}{V - b} - \frac{a}{V^2}$$

Next, we write the partial derivatives of P with respect to V and T as follows:

$$\left(\frac{\partial P}{\partial V}\right)_T = -\frac{RT}{(V - b)^2} + \frac{2a}{V^3} \quad \text{and} \quad \left(\frac{\partial P}{\partial T}\right)_V = \frac{R}{V - b}$$

Substituting these expressions in Equation 2, we obtain the total differential, dP, for a van der Waals gas:

$$dP = \left[-\frac{RT}{(V - b)^2} + \frac{2a}{V^3}\right] dV + \frac{R}{V - b} dT \tag{3}$$

Total differentials are either *exact* or *inexact*, with important differences. A differential of the type

$$dz = M(x, y)dx + N(x, y)dy$$

is said to be an exact differential if the following condition is satisfied:

$$\left(\frac{\partial M}{\partial y}\right)_x = \left(\frac{\partial N}{\partial x}\right)_y$$

This test is known as Euler's theorem (after the Swiss mathematician Leonhard Euler, 1707–1783). Suppose that we have a function given by

$$dz = (y^2 + 3x)dx + e^x dy$$

where $M(x, y) = y^2 + 3x$ and $N(x, y) = e^x$. Applying Euler's theorem, we write

$$\left(\frac{\partial M}{\partial y}\right)_x = \left[\frac{\partial(y^2 + 3x)}{\partial y}\right]_x = 2y$$

and

$$\left(\frac{\partial N}{\partial x}\right)_y = \left(\frac{\partial e^x}{\partial x}\right)_y = e^x$$

Therefore, dz is not an exact differential.

On the other hand, dP in Equation 3 is an exact differential because

$$\left[\frac{\partial\left(-\frac{RT}{(V-b)^2} + \frac{2a}{V^3}\right)}{\partial T}\right]_V = -\frac{R}{(V-b)^2}$$

and

$$\left[\frac{\partial\left(\frac{R}{V-b}\right)}{\partial V}\right]_T = -\frac{R}{(V-b)^2}$$

The significance of exact differentials is that if df is an exact differential (where f is a function of x and y), then the value of the following integral depends only on the limits of integration; that is,

$$\int_1^2 df = f_2 - f_1$$

However, if $đf$ is an inexact differential, then

$$\int_1^2 đf \neq f_2 - f_1$$

Note that we have used the symbol $đf$ to denote an inexact differential. Unless the functional relationship between the variables x and y is known, the integral $\int_1^2 đf$ cannot be carried out. We saw earlier that dU and dH are exact differentials whereas $đw$ and $đq$ are inexact differentials. This means that the amount of work done or heat exchanged in a process depends on the path and not just on the initial and final states of the system. The important conclusion is that if any thermodynamic function X is a state function, then dX is an exact differential.

Exercise 1

For a given amount of an ideal gas, we can write $V = f(P, T)$. Prove that dV is an exact differential.

Exercise 2

Using the result from Exercise 1, show that $đw$, expressed as $đw = -PdV$, is an inexact differential.

Suggestions for Further Reading

BOOKS

Hanson, R. M., and S. Green, *Introduction to Molecular Thermodynamics*, University Science Books, Sausalito, CA, 2008.

Klotz, I. M., and R. M. Rosenberg, *Chemical Thermodynamics: Basic Theory and Methods*, 5th ed., John Wiley & Sons, New York, 1994.

Levine, I. N., *Physical Chemistry*, 6th ed., McGraw-Hill, New York, 2009.

McQuarrie, D. A., and J. D. Simon, *Molecular Thermodynamics*, University Science Books, Sausalito, CA, 1999.

Rock, P. A., *Chemical Thermodynamics*, University Science Books, Mill Valley, CA, 1983.

ARTICLES

"What is Heat," F. J. Dyson, *Sci. Am.* September 1954.

"Perpetual Motion Machines," S. W. Angrist, *Sci. Am.* January 1968.

"The Definition of Heat," T. B. Tripp, *J. Chem. Educ.* **53**, 782 (1976).

"Conversion of Standard (1 atm) Thermodynamic Data to the New Standard-State Pressure, 1 bar (10^5 Pa)," *Bull. Chem. Thermodynamics* **25**, 523 (1982).

"Heat, Work, and Metabolism," J. N. Spencer, *J. Chem. Educ.* **62**, 571 (1985).

"Conversion of Standard Thermodynamic Data to the New Standard-State Pressure," R. D. Freeman, *J. Chem. Educ.* **62**, 681 (1985).

"General Definitions of Work and Heat in Thermodynamic Processes," E. A. Gislason and N. C. Craig, *J. Chem. Educ.* **64**, 660 (1987).

"Power From the Sea," T. R. Penney and P. Bharathan, *Sci. Am.* January 1987.

"Simplification of Some Thermochemical Calculations," E. R. Boyko and J. F. Belliveau, *J. Chem. Educ.* **67**, 743 (1990).

"Understanding the Language: Problem Solving and the First Law of Thermodynamics," M. Hamby, *J. Chem. Educ.* **67**, 923 (1990).

"Standard Enthalpies of Formation of Ions in Solution," T. Solomon, *J. Chem. Educ.* **68**, 41 (1991).
"Why There's Frost on the Pumpkin," W. H. Corkern and L. H. Holmes, Jr., *J. Chem. Educ.* **68**, 825 (1991).
"Bond Energies and Enthalpies," R. S. Treptow, *J. Chem. Educ.* **72**, 497 (1995).
"Thermochemistry," P. A. G. O'Hare, *Encyclopedia of Applied Physics*, Trigg, G. L., Ed., VCH Publishers, New York (1997), Vol. 21, p. 265.
"The Thermodynamics of Drunk Driving," R. Q. Thompson, *J. Chem. Educ.* **74**, 532 (1997).
"Heat Capacity, Body Temperature, and Hypothermia," D. R. Kimbrough, *J. Chem. Educ.* **75**, 48 (1998).
"How Thermodynamic Data and Equilibrium Constants Changed When the Standard-State Pressure Became 1 Bar," R. S. Treptow, *J. Chem. Educ.* **76**, 212 (1999).
"Stories to Make Thermodynamics and Related Subjects More Palatable," L. S. Bartell, *J. Chem. Educ.* **78**, 1059 (2001).
"First Law of Thermodynamics: Irreversible and Reversible Processes," E. A. Gislason and N. C. Craig, *J. Chem. Educ.* **79**, 193 (2002).
"H is for Enthalpy," H. C. Van Ness, *J. Chem. Educ.* **80**, 486 (2003).
"A Close Look at Temperature during the Free Expansion of a Dilute Monatomic Ideal Gas," D. Keeports, *Chem. Educator* **10**, 250 (2005).
"Mysteries of the First and Second law of Thermodynamics," R. Battino, *J. Chem. Educ.* **84**, 753 (2007).
"Effect of a Dynamic Learning Tutorial on Undergraduate Students' Understanding of Heat and the First Law of Thermodynamics," J. Barbera and C. E. Wieman, *Chem. Educator* [Online] **14**, 45 (2009) DOI 10.1333/s00897092193a.
"Heat of Combustion and GC-MS of Regular, Regular-Plus and Premium Gasoline: An Undergraduate Experiment," A. Gaquere-Parker, K. Lawson, M. Logue, K. Sutton, C. Richardson, and S. Gant, *Chem. Educator* [Online] **16**, 310 (2011) DOI 10.1333/s00897112401a.

Problems

Work and Heat

3.1 Explain the term *state function*. Which of the following are state functions? P, V, T, w, q.

3.2 What is heat? How does heat differ from thermal energy? Under what condition is heat transferred from one system to another?

3.3 Show that 1 L atm = 101.3 J.

3.4 A 7.24-g sample of ethane occupies 4.65 L at 294 K. **(a)** Calculate the work done when the gas expands isothermally against a constant external pressure of 0.500 atm until its volume is 6.87 L. **(b)** Calculate the work done if the same expansion occurs reversibly.

3.5 A 19.2-g quantity of dry ice (solid carbon dioxide) is allowed to sublime (evaporate) in an apparatus like the one shown in Figure 3.1. Calculate the expansion work done against a constant external pressure of 0.995 atm and at a constant temperature of 22°C. Assume that the initial volume of dry ice is negligible and that CO_2 behaves like an ideal gas.

3.6 Calculate the work done by the reaction

$$\mathrm{Zn}(s) + \mathrm{H_2SO_4}(aq) \rightarrow \mathrm{ZnSO_4}(aq) + \mathrm{H_2}(g)$$

when 1.0 mole of hydrogen gas is collected at 273 K and 1.0 atm. (Neglect volume changes other than the change in gas volume.)

First Law of Thermodynamics

3.7 A truck traveling 60 kilometers per hour is brought to a complete stop at a traffic light. Does this change in velocity violate the law of conservation of energy?

3.8 Some driver's test manuals state that the stopping distance quadruples as the velocity doubles. Justify this statement by using mechanics and thermodynamic arguments.

3.9 Provide a first law analysis for each of the following cases: **(a)** When a bicycle tire is inflated with a hand pump, the temperature inside rises. You can feel the warming effect at the valve stem. **(b)** Artificial snow is made by quickly releasing a mixture of compressed air and water vapor at about 20 atm from a snow-making machine to the surroundings.

3.10 An ideal gas is compressed isothermally by a force of 85 newtons acting through 0.24 meter. Calculate the values of ΔU and q.

3.11 Calculate the internal energy of 2 moles of argon gas (assuming ideal behavior) at 298 K. Suggest two ways to increase its internal energy by 10 J.

3.12 A thermos bottle containing milk is shaken vigorously. Consider the milk as the system. **(a)** Will the temperature rise as a result of the shaking? **(b)** Has heat been added to the system? **(c)** Has work been done on the system? **(d)** Has the system's internal energy changed?

ΔU and ΔH

3.13 A 1.00-mole sample of ammonia at 14.0 atm and 25°C in a cylinder fitted with a movable piston expands against a constant external pressure of 1.00 atm. At equilibrium, the pressure and volume of the gas are 1.00 atm and 23.5 L, respectively. **(a)** Calculate the final temperature of the sample. **(b)** Calculate the values of q, w, and ΔU for the process.

3.14 An ideal gas is compressed isothermally from 2.0 atm and 2.0 L to 4.0 atm and 1.0 L. Calculate the values of ΔU and ΔH if the process is carried out **(a)** reversibly and **(b)** irreversibly.

3.15 Explain the energy changes at the molecular level when liquid acetone is converted to vapor at its boiling point.

3.16 A piece of potassium metal is added to water in a beaker. The reaction that takes place is

$$2K(s) + 2H_2O(l) \rightarrow 2KOH(aq) + H_2(g)$$

Predict the signs of w, q, ΔU, and ΔH.

3.17 At 373.15 K and 1 atm, the molar volumes of liquid water and steam are 1.88×10^{-5} m^3 and 3.06×10^{-2} m^3, respectively. Given that the heat of vaporization of water is 40.79 kJ mol^{-1}, calculate the values of ΔH and ΔU for 1 mole in the following process:

$$H_2O(l, 373.15\text{ K}, 1\text{ atm}) \rightarrow H_2O(g, 373.15\text{ K}, 1\text{ atm})$$

3.18 Consider a cyclic process involving a gas. If the pressure of the gas varies during the process but returns to the original value at the end, is it correct to write $\Delta H = q_P$?

3.19 Calculate the value of ΔH when the temperature of 1 mole of a monatomic gas is increased from 25°C to 300°C.

3.20 One mole of an ideal gas undergoes an isothermal expansion at 300 K from 1.00 atm to a final pressure while performing 200 J of expansion work. Calculate the final pressure of the gas if the external pressure is 0.20 atm.

Heat Capacities

3.21 A 6.22-kg piece of copper metal is heated from 20.5°C to 324.3°C. Given that the specific heat of Cu is 0.385 J g^{-1} °C^{-1}, calculate the heat absolved (in kJ) by the metal.

3.22 A 10.0-g sheet of gold with a temperature of 18.0°C is laid flat on a sheet of iron that weighs 20.0 g and has a temperature of 55.6°C. Given that the specific heats of Au and Fe are 0.129 J g^{-1} °C^{-1} and 0.444 J g^{-1} °C^{-1}, respectively, what is the final temperature of the combined metals? Assume that no heat is lost to the surroundings. (*Hint:* The heat gained by the gold must be equal to the heat lost by the iron.)

3.23 It takes 330 joules of energy to raise the temperature of 24.6 g of benzene from 21.0°C to 28.7°C at constant pressure. What is the molar heat capacity of benzene at constant pressure?

3.24 The molar heat of vaporization for water is 44.01 kJ mol^{-1} at 298 K and 40.79 kJ mol^{-1} at 373 K. Give a qualitative explanation of the difference in these two values.

3.25 The constant-pressure molar heat capacity of nitrogen is given by the expression

$$\overline{C}_P = (27.0 + 5.90 \times 10^{-3}\, T/\text{K} - 0.34 \times 10^{-6}\, T^2/\text{K}^2)\text{ J K}^{-1}\text{ mol}^{-1}$$

Calculate the value of ΔH for heating 1 mole of nitrogen from 25.0°C to 125°C.

3.26 The heat capacity ratio (γ) for a gas with the molecular formula X_2Y is 1.38. What can you deduce about the structure of the molecule?

3.27 One way to measure the heat capacity ratio (γ) of a gas is to measure the speed of sound in the gas (c), which is given by

$$c = \left(\frac{\gamma RT}{\mathcal{M}}\right)^{1/2}$$

where $\mathcal{M}$ is the molar mass of the gas. Calculate the speed of sound in helium at 25°C.

3.28 Which of the following gases has the largest $\overline{C}_V$ value at 298 K? He, N_2, CCl_4, HCl.

3.29 (**a**) For the most efficient use, refrigerator freezer compartments should be fully packed with food. What is the thermochemical basis for this recommendation? (**b**) Starting at the same temperature, tea and coffee remain hot longer in a thermal flask than soup. Explain.

3.30 In the nineteenth century, two scientists named Dulong and Petit noticed that the product of the molar mass of a solid element and its specific heat is approximately 25 J °C^{-1}. This observation, now called Dulong and Petit's law, was used to estimate the specific heat of metals. Verify the law for aluminum (0.900 J g^{-1} °C^{-1}), copper (0.385 J g^{-1} °C^{-1}), and iron (0.444 J g^{-1} °C^{-1}). The law does not apply to one of the metals. Which one is it? Why?

Gas Expansion

3.31 The following diagram represents the P–V changes of a gas. Write an expression for the total work done.

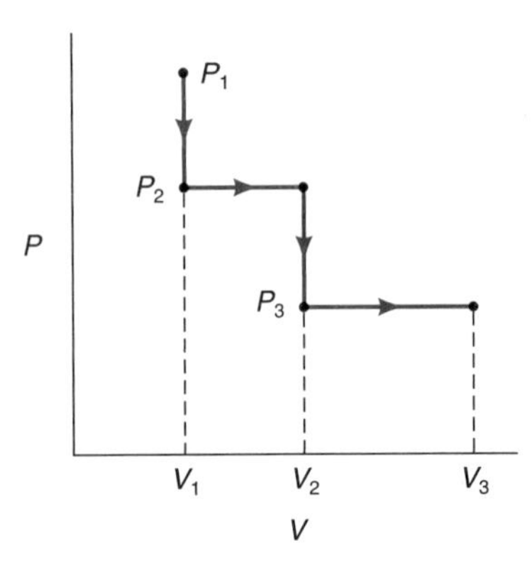

3.32 The equation of state for a certain gas is given by $P[(V/n) - b] = RT$. Obtain an expression for the maximum work done by the gas in a reversible isothermal expansion from V_1 to V_2.

3.33 Calculate the values of q, w, ΔU, and ΔH for the reversible adiabatic expansion of 1 mole of a monatomic ideal gas from 5.00 m^3 to 25.0 m^3. The temperature of the gas is initially 298 K.

3.34 A quantity of 0.27 mole of neon is confined in a container at 2.50 atm and 298 K and then allowed to expand adiabatically under two different conditions: (**a**) reversibly to 1.00 atm and (**b**) against a constant pressure of 1.00 atm. Calculate the final temperature in each case.

3.35 One mole of an ideal monatomic gas initially at 300 K and a pressure of 15.0 atm expands to a final pressure of 1.00 atm. The expansion can occur via any one of four different paths: (**a**) isothermal and reversible. (**b**) isothermal and irreversible. (**c**) adiabatic and reversible, and (**d**) adiabatic and irreversible. In irreversible processes, the expansion occurs against an external pressure of 1.00 atm. For each case, calculate the values of q, w, ΔU, and ΔH.

Calorimetry

3.36 A 0.1375-g sample of magnesium is burned in a constant-volume bomb calorimeter that has a heat capacity of 1769 J °C^{-1}. The calorimeter contains exactly 300 g of water, and the temperature increases by 1.126°C. Calculate the heat given off by the burning magnesium, in kJ g^{-1} and in kJ mol^{-1}. The specific heat of water is 4.184 J g^{-1} °C^{-1}.

3.37 The enthalpy of combustion of benzoic acid (C_6H_5COOH) is commonly used as the standard for calibrating constant-volume bomb calorimeters; its value has been accurately determined to be −3226.7 kJ mol^{-1}. **(a)** When 0.9862 g of benzoic acid was oxidized, the temperature rose from 21.84°C to 25.67°C. What is the heat capacity of the calorimeter? **(b)** In a separate experiment, 0.4654 g of glucose ($C_6H_{12}O_6$) was oxidized in the same calorimeter, and the temperature rose from 21.22°C to 22.28°C. Calculate the enthalpy of combustion of glucose, the value of $\Delta_r U$ for the combustion, and the molar enthalpy of formation of glucose.

3.38 A quantity of 2.00×10^2 mL of 0.862 *M* HCl is mixed with 2.00×10^2 mL of 0.431 *M* $Ba(OH)_2$ in a constant-pressure calorimeter that has a heat capacity of 453 J °C^{-1}. The initial temperature of the HCl and $Ba(OH)_2$ solutions is the same at 20.48°C. For the process

$$H^+(aq) + OH^-(aq) \rightarrow H_2O(l)$$

the heat of neutralization is −56.2 kJ mol^{-1}. What is the final temperature of the mixed solution?

3.39 When 1 mole of naphthalene ($C_{10}H_8$) is completely burned in a constant-volume bomb calorimeter at 298 K, 5150 kJ of heat is evolved. Calculate the values of $\Delta_r U$ and $\Delta_r H$ for the reaction.

Thermochemistry

3.40 Consider the following reaction:

$$2CH_3OH(l) + 3O_2(g) \rightarrow 4H_2O(l) + 2CO_2(g) \qquad \Delta_r H° = -1452.8 \text{ kJ mol}^{-1}$$

What is the value of $\Delta_r H°$ if **(a)** the equation is multiplied throughout by 2. **(b)** the direction of the reaction is reversed so that the products become the reactants and vice versa. **(c)** water vapor instead of liquid water is the product?

3.41 Which of the following standard enthalpy of formation values is not zero at 25°C? Na(*s*), Ne(*g*), $CH_4(g)$, $S_8(s)$, Hg(*l*), H(*g*).

3.42 The standard enthalpies of formation of ions in aqueous solution are obtained by arbitrarily assigning a value of zero to H^+ ions; that is, $\Delta_f \overline{H}°[H^+(aq)] = 0$. **(a)** For the following reaction,

$$HCl(g) \rightarrow H^+(aq) + Cl^-(aq) \qquad \Delta_r H° = -74.9 \text{ kJ mol}^{-1}$$

calculate the value of $\Delta_f \overline{H}°$ for the Cl^- ions. **(b)** The standard enthalpy of neutralization between a HCl solution and a NaOH solution is −56.2 kJ mol^{-1}. Calculate the standard enthalpy of formation of the hydroxide ion at 25°C.

3.43 Determine the amount of heat (in kJ) given off when 1.26×10^4 g of ammonia is produced according to the equation

$$N_2(g) + 3H_2(g) \rightarrow 2NH_3(g) \qquad \Delta_r H° = -92.6 \text{ kJ mol}^{-1}$$

Assume the reaction takes place under standard-state conditions at 25°C.

3.44 When 2.00 g of hydrazine decomposed under constant-pressure conditions, 7.00 kJ of heat were transferred to the surroundings:

$$3N_2H_4(l) \rightarrow 4NH_3(g) + N_2(g)$$

What is the $\Delta_r H°$ value for the reaction?

3.45 Consider the reaction

$$N_2(g) + 3H_2(g) \rightarrow 2NH_3(g) \qquad \Delta_r H° = -92.6 \text{ kJ mol}^{-1}$$

If 2.0 moles of N_2 react with 6.0 moles of H_2 to form NH_3, calculate the work done (in joules) against a pressure of 1.0 atm at 25°C. What is the value of $\Delta_r U$ for this reaction? Assume the reaction goes to completion.

3.46 The standard enthalpies of combustion of fumaric acid and maleic acids (to form carbon dioxide and water) are -1336.0 kJ mol^{-1} and -1359.2 kJ mol^{-1}, respectively. Calculate the enthalpy of the following isomerization process:

```
HOOC      COOH            H       COOH
    \    /                 \     /
     C=C        ⟶           C=C
    /    \                 /     \
   H      H            HOOC       H

  Maleic acid           Fumaric acid
```

3.47 From the reaction

$$C_{10}H_8(s) + 12O_2(g) \rightarrow 10CO_2(g) + 4H_2O(l) \qquad \Delta_r H° = -5153.0 \text{ kJ mol}^{-1}$$

and the enthalpies of formation of CO_2 and H_2O (see Appendix B), calculate the enthalpy of formation of naphthalene ($C_{10}H_8$).

3.48 The standard molar enthalpy of formation of molecular oxygen at 298 K is zero. What is its value at 315 K? (*Hint:* Look up the $\overline{C}_P$ value in Appendix B.)

3.49 Which of the following substances has a nonzero $\Delta_f \overline{H}°$ value at 25°C? Fe(s), $I_2(l)$, $H_2(g)$, Hg(l), $O_2(g)$, C(graphite).

3.50 The hydrogenation of ethylene is

$$C_2H_4(g) + H_2(g) \rightarrow C_2H_6(g)$$

Calculate the change in the enthalpy of hydrogenation from 298 K to 398 K. The $\overline{C}_P^{\circ}$ values are C_2H_4: 43.6 J K^{-1} mol^{-1} and C_2H_6: 52.7 J K^{-1} mol^{-1}.

3.51 Use the data in Appendix B to calculate the value of $\Delta_r H°$ for the following reaction at 298 K:

$$N_2O_4(g) \rightarrow 2NO_2(g)$$

What is its value at 350 K? State any assumptions used in your calculation.

3.52 Calculate the standard enthalpy of formation for diamond, given that

$$\text{C(graphite)} + O_2(g) \rightarrow CO_2(g) \qquad \Delta_r H° = -393.5 \text{ kJ mol}^{-1}$$

$$\text{C(diamond)} + O_2(g) \rightarrow CO_2(g) \qquad \Delta_r H° = -395.4 \text{ kJ mol}^{-1}$$

3.53 Photosynthesis produces glucose, $C_6H_{12}O_6$, and oxygen from carbon dioxide and water:

$$6CO_2 + 6H_2O \rightarrow C_6H_{12}O_6 + 6O_2$$

(a) How would you determine the $\Delta_r H°$ value for this reaction experimentally? **(b)** Solar radiation produces approximately 7.0×10^{14} kg glucose per year on Earth. What is the corresponding change in the $\Delta_r H°$ value?

3.54 From the following heats of combustion,

$$CH_3OH(l) + \tfrac{3}{2}O_2(g) \rightarrow CO_2(g) + 2H_2O(l) \qquad \Delta_r H° = -726.4 \text{ kJ mol}^{-1}$$

$$C(\text{graphite}) + O_2(g) \rightarrow CO_2(g) \qquad \Delta_r H° = -393.5 \text{ kJ mol}^{-1}$$

$$H_2(g) + \tfrac{1}{2}O_2(g) \rightarrow H_2O(l) \qquad \Delta_r H° = -285.8 \text{ kJ mol}^{-1}$$

calculate the enthalpy of formation of methanol (CH_3OH) from its elements:

$$C(\text{graphite}) + 2H_2(g) + \tfrac{1}{2}O_2(g) \rightarrow CH_3OH(l)$$

3.55 The standard enthalpy change for the following reaction is 436.4 kJ mol^{-1}:

$$H_2(g) \rightarrow H(g) + H(g)$$

Calculate the standard enthalpy of formation of atomic hydrogen (H).

3.56 Calculate the difference between the values of $\Delta_r H°$ and $\Delta_r U°$ for the oxidation of glucose at 298 K:

$$C_6H_{12}O_6(s) + 6O_2(g) \rightarrow 6CO_2(g) + 6H_2O(l)$$

3.57 Alcoholic fermentation is the process in which carbohydrates are broken down into ethanol and carbon dioxide. The reaction is very complex and involves a number of enzyme-catalyzed steps. The overall change is

$$C_6H_{12}O_6(s) \rightarrow 2C_2H_5OH(l) + 2CO_2(g)$$

Calculate the standard enthalpy change for this reaction, assuming that the carbohydrate is glucose.

Bond Enthalpy

3.58 **(a)** Explain why the bond enthalpy of a molecule is always defined in terms of a gas-phase reaction. **(b)** The bond dissociation enthalpy of F_2 is 158.8 kJ mol^{-1}. Calculate the value of $\Delta_f \overline{H}°$ for F(g).

3.59 From the molar enthalpy of vaporization of water at 373 K and the bond dissociation enthalpies of H_2 and O_2 (see Table 3.4), calculate the average O–H bond enthalpy in water, given that

$$H_2(g) + \tfrac{1}{2}O_2(g) \rightarrow H_2O(l) \qquad \Delta_r H° = -285.8 \text{ kJ mol}^{-1}$$

3.60 Use the bond enthalpy values in Table 3.4 to calculate the enthalpy of combustion for ethane,

$$2C_2H_6(g) + 7O_2(g) \rightarrow 4CO_2(g) + 6H_2O(l)$$

Compare your result with that calculated from the enthalpy of formation values of the products and reactants listed in Appendix B.

Additional Problems

3.61 A 2.10-mole sample of crystalline acetic acid, initially at 17.0°C, is allowed to melt at 17.0°C and is then heated to 118.1°C (its normal boiling point) at 1.00 atm. The sample is allowed to vaporize at 118.1°C and is then rapidly quenched to 17.0°C, so that it recrystallizes. Calculate the value of $\Delta_r H°$ for the total process as described.

3.62 Predict whether the values of q, w, ΔU, and ΔH are positive, zero, or negative for each of the following processes: **(a)** melting of ice at 1 atm and 273 K. **(b)** melting of solid cyclohexane at 1 atm and the normal melting point. **(c)** reversible isothermal expansion of an ideal gas. **(d)** reversible adiabatic expansion of an ideal gas.

3.63 Einstein's special relativity equation is $E = mc^2$, where E is energy, m is mass, and c is the velocity of light. Does this equation invalidate the law of conservation of energy, and hence the first law of thermodynamics?

3.64 The convention of arbitrarily assigning a zero enthalpy value to all the (most stable) elements in the standard state and (usually) 298 K is a convenient way of dealing with the enthalpy changes of chemical processes. This convention does not apply to one kind of process, however. What process is it? Why?

3.65 Two moles of an ideal gas are compressed isothermally at 298 K from 1.00 atm to 200 atm. Calculate the values of q, w, ΔU, and ΔH for the process if it is carried out **(a)** reversibly and **(b)** by applying an external pressure of 300 atm.

3.66 The fuel value of hamburger is approximately 3.6 kcal g^{-1}. If a person eats 1 pound of hamburger for lunch and if none of the energy is stored in his body, estimate the amount of water that would have to be lost in perspiration to keep his body temperature constant. (1 lb = 454 g.)

3.67 A quantity of 4.50 g of CaC_2 is reacted with an excess of water at 298 K and atmospheric pressure:

$$CaC_2(s) + 2H_2O(l) \rightarrow Ca(OH)_2(aq) + C_2H_2(g)$$

Calculate the work done in joules by the acetylene gas against the atmospheric pressure.

3.68 An oxyacetylene flame is often used in the welding of metals. Estimate the flame temperature produced by the reaction

$$2C_2H_2(g) + 5O_2(g) \rightarrow 4CO_2(g) + 2H_2O(g)$$

Assume the heat generated from this reaction is all used to heat the products. (*Hint:* First calculate the value of $\Delta_r H°$ for the reaction. Next, look up the heat capacities of the products. Assume the heat capacities are temperature independent.)

3.69 The $\Delta_f \overline{H}°$ values listed in Appendix B all refer to 1 bar and 298 K. Suppose that a student wants to set up a new table of $\Delta_f \overline{H}°$ values at 1 bar and 273 K. Show how she should proceed on the conversion, using acetone as an example.

3.70 The enthalpies of hydrogenation of ethylene and benzene have been determined at 298 K:

$$C_2H_4(g) + H_2(g) \rightarrow C_2H_6(g) \qquad \Delta_r H° = -132 \text{ kJ mol}^{-1}$$

$$C_6H_6(g) + 3H_2(g) \rightarrow C_6H_{12}(g) \qquad \Delta_r H° = -246 \text{ kJ mol}^{-1}$$

What would be the enthalpy of hydrogenation for benzene if it contained three isolated, unconjugated double bonds? How would you account for the difference between the calculated value based on this assumption and the measured value?

3.71 The molar enthalpies of fusion and vaporization of water are 6.01 kJ mol^{-1} and 44.01 kJ mol^{-1} (at 298 K), respectively. From these values, estimate the molar enthalpy of sublimation of ice.

3.72 The standard enthalpy of formation at 298 K of $HF(aq)$ is −320.1 kJ mol^{-1}; $OH^-(aq)$, −229.6 kJ mol^{-1}; $F^-(aq)$, −329.11 kJ mol^{-1}; and $H_2O(l)$, −285.8 kJ mol^{-1}. **(a)** Calculate the enthalpy of neutralization of $HF(aq)$,

$$HF(aq) + OH^-(aq) \rightarrow F^-(aq) + H_2O(l)$$

(b) Using the value of −55.83 kJ mol^{-1} as the enthalpy change from the reaction

$$H^+(aq) + OH^-(aq) \rightarrow H_2O(l)$$

calculate the enthalpy change for the dissociation

$$HF(aq) \rightarrow H^+(aq) + F^-(aq)$$

3.73 It was stated in the chapter that for reactions in condensed phases, the difference between the values of $\Delta_r H$ and $\Delta_r U$ is usually negligibly small. This statement holds for processes carried out under atmospheric conditions. For certain geochemical processes, however, the external pressures may be so great that $\Delta_r H$ and $\Delta_r U$ values can differ by a significant amount. A well-known example is the slow conversion of graphite to diamond under Earth's surface. Calculate the value of the quantity $(\Delta_r H - \Delta_r U)$ for the conversion of 1 mole of graphite to 1 mole of diamond at a pressure of 50,000 atm. The densities of graphite and diamond are 2.25 g cm^{-3} and 3.52 g cm^{-3}, respectively.

3.74 Metabolic activity in the human body releases approximately 1.0×10^4 kJ of heat per day. Assuming the body is 50 kg of water, how fast would the body temperature rise if it were an isolated system? How much water must the body eliminate as perspiration to maintain the normal body temperature (98.6°F)? Comment on your results. The heat of vaporization of water may be taken as 2.41 kJ g^{-1}.

3.75 An ideal gas in a cylinder fitted with a movable piston is adiabatically compressed from V_1 to V_2. As a result, the temperature of the gas rises. Explain what causes the temperature of the gas to rise.

3.76 Calculate the fraction of the enthalpy of vaporization of water used for the expansion of steam at its normal boiling point.

3.77 From a thermochemical point of view, explain why a carbon dioxide fire extinguisher should not be used on a magnesium fire.

3.78 Calculate the internal energy of a Goodyear blimp filled with helium gas at 1.2×10^5 Pa (compared to the empty blimp). The volume of the inflated blimp is 5.5×10^3 m^3. If all the internal energy were used to heat 10.0 tons of copper at 21°C, calculate the final temperature of the metal. (*Hint:* 1 ton = 9.072×10^5 g.)

3.79 Without referring to the chapter, state the conditions for each of the following equations: **(a)** $\Delta H = \Delta U + P\Delta V$. **(b)** $C_P = C_V + nR$. **(c)** $\gamma = \frac{5}{3}$. **(d)** $P_1V_1^\gamma = P_2V_2^\gamma$. **(e)** $w = n\bar{C}_V(T_2 - T_1)$. **(f)** $w = -P\Delta V$. **(g)** $w = -nRT\ln(V_2/V_1)$. **(h)** $dH = dq$.

3.80 The combustion of what volume of ethane (C_2H_6), measured at 23.0°C and 752 mmHg, would be required to heat 855 g of water from 25.0°C to 98.0°C?

3.81 Construct a table with the headings q, w, ΔU, and ΔH. For each of the following processes, deduce whether each of the quantities listed is positive (+), negative (−), or zero (0). **(a)** Freezing of acetone at 1 atm and its normal melting point. **(b)** Irreversible isothermal expansion of an ideal gas. **(c)** Adiabatic compression of an ideal gas. **(d)** Reaction of sodium with water. **(e)** Boiling of liquid ammonia at its normal boiling point. **(f)** Irreversible adiabatic expansion of a gas against an external pressure. **(g)** Reversible isothermal compression of an ideal gas. **(h)** Heating of a gas at constant volume. **(i)** Freezing of water at 0°C.

3.82 State whether each of the following statements is true or false: **(a)** $\Delta U \approx \Delta H$ except for gases or high-pressure processes. **(b)** In gas compression, a reversible process does maximum work. **(c)** ΔU is a state function. **(d)** $\Delta U = q + w$ for an open system. **(e)** C_V is temperature independent for gases. **(f)** The internal energy of a real gas depends only on temperature.

3.83 Show that $(\partial C_V/\partial V)_T = 0$ for an ideal gas.

3.84 Calculate the work done during the isothermal, reversible expansion of a van der Waals gas. Account physically for the way in which the coefficients a and b appear in the final expression. *Hint*: You need to apply the Taylor series expansion:

$$\ln(1 - x) = -x - \frac{x^2}{2} \cdots \quad \text{for } |x| \ll 1$$

to the expression $\ln(V - nb)$. Recall that the a term represents attraction and the b term repulsion.

3.85 Show that for the adiabatic reversible expansion of an ideal gas,

$$T_1^{C_V/R} V_1 = T_2^{C_V/R} V_2$$

3.86 A 4.0-L ideal gas initially at 2.0 atm and 300 K is isothermally compressed to a final volume of 2.0 L. Calculate the work done if the process is carried out **(a)** reversibly and **(b)** irreversibly. Support your answers with a graphical illustration of the processes.

3.87 Suppose that 0.0500-mole of an ideal monatomic gas undergoes the reversible cyclic process shown below. Calculate w, q, and ΔU for each step and for the complete cycle.

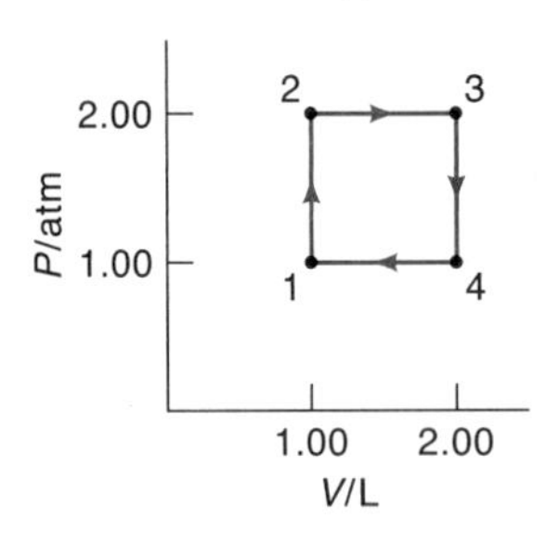

3.88 In acid–base theory, the species $H^+(aq)$ and $H_3O^+(aq)$ are usually treated as the same. This is not the case, however, in thermochemistry. What are the values for $\Delta \overline{H}_f^\circ[H^+(aq)]$ and $\Delta \overline{H}_f^\circ[H_3O^+(aq)]$?

3.89 **(a)** Derive an expression for the work done (w) in a reversible isothermal expansion of a van der Waals gas from V_1 to V_2. **(b)** Calculate w if 2.0 moles of Ne undergo such an expansion from 0.50 L to 1.0 L at 298 K. **(c)** Compare the result in **(b)** with that of an ideal gas. Account for the difference.

CHAPTER 4

The Second Law of Thermodynamics

Humpty Dumpty sat on a wall
Humpty Dumpty had a great fall
All the King's horses and all the King's men
Couldn't put Humpty together again.

As we saw in Chapter 3, the first law of thermodynamics specifies that
neither created nor destroyed, but flows from one part of the universe
is converted from one form into another. The total amount of energy i
remains constant. Despite its immense value in the study of the energet
reactions, the first law does have a major limitation: it cannot predict t
change. It helps us to do the bookkeeping of energy balance, such as th
heat released, work done, and so forth, but it says nothing about wheth
process can indeed occur. For this kind of information we must turn to
of thermodynamics.

In this chapter we introduce a new thermodynamic function, call
which is central to both the second law and the third law of thermo
shall see that changes in entropy, ΔS, provide the necessary criterion
the direction of any process. As a prelude to discussing the entropy fu
take a look at spontaneous processes that occur on their own, under
conditions.

4.1 Spontaneous Processes

A lump of sugar dissolves in a cup of coffee, an ice cube melts in your hand, and a struck match burns in air—we witness so many of these *spontaneous* processes in everyday life that it is almost impossible to list them all. The interesting aspect of a spontaneous process is that the reverse process never happens under the same set of conditions. A leaf lying on the ground will not rise into the air on its own and return to the branch from which it came. Run backwards, a movie of a baseball smashing a window to pieces is funny because everyone knows that it is an impossible event. Ice melts at 20°C and 1 atm, but liquid water at the same temperature and pressure will not spontaneously turn into ice. But why not? Surely we can demonstrate that any of the changes just described (and countless more) can occur in *either* direction in accord with the first law of thermodynamics; yet, in fact, each process occurs in only one direction. After many observations we can conclude that processes occurring spontaneously in one direction cannot also take place spontaneously in the opposite direction; otherwise, nothing would ever happen (Figure 4.1).

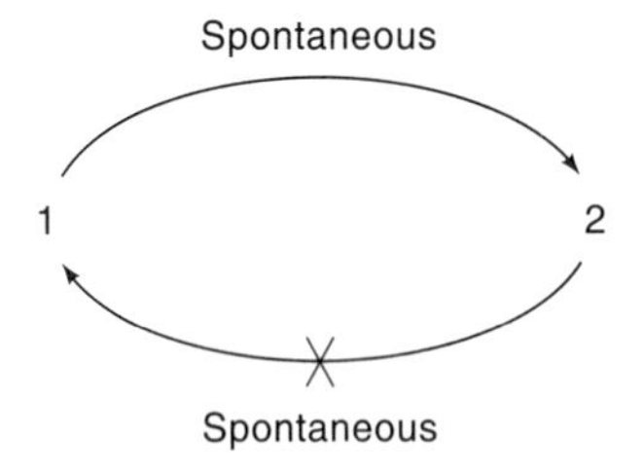

Figure 4.1
If the change from state 1 to 2 occurs spontaneously, then the reverse step, that is, 2 to 1, cannot also be a spontaneous process.

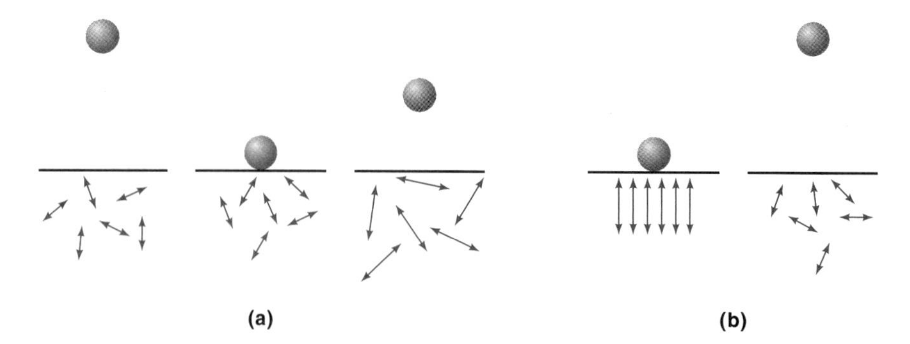

Figure 4.2
(a) A spontaneous process. A falling ball strikes the floor and loses some of its kinetic energy to the molecules in the floor. As a result, the ball does not bounce quite as high and the floor heats up a little. The length of arrows indicates the relative amplitude of molecular vibration. (b) An impossible event. A ball resting on the floor cannot spontaneously rise into the air by absorbing thermal energy from the floor.

Why can't the reverse of a spontaneous process occur by itself? Consider a rubber ball held at some distance above the floor. When the ball is released, it falls. The impact between the ball and the floor causes the ball to bounce upward, and when it has reached a certain height, it repeats its downward motion. In the process of falling, the potential energy of the ball is converted to kinetic energy. Experience tells us that after every bounce the ball does not rise quite as high as before (Figure 4.2a). The reason is that the collision between the ball and the floor is inelastic, so that upon each impact some of the ball's kinetic energy is dissipated among the molecules in the floor. After each bounce, the floor becomes a little bit hotter.* This intake of energy increases the rotational and vibrational motions of the molecules in the floor. Eventually, the ball comes to a complete rest because its kinetic energy is totally lost to the floor. To describe this process in another way, we say that the original potential energy of the ball, through its conversion to kinetic energy, is degraded into heat.

Now let us consider what would be necessary for the reverse process to occur on its own; that is, a ball sitting on the floor spontaneously rises to a certain height in the air by absorbing heat from the floor. That such a process will not violate the first law is clear. If the mass of the ball is m and the height above the floor to which it rises is h, we have

$$\text{energy extracted from the floor} = mgh$$

where g is acceleration due to gravity. The thermal energy of the floor is random molecular motion. To impart an amount of energy large enough to raise the ball from the floor, most of the molecules would have to line up under the ball and vibrate in phase with one another, as shown in Figure 4.2b. At the instant the ball leaves the floor, all the atoms in these molecules must be moving upward for proper energy transfer.

* Actually, the temperature of the ball and the surrounding air also rises slightly after each impact. But here we are concerned only with what happens to the floor.

It is conceivable for several million molecules to execute this kind of synchronized motion, but because of the magnitude of energy transfer, the number of molecules involved would have to be truly enormous, perhaps on the order of Avogadro's number, or 6×10^{23}. Given the random nature of molecular motion, this is such an improbable event that it is virtually impossible. Indeed, no one has ever witnessed the spontaneous rising of a ball from the floor, and we can safely conclude that no one ever will.

Thinking about the improbability of a ball spontaneously rising upward from the floor helps us understand the nature of many spontaneous processes. Consider the familiar example of a gas in a cylinder fitted with a movable piston. If the pressure of the gas is greater than the external pressure, then the gas will expand until the internal and external pressures are equal. This is a spontaneous process. What would it take for the gas to contract spontaneously once mechanical equilibrium is reached? Most of the gas molecules would have to move away from the piston and toward other parts of the cylinder at the same time. Now, at any given moment many molecules are indeed doing this, but we will never find 6×10^{23} molecules engaged in such a unidirectional motion because molecular translational motion is totally random. By the same token, a metal bar at a uniform temperature will not suddenly become hotter at one end and colder at the other. To establish this temperature gradient, the thermal motion due to collisions between randomly vibrating atoms would have to decrease at one end and rise at the other—a highly improbable event.

Let us look at this problem from a different perspective and ask what changes accompany a spontaneous process. It seems logical to assume that all spontaneous processes occur in such a way as to decrease the energy of the system. This assumption helps us explain why things fall downward, why springs unwind, and so forth. But a change of energy alone is not enough to predict whether a process will be spontaneous. For example, in Chapter 3 we saw that the expansion of an ideal gas against vacuum does not result in a change in its internal energy. Yet the process is spontaneous. When ice melts spontaneously at 20°C to form liquid water, the internal energy of the system actually *increases*. In fact, many endothermic physical and chemical processes are spontaneous and many exothermic processes are not. If energy change cannot be used to indicate the direction of a spontaneous process, then we need another thermodynamic function to help us. This function turns out to be entropy (S).

4.2 Entropy

Our discussion of spontaneous processes is based on macroscopic events. Referring to our earlier case of a contracting gas, we can imagine several million molecules moving in a particular direction simultaneously, but certainly not an Avogadro's number of molecules. When the number of molecules is as enormous as 6×10^{23}, roughly the same number of molecules are moving in every direction. There is no reason why, in the absence of an external influence, all of the molecules would choose a specific direction at the same time. In trying to understand spontaneous processes, we should therefore focus our attention on the *statistical* behavior of a very large number of molecules, not on the motion of just a few of them. In this section we shall derive a statistical definition of entropy and then define entropy in terms of thermodynamic quantities.

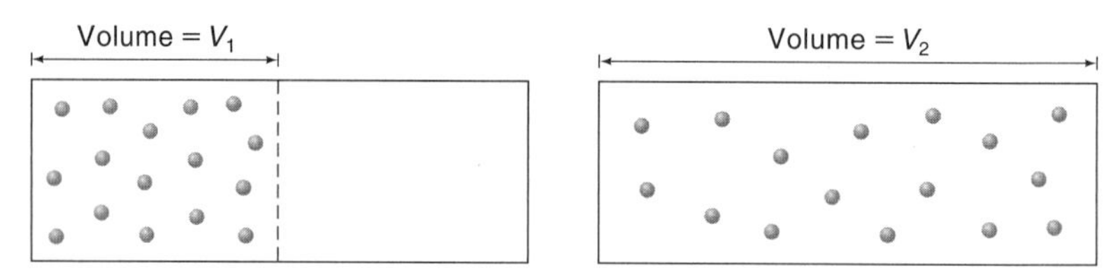

Figure 4.3
Schematic diagram showing N helium atoms occupying volumes V_1 and V_2 of a container.

Statistical Definition of Entropy

For a cylinder containing helium atoms, depicted in Figure 4.3, the probability of finding any one He atom in the entire volume of the cylinder is 1, because all He atoms are known to be inside the cylinder. On the other hand, the probability of finding a helium atom in half of the volume of the cylinder is only $\frac{1}{2}$. If the number of He atoms is increased to two, the probability of finding both of them in the whole volume is still 1, but that of finding both of them in half the volume becomes $\left(\frac{1}{2}\right)\left(\frac{1}{2}\right)$ or $\frac{1}{4}$.* Because $\frac{1}{4}$ is an appreciable quantity, it would not be surprising to find both He atoms in the same region at a given time. However, it is not difficult to see that as the number of He atoms increases, the probability (p) of finding all of them in half the volume becomes progressively smaller:

$$p = \left(\tfrac{1}{2}\right)\left(\tfrac{1}{2}\right)\left(\tfrac{1}{2}\right)\ldots$$
$$= \left(\tfrac{1}{2}\right)^N$$

where N is the total number of atoms present. If $N = 100$, we have

$$p = \left(\tfrac{1}{2}\right)^{100} = 8 \times 10^{-31}$$

Put in perspective, this probability is less than that for the production of Shakespeare's complete works 15 quadrillion times in succession without a single error by a tribe of wild monkeys randomly pounding on computer keyboards.

If N is of the order of 6×10^{23}, the probability becomes $\left(\frac{1}{2}\right)^{6\times 10^{23}}$, a quantity so small that for all practical purposes it can be regarded as zero. From the results of these simple calculations comes a most important message. If initially we had compressed all the He atoms into half the volume, and allowed the gas to expand on its own, we would find that eventually the atoms would be evenly distributed over the entire volume because this situation corresponds to the most probable state. Thus, the direction of spontaneous change is from a smaller volume to a larger one, or from a state with low probability of occurring to one of maximum probability.

Now that we know how to predict the direction of a spontaneous change in terms of probabilities of the initial and final states, it may seem appropriate to treat entropy as being directly proportional to probability as $S = kp$, where k is a proportionality constant. But this expression is invalid for the following reason. Entropy, like U and H, is an extensive property. Consequently, doubling the number of molecules would lead to a twofold increase in the entropy of the system. As we just saw, however, the probability that two independent events will both happen is the product of the probabilities for each

* The probability of both events occurring is a *product* of the probabilities of two independent events. We assume that the He gas behaves ideally so that the presence of one He atom in half the volume does not affect the presence of another He atom in the same region in any way.

event. Hence, changing from one to two molecules gives us p^2. Thus, the increase in entropy (from S to $2S$) and decrease in probability (from p to p^2) are not related to each other as predicted by the simple equation given above. A way out of this dilemma is to express entropy as a natural logarithmic function of probability as follows:

Because $0 \leq p \leq 1$, we have $\ln p < 0$ and $p^2 < p$.

$$S = k_B \ln p + a \tag{4.1}$$

where k_B is the Boltzmann constant (1.381×10^{-23} J K^{-1}) and a is a constant of unknown value. Because the quantity $\ln p$ is dimensionless, the units of entropy are J K^{-1}. While it is tempting to use Equation 4.1 to evaluate absolute entropies, this is not possible because we cannot determine the value of the constant a. However, we can use Equation 4.1 to calculate changes in entropy when a system changes from an initial state 1 to a final state 2. Because entropy is a state function (it depends only on the probability of a state occurring and not on the manner in which the state is created), the change in entropy ΔS for the $1 \rightarrow 2$ process is

$$\begin{aligned}\Delta S &= S_2 - S_1 \\ &= (k_B \ln p_2 + a) - (k_B \ln p_1 + a) \\ &= k_B \ln \frac{p_2}{p_1} \end{aligned} \tag{4.2}$$

Applying Equation 4.2 to the situation described in Figure 4.3, we see that $p_2 = 1$ and $p_1 = (\frac{1}{2})^N$ and the change in entropy is given by

$$\Delta S = k_B \ln \left[\frac{1}{(\frac{1}{2})^N}\right] = k_B \ln 2^N = Nk_B \ln 2$$

Using the relation $N = nN_A$, where n is the number of moles and N_A is Avogardro's constant, the above equation becomes

$$\Delta S = nN_A k_B \ln 2 = nR \ln 2$$

where

$$k_B = \frac{R}{N_A} = \frac{8.314 \text{ J K}^{-1}\text{ mol}^{-1}}{6.022 \times 10^{23}\text{ mol}^{-1}} = 1.381 \times 10^{-23}\text{ J K}^{-1}$$

We see in Figure 4.3 that $V_2/V_1 = 2$. Thus, in general when a gas expands from V_1 to V_2, the change in entropy is given by

$$\Delta S = nR \ln \frac{V_2}{V_1} \tag{4.3}$$

It is important to remember that Equation 4.3 holds only for an isothermal expansion because the entropy of a system is also affected by changes in temperature. Furthermore, we do not have to specify the manner in which the expansion was brought about; that is, reversible or irreversible because S is a state function.

EXAMPLE 4.1

Calculate the entropy change when 2.0 moles of an ideal gas are allowed to expand isothermally from an initial volume of 1.5 L to 2.4 L. Estimate the probability that the gas will contract spontaneously from the final volume to the initial one.

ANSWER

From Equation 4.3 we write

$$\Delta S = nR \ln \frac{V_2}{V_1}$$

$$\Delta S = (2.0 \text{ mol})(8.314 \text{ J K}^{-1} \text{ mol}^{-1}) \ln \frac{2.4 \text{ L}}{1.5 \text{ L}}$$

$$= 7.8 \text{ J K}^{-1}$$

To estimate the probability for spontaneous contraction, we note that this process must be accompanied by a *decrease* in entropy equal to -7.8 J K^{-1}. Because the process is now defined as $2 \rightarrow 1$, we have, from Equation 4.2,

$$\Delta S = k_B \ln \frac{p_1}{p_2}$$

$$-7.8 \text{ J K}^{-1} = (1.381 \times 10^{-23} \text{ J K}^{-1}) \ln \frac{p_1}{p_2}$$

$$\ln \frac{p_1}{p_2} = -5.7 \times 10^{23}$$

or

$$\frac{p_1}{p_2} = e^{-5.7 \times 10^{23}}$$

COMMENT

This exceedingly small ratio means that there is no possibility for the process to occur by itself. This does not mean that the gas cannot be compressed from 2.4 L to 1.5 L, but it must be done with the aid of an external force.

Thermodynamic Definition of Entropy

Equation 4.1 is a statistical formulation of entropy; defining entropy in terms of probability provides us with a molecular interpretation. In general, however, this equation is not used for calculating changes in entropy. Calculating p for complex systems, those in which chemical reactions occur, for example, is too difficult. Entropy changes can be conveniently measured from changes of other thermodynamic quantities such as ΔH. In Section 3.5 we saw that the heat absorbed by an ideal gas in an isothermal, reversible expansion is given by

$$q_{\text{rev}} = nRT \ln \frac{V_2}{V_1}$$

or

$$\frac{q_{\text{rev}}}{T} = nR \ln \frac{V_2}{V_1}$$

Because the right-hand side of the equation above is equal to ΔS (see Equation 4.3), we have

$$\Delta S = \frac{q_{\text{rev}}}{T} \tag{4.4}$$

In words, Equation 4.4 says that the entropy change of a system in a reversible process is given by the heat absorbed divided by the temperature at which the process occurs. For an infinitesimal process, we can write

$$dS = \frac{dq_{\text{rev}}}{T} \tag{4.5}$$

We write dq_{rev} rather than $đq_{\text{rev}}$ because the path is defined.

Both Equations 4.4 and 4.5 are the thermodynamic definition of entropy. Although these equations were derived for the expansion of gases, they are applicable to any type of process at constant temperature. It is important to note that the definition holds only for a reversible process, as the subscript "rev" indicates. While S is a path-independent state function, q is not and we must specify the reversible path in defining entropy. If the expansion were irreversible, then the work done by the gas on the surroundings would be less and so would the heat absorbed by the gas from the surroundings, that is, $q_{\text{irrev}} < q_{\text{rev}}$. Although the entropy change would be the same, that is, $\Delta S_{\text{rev}} = \Delta S_{\text{irrev}} = \Delta S$, we would have $\Delta S > q_{\text{irrev}}/T$. We shall return to this point in Section 4.4.

4.3 The Carnot Heat Engine

For a more rigorous thermodynamic definition of entropy, we can analyze the performance of a Carnot heat engine, named for the French engineer Sadi Carnot (1796–1832). A heat engine converts heat to mechanical work. Heat engines play an essential role in our technological society; they include steam locomotives, steam turbines that generate electricity, and the internal combustion engines in automobiles. The Carnot heat engine is an idealized model for the operation of any heat engine. Incorporated in this model is the concept of thermodynamic efficiency, which is of great importance in the study of industrial, chemical, and biological processes.

For our purpose, the Carnot heat engine can be represented by one mole of an ideal gas in a cylinder fitted with a movable, frictionless piston that allows P–V work to be done on and by the gas. Figure 4.4 shows the relationships of the engine to its thermal and mechanical surroundings. A complete cycle of the engine includes four steps.

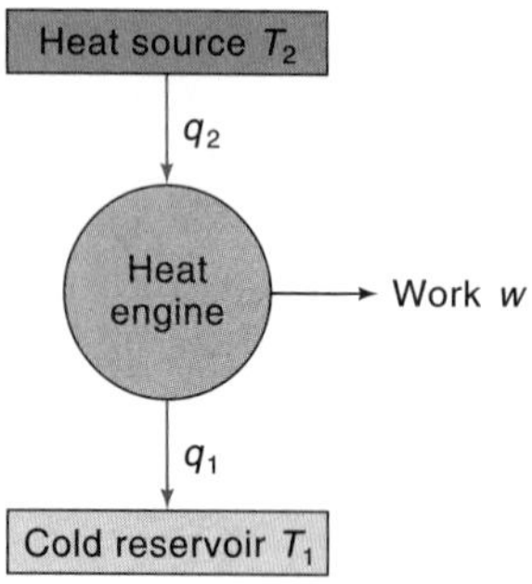

Figure 4.4
A heat engine draws heat from a heat source to do work on the surroundings and discharges some of the heat to a cold reservoir.

STEP 1. The gas at temperature T_2 absorbs heat q_2 from the heat source and expands isothermally and reversibly from V_1 to V_2. The changes are:

$$\Delta U = 0 \qquad \text{(isothermal process, ideal gas)}$$

$$\text{work done} = w_2 = -RT_2 \ln \frac{V_2}{V_1} \qquad \text{(Equation 3.5)}$$

$$\text{heat absorbed} = q_2 = RT_2 \ln \frac{V_2}{V_1} \qquad (q_2 = -w_2 \text{ from the first law})$$

STEP 2. The gas expands adiabatically and reversibly from V_2 to V_3. In the process, the gas temperature drops from T_2 to T_1. The changes are:

$$q = 0 \qquad \text{(adiabatic process)}$$

$$\text{work done} = \Delta U = \overline{C}_V(T_1 - T_2) \qquad \text{(Equation 3.25; assume } \overline{C}_V \text{ to be temperature independent)}$$

STEP 3. The gas is compressed isothermally and reversibly from V_3 to V_4. The heat released is transferred to the cold reservoir at T_1. The changes are:

$$\Delta U = 0 \qquad \text{(isothermal process, ideal gas)}$$

$$\text{work done} = w_1 = -RT_1 \ln \frac{V_4}{V_3} \qquad \text{(Equation 3.5)}$$

$$\text{heat absorbed} = q_1 = RT_1 \ln \frac{V_4}{V_3} \qquad (q_1 = -w_1)$$

STEP 4. The gas is compressed adiabatically and reversibly from V_4 to V_1. As a result, the temperature of the gas increases from T_1 to T_2. The changes are:

$$q = 0 \qquad \text{(adiabatic process)}$$

$$\text{work done} = \Delta U = \overline{C}_V(T_2 - T_1) \qquad \text{(Equation 3.25; assume } \overline{C}_V \text{ to be temperature independent)}$$

Figure 4.5 illustrates these four steps, called the *Carnot cycle*, in the operation of a Carnot heat engine.

Summarizing the calculations for the entire Carnot cycle, we write

$$\Delta U(\text{cycle}) = 0 \qquad (U \text{ is a state function})$$

$$q(\text{cycle}) = q_2 + q_1 \qquad (q_2 \text{ is positive and } q_1 \text{ is negative})$$

$$\begin{aligned} w(\text{cycle}) &= -RT_2 \ln \frac{V_2}{V_1} + \overline{C}_V(T_1 - T_2) - RT_1 \ln \frac{V_4}{V_3} + \overline{C}_V(T_2 - T_1) \\ &= -RT_2 \ln \frac{V_2}{V_1} - RT_1 \ln \frac{V_4}{V_3} \end{aligned}$$

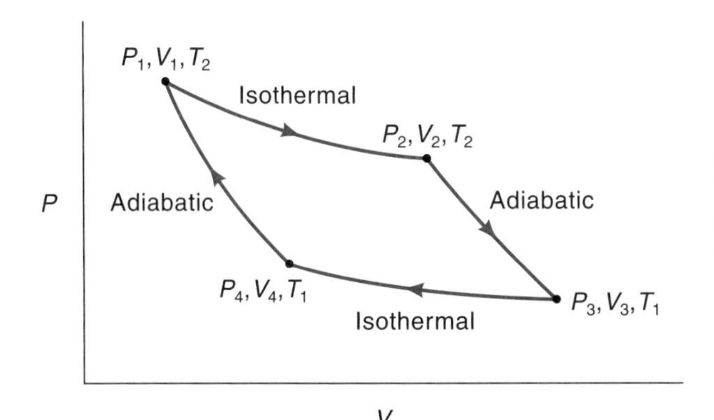

Figure 4.5
The four steps of the Carnot cycle. The enclosed area represents the work done by the heat engine on the surroundings. $T_2 > T_1$.

Next we find a relationship between V_1, V_2, V_3, and V_4. The P–V relations for isothermal and adiabatic processes are

$$P_1V_1 = P_2V_2 \qquad P_3V_3 = P_4V_4 \qquad \text{(Boyle's law)}$$

$$P_2V_2^\gamma = P_3V_3^\gamma \qquad P_1V_1^\gamma = P_4V_4^\gamma \qquad \text{(Equation 3.24)}$$

Taking the ratios of the adiabatic relations, we write

$$\frac{P_2V_2^\gamma}{P_1V_1^\gamma} = \frac{P_3V_3^\gamma}{P_4V_4^\gamma}$$

or

$$\frac{P_2V_2}{P_1V_1} \times \frac{V_2^{\gamma-1}}{V_1^{\gamma-1}} = \frac{P_3V_3}{P_4V_4} \times \frac{V_3^{\gamma-1}}{V_4^{\gamma-1}}$$

so that

$$\left(\frac{V_2}{V_1}\right)^{\gamma-1} = \left(\frac{V_3}{V_4}\right)^{\gamma-1}$$

Therefore,

$$\frac{V_2}{V_1} = \frac{V_3}{V_4}$$

The net work done during the cycle can now be written as

$$w(\text{cycle}) = -R(T_2 - T_1)\ln\frac{V_2}{V_1} \tag{4.6}$$

and the heat absorbed from the heat reservoir and discharged to the cold reservoir are

$$q_2 = RT_2 \ln\frac{V_2}{V_1} \tag{4.7}$$

$$q_1 = RT_1 \ln\frac{V_4}{V_3} = -RT_1 \ln\frac{V_2}{V_1} \tag{4.8}$$

and thus

$$q(\text{cycle}) = R(T_2 - T_1)\ln\frac{V_2}{V_1} = -w(\text{cycle})$$

Thermodynamic Efficiency

Having obtained expressions for the net work done and heat absorbed, we can determine the efficiency of the heat engine. Keeping in mind that efficiency represents a ratio of output to input, we can write the efficiency η for a heat engine as follows:

$$\begin{aligned}\eta &= \frac{\text{net work done by heat engine}}{\text{heat absorbed by engine}}\\ &= \frac{|w|}{q_2}\\ &= \frac{R(T_2 - T_1)\ln(V_2/V_1)}{RT_2\ln(V_2/V_1)}\\ &= \frac{T_2 - T_1}{T_2} = 1 - \frac{T_1}{T_2} \qquad (4.9)\end{aligned}$$

Note that in arriving at the above equation, we have used the magnitude but not the sign of w because it is a negative quantity by our convention (the net work is work done by the gas on the surroundings). Equation 4.9 defines the thermodynamic efficiency of all heat engines. It is given by the difference in the temperatures of the heat source and the cold reservoir, divided by the temperature of the heat source. In practice, thermodynamic efficiency can never be 1 (or 100%) because T_1 cannot be zero and T_2 cannot be infinite.*

The absolute zero of temperature can never be attained in practice.

* We can never convert heat totally into work; some of it escapes into the surroundings as waste heat.

EXAMPLE 4.2

At a power plant, superheated steam at 560°C is used to drive a turbine for electricity generation. The steam is discharged to a cooling tower at 38°C. Calculate the maximum efficiency of this process.

ANSWER

First we convert the temperatures to kelvins: $T_2 = 833$ K and $T_1 = 311$ K. From Equation 4.9, we write

$$\begin{aligned}\eta &= \frac{T_2 - T_1}{T_2}\\ &= \frac{833\text{ K} - 311\text{ K}}{833\text{ K}}\\ &= 0.63 \text{ or } 63\%\end{aligned}$$

COMMENT

In reality, friction, heat loss, and other complications reduce the maximum efficiency of a steam turbine to about 40%. Therefore, for every ton of coal burned at a power plant, 0.40 ton generates electricity and the rest of it heats the surroundings!

The Entropy Function

As mentioned earlier, the other important finding in the analysis of the Carnot cycle concerns entropy. We have seen that for this cyclic process, the overall change in U is zero but not so for work and heat because they are not state functions. If we examine the q/T ratio, however, we find that

$$\frac{q_2}{T_2} + \frac{q_1}{T_1} = \frac{RT_2 \ln (V_2/V_1)}{T_2} + \frac{-RT_1 \ln (V_2/V_1)}{T_1}$$
$$= 0$$

In fact, for a heat engine performing a cyclic process involving any number of steps, it is true that the sum of the q/T terms is zero; that is, for i steps

$$\sum_i \frac{q_i}{T_i} = 0$$

This result is remarkable, for it suggests that there is a state function, S, defined by

$$\Delta S = \frac{q_{\text{rev}}}{T}$$

which is the same as Equation 4.4. For an infinitesimal process,

$$dS = \frac{dq_{\text{rev}}}{T}$$

The subscript "rev" indicates that the process is reversible. For a cyclic process, then, we can write

$$\sum_i \frac{q_i}{T_i} = \sum_i \Delta S_i = 0$$

where ΔS_i is the entropy change for the ith step. The change in S for the cycle is zero because it is a state function.

Refrigerators, Air Conditioners, and Heat Pumps

Daily experience tells us that heat flows spontaneously from a hot body to a cold one. If work is applied, however, heat can be made to flow in the reverse direction, against its natural tendency. The three familiar devices that reverse the direction of heat flow are refrigerators, air conditioners, and heat pumps. Figure 4.6 shows a Carnot engine operating in reverse. Work is done to extract an amount of heat, q_1, from a cold reservoir, and an amount of heat, q_2, is deposited into the heat source. As in the Carnot cycle, the law of conservation of energy requires that (see Equations 4.6, 4.7, and 4.8)

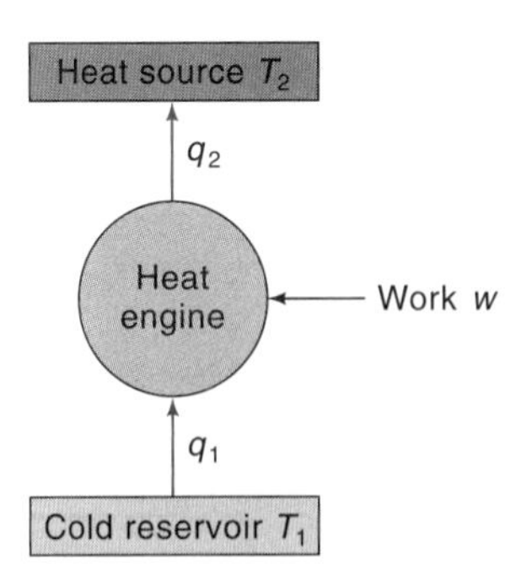

Figure 4.6
A Carnot engine run in the reverse direction. Refrigerators, air conditioners, and heat pumps perform work w to remove heat q_1 from a cold reservoir and deposit heat q_2 into the heat source.

$$-q_2 = q_1 + w$$

Refrigerators. Figure 4.7 shows a schematic diagram of a refrigerator. A gaseous substance, called a *refrigerant*, circulates in a closed system. In a complete cycle, the substance is compressed, cooled, and then expanded. As it expands inside the refrigerator, the refrigerant absorbs heat from the food. When it travels outside of the refrigerator, the refrigerant is compressed and heat is given off into the surroundings. For this reason, the outside surface (usually the sides and the back) of a working refrigerator feels warm.

Hydrofluorocarbons (HFCs) have replaced chlorofluorocarbons (CFCs) as the most commonly used refrigerants.

The performance of a refrigerator is rated by its *coefficient of performance* (COP), which is the ratio of the heat extracted from the cold reservoir to the amount of work applied:

$$\text{COP} = \frac{q_1}{w} \qquad \text{(both } q_1 \text{ and } w \text{ are positive)} \tag{4.10}$$

Our next step is to derive an equation relating the COP to the temperatures of the heat source (T_2) and the cold reservoir (T_1). From Equation 4.9, we have

$$\frac{w}{|q_2|} = 1 - \frac{T_1}{T_2} \qquad (q_2 \text{ is negative}) \tag{4.11}$$

Therefore,

$$\frac{T_1}{T_2} = 1 - \frac{w}{|q_2|} = \frac{|q_2| - w}{|q_2|} = \frac{q_1}{|q_2|} \tag{4.12}$$

Inverting the last and first terms in Equation 4.12, we get

$$\frac{|q_2|}{q_1} = \frac{T_2}{T_1}$$

or

$$\frac{|q_2| - q_1}{q_1} = \frac{w}{q_1} = \frac{T_2 - T_1}{T_1}$$

Finally, we arrive at the result:

$$\text{COP} = \frac{q_1}{w} = \frac{T_1}{T_2 - T_1} \tag{4.13}$$

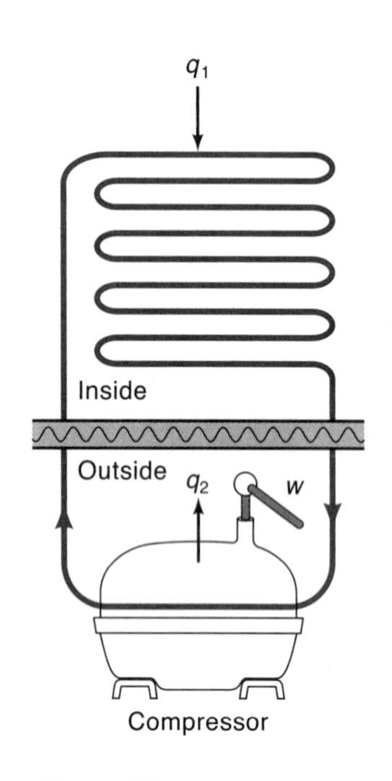

Figure 4.7
How a refrigerator works. Expansion coils are located inside the refrigerator and a compressor on the outside. When the refrigerant gas is compressed, its temperature rises, and heat (q_2) flows from the gas to the surroundings. After cooling, the compressed gas passes into the refrigerator, where it expands and its temperature drops. Heat (q_1) flows from the food to the gas in the coils. The now warm, uncompressed gas reenters the compressor and the cycle is repeated.

Note that because the derivation was based on an ideal gas working reversibly. Equation 4.13 gives the *maximum* value of the COP. For example, if the refrigerator is set at 273 K and the room temperature is 293 K, then COP $\approx$ 14. In reality, the COP values of commercially available refrigerators are in the range of only 2–6 because these devices are not working under ideal conditions.

Air Conditioners. An air conditioner operates much like a refrigerator. In this case, the room itself becomes the cold reservoir, and the outdoors is the heat source. The performance of an air conditioner is also rated by its COP value.

Heat Pumps. In principle, there is no difference between a heat pump and a refrigerator (or an air conditioner). In practice, however, we use a heat pump to warm a room rather than to lower its temperature. The word *pump*, which applies to all three devices, describes the transport of heat from a cold reservoir to a heat source. The process is akin to pumping water up a tower against Earth's gravity.

The advantage of a heat pump over a conventional electric heater, say, can be demonstrated with the following analysis. Suppose 500 J of energy is available to heat a room. If this amount of energy is channeled to an electric heater, the heater would deliver 500 J of heat to the room. On the other hand, a heat pump uses energy to do the work of pumping heat, q_1, from the cooler outdoors into the warmer room and delivers to the inside an amount of energy, $-q_2 = q_1 + w$. Thus, a heat pump puts more than 500 J of heat into the room. The performance of a heat pump is also rated by a coefficient of performance value. Because the function of a heat pump is to deliver heat, however, the COP is the ratio of the heat delivered, q_2, to the work done,

$$\text{COP} = \frac{|q_2|}{w} \tag{4.14}$$

Following the same procedure we used for refrigerators and air conditioners, we can show that

$$\text{COP} = \frac{T_2}{T_2 - T_1} \tag{4.15}$$

A comparison of Equation 4.15 with Equation 4.13 shows that, for the same values of T_1 and T_2, the COP value of a heat pump is greater than that of a refrigerator.

EXAMPLE 4.3

How much work must be done by a heat pump to deliver 5000 J of heat into a house maintained at 22°C when the outdoor temperature is (a) 5°C and (b) −10°C?

ANSWER

First, we need to derive an expression relating w to the outdoor (T_1) and indoor (T_2) temperatures. From energy conservation and Equation 4.12,

$$w = |q_2| - q_1 = |q_2| - |q_2|\left(\frac{T_1}{T_2}\right) = |q_2|\left(1 - \frac{T_1}{T_2}\right)$$

(a) $$w = 5000\text{ J}\left(1 - \frac{278\text{ K}}{295\text{ K}}\right) = 288\text{ J}$$

(b) $$w = 5000\text{ J}\left(1 - \frac{263\text{ K}}{295\text{ K}}\right) = 542\text{ J}$$

COMMENT

In both cases, we see that the amount of work done is considerably less than the heat delivered to the inside of the house. As expected, it takes more work to deliver the same amount of heat when the outdoor temperature is lower (case b).

4.4 The Second Law of Thermodynamics

So far, our discussion of entropy changes has focused on the system. For a proper understanding of entropy, we must also examine what happens to the surroundings. Because of its size and the amount of material it contains, the surroundings can be thought of as an infinitely large reservoir. Therefore, the exchange of heat and work between a system and its surroundings alters the properties of the surroundings by only an infinitesimal amount. Because infinitesimal changes are characteristic of reversible processes, it follows that *any* process has the same effect on the surroundings as a reversible process. Thus, regardless of whether a process is reversible or irreversible with respect to the system, we can write the heat change in the surroundings as

$$(dq_{\text{surr}})_{\text{rev}} = (dq_{\text{surr}})_{\text{irrev}} = dq_{\text{surr}}$$

For this reason, we shall not bother to specify the path for dq_{surr}. The change in entropy of the surroundings is

$$dS_{\text{surr}} = \frac{dq_{\text{surr}}}{T_{\text{surr}}}$$

and for a finite isothermal process, that is, for a process that can be studied in the laboratory,

$$\Delta S_{\text{surr}} = \frac{q_{\text{surr}}}{T_{\text{surr}}}$$

Returning to the isothermal expansion of an ideal gas, we saw earlier that the heat absorbed from the surroundings during a reversible process is $nRT_{\text{sys}} \ln(V_2/V_1)$, where T_{sys} is the temperature of the system. Because the system is at thermal equilibrium with its surroundings throughout the process, $T_{\text{sys}} = T_{\text{surr}} = T$. The heat lost by the surroundings to the system is therefore $-nRT \ln(V_2/V_1)$, and the corresponding change in entropy is

$$\Delta S_{\text{surr}} = \frac{q_{\text{surr}}}{T}$$

The total change in the entropy of the universe (system plus surroundings), ΔS_{univ}, is given by

$$\begin{aligned}
\Delta S_{\text{univ}} &= \Delta S_{\text{sys}} + \Delta S_{\text{surr}} \\
&= \frac{q_{\text{sys}}}{T} + \frac{q_{\text{surr}}}{T} \\
&= \frac{nRT \ln(V_2/V_1)}{T} + \frac{[-nRT \ln(V_2/V_1)]}{T} = 0
\end{aligned}$$

Thus, for a reversible process, the total change in the entropy of the universe is equal to zero.

Now let us consider what happens if the expansion is irreversible. In the extreme case, we can assume that the gas is expanding against a vacuum. Again, the change in the entropy of the system is given by $\Delta S_{sys} = nR \ln(V_2/V_1)$ because S is a state function. Because no work is done in this process, however, no heat is exchanged between the system and the surroundings. Therefore, we have $q_{surr} = 0$ and $\Delta S_{surr} = 0$. The change in the entropy of the universe is now given by

$$\Delta S_{univ} = \Delta S_{sys} + \Delta S_{surr}$$

$$= nR \ln\left(\frac{V_2}{V_1}\right) > 0$$

Combining these two expressions for ΔS_{univ}, we obtain

$$\Delta S_{univ} = \Delta S_{sys} + \Delta S_{surr} \geq 0 \qquad (4.16)$$

where the equality sign applies to a reversible process and the greater-than (>) sign applies to an irreversible (i.e., spontaneous) process. Equation 4.16 is the mathematical statement of the *second law of thermodynamics.* In words, the second law may be stated as follows: *The entropy of an isolated system increases in an irreversible process and remains unchanged in a reversible process. It can never decrease.* (See Appendix 4.1 on p. 165 for a list of equivalent statements of the second law.) Thus, either ΔS_{sys} or ΔS_{surr} can be a negative quantity for a particular process, but their sum can never be less than zero.

Just thinking about entropy increases its value in the universe.

EXAMPLE 4.4

A quantity of 0.50 mole of an ideal gas at 20°C expands isothermally against a constant pressure of 2.0 atm from 1.0 L to 5.0 L. Calculate the values of ΔS_{sys}, ΔS_{surr} and ΔS_{univ}.

ANSWER

From the initial conditions, we can show that the pressure of the gas is 12 atm. First we calculate the value of ΔS_{sys}. Noting that the process is isothermal and ΔS_{sys} is the same whether the process is reversible or irreversible, we write, from Equation 4.3,

$$\Delta S_{sys} = nR \ln \frac{V_2}{V_1}$$

$$= (0.50 \text{ mol})(8.314 \text{ J K}^{-1} \text{ mol}^{-1}) \ln \frac{5.0 \text{ L}}{1.0 \text{ L}}$$

$$= 6.7 \text{ J K}^{-1}$$

To calculate the value of ΔS_{surr}, we first determine the work done in the irreversible gas expansion:

$$w = -P\Delta V$$
$$= -(2.0 \text{ atm})(5.0 - 1.0)\text{L}$$
$$= -8.0 \text{ L atm}$$
$$= -810 \text{ J} \quad (1 \text{ L atm} = 101.3 \text{ J})$$

Because $\Delta U = 0$, $q = -w = +810$ J. The heat lost by the surroundings must then be -810 J. The change in entropy of the surroundings is given by

$$\Delta S_{surr} = \frac{q_{surr}}{T}$$
$$= \frac{-810 \text{ J}}{293 \text{ K}}$$
$$= -2.8 \text{ J K}^{-1}$$

Finally, from Equation 4.16,

$$\Delta S_{univ} = \Delta S_{sys} + \Delta S_{surr}$$
$$= 6.7 \text{ J K}^{-1} - 2.8 \text{ J K}^{-1}$$
$$= 3.9 \text{ J K}^{-1}$$

COMMENT

The result shows that the process is spontaneous, which is what we would expect given the initial pressure of the gas.

4.5 Entropy Changes

Having learned the statistical and thermodynamic definitions of entropy and examined the second law of thermodynamics, we are now ready to study how various processes affect the entropy of the system. We have already seen that the entropy change for the reversible, isothermal expansion of an ideal gas is given by $nR \ln(V_2/V_1)$. In this section, we shall consider several other examples of entropy changes.

Entropy Change due to Mixing of Ideal Gases

Figure 4.8 shows a container in which n_A moles of ideal gas A at T, P, and V_A are separated by a partition from n_B moles of ideal gas B at T, P, and V_B. When the partition

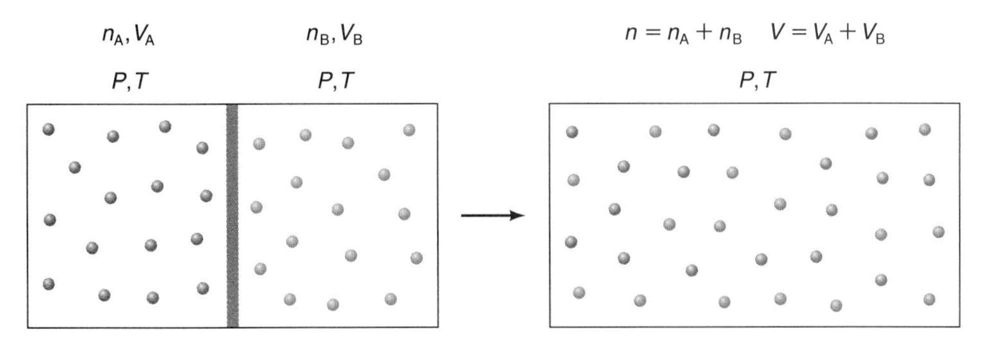

Figure 4.8
The mixing of two ideal gases at the same temperature and pressure leads to an increase in entropy.

is removed, the gases mix spontaneously and the entropy of the system increases. To calculate the entropy of mixing, $\Delta_{mix}S$, we can treat the process as two separate, isothermal gas expansions.

$$\text{For gas A} \qquad \Delta S_A = n_A R \ln \frac{V_A + V_B}{V_A}$$

$$\text{For gas B} \qquad \Delta S_B = n_B R \ln \frac{V_A + V_B}{V_B}$$

Therefore,

$$\Delta_{mix}S = \Delta S_A + \Delta S_B = n_A R \ln \frac{V_A + V_B}{V_A} + n_B R \ln \frac{V_A + V_B}{V_B}$$

According to Avogadro's law, volume is directly proportional to the number of moles of the gas at constant T and P, and so the above equation can be written as

$$\begin{aligned} \Delta_{mix}S &= n_A R \ln \frac{n_A + n_B}{n_A} + n_B R \ln \frac{n_A + n_B}{n_B} \\ &= -n_A R \ln \frac{n_A}{n_A + n_B} - n_B R \ln \frac{n_B}{n_A + n_B} \\ &= -n_A R \ln x_A - n_B R \ln x_B \\ &= -R(n_A \ln x_A + n_B \ln x_B) \end{aligned} \qquad (4.17)$$

where x_A and x_B are the mole fractions of A and B, respectively. Because $x < 1$, it follows that $\ln x < 0$ and that the right side of Equation 4.17 is a positive quantity, which is consistent with the spontaneous nature of the process.*

* Because A and B are ideal gases, there are no intermolecular forces between the molecules, so no heat change results from the mixing. Consequently, the change in entropy of the surroundings is zero, and the direction of the process is solely dependent on the change in entropy of the system.

Entropy Changes due to Phase Transitions

The melting of ice is a familiar phase change. At 0°C and 1 atm, ice and liquid water are in equilibrium. Under these conditions, heat is absorbed reversibly by the ice during the melting process. Furthermore, because this is a constant-pressure process, the heat absorbed is equal to the enthalpy change of the system, so that $q_{\text{rev}} = \Delta_{\text{fus}}H$, where $\Delta_{\text{fus}}H$ is called the *heat* or *enthalpy of fusion*. Note that because H is a state function, it is no longer necessary to specify the path, and the melting process need not be carried out reversibly. The entropy of fusion, $\Delta_{\text{fus}}S$, is given by

$$\Delta_{\text{fus}}S = \frac{\Delta_{\text{fus}}H}{T_{\text{f}}} \tag{4.18}$$

where T_{f} is the fusion or melting point (273 K for ice). Similarly, we can write the entropy of vaporization, $\Delta_{\text{vap}}S$, as

$$\Delta_{\text{vap}}S = \frac{\Delta_{\text{vap}}H}{T_{\text{b}}} \tag{4.19}$$

where $\Delta_{\text{vap}}H$ and T_{b} are the enthalpy of vaporization and boiling point of the liquid, respectively.

EXAMPLE 4.5

The molar enthalpies of fusion and vaporization of water are 6.01 kJ mol^{-1} and 40.79 kJ mol^{-1}, respectively. Calculate the entropy changes for the fusion and vaporization of 1 mole of water at its normal melting point and boiling point.

ANSWER

From Equation 4.18, we write

$$\Delta_{\text{fus}}\overline{S} = \frac{\Delta_{\text{fus}}\overline{H}}{T_{\text{f}}}$$

$$= \frac{6.01 \times 1000 \text{ J mol}^{-1}}{273 \text{ K}} = 22.0 \text{ J K}^{-1} \text{ mol}^{-1}$$

and, from Equation 4.19,

$$\Delta_{\text{vap}}\overline{S} = \frac{\Delta_{\text{vap}}\overline{H}}{T_{\text{b}}}$$

$$= \frac{40.79 \times 1000 \text{ J mol}^{-1}}{373 \text{ K}} = 109.3 \text{ J K}^{-1} \text{ mol}^{-1}$$

COMMENTS

(1) Because entropy values are generally quite small, we express them as J K^{-1} mol^{-1} rather than kJ K^{-1} mol^{-1}. (2) In both melting and vaporization, we find an increase in entropy. This result may seem strange because ice is in equilibrium with water at 273 K, and water is in equilibrium with its vapor at 373 K, so we might expect the change ins entropy in each case to be zero. However, the entropy changes we have calculated refer only to the system. Because heat is absorbed reversibly in an equilibrium process, the changes in entropy in the surroundings for the fusion and vaporization are given by $-\Delta_{fus}\overline{H}/T_f$ and $-\Delta_{vap}\overline{H}/T_b$, respectively. Therefore, the total change in entropy of the universe is zero in each case. (3) Our calculation shows, and indeed it is true in general, that $\Delta_{vap}\overline{S}$ is greater than $\Delta_{fus}\overline{S}$ for the same substance. Because both solids and liquids are condensed phases, they possess considerable structure, or order. Consequently, the solid-to-liquid transition results in a relatively small increase in molecular disorder. On the other hand, the arrangement of molecules in the gaseous state is completely random, so that the liquid-to-vapor process is accompanied by a large increase in disorder.

Table 4.1 lists the molar entropies of vaporization of several liquids. The interesting feature of this table is that these different liquids have approximately the same molar entropy of vaporization (about 88 J K^{-1} mol^{-1}). This empirical observation is known as *Trouton's rule*. The molecular interpretation of this phenomenon is that most liquids and almost all gases have similar structures so that a comparable amount of disorder is generated upon vaporization; hence, the $\Delta_{vap}\overline{S}$ values are similar. In Table 4.1, the exceptions are water and ethanol, which have noticeably larger $\Delta_{vap}\overline{S}$ values. The reason is that these liquids have a more ordered structure because of hydrogen bonding, and so greater disorder results at their boiling points.

Table 4.1
The Boiling Point (T_b), Molar Enthalpy of Vaporization ($\Delta_{vap}\overline{H}$), and Molar Entropy of Vaporization ($\Delta_{vap}\overline{S}$) of Several Substances

Substance	T_b/K	$\Delta_{vap}\overline{H}$/kJ mol^{-1}	$\Delta_{vap}\overline{S}$/J K^{-1} mol^{-1}
Bromine (Br_2)	331.8	30.0	90.4
Ethanol (C_2H_5OH)	351.3	39.3	111.9
Diethyl ether ($C_2H_5OC_2H_5$)	307.6	26.0	84.5
Hexane (C_6H_{12})	341.7	28.9	84.6
Mercury (Hg)	630	59.0	93.7
Methane (CH_4)	109	9.2	84.4
Water (H_2O)	373	40.79	109.4

Entropy Change due to Heating

When the temperature of a system is raised from T_1 to T_2, its entropy also increases. This correlation is due to the intake of energy, which promotes molecules to higher translational, rotational, and vibrational energy levels and increases disorder at the molecular level. We can calculate this entropy increase as follows. Let S_1 and S_2 be the entropies of the system in states 1 and 2 (characterized by T_1 and T_2). If heat is transferred reversibly to the system, then the increase in entropy for an infinitesimal amount of heat transfer is given by Equation 4.5:

$$dS = \frac{dq_{\text{rev}}}{T}$$

The entropy at T_2 is given by

$$S_2 = S_1 + \int_{T_1}^{T_2} \frac{dq_{\text{rev}}}{T}$$

If we define this to be a constant-pressure process, as is usually the case, then $dq_{\text{rev}} = dH$, so that

$$S_2 = S_1 + \int_{T_1}^{T_2} \frac{dH}{T}$$

From Equation 3.18, we have $dH = C_P dT$, and so we can write

Remember that $\int \frac{dx}{x} = \int d\ln x$.

$$S_2 = S_1 + \int_{T_1}^{T_2} \frac{C_P}{T} dT = S_1 + \int_{T_1}^{T_2} C_P d\ln T \qquad (4.20)$$

If the temperature range is small, we can assume that C_P is independent of temperature. Then Equation 4.20 becomes

$$S_2 = S_1 + C_P \ln \frac{T_2}{T_1} \qquad (4.21)$$

and the increase in entropy, ΔS, as a result of heating is

$$\Delta S = S_2 - S_1 = C_P \ln \frac{T_2}{T_1}$$

$$= n\overline{C}_P \ln \frac{T_2}{T_1} \qquad (4.22)$$

EXAMPLE 4.6

At constant pressure, 200.0 g of water is heated from 10°C to 20°C. Calculate the increase in entropy for this process. The molar heat capacity of water at constant pressure is 75.3 J K^{-1} mol^{-1}.

ANSWER

The number of moles of water present is 200.0 g/18.02 g mol^{-1} = 11.1 mol. The increase in entropy, according to Equation 4.22, is given by

$$\Delta S = n\overline{C}_P \ln \frac{T_2}{T_1}$$

$$= (11.1 \text{ mol})(75.3 \text{ J K}^{-1} \text{ mol}^{-1}) \ln \frac{293 \text{ K}}{283 \text{ K}}$$

$$= 29.0 \text{ J K}^{-1}$$

COMMENT

For this calculation we have assumed that $\overline{C}_P$ is independent of temperature and that water does not expand when heated, so that no work is done.

Suppose that the heating of water in Example 4.6 had been carried out irreversibly (as is the case in practice), say with a Bunsen burner. What would be the increase in entropy? We note that regardless of the path, the initial and final states are the same, that is, 200 g of water heated from 10°C to 20°C. Therefore, the integral on the right-hand side of Equation 4.20 gives ΔS for the irreversible heating. This conclusion follows from the fact that ΔS depends only on T_1 and T_2 and not on the path. Thus, ΔS for this process is 29.0 J K^{-1}, whether the heating is done reversibly or irreversibly.

EXAMPLE 4.7

Supercooled water is liquid water that has been cooled below its normal freezing point. This state is thermodynamically unstable and has a tendency to spontaneously freeze into ice. Suppose we have 2.00 moles of supercooled water turning into ice at −10°C and 1.0 atm. Calculate ΔS_{sys}, ΔS_{surr}, and ΔS_{univ} for this process. The $\overline{C}_P$ of water and ice for the temperature range between 0°C and −10°C are 75.3 J K^{-1} mol^{-1} and 37.7 J K^{-1} mol^{-1}, respectively.

ANSWER

First we note that a change in phase is reversible only at the temperature at which the two phases are at equilibrium. Because supercooled water at −10°C and ice at −10°C are not at equilibrium, the freezing process is not reversible. To calculate ΔS_{sys}, we devise a series of reversible steps by which supercooled water at −10°C is converted to ice at −10°C, shown in Figure 4.9.

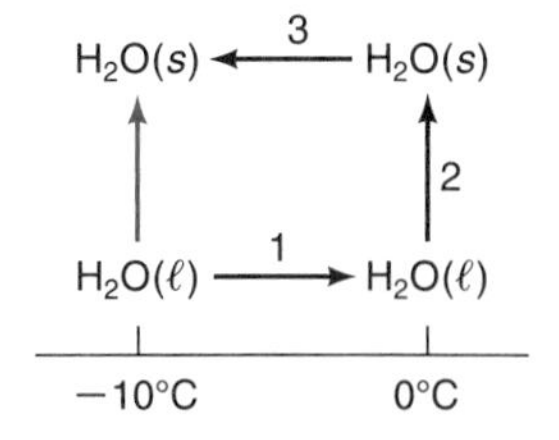

Figure 4.9
The spontaneous freezing of the super-cooled water (red arrow) at −10°C can be broken down into three reversible paths (1, 2, and 3).

STEP 1. Reversible heating of supercooled water at −10°C to 0°C.

$$\begin{array}{ccc} H_2O(l) & \rightarrow & H_2O(l) \\ -10°C & & 0°C \end{array}$$

From Equation 4.22

$$\Delta S_1 = n\overline{C}_P \ln \frac{T_2}{T_1}$$

$$= (2.00 \text{ mol})(75.3 \text{ J K}^{-1} \text{ mol}^{-1}) \ln \frac{273 \text{ K}}{263 \text{ K}}$$

$$= 5.62 \text{ J K}^{-1}$$

STEP 2. Water freezes into ice at 0°C.

$$\begin{array}{ccc} H_2O(l) & \rightarrow & H_2O(s) \\ 0°C & & 0°C \end{array}$$

Following the procedure in Example 4.5, we write

$$\Delta S_2 = n\Delta_{\text{fus}}\overline{S}$$

$$= -(2.00 \text{ mol})(22.0 \text{ J K}^{-1} \text{ mol}^{-1})$$

$$= -44.0 \text{ J K}^{-1}$$

STEP 3. Reversible cooling of ice from 0°C to −10°C.

$$\begin{array}{ccc} H_2O(s) & \rightarrow & H_2O(s) \\ 0°C & & -10°C \end{array}$$

Again from Equation 4.22,

$$\Delta S_3 = n\overline{C}_P \ln \frac{T_2}{T_1}$$

$$= (2.00 \text{ mol})(37.7 \text{ J K}^{-1} \text{ mol}^{-1}) \ln \frac{263 \text{ K}}{273 \text{ K}}$$

$$= -2.81 \text{ J K}^{-1}$$

Finally,

$$\Delta S_{\text{sys}} = \Delta S_1 + \Delta S_2 + \Delta S_3$$

$$= (5.62 - 44.0 - 2.81) \text{ J K}^{-1}$$

$$= -41.2 \text{ J K}^{-1}$$

To calculate ΔS_{surr}, we first determine the heat change in the surroundings for each of the above steps.

STEP 1. Heat gained by the supercooled water = heat lost by the surroundings, given by

$$(q_{surr})_1 = -n\overline{C}_P \Delta T$$

$$= -(2.00 \text{ mol})(75.3 \text{ J K}^{-1} \text{ mol}^{-1})(10 \text{ K})$$

$$= -1.5 \times 10^3 \text{ J}$$

STEP 2. When the water freezes at 0°C, heat is given off to the surroundings. Using the data in Example 4.5,

$$(q_{surr})_2 = n\Delta_{fus}\overline{H}$$

$$= (2.00 \text{ mol})(6010 \text{ J mol}^{-1})$$

$$= 1.20 \times 10^4 \text{ J}$$

STEP 3. Cooling ice from 0°C to −10°C releases heat to the surroundings equal to

$$(q_{surr})_3 = n\overline{C}_P \Delta T$$

$$= (2.00 \text{ mol})(37.7 \text{ J K}^{-1} \text{ mol}^{-1})(10 \text{ K})$$

$$= 754 \text{ J}$$

The total heat change is given by

$$(q_{surr})_{total} = (q_{surr})_1 + (q_{surr})_2 + (q_{surr})_3$$

$$= (-1.5 \times 10^3 + 1.20 \times 10^4 + 754) \text{ J}$$

$$= 1.12 \times 10^4 \text{ J}$$

Recall that $\Delta S_{surr} = q_{surr}/T$ regardless of whether the process is reversible or irreversible. Therefore, the change in entropy at −10°C is

$$\Delta S_{surr} = \frac{1.12 \times 10^4 \text{ J}}{263 \text{ K}}$$

$$= 42.6 \text{ J K}^{-1}$$

Finally,

$$\Delta S_{univ} = \Delta S_{sys} + \Delta S_{surr}$$

$$= -41.2 \text{ J K}^{-1} + 42.6 \text{ J K}^{-1}$$

$$= 1.4 \text{ J K}^{-1}$$

COMMENT

The result ($\Delta S_{univ} > 0$) confirms the statement that supercooled water is unstable and will spontaneously freeze on standing. Note that in this process, the entropy of the system decreases because water is converted to ice. However, the heat released to the surroundings results in an increase in the value of ΔS_{surr} that is greater (in magnitude) than ΔS_{sys}, so the ΔS_{univ} value is positive.

4.6 The Third Law of Thermodynamics

In thermodynamics we are normally interested only in changes in properties, such as ΔU and ΔH. Although there is no known way of measuring the absolute values of internal energy and enthalpy, it is possible to find the absolute entropy of a substance. As written, Equation 4.20 enables us to measure the change in entropy over a suitable temperature range between T_1 and T_2. But suppose we set the lower temperature to absolute zero, that is, $T_1 = 0$ K, and call the upper temperature (T_2) T. Equation 4.20 now becomes

$$\Delta S = S_T - S_0 = \int_0^T \frac{C_P}{T} dT \tag{4.23}$$

The entropy of a substance at any temperature T is equal to the sum of the contributions from 0 K to the specified temperature. We can measure heat capacity as a function of temperature to evaluate the integral in Equation 4.23. Additionally, we must also include any entropy changes due to phase transitions. But this approach presents two obstacles. First, what is the entropy of a substance at absolute zero; that is, what is S_0? Second, how do we account for that part of the contribution to the total entropy that lies between absolute zero and the lowest temperature at which measurements are feasible?

The third law of thermodynamics helps us solve the first obstacle. This law states that *every substance has a finite positive entropy, but at the absolute zero of temperature the entropy may become zero, and it does in the case of a pure, perfect crystalline substance.* Mathematically the third law can be expressed as

$$\lim_{T \to 0\,\text{K}} S = 0 \qquad \text{(pure, perfect crystalline substance)} \tag{4.24}$$

At temperatures above absolute zero, thermal motion contributes to the entropy of the substance so that its entropy is no longer zero, even if it is pure and remains perfectly crystalline. The significance of the third law is that it enables us to calculate the absolute values of entropies, discussed below.

Third-Law or Absolute Entropies

The third law of thermodynamics enables us to measure the entropy of a substance at temperature T. For a perfect crystalline substance, $S_0 = 0$ so that Equation 4.23 becomes

Contributions to S due to phase transitions must also be included in Equation 4.25

$$S_T = \int_0^T \frac{C_P}{T} dT = \int_0^T C_P d\ln T \tag{4.25}$$

Now we can measure the heat capacity over the desired temperature range. For very low temperatures ($\leq$ 15 K), where such measurements are difficult to carry out, we can use Debye's theory of heat capacity [after the Dutch-American physicist Peter Debye (1884–1966)]:

$$C_P = aT^3 \tag{4.26}$$

where a is a constant for a given substance. The entropy change over this small temperature range is

$$\Delta S = \int_0^T \frac{aT^3}{T} dT = \int_0^T aT^2 dt$$

Keep in mind that Equation 4.26 is applicable only near absolute zero. Table 4.2 lists the various steps taken in measuring the standard molar entropy ($\overline{S}°$) of HCl gas at 298.15 K. In applying Equation 4.25, we must keep in mind that it holds only for a perfectly ordered substance at 0 K. Figure 4.10 shows a plot of $S°$ versus temperature for substances in general.

Entropies calculated by using Equation 4.25 are called third-law or absolute entropies because these values are not based on some reference state. Table 4.3 lists the absolute molar entropies of a number of common elements and compounds at 298 K. More data are given in Appendix B. Note that because these are absolute values, we omit the Δ sign and the subscript "f" for $\overline{S}°$, but we retain them for the molar standard enthalpies of formation ($\Delta_f\overline{H}°$).

For convenience, we shall often omit "absolute" in referring to the standard molar entropy values.

Table 4.2
Determining Molar Entropy of HCl at 298.15 K[a]

Contribution	$\overline{S}_T°$/J K^{-1} mol^{-1}
1. Extrapolation from 0 to 16 K (Equation 4.26)	1.3
2. $\int \overline{C}_P d \ln T$ for solid I from 16 K to 98.36 K	29.5
3. Phase transition at 98.36 K, solid I $\rightarrow$ solid II, $\Delta\overline{H}/T$ = 1190 J mol^{-1}/98.36 K	12.1
4. $\int \overline{C}_P d\ln T$ for solid II from 98.36 K to 158.91 K	21.1
5. Fusion, 1992 J mol^{-1}/158.91 K	12.6
6. $\int \overline{C}_P d\ln T$ for liquid from 158.91 K to 188.07 K	9.9
7. Vaporization, $\frac{16,150 \text{ J mol}^{-1}}{188.07 \text{ K}}$	85.9
8. $\int \overline{C}_P d\ln T$ for gas from 188.07 K to 298.15 K	13.5
	$\overline{S}_{298.15}°$ = 185.9[b]

[a] Moore, W. J. *Physical Chemistry*, 4th ed., © 1972 Prentice-Hall, Englewood Cliffs, NJ.

[b] This value differs slightly from that in Table 4.3 and Appendix B because the standard state is 1 atm rather than 1 bar.

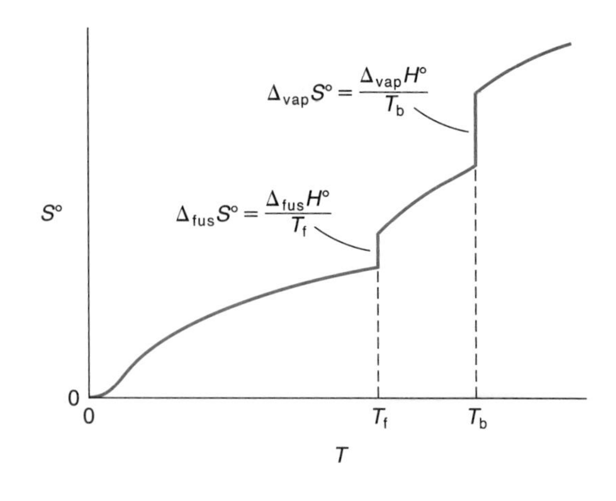

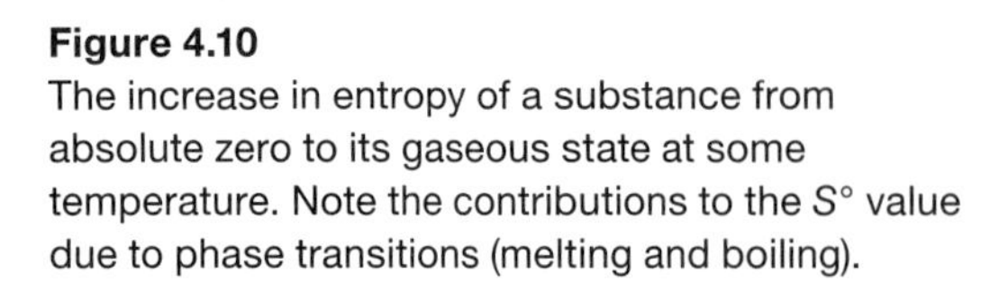

Figure 4.10
The increase in entropy of a substance from absolute zero to its gaseous state at some temperature. Note the contributions to the $S°$ value due to phase transitions (melting and boiling).

Table 4.3
Standard Molar Entropies at 298 K and 1 Bar for Some Inorganic and Organic Substances

Substance	$\overline{S}°$/J K^{-1} mol^{-1}	Substance	$\overline{S}°$/J K^{-1} mol^{-1}
C(graphite)	5.7	$CH_4(g)$	186.2
C(diamond)	2.4	$C_2H_6(g)$	229.5
$CO(g)$	197.9	$C_3H_8(g)$	269.9
$CO_2(g)$	213.6	$C_2H_2(g)$	200.8
$HF(g)$	173.5	$C_2H_4(g)$	219.5
$HCl(g)$	186.5	$C_6H_6(l)$	172.8
$HBr(g)$	198.7	$CH_3OH(l)$	126.8
$HI(g)$	206.3	$C_2H_5OH(l)$	161.0
$H_2O(g)$	188.7	$CH_3CHO(l)$	160.2
$H_2O(l)$	69.9	$HCOOH(l)$	129.0
$NH_3(g)$	192.5	$CH_3COOH(l)$	159.8
$NO(g)$	210.6	$C_6H_{12}O_6(s)$	210.3
$NO_2(g)$	240.5	$C_{12}H_{22}O_{11}(s)$	360.2
$N_2O_4(g)$	304.3		
$N_2O(g)$	220.0		
$O_2(g)$	205.0		
$O_3(g)$	237.7		
$SO_2(g)$	248.5		

Entropy of Chemical Reactions

We are now ready to calculate the entropy change that occurs in a chemical reaction. As for the enthalpy of reaction (see Equation 3.31), the change in entropy for the hypothetical reaction

$$a\text{A} + b\text{B} \rightarrow c\text{C} + d\text{D}$$

is given by the equation

$$\Delta_r S^\circ = c\overline{S}^\circ(\text{C}) + d\overline{S}^\circ(\text{D}) - a\overline{S}^\circ(\text{A}) - b\overline{S}^\circ(\text{B})$$

$$= \Sigma v\overline{S}^\circ(\text{products}) - \Sigma v\overline{S}^\circ(\text{reactants}) \quad (4.27)$$

When the stoichiometric coefficient is one, it is not shown in Equation 4.27.

where v represents the stoichiometric coefficient. This procedure is illustrated in the following example.

EXAMPLE 4.8

Calculate the value of the standard molar entropy changes for the following reactions at 298 K:

(a) $CaCO_3(s) \rightarrow CaO(s) + CO_2(g)$

(b) $2H_2(g) + O_2(g) \rightarrow 2H_2O(l)$

(c) $N_2(g) + O_2(g) \rightarrow 2NO(g)$

ANSWER

The entropy change of a reaction is given by Equation 4.27. From the S° values listed in Appendix B, we write

(a) $$\Delta_r S^\circ = [\overline{S}^\circ(\text{CaO}) + \overline{S}^\circ(\text{CO}_2)] - \overline{S}^\circ(\text{CaCO}_3)$$

$$= (39.8 + 213.6)\text{ J K}^{-1}\text{ mol}^{-1} - 92.9\text{ J K}^{-1}\text{ mol}^{-1}$$

$$= 160.5\text{ J K}^{-1}\text{ mol}^{-1}$$

(b) $$\Delta_r S^\circ = (2)\overline{S}^\circ(\text{H}_2\text{O}) - [(2)\overline{S}^\circ(\text{H}_2) + \overline{S}^\circ(\text{O}_2)]$$

$$= (2)(69.9\text{ J K}^{-1}\text{ mol}^{-1}) - [(2)(130.6) + (205.0)]\text{ J K}^{-1}\text{ mol}^{-1}$$

$$= -326.4\text{ J K}^{-1}\text{ mol}^{-1}$$

(c) $$\Delta_r S^\circ = (2)\overline{S}^\circ(\text{NO}) - [\overline{S}^\circ(\text{N}_2) + \overline{S}^\circ(\text{O}_2)]$$

$$= (2)(210.6\text{ J K}^{-1}\text{ mol}^{-1}) - (191.5 + 205.0)\text{ J K}^{-1}\text{ mol}^{-1}$$

$$= 24.7\text{ J K}^{-1}\text{ mol}^{-1}$$

COMMENT

The results are consistent with our expectation that reactions producing a net increase in the number of gas molecules are accompanied by an appreciable increase in entropy as in reaction (a), whereas the reverse holds true for reaction (b). In (c), there is no change in the total number of gas molecules; therefore, the change in entropy is relatively small. Note that all the entropy changes apply to the system.

EXAMPLE 4.9

Calculate the values of ΔS_{sys}, ΔS_{surr}, and ΔS_{univ} for the synthesis of ammonia at 25°C:

$$N_2(g) + 3H_2(g) \rightarrow 2NH_3(g) \qquad \Delta_r H° = -92.6 \text{ kJ mol}^{-1}$$

ANSWER

Using the data in Appendix B and following the procedure in Example 4.8, we find that the entropy change for the reaction is $-198 \text{ J K}^{-1} \text{ mol}^{-1}$, which is the value for ΔS_{sys}. To calculate the value of ΔS_{surr}, we note that the system is in thermal equilibrium with the surroundings. Because $\Delta H_{surr} = -\Delta H_{sys}$, we write

$$\Delta S_{surr} = \frac{\Delta H_{surr}}{T}$$

$$= \frac{-(-92.6 \times 1000) \text{ J mol}^{-1}}{298 \text{ K}} = 311 \text{ J K}^{-1} \text{ mol}^{-1}$$

The change in entropy for the universe is

$$\Delta S_{univ} = \Delta S_{sys} + \Delta S_{surr}$$

$$= -198 \text{ J K}^{-1} \text{ mol}^{-1} + 311 \text{ J K}^{-1} \text{ mol}^{-1}$$

$$= 113 \text{ J K}^{-1} \text{ mol}^{-1}$$

COMMENT

Because the ΔS_{univ} value is positive, we predict that the reaction is spontaneous at 25°C. Recall that just because a reaction is spontaneous does not mean that it will occur at an observable rate. The synthesis of ammonia is, in fact, extremely slow at room temperature. Thermodynamics can tell us whether a reaction will occur spontaneously under specific conditions, but it does not say how fast it will occur. Reaction rates are the subject of chemical kinetics (Chapter 15).

4.7 The Meaning of Entropy

At this point we have defined entropy statistically and thermodynamically. With the third law of thermodynamics, it is possible to determine the absolute entropy of substances. We have seen some examples of entropy changes in physical processes and chemical reactions. But what is entropy, really?

Frequently entropy is described as a measure of disorder or randomness. The higher the disorder, the greater the entropy of the system. While useful, these terms must be used with caution because they are subjective concepts.* On the other hand, relating entropy to probability makes more sense because probability is a quantitative concept. Earlier we saw how gas expansion was viewed in terms of probability. In a spontaneous process, a system goes from a less probable state to a more probable one. The corresponding change in entropy is calculated using Equation 4.2.

Our next step is to interpret entropy at the molecular level. In thermodynamics, the state of a macroscopic system, or *macrostate*, is described by properties such as P, V, T, n, U, and H. A microscopic system or *microstate*, on the other hand, is a condition in which we have specified some or all of the variables associated with the individual particles—the atoms and molecules—of the system. But we normally are not interested in the individual microstates of bulk matter because there are so many particles. Instead, we develop the variables of the macroscopic state such as temperature, pressure, volume, energy, enthalpy, and so on, which are based on the average properties of the microstates.

One way to think of microstates is to relate them to the number of ways the overall state of a system can be constructed. The number of ways the state of a system can arise determines the probability of its occurrence. Let us begin with the throwing of dice to show the relationship between microstates and macrostates. If we have only one die, there are six microstates given by 1, 2, 3, 4, 5, and 6. And these are the same as the macrostates because there is only one way to achieve each number. The situation is different for two dice. Here the macrostates are defined by the sums of both dice, which are 2, 3, 4, …, 10, 11, 12. Figure 4.11 shows that the 36 (6×6) microstates can be grouped into 11 macrostates. The most probable macrostate is the number 7 because it can be made up by six different microstates. Put another way, the probability of getting 7 by throwing a pair of dice is 6/36. On the other hand, the macrostate 2 (or 12) has only one microstate and the probability of getting this number is 1/36. An important observation should be made here: the probability of getting any of the 36 microstates is the same; we are more likely to hit 7 rather than 2 simply because there are more ways to form the number 7 than 2.

With four dice there are 1296 ($6\times6\times6\times6$) microstates ranging from 4 to 24, giving rise to a total of 21 macrostates. The probability of getting the number 4 is 1/1296 (one of the two least probable macrostates; the other is number 24), while that of getting 14 (the most probable macrostate) is 146/1296. As the number of dice increases further, we find the number of microstates corresponding to the most probable macrostate increases rapidly. When the number of dice approaches Avogadro's number, the most probable macrostate will have such an overwhelming number of microstates compared to all other macrostates that we will always hit that number whenever we throw that many dice (Figure 4.12).

---

* See D. F. Styer, *Am. J. Phys.* **68**, 1090 (2000) and F. L. Lambert, *J. Chem. Educ.* **79**, 187 (2002).

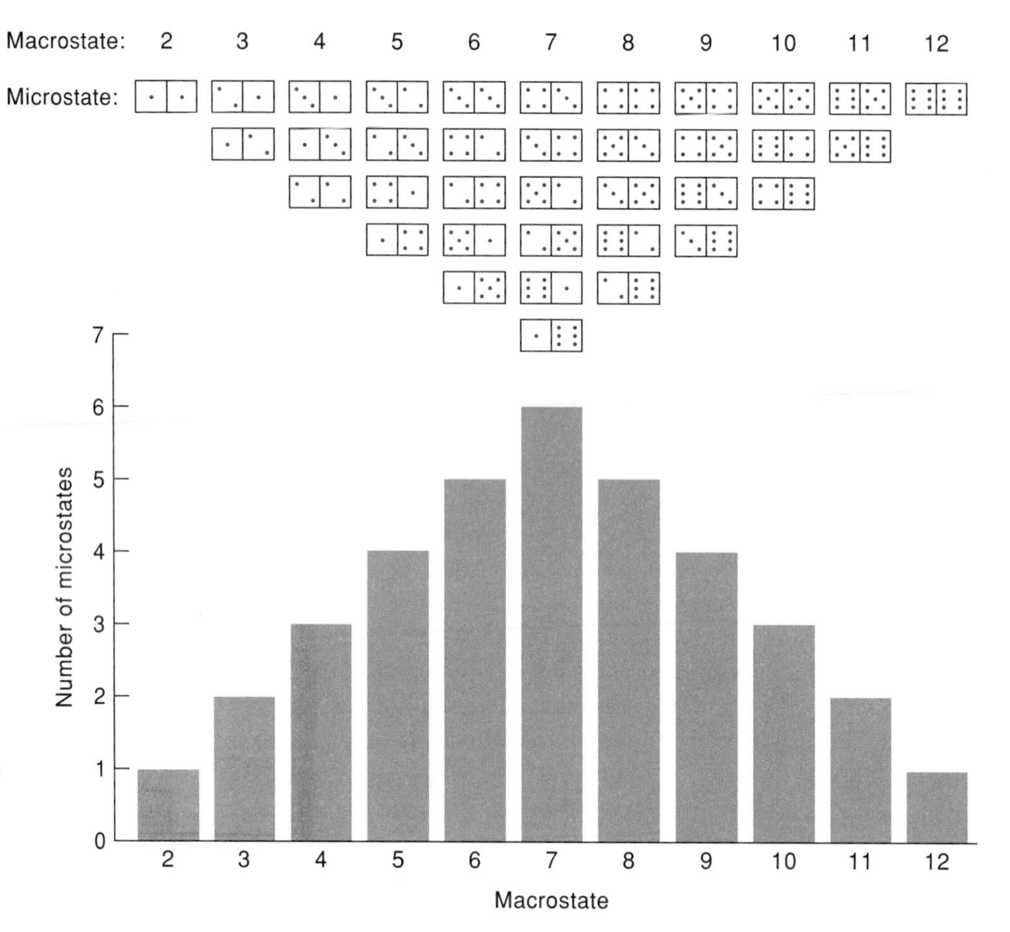

Figure 4.11
The microstates arising from the combination of two dice and the corresponding macrostates.

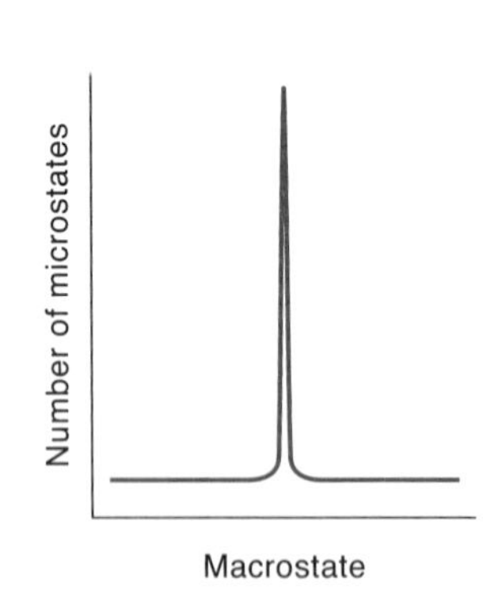

Figure 4.12
For an Avogadro's number of dice, the most probable macrostate has an overwhelmingly large number of microstates compared to other macrostates.

We now turn to a molecular system. Suppose we have three identical, noninteracting molecules distributed over energy levels with the total energy of the system restricted to three units. How many ways can this be accomplished? Although the molecules are identical, they can be distinguished from one another by their locations (e.g., if they occupy different lattice points in a crystal). We see that there are 10 ways (10 microstates) to distribute the molecules that make up three distinct *distributions* (three macrostates), designated I, II, and III, as shown in Figure 4.13. Not all the macrostates are equally probable—macrostate state II is six times as probable as I, and twice as probable as III.

The general formula for calculating the number of microstates in a distribution (W) is

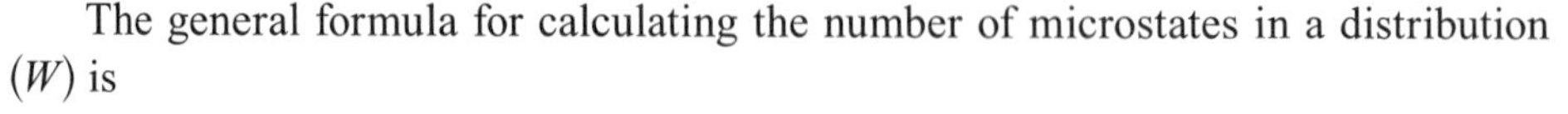

$$W = \frac{N!}{n_1!n_2!\ldots} \tag{4.28}$$

where N is the total number of molecules present, and n_1, n_2, ... are the number of molecules in the lowest level, second lowest level, and so on. The symbol ! is called "factorial." $N!$ is given by

$$N! = N(N - 1)(N - 2)\ldots1$$

and by definition, $0! = 1$. Equation 4.28 can be simplified as

$$W = \frac{N!}{\Pi_i n_i!} \tag{4.29}$$

where Π_i is the product for all values of i of the distributions.

Applying Equation 4.29 to our system, we write

$$W_{\text{I}} = \frac{3!}{3!} = \frac{3 \times 2 \times 1}{3 \times 2 \times 1} = 1$$

$$W_{\text{II}} = \frac{3!}{1!1!1!} = \frac{3 \times 2 \times 1}{1 \times 1 \times 1} = 6$$

$$W_{\text{III}} = \frac{3!}{2!1!} = \frac{3 \times 2 \times 1}{2 \times 1 \times 1} = 3$$

An empty level contributes 0! and therefore does not affect the value of W.

Consider a macroscopic system at equilibrium containing N particles (where N may be on the order of Avogadro's number) distributed among many, many energy, levels. As in the case of throwing dice, there is only one distribution or macrostate with an overwhelming number of microstates compared to other macrostates. An important underlying assumption here is called the *principle of equal a priori probability*, which states that all microstates are equally probable. This means that it is just as probable to find the system in a specific microstate that belongs to an improbable macrostate as it is to find the same system in a microstate that belongs to a probable macrostate. Because there are so many more microstates that belong to the most probable macrostate, we will *always* find the system with this distribution.

A priori means without influence of experience.

From this discussion we can say the following about entropy. Entropy is related to the *distribution* or spread of energy among the available molecular energy levels.

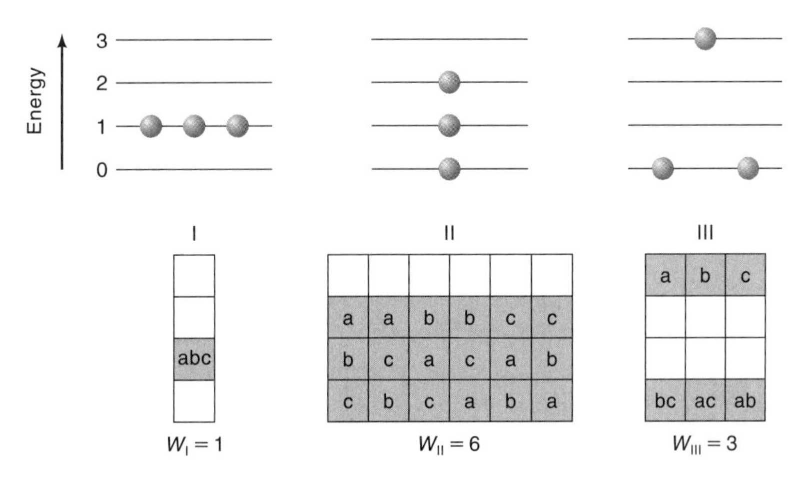

Figure 4.13
Arrangement of three molecules among energy levels with a total energy of three units.

At thermal equilibrium we always find the system in the most probable macrostate, which has the largest number of microstates and the most probable distribution of energy. Keep in mind that the greater the number of energy levels that have significant occupation, the larger the entropy is. It follows, therefore, that entropy of a system is a maximum at equilibrium because W itself is a maximum.

Ludwig Boltzmann's gravestone in Vienna, Austria, is inscribed with his equation. (Photo courtesy of John Simon, reprinted with permission.)

Previously Equation 4.1 was used to relate entropy to probability. Because the probability of a specific macrostate of a system is proportional to the number of microstates that makes up that state, it is more proper to express entropy in terms of W:

$$S = k_{\mathrm{B}} \ln W \tag{4.30}$$

Equation 4.30 is the Boltzmann equation that defines entropy statistically. It is not used to calculate entropy in general, however, because we do not know what W is for macroscopic systems. As mentioned earlier, entropy values are usually determined by calorimetric methods. Nevertheless, the molecular interpretation enables us to gain a better understanding of the nature of entropy and its changes. Note that Equation 4.30 can also be used to illustrate the third law of thermodynamics. If we have a perfect crystalline substance at the absolute zero, then there can only be one particular arrangement of atoms or molecules and hence only one microstate. Consequently, $W = 1$ and

$$S_0 = k_{\mathrm{B}} \ln W = k_{\mathrm{B}} \ln 1 = 0$$

We now present a molecular interpretation of the entropy changes discussed earlier.

Isothermal Gas Expansion

In an expansion, the gas molecules move in a larger volume. As we shall see in Chapter 10, the translational kinetic energy of a molecule is quantized and the energy of any particular level is *inversely* proportional to the dimension of the container. It follows, therefore, that in the larger volume the levels become more closely spaced and hence more accessible for distributing energy. Consequently, more energy levels will be occupied, resulting in an increased number of microstates corresponding to the most probable macrostate and hence an increase in entropy.

Isothermal Mixing of Gases

The mixing of two gases at constant temperature can be treated as two separate gas expansions. Again, we predict an increase in entropy.

Heating

When the temperature of a substance is raised, the energy input is used to promote the molecular motions (translational, rotational, and vibrational) from the low-lying levels to higher ones. The result is an increase in the occupancy among the molecular energy levels and hence the number of microstates. Consequently, there will be an increase in entropy. This is what happens for heating at constant volume. If heating is carried out at constant pressure, there will be an additional contribution to entropy due to expansion. The difference between constant-volume and constant-pressure conditions is significant only if the substance is a gas.

Phase Transitions

In a solid, molecules can usually vibrate only about their lattice points with restricted rotational motion and no translational motion. At the melting point, the molecules enter the liquid phase and will have more freedom in rotational and translational motions. This phase transition results in an increase in the microstates and hence also the entropy of the system. At the boiling point there will be a more pronounced enhancement in molecular motions from the condensed phase to the unrestricted motions in free space. The corresponding increase in microstates will be considerably larger and so will be the entropy increase.

Chemical Reactions

Referring to Example 4.9, we see that the synthesis of ammonia from nitrogen and hydrogen results in a net loss of two moles of gases per reaction unit. The decrease in molecular motions is reflected in fewer microstates so we would expect to see a decrease in entropy of the system. Because the reaction is exothermic, the heat released energizes the motions of the surrounding air molecules. The increase in the microstates of the air molecules leads to an increase in the entropy of the surroundings, which outweighs the decrease in entropy of the system, so the reaction is spontaneous. Keep in mind that predicting entropy changes becomes less certain for reactions involving condensed phases or in cases where there are no changes in the number of gaseous components.

4.8 Residual Entropy

As we have seen, absolute entropy values can be determined by applying the third law of thermodynamics. It turns out that absolute entropies can also be calculated using methods of statistical thermodynamics (see Chapter 20), based on spectroscopic data at 298 K. Generally these two approaches yield comparable results. In some cases, however, the absolute entropy determined experimentally with the third law is found to be less than the calculated value. This difference is due to *residual entropy*, or the "greater-than-zero" entropy value, of a substance at 0 K.

To see how residual entropy can arise, let us consider a carbon monoxide crystal at absolute zero. Carbon monoxide has a small dipole moment ($\mu = 0.12$ D), and there is just a slight charge separation in the molecule. Consequently, molecules may be randomly oriented as shown in Figure 4.14. If there is no preferred orientation, then for one molecule there are two, or 2^1, choices. For two molecules there are four, or 2^2, choices, and for n moles of CO there are 2^{nN_A} choices. Recalling that $R = k_B N_A$, we can calculate the residual entropy (S_0) for 1 mole of CO molecules from the Boltzmann equation:

$$
\begin{aligned}
S_0 &= k_B \ln W \\
&= k_B \ln 2^{nN_A} \\
&= nR \ln 2 \qquad (k_B N_A = R) \\
&= (1 \text{ mol})(8.314 \text{ J K}^{-1} \text{ mol}^{-1}) \ln 2 \\
&= 5.8 \text{ J K}^{-1}
\end{aligned}
$$

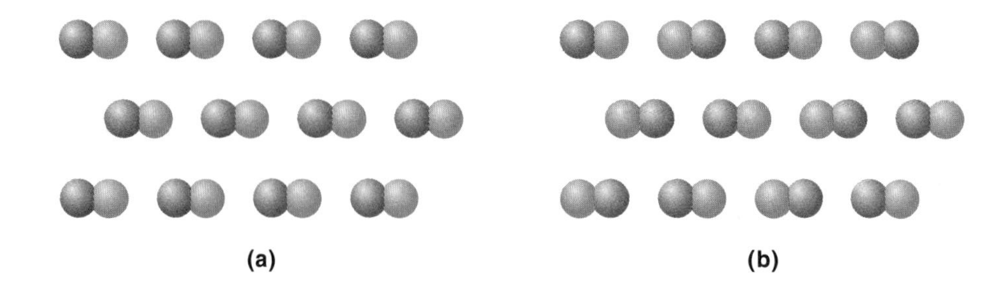

Figure 4.14
(a) Perfect arrangement of carbon monoxide molecules in the crystal; $S_0 = 0$ at 0 K.
(b) Imperfect arrangement of carbon monoxide molecules in the crystal; $S_0 > 0$ at 0 K.
Color codes: gray is C and red is O.

The molar entropy of CO at 298 K as calculated by statistical thermodynamics is 4.2 J K^{-1} mol^{-1} greater than the third-law entropy, a difference that suggests that there is a residual entropy at absolute zero. The fact that this discrepancy is smaller than 5.8 J K^{-1} mol^{-1} means that the orientation of the CO molecules in the crystal is not totally random.

Another example of a substance exhibiting residual entropy is nitrous oxide (N_2O) with a dipole moment of 0.166 D. In this case the calculated value is greater than the third-law value by 5.8 J K^{-1} mol^{-1}. Nitrous oxide is a linear molecule with the atomic arrangement NNO. Energetically it makes no difference whether the orientation of the molecules is NNO or ONN. As in the CO case, the Boltzmann equation predicts a residual molar entropy of 5.8 J K^{-1} mol^{-1}, which agrees well with the experimental result.

Finally we consider water. The molar entropy calculated from statistical thermodynamics and spectroscopic measurements on water vapor at 298 K is 188.7 J K^{-1} mol^{-1}, whereas the value determined by the third-law method is 185.3 J K^{-1} mol^{-1}. The difference, 3.4 J K^{-1} mol^{-1}, is attributed to the residual entropy of ice at 0 K. The American chemist Linus Pauling (1901–1994) provided an accurate explanation of this result based on geometric considerations. An ice crystal made up of N H_2O molecules contains $2N$ H atoms. Each of the H atoms can be in one of two positions: close to an O atom (as in an O—H sigma bond) or far from it (as in an O···H hydrogen bond). Therefore, there are 2^{2N} arrangements in all. Table 4.4 summarizes the $2^4 = 16$

Pauling is the only winner of two unshared Nobel Prizes: Chemistry in 1954 and Peace in 1962.

Table 4.4
Arrangements of Four H atoms Around an O Atom in Ice

Description	Chemical Species	Number of Equivalent Arrangements
All H atoms are sigma-bonded to the O atom	H_4O^{2+}	1
Three H atoms are sigma-bonded and one H is hydrogen-bonded to the O atom	H_3O^+	4
Two H atoms are sigma-bonded and two H atoms are hydrogen-bonded to the O atom	H_2O	6
One H is sigma-bonded and three H atoms are hydrogen-bonded to the O atom	OH^-	4
All four H atoms are hydrogen-bonded to the O atom	O^{2-}	1

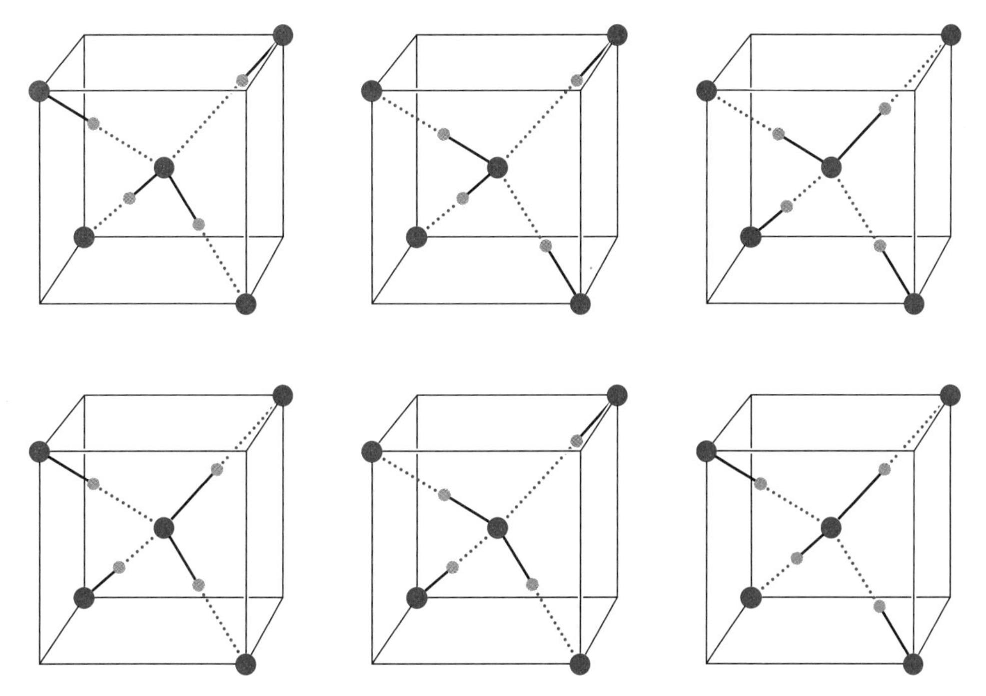

Figure 4.15
The six equivalent arrangements of four H atoms about a central O atom. Note that in each case the O atom is tetrahedrally bonded to the H atoms by two sigma bonds and two hydrogen bonds.

different ways of positioning 4 H atoms around an O atom in ice. Ten of these arrangements would leave positive or negative charges in the lattice because there would be more or fewer than two H atoms bonded to the O atom. These charge separations are energetically unfavorable and do not occur at 0 K. The remaining six arrangements, shown in Figure 4.15, are acceptable. Consequently, the 2^{2N} ways of arranging the H atoms about N O atoms must be reduced by a factor $(6/16)^N$; that is,

$$W = 2^{2N} \times \left(\frac{6}{16}\right)^N = \left(\frac{3}{2}\right)^N$$

For n moles of H_2O molecules, $N = nN_A$ and the residual entropy is given by

$$\begin{aligned}\overline{S}_0 &= k_B \ln \left(\frac{3}{2}\right)^{nN_A} \\ &= (8.314 \text{ J K}^{-1} \text{ mol}^{-1}) \ln \left(\frac{3}{2}\right) \quad (n = 1 \text{ mol and } k_B N_A = R) \\ &= 3.4 \text{ J K}^{-1} \text{ mol}^{-1}\end{aligned}$$

which is in excellent agreement with experimental results.

Finally we note that residual entropy does not exist in the majority of substances. Nevertheless, when it occurs, it provides us with a way of studying molecular arrangement in a crystal at the unattainable temperature of absolute zero.

Key Equations

$\Delta S = nR \ln \frac{V_2}{V_1}$	(Entropy change due to isothermal gas expansion)	(4.3)
$\Delta S = \frac{q_{\text{rev}}}{T}$	(Thermodynamic definition of entropy)	(4.4)
$dS = \frac{dq_{\text{rev}}}{T}$	(Thermodynamic definition of entropy)	(4.5)
$\eta = \frac{T_2 - T_1}{T_2} = 1 - \frac{T_1}{T_2}$	(Thermodynamic efficiency)	(4.9)
$\Delta S_{\text{univ}} = \Delta S_{\text{sys}} + \Delta S_{\text{surr}} \geq 0$	(Second law of thermodynamics)	(4.16)
$\Delta_{\text{mix}} S = -R(n_{\text{A}} \ln x_{\text{A}} + n_{\text{B}} \ln x_{\text{B}})$	(Entropy change due to isothermal mixing)	(4.17)
$\Delta_{\text{fus}} S = \frac{\Delta_{\text{fus}} H}{T_{\text{f}}}$	(Entropy change due to melting)	(4.18)
$\Delta_{\text{vap}} S = \frac{\Delta_{\text{vap}} H}{T_{\text{b}}}$	(Entropy change due to vaporization)	(4.19)
$\Delta S = S_2 - S_1 = n\overline{C}_P \ln \frac{T_2}{T_1}$	(Entropy change due to heating at constant pressure)	(4.22)
$\lim_{T \to 0\,\text{K}} S = 0$	(Third law of thermodynamics)	(4.24)
$S_T = \int_0^T \frac{C_P}{T} dT = \int_0^T C_P d\ln T$	(Third-law entropy)	(4.25)
$\Delta_{\text{r}} S^\circ = \Sigma \nu \overline{S}^\circ(\text{products}) - \Sigma \nu \overline{S}^\circ(\text{reactants})$	(Standard entropy of reaction)	(4.26)
$S = k_{\text{B}} \ln W$	(The Boltzmann equation)	(4.30)

APPENDIX 4.1

Statements of the Second Law of Thermodynamics

The second law of thermodynamics is one of the most important physical laws. Over the years, this law has been studied and applied by scientists in vastly different disciplines. Not surprisingly, there are many different (but equivalent) ways of stating the second law. Listed below are some of the more familiar statements.

1. "It is impossible for a self-acting machine, unaided by any external agency, to convey heat from one body to another at a higher temperature, or heat cannot by itself pass from a colder body to a warmer one." (Rudolf Julius Clausius, 1822–1888; German physicist.)
2. "The energy of the universe is constant, the entropy increases toward a maximum." (Clausius)
3. "Entropy is time's arrow." (Sir Arthur Stanley Eddington, 1882–1944; English mathematician and astrophysicist.)
4. "The state of maximum entropy is the most stable state for an isolated system." (Enrico Fermi, 1901–1954; Italian physicist.)
5. "Every system which is left to itself will, on average, change toward a condition of maximum probability." (Gilbert Newton Lewis, 1875–1946; American chemist.)
6. "In any irreversible process the total entropy of all bodies concerned is increased." (Lewis)
7. "When any actual process occurs it is impossible to invent a means of restoring every system concerned to its original condition." (Lewis)
8. "The second law of thermodynamics has as much truth as saying that, if you poured a glass of water into the ocean, it would not be possible to get the same glass of water back again." (James Clerk Maxwell)
9. "It is impossible in any way to diminish the entropy of a system of bodies without thereby leaving behind changes in other bodies." (Max Planck, 1858–1947; German physicist.)
10. "The change of mechanical work into heat may be complete, but on the contrary that of heat into work can never be complete, since whenever a certain quantity of heat is transformed into work, another quantity of heat must undergo a corresponding and compensating change." (Planck)

Suggestions for Further Reading

BOOKS

Bent, H. A., *The Second Law*, Oxford University Press, New York, 1965.

Fenn, J. B., *Engines, Energy, and Entropy*, W. H. Freeman, New York, 1982.

Hanson, R. M., and S. Green, *Introduction to Molecular Thermodynamics*, University Science Books, Sausalito, CA, 2008.

Klotz, I. M., and R. M. Rosenberg, *Chemical Thermodynamics: Basic Theory and Methods,* 5th ed., John Wiley and Sons, New York, 1994.

von Baeyer, H. C., *Warmth Disperses and Time Passes*, Random House, New York, 1998.

McQuarrie, D. A., and J. D. Simon, *Molecular Thermodynamics*, University Science Books, Sausalito, CA, 1999.

ARTICLES

The Second Law of Thermodynamics (General)

"The Second Law of Thermodynamics," H. A. Bent, *J. Chem. Educ.* **39**, 491 (1962).

"The Use and Misuse of the Laws of Thermodynamics," M. L. McGlashan, *J. Chem. Educ.* **43**, 226 (1966).

"The Scope and Limitations of Thermodynamics," K. G. Denbigh, *Chem. Brit.* **4**, 339 (1968).

"Conversion of Standard Thermodynamic Data to the New Standard-State Pressure," R. D. Freeman, *J. Chem. Educ.* **62**, 681 (1985).

"Student Misconceptions in Thermodynamics," M. F. Granville, *J. Chem. Educ.* **62**, 847 (1985).

"How Thermodynamic Data and Equilibrium Constants Changed When the Standard-State Pressure Became 1 Bar," R. S. Treptow, *J. Chem. Educ.* **76**, 212 (1999).

"Revitalizing Modern Thermodynamics Instruction within an Old Technique: A State Functions Table," P. A. Molina, *Chem. Educator* [Online] **12**, 137 (2007) DOI 10.1333/s00897072026a.

"The Long Arm of the Second Law," J. M. Rubi, *Sci. Am.* November 2008.

Entropy

"Chance," A. J. Ayer, *Sci. Am.* October 1965.

"Maxwell's Demon," W. Ehrenberg, *Sci. Am.* November 1967.

"States, Indistinguishability, and the Formula $S = k \ln W$ in Thermodynamics," J. Braunstein, *J. Chem. Educ.* **46**, 719 (1969).

"Temperature-Entropy Diagrams," A. Wood, *J. Chem. Educ.* **47**, 285 (1970).

"The Arrow of Time," D. Layzer, *Sci. Am.* December 1975.

"Negative Absolute Temperature," W. G. Proctor, *Sci. Am.* August 1978.

"Reversibility and Returnability," J. A. Campbell, *J. Chem. Educ.* **57**, 345 (1980).

"Entropy and Unavailable Energy," J. N. Spencer and E. S. Holmboe, *J. Chem. Educ.* **60**, 1018 (1983).

"A Simple Method for Showing that Entropy is a Function of State," P. Djurdjevic and I. Gutman, *J. Chem. Educ.* **65**, 399 (1985).

"Entropy: Conceptual Disorder," J. P. Low, *J. Chem. Educ.* **65**, 403 (1988).

"Entropy Analyses of Four Familiar Processes," N. C. Craig, *J. Chem. Educ.* **65**, 760 (1988).

"Order and Disorder and Entropies of Fusion," D. F. R. Gilson, *J. Chem. Educ.* **69**, 23 (1992).

"Periodic Trends for the Entropy of Elements," T. Thoms, *J. Chem. Educ.* **72**, 16 (1995).

"Entropy Diagrams," N. C. Craig, *J. Chem. Educ.* **73**, 716 (1996).

"Thermodynamics and Spontaneity," R. S. Ochs, *J. Chem. Educ.* **73**, 952 (1996).

"*S* is for Entropy. *U* is for Energy. What Was Clausius Thinking?" I. K. Howard, *J. Chem. Educ.* **78**, 505 (2001).
"Disorder—A Cracked Crutch for Supporting Entropy Discussions," F. L. Lambert, *J. Chem. Educ.* **79**, 187 (2002).
"Entropy Is Simple, Qualitatively," F. L. Lambert, *J. Chem. Educ.* **79**, 1241 (2002).
"Entropy Explained: The Origin of Some Simple Trends," L. Watson and O. Eisenstein, *J. Chem. Educ.* **79**, 1269 (2002).
"Campbell's Rule for Estimating Entropy Changes in Gas-Producing and Gas-Consuming Reactions and Related Generalizations about Entropies and Enthalpies," N. C. Craig, *J. Chem. Educ.* **80**, 1432 (2003).
"Understanding Entropy with the Boltzmann Formula," F. Hynne, *Chem. Educator* [Online] **9**, 74 (2004) DOI 10.1333/s00897040755a.
"From Microstates to Thermodynamics," F. Hynne, *Chem. Educator* [Online] **9**, 262 (2004) DOI 10.1333/s00897040827a.
"Regarding Entropy Analysis," R. M. Hanson, *J. Chem. Educ.* **82**, 839 (2005).
"Mysteries of the First and Second Laws of Thermodynamics," R. Battino, *J. Chem. Educ.* **84**, 753 (2007). [Also see *J. Chem. Educ.* **86**, 31 (2009).]
"Rescaling Temperature and Entropy," J. Olmsted III, *J. Chem. Educ.* **87**, 1195 (2010).
"Entropy: Order or Information," A. Ben-Naim, *J. Chem. Educ.* **88**, 594 (2011).
"The Statistical Interpretation of Classical Thermodynamic Heating and Expansion Processes," S. F. Cartier, *J. Chem. Educ.* **88**, 1531 (2011).
"Demons, Entropy, and the Quest for Absolute Zero," M. G. Raizen, *Sci. Am.* March 2011.

Heat Engines, Heat Pumps, and Thermodynamic Efficiencies
"Heat Pumps," J. F. Sandfoot, *Sci. Am.* May 1951.
"The Conversion of Energy," C. M. Summers, *Sci. Am.* September 1971.
"Energy Conversion: Better Living Through Thermodynamics," W. D. Metz, *Science* **188**, 820 (1975).
"Heat-Fall and Entropy," J. P. Lowe, *J. Chem. Educ.* **59**, 353 (1982).
"Demons, Engines, and the Second Law," C. H. Bennett, *Sci. Am.* November, 1987.
"The Conversion of Chemical Energy," D. J. Wink, *J. Chem. Educ.* **69**, 109 (1992).
"Steam Engines," S. Luchter, *Encyclopedia of Applied Physics,* Trigg, G. L., Ed., VCH Publishers, New York (1997), Vol. 19, p. 563.
"A Simple Approach to Heat Engine Efficiency," C. Salter, *J. Chem. Educ.* **77**, 127 (2000).
"Carnot Cycle Revisited," E. Peacock-López, *Chem. Educator* [Online] **7**, 127 (2002) DOI 10.1333/s00897020555a.
"Diesels Come Clean," S. Ashley, *Sci. Am.* March 2007.
"Warming *and* Cooling," M. Fischetti, *Sci. Am.* August 2008.
"Getting the Most from Energy," T. R. Casten and P. F. Schewe, *Am. Scientist* Jan–Feb 2009.

The Third Law of Thermodynamics and Residual Entropy
"The Third Law of Thermodynamics, the Unattainability of Absolute Zero, and Quantum Mechanics," E. M. Loebl, *J. Chem. Educ.* **37**, 361 (1960).
"Ice," L. K. Runnels, *Sci. Am.* December 1966.
"The Third Law of Thermodynamics and the Residual Entropy of Ice," M. M. Julian, F. H. Stillinger, and R. R. Festa, *J. Chem. Educ.* **60**, 65 (1983).
"Demons, Entropy, and the Quest for Absolute Zero," M. G. Raizen, *Sci. Am.* March 2011.

Problems

Probability

4.1 Determine the probability that all the molecules of a gas will be found in one half of a container when the gas consists of **(a)** 1 molecule, **(b)** 20 molecules, and **(c)** 2 million molecules.

4.2 Suppose that your friend told you of the following extraordinary event. A block of metal weighing 500 g was seen rising spontaneously from the table on which it was resting to a height of 1.00 cm above the table. He stated that the metal had absorbed thermal energy from the table that was then used to raise itself against gravitational pull. **(a)** Does this process violate the first law of thermodynamics? **(b)** How about the second law? Assume that the room temperature was 298 K and that the table was large enough that its temperature was unaffected by this transfer of energy. (*Hint:* First calculate the decrease in entropy as a result of this process and then estimate the probability for the occurrence of such a process. The acceleration due to gravity is 9.81 m s^{-2}.)

The Carnot Heat Engine

4.3 Compare the generation of electricity by a hydroelectric plant to the use of a heat engine. Which method is more efficient? Why?

4.4 Convert the P–V diagram for the Carnot cycle to a T–S diagram. What is the area of the enclosed portion?

4.5 The internal engine of a 1200-kg car is designed to run on octane (C_8H_{18}), whose enthalpy of combustion is 5510 kJ mol^{-1}. If the car is moving up a slope, calculate the maximum height (in meters) to which the car can be driven on 1.0 gallon of the fuel. Assume that the engine cylinder temperature is 2200°C and the exit temperature is 760°C, and neglect all forms of friction. The mass of 1 gallon of fuel is 3.1 kg. [*Hint:* The work done in moving the car over a vertical distance is mgh, where m is the mass of the car in kg, g the acceleration due to gravity (9.81 m s^{-2}), and h the height in meters.]

4.6 A heat engine operates between 210°C and 35°C. Calculate the minimum amount of heat that must be withdrawn from the hot source to obtain 2000 J of work.

The Second Law of Thermodynamics

4.7 Comment on the statement: "Even thinking about entropy increases its value in the universe."

4.8 One of the many statements of the second law of thermodynamics is: Heat cannot flow from a colder body to a warmer one without external aid. Assume two systems, 1 and 2, at T_1 and T_2 ($T_2 > T_1$). Show that if a quantity of heat q did flow spontaneously from 1 to 2, the process would result in a decrease in entropy of the universe. (You may assume that the heat flows very slowly so that the process can be regarded as reversible. Assume also that the loss of heat by system 1 and the gain of heat by system 2 do not affect T_1 and T_2.)

4.9 A ship sailing in the Indian Ocean takes the warmer surface water at 28°C to run a heat engine that powers the ship and discharges the used water back to the surface of the sea. Does this scheme violate the second law of thermodynamics? If so, what change would you make to make it work?

4.10 Molecules of a gas at any temperature T above the absolute zero are in constant motion. Does this "perpetual motion" violate the laws of thermodynamics?

4.11 According to the second law of thermodynamics, the entropy of an irreversible process in an isolated system must always increase. On the other hand, it is well known that the entropy of living systems remains small. (For example, the synthesis of highly complex protein molecules from individual amino acids is a process that leads to a decrease in entropy.) Is the second law invalid for living systems? Explain.

4.12 On a hot summer day, a person tries to cool himself by opening the door of a refrigerator. Is this a wise action, thermodynamically speaking?

Entropy Changes

4.13 The molar heat of vaporization of ethanol is 39.3 kJ mol^{-1}, and the boiling point of ethanol is 78.3°C. Calculate the value of $\Delta_{vap}S$ for the vaporization of 0.50 mole of ethanol.

4.14 Calculate the values of ΔU, ΔH, and ΔS for the following process:

1 mole of liquid water at 25°C and 1 atm	$\rightarrow$	1 mole of steam at 100°C and 1 atm

The molar heat of vaporization of water at 373 K is 40.79 kJ mol^{-1}, and the molar heat capacity of water is 75.3 J K^{-1} mol^{-1}. Assume the molar heat capacity to be temperature independent and ideal-gas behavior.

4.15 Calculate the value of ΔS in heating 3.5 moles of a monatomic ideal gas from 50°C to 77°C at constant pressure.

4.16 A quantity of 6.0 moles of an ideal gas is reversibly heated at constant volume from 17°C to 35°C. Calculate the entropy change. What would be the value of ΔS if the heating were carried out irreversibly?

4.17 One mole of an ideal gas is first, heated at constant pressure from T to $3T$ and second, cooled back to T at constant volume. (**a**) Derive an expression for ΔS for the overall process. (**b**) Show that the overall process is equivalent to an isothermal expansion of the gas at T from V to $3V$, where V is the original volume. (**c**) Show that the value of ΔS for the process in (**a**) is the same as that for (**b**).

4.18 A quantity of 35.0 g of water at 25.0°C (called A) is mixed with 160.0 g of water at 86.0°C (called B). (**a**) Calculate the final temperature of the system, assuming that the mixing is carried out adiabatically. (**b**) Calculate the entropy change of A, B, and the entire system.

4.19 The heat capacity of chlorine gas is given by

$$\overline{C}_P = (31.0 + 0.008T/\text{K})\ \text{J K}^{-1}\ \text{mol}^{-1}$$

Calculate the entropy change when 2 moles of gas are heated from 300 K to 400 K at constant pressure.

4.20 A sample of helium (He) gas initially at 20°C and 1.0 atm is expanded from 1.2 L to 2.6 L and simultaneously heated to 40°C. Calculate the entropy change for the process.

4.21 One of the early experiments in the development of the atomic bomb was to demonstrate that ^{235}U and not ^{238}U is the fissionable isotope. A mass spectrometer was employed to separate $^{235}UF_6$ from $^{238}UF_6$. Calculate the value of ΔS for the separation of 100 mg of the mixture of gas, given that the natural abundances of ^{235}U and ^{238}U are 0.72% and 99.28%, respectively, and that of ^{19}F is 100%.

4.22 One mole of an ideal gas at 298 K expands isothermally from 1.0 L to 2.0 L (**a**) reversibly and (**b**) against a constant external pressure of 12.2 atm. Calculate the values of ΔS_{sys}, ΔS_{surr}, and ΔS_{univ} in both cases. Are your results consistent with the nature of the processes?

4.23 The absolute molar entropies of O_2 and N_2 and 205 J K^{-1} mol^{-1} and 192 J K^{-1} mol^{-1}, respectively, at 25°C. What is the entropy of a mixture made up of 2.4 moles of O_2 and 9.2 moles of N_2 at the same temperature and pressure?

4.24 A quantity of 0.54 mole of steam initially at 350°C and 2.4 atm undergoes a cyclic process for which $q = -74$ J. Calculate the value of ΔS for the process.

4.25 Predict whether the entropy change is positive or negative for each of the following reactions at 298 K:

(**a**) $4Fe(s) + 3O_2(g) \rightarrow 2Fe_2O_3(s)$

(**b**) $O(g) + O(g) \rightarrow O_2(g)$

(**c**) $NH_4Cl(s) \rightarrow NH_3(g) + HCl(g)$

(**d**) $H_2(g) + Cl_2(g) \rightarrow 2HCl(g)$

4.26 Use the data in Appendix B to calculate the values of $\Delta_r S°$ of the reactions listed in the previous problem.

4.27 A quantity of 0.35 mole of an ideal gas initially at 15.6°C is expanded from 1.2 L to 7.4 L. Calculate the values of w, q, ΔU, and ΔS if the process is carried out (**a**) isothermally and reversibly, and (**b**) isothermally and irreversibly against an external pressure of 1.0 atm.

4.28 One mole of an ideal gas is isothermally expanded from 5.0 L to 10 L at 300 K. Compare the entropy changes for the system, surroundings, and the universe if the process is carried out (**a**) reversibly, and (**b**) irreversibly against an external pressure of 2.0 atm.

4.29 The heat capacity of hydrogen may be represented by

$$\overline{C}_P = (1.554 + 0.0022T/\mathrm{K})\ \mathrm{J\ K^{-1}\ mol^{-1}}$$

Calculate the entropy changes for the system, surroundings, and the universe for the (**a**) reversible heating, and (**b**) irreversible heating of 1.0 mole of hydrogen from 300 K to 600 K. [*Hint:* In (**b**), assume the surroundings to be at 600 K.]

4.30 Consider the reaction

$$N_2(g) + O_2(g) \rightarrow 2NO(g)$$

Calculate the values of $\Delta_r S°$ for the reaction mixture, surroundings, and the universe at 298 K. Why is your result reassuring to Earth's inhabitants?

The Third Law of Thermodynamics and Residual Entropy

4.31 The $\Delta_f\overline{H}°$ values can be negative, zero, or positive, but the $\overline{S}°$ values can be only zero or positive. Explain.

4.32 Choose the substance with the greater molar entropy in each of the following pairs: (**a**) $H_2O(l)$, $H_2O(g)$, (**b**) $NaCl(s)$, $CaCl_2(s)$, (**c**) N_2 (0.1 atm), N_2 (1 atm), (**d**) C (diamond), C (graphite), (**e**) $O_2(g)$, $O_3(g)$, (**f**) ethanol (C_2H_5OH), dimethly ether (CH_3OCH_3), (**g**) $N_2O_4(g)$, $2NO_2(g)$, (**h**) $Fe(s)$ at 298 K, $Fe(s)$ at 398 K. (Unless otherwise stated, assume the temperature is 298 K.)

4.33 A chemist found a discrepancy between the third law entropy and the calculated entropy from statistical thermodynamics for a compound. (**a**) Which value is larger? (**b**) Suggest two reasons that may give rise to this discrepancy.

4.34 Calculate the molar residual entropy of a solid in which the molecules can adopt (**a**) three, (**b**) four, and (**c**) five orientations of equal energy at absolute zero.

4.35 Account for the measured residual entropy of 10.1 J K^{-1} mol^{-1} for the CH_3D molecule.

4.36 Explain why the value of $\overline{S}°$ (graphite) is greater than that of $\overline{S}°$ (diamond) at 298 K (see Appendix B). Would this inequality hold at 0 K?

Additional Problems

4.37 Entropy has sometimes been described as "time's arrow" because it is the property that determines the forward direction of time. Explain.

4.38 State the condition(s) under which the following equations can be applied: (**a**) $\Delta S = \Delta H/T$, (**b**) $S_0 = 0$, (**c**) $dS = C_P dT/T$, (**d**) $dS = dq/T$.

4.39 Without referring to any table, predict whether the entropy change is positive, nearly zero, or negative for each of the following reactions:

(**a**) $N_2(g) + O_2(g) \rightarrow 2NO(g)$

(**b**) $2Mg(s) + O_2(g) \rightarrow 2MgO(s)$

(**c**) $2H_2O_2(l) \rightarrow 2H_2O(l) + O_2(g)$

(**d**) $H_2(g) + CO_2(g) \rightarrow H_2O(g) + CO(g)$

4.40 Calculate the entropy change when neon at 25°C and 1.0 atm in a container of volume 0.780 L is allowed to expand to 1.25 L and is simultaneously heated to 85°C. Assume ideal behavior. (*Hint:* Because S is a state function, you can first calculate the value of ΔS for expansion and then calculate the value of ΔS for heating at constant final volume.)

4.41 A reversible refrigerator has a coefficient of performance of 4.0. How much work must be done to freeze 1.0 kg of water initially at 0°C?

4.42 One mole of an ideal monatomic gas is compressed from 2.0 atm to 6.0 atm while being cooled from 400 K to 300 K. Calculate the values of ΔU, ΔH, and ΔS for the process.

4.43 The three laws of thermodynamics are sometimes stated colloquially as follows: First law: You cannot get something for nothing; Second law: The best you can do is get even; Third law: You cannot get even. Provide a scientific basis for each of these statements. (*Hint:* One consequence of the third law is that it is impossible to attain the absolute zero of temperature.)

4.44 Use the following data to determine the normal boiling point, in kelvins, of mercury. What assumptions must you make to do the calculation?

$$\text{Hg}(l): \quad \Delta_f\overline{H}^\circ = 0 \text{ (by definition)}$$
$$\overline{S}^\circ = 75.9 \text{ J K}^{-1} \text{ mol}^{-1}$$
$$\text{Hg}(g): \quad \Delta_f\overline{H}^\circ = 60.78 \text{ kJ mol}^{-1}$$
$$\overline{S}^\circ = 175.0 \text{ J K}^{-1} \text{ mol}^{-1}$$

4.45 Referring to Trouton's rule, explain why the ratio $\Delta_{vap}\overline{H}/T_b$ is considerably smaller than 90 J K^{-1} mol^{-1} for liquid HF.

4.46 Give a detailed example of each of the following, with an explanation: **(a)** a thermodynamically spontaneous process; **(b)** a process that would violate the first law of thermodynamics; **(c)** a process that would violate the second law of thermodynamics; **(d)** an irreversible process; **(e)** an equilibrium process.

4.47 In the reversible adiabatic expansion of an ideal gas, there are two contributions to entropy changes: the expansion of the gas and the cooling of the gas. Show that these two contributions are equal in magnitude but opposite in sign. Show also that for an irreversible adiabatic gas expansion, these two contributions are no longer equal in magnitude. Predict the sign of ΔS.

4.48 A refrigerator set at 0°C discharges heat into the kitchen at 20°C. **(a)** How much work would be required to freeze 500 mL of water (about an ice tray's volume)? **(b)** How much heat would be discharged during this process? (The molar enthalpy of fusion of water is 6.01 kJ mol^{-1}, and the refrigerator operates at 35% efficiency.)

4.49 Superheated water is water heated above 100°C without boiling. As for supercooled water (see Example 4.7), superheated water is thermodynamically unstable. Calculate the values of ΔS_{sys}, ΔS_{surr}, and ΔS_{univ} when 1.5 moles of superheated water at 110°C and 1.0 atm are converted to steam at the same temperature and pressure. (The molar enthalpy of vaporization of water is 40.79 kJ mol^{-1}, and the molar heat capacities of water and steam in the temperature range 100–110°C are 75.5 J K^{-1} mol^{-1} and 34.4 J K^{-1} mol^{-1}, respectively.

4.50 Toluene (C_7H_8) has a dipole moment, whereas benzene (C_6H_6) is nonpolar:

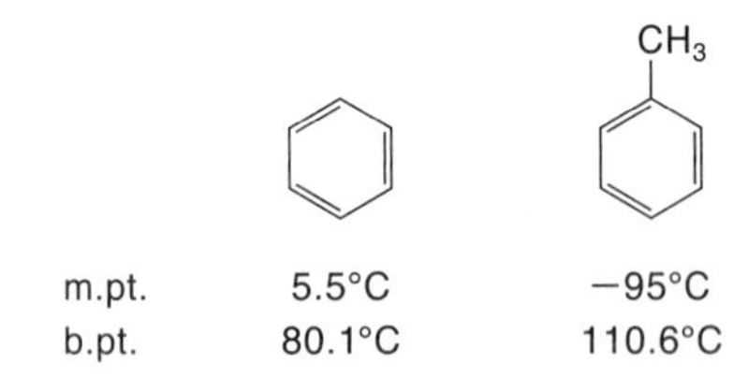

m.pt.	5.5°C	−95°C
b.pt.	80.1°C	110.6°C

Explain why, contrary to our expectation, benzene melts at a much higher temperature than toluene. Why is the boiling point of toluene higher than that of benzene?

4.51 Comment on the correctness of the analogy sometimes used to relate a student's dormitory room becoming disorderly and untidy to an increase in entropy.

4.52 In Section 4.2 we saw that the probability of finding all 100 helium atoms in one half of the cylinder is 8×10^{-31} (see Figure 4.3). Assuming that the age of the universe is 14 billion years, calculate the time in seconds during which this event can be observed.

4.53 Calculate the entropy of a system containing 2.0 moles of Ar and 3.0 moles of Xe at 298 K.

4.54 Two moles of argon gas initially at 300 K and 2.00 L are heated to 400 K and the volume changed to 6.00 L. Calculate ΔS for the process. Assume ideal behavior and $\overline{C}_V$ to be independent of temperature.

4.55 Calculate the ΔS and ΔS_{univ} for each step of the Carnot cycle.

4.56 A certain reaction is spontaneous at 72°C. If the enthalpy change for the reaction is 19 kJ mol^{-1}, what is the minimum value of ΔS for the reaction?

4.57 Suppose we have eight distinguishable particles that have a total energy equal to six units (see Figure 4.13). The distributions for this system are labeled as (n_0,n_1,n_2,n_3). Calculate the number of microstates (W) for each of the following distributions: **(a)** (6,0,0,2), **(b)** (5,1,1,1), **(c)** (4,2,2,0), and **(d)** (2,6,0,0).

4.58 Calculate the number of microstates (W) when 10 molecules are equally distributed among five energy levels. What is the value of W if one molecule is removed from one state and added to another?

4.59 A useful estimate of the enormity of the number of microstates associated with the most probable macrostate for a macroscopic system is the Poincaré recurrence time, which is the average time it takes for a system to return to any microstate that it has once occupied. The magnitude of Poincaré recurrence time can be appreciated by doing the following exercise. Consider a deck of cards dealt out in any specific order (both in numbers and in suits) as defining a microstate. **(a)** How many microstates are there? **(b)** If we could shuffle the cards and deal one per second, how long would we have to wait before a duplicate of that order is dealt again?

CHAPTER 5

Gibbs and Helmholtz Energies and Their Applications

Mathematics is a language.

Josiah Willard Gibbs*

Having introduced the entropy function and the second and third laws of thermodynamics in the last chapter, we now have essentially all the tools we need for studying chemical and physical processes. However, to help us focus on the system and on specific practical conditions in this chapter, we shall develop two functions that form the basis of chemical thermodynamics: Gibbs energy and Helmholtz energy.

5.1 Gibbs and Helmholtz Energies

With the first law of thermodynamics to take care of energy balance and the second law to help us decide which processes can occur spontaneously, we might reasonably expect that we have enough thermodynamic quantities to deal with any situation. Although this expectation is true in principle, in practice the equations we have derived so far are not the most convenient to apply. For example, to use the second law equation developed in Chapter 4 (Equation 4.16), we must calculate the entropy change in both the system and the surroundings. Because we are generally interested only in what happens in the system and are not concerned with events in the surroundings, it would be simpler if we could establish criteria for equilibrium and spontaneity in terms of the change in a certain thermodynamic function of the system and not of the entire universe, as for ΔS_{univ}.

Consider a system in thermal equilibrium with its surroundings at temperature T. A process occurring in the system results in the transfer of an infinitesimal amount of heat, dq, from the system to the surroundings. Thus, we have $-dq_{\text{sys}} = dq_{\text{surr}}$. The total change in entropy, according to Equation 4.16, is

$$\begin{aligned} dS_{\text{univ}} &= dS_{\text{sys}} + dS_{\text{surr}} \geq 0 \\ &= dS_{\text{sys}} + \frac{dq_{\text{surr}}}{T} \geq 0 \\ &= dS_{\text{sys}} - \frac{dq_{\text{sys}}}{T} \geq 0 \end{aligned}$$

* Gibbs made this brief statement at the end of a prolonged debate of the Yale faculty on whether studying classical language or mathematics is the better discipline for students.

Note that every quantity on the right side of the above equation refers to the system. If the process takes place at constant pressure, then $dq_{sys} = dH_{sys}$, or

$$dS_{sys} - \frac{dH_{sys}}{T} \geq 0$$

The reversal of the inequality sign follows from the fact that if $x > 0$ then $-x < 0$.

Multiplying the equation above by $-T$, we obtain

$$dH_{sys} - TdS_{sys} \leq 0$$

We now define a function, called *Gibbs energy** (after the American physicist Josiah Willard Gibbs, 1839–1903), G, as

$$G = H - TS \tag{5.1}$$

From Equation 5.1 we see that because H, T, and S are all state functions, G is also a state function. Further, like enthalpy, G has the units of energy.

At constant temperature, the change in the Gibbs energy of the system in an infinitesimal process is given by

$$dG_{sys} = dH_{sys} - TdS_{sys}$$

We can apply dG_{sys} as a criterion for equilibrium and spontaneity as follows:

$$dG_{sys} \leq 0 \tag{5.2}$$

where the $<$ sign denotes a spontaneous process and the equality sign denotes equilibrium at constant temperature and pressure.

Unless otherwise indicated, from now on we shall consider only the system in our discussion of Gibbs energy changes. For this reason, the subscript "sys" will be omitted for simplicity. For a finite isothermal process $1 \rightarrow 2$, the change of Gibbs energy is given by

$$\Delta G = \Delta H - T\Delta S \tag{5.3}$$

and the conditions of equilibrium and spontaneity at constant temperature and pressure are given by

$$\Delta G = G_2 - G_1 = 0 \qquad \text{system at equilibrium}$$

$$\Delta G = G_2 - G_1 < 0 \qquad \text{spontaneous process from 1 to 2}$$

If ΔG is negative, the process is said to be *exergonic* (from the Greek word for "work producing"); if positive, the process is *endergonic* (work consuming). Note that pressure must be constant to set $q = \Delta H$, and temperature must be constant to derive Equation 5.3. In general, we can replace q with ΔH only if pressure is constant

* Gibbs energy was previously called *Gibbs free energy* or just *free energy*. However, IUPAC (the International Union of Pure and Applied Chemistry) has recommended that the *free* be dropped. The same recommendation applies to Helmholtz energy, to be discussed shortly.

throughout the process. Because G is a state function, however, ΔG is independent of path. Therefore, Equation 5.3 applies to any process as long as the temperature and pressure are the same in the initial and final states.

Gibbs energy is useful because it incorporates both enthalpy and entropy. In some reactions, the enthalpy and entropy contributions reinforce each other. For example, if ΔH is negative (an exothermic reaction) and ΔS is positive (more molecular disorder), then $(\Delta H - T\Delta S)$ or ΔG is a negative quantity, and the process is favored from left to right. In other reactions, enthalpy and entropy may work against each other; that is, ΔH and $(-T\Delta S)$ have different signs. In such cases, the sign of ΔG is determined by the *magnitudes* of ΔH and $T\Delta S$. If $|\Delta H| \gg |T\Delta S|$, then the reaction is said to be enthalpy driven because the sign of ΔG is predominantly determined by ΔH. Conversely, if $|T\Delta S| \gg |\Delta H|$, then the process is an entropy-driven one. Table 5.1 shows how positive and negative values of ΔH and ΔS affect ΔG at different temperatures.

A similar thermodynamic function can be derived for processes in which the temperature and volume are kept constant. *Helmholtz energy* (after the German physiologist and physicist Hermann Ludwig Helmholtz, 1821–1894), A, is defined as

$$A = U - TS \tag{5.4}$$

where all the terms refer to the system. Like G, A is a state function and has the units of energy. Following the same procedure described above for Gibbs energy, we can show that at constant temperature and volume, the criteria for equilibrium and spontaneity are given by

$$dA_{\text{sys}} \leq 0 \tag{5.5}$$

Omitting the subscript for system, we have, for a finite process

$$\Delta A = \Delta U - T\Delta S \tag{5.6}$$

Table 5.1
Factors Affecting ΔG of a Reaction[a]

ΔH	ΔS	ΔG	Example
+	+	Positive at low temperatures; negative at high temperatures. Reaction spontaneous in the forward direction at high temperatures and spontaneous in the reverse direction at low temperatures.	$2HgO(s) \rightarrow 2Hg(l) + O_2(g)$
+	−	Positive at all temperatures. Reaction spontaneous in the reverse direction at all temperatures.	$3O_2(g) \rightarrow 2O_3(g)$
−	+	Negative at all temperatures. Reaction spontaneous in the forward direction at all temperatures.	$2H_2O_2(l) \rightarrow 2H_2O(l) + O_2(g)$
−	−	Negative at low temperatures; positive at high temperatures. Reaction spontaneous at low temperatures; tends to reverse at high temperatures.	$NH_3(g) + HCl(g) \rightarrow NH_4Cl(s)$

[a]Assuming both ΔH and ΔS are independent of temperature.

5.2 The Meaning of Helmholtz and Gibbs Energies

Equations 5.3 and 5.6 provide us with extremely useful criteria for dealing with the direction of spontaneous changes and the nature of chemical and physical equilibria. In addition, these thermodynamic functions enable us to determine the amount of work that can be done in a given process.

Helmholtz Energy

For an infinitesimal process at constant temperature, Equation 5.4 takes the form

$$dA = dU - TdS$$

For a reversible change, $dq_{\text{rev}} = TdS$ (see Equation 4.5), so that the above equation becomes

$$dA = dU - dq_{\text{rev}}$$

Applying the first law of thermodynamics (see Equation 3.7), we write

$$\begin{aligned} dA &= dq_{\text{rev}} + dw_{\text{rev}} - dq_{\text{rev}} \\ &= dw_{\text{rev}} \end{aligned} \tag{5.7}$$

or, for a finite process,

$$\Delta A = w_{\text{rev}} \tag{5.8}$$

If $\Delta A < 0$, the process will occur spontaneously, and w_{rev} represents the work that can be done *by* the system on the surroundings if this change is carried out reversibly. (Note that w_{rev} in this case is a negative quantity in accord with our convention.) Furthermore, w_{rev} is the maximum work that can be obtained.

One way to apply Equation 5.6 is to calculate the value of ΔA for the mixing of two ideal gases 1 and 2 at constant T and V. Because this is an isothermal process, $\Delta U = 0$ and, from Equation 4.17, the entropy of the mixing is given by $-R(n_1 \ln x_1 + n_2 \ln x_2)$. Therefore,

$$\begin{aligned} \Delta A &= \Delta U - T\Delta S \\ &= RT(n_1 \ln x_1 + n_2 \ln x_2) \end{aligned}$$

which is a negative quantity because $x < 1$ so that $\ln x < 0$. This result is consistent with our knowledge that the isothermal mixing of two ideal gases is a spontaneous process. The following example shows how to use Equation 5.6 to determine the maximum work that can be extracted from a chemical reaction.

EXAMPLE 5.1

Consider the metabolism of glucose to water and carbon dioxide at 25°C:

$$C_6H_{12}O_6(s) + 6O_2(g) \rightarrow 6CO_2(g) + 6H_2O(l)$$

The following changes for the combustion of 1 mole of glucose are obtained from calorimetric measurements and Appendix B: $\Delta_r U = -2801.3$ kJ mol^{-1} and $\Delta_r S = 260.7$ J K^{-1} mol^{-1}. How much of the energy change can be extracted as work?

ANSWER

To calculate the change in Helmholtz energy using Equation 5.6, we write

$$\begin{aligned}\Delta_r A &= \Delta_r U - T\Delta_r S \\ &= -2801.3 \text{ kJ mol}^{-1} - (298 \text{ K})\left(\frac{260.7 \text{ J K}^{-1} \text{ mol}^{-1}}{1000 \text{ J/kJ}}\right) \\ &= -2879.0 \text{ kJ mol}^{-1}\end{aligned}$$

COMMENT

The result shows that the maximum work theoretically obtainable from the biological breakdown of glucose (2879.0 kJ mol^{-1}) is actually greater than the internal energy change (2801.3 kJ mol^{-1}). The reason is that the process is accompanied by an increase in entropy, which contributes to the maximum work. In practice, only a portion of this work is converted to useful biological activity.

Gibbs Energy

To show the relationship between the change in Gibbs energy and work, we start with the definition of G:

$$G = H - TS$$

For an infinitesimal process

$$dG = dH - TdS - SdT$$

Now, because

$$H = U + PV$$

$$dH = dU + PdV + VdP$$

According to the first law of thermodynamics,

$$dU = đq + đw$$

and

Here we consider only P–V type work.

$$dU = đq - PdV$$

For a reversible process,

$$dq_{\text{rev}} = TdS$$

so that

$$dU = TdS - PdV \tag{5.9}$$

and

$$\begin{aligned} dH &= (TdS - PdV) + PdV + VdP \\ &= TdS + VdP \end{aligned}$$

Finally, we have

$$\begin{aligned} dG &= (TdS + VdP) - TdS - SdT \\ &= VdP - SdT \end{aligned} \tag{5.10}$$

Equation 5.9 incorporates the first and second laws, whereas Equation 5.10 shows the dependence of G on pressure and temperature. Both are important, fundamental equations of thermodynamics. (More thermodynamic relationships are derived in Appendix 5.1 on p. 200.)

Equation 5.10 holds for a process in which only expansion work occurs. If, in addition to expansion work, another type of work is done, we must take that into account. For example, for a redox reaction in an electrochemical cell that generates electrons and does electrical work (w_{el}), Equation 5.9 is modified to be

$$dU = TdS - PdV + dw_{\text{el}}$$

and therefore

$$dG = VdP - SdT + dw_{\text{el}}$$

where the subscript "el" denotes electrical. At constant P and T, we have

$$dG = dw_{\text{el, rev}}$$

and for a finite change

$$\Delta G = w_{\text{el, rev}} = w_{\text{el, max}} \tag{5.11}$$

This derivation shows that ΔG is the maximum nonexpansion work we can obtain for a reversible process at constant P and T. We shall make use of Equation 5.11 when we discuss electrochemistry in Chapter 9.

EXAMPLE 5.2

Referring to Example 4.5, calculate the value of ΔG for the melting of ice at (a) 0°C, (b) 10°C, and (c) −10°C. The molar enthalpy and entropy of fusion of water are 6.01 kJ mol^{-1} and 22.0 J K^{-1} mol^{-1}, respectively, and are assumed to be temperature independent.

ANSWER

We need Equation 5.3 for all three cases.
(a) At the normal melting point of ice,

$$\begin{aligned}\Delta G &= \Delta H - T\Delta S\\ &= 6.01\text{ kJ} - 273\text{ K}\left(\frac{22.0\text{ J K}^{-1}}{1000\text{ J/kJ}}\right)\\ &= 0\end{aligned}$$

(b) At 10°C,

$$\begin{aligned}\Delta G &= 6.01\text{ kJ} - 283\text{ K}\left(\frac{22.0\text{ J K}^{-1}}{1000\text{ J/kJ}}\right)\\ &= -0.22\text{ kJ}\end{aligned}$$

(c) At −10°C,

$$\begin{aligned}\Delta G &= 6.01\text{ kJ} - 263\text{ K}\left(\frac{22.0\text{ J K}^{-1}}{1000\text{ J/kJ}}\right)\\ &= 0.22\text{ kJ}\end{aligned}$$

COMMENT

These results are consistent with our knowledge that at 0°C the system is at equilibrium and $\Delta G = 0$ [case (a)]; at 10°C, ice melts spontaneously and $\Delta G < 0$ [case (b)]; and at −10°C, ice will not melt spontaneously and $\Delta G > 0$ [case (c)].

EXAMPLE 5.3

In a fuel cell, natural gases such as methane undergo the same redox reaction as in the combustion process to produce carbon dioxide and water and generate electricity (see Section 9.5). Calculate the maximum electrical work that can be obtained from 1 mole of methane at 25°C and constant pressure.

ANSWER

The reaction is

$$CH_4(g) + 2O_2(g) \rightarrow CO_2(g) + 2H_2O(l)$$

From the $\Delta_f\overline{H}°$ and $\overline{S}°$ values in Appendix B, we find that $\Delta_r H = -890.3$ kJ mol^{-1} and $\Delta_r S = -242.8$ J K^{-1} mol^{-1}. Therefore, from Equation 5.3

$$\Delta_r G = -890.3 \text{ kJ mol}^{-1} - 298 \text{ K}\left(\frac{-242.8 \text{ J K}^{-1}\text{ mol}^{-1}}{1000 \text{ J/kJ}}\right)$$

$$= -818.0 \text{ kJ mol}^{-1}$$

From Equation 5.11 we write

$$w_{\text{el, max}} = -818.0 \text{ kJ mol}^{-1}$$

Thus, the maximum electrical work the system can do on the surroundings is equal to 818.0 kJ per mole of CH_4 reacted.

COMMENTS

Two points of interest: First, because the reaction results in a decrease in entropy, the electrical work done is *less* than the heat generated. This is the price we pay for order. Second, if the enthalpy of combustion were used to do work in a heat engine, then the efficiency of the heat-to-work conversion would be limited by Equation 4.9. In principle, 100% of Gibbs energy released in a fuel cell can be converted to work because the cell is not a heat engine and therefore is not governed by the second law restrictions.

Finally, note that constant-pressure conditions are more common than constant-volume conditions; therefore, Gibbs energy is more frequently used than Helmholtz energy.

5.3 Standard Molar Gibbs Energy of Formation ($\Delta_f\overline{G}°$)

As for enthalpy, we cannot measure the absolute value of Gibbs energies, and so for conveniece, we assign a value of zero to the standard molar Gibbs energy of formation of an element in its most stable allotropic form at 1 bar and 298 K. Again using the combustion of graphite as an example (see Section 3.7):

$$C(\text{graphite}) + O_2(g) \rightarrow CO_2(g)$$

If the reaction is carried out with reactants at 1 bar being converted to products at 1 bar, then the standard molar Gibbs energy change, $\Delta_r G°$, for the reaction is

$$\Delta_r G° = \Delta_f\overline{G}°(CO_2) - \Delta_f\overline{G}°(\text{graphite}) - \Delta_f\overline{G}°(O_2)$$
$$= \Delta_f\overline{G}°(CO_2)$$

or

$$\Delta_f\overline{G}°(CO_2) = \Delta_r G°$$

because the $\Delta_f\overline{G}°$ values for both graphite and O_2 are zero. To determine the value of $\Delta_r G°$, we use Equation 5.3:

$$\Delta_r G° = \Delta_r H° - T\Delta_r S°$$

In Chapter 3 (p. 103), we saw that $\Delta_r H° = -393.5 \text{ kJ mol}^{-1}$. To find the value of $\Delta_r S°$, we use Equation 4.27 and the data in Appendix B:

$$\Delta_r S° = \overline{S}°(CO_2) - \overline{S}°(\text{graphite}) - S°(O_2)$$
$$= (213.6 - 5.7 - 205.0) \text{ J K}^{-1} \text{ mol}^{-1}$$
$$= 2.9 \text{ J K}^{-1} \text{ mol}^{-1}$$

Thus,

$$\Delta_r G° = -393.5 \text{ kJ mol}^{-1} - 298 \text{ K}\left(\frac{2.9 \text{ J K}^{-1} \text{ mol}^{-1}}{1000 \text{ J/kJ}}\right)$$
$$= -394.4 \text{ kJ mol}^{-1}$$

Finally, we arrive at the result:

$$\Delta_f\overline{G}°(CO_2) = -394.4 \text{ kJ mol}^{-1}$$

In this manner, we can determine the $\Delta_f\overline{G}°$ values of most substances. Table 5.2 lists the $\Delta_f\overline{G}°$ values for a number of common inorganic and organic substances (a more extensive listing is given in Appendix B).

In general, $\Delta_r G°$ for a reaction of the type

$$a\text{A} + b\text{B} \rightarrow c\text{C} + d\text{D}$$

is given by

$$\Delta_r G° = c\Delta_f\overline{G}°(\text{C}) + d\Delta_f\overline{G}°(\text{D}) - a\Delta_f\overline{G}°(\text{A}) - b\Delta_f\overline{G}°(\text{B})$$
$$= \Sigma\nu\Delta_f\overline{G}°(\text{products}) - \Sigma\nu\Delta_f\overline{G}°(\text{reactants}) \quad (5.12)$$

When the stoichiometric coefficient is one, it is not shown in Equation 5.12.

where ν is the stoichiometric coefficient. In later chapters, we shall see that $\Delta_r G°$ can also be obtained from the equilibrium constant and electrochemical measurements.

Table 5.2
Standard Molar Gibbs Energies of Formation at 1 Bar and 298 K for Some Inorganic and Organic Substances

Substance	$\Delta_f\overline{G}°$/kJ mol^{-1}	Substance	$\Delta_f\overline{G}°$/kJ mol^{-1}
C(graphite)	0	$CH_4(g)$	−50.79
C(diamond)	2.87	$C_2H_6(g)$	−32.9
$CO(g)$	−137.3	$C_3H_8(g)$	−23.5
$CO_2(g)$	−394.4	$C_2H_2(g)$	209.2
$HF(g)$	−275.4	$C_2H_4(g)$	68.12
$HCl(g)$	−95.3	$C_6H_6(l)$	124.5
$HBr(g)$	−53.4	$CH_3OH(l)$	−166.3
$HI(g)$	1.7	$C_2H_5OH(l)$	−174.2
$H_2O(g)$	−228.6	$CH_3CHO(l)$	−128.1
$H_2O(l)$	−237.2	$HCOOH(l)$	−361.4
$NH_3(g)$	−16.6	$CH_3COOH(l)$	−389.9
$NO(g)$	86.7	$C_6H_{12}O_6(s)$	−910.6
$NO_2(g)$	51.84	$C_{12}H_{22}O_{11}(s)$	−1544.3
$N_2O_4(g)$	98.3		
$N_2O(g)$	103.6		
$O_3(g)$	163.4		
SO_2	−300.1		
SO_3	−370.4		

Because the Gibbs energy change is made up of two parts—enthalpy and temperature times entropy—comparing their contributions to $\Delta_r G°$ in a process is instructive. Figure 5.1 shows these contributions on a vector diagram, using the following combustion data from the previous section:

$C_6H_{12}O_6$* $\Delta_r H° = -2801.3$ kJ mol^{-1} $-T\Delta_r S° = -77.7$ kJ mol^{-1} $\Delta_r G° = -2879.0$ kJ mol^{-1}

CH_4 $\Delta_r H° = -890.3$ kJ mol^{-1} $-T\Delta_r S° = 72.3$ kJ mol^{-1} $\Delta_r G° = -818.0$ kJ mol^{-1}

A question arises here: If the combustions of glucose and methane are as spontaneous as the large negative $\Delta G°$ values indicate, how can these substances be kept in air for long periods of time without any apparent change? Herein lies the limitation

*The glucose data are from Example 5.1. Because there is no change in the number of moles of gases, $\Delta n = 0$, and so $\Delta_r U° = \Delta_r H°$ and $\Delta_r A° = \Delta_r G°$.

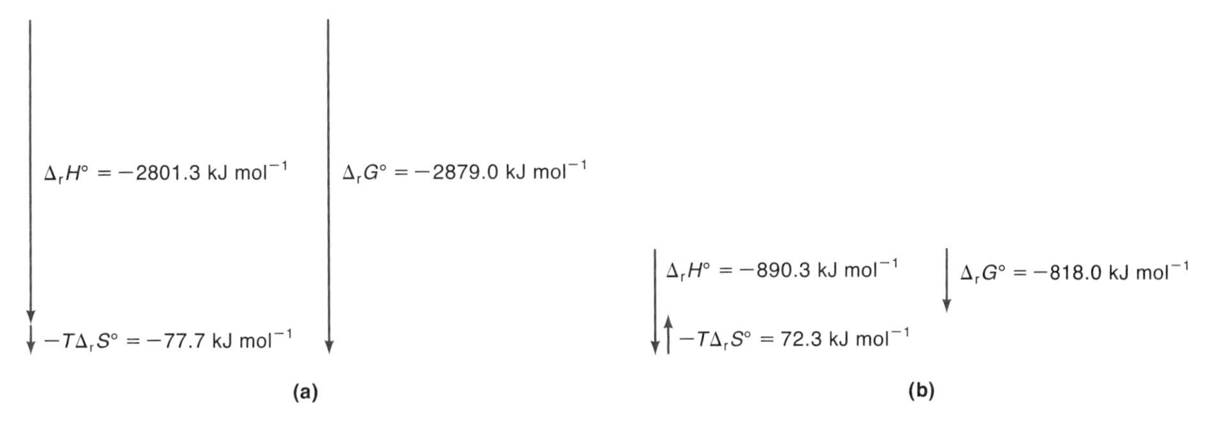

Figure 5.2
Vector diagrams shows the changes of $\Delta_r H^\circ$, $-T\Delta_r S^\circ$, and $\Delta_r G^\circ$ at 298 K for the combustion of (a) glucose and (b) methane.

of thermodynamics, for it tells us only the *direction* in which a reaction will go and nothing about the *rate* at which it occurs. For any reaction to start, the reactants must first acquire sufficient energy to get over the activation energy barrier. Glucose (or methane) molecules in a container at room temperature lack this energy and are therefore perfectly stable. More will be said about this topic in Chapter 15.

5.4 Dependence of Gibbs Energy on Temperature and Pressure

Because Gibbs energy plays such a central role in chemical thermodynamics, understanding its properties is important. Equation 5.10 shows that it is a function of both pressure and temperature. Here, we shall see how the value of G changes with each of these variables.

Dependence of G on Temperature

We start with Equation 5.10:

$$dG = VdP - SdT$$

At constant pressure, this equation becomes

$$dG = -SdT$$

so that the variation of G with respect to T at constant pressure is given by

$$\left(\frac{\partial G}{\partial T}\right)_P = -S \tag{5.13}$$

Equation 5.1 now becomes

$$G = H + T\left(\frac{\partial G}{\partial T}\right)_P$$

Dividing the above equation by T^2 and rearranging, we obtain

$$-\frac{G}{T^2} + \frac{1}{T}\left(\frac{\partial G}{\partial T}\right)_P = -\frac{H}{T^2}$$

The left side of the above equation is the partial derivative of G/T with respect to T at contant pressure; that is,

$$\left[\frac{\partial\left(\frac{G}{T}\right)}{\partial T}\right]_P = -\frac{G}{T^2} + \frac{1}{T}\left(\frac{\partial G}{\partial T}\right)_P$$

Therefore,

$$\left[\frac{\partial\left(\frac{G}{T}\right)}{\partial T}\right]_P = -\frac{H}{T^2} \quad (5.14)$$

Equation 5.14 is known as the Gibbs-Helmholtz equation. When applied to a finite process, G and H become ΔG and ΔH so that the equation becomes

$$\left[\frac{\partial\left(\frac{\Delta G}{T}\right)}{\partial T}\right]_P = -\frac{\Delta H}{T^2} \quad (5.15)$$

Equation 5.15 is important because it relates the temperature dependence of the Gibbs energy change, and hence the position of equilibrium, to the enthalpy change. We shall return to this equation in Chapter 8.

Dependence of *G* on Pressure

To see how the Gibbs energy depends on pressure, we again employ Equation 5.10. At constant temperature,

$$dG = VdP$$

or

$$\left(\frac{\partial G}{\partial P}\right)_T = V \quad (5.16)$$

Because volume must be a positive quantity, Equation 5.16 says that the Gibbs energy of a system always increases with pressure at constant temperature. We are interested in how the value of G increases when the pressure of the system increases from P_1 to P_2. We can write the change in G, ΔG, as the system goes from state 1 to state 2 as

$$\Delta G = \int_1^2 dG = G_2 - G_1 = \int_{P_1}^{P_2} VdP$$

For an ideal gas, $V = nRT/P$ so that

$$\Delta G = G_2 - G_1 = \int_{P_1}^{P_2} \frac{nRT}{P} dP$$

$$= nRT \ln \frac{P_2}{P_1} \qquad (5.17)$$

If we set $P_1 = 1$ bar (the standard state), we can replace G_1 with the symbol for the standard state, $G°$, G_2 by G, and P_2 by P. Equation 5.17 now becomes

$$G = G° + nRT \ln \frac{P}{1 \text{ bar}}$$

Expressed in molar quantities,

$$\overline{G} = \overline{G}° + RT \ln \frac{P}{1 \text{ bar}} \qquad (5.18)$$

where $\overline{G}$ depends on both temperature and pressure, and $\overline{G}°$ is a function of temperature only. Equation 5.18 relates the molar Gibbs energy of an ideal gas to its pressure. Later, we shall see a similar equation relating the Gibbs energy of a substance to its concentration in a mixture.

EXAMPLE 5.4

A 0.590-mol sample of an ideal gas initially at 300 K and 1.50 bar is compressed isothermally to a final pressure of 6.90 bar. Calculate the change in Gibbs energy for this process.

ANSWER

From Equation 5.17, we write

$$P_1 = 1.50 \text{ bar} \qquad P_2 = 6.90 \text{ bar}$$

so that

$$\Delta G = nRT \ln \frac{P_2}{P_1}$$

$$= (0.590 \text{ mol})(8.314 \text{ J K}^{-1} \text{ mol}^{-1})(300 \text{ K}) \ln \frac{6.90 \text{ bar}}{1.50 \text{ bar}}$$

$$= 2.25 \times 10^3 \text{ J}$$

Thus far, we have focused on gases in discussing the dependence of G on pressure. Because the volume of a liquid or a solid is practically independent of applied pressure, we write

$$G_2 - G_1 = \int_{P_1}^{P_2} VdP$$

$$= V(P_2 - P_1) = V\Delta P$$

or

$$G_2 = G_1 + V\Delta P$$

The volume, V, is treated as a constant and may be taken outside the integral. In general, the Gibbs energies of liquids and solids are much less dependent on pressure so that the variation of G with P can be ignored, except when dealing with geological processes in Earth's interior or specially created high-pressure conditions in the laboratory.

5.5 Gibbs Energy and Phase Equilibria

In this section, we shall see how Gibbs energy can be applied to the study of phase equilibria. A *phase* is a homogeneous part of a system that is in contact with other parts of the system but separated from them by a well-defined boundary. Examples of phase equilibria are physical processes such as freezing and boiling. In Chapter 8, we shall apply Gibbs energy to the study of chemical equilibria. Our discussion here is restricted to one-component systems.

Consider that at some temperature and pressure, two phases, say solid and liquid, of a one-component system are in equilibrium. How do we formulate this condition? We might be tempted to equate the Gibbs energies as follows:

$$G_{\text{solid}} = G_{\text{liquid}}$$

But this formulation will not hold, for it is possible to have a small ice cube floating in an ocean of water at 0°C, and yet the Gibbs energy of water is much larger than that of the ice cube. Instead, we must insist that the Gibbs energy *per mole* (or the molar Gibbs energy) of the substance, an intensive property, be the same in both phases at equilibrium because intensive quantities are independent of the amount present:

$$\overline{G}_{\text{solid}} = \overline{G}_{\text{liquid}}$$

If external conditions (temperature or pressure) were altered so that $\overline{G}_{\text{solid}} > \overline{G}_{\text{liquid}}$, then some solid would melt because

$$\Delta G = \overline{G}_{\text{liquid}} - \overline{G}_{\text{solid}} < 0$$

On the other hand, if $\overline{G}_{\text{solid}} < \overline{G}_{\text{liquid}}$, then some liquid would freeze spontaneously.

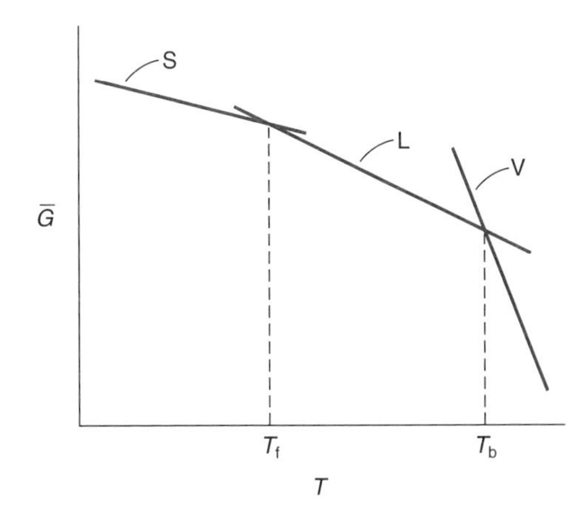

Figure 5.2
Dependence of molar Gibbs energy on temperature for the solid (S), liquid (L), and vapor (V) phases of a substance at constant pressure. The phase with the lowest $\overline{G}$ is the most stable phase at that temperature. The intercept of the vapor and liquid lines gives the boiling point (T_b) and that between the liquid and solid lines gives the melting point (T_f).

Next, let us see how the molar Gibbs energies of solid, liquid, and vapor depend on temperature and pressure. Equation 5.13 expressed in molar quantities becomes

$$\left(\frac{\partial \overline{G}}{\partial T}\right)_P = -\overline{S}$$

Because the entropy of a substance in any phase is always positive, a plot of $\overline{G}$ versus T at constant pressure gives us a line with a negative slope. For the three phases of a single substance, we have*

$$\left(\frac{\partial \overline{G}_{\text{solid}}}{\partial T}\right)_P = -\overline{S}_{\text{solid}} \qquad \left(\frac{\partial \overline{G}_{\text{liquid}}}{\partial T}\right)_P = -\overline{S}_{\text{liquid}} \qquad \left(\frac{\partial \overline{G}_{\text{vapor}}}{\partial T}\right)_P = -\overline{S}_{\text{vapor}}$$

At any temperature, the molar entropies of a substance decrease in the order

$$\overline{S}_{\text{vapor}} \gg \overline{S}_{\text{liquid}} > \overline{S}_{\text{solid}}$$

These differences are reflected in the slopes of lines shown in Figure 5.2. At high temperatures, the vapor phase is the most stable, because it has the lowest molar Gibbs energy. As temperature decreases, however, liquid becomes the stable phase, and finally, at even lower temperatures, solid becomes the most stable phase. The intercept between the vapor and liquid lines is the point at which these two phases are in equilibrium, that is, $\overline{G}_{\text{vapor}} = \overline{G}_{\text{liquid}}$. The corresponding temperature is T_b, the boiling point. Similarly, solid and liquid coexist in equilibrium at the temperature T_f, the melting (or fusion) point.

How does an increase in pressure affect phase equilibria? In the previous section, we saw that the Gibbs energy of a substance always increases with pressure (see Equation 5.16). Further, for a given change in pressure, the increase is greatest

* Although we use the terms *gas* and *vapor* interchangeably, strictly speaking, there is a difference. A gas is a substance that is normally in the gaseous state at ordinary temperatures and pressures; a vapor is the gaseous form of any substance that is a liquid or a solid at normal temperatures and pressures. Thus, at 25°C and 1 atm, we speak of water vapor and oxygen gas.

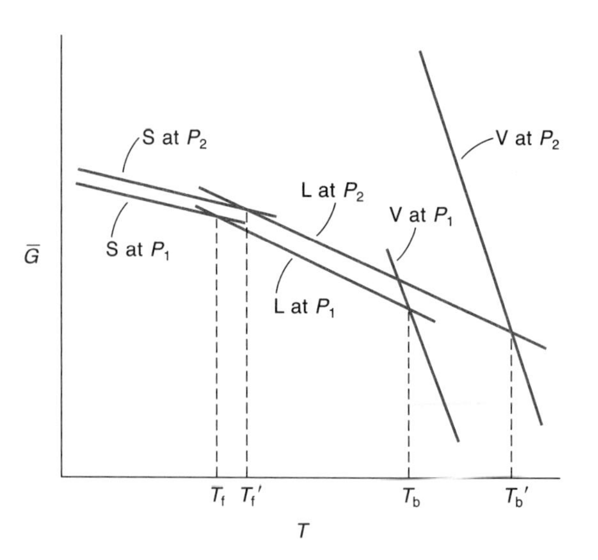

Figure 5.3
Pressure dependence of molar Gibbs energy. For the majority of substances (water being the important exception), an increase in pressure leads to an increase in both the melting point and the boiling point. (Here we have $P_2 > P_1$.)

for vapors, much less for liquids and solids. This result follows from Equation 5.16, expressed in molar quantities

$$\left(\frac{\partial \overline{G}}{\partial P}\right)_T = \overline{V}$$

The molar volume of a vapor is normally about a thousand times greater than that for a liquid or a solid.

Figure 5.3 shows the increases in the value of $\overline{G}$ for the three phases as the pressure increases from P_1 to P_2. We see that both T_f and T_b shift to higher values, but the shift in T_b is greater because of the larger increase in the value of $\overline{G}$ for the vapor. Thus, in general, an increase in external pressure will raise both the melting point and boiling point of a substance. Although not shown in Figure 5.3, the reverse also holds true; that is, decreasing the pressure will lower both the melting point and boiling point. Keep in mind that our conclusion about the effect of pressure on melting point is based on the assumption that the molar volume of liquid is greater than that of solid. This assumption is true of most, but not all, substances. A key exception is water. The molar volume of ice is actually greater than that of liquid water, accounting for the fact that ice floats on water. For water, then, an increase in pressure will *lower* the melting point. More will be said about this property of water later.

The Clapeyron and the Clausius-Clapeyron Equations

We shall now derive some useful, general relations for the quantitative understanding of phase equilibria. Consider a substance that exists in two phases, α and β. The condition for equilibrium at constant temperature and pressure is that

$$\overline{G}_\alpha = \overline{G}_\beta$$

so that

$$d\overline{G}_\alpha = d\overline{G}_\beta$$

To establish the relationship of dT to dP in the change that links these two phases, we have, from Equation 5.10,

$$d\overline{G}_\alpha = \overline{V}_\alpha dP - \overline{S}_\alpha dT = d\overline{G}_\beta = \overline{V}_\beta dP - \overline{S}_\beta dT$$

$$(\overline{S}_\beta - \overline{S}_\alpha)dT = (\overline{V}_\beta - \overline{V}_\alpha)dP$$

or

$$\frac{dP}{dT} = \frac{\Delta\overline{S}}{\Delta\overline{V}}$$

where $\Delta\overline{V}$ and $\Delta\overline{S}$ are the change in molar volume and molar entropies for the $\alpha \rightarrow \beta$ phase transition, respectively. Because $\Delta\overline{S} = \Delta\overline{H}/T$ at equilibrium, the above equation becomes

$$\frac{dP}{dT} = \frac{\Delta\overline{H}}{T\Delta\overline{V}} \tag{5.19}$$

where T is the phase transition temperature (it may be the melting point or the boiling point or any other temperature at which the two phases can coexist in equilibrium). Equation 5.19 is known as the Clapeyron equation (after the French engineer Benoit-Paul-Émile Clapeyron, 1799–1864), This simple expression gives us the ratio of the change in pressure to the change in temperature in terms of some readily measurable quantities such as molar volume and molar enthalpy change for the process. It applies to fusion, vaporization, and sublimation as well as to equilibria between two allotropic forms, such as graphite and diamond.

The Clapeyron equation can be expressed in a convenient approximate form for vaporization and sublimation equilibria. In these cases, the molar volume of the vapor is so much greater than that for the condensed phase, we can write

$$\Delta_{\text{vap}}\overline{V} = \overline{V}_{\text{vap}} - \overline{V}_{\text{condensed}} \approx \overline{V}_{\text{vap}}$$

Further, if we assume ideal-gas behavior, then

$$\Delta_{\text{vap}}\overline{V} = \overline{V}_{\text{vap}} = \frac{RT}{P}$$

Substitution for $\Delta_{\text{vap}}\overline{V}$ in Equation 5.19 yields

$$\frac{dP}{dT} = \frac{P\Delta_{\text{vap}}\overline{H}}{RT^2}$$

or

$$\frac{dP}{P} = d\ln P = \frac{\Delta_{\text{vap}}\overline{H}\,dT}{RT^2} \tag{5.20}$$

Equation 5.20 is known as the Clausius-Clapeyron equation (after Clapeyron and the German physicist Rudolf Julius Clausius, 1822–1888). Integrating Equation 5.20 between limits of P_1, T_1 and P_2, T_2, we obtain

$$\int_{P_1}^{P_2} d \ln P = \ln \frac{P_2}{P_1} = \frac{\Delta_{\text{vap}}\overline{H}}{R} \int_{T_1}^{T_2} \frac{dT}{T^2} = -\frac{\Delta_{\text{vap}}\overline{H}}{R}\left(\frac{1}{T_2} - \frac{1}{T_1}\right)$$

or

$$\ln \frac{P_2}{P_1} = \frac{\Delta_{\text{vap}}\overline{H}}{R} \frac{(T_2 - T_1)}{T_1 T_2} \tag{5.21}$$

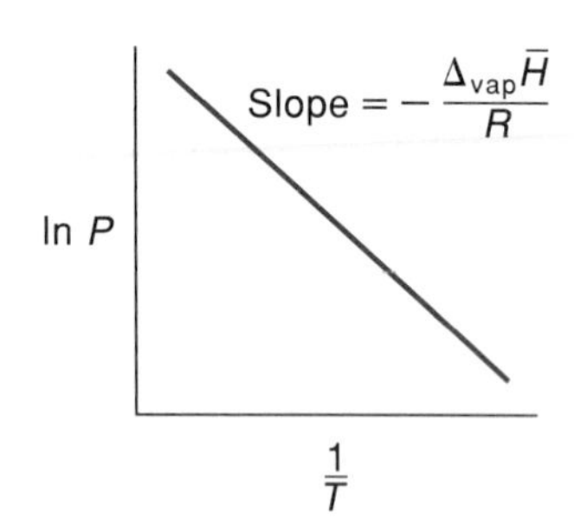

Figure 5.4
Plot of ln P versus $1/T$ to determine $\Delta_{\text{vap}}\overline{H}$ of a liquid.

We assume that $\Delta_{\text{vap}}\overline{H}$ is independent of temperature. If we had carried out an indefinite integral (integration without the limits), we could express ln P as a function of temperature as follows:

$$\ln P = -\frac{\Delta_{\text{vap}}\overline{H}}{RT} + C \tag{5.22}$$

where C is a constant. Thus, a plot of ln P versus $1/T$ gives a straight line whose slope (which is negative) is equal to $-\Delta_{\text{vap}}\overline{H}/R$ (Figure 5.4).

Phase Diagrams

At this point, we are ready to examine the phase equilibria of some familiar systems. The conditions at which a system exists as a solid, liquid, or vapor are conveniently summarized in a *phase diagram*, which is a plot of pressure versus temperature. We shall consider the phase equilibria of water and carbon dioxide.

Water. Figure 5.5 shows the phase diagram of water, where S, L, and V represent regions in which only one phase (solid, liquid, or vapor) can exist. Along any one curve, however, the two corresponding phases can coexist. The slope of any curve is given

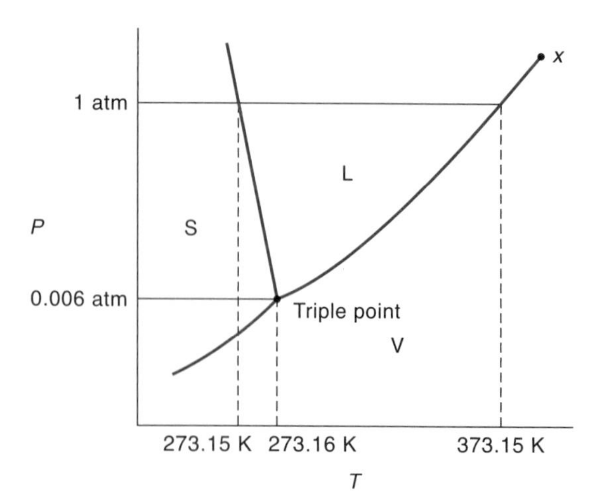

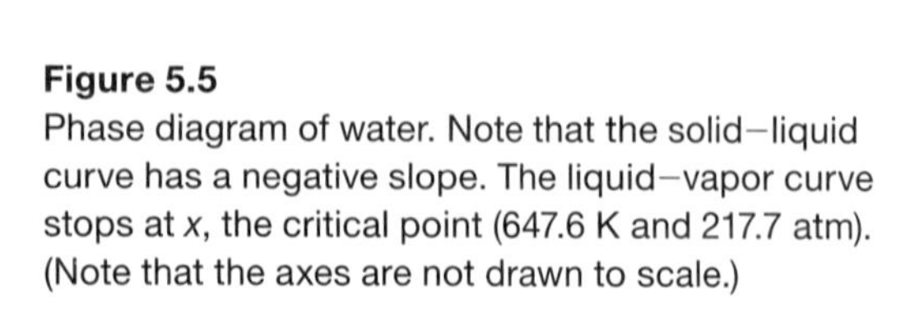
Figure 5.5
Phase diagram of water. Note that the solid–liquid curve has a negative slope. The liquid–vapor curve stops at x, the critical point (647.6 K and 217.7 atm). (Note that the axes are not drawn to scale.)

by dP/dT. For example, the curve separating regions L and V shows how the vapor pressure of water varies with respect to temperature. At 373.15 K, its vapor pressure is 1 atm, and these conditions mark the normal boiling point of water. Note that the L–V curve stops abruptly at the critical point, beyond which the liquid phase cannot exist. The normal freezing point of water (or melting point of ice) is similarly defined by the S–L curve at 1 atm, which is 273.15 K. Finally, all three phases can coexist at only one point called the triple point; for water, the triple point is at $T = 273.16$ K and $P = 0.006$ atm.

EXAMPLE 5.5

Calculate the slope of the S–L curve at 273.15 K in atm K^{-1}, given that $\Delta_{\text{fus}}\overline{H} = 6.01$ kJ mol^{-1}, $\overline{V}_{\text{L}} = 0.0180$ L mol^{-1}, and $\overline{V}_{\text{S}} = 0.0196$ L mol^{-1}.

ANSWER

We need the Clapeyron equation (Equation 5.19)

$$\frac{dP}{dT} = \frac{\Delta_{\text{fus}}\overline{H}}{T_{\text{f}}\Delta_{\text{fus}}\overline{V}}$$

Using the conversion factor 1 J = 9.87×10^{-3} L atm, we obtain

$$\frac{dP}{dT} = \frac{(6010 \text{ J mol}^{-1})(9.87 \times 10^{-3} \text{ L atm J}^{-1})}{(273.15 \text{ K})(0.0180 - 0.0196) \text{ L mol}^{-1}}$$

$$= -136 \text{ atm K}^{-1}$$

COMMENTS

(1) Because the molar volume of liquid water is smaller than that for ice, the slope is negative, as shown in Figure 5.5. Furthermore, because the quantity $(\overline{V}_{\text{L}} - \overline{V}_{\text{S}})$ is small, the slope is also quite steep. (2) An interesting result is obtained by calculating the quantity dT/dP, which gives the change (decrease) in melting point as a function of pressure. We find that $dT/dP = -7.35 \times 10^{-3}$ K atm^{-1}, which means that the melting point of ice decreases by 7.35×10^{-3} K whenever the pressure increases by 1 atm. This effect helps make ice skating possible. The weight of a skater exerts considerable pressure on the ice (of the order of 500 atm) because of the small area of the blades. As ice melts, the film of water formed between skates and ice acts as a lubricant to facilitate movement over the ice. However, more detailed studies indicate that the frictional heat generated between the skates and ice is the main reason for ice melting.

EXAMPLE 5.6

The following data show the variation of the vapor pressure of water as a function of temperature:

P/mmHg	17.54	31.82	55.32	92.51	149.38	233.7
t/°C	20	30	40	50	60	70

Determine the molar enthalpy of vaporization for water.

ANSWER

We need Equation 5.22. The first step is to convert the data into a suitable form for plotting:

$\ln P$	2.865	3.460	4.013	4.527	5.007	5.454
K/T	3.41×10^{-3}	3.30×10^{-3}	3.19×10^{-3}	3.10×10^{-3}	3.00×10^{-3}	2.92×10^{-3}

Figure 5.6 shows the plot of ln P versus $1/T$. From the measured slope, we have

$$-5090 \text{ K} = -\frac{\Delta_{\text{vap}}\overline{H}}{R}$$

or

$$\Delta_{\text{vap}}\overline{H} = (8.314 \text{ J K}^{-1} \text{ mol}^{-1})(5090 \text{ K})$$

$$= 42.3 \text{ kJ mol}^{-1}$$

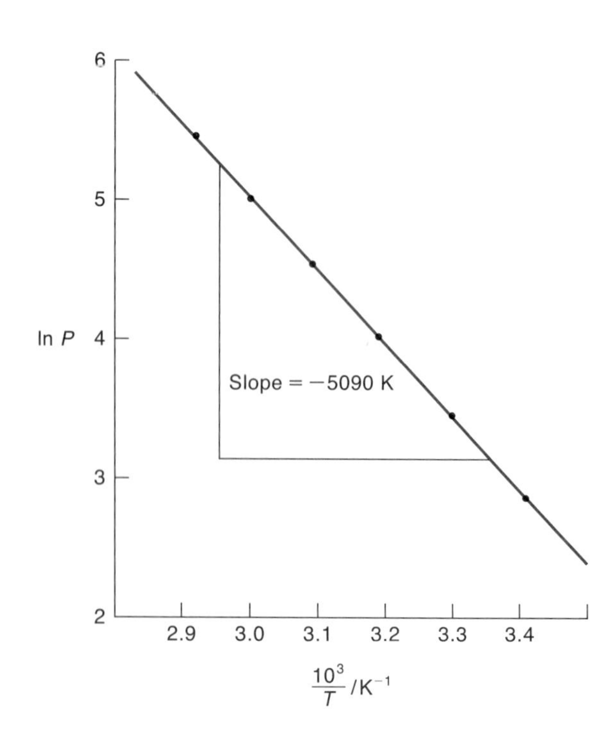

Figure 5.6
Plot of ln P versus $1/T$. The slope of the line is given by $-\Delta_{\text{vap}}\overline{H}/R$.

COMMENTS

(1) The molar heat of vaporization for water measured at its normal boiling point is 40.79 kJ mol^{-1}. Because $\Delta_{vap}\overline{H}°$ does depend on temperature to a certain extent, however, our graphically determined value is taken to be the average value between 20°C and 80°C. (2) In plotting the data, the quantity ln P is dimensionless. Note that the same slope is obtained whether we express the pressure as mmHg or as atm, for the following reason. The relationship between atm and mmHg is

$$P' = CP$$

where P' is the pressure expressed in atm, P is the pressure expressed in mmHg, and C is a conversion factor. For pressures expressed in atm, the slope of the line is given by

$$\text{slope} = \frac{(\ln P_2' - \ln P_1')}{1/T_2 - 1/T_1} = \frac{\ln CP_2 - \ln CP_1}{1/T_2 - 1/T_1}$$

$$= \frac{\ln P_2 - \ln P_1}{1/T_2 - 1/T_1}$$

Thus, whether the pressure is in atm (P_1' and P_2') or in mmHg (P_1 and P_2), the slope is the same.

Carbon dioxide. Figure 5.7 shows the phase diagram for carbon dioxide. The main difference between this diagram and that for water is that the S–L curve for CO_2 has a positive slope. This follows from the fact that because $\overline{V}_{liq} > \overline{V}_{solid}$, the quantity on the right side of Equation 5.19 is positive and therefore so is dP/dT. Note that liquid CO_2 is not stable at pressures lower than 5 atm. For this reason, solid CO_2 is called "dry

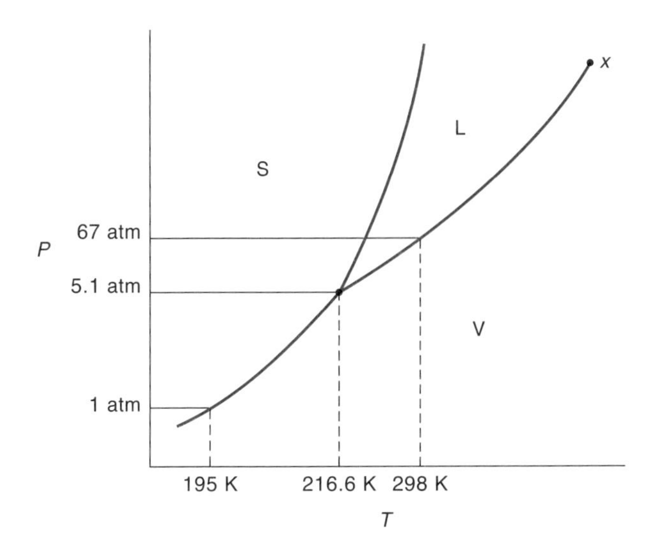

Figure 5.7
Phase diagram of carbon dioxide. Note that the solid–liquid curve has a positive slope, which is true of most substances. The liquid–vapor curve stops at x, the critical point (304.2 K and 73.0 atm). (Note that the axes are not drawn to scale.)

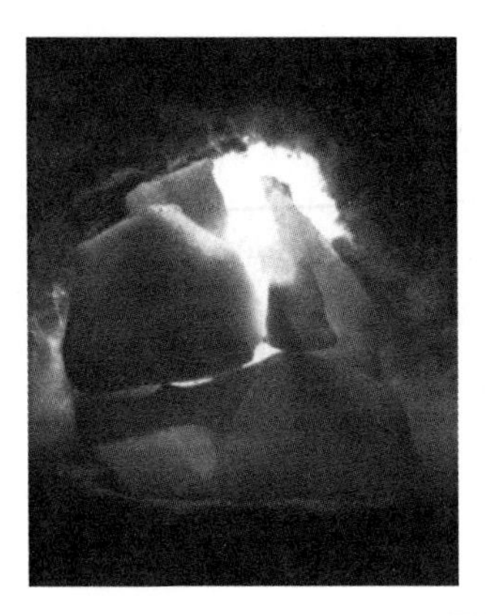

Figure 5.8
At 1 atm, solid carbon dioxide cannot melt; it can only sublime.

ice"—under atmospheric conditions it does not melt; it can only sublime. Furthermore, it looks like ice (Figure 5.8). Liquid CO_2 does exist at room temperature, but it is normally confined to a metal cylinder under a pressure of 67 atm!

The Gibbs Phase Rule

To conclude our discussion of phase equilibria, let us consider a useful rule that was derived by Gibbs (for a derivation, see Appendix 5.2 on p. 203):

$$f = c - p + 2 \tag{5.23}$$

where c is the number of components and p is the number of phases present in a system. The *degree of freedom*, f, gives the number of intensive variables (pressure, temperature, and composition) that can be changed independently without disturbing the number of phases in equilibrium. For example, in a single-component, single-phase system ($c = 1, p = 1$), say, a gas in a container, the pressure and temperature of the gas may be changed independently without changing the number of phases, so $f = 2$, or the system has two degrees of freedom.

The degree of freedom term used here has a different meaning from that used in Chapter 2 for molecular motions.

Now let us apply the phase rule to water ($c = 1$). Figure 5.5 shows that in the pure phase region (S, L, or V) we have $p = 1$ and $f = 2$, meaning that the pressure can be varied independently of temperature (two degrees of freedom). Along each of the S–L, L–V, or S–V boundaries, however, $p = 2$, and $f = 1$. Thus, for every value of P, there can be only one specific value of T and vice versa (one degree of freedom). Finally, at the triple point, $p = 3$ and $f = 0$ (no degree of freedom). Under these conditions, the system is totally fixed, and no variation of either the pressure or the temperature is possible. Such a system is said to be *invariant* and is represented by a point in a plot of pressure versus temperature.

5.6 Thermodynamics of Rubber Elasticity

In this section, we shall see an application of thermodynamic functions to a system other than gases—the familiar rubber band.

Natural rubber is poly-*cis*-isoprene and has the following repeating monomeric unit:

$$\left[-CH_2(CH_3)C{=}CH{-}CH_2- \right]_n$$

where n is in the hundreds. The characteristic property of rubber is its elasticity. It can be stretched up to 10 times its length, and, if released, will return to its original size. This behavior is due to the flexibility of rubber's long-chain molecules. In the bulk state, rubber is a tangle of polymeric chains, and if the external force is strong enough, individual chains will slip past one another, causing the rubber to lose most of its elasticity. In 1839, the American chemist Charles Goodyear (1800–1860) discovered that natural rubber could be cross-linked with sulfur to prevent chain slippage in a process called *vulcanization*. As Figure 5.9 shows, rubber in the unstretched state has many possible conformations and hence a greater entropy than the stretched state, which has relatively few conformations and a lower entropy.

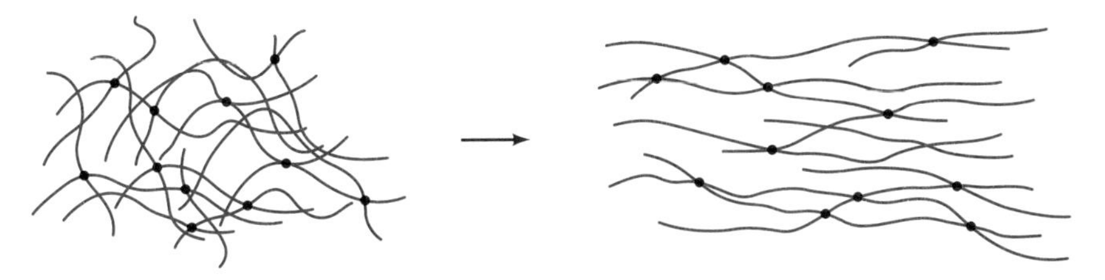

Figure 5.9
Unstretched rubber (left) has many more conformations than stretched rubber (right). The long chains of vulcanized rubber molecules are held together by sulfur linkages (represented by dots) to prevent slippage.

When a rubber band is stretched elastically by a force, f, the differential work done, dw, is given by two terms:

$$dw = f dl - P dV \tag{5.24}$$

The first term is force times the extension. The second term is small, however, and can usually be ignored. (The rubber band becomes thinner when stretched, but it also gets longer so that the change in volume, dV, is negligible.) If the rubber band is stretched slowly, the restoring force is equal to the applied force at every stage and we can therefore assume the process to be reversible. We saw earlier that for a process at constant volume and temperature, the maximum work done is equal to the change in Helmholtz energy:

$$dw_{\text{rev}} = dw_{\text{max}} = dA$$

or

$$dA = f dl$$

We can now express the restoring force in terms of the Helmholtz energy as

$$f = \left(\frac{\partial A}{\partial l}\right)_T \tag{5.25}$$

From Equation 5.4,

$$A = U - TS$$

we find the variation of A with extension l to be

$$\left(\frac{\partial A}{\partial l}\right)_T = \left(\frac{\partial U}{\partial l}\right)_T - T\left(\frac{\partial S}{\partial l}\right)_T \tag{5.26}$$

Substituting Equation 5.25 into 5.26 we obtain

$$f = \left(\frac{\partial U}{\partial l}\right)_T - T\left(\frac{\partial S}{\partial l}\right)_T \tag{5.27}$$

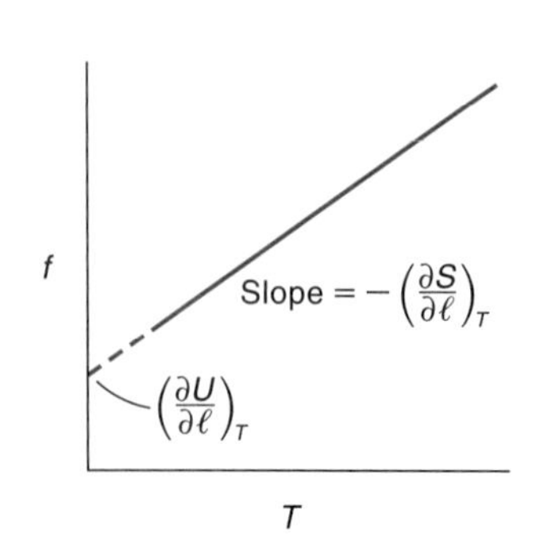

Figure 5.10
Plot of restoring force (f) in a rubber band versus temperature (T).

Equation 5.27 shows that there are two contributions to the restoring force—one from energy change with extension and the other from entropy change.*

A plot of f versus T is shown in Figure 5.10. Note that the line has a positive slope, which means that $(\partial S/\partial l)_T$ is negative. This result is consistent with the notion that the polymer molecules become more ordered in the stretched state, leading to a decrease in entropy. Experimental results also show that the $(\partial U/\partial l)_T$ term (the intercept on the y axis) is 5 to 10 times smaller than the $(\partial S/\partial l)_T$ term. The reason is that intermolecular forces between hydrocarbon molecules are relatively small so that the internal energy of the rubber band does not vary appreciably with extension. Therefore, the predominant contribution to the restoring force is entropy, not energy. When a stretched rubber band snaps back to its original position, the process is largely driven by an increase in entropy!

Finally, note the analogy between the stretching of a rubber band and the compression of a gas. If the rubber and the gas behave ideally, then

$$\left(\frac{\partial U}{\partial l}\right)_T = 0 \quad \text{(rubber)} \qquad \text{and} \qquad \left(\frac{\partial U}{\partial V}\right)_T = 0 \quad \text{(gas)}$$

Ideal behavior for rubber means that intermolecular forces are independent of conformations, whereas no intermolecular forces are present in an ideal gas. Similarly, the entropy of a rubber band decreases on stretching at constant temperature, just as the entropy of a gas decreases when it is compressed isothermally.

---

* For an experimental measurement of the restoring force of a stretched rubber band, see J. P. Byrne, *J. Chem. Educ.* **71**, 531 (1994).

Key Equations

$G = H - TS$	(Definition of Gibbs energy)	(5.1)
$dG_{sys} \leq 0$	(At constant T and P)	(5.2)
$\Delta G = \Delta H - T\Delta S$	(At constant T and P)	(5.3)
$A = U - TS$	(Definition of Helmholtz energy)	(5.4)
$dA_{sys} \leq 0$	(At constant T and V)	(5.5)
$\Delta A = \Delta U - T\Delta S$	(At constant T and V)	(5.6)
$\Delta A = w_{rev}$	(Relating ΔA to maximum work)	(5.8)
$dU = TdS - PdV$	(Combining first and second laws of thermodynamics)	(5.9)
$dG = VdP - SdT$	(Dependence of G on T and P)	(5.10)
$\Delta G = w_{el,\,max}$	(Relating ΔG to electrical work)	(5.11)
$\Delta_r G^\circ = \Sigma v \Delta_f \overline{G}^\circ(\text{products}) - \Sigma v \Delta_f \overline{G}^\circ(\text{reactants})$	(Standard Gibbs energy change of a reaction)	(5.12)
$\left[\frac{\partial\left(\frac{\Delta G}{T}\right)}{\partial T}\right]_P = -\frac{\Delta H}{T^2}$	(Gibbs-Helmholtz equation)	(5.15)
$\Delta G = nRT \ln \frac{P_2}{P_1}$	(Change in G of an ideal gas due to change in P)	(5.17)
$\overline{G} = \overline{G}^\circ + RT \ln \frac{P}{1\text{ bar}}$	(Molar Gibbs energy of a gas)	(5.18)
$\frac{dP}{dT} = \frac{\Delta \overline{H}}{T\Delta \overline{V}}$	(Clapeyron equation)	(5.19)
$\ln \frac{P_2}{P_1} = \frac{\Delta_{vap}\overline{H}}{R} \frac{(T_2 - T_1)}{T_1 T_2}$	(Clausius-Clapeyron equation)	(5.21)
$\ln P = -\frac{\Delta_{vap}\overline{H}}{RT} + C$	(Clausius-Clapeyron equation)	(5.22)
$f = c - p + 2$	(The Gibbs phase rule)	(5.23)
$f = \left(\frac{\partial U}{\partial l}\right)_T - T\left(\frac{\partial S}{\partial l}\right)_T$	(Restoring force of a stretched rubber band)	(5.27)

APPENDIX 5.1

Some Thermodynamic Relationships

In this appendix, we shall derive a number of fundamental thermodynamic relationships.

From Equations 5.9 and 5.10, we have

$$dU = -PdV + TdS \tag{1}$$

$$dG = VdP - SdT \tag{2}$$

We can obtain analogous expressions for dH and dA as follows. Starting with the definition of H ($H = U + PV$) and using Equation 1, we write

$$\begin{aligned} dH &= dU + PdV + VdP \\ &= -PdV + TdS + PdV + VdP \\ &= VdP + TdS \end{aligned} \tag{3}$$

and from the definition of A ($A = U - TS$),

$$\begin{aligned} dA &= dU - TdS - SdT \\ &= -PdV + TdS - TdS - SdT \\ &= -PdV - SdT \end{aligned} \tag{4}$$

From Equation 1, we see that U is a function of V and S, so we can write the exact differential dU as (see Appendix 3.1 on p. 116)

$$dU = \left(\frac{\partial U}{\partial V}\right)_S dV + \left(\frac{\partial U}{\partial S}\right)_V dS \tag{5}$$

Comparing the coefficients for dV and dS in Equations 1 and 5, we arrive at the results,

$$\left(\frac{\partial U}{\partial V}\right)_S = -P \tag{6}$$

$$\left(\frac{\partial U}{\partial S}\right)_V = T \tag{7}$$

By a similar procedure, we proceed as follows for dH:

$$dH = \left(\frac{\partial H}{\partial P}\right)_S dP + \left(\frac{\partial H}{\partial S}\right)_P dS \tag{8}$$

Comparing the coefficients for dP and dS in Equations 3 and 8, we get

$$\left(\frac{\partial H}{\partial P}\right)_S = V \tag{9}$$

$$\left(\frac{\partial H}{\partial S}\right)_P = T \tag{10}$$

From Equation 4,

$$dA = \left(\frac{\partial A}{\partial V}\right)_T dV + \left(\frac{\partial A}{\partial T}\right)_V dT \tag{11}$$

Comparison of the coefficients for dV and dT in Equations 4 and 11 gives

$$\left(\frac{\partial A}{\partial V}\right)_T = -P \tag{12}$$

$$\left(\frac{\partial A}{\partial T}\right)_V = -S \tag{13}$$

From Equation 2,

$$dG = \left(\frac{\partial G}{\partial P}\right)_T dP + \left(\frac{\partial G}{\partial T}\right)_P dT \tag{14}$$

Comparing the coefficients for dP and dT in Equations 2 and 14, we obtain

$$\left(\frac{\partial G}{\partial P}\right)_T = V \tag{15}$$

$$\left(\frac{\partial G}{\partial T}\right)_P = -S \tag{16}$$

From Equations 6, 9, 12, 13, 15, and 16, we see how the four thermodynamic functions (U, H, A, and G) vary with pressure, volume, and temperature under different conditions.

The Maxwell Relations

In the 1870s, Maxwell derived a set of equations that relate S to P, V, and T. Starting with Equation 1 and realizing that dU is an exact differential, we apply Euler's theorem (see Appendix 3.1) as follows:

$$-\left(\frac{\partial P}{\partial S}\right)_V = \left(\frac{\partial T}{\partial V}\right)_S \tag{17}$$

Similarly, from Equation 3,

$$\left(\frac{\partial V}{\partial S}\right)_P = \left(\frac{\partial T}{\partial P}\right)_S \tag{18}$$

from Equation 4,

$$\left(\frac{\partial P}{\partial T}\right)_V = \left(\frac{\partial S}{\partial V}\right)_T \tag{19}$$

and from Equation 2,

$$\left(\frac{\partial V}{\partial T}\right)_P = -\left(\frac{\partial S}{\partial P}\right)_T \tag{20}$$

The usefulness of the equations derived above is that they provide relations among the variables that may not appear obvious to us and that they enable us to study the dependence of thermodynamic functions on P, V, and T. For example, $(\partial S/\partial V)_T$ is not an easy quantity to measure. According to Equation 19, however, we see that this quantity is equal to $(\partial P/\partial T)_V$, which is the change in pressure of a system with temperature at constant volume, a much easier quantity to determine experimentally. Some of these relationships also enable us to derive the *thermodynamic equations of state*, which is an equation that describes the dependence of the internal energy of a system with volume at constant temperature (see Problem 5.41).

APPENDIX 5.2

Derivation of the Gibbs Phase Rule

Before we derive the Gibbs phase rule, it is useful to first define the term *component* (c). The number of components of a system is the minimum number of composition variables necessary to describe all possible variations in the composition of the system. It is represented by

$$c = s - r \tag{1}$$

where s is the total number of constituents (atoms, molecules, or ions) present, and r is the number of restrictive conditions, which is the number of algebraic relationships among the composition variables. The following examples help clarify Equation 1.

Example 1 Pure water. The number of constituents is one ($s = 1$), and because there is no reaction (ignoring water autodissociation), $r = 0$ so that $c = 1$, and we have a one-component system.

Example 2 A mixture of ethanol and water. Here, we have two constituents ($s = 2$). There is no reaction between ethanol and water so $r = 0$ and $c = 2$, giving us a two-component system.

Example 3 Ammonium chloride in equilibrium with ammonia and hydrogen chloride. We assume that, initially, only NH_4Cl was present.

$$NH_4Cl(s) \rightleftharpoons NH_3(g) + HCl(g)$$

This system has three constituents ($s = 3$) and two algebraic relationships among the constituents: the equilibrium constant

$$K = [NH_3][HCl]$$

and

$$[NH_3] = [HCl]$$

Therefore, $r = 2$, and we have

$$c = 3 - 2 = 1$$

or a one-component system.

The Gibbs Phase Rule

Suppose we have a system containing c components in p phases at equilibrium. How many concentration terms do we need to know to define the system completely in any one phase? If only one component is present ($c = 1$), no concentration terms are needed. For two or more components, we need know only $(c - 1)$ concentration terms. For example, if c = 2, then knowing the mole fraction of only one component will define both concentration terms. Thus, $p(c - 1)$ terms are required to define the concentration terms completely in p phases. Keeping in mind that pressure and temperature are two additional variables at our disposal, we write

$$\text{total number of independent intensive variables} = p(c - 1) + 2 \tag{2}$$

Because the molar Gibbs energy of a component is the same in all the phases (α, β, γ, ...) at equilibrium, the number of independent equations for this condition for each component is thus $(p - 1)$.* For example, if one component is present in two phases, the condition for equilibrium is

$$\overline{G}_\alpha = \overline{G}_\beta \quad \text{(one equation)}$$

and for one component in three phases,

$$\overline{G}_\alpha = \overline{G}_\beta$$

and

$$\overline{G}_\beta = \overline{G}_\gamma \quad \text{(two equations)}$$

(Note that $\overline{G}_\alpha = \overline{G}_\gamma$ is not an independent equation.) Each equation reduces our freedom to vary the total variables in Equation 2 by one. For c components, $c(p - 1)$ equations must be satisfied. Finally, the difference between the total number of intensive variables and total number of independent equations gives us the number of variables that can be varied independently without changing the number of phases present in equilibrium, or the degree of freedom (f):

$$\begin{aligned} f &= [p(c - 1) + 2] - c(p - 1) \\ &= c - p + 2 \end{aligned} \tag{3}$$

which is Equation 5.23.

* The proper derivation of the phase rule requires the use of partial molar Gibbs energy or chemical potential (see Chapter 6). Our result here applies only to a one-component system because we use molar Gibbs energy, but it can be generalized for a multicomponent system.

Suggestions for Further Reading

BOOKS

Bent, H. A., *The Second Law*, Oxford University Press, New York, 1965.

Hanson, R. M., and S. Green, *Introduction to Molecular Thermodynamics*, University Science Books, Sausalito, CA, 2008.

Klotz, I. M., and R. M. Rosenberg, *Chemical Thermodynamics: Basic Theory and Methods*, 5th ed., John Wiley and Sons, New York, 1994.

McQuarrie, D. A., and J. Simon, *Molecular Thermodynamics*, University Science Books, Sausalito, CA, 1999.

Rock, P. A., *Chemical Thermodynamics*, University Science Books, Sausalito, CA, 1983.

ARTICLES

General

"The Synthesis of Diamond," H. Hall, *J. Chem. Educ.* **38**, 484 (1961).

"The Second Law of Thermodynamics," H. A. Bent, *J. Chem. Educ.* **39**, 491 (1962).

"The Use and Misuse of the Laws of Thermodynamics," M. L. McGlashan, *J. Chem. Educ.* **43**, 226 (1966).

"The Scope and Limitations of Thermodynamics," K. G. Denbigh, *Chem. Brit.* **4**, 339 (1968).

"Thermodynamics of Hard Molecules," L. K. Runnels, *J. Chem. Educ.* **47**, 742 (1970).

"The Thermodynamic Transformation of Organic Chemistry," D. E. Stull, *Am. Sci.* **54**, 734 (1971).

"Introduction to the Thermodynamics of Biopolymer Growth," C. Kittel, *Am. J. Phys.* **40**, 60 (1972).

"High Pressure Synthetic Chemistry," A. P. Hagen, *J. Chem. Educ.* **55**, 620 (1978).

"Reversibility and Returnability," J. A. Campbell, *J. Chem. Educ.* **57**, 345 (1980).

"Student Misconceptions in Thermodynamics," M. F. Granville, *J. Chem. Educ.* **62**, 847 (1985).

"The True Meaning of Isothermal," D. Fain, *J. Chem Educ.* **65**, 187 (1988).

"The Conversion of Chemical Energy," D. J. Wink, *J. Chem. Educ.* **69**, 109 (1992).

"The Thermodynamics of Home-Made Ice Cream," D. L. Gibbon, K. Kennedy, N. Reading, and M. Quierox, *J. Chem. Educ.* **69**, 658 (1992).

"Spontaneity, Accessibility, Irreversibility, 'Useful Work': The Availability Function, the Helmholtz Function, and the Gibbs Function," R. J. Tykodi, *J. Chem. Educ.* **72**, 103 (1995).

"The Gibbs Function Controversy," S. E. Wood and R. Battino, *J. Chem. Educ.* **73**, 408 (1996).

"How Thermodynamic Data and Equilibrium Constants Changed When the Standard-State Pressure Became 1 Bar," R. S. Treptow, *J. Chem. Educ.* **76**, 212 (1999).

"Revitalizing Modern Thermodynamics Instruction with an Old Technique: A State Functions Table," P. A. Molina, *Chem. Educator* [Online] **12**, 137 (2007) DOI 10.1333/s00897072026a.

Phase Equilibria

"The Triple Point of Water," F. L. Swinton, *J. Chem. Educ.* **44**, 541 (1967).

"Subtleties of Phenomena Involving Ice-Water Equilibria," L. F. Loucks, *J. Chem. Educ.* **63**, 115 (1986). Also see *J. Chem. Educ.* **65**, 186 (1988).

"Reappearing Phases," J. Walker and C. A. Vanse, *Sci. Am.* May 1987.

"Supercritical Phase Transitions at Very High Pressure," K. M. Scholsky, *J. Chem. Educ.* **66**, 989 (1989).
"The Direct Relation Between Altitude and Boiling Point," B. L. Earl, *J. Chem. Educ.* **67**, 45 (1990).
"Ice Under Pressure," R. Chang and J. F. Skinner, *J. Chem. Educ.* **67**, 789 (1990).
"The Critical Point and the Number of Degrees of Freedom," R. Battino, *J. Chem. Educ.* **68**, 276 (1991).
"Phase Diagrams of One-Component Systems," G. D. Peckham and I. J. McNaught, *J. Chem. Educ.* **70**, 560 (1993).
"Description of Regions in Two-Component Phase Diagrams," R. M. Rosenberg, *J. Chem. Educ.* **76**, 223 (1999).
"The Gibbs Phase Rule Revisited: Interrelationships between Components and Phases," J. S. Alper, *J. Chem. Educ.* **76**, 1567 (1999).
"Melting Below Zero," J. S. Wellaufer and J. G. Dash, *Sci. Am.* February 2000.
"On the Clausius-Clapeyron Vapor Pressure Equation," S. Velasco, F. L. Román, and J. A. White, *J. Chem. Educ.* **86**, 106 (2009).
"Nanotechnology Provides a New Perspective on Chemical Thermodynamics," R. G. Haverkamp, *J. Chem. Educ.* **86**, 50 (2009).
"Complexities of One-Component Phase Diagrams" A. Ciccioli and L. Glasser, *J. Chem. Educ.* **88**, 586 (2011).

Problems

ΔG and ΔA

5.1 A quantity of 0.35 mole of an ideal gas initially at 15.6°C is expanded from 1.2 L to 7.4 L. Calculate the values of w, q, ΔU, ΔS, and ΔG if the process is carried out **(a)** isothermally and reversibly, and **(b)** isothermally and irreversibly against an external pressure of 1.0 atm.

5.2 At one time, the domestic gas used for cooking, called "water gas," was prepared as follows:

$$H_2O(g) + C(\text{graphite}) \rightarrow CO(g) + H_2(g)$$

From the thermodynamic quantities listed in Appendix B, predict whether this reaction will occur at 298 K. If not, at what temperature will the reaction occur? Assume $\Delta_r H°$ and $\Delta_r S°$ are temperature independent.

5.3 Use the values listed in Appendix B to calculate the value of $\Delta_r G°$ for the following alcohol fermentation:

$$\text{glucose}(aq) \rightarrow 2C_2H_5OH(l) + 2CO_2(g)$$

$(\Delta_f \overline{G}°[\text{glucose}(aq)] = -914.5 \text{ kJ mol}^{-1})$

5.4 Without referring to Appendix B, calculate the quantity $(\Delta_r G° - \Delta_r A°)$ for the following reaction at 298 K:

$$C(s) + CO_2(g) \rightarrow 2CO(g)$$

Assume ideal-gas behavior.

5.5 As an approximation, we can assume that proteins exist either in the native (or physiologically functioning) state and the denatured state. The standard molar enthalpy and entropy of the denaturation of a certain protein are 512 kJ mol^{-1} and 1.60 kJ K^{-1} mol^{-1}, respectively. Comment on the signs and magnitudes of these quantities, and calculate the temperature at which the denaturation becomes spontaneous.

5.6 Certain bacteria in the soil obtain the necessary energy for growth by oxidizing nitrite to nitrate:

$$2NO_2^-(aq) + O_2(g) \rightarrow 2NO_3^-(aq)$$

Given that the standard Gibbs energies of formation of NO_2^- and NO_3^- are -34.6 kJ mol^{-1} and -110.5 kJ mol^{-1}, respectively, calculate the amount of Gibbs energy released when 1 mole of NO_2^- is oxidized to 1 mole of NO_3^-.

5.7 Consider the synthesis of urea according to the equation

$$CO_2(g) + 2NH_3(g) \rightarrow (NH_2)_2CO(s) + H_2O(l)$$

From the data listed in Appendix B, calculate the value of $\Delta_r G°$ for the reaction at 298 K. Assuming ideal gas behavior, calculate the value of $\Delta_r G$ for the reaction at a pressure of 10.0 bar. The $\Delta_f \overline{G}°$ of urea is -197.15 kJ mol^{-1}.

5.8 This problem involves the synthesis of diamond from graphite:

$$\text{C(graphite)} \rightarrow \text{C(diamond)}$$

(a) Calculate the values of $\Delta_r H°$ and $\Delta_r S°$ for the reaction. Will the conversion occur spontaneously at 25°C or any other temperature? **(b)** From density measurements, the molar volume of graphite is found to be 2.1 cm^3 greater than that of diamond. Can the conversion of graphite to diamond be brought about at 25°C by applying pressure on graphite? If so, estimate the pressure at which the process becomes spontaneous. [*Hint:* Starting from Equation 5.16, derive the equation $\Delta G = (\overline{V}_{\text{diamond}} - \overline{V}_{\text{graphite}})\Delta P$ for a constant-temperature process. Next, calculate the ΔP value that would lead to the necessary decrease in Gibbs energy.]

5.9 How do the requirements that T and V are constant enter the derivation for $\Delta A_{\text{sys}} < 0$ for a spontaneous process?

5.10 From the standard molar enthalpy of combustion of benzene at 298 K, calculate the value of $\Delta_r A°$ for the process. Compare the value of $\Delta_r A°$ with that of $\Delta_r H°$. Comment on the difference.

5.11 A student placed 1 g of each of three compounds A, B, and C in a container and found that no change had occurred after one week. Offer possible explanations for the lack of reaction. Assume that A, B, and C are totally miscible liquids.

5.12 Predict the signs of ΔH, ΔS, and ΔG of the system for the following processes at 1 atm: **(a)** ammonia melts at −60°C, **(b)** ammonia melts at −77.7°C, **(c)** ammonia melts at −100°C. (The normal melting point of ammonia is −77.7°C.)

5.13 Crystallization of sodium acetate from a supersaturated solution occurs spontaneously. What can you deduce about the signs of ΔS and ΔH?

5.14 A student looked up the $\Delta_f \overline{G}°$, $\Delta_f \overline{H}°$, and $\overline{S}°$ values for CO_2 in Appendix B. Plugging these values into Equation 5.3, he found that $\Delta_f \overline{G}° \neq \Delta_f \overline{H}° - T\overline{S}°$ at 298 K. What is wrong with his approach?

5.15 A certain reaction is spontaneous at 72°C. If the enthalpy change for the reaction is 19 kJ, what is the *minimum* value of $\Delta_r S$ (in joules per kelvin) for the reaction?

5.16 A certain reaction is known to have a $\Delta_r G°$ value of −122 kJ. Will the reaction necessarily occur if the reactants are mixed together?

Phase Equilibria

5.17 The vapor pressure of mercury at various temperatures has been determined as follows:

T/K	P/mmHg
323	0.0127
353	0.0888
393.5	0.7457
413	1.845
433	4.189

Calculate the value of $\Delta_{\text{vap}}\overline{H}$ for mercury.

5.18 The pressure exerted on ice by a 60.0-kg skater is about 300 atm. Calculate the depression in freezing point. The molar volumes are: $\overline{V}_L = 0.0180 \text{ L mol}^{-1}$ and $\overline{V}_S = 0.0196 \text{ L mol}^{-1}$.

5.19 Use the phase diagram of water (Figure 5.5) to predict the direction for the following changes: **(a)** at the triple point of water, temperature is lowered at constant pressure, and **(b)** somewhere along the S–L curve of water, pressure is increased at constant temperature.

5.20 Use the phase diagram of water (Figure 5.5) to predict the dependence of the freezing and boiling points of water on pressure.

5.21 Consider the following system at equilibrium

$$CaCO_3(s) \rightleftharpoons CaO(s) + CO_2(g)$$

How many phases are present?

5.22 Below is a rough sketch of the phase diagram of carbon. **(a)** How many triple points are there, and what are the phases that can coexist at each triple point? **(b)** Which has a higher density, graphite or diamond? **(c)** Synthetic diamond can be made from graphite. Using the phase diagram, how would you go about making diamond?

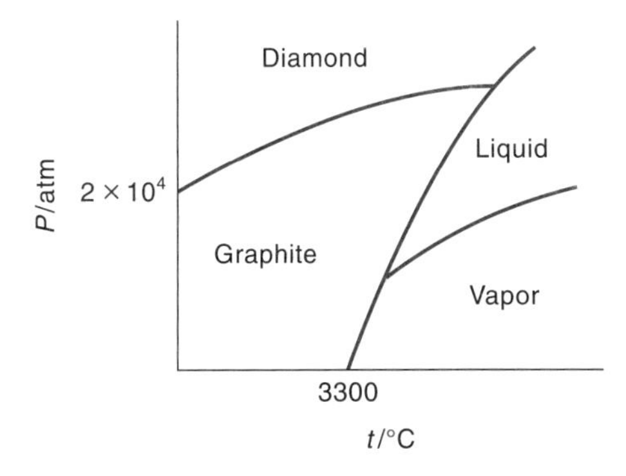

5.23 What is wrong with the following phase diagram for a one-component system?

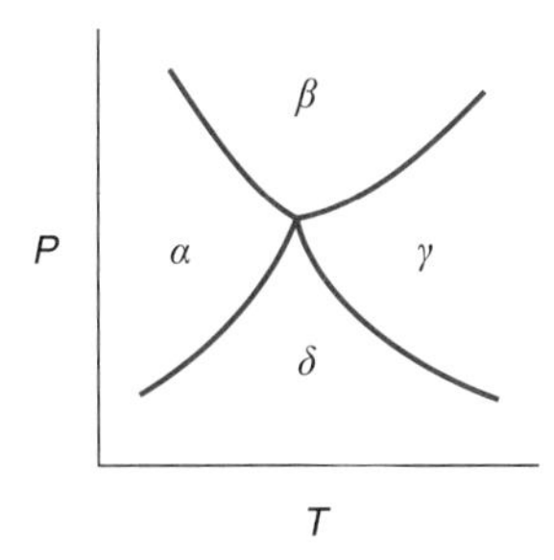

5.24 The plot in Figure 5.4 is no longer linear at high temperatures. Explain.

5.25 Pike's Peak in Colorado is approximately 4300 m above sea level (10°C). What is the boiling point of water at the summit? (*Hint:* See Equation 2.24. The molar mass of air is 29.0 g mol^{-1}, and $\Delta_{vap}\overline{H}$ for water is 40.79 kJ mol^{-1}.)

5.26 The normal boiling point of ethanol is 78.3°C, and its molar enthalpy of vaporization is 39.3 kJ mol^{-1}. What is its vapor pressure at 30°C?

5.27 Calculate the number of components present in each of the following situations:

(a) Water, including autodissociation into H^+ and OH^- ions.

(b) Consider the following reaction in a closed container:

$$2NH_3(g) \rightleftharpoons N_2(g) + 3H_2(g)$$

(i) All three gases were present initially in arbitrary amounts, but the temperature is too low for the reaction to occur.

(ii) Same as **(i)**, but the temperature is raised sufficiently to allow for the equilibrium to be established.

(iii) Initially only NH_3 was present. The system is then allowed to reach equilibrium.

Additional Problems

5.28 Give the conditions under which each of the following equations may be applied.

(a) $dA \leq 0$ (for equilibrium and spontaneity)

(b) $dG \leq 0$ (for equilibrium and spontaneity)

(c) $\ln \frac{P_2}{P_1} = \frac{\Delta \bar{H}}{R} \frac{(T_2 - T_1)}{T_1 T_2}$

(d) $\Delta G = nRT \ln \frac{P_2}{P_1}$

5.29 When ammonium nitrate is dissolved in water, the solution becomes colder. What conclusion can you draw about $\Delta S°$ for the process?

5.30 Protein molecules are polypeptide chains made up of amino acids. In their physiologically functioning or native state, these chains fold in a unique manner such that the nonpolar groups of the amino acids are usually buried in the interior region of the proteins, where there is little or no contact with water. When a protein denatures, the chain unfolds so that these nonpolar groups are exposed to water. A useful estimate of the changes of the thermodynamic quantities as a result of denaturation is to consider the transfer of a hydrocarbon such as methane (a nonpolar substance) from an inert solvent (such as benzene or carbon tetrachloride) to the aqueous environment:

(a) $CH_4(\text{inert solvent}) \rightarrow CH_4(g)$

(b) $CH_4(g) \rightarrow CH_4(aq)$

If the values of $\Delta H°$ and $\Delta G°$ are approximately 2.0 kJ mol^{-1} and -14.5 kJ mol^{-1}, respectively, for **(a)** and -13.5 kJ mol^{-1} and 26.5 kJ mol^{-1}, respectively, for **(b)**, then calculate the values of $\Delta H°$ and $\Delta G°$ for the transfer of 1 mole of CH_4 according to the equation

$$CH_4(\text{inert solvent}) \rightarrow CH_4(aq)$$

Comment on your results. Assume $T = 298$ K.

5.31 Find a rubber band that is about 0.5 cm wide. Quickly stretch the rubber band and then press it against your lips. You will feel a slight warming effect. Next, reverse the process. Stretch a rubber band and hold it in position for a few seconds. Then quickly release the tension and press the rubber band against your lips. This time you will feel a slight cooling effect. Use Equation 5.3 to present a thermodynamic analysis of this behavior.

5.32 A rubber band is stretched vertically by tying a weight to one end and attaching the other end to a ring stand. When heated with a hot-air blower, the rubber band shrinks slightly in length. Account for this observation.

5.33 Hydrogenation reactions are facilitated by the use of a transition metal catalyst, such as Ni or Pt. Predict the signs of $\Delta_r H$, $\Delta_r S$, and $\Delta_r G$ when hydrogen gas is adsorbed onto the surface of nickel metal.

5.34 A sample of supercooled water freezes at −10°C. What are the signs of ΔH, ΔS, and ΔG for this process? All the changes refer to the system.

5.35 The boiling point of benzene is 80.1°C. Estimate **(a)** its $\Delta_{vap}\overline{H}$ value and **(b)** its vapor pressure at 74°C. (*Hint:* Use Trouton's rule on p. 147.)

5.36 A chemist has synthesized a hydrocarbon compound (C_xH_y). Briefly describe what measurements are needed to determine the values of $\Delta_f\overline{H}^\circ$, $\overline{S}^\circ$, and $\Delta_f\overline{G}^\circ$ of the compound.

5.37 Calculate ΔA and ΔG for the vaporization of 2.00 moles of water at 100°C and 1.00 atm. The molar volume of $H_2O(l)$ at 100°C is 0.0188 L mol^{-1}. Assume ideal gas behavior.

5.38 A person heated water in a cup in a microwave oven for tea. After removing the cup from the oven, she added a tea bag to the hot water. To her surprise, the water started to boil violently. Explain what happened.

5.39 Consider the reversible isothermal compression of 0.45 mole of helium gas from 0.50 atm and 22 L to 1.0 atm at 25°C. **(a)** Calculate the values of w, ΔU, ΔH, ΔS, and ΔG for the process. **(b)** Can you use the sign of ΔG to predict whether the process is spontaneous? Explain. **(c)** What is the maximum work that can be done for the compression process? Assume ideal-gas behavior.

5.40 The molar entropy of argon (Ar) is given by

$$\overline{S}^\circ = (36.4 + 20.8 \ln T/\mathrm{K})\ \mathrm{J\ K^{-1}\ mol^{-1}}$$

Calculate the change in Gibbs energy when 1.0 mole of Ar is heated at constant pressure from 20°C to 60°C. (*Hint:* Use the relation $\int \ln x\, dx = x \ln x - x$.)

5.41 Derive the thermodynamic equation of state:

$$\left(\frac{\partial U}{\partial V}\right)_T = -P + T\left(\frac{\partial P}{\partial T}\right)_V$$

Apply the equation to **(a)** an ideal gas and **(b)** a van der Waals gas. Comment on your results. (*Hint:* See Appendix 5.1 for thermodynamic relationships.)

CHAPTER 6

Nonelectrolyte Solutions

If a solution of iodine in benzene is cooled, the red color deepens, while if it is warmed, the color approaches the violet of iodine vapor, indicating that the solvation decreases with rising temperature, as would be expected.

—J. H. Hildebrand and C. A. Jenks*

The study of solutions is of great importance because many interesting and useful chemical and biological processes occur in liquid solutions. Generally, a solution is defined as a homogeneous mixture of two or more components that form a single phase. Most solutions are liquids, although gas solutions (e.g., air) and solid solutions (e.g., solder) also exist. This chapter is devoted to the thermodynamic study of ideal and nonideal solutions of nonelectrolytes—solutions that do not contain ionic species—and the colligative properties of these solutions.

6.1 Concentration Units

Any quantitative study of solutions requires that we know the amount of solute dissolved in a solvent or the concentration of the solution. Chemists employ several different concentration units, each one having advantages and limitations. The use of the solution generally determines how we express its concentration. In this section, we shall define four concentration units: percent by weight, mole fraction, molarity, and molality.

Percent by Weight

The percent by weight (also called percent by mass) of a solute in a solution is defined as

$$\text{percent by weight} = \frac{\text{weight of solute}}{\text{weight of solute} + \text{weight of solvent}} \times 100\%$$

$$= \frac{\text{weight of solute}}{\text{weight of solution}} \times 100\% \qquad (6.1)$$

* J. H. Hildebrand and C. A. Jenks, *J. Am. Chem. Soc.* **42**, 2180 (1920).

Mole Fraction (x)

The concept of mole fraction was introduced in Section 1.7. We define the mole fraction of a component i of a solution, x_i, as

$$x_i = \frac{\text{number of moles of component } i}{\text{number of moles of all components}}$$

$$= \frac{n_i}{\sum_i n_i} \tag{6.2}$$

The mole fraction has no units.

Molarity (M)

Molarity is defined as the number of moles of solute dissolved in 1 liter of solution, that is,

$$\text{molarity} = \frac{\text{number of moles of solute}}{\text{liters of solution}} \tag{6.3}$$

Thus, molarity has the units moles per liter (mol L^{-1}). By convention, we use square brackets [ ] to represent molarity.

Molality (m)

Molality is defined as the number of moles of solute dissolved in 1 kg (1000 g) of solvent, that is,

$$\text{molality} = \frac{\text{number of moles of solute}}{\text{mass of solvent in kg}} \tag{6.4}$$

Thus, molality has the units of moles per kg of solvent (mol kg^{-1}).

We shall now compare the usefulness of these four concentration terms. Percent by weight has the advantage that we do not need to know the molar mass of the solute. This unit is useful to biochemists, who frequently work with macromolecules either of unknown molar mass or of unknown purity. (A common unit for protein and DNA solutions is mg mL^{-1}, or mg per milliliter.) Furthermore, the percent by weight of a solute in a solution is independent of temperature, because it is defined in terms of weight. Mole fractions are useful for calculating partial pressures of gases (see Section 1.7) and in the study of vapor pressures of solutions (to be introduced later). Molarity is one of the most commonly employed concentration units. The advantage of using molarity is that it is generally easier to measure the volume of a solution using precisely calibrated volumetric flasks than to weigh the solvent. Its main drawback is that it is temperature dependent, because the volume of a solution usually increases with increasing temperature. Another drawback is that molarity does not tell us the amount of solvent present. Molality, on the other hand, is temperature independent because it is a ratio of the number of moles of solute to the weight of the solvent. For this reason, molality is the preferred concentration unit in studies that involve changes in temperature, as in some of the colligative properties of solutions (see Section 6.7).

6.2 Partial Molar Quantities

The extensive properties of a one-component system at a constant temperature and pressure depend only on the amount of the system present. For example, the volume of water depends on the quantity of water present. If the volume is expressed as a molar quantity, however, it is an intensive property. Thus, the molar volume of water at 1 atm and 298 K is 0.018 L mol^{-1}, no matter how little or how much water is present. For solutions, the criteria are different. A solution, by definition, contains at least two components. The extensive properties of a solution depend on temperature, pressure, and the composition of the solution. In discussing the properties of any solution, we cannot employ molar quantities; instead, we must use *partial molar quantities.* Perhaps the easiest partial molar quantity to understand is *partial molar volume*, described below.

Partial Molar Volume

The molar volumes of water and ethanol at 298 K are 0.018 L and 0.058 L, respectively. If we mix half a mole of each liquid, we might expect the combined volume to be the sum of 0.018 L/2 and 0.058 L/2, or 0.038 L. Instead, we find the volume to be only 0.036 L. The shrinkage of the volume is the result of unequal intermolecular interaction between unlike molecules. Because the forces of attraction between water and ethanol molecules are greater than those between water molecules and between ethanol molecules, the total volume is less than the sum of the individual volumes. If the intermolecular forces are weaker, then expansion will occur and the final volume will be greater than the sum of individual volumes. Only if the interactions between like and unlike molecules are the same will volume be additive. If the final volume is equal to the sum of the separate volumes, the solution is called an *ideal* solution. Figure 6.1 shows the total volume of a water–ethanol solution as a function of their mole fractions. In a real (nonideal) solution, the partial molar volume of each component is affected by the presence of the other components.

Financially, this shrinkage in volume has a detrimental effect on bartenders.

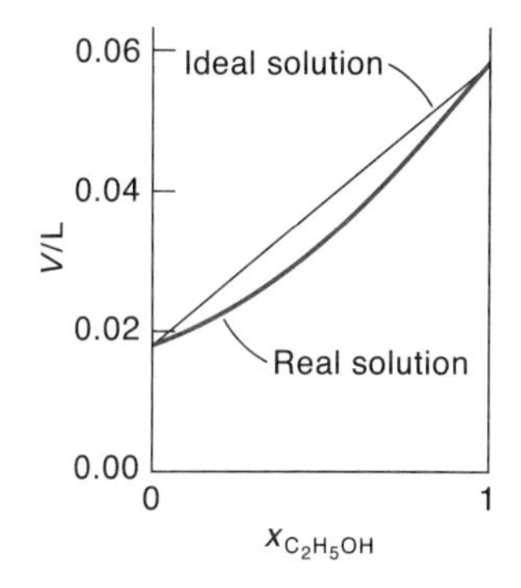

Figure 6.1
Total volume of a water–ethanol mixture as a function of the mole fraction of ethanol. At any concentration, the sum of the number of moles is 1. The straight line represents the variation of volume with mole fraction for an ideal solution. The curve represents the actual variation. Note that at $x_{C_2H_5OH} = 0$, the volume corresponds to that of the molar volume of water, and at $x_{C_2H_5OH} = 1$, V is the molar volume of ethanol.

At constant temperature and pressure, the volume of a solution is a function of the number of moles of different substances present; that is,

$$V = V(n_1, n_2, \ldots)$$

For a two-component system the total differential, dV, is given by

$$dV = \left(\frac{\partial V}{\partial n_1}\right)_{T,P,n_2} dn_1 + \left(\frac{\partial V}{\partial n_2}\right)_{T,P,n_1} dn_2$$

$$= \overline{V}_1 dn_1 + \overline{V}_2 dn_2 \tag{6.5}$$

where $\overline{V}_1$ and $\overline{V}_2$ are the partial molar volumes of components 1 and 2. The partial molar volume $\overline{V}_1$, for example, tells us the rate of change in volume with number of moles of component 1, at constant T, P, and component 2. Alternatively, $\overline{V}_1$ can be viewed as the increase in volume resulting from the addition of 1 mole of component 1 to a very large volume of solution so that its concentration remains unchanged. The quantity $\overline{V}_2$ can be similarly interpreted. Equation 6.5 can be integrated to give

$$V = n_1\overline{V}_1 + n_2\overline{V}_2 \tag{6.6}$$

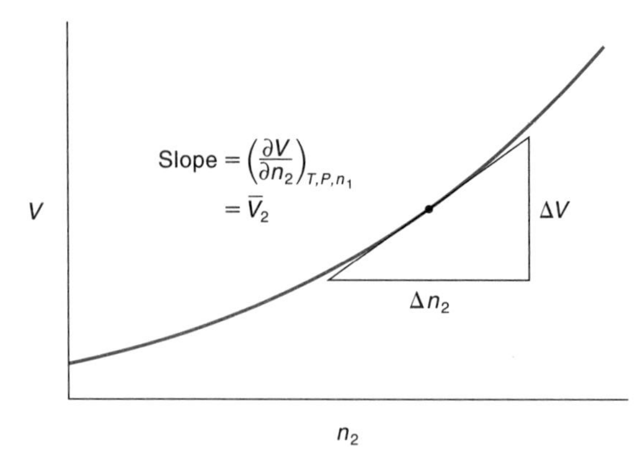

Figure 6.2
Determination of partial molar volume. The volume of a two-component solution is measured as a function of the number of moles, n_2, of component 2. The slope at a particular value n_2 gives the partial molar volume, $\overline{V}_2$, at that concentration while holding temperature, pressure, and number of moles of component 1 constant.

This equation enables us to calculate the volume of the solution by summing the products of the number of moles and the partial molar volume of each component (see Problem 6.55).

Figure 6.2 suggests a way of measuring partial molar volumes. Consider a solution composed of substances 1 and 2. To measure $\overline{V}_2$, we prepare a series of solutions at certain T and P, all of which contain a fixed number of moles of component 1 (i.e., n_1 is fixed) but different amounts of n_2. When we plot the measured volumes, V, of the solutions against n_2, the slope of the curve at a particular composition of 2 gives $\overline{V}_2$ for that composition. Once $\overline{V}_2$ has been measured, $\overline{V}_1$ at the same composition can be calculated using Equation 6.6:

$$\overline{V}_1 = \frac{V - n_2\overline{V}_2}{n_1}$$

Figure 6.3 shows the partial molar volumes of ethanol and water in an ethanol–water solution. Note that whenever the partial molar volume of one component rises, that of the other component falls. This relationship is a characteristic of *all* partial molar quantities.

Partial Molar Gibbs Energy

Partial molar quantities permit us to express the total extensive properties, such as volume, energy, enthalpy, and Gibbs energy, of a solution of any composition. The partial molar Gibbs energy of the ith component in solution $\overline{G}_i$ is given by

$$\overline{G}_i = \left(\frac{\partial G}{\partial n_i}\right)_{T,P,n_j} \tag{6.7}$$

where n_j represents the number of moles of all other components present. Again we can think of $\overline{G}_i$ as the coefficient that gives the increase in the Gibbs energy of the

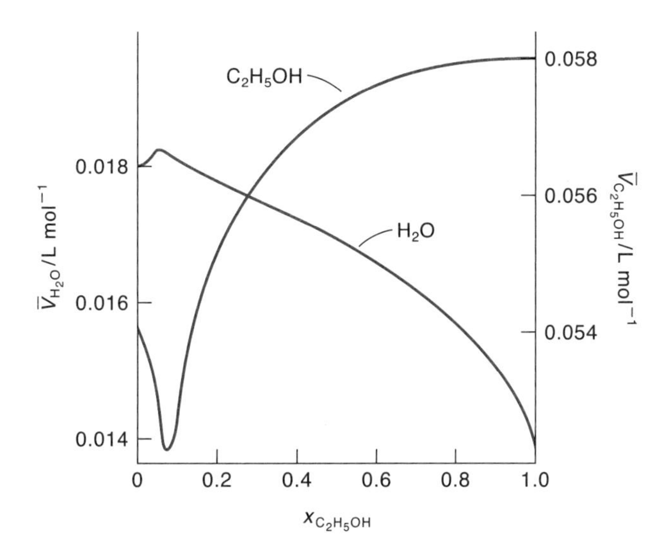

Figure 6.3
The partial molar volumes of water and ethanol as a function of the mole fraction of ethanol. Note the different scales for water (left) and ethanol (right).

solution upon the addition of 1 mole of component i at constant temperature and pressure to a large amount of solution of specified concentration. Partial molar Gibbs energy is also called the *chemical potential* (μ), so we can write

$$\overline{G}_i = \mu_i \tag{6.8}$$

The expression for the total Gibbs energy of a two-component solution is similar to Equation 6.6 for volumes:

$$G = n_1\mu_1 + n_2\mu_2 \tag{6.9}$$

The chemical potential provides a criterion for equilibrium and spontaneity for a multicomponent system, just as molar Gibbs energy does for a single-component system. Consider the transfer of dn_i moles of component i from some initial state A, where its chemical potential is μ_i^{A}, to some final state B, where its chemical potential is μ_i^{B}. For a process carried out at constant temperature and pressure, the change in the Gibbs energy, dG, is given by

$$\begin{aligned} dG &= \mu_i^{\mathrm{B}} dn_i - \mu_i^{\mathrm{A}} dn_i \\ &= (\mu_i^{\mathrm{B}} - \mu_i^{\mathrm{A}}) dn_i \end{aligned}$$

If $\mu_i^{\mathrm{B}} < \mu_i^{\mathrm{A}}$, $dG < 0$, and transfer of dn_i moles from A to B will be a spontaneous process; if $\mu_i^{\mathrm{B}} > \mu_i^{\mathrm{A}}$, $dG > 0$, and the process will be spontaneous from B to A. As we shall see later, the transfer can be from one phase to another or from one state of chemical combination to another. The transfer can be transport by diffusion, evaporation, sublimation, condensation, crystallization, solution formation, or chemical reaction. Regardless of the nature of the process, in each case the transfer proceeds from a higher μ_i value to a lower μ_i value. This characteristic explains the name *chemical potential.* In mechanics, the direction of spontaneous change always takes the system from a higher potential-energy state to a lower one. In thermodynamics, the situation is not quite so simple because we have to consider both energy and entropy factors. Nevertheless, we know that at constant temperature and pressure, the direction of a spontaneous change is always toward a decrease in the system's Gibbs energy. Thus, the role Gibbs energy plays in thermodynamics is analogous to that of potential energy in mechanics. This is the reason that molar Gibbs energy or, more commonly, partial molar Gibbs energy, is called the chemical potential.

6.3 Thermodynamics of Mixing

The formation of solutions is governed by the principles of thermodynamics. In this section, we shall discuss the changes in thermodynamic quantities that result from mixing. In particular, we shall focus on gases.

Equation 6.9 gives the dependence of the Gibbs energy of a system on its composition. The spontaneous mixing of gases is accompanied by a change in composition; consequently, the system's Gibbs energy decreases. In Section 5.4, we obtained an expression for the molar Gibbs energy of an ideal gas (Equation 5.18):

$$\overline{G} = \overline{G}^\circ + RT \ln \frac{P}{1 \text{ bar}}$$

In a mixture of ideal gases, the chemical potential of the ith component is given by

$$\mu_i = \mu_i^\circ + RT \ln \frac{P_i}{1 \text{ bar}} \tag{6.10}$$

where P_i is the partial pressure of component i in the mixture and μ_i° is the standard chemical potential of component i when its partial pressure is 1 bar. Now consider the mixing of n_1 moles of gas 1 at temperature T and pressure P with n_2 moles of gas 2 at the same T and P. Before mixing, the total Gibbs energy of the system is given by Equation 6.9, where chemical potentials are the same as molar Gibbs energies,

For simplicity, we omit the term "1 bar." Note that the resulting P values are dimensionless.

$$G = n_1\overline{G}_1 + n_2\overline{G}_2 = n_1\mu_1 + n_2\mu_2$$

$$G_{\text{initial}} = n_1(\mu_1^\circ + RT \ln P) + n_2(\mu_2^\circ + RT \ln P)$$

After mixing, the gases exert partial pressures P_1 and P_2, where $P_1 + P_2 = P$, and the Gibbs energy is*

$$G_{\text{final}} = n_1(\mu_1^\circ + RT \ln P_1) + n_2(\mu_2^\circ + RT \ln P_2)$$

* Note that $P_1 + P_2 = P$ only if there is no change in volume as a result of mixing; that is, $\Delta_{\text{mix}}V = 0$. This condition holds for ideal solutions.

The Gibbs energy of mixing, $\Delta_{\text{mix}}G$, is given by

$$\begin{aligned}\Delta_{\text{mix}}G &= G_{\text{final}} - G_{\text{initial}} \\ &= n_1RT\ln\frac{P_1}{P} + n_2RT\ln\frac{P_2}{P} \\ &= n_1RT\ln x_1 + n_2RT\ln x_2\end{aligned}$$

where $P_1 = x_1P$ and $P_2 = x_2P$, and x_1 and x_2 are the mole fractions of 1 and 2, respectively. (The standard chemical potential, $\mu°$, is the same in the pure state and in the mixture.) Further, from the relations

$$x_1 = \frac{n_1}{n_1 + n_2} = \frac{n_1}{n} \quad \text{and} \quad x_2 = \frac{n_2}{n_1 + n_2} = \frac{n_2}{n}$$

where n is the total number of moles, we have

$$\Delta_{\text{mix}}G = nRT(x_1\ln x_1 + x_2\ln x_2) \tag{6.11}$$

Because both x_1 and x_2 are less than unity, $\ln x_1$ and $\ln x_2$ are negative quantities, and hence so is $\Delta_{\text{mix}}G$. This result is consistent with our expectation that the mixing of gases is a spontaneous process at constant T and P.

Now we can calculate other thermodynamic quantities of mixing. From Equation 5.10, we see that at constant pressure

$$\left(\frac{\partial G}{\partial T}\right)_P = -S$$

Thus, the entropy of mixing is obtained by differentiating Equation 6.11 with respect to temperature at constant pressure:

$$\begin{aligned}\left(\frac{\partial \Delta_{\text{mix}}G}{\partial T}\right)_P &= nR(x_1 \ln x_1 + x_2 \ln x_2) \\ &= -\Delta_{\text{mix}}S\end{aligned}$$

or

$$\Delta_{\text{mix}}S = -nR(x_1 \ln x_1 + x_2 \ln x_2) \tag{6.12}$$

This result is equivalent to Equation 4.17. The minus sign in Equation 6.12 makes $\Delta_{\text{mix}}S$ a positive quantity, in accord with a spontaneous process. The enthalpy of mixing is given by rearranging Equation 5.3:

$$\begin{aligned}\Delta_{\text{mix}}H &= \Delta_{\text{mix}}G + T\Delta_{\text{mix}}S \\ &= 0\end{aligned}$$

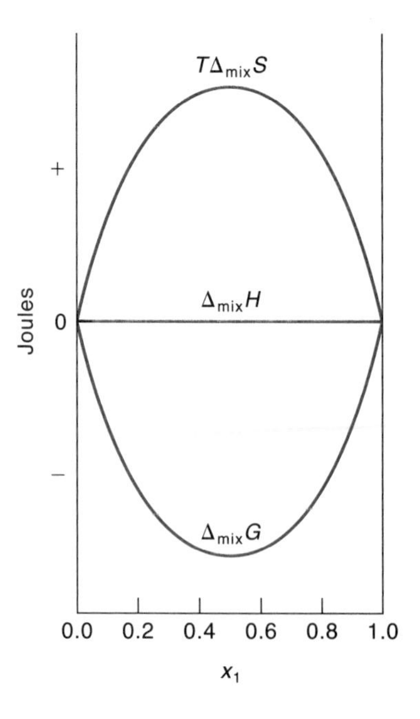

Figure 6.4
Plots of $T\Delta_{mix}S$, $\Delta_{mix}H$, and $\Delta_{mix}G$ as a function of composition x_1 for the mixing of two components to form an ideal solution.

This result is not surprising, because molecules of ideal gases do not interact with one another, so no heat is absorbed or produced as a result of mixing. Figure 6.4 shows the plots of $\Delta_{mix}G$, $T\Delta_{mix}S$, and $\Delta_{mix}H$ for a two-component system as a function of composition. Note that both the maximum (for $T\Delta_{mix}S$) and the minimum (for $\Delta_{mix}G$) occur at $x_1 = 0.5$. This result means that we achieve maximum disorder by mixing equimolar amounts of gases and that the Gibbs energy of mixing reaches a minimum at this point (see Problem 6.57).

Reversing the process for a two-component solution of equal mole fractions leads to an increase in Gibbs energy and a decrease in entropy of the system, so energy must be supplied to the system from the surroundings. Initially, at $x_1 \approx x_2$, the $\Delta_{min}G$ and $T\Delta_{mix}S$ curves are fairly flat (see Figure 6.4), and separation can be carried out easily. However, as the solution becomes progressively richer in one component, say 1, the curves become very steep. Then, a considerable amount of energy input is needed to separate component 2 from 1. This difficulty is encountered, for example, in the attempt to clean up a lake contaminated by small amounts of undesirable chemicals. The same consideration applies to the purification of compounds. Preparing most compounds in 95% purity is relatively easy, but much more effort is required to attain 99% or higher purity, which is needed, for example, for the silicon crystals used in solid-state electronics.

As another example, let us explore the possibility of mining gold from the oceans. Estimates are that there is approximately 4×10^{-12} g of gold/mL of seawater. This amount may not seem like much, but when we multiply it by the total volume of ocean water, 1.5×10^{21} L, we find the amount of gold present to be 6×10^{12} g or 7 million tons, which should satisfy anybody. Unfortunately, not only is the concentration of gold in seawater very low, but gold is also just one of some 60 different elements in the ocean. Separating one pure component initially present in a very low concentration in seawater (i.e., starting at the steep portions of the curves in Figure 6.4) would be a very formidable (and expensive) undertaking indeed.

EXAMPLE 6.1

Calculate the Gibbs energy and entropy of mixing 1.6 moles of argon at 1 atm and 25°C with 2.6 moles of nitrogen at 1 atm and 25°C. Assume ideal behavior.

ANSWER

The mole fractions of argon and neon are

$$x_{Ar} = \frac{1.6}{1.6 + 2.6} = 0.38 \quad x_{N_2} = \frac{2.6}{1.6 + 2.6} = 0.62$$

From Equation 6.11,

$$\begin{aligned}\Delta_{mix}G &= nRT(x_1 \ln x_1 + x_2 \ln x_2) \\ &= (4.2 \text{ mol})(8.314 \text{ J K}^{-1} \text{ mol}^{-1})(298 \text{ K})[(0.38) \ln 0.38 + (0.62) \ln 0.62] \\ &= -6.9 \text{ kJ}\end{aligned}$$

Because $\Delta_{mix}S = \Delta_{mix}G/T$, we write

$$\begin{aligned}\Delta_{mix}S &= -\frac{-6.9 \times 10^3 \text{ J}}{298 \text{ K}} \\ &= 23 \text{ J K}^{-1}\end{aligned}$$

COMMENT

In this example, the gases are at the same temperature and pressure when they are mixed. If the initial pressures of the gases differ, then there will be two contributions to $\Delta_{mix}G$: the mixing itself and the changes in pressure. Problem 6.58 illustrates this situation.

6.4 Binary Mixtures of Volatile Liquids

The results obtained in Section 6.3 for mixtures of gases also apply to ideal liquid solutions. For the study of solutions, we need to know how to express the chemical potential of each component. We shall consider a solution containing two volatile liquids, that is, liquids with easily measurable vapor pressures.

Let us start with a liquid in equilibrium with its vapor in a closed container. Because the system is at equilibrium, the chemical potentials of the liquid phase and the vapor phase must be the same, that is,

$$\mu^*(l) = \mu^*(g)$$

where the asterisk denotes a pure component. Further, from the expression for $\mu^*(g)$ for an ideal gas, we can write†

$$\mu^*(l) = \mu^*(g) = \mu^\circ(g) + RT \ln \frac{P^*}{1 \text{ bar}} \tag{6.13}$$

where $\mu^\circ(g)$ is the standard chemical potential at $P^* = 1$ bar. For a two-component solution at equilibrium with its vapor, the chemical potential of each component is still the same in the two phases. Thus, for component 1 we write

$$\mu_1(l) = \mu_1(g) = \mu_1^\circ(g) + RT \ln \frac{P_1}{1 \text{ bar}} \tag{6.14}$$

where P_1 is the partial pressure. Because $\mu^\circ(g) = \mu_1^\circ(g)$, we can combine the previous two equations to get

$$\begin{aligned}\mu_1(l) &= \mu_1^\circ(g) + RT \ln \frac{P_1}{1 \text{ bar}} \\ &= \mu_1^*(l) - RT \ln \frac{P_1^*}{1 \text{ bar}} + RT \ln \frac{P_1}{1 \text{ bar}} \\ &= \mu_1^*(l) + RT \ln \frac{P_1}{P_1^*}\end{aligned} \tag{6.15}$$

Thus, the chemical potential of component 1 in solution is expressed in terms of the chemical potential of the liquid in the pure state and the vapor pressures of the liquid in solution and in the pure state.

Raoult's Law

The French chemist François Marie Raoult (1830–1901) found that for some solutions, the ratio P_1/P_1^* in Equation 6.15 is equal to the mole fraction of component 1, that is,

$$\frac{P_1}{P_1^*} = x_1$$

or

$$P_1 = x_1 P_1^* \tag{6.16}$$

Equation 6.16 is known as *Raoult's law*, which states that the vapor pressure of a component of a solution is equal to the product of its mole fraction and the vapor pressure of the pure liquid. Substituting Equation 6.16 into Equation 6.15, we obtain

$$\mu_1(l) = \mu_1^*(l) + RT \ln x_1 \tag{6.17}$$

† This equation follows from Equation 5.18. For a pure component, the chemical potential is equal to the molar Gibbs energy.

We see that in a pure liquid ($x_1 = 1$ and $\ln x_1 = 0$), $\mu_1(l) = \mu_1^*(l)$. Solutions that obey Raoult's law are called *ideal solutions*. An example of a nearly ideal solution is the benzene–toluene system. Figure 6.5 shows a plot of the vapor pressures versus the mole fraction of benzene.

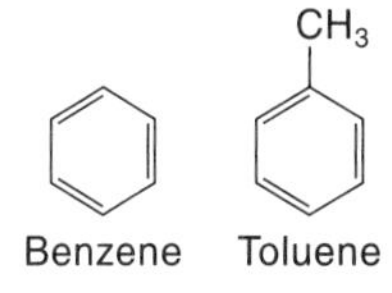

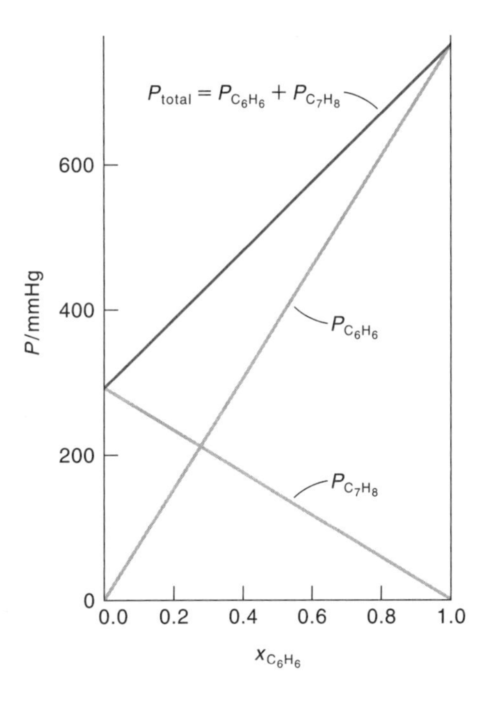

Figure 6.5
Total vapor pressure of the benzene–toluene mixture as a function of the benzene mole fraction at 80.1°C.

EXAMPLE 6.2

Liquids A and B form an ideal solution. At 45°C, the vapor pressures of pure A and pure B are 66 torr and 88 torr, respectively. Calculate the composition of the vapor in equilibrium with a solution containing 36 mole percent A at this temperature.

ANSWER

Because $x_A = 0.36$ and $x_B = 1 - 0.36 = 0.64$, we have, according to Raoult's law

$$P_A = x_A P_A^* = 0.36(66 \text{ torr}) = 23.8 \text{ torr}$$

$$P_B = x_B P_B^* = 0.64(88 \text{ torr}) = 56.3 \text{ torr}$$

The total vapor pressure, P_T, is given by

$$P_T = P_A + P_B = 23.8 \text{ torr} + 56.3 \text{ torr} = 80.1 \text{ torr}$$

Finally, the mole fractions of A and B in the vapor phase, x_A^v and x_A^v, are given by

$$x_A^v = \frac{P_A}{P_T} = \frac{23.8 \text{ torr}}{80.1 \text{ torr}} = 0.30$$

and

$$x_B^v = \frac{P_B}{P_T} = \frac{56.3 \text{ torr}}{80.1 \text{ torr}} = 0.70$$

In an ideal solution, all intermolecular forces are equal, whether the molecules are alike or not. The benzene–toluene system approximates this requirement because benzene and toluene molecules have similar shapes and electronic structures. For an ideal solution, we have both $\Delta_{mix}H = 0$ and $\Delta_{mix}V = 0$. Most solutions do *not* behave ideally, however. Figure 6.6 shows the positive and negative deviations from Raoult's law. The positive deviation (Figure 6.6a) corresponds to the case in which the intermolecular forces between unlike molecules are weaker than those between like molecules, and

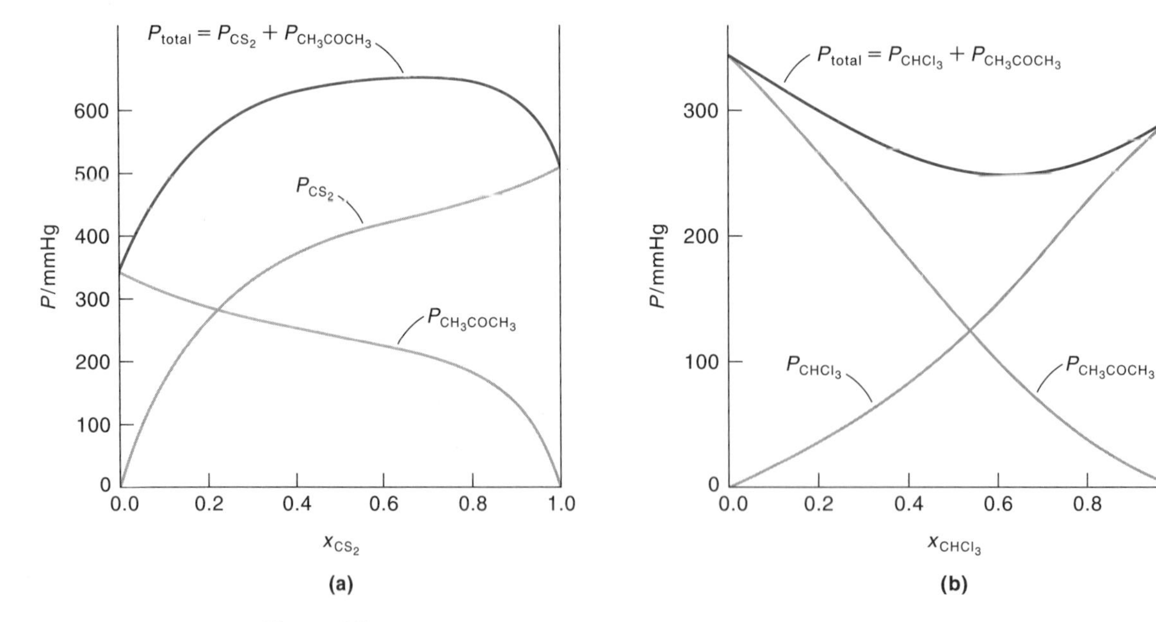

Figure 6.6
Deviations from Raoult's law by nonideal solutions. (a) Positive deviation from Raoult's law: carbon disulfide–acetone system at 35.2°C. (b) Negative deviation: chloroform–acetone system at 25.2°C. (From Hildebrand, J. and R. Scott, *The Solubility of Nonelectrolytes* © 1950, Litton Educational Publishing. Reprinted by permission of Van Nostrand Reinhold Company.)

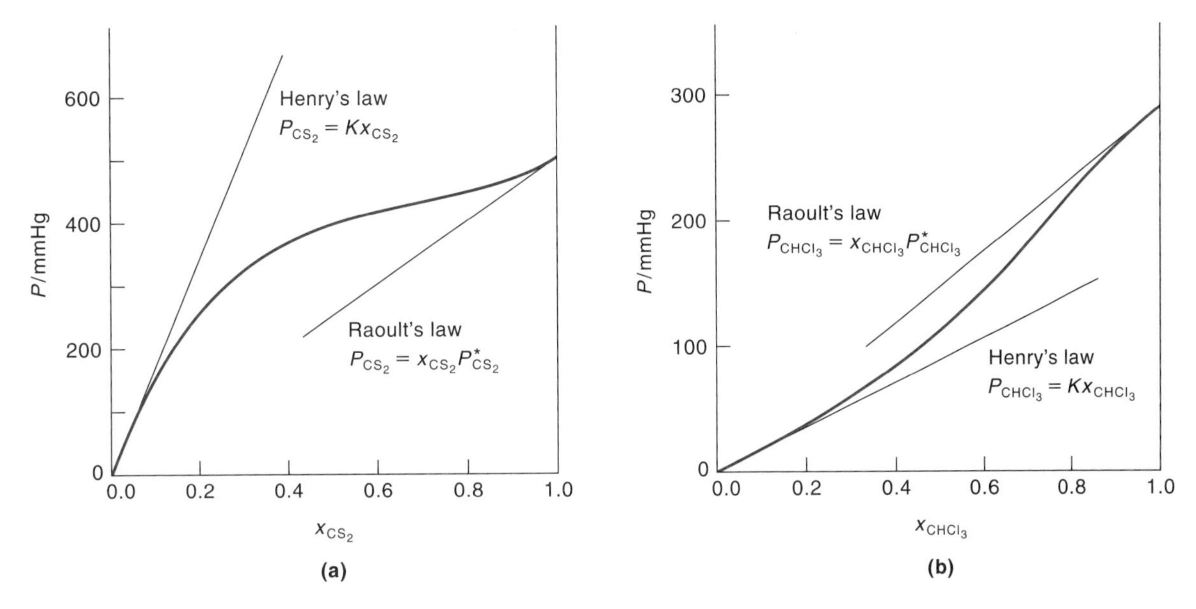

Figure 6.7
Diagrams showing regions over which Raoult's law and Henry's law are applicable for a two-component system (see Figure 6.6) The Henry's law constants can be obtained from the intercepts on the *y* (pressure) axis.

there is a greater tendency for these molecules to leave the solution than in the case of an ideal solution. Consequently, the vapor pressure of the solution is greater than the sum of the vapor pressures for an ideal solution. Just the opposite holds for a negative deviation from Raoult's law (Figure 6.6b). In this case, unlike molecules attract each other more strongly than they do their own kind, and the vapor pressure of the solution is less than the sum of the vapor pressures for an ideal solution.

Henry's Law

When one solution component is present in excess (this component is called the solvent), its vapor pressure is quite accurately described by Equation 6.16. The regions where Raoult's law is applicable are shown for the carbon disulfide–acetone system in Figure 6.7. In contrast, the vapor pressure of the component present in a small amount (this component is called the solute) does not vary with the composition of the solution, as predicted by Equation 6.16. Still, the vapor pressure of the solute varies with concentration in a linear manner:

There is no sharp distinction between solvent and solute. Where applicable, we shall call component 1 the solvent and component 2 the solute.

$$P_2 = Kx_2 \tag{6.18}$$

Equation 6.18 is known as *Henry's law* (after the English chemist William Henry, 1775–1836), where K, the Henry's law constant, has units of pressure. Henry's law relates the mole fraction of the solute to its partial (vapor) pressure. Alternatively, Henry's law can be expressed as

$$P_2 = K'm \tag{6.19}$$

Table 6.1
Henry's Law Constants for Some Gases in Water at 298 K

Gas	K/torr	K'/atm mol^{-1} kg H_2O
H_2	5.54×10^7	1311
He	1.12×10^8	2649
Ar	2.80×10^7	662
N_2	6.80×10^7	1610
O_2	3.27×10^7	773
CO_2	1.24×10^6	29.3
H_2S	4.27×10^5	10.1

where m is the molality of the solution and the constant K' now has the units atm mol^{-1} kg of the solvent. Table 6.1 lists the values of K and K' for several gases in water at 298 K.

Henry's law is normally associated with solutions of gases in liquids, although it is equally applicable to solutions containing nongaseous volatile solutes. It has great practical importance in chemical and biological systems and therefore merits further discussion. The effervescence observed when a soft drink or champagne bottle is opened is a nice demonstration of the decrease in gas—mostly CO_2—solubility as its partial pressure is lowered. The emboli (gas bubbles in the bloodstream) suffered by deep-sea divers who rise to the surface too rapidly also illustrate Henry's law. At a point some 40 m below the surface of seawater, the total pressure is about 6 atm. The solubility of nitrogen in the blood plasma is then about 0.8×6 atm/1610 atm mol^{-1} kg H_2O, or 3.0×10^{-3} mol (kg H_2O)$^{-1}$, which is six times the solubility at sea level. If the diver swims upward too rapidly, dissolved nitrogen gas will start boiling off. The mildest result is dizziness; the most serious, death.* Because helium is less soluble in the blood plasma than nitrogen is, it is the preferred gas for diluting oxygen gas for use in deep-sea diving tanks.

There are several types of deviations from Henry's law. First, as mentioned earlier, the law holds only for dilute solutions. Second, if the dissolved gas interacts chemically with the solvent, then the solubility can be greatly enhanced. Gases such as CO_2, H_2S, NH_3, and HCl all have high solubilities in water because they react with the solvent. The third type of deviation is illustrated by the dissolution of oxygen in blood. Normally, oxygen is only sparingly soluble in water (see Table 6.1), but its solubility increases dramatically if the solution contains hemoglobin or myoglobin.

* For other interesting illustrations of Henry's law, see T. C. Loose, *J. Chem. Educ.* **48**, 154 (1971), W. J. Ebel, *J. Chem. Educ.* **50**, 559 (1973), and D. R. Kimbrough, *J. Chem. Educ.* **76**, 1509 (1999).

EXAMPLE 6.3

Calculate the molal solubility of carbon dioxide in water at 298 K and a CO_2 pressure of 3.3×10^{-4} atm, which corresponds to the partial pressure of CO_2 in air.

ANSWER

The mole fraction of solute (carbon dioxide) is given by Equation 6.18:

$$x_{CO_2} = \frac{P_{CO_2}}{K}$$

Because the number of moles of CO_2 dissolved in 1000 g of water is small, we can approximate the mole fraction as follows:

$$x_{CO_2} = \frac{n_{CO_2}}{n_{CO_2} + n_{H_2O}} \approx \frac{n_{CO_2}}{n_{H_2O}}$$

so that

$$n_{CO_2} = \frac{P_{CO_2} n_{H_2O}}{K}$$

Finally, looking up the K value of CO_2 in Table 6.1 we write

$$n_{CO_2} = (3.3 \times 10^{-4} \times 760) \text{ torr} \times \frac{1000 \text{ g}}{18.01 \text{ g mol}^{-1}} \times \frac{1}{1.24 \times 10^6 \text{ torr}}$$

$$= 1.12 \times 10^{-5} \text{ mol}$$

Because this is the number of moles of CO_2 in 1000 g or 1 kg of H_2O, the molality is 1.12×10^{-5} mol (kg H_2O)$^{-1}$.

Alternatively, we can use Equation 6.19 and proceed as follows:

$$m = \frac{P_{CO_2}}{K'}$$

$$= \frac{3.3 \times 10^{-4} \text{ atm}}{29.3 \text{ atm mol}^{-1} \text{ kg H}_2\text{O}} = 1.12 \times 10^{-5} \text{ mol (kg H}_2\text{O)}^{-1}$$

COMMENT

Carbon dioxide dissolved in water is converted to carbonic acid, which causes water that is exposed to air for a long period of time to become acidic.

6.5 Real Solutions

As pointed out in Section 6.4, most solutions do not behave ideally. One problem that immediately arises in dealing with nonideal solutions is how to write the chemical potentials for the solvent and solute components.

The Solvent Component

Let us look at the solvent component first. As we saw earlier, the chemical potential of the solvent in an ideal solution is given by (see Equation 6.17)

$$\mu_1(l) = \mu_1^*(l) + RT \ln x_1$$

where $x_1 = P_1/P_1^*$ and P_1^* is the equilibrium vapor pressure of pure component 1 at T. The standard state is the pure liquid and is attained when $x_1 = 1$. For a nonideal solution, we write

$$\mu_1(l) = \mu_1^*(l) + RT \ln a_1 \tag{6.20}$$

where a_1 is the *activity* of the solvent. Nonideality is the consequence of unequal intermolecular forces between solvent–solvent and solvent–solute molecules. Therefore, the extent of nonideality depends on the composition of solution, and the activity of the solvent plays the role of "effective" concentration. The solvent's activity can be expressed in terms of vapor pressure as

$$a_1 = \frac{P_1}{P_1^*} \tag{6.21}$$

Table 6.2
Activity of Water in Water–Urea Solutions at 273 K[a]

Molality of urea, m_2	Mole fraction of water, x_1	Vapor pressure of water, P_1/atm	Activity of water, a_1	Activity coefficient of water, γ_1
0	1.000	6.025×10^{-3}	1.000	1.000
1	0.982	5.933×10^{-3}	0.985	1.003
2	0.965	5.846×10^{-3}	0.970	1.005
4	0.933	5.672×10^{-3}	0.942	1.010
6	0.902	5.501×10^{-3}	0.913	1.012
10	0.847	5.163×10^{-3}	0.857	1.012

[a] Data from National Research Council. *International Critical Tables of Numerical Data: Physics, Chemistry, and Technology*, Vol. 3. © 1928, McGraw-Hill. Used by permission of McGraw-Hill. Note that the solute (urea) is nonvolatile.

where P_1 is the partial vapor pressure of component 1 over the (nonideal) solution. Activity is related to concentration (mole fraction) as follows:

$$a_1 = \gamma_1 x_1 \tag{6.22}$$

where γ_1 is the *activity coefficient.* Equation 6.20 can now be written as

$$\mu_1(l) = \mu_1^*(l) + RT \ln \gamma_1 + RT \ln x_1 \tag{6.23}$$

The value of γ_1 is a measure of the deviation from ideality. In the limiting case, where $x_1 \to 1$, $\gamma_1 \to 1$ and activity and the mole fraction are identical. This condition also holds for an ideal solution at all concentrations.

Equation 6.21 provides a way of obtaining the activity of the solvent. By measuring P_1 of the solvent vapor over a range of concentrations, we can calculate the value of a_1 at each concentration if P_1^* is known.† Table 6.2 shows the activity of water in various water–urea solutions.

The Solute Component

We now come to the solute. Ideal solutions in which both components obey Raoult's law over the entire concentration range are rare. In dilute, nonideal solutions in which there is no chemical interaction, the solvent obeys Raoult's law, and the solute obeys Henry's law.‡ Such solutions are sometimes called "ideal dilute solutions." If the solution were ideal, the chemical potential of the solute is also given by Raoult's law

$$\begin{aligned}\mu_2(l) &= \mu_2^*(l) + RT \ln x_2 \\ &= \mu_2^*(l) + RT \ln \frac{P_2}{P_2^*}\end{aligned}$$

In an ideal dilute solution, Henry's law applies. That is $P_2 = Kx_2$, so that

$$\begin{aligned}\mu_2(l) &= \mu_2^*(l) + RT \ln \frac{K}{P_2^*} + RT \ln x_2 \\ &= \mu_2^\circ(l) + RT \ln x_2 \end{aligned} \tag{6.24}$$

where $\mu_2^\circ(l) = \mu_2^*(l) + RT \ln (K/P_2^*)$. Although Equation 6.24 seems to take the same form as Equation 6.17, there is an important difference, which lies in the choice of standard state. According to Equation 6.24, the standard state is defined as the pure solute, attained by setting $x_2 = 1$. Equation 6.24 holds only for dilute solutions, however. How can these two conditions be met simultaneously? The simple way out of this dilemma is to recognize that standard states are often hypothetical states, not physically realizable. Thus, the standard state of solute defined by Equation 6.24 is

† To obtain the value of P_1, we must measure the total pressure, P, and also analyze the composition of the mixture. Then we can calculate partial pressure P_1 using Equation 1.6; that is, $P_1 = x_1^{\text{v}} P$, where x_1^{v} is the mole fraction of the solvent in the vapor phase.

‡ For ideal solutions, Raoult's law and Henry's law become identical; that is, $P_2 = Kx_2 = P_2^* x_2$.

the hypothetical pure component 2 with a vapor pressure equal to K (when $x_2 = 1$, $P_2 = K$). In a sense, this is an "infinite dilution state of unit mole fraction"; that is, it is infinitely dilute with respect to component 1, the solvent, with the solute at unit mole fraction. For nonideal solutions in general (beyond the dilute solution limit), Equation 6.24 is modified to

$$\mu_2(l) = \mu_2^\circ(l) + RT \ln a_2 \tag{6.25}$$

where a_2 is the activity of the solute. As in the case of the solvent component, we have $a_2 = \gamma_2 x_2$, where γ_2 is the activity coefficient of the solute. Here, we have $a_2 \rightarrow x_2$ or $\gamma_2 \rightarrow 1$ as $x_2 \rightarrow 0$. Henry's law is now given by

$$P_2 = Ka_2 \tag{6.26}$$

Concentrations are usually expressed in molalities (or molarities) instead of mole fractions. In molality, Equation 6.24 takes the form

$$\mu_2(l) = \mu_2^\circ(l) + RT \ln \frac{m_2}{m^\circ} \tag{6.27}$$

where $m^\circ = 1$ mol kg^{-1} so that the ratio m_2/m° is dimensionless. Here, the standard state is defined as a state at unit molality but in which the solution is behaving ideally. Again, this standard state is a hypothetical state, not attainable in practice (Figure 6.8). For nonideal solutions, Equation 6.27 is rewritten as

$$\mu_2(l) = \mu_2^\circ(l) + RT \ln a_2 \tag{6.28}$$

where $a_2 = \gamma_2(m_2/m^\circ)$. In the limiting case of $m_2 \rightarrow 0$, we have $a_2 \rightarrow m_2/m^\circ$ or $\gamma_2 \rightarrow 1$ (see Figure 6.8b).

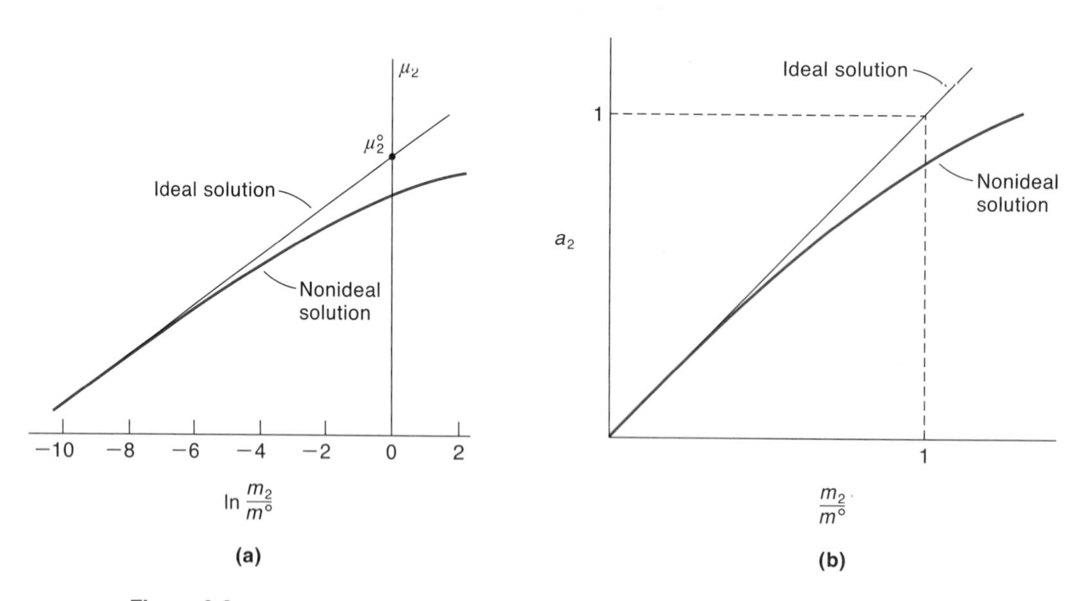

Figure 6.8
(a) Chemical potential of a solute plotted against the logarithm of molality for a nonideal solution. (b) Activity of a solute as a function of molality for a nonideal solution. The standard state is at $m_2/m^\circ = 1$.

Keep in mind that although Equations 6.24 and 6.27 were derived using Henry's law, they are applicable to any solute, whether or not it is volatile. These expressions are useful in discussing the colligative properties of solutions (see Section 6.7), and, as we shall see in Chapter 8, in deriving the equilibrium constant.

6.6 Phase Equilibria of Two-Component Systems

Before we study colligative properties, let us first apply the phase diagram and the Gibbs phase rule to the properties of solutions containing two components.

Distillation

The separation of two volatile liquid components is usually accomplished by fractional distillation, which has many applications in the laboratory and in industrial processes. To use this procedure, we must understand how pressure and temperature affect the vapor–liquid equilibrium of binary liquid mixtures.

Pressure–Composition Diagram. We begin by constructing phase diagrams that show the vapor pressure of a solution as a function of mole fraction and as a function of the composition of the vapor in equilibrium with the solution. Using the ideal benzene–toluene solution as an example, we can express the vapor pressures of both components in terms of Raoult's law:

$$P_b = x_b P_b^* \quad \text{and} \quad P_t = x_t P_t^*$$

where x_b and x_t are the mole fractions of benzene and toluene in solution, respectively, and the asterisk denotes a pure component. The total pressure, P, is given by

$$\begin{aligned} P &= P_b + P_t \\ &= x_b P_b^* + x_t P_t^* \\ &= x_b P_b^* + (1 - x_b) P_t^* \\ &= P_t^* + (P_b^* - P_t^*) x_b \end{aligned} \tag{6.29}$$

Figure 6.9a shows a plot of P versus x_b, which is a straight line.

Next we want to express P in terms of x_b^v, which is the mole fraction of benzene in the vapor phase. According to Equation 1.6, the mole fraction of benzene in the vapor phase is given by

$$x_b^v = \frac{P_b}{P} = \frac{x_b P_b^*}{P_t^* + (P_b^* - P_t^*) x_b}$$

This equation can be solved for x_b to obtain the following expression for the mole fraction of benzene in solution that corresponds to its mole fraction x_b^v in the vapor phase at equilibrium:

$$x_b = \frac{x_b^v P_t^*}{P_b^* - (P_b^* - P_t^*) x_b^v} \tag{6.30}$$

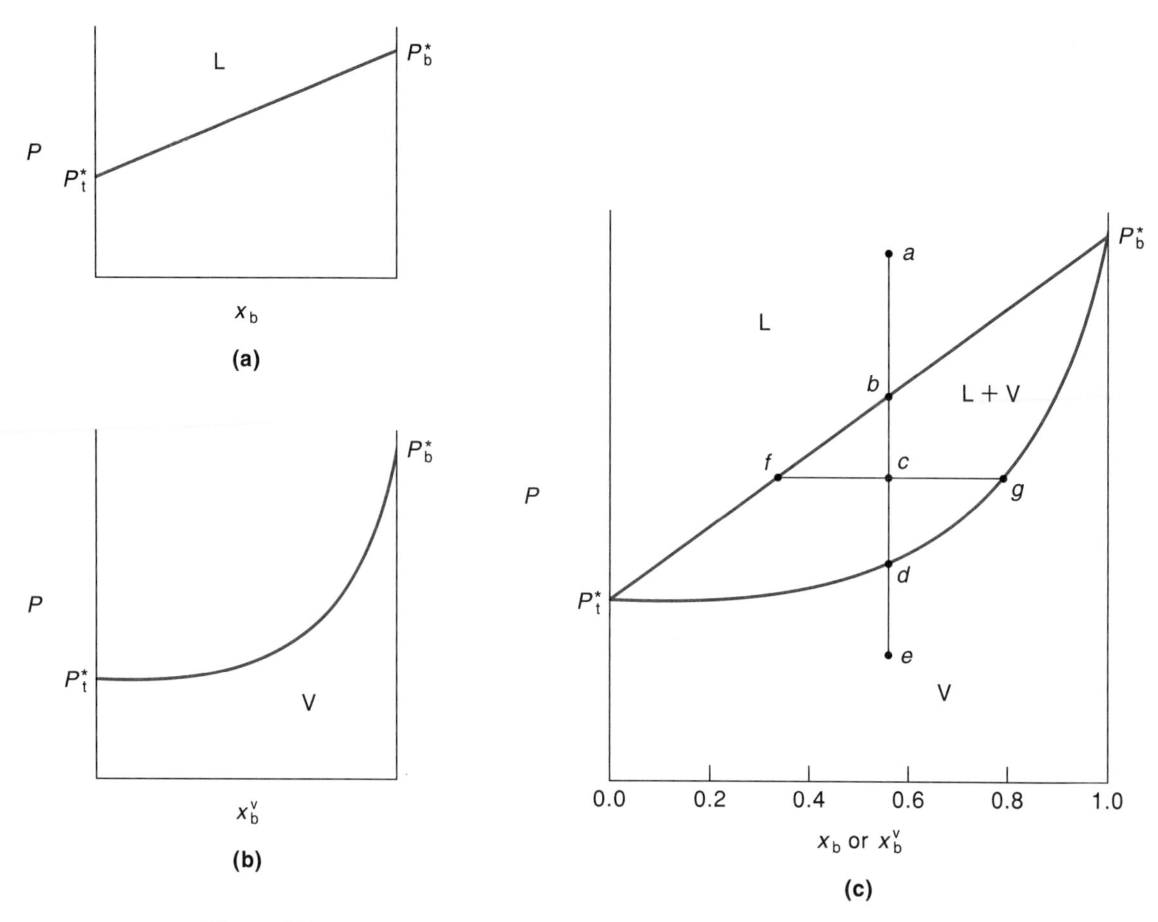

Figure 6.9
Pressure versus composition liquid (L)–vapor (V) phase diagram for the benzene–toluene solution at 23°C. (a) Plot of vapor pressure versus the mole fraction of a benzene in solution. (b) Plot of vapor pressure versus the mole fraction of benzene in the vapor phase. (c) Combined plots of (a) and (b). Above the straight line, the system is entirely in the liquid state; below the curve, the system is entirely in the vapor state; both liquid and vapor are present in the enclosed area. At point *c*, the liquid composition is $x_b = 0.30$ and the vapor composition is $x_b^v = 0.80$.

Again, referring to Equation 1.6 and Raoult's law,

$$P_b = x_b^v P = x_b P_b^*$$

and using Equation 6.30, we write

$$P = \frac{x_b P_b^*}{x_b^v} = \frac{P_b^* P_t^*}{P_b^* - (P_b^* - P_t^*)x_b^v} \tag{6.31}$$

Figure 6.9b shows a plot of P versus x_b^v.

The combined plots of Equations 6.29 and 6.31 are shown in Figure 6.9c. Above the straight line, the system is in the liquid state. (This result is consistent with our expectation that at high pressures, liquid is the more stable phase.) Below the curve, that is, at low pressures, the system exists as a vapor. At a point between the two lines (i.e., a point in the enclosed area) both liquid and vapor are present.

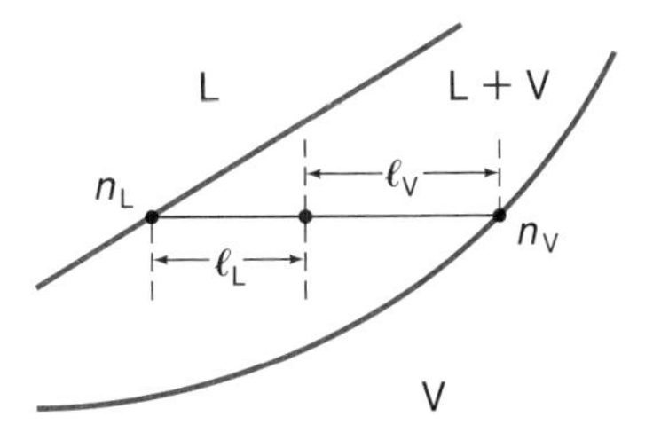

Figure 6.10
The lever rule gives the ratio of moles present in each phase of a two-phase system as $n_L\ell_L = n_V\ell_V$, where ℓ_L and ℓ_V are the distances from the point corresponding to the system's overall mole fraction to the endpoints of the tie line.

Now suppose we begin with the benzene–toluene system initially in the liquid phase at point *a* in Figure 6.9c. According to the Gibbs phase rule (see p. 196), we have a two-component, one-phase system, so that the degrees of freedom are

$$f = c - p + 2$$
$$= 2 - 1 + 2 = 3$$

At constant temperature and a fixed mole fraction, we are left with one degree of freedom. Therefore, we can lower the pressure at constant composition and temperature (from point *a*) until point *b* is reached. At this point, the liquid can exist in equilibrium with its vapor. (This is the point at which the liquid just begins to vaporize.) Further decreasing the pressure takes us to point *c*. The horizontal "tie line" indicates that the composition (in mole fraction) of the liquid is *f* and that of the vapor is *g* at this point. The relative amounts of liquid and vapor present in equilibrium are given by the lever rule (Figure 6.10):

$$n_L\ell_L = n_V\ell_V \tag{6.32}$$

where n_L and n_V are the number of moles of liquid and vapor, respectively, and ℓ_L and ℓ_V are the distances defined in Figure 6.10. At point *c*, we have a two-component, two-phase system so that the total degrees of freedom are

$$f = 2 - 2 + 2$$
$$= 2$$

At constant temperature, then, we have only one degree of freedom, so that if we select a particular pressure, the compositions of the liquid and vapor must be fixed (as shown by the tie line).

Figure 6.11 suggests a way to separate benzene from toluene at a fixed temperature. We can lower the pressure on a solution until it begins to vaporize (point *a*). At this point, the mole fractions are $x_b = 0.2$ and $x_t = 1 - 0.2 = 0.8$, respectively. The composition of the vapor in equilibrium with the solution (point *b*) has the following values: $x_b^V = 0.5$ and $x_t^V = 0.5$. Thus, the vapor phase is richer in benzene than is the liquid phase. Now, if we condense the vapor ($b \rightarrow c$) and re-evaporate the liquid ($c \rightarrow d$), the mole fraction of benzene will be even higher in the vapor phase. Repeating the process at constant temperature will eventually lead to a quantitative separation of benzene and toluene.

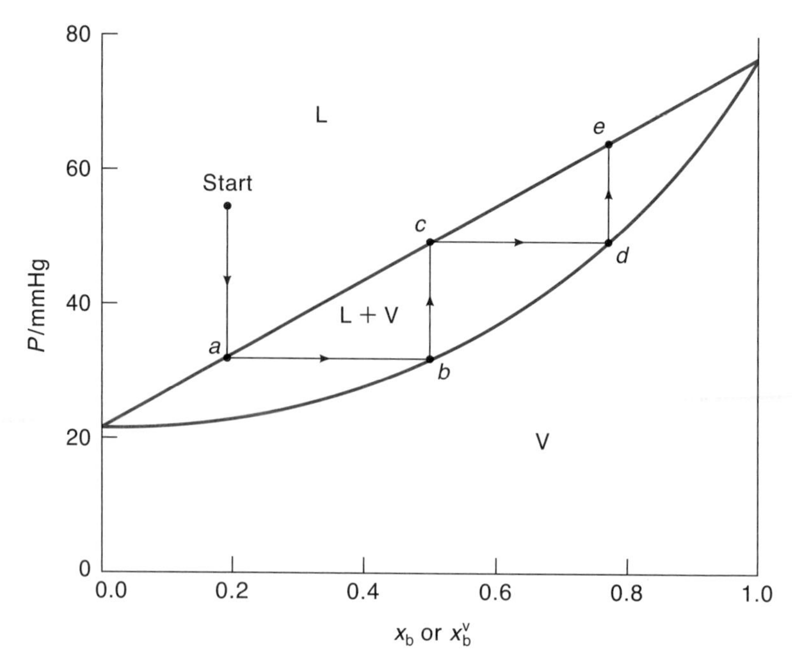

Figure 6.11
Pressure–composition diagram of the benzene–toluene system at 23°C.

Temperature–Composition Diagram. In practice, distillation is carried out more conveniently at constant pressure than at constant temperature; therefore, we need to examine the *temperature–composition* or *boiling-point diagram*. The relationship between temperature and composition is complex and is usually determined experimentally.

Referring again to the benzene–toluene system and comparing Figure 6.12 with Figure 6.9, we see that the liquid and vapor regions are inverted and that neither the liquid nor the vapor curve is a straight line. The more volatile component, benzene, has a higher vapor pressure and hence a lower boiling point. Figure 6.9 shows that at a constant temperature, liquid is the stable phase at high pressures. Similarly, Figure 6.12 tells us that at a constant pressure, the stable phase at low temperatures is also the liquid. During fractional distillation, the solution is heated until it starts to evaporate ($a \rightarrow b$). The vapor, which is richer in benzene, is condensed ($b \rightarrow c$) and then evaporated ($c \rightarrow d$). Repeating the process will eventually separate the two components completely. This procedure is called *fractional distillation*. Each vaporization and condensation step is called a *theoretical plate*.

In the laboratory, chemists use an apparatus like that shown in Figure 6.13 to separate volatile liquids. The round-bottomed flask containing the benzene–toluene solution is fitted with a long column packed with small glass beads. When the solution boils, the vapor condenses on the beads in the lower portion of the column, and the liquid falls back into the distilling flask. As time goes on, the beads gradually heat up, allowing the vapor to move upward slowly. In essence, the packing material gives the column many theoretical plates and causes the benzene–toluene mixture to be subjected continuously to numerous vaporization–condensation steps. At each step, the composition of the vapor in the column will be richer in the more volatile, or lower boiling-point, component (in this case, benzene). The vapor that rises to the top

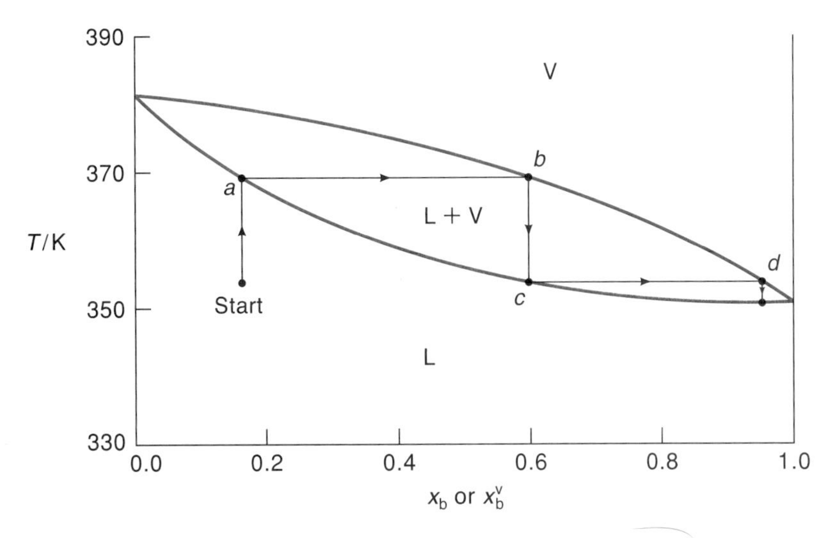

Figure 6.12
Temperature–composition diagram of the benzene–toluene system at 1 atm. The boiling points of benzene and toluene are 80.1°C and 110.6°C, respectively.

of the column is essentially pure benzene, which is then condensed and collected in a receiving flask. Petroleum refining employs a similar approach. Crude oil is a complex mixture of thousands of compounds. By heating and condensing the crude oil up a distillation column that may be 80 meters high and contain hundreds of theoretical plates, workers can separate fractions according to their boiling-point ranges.

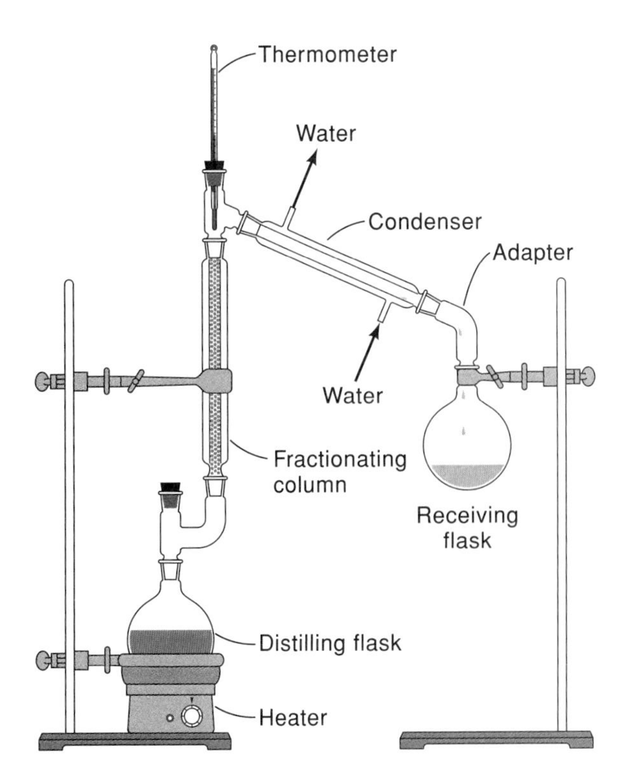

Figure 6.13
A laboratory setup for fractional distillation. The fractioning column contains many small glass beads, which act as theoretical plates for the condensation–evaporation steps.

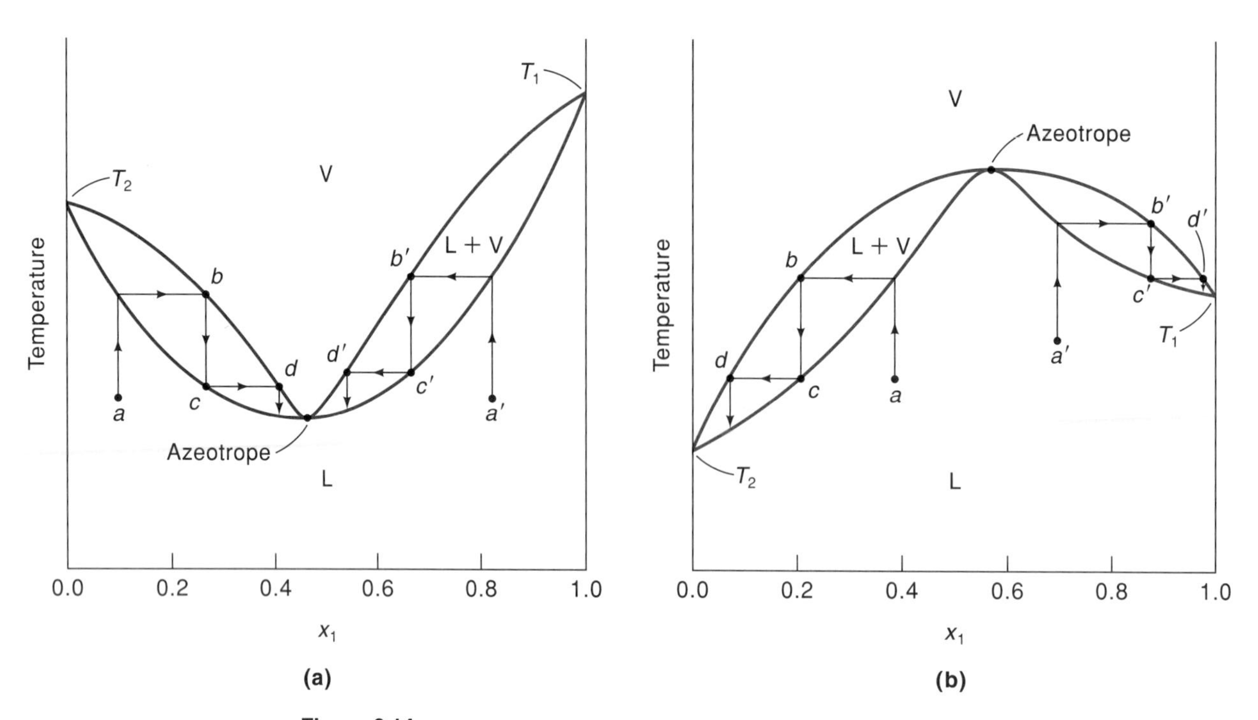

Figure 6.14
Azeotropes: (a) minimum boiling point, (b) maximum boiling point.

Azeotropes. Because most solutions are nonideal, their experimentally determined temperature–composition diagrams are more complicated than the one shown in Figure 6.12. If the system exhibits a positive deviation from Raoult's law, the curve will show a minimum boiling point. Conversely, a negative deviation from Raoult's law will result in a maximum boiling point (Figure 6.14). Examples of the former are acetone–carbon disulfide, ethanol–water, and *n*-propanol–water. Systems that show a maximum boiling point are less common; among the known examples are acetone–chloroform and hydrochloric acid–water. In every case, the mixture cannot be completely separated into pure components by simple fractional distillation.

Consider the following steps, shown in Figure 6.14a. The solution is heated at a certain composition denoted by point a. The condensed vapor ($b \rightarrow c$) becomes richer in component 1, while the solution remaining in the pot becomes richer in component 2. Consequently, the point representing the composition of the solution in the pot on the liquid curve will move toward the left as the distillation proceeds and the boiling point rises. Boiling the solution, condensing the vapor, and again boiling the condensed vapor results in a distillate that has the same composition as the solution in the pot. Such a distillate is known as an *azeotrope* (from the Greek word meaning "to boil unchanged"). The boiling point of the solution remaining in the pot will eventually reach that of pure component 2 (i.e., T_2). Once the azeotrope distillate has been produced, further distillation will result in no further separation, and it will boil at a constant temperature. If we started at point a' and went through the same evaporation–condensation steps, the vapor would become richer in component 2 until it formed the same azeotrope, while the boiling point of the solution would approach that of component 1. The maximum boiling point system (Figure 6.14b) can be similarly explained except that a pure component appears in the distillate, and the azeotrope appears in the pot.

Table 6.3
Variation of Boiling Point and Composition of the HCl-Water Azeotrope with Pressure[a]

P/torr	Composition (percent by weight HCl)	Boiling point/°C
760	20.222	108.584
700	20.360	106.424
500	20.916	97.578

[a] W. D. Bonner and R. E. Wallace, *J. Am. Chem. Soc.* **52**, 1747 (1930).

Although an azeotrope behaves in distillation as if it were a single component, that can easily be demonstrated not to be the case. As the data in Table 6.3 show, the composition of the azeotrope depends on the pressure.

Solid–Liquid Equilibria

If a liquid solution of two substances is cooled to a sufficiently low temperature, a solid will form. This temperature is the solution's freezing point, which depends on the composition of the solution. As we shall see in the next section, the freezing point of a solution is always lower than that of the solvent.

Let us consider a two-component system made of antimony (Sb) and lead (Pb). The solid–liquid phase diagram for this system is shown in Figure 6.15. This phase diagram is constructed by measuring the melting points of a series of solutions of

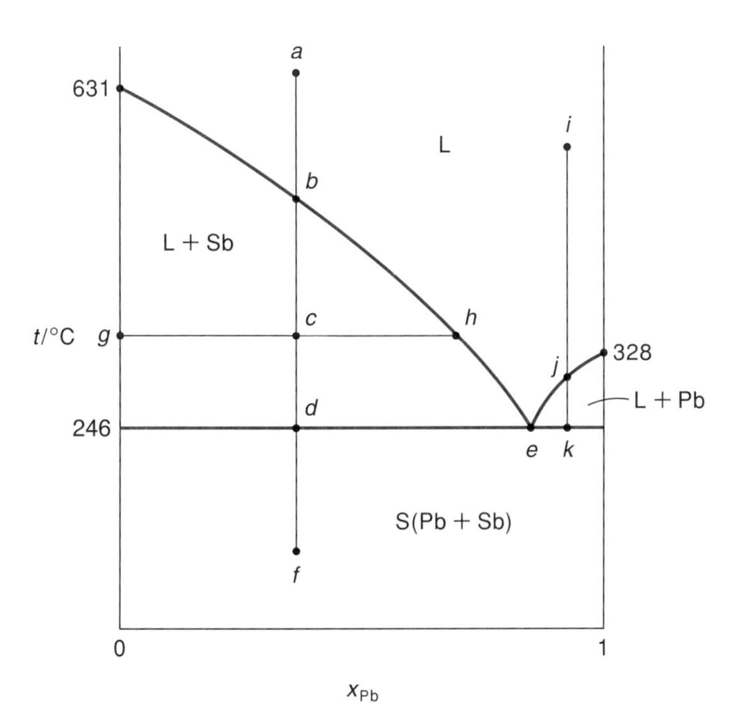

Figure 6.15
The solid–liquid phase diagram of the lead–antimony system. The eutectic point is at *e*.

different compositions at constant pressure. The asymmetric, V-shaped curve is the freezing point curve, above which the system is liquid. (The melting points of Pb and Sb are 328°C and 631°C, respectively.) Consider what happens when the solution at point a is cooled at constant pressure. When point b is reached, the solution begins to freeze, and the solid that separates from the solution is pure Sb. As the temperature is lowered further, more Sb freezes, and the solution becomes progressively richer in Pb. The composition of the solution at point c, for example, is given by drawing the tie line gch. As this point, the composition of the solution is given by the projection of the vertical line from point h onto the x axis. If we continue to lower the temperature of the solution, eventually we reach point d. At this temperature, the composition of the solution is given by point e.

Now suppose we start to cool the solution at point i. At point j, the solution begins to freeze, and solid Pb forms. On further cooling, eventually we reach point k. At this point, the composition of the solution is again given by point e; therefore, this is the point at which the liquid is in equilibrium with both solids. Point e is called the *eutectic point*. The eutectic point has the following significance: (1) It represents the lowest temperature at which a liquid solution can exist, and (2) at the eutectic point, the solid that separates from the solution has the *same* composition as the solution. In this respect, a solution that has eutectic composition behaves like a pure compound. Because the composition of a eutectic mixture depends on external pressure, however, we can readily distinguish its behavior from that of a pure liquid by studying the freezing phenomenon at different pressures.

The eutectic point is at 246°C.

At the eutectic point, we have two components (Pb and Sb) and three phases (solid Pb, solid Sb, and solution), so the degrees of freedom, f, is given by

$$\begin{aligned} f &= c - p + 2 \\ &= 2 - 3 + 2 \\ &= 1 \end{aligned}$$

This single degree of freedom, however, is used to specify the pressure. Consequently, at the eutectic point, neither the temperature nor the composition can vary.

A familiar eutectic mixture is the solder used in constructing electronic circuit boards. Solder is about 33% lead and 67% tin and melts at 183°C. (Tin melts at 232°C.)

6.7 Colligative Properties

General properties of solutions include vapor-pressure lowering, boiling-point elevation, freezing-point depression, and osmotic pressure. These properties are commonly referred to as *colligative*, or *collective*, *properties* because they are bound together through their common origin. Colligative properties depend only on the number of solute molecules present, not on the size or molar mass of the molecules. To derive equations describing these phenomena, we shall make three important assumptions: (1) The solutions are ideal dilute, so that the solvent obeys Raoult's law; (2) the solutions are dilute; and (3) the solutions contain nonelectrolytes. As usual, we shall consider only a two-component system.

Vapor-Pressure Lowering

Consider a solution that contains a solvent 1 and a *nonvolatile* solute 2, such as a solution of sucrose in water. Because the solution is ideal dilute, Raoult's law applies:

$$P = x_1 P_1^*$$

Because $x_1 = 1 - x_2$, the equation above becomes

$$P_1 = (1 - x_2) P_1^*$$

Rearranging this equation gives

$$P_1^* - P_1 = \Delta P = x_2 P_1^* \tag{6.33}$$

where ΔP, the decrease in vapor pressure from that of the pure solvent, is directly proportional to the mole fraction of the solute.

Why does the vapor pressure of a solution fall in the presence of a solute? We may be tempted to suggest that it does because of the modification of intermolecular forces. But this idea is not right because vapor-pressure lowering occurs even in ideal solutions, in which there is no difference between solute–solvent and solvent–solvent interactions. A more convincing explanation is provided by the entropy effect. When a solvent evaporates, the entropy of the universe increases because the entropy of any substance in the gaseous state is greater than that in the liquid state (at the same temperature). As we saw in Section 6.3, the solution process itself is accompanied by an increase in entropy. This result means that there is an extra degree of randomness, or disorder, in a solution that was not present in the pure solvent. Therefore, the evaporation of solvent from a solution will result in a *smaller* increase in entropy. Consequently, the solvent has less of a tendency to leave the solution, and the solution will have a lower vapor pressure than the pure solvent (see Problem 6.40).

Boiling-Point Elevation

The boiling point of a solution is the temperature at which its vapor pressure is equal to the external pressure. The previous discussion might lead you to expect that because the addition of a nonvolatile solute lowers the vapor pressure, it should also raise the boiling point of a solution. This effect is indeed the case.

For a solution containing a *nonvolatile* solute, the boiling-point elevation originates in the change in the chemical potential of the solvent due to the presence of the solute. From Equation 6.17, we can see that the chemical potential of the solvent in a solution is less than the chemical potential of the pure solvent by an amount equal to $RT \ln x_1$. How this change affects the boiling point of the solution can be seen from Figure 6.16. The solid lines refer to the pure solvent. Because the solute is nonvolatile, it does not vaporize; therefore, the curve for the vapor phase is the same as that for the pure vapor. On the other hand, because the liquid contains a solute, the chemical potential of the solvent decreases (see the dashed curve). The points where the curve for the vapor intersects the curves for the liquids (pure and solution) correspond to the boiling points of the pure solvent and the solution, respectively. We see that the boiling point of the solution (T_b') is higher than that of the pure solvent (T_b).

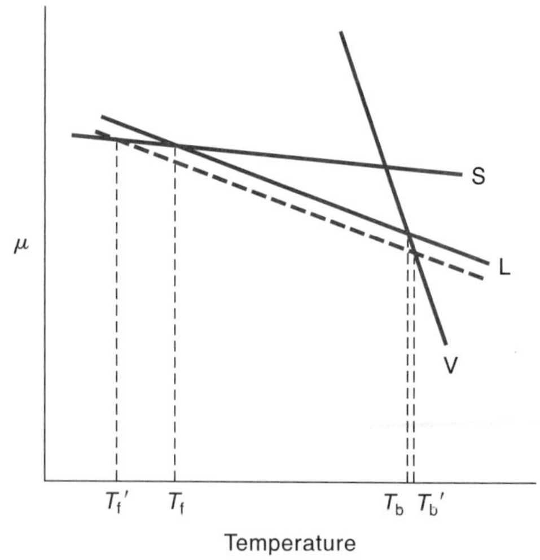

Figure 6.16
Plot of chemical potentials versus temperature to illustrate colligative properties. The dashed red line denotes the solution phase. T_b and T_b' are the boiling points of the solvent and the solution, and T_f and T_f' are the freezing points of the solvent and solution, respectively.

We now turn to a quantitative treatment of the boiling-point elevation phenomenon. At the boiling point, the solvent vapor is in equilibrium with the solvent in solution, so that

$$\mu_1(g) = \mu_1(l) = \mu_1^*(l) + RT \ln x_1$$

or

$$\Delta\mu_1 = \mu_1(g) - \mu_1^*(l) = RT \ln x_1 \tag{6.34}$$

where $\Delta\mu_1$ is the Gibbs energy change associated with the evaporation of 1 mole of solvent from the solution at temperature T, its boiling point. Thus, we can write $\Delta\mu_1 = \Delta_{\text{vap}}\overline{G}$. Dividing Equation 6.34 by T, we obtain

$$\frac{\Delta_{\text{vap}}\overline{G}}{T} = \frac{\mu_1(g) - \mu_1^*(l)}{T} = R \ln x_1$$

From the Gibbs–Helmholz equation (Equation 5.15), we write

$$\frac{d(\Delta G/T)}{dT} = -\frac{\Delta H}{T^2} \quad \text{(at constant } P\text{)}$$

or

$$\frac{d(\Delta_{\text{vap}}\overline{G}/T)}{dT} = \frac{-\Delta_{\text{vap}}\overline{H}}{T^2} = R\frac{d(\ln x_1)}{dT}$$

where $\Delta_{\text{vap}}\overline{H}$ is the molar enthalpy of vaporization of the solvent from the solution. Because the solution is dilute, $\Delta_{\text{vap}}\overline{H}$ is taken to be the same as the molar enthalpy of vaporization of the pure solvent. Rearranging the last equation gives

$$d \ln x_1 = \frac{-\Delta_{\text{vap}}\overline{H}}{RT^2} dT \tag{6.35}$$

To find the relationship between x_1 and T, we integrate Equation 6.35 between the limits T_b' and T_b, the boiling points of the solution and pure solvent, respectively. Because the mole fraction of the solvent is x_1 at T_b' and 1 at T_b, we write

$$\int_{\ln 1}^{\ln x_1} d\ln x_1 = \int_{T_b}^{T_b'} \frac{-\Delta_{\text{vap}}\overline{H}}{RT^2}\,dT$$

or

$$\begin{aligned}\ln x_1 &= \frac{\Delta_{\text{vap}}\overline{H}}{R}\left(\frac{1}{T_b'} - \frac{1}{T_b}\right)\\ &= \frac{-\Delta_{\text{vap}}\overline{H}}{R}\left(\frac{T_b' - T_b}{T_b'T_b}\right)\\ &= \frac{-\Delta_{\text{vap}}\overline{H}}{R}\frac{\Delta T}{T_b^2} \qquad (6.36)\end{aligned}$$

where $\Delta T = T_b' - T_b$. Two assumptions were used to obtain Equation 6.36, both of which are based on the fact that T_b' and T_b differ only by a small amount (a few degrees). First, we assumed $\Delta_{\text{vap}}\overline{H}$ to be temperature independent and second, $T_b' \approx T_b$, so that $T_b'T_b \approx T_b^2$.

Equation 6.36 gives the boiling-point elevation, ΔT, in terms of the concentration of the solvent (x_1). By custom, however, we express the concentration in terms of the amount of solute present, so we write

$$\ln x_1 = \ln(1 - x_2) = \frac{-\Delta_{\text{vap}}\overline{H}}{R}\frac{\Delta T}{T_b^2}$$

where*

$$\begin{aligned}\ln(1 - x_2) &= -x_2 - \frac{x_2^2}{2} - \frac{x_2^3}{3}\cdots\\ &\approx -x_2 \quad (x_2 \ll 1)\end{aligned}$$

We now have

$$\Delta T = \frac{RT_b^2}{\Delta_{\text{vap}}\overline{H}}x_2$$

To convert the mole fraction x_2 into a more practical concentration unit, such as molality (m_2), we write

$$x_2 = \frac{n_2}{n_1 + n_2} \approx \frac{n_2}{n_1} = \frac{n_2}{w_1/\mathcal{M}_1} \quad (n_1 \gg n_2)$$

* This expansion is known as Maclaurin's series. You can verify this relationship by employing a small numerical value for x_2 (≤ 0.1).

where w_1 is the mass of the solvent in kg and $\mathscr{M}_1$ is the molar mass of the solvent in kg mol^{-1}, respectively. Because n_2/w_1 gives the molality of the solution, m_2, it follows that $x_2 = \mathscr{M}_1 m_2$ and thus

$$\Delta T = \frac{RT_b^2 \mathscr{M}_1}{\Delta_{vap}\overline{H}} m_2 \tag{6.37}$$

Note that all the quantities in the first term on the right of Equation 6.37 are constants for a given solvent, and so we have

$$K_b = \frac{RT_b^2 \mathscr{M}_1}{\Delta_{vap}\overline{H}} \tag{6.38}$$

where K_b is called the *molal boiling-point-elevation constant*. The units of K_b are K mol^{-1} kg. Finally,

$$\Delta T = K_b m_2 \tag{6.39}$$

The advantage of using molality, as mentioned in Section 6.1, is that it is independent of temperature and thus is suitable for boiling-point elevation studies.

Figure 6.17 shows the phase diagrams of pure water and an aqueous solution. Upon the addition of a nonvolatile solute, the vapor pressure of the solution decreases at every temperature. Consequently, the boiling point of the solution at 1 atm will be greater than 373.15 K.

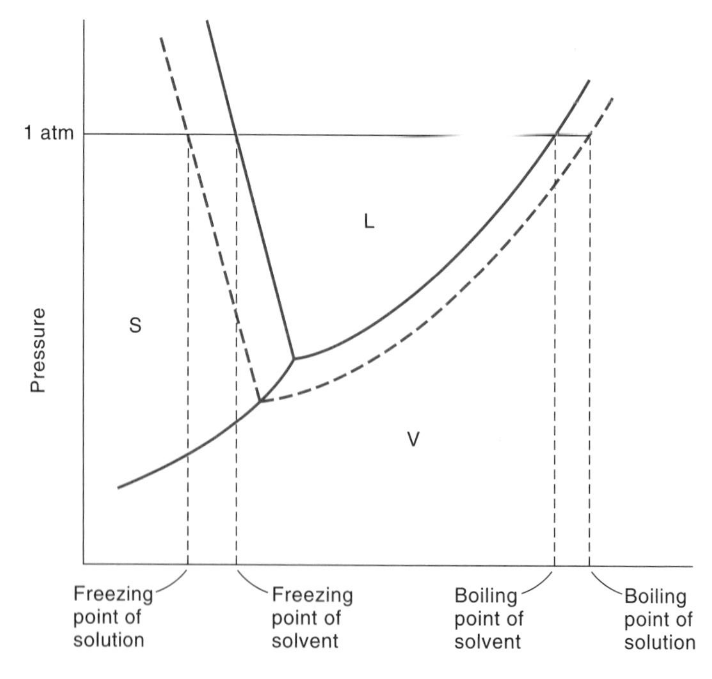

Figure 6.17
Phase diagrams of pure water (solid red lines) and of water in an aqueous solution containing a nonvolatile solid (dashed red lines). (Note that the axes are not drawn to scale.)

Freezing-Point Depression

A nonchemist may be forever unaware of the boiling-point-elevation phenomenon, but any casual observer living in a cold climate witnesses an illustration of freezing-point depression: ice on winter roads and sidewalks melts readily when sprinkled with salt.* This method of thawing depresses the freezing point of water.

The thermodynamic analysis of freezing-point depression is similar to that of boiling-point elevation. If we assume that when a solution freezes, the solid that separates from the solution contains only the solvent, then the curve for the chemical potential of the solid does not change (see Figure 6.16). Consequently, the solid curve for the solid and the dashed curve for the solvent in solution now intersect at a point (T_f') *below* the freezing point of the pure solvent (T_f). By following the same procedure as that for the boiling-point elevation, we can show that the drop in freezing point ΔT (i.e., $T_f - T_f'$, where T_f and T_f' are the freezing points of the pure solvent and solution, respectively) is

$$\Delta T = K_f m_2 \tag{6.40}$$

where K_f is the *molal freezing-point-depression constant* given by

$$K_f = \frac{RT_f^2 \mathcal{M}_1}{\Delta_{fus}\overline{H}} \tag{6.41}$$

where $\Delta_{fus}\overline{H}$ is the molar enthalpy of fusion of the solvent.

The freezing-point-depression phenomenon can also be understood by studying Figure 6.17. At 1 atm, the freezing point of solution lies at the intersection point of the dashed curve (between the solid and liquid phases) and the horizontal line at 1 atm. It is interesting that whereas the solute must be nonvolatile in the boiling-point-elevation case, no such restriction applies to lowering the freezing point. A proof of this statement is the use of ethanol (b.p. = 351.65 K) as an antifreeze.

Either Equation 6.39 or 6.40 can be used to determine the molar mass of a solute. In general, the freezing-point-depression experiment is much easier to carry out. It is commonly employed in measuring the molar mass of compounds. Table 6.4 lists the K_b and K_f values for several common solvents.

The freezing-point-depression phenomenon has many examples in everyday life and in biological systems. As mentioned above, salts, such as sodium chloride and calcium chloride, are used to melt ice on roads and sidewalks. The organic compound ethylene glycol [$CH_2(OH)CH_2(OH)$] is the common automobile antifreeze. It is also employed to de-ice airplanes. In recent years, there has been much interest in understanding how certain species of fish manage to survive in the ice-cold waters of the polar oceans. The freezing point of seawater is approximately −1.9°C, which is the temperature of seawater surrounding an iceberg. A depression in freezing point of 1.9 degrees corresponds to a concentration of one molal, which is much too high for proper physiological function; for example, it alters osmotic balance (see section below on Osmotic Pressure). Besides dissolved salts and other substances that can

* The salt employed is usually sodium chloride, which attacks cement and is harmful to many plants. Also see "Freezing Ice Cream and Making Caramel Topping," J. O. Olson and L. H. Bowman, *J. Chem. Educ.* **53**, 49 (1976).

Table 6.4
Molal Boiling-Point-Elevation and Molal Freezing-Point-Depression Constants of Some Common Solvents

Solvent	K_b/K mol^{-1} kg	K_f/K mol^{-1} kg
H_2O	0.51	1.86
C_2H_5OH	1.22	1.99
C_6H_6	2.53	5.12
$CHCl_3$	3.63	4.68
CH_3COOH	2.93	3.90
CCl_4	5.03	29.8

lower the freezing point colligatively, a special class of proteins resides in the blood of polar fishes that has some kind of protective effect. These proteins contain both amino acid and sugar units and are called glycoprotetins. The concentration of glycoproteins in the fishes' blood is quite low (approximately 4×10^{-4} m), so their action cannot be explained by colligative properties. The belief is that the glycoproteins have the ability to adsorb onto the surface of each tiny ice crystal as soon as it begins to form, thus preventing it from growing to a size that would cause biological damage. Consequently, the freezing point of blood in these fishes is below −2°C.

EXAMPLE 6.4

For a solution of 45.20 g of sucrose ($C_{12}H_{22}O_{11}$) in 316.0 g of water, calculate (a) the boiling point, and (b) the freezing point.

ANSWER

(a) Boiling point: $K_b = 0.51$ K mol^{-1} kg, and the molality of the solution is given by

$$m_2 = \frac{(45.20\text{ g})(1000\text{ g/1 kg})}{(342.3\text{ g mol}^{-1})(316.0\text{ g})} = 0.418\text{ mol kg}^{-1}$$

From Equation 6.39

$$\begin{aligned}\Delta T &= K_b m_2 \\ &= (0.51\text{ K mol}^{-1}\text{ kg})(0.418\text{ mol kg}^{-1}) \\ &= 0.21\text{ K}\end{aligned}$$

Thus, the solution will boil at (373.15 + 0.21) K, or 373.36 K.

(b) Freezing point: From Equation 6.40,

$$\Delta T = K_f m_2$$
$$= (1.86 \text{ K mol}^{-1} \text{ kg})(0.418 \text{ mol kg}^{-1})$$
$$= 0.78 \text{ K}$$

Thus, the solution will freeze at (273.15 − 0.78) K, or 272.37 K.

COMMENT

For aqueous solutions of equal concentrations, the depression in freezing point is always greater than the corresponding elevation in boiling point. The reason can be seen by comparing the following two expressions from Equations 6.38 and 6.41:

$$K_b = \frac{RT_b^2 \mathcal{M}_1}{\Delta_{vap}\overline{H}} \qquad K_f = \frac{RT_f^2 \mathcal{M}_1}{\Delta_{fus}\overline{H}}$$

Although $T_b > T_f$, $\Delta_{vap}\overline{H}$ for water is 40.79 kJ mol^{-1}, whereas $\Delta_{fus}\overline{H}$ is only 6.01 kJ mol^{-1}. The large value of $\Delta_{vap}\overline{H}$ in the denominator is what causes K_b and hence ΔT to be smaller.

Osmotic Pressure

The phenomenon of *osmosis* is illustrated in Figure 6.18. The left compartment of the apparatus contains pure solvent; the right compartment contains a solution. The two compartments are separated by a *semipermeable membrane* (e.g., a cellophane membrane), one that permits the solvent molecules to pass through but does not permit the movement of solute molecules from right to left. Practically speaking, then, this system has two different phases. At equilibrium, the height of the solution in the tube on the right is greater than that of the pure solvent in the left tube by h. This excess hydrostatic pressure is called the *osmotic pressure*. We can now derive an expression for osmotic pressure as follows.

Let μ_1^L and μ_1^R be the chemical potential of the solvent in the left and right compartments, respectively. Initially, before equilibrium is established, we have

$$\mu_1^L = \mu_1^* + RT \ln x_1$$
$$= \mu_1^* \qquad (x_1 = 1)$$

and

$$\mu_1^R = \mu_1^* + RT \ln x_1 \qquad (x_1 < 1)$$

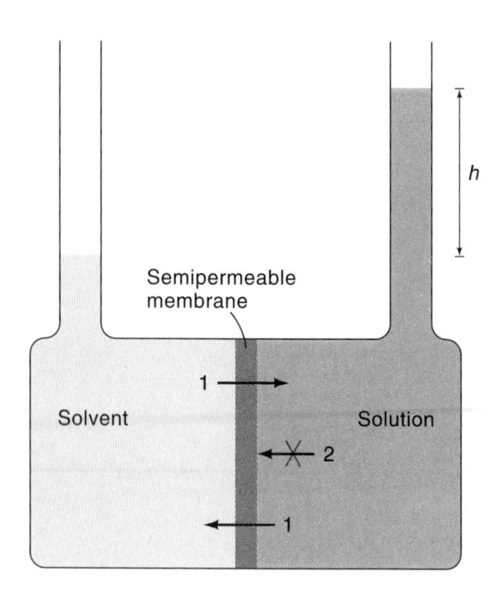

Figure 6.18
Apparatus demonstrating the osmotic pressure phenomenon. Here 1 represents the solvent and 2 the solute molecules.

Thus,

$$\mu_1^{\mathrm{L}} = \mu_1^* > \mu_1^{\mathrm{R}} = \mu_1^* + RT \ln x_1$$

Note that μ_1^{L} is the same as the standard chemical potential for the pure solvent, μ_1^*, and the inequality sign denotes that $RT \ln x_1$ is a negative quantity. Consequently, more solvent molecules, on the average, will move from left to right across the membrane. The process is spontaneous because the dilution of the solution in the right compartment by solvent leads to a decrease in the Gibbs energy and an increase in entropy. Equilibrium is finally reached when the flow of solvent is exactly balanced by the hydrostatic pressure difference in the two side tubes. This extra pressure increases the chemical potential of the solvent in solution, μ_1^{R}. From Equation 5.16, we know that

$$\left(\frac{\partial G}{\partial P}\right)_T = V$$

We can write a similar equation for the variation of the chemical potential with pressure at constant temperature. Thus, for the solvent component in the right compartment,

$$\left(\frac{\partial \mu_1^{\mathrm{R}}}{\partial P}\right)_T = \overline{V}_1 \tag{6.42}$$

where $\overline{V}_1$ is the partial molar volume of the solvent. For a dilute solution, $\overline{V}_1$ is approximately equal to $\overline{V}$, the molar volume of the pure solvent. The increase in the chemical potential of the solvent in the solution compartment ($\Delta\mu_1^{\mathrm{R}}$) when the pressure increases from P, the external atmospheric pressure, to $(P + \Pi)$ is given by

$$\Delta\mu_1^{\mathrm{R}} = \int_P^{P+\Pi} \overline{V}\, dP = \overline{V}\Pi$$

Note that $\overline{V}$ is treated as a constant because the volume of a liquid changes little with pressure. The Greek letter Π represents the osmotic pressure. The term *osmotic pressure of a solution* refers to the pressure that must be applied to the solution to increase the chemical potential of the solvent to the value of its pure liquid under atmospheric pressure.

At equilibrium, the following relations must hold:

$$\mu_1^{\mathrm{L}} = \mu_1^{\mathrm{R}} = \mu_1^* + RT \ln x_1 + \Pi\overline{V}$$

Because $\mu_1^{\mathrm{L}} = \mu_1^*$, we have

$$\Pi\overline{V} = -RT \ln x_1 \tag{6.43}$$

To relate Π to the concentration of the solute, we take the following steps. From the procedure employed for boiling-point elevation:

$$\begin{aligned} -\ln x_1 &= -\ln (1 - x_2) \\ &\approx x_2 \quad (x_2 \ll 1) \end{aligned}$$

Furthermore,

$$x_2 = \frac{n_2}{n_1 + n_2} \approx \frac{n_2}{n_1} \quad (n_1 \gg n_2)$$

where n_1 and n_2 are the number of moles of solvent and solute, respectively. Equation 6.43 now becomes

$$\begin{aligned} \Pi\overline{V} &= RTx_2 \\ &= RT\left(\frac{n_2}{n_1}\right) \end{aligned} \tag{6.44}$$

Substituting $\overline{V} = V/n_1$ into Equation 6.44, we get

$$\Pi V = n_2 RT \tag{6.45}$$

If V is in liters, then

$$\begin{aligned} \Pi &= \frac{n_2}{V} RT \\ &= MRT \end{aligned} \tag{6.46}$$

Equation 6.46 also applies to two similar solutions that have different concentrations.

where M is the molarity of the solution. Note that molarity is a convenient concentration unit here, because osmotic pressure measurements are normally made at constant temperature. Alternatively, we can rewrite Equation 6.46 as

$$\Pi = \frac{c_2}{\mathcal{M}_2} RT \tag{6.47}$$

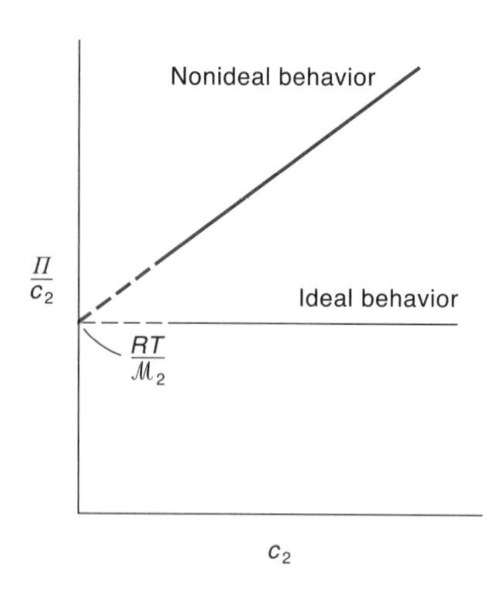

Figure 6.19
Determination of the molar mass of a solute by osmotic pressure measurements for an ideal and a nonideal solution. The intercept on the y axis (as $c_2 \rightarrow 0$) gives the correct value for the molar mass.

or

$$\frac{\Pi}{c_2} = \frac{RT}{\mathcal{M}_2} \tag{6.48}$$

where c_2 is the concentration of the solute in g L^{-1} of the solution and $\mathcal{M}_2$ is the molar mass of the solute. Equation 6.48 provides a way to determine molar masses of compounds from osmotic pressure measurements.

Equation 6.48 is derived by assuming ideal behavior, so it is desirable to measure Π at several different concentrations and extrapolate to zero concentration for molar mass determination (Figure 6.19). For a nonideal solution, the osmotic pressure at *any* concentration, c_2, is given by

Compare Equation 6.49 with Equation 1.10

$$\frac{\Pi}{c_2} = \frac{RT}{\mathcal{M}_2}(1 + Bc_2 + Cc_2^2 + Dc_2^3 + \cdots) \tag{6.49}$$

where B, C, and D are called the second, third, and fourth virial coefficients, respectively. The magnitude of the virial coefficients is such that $B \gg C \gg D$. In dilute solutions, we need be concerned only with the second virial coefficient. For an ideal solution, the second and higher virial coefficients are all equal to zero, so Equation 6.49 reduces to Equation 6.48.

Even though osmosis is a well-studied phenomenon, the mechanism involved is not always clearly understood. In some cases, a semipermeable membrane may act as a molecular sieve, allowing smaller solvent molecules to pass through while blocking larger solute molecules. In other cases, osmosis may be caused by the higher solubility of the solvent in the membrane than the solute. Each system must be studied individually. The previous discussion illustrates both the usefulness and limitation of thermodynamics. We have derived a convenient equation relating the molar mass of the solute to an experimentally measurable quantity—the osmotic pressure—simply in terms of the chemical potential difference. Because thermodynamics is not based on any specific model, however, Equation 6.47 tells us nothing about the mechanism of osmosis.

EXAMPLE 6.5

Consider the following arrangement, in which a solution containing 20 g of hemoglobin in 1 liter of the solution is placed in the right compartment, and pure water is placed in the left compartment (see Figure 6.18). At equilibrium, the height of the water in the right column is 77.8 mm in excess of the height of the solvent in the left column. What is the molar mass of hemoglobin? The temperature of the system is constant at 298 K.

ANSWER

To determine the molar mass of hemoglobin, we first need to calculate the osmotic pressure of the solution. We start by writing

$$\text{pressure} = \frac{\text{force}}{\text{area}}$$

$$= \frac{Ah\rho g}{A} = h\rho g$$

where A is the cross-sectional area of the tube, h the excess liquid height in the right column, ρ the density of the solution, and g the acceleration due to gravity. The parameters are

$$h = 0.0778 \text{ m}$$

$$\rho = 1 \times 10^3 \text{ kg m}^{-3}$$

$$g = 9.81 \text{ m s}^{-2}$$

(We have assumed that the density of the dilute solution is the same as that of water.) The osmotic pressure in pascals (N m^{-2}) is given by

$$\Pi = 0.0778 \text{ m} \times 1 \times 10^3 \text{ kg m}^{-3} \times 9.81 \text{ m s}^{-2}$$

$$= 763 \text{ kg m}^{-1} \text{ s}^{-2}$$

$$= 763 \text{ N m}^{-2}$$

From Equation 6.47,

$$\mathcal{M}_2 = \frac{c_2}{\Pi} RT$$

$$= \frac{(20 \text{ kg m}^{-3})(8.314 \text{ J K}^{-1} \text{ mol}^{-1})(298 \text{ K})}{763 \text{ N m}^{-2}} \times \frac{1 \text{ N m}}{1 \text{ J}}$$

$$= 65 \text{ kg mol}^{-1}$$

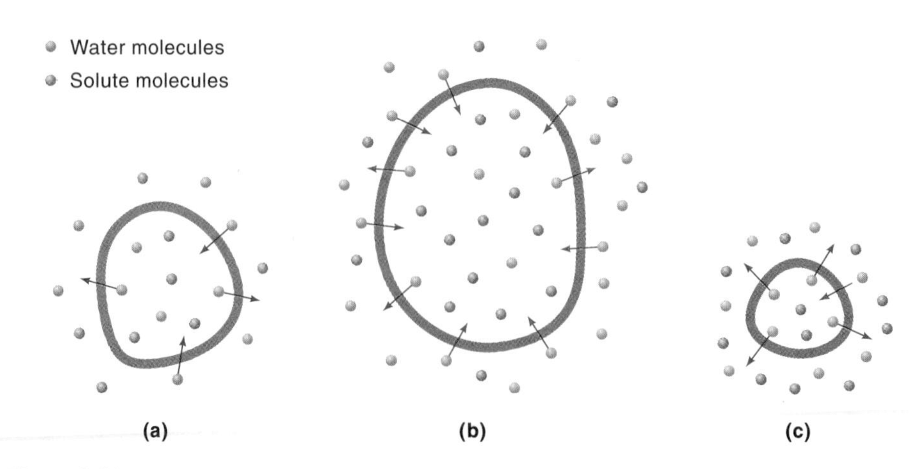

Figure 6.20
A cell in (a) an isotonic solution, (b) a hypotonic solution, and (c) a hypertonic solution. The cell remains unchanged in (a), swells in (b) , and shrinks in (c).

High-resolution mass spectrometry is now a common and convenient method for determining the molar mass of macromolecules.

Example 6.5 shows that osmotic pressure measurement is a more sensitive method to determine molar mass than the boiling-point-elevation and freezing-point-depression techniques, because 7.8 cm is an easily measurable height. On the other hand, the same solution will lead to an elevation in boiling point of approximately 1.6×10^{-4} °C and a depression in freezing point of 5.8×10^{-4} °C, which are too small to measure accurately. Most proteins are less soluble than hemoglobin. Nevertheless, their molar masses can often be determined by osmotic pressure measurements. A disadvantage of osmotic pressure measurements is that the time to reach equilibrium may be hours or days.

Many examples of the osmotic-pressure phenomenon are found in chemical and biological systems. If two solutions are of equal concentration, and hence have the same osmotic pressure, they are said to be *isotonic*. For two solutions of unequal osmotic pressures, the more concentrated solution is said to be *hypertonic*, and the less concentrated solution is said to be *hypotonic* (Figure 6.20). To study the contents of red blood cells, which are protected from the outside environment by a semipermeable membrane, biochemists employ a technique called *hemolysis*. They place the red blood cells in a hypotonic solution, which causes water to move into the cell. The cells swell and eventually burst, releasing hemoglobin and other protein molecules. When a cell is placed in a hypertonic solution, on the other hand, the intracellular water tends to move out of the cell by osmosis to the more concentrated, surrounding solution. This process, known as *crenation*, causes the cell to shrink and eventually cease functioning.

The mammalian kidney is a particularly effective osmotic device. Its main function is to remove metabolic waste products and other impurities from the bloodstream by osmosis to the more concentrated urine outside through a semipermeable membrane. Biologically important ions (such as Na^+ and Cl^-) lost in this manner are then actively pumped back into the blood through the same membrane. The loss of water through the kidney is controlled by the antidiuretic hormone (ADH), which is secreted into the blood by the hypothalamus and posterior pituitary gland. When little or no ADH is secreted, large amounts of water (perhaps 10 times normal) pass into the urine each day. On the other hand, when large quantities of ADH are present in the blood, the permeability of water through the membrane decreases so that the volume of urine formed may be as little as one-half the normal amount. Thus, the kidney–ADH combination controls the rate of loss of both water and other small waste molecules.

The chemical potential of water within the body fluids of freshwater fishes is lower than that in their environment, so they are able to draw in water by osmosis through their gill membranes. Surplus water is excreted as urine. An opposite process occurs for the marine teleost fishes. They lose body water to the more concentrated environment by osmosis across the gill membranes. To balance the loss, they drink seawater.

Osmotic pressure is also the major mechanism for water rising upward in plants. The leaves of trees constantly lose water to their surroundings, a process called *transpiration*, so the solute concentrations in leaf fluids increase. Water is then pushed up through the trunks and branches by osmotic pressure, which, to reach the tops of the tallest trees, can be as high as 10 to 15 atm.* Leaf movement is an interesting phenomenon that may also be related to osmotic pressure. The belief is that some processes can increase salt concentration in leaf cells in the presence of light. Osmotic pressure rises and cells become enlarged and turgid, causing the leaves to orient toward light.

Reverse Osmosis. A related phenomenon to osmosis is called *reverse osmosis*. If we apply pressure greater than the equilibrium osmotic pressure to the solution compartment shown in Figure 6.18, pure solvent will flow from the solution to the solvent compartment. This reversal of the osmotic process results in the unmixing of the solution components. An important application of reverse osmosis is the desalination of water. Several techniques discussed in this chapter are suitable, at least in principle, for obtaining pure water from the sea. For example, either distilling or freezing seawater would achieve the goal. However, these processes involve a phase change from liquid to vapor or liquid to solid and so would require considerable energy input to maintain. Reverse osmosis is more appealing, for it does not involve a phase change and is economically sound for large amounts of water.† Seawater, which is approximately 0.7 *M* in NaCl, has an estimated osmotic pressure of 30 atm. For a 50% recovery of pure water from the sea, an additional 60 atm would have to be applied on the seawater-side compartment to cause reverse osmosis. The success of large-scale desalination depends on the selection of a suitable membrane that is permeable to water but not to dissolved salts and that can withstand the high pressure over long periods of time.

---

* *See* "Entropy Makes Water Run Uphill—in Trees," P. E. Steveson, *J. Chem. Educ.* **48**, 837 (1971).

† *See* "Desalination of Water by Reverse Osmosis," C. E. Hecht, *J. Chem. Educ.* **44**, 53 (1967).

Key Equations

$V = n_1\overline{V}_1 + n_2\overline{V}_2$	(Volume of a solution in terms of partial molar volumes)	(6.6)
$\overline{G}_i = \mu_i = \left(\frac{\partial G}{\partial n_i}\right)_{T,P,n_j}$	(Definition of chemical potential)	(6.7)
$\Delta_{\text{mix}}G = nRT(x_1 \ln x_1 + x_2 \ln x_2)$	(Gibbs energy of mixing)	(6.11)
$\Delta_{\text{mix}}S = -nR(x_1 \ln x_1 + x_2 \ln x_2)$	(Entropy of mixing)	(6.12)
$P_1 = x_1P_1^*$	(Raoult's law)	(6.16)
$\mu_1(l) = \mu_1^*(l) + RT \ln x_1$	(Chemical potential of solvent in an ideal solution)	(6.17)
$P_2 = Kx_2$	(Henry's law)	(6.18)
$P_2 = K'm$	(Henry's law)	(6.19)
$a_1 = \frac{P_1}{P_1^*}$	(Activity of solvent)	(6.21)
$a_1 = \gamma_1 x_1$	(Definition of activity coefficient)	(6.22)
$\mu_2(l) = \mu_2^\circ(l) + RT \ln a_2$	(Chemical potential of solute in a real solution)	(6.25)
$\mu_2(l) = \mu_2^\circ(l) + RT \ln \frac{m_2}{m^\circ}$	(Chemical potential of solute in an ideal solution)	(6.27)
$\Delta P = x_2P_1^*$	(Vapor-pressure lowering)	(6.33)
$\Delta T = K_b m_2$	(Boiling-point elevation)	(6.39)
$\Delta T = K_f m_2$	(Freezing-point depression)	(6.40)
$\Pi = MRT$	(Osmotic pressure)	(6.46)

Suggestions for Further Reading

ARTICLES

General

"Ideal Solutions," W. A. Oates, *J. Chem. Educ.* **46**, 501 (1969).
"Standard States of Real Solutions," A. Lainez and G. Tardajos, *J. Chem. Educ.* **62**, 678 (1985).
"Thermodynamics of Mixing of Ideal Gases: A Persistent Pitfall," E. F. Meyer, *J. Chem. Educ.* **64**, 676 (1987).
"Understanding Chemical Potential," M. P. Tarazona and E. Saiz, *J. Chem. Educ.* **72**, 882 (1995).
"Thermodynamics of Mixing of Real Gases," S. Sattar, *J. Chem. Educ.* **77**, 1361 (2000).
"Applying Chemical Potential and Partial Pressure Concepts To Understanding the Spontaneous Mixing of Helium and Air in a Helium-Inflated Balloon," J.-Y. Lee, H.-S. Yoo, J. S. Park, K.-J. Hwang, and J. S. Kim, *J. Chem. Educ.* **82**, 288 (2005).
"An Undergraduate Experiment Using Differential Scanning Calorimetry: A Study of the Thermal Properties of a Binary Eutectic Alloy of Tin and Lead," R. P. D'Amelia, D. Clark, and W. Nirode, *J. Chem. Educ.* **89**, 548 (2012).

Phase Equilibria

"The Mechanism of Vapor Pressure Lowering," K. J. Mysels, *J. Chem. Educ.* **32**, 179 (1955).
"Deviations from Raoult's Law," M. L. McGlashan, *J. Chem. Educ.* **40**, 516 (1963).
"Removal of an Assumption in Deriving the Phase Change Formula $T = K \cdot m$," F. E. Schubert, *J. Chem. Educ.* **56**, 259 (1979).
"Reappearing Phases," J. S. Walker and C. A. Vause, *Sci. Am.* May 1987.
"The Direct Relation Between Altitude and Boiling Point," L. Earl, *J. Chem. Educ.* **67**, 45 (1990).
"Henry's Law: A Historical View," J. J. Carroll, *J. Chem. Educ.* **70**, 91 (1993).
"Phase Diagrams for Aqueous Systems," R. E. Treptow, *J. Chem. Educ.* **70**, 616 (1993).
"Journey Around a Phase Diagram," N. K. Kildahl, *J. Chem. Educ.* **71**, 1052 (1994).
"Henry's Law and Noisy Knuckles," D. R. Kimbrough, *J. Chem. Educ.* **76**, 1509 (1999).
"Thermodynamics of Water Superheated in the Microwave Oven," B. H. Erne, *J. Chem. Educ.* **77**, 1309 (2000).
"From Chicken Breath to the Killer Lake of Cameroon: Uniting Seven Interesting Phenomena with a Single Chemical Underpinning," R. DeLorenzo, *J. Chem. Educ.* **78**, 191 (2001).
"Henry's Law: A Retrospective," R. M. Rosenberg, *J. Chem. Educ.* **81**, 1647 (2004).
"Determining the Pressure Inside an Unopened Carbonated Beverage," H. de Grys, *J. Chem. Educ.* **82**, 116 (2005).

Colligative Properties

"The Kidney," H. W. Smith, *Sci. Am.* January 1953.
"Desalting Water by Freezing," A. E. Snyder, *Sci. Am.* December 1962.
"Desalination of Water by Reverse Osmosis," C. E. Hecht, *J. Chem. Educ.* **44**, 53 (1967).
"Demonstrating Osmotic and Hydrostatic Pressures in Blood Capillaries," J. W. Ledbetter, Jr., and H. D. Jones, *J. Chem. Educ.* **44**, 362 (1967).
"Reverse Osmosis," M. J. Suess, *J. Chem. Educ.* **48**, 190 (1971).
"Desalination," R. F. Probstein, *Am. Sci.* **61**, 280 (1973).
"Colligative Properties," F. Rioux, *J. Chem. Educ.* **50**, 490 (1973).

"Osmotic Pressure in the Physics Course for Students of the Life Sciences," R. K. Hobbie, *Am. J. Phys.* **42**, 188 (1974).
"A Biological Antifreeze," R. E. Feeney, *Am. Sci.* **62**, 172 (1974).
"Colligative Properties of a Solution," H. T. Hammel, *Science* **192**, 748 (1976).
"Antarctic Fishes," J. T. Eastman and A. C. DeVries, *Sci. Am.* November 1986.
"The Freezing Point Depression Law in Physical Chemistry," H. F. Franzen, *J. Chem. Educ.* **65**, 1077 (1988).
"Regulating Cell Volume," F. Lang and S. Waldegger, *Am. Sci.* **85**, 456 (1997).
"Transporting Water in Plants," M. J. Canny, *Am. Sci.* **86**, 152 (1998).
"An After-Dinner Trick," *J. Chem. Educ.* **79**, 480A (2002).
"A Greener Approach for Measuring Colligative Properties," S. M. McCarthy and S. W. Gordon-Wylie, *J. Chem. Educ.* **82**, 116 (2005).
"Fresh From the Sea," M. Fischetti, *Sci Am.* September 2007.

Problems

Concentration Units

6.1 How many grams of water must be added to 20.0 g of urea to prepare a 5.00% aqueous urea solution by weight?

6.2 What is the molarity of a 2.12 mol kg^{-1} aqueous sulfuric acid solution? The density of this solution is 1.30 g cm^{-3}.

6.3 Calculate the molality of a 1.50 *M* aqueous ethanol solution. The density of the solution is 0.980 g cm^{-3}.

6.4 The concentrated sulfuric acid we use in the laboratory is 98.0% sulfuric acid by weight. Calculate the molality and molarity of concentrated sulfuric acid if the density of the solution is 1.83 g cm^{-3}.

6.5 Convert a 0.25 mol kg^{-1} sucrose solution into percent by weight. The density of the solution is 1.2 g cm^{-3}.

6.6 For dilute aqueous solutions in which the density of the solution is roughly equal to that of the pure solvent, the molarity of the solution is equal to its molality. Show that this statement is correct for a 0.010 *M* aqueous urea $[(NH_2)_2CO]$ solution.

6.7 The blood sugar (glucose) level of a diabetic patient is approximately 0.140 g of glucose/100 mL of blood. Every time the patient ingests 40 g of glucose, her blood glucose level rises to approximately 0.240 g/100 mL of blood. Calculate the number of moles of glucose per milliliter of blood and the total number of moles and grams of glucose in the blood before and after consumption of glucose. (Assume that the total volume of blood in her body is 5.0 L.)

6.8 The strength of alcoholic beverages is usually described in terms of "proof," which is defined as twice the percentage by volume of ethanol. Calculate the number of grams of alcohol in 2 quarts of 75-proof gin. What is the molality of the gin? (The density of ethanol is 0.80 g cm^{-3} and 1 quart = 0.946 L.)

Thermodynamics of Mixing

6.9 Liquids A and B form a nonideal solution. Provide a molecular interpretation for each of the following situations: $\Delta_{mix}H > 0$, $\Delta_{mix}H < 0$, $\Delta_{mix}V > 0$, $\Delta_{mix}V < 0$.

6.10 Calculate the changes in entropy for the following processes: **(a)** mixing of 1 mole of nitrogen and 1 mole of oxygen, and **(b)** mixing of 2 moles of argon, 1 mole of helium, and 3 moles of hydrogen. Both **(a)** and **(b)** are carried out under conditions of constant temperature (298 K) and constant pressure. Assume ideal behavior.

6.11 At 25°C and 1 atm pressure, the absolute third-law entropies of methane and ethane are 186.19 J K^{-1} mol^{-1} and 229.49 J K^{-1} mol^{-1}, respectively, in the gas phase. Calculate the absolute third-law entropy of a "solution" containing 1 mole of each gas. Assume ideal behavior.

Henry's Law

6.12 Prove the statement that an alternative way to express Henry's law of gas solubility is to say that the volume of gas that dissolves in a fixed volume of solution is independent of pressure at a given temperature.

6.13 A miner working 900 ft below the surface had a soft drink beverage during the lunch break. To his surprise, the drink seemed very flat (i.e., not much effervescence was observed upon removing the cap). Shortly after lunch, he took the elevator up to the surface. During the trip up, he felt a great urge to belch. Explain.

6.14 The Henry's law constant of oxygen in water at 25°C is 773 atm mol^{-1} kg of water. Calculate the molality of oxygen in water under a partial pressure of 0.20 atm. Assuming that the solubility of oxygen in blood at 37°C is roughly the same as that in water at 25°C, comment on the prospect for our survival without hemoglobin molecules. (The total volume of blood in the human body is about 5 L.)

6.15 The solubility of N_2 in blood at 37°C and a partial pressure of 0.80 atm is 5.6×10^{-4} mol L^{-1}. A deep-sea diver breathes compressed air with a partial pressure of N_2 equal to 4.0 atm. Assume that the total volume of blood in the body is 5.0 L. Calculate the amount of N_2 gas released (in liters) when the diver returns to the surface of water, where the partial pressure of N_2 is 0.80 atm.

Chemical Potential and Activity

6.16 Which of the following has a higher chemical potential? If neither, answer "same." **(a)** $H_2O(s)$ or $H_2O(l)$ at water's normal melting point, **(b)** $H_2O(s)$ at −5°C and 1 bar or $H_2O(l)$ at −5°C and 1 bar, **(c)** benzene at 25°C and 1 bar or benzene in a 0.1 *M* toluene solution in benzene at 25°C and 1 bar.

6.17 A solution of ethanol and *n*-propanol behaves ideally. Calculate the chemical potential of ethanol in solution relative to that of pure ethanol when its mole fraction is 0.40 at its boiling point (78.3°C).

6.18 Derive the Gibbs phase rule (Equation 5.23) in terms of chemical potentials.

6.19 The following data give the pressures for carbon disulfide–acetone solutions at 35.2°C. Calculate the activity coefficients of both components based on deviations from Raoult's law and Henry's law. (*Hint:* First determine Henry's law constants graphically.)

x_{CS_2}	0	0.20	0.45	0.67	0.83	1.00
P_{CS_2}/torr	0	272	390	438	465	512
$P_{C_3H_6O}$/torr	344	291	250	217	180	0

6.20 A solution is made up by dissolving 73 g of glucose ($C_6H_{12}O_6$; molar mass 180.2 g) in 966 g of water. Calculate the activity coefficient of glucose in this solution if the solution freezes at −0.66°C.

6.21 A certain dilute solution has an osmotic pressure of 12.2 atm at 20°C. Calculate the difference between the chemical potential of the solvent in the solution and that of pure water. Assume that the density is the same as that of water. (*Hint:* Express the chemical potential in terms of mole fraction, x_1, and rewrite the osmotic pressure equation as $\Pi V = n_2 RT$, where n_2 is the number of moles of the solute and $V = 1$ L.)

6.22 At 45°C, the vapor pressure of water is 65.76 mmHg for a glucose solution in which the mole fraction of glucose is 0.080. Calculate the activity and activity coefficient of the water in the solution. The vapor pressure of pure water at 45°C is 71.88 mmHg.

6.23 Consider a binary liquid mixture of A and B, where A is volatile and B is nonvolatile. The composition of the solution in terms of mole fraction is $x_A = 0.045$ and $x_B = 0.955$. The vapor pressure of A from the mixture is 5.60 mmHg, and that of pure A is 196.4 mmHg at the same temperature. Calculate the activity coefficient of A at this concentration.

Colligative Properties

6.24 List the important assumptions in the derivation of Equation 6.39.

6.25 Liquids A (b.p. = T_A°) and B (b.p. = T_B°) form an ideal solution. Predict the range of boiling points of solutions formed by mixing different amounts of A and B.

6.26 A mixture of ethanol and *n*-propanol behaves ideally at 36.4°C. **(a)** Determine graphically the mole fraction of *n*-propanol in a mixture of ethanol and *n*-propanol that boils at 36.4°C and 72 mmHg. **(b)** What is the total vapor pressure over the mixture at 36.4°C when the mole fraction of *n*-propanol is 0.60? **(c)** Calculate the composition of the vapor in **(b)**. (The equilibrium vapor pressures of ethanol and *n*-propanol at 36.4°C are 108 mmHg and 40.0 mmHg, respectively.)

6.27 Two beakers, 1 and 2, containing 50 mL of 0.10 *M* urea and 50 mL of 0.20 *M* urea, respectively, are placed under a tightly sealed bell jar at 298 K. Calculate the mole fraction of urea in the solutions at equilibrium. Assume ideal behavior. (*Hint:* Use Raoult's law and note that at equilibrium, the mole fraction of urea is the same in both solutions.)

6.28 At 298 K, the vapor pressure of pure water is 23.76 mmHg and that of seawater is 22.98 mmHg. Assuming that seawater contains only NaCl, estimate its concentration. (*Hint:* Sodium chloride is a strong electrolyte.)

6.29 Trees in cold climates may be subjected to temperatures as low as −60°C. Estimate the concentration of an aqueous solution in the body of the tree that would remain unfrozen at this temperature. Is this a reasonable concentration? Comment on your result.

6.30 Explain why jams and honey can each be stored under atmospheric conditions for long periods of time without spoilage.

6.31 Provide a molecular interpretation for the positive and negative deviations in the boiling-point curves and the formation of azeotropes.

6.32 The freezing-point-depression measurement of benzoic acid in acetone yields a molar mass of 122 g; the same measurement in benzene gives a value of 244 g. Account for this discrepancy. (*Hint:* Consider solvent–solute and solute–solute interactions.)

6.33 A common antifreeze for car radiators is ethylene glycol, $CH_2(OH)CH_2(OH)$. How many milliliters of this substance would you add to 6.5 L of water in the radiator if the coldest day in winter is −20°C? Would you keep this substance in the radiator in the summer to prevent the water from boiling? (The density and boiling point of ethylene glycol are 1.11 g cm^{-3} and 470 K, respectively.)

6.34 For intravenous injections, great care is taken to ensure that the concentration of solutions to be injected is comparable to that of blood plasma. Why?

6.35 The tallest trees known are the redwoods in California. Assuming the height of a redwood to be 105 m (about 350 ft), estimate the osmotic pressure required to push water up from the roots to the treetop.

6.36 A mixture of liquids A and B exhibits ideal behavior. At 84°C, the total vapor pressure of a solution containing 1.2 moles of A and 2.3 moles of B is 331 mmHg. Upon the addition of another mole of B to the solution, the vapor pressure increases to 347 mmHg. Calculate the vapor pressures of pure A and B at 84°C.

6.37 Fish breathe the dissolved air in water through their gills. Assuming the partial pressures of oxygen and nitrogen in air to be 0.20 atm and 0.80 atm, respectively, calculate the mole fractions of oxygen and nitrogen in water at 298 K. Comment on your results.

6.38 Liquids A (molar mass 100 g mol^{-1}) and B (molar mass 110 g mol^{-1}) form an ideal solution. At 55°C, A has a vapor pressure of 95 mmHg and B a vapor pressure of 42 mmHg. A solution is prepared by mixing equal weights of A and B. (**a**) Calculate the mole fraction of each component in the solution. (**b**) Calculate the partial pressures of A and B over the solution at 55°C. (**c**) Suppose that some of the vapor described in (**b**) is condensed to a liquid. Calculate the mole fraction of each component in this liquid and the vapor pressure of each component above this liquid at 55°C.

6.39 Lysozyme extracted from chicken egg white has a molar mass of 13,930 g mol^{-1}. Exactly 0.1 g of this protein is dissolved in 50 g of water at 298 K. Calculate the vapor pressure lowering, the depression in freezing point, the elevation of boiling point, and the osmotic pressure of this solution. The vapor pressure of pure water at 298 K is 23.76 mmHg.

6.40 The following argument is frequently used to explain the fact that the vapor pressure of the solvent is lower over a solution than over the pure solvent and that lowering is proportional to the concentration. A dynamic equilibrium exists in both cases, so that the rate at which molecules of solvent evaporate from the liquid is always equal to that at which they condense. The rate of condensation is proportional to the partial pressure of the vapor, whereas that of evaporation is unimpaired in the pure solvent but is impaired by solute molecules in the surface of the solution. Hence the rate of escape is reduced in proportion to the concentration of the solute, and maintenance of equilibrium requires a corresponding lowering of the rate of condensation and therefore of the partial pressure of the vapor phase. Explain why this argument is incorrect. [*Source:* K. J. Mysels, *J. Chem. Educ.* **32**, 179 (1955).]

6.41 A compound weighing 0.458 g is dissolved in 30.0 g of acetic acid. The freezing point of the solution is found to be 1.50 K below that of the pure solvent. Calculate the molar mass of the compound.

6.42 Two aqueous urea solutions have osmotic pressures of 2.4 atm and 4.6 atm, respectively, at a certain temperature. What is the osmotic pressure of a solution prepared by mixing equal volumes of these two solutions at the same temperature?

6.43 A forensic chemist is given a white powder for analysis. She dissolves 0.50 g of the substance in 8.0 g of benzene. The solution freezes at 3.9°C. Can the chemist conclude that the compound is cocaine ($C_{17}H_{21}NO_4$)? What assumptions are made in the analysis? The freezing point of benzene is 5.5°C.

6.44 "Time-release" drugs have the advantage of releasing the drug to the body at a constant rate so that the drug concentration at any time is not high enough to have harmful side effects or so low as to be ineffective. A schematic diagram of a pill that works on this basis is shown below. Explain how it works.

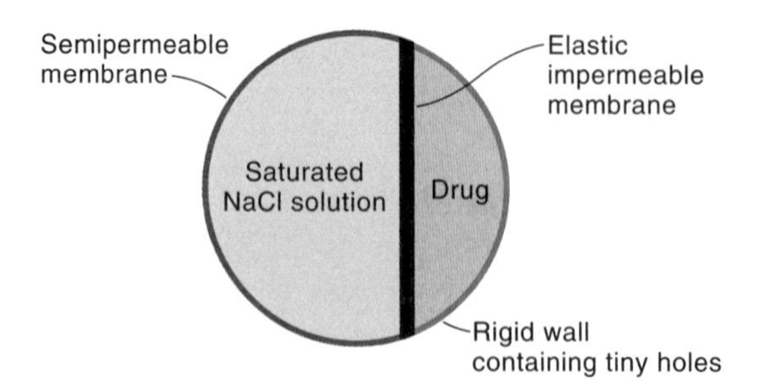

6.45 A nonvolatile organic compound, Z, was used to make up two solutions. Solution A contains 5.00 g of Z dissolved in 100 g of water, and solution B contains 2.31 g of Z dissolved in 100 g of benzene. Solution A has a vapor pressure of 754.5 mmHg at the normal boiling point of water, and solution B has the same vapor pressure at the normal boiling point of benzene. Calculate the molar mass of Z in solutions A and B, and account for the difference.

6.46 Acetic acid is a polar molecule that can form hydrogen bonds with water molecules. Therefore, it has a high solubility in water. Yet acetic acid is also soluble in benzene (C_6H_6), a nonpolar solvent that lacks the ability to form hydrogen bonds. A solution of 3.8 g of CH_3COOH in 80 g C_6H_6 has a freezing point of 3.5°C. Calculate the molar mass of the solute, and suggest what its structure might be. (*Hint:* Acetic acid molecules can form hydrogen bonds among themselves.)

6.47 At 85°C, the vapor pressure of A is 566 torr and that of B is 250 torr. Calculate the composition of a mixture of A and B that boils at 85°C when the pressure is 0.60 atm. Also, calculate the composition of the vapor mixture. Assume ideal behavior.

6.48 Comment on whether each of the following statements is true or false, and briefly explain your answer: **(a)** If one component of a solution obeys Raoult's law, then the other component must also obey the same law. **(b)** Intermolecular forces are small in ideal solutions. **(c)** When 15.0 mL of an aqueous 3.0 *M* ethanol solution is mixed with 55.0 mL of an aqueous 3.0 *M* ethanol solution, the total volume is 70.0 mL.

6.49 Liquids A and B form an ideal solution at a certain temperature. The vapor pressures of pure A and B are 450 torr and 732 torr, respectively, at this temperature. **(a)** A sample of the solution's vapor is condensed. Given that the original solution contains 3.3 moles of A and 8.7 moles of B, calculate the composition of the condensate in mole fractions. **(b)** Suggest a method for measuring the partial pressures of A and B at equilibrium.

6.50 Nonideal solutions are the result of unequal intermolecular forces between components. Based on this knowledge, comment on whether a racemic mixture of a liquid compound would behave as an ideal solution.

6.51 Calculate the molal boiling-point elevation constant (K_b) for water. The molar enthalpy of vaporization of water is 40.79 kJ mol^{-1} at 100°C.

6.52 Explain the following phenomena. **(a)** A cucumber placed in concentrated brine (saltwater) shrivels into a pickle. **(b)** A carrot placed in fresh water swells in volume.

Additional Problems

6.53 Calculate the change in the Gibbs energy at 37°C when the human kidneys secrete 0.275 mole of urea per kilogram of water from blood plasma to urine if the molal concentrations of urea in blood plasma and urine are 0.005 mol kg^{-1} and 0.326 mol kg^{-1}, respectively.

6.54 **(a)** Which of the following expressions is incorrect as a representation of the partial molar volume of component A in a two-component solution? Why? How would you correct it?

$$\left(\frac{\partial V_m}{\partial n_A}\right)_{T,P,n_B} \qquad \left(\frac{\partial V_m}{\partial x_A}\right)_{T,P,x_B}$$

(b) Given that the molar volume of this mixture (V_m) of A and B is given by

$$V_m = [0.34 + 3.6x_Ax_B + 0.4x_B(1 - x_A)]\text{ L mol}^{-1}$$

derive an expression for the partial molar volume for A and calculate its value at $x_A = 0.20$.

6.55 The partial molar volumes for a benzene–carbon tetrachloride solution at 25°C at a mole fraction of 0.5 are: $\overline{V}_b = 0.106\ \text{L mol}^{-1}$ and $\overline{V}_c = 0.100\ \text{L mol}^{-1}$, respectively, where the subscripts b and c denote C_6H_6 and CCl_4. **(a)** What is the volume of a solution made up of one mole of each? **(b)** Given that the molar volumes are: $C_6H_6 = 0.089\ \text{L mol}^{-1}$ and $CCl_4 = 0.097\ \text{L mol}^{-1}$, what is the change in volume on mixing 1 mole each of C_6H_6 and CCl_4? **(c)** What can you deduce about the nature of intermolecular forces between C_6H_6 and CCl_4?

6.56 The osmotic pressure of poly(methyl methacrylate) in toluene has been measured at a series of concentrations at 298 K. Determine graphically the molar mass of the polymer.

Π/atm	8.40×10^{-4}	1.72×10^{-3}	2.52×10^{-3}	3.23×10^{-3}	7.75×10^{-3}
c/g L^{-1}	8.10	12.31	15.00	18.17	28.05

6.57 Benzene and toluene form an ideal solution. Prove that to achieve the maximum entropy of mixing, the mole fraction of each component must be 0.5.

6.58 Suppose 2.6 moles of He at 0.80 atm and 25°C are mixed with 4.1 moles of Ne at 2.7 atm and 25°C. Calculate the Gibbs energy change for the process. Assume ideal behavior.

6.59 Two beakers are placed in a closed container. Beaker A initially contains 0.15 mole of naphthalene ($C_{10}H_8$) in 100 g of benzene (C_6H_6) and beaker B initially contains 31 g of an unknown compound dissolved in 100 g of benzene. At equilibrium, beaker A is found to have lost 7.0 g. Assuming ideal behavior, calculate the molar mass of the unknown compound. State any assumptions made.

6.60 As an after dinner party trick, the host brings out a glass of water with an ice cube floating on top and a thread. Then he asks the guests to remove the ice cube with the thread, but they are not allowed to tie a loop around the ice cube. Describe how the guests might accomplish this task.

6.61 A student carried out the following procedure to measure the pressure of carbon dioxide in a carbonated soft drink bottle. First, she weighed the bottle (853.5 g). Next, she carefully removed the cap to let the CO_2 gas escape. Finally, she measured the volume of the soft drink (452.4 mL). Given that the Henry's law constant for CO_2 in water at 25°C is $3.4 \times 10^{-2}\ \text{mol L}^{-1}\ \text{atm}^{-1}$, calculate the pressure of CO_2 in the original bottle. List the sources of errors.

6.62 **(a)** Derive the equation relating the molality (m) of a solution to its molarity (M)

$$m = \frac{M}{d - \dfrac{M\mathcal{M}}{1000}}$$

where d is the density of the solution (g/mL) and $\mathcal{M}$ is the molar mass of the solute (g/mol). (*Hint:* Start by expressing the solvent in kilograms in terms of the difference between the mass of the solution and the mass of the solute.) **(b)** Show that, for dilute aqueous solutions, m is approximately equal to M.

6.63 At 298 K, the osmotic pressure of a glucose solution is 10.50 atm. Calculate the freezing point of the solution. The density of the solution is 1.16 g/mL.

6.64 The mole fractions of dry air are approximately 21% O_2 and 79% N_2. Calculate the masses of these two gases dissolved in 1000 g of water at 25°C and 1 atm.

CHAPTER 7

Electrolyte Solutions

And when the rain has wet the kite and twine, so that it can conduct the electric fire freely, you will find it stream out plentifully from the key on the approach of your knuckle.

—Benjamin Franklin*

All biological and many chemical systems are aqueous solutions that contain various ions. The rates of many reactions are highly dependent on the type and concentration of ions present. It is important to have a clear understanding of the behavior of ions in solution.

An electrolyte is a substance that, when dissolved in a solvent (usually water), produces a solution that will conduct electricity. An electrolyte can be an acid, a base, or a salt. In this chapter, we shall consider ionic conductance, ionic dissociation, the thermodynamics of ions in solution, and the theory and colligative properties of electrolyte solutions.

7.1 Electrical Conduction in Solution

Some Basic Definitions

The ability of an electrolyte to conduct electricity provides us with a simple and direct means of studying ionic behavior in solution. We begin with a few basic definitions.

Ohm's Law. Ohm's law (after the German physicist George Simon Ohm, 1787–1854) states that the current (I) flowing through a particular medium is directly proportional to the voltage or the electrical potential difference (V) across the medium and indirectly proportional to the resistance (R) of the medium. Thus,

$$I = \frac{V}{R} \tag{7.1}$$

where I is in amperes (A), V in volts (V), and R in ohms (Ω).

---

* Labaree. L. W., et al., Eds., *The Papers of Benjamin Franklin*, Yale University Press, New Haven, CT, 1961. Vol. 4, p.367. Used by permission.

Resistance (R). The *resistance* across a particular medium depends on the geometry of the medium; it is directly proportional to the length (l) and inversely proportional to the cross section area (A) of the medium. Thus,

$$R \propto \frac{l}{A}$$

$$= \rho \frac{l}{A} \tag{7.2}$$

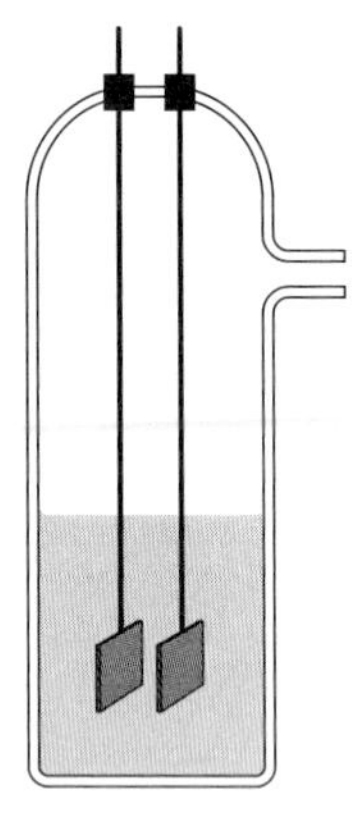

Figure 7.1
A conductance cell. The electrodes are made of platinum.

where the proportionality constant ρ is called the *specific resistance*, or *resistivity*. Because the units of R are ohms (Ω), of l centimeters or meters, and of A square centimeters or square meters, the units of ρ are Ω cm or Ω m. Resistivity is a property characteristic of the material comprising the medium.

Conductance (C). *Conductance* is the reciprocal of resistance, that is,

$$C = \frac{1}{R} = \frac{1}{\rho}\frac{A}{l} = \kappa \frac{A}{l} \tag{7.3}$$

where κ is the *specific conductance*, or *conductivity* equal to $1/\rho$. The SI unit for conductance is siemens (S) (after the German physicist Werner von Siemens, 1816–1892), where $1\ \text{S} = 1\ \Omega^{-1}$. Conductivity has the units $\Omega^{-1}\ \text{cm}^{-1}$ or $\Omega^{-1}\ \text{m}^{-1}$.

The historical unit of electric conductance was the mho ("ohm" backwards).

A typical conductance cell* is shown in Figure 7.1. The conductance is given by Equation 7.3. The ratio l/A, called the *cell constant*, is the same for all solutions. Here, A is the area of the electrodes and l the distance of separation between the electrodes. In practice, the cell constant is calibrated by measuring the conductance of a standard solution of potassium chloride with a known value of κ.

* It is customary to quote the conductance rather than the resistance of an electrolyte solution.

EXAMPLE 7.1

The conductance of a solution is $0.689\ \Omega^{-1}$. Calculate the specific conductance if the cell constant is $0.255\ \text{cm}^{-1}$.

ANSWER

From Equation 7.3,

$$\kappa = C\frac{l}{A}$$

$$= 0.689\ \Omega^{-1} \times 0.255\ \text{cm}^{-1}$$

$$= 0.176\ \Omega^{-1}\ \text{cm}^{-1}$$

Although the specific conductance can be measured easily (from the known cell constant and the experimentally determined conductance), it is not the best value to use in studying the conduction process in electrolyte solutions. The specific conductances of solutions of different concentrations, for example, will differ simply because a given volume of the solutions will contain different numbers of ions. For this reason, expressing the conductance as a molar quantity is preferable. We define the *molar conductance* (Λ) as

$$\Lambda = \frac{\kappa}{c} \tag{7.4}$$

where c is the molar concentration of the solution. The SI units for Λ are $\Omega^{-1}\ \text{mol}^{-1}\ \text{m}^2$, although it is often more conveniently expressed as $\Omega^{-1}\ \text{mol}^{-1}\ \text{cm}^2$.

EXAMPLE 7.2

The conductance of a cell containing an aqueous 0.0560 *M* KCl solution is $0.0239\ \Omega^{-1}$. When the same cell is filled with an aqueous 0.0836 *M* NaCl solution, its conductance is $0.0285\ \Omega^{-1}$. Given that the molar conductance of KCl is $1.345 \times 10^2\ \Omega^{-1}\ \text{mol}^{-1}\ \text{cm}^2$, calculate the molar conductance of the NaCl solution.

ANSWER

We need the cell constant. Our first step is to calculate the specific conductance of the KCl solution. From Equation 7.4,

$$\begin{aligned}\kappa &= \Lambda c \\ &= 1.345 \times 10^2\ \Omega^{-1}\ \text{mol}^{-1}\ \text{cm}^2 \times \frac{0.0560\ \text{mol}}{1\ \text{L}} \times \frac{1\ \text{L}}{1000\ \text{cm}^3} \\ &= 7.53 \times 10^{-3}\ \Omega^{-1}\ \text{cm}^{-1}\end{aligned}$$

Next, from Equation 7.3 we obtain the cell constant

$$\begin{aligned}\frac{l}{A} = \frac{\kappa}{C} &= \frac{7.53 \times 10^{-3}\ \Omega^{-1}\ \text{cm}^{-1}}{0.0239\ \Omega^{-1}} \\ &= 0.315\ \text{cm}^{-1}\end{aligned}$$

The specific conductance of the NaCl solution is obtained by rearranging Equation 7.3:

$$\begin{aligned}\kappa &= \frac{l}{A}C \\ &= (0.315\ \text{cm}^{-1})(0.0285\ \Omega^{-1}) \\ &= 8.98 \times 10^{-3}\ \Omega^{-1}\ \text{cm}^{-1}\end{aligned}$$

Finally, the molar conductance of the NaCl solution is given by (see Equation 7.4)

$$\Lambda = \frac{\kappa}{c} = \frac{8.98 \times 10^{-3}\ \Omega^{-1}\ \text{cm}^{-1}}{0.0836\ \text{mol L}^{-1}} \times \frac{1000\ \text{cm}^3}{1\ \text{L}}$$

$$= 1.07 \times 10^2\ \Omega^{-1}\ \text{mol}^{-1}\ \text{cm}^2$$

Strong electrolytes are substances that are completely dissociated into ions in solution.

Looking at Equation 7.4, we might expect Λ to be independent of the concentration of the solution. (κ is directly proportional to concentration, but κ/c should be constant for a given substance.) Careful measurements show that this is not the case, however. The German chemist Friedrich Wilhelm Georg Kohlrausch (1840–1910) discovered the following relationship between molar conductance and concentration for strong electrolytes at a particular temperature:

$$\Lambda = \Lambda_0 - B\sqrt{c} \tag{7.5}$$

where Λ_0 is the molar conductance at infinite dilution; that is, $\Lambda \rightarrow \Lambda_0$ as $c \rightarrow 0$, and B is a positive constant for a given electrolyte. Thus, Λ_0 can be readily obtained by plotting Λ versus $\sqrt{c}$ and extrapolating to zero concentration (Figure 7.2). This method is unsatisfactory for weak electrolytes because of the steepness of their curves at low concentrations (see plot for CH_3COOH).

Table 7.1 shows the values of Λ_0 for several electrolytes. An interesting pattern emerges when we examine the difference in Λ_0 for two electrolytes containing the same cation or anion. For example,

$$\Lambda_0^{\text{KCl}} - \Lambda_0^{\text{NaCl}} = 23.4\ \Omega^{-1}\ \text{mol}^{-1}\ \text{cm}^2$$

$$\Lambda_0^{\text{KNO}_3} - \Lambda_0^{\text{NaNO}_3} = 23.4\ \Omega^{-1}\ \text{mol}^{-1}\ \text{cm}^2$$

This same difference and similar observations led Kohlrausch to suggest that molar conductance at infinite dilution can be broken down into two contributions, one from the anion and the other from the cation:

$$\Lambda_0 = \nu_+\lambda_0^+ + \nu_-\lambda_0^- \tag{7.6}$$

where λ_0^+ and λ_0^- are the molar ionic conductances at infinite dilution, and ν_+ and ν_- are the number of cations and anions in the formula. Equation 7.6 is known as *Kohlrausch's law of independent migration*. It means that molar conductance at infinite dilution is made up of independent contributions from the cationic and anionic species. We can now see why the same value was obtained in the example above, because

Note that these are all 1:1 electrolytes, so $\nu_+ = \nu_- = 1$.

$$\Lambda_0^{\text{KCl}} - \Lambda_0^{\text{NaCl}} = \lambda_0^{\text{K}^+} + \lambda_0^{\text{Cl}^-} - \lambda_0^{\text{Na}^+} - \lambda_0^{\text{Cl}^-} = \lambda_0^{\text{K}^+} - \lambda_0^{\text{Na}^+}$$

and

$$\Lambda_0^{\text{KNO}_3} - \Lambda_0^{\text{NaNO}_3} = \lambda_0^{\text{K}^+} + \lambda_0^{\text{NO}_3^-} - \lambda_0^{\text{Na}^+} - \lambda_0^{\text{NO}_3^-} = \lambda_0^{\text{K}^+} - \lambda_0^{\text{Na}^+}$$

Table 7.2 lists the molar conductances of a number of ions.

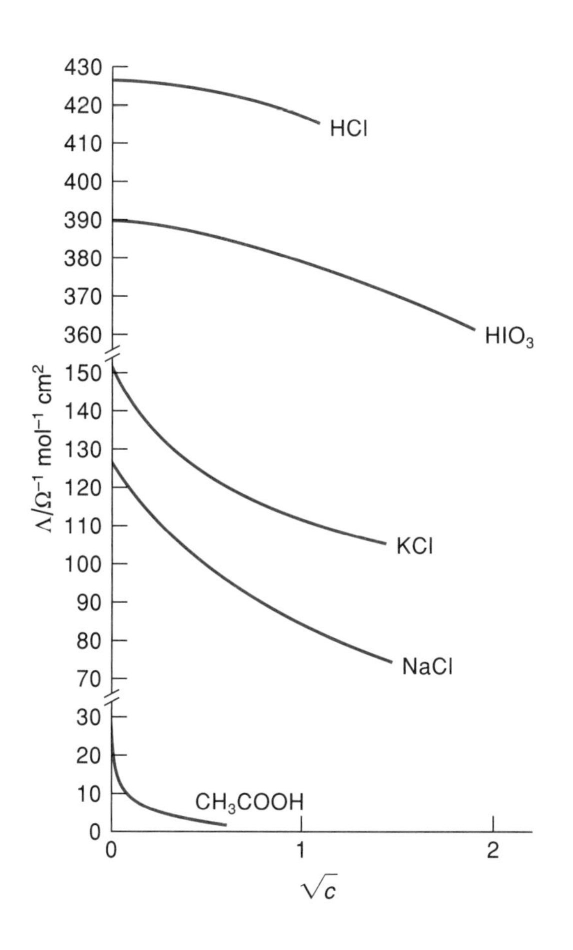

Figure 7.2
Plots of molar conductance versus the square root of concentration (mol L^{-1}) for several electrolytes.

Table 7.1
Molar Conductance at Infinite Dilution for Some Electrolytes in Water at 298 K[a]

Electrolyte	Λ_0/Ω^{-1} mol^{-1} cm^2
HCl	426.16
CH_3COOH	390.71
LiCl	115.03
NaCl	126.45
AgCl	137.20
KCl	149.85
$LiNO_3$	110.14
$NaNO_3$	121.56
KNO_3	144.96
$CuSO_4$	267.24
CH_3COONa	91.00

[a] To express Λ_0 as Ω^{-1} mol^{-1} m^2, multiply each number by 10^{-4}. Thus, Λ_0 for HCl is 426.16 Ω^{-1} mol^{-1} cm^2 or 4.2616×10^{-2} Ω^{-1} mol^{-1} m^2.

Table 7.2
Molar Ionic Conductance and Ionic Mobility of Some Common Ions at 298 K

Ion	λ_0[a] / Ω^{-1} mol^{-1} cm^2	Ionic mobility[b] / 10^{-4} cm^2 s^{-1} V^{-1}	Ionic radius/Å
H^+	349.81	36.3	
Li^+	38.68	4.01	0.60
Na^+	50.10	5.19	0.95
K^+	73.50	7.62	1.33
Rb^+	77.81	7.92	1.48
Cs^+	77.26	7.96	1.69
NH_4^+	73.5	7.62	
Mg^{2+}	106.1	5.50	0.65
Ca^{2+}	119.0	6.17	0.99
Ba^{2+}	127.3	6.59	1.35
Cu^{2+}	107.2	5.56	0.72
OH^-	198.3	20.50	
F^-	55.4	5.74	1.36
Cl^-	76.35	7.91	1.81
Br^-	78.14	8.10	1.95
I^-	76.88	7.95	2.16
NO_3^-	71.46	7.41	
HCO_3^-	44.50	4.61	
CH_3COO^-	40.90	4.24	
SO_4^{2-}	160.0	8.29	

[a] From Robinson, R. A., and R. H. Stokes, *Electrolyte Solutions*, Academic Press, New York, 1959. Used by permission.

[b] From Adamson, A. W. *A Textbook of Physical Chemistry*, Academic Press, New York, 1973. Used by permission.

Degree of Dissociation

At a certain concentration, an electrolyte may be only partially dissociated. At infinite dilution, any electrolyte, weak or strong, is completely dissociated. In 1887, the Swedish chemist Svante August Arrhenius (1859–1927) suggested that the *degree of dissociation* (α) of an electrolyte can be calculated by the simple relation

$$\alpha = \frac{\Lambda}{\Lambda_0} \tag{7.7}$$

where Λ is the molar conductance at a particular concentration to which α refers. Using Equation 7.7, the German chemist Wilhelm Ostwald (1853–1932) showed how one can measure the dissociation constant of an acid. Consider a weak acid HA of concentration c (mol L^{-1}). At equilibrium, we have

$$\begin{array}{ccc} \mathrm{HA} & \rightleftharpoons \mathrm{H^+} & + \ \mathrm{A^-} \\ c(1-\alpha) & c\alpha & c\alpha \end{array}$$

where α is the fraction of HA dissociated. The dissociation constant, K_a, is given by

$$K_a = \frac{[\mathrm{H^+}][\mathrm{A^-}]}{[\mathrm{HA}]} = \frac{c^2\alpha^2}{c(1-\alpha)} = \frac{c\alpha^2}{(1-\alpha)}$$

Using the expression for α in Equation 7.7, we obtain

$$K_a = \frac{c\Lambda^2}{\Lambda_0(\Lambda_0 - \Lambda)} \tag{7.8}$$

Equation 7.8 can be rearranged to give

$$\frac{1}{\Lambda} = \frac{1}{K_a\Lambda_0^2}(\Lambda c) + \frac{1}{\Lambda_0} \tag{7.9}$$

Equation 7.9 is known as the *Ostwald dilution law*. Thus, the value of K_a can be obtained either directly from Equation 7.8 or more accurately from Equation 7.9 by plotting $1/\Lambda$ versus Λc (Figure 7.3). Note that Λ_0 must be known to apply Equation 7.8 alone.

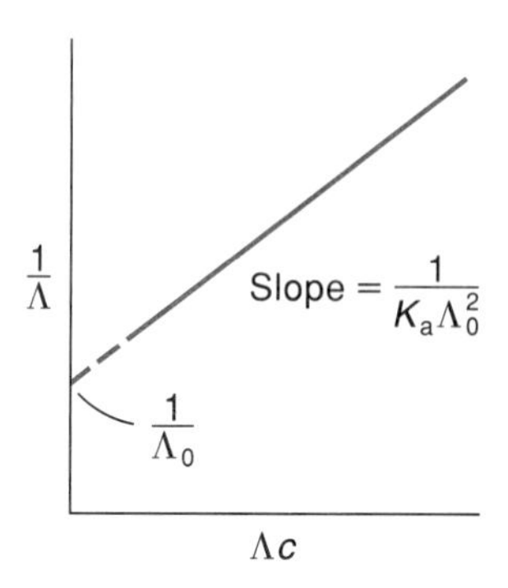

Figure 7.3
Graphical determination of K_a according to Equation 7.9.

EXAMPLE 7.3

The molar conductance of an aqueous acetic acid solution at concentration 0.10 *M* is 5.2 Ω^{-1} mol^{-1} cm^2 at 298 K. Calculate the dissociation constant of acetic acid.

ANSWER

From Table 7.1, $\Lambda_0 = 390.71\ \Omega^{-1}\ \mathrm{mol^{-1}\ cm^2}$, and from Equation 7.8 we write

$$K_a = \frac{(0.10\ \mathrm{mol\ L^{-1}})(5.2\ \Omega^{-1}\ \mathrm{mol^{-1}\ cm^2})^2}{(390.71\ \Omega^{-1}\ \mathrm{mol^{-1}\ cm^2})(390.71 - 5.2)\Omega^{-1}\ \mathrm{mol^{-1}\ cm^2}}$$

$$= 1.8 \times 10^{-5}\ \mathrm{mol\ L^{-1}}$$

Ionic Mobility

The molar conductance of a solution depends on the ease of ionic movement. Ionic velocity is not a constant, however, because it depends on the strength of the electric field (E).* On the other hand, *ionic mobility* (u), defined as the ionic velocity per unit electric field, is a constant. Thus, the ionic mobility of a cation, u_+, is given by

$$u_+ = \frac{\upsilon_+}{E} \tag{7.10}$$

where υ_+ is the velocity of the cation moving in an electric field of strength E. Ionic mobility has the units $(\text{cm s}^{-1})/\text{V cm}^{-1}$, or $\text{cm}^2\ \text{s}^{-1}\ \text{V}^{-1}$; it is related to the molar ionic conductance at infinite dilution as follows:

$$u_+ = \frac{\lambda_0^+}{F} \quad \text{and} \quad u_- = \frac{\lambda_0^-}{F} \tag{7.11}$$

where F is the Faraday constant (after the English chemist and physicist Michael Faraday, 1791–1867).†

Table 7.2 lists the ionic mobilities of various ions at 298 K. The ionic mobilities of H^+ and OH^- ions are much higher than those of other ions. These high values are due to hydrogen bonding. In water, the proton is hydrated, and its movement can be represented as follows:

```
       H+     H               H       H              H      H+
        \      \             /         \            /      /
 --->    O—H·····O—H·····O      --->    O·····H—O·····H—O       --->
        /                    \         /                   \
       H                      H       H                     H
```

Similarly, the movement of the hydroxide ion is

```
        H          H       H          H       H           H
         \        /       /            \       \         /
 <---    _O·····H—O·····H—O     <---    O—H·····O—H·····O_     <---
```

In each case, the ion can move along an extended hydrogen-bond network, resulting in very high mobility.

Ionic mobility is utilized in electrophoresis, a technique for purifying and identifying proteins and nucleic acids.

---

* For example, if the potential drop between two electrodes separated by 2.0 cm in an electrolytic cell is 5.0 V, then the electric field is 5.0 V/2.0 cm. or 2.5 V cm^{-1}.

† The charge carried by 1 mole of electrons is called the faraday (F), given by 96,485 coulombs. In most calculations, we shall round off the quantity to 96,500 C mol^{-1}.

EXAMPLE 7.4

The mobility of a chloride ion in water at 25°C is 7.91×10^{-4} cm^2 s^{-1} V^{-1}. (a) Calculate the molar conductance of the ion at infinite dilution. (b) How long will it take for the ion to travel between two electrodes separated by 4.0 cm if the electric field is 20 V cm^{-1}?

ANSWER

From Equation 7.11,

$$\lambda_0^- = Fu_-$$

$$= (96500 \text{ C mol}^{-1})(7.91 \times 10^{-4} \text{ cm}^2 \text{ s}^{-1} \text{ V}^{-1})$$

$$= 76.3 \text{ C s}^{-1} \text{ V}^{-1} \text{ mol}^{-1} \text{ cm}^2$$

$$= 76.3 \ \Omega^{-1} \text{ mol}^{-1} \text{ cm}^2$$

because 1 C s^{-1} = 1 A and A/V = Ω^{-1} (Ohm's law).
(b) First, we calculate the ionic velocity, given by Equation 7.10, for an anion

$$v_- = Eu_-$$

$$= (20 \text{ V cm}^{-1})(7.91 \times 10^{-4} \text{ cm}^2 \text{ s}^{-1} \text{ V}^{-1})$$

$$= 1.58 \times 10^{-2} \text{ cm s}^{-1}$$

Next, we write

$$\text{time} = \frac{\text{distance}}{\text{velocity}}$$

$$= \frac{4.0 \text{ cm}}{1.58 \times 10^{-2} \text{ cm s}^{-1}}$$

$$= 2.5 \times 10^2 \text{ s}$$

$$= 4.2 \text{ min}$$

Applications of Conductance Measurements

Accurate conductance measurements are easy to make and have many applications. Two examples are described below.

Acid–Base Titration. As mentioned earlier, the conductances of H^+ and OH^- are considerably higher than those of other cations and anions. By plotting the conductance of a HCl solution as a function of NaOH solution added, we obtain a titration

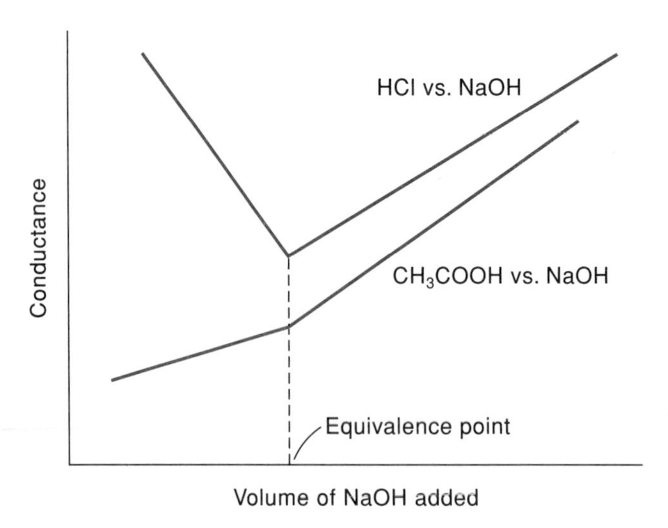

Figure 7.4
Acid–base titration monitored by conductance measurements. Note the difference between a strong acid–base titration (HCl vs. NaOH) and a weak acid–strong base titration (CH_3COOH vs. NaOH).

curve such as the one shown in Figure 7.4. Initially, the conductance of the solution falls, because H^+ ions are replaced by Na^+ ions, which have a lower ionic conductance. This trend continues until the equivalence point is reached. Beyond this point, the conductance begins to rise due to the excess of OH^- ions. If the acid is a weak electrolyte, such as acetic acid, the slope of the first part of the curve is much less steep—the conductance actually increases right from the beginning—and there is more uncertainty in determining the equivalence point.

Solubility Determination. We have seen how the dissociation constant of acetic acid can be obtained from conductance measurements. The same procedure can be applied to determine the solubility of a sparingly soluble salt. Suppose that we are interested in the molar solubility (mol L^{-1}) and the solubility product of AgCl (a 1:1 electrolyte) in water at 298 K. From Equation 7.4, we write

$$\Lambda = \frac{\kappa}{c} = \frac{\kappa}{S}$$

where S is the molar solubility in mol L^{-1}. Because AgCl is an insoluble salt, the concentration of the solution is low, so we can assume that $\Lambda \approx \Lambda_0$. Thus,

$$S = \frac{\kappa}{\Lambda} \approx \frac{\kappa}{\Lambda_0}$$

Experimentally, the specific conductance of a saturated AgCl solution is $1.86 \times 10^{-6}\ \Omega^{-1}\ cm^{-1}$. Because water is a weak electrolyte, however, we must take out the contribution due to water itself (κ for water is $6.0 \times 10^{-8}\ \Omega^{-1}\ cm^{-1}$). Thus,

$$\kappa(\text{AgCl}) = (1.86 \times 10^{-6}) - (6.0 \times 10^{-8}) = 1.8 \times 10^{-6}\ \Omega^{-1}\ \text{cm}^{-1}$$

According to Table 7.1, $\Lambda_0 = 137.2\ \Omega^{-1}\ \text{mol}^{-1}\ \text{cm}^2$ for AgCl. Finally, we have

$$S = \frac{\kappa}{\Lambda_0} = \frac{1.8 \times 10^{-6}\ \Omega^{-1}\ \text{cm}^{-1}}{137.2\ \Omega^{-1}\ \text{mol}^{-1}\ \text{cm}^2} \times \frac{1000\ \text{cm}^3}{1\ \text{L}}$$
$$= 1.3 \times 10^{-5}\ \text{mol L}^{-1}$$

The solubility product, K_{sp}, for AgCl is given by

$$K_{sp} = [\text{Ag}^+][\text{Cl}^-] = (1.3 \times 10^{-5})(1.3 \times 10^{-5})$$
$$= 1.7 \times 10^{-10}$$

7.2 A Molecular View of the Solution Process

Why does NaCl dissolve in water and not in benzene? We know that NaCl is a stable compound in which the Na^+ and Cl^- ions are held by electrostatic forces in the crystal lattice. For NaCl to enter the aqueous environment, the strong attractive forces must somehow be overcome. The dissolution of NaCl in water presents two questions: How do the ions interact with water molecules, and how do they interact with one another?

Water is a good solvent for ionic compounds because it is a polar molecule and therefore can stabilize the ions through ion-dipole interaction that results in hydration. Generally, smaller ions can be hydrated more effectively than larger ions. A small ion contains a more concentrated charge, which leads to greater electrostatic interaction with the polar water molecules.* Figure 7.5 shows a schematic diagram of hydration. Because a different number of water molecules surrounds each type of ion, we speak of the *hydration number* of an ion. This number is directly proportional to the charge and inversely proportional to the size of the ion. Note that water in the "hydration sphere" and bulk water molecules have different properties,† which can be distinguished by spectroscopic techniques such as nuclear magnetic resonance. There is a dynamic equilibrium between the two types of molecules. Depending on the ion, the mean lifetime of a H_2O molecule in the hydration sphere can vary tremendously. For example, consider the mean lifetime of H_2O in the hydration sphere for the following ions: Br^-, 10^{-11} s; Na^+, 10^{-9} s; Cu^{2+}, 10^{-7} s; Fe^{2+}, 10^{-5} s; Al^{3+}, 7 s; and Cr^{3+}, 1.5×10^5 s, or 42 h.

---

* According to electrostatic theory, the electric field at the surface of a charged sphere of radius r is proportional to ze/r^2, where z is the number of charges and e is the electronic charge.

† Water molecules in the hydration sphere of an ion do not exhibit individual translational motion. They move with the ion as a whole.

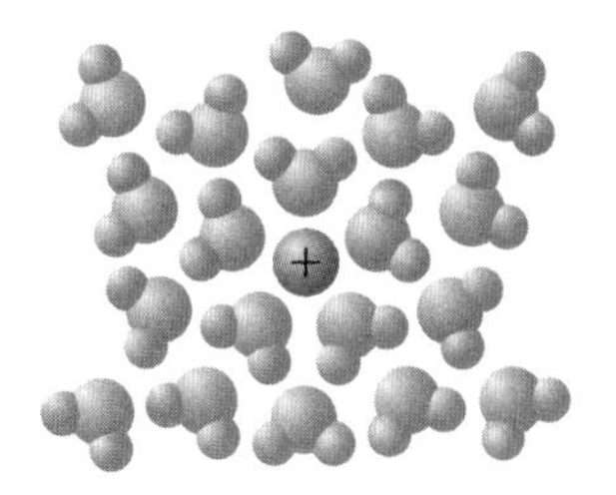

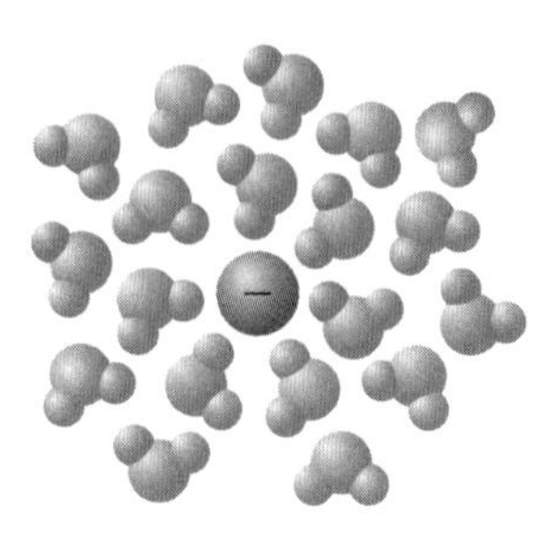

Figure 7.5
Hydration of a cation and an anion. In general, each cation and each anion has a specific number of water molecules associated with it in the hydration sphere.

The ion–dipole interaction (see Chapter 17) between dissolved ions and water molecules can affect several bulk properties of water. Small and/or multicharged ions such as Li^+, Na^+, Mg^{2+}, Al^{3+}, Er^{3+}, OH^-, and F^- are often called *structure-making ions*. The strong electric fields exerted by these ions can polarize water molecules, producing additional order beyond the first hydration layer. This interaction increases the solution's viscosity. On the other hand, large monovalent ions such as K^+, Rb^+, Cs^+, NH_4^+, Cl^-, NO_3^-, and ClO_4^- are *structure-breaking ions*. Because of their diffuse surface charges and hence weak electric fields, these ions are unable to polarize water molecules beyond the first layer of hydration. Consequently, the viscosities of solutions containing these ions are usually lower than that of pure water.

The effective radii of hydrated ions in solution can be appreciably greater than their crystal or ionic radii. For example, the radii of the hydrated Li^+, Na^+, and K^+ ions are estimated to be 3.66 Å, 2.80 Å, and 1.87 Å, respectively, although the ionic radii actually increase from Li^+ to K^+. We might expect the mobility of an ion to be inversely proportional to its hydrated radius. Table 7.2 nicely demonstrates the truth of this expectation. The high ionic mobility of the proton, which we would expect to be strongly hydrated because of its very small size, stems from the rapid movement of the H^+ ions via hydrogen bonds.

We now turn to the other question raised earlier: How do ions interact with one another? According to Coulomb's law (after the French physicist Charles Augustin de Coulomb, 1736–1806), the force (F) between Na^+ and Cl^- ions in a vacuum is given by

$$F = \frac{q_{Na^+}q_{Cl^-}}{4\pi\varepsilon_0 r^2} \tag{7.12}$$

where ε_0 is the *permittivity of the vacuum* (8.854×10^{-12} C^2 N^{-1} m^{-2}), q_{Na^+} and q_{Cl^-} are the charges on the ions, and r is the distance of separation. The factor $4\pi\varepsilon_0$ is present as a result of using SI units. In the polar medium of water, as Figure 7.6 shows, the dipolar molecules align themselves with their positive ends facing the negative charge and the negative ends facing the positive charge. This arrangement reduces the effective charge at the positive and negative charge centers by a factor of $1/\varepsilon$, where ε is the *dielectric constant* of the medium (see Appendix 7.1 on p. 295). Therefore, in any medium other than a vacuum, Equation 7.12 takes the form

The dielectric constant of vacuum is 1.

$$F = \frac{q_{Na^+}q_{Cl^-}}{4\pi\varepsilon_0\varepsilon r^2} \tag{7.13}$$

Table 7.3 lists the dielectric constants of several solvents. Keep in mind that ε always decreases with increasing temperature. For example, at 343 K, the dielectric constant

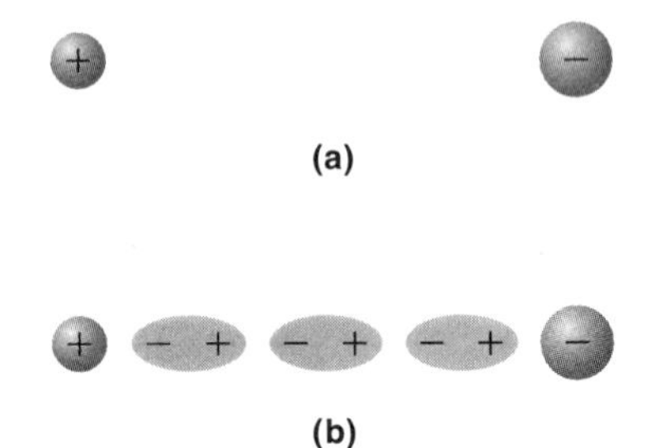

Figure 7.6
(a) Separation of a cation and an anion in a vacuum. (b) Separation of the same ions in water. The alignment of polar water molecules is exaggerated. Because of thermal motion, the polar molecules are only partially aligned. Nevertheless, this arrangement reduces the electric field and hence the attractive force between the ions.

Table 7.3
Dielectric Constants of Some Pure Liquids at 298 K

Liquid	Dielectric constant, ε[a]
H_2SO_4	101
H_2O	78.54
$(CH_3)_2SO$ (dimethylsulfoxide)	49
$C_3H_8O_3$ (glycerol)	42.5
CH_3NO_2 (nitromethane)	38.6
$HOCH_2CH_2OH$ (ethylene glycol)	37.7
CH_3CN (acetonitrile)	36.2
CH_3OH	32.6
C_2H_5OH	24.3
CH_3COCH_3 (dimethyl ether)	20.7
CH_3COOH	6.2
C_6H_6	4.6
$C_2H_5OC_2H_5$ (diethyl ether)	4.3
CS_2	2.6

[a] The dielectric constant is a dimensionless quantity.

of water is reduced to about 64. Water's large dielectric constant is what reduces the attractive force between the Na^+ and Cl^- ions and enables them to separate in solution.

The dielectric constant of a solvent also determines the "structure" of ions in solution. To maintain electrical neutrality in solution, an anion must be near a cation, and vice versa. Depending on the proximity of these two ions, we can think of them either as "free" ions or as "ion pairs." Each free ion is surrounded by at least one and perhaps several layers of water molecules. In an ion pair, the cation and anion are close to each other, and few or no solvent molecules are between them. Generally, free ions and ion pairs are thermodynamically distinguishable species that have quite different chemical reactivities. For dilute 1:1 aqueous electrolyte solutions, such as NaCl, ions are believed to be in the free-ion form. In higher-valence electrolytes, such as $CaCl_2$ and Na_2SO_4, the formation of ion pairs is indicated by conductance measurements, for a neutral ion pair cannot conduct electricity. Two opposing factors determine whether we have free ions or ion pairs in solution: the potential energy of attraction between the cation and anion, and the kinetic or thermal energy, of the order of k_BT, for individual ions.

We can now understand easily why NaCl does not dissolve in benzene. A nonpolar molecule, benzene does not solvate Na^+ and Cl^- ions effectively. In addition, its small dielectric constant means that the cations and anions will have little tendency to enter the solution as separate ions.

EXAMPLE 7.5

Calculate the force in newtons between a pair of Na^+ and Cl^- ions separated by exactly 1 nm (10 Å) in (a) a vacuum and (b) water at 25°C. The charges on the Na^+ and Cl^- ions are 1.602×10^{-19} C and -1.602×10^{-19} C, respectively.

ANSWER

From Equations 7.12 and 7.13 and Table 7.3, we proceed as follows.

(a)

$$F = \frac{(1.602 \times 10^{-19}\ \text{C})(-1.602 \times 10^{-19}\ \text{C})}{4\pi(8.854 \times 10^{-12}\ \text{C}^2\ \text{N}^{-1}\ \text{m}^{-2})(1)(1 \times 10^{-9}\ \text{m})^2}$$

$$= -2.31 \times 10^{-10}\ \text{N}$$

(b)

$$F = \frac{(1.602 \times 10^{-19}\ \text{C})(-1.602 \times 10^{-19}\ \text{C})}{4\pi(8.854 \times 10^{-12}\ \text{C}^2\ \text{N}^{-1}\ \text{m}^{-2})(78.54)(1 \times 10^{-9}\ \text{m})^2}$$

$$= -2.94 \times 10^{-12}\ \text{N}$$

COMMENT

As expected, the attractive force between the ions is reduced by a factor of about 80 from vacuum to the aqueous environment. The negative sign convention for F denotes attraction.

7.3 Thermodynamics of Ions in Solution

In this section, we shall briefly examine the thermodynamic parameters of the solution process involving ionic compounds and the thermodynamic functions of the formation of ions in aqueous solution.

The constant-pressure dissolution of NaCl can be represented by

$$\begin{array}{ccc} & \overset{1}{\nearrow} & Na^+(g) + Cl^-(g) \\ NaCl(s) & & \downarrow 2 \\ & \underset{3}{\searrow} & Na^+(aq) + Cl^-(aq) \end{array}$$

The enthalpy change for process 1 corresponds to the energy required to separate one mole of ions from the crystal lattice to an infinite distance. This energy is called the *lattice energy* (U_0). The enthalpy change for process 3 is the enthalpy of solution, $\Delta_{\text{soln}}\overline{H}°$, which is the heat absorbed or released when NaCl dissolves in a large amount of water. The heat or enthalpy of hydration, $\Delta_{\text{hydr}}\overline{H}°$, for process 2 is given by Hess's law:

$$\Delta_{\text{hydr}}\overline{H}° = \Delta_{\text{soln}}\overline{H}° - U_0$$

The quantity $\Delta_{soln}\overline{H}°$ is experimentally measurable; the value of U_0 can be estimated if the structure of the lattice is known. For NaCl, we have $U_0 = 787$ kJ mol^{-1} and $\Delta_{soln}\overline{H}° = 3.8$ kJ mol^{-1}, so that

$$\Delta_{hydr}\overline{H}° = 3.8 - 787 = -783 \text{ kJ mol}^{-1}$$

Thus, the hydration of Na^+ and Cl^- ions by water releases a large amount of heat.

The enthalpy of hydration obtained above comes from both ions together. We often want to know the value of individual ions. In reality, we cannot study them separately, but their values can be obtained as follows. The enthalpy of hydration for the process

$$H^+(g) \rightarrow H^+(aq)$$

has been reliably estimated by theoretical methods as -1089 kJ mol^{-1}. Using this value as a starting point, we can calculate the $\Delta_{hydr}\overline{H}°$ values for individual anions such as F^-, Cl^-, Br^-, and I^- (from data on HF, HCl, HBr, and HI), and in turn obtain $\Delta_{hydr}\overline{H}°$ values for Li^+, Na^+, K^+, and other cations (from data on alkali metal halides). Table 7.4 lists the standard $\Delta_{hydr}\overline{H}°$ values for a number of ions. All the $\Delta_{hydr}\overline{H}°$ values are negative because the hydration of a gaseous ion is an exothermic process. Furthermore, there is a correlation between ionic charge/radius and hydration enthalpy. The values of $\Delta_{hydr}\overline{H}°$ are larger (more negative) for smaller ions than for large ions of the same charge. A smaller ion has a more concentrated charge and can interact more strongly with water molecules. Ions bearing higher charges also have larger $\Delta_{hydr}\overline{H}°$ values.

The other quantity of interest is the entropy of hydration, $\Delta_{hydr}\overline{S}°$. The hydration process results in considerable ordering of water molecules around each ion, so that $\Delta_{hydr}\overline{S}°$ also is a negative quantity. As Table 7.4 shows, the variation in standard $\Delta_{hydr}\overline{S}°$ with ionic radius closely corresponds to that for $\Delta_{hydr}\overline{H}°$. Finally, note that there are two contributions to the entropy of solution, $\Delta_{soln}\overline{S}°$. The first is the hydration process, which results in a decrease in entropy. The other is the entropy gained when the solid breaks up into ions, which can move freely in solution. The sign of $\Delta_{soln}\overline{S}°$ depends on the magnitudes of these opposing factors.

The entropy of hydration for $H^+(g)$ is estimated to be -109 J K^{-1} mol^{-1}.

Enthalpy, Entropy, and Gibbs Energy of Formation of Ions in Solution

Because ions cannot be studied separately, we cannot measure the standard molar enthalpy of formation, $\Delta_f\overline{H}°$, of an individual ion. To get around this problem, we arbitrarily assign a zero value to the formation of the hydrogen ion—that is, $\Delta_f\overline{H}°[H^+(aq)] = 0$—and then evaluate the $\Delta_f\overline{H}°$ values of other ions relative to this scale. Consider the following reaction:

$$\tfrac{1}{2}H_2(g) + \tfrac{1}{2}Cl_2(g) \rightarrow H^+(aq) + Cl^-(aq) \quad \Delta_r H° = -167.2 \text{ kJ mol}^{-1}$$

The standard enthalpy of the reaction, which is an experimentally measurable quantity, can be expressed as

$$\Delta_r H° = \Delta_f\overline{H}°[H^+(aq)] + \Delta_f\overline{H}°[Cl^-(aq)] - (\tfrac{1}{2})(0) - (\tfrac{1}{2})(0)$$

Table 7.4
Thermodynamic Values for the Hydration of Gaseous Ions at 298 K

Ion	$\frac{-\Delta_{hydr}\overline{H}^\circ}{\text{kJ mol}^{-1}}$	$\frac{-\Delta_{hydr}\overline{S}^\circ}{\text{J K}^{-1}\text{ mol}^{-1}}$	Ionic radius/Å
H^+	1089	109	—
Li^+	520	119	0.60
Na^+	405	89	0.95
K^+	314	51	1.33
Ag^+	468	94	1.26
Mg^{2+}	1926	268	0.65
Ca^{2+}	1579	209	0.99
Ba^{2+}	1309	159	1.35
Mn^{2+}	1832	243	0.80
Fe^{2+}	1950	272	0.76
Cu^{2+}	2092	259	0.72
Fe^{3+}	4355	460	0.64
F^-	506	151	1.36
Cl^-	378	96	1.81
Br^-	348	80	1.95
I^-	308	60	2.16

so that

$$\Delta_r H^\circ = \Delta_f \overline{H}^\circ[\text{Cl}^-(aq)]$$

or

$$\Delta_f \overline{H}^\circ[\text{Cl}^-(aq)] = -167.2 \text{ kJ mol}^{-1}$$

Once the value of $\Delta_f \overline{H}^\circ[\text{Cl}^-(aq)]$ has been determined, we can measure the $\Delta_r H^\circ$ of the reaction

$$\text{Na}(s) + \tfrac{1}{2}\text{Cl}_2(g) \rightarrow \text{Na}^+(aq) + \text{Cl}^-(aq)$$

from which we can determine the value of $\Delta_f \overline{H}^\circ[\text{Na}^+(aq)]$ and so on.

Table 7.5 lists the $\Delta_f \overline{H}^\circ$ values of some common cations and anions. Two points are worth noting about this table. First, for aqueous solutions, the standard state at

Table 7.5
Thermodynamic Data for Aqueous Ions at 1 bar and 298 K

Ion	$\Delta_f\overline{H}°$/kJ mol^{-1}	$\Delta_f\overline{G}°$/kJ mol^{-1}	$\overline{S}°$/J K^{-1} mol^{-1}
H^+	0	0	0
Li^+	−278.5	−293.8	14.23
Na^+	−239.7	−261.9	50.9
K^+	−252.4	−283.3	102.5
Mg^{2+}	−466.9	−454.8	−138.1
Ca^{2+}	−542.8	−553.6	−53.1
Fe^{2+}	−89.1	−78.9	−137.7
Zn^{2+}	−153.9	−147.2	−112.1
Fe^{3+}	−48.5	−4.7	−315.9
OH^-	−229.6	−157.3	−10.75
F^-	−329.1	−276.5	−13.8
Cl^-	−167.2	−131.2	56.5
Br^-	−121.6	−104.0	82.4
I^-	−55.2	−51.57	111.3
CO_3^{2-}	−677.1	−527.8	−56.9
NO_3^-	−206.6	−110.5	146.4
PO_4^{3-}	−1277.4	−1018.7	−221.8

298 K is a hypothetical state defined as the ideal solution of unit molality at 1 bar pressure, in which the activity of the solute (the ion) is unity. The ion thus has the properties it would possess in an infinitely dilute solution, in which interactions between the ions are negligible. Second, all the $\Delta_f\overline{H}°$ values are *relative* values based on the $\Delta_f\overline{H}°[H^+(aq)] = 0$ scale.

We can determine the standard molar Gibbs energy of formation of ions and standard molar entropy of ions at 298 K in a similar fashion, that is, by arbitrarily assigning zero values to $\Delta_f\overline{G}°[H^+(aq)]$ and $\overline{S}°[H^+(aq)]$. These values are also listed in Table 7.5. Because the entropy values of ions in aqueous solution are relative to that of the H^+ ion, they may be either positive or negative. For example, the entropy of $Ca^{2+}(aq)$ is −53.1 J K^{-1} mol^{-1}, and that of $NO_3^-(aq)$ is 146.4 J K^{-1} mol^{-1}. The magnitude and sign of these entropies are influenced by the extent to which they can order the water molecules around themselves in solution, compared with $H^+(aq)$. Small, highly charged ions have negative entropy values, whereas large, singly charged ions have positive entropy values.

EXAMPLE 7.6

Use the standard enthalpy of the reaction

$$\mathrm{Na}(s) + \tfrac{1}{2}\mathrm{Cl_2}(g) \rightarrow \mathrm{Na^+}(aq) + \mathrm{Cl^-}(aq) \qquad \Delta_r H^\circ = -406.9\ \mathrm{kJ\ mol^{-1}}$$

to calculate the value of $\Delta_f \overline{H}^\circ[\mathrm{Na^+}(aq)]$.

ANSWER

The standard enthalpy of reaction is given by

$$\Delta_r H^\circ = \Delta_f \overline{H}^\circ[\mathrm{Na^+}(aq)] + \Delta_f \overline{H}^\circ[\mathrm{Cl^-}(aq)] - (0) - (\tfrac{1}{2})(0)$$

$$-406.9\ \mathrm{kJ\ mol^{-1}} = \Delta_f \overline{H}^\circ[\mathrm{Na^+}(aq)] - 167.2\ \mathrm{kJ\ mol^{-1}}$$

so

$$\Delta_f \overline{H}^\circ[\mathrm{Na^+}(aq)] = -239.7\ \mathrm{kJ\ mol^{-1}}$$

7.4 Ionic Activity

Our next task is to learn to write chemical potentials of electrolytes in solution. First, we shall discuss ideal electrolyte solutions in which the concentrations are expressed on the molality scale.

For an ideal NaCl solution, the chemical potential, μ_{NaCl}, is given by

$$\mu_{\mathrm{NaCl}} = \mu_{\mathrm{Na^+}} + \mu_{\mathrm{Cl^-}} \tag{7.14}$$

Because cations and anions cannot be studied individually, $\mu_{\mathrm{Na^+}}$ and $\mu_{\mathrm{Cl^-}}$ are not measurable. Nevertheless, we can express the chemical potentials of the cation and anion as

Each m term is divided by m° [1 mol (kg H_2O)$^{-1}$], so the logarithmic term is dimensionless.

$$\mu_{\mathrm{Na^+}} = \mu^\circ_{\mathrm{Na^+}} + RT \ln m_{\mathrm{Na^+}}$$

$$\mu_{\mathrm{Cl^-}} = \mu^\circ_{\mathrm{Cl^-}} + RT \ln m_{\mathrm{Cl^-}}$$

where $\mu^\circ_{\mathrm{Na^+}}$ and $\mu^\circ_{\mathrm{Cl^-}}$ are the standard chemical potentials of the ions. Equation 7.14 can now be written as

$$\mu_{\mathrm{NaCl}} = \mu^\circ_{\mathrm{NaCl}} + RT \ln m_{\mathrm{Na^+}} m_{\mathrm{Cl^-}}$$

where

$$\mu^\circ_{\text{NaCl}} = \mu^\circ_{\text{Na}^+} + \mu^\circ_{\text{Cl}^-}$$

In general, a salt with the formula $M_{\nu_+}X_{\nu_-}$ dissociates as follows:

$$M_{\nu_+}X_{\nu_-} \rightleftharpoons \nu_+ M^{z_+} + \nu_- X^{z_-}$$

where ν_+ and ν_- are the numbers of cations and anions per unit and z_+ and z_- are the numbers of charges on the cation and anion, respectively. For NaCl, $\nu_+ = \nu_- = 1$, $z_+ = +1$, $z_- = -1$. For $CaCl_2$, $\nu_+ = 1$, $\nu_- = 2$, $z_+ = +2$, and $z_- = -1$. The chemical potential is given by

$$\mu = \nu_+\mu_+ + \nu_-\mu_- \tag{7.15}$$

where

$$\mu_+ = \mu^\circ_+ + RT \ln m_+$$

Recall that each m term is divided by m°.

and

$$\mu_- = \mu^\circ_- + RT \ln m_-$$

The molalities of the cation and anion are related to the molality of the salt originally dissolved in solution, m, as follows:

$$m_+ = \nu_+ m \qquad m_- = \nu_- m$$

Substituting the expressions for μ_+ and μ_- into Equation 7.15 yields

$$\mu = (\nu_+\mu^\circ_+ + \nu_-\mu^\circ_-) + RT \ln m_+^{\nu_+} m_-^{\nu_-} \tag{7.16}$$

We define *mean ionic molality* ($m_\pm$) as a geometric mean (see Appendix A) of the individual ionic molalities

$$m_\pm = (m_+^{\nu_+} m_-^{\nu_-})^{1/\nu} \tag{7.17}$$

where $\nu = \nu_+ + \nu_-$, and Equation 7.16 becomes

$$\mu = (\nu_+\mu^\circ_+ + \nu_-\mu^\circ_-) + \nu RT \ln m_\pm \tag{7.18}$$

Mean ionic molality can also be expressed in terms of the molality of the solution, m. Because $m_+ = \nu_+ m$ and $m_- = \nu_- m$, we have

$$m_\pm = [(\nu_+ m)^{\nu_+}(\nu_- m)^{\nu_-}]^{1/\nu}$$

$$= m[(\nu_+^{\nu_+})(\nu_-^{\nu_-})]^{1/\nu} \tag{7.19}$$

EXAMPLE 7.7

Write the expression for the chemical potential of $Mg_3(PO_4)_2$ in terms of the molality of the solution.

ANSWER

For $Mg_3(PO_4)_2$, we have $\nu_+ = 3$, $\nu_- = 2$, and $\nu = 5$. The mean ionic molality is

$$m_\pm = (m_+^3 m_-^2)^{1/5}$$

and the chemical potential is given by

$$\mu_{Mg_3(PO_4)_2} = \mu^\circ_{Mg_3(PO_4)_2} + 5RT \ln m_\pm$$

From Equation 7.19,

$$m_\pm = m(3^3 \times 2^2)^{1/5}$$
$$= 2.55m$$

so that

$$\mu_{Mg_3(PO_4)_2} = \mu^\circ_{Mg_3(PO_4)_2} + 5RT \ln 2.55m$$

Unlike nonelectrolyte solutions, most electrolyte solutions behave nonideally. The reason is as follows. The intermolecular forces between uncharged species generally depend on $1/r^7$, where r is the distance of separation; a 0.1-m nonelectrolyte solution is considered ideal for most practical purposes. But Coulomb's law has a $1/r^2$ dependence (Figure 7.7). This dependence means that even in quite dilute solutions (e.g., 0.05 m), the electrostatic forces exerted by ions on one another are enough to cause a deviation from ideal behavior. Thus, in the vast majority of cases, we must replace molality with activity. By analogy to the mean ionic molality, we define the *mean ionic activity* ($a_\pm$) as

$$a_\pm = (a_+^{\nu_+} a_-^{\nu_-})^{1/\nu} \tag{7.20}$$

where a_+ and a_- are the activities of the cation and anion, respectively. The mean ionic activity and mean ionic molality are related by the *mean ionic activity coefficient*, $\gamma_\pm$; that is,

$$a_\pm = \gamma_\pm m_\pm \tag{7.21}$$

where

$$\gamma_\pm = (\gamma_+^{\nu_+} \gamma_-^{\nu_-})^{1/\nu} \tag{7.22}$$

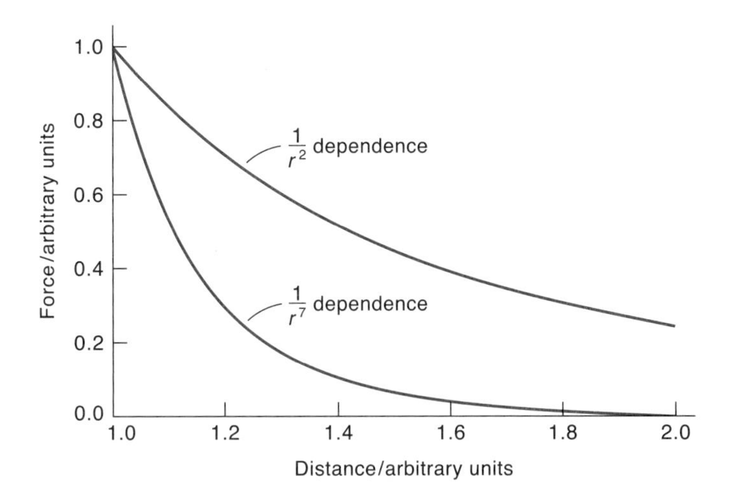

Figure 7.7
Comparison of dependence of attractive forces on distance, r: electrostatic forces ($1/r^2$) between ions and van der Waals forces ($1/r^7$) between molecules.

The chemical potential of a nonideal electrolyte solution is given by

$$\begin{aligned}\mu &= (\nu_+\mu_+^\circ + \nu_-\mu_-^\circ) + \nu RT \ln a_\pm \\ &= (\nu_+\mu_+^\circ + \nu_-\mu_-^\circ) + RT \ln a_\pm^\nu \\ &= (\nu_+\mu_+^\circ + \nu_-\mu_-^\circ) + RT \ln a \qquad (7.23)\end{aligned}$$

where the activity of the electrolyte, a, is related to its mean ionic activity by

$$a = a_\pm^\nu$$

Experimental values of $\gamma_\pm$ can be obtained from freezing-point depression and osmotic-pressure measurements* or electrochemical studies (see Chapter 9). Hence, the value of $a_\pm$ can be calculated from Equation 7.21. In the limiting case of infinite dilution ($m \to 0$), we have

$$\lim_{m \to 0} \gamma_\pm = 1$$

Figure 7.8 shows the plots of $\gamma_\pm$ versus m for several electrolytes. At very low concentrations, $\gamma_\pm$ approaches unity for all types of electrolytes. As the concentrations of electrolytes increase, deviations from ideality occur. The variation of $\gamma_\pm$ with concentration for dilute solutions can be explained by the Debye–Hückel theory, which is discussed next.

* Interested readers should consult the standard physical chemistry texts listed in Chapter 1 for details of $\gamma_\pm$ measurements.

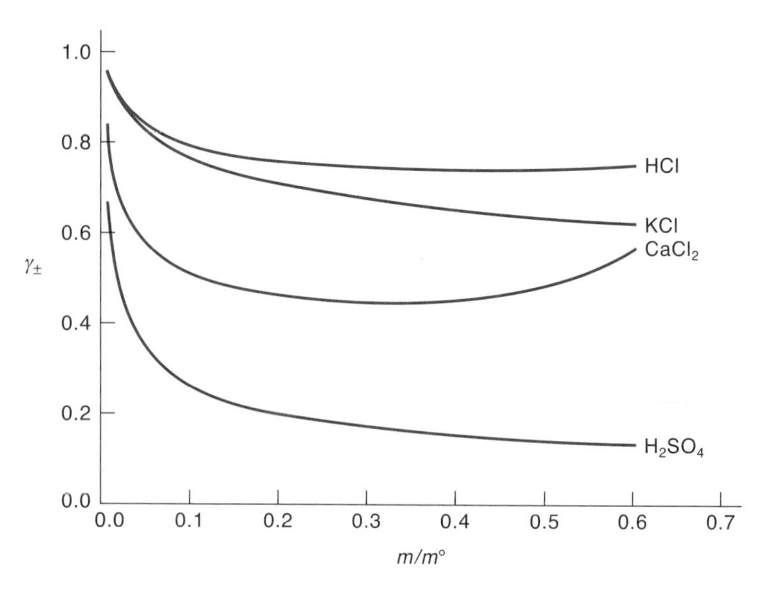

Figure 7.8
Plots of mean activity coefficient, $\gamma_\pm$ versus molality, m, for several electrolytes. At infinite dilution ($m \rightarrow 0$), the mean activity coefficient approaches unity.

EXAMPLE 7.8

Write expressions for the activities (a) of KCl, Na_2CrO_4, and $Al_2(SO_4)_3$ in terms of their molalities and mean ionic activity coefficients.

ANSWER

We need the relations $a = a_\pm^\nu = (\gamma_\pm m_\pm)^\nu$.

KCl: $\nu = 1 + 1 = 2;\ m_\pm = (m^2)^{1/2} = m$

Therefore, $a_{KCl} = m^2\gamma_\pm^2$

Na_2CrO_4: $\nu = 2 + 1 = 3;\ m_\pm = [(2m)^2(m)]^{1/3} = 4^{1/3}m$

Therefore, $a_{Na_2CrO_4} = 4m^3\gamma_\pm^3$

$Al_2(SO_4)_3$: $\nu = 2 + 3 = 5;\ m_\pm = [(2m)^2(3m)^3]^{1/5} = 108^{1/5}m$

Therefore, $a_{Al_2(SO_4)_3} = 108m^5\gamma_\pm^5$

7.5 Debye–Hückel Theory of Electrolytes

Our treatment of deviations from ideality by electrolyte solutions has been empirical: Using the ionic activities obtained from the activity coefficient and the known concentration, we calculate chemical potential, the equilibrium constant, and other properties. Missing in this approach is a physical interpretation of ionic behavior in solution. In 1923, Debye and the German chemist Erich Hückel (1896–1980) put forward a quantitative theory that has greatly advanced our knowledge of electrolyte solutions.

Based on a rather simple model, the Debye–Hückel theory enables us to calculate the value of $\gamma_\pm$ from the properties of the solution.

The mathematical details of Debye's and Hückel's treatment are too complex to present here. (Interested readers should consult the standard physical chemistry texts listed in Chapter 1.) Instead, we shall discuss the underlying assumptions and final results. Debye and Hückel began by assuming the following: (1) electrolytes are completely dissociated into ions in solution; (2) the solutions are dilute, with a concentration of 0.01 *m* or lower; and (3) on average, each ion is surrounded by ions of opposite charge, forming an *ionic atmosphere* (Figure 7.9a). Working from these assumptions, Debye and Hückel calculated the average electric potential at each ion caused by the presence of other ions in the ionic atmosphere. The Gibbs energy of the ions was then related to the activity coefficient of the individual ion. Because neither γ_+ nor γ_- could be measured directly, the final result is expressed in terms of the mean ionic activity coefficient of the electrolyte as follows:

$$\log \gamma_\pm = -0.509|z_+z_-|\sqrt{I} \tag{7.24}$$

where the | | signs denote the magnitude but not the sign of the product z_+z_-. Thus, for $CuSO_4$, we have $z_+ = 2$ and $z_- = -2$, but $|z_+z_-| = 4$. The quantity I, called the *ionic strength*, is defined as follows:

$$I = \tfrac{1}{2}\sum_i m_i z_i^2 \tag{7.25}$$

where m_i and z_i are the molality and the charge of the ith ion in the electrolyte, respectively. This quantity was first introduced by the American chemist Gilbert Newton Lewis (1875–1946), who noted that nonideality observed in electrolye solutions primarily stems from the *total* concentration of charges present rather than from the chemical nature of the individual ionic species. Equation 7.25 enables us to express the ionic concentrations for all types of electrolytes on a common basis so that we need not sort out the charges on the individual ions. Equation 7.24 is known as the *Debye–Hückel limiting law.* As written, it applies to an aqueous electrolyte solution at 298 K. Note that the right side of Equation 7.24 is dimensionless. Thus, the I term is assumed to be divided by $m°[1 \text{ mol (kg } H_2O)^{-1}]$.

The word "limiting" means that the law applies only to solutions in the limit of dilute concentrations.

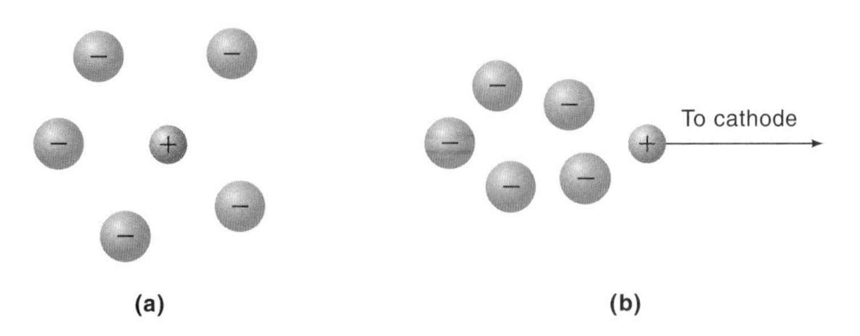

Figure 7.9
(a) Simplified presentation of an ionic atmosphere surrounding a cation in solution. (b) In a conductance measurement, the movement of a cation toward the cathode is retarded by the electric field exerted by the ionic atmosphere left behind.

EXAMPLE 7.9

Calculate the mean activity coefficient ($\gamma_\pm$) of a 0.010 m aqueous solution of $CuSO_4$ at 298 K.

ANSWER

The ionic strength of the solution is given by Equation 7.25:

$$I = \tfrac{1}{2}[(0.010\ m) \times 2^2 + (0.010\ m) \times (-2)^2]$$

$$= 0.040\ m$$

From Equation 7.26,

$$\log \gamma_\pm = -0.509(2 \times 2)\sqrt{0.040}$$

$$= -0.407$$

or

$$\gamma_\pm = 0.392$$

Experimentally, $\gamma_\pm$ is found to be 0.41 at the same concentration.

Two points are worth noting in applying Equation 7.24. First, in a solution containing several electrolytes, *all* the ions in solution contribute to the ionic strength, but z_+ and z_- refer only to the ionic charges of the particular electrolyte for which $\gamma_\pm$ is being calculated. Second, Equation 7.24 can be used to calculate the ionic activity coefficient of individual cations or anions. Thus, for the ith ion, we write

$$\log \gamma_i = -0.509 z_i^2 \sqrt{I} \tag{7.26}$$

where z_i is the charge of the ion. The γ_+ and γ_- values calculated this way are related to $\gamma_\pm$ according to Equation 7.22.

Figure 7.10 shows calculated and measured values of log $\gamma_\pm$ at various ionic strengths. We can see Equation 7.24 holds quite well for dilute solutions but must be modified to account for the drastic deviations that occur at high concentrations of electrolytes. Several improvements and modifications have been applied to this equation for treating more concentrated solutions.

The generally good agreement between experimentally determined $\gamma_\pm$ values and those calculated using the Debye–Hückel theory provides strong support for the existence of an ionic atmosphere in solution. The model can be tested by taking a conductance measurement in a very strong electric field. In reality, ions do not move in a straight line toward the electrodes in a conductance cell but move along

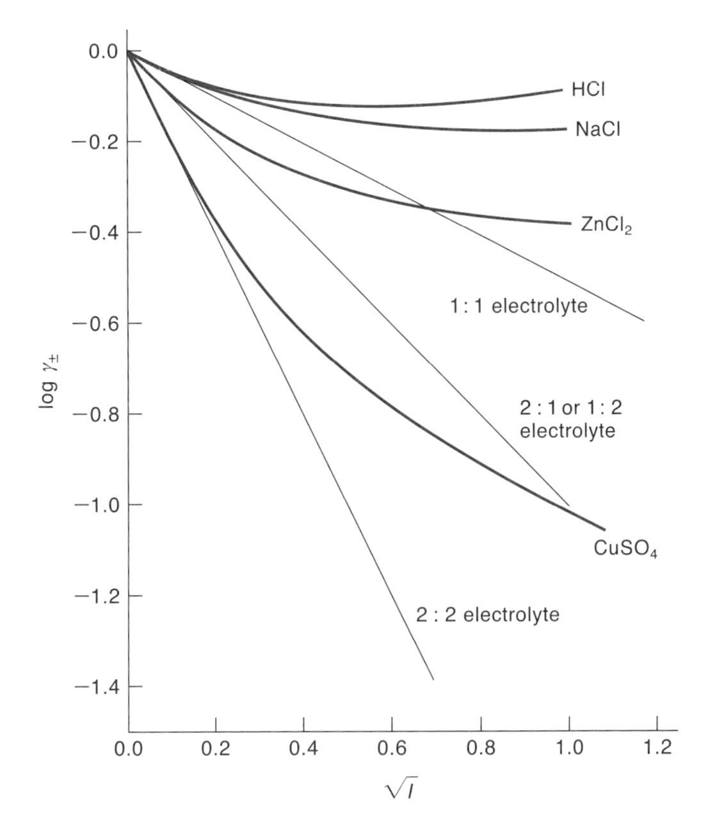

Figure 7.10
Plots of log $\gamma_\pm$ versus $\sqrt{I}$ for several electrolytes. The straight lines are those predicted by Equation 7.24.

a zigzag path. Microscopically, the solvent is not a continuous medium. Each ion actually "jumps" from one solvent hole to another, and as the ion moves across the solution, its ionic atmosphere is repeatedly being destroyed and formed again. The formation of an ionic atmosphere does not occur instantaneously but requires a finite amount of time, called the *relaxation time*, which is approximately 10^{-7} s in a 0.01-m solution. Under normal conditions of conductance measurement, the velocity of an ion is sufficiently slow so that the electrostatic force exerted by the atmosphere on the ion tends to retard its motion and hence to decrease the conductance (see Figure 7.9b). If the conductance measurement were carried out at a very strong electric field (approximately 2×10^5 V cm^{-1}), the ionic velocity would be approximately 10 cm s^{-1}. The radius of the ionic atmosphere in a 0.01-m solution is approximately 5 Å, or 5×10^{-8} cm, so that the time required for the ion to move out of the atmosphere is 5×10^{-8} cm/(10 cm s^{-1}), or 5×10^{-9} s, which is considerably shorter than the relaxation time. Consequently, the ion can move through the solution free of the retarding influence of the ionic atmosphere. The free movement leads to a marked increase in conductance. This phenomenon is called the *Wien effect*, after the German physicist Wilhelm Wien (1864–1928), who first performed the experiment in 1927. The Wien effect is one of the strongest pieces of evidence for the existence of an ionic atmosphere.

The Salting-In and Salting-Out Effects

The Debye–Hückel limiting law can be applied to study the solubility of proteins. The solubility of a protein in an aqueous solution depends on the temperature, pH, dielectric constant, ionic strength, and other characteristics of the medium. In this section, however, we shall focus on the influence of ionic strength.

Let us first investigate the effect of ionic strength on the solubility of an inorganic compound, AgCl. The solubility equilibrium is

$$\mathrm{AgCl}(s) \rightleftharpoons \mathrm{Ag}^+(aq) + \mathrm{Cl}^-(aq)$$

The *thermodynamic* solubility product for the process, K_{sp}°, is given by

$$K_{\mathrm{sp}}^{\circ} = a_{\mathrm{Ag}^+} a_{\mathrm{Cl}^-}$$

The ionic activities are related to ionic concentrations as follows:

$$a_+ = \gamma_+ m_+ \quad \text{and} \quad a_- = \gamma_- m_-$$

so that

$$\begin{aligned} K_{\mathrm{sp}}^{\circ} &= \gamma_{\mathrm{Ag}^+} m_{\mathrm{Ag}^+} \gamma_{\mathrm{Cl}^-} m_{\mathrm{Cl}^-} \\ &= \gamma_{\mathrm{Ag}^+} \gamma_{\mathrm{Cl}^-} K_{\mathrm{sp}} \end{aligned}$$

where $K_{\mathrm{sp}} = m_{\mathrm{Ag}^+} m_{\mathrm{Cl}^-}$ is the *apparent* solubility product. The difference between the thermodynamic and apparent solubility products is as follows. As we can see, the apparent solubility product is expressed in molalities (or some other concentration unit). We can readily calculate this quantity if we know the amount of AgCl dissolved in a known amount of water to produce a saturated solution. Because of electrostatic forces, however, the dissolved ions are under the influence of their immediate neighbors. Consequently, the actual or effective number of ions is not the same as that calculated from the concentration of the solution. For example, if a cation forms a tight ion pair with an anion, then the actual number of species in solution, from a thermodynamic perspective, is one and not, as we would expect, two. This is the reason for replacing concentration with activity, which is the effective concentration. Thus, the thermodynamic solubility product represents the true value of the solubility product, which generally differs from the apparent solubility product. Because

$$\gamma_{\mathrm{Ag}^+} \gamma_{\mathrm{Cl}^-} = \gamma_{\pm}^2$$

we write

$$K_{\mathrm{sp}}^{\circ} = \gamma_{\pm}^2 K_{\mathrm{sp}}$$

Taking the logarithm of both sides and rearranging, we obtain

$$-\log \gamma_{\pm} = \log\left(\frac{K_{\mathrm{sp}}}{K_{\mathrm{sp}}^{\circ}}\right)^{1/2} = 0.509|z_+ z_-|\sqrt{I}$$

The last equality in the above equation is the Debye–Hückel limiting law. The solubility product can be directly related to the molar solubility (S); for a 1:1 electrolyte,

$$(K_{sp})^{1/2} = S \quad \text{and} \quad (K^\circ_{sp})^{1/2} = S^\circ$$

where S and S° are the apparent and thermodynamic molar solubilities in mol L^{-1}. Finally, we obtain the following equation relating the solubility of an electrolyte to the ionic strength of the solution:

$$\log \frac{S}{S^\circ} = 0.509|z_+z_-|\sqrt{I} \tag{7.27}$$

Note that the value of S° can be determined by plotting log S versus $\sqrt{I}$. The intercept on the log S axis ($I = 0$) gives log S°, and hence S°.

If AgCl is dissolved in water, its solubility (S) is 1.3×10^{-5} mol L^{-1}. If it is dissolved in a KNO_3 solution, according to Equation 7.27, its solubility is greater because of the solution's increase in ionic strength. In a KNO_3 solution, the ionic strength is a sum of two concentrations, one from AgCl and the other from KNO_3. The increase in solubility, caused by the increase in ionic strength, is called the *salting-in effect.*

Equation 7.27 holds up only to a certain value of ionic strength. As the ionic strength of a solution increases further, it must be replaced by the following expression:

$$\log \frac{S}{S^\circ} = -K'I \tag{7.28}$$

where K' is a positive constant whose value depends on the nature of the solute and on the electrolyte present. The larger the solute molecule, the greater the value of K' is. Equation 7.28 tells us that the ratio of the solubilities in the region of high ionic strength actually decreases with I (note the negative sign). The decrease in solubility with increasing ionic strength of the solution is called the *salting-out effect.* This phenomenon can be explained in terms of hydration. Recall that hydration is the process that stabilizes ions in solution. At high salt concentrations, the availability of water molecules decreases, and so the solubility of ionic compounds also decreases. The salting-out effect is particularly noticeable with proteins, whose solubility in water is sensitive to ionic strength because of their large surface areas. Combining Equations 7.27 and 7.28, we have the approximate equation

$$\log \frac{S}{S^\circ} = 0.509|z_+z_-|\sqrt{I} - K'I \tag{7.29}$$

Equation 7.29 is applicable over a wider range of ionic strengths.

Figure 7.11 shows how the ionic strength of various inorganic salts affects the solubility of horse hemoglobin. As we can see, the protein exhibits a salting-in region at low ionic strengths.* As I increases, the curve goes through a maximum and eventually the slope becomes negative, indicating that the solubility decreases with increasing ionic strength. In this region, the second term in Equation 7.29 predominates. This trend is most pronounced for salts such as Na_2SO_4 and $(NH_4)_2SO_4$.

---

* When I is less than unity, $\sqrt{I} > I$. Thus, at low ionic strengths, the first term in Equation 7.29 predominates.

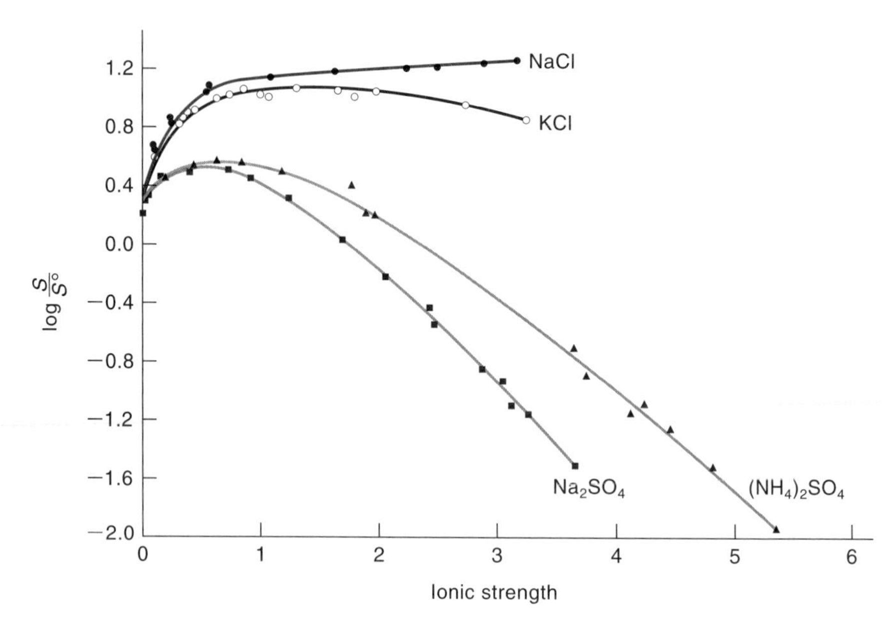

Figure 7.11
Plots of log($S/S°$) versus ionic strength for horse hemoglobin in the presence of various inorganic salts. Note that when $I = 0$, all the curves converge to the same point on the log($S/S°$) axis at zero and $S = S°$. [From Cohn, E., and J. Edsall, *Proteins, Amino Acids and Peptides*, © Litton Educational Publishing, 1943. Reprinted by permission of Van Nostrand Reinhold, New York.]

The practical value of the salting-out effect is that it enables us to precipitate proteins from solutions. In addition, the effect can also be used to purify proteins. Figure 7.12 shows the range of the salting-out phenomenon for several proteins in the presence of ammonium sulfate. Although the solubility of proteins is sensitive to the degree of hydration, the strength of binding of water molecules is not the same for all proteins. The relative solubility of different proteins at a particular ionic strength provides a means for selective precipitation. The point is that although higher ionic strengths are needed to salt out proteins, precipitation occurs over a small range of ionic strength, providing sharp separations.

7.6 Colligative Properties of Electrolyte Solutions

The colligative properties of an electrolyte solution are influenced by the number of ions present in solution. For example, we expect the aqueous freezing-point depression caused by a 0.01-m solution of NaCl to be twice that effected by a 0.01-m sucrose solution, assuming complete dissociation of the former. For incompletely dissociated salts, the relationship is more complicated, but understanding it provides us with another way to measure an electrolyte's degree of dissociation.

Let us define a factor i, called the van't Hoff factor (after the Dutch chemist Jacobus Hendricus van't Hoff, 1852–1911), as follows:

$$i = \frac{\text{actual number of particles in solution at equilibrium}}{\text{number of particles in solution before dissociation}} \tag{7.30}$$

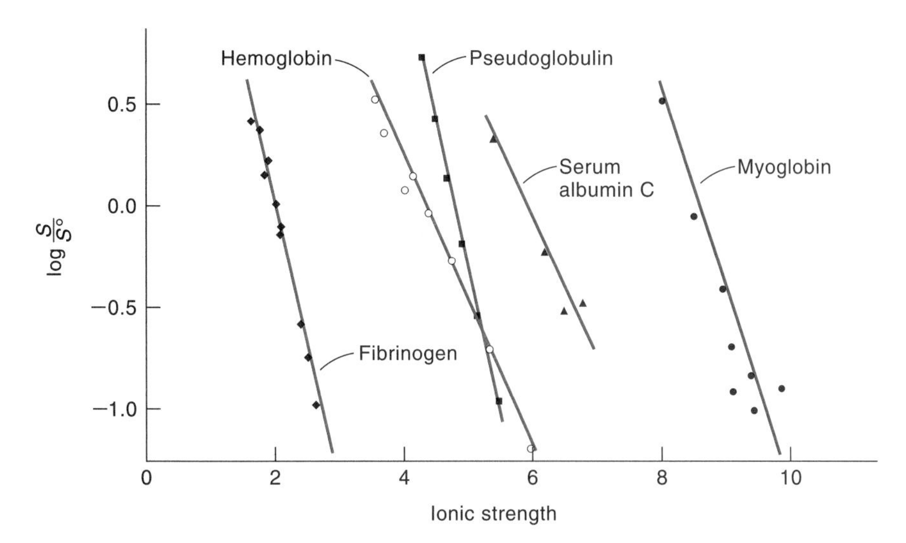

Figure 7.12
Plots of ($S/S°$) versus ionic strength for several proteins in aqueous ammonium sulfate, demonstrating the salting-out effect. [From Cohn, E. J., *Chem. Rev.* **19**, 241 (1936). Used by permission of Williams & Wilkins, Baltimore.]

If a solution contains N units of an electrolyte, and if α is the degree of dissociation,

$$\begin{array}{ccccc} \mathrm{M}_{\nu_+}\mathrm{X}_{\nu_-} & \rightleftharpoons & \nu_+\mathrm{M}^{z_+} & + & \nu_-\mathrm{X}^{z_-} \\ N(1-\alpha) & & N\nu_+\alpha & & N\nu_-\alpha \end{array}$$

there will be $N(1 - \alpha)$ undissociated units and $(N\nu_+\alpha + N\nu_-\alpha)$, or $N\nu\alpha$ ions in solution at equilibrium, where $\nu = \nu_+ + \nu_-$. We can now write the van't Hoff factor as

$$i = \frac{N(1-\alpha) + N\nu\alpha}{N} = 1 - \alpha + \nu\alpha$$

and

$$\alpha = \frac{i-1}{\nu-1} \tag{7.31}$$

For strong electrolytes, i is approximately equal to the number of ions formed from each unit of the electrolyte; for example, $i \approx 2$ for NaCl and $CuSO_4$; $i \approx 3$ for K_2SO_4 and $BaCl_2$, and so on. The value of i decreases with increasing concentration of the solution, which is attributed to the formation of ion pairs.

The presence of ion pairs will also affect the colligative properties because it decreases the number of free particles in solution. In general, formation of ion pairs is most pronounced between highly charged cations and anions and in media of low dielectric constants. In an aqueous solution of $Ca(NO_3)_2$, for example, the Ca^{2+} and NO_3^- ions form ion pairs as follows:

$$Ca^{2+}(aq) + NO_3^-(aq) \rightleftharpoons Ca(NO_3)^+(aq)$$

The equilibrium constants for such ion pairing are not accurately known, however, thus making calculations of colligative properties of electrolyte solutions difficult. A recent study* showed that deviation from colligative properties for many electrolyte solutions is not the result of ion pair formation, but rather of hydration. In an electrolyte solution the cations and anions can tie up large numbers of water molecules in their hydration spheres, thus reducing the number of free water molecules in the bulk solvent. Deviations disappear when the correct number of water molecules in the hydration sphere is subtracted from the total water solvent molecules in calculating the concentrations (molality or molarity) of the solution.

EXAMPLE 7.10

The osmotic pressures of a 0.01 *m* solution of $CaCl_2$ and a 0.01 *m* sucrose solution at 298 K are 0.605 atm and 0.224 atm, respectively. Calculate the van't Hoff factor and the degree of dissociation for $CaCl_2$. Assume ideal behavior.

ANSWER

As far as osmotic pressure measurements are concerned, the main difference between calcium chloride and sucrose is that only $CaCl_2$ can dissociate into ions (Ca^{2+} and Cl^-). Otherwise, equal concentrations of $CaCl_2$ and sucrose solutions would have the same osmotic pressure. Because the osmotic pressure of a solution is directly proportional to the number of particles present, we can calculate the van't Hoff factor for the $CaCl_2$ solution as follows. From Equation 7.30,

$$i = \frac{0.605 \text{ atm}}{0.224 \text{ atm}} = 2.70$$

Because for $CaCl_2$, $\nu_+ = 1$, and $\nu_- = 2$, we have $\nu = 3$ and

$$\alpha = \frac{2.70 - 1}{3 - 1} = 0.85$$

Finally, we note that the equations used to determine the colligative properties of nonelectrolyte solutions (Equations 6.39, 6.40, and 6.46) must be modified for electrolyte solutions as follows:

$$\Delta T = K_b(im_2) \qquad (7.32)$$

$$\Delta T = K_f(im_2) \qquad (7.33)$$

$$\Pi = iMRT \qquad (7.34)$$

Ideal behavior is assumed for the electrolyte solutions, so we use concentrations instead of activities.

* A. A. Zavitsas, *J. Phys. Chem.* **105**, 7805 (2001).

The Donnan Effect

The Donnan effect (after the British chemist Frederick George Donnan, 1870–1956) has its starting point in the treatment of osmotic pressure. It describes the equilibrium distribution of small diffusible ions on the two sides of a membrane that is freely permeable to these ions but impermeable to macromolecular ions, in the presence of a macromolecular electrolyte on one side of the membrane.

Suppose that a cell is separated into two parts by a semipermeable membrane that allows the diffusion of water and small ions but not protein molecules. Let us consider the following three cases.

Case 1. The protein solution is placed in the left compartment; water is placed in the right compartment. We assume that the protein molecules are neutral species.* Let the concentration of the protein solution be c (mol L^{-1}) so that the osmotic pressure of the solution, according to Equation 6.46, is given by

$$\Pi_1 = cRT$$

Thus, from a measurement of the osmotic pressure, we can readily determine the molar mass of the protein molecule.

Case 2. In this case, the protein is the anion of the sodium salt, Na^+P^-, which we assume to be a strong electrolyte. Again, the protein solution of concentration c is placed in the left compartment, and pure water is in the right compartment. To maintain electrical neutrality, all of the Na^+ ions remain in the left compartment; the osmotic pressure of the solution now becomes

$$\Pi_2 = (c + c)RT = 2cRT$$

Because $\Pi_2 = 2\Pi_1$, it follows that the molar mass determined in this case will be only half of the true molar mass. (From Equation 6.47 we have $\mathcal{M}_2 = c_2RT/\Pi$, so that doubling Π would decrease the value of $\mathcal{M}_2$ by half.) In practice, the situation is actually much worse, because the protein ion may bear as many as 20 or 30 net negative (or positive) charges. In the early days of protein-molar-mass determination by osmotic pressure, disastrously poor results were obtained before the dissociation process was recognized and efforts made to correct for it (see Appendix 7.2 on p. 298).

Case 3. We start with an arrangement similar to that discussed in Case 2 and then add NaCl (of concentration b in mol L^{-1}) to the right compartment (Figure 7.13a). At equilibrium, a certain amount, x (mol L^{-1}), of Na^+ and Cl^- ions has diffused through the membrane from right to left, creating a final state shown in Figure 7.13b. Both sides of the membrane must be electrically neutral: in each compartment the number of cations equals the number of anions. The condition of equilibrium enables us to equate the chemical potentials of NaCl in the two compartments as follows (see Equation 7.23):

$$(\mu_{\text{NaCl}})^{\text{L}} = (\mu_{\text{NaCl}})^{\text{R}}$$

* Proteins are ampholytes, that is, they possess both acidic and basic properties. Depending on the pH of the medium, a protein can exist as an anion, a cation, or a neutral species.

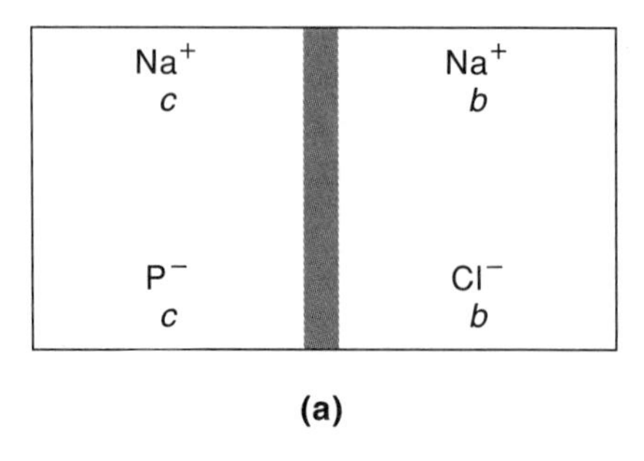

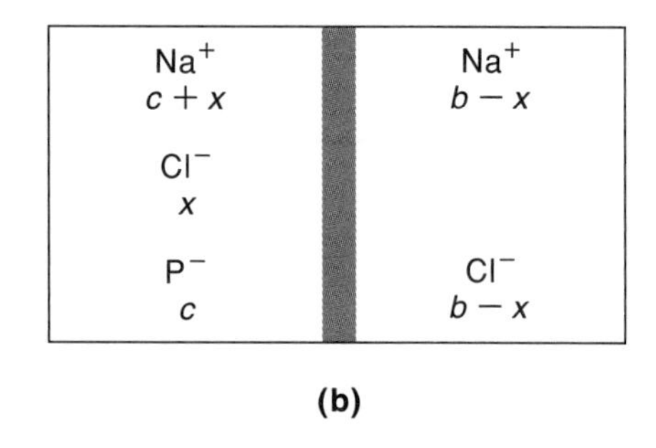

Figure 7.13
Schematic representation of the Donnan effect. (a) Before diffusion has begun. (b) At equilibrium. The membrane separating the left and right compartments is permeable to all but the P^- ions. The volumes in the two compartments are equal and assumed to remain constant.

or

$$(\mu^\circ + 2RT \ln a_\pm)^{\mathrm{L}}_{\mathrm{NaCl}} = (\mu^\circ + 2RT \ln a_\pm)^{\mathrm{R}}_{\mathrm{NaCl}}$$

Because μ°, the standard chemical potential, is the same on both sides, we obtain

$$(a_\pm)^{\mathrm{L}}_{\mathrm{NaCl}} = (a_\pm)^{\mathrm{R}}_{\mathrm{NaCl}}$$

From Equation 7.20,

$$(a_{\mathrm{Na^+}} a_{\mathrm{Cl^-}})^{\mathrm{L}} = (a_{\mathrm{Na^+}} a_{\mathrm{Cl^-}})^{\mathrm{R}}$$

If the solutions are dilute, then the ionic activities may be replaced by the corresponding concentrations, that is, $a_{\mathrm{Na^+}} = [\mathrm{Na^+}]$ and $a_{\mathrm{Cl^-}} = [\mathrm{Cl^-}]$. Hence,

$$([\mathrm{Na^+}][\mathrm{Cl^-}])^{\mathrm{L}} = ([\mathrm{Na^+}][\mathrm{Cl^-}])^{\mathrm{R}}$$

or

$$(c + x)x = (b - x)(b - x)$$

Table 7.6
The Donnan Effect and Osmotic Pressure

Initial concentration		Equilibrium concentration		
Left compartment	Right compartment	Left compartment		
$c = [\mathrm{Na^+}] = [\mathrm{P^-}]$	$b = [\mathrm{Na^+}] = [\mathrm{Cl^-}]$	$(c + x) = [\mathrm{Na^+}]$	$c = [\mathrm{P^-}]$	$x = [\mathrm{Cl^-}]$
0.1	0	0.1	0.1	0
0.1	0.01	0.1008	0.1	0.00083
0.1	0.1	0.1333	0.1	0.0333
0.1	1.0	0.576	0.1	0.476
0.1	10.0	5.075	0.1	4.975
$[\mathrm{P}] = 0.1$	0	0	$[\mathrm{P}] = 0.1$	0

Solving for x, we obtain

$$x = \frac{b^2}{c + 2b} \tag{7.35}$$

Equation 7.35 says that the amount of NaCl, x, that diffuses from right to left is inversely proportional to the concentration of the nondiffusible ion (P^-), c, in the left compartment. This unequal distribution of the diffusible ions (Na^+ and Cl^-) in the two compartments is the result of the Donnan effect.

In this case, then, the osmotic pressure of the protein solution is determined by the *difference* between the number of particles in the left compartment and the number in the right compartment. We write

$$\Pi_3 = [\underbrace{(c + c + x + x)}_{\text{left compartment}} - \underbrace{2(b - x)}_{\text{right compartment}}]RT = (2c + 4x - 2b)RT$$

From Equation 7.35,

$$\Pi_3 = \left(2c + \frac{4b^2}{c + 2b} - 2b\right)RT = \left(\frac{2c^2 + 2cb}{c + 2b}\right)RT$$

Two limiting cases may be applied to the equation above. If $b \ll c$, $\Pi_3 = 2cRT$, which gives the same result as Case 2. On the other hand, if $b \gg c$, $\Pi_3 = cRT$, which is identical to Case 1. The important conclusion we reach is that the presence of NaCl in the right compartment decreases the osmotic pressure of the protein solution compared to Case 2 and therefore minimizes the Donnan effect. When a very large amount of NaCl is present, the Donnan effect can be effectively eliminated. In general, we have $\Pi_1 \leq \Pi_3 \leq \Pi_2$. Because proteins are usually studied in buffer solutions that contain ionic species, the osmotic pressure measured will be less than in the case in which pure water is the solvent. Table 7.6 shows the Donnan effect for the NaCl example at several concentrations and the corresponding osmotic pressure at 298 K.

Equilibrium concentration		Percent NaCl transferred from right to left	Osmotic pressure/atm	Case
Right compartment				
$(b - x) = [Na^+]$	$(b - x) = [Cl^-]$			
0	0	0	4.90	2
0.00917	0.00917	8.3	4.48	3
0.0667	0.0667	33.3	3.26	3
0.524	0.524	47.6	2.56	3
5.025	5.025	49.75	2.46	3
0	0	0	2.45	1

Another way to eliminate the Donnan effect is to choose a pH at which the protein has no net charge, called the *isoelectric point.* At this pH, the distribution of any diffusible ion will always be equal in both compartments. This method is difficult to apply because most proteins are least soluble at their isoelectric points.

The discussion of the Donnan effect was simplified by assuming ideal behavior and no change in either the pH or the volume of the solution. In addition, for simplicity, we used a common diffusible ion, Na^+, in deriving Equation 7.35.

Understanding the Donnan effect is essential in studying the distribution of ions across the membranes of living organisms and membrane potentials (see Chapter 9).

Key Equations

$$\Lambda = \frac{\kappa}{c}$$ (Molar conductance) (7.4)

$$\Lambda_0 = \nu_+\lambda_0^+ + \nu_-\lambda_0^-$$ (Kohlrausch's law of independent migration) (7.6)

$$\frac{1}{\Lambda} = \frac{1}{K_a\Lambda_0^2}(\Lambda c) + \frac{1}{\Lambda_0}$$ (Ostwald dilution law) (7.9)

$$F = \frac{q_A q_B}{4\pi\varepsilon_0 r^2}$$ (Coulomb's law) (7.12)

$$m_\pm = (m_+^{\nu_+} m_-^{\nu_-})^{1/\nu}$$ (Mean ionic molality) (7.17)

$$a_\pm = (a_+^{\nu_+} a_-^{\nu_-})^{1/\nu}$$ (Mean ionic activity) (7.20)

$$a_\pm = \gamma_\pm m_\pm$$ (Definition of $\gamma_\pm$) (7.21)

$$\gamma_\pm = (\gamma_+^{\nu_+} \gamma_-^{\nu_-})^{1/\nu}$$ (Mean ionic activity coefficient) (7.22)

$$\log \gamma_\pm = -0.509|z_+z_-|\sqrt{I}$$ (Debye–Hückel limiting law) (7.24)

$$I = \tfrac{1}{2}\sum_i m_i z_i^2$$ (Ionic strength) (7.25)

$$\log \frac{S}{S^\circ} = 0.509|z_+z_-|\sqrt{I}$$ (Salting-in effect) (7.27)

$$\log \frac{S}{S^\circ} = -K'I$$ (Salting-out effect) (7.28)

Notes on Electrostatics

An electric charge (q_A) is said to produce an *electric field* (E) in the space around itself. This field exerts a force on any charge (q_B) within that space. According to Coulomb's law, the potential energy (V) between these two charges separated by distance r in a vacuum is given by

Potential energy V has units of energy Joule (J) or electron volt (eV); 1 eV = 1.602×10^{-19} J.

$$V = \frac{q_A q_B}{4\pi\varepsilon_0 r} \quad (1)$$

and the electrostatic force, F, between the charges is

$$F = \frac{q_A q_B}{4\pi\varepsilon_0 r^2} \quad (2)$$

where ε_0 is the permittivity of the vacuum (see p. 272). The electric field is the electrostatic force on a unit positive charge. Thus, the electric field at q_B due to q_A is F divided by q_B, or

$$E = \frac{q_A q_B}{4\pi\varepsilon_0 r^2 q_B} = \frac{q_A}{4\pi\varepsilon_0 r^2} \quad (3)$$

Note that E is a vector and is directed away from q_A toward q_B. Its units are V m^{-1} or V cm^{-1}.

Another important property of the electric field is its *electric potential*, ϕ, which is the potential energy of a unit positive charge in the electric field. Its units are J/C or V (1 J = 1 C $\times$ 1 V). A unit positive charge in an electric field E experiences a force equal in magnitude to E. When the charge is moved through a certain distance, dr, the potential energy change is equal to $qEdr$ or Edr because q = 1 C. Because the repulsive potential energy increases as the unit positive charge comes closer to the positive charge q_A that generates the electric field, the change in potential energy, $d\phi$, is $-Edr$. (The negative sign ensures that as dr decreases, $-Edr$ is a positive quantity, signifying the increase in repulsion between the two positive charges.) The electric potential at a certain point at a distance r from the charge q_A is the potential energy change that occurs in bringing the unit positive charge from infinity to distance r from the charge:

$$\phi = -\int_{r=\infty}^{r=r} Edr = -\int_{r=\infty}^{r=r} \frac{q_A}{4\pi\varepsilon_0 r^2} dr = \frac{q_A}{4\pi\varepsilon_0 r} \quad (4)$$

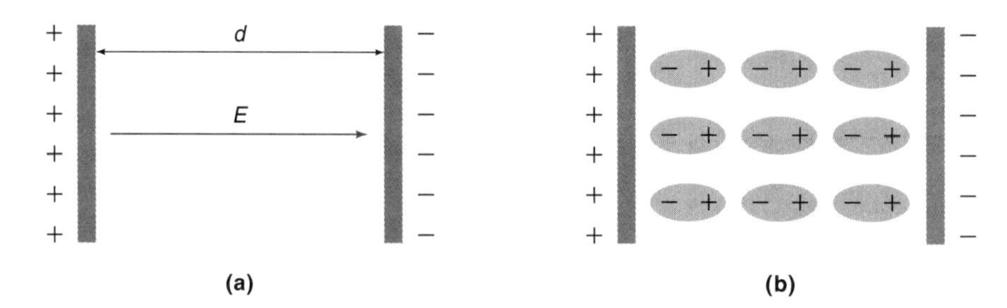

Figure 7.14
(a) The charge separation of a capacitor. The electric field, E, is directed from the positive plate to the negative plate, separated by distance d. With a vacuum between the plates, the dielectric constant is ε_0. (b) Orientation of the dipoles of a dielectric in a capacitor. The degree of orientation is exaggerated. The dielectric material decreases the electric field between the capacitor plates. The dielectric constant of the medium is ε.

Note that $\phi = 0$ at $r = \infty$. From Equation 4, we can define the electric potential difference between points 1 and 2 in an electric field as the work done in bringing a unit charge from 1 to 2; that is,

$$\Delta\phi = \phi_2 - \phi_1 \tag{5}$$

This difference is commonly referred to as the voltage between points 1 and 2.

Dielectric Constant (ε) and Capacitance (C)

When a nonconducting substance (called a *dielectric*) is placed between two flat, parallel metal plates with opposite charges that are equal in magnitude (called a *capacitor*), the substance becomes polarized. The reason is that the electric field of the plates either orients the permanent dipoles of the dielectric or induces dipole moments, as shown in Figure 7.14. The *dielectric constant* of the substance is defined as

$$\varepsilon = \frac{E_0}{E} \tag{6}$$

where E_0 and E are the electric fields in the space between the plates of the capacitor in the absence of a dielectric (a vacuum) and presence of a dielectric, respectively. Keep in mind that the orientation of the dipoles (or induced dipoles) reduces the electric field between the capacitor plates so that $E < E_0$ and $\varepsilon > 1$. For ions in aqueous solution, this decrease in electric field reduces the attraction between the cation and the anion (see Figure 7.6).

The *capacitance* (C) of a capacitor measures its ability to hold charges for a given electric potential difference between the plates; that is, it is given by the ratio of charge to potential difference. The capacitances of a capacitor, when the space between the plates is filled with a dielectric (C) and with a vacuum (C_0), are given respectively by

$$C = \frac{Q}{\Delta\phi} \tag{7}$$

and

$$C_0 = \frac{Q}{\Delta\phi_0} \tag{8}$$

Because $\Delta\phi = Ed$, where d is the distance between the plates, Equation 6 can also be written as

$$\varepsilon = \frac{(\Delta\phi_0/d)}{(\Delta\phi/d)} = \frac{(Q/C_0)}{(Q/C)} = \frac{C}{C_0} \tag{9}$$

Capacitance is an experimentally measurable quantity, so the dielectric constant of a substance can be determined. It has the unit farad (F), where 1 F = 1 C/V. Note that the ratio C/C_0 in Equation 9 makes ε a dimensionless quantity.

APPENDIX 7.2

The Donnan Effect Involving Proteins Bearing Multiple Charges

For a protein at a pH other than its isoelectric point, at which it will possess either a net positive or net negative charge, an additional factor must be considered; that is, the counterions needed to maintain charge neutrality. In the chapter, we considered the simple case where the protein bears only one negative charge. Here we deal with the situation in which a protein bears a number of negative charges (z). We assume that the protein, $Na_z^+P^{z-}$, is a strong electrolyte so that

$$Na_z^+P^{z-} \longrightarrow zNa^+ + P^{z-}$$

Referring to Figure 7.13, we shall consider two cases.

Case 1

The protein solution is placed in the left compartment and water in the right compartment. The osmotic pressure of the solution (Π_1) is given by

$$\Pi_1 = (z + 1)cRT$$

where c is the concentration (molarity) of the protein solution. Because z is typically of the order of 30, using this arrangement to determine the molar mass of the protein yields a value that is only $\frac{1}{30}$ of the true value.

Case 2

Again the protein solution is placed in the left compartment, but a NaCl solution is placed in the right compartment. The requirement that the chemical potential of a component be the same throughout the system applies to the NaCl as well as to the water. To attain equilibrium. NaCl will move from the right to the left compartment. We can calculate the actual amount of NaCl that is transported. The initial molar concentration of $Na_z^+P^{z-}$ is c, and that of NaCl is b. At equilibrium, the concentrations are

$$[P^{z-}]^L = c \quad [Na^+]^L = (zc + x) \quad [Cl^-]^L = x$$

and

$$[Na^+]^R = (b - x) \quad [Cl^-]^R = (b - x)$$

where x is the amount of NaCl transported from right to left.

Because $(\mu_{NaCl})^L = (\mu_{NaCl})^R$ at equilibrium and for dilute solutions, we replace activities with concentrations so that

$$([Na^+][Cl^-])^L = ([Na^+][Cl^-])^R$$

or

$$(zc + x)(x) = (b - x)(b - x)$$

$$x = \frac{b^2}{zc + 2b}$$

The osmotic pressure (Π_2), which is proportional to the difference in solute concentration between the two sides, is now given by

$$\Pi_2 = [\underbrace{(c + zc + x + x)}_{\text{left compartment}} - \underbrace{(b - x + b - x)}_{\text{right compartment}}]RT$$

or

$$\Pi_2 = (c + zc - 2b + 4x)RT$$

Substituting for x, we obtain

$$\Pi_2 = \left(c + zc - 2b + \frac{4b^2}{zc + 2b}\right)RT$$

$$= \frac{zc^2 + 2cb + z^2c^2}{zc + 2b}RT \qquad (1)$$

Equation 1 was derived assuming no change in either the pH or the volume of the solutions. Two limiting cases follow.

If $b \ll zc$, (the salt concentration is much less than the protein concentration), then

$$\Pi_2 = \frac{zc^2 + z^2c^2}{zc}RT = (zc + c)RT$$

$$= (z + 1)cRT$$

$$= \Pi_1$$

If $b \gg z^2c$ (the salt concentration is much greater than the protein concentration),* then

$$\Pi_2 = \frac{2cb}{2b}RT = cRT \qquad (2)$$

In this limiting case, the osmotic pressure approaches that of the pure isoelectric protein. In effect, the added salt reduces (and at high enough salt concentrations, eliminates) the Donnan effect. Under these conditions, the molar mass determined by osmotic pressure measurement would correspond closely to the true value.

---

* In practice, $c \leq 1 \times 10^{-4}$ M, $z \leq 30$, so that $z^2c \lesssim 0.1$ M. Thus, for this limiting case to hold, the concentration of the added salt should be about 1 M.

Suggestions for Further Reading

BOOKS

Hunt, J. P., *Metal Ions in Aqueous Solution,* W. A. Benjamin, Menlo Park, CA, 1963.

Pass, G., *Ions in Solution*, Clarendon Press, Oxford, 1973.

Robinson, R. A., and R. H. Stokes, *Electrolyte Solutions*, 2nd ed. Academic Press, New York, 1959.

Tombs, M. P., and A. R. Peacocke, *The Osmotic Pressure of Biological Macromolecules*, Clarendon Press, New York, 1975.

Wright, M. R., *An Introduction to Aqueous Electrolyte Solutions*, John Wiley & Sons, New York, 2007

ARTICLES

"Ion Pairs and Complexes: Free Energies, Enthalpies, and Entropies," J. E. Prue, *J. Chem. Educ.* **46**, 12 (1969).

"On Squid Axons, Frog Skins, and the Amazing Uses of Thermodynamics," W. H. Cropper, *J. Chem. Educ.* **48**, 182 (1971).

"Osmotic Pressure in the Physics Course for Students of the Life Sciences," R. K. Hobbie, *Am. J. Phys.* **42**, 188 (1974).

"Electrolyte Theory and SI Units," R. I. Holliday, *J. Chem. Educ.* **53**, 21 (1976).

"The Motion of Ions in Solution Under the Influence of an Electric Field," C. A. Vincent, *J. Chem. Educ.* **53**, 490 (1976).

"Thermodynamics of Ion Solvation and Its Significance in Various Systems," C. M. Criss and M. Salomon, *J. Chem. Educ.* **53**, 763 (1976).

"The Donnan Equilibrium and Osmotic Pressure," R. Chang and L. J. Kaplan, *J. Chem. Educ.* **54**, 218 (1977).

"Ionic Hydration Enthalpies," D. W. Smith, *J Chem. Educ.* **54**, 540 (1977).

"Paradox of the Activity Coefficient $\gamma_{\pm}$," E-I Ochiai, *J. Chem. Educ.* **67**, 489 (1990).

"Standard Enthalpies of Formation of Ions in Solution," T. Solomon, *J. Chem. Educ.* **68**, 41 (1991).

"Determination of the Thermodynamic Solubility Product, K°_{sp}, of PbI_2 Assuming Nonideal Behavior," D. B. Green, G. Rechtsteiner, and A. Honodel, *J. Chem. Educ.* **73**, 789 (1996).

"Properties of Water Solutions of Electrolytes and Nonelectrolytes," A. A. Zavitsas, *J. Phys. Chem.* B, **105**, 7805 (2001).

"The Definition and Unit of Ionic Strength," T. Solomon, *J. Chem. Educ.* **78**, 1691 (2001)

"The Conductivity of Strong Electrolytes: A Computer Simulation in LabVIEW," A. Belletti, R. Borromei, and G. Ingletto, *Chem. Educator* [Online] **13**, 224 (2008) DOI 10.1333/s00897082144a.

Problems

Ionic Conductance

7.1 The resistance of a 0.010 M NaCl solution is 172 Ω. If the molar conductance of the solution is 153 Ω^{-1} mol^{-1} cm^2, what is the cell constant?

7.2 Using the cell described in Problem 7.1, a student determined the resistance of a 0.086 M KCl solution to be 20.4 Ω. Calculate the molar conductance of this solution.

7.3 The cell constant (l/A) of a conductance cell is 388.1 m^{-1}. At 25°C, the resistance of a 4.8×10^{-4} mol L^{-1} aqueous solution of sodium chloride is 6.4×10^4 Ω and that of a sample of water is 7.4×10^6 Ω. Calculate the molar conductance of the NaCl in solution at this concentration.

7.4 Given that the measurement of Λ_0 for weak electrolytes is generally difficult to obtain, how would you deduce the value of Λ_0 for CH_3COOH from the data listed in Table 7.1? (*Hint:* Consider CH_3COONa, HCl, and NaCl.)

7.5 A simple way to determine the salinity of water is to measure its conductivity and assume that the conductivity is entirely due to sodium chloride. In a particular experiment, the resistance of a sample solution is found to be 254 Ω. The resistance of a 0.050 M KCl solution measured in the same cell is 467 Ω. Estimate the concentration of NaCl in the solution. (*Hint:* First derive an equation relating R to Λ and c and then use Λ_0 values for Λ.)

7.6 A conductance cell consists of two electrodes, each with an area of 4.2×10^{-4} m^2, separated by 0.020 m. The resistance of the cell when filled with a 6.3×10^{-4} M KNO_3 solution is 26.7 Ω. What is the molar conductance of the solution?

7.7 Referring to Figure 7.4, explain why the slope of conductance versus volume of NaOH added rises right at the start if the acid employed in the titration is weak.

Solubility

7.8 Calculate the solubility of $BaSO_4$ (in g L^{-1}) in **(a)** water and **(b)** a 6.5×10^{-5} M $MgSO_4$ solution. The solubility product of $BaSO_4$ is 1.1×10^{-10}. Assume ideal behavior.

7.9 The thermodynamic solubility product of AgCl is 1.6×10^{-10}. What is $[Ag^+]$ in **(a)** a 0.020 M KNO_3 solution and **(b)** a 0.020 M KCl solution?

7.10 Referring to Problem 7.9, calculate the value of $\Delta G°$ for the process

$$AgCl(s) \rightleftharpoons Ag^+(aq) + Cl^-(aq)$$

to yield a saturated solution at 298 K. (*Hint:* Use the well-known equation $\Delta G° = -RT \ln K$.)

7.11 The apparent solubility products of CdS and CaF_2 at 25°C are 3.8×10^{-29} and 4.0×10^{-11}, respectively. Calculate the solubility (g/100 g of solution) of these compounds.

7.12 Oxalic acid, $(COOH)_2$, is a poisonous compound present in many plants and vegetables, including spinach. Calcium oxalate is only slightly soluble in water ($K_{sp} = 3.0 \times 10^{-9}$ at 25°C) and its ingestion can result in kidney stones. Calculate **(a)** the apparent and thermodynamic solubility of calcium oxalate in water, and **(b)** the concentrations of calcium and oxalate ions in a 0.010 M $Ca(NO_3)_2$ solution. Assume ideal behavior in **(b)**.

Ionic Activity

7.13 Express the mean activity, mean activity coefficient, and mean molality in terms of the individual ionic quantities (a_+, a_-, γ_+, γ_-, m_+, and m_-) for the following electrolytes: KI, $SrSO_4$, $CaCl_2$, Li_2CO_3, $K_3Fe(CN)_6$, and $K_4Fe(CN)_6$. [*Hint*: $Fe(CN)_6^{4-}$ is a complex ion.]

7.14 Calculate the ionic strength and the mean activity coefficient for the following solutions at 298 K: **(a)** 0.10 *m* NaCl, **(b)** 0.010 *m* $MgCl_2$, and **(c)** 0.10 *m* $K_4Fe(CN)_6$.

7.15 The mean activity coefficient of a 0.010 *m* H_2SO_4 solution is 0.544. What is its mean ionic activity?

7.16 A 0.20 *m* $Mg(NO_3)_2$ solution has a mean ionic activity coefficient of 0.13 at 25°C. Calculate the mean molality, the mean ionic activity, and the activity of the compound.

Debye–Hückel Limiting Law

7.17 The Debye–Hückel limiting law is more reliable for 1:1 electrolytes than for 2:2 electrolytes. Explain.

7.18 The size of the ionic atmosphere, called the Debye radius, is $1/\kappa$, where κ is given by

$$\kappa = \left(\frac{e^2 N_A}{\varepsilon_0 \varepsilon k_B T}\right)^{1/2} \sqrt{I}$$

where e is the electronic charge, N_A Avogadro's constant, ε_0 the permittivity of vacuum ($8.854 \times 10^{-12}\ C^2\ N^{-1}\ m^{-2}$), ε the dielectric constant of the solvent, k_B the Boltzmann constant, T the absolute temperature, and I the ionic strength (see the physical chemistry texts listed in Chapter 1). Calculate the Debye radius in a 0.010 *m* aqueous Na_2SO_4 solution at 25°C.

7.19 Explain why it is preferable to take the geometric mean rather than the arithmetic mean when defining mean activity, mean molality, and mean activity coefficient.

Colligative Properties

7.20 The freezing-point depression of a 0.010 *m* acetic acid solution is 0.0193 K. Calculate the degree of dissociation for acetic acid at this concentration.

7.21 A 0.010 *m* aqueous solution of the ionic compound $Co(NH_3)_5Cl_3$ has a freezing point depression of 0.0558 K. What can you conclude about its structure? Assume the compound is a strong electrolyte.

7.22 The osmotic pressure of blood plasma is approximately 7.5 atm at 37°C. Estimate the total concentration of dissolved species and the freezing point of blood plasma.

7.23 Calculate the ionic strength of a 0.0020 *m* aqueous solution of $MgCl_2$ at 298 K. Use the Debye–Hückel limiting law to estimate **(a)** the activity coefficients of the Mg^{2+} and Cl^- ions in this solution and **(b)** the mean ionic activity coefficients of these ions.

7.24 Referring to Figure 7.13, calculate the osmotic pressure for the following cases at 298 K: **(a)** The left compartment contains 200 g of hemoglobin in 1 liter of solution; the right compartment contains pure water. **(b)** The left compartment contains the same hemoglobin solution as in part **(a)**, and the right compartment initially contains 6.0 g of NaCl in 1 liter of solution. Assume that the pH of the solution is such that the hemoglobin molecules are in the $Na^+\ Hb^-$ form. (The molar mass of hemoglobin is 65,000 g mol^{-1}.)

Additional Problems

7.25 From the following data, calculate the heat of solution for KI:

	NaCl	NaI	KCl	KI
Lattice energy/kJ mol^{-1}	787	700	716	643
Heat of solution/kJ mol^{-1}	3.8	−5.1	17.1	?

7.26 From the data in Table 7.2, determine Λ_0 for H_2O. Given that the specific conductance (κ) for water is $5.7 \times 10^{-8}\ \Omega^{-1}\ cm^{-1}$, calculate the ion product (K_w) of water at 298 K.

7.27 In this chapter (see Figures 7.2 and 7.11) and in Chapter 6 (see Π measurements in Figure 6.19), we extrapolated concentration-dependent values to zero solute concentration. Explain what these extrapolated values mean physically and why they differ from the value obtained for the pure solvent.

7.28 (**a**) The root cells of plants contain a solution that is hypertonic in relation to water in the soil. Thus, water can move into the roots by osmosis. Explain why salts (NaCl and $CaCl_2$) spread on roads to melt ice can be harmful to nearby trees. (**b**) Just before urine leaves the human body, the collecting ducts in the kidney (which contain the urine) pass through a fluid whose salt concentration is considerably greater than is found in the blood and tissues. Explain how this action helps conserve water in the body.

7.29 A very long pipe is capped at one end with a semipermeable membrane. How deep (in meters) must the pipe be immersed into the sea for fresh water to begin passing through the membrane? Assume seawater is at 20°C and treat it as a 0.70 *M* NaCl solution. The density of seawater is 1.03 g cm^{-3}.

7.30 (**a**) Using the Debye–Hückel limiting law, calculate the value of $\gamma_\pm$ for a 2.0×10^{-3} *m* Na_3PO_4 solution at 25°C. (**b**) Calculate the values of γ_+ and γ_- for the Na_3PO_4 solution, and show that they give the same value for $\gamma_\pm$ as that obtained in (**a**).

CHAPTER 8

Chemical Equilibrium

When a system is in chemical equilibrium, a change in one of the parameters of the equilibrium produces a shift in such a direction that, were no other factors involved in this shift, it would lead to a change of opposite sign in the parameter considered.

—Henri-Louis Le Châtelier

Turning from our discussion of nonelectrolyte and electrolyte solutions and physical equilibria in Chapters 6 and 7, we shall focus in this chapter on chemical equilibrium in gaseous and condensed phases. Equilibrium is a state in which there are no observable changes over time; at equilibrium, the concentrations of reactants and products in a chemical reaction remain constant. Much activity occurs at the molecular level, however, because reactant molecules continue to form product molecules while product molecules react to yield reactant molecules. This process is an example of dynamic equilibrium. The laws of thermodynamics help us predict equilibrium composition under different reaction conditions.

8.1 Chemical Equilibrium in Gaseous Systems

In this section, we shall derive an expression relating the Gibbs energy change for a reaction in the gas phase to the concentrations of the reacting species and temperature. We first consider the case in which all gases exhibit ideal behavior.

Ideal Gases

Examples of the simplest type of chemical equilibrium,

$$A(g) \rightleftharpoons B(g)$$

are cis–trans isomerization, racemization, and the cyclopropane-ring-opening reaction to form propene. The progress of the reaction can be monitored by the quantity ξ (Greek letter xi), called the *extent of reaction*. When an infinitesimal amount of A is converted to B, the change in A is $dn_A = -d\xi$ and that in B is $dn_B = +d\xi$ where dn denotes the change in number of moles. The change in Gibbs energy for this transformation at constant T and P is given by

$$\begin{aligned} dG &= \mu_A dn_A + \mu_B dn_B \\ &= -\mu_A d\xi + \mu_B d\xi \\ &= (\mu_B - \mu_A) d\xi \end{aligned} \tag{8.1}$$

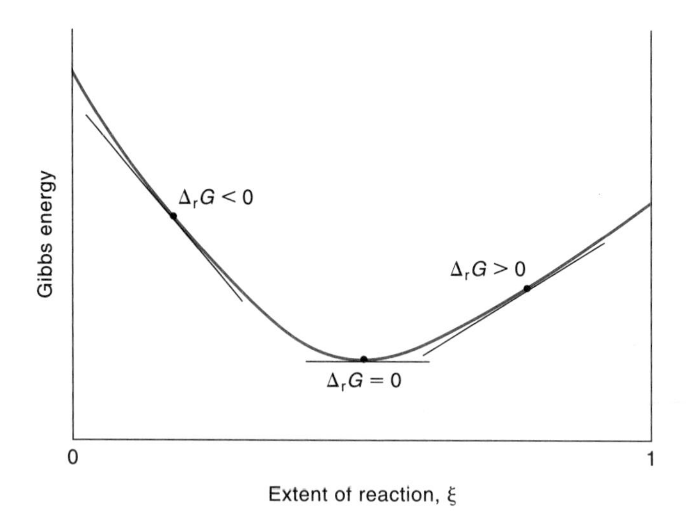

Figure 8.1
A plot of Gibbs energy versus extent of reaction. For a reacting system at equilibrium, the slope of the curve is zero.

where μ_A and μ_B are the chemical potentials of A and B, respectively. Equation 8.1 can be rearranged to give

$$\left(\frac{\partial G}{\partial \xi}\right)_{T,P} = \mu_B - \mu_A \tag{8.2}$$

For simplicity, we use the units kJ mol^{-1} rather than kJ (mol reaction)$^{-1}$.

The quantity $(\partial G/\partial \xi)_{T,P}$ is represented by $\Delta_r G$, which is the change in Gibbs energy per mole of reaction; it has the units kJ mol^{-1}.

During the reaction, the chemical potentials vary with composition. The reaction proceeds in the direction of decreasing G, that is, $(\partial G/\partial \xi)_{T,P} < 0$. Therefore, the forward reaction (A → B) is spontaneous when $\mu_A > \mu_B$ whereas the reverse reaction (B → A) is spontaneous when $\mu_B > \mu_A$. At equilibrium, $\mu_A = \mu_B$ so that

$$\left(\frac{\partial G}{\partial \xi}\right)_{T,P} = 0$$

Figure 8.1 shows a plot of Gibbs energy versus extent of reaction. At constant T and P, we have

$\Delta_r G < 0$ Forward reaction is spontaneous

$\Delta_r G > 0$ Reverse reaction is spontaneous

$\Delta_r G = 0$ Reacting system is at equilibrium

Let us now consider a more complicated case:

$$a\text{A}(g) \rightleftharpoons b\text{B}(g)$$

where a and b are stoichiometric coefficients. According to Equation 6.10, the chemical potential of the ith component in a mixture, assuming ideal behavior, is given by

$$\mu_i = \mu_i^\circ + RT \ln \frac{P_i}{P^\circ}$$

where P_i is the partial pressure of component i in the mixture, μ_i° is the standard chemical potential of component i, and $P^\circ = 1$ bar. Therefore, we can write

$$\mu_A = \mu_A^\circ + RT \ln \frac{P_A}{P^\circ} \tag{8.3a}$$

$$\mu_B = \mu_B^\circ + RT \ln \frac{P_B}{P^\circ} \tag{8.3b}$$

The Gibbs energy change for the reaction, $\Delta_r G$, can be expressed as

$$\Delta_r G = b\mu_B - a\mu_A \tag{8.4}$$

Substituting the expressions in Equation 8.3 into Equation 8.4, we get

$$\Delta_r G = b\mu_B^\circ - a\mu_A^\circ + bRT \ln \frac{P_B}{P^\circ} - aRT \ln \frac{P_A}{P^\circ} \tag{8.5}$$

The standard Gibbs energy change of the reaction, $\Delta_r G^\circ$, is just the difference between the standard Gibbs energies of products and reactants, that is,

$$\Delta_r G^\circ = b\mu_B^\circ - a\mu_A^\circ$$

Therefore, we can write Equation 8.5 as

$$\Delta_r G = \Delta_r G^\circ + RT \ln \frac{(P_B/P^\circ)^b}{(P_A/P^\circ)^a} \tag{8.6}$$

By definition, $\Delta_r G = 0$ at equilibrium, so Equation 8.6 becomes

$$0 = \Delta_r G^\circ + RT \ln \frac{(P_B/P^\circ)^b}{(P_A/P^\circ)^a}$$

$$0 = \Delta_r G^\circ + RT \ln K_P$$

or

$$\Delta_r G^\circ = -RT \ln K_P \tag{8.7}$$

K_P, the equilibrium constant (where the subscript P denotes that concentrations are expressed in pressures), is given by

$$K_P = \frac{(P_B/P^\circ)^b}{(P_A/P^\circ)^a} = \frac{P_B^b}{P_A^a}(P^\circ)^{a-b} \tag{8.8}$$

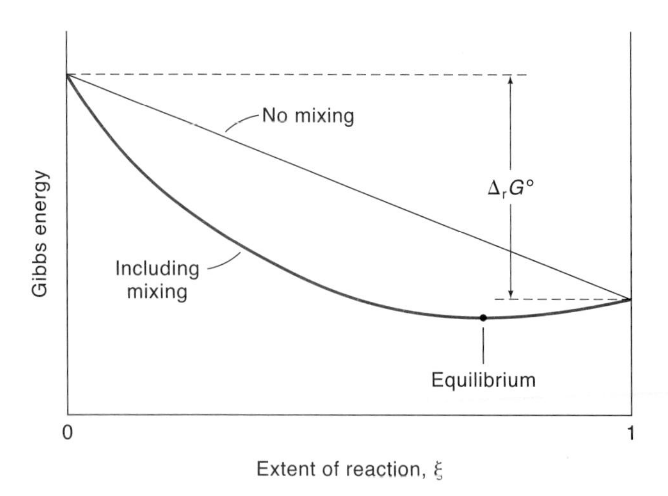

Figure 8.2
Total Gibbs energy versus extent of reaction for the $aA(g) \rightleftharpoons bB(g)$ reaction assuming $\Delta_r G° < 0$. At equilibrium, product is favored over reactant. Note that the equilibrium point, which is at the minimum of Gibbs energy, is a compromise between $\Delta_r G°$ and the Gibbs energy of mixing.

Equation 8.7 is one of the most important and useful equations in chemical thermodynamics. It relates the equilibrium constant, K_P, and the standard Gibbs energy change of the reaction, $\Delta_r G°$, in a remarkably simple fashion. Keep in mind that at a given temperature, $\Delta_r G°$ is a constant whose value depends only on the nature of the reactants and products and temperature. In our example, $\Delta_r G°$ is the standard Gibbs energy change when reactant A at 1 bar pressure and temperature T is converted to product B at 1 bar pressure and the same temperature per mole of reaction, as shown above. Figure 8.2 shows the Gibbs energy versus the extent of reaction for $\Delta_r G° < 0$. When there is no mixing of reactants with products, the Gibbs energy decreases linearly as the reaction progresses, and eventually the reactants will be completely converted to products. As Equation 6.11 shows, however, $\Delta_{mix}G$ is a negative quantity; therefore, the Gibbs energy for the actual path will be lower than that for the nonmixing case. Consequently, the equilibrium point, which is at the *minimum* of the Gibbs energy, is a compromise between these two opposing tendencies, that is, the conversion of reactants to products and the mixing of products with reactants.

Equation 8.7 tells us that if we know the value of $\Delta_r G°$, we can calculate the equilibrium constant K_P, and vice versa. The standard Gibbs energy of a reaction is just the difference between the standard Gibbs energies of formation ($\Delta_f \overline{G}°$) of the products and the reactants, discussed in Section 5.3. Thus, once we have defined a reaction, we can usually calculate the equilibrium constant from the $\Delta_f \overline{G}°$ values listed in Appendix B and Equation 8.7. Note that these values all refer to 298 K. Later in this chapter (Section 8.5), we shall learn how to calculate the value of K_P at another temperature if its value at 298 K is known.

Finally, note that the equilibrium constant is a function of temperature alone (because $\mu°$ depends only on temperature) and is dimensionless. This follows from the fact that in the expression for K_P, each pressure term is divided by its standard-state value of 1 bar, which cancels the pressure unit but does not alter P numerically.

EXAMPLE 8.1

From the thermodynamic data listed in Appendix B, calculate the equilibrium constant for this reaction at 298 K:

$$N_2(g) + 3H_2(g) \rightleftharpoons 2NH_3(g)$$

ANSWER

The equilibrium constant for the equation is given by

$$K_P = \frac{(P_{NH_3}/P^\circ)^2}{(P_{N_2}/P^\circ)(P_{H_2}/P^\circ)^3}$$

To calculate the value of K_P, we need Equation 8.7 and the value of $\Delta_r G^\circ$. From Equation 5.12 and Appendix B, we have

$$\begin{aligned}\Delta_r G^\circ &= 2\Delta_f \overline{G}^\circ(NH_3) - \Delta_f \overline{G}^\circ(N_2) - 3\Delta_f \overline{G}^\circ(H_2)\\ &= (2)(-16.6 \text{ kJ mol}^{-1}) - (0) - (3)(0)\\ &= -33.2 \text{ kJ mol}^{-1}\end{aligned}$$

From Equation 8.7,

$$-33{,}200 \text{ J mol}^{-1} = -(8.314 \text{ J K}^{-1} \text{ mol}^{-1})(298 \text{ K}) \ln K_P$$

$$\ln K_P = 13.4$$

or

$$K_P = 6.6 \times 10^5$$

COMMENT

Note that if the reaction were written as

$$\tfrac{1}{2}N_2(g) + \tfrac{3}{2}H_2(g) \rightleftharpoons NH_3(g)$$

the value of $\Delta_r G^\circ$ would be -16.6 kJ mol^{-1}, and the equilibrium constant would be calculated as follows:

$$K_P = \frac{(P_{NH_3}/P^\circ)}{(P_{N_2}/P^\circ)^{1/2}(P_{H_2}/P^\circ)^{3/2}} = 8.1 \times 10^2$$

Thus, whenever we multiply a balanced equation throughout by a factor n, we change the equilibrium constant K_P to K_P^n. Here $n = \frac{1}{2}$, so we have changed K_P to $K_P^{1/2}$.

A Closer Look at Equation 8.7

The change in standard Gibbs energy, $\Delta_r G°$, is generally not equal to zero. According to Equation 8.7, if $\Delta_r G°$ is negative, the equilibrium constant must be greater than unity; in fact, the more negative $\Delta_r G°$ is at a given temperature, the larger K_P is. The reverse holds true if $\Delta_r G°$ is a positive number. Here the equilibrium constant is less than unity. Of course, just because $\Delta_r G°$ is positive does not mean that no reaction will take place. For example, if $\Delta_r G° = 10 \text{ kJ mol}^{-1}$ and $T = 298$ K, then $K_P = 0.018$. While 0.018 is a small number compared to unity, an appreciable amount of products can still be obtained at equilibrium if we use large quantities of reactants for the reaction. There is also the special case in which $\Delta_r G° = 0$, which corresponds to a K_P of unity, meaning that the products and reactants are equally favored at equilibrium.

It is instructive to look at the factors that affect $\Delta_r G°$ and hence K_P. From Equation 5.3 we have

$$\Delta_r G° = \Delta_r H° - T\Delta_r S°$$

so the equilibrium constant at temperature T is governed by two terms: the change in enthalpy and temperature times the change in entropy. For many exothermic reactions ($\Delta_r H° < 0$) at room temperature and below, the first term on the right side of the above equation dominates. This means that K_P is greater than one, so products are favored over reactants. For an endothermic reaction ($\Delta_r H° > 0$), the equilibrium composition will favor products only if $\Delta_r S° > 0$ and the reaction is run at a high temperature. Consider the following reaction in which the A $\rightarrow$ B step is endothermic

$$\text{A} \rightleftharpoons \text{B}$$

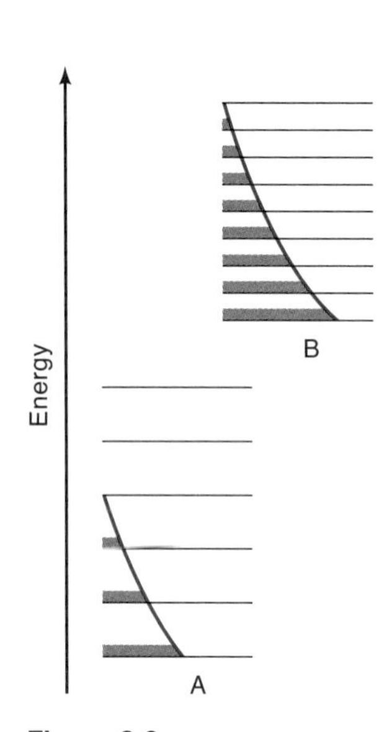

Figure 8.3
The occupancy of energy levels in B is greater than than in A. Consequently, there are more microstates in B and the entropy of B is greater than that of A. At equilibrium, B dominates even though the A $\rightarrow$ B reaction is endothermic.

As Figure 8.3 shows, the energy levels of A are below those of B, so the conversion of A to B is energetically unfavorable. This is, in fact, the nature of all endothermic reactions. But because the energy levels of B are more closely spaced together, the Boltzmann distribution law (see Equation 2.33) tells us that the population spread of the B molecules over the energy levels is greater than that of the A molecules. Consequently $\Delta_r S° > 0$ because the entropy of B is greater than that of A. At a sufficiently high temperature, then, the $T\Delta_r S°$ term will outweigh the $\Delta_r H°$ term in magnitude and $\Delta_r G°$ will be a negative quantity.

As an illustration of the relative importance of $\Delta_r H°$ versus $T\Delta_r S°$, let's consider the thermal decomposition of limestone or chalk ($CaCO_3$):

$$CaCO_3(s) \rightleftharpoons CaO(s) + CO_2(g) \qquad \Delta_r H° = 177.8 \text{ kJ mol}^{-1}$$

Using the data in Appendix B we can show that $\Delta_r S° = 160.5 \text{ J K}^{-1} \text{ mol}^{-1}$. At 298 K, we have

$$\begin{aligned}\Delta_r G° &= 177.8 \text{ kJ mol}^{-1} - (298 \text{ K})(160.5 \text{ J K}^{-1} \text{ mol}^{-1})(1 \text{ kJ}/1000 \text{ J}) \\ &= 130.0 \text{ kJ mol}^{-1}\end{aligned}$$

Because $\Delta_r G°$ is a large positive quantity, we conclude that the reaction is not favored for product formation at 298 K. Indeed, the equilibrium pressure of CO_2 is so low at room temperature that it cannot be measured. In order to make $\Delta_r G°$ negative, we first have to find the temperature at which $\Delta_r G°$ is zero; that is,

$$0 = \Delta_r H° - T\Delta_r S°$$

or

$$T = \frac{\Delta_r H^\circ}{\Delta_r S^\circ}$$

$$= \frac{(177.8 \text{ kJ mol}^{-1})(1000 \text{ J}/1 \text{ kJ})}{160.5 \text{ J K}^{-1} \text{ mol}^{-1}}$$

$$= 1108 \text{ K or } 835°\text{C}$$

At a temperature higher than 835°C, $\Delta_r G^\circ$ becomes negative, indicating that the reaction now favors the formation of CaO and CO_2. For example, at 840°C, or 1113 K,

$$\Delta_r G^\circ = \Delta_r H^\circ - T\Delta_r S^\circ$$

$$= 177.8 \text{ kJ mol}^{-1} - (1113 \text{ K})(160.5 \text{ J K}^{-1} \text{ mol}^{-1})\left(\frac{1 \text{ kJ}}{1000 \text{ J}}\right)$$

$$= -0.8 \text{ kJ mol}^{-1}$$

Two points are worth making about such a calculation. First, we used the $\Delta_r H^\circ$ and $\Delta_r S^\circ$ values at 25°C to calculate changes that occur at a much higher temperature. Because both $\Delta_r H^\circ$ and $\Delta_r S^\circ$ change with temperature, this approach will not give us an accurate value of $\Delta_r G^\circ$, but it is good enough for "ballpark" estimates. Second, we should not be misled into thinking that nothing happens below 835°C and that at 835°C $CaCO_3$ suddenly begins to decompose. Far from it. The fact that $\Delta_r G^\circ$ is a positive value at some temperature below 835°C does not mean that no CO_2 is produced, but rather that the pressure of the CO_2 gas formed at that temperature will be below 1 bar (its standard-state value). The significance of 835°C is that this is the temperature at which the equilibrium pressure of CO_2 reaches 1 bar. Above 835°C, the equilibrium pressure of CO_2 exceeds 1 bar.

A Comparison of $\Delta_r G^\circ$ with $\Delta_r G$

Suppose we start a gaseous reaction with all the reactants in their standard states (i.e., all at 1 bar). As soon as the reaction starts, the standard-state condition no longer exists for the reactants or the products because their pressures are different from 1 bar. Under conditions that are not standard state, we must use $\Delta_r G$ rather than $\Delta_r G^\circ$ to predict the direction of a reaction.

We can use the change in standard Gibbs energy ($\Delta_r G^\circ$) and Equation 8.6 to find $\Delta_r G$. The value of $\Delta_r G$ is determined by two terms: $\Delta_r G^\circ$ and a concentration dependent term. At a given temperature the value of $\Delta_r G^\circ$ is fixed, but we can change the value of $\Delta_r G$ by adjusting the partial pressures of the gases. Although the quotient composed of the pressures of reactants and products has the form of an equilibrium constant, it is *not* equal to the equilibrium constant unless P_A and P_B are the partial pressures at equilibrium. In general, we can rewrite Equation 8.6 as

The sign of $\Delta_r G$ and not that of $\Delta_r G^\circ$ determines the direction of reaction spontaneity.

$$\Delta_r G = \Delta_r G^\circ + RT \ln Q \tag{8.9}$$

where Q is the *reaction quotient* and $Q \neq K_P$ unless $\Delta_r G = 0$. The usefulness of Equation 8.6 or Equation 8.9 is that it tells us the direction of a spontaneous change if the concentrations of the reacting species are known. If $\Delta_r G^\circ$ is a large positive or a large negative number (say, 50 kJ mol^{-1} or more), then the direction of the reaction (or the sign of $\Delta_r G$) is primarily determined by $\Delta_r G°$ alone, unless either the reactants or the products are present in a much larger amount so that the $RT \ln Q$ term in Equation 8.9 is comparable to $\Delta_r G°$ in magnitude but opposite in sign. If $\Delta_r G°$ is a small number, either positive or negative (say, 10 kJ mol^{-1} or less), then the reaction can go either way.*

---

* Alternatively, we can determine the direction of a reaction by comparing Q with K_P. From Equations 8.7 and 8.9 we can show that $\Delta_r G = RT \ln(Q/K_P)$. Therefore, if $Q < K_P$, $\Delta_r G$ is negative and the reaction will proceed in the forward direction (left to right). If $Q > K_P$, $\Delta_r G$ is positive. Here, the reaction will proceed in the reverse direction (right to left).

EXAMPLE 8.2

The equilibrium constant (K_P) for the reaction

$$N_2O_4(g) \rightleftharpoons 2NO_2(g)$$

is 0.113 at 298 K, which corresponds to a standard Gibbs energy change of 5.40 kJ mol^{-1}. In a certain experiment, the initial pressures are $P_{NO_2} = 0.122$ bar and $P_{N_2O_4} = 0.453$ bar. Calculate $\Delta_r G$ for the reaction at these pressures and predict the direction of the net reaction.

ANSWER

To determine the direction of the net reaction, we need to calculate the Gibbs energy change under nonstandard-state conditions ($\Delta_r G$) using Equation 8.9 and the given $\Delta_r G°$ value. Note that the partial pressures are expressed as dimensionless quantities in the reaction quotient Q because each pressure is divided by its standard-state value of 1 bar.

$$\Delta_r G = \Delta_r G° + RT \ln Q$$

$$= \Delta_r G° + RT \ln \frac{P_{NO_2}^2}{P_{N_2O_4}}$$

$$= 5.40 \times 10^3 \text{ J mol}^{-1} + (8.314 \text{ J K}^{-1} \text{ mol}^{-1})(298 \text{ K}) \times \ln \frac{(0.122)^2}{0.453}$$

$$= 5.40 \times 10^3 \text{ J mol}^{-1} - 8.46 \times 10^3 \text{ J mol}^{-1}$$

$$= -3.06 \times 10^3 \text{ J mol}^{-1} = -3.06 \text{ kJ mol}^{-1}$$

Because $\Delta_r G < 0$, the net reaction proceeds from left to right to reach equilibrium.

COMMENT

Note that although $\Delta_r G° > 0$, the reaction can be made initially to favor product formation by having a small concentration (pressure) of the product compared to that of the reactant.

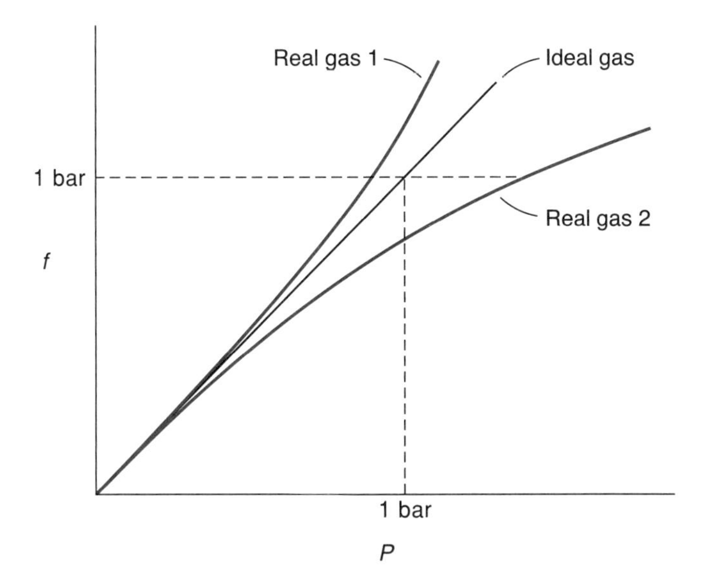

Figure 8.4
Fugacity (f) versus pressure (P) for real and ideal gases. The standard state is the gas at a pressure of 1 bar in a hypothetical state in which it exhibits ideal gas behavior.

Real Gases

What form does the equilibrium constant take for real gases? As we saw in Chapter 1, the behavior of real gases cannot be described by the ideal-gas equation and requires a more accurate equation of state, such as the van der Waals equation. However, if we tried to calculate the value of P using the van der Waals equation for every gas and substituted this quantity in the equilibrium constant expression, the final form would be very unwieldy. Instead, we adopt a simpler procedure analogous to the use of activity for concentration in Chapter 6. For real gases, we define a new variable called *fugacity* (f) to replace partial pressure. Equation 8.3 applies only to an ideal gas. For a real gas, we must write

$$\mu = \mu^\circ + RT \ln \frac{f}{P^\circ} \tag{8.10}$$

Fugacity has the same dimensions as pressure. Figure 8.4 shows the variation of the fugacity of a gas with its pressure. At low pressure, the gas behaves ideally and fugacity equals pressure, but deviation occurs when the pressure increases. As we can see, the fugacity of an ideal gas is 1 bar when the pressure is 1 bar. The standard state for the fugacity of a real gas is defined as the state in which the fugacity would be 1 bar if the gas behaved ideally from low pressures to 1 bar pressure. In general, we have

$$\lim_{P \to 0} f = P$$

The *fugacity coefficient*, γ, is given by

$$\gamma = \frac{f}{P} \tag{8.11}$$

$\gamma < 1$ indicates that the attractive intermolecular forces are dominant. Conversely, $\gamma > 1$ means repulsive intermolecular forces are dominant.

and

$$\lim_{P \to 0} \gamma = 1$$

Although pressure is directly measurable, fugacity can only be calculated. The relationship between fugacity and pressure is discussed in Appendix 8.1 (p. 335).

Starting with Equation 8.10, we can derive an equilibrium constant, K_f (where the subscript f denotes fugacity), for the hypothetical reaction ($a\text{A} \rightleftharpoons b\text{B}$) discussed earlier:

$$K_f = \frac{(f_B/1 \text{ bar})^b}{(f_A/1 \text{ bar})^a} \tag{8.12}$$

Because $f = \gamma P$, Equation 8.12 can be rewritten as

$$K_f = \frac{\gamma_B^b (P_B/1 \text{ bar})^b}{\gamma_A^a (P_A/1 \text{ bar})^a} = K_\gamma K_P \tag{8.13}$$

where K_γ is given by γ_B^b/γ_A^a, and K_P is given by $(P_B^b/P_A^a)(1 \text{ bar})^{a-b}$. K_f, as defined by Equation 8.12 or Equation 8.13, is called the *thermodynamic equilibrium constant* and gives the exact result, whereas K_P is called the *apparent equilibrium constant* because its value is not constant but depends on pressure at a given temperature. For an ideal gas reaction, K_f is equal to K_P. As an example of the difference between these two quantities, consider the synthesis of ammonia from nitrogen and hydrogen at high temperature and pressure:

$$\tfrac{1}{2}N_2(g) + \tfrac{3}{2}H_2(g) \rightleftharpoons NH_3(g)$$

Table 8.1 shows the apparent and thermodynamic equilibrium constants for this reaction at different pressures. At low total pressures, K_P is quite close to K_f, indicating that gases behave fairly ideally. Appreciable deviations occur as the total pressure exceeds 300 bar. On the other hand, the thermodynamic equilibrium constant remains relatively constant with changing pressure. The small variations in K_f are due to uncertainties in determining the fugacities of the gases.

Table 8.1
Equilibrium Constants for the Reaction $\tfrac{1}{2}N_2(g) + \tfrac{3}{2}H_2(g) \rightleftharpoons NH_3(g)$ at 450°C[a]

Total pressure/bar	P_{NH_3}/bar	P_{N_2}/bar	P_{H_2}/bar	K_P	K_γ	$K_f(K_PK_\gamma)$
10.2	0.204	2.30	7.67	0.0064	0.994	0.0064
30.3	1.76	6.68	21.9	0.0066	0.975	0.0064
50.6	4.65	10.7	35.2	0.0068	0.95	0.0065
101.0	16.6	19.4	65.0	0.0072	0.89	0.0064
302.8	108	42.8	152	0.0088	0.70	0.0062
606	326	56.5	223	0.0130	0.50	0.0065

[a] Data from A. J. Larson, *J. Am. Chem. Soc.* **46**, 367 (1924).

8.2 Reactions in Solution

The treatment of chemical equilibrium in solution is analogous to that in the gas phase, although we normally express concentrations of reacting species in solution in molality or molarity. We again start with a hypothetical reaction, this time in solution:

$$a\mathrm{A} \rightleftharpoons b\mathrm{B}$$

where A and B are the nonelectrolyte solutes. Assuming ideal behavior and expressing the solute concentrations in molalities, we have, from Equation 6.27,

$$\mu_{\mathrm{A}} = \mu_{\mathrm{A}}^{\circ} + RT \ln \frac{m_{\mathrm{A}}}{m^{\circ}}$$

where m° represents 1 mol (kg solvent)$^{-1}$. We can write a similar equation for B. Following the same procedure as that employed for ideal gases in Section 8.1, we arrive at the standard Gibbs energy change:

$$\Delta_{\mathrm{r}}G^{\circ} = -RT \ln K_m \qquad (8.14)$$

where

$$K_m = \frac{(m_{\mathrm{B}}/m^{\circ})^b}{(m_{\mathrm{A}}/m^{\circ})^a}$$

If we express the solute concentrations in molarities, the equilibrium constant takes the form

$$K_{\mathrm{c}} = \frac{([\mathrm{B}]/1\ M)^b}{([\mathrm{A}]/1\ M)^a}$$

where the square brackets signify mol L^{-1}. Again, both K_m and K_{c} are dimensionless quantities, because each concentration term is divided by its standard-state value (1 m or 1 M). For nonequilibrium reactions, the Gibbs energy change is given by

$$\Delta_{\mathrm{r}}G = \Delta_{\mathrm{r}}G^{\circ} + RT \ln Q$$

where Q is the reaction quotient.

For nonideal solutions, we must replace concentrations with activities. From Equation 6.25, we write the chemical potential of the ith component as

$$\mu_i = \mu_i^{\circ} + RT \ln a_i$$

The substitution of activity for concentration is analogous to the substitution of fugacity for pressure. Starting with the above chemical potential expression, we obtain the thermodynamic equilibrium constant, K_a:

$$K_a = \frac{a_{\mathrm{B}}^b}{a_{\mathrm{A}}^a} \qquad (8.15)$$

Because $a = \gamma m$, Equation 8.15 can be written as

$$K_a = \frac{\gamma_B^b}{\gamma_A^a} \times \frac{(m_B/m^\circ)^b}{(m_A/m^\circ)^a} = K_\gamma K_m \tag{8.16}$$

where K_γ is given by (γ_B^b/γ_A^a) and K_m, the apparent equilibrium constant for a nonideal solution reaction, is given by $(m_B^b/m_A^a)(m^\circ)^{a-b}$.

8.3 Heterogeneous Equilibria

So far, we have concentrated on homogeneous equilibria, that is, reactions that occur in a single phase. Here, we shall discuss heterogeneous equilibria in which the reactants and products are present in more than one phase.

Consider the thermal decomposition of calcium carbonate in a closed system:

$$CaCO_3(s) \rightleftharpoons CaO(s) + CO_2(g)$$

The two solids and one gas constitute three separate phases. We might write the equilibrium constant of this reaction as

$$K_c' = \frac{[CaO][CO_2]}{[CaCO_3]}$$

By common practice, however, we do not include the concentrations of solids in the equilibrium-constant expression. The concentration of any pure solid is the ratio of the total number of moles present in the solid divided by the volume of the solid. If part of the solid is removed, the number of moles of the solid will decrease, but so will its volume. The same holds true for the addition of more solid substance—increasing the number of moles of the solid results in an increase in volume. For this reason, the ratio of moles to volume always remains unchanged. Thus, the amount of CO_2 and CaO produced is always the same, regardless of the amount of $CaCO_3$ employed in the beginning, as long as some of the solid is present at equilibrium. The equilibrium-constant expression given above can now be arranged as follows:

$$\frac{[CaCO_3]}{[CaO]} K_c' = [CO_2]$$

Because both $[CaCO_3]$ and $[CaO]$ are constants, every term on the left side is a constant, and we write

$$K_c = [CO_2]$$

where K_c, the "new" equilibrium constant, is given by $[CaCO_3]K_c'/[CaO]$. More conveniently, we can measure the pressure of CO_2 and obtain

$$K_P = P_{CO_2}$$

Both K_c and K_P are dimensionless because $[CO_2]$ is divided by its standard-state value of 1 M, and P_{CO_2} is divided by 1 bar. The equilibrium constants K_c and K_P are related to each other in a simple manner (see Problem 8.1).

Heterogeneous equilibria are more easily handled if we write the thermodynamic equilibrium constant instead of the apparent equilibrium constant. Replacing concentrations with activities, we have

$$K_a = \frac{a_{\mathrm{CaO}} a_{\mathrm{CO_2}}}{a_{\mathrm{CaCO_3}}}$$

By convention, the activities of pure solids (and pure liquids) in their standard states (i.e., at 1 bar) are equal to unity; that is, $a_{\mathrm{CaO}} = 1$ and $a_{\mathrm{CaCO3}} = 1$. For reactions carried out under moderate pressure conditions, we can assume that their values do not change appreciably from unity and write the equilibrium constant in terms of the fugacity of the CO_2 gas:

$$K_a = \frac{f_{\mathrm{CO_2}}}{1 \text{ bar}}$$

or, assuming ideal behavior,

$$K_P = \frac{P_{\mathrm{CO_2}}}{1 \text{ bar}}$$

where the fugacity and pressure are in bars. Figure 8.5 shows the equilibrium pressure of CO_2 for the decomposition of $CaCO_3$ as a function of temperature.

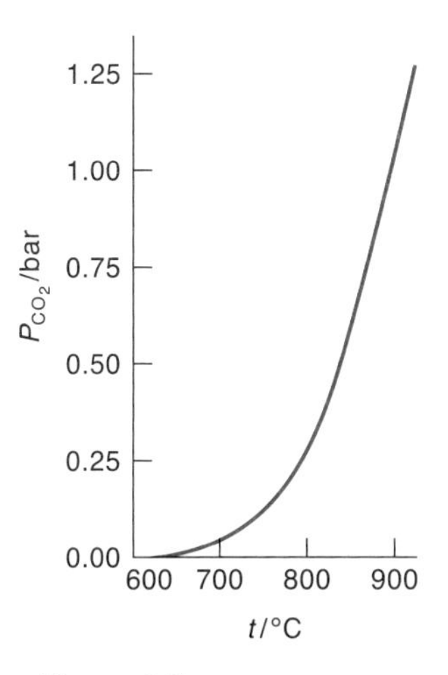

Figure 8.5
A plot of the equilibrium pressure (P) of CO_2 over CaO and $CaCO_3$ as a function of temperature (t).

EXAMPLE 8.3

Calculate the equilibrium constant for the following reaction at 298 K, using the data listed in Appendix B:

$$2\mathrm{H_2}(g) + \mathrm{O_2}(g) \rightleftharpoons 2\mathrm{H_2O}(l)$$

ANSWER

The thermodynamic equilibrium constant is given by

$$K_a = \frac{a^2_{\mathrm{H_2O}}}{(f_{\mathrm{H_2}}/1 \text{ bar})^2 (f_{\mathrm{O_2}}/1 \text{ bar})} = \frac{1}{f^2_{\mathrm{H_2}} f_{\mathrm{O_2}}} (1 \text{ bar})^3$$

Assuming ideal behavior, we write

$$K_P = \frac{1}{P^2_{\mathrm{H_2}} P_{\mathrm{O_2}}} (1 \text{ bar})^3$$

The standard Gibbs energy change for the reaction is given by

$$\begin{aligned}\Delta_r G^\circ &= 2\Delta_f \overline{G}^\circ(\mathrm{H_2O}) - 2\Delta_f \overline{G}^\circ(\mathrm{H_2}) - \Delta_f \overline{G}^\circ(\mathrm{O_2}) \\ &= (2)(-237.2 \text{ kJ mol}^{-1}) - 2(0) - (0) \\ &= -474.4 \text{ kJ mol}^{-1}\end{aligned}$$

Finally, from Equation 8.7,

$$-474.4 \times 10^3 \text{ J mol}^{-1} = -(8.314 \text{ J K}^{-1} \text{ mol}^{-1})(298 \text{ K}) \ln K_P$$

$$K_P = 1.4 \times 10^{83}$$

The very large K_P value indicates that the reaction virtually goes essentially to completion.

Solubility Equilibria

Another example of heterogeneous equilibria is that between a sparingly soluble salt and its ions in a saturated solution. For example, when silver chloride dissolves in water, the following equilibrium is established:

$$AgCl(s) \rightleftharpoons Ag^+(aq) + Cl^-(aq)$$

The equilibrium constant for the process expressed in activities is

$$K_a = \frac{a_{Ag^+} a_{Cl^-}}{a_{AgCl}}$$

Because the activity of the solid AgCl is 1, the above equation can be rewritten as

$$K_a = K_{sp} = a_{Ag^+} a_{Cl^-}$$

where K_{sp} is called the *solubility product.* Note that because the concentrations of the ions are usually quite low, it is safe to use concentrations rather than activities in calculating K_{sp}.

EXAMPLE 8.4

The concentrations of the Ag^+ and Cl^- ions in a saturated AgCl solution are both 1.27×10^{-5} *M* at 25°C. Calculate and K_{sp} and $\Delta_r G°$ for the process:

$$AgCl(s) \rightleftharpoons Ag^+(aq) + Cl^-(aq)$$

ANSWER

From the expression for the solubility product for AgCl and using concentrations instead of activities, we write

$$\begin{aligned} K_{sp} &= [Ag^+][Cl^-] \\ &= (1.27 \times 10^{-5})(1.27 \times 10^{-5}) \\ &= 1.61 \times 10^{-10} \end{aligned}$$

From Equation 8.14,

$$\begin{aligned}\Delta_r G^\circ &= -RT \ln K_{sp} \\ &= -(8.314 \text{ J K}^{-1} \text{ mol}^{-1})(298 \text{ K}) \ln 1.61 \times 10^{-10} \\ &= 5.59 \times 10^4 \text{ J mol}^{-1} \\ &= 55.9 \text{ kJ mol}^{-1}\end{aligned}$$

The fact that $\Delta_r G^\circ$ is a large positive quantity means that the process favors the formation of reactant at equilibrium. This result is consistent with the low solubility of AgCl.

8.4 Multiple Equilibria and Coupled Reactions

The reactions we have considered so far are all relatively straightforward. A more complicated situation is one in which the product molecules in one equilibrium process are involved in a second equilibrium process:

$$\text{(1)}\quad \text{A} + \text{B} \rightleftharpoons \text{C} + \text{D}$$
$$\text{(2)}\quad \text{C} \rightleftharpoons \text{E}$$

One of the products formed in Reaction (1) reacts further to form product E. The equilibrium constant for Reaction (1), K_1, is given by

$$K_1 = \frac{[\text{C}][\text{D}]}{[\text{A}][\text{B}]}$$

Similarly, for Reaction (2) we have

$$K_2 = \frac{[\text{E}]}{[\text{C}]}$$

The overall reaction is given by the sum of Reactions (1) and (2):

$$\text{(1)}\quad \text{A} + \text{B} \rightleftharpoons \text{C} + \text{D}$$
$$\text{(2)}\quad \text{C} \rightleftharpoons \text{E}$$
$$\overline{\text{(3)}\quad \text{A} + \text{B} \rightleftharpoons \text{D} + \text{E}}$$

and the corresponding equilibrium constant, K_3, is given by

$$K_3 = \frac{[\text{D}][\text{E}]}{[\text{A}][\text{B}]}$$

We obtain the same expression if we take the product of K_1 and K_2:

$$K_1K_2 = \frac{[C][D]}{[A][B]} \times \frac{[E]}{[C]}$$
$$= \frac{[D][E]}{[A][B]}$$

Therefore,

$$K_3 = K_1K_2$$

An important statement can be made about multiple equilibria: *If a reaction can be expressed as the sum of two or more reactions, the equilibrium constant for the overall reaction is given by the product of the equilibrium constants of the individual reactions.*

Among the many examples of multiple equilibria is the dissociation of a diprotic acid in aqueous solution. The following equilibrium constants have been determined for carbonic acid (H_2CO_3) at 25°C:

$$H_2CO_3 \rightleftharpoons H^+ + HCO_3^- \qquad K_1 = \frac{[H^+][HCO_3^-]}{[H_2CO_3]} = 4.2 \times 10^{-7}$$

$$HCO_3^- \rightleftharpoons H^+ + CO_3^{2-} \qquad K_2 = \frac{[H^+][CO_3^{2-}]}{[HCO_3^-]} = 4.8 \times 10^{-11}$$

The overall reaction is the sum of these two reactions

$$H_2CO_3 \rightleftharpoons 2H^+ + CO_3^{2-}$$

and the corresponding equilibrium constant is given by

$$K_3 = \frac{[H^+]^2[CO_3^{2-}]}{[H_2CO_3]}$$

or

$$K_3 = K_1K_2$$
$$= (4.2 \times 10^{-7})(4.8 \times 10^{-11})$$
$$= 2.0 \times 10^{-17}$$

We can also relate the standard Gibbs energy change for the overall reaction to the individual reactions. Because

$$-RT\ln K_3 = -RT\ln K_1 - RT\ln K_2$$

it follows that

$$\Delta_r G_3^\circ = \Delta_r G_1^\circ + \Delta_r G_2^\circ$$

Principle of Coupled Reactions

Many chemical and biological reactions are endergonic ($\Delta_r G° > 0$) and therefore the products are not favored at equilibrium. However, it is possible in some cases to carry out these reactions to an appreciable extent by coupling them with an exergonic ($\Delta_r G° < 0$) reaction. Let us consider a chemical process for the extraction of copper from its ore, Cu_2S. As the following equation shows, heating the ore alone will not yield much copper because $\Delta_r G°$ for the reaction is a large positive quantity:

$$Cu_2S(s) \rightarrow 2Cu(s) + S(s) \qquad \Delta_r G° = 86.2 \text{ kJ mol}^{-1}$$

But, if we couple the thermal decomposition of Cu_2S with the oxidation of sulfur to sulfur dioxide, the outcome changes dramatically :

$$Cu_2S(s) \rightarrow 2Cu(s) + S(s) \qquad \Delta_r G° = 86.2 \text{ kJ mol}^{-1}$$

$$S(s) + O_2(g) \rightarrow SO_2(g) \qquad \Delta_r G° = -300.1 \text{ kJ mol}^{-1}$$

$$\text{Overall: } Cu_2S(s) + O_2(g) \rightarrow 2Cu(s) + SO_2(g) \qquad \Delta_r G° = -213.9 \text{ kJ mol}^{-1}$$

The price we pay for this coupled reaction is acid rain due to the formation of SO_2.

The Gibbs energy change for the overall reaction is the sum of the Gibbs energy changes of the two reactions. Because the negative Gibbs energy change for the oxidation of sulfur is considerably larger than the positive Gibbs energy change for the decomposition of Cu_2S, the overall reaction has a large negative Gibbs energy change and therefore favors the formation of Cu. Figure 8.6 shows a mechanical analog of a coupled reaction.

Coupled reactions play a crucial role in biological systems. For example, in the human body, food molecules, represented by glucose ($C_6H_{12}O_6$), are converted to carbon dioxide and water during metabolism with a substantial release of Gibbs energy:

$$C_6H_{12}O_6 + 6O_2 \rightarrow 6CO_2 + 6H_2O \qquad \Delta_r G° = -2880 \text{ kJ mol}^{-1}$$

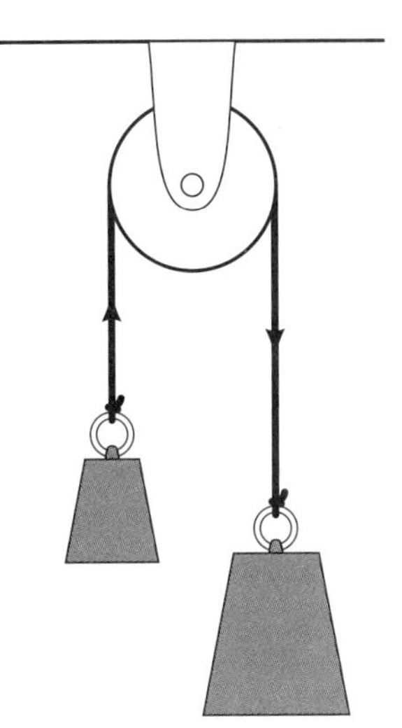

Figure 8.6
A mechanical analog for coupled reactions. Normally weights fall downward under the influence of gravity (a spontaneous process). However, it is possible to make the smaller weight move upwards (a nonspontaneous process) by coupling it with the falling of a larger weight. Overall the process is still spontaneous. Similarly, a reaction with a large negative $\Delta_r G°$ can cause another reaction with a smaller positive $\Delta_r G°$ to proceed in its nonspontaneous direction.

In a living cell, the reaction does not take place in a single step (as burning glucose in air would); rather, the glucose molecule is broken down with the aid of enzymes in a series of steps. Much of the Gibbs energy released along the way is used to synthesize adenosine triphosphate (ATP) from adenosine diphosphate (ADP) and phosphoric acid:

$$\text{ADP} + H_3PO_4 \rightarrow \text{ATP} + H_2O \qquad \Delta_r G^\circ = +31 \text{ kJ mol}^{-1}$$

The function of ATP is to store Gibbs energy until it is needed by cells. Under appropriate conditions, ATP undergoes hydrolysis to give ADP and phosphoric acid, with a release of 31 kJ mol^{-1} of Gibbs energy, which can be used to promote energetically unfavorable reactions, such as protein synthesis.

Proteins are polymers of amino acids. The stepwise synthesis of a protein molecule involves the joining of individual amino acids. Consider the formation of the dipeptide (a two-amino-acid unit) alanylglycine from alanine and glycine:

$$\text{Alanine} + \text{Glycine} \rightarrow \text{Alanylglycine} \qquad \Delta_r G^\circ = +29 \text{ kJ mol}^{-1}$$

This reaction does not favor the formation of product, so only a small amount of the dipeptide will be formed at equilibrium. However, with the aid of an enzyme, the reaction is coupled to the hysrolysis of ATP as follows:

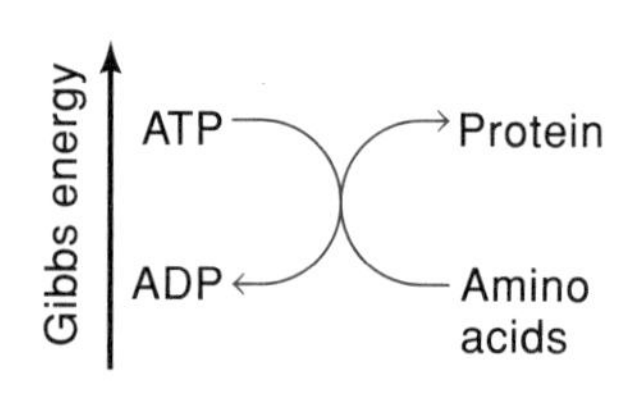

Figure 8.7
Schematic representation of the Gibbs energy changes that occur in protein synthesis.

$$\text{ATP} + H_2O + \text{Alanine} + \text{Glycine} \rightarrow \text{ADP} + H_3PO_4 + \text{Alanylglycine}$$

The overall Gibbs energy change is given by $\Delta_r G^\circ = (-31 + 29)$ kJ mol^{-1} or -2 kJ mol^{-1}, which means that the coupled reaction now favors the formation of products (Figure 8.7).

Finally note that coupled reactions may or may not involve multiple equilibria; that is, the products of one reaction may not take part in the other reaction. In these cases, the coupling mechanism is made possible with the aid of enzymes.

8.5 The Influence of Temperature, Pressure, and Catalysts on the Equilibrium Constant

The aim of many industrial processes is to obtain optimal yield in the shortest period of time, and if possible, under moderate conditions to reduce cost. Therefore, the study of how the equilibrium constant is influenced by external parameters such as temperature or pressure has great practical importance. Here, we shall consider the influence of temperature, pressure, and the use of a catalyst on the equilibrium constant.

The Effect of Temperature

Although Equation 8.7 relates the change in the standard Gibbs energy to the equilibrium constant at *any* temperature, calculating the K_P value at 298 K is usually the most convenient because of the availability of thermodynamic data. In practice, however, a reaction may be run at temperatures other than 298 K, and so we either have to know the $\Delta_r G^\circ$ values at that particular temperature or find some other way to calculate the

K_P value. The question we ask is: If we know the equilibrium constant of a reaction K_1 at temperature T_1, can we calculate the equilibrium constant of the same reaction K_2 at temperature T_2 The answer is yes.

A very useful equation relating the equilibrium constant to temperature can be derived as follows. Substituting Equation 8.7 into the Gibbs–Helmholtz equation (Equation 5.15) written for the changes in the standard state, we obtain

$$\left[\frac{\partial\left(\frac{\Delta_r G^\circ}{T}\right)}{\partial T}\right]_P = -\frac{\Delta_r H^\circ}{T^2}$$

$$\left[\frac{\partial\left(\frac{-RT\ln K}{T}\right)}{\partial T}\right]_P = -\frac{\Delta_r H^\circ}{T^2}$$

$$\left(\frac{\partial \ln K}{\partial T}\right)_P = \frac{\Delta_r H^\circ}{RT^2} \tag{8.17}$$

Equation 8.17 is known as the *van't Hoff equation*. Assuming $\Delta_r H^\circ$ to be temperature independent, this equation can be integrated to give

$$\ln\frac{K_2}{K_1} = \frac{\Delta_r H^\circ}{R}\left(\frac{1}{T_1} - \frac{1}{T_2}\right)$$

$$= \frac{\Delta_r H^\circ}{R}\left(\frac{T_2 - T_1}{T_1 T_2}\right) \tag{8.18}$$

Because

$$\ln K = -\frac{\Delta_r G^\circ}{RT}$$

and

$$\Delta_r G^\circ = \Delta_r H^\circ - T\Delta_r S^\circ$$

it follows that

$$\ln K = -\frac{\Delta_r H^\circ}{RT} + \frac{\Delta_r S^\circ}{R} \tag{8.19}$$

Equation 8.19 is another form of the van't Hoff equation in which $\Delta_r S^\circ$ is the standard entropy change of the reaction.* Thus, by plotting $\ln K$ versus $1/T$, we obtain a straight line with a slope equal to $-\Delta_r H^\circ/R$ and the intercept on the ordinate gives $\Delta_r S^\circ/R$ (Figure 8.8). This is a convenient way to determine the values of $\Delta_r H^\circ$ and

* If $\Delta_r H^\circ$ is independent of temperature, then the change in heat capacities between the products and reactants is zero, which means that $\Delta_r S^\circ$ is also independent of temperature. The term $\Delta_r S^\circ/R$ is the integration constant for the indefinite integral of Equation 8.17.

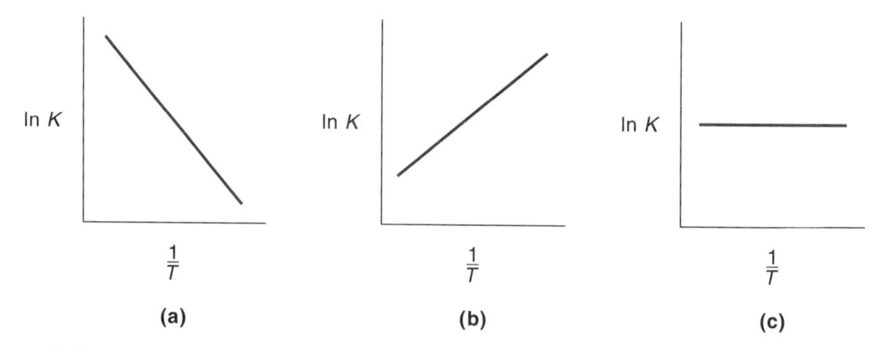

Figure 8.8
Graph of ln K versus $1/T$ for the van't Hoff equation. The slope is equal to $-\Delta_r H°/R$ and the intercept on the ordinate (ln K) is equal to $\Delta_r S°/R$. (a) $\Delta_r H° > 0$, (b) $\Delta_r H° < 0$, (c) $\Delta_r H° = 0$. In all cases, $\Delta_r H°$ is assumed to be temperature independent.

$\Delta_r S°$ for a reaction. Note that this graph will yield a straight line only if both $\Delta_r H°$ and $\Delta_r S°$ are independent of temperature. For relatively small temperature ranges (say 50 K or less), this method provides a fairly good approximation.

Some interesting observations can be made from Equation 8.18. If the reaction is endothermic in the forward direction (i.e., $\Delta_r H°$ is positive) and if we assume that $T_2 > T_1$, then the quantity on the left side of Equation 8.18 is positive, which means that $K_2 > K_1$. On the other hand, if the reaction is exothermic in the forward direction (i.e., $\Delta_r H°$ is negative), then the quantity on the left side of Equation 8.18 is negative, or $K_2 < K_1$. Thus, we can conclude that an increase in temperature shifts the equilibrium from left to right (favors the formation of products) in an endothermic reaction and shifts the equilibrium from right to left (favors the formation of reactants) in an exothermic reaction. These results are consistent with *Le Châtelier's principle* (after the French chemist Henri Louis Le Châtelier, 1850–1936), which states that if an external stress is applied to a system at equilibrium, the system will adjust itself in such a way as to partially offset the stress as it tries to re-establish equilibrium. In this case the "stress" is the change in temperature.

EXAMPLE 8.5

The equilibrium constant for the gas-phase dissociation of molecular iodine,

$$I_2(g) \rightleftharpoons 2I(g)$$

has been measured at the following temperatures:

T/K	872	973	1073	1173
K_P	1.8×10^{-4}	1.8×10^{-3}	1.08×10^{-2}	0.0480

Determine graphically the values of $\Delta_r H°$ and $\Delta_r S°$ for the reaction.

ANSWER

We need Equation 8.19. First, we construct the table,

$1/T/\mathrm{K}^{-1}$	1.15×10^{-3}	1.03×10^{-3}	9.32×10^{-4}	8.53×10^{-4}
$\ln K_P$	-8.62	-6.32	-4.53	-3.04

Next, we plot $\ln K_P$ versus $1/T$ (Figure 8.9). The points fall on a straight line whose equation is $\ln K_P = -1.875 \times 10^4\ \mathrm{K}/T + 12.954$. Assuming that $\Delta_r H°$ is temperature independent, we have

$$-\frac{\Delta_r H°}{R} = -1.875 \times 10^4\ \mathrm{K}$$

or

$$\Delta_r H° = 1.56 \times 10^2\ \mathrm{kJ\ mol^{-1}}$$

The intercept on the ordinate is equal to $\Delta_r S°/R$, so

$$\Delta_r S° = 12.954(8.314\ \mathrm{J\ K^{-1}\ mol^{-1}})$$

$$= 108\ \mathrm{J\ K^{-1}\ mol^{-1}}$$

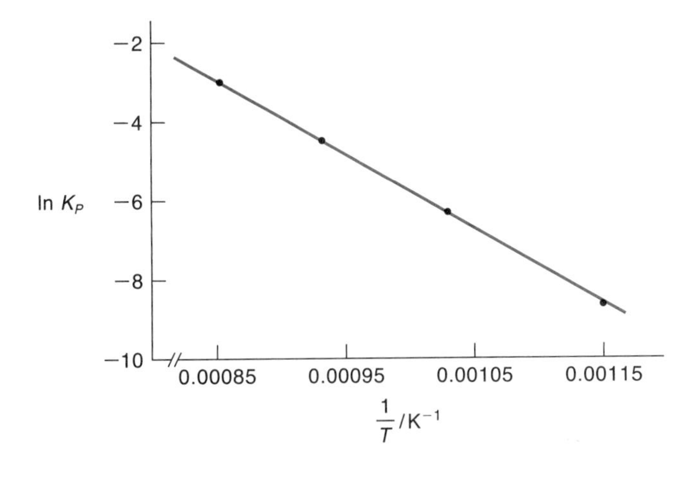

Figure 8.9
The van't Hoff plot for the dissociation of iodine vapor.

The Effect of Pressure

The question of whether the equilibrium constant changes when the pressure changes can be answered by examining Equation 8.7, which relates the equilibrium constant K_P to $\Delta_r G°$. Because $\Delta_r G°$ is defined for the reacting species at a specific pressure (1 bar), it does not vary when the pressure of the experiment is changed. However, the apparent equilibrium constant, K_P, is independent of pressure only if the reaction involves ideal gases. For a reaction involving real gases, the apparent equilibrium constant does depend on pressure. On the other hand, the thermodynamic equilibrium constant, K_f, is independent of pressure for reactions involving either ideal or real gases. For reactions involving ideal gases, $K_f = K_P$, so at constant temperature T we write,

$$\left(\frac{\partial K_P}{\partial P}\right)_T = 0$$

The fact that K_P is not affected by pressure does not mean that the amounts of various gases at equilibrium do not change with pressure. To illustrate this point, let us consider the ideal gas-phase reaction at equilibrium:

$$\begin{array}{ccc} A(g) & \rightleftharpoons & 2B(g) \\ n(1-\alpha) & & 2n\alpha \end{array}$$

where n is the number of moles of A originally present and α is the fraction of the A molecules dissociated. The total number of moles of molecules present at equilibrium is $n(1 + \alpha)$, and the mole fractions of A and B are

$$x_A = \frac{n(1-\alpha)}{n(1+\alpha)} = \frac{(1-\alpha)}{(1+\alpha)}$$

$$x_B = \frac{2n\alpha}{n(1+\alpha)} = \frac{2\alpha}{(1+\alpha)}$$

and the partial pressures of A and B are

$$P_A = \frac{(1-\alpha)}{(1+\alpha)}P \quad \text{and} \quad P_B = \frac{2\alpha}{(1+\alpha)}P$$

where P is the total pressure of the system. The equilibrium constant is given by

For simplicity, we omit the $P°$ terms in K_P.

$$K_P = \frac{P_B^2}{P_A} = \left(\frac{2\alpha}{1+\alpha}P\right)^2 \bigg/ \frac{(1-\alpha)}{(1+\alpha)}P$$

$$= \left(\frac{4\alpha^2}{1-\alpha^2}\right)P$$

Rearranging the last equation, we obtain

$$\alpha = \sqrt{\frac{K_P}{K_P + 4P}}$$

Because K_P is a constant, the value of α depends only on P. If P is large, α is small; if P is small, α is large. These predictions again remind us of Le Châtelier's principle. If the stress applied to the system is an increase in pressure, then the equilibrium will shift to the side that produces fewer molecules, which is right to left in our example, and hence a decreases. The reverse holds true for a decrease in pressure.

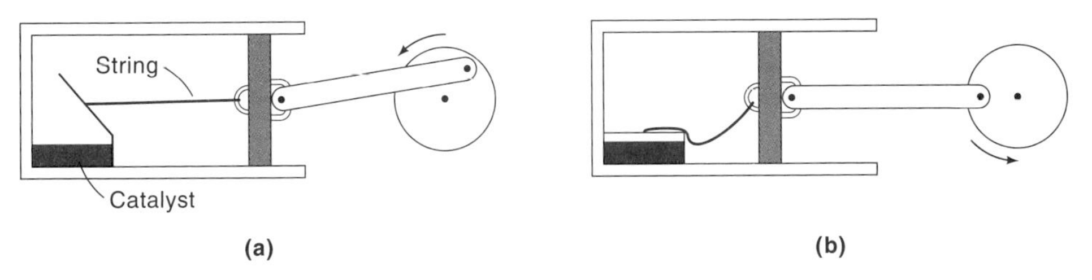

Figure 8.10
A perpetual-motion machine operating with a hypothetical catalyst capable of shifting the equilibrium position of a gaseous reaction in only one direction.

The Effect of a Catalyst

By definition, a catalyst is a substance that can speed up the rate of a reaction without itself being used up. Will the addition of a catalyst to a reacting system at equilibrium shift the equilibrium in a particular direction? To answer this question, let us again consider the gas-phase equilibrium,

$$A(g) \rightleftharpoons 2B(g)$$

as a thought experiment. Suppose a catalyst exists that favors the reverse reaction ($2B \longrightarrow A$) but not the forward reaction ($A \longrightarrow 2B$). We can then construct an apparatus such as that shown in Figure 8.10. A small box is placed inside a cylinder fitted with a movable piston. The lid of the box is connected to the piston by a piece of string, so that the cover can be closed and opened by the movement of the piston. We start with the equilibrium mixture of gases A and B in the cylinder and then add the catalyst to the box (Figure 8.10a). Immediately, the equilibrium shifts toward the formation of A. Because two molecules of B are consumed to produce one molecule of A, this step decreases the total number of molecules present in the cylinder and hence the internal gas pressure. Consequently, the piston will be pushed inward by the external pressure until the internal and external pressures are again balanced. At this stage, the box lid drops (Figure 8.10b). Without the catalyst, the gases gradually return to their original concentrations, and the increase in the number of molecules as a result of the formation of B pushes the piston from left to right until the cover is lifted. The whole process will then repeat itself.

You may have noticed something strange about the whole arrangement. The piston can be made to perform work even though there is no energy input or net consumption of chemicals. Such a device is known as a perpetual-motion machine. Construction of a perpetual-motion machine is impossible, however, because it violates the laws of thermodynamics. The reason such a machine can never be realized is that no catalysts can speed up the rate of a reaction in one direction without affecting the rate of the reverse reaction in a similar manner. The important conclusion from this simple illustration is that a catalyst cannot shift the position of an equilibrium.* For a reaction mixture that is not at equilibrium, a catalyst will increase both the forward and reverse rate processes so that the reaction will reach equilibrium sooner, but the same equilibrium state will be attained eventually even if no catalyst is present.

* The fact that a catalyst must enhance both the forward and reverse rates of a reaction is predicted by the principle of microscopic reversibility (see Chapter 15).

8.6 Binding of Ligands and Metal Ions to Macromolecules

The interaction of small molecules (ligands) with proteins and with specific receptor sites on membrane surfaces is one of the most extensively studied biochemical phenomena. Examples of these reversible interactions include the binding and release of protons by the acidic and basic groups in proteins, the association of cations such as Mg^{2+} and Ca^{2+} with proteins and nucleic acids, antibody–antigen reactions, and the reversible binding of oxygen by myoglobin and hemoglobin. These processes are closely related to the combination of enzymes with substrates and inhibitors.

In this section, we shall apply the equilibrium treatment to the study of the binding of ligands and metal ions to macromolecules in solution. We concentrate on two cases: one in which the macromolecule possesses one binding site per molecule, and another in which there are *n equivalent* and *independent* binding sites per molecule. Because our approach here is strictly thermodynamic, we need not discuss the structure of macromolecules and the nature of covalent and other intermolecular forces responsible for the binding.

One Binding Site per Macromolecule

This is the simplest case, in which one site of a macromolecule, P, binds one molecule (or ion) of a ligand, L. The reaction can be represented as

$$\mathrm{P} + \mathrm{L} \rightleftharpoons \mathrm{PL}$$

The equilibrium constant for this association reaction, K_a, is

We assume ideal behavior and use concentrations instead of activities.

$$K_a = \frac{[\mathrm{PL}]}{[\mathrm{P}][\mathrm{L}]}$$

Frequently, working with the dissociation constant, K_d, is more convenient:

$$K_d = \frac{[\mathrm{P}][\mathrm{L}]}{[\mathrm{PL}]} \tag{8.20}$$

The smaller the value of K_d, the "tighter" the complex PL is. These equilibrium constants are related by the simple relation $K_a K_d = 1$.

To determine the value of K_d, we first define a quantity Y, called the *fractional saturation of sites*, such that

$$Y = \frac{\text{concentration of L bound to P}}{\text{total concentration of all forms of P}}$$

$$= \frac{[\mathrm{PL}]}{[\mathrm{P}] + [\mathrm{PL}]} \tag{8.21}$$

The value of Y ranges from zero, when $[\mathrm{PL}] = 0$, to 1, when $[\mathrm{P}] = 0$. For example, when $Y = 0.5$, half of the P molecules are complexed with L and half of the P molecules are in the free form, so that $[\mathrm{P}] = [\mathrm{PL}]$ and $[\mathrm{L}] = K_d$. To determine the value of K_d, we first rearrange Equation 8.20 to give

$$[\mathrm{PL}] = \frac{[\mathrm{P}][\mathrm{L}]}{K_d}$$

Substituting the expression for $[\mathrm{PL}]$ in Equation 8.21, we obtain

$$Y = \frac{[\mathrm{P}][\mathrm{L}]/K_d}{[\mathrm{P}] + [\mathrm{P}][\mathrm{L}]/K_d}$$

$$= \frac{[\mathrm{L}]}{[\mathrm{L}] + K_d} \qquad (8.22)$$

Keep in mind that $[\mathrm{L}]$ is the concentration of free ligands at equilibrium. The value of Y (see Equation 8.21) can be determined in the following way. Initially a known concentration of L, $[\mathrm{L}]_0$, is added to a known concentration of P, $[\mathrm{P}]_0$. At equilibrium, we can measure either $[\mathrm{PL}]$ or $[\mathrm{L}]$ because $[\mathrm{L}]_0$ is known at the outset, and mass balance requires that $[\mathrm{L}] + [\mathrm{PL}] = [\mathrm{L}]_0$. The value of $[\mathrm{P}]$ need not be measured because $[\mathrm{P}] + [\mathrm{PL}] = [\mathrm{P}]_0$, so that $[\mathrm{P}] = [\mathrm{P}]_0 - [\mathrm{PL}]$. We shall soon discuss an experimental procedure for determining $[\mathrm{L}]$ and $[\mathrm{PL}]$.

By taking the reciprocal of Equation 8.22, we get

$$\frac{1}{Y} = 1 + \frac{K_d}{[\mathrm{L}]} \qquad (8.23)$$

A plot of $1/Y$ versus $1/[\mathrm{L}]$ gives a straight line of slope K_d. Alternatively, Equation 8.22 can be rearranged to give

$$\frac{Y}{[\mathrm{L}]} = \frac{1}{K_d} - \frac{Y}{K_d} \qquad (8.24)$$

In this case, a plot of $Y/[\mathrm{L}]$ versus Y gives a straight line whose slope is $-1/K_d$ (Figure 8.11).

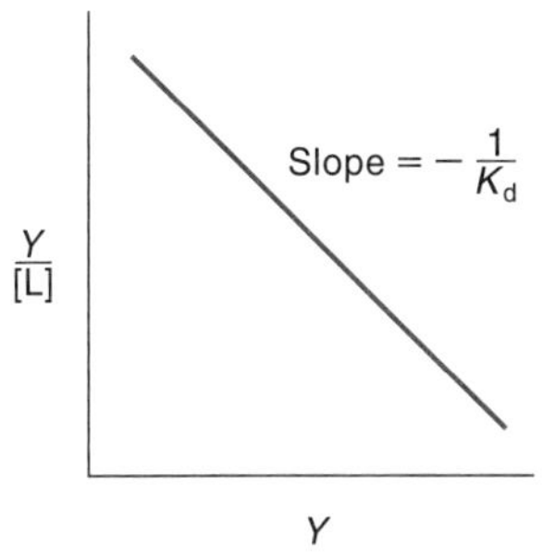

Figure 8.11
A plot of $Y/[\mathrm{L}]$ versus Y according to Equation 8.24.

n Equivalent Binding Sites per Macromolecule

Now consider the case in which a macromolecule has n equivalent sites; that is, each binding site has the same K_d value, regardless of whether other sites on the same molecule are occupied. We shall take $n = 2$ first and then generalize the result to cases for which $n > 2$.

If a macromolecule has two equivalent binding sites, there will be two binding equilibria:

$$\mathrm{P} + \mathrm{L} \rightleftharpoons \mathrm{PL} \qquad K_1 = \frac{[\mathrm{P}][\mathrm{L}]}{[\mathrm{PL}]}$$

$$\mathrm{PL} + \mathrm{L} \rightleftharpoons \mathrm{PL}_2 \qquad K_2 = \frac{[\mathrm{PL}][\mathrm{L}]}{[\mathrm{PL}_2]}$$

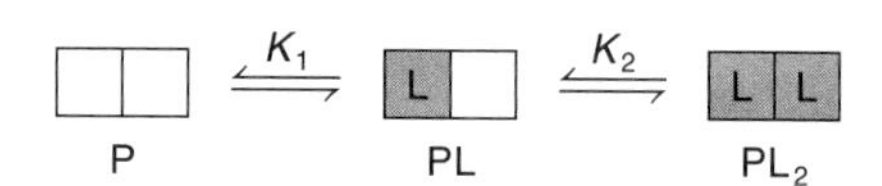

Figure 8.12
The successive bindings of ligand L to a macromolecule that has two equivalent binding sites.

where K_1 and K_2 are the dissociation constants (Figure 8.12). This time, we define Y such that

$$Y = \frac{\text{concentration of L bound to P}}{\text{total concentration of all forms of P}}$$

$$= \frac{[PL] + 2[PL_2]}{[P] + [PL] + [PL_2]} \tag{8.25}$$

Note that because PL_2 has two molecules of L bound to one molecule of P, its concentration must be multiplied by 2. Substituting the relations

$$[PL] = \frac{[P][L]}{K_1} \quad \text{and} \quad [PL_2] = \frac{[P][L]}{K_2}$$

into Equation 8.25, we write

$$Y = \frac{[P][L]/K_1 + 2[P][L]^2/K_1K_2}{[P] + [P][L]/K_1 + [P][L]^2/K_1K_2}$$

$$= \frac{[L]/K_1 + 2[L]^2/K_1K_2}{1 + [L]/K_1 + [L]^2/K_1K_2} \tag{8.26}$$

At first, K_1 might appear to be equal to K_2 because both sites are independent and have the same dissociation constant. But this is not the case because of certain statistical factors. Let K be the dissociation constant at a *given* site (in this case, K is called the *intrinsic dissociation constant*). Then $2K_1 = K$ because there are two ways for L to be attached to P and one way for it to detach from PL. (K_1 would be equal to K if there were only one site present.) After an L is bound, there is one way for L to be attached to PL and two ways for it to detach from PL_2, so $K_2 = 2K$. The general relationship between the ith dissociation constant K_i and K is given by

$$K_i = \left(\frac{i}{n - i + 1}\right)K \tag{8.27}$$

For our example $i = 1, 2$ and $n = 2$, so from Equation 8.27,

$$K_1 = \frac{K}{2} \quad \text{and} \quad K_2 = 2K$$

Thus, purely on the basis of statistical analysis, we find that the second dissociation constant is four times as large as the first dissociation constant, that is, $K_2 = 4K_1$ (see Appendix 8.2 on p. 338). Note that K is the geometric mean of the individual dissociation constants, that is,

$$K = \sqrt{K_1K_2} \tag{8.28}$$

Equation 8.26 can now be written as

$$Y = \frac{2[\mathrm{L}]/K + 2[\mathrm{L}]^2/K^2}{1 + 2[\mathrm{L}]/K + [\mathrm{L}]^2/K^2}$$

$$= \frac{(2[\mathrm{L}]/K)(1 + [\mathrm{L}]/K}{(1 + [\mathrm{L}]/K)^2} = \frac{2[\mathrm{L}]}{[\mathrm{L}] + K} \qquad (8.29)$$

Equation 8.29 is the result obtained for two equivalent sites.

In general, for n equivalent sites, we have*

$$Y = \frac{n[\mathrm{L}]}{[\mathrm{L}] + K} \qquad (8.30)$$

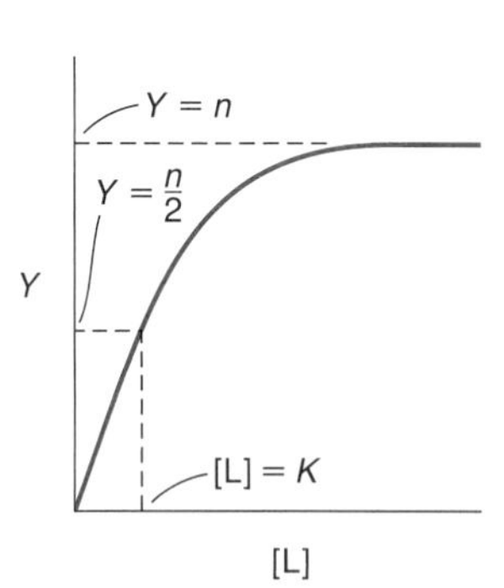

Figure 8.13
A plot of fractional saturation (Y) versus ligand concentration ([L]).

Equation 8.30 can be rearranged into several forms suitable for graphing. We shall look at three of the most common procedures.

1. The Direct Plot. Figure 8.13 shows a graph of Y versus the ligand concentration. A direct plot of this type yields a hyperbolic curve, which is characteristic of simple binding (i.e., the binding sites are all equivalent and noninteracting). When $[\mathrm{L}] = K$, we have $Y = n/2$, and at very high ligand concentrations, we can assume that $[\mathrm{L}] \gg K$, so that $Y = n$. The direct plot is generally not very useful in determining the values of n and K, however, because determining n at very high ligand concentrations is often difficult and we need to know the value of n to determine that of K.

2. The Double Reciprocal Plot. By taking the reciprocal of each side of Equation 8.30, we obtain

$$\frac{1}{Y} = \frac{1}{n} + \frac{K}{n[\mathrm{L}]} \qquad (8.31)$$

Thus, a plot of $1/Y$ versus $1/[\mathrm{L}]$ gives a straight line of slope K/n and an intercept on the ordinate of $1/n$ (Figure 8.14). This plot is known as the Hughes–Klotz plot.

Figure 8.14
A plot of $1/Y$ versus $1/[\mathrm{L}]$. This graph is also known as the Hughes–Klotz plot.

3. The Scatchard Plot. Starting with Equation 8.30, we have

$$Y[\mathrm{L}] + KY = n[\mathrm{L}]$$

$$\frac{Y[\mathrm{L}]}{K} + Y = \frac{n[\mathrm{L}]}{K}$$

$$Y = \frac{n[\mathrm{L}]}{K} - \frac{Y[\mathrm{L}]}{K}$$

or

$$\frac{Y}{[\mathrm{L}]} = \frac{n}{K} - \frac{Y}{K} \qquad (8.32)$$

* For n equivalent sites, we have $K = (K_1K_2K_3 \cdots K_n)^{1/n}$.

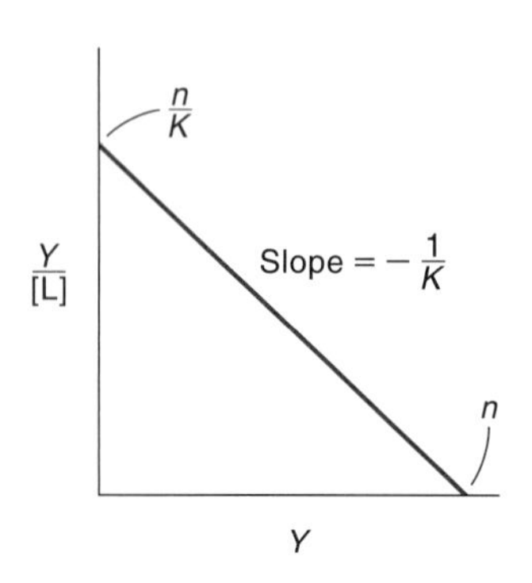

Figure 8.15
A plot of $Y/[L]$ versus Y. This graph is also known as the Scatchard plot.

Equation 8.32 is known as the Scatchard equation (after the American chemist George Scatchard, 1892–1973). Thus, plotting $Y/[L]$ versus Y gives a straight line of slope $-1/K$ and an intercept on the abscissa of n (Figure 8.15).

Equilibrium Dialysis

Having discussed the theoretical aspects of the binding of ligands to macromolecules, we now look at one particular experimental method for determining the values of n and K—*equilibrium dialysis.*

Dialysis is the process of exchanging small ions and other solute molecules from a protein solution through a semipermeable membrane. Suppose that in an experiment, we have precipitated hemoglobin from a solution by using the salting out technique with ammonium sulfate (see Section 7.5). The protein can be freed of the $(NH_4)_2SO_4$ salt as follows. The precipitate is dissolved in water or more typically, in a buffer solution. The protein solution is then placed in a cellophane bag, which, in turn, is immersed in a beaker containing the same buffer (Figure 8.16). Because both the NH_4^+ and SO_4^{2-} ions are small enough to diffuse through the membrane but the protein molecules are not, the ions in the bag will begin to enter into the outside solution, where the chemical potentials are lower:

$$(\mu_{NH_4^+})_{inside} > (\mu_{NH_4^+})_{outside}$$

$$(\mu_{SO_4^{2-}})_{inside} > (\mu_{SO_4^{2-}})_{outside}$$

The flow of ions out of the bag continues until the chemical potentials of the cation and anion outside of the bag equal those on the inside, and an equilibrium is established. If desired, all the $(NH_4)_2SO_4$ can be removed by continually changing the buffer solution in the beaker.

The procedure described above may be reversed to study the binding of ions or small ligands to proteins. In this case, we begin by placing the protein without ligands in a buffer solution (called phase 1) inside the cellophane bag, which is then immersed

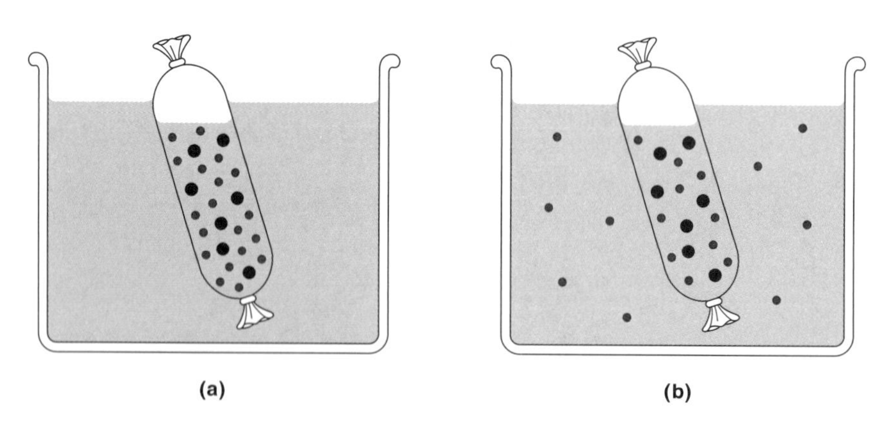

Figure 8.16
A dialysis experiment. The small dots denote ligands and the larger dots proteins. (a) At the start of dialysis. (b) At equilibrium, some of the small ions have diffused out of the cellophane bag. By repeatedly replacing the buffer solution in the beaker, it is possible to remove all of the small ions not bound to protein.

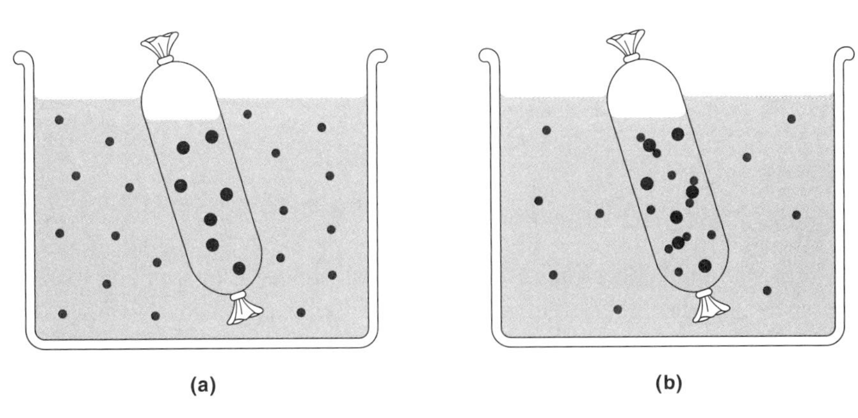

Figure 8.17
Equilibrium dialysis. The small dots denote ligands and the large dots proteins. (a) Initially, a cellophane bag containing a protein solution is immersed in a buffer solution containing the ligand molecules. (b) At equilibrium, some of the ligand molecules have diffused into the bag and are bound to the protein molecules.

in a similar buffer solution (called phase 2) that contains the ligands (L) of a known concentration. At equilibrium, the chemical potentials of the free (unbound) ligands (μ_L) in both phases must be the same (Figure 8.17), so

$$(\mu_\mathrm{L})_1^{\text{unbound}} = (\mu_\mathrm{L})_2^{\text{unbound}} \tag{8.33}$$

or

$$(\mu^\circ + RT \ln a_\mathrm{L})_1^{\text{unbound}} = (\mu^\circ + RT \ln a_\mathrm{L})_2^{\text{unbound}} \tag{8.34}$$

Because the standard chemical potential, μ°, is the same on both sides, we have

$$(a_\mathrm{L})_1^{\text{unbound}} = (a_\mathrm{L})_2^{\text{unbound}} \tag{8.35}$$

If the solutions are dilute, we can replace activities with concentrations so that

$$[\mathrm{L}]_1^{\text{unbound}} = [\mathrm{L}]_2^{\text{unbound}} \tag{8.36}$$

However, the total concentration of the ligands inside the bag is given by

$$[\mathrm{L}]_1^{\text{total}} = [\mathrm{L}]_1^{\text{bound}} + [\mathrm{L}]_1^{\text{unbound}} \tag{8.37}$$

Therefore, the concentration of ligands bound to protein molecules is

$$[\mathrm{L}]_1^{\text{bound}} = [\mathrm{L}]_1^{\text{total}} - [\mathrm{L}]_1^{\text{unbound}} \tag{8.38}$$

The first quantity on the right side of Equation 8.38 can be determined by analyzing the solution in the bag *after* its removal from the beaker; the second quantity is obtained by measuring the concentration of the ligands in the solution remaining in the beaker. (Remember that the concentrations of the unbound ligands are equal in phases 1 and 2.) We can now see how both the intrinsic dissociation constant K and the number of binding sites n can be obtained by using the equilibrium dialysis technique. Suppose that we start with a protein solution of known concentration in phase 1 and a ligand solution of known concentration in phase 2. At equilibrium, the quantity Y is given by (see Equation 8.25)

$$Y = \frac{[\mathrm{L}]_1^{\mathrm{bound}}}{[\mathrm{P}]^{\mathrm{total}}} \tag{8.39}$$

where $[\mathrm{P}]^{\mathrm{total}}$ is the concentration of the original protein solution. The experiment can be repeated by using different concentrations for the protein and ligand solutions, and the values of K and n can be determined from either a Hughes–Klotz plot or a Scatchard plot. Keep in mind that the quantity $[\mathrm{L}]$ in Equations 8.31 and 8.32 refers to the concentration of unbound ligands at equilibrium, which is $[\mathrm{L}]_1^{\mathrm{unbound}}$ in our case.

The equilibrium dialysis technique has been successfully employed to bind drugs, hormones, and other small molecules to proteins and nucleic acids. Note that we have implicitly assumed that the ligand is a nonelectrolyte, so the concentrations of unbound ligands are equal in phases 1 and 2 at equilibrium. If the ligand were an electrolyte, we would have to apply the Donnan effect in treating the dialysis data.

Key Equations

$\Delta_r G^\circ = -RT \ln K_P$	(Relation between $\Delta_r G^\circ$ and K_P)	(8.7)
$\Delta_r G = \Delta_r G^\circ + RT \ln Q$	(Gibbs energy change of a reaction)	(8.9)
$\gamma = \frac{f}{P}$	(Fugacity coefficient)	(8.11)
$\ln \frac{K_2}{K_1} = \frac{\Delta_r H^\circ}{R}\left(\frac{T_2 - T_1}{T_1 T_2}\right)$	(van't Hoff equation)	(8.18)
$\ln K = -\frac{\Delta_r H^\circ}{RT} + \frac{\Delta_r S^\circ}{R}$	(van't Hoff equation)	(8.19)

APPENDIX 8.1

The Relationship Between Fugacity and Pressure

As mentioned in Section 8.1, fugacity is a quantity devised to represent the pressure of real gases, especially in the study of chemical equilibria involving gases at high pressure. For simplicity, we shall derive the relationship between fugacity and pressure for a pure gas, although the result can be extended readily to partial pressures in a mixture.

The Gibbs energy for an ideal gas is given by (see Equation 5.18):

$$\overline{G}_{\text{ideal}} = \overline{G}^\circ + RT \ln \frac{P}{1 \text{ bar}} \tag{1}$$

Gilbert Lewis suggested that for a real gas, we can use an analogous equation:

$$\overline{G}_{\text{real}} = \overline{G}^\circ + RT \ln \frac{f}{1 \text{ bar}} \tag{2}$$

where f is the fugacity of the gas. Fugacity has the same unit as pressure, but it cannot be measured experimentally. As gas pressure approaches zero, the gas approaches ideal behavior, and fugacity and pressure are one and the same quantity.

Our next step is to find ways to calculate the fugacity of a gas at different temperatures and pressures. For an ideal gas in the standard state, $P = 1$ bar. Real gases at 1 bar pressure are not in their standard state, but the difference is small, and so we assume that the values of $\overline{G}^\circ$ in Equations 1 and 2 are the same (see Figure 8.4). We can now write

$$\overline{G}_{\text{real}} - \overline{G}_{\text{ideal}} = RT \ln \frac{f}{P} \tag{3}$$

To evaluate $(\overline{G}_{\text{real}} - \overline{G}_{\text{ideal}})$, we proceed as follows. Equation 5.16, in molar quantities, becomes

$$\left(\frac{\partial \overline{G}}{\partial P}\right)_T = \overline{V} \tag{4}$$

or

$$d\overline{G} = \overline{V} dP$$

This equation applies to all gases, real and ideal. For an ideal gas,

$$d\overline{G}_{\text{ideal}} = \overline{V}_{\text{ideal}} dP$$

$$= \left(\frac{RT}{P}\right) dP \tag{5}$$

because $P\overline{V}_{\text{ideal}} = RT$ and

$$d\overline{G}_{\text{real}} = \overline{V}_{\text{real}} dP \tag{6}$$

The molar volume of a real gas, $\overline{V}_{\text{real}}$, can be measured directly or calculated using an equation of state for real gases. Thus,

$$d(\overline{G}_{\text{real}} - \overline{G}_{\text{ideal}}) = \left(\overline{V}_{\text{real}} - \frac{RT}{P}\right) dP$$

Integration between $P = 0$, where $\overline{G}_{\text{real}} = \overline{G}_{\text{ideal}}$ (when pressure equals fugacity) and P, the pressure of interest, yields

$$\int_0^P d(\overline{G}_{\text{real}} - \overline{G}_{\text{ideal}}) = \overline{G}_{\text{real}} - \overline{G}_{\text{ideal}}$$

$$= \int_0^P \left[\overline{V}_{\text{real}} - \left(\frac{RT}{P}\right)\right] dP \tag{7}$$

Substituting Equation 3 into Equation 7, we obtain

$$\ln \frac{f}{P} = \frac{1}{RT} \int_0^P \left[\overline{V}_{\text{real}} - \left(\frac{RT}{P}\right)\right] dP$$

$$= \int_0^P \left(\frac{\overline{V}_{\text{real}}}{RT} - \frac{1}{P}\right) dP \tag{8}$$

or

$$\ln f = \ln P + \int_0^P \left(\frac{\overline{V}_{\text{real}}}{RT} - \frac{1}{P}\right) dP \tag{9}$$

The integral in Equation 9 can be evaluated graphically if we know $\overline{V}_{\text{real}}$ as a function of pressure at T. Alternatively, if $\overline{V}_{\text{real}}$ can be expressed as a function of pressure by an equation of state, the integral can be evaluated analytically. The following example illustrates the second approach.

EXAMPLE 1

A satisfactory equation of state for ammonia is the modified van der Waals equation (for a so-called hard-sphere gas) $P(\overline{V} - b) = RT$, where b is 0.0379 L mol^{-1}. Calculate the fugacity of the gas when the pressure is 50 atm at 298 K.

ANSWER

Because $\overline{V}_{\text{real}} = (RT/P) + b$, Equation 8 can be written as

$$\ln \frac{f}{P} = \int_0^{50\text{ atm}} \left(\frac{b}{RT}\right) dP$$

$$= \frac{(0.0379\text{ L mol}^{-1})(50\text{ atm})}{(0.08206\text{ L atm K}^{-1}\text{ mol}^{-1})(298\text{ K})} = 0.0775$$

Therefore,

$$f = Pe^{0.0775}$$

$$= (50\text{ atm})(1.08)$$

$$= 54\text{ atm}$$

COMMENT

Because the pressure is high, we do not expect the ammonia gas to behave ideally. The fugacity coefficient, γ, or f/P, is 54/50, or 1.1.

APPENDIX 8.2

The Relationships Between K_1 and K_2 and the Intrinsic Dissociation Constant K

Consider a macromolecule (P) with two equivalent and independent binding sites. Suppose that these two sites are distinguishable so we can label them (1) and (2). Because these sites are equivalent, the equilibrium constant for the dissociation of ligand (L), which we call the *intrinsic dissociation constant (K)*, is the same for the following two processes:

$$\mathrm{P} + \mathrm{L} \rightleftharpoons \mathrm{PL}^{(1)} \qquad K = \frac{[\mathrm{P}][\mathrm{L}]}{[\mathrm{PL}^{(1)}]}$$

$$\mathrm{P} + \mathrm{L} \rightleftharpoons \mathrm{PL}^{(2)} \qquad K = \frac{[\mathrm{P}][\mathrm{L}]}{[\mathrm{PL}^{(2)}]}$$

The total concentration of PL is given by

$$\begin{aligned} [\mathrm{PL}] &= [\mathrm{PL}^{(1)}] + [\mathrm{PL}^{(2)}] \\ &= \frac{[\mathrm{P}][\mathrm{L}]}{K} + \frac{[\mathrm{P}][\mathrm{L}]}{K} \\ &= \frac{2[\mathrm{P}][\mathrm{L}]}{K} \end{aligned} \tag{1}$$

We can also express the binding equilibrium as

$$\mathrm{P} + \mathrm{L} \overset{K_1}{\rightleftharpoons} \mathrm{PL} \quad \text{(any site)}$$

where

$$K_1 = \frac{[\mathrm{P}][\mathrm{L}]}{[\mathrm{PL}]} \tag{2}$$

Substituting Equation 1 into 2, we get

$$K_1 = \frac{[\mathrm{P}][\mathrm{L}]}{2[\mathrm{P}][\mathrm{L}]/K} = \frac{K}{2} \tag{3}$$

Returning to the definition for the intrinsic dissociation constant, let us assume that the ligands bind sites (1) and (2) sequentially, so we can write

$$P + L \rightleftharpoons PL^{(1)} \qquad K = \frac{[P][L]}{[PL^{(1)}]}$$

$$PL^{(1)} + L \rightleftharpoons PL_2^{(1)(2)} \qquad K = \frac{[PL^{(1)}][L]}{[PL_2^{(1)(2)}]}$$

$$\text{Overall:} \quad P + 2L \rightleftharpoons PL_2^{(1)(2)}$$

The equilibrium (dissociation) constant for the overall reaction is the product of the equilibrium constants for the individual steps, that is,

$$K_{\text{overall}} = K^2 = \frac{[P][L]^2}{[PL_2^{(1)(2)}]} \tag{4}$$

or

$$[PL_2^{(1)(2)}] = \frac{[P][L]^2}{K^2} \tag{5}$$

To derive the relationship between K_2 and K, we write

$$PL(\text{any site}) + L \overset{K_2}{\rightleftharpoons} PL_2^{(1)(2)}$$

where

$$K_2 = \frac{[PL][L]}{[PL_2^{(1)(2)}]} \tag{6}$$

Substituting Equations 1 and 5 into 6, we arrive at the results

$$K_2 = \frac{(2[P][L]/K)[L]}{[P][L]^2/K^2} = 2K \tag{7}$$

and

$$K_2 = 4K_1 \tag{8}$$

Suggestions for Further Reading

BOOKS

Hanson, R. M., and S. Green, *Introduction to Molecular Thermodynamics*, University Science Books, Sausalito, CA, 2008.

Klotz, I. M., and R. M. Rosenberg, *Chemical Thermodynamics*, 6th ed., John Wiley & Sons, New York, 2000.

McQuarrie, D. A., and J. Simon, *Molecular Thermodynamics*, University Science Books, Sausalito, CA, 1999.

Rock, P., *Chemical Thermodynamics*, University Science Books, Sausalito, CA, 1983.

ARTICLES

General

"Some Observations Concerning the van't Hoff Equation," J. T. MacQueen, *J. Chem. Educ.* **44**, 755 (1967).

"High Altitude Acclimatization," R. C. Plumb, *J. Chem. Educ.* **48**, 75 (1971).

"Effect of Ionic Strength on Equilibrium Constants," M. D. Seymour and Q. Fernando, *J. Chem. Educ.* **54**, 225 (1977).

"Clarifying the Concept of Equilibrium in Chemically Reacting Systems," W. F. Harris, *J. Chem. Educ.* **59**, 1034 (1982).

"On the Dynamic Nature of Chemical Equilibrium," M. L. Hernandez and J. M. Alvarino, *J. Chem. Educ.* **60**, 930 (1983).

"Le Châtelier's Principle—a Redundant Principle?" R. T. Allsop and N. H. George, *Educ. Chem.* **21**, 82 (1984).

"Le Châtelier's Principle and the Law of van't Hoff," J. Gold and V. Gold, *Educ. Chem.* **22**, 82 (1985).

"A Better Way of Dealing with Chemical Equilibrium," R. J. Tykodi, *J. Chem. Educ.* **63**, 582 (1986).

"The Effect of Temperature and Pressure on Equilibria: A Derivation of the van't Hoff Rules," H. R. Kemp, *J. Chem. Educ.* **64**, 482 (1987).

"Entropy of Mixing and Homogeneous Equilibrium," S. R. Logan, *Educ. Chem.* **25**, 44 (1988).

"Equilibrium, Free Energy, and Entropy: Rates and Differences," J. J. MacDonald, *J. Chem. Educ.* **67**, 380 (1990).

"Equilibria and $\Delta G°$," J. J. MacDonald, *J. Chem. Educ.* **67**, 745 (1990).

"Chemical Equilibrium," A. A. Gordus, *J. Chem. Educ.* **68**, 138, 215, 291, 397, 566, 656, 759, 927 (1991).

"Reaction Thermodynamics: A Flawed Derivation," F. M. Horuack, *J. Chem. Educ.* **69**, 112 (1992).

"The Conversion of Chemical Energy," D. J. Wink, *J. Chem. Educ.* **69**, 264 (1992).

"Practical Calculation of the Equilibrium Constant and the Enthalpy of Reaction at Different Temperature," K. Anderson, *J. Chem. Educ.* **71**, 474 (1994).

"Teaching Chemical Equilibrium and Thermodynamics in Undergraduate General Chemistry Classes," A. C. Banerjee, *J. Chem. Educ.* **72**, 879 (1995). Also see *J. Chem. Educ.* **73**, A261 (1996).

"Thermodynamics and Spontaneity," R. S. Ochs, *J. Chem. Educ.* **73**, 952 (1996).

"Free Energy Versus Extent of Reaction," R. S. Treptow, *J. Chem. Educ.* **73**, 51 (1996). Also see *J. Chem. Educ.* **74**, 22 (1997).

"The Iron Blast Furnace: A Study in Chemical Thermodynamics," R. S. Treptow and L.

Jean, *J. Chem. Educ.* **75**, 43 (1998).
"A Mechanical Analogue for Chemical Potential, Extent of Reaction, and the Gibbs Energy," S. V. Glass and R. L. DeKock, *J. Chem. Educ.* **75**, 190 (1998).
"The Temperature Dependence of ΔG° and the Equilibrium Constant, K_{eq}; Is There a Paradox?" F. H. Chappie, *J. Chem. Educ.* **75**, 342 (1998).
"How Thermodynamic Data and Equilibrium Constants Changed When the Standard-State Pressure Became 1 Bar," R. S. Treptow, *J. Chem. Educ.* **76**, 212 (1999).
"The Use of Extent of Reaction in Introductory Courses," S. G. Canagaratna, *J. Chem. Educ.* **77**, 52 (2000).
"Illustrating Chemical Concepts with Coin Flipping," R. Chang and J. W. Thoman, Jr., *Chem. Educator* [Online] **6**, 360 (2001) DOI 10.1007/s00897010524a.
"Determining the Enthalpy, Free Energy, and Entropy for the Solubility of Salicylic Acid with the van't Hoff Equation: A Spectrophotometric Determination of of K_{eq}," J. C. Barreto, T. Dubetz, D. W. Brown, P. D. Barreto, C. M. Coates, and A. Cobb, *Chem. Educator* [Online] **12**, 18 (2007) DOI 10.1333/s00897072002a.

Fugacity
"The Fugacity of van der Waals Gas," J. S. Winn, *J. Chem. Educ.* **65**, 772 (1988).
"An Alternative View of Fugacity," L. L. Combs, *J. Chem. Educ.* **69**, 218 (1992).
"Fugacity—More Than a Fake Pressure," M. C. A. Donkersloot, *J. Chem. Educ.* **69**, 290 (1992).
"Fugacity and Activity in a Nutshell," J. D. Ramshaw, *J. Chem. Educ.* **72**, 601 (1995).
"An Alternate Method of Introducing Fugacity," M. Jemal, *Chem. Educator* [Online] **4**, 1 (1999) DOI 10.1007/s00897990284a.

Binding Equilibria
"Equilibrium Dialysis," S. A. Katz, C. Parfitt, and R. Purdy, *J. Chem. Educ.* **47**, 721 (1970).
"Spectroscopic Determination of Protein-Ligand Binding Constants," A. Orstan and J. F. Wojcik, *J. Chem. Educ.* **64**, 814 (1987).
"Dialysis," R. H. Barth, *Encyclopedia of Applied Physics*, Trigg, G. L., Ed., VCH Publishers, New York (1992), Vol. 4, p. 533.
"Analysis of Receptor-Ligand Interactions," A. D. Atlie and R. T. Raines, *J. Chem. Educ.* **72**, 119 (1995).
"A Thermodynamic Study of Azide Binding to Myoglobin," A. T. Marcoline and T. E. Elgren, *J. Chem. Educ.* **75**, 1622 (1998).

Problems

Chemical Equilibrium

8.1 Equilibrium constants of gaseous reactions can be expressed in terms of pressures only (K_P), concentrations only (K_c), or mole fractions only (K_x). For the hypothetical reaction

$$aA(g) \rightleftharpoons bB(g)$$

derive the following relationships: **(a)** $K_P = K_c(RT)^{\Delta n}(P^\circ)^{-\Delta n}$ and **(b)** $K_P = K_x P^{\Delta n}(P^\circ)^{-\Delta n}$, where Δn is the difference in the number of moles of products and reactants, and P is the total pressure of the system. Assume ideal-gas behavior.

8.2 At 1024°C, the pressure of oxygen gas from the decomposition of copper(II) oxide (CuO) is 0.49 bar:

$$4CuO(s) \rightleftharpoons 2Cu_2O(s) + O_2(g)$$

(a) What is the value of K_P for the reaction? **(b)** Calculate the fraction of CuO that will decompose if 0.16 mole of it is placed in a 2.0-L flask at 1024°C. **(c)** What would the fraction be if a 1.0-mole sample of CuO were used? **(d)** What is the smallest amount of CuO (in moles) that would establish the equilibrium?

8.3 Gaseous nitrogen dioxide is actually a mixture of nitrogen dioxide (NO_2) and dinitrogen tetroxide (N_2O_4). If the density of such a mixture is 2.3 g L^{-1} at 74°C and 1.3 atm, calculate the partial pressures of the gases and the value of K_P for the dissociation of N_2O_4.

8.4 About 75% of the hydrogen produced for industrial use is produced by the *steam-reforming* process. This process is carried out in two stages called primary and secondary reforming. In the primary stage, a mixture of steam and methane at about 30 atm is heated over a nickel catalyst at 800°C to give hydrogen and carbon monoxide:

$$CH_4(g) + H_2O(g) \rightleftharpoons CO(g) + 3H_2(g) \qquad \Delta_r H^\circ = 206 \text{ kJ mol}^{-1}$$

The secondary stage is carried out at about 1000°C, in the presence of air, to convert the remaining methane to hydrogen:

$$CH_4(g) + \tfrac{1}{2}O_2(g) \rightleftharpoons CO(g) + 2H_2(g) \qquad \Delta_r H^\circ = 35.7 \text{ kJ mol}^{-1}$$

(a) What conditions of temperature and pressure would favor the formation of products in both the primary and secondary stages? **(b)** The equilibrium constant, K_c, for the primary stage is 18 at 800°C. **(i)** Calculate the value of K_P for the reaction. **(ii)** If the partial pressures of methane and steam were both 15 atm at the start, what would the pressures of all the gases be at equilibrium?

8.5 Consider the reaction

$$PCl_5(g) \rightleftharpoons PCl_3(g) + Cl_2(g)$$

for which $K_P = 1.05$ at 250°C. A quantity of 2.50 g of PCl_5 is placed in an evacuated flask of volume 0.500 L and heated to 250°C. **(a)** Calculate the pressure of PCl_5 if it did not dissociate.

(**b**) Calculate the partial pressure of PCl_5 at equilibrium. (**c**) What is the total pressure at equilibrium? (**d**) What is the degree of dissociation of PCl_5? (The degree of dissociation is given by the fraction of PCl_5 that has undergone dissociation.)

8.6 The vapor pressure of mercury is 0.002 mmHg at 26°C. (**a**) Calculate the values of K_c and K_P for the process $Hg(l) \rightleftharpoons Hg(g)$. (**b**) A chemist breaks a thermometer and spills mercury onto the floor of a laboratory measuring 6.1 m long, 5.3 m wide, and 3.1 m high. Calculate the mass of mercury (in grams) vaporized at equilibrium and the concentration of mercury vapor in mg m^{-3}. Does this concentration exceed the safety limit of 0.05 mg m^{-3}? (Ignore the volume of furniture and other objects in the laboratory.)

8.7 A quantity of 0.20 mole of carbon dioxide was heated to a certain temperature with an excess of graphite in a closed container until the following equilibrium was reached:

$$C(s) + CO_2(g) \rightleftharpoons 2CO(g)$$

Under these conditions, the average molar mass of the gases was 35 g mol^{-1}. (**a**) Calculate the mole fractions of CO and CO_2. (**b**) What is the value of K_P if the total pressure is 11 atm? (*Hint:* The average molar mass is the sum of the products of the mole fraction of each gas times its molar mass.)

van't Hoff Equation

8.8 Consider the thermal decomposition of $CaCO_3$:

$$CaCO_3(s) \rightleftharpoons CaO(s) + CO_2(g)$$

The equilibrium vapor pressures of CO_2 are 22.6 mmHg at 700°C and 1829 mmHg at 950°C. Calculate the standard enthalpy of the reaction.

8.9 Consider the following reaction:

$$CO_2(g) + H_2(g) \rightleftharpoons CO(g) + H_2O(g)$$

The equilibrium constant is 0.534 at 960 K and 1.571 at 1260 K. What is the enthalpy of the reaction?

8.10 The vapor pressure of dry ice (solid CO_2) is 672.2 torr at −80°C and 1486 torr at −70°C. Calculate the molar heat of sublimation of CO_2.

8.11 Nitric oxide (NO) from car exhaust is a primary air pollutant. Calculate the equilibrium constant for the reaction

$$N_2(g) + O_2(g) \rightleftharpoons 2NO(g)$$

at 25°C using the data listed in Appendix B. Assume that both $\Delta_r H°$ and $\Delta_r S°$ are temperature independent. Calculate the equilibrium constant at 1500°C, which is the typical temperature inside the cylinders of a car's engine after it has been running for some time.

$\Delta G°$ and K

8.12 Calculate the value of $\Delta_r G°$ for each of the following equilibrium constants: 1.0×10^{-4}, 1.0×10^{-2}, 1.0, 1.0×10^{2}, and 1.0×10^{4} at 298 K.

8.13 Use the data listed in Appendix B to calculate the equilibrium constant, K_P, for the synthesis of HCl at 298 K:

$$H_2(g) + Cl_2(g) \rightleftharpoons 2HCl(g)$$

What is the value of K_P if the equilibrium is expressed as

$$\tfrac{1}{2}H_2(g) + \tfrac{1}{2}Cl_2(g) \rightleftharpoons HCl(g)$$

8.14 The dissociation of N_2O_4 into NO_2 is 16.7% complete at 298 K and 1 atm:

$$N_2O_4(g) \rightleftharpoons 2NO_2(g)$$

Calculate the equilibrium constant and the standard Gibbs energy change for the reaction. [*Hint:* Let α be the degree of dissociation and show that $K_P = 4\alpha^2 P/(1 - \alpha^2)$, where P is the total pressure.]

8.15 The standard Gibbs energies of formation of gaseous *cis*- and *tran*-2-butene are 67.15 kJ mol^{-1} and 64.10 kJ mol^{-1}, respectively. Calculate the ratio of equilibrium pressures of the gaseous isomers at 298 K.

8.16 Consider the decomposition of calcium carbonate:

$$MgCO_3(s) \rightleftharpoons MgO(s) + CO_2(g)$$

(a) Write an equilibrium constant expression (K_P) for the reaction. **(b)** The rate of decomposition is slow until the partial pressure of carbon dioxide is equal to 1 bar. Calculate the temperature at which the decomposition becomes spontaneous. Assume that $\Delta_r H°$ and $\Delta_r S°$ are temperature independent. Use the data in Appendix B for your calculation.

8.17 Use the data in Appendix B to calculate the equilibrium constant (K_P) for the following reaction at 25°C:

$$2SO_2(g) + O_2(g) \rightleftharpoons 2SO_3(g)$$

Calculate K_P for the reaction at 60°C **(a)** using the van't Hoff equation (Equation 8.17); **(b)** using the Gibbs–Helmholtz equation (Equation 5.15) to find $\Delta_r G°$ at 60°C and hence K_P at the same temperature; and **(c)** using $\Delta_r G° = \Delta_r H° - T\Delta_r S°$ to find $\Delta_r G°$ at 60°C and hence K_P at the same temperature. State the approximations employed in each case and compare your results. (*Hint:* From Equation 5.15, you can derive the relationship

$$\frac{\Delta_r G_2}{T_2} - \frac{\Delta_r G_1}{T_1} = \Delta_r H\left(\frac{1}{T_2} - \frac{1}{T_1}\right)$$

Le Châtelier's Principle

8.18 Consider the reaction

$$2NO_2(g) \rightleftharpoons N_2O_4(g) \qquad \Delta_r H° = -58.04 \text{ kJ mol}^{-1}$$

Predict what happens to the system at equilibrium if **(a)** the temperature is raised, **(b)** the pressure on the system is increased, **(c)** an inert gas is added to the system at constant pressure, **(d)** an inert gas is added to the system at constant volume, and **(e)** a catalyst is added to the system.

8.19 Referring to Problem 8.14, calculate the degree of dissociation of N_2O_4 if the total pressure is 10 atm. Comment on your result.

8.20 At a certain temperature, the equilibrium pressures of NO_2 and N_2O_4 are 1.6 bar and 0.58 bar, respectively. If the volume of the container is doubled at constant temperature, what would be the partial pressures of the gases when equilibrium is re-established?

8.21 Eggshells are composed mostly of calcium carbonate ($CaCO_3$) formed by the reaction

$$Ca^{2+}(aq) + CO_3^{2-}(aq) \rightleftharpoons CaCO_3(s)$$

The carbonate ions are supplied by carbon dioxide produced during metabolism. Explain why eggshells are thinner in the summer when the rate of panting by chickens is greater. Suggest a remedy for this situation.

8.22 Photosynthesis can be represented by

$$6CO_2(g) + 6H_2O(l) \rightleftharpoons C_6H_{12}O_6(s) + 6O_2(g) \qquad \Delta_r H° = 2801 \text{ kJ mol}^{-1}$$

Explain how the equilibrium would be affected by the following changes: **(a)** the partial pressure of CO_2 is increased, **(b)** O_2 is removed from the mixture, **(c)** $C_6H_{12}O_6$ (glucose) is removed from the mixture, **(d)** more water is added, **(e)** a catalyst is added, **(f)** the temperature is decreased, and **(g)** more sunlight shines on the plants.

8.23 When a gas was heated at atmospheric pressure and 25°C, its color deepened. Heating above 150°C caused the color to fade, and at 550°C the color was barely detectable. At 550°C, however, the color was partially restored by increasing the pressure of the system. Which of the following scenarios best fits the above description? Justify your choice. **(a)** A mixture of hydrogen and bromine, **(b)** pure bromine, **(c)** a mixture of nitrogen dioxide and dinitrogen tetroxide. (*Hint:* Bromine is reddish, and nitrogen dioxide is brown. The other gases are colorless.)

8.24 Industrially, sodium metal is obtained by electrolyzing molten sodium chloride. The reaction at the cathode is $Na^+ + e^- \rightarrow Na$. We might expect that potassium metal could also be prepared by electrolyzing molten potassium chloride. Potassium metal is soluble in molten potassium chloride, however, and is therefore hard to recover. Furthermore, potassium vaporizes readily at the operating temperature, creating hazardous conditions. Instead, potassium is prepared by the distillation of molten potassium chloride in the presence of sodium vapor at 892°C:

$$Na(g) + KCl(l) \rightleftharpoons NaCl(l) + K(g)$$

Considering that potassium is a stronger reducing agent than sodium, explain why this approach works. (The boiling points of sodium and potassium are 892°C and 770°C, respectively.)

8.25 People living at high altitudes have higher hemoglobin content in their red blood cells than those living near sea level. Explain.

Binding Equilibria

8.26 Derive Equation 8.23 from 8.21.

8.27 The calcium ion binds to a certain protein to form a 1:1 complex. The following data were obtained in an experiment:

Total $Ca^{2+}/\mu M$	60	120	180	240	480
Ca^{2+} bound to Protein$/\mu M$	31.2	51.2	63.4	70.8	83.4

Determine graphically the dissociation constant of the Ca^{2+}–protein complex. The protein concentration was kept at 96 μM for each run. ($1\ \mu M = 1 \times 10^{-6}\ M$.)

8.28 An equilibrium dialysis experiment showed that the concentrations of the free ligand, bound ligand, and protein are $1.2 \times 10^{-5}\ M$, $5.4 \times 10^{-6}\ M$, and $4.9 \times 10^{-6}\ M$, respectively. Calculate the dissociation constant for the reaction PL $\rightleftharpoons$ P + L. Assume there is one binding site per protein molecule.

Additional Problems

8.29 List two important differences between a steady state and an equilibrium state.

8.30 Based on the material covered so far in the text, describe as many ways as you can for calculating the $\Delta_r G°$ value of a process.

8.31 The solubility of *n*-heptane in water is 0.050 g per liter of solution at 25°C. What is the Gibbs energy change for the hypothetical process of dissolving *n*-heptane in water at a concentration of 2.0 g L^{-1} at the same temperature? (*Hint:* First calculate the value of $\Delta_r G°$ from the equilibrium process and then the $\Delta_r G$ value using Equation 8.6.)

8.32 In physical chemistry, the standard state for a solution is 1 *M*. In biological systems, however, we define the standard state as $1 \times 10^{-7}\ M$ because the physiological pH is about 7. Consequently, the change in the standard Gibbs energy according to these two conventions will be different involving uptake or release of H^+ ions, depending on which convention is used. We will therefore replace $\Delta_r G°$ with $\Delta_r G°'$, where the prime denotes that it is the standard Gibbs energy change for a biological process. **(a)** Consider the reaction

$$A + B \rightarrow C + xH^+$$

where x is a stoichiometric coefficient. Derive a relation between $\Delta_r G°$ and $\Delta_r G°'$, keeping in mind that $\Delta_r G$ is the same for a process regardless of which convention is used. Repeat the derivation for the reverse process:

$$C + xH^+ \rightarrow A + B$$

(b) NAD^+ and NADH are the oxidized and reduced forms of nicotinamide adenine dinucleotide, two key compounds in the metabolic pathways. For the oxidation of NADH,

$$NADH + H^+ \rightarrow NAD^+ + H_2$$

$\Delta_r G°$ is -21.8 kJ mol^{-1} at 298 K. Calculate $\Delta_r G°'$. Also calculate $\Delta_r G$ using both the chemical and biological conventions when $[NADH] = 1.5 \times 10^{-2}\ M$, $[H^+] = 3.0 \times 10^{-5}\ M$, $[NAD^+] = 4.6 \times 10^{-3}\ M$, and $P_{H_2} = 0.010$ bar.

8.33 Referring to Problem 8.32, we see that the quantity $\Delta_r G°'$ can also be applied to reactions involving the uptake and liberation of gases such as O_2 and CO_2. In these cases, the biochemical standard states are $P_{O_2} = 0.2$ bar and $P_{CO_2} = 0.0004$ bar, where 0.2 bar and 0.0004 bar are the partial pressures of O_2 and CO_2 in air, respectively. **(a)** Consider the reaction

$$A(aq) + B(aq) \rightarrow C(aq) + CO_2(g)$$

where A, B, C are molecular species. Derive a relation between $\Delta_r G°$ and $\Delta_r G°'$ for this reaction at 310 K. **(b)** The binding of oxygen to hemoglobin (Hb) is quite complex, but for our purpose here we can represent the reaction as

$$Hb(aq) + O_2(g) \rightarrow HbO_2(aq)$$

If the value of $\Delta_r G°$ for the reaction is -11.2 kJ mol^{-1} at 20°C, calculate the value of $\Delta_r G°'$ for the reaction.

8.34 The K_{sp} value of AgCl is 1.6×10^{-10} at 25°C. What is its value at 60°C?

8.35 Many hydrocarbons exist as structural isomers, which are compounds that have the same molecular formula but different structures. For example, both butane and isobutane have the same molecular formula: C_4H_{10}. Calculate the mole percent of these molecules in an equilibrium mixture at 25°C, given that the standard Gibbs energy of formation of butane is -15.9 kJ mol^{-1} and that of isobutane is -18.0 kJ mol^{-1}. Does your result support the notion that straight-chain hydrocarbons (i.e., hydrocarbons in which the C atoms are joined in a line) are less stable than branch-chain hydrocarbons?

8.36 Consider the equilibrium system $3A \rightleftharpoons B$. Sketch the change in the concentrations of A and B with time for the following situations: **(a)** initially only A is present; **(b)** initially only B is present; and **(c)** initially both A and B are present (with A in higher concentration). In each case, assume that the concentration of B is higher than that of A at equilibrium.

8.37 From the following reactions at 25°C:

$$\text{fumarate}^{2-} + NH_4^+ \rightarrow \text{aspartate}^- \qquad \Delta_r G° = -36.7 \text{ kJ mol}^{-1}$$

$$\text{fumarate}^{2-} + H_2O \rightarrow \text{malate}^{2-} \qquad \Delta_r G° = -2.9 \text{ kJ mol}^{-1}$$

calculate the standard Gibbs energy change and the equilibrium constant for the following reaction:

$$\text{malate}^{2-} + NH_4^+ \rightarrow \text{aspartate}^- + H_2O$$

8.38 A polypeptide can exist in either the helical or random coil forms. The equilibrium constant for the equilibrium reaction of the helix to the random coil transition is 0.86 at 40°C and 0.35 at 60°C. Calculate the values of $\Delta_r H°$ and $\Delta_r S°$ for the reaction.

8.39 A student placed a few ice cubes in a drinking glass with water. A few minutes later she noticed that some of the ice cubes were fused together. Explain what happened.

8.40 A 14.6-g sample of ammonia is placed in a closed 4.00-L flask and heated to 375°C. Calculate the concentrations of all the gases in molarities when equilibrium is established. The equilibrium constant K_c for the reaction $2NH_3(g) \rightleftharpoons N_2(g) + 3H_2(g)$ is 0.83 at 375°C.

8.41 A quantity of 1.0 mole of N_2O_4 was introduced into an evacuated bulb and allowed to attain equilibrium at a certain temperature

$$N_2O_4(g) \rightleftharpoons 2NO_2(g)$$

The average molar mass of the reacting mixture was 70.6 g mol^{-1}. **(a)** Calculate the mole fractions of the gases. **(b)** Calculate K_P for the reaction if the total pressure was 1.2 atm. **(c)** What would be the mole fractions if the pressure were increased to 4.0 atm by reducing the volume at the same temperature?

8.42 The equilibrium constant (K_P) for the reaction

$$C(s) + CO_2(g) \rightleftharpoons 2CO(g)$$

is 1.9 at 727°C. What total pressure must be applied to the reacting system to obtain 0.012 mole of CO_2 and 0.025 mole of CO?

8.43 The following table gives the equilibrium constant for the reaction at various temperatures

$$2NO(g) + O_2(g) \rightleftharpoons 2NO_2(g)$$

K_P	138	5.12	0.436	0.0626	0.0130
T/K	600	700	800	900	1000

Determine graphically the $\Delta_r H°$ for the reaction.

8.44 Consider the following reaction at a certain temperature

$$A_2 + B_2 \rightleftharpoons 2AB$$

The mixing of 1 mole of A_2 with 3 moles of B_2 gives rise to x mole of AB at equilibrium. The addition of 2 more moles of A_2 produces another x mole of AB. What is the equilibrium constant for the reaction?

8.45 At a certain temperature, the equilibrium partial pressures are $P_{NH_3} = 321.6$ atm, $P_{N_2} = 69.6$ atm, and $P_{H_2} = 208.8$ atm, respectively. **(a)** Calculate the value of K_P, for the reaction described in Example 8.1. **(b)** Calculate the thermodynamic equilibrium constant if $\gamma_{NH_3} = 0.782$, $\gamma_{N_2} = 1.266$, and $\gamma_{H_2} = 1.243$.

8.46 Iodine is sparingly soluble in water but much more so in carbon tetrachloride. The equilibrium constant, also called the partition coefficient, for the distribution of I_2 between these two phases

$$I_2(aq) \rightleftharpoons I_2(CCl_4)$$

is 83 at 20°C. **(a)** A student adds 0.030 L of CCl_4 to 0.200 L of an aqueous solution containing 0.032 g I_2. The mixture is shaken and the two phases are then allowed to separate. Calculate the fraction of I_2 remaining in the aqueous phase. **(b)** The student now repeats the extraction of I_2 with another 0.030 L of CCl_4. Calculate the fraction of I_2 from the original solution that remains in the aqueous phase. **(c)** Compare the result in **(b)** with a single extraction using 0.060 L of CCl_4. Comment on the difference.

8.47 Consider the following equilibrium system

$$N_2O_4(g) \rightleftharpoons 2NO_2(g) \qquad \Delta_r H° = 58.0 \text{ kJ mol}^{-1}$$

(a) If the volume of the reacting system is changed at constant temperature, describe what a plot of P versus $1/V$ would look like for the system. **(b)** If the temperature of the reacting system is changed at constant pressure, describe what a plot of V versus T would look like for the system.

8.48 Use the appropriate equation in this chapter to estimate the vapor pressure of water at 60°C.

8.49 The equilibrium constant (K_P) for the reaction

$$I_2(g) \rightleftharpoons 2I(g)$$

is 1.8×10^{-4} at 872 K and 0.048 at 1173 K. From these data, calculate the bond enthalpy of I_2.

CHAPTER 9

Electrochemistry

Prometheus, they say, brought fire to the service of mankind; electricity we owe to Faraday.

—Sir William Lawrence Bragg

Electrochemical reactions reverse the action of electrolysis. Whereas electrolysis converts electrical energy to chemical energy, electrochemical reactions convert chemical energy directly to electrical energy. There is a convenient difference between electrochemical reactions and chemical reactions: the Gibbs energy change for an electrochemical reaction is equivalent to the maximum electrical work done, which can be measured readily.

In this chapter, we shall discuss the basic principles of electrochemistry and their applications, including membrane potentials.

9.1 Electrochemical Cells

When a piece of zinc metal is placed in a $CuSO_4$ solution, two things happen. Some of the zinc metal enters the solution as Zn^{2+} ions and some of the Cu^{2+} ions are converted to metallic copper at the electrode. This spontaneous redox reaction is represented by

$$\mathrm{Zn}(s) + \mathrm{Cu}^{2+}(aq) \rightarrow \mathrm{Zn}^{2+}(aq) + \mathrm{Cu}(s)$$

In time, the blue of the $CuSO_4$ solution fades. Similarly, if a piece of copper wire is placed in a $AgNO_3$ solution, silver metal is deposited on the copper wire, and the solution gradually turns blue due to the presence of the hydrated Cu^{2+} ions. In each case, nothing will happen if we exchange the roles of the metals involved.

Now, suppose zinc and copper metals are placed in two separate compartments containing $ZnSO_4$ and $CuSO_4$ solutions, respectively, as shown in Figure 9.1. These solutions are connected by a *salt bridge*, a tube that contains an inert electrolyte solution, such as NH_4NO_3 or KCl. This solution is kept from flowing into the compartments by either a sintered disc on each end of the tube or a gelatinous material, such as agar-agar, mixed with the electrolyte solution. When the two electrodes are connected by a piece of metal wire, electrons will flow from the zinc electrode to the copper electrode through the external wire. At the same time, zinc will dissolve in the left compartment, forming Zn^{2+} ions, and Cu^{2+} ions will be converted to metallic copper at the copper electrode. The purpose of the salt bridge is to complete the electrical circuit between the two solutions and to facilitate the movement of ions from one compartment to the other.

Agar-agar is a polysaccharide.

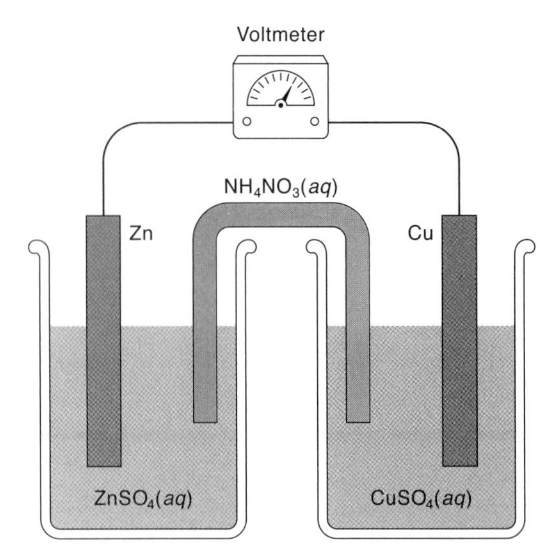

Figure 9.1
Schematic diagram of a galvanic cell. Electrons flow externally from the zinc electrode to the copper electrode. In solution, the anions (SO_4^{2-} and NO_3^-) move toward the zinc anode while the cations (Zn^{2+}, Cu^{2+}, and NH_4^+) move toward the copper cathode.

The setup described above is known as the *Daniell cell*, a type of *galvanic* or *voltaic cell*. The operation of galvanic cells is based on oxidation–reduction, or *redox*, reactions. For the zinc-copper cell, the redox reactions can be expressed in terms of two *half-cell reactions* at the electrodes:

$$\text{Anode:} \qquad Zn(s) \rightarrow Zn^{2+}(aq) + 2e^-$$

$$\text{Cathode:} \quad Cu^{2+}(aq) + 2e^- \rightarrow Cu(s)$$

The zinc electrode is called the *anode*, where oxidation (loss of electrons) takes place; the copper electrode is called the *cathode*, where reduction (gain of electrons) takes place. The *cell diagram* for the Daniell cell is given by

$$Zn(s)|ZnSO_4(1.00\ M)||CuSO_4(1.00\ M)|Cu(s)$$

The single vertical line represents a phase boundary. The double vertical lines denote the salt bridge. By convention, the anode is written first, to the left of the double lines, and the other components appear in the order in which we would encounter them in moving from the anode to the cathode. The concentrations of the solutions are usually indicated in the cell diagram.

The fact that electrons flow from the anode to the cathode means that there is an electrical potential difference between the electrodes. This difference is called the cell voltage, or cell potential. Another common term for cell voltage is the electromotive force or emf (E), which, despite its name, is a measure of voltage, not force. For the Daniell cell, $E = 1.104$ V at 298 K and equal molar concentrations of $CuSO_4$ and $ZnSO_4$.

The emf of a cell is usually measured with a potentiometer (Figure 9.2). A cell S, which has a larger emf than that of any other cell to be measured, is connected to the ends of a uniform wire (AB) of high resistance. The cell under study, X, is connected to A and a galvanometer, which detects the flow of electric current. In a typical experiment, the sliding contact—at the arrow—is moved along AB to a point C, at which the galvanometer registers zero current flow. At this point, the potential from cell S across AC is exactly balanced by the emf of cell X, E_X. The procedure is repeated with cell W, whose emf, E_W, is accurately known. If the new balance point is C′, we write

$$\frac{E_X}{E_W} = \frac{AC}{AC'}$$

Normally, W is a Weston cell ($E_W = 1.018$ V at 298 K), so that

$$E_X = 1.018 \text{ V} \times \frac{AC}{AC'}$$

Accurate measurements of cell emfs are essential for calculating thermodynamic quantities of electrochemical reactions.

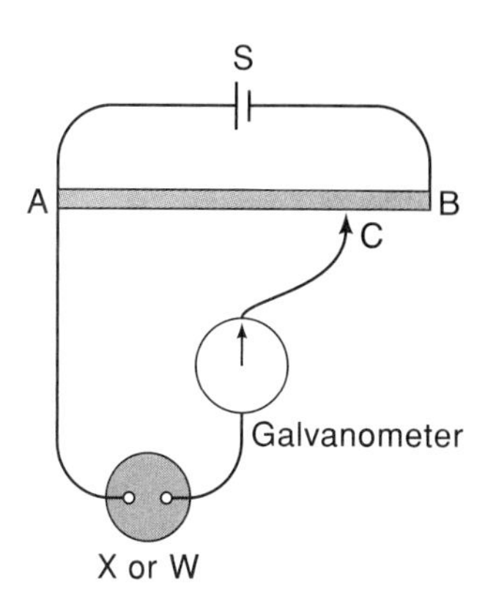

Figure 9.2
Measurement of the emf of a cell using a potentiometer.

9.2 Single-Electrode Potential

Just as monitoring the activity of a single ion is impossible, so is measuring the potential of a single electrode. Any complete circuit must contain two electrodes. By convention, we measure the potential of all electrodes in reference to the *standard hydrogen electrode*, or SHE (Figure 9.3). The potential of the SHE at 298 K, 1 bar H_2 pressure, and 1 *M* H^+ concentration (more correctly, unit activity) is arbitrarily set to be zero; that is,

$$H^+(1\ M) + e^- \rightleftharpoons \tfrac{1}{2}H_2(1 \text{ bar}) \qquad E° = 0 \text{ V}$$

The double arrows mean that the SHE can act as either a cathode or an anode. The measured emf, then, is the potential of the other electrode. Note that we do not need to employ the standard hydrogen electrode for all measurements. As we shall see, it is more convenient to use other electrodes that have been calibrated against the SHE to measure the standard reduction potentials of still other electrodes.

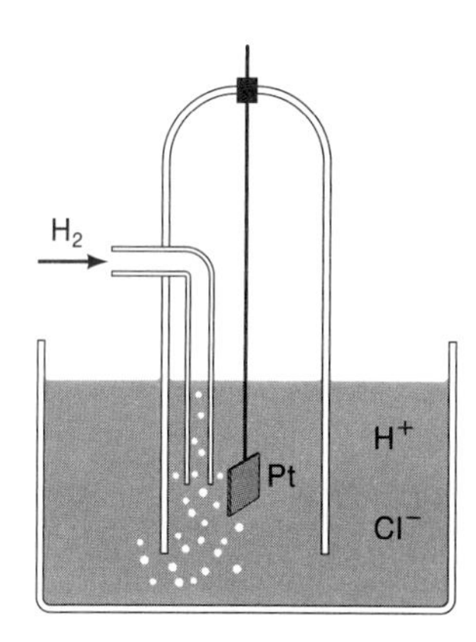

Figure 9.3
Schematic diagram of a hydrogen gas electrode. Hydrogen gas is bubbled into a solution containing H^+ ions. The half-cell redox reaction takes place on the platinum metal immersed in the solution.

Table 9.1 lists the *standard reduction potentials*, $E°$, for some common half-cell reactions. The more positive the reduction potential, the greater the strength of the oxidizing agent. Thus, F_2 is the strongest oxidizing agent because it has the greatest tendency to pick up electrons, and F^- is the weakest reducing agent. The weakest oxidizing agent is Li^+, which makes lithium metal the most powerful reducing agent. The standard reduction potentials are measured at 298 K for aqueous solutions in which the concentration of each dissolved species is at 1 *M* and the gas is at 1 bar.

Keep in mind that electrode potential is an intensive property, the value of which depends only on the type of substance, concentration, and temperature, and not on the size of the electrode or on the amount of solution present. Furthermore, the half-cell reactions are reversible. Depending on the conditions, any electrode can act either as

Table 9.1
Standard Reduction Potentials, $E°$, for Half-Cells at 298 K (pH = 0)[a]

Electrode	Electrode reaction	$E°$/V
$Pt\|F_2\|F^-$	$F_2(g) + 2e^- \rightarrow 2F^-$	+2.87
$Pt\|Co^{3+}, Co^{2+}$	$Co^{3+} + e^- \rightarrow Co^{2+}$	+1.92
$Pt\|Ce^{4+}, Ce^{3+}$	$Ce^{4+} + e^- \rightarrow Ce^{3+}$	+1.72
$Pt\|MnO_4^-, Mn^{2+}$	$MnO_4^- + 8H^+ + 5e^- \rightarrow Mn^{2+} + 4H_2O$	+1.507
$Pt\|Mn^{3+}, Mn^{2+}$	$Mn^{3+} + e^- \rightarrow Mn^{2+}$	+1.54
$Au\|Au^{3+}$	$Au^{3+} + 3e^- \rightarrow Au$	+1.498
$Pt\|Cl_2\|Cl^-$	$Cl_2(g) + 2e^- \rightarrow 2Cl^-$	+1.36
$Pt\|Cr_2O_7^{2-}, Cr^{3+}$	$Cr_2O_7^{2-} + 14H^+ + 6e^- \rightarrow 2Cr^{3+} + 7H_2O$	+1.23
$Pt\|Tl^{3+}, Tl^+$	$Tl^{3+} + 2e^- \rightarrow Tl^+$	+1.252
$Pt\|O_2, H_2O$	$O_2(g) + 4H^+ + 4e^- \rightarrow 2H_2O$	+1.229
$Pt\|Br_2, Br^-$	$Br_2 + 2e^- \rightarrow 2Br^-$	+1.087
$Pt\|Hg^{2+}, Hg_2^{2+}$	$2Hg^{2+} + 2e^- \rightarrow Hg_2^{2+}$	+0.92
$Hg\|Hg^{2+}$	$Hg^{2+} + 2e^- \rightarrow Hg$	+0.851
$Ag\|Ag^+$	$Ag^+ + e^- \rightarrow Ag$	+0.800
$Pt\|Fe^{3+}, Fe^{2+}$	$Fe^{3+} + e^- \rightarrow Fe^{2+}$	+0.771
$Pt\|I_2, I^-$	$I_2 + 2e^- \rightarrow 2I^-$	+0.536
$Pt\|O_2, OH^-$	$O_2(g) + 2H_2O + 4e^- \rightarrow 4OH^-$	+0.401
$Pt\|Fe(CN)_6^{3-}, Fe(CN)_6^{4-}$	$Fe(CN)_6^{3-} + e^- \rightarrow Fe(CN)_6^{4-}$	+0.36
$Cu\|Cu^{2+}$	$Cu^{2+} + 2e^- \rightarrow Cu$	+0.342
$Pt\|Hg\|Hg_2Cl_2\|Cl^-$	$Hg_2Cl_2 + 2e^- \rightarrow 2Hg + 2Cl^-$	+0.268
$Ag\|AgCl\|Cl^-$	$AgCl + e^- \rightarrow Ag + Cl^-$	+0.222
$Pt\|Sn^{4+}, Sn^{2+}$	$Sn^{4+} + 2e^- \rightarrow Sn^{2+}$	+0.151

[a] Data mostly from *CRC Handbook of Chemistry and Physics*, 94th ed., Taylor-Francis/CRC Press, Boca Raton, FL, 2013.

Electrode	Electrode reaction	$E°$/V
$Pt\|Cu^{2+}, Cu^{+}$	$Cu^{2+} + e^{-} \rightarrow Cu^{+}$	+0.153
$Ag\|AgBr\|Br^{-}$	$AgBr + e^{-} \rightarrow Ag + Br^{-}$	+0.0713
$Pt\|H_2\|H^{+}$	$2H^{+} + 2e^{-} \rightarrow H_2$	0.0
$Pb\|Pb^{2+}$	$Pb^{2+} + 2e^{-} \rightarrow Pb$	−0.126
$Sn\|Sn^{2+}$	$Sn^{2+} + 2e^{-} \rightarrow Sn$	−0.138
$Co\|Co^{2+}$	$Co^{2+} + 2e^{-} \rightarrow Co$	−0.277
$Tl\|Tl^{+}$	$Tl^{+} + e^{-} \rightarrow Tl$	−0.336
$Pb\|PbSO_4\|SO_4^{2-}$	$PbSO_4 + 2e^{-} \rightarrow Pb + SO_4^{2-}$	−0.359
$Cd\|Cd^{2+}$	$Cd^{2+} + 2e^{-} \rightarrow Cd$	−0.403
$Pt\|Cr^{3+}, Cr^{2+}$	$Cr^{3+} + e^{-} \rightarrow Cr^{2+}$	−0.41
$Fe\|Fe^{2+}$	$Fe^{2+} + 2e^{-} \rightarrow Fe$	−0.447
$Zn\|Zn^{2+}$	$Zn^{2+} + 2e^{-} \rightarrow Zn$	−0.762
$Pt\|H_2O\|H_2, OH^{-}$	$2H_2O + 2e^{-} \rightarrow H_2 + 2OH^{-}$	−0.828
$Mn\|Mn^{2+}$	$Mn^{2+} + 2e^{-} \rightarrow Mn$	−1.180
$Al\|Al^{3+}$	$Al^{3+} + 3e^{-} \rightarrow Al$	−1.662
$Mg\|Mg^{2+}$	$Mg^{2+} + 2e^{-} \rightarrow Mg$	−2.372
$Na\|Na^{+}$	$Na^{+} + e^{-} \rightarrow Na$	−2.714
$Ca\|Ca^{2+}$	$Ca^{2+} + 2e^{-} \rightarrow Ca$	−2.868
$Sr\|Sr^{2+}$	$Sr^{2+} + 2e^{-} \rightarrow Sr$	−2.899
$Ba\|Ba^{2+}$	$Ba^{2+} + 2e^{-} \rightarrow Ba$	−2.905
$K\|K^{+}$	$K^{+} + e^{-} \rightarrow K$	−2.931
$Li\|Li^{+}$	$Li^{+} + e^{-} \rightarrow Li$	−3.05

an anode or as a cathode. When we reverse a half-cell reaction, the numerical value of $E°$ remains the same, but its sign changes. For example,

$$Sr^{2+}(aq) + 2e^- \rightarrow Sr(s) \qquad E° = -2.899\ V$$

and

$$Sr(s) \rightarrow Sr^{2+}(aq) + 2e^- \qquad E° = 2.899\ V$$

The standard electrode potential for any electrochemical cell can be readily obtained from the data in Table 9.1. By convention, the emf of a galvanic cell ($E°$) is given by

$$E° = E°_{cathode} - E°_{anode} \tag{9.1}$$

where both $E°_{cathode}$ and $E°_{anode}$ refer to the standard reduction potential. Using the Daniell cell as an example, we write

$$\begin{aligned} &\text{Anode:} \quad Zn(s) \rightarrow Zn^{2+}(aq) + 2e^- \\ &\text{Cathode:} \quad Cu^{2+}(aq) + 2e^- \rightarrow Cu(s) \\ &\text{Overall:} \quad Zn(s) + Cu^{2+}(aq) \rightarrow Zn^{2+}(aq) + Cu(s) \end{aligned}$$

Thus, we calculate the emf of the cell as

$$\begin{aligned} E° &= 0.342\ V - (-0.762\ V) \\ &= 1.104\ V \end{aligned}$$

Finally, note that as the electrode reactions progress, the concentrations in both the anode and the cathode departments change, and the solutes are no longer in their standard-state concentrations. Therefore, the emf of a cell refers only to the initial measured value.

9.3 Thermodynamics of Electrochemical Cells

For us to relate the electrochemical energy associated with a cell to $\Delta_r G$, the cell must behave reversibly in the following manner. If we apply an external potential that is exactly equal but opposite to that of the cell, no reaction occurs within the cell. An infinitesimal decrease or increase in the external potential would lead to either the normal or the reverse cell reaction. Any cell that satisfies these conditions is called a *reversible cell*. The situation described here is analogous to the reversible expansion of a gas discussed in Section 3.1. Under normal conditions, cells never operate reversibly; if they did, no current would ever flow through them. By measuring the emf of a cell, however, we learn what it would take to reverse the reaction. Indeed, at the balance point, the net current flow is zero, as discussed earlier (see Figure 9.2).

Consider an electrochemical cell reaction in which electrons are transferred from one electrode to the other (i.e., from the anode to the cathode). The quantity of charge per mole of reaction, Q (in coulombs), is given by

$$Q = \nu F$$

where ν is the stoichiometric coefficient and F is the Faraday constant, which is the charge carried by 1 mole of electrons; that is,

$$\begin{aligned}\text{Faraday constant} &= \text{charge of electron} \times \text{number of electrons per mole}\\ &= 1.6022 \times 10^{-19}\ \text{C} \times 6.022 \times 10^{23}\ \text{mol}^{-1}\\ &= 96{,}485\ \text{C mol}^{-1}\end{aligned}$$

As mentioned earlier, except for very accurate work, we round the value of the Faraday constant to 96,500 C mol^{-1}. In principle, the total electric current generated by an electrochemical cell can be used for work. The amount of electrical work done is given by the product of potential and charge, $-\nu FE$, where the potential is the emf E (in volts), charge (in coulombs) is given by νF, and units are related by 1 J = 1 V × 1 C. The negative sign indicates that work is done by the cell on the surroundings, in keeping with the convention established in Section 3.1.

For a reversible cell at a given temperature and pressure, $-\nu FE$ is the maximum work done, which is equal to the decrease in the Gibbs energy of the system (see Equation 5.11):

Electrochemical measurements provide the most direct determination of $\Delta_r G$ (or $\Delta_r G°$) for a process.

$$\Delta_r G = -\nu FE \tag{9.2}$$

where $\Delta_r G$ is the difference between the Gibbs energy of the products and that of the reactants. Equation 9.2 may be rewritten as

$$E = \frac{-\Delta_r G}{\nu F} \tag{9.3}$$

Recall that at constant temperature and pressure, the criterion for a reaction to proceed spontaneously in the direction written is that $\Delta_r G$ must be a negative value. According to Equation 9.3, this condition corresponds to a positive emf E. Thus, when E is positive for an electrochemical cell, the cell is galvanic, and the cell reaction proceeds as written. When E for an electrochemical cell reaction as written is negative, the forward reaction is electrolytic. An external voltage larger than E must be supplied to the cell to sustain the nonspontaneous process, which is electrolysis.

A special case of Equation 9.3 is a cell in which all reactants and products are in their standard states. In this case, the emf is called the standard emf $E°$, which is related to the change in the standard Gibbs energy as follows:

$$E° = \frac{-\Delta_r G°}{\nu F} \tag{9.4}$$

Because $\Delta_r G°$ is related to the equilibrium constant by Equation 8.7, we have

$$E° = \frac{RT \ln K}{\nu F} \tag{9.5}$$

or

$$K = e^{\nu FE°/RT} \tag{9.6}$$

Thus, a knowledge of $E°$ enables us to calculate the equilibrium constant of the redox cell reaction.

EXAMPLE 9.1

Predict whether the following reaction would occur spontaneously under standard-state conditions, and calculate the equilibrium constant at 25°C:

$$\mathrm{Sn}(s) + 2\mathrm{Ag}^+(aq) \rightleftharpoons \mathrm{Sn}^{2+}(aq) + 2\mathrm{Ag}(s)$$

ANSWER

The two half-reactions for the overall process are

$$\text{Oxidation:} \quad \mathrm{Sn}(s) \rightarrow \mathrm{Sn}^{2+}(aq) + 2e^-$$

$$\text{Reduction:} \quad 2[\mathrm{Ag}^+(aq) + e^- \rightarrow \mathrm{Ag}(s)]$$

$$\text{Overall:} \quad \mathrm{Sn}(s) + 2\mathrm{Ag}^+(aq) \rightarrow \mathrm{Sn}^{2+}(aq) + 2\mathrm{Ag}(s)$$

In Table 9.1, we find the standard reduction potentials for these half-reactions and use Equation 9.1 to calculate the $E°$ value for the reaction

$$E° = 0.800 \text{ V} - (-0.138 \text{ V})$$
$$= 0.938 \text{ V}$$

Because $E°$ is a positive quantity, the reaction is spontaneous under standard-state conditions.

Because $v = 2$ (two electrons are transferred in the overall reaction), from Equation 9.6, we write

$$K = \exp\left[\frac{(2)(96500 \text{ C mol}^{-1})(0.938 \text{ V})}{(8.314 \text{ J K}^{-1} \text{ mol}^{-1})(298 \text{ K})}\right]$$
$$= 5.4 \times 10^{31}$$

An alternative way to show the spontaneity of the reaction is to write

$$\Delta_r G° = -vFE°$$
$$= -(2)(96500 \text{ C mol}^{-1})(0.938 \text{ V})$$
$$= -1.81 \times 10^5 \text{ C V mol}^{-1}$$
$$= -181 \text{ kJ mol}^{-1}$$

The large negative $\Delta_r G°$ value indicates that the reaction is spontaneous under standard-state conditions.

EXAMPLE 9.2

Based on the following electrode potentials,

$$Fe^{2+}(aq) + 2e^- \rightarrow Fe(s) \qquad (1) \qquad E_1^\circ = -0.447\ \text{V}$$

and

$$Fe^{3+}(aq) + e^- \rightarrow Fe^{2+}(aq) \qquad (2) \qquad E_2^\circ = 0.771\ \text{V}$$

calculate the standard reduction potential for the half-reaction

$$Fe^{3+}(aq) + 3e^- \rightarrow Fe(s) \qquad (3) \qquad E_3^\circ = ?$$

ANSWER

It might appear that because the sum of the first two half-reactions gives Equation 3, E_3° is given by $E_1^\circ + E_2^\circ$, or 0.324 V. This is *not* the case, however. The reason is that emf is not an extensive property, so we cannot set $E_3^\circ = E_1^\circ + E_2^\circ$. On the other hand, the Gibbs energy is an extensive property, so we can add the separate Gibbs energy changes to get the overall Gibbs energy change; that is,

$$\Delta_r G_3^\circ = \Delta_r G_1^\circ + \Delta_r G_2^\circ$$

Substituting the relationship $\Delta_r G^\circ = -\nu F E^\circ$, we obtain

$$\nu_3 F E_3^\circ = \nu_1 F E_1^\circ + \nu_2 F E_2^\circ$$

or

$$E_3^\circ = \frac{\nu_1 E_1^\circ + \nu_2 E_2^\circ}{\nu_3} \qquad (\nu_1 = 2, \nu_2 = 1, \nu_3 = 3)$$

$$= \frac{(2)(-0.447\ \text{V}) + (1)(0.771\ \text{V})}{3}$$

$$= -0.041\ \text{V}$$

Equation 9.4 enables us to determine the E° values of the alkali metals and certain alkaline earth metals that react with water. Suppose we want to determine the standard reduction potential of lithium:

$$Li^+(aq) + e^- \rightarrow Li(s) \qquad E^\circ = ?$$

Although we cannot place a lithium electrode in water because lithium reacts with water to form hydrogen gas and lithium hydroxide, we can imagine the following electrochemical process:

$$Li^+(aq) + \tfrac{1}{2}H_2(g) \rightarrow Li(s) + H^+(aq)$$

This is not a spontaneous reaction.

in which lithium ions are reduced at the lithium electrode, and H_2 molecules are oxidized at the hydrogen electrode. From data in Appendix B, we can calculate the $\Delta_r H^\circ$ and $\Delta_r S^\circ$ values as follows:

$$\begin{aligned}\Delta_r H^\circ &= \Delta_f \overline{H}^\circ[\text{Li}(s)] + \Delta_f \overline{H}^\circ[\text{H}^+(aq)] - \Delta_f \overline{H}^\circ[\text{Li}^+(aq)] - (\tfrac{1}{2})\Delta_f \overline{H}^\circ[\text{H}_2(g)] \\ &= (0) + (0) - (-278.5 \text{ kJ mol}^{-1}) - (\tfrac{1}{2})(0) \\ &= 278.5 \text{ kJ mol}^{-1}\end{aligned}$$

$$\begin{aligned}\Delta_r S^\circ &= \overline{S}^\circ[\text{Li}(s)] + \overline{S}^\circ[\text{H}^+(aq)] - \overline{S}^\circ[\text{Li}^+(aq)] - (\tfrac{1}{2})\overline{S}^\circ[\text{H}_2(g)] \\ &= 28.03 \text{ J K}^{-1}\text{ mol}^{-1} + (0) - (14.23 \text{ J K}^{-1}\text{ mol}^{-1}) - (\tfrac{1}{2})(130.6 \text{ J K}^{-1}\text{ mol}^{-1}) \\ &= -51.5 \text{ J K}^{-1}\text{ mol}^{-1}\end{aligned}$$

Thus, at 298 K,

$$\begin{aligned}\Delta_r G^\circ &= \Delta_r H^\circ - T\Delta_r S^\circ \\ &= 278.5 \text{ kJ mol}^{-1} - (298 \text{ K})\left(\frac{-51.5 \text{ J K}^{-1}\text{ mol}^{-1}}{1000 \text{ J/kJ}}\right) \\ &= 293.8 \text{ kJ mol}^{-1}\end{aligned}$$

Finally, from Equation 9.4

Note that we are using a more accurate value for the Faraday constant here.

$$\begin{aligned}E^\circ &= \frac{-\Delta_r G^\circ}{\nu F} \\ &= \frac{-293.8 \times 1000 \text{ J mol}^{-1}}{96485 \text{ C mol}^{-1}} \\ &= -3.05 \text{ V}\end{aligned}$$

By the same procedure, we can calculate the E° values of other reactive metals and fluorine (F_2), which also reacts with water (see Problem 9.32).

The Nernst Equation

We can now derive an equation relating the emf of a cell to variables such as temperature and the concentrations of reacting species. The Gibbs energy change for the cell reaction

$$a\text{A} + b\text{B} \rightarrow c\text{C} + d\text{D}$$

is given by (see Equation 8.6)

$$\Delta_r G = \Delta_r G^\circ + RT \ln \frac{a_\text{C}^c a_\text{D}^d}{a_\text{A}^a a_\text{B}^b}$$

where a denotes activity. Dividing this equation throughout by $-\nu F$ and using both Equations 9.3 and 9.4 we obtain

$$E = E^\circ - \frac{RT}{\nu F} \ln \frac{a_C^c a_D^d}{a_A^a a_B^b} \tag{9.7}$$

Equation (9.7) is known as the *Nernst equation*, after the German chemist Walther Hermann Nernst (1864–1941). Here, E is the observed emf of the cell and E° is the standard emf of the cell, that is, the emf when all the reactants and products are in their unit activity standard states. At equilibrium, $E = 0$, so

$$E^\circ = \frac{RT}{\nu F} \ln K = \frac{-\Delta_r G^\circ}{\nu F}$$

Because most electrochemical cells operate at or near room temperature, we can evaluate the quantity RT/F by taking $R = 8.314$ J K^{-1} mol^{-1}, $T = 298$ K, and $F = 96{,}500$ C mol^{-1}. Therefore,

$$\frac{(8.314 \text{ J K}^{-1} \text{ mol}^{-1})(298 \text{ K})}{96500 \text{ C mol}^{-1}} = 0.0257 \text{ J C}^{-1}$$

$$= 0.0257 \text{ V}$$

Finally, Equation 9.7 can be expressed as

$$E = E^\circ - \frac{0.0257 \text{ V}}{\nu} \ln \frac{a_C^c a_D^d}{a_A^a a_B^b} \tag{9.8}$$

EXAMPLE 9.3

Predict whether the following reaction would proceed spontaneously as written at 298 K:

$$Cd(s) + Fe^{2+}(aq) \rightarrow Cd^{2+}(aq) + Fe(s)$$

given that $[Cd^{2+}] = 0.15$ *M* and $[Fe^{2+}] = 0.68$ *M*.

ANSWER

The half-cell reactions are

Anode: $Cd(s) \rightarrow Cd^{2+}(aq) + 2e^-$

Cathode: $Fe^{2+}(aq) + 2e^- \rightarrow Fe(s)$

From Equation 9.1 and Table 9.1, we write

$$E^\circ = -0.447 \text{ V} - (-0.403 \text{ V})$$

$$= -0.044 \text{ V}$$

Assuming ideal behavior and noting that the activities of the solids are unity, the Nernst equation for this reaction is

$$E = -0.044\ \text{V} - \frac{0.0257\ \text{V}}{2}\ln\frac{[\text{Cd}^{2+}]}{[\text{Fe}^{2+}]}$$

$$= -0.044\ \text{V} - \frac{0.0257\ \text{V}}{2}\ln\frac{0.15\ M}{0.68\ M}$$

$$= -0.025\ \text{V}$$

Because E is negative, the reaction is not spontaneous as written, and the cell reaction must be

$$\text{Fe}(s) + \text{Cd}^{2+}(aq) \rightarrow \text{Fe}^{2+}(aq) + \text{Cd}(s)$$

At what ratio of $[\text{Cd}^{2+}]$ to $[\text{Fe}^{2+}]$ does the reaction in Example 9.3 become spontaneous as written? To find out, we first set E equal to zero, which corresponds to the equilibrium situation, and we write the Nernst equation as

$$0 = -0.044\ \text{V} - \frac{0.0257\ \text{V}}{2}\ln\frac{[\text{Cd}^{2+}]}{[\text{Fe}^{2+}]}$$

or

$$\frac{[\text{Cd}^{2+}]}{[\text{Fe}^{2+}]} = 0.033 = K$$

Thus, the reaction will be spontaneous if the ratio $[\text{Cd}^{2+}]/[\text{Fe}^{2+}]$ is smaller than 0.033, so that E is positive.

Temperature Dependence of EMF

Thermodynamic values for a cell reaction can be obtained from the temperature dependence of the emf. Starting with

$$\Delta_r G^\circ = -\nu F E^\circ$$

and differentiating $\Delta_r G^\circ$ with respect to temperature at constant pressure, we get

$$\left(\frac{\partial \Delta_r G^\circ}{\partial T}\right)_P = -\nu F\left(\frac{\partial E^\circ}{\partial T}\right)_P$$

Equation 5.13 when expressed in terms of changes in G and S, is given by

$$\left(\frac{\partial \Delta_r G^\circ}{\partial T}\right)_P = -\Delta_r S^\circ$$

so that

$$\Delta_r S^\circ = \nu F\left(\frac{\partial E^\circ}{\partial T}\right)_P \tag{9.9}$$

Thus, from the variation of E° with temperature,* we can measure the standard entropy change of a cell reaction. Suppose that we want to determine the value of $(\partial E^\circ/\partial T)_P$ for the Daniel cell. The simplest way is to set $[Zn^{2+}] = 1.00\ M$ and $[Cu^{2+}] = 1.00\ M$ (the standard states) and then measure the emf of the cell at several different temperatures. Once $\Delta_r S^\circ$ and $\Delta_r G^\circ$ are known at a certain temperature, we can calculate the value of $\Delta_r H^\circ$ as follows:

$$\Delta_r G^\circ = \Delta_r H^\circ - T\Delta_r S^\circ$$

or

$$\Delta_r H^\circ = \Delta_r G^\circ + T\Delta_r S^\circ$$

$$= -\nu F E^\circ + \nu F T\left(\frac{\partial E^\circ}{\partial T}\right)_P \tag{9.10}$$

In general, $\Delta_r H^\circ$ and $\Delta_r S^\circ$ are roughly temperature independent (for a range of 50 K or smaller), but $\Delta_r G^\circ$, as we know, varies with temperature. Note that Equation 9.10 provides us with a noncalorimetric way of determining the enthalpy change of a reaction.

9.4 Types of Electrodes

Depending on the type of redox reactions that occur in an electrochemical cell, there are many different types of electrodes in use. The following are a few examples.

Metal Electrodes

A metal electrode consists of a piece of metal immersed in a solution containing cations of the same metal. The electrode reactions are

$$M^{z+}(aq) + ze^- \rightleftharpoons M(s)$$

where z is the number of positive charges on the metal ion. The Daniell cell employs metal electrodes (Zn and Cu). As shown earlier, metals that react with water, such as the alkali metals and certain alkaline-earth metals (Ca, Sr, and Ba), cannot be used as electrodes. For these metals, standard reduction potentials (see Table 9.1) are obtained by first determining the $\Delta_r H^\circ$ and $\Delta_r S^\circ$ values for the cell reaction, which enable us to calculate the $\Delta_r G^\circ$ value at 298 K and finally the value of E°, according to Equation 9.4.

* The temperature dependence of the emf of most car batteries is generally quite small, of the order of 5×10^{-4} V K^{-1}, which is insufficient to explain why cars will not start on a cold morning. For an interesting explanation of the real cause, see L. K. Nash, *J. Chem. Educ.* **47**, 382 (1970).

Gas Electrodes

An example of a gas electrode is the standard hydrogen electrode discussed earlier (see Figure 9.3) and is represented by $Pt|H_2(g)|H^+(aq)$. The inert platinum metal has a dual purpose: it acts as a catalyst in the decomposition of H_2 into atomic hydrogen (or the recombination of H atoms to form H_2) and as an electrical conductor to the external circuit. The electrode reactions are

$$2H^+(aq) + 2e^- \rightleftharpoons H_2(g)$$

Other gas electrodes include the chlorine electrode and the oxygen electrode.

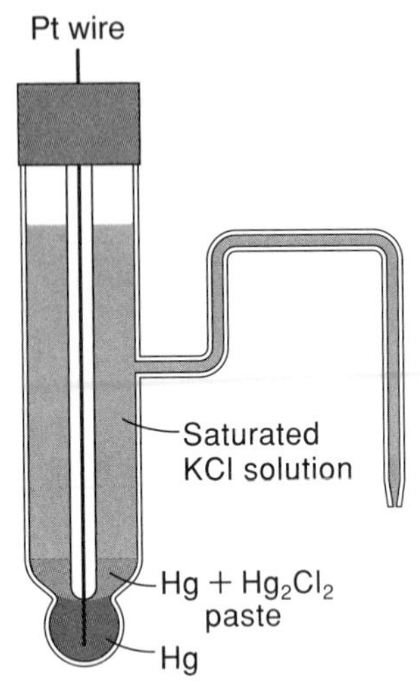

Figure 9.4
A saturated calomel electrode. The end portion of the vertical sidearm contains agar-agar to prevent the flow of the KCl solution into the cell compartment (not shown).

Metal-Insoluble Salt Electrodes

A metal-insoluble salt electrode can be prepared by coating a piece of metal with an insoluble salt of the same metal. The electrode is immersed in a solution containing the anion of the salt. A common example is the silver–silver chloride electrode, $Ag(s)|AgCl(s)|Cl^-(aq)$, where the Cl^- ions are supplied by KCl or HCl. The electrode reactions are

$$AgCl(s) + e^- \rightleftharpoons Ag(s) + Cl^-(aq)$$

Perhaps the most common electrode of this type is the *calomel electrode*, shown in Figure 9.4. The electrode consists of metallic mercury in contact with calomel, or mercury(I) chloride (Hg_2Cl_2), which is in contact with a solution containing Cl^- ions (from KCl). The electrode is represented as $Pt(s)|Hg(l)|Hg_2Cl_2(s)|Cl^-(aq)$, and the electrode reactions are

$$Hg_2Cl_2(s) + 2e^- \rightleftharpoons 2Hg(l) + 2Cl^-(aq)$$

If the solution is saturated with KCl, as is often the case, the electrode is called a *saturated calomel electrode*. For this type of electrode, the concentration of the Cl^- ion is fixed at a given temperature. The saturated calomel electrode is useful as a reference electrode in electrochemical work. Once it has been calibrated against the SHE, it can be used to determine the standard reduction potentials of many other electrodes.

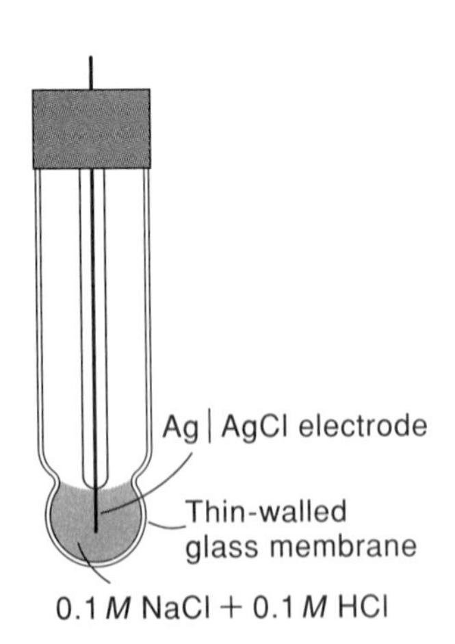

Figure 9.5
A glass electrode.

The Glass Electrode

One of the most widely used electrodes, the glass electrode is an example of an *ion-selective electrode*, because it is specific for H^+ ions. The essential features of a glass electrode are shown in Figure 9.5. The electrode consists of a very thin bulb or membrane made of a special type of glass that is permeable to H^+ ions. An Ag|AgCl electrode is immersed in a buffer solution (constant pH) containing Cl^- ions. When the electrode is placed in a solution whose pH is different from that of the buffer solution, the potential difference that develops between the two sides is a measure of the difference in the two pH values.*

* See R. A. Durst, *J. Chem. Educ.* **44**, 175 (1967) and M. Dole, *ibid*, **57**, 134 (1980) for a detailed discussion of the glass electrode.

Ion-Selective Electrodes

There are a variety of electrodes besides the glass electrode that are specific for cations such as Li^+, Na^+, K^+, Ca^{2+}, NH_4^+, Ag^+, and Cu^{2+} and for anions such as the halides, S^{2-}, and CN^-. Because these specific electrodes lend themselves to easy and accurate emf measurements, they are handy analytical tools in fields ranging from medicine to environmental studies. The operational details of these electrodes will not be given here. The interested reader should consult the references listed at the end of the chapter.

9.5 Types of Electrochemical Cells

The galvanic cell discussed earlier is one of several types of electrochemical cells in use. Two examples we shall consider here are concentration cells and fuel cells.

Concentration Cells

A concentration cell contains electrodes made of the same metal and solutions containing the same ions but at different concentrations. An example is the $ZnSO_4$ concentration cell:

$$\mathrm{Zn}(s)|\mathrm{ZnSO_4}(0.10\ M)||\mathrm{ZnSO_4}(1.0\ M)|\mathrm{Zn}(s)$$

The electrode reactions are given by

$$\begin{aligned} &\text{Anode:} \quad \mathrm{Zn}(s) \rightarrow \mathrm{Zn^{2+}}(0.10\ M) + 2e^- \\ &\text{Cathode:} \quad \mathrm{Zn^{2+}}(1.0\ M) + 2e^- \rightarrow \mathrm{Zn}(s) \\ \hline &\text{Overall:} \quad \mathrm{Zn^{2+}}(1.0\ M) \rightarrow \mathrm{Zn^{2+}}(0.10\ M) \end{aligned}$$

Note that the cathode compartment in a concentration cell always contains the more concentrated solution because it has a greater tendency to accept electrons.

This is a dilution process, so as the electrode reactions progress, the anode compartment becomes more concentrated in Zn^{2+} ions while the cathode department becomes more diluted in Zn^{2+} ions. Eventually, when the concentrations in the two compartments are the same, the cell ceases to function. The initial emf of the cell at 298 K is

$$\begin{aligned} E &= E^\circ - \frac{RT}{\nu F}\ln\frac{[\mathrm{Zn^{2+}}]_{\mathrm{dil}}}{[\mathrm{Zn^{2+}}]_{\mathrm{conc}}} \\ &= 0 - \frac{0.0257\ \mathrm{V}}{2}\ln\frac{0.10\ M}{1.0\ M} \\ &= 0.030\ \mathrm{V} \end{aligned}$$

Note that because the same electrode is used in the cell, E° is zero in the Nernst equation. The concentration cell generally produces small emfs and has no practical use. The concept of its operation is important in the study of membrane potentials, however, as we shall see later.

Fuel Cells

Fossil fuel is a major source of energy at present. Unfortunately, the combustion of fossil fuel is a highly irreversible process, and so its thermodynamic efficiency is low. Fuel cells, however, can make combustion largely reversible by converting a greater amount of chemical energy to useful work. Furthermore, they do not operate like a heat engine and therefore are not subject to the same kind of thermodynamic limitations in energy conversion (see Equation 4.9).

Consider the hydrogen–oxygen fuel cell, the simplest example. Such a cell consists of an electrolyte solution, such as sulfuric acid or sodium hydroxide, and two inert electrodes. Hydrogen and oxygen gases are bubbled through the anode and cathode compartments, where the following reactions take place:

$E°$ of the cell is 1.229 V at 298 K.

$$\text{Anode:}\quad H_2(g) + 2OH^-(aq) \rightarrow 2H_2O(l) + 2e^-$$

$$\text{Cathode:}\quad \tfrac{1}{2}O_2(g) + H_2O(l) + 2e^- \rightarrow 2OH^-(aq)$$

$$\text{Overall:}\quad H_2(g) + \tfrac{1}{2}O_2(g) \rightarrow H_2O(l)$$

As you can see, the overall reaction is the same as the combustion of hydrogen gas in air. A potential difference is established between the two electrodes, and electrons flow from the anode to the cathode through the wire connecting the two electrodes.

The function of the electrodes is twofold. First, the anode acts as a source, the cathode as a sink, of electrons. Second, the electrodes provide the necessary surface for the initial decomposition of the molecules into atomic species. They are *electrocatalysts*. Metals such as platinum, iridium, and rhodium are good electrocatalysts.

Another type of fuel cell is the propane–oxygen fuel cell shown in Figure 9.6. The half-cell reactions are

$$\text{Anode:}\quad C_3H_8(g) + 6H_2O(l) \rightarrow 3CO_2(g) + 20H^+(aq) + 20e^-$$

$$\text{Cathode:}\quad 5O_2(g) + 20H^+(aq) + 20e^- \rightarrow 10H_2O(l)$$

$$\text{Overall:}\quad C_3H_8(g) + 5O_2(g) \rightarrow 3CO_2(g) + 4H_2O(l)$$

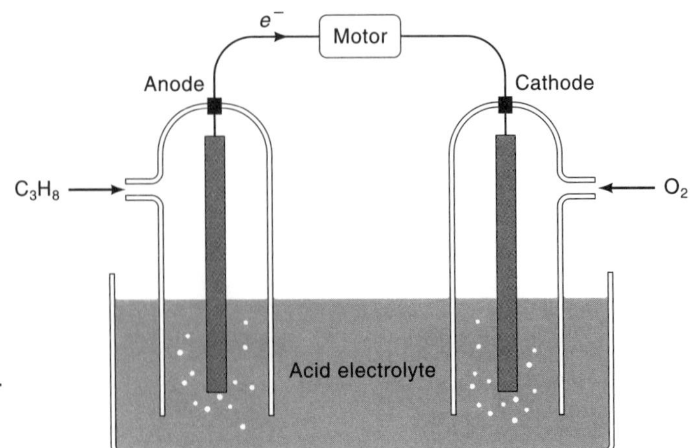

Figure 9.6
Schematic diagram of a propane–oxygen fuel cell.

The overall reaction is identical to burning propane in oxygen. With proper design, the efficiency of a propane-oxygen fuel cell may be as high as 70%, about twice that of an internal combustion engine. In addition, fuel cells generate electricity without the noise, vibration, heat transfer, and other problems normally associated with conventional power plants. The advantages are so attractive that fuel cells are most likely to become operational on a large scale in the near future. At present, much effort is being spent in search of suitable electrocatalysts for various gases.

9.6 Applications of EMF Measurements

We shall now discuss two important applications of emf measurements.

Determination of Activity Coefficients

Emf measurements provide one of the most convenient and accurate methods for determining the activity coefficient of ions. As an example, let us consider the following cell arrangement:

$$\mathrm{Pt}|\mathrm{H_2}(1\ \mathrm{bar})|\mathrm{HCl}(m)|\mathrm{AgCl}(s)|\mathrm{Ag}$$

The overall reaction for the cell is

$$\tfrac{1}{2}\mathrm{H_2}(g) + \mathrm{AgCl}(s) \rightarrow \mathrm{Ag}(s) + \mathrm{H^+}(aq) + \mathrm{Cl^-}(aq)$$

and the emf of the cell at 298 K is given by

$$E = E^\circ - 0.0257\ \mathrm{V}\ \ln \frac{a_{\mathrm{H^+}} a_{\mathrm{Cl^-}} a_{\mathrm{Ag}}}{f_{\mathrm{H_2}}^{1/2}\, a_{\mathrm{AgCl}}}$$

Because both Ag and AgCl are solids, their activities are unity; the fugacity of hydrogen gas at 1 bar is also approximately unity, so the preceding equation reduces to

$$E = E^\circ - 0.0257\ \mathrm{V}\ \ln a_{\mathrm{H^+}} a_{\mathrm{Cl^-}}$$

From Equation 7.21, we find that

$$\begin{aligned} a_{\mathrm{H^+}} a_{\mathrm{Cl^-}} &= \gamma_\pm^2 m_\pm^2 \\ &= \gamma_\pm^2 m^2 \end{aligned}$$

For a 1:1 electrolyte such as HCl, $m_\pm = m$. Thus, the emf of the cell can be expressed as

$$\begin{aligned} E &= E^\circ - 0.0257\ \mathrm{V}\ \ln(\gamma_\pm m)^2 \\ &= E^\circ - 0.0514\ \mathrm{V}\ \ln m - 0.0514\ \mathrm{V}\ \ln \gamma_\pm \end{aligned}$$

The above equation can be rearranged to give

$$E + 0.0514\ \mathrm{V}\ \ln m = E^\circ - 0.0514\ \mathrm{V}\ \ln \gamma_\pm$$

By measuring E over a range of molalities of HCl, the quantity (E + 0.0514 V ln m) can be calculated at various molalities. If we plot (E + 0.0514 V ln m) against m and extrapolate to zero m, we determine the value of $E°$ because at zero m, $\gamma_\pm$ is unity, so ln $\gamma_\pm = 0$. Once we know the value of $E°$, the mean activity coefficient can be found at a particular value of m (see Problem 9.34).

Determination of pH

Determining pH from emf measurements is a standard technique. Because using the hydrogen electrode itself for this purpose is impractical, a combination of the glass electrode and saturated calomel electrode provides the most suitable arrangement. Consider the cell (also see Figure 9.4)

$$\underbrace{\mathrm{Ag}(s)|\mathrm{AgCl}(s)|\mathrm{HCl}(aq),\ \mathrm{NaCl}(aq)}_{\text{glass electrode}}\ \underbrace{|\mathrm{HCl}(aq)|}_{\text{solution of unknown pH}}\ \underbrace{\mathrm{Cl}^-(aq)|\mathrm{Hg_2Cl_2}(s)|\mathrm{Hg}(l)|\mathrm{Pt}(s)}_{\text{saturated calomel electrode}}$$

The overall emf E for this arrangement at 298 K is given by

We change from ln to log because of the definition of pH.

$$E = E_{\mathrm{ref}} - 0.0591\ \mathrm{V}\log a_{\mathrm{H^+}}$$
$$= E_{\mathrm{ref}} + 0.0591\ \mathrm{V\ pH}$$

where pH $= -\log a_{\mathrm{H^+}}$ and E_{ref} is the standard electrode potential difference between the glass electrode and the calomel electrode. In practice, we can replace $a_{\mathrm{H^+}}$ with $[\mathrm{H^+}]$, except in very precise work. Rearranging the above equation, we get

$$\mathrm{pH} = \frac{E - E_{\mathrm{ref}}}{0.0591\ \mathrm{V}}$$

We can determine the value of E_{ref} by measuring E for a number of solutions of accurately known pH. Once E_{ref} is known, the combination of the glass electrode and the calomel electrode can be used to find the pH of other solutions from values of E. The practical arrangement of this combination is called a pH meter.

9.7 Membrane Potential

Electrical potentials exist across the membranes of various kinds of cells. Some cells, such as nerve cells and muscle cells, are said to be excitable because they are capable of transmitting a change of potential along their membranes. In this section, we shall briefly discuss the nature of membrane potentials.

A human nerve cell consists of a cell body and a single long fiber extension approximately 10^{-5} to 10^{-3} cm in diameter, called the *axon*, which transmits impulses from the cell body to the adjacent nerve cell (Figure 9.7). Table 9.2 shows the ion distribution of a typical nerve cell. The membrane of the axon is similar in structure to other cell membranes and similar in composition to the fluid in the cell body. The electrical potential established by the difference in ionic concentrations across the membrane is known as the *membrane potential.*

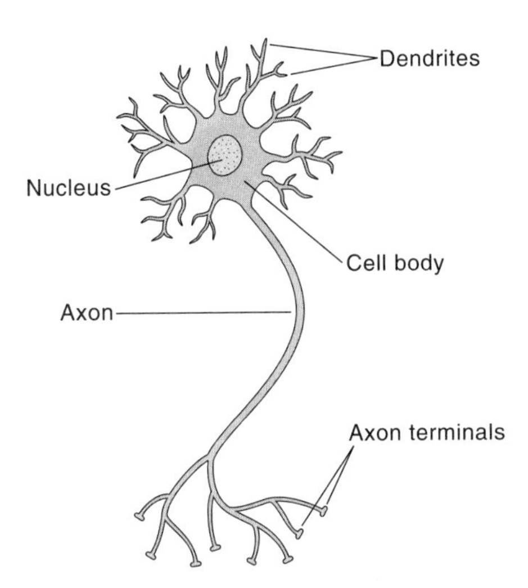

Figure 9.7
Schematic diagram of a neuron (nerve cell), made up of a cell body, an axon, and dendrites. The dendrites carry unidirectional nerve impulses from other neurons toward the cell body. The axon transmits impulses to the adjacent neurons.

To understand how membrane potentials arise, let us consider the simple chemical systems shown in Figure 9.8. Figure 9.8a depicts two KCl solutions, both at a concentration of 0.01 *M*, separated by a membrane permeable to K^+ but not to Cl^- ions, so K^+ diffuses across the membrane without its counter ion (Cl^-). Because the concentration in the two compartments is the same, the net transport of K^+ ions in either direction is zero and so is the electric potential across the membrane. The arrangement in Figure 9.8b shows that the concentration in the left compartment is 10 times that in the right compartment. In this case, more K^+ will diffuse from left to right, producing an increase in positive charge on the right and establishing a potential difference across the membrane. The movement of K^+ ions ceases when the extra positive charges in the right compartment begin repelling additional positive charges, while the electrostatic attraction of the excess negative charges in the left compartment holds K^+ ions there. The potential difference due to charge separation across the membrane at

Table 9.2
Distribution of Major Ions on Opposite Sides of the Membrane of a Typical Nerve Cell

	Concentration/m*M*	
Ion	Intracellular	Extracellular
Na^+	15	150
K^+	150	5
Cl^-	10	110

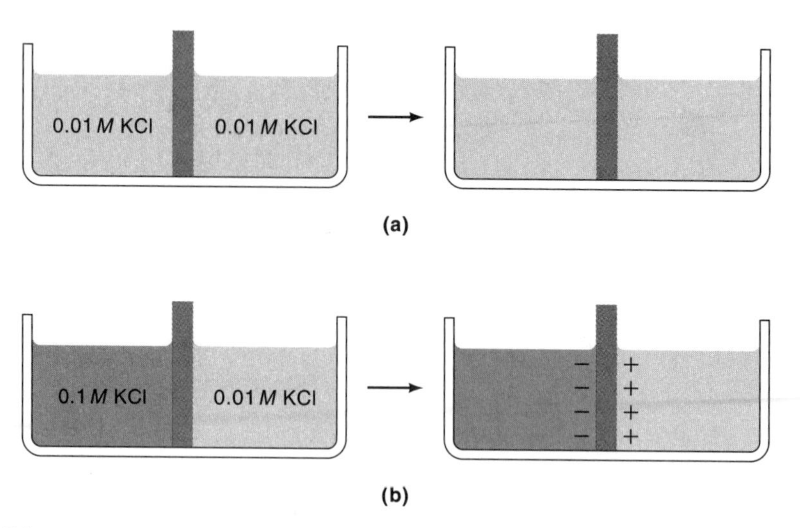

Figure 9.8
Two compartments are separated by a membrane permeable only to K^+ ions. (a) Because the concentrations in the two compartments are equal, there is no net flow of ions across the membrane, and no electrical potential. (b) A difference in concentration causes K^+ ions to move from the left compartment to the right one. At equilibrium, an electric potential is established across the membrane due to an accumulation of negative charges on the left side and positive charges on the right. Only a small fraction of the K^+ ions take part in establishing the membrane potential.

equilibrium is the equilibrium membrane potential, or simply the membrane potential, of K^+ ions.*

We can calculate the membrane potential for K^+ ions as follows. The Nernst equation for a single type of ion at 298 K is

$$E_{K^+} = E^\circ_{K^+} - \frac{0.0257\text{ V}}{\nu}\ln[K^+]$$

By convention, the electric potential inside a nerve cell (or other living cells), E_{in}, is expressed relative to the potential outside the cell, E_{ex}. That is, the membrane potential is given as $E_{in} - E_{ex}$. Because $\nu = 1$, we write the membrane potential of the K^+ ions, ΔE_{K^+}, as

$$\Delta E_{K^+} = E_{K^+,in} - E_{K^+,ex} = 0.0257\text{ V}\ln\frac{[K^+]_{ex}}{[K^+]_{in}}$$

From Table 9.2, we obtain

$$\Delta E_{K^+} = 0.0257\text{ V}\ln\frac{5\text{ m}M}{150\text{ m}M} = -8.7\times10^{-2}\text{ V} = -87\text{ mV}$$

Using an arrangement like the one shown in Figure 9.9, we find that the membrane potential of a nerve cell is only about −70 mV. One reason for the discrepancy is that

* If the membrane is impermeable to certain ions (Cl^- in this case), then their presence will have no effect on the membrane potential.

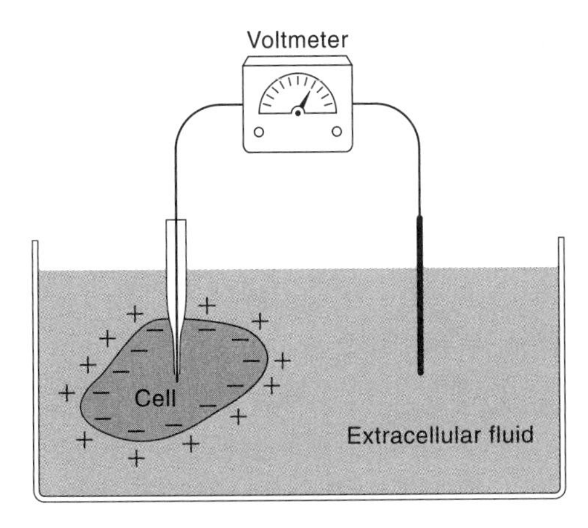

Figure 9.9
Arrangement for measuring the membrane potential of a cell.

there is also a membrane potential due to the presence of Na^+ ions. Because the concentration of Na^+ ions is much higher outside the cell, the movement of Na^+ ions into the cell makes the inside more positive. Referring again to Table 9.2, we write

$$\Delta E_{Na^+} = 0.0257 \text{ V} \ln \frac{[Na^+]_{ex}}{[Na^+]_{in}}$$

$$= 0.0257 \text{ V} \ln \frac{150 \text{ m}M}{15 \text{ m}M} = 5.9 \times 10^{-2} \text{ V} = 59 \text{ mV}$$

Because the membrane is much more permeable to K^+ ions than to Na^+ ions, however, the measured potential is closer to the membrane potential of K^+ ions.

Recall that the experimentally measured membrane potential is not equal to the K^+ membrane potential because some Na^+ ions continually move into the cell, canceling the effect of K^+ ions that are simultaneously moving out. If such net ion movements occur, why does the concentration of intracellular Na^+ not progressively increase and that of intracellular K^+ progressively decrease? The reason is that there is a specific membrane protein called $Na^+ - K^+$ ATPase that transports Na^+ out of the cell and K^+ into it using energy from ATP hydrolysis.

The Goldman Equation

The Nernst equation enables us to calculate membrane potential for only one ionic species at a time; it is not applicable for several types of ions that are distributed in unequal concentrations across the same membrane. To calculate membrane potential in such cases, we must employ the Goldman equation (after the American biophysicist David Eliot Goldman, 1910–1998), a generalization of the Nernst equation that has been extended to include the relative permeability of each ionic species. Applied to a nerve cell at 298 K, the Goldman equation takes the form

$$E_m = 0.0257 \text{ V} \ln \frac{[K^+]_{ex} P_{K^+} + [Na^+]_{ex} P_{Na^+} + [Cl^-]_{ex} P_{Cl^-}}{[K^+]_{in} P_{K^+} + [Na^+]_{in} P_{Na^+} + [Cl^-]_{in} P_{Cl^-}} \quad (9.11)$$

P_{ion} has units of meters per second, but we are usually more interested in the relative values of P_{ion}.

where E_m is the membrane potential and P_{ion} is the permeability of the membrane to an ion. Nerve cell membranes in the resting (unperturbed) state are about 100 times more permeable to K^+ ions than to Na^+ ions and are nearly impermeable to Cl^- ions, that is, $P_{Cl^-} \approx 0$. Under these conditions, Equation 9.11 becomes

$$E_m = 0.0257 \text{ V} \ln \frac{[K^+]_{ex} P_{K^+} + [Na^+]_{ex} P_{Na^+}}{[K^+]_{in} P_{K^+} + [Na^+]_{in} P_{Na^+}}$$

$$= 0.0257 \text{ V} \ln \frac{[K^+]_{ex} P_{K^+}/P_{Na^+} + [Na^+]_{ex}}{[K^+]_{in} P_{K^+}/P_{Na^+} + [Na^+]_{in}} \quad (9.12)$$

Because $P_{K^+}/P_{Na^+} \approx 100$, we write

$$E_m = 0.0257 \text{ V} \ln \frac{5 \text{ m}M \times 100 + 150 \text{ m}M}{150 \text{ m}M \times 100 + 15 \text{ m}M}$$

$$= -81 \text{ mV}$$

This value is closer to the experimentally determined membrane potential.

The Action Potential

If a nerve cell is stimulated electrically, chemically, or mechanically, the cell membrane becomes much more permeable to Na^+ ions than to K^+ ions so $P_{K^+}/P_{Na^+} \approx 0.17$. The permeability of the membrane to K^+ ions does not change very much at first but there is a 600-fold increase in the permeability to Na^+ ions. Nerve-cell stimulation causes a small fraction of the Na^+ ions to rush into the cell, resulting in change in the membrane potential (the membrane is said to be *depolarized*). From Equation 9.12, we write

$$E_m = 0.0257 \text{ V} \ln \frac{5 \text{ m}M \times 0.17 + 150 \text{ m}M}{150 \text{ m}M \times 0.17 + 15 \text{ m}M}$$

$$= 34 \text{ mV}$$

During a very short time (less than 1 ms), the membrane potential changes from −70 mV to about 35 mV (inside positive) and then rapidly returns to its original value (Figure 9.10). The sudden spike in the membrane potential is called the *action potential.*

What causes the membrane potential to return so quickly to its resting value? Two factors are involved. First, the increased Na^+ permeability is rapidly turned off after the initial influx of Na^+ ions into the cell. Second, the membrane's permeability to K^+ ions increases relative to its resting value over a short time (about 1 ms). For this reason, the membrane potential actually dips below −70 mV initially before it returns to its normal value (see Figure 9.10) at which point the cell is ready to "fire" again when it receives another signal. The small number of excess Na^+ ions present in the cell are eventually pumped out of the cell.

The events that give rise to an action potential occur in and around a small area of the nerve cell membrane. The action potential is then propagated along the axon of a neuron. Looking at Figure 9.7, you may be tempted to assume that axons act like electrical cables. After all, the axon does have a cablelike structure. It has a core of electrolytic solution, and it is surrounded by a membrane that acts as an electrical

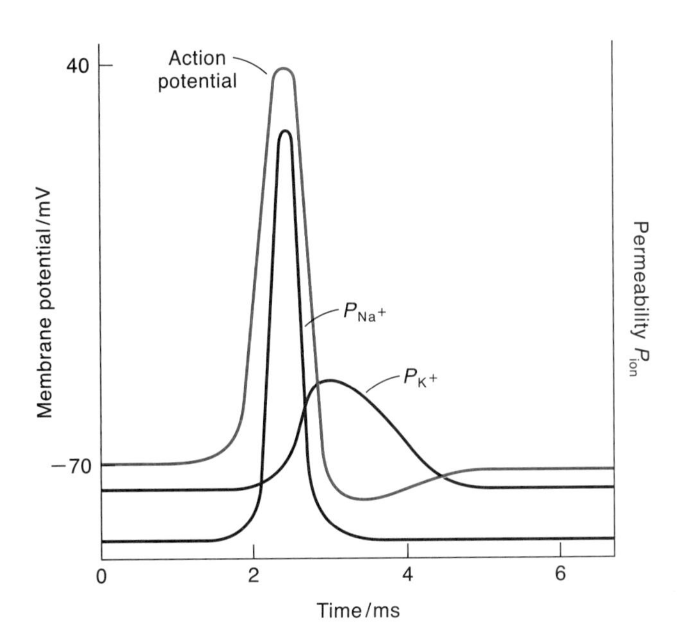

Figure 9.10
The rise and fall of an action potential and the changes in membrane permeability to Na^+ and K^+ ions during this event.

insulator. The resistance of axoplasm (the cytoplasm within the axon), however, is some 100 million times greater than that of copper of the same size. Therefore, the axon would be a relatively poor electrical conductor. Yet we know that when an action potential is generated at a particular site on a neuron, it moves rapidly and without any decrease in magnitude along the axon. Figure 9.11 shows the mechanism of action-potential propagation. The depolarizing influx of Na^+ ions at the immediate site of the action potential causes the membrane potential in adjacent regions to depolarize slowly. When this slow depolarization has pushed the potential of the adjacent membrane beyond a certain value, called the *threshold potential*,* the membrane's permeability to Na^+ ions increases drastically and there is an influx of Na^+ ions into the cell that is greater than the efflux of K^+ ions. Consequently, the potential becomes more positive, and an action potential is generated at this site. This event, in turn, causes a slow depolarization at an adjacent site farther down from the original site, and so on. In this manner, the action potential moves down the neuron without any decrease in magnitude. The speed at which the fastest action potentials in human nerves move along their axon is about 30 m s^{-1}.

The action potential travels along the axon until it reaches either a *synaptic junction* (the connection between nerve cells) or a *neuromuscular junction* (the connection between a nerve cell and a muscle cell). The arrival of an action potential at a synapse triggers the release of a *neurotransmitter*, which is a small, diffusible molecule such as acetylcholine present in the synaptic vesicles. The acetylcholine molecules then diffuse to the postsynaptic membranes, where they produce a large change in the

* The threshold potential is the potential at which gradual depolarization is replaced by explosive depolarization. It is about 20 to 40 mV more positive than the resting membrane potential; that is, it is somewhere between −30 mV and −50 mV.

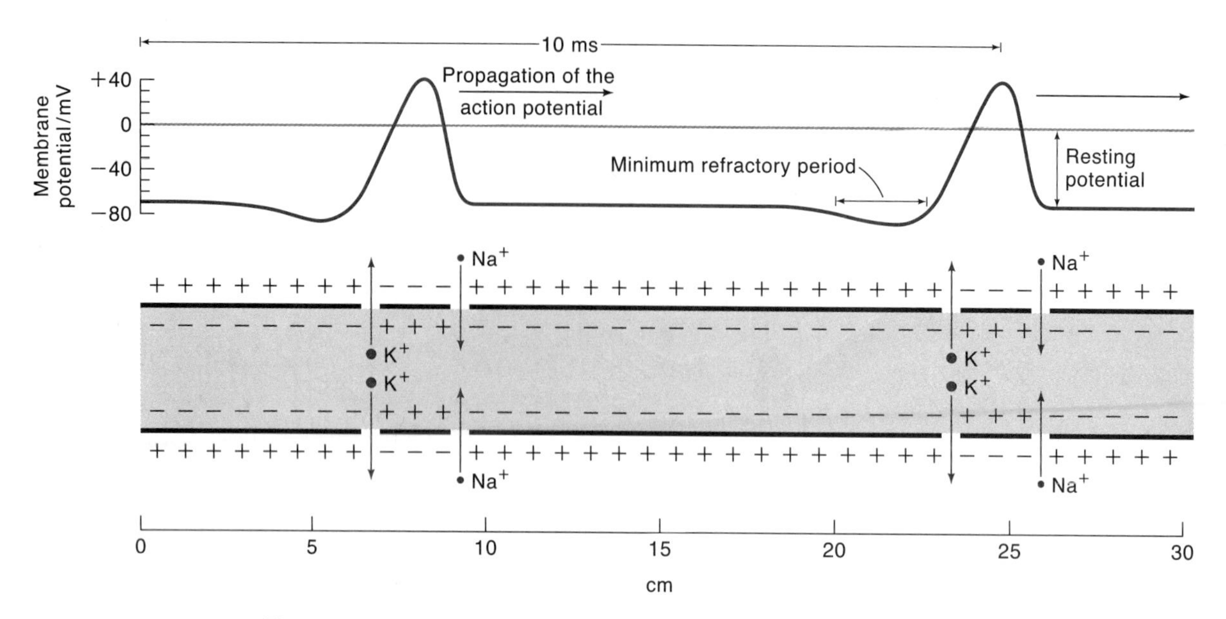

Figure 9.11
Propagation of nerve impulses along the axon coincides with a localized influx of Na^+ ions followed by an outflux of K^+ ions through channels that are "gated," or controlled, by voltage changes across the axon membrane. The electrical event that sends a nerve impulse traveling down the axon normally originates in the cell body. The impulse begins with a slight depolarization, or reduction in the negative potential, across the membrane of the axon where it leaves the cell body. The slight voltage shift opens some of the sodium channels, shifting the voltage still further. The influx of Na^+ ions accelerates until the inner surface of the membrane is locally positive. The voltage reversal closes the sodium channel and opens the potassium channel. The outflux of K^+ ions quickly restores the negative potential. The voltage reversal, known as the action potential, propagates itself down the axon. After a brief refractory period, a second impulse can follow. (A refractory period is the period during and following an action potential in which an excitable membrane cannot be re-excited.) The impulse-propagation speed is that measured in the giant axon of the squid.

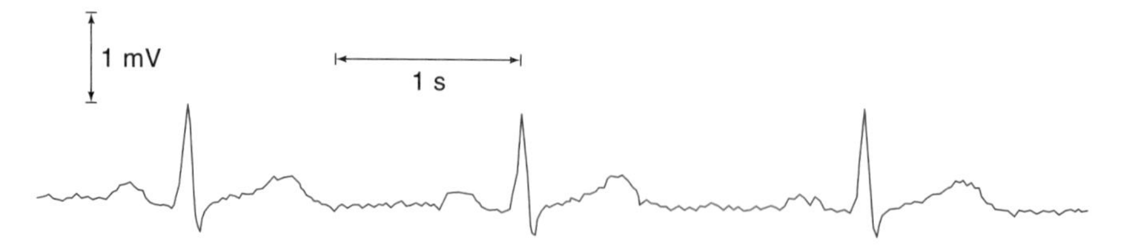

Figure 9.12
A printout of a person's EKG. The appearance is more complicated than the action potential shown in Figure 9.11 due to the depolarization and repolarization in the atria and ventricles. Because measurements are made on the skin, the magnitude of the action potential is much smaller than that at the heart.

permeability of the membranes. The conductance of both Na^+ and K^+ ions increases markedly, resulting in a large inward current of Na^+ ions and a small outward current of K^+ ions. The inward flow of Na^+ ions again depolarizes the postsynaptic membrane and triggers an action potential in the adjacent axon. Finally, acetylcholine is hydrolyzed to acetate and choline by the enzyme acetylcholinesterase as follows:

$$\underset{\text{Acetylcholine}}{CH_3-\overset{\overset{\displaystyle O}{\|}}{C}-O-CH_2-CH_2-\overset{+}{N}(CH_3)_3} + H_2O \longrightarrow \underset{\text{Choline}}{HO-CH_2-CH_2-\overset{+}{N}(CH_3)_3} + CH_3COO^- + H^+$$

In a similar manner, an action potential generated in a nerve cell can be transmitted to a muscle cell. In the muscle cell of the heart, a large action potential is generated during each heartbeat. This potential produces enough current to be detected by placing electrodes on the chest. After amplification, the signals can be recorded either on a moving chart or displayed on an oscilloscope. The record, called an *electrocardiogram* (ECG, also known as EKG, where K is from the German *kardio*, heart), is of great value in diagnosing heart diseases (Figure 9.12).

Key Equations

$E° = E°_{\text{cathode}} - E°_{\text{anode}}$	(Standard emf of a cell)	(9.1)
$\Delta_r G = -\nu FE$	(Relating $\Delta_r G$ to the emf of a cell)	(9.2)
$E° = \dfrac{-\Delta_r G°}{\nu F}$	(Relating $\Delta_r G°$ to the standard emf of a cell)	(9.4)
$E° = \dfrac{RT \ln K}{\nu F}$	(Relating $E°$ to the equilibrium constant)	(9.5)
$E = E° - \dfrac{RT}{\nu F} \ln \dfrac{a_C^c a_D^d}{a_A^a a_B^b}$	(The Nernst equation)	(9.7)
$E = E° - \dfrac{0.0257\ \text{V}}{\nu} \ln \dfrac{a_C^c a_D^d}{a_A^a a_B^b}$	(The Nernst equation at 298 K)	(9.8)
$\Delta_r S° = -\nu F\left(\dfrac{\partial E°}{\partial T}\right)_P$	(Relating $\Delta_r S°$ to the temperature coefficient of $E°$)	(9.9)
$\Delta_r H° = -\nu FE° + \nu FT\left(\dfrac{\partial E°}{\partial T}\right)_P$	(Standard enthalpy change of an electrochemical reaction)	(9.10)

Suggestions for Further Reading

BOOKS

Brett, C. M. A., and A. M. Oliveira Brett, *Electrochemistry: Principles, Methods, and Applications*, Oxford University Press, New York, 1993.

Compton, R. G., and G. H. W. Sanders, *Electrode Potentials*, Oxford Science Publications, New York, 1996.

Rieger, P. H., *Electrochemistry*, Prentice-Hall, Englewood Cliffs, NJ, 1987.

Sawyer, D. T., A. Sobkowiak, and J. L. Roberts, Jr., *Electrochemistry for Chemists*, John Wiley & Sons, New York, 1995.

ARTICLES

General

"Fuel Cells—Electrochemical Converters of Chemical to Electrical Energy," J. Weissbart, *J. Chem. Educ.* **38**, 267 (1961).

"Mechanisms of Oxidation-Reduction Reactions," H. Taube, *J. Chem. Educ.* **45**, 452 (1968).

"Fuel Cells—Present and Future," D. P. Gregory, *Chem. Brit.* **5**, 308 (1969).

"Equivalence Point Potential in Redox Titrations," A. H. A. Heyn, *J. Chem. Educ.* **47**, 240 (1970).

"Thermodynamic Parameters from an Electrochemical Cell," C. A. Vincent, *J. Chem. Educ.* **47**, 365 (1970).

"Electrochemical Principles Involved in a Fuel Cell," A. K. Vijh, *J. Chem. Educ.* **47**, 680 (1970).

"Ion-Selective Electrodes in Science, Medicine and Technology," R. A. Durst, *Am. Sci.* **59**, 353 (1971).

"Electrochemical Cells for Space Power," R. M. Lawrence and W. H. Bowman *J. Chem. Educ.* **48**, 359 (1971).

"On the Relationship Between Cell Potential and Half-Cell Reactions," D. N. Bailey, A. Moe, Jr., and J. N. Spencer, *J. Chem. Educ.* **53**, 77 (1976).

"Dental Filling Discomforts Illustrates the Electrochemical Potential of Metals," R. E. Treptow, *J. Chem. Educ.* **55**, 189 (1978).

"Ion and Bio-Selective Membrane Electrodes," G. A. Rechnitz, *J. Chem. Educ.* **60**, 282 (1983).

"Cathodes, Terminals, and Signs," J. J. MacDonald, *Educ. Chem.* **25**, 52 (1988).

"Alleviating the Common Confusion caused by Polarity in Electrochemistry," P. J. Morgan, *J. Chem. Educ.* **66**, 912 (1989).

"Electrochemical Measurements in General Chemistry Lab Using a Student-Constructed Ag-AgCl Reference Electrode," M. K. Ahn, D. J. Reuland, K. D. Chadd, *J. Chem. Educ.* **69**, 74 (1992).

"The Nernst Equation," A. S. Feiner and A. J. McEvoy, *J. Chem. Educ.* **71**, 493 (1994).

"Tendency of Reaction, Electrochemistry, and Units," R. L. DeKock, *J. Chem. Educ.* **73**, 995 (1996).

"Electric Potential Distribution in an Electrochemical Cell," P. Millet, *J. Chem. Educ.* **73**, 956 (1996).

"Students' Misconceptions in Electrochemistry," M. J. Sanger and T. J. Greenbowe, *J. Chem. Educ.* **74**, 819 (1997).

"The Future of Fuel Cells," (Three articles), *Sci. Am.* July (1999).

"Textbook Error: Short Circuiting an Electrochemical Cell," J. M. Bonicamp and R. W. Clark, *J. Chem. Educ.* **84**, 731 (2007).

"Gassing Up With Hydrogen," S. Satayapal, J. Petrovic, and G. Thomas, *Sci. Am.* April 2007.
"Insights Obtained Through the Study of a Concentration Cell," R. Toomey, E. DePierro, and F. Garafalo, *Chem. Educator* [Online] **12**, 67 (2007) DOI 10.1333/s00897072008a.
"Use of a Commercial Silver-Silver Chloride Electrode for the Measurement of Cell Potentials to Determine Mean Ionic Activity Coefficients," C. A. Kauffman, A. L. Muza, M. W. Porambo, and A. L. Marsh, *Chem. Educator* [Online] **15**, 178 (2010) DOI 10.1007/s00897102272a.

Bioelectrochemistry

"The Nerve Axon," P. F. Baker, *Sci. Am.* March 1966.
"Biogalvanic Cells," W. D. Hobey, *J. Chem. Educ.* **49**, 413 (1972).
"Electron Transfer in Chemical and Biological Systems," N. Sutin, *Chem. Brit.* **8**, 148 (1972).
"Neurotransmitters," I. Axelrod, *Sci. Am.* June 1974.
"Electrochemistry in Organisms. Electron Flow and Power Output," T. P. Chirpith. *J. Chem. Educ.* **52**, 99 (1975).
"Membrane Electrode Probes for Biological Systems," G. A. Rechnitz, *Science* **190**, 234 (1975).
"Chemistry and Nerve Conduction," K. A. Rubinson, *J. Chem. Educ.* **54**, 345 (1977).
"Ion Channels in the Nerve-Cell Membrane," R. D. Keynes, *Sci. Am.* March 1979.
"The Neuron," C. F. Stevens, *Sci. Am.* September 1979.
"The Transport of Substances in Nerve Cells," J. H. Schwartz, *Sci. Am.* April 1980.
"Davy's Electrochemistry: Nature's Protochemistry," P. Mitchell. *Chem. Brit.* **17**, 14 (1981).
"The Release of Acetylcholine," Y. Dunant and M. Israel, *Sci. Am.* April 1985.
"Bio-Electrochemistry," H. A. O. Hill, *Pure Appl. Chem.* **59**, 743 (1987).
"Classical Neurotransmitters and Their Significance within the Nervous System," A. Veca and J. H. Dreisbach, *J. Chem. Educ.* **65**, 108 (1988).

Problems

Emf of Electrochemical Cells

9.1 Calculate the standard emf for the following reaction at 298 K:

$$Fe(s) + Tl^{3+} \rightarrow Fe^{2+} + Tl^{+}$$

9.2 Calculate the emf of the Daniell cell at 298 K when the concentrations of $CuSO_4$ and $ZnSO_4$ are 0.50 *M* and 0.10 *M*, respectively. What would the emf be if activities were used instead of concentrations? (The $\gamma_\pm$ values for $CuSO_4$ and $ZnSO_4$ at their respective concentrations are 0.068 and 0.15, respectively.)

9.3 The half-reaction at an electrode is

$$Al^{3+}(aq) + 3e^- \rightarrow Al(s)$$

Calculate the number of grams of aluminum that can be produced by passing 1.00 faraday through the electrode.

9.4 Consider a Daniell cell operating under non-standard-state conditions. Suppose that the cell's reaction is multiplied by 2. What effect does this have on each of the following quantities in the Nernst equation? **(a)** E, **(b)** $E°$, **(c)** Q, **(d)** $\ln Q$, and **(e)** ν

9.5 A student is given two beakers in the laboratory. One beaker contains a solution that is 0.15 *M* in Fe^{3+} and 0.45 *M* in Fe^{2+}, and the other beaker contains a solution that is 0.27 *M* in I^- and 0.050 *M* in I_2. A piece of platinum wire is dipped into each solution. **(a)** Calculate the potential of each electrode relative to a standard hydrogen electrode at 25°C. **(b)** Predict the chemical reaction that will occur when these two electrodes are connected and a salt bridge is used to join the two solutions together.

9.6 From the standard reduction potentials listed in Table 9.1 for $Cu^{2+}|Cu$ and $Pt|Cu^{2+}, Cu^{+}$, calculate the standard reduction potential for $Cu^{+}|Cu$.

Thermodynamics of Electrochemical Cells and the Nernst Equation

9.7 Complete the following table, indicating in the third column whether the cell reaction is spontaneous:

E	$\Delta_r G$	Cell Reaction
+		
	+	
0		

9.8 Calculate the values of $E°$, $\Delta_r G°$, and K for the following reactions at 25°C:

(a) $Zn + Sn^{4+} \rightleftharpoons Zn^{2+} + Sn^{2+}$

(b) $Cl_2 + 2I^- \rightleftharpoons 2Cl^- + I_2$

(c) $5Fe^{2+} + MnO_4^- + 8H^+ \rightleftharpoons Mn^{2+} + 4H_2O + 5Fe^{3+}$

9.9 The equilibrium constant for the reaction

$$\mathrm{Sr} + \mathrm{Mg}^{2+} \rightleftharpoons \mathrm{Sr}^{2+} + \mathrm{Mg}$$

is 6.56×10^{17} at 25°C. Calculate the value of $E°$ for a cell made up of the $Sr|Sr^{2+}$ and $Mg|Mg^{2+}$ half-cells.

9.10 Consider a concentration cell consisting of two hydrogen electrodes. At 25°C, the cell emf is found to be 0.0267 V. If the pressure of hydrogen gas at the anode is 4.0 bar, what is the pressure of hydrogen gas at the cathode?

9.11 An electrochemical cell consists of a half-cell in which a piece of platinum wire is dipped into a solution that is 2.0 M in KBr and 0.050 M in Br_2. The other half-cell consists of magnesium metal immersed in a 0.38 M Mg^{2+} solution. **(a)** Which electrode is the anode and which is the cathode? **(b)** What is the emf of the cell? **(c)** What is the spontaneous cell reaction? **(d)** What is the equilibrium constant of the cell reaction? Assume that the temperature is 25°C.

9.12 From the standard reduction potentials listed in Table 9.1 for $Sn^{2+}|Sn$ and $Pb^{2+}|Pb$, calculate the ratio of $[Sn^{2+}]$ to $[Pb^{2+}]$ at equilibrium at 25°C and the $\Delta_r G°$ value for the cell reaction.

9.13 Consider the following cell:

$$\mathrm{Ag}(s)|\mathrm{AgCl}(s)|\mathrm{NaCl}(aq)|\mathrm{Hg_2Cl_2}(s)|\mathrm{Hg}(l)|\mathrm{Pt}(s)$$

(a) Write the half-cell reactions. **(b)** The standard emfs of the cell at several temperatures are as follows:

T/K	291	298	303	311
$E°$/mV	43.0	45.4	47.1	50.1

Calculate the values of $\Delta_r G°$, $\Delta_r S°$, and $\Delta_r H°$ for the reaction at 298 K.

9.14 Calculate the emf of the following concentration cell at 298 K:

$$\mathrm{Mg}(s)|\mathrm{Mg}^{2+}(0.24\ M)||\mathrm{Mg}^{2+}(0.53\ M)|\mathrm{Mg}(s)$$

9.15 An electrochemical cell consists of a silver electrode in contact with 346 mL of 0.100 M $AgNO_3$ solution and a magnesium electrode in contact with 288 mL of 0.100 M $Mg(NO_3)_2$ solution. **(a)** Calculate the value of E for the cell at 25°C. **(b)** A current is drawn from the cell until 1.20 g of silver have been deposited at the silver electrode. Calculate the value of E for the cell at this stage of operation.

Membrane Potentials

9.16 Describe an experiment that would show that the nerve cell membrane is much more permeable to K^+ than to Na^+.

9.17 A membrane permeable only to K^+ ions is used to separate the following two solutions:

α $[\mathrm{KCl}] = 0.10\ M$ $\quad$ $[\mathrm{NaCl}] = 0.050\ M$

β $[\mathrm{KCl}] = 0.050\ M$ $\quad$ $[\mathrm{NaCl}] = 0.10\ M$

Calculate the membrane potential at 25°C, and determine which solution has the more negative potential.

9.18 Referring to Figure 9.8b, carry out the following operations: **(a)** Calculate the membrane potential due to K^+ ions at 25°C. **(b)** Given that biological membranes typically have a capacitance of approximately 1 μF cm^{-2}, calculate the charge in coulombs on a unit area (1 cm^2) of the membrane. (See Appendix 7.1 for units of capacitance.) **(c)** Convert the charge in **(b)** to number of K^+ ions. **(d)** Compare the result in **(c)** with the number of K^+ ions in 1 cm^3 of the solution in the left compartment. What can you conclude about the relative number of K^+ ions needed to establish the membrane potential?

Additional Problems

9.19 Look up the values of $E°$ for the following half-cell reactions:

$$Ag^+ + e^- \rightarrow Ag$$

$$AgBr + e^- \rightarrow Ag + Br^-$$

Describe how you would use these values to determine the solubility product (K_{sp}) of AgBr at 25°C.

9.20 A well-known organic redox system is the quinone–hydroquinone couple. In an aqueous solution at a pH below 8, we have

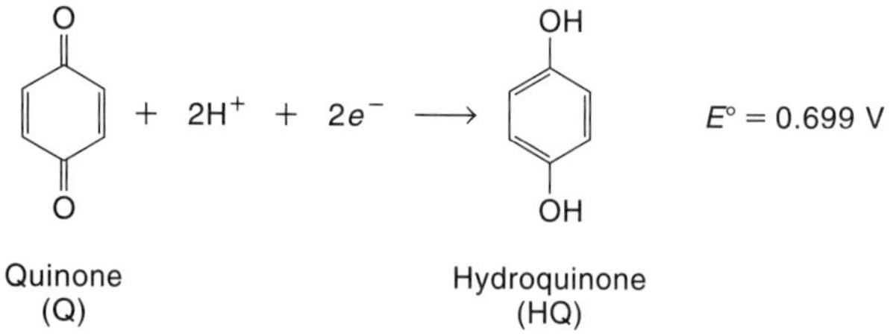

O, O (Quinone (Q)) + $2H^+$ + $2e^-$ ⟶ OH, OH (Hydroquinone (HQ)) $E° = 0.699$ V

Quinone
(Q)

Hydroquinone
(HQ)

This system can be prepared by dissolving quinhydrone, QH (a complex consisting of equimolar amounts of Q and HQ), in water. A quinhydrone electrode can be constructed by immersing a piece of platinum wire in a quinhydrone solution. **(a)** Derive an expression for the electrode potential of this couple in terms of $E°$ and the hydrogen-ion concentration. **(b)** When the quinone–hydroquinone couple is joined to a saturated calomel electrode, the emf of the cell is found to be 0.18 V. In this arrangement, the saturated calomel electrode acts as the anode. Calculate the pH of the quinhydrone solution. Assume the temperature is 25°C.

9.21 One way to prevent a buried iron pipe from rusting is to connect it with a piece of wire to a magnesium or zinc rod. What is the electrochemical principle for this action?

9.22 Aluminum has a more negative standard reduction potential than iron. Yet aluminum does not form rust or corrode as easily as iron. Explain.

9.23 Given that the $\Delta_r S°$ value for the Daniell cell is −21.7 J K^{-1} mol^{-1}, calculate the temperature coefficient $(\partial E°/\partial T)_P$ of the cell and the emf of the cell at 80°C.

9.24 For years it was not clear whether mercury(I) ions existed in solution as Hg^+ or as Hg_2^{2+}. To distinguish between these two possibilities, we could set up the following system:

$$Hg(l)|\text{soln A}||\text{soln B}|Hg(l)$$

where solution A contained 0.263 g mercury(I) nitrate per liter and solution B contained 2.63 g mercury(I) nitrate per liter. If the measured emf of such a cell is 0.0289 V at 18°C, what can you deduce about the nature of the mercury ions?

9.25 Given the following standard reduction potentials, calculate the ion-product K_w value ($[H^+][OH^-]$) at 25°C:

$$2H^+(aq) + 2e^- \rightarrow H_2(g) \qquad E° = 0.00 \text{ V}$$

$$2H_2O(l) + 2e^- \rightarrow H_2(g) + 2OH^-(aq) \qquad E° = -0.828 \text{ V}$$

9.26 Given that

$$2Hg^{2+}(aq) + 2e^- \rightarrow Hg_2^{2+}(aq) \qquad E° = 0.920 \text{ V}$$

$$Hg_2^{2+}(aq) + 2e^- \rightarrow 2Hg(l) \qquad E° = 0.797 \text{ V}$$

calculate the values of $\Delta_r G°$ and for the following process at 25°C:

$$Hg_2^{2+}(aq) \rightarrow Hg^{2+}(aq) + Hg(l)$$

(The above reaction is an example of a *disproportionation reaction*, in which an element in one oxidation state is both oxidized and reduced.)

9.27 The magnitudes of the standard electrode potentials of two metals, X and Y, are

$$X^{2+} + 2e^- \rightarrow X \qquad |E°| = 0.25 \text{ V}$$

$$Y^{2+} + 2e^- \rightarrow Y \qquad |E°| = 0.34 \text{ V}$$

where the || notation denotes that only the magnitude (but *not* the sign) of the $E°$ value is shown. When the half-cells of X and Y are connected, electrons flow from X to Y. When X is connected to a SHE, electrons flow from X to SHE. **(a)** Which value of $E°$ is positive and which is negative? **(b)** What is the standard emf of a cell made up of X and Y?

9.28 An electrochemical cell is constructed as follows. One half-cell consists of a platinum wire immersed in a solution containing 1.0 *M* Sn^{2+} and 1.0 *M* Sn^{4+}, and the other half-cell has a thallium rod immersed in a solution of 1.0 *M* Tl^+. **(a)** Write the half-cell reactions and the overall reaction. **(b)** What is the equilibrium constant at 25°C? **(c)** What is the cell voltage if the Tl^+ concentration is increased tenfold?

9.29 Given the standard reduction potential for Au^{3+} in Table 9.1 and

$$Au^+(aq) + e^- \rightarrow Au(s) \qquad E° = 1.69 \text{ V}$$

answer the following questions. **(a)** Why does gold not tarnish in air? **(b)** Will the following disproportionation occur spontaneously?

$$3Au^+(aq) \rightarrow Au^{3+}(aq) + 2Au(s)$$

(c) Predict the reaction between gold and fluorine gas.

9.30 Consider the Daniell cell shown in Figure 9.1. In the diagram, the anode appears to be negative and the cathode positive (electrons are flowing from the anode to the cathode). Yet the anions in solution are moving toward the anode, which must therefore seem positive to the anions. Because the anode cannot simultaneously be negative and positive, give an explanation for this apparently contradictory situation.

9.31 Calculate the pressure of H_2 (in bar) required to maintain equilibrium with respect to the following reaction at 25°C:

$$\mathrm{Pb}(s) + 2\mathrm{H}^+(aq) \rightleftharpoons \mathrm{Pb}^{2+}(aq) + \mathrm{H}_2(g)$$

given that $[\mathrm{Pb}^{2+}] = 0.035\ M$ and the solution is buffered at pH 1.60.

9.32 Use the data in Appendix B and the convention that $\Delta_f\overline{G}°[\mathrm{H}^+(aq)] = 0$ to determine the standard reduction potentials for sodium and fluorine. (Like sodium, fluorine also reacts violently with water.)

9.33 Use the data in Table 9.1 to determine the value of $\Delta_f\overline{G}°$ for $\mathrm{Fe}^{2+}(aq)$.

9.34 Consider the following cell:

$$\mathrm{Pt}|\mathrm{H}_2(1\ \mathrm{bar})|\mathrm{HCl}(m)|\mathrm{AgCl}(s)|\mathrm{Ag}$$

At 25°C, the emf values at various molalities are given by

$m/(\mathrm{mol\ kg^{-1}})$	0.124	0.0539	0.0256	0.0134	0.00914	0.00562	0.00322
E/V	0.342	0.382	0.418	0.450	0.469	0.493	0.521

(**a**) Determine the value of $E°$ graphically. Compare your value of $E°$ with that listed in Table 9.1. (**b**) Calculate the mean activity coefficient ($\gamma_\pm$) for HCl at 0.124 m.

9.35 A concentration cell ceases to operate when the concentrations of the two cell compartments are equal. At this stage, is it possible to generate an emf from the cell by adjusting another parameter without changing the concentrations? Explain.

9.36 Consider the following reaction:

$$\mathrm{Mg}(s) + 2\mathrm{AgNO_3}(aq) \rightarrow \mathrm{Mg(NO_3)_2}(aq) + 2\mathrm{Ag}(s)$$

Describe how you would measure the $\Delta_rG°$, $\Delta_rH°$, and $\Delta_rS°$ of the reaction (**a**) thermochemically and (**b**) electrochemically. Compare the two methods.

9.37 Compare the efficiencies of a hydrogen–oxygen fuel cell operating at 300 K with the maximum work available through a reversible heat engine operating between 300 K (cold reservoir) and 600 K (heat source). (*Hint:* Calculate the $\Delta_rH°$ for the combustion of hydrogen and use Equation 4.9.)

9.38 Calculate the $E°$ value for the propane–oxygen fuel cell at 298 K. The $\Delta_f\overline{G}°$ value for propane is $-23.49\ \mathrm{kJ\ mol^{-1}}$.

9.39 A galvanic cell using $\mathrm{Mg/Mg^{2+}}$ and $\mathrm{Cu/Cu^{2+}}$ half-cells operates under standard-state conditions at 25°C and each compartment has a volume of 218 mL. The cell delivers 0.22 A for 31.6 h. (**a**) How many grams of Cu are deposited? (**b**) What is the $[\mathrm{Mg^{2+}}]$? Assume constant volumes.

9.40 Consider a concentration cell that consists of $\mathrm{Co/Co^{2+}}$ compartments of concentrations 0.10 M and 2.0 M, respectively. (**a**) Calculate the E_{cell} at 25°C. (**b**) What are the concentrations in the compartments when E_{cell} drops to 0.020 V? Assume volumes remain constant at 1.00 L in each compartment.

CHAPTER 10

Quantum Mechanics

Today I have made a discovery as important as that of Newton.
—Max Planck speaking to his son in 1900*

Up to this point we have focused mainly on the bulk properties of matter. Thermodynamics provides important information regarding chemical processes, but it does not explain what takes place at the molecular level. Now we shall take a closer look at the properties of atoms and molecules. To do so, we need to be familiar with quantum mechanics. At the end of the nineteenth century, there were new experimental results that could not be explained by the so-called classical theories of physics. In 1900, the German physicist Max Planck proposed the quantum theory to explain one of these experiments. In this chapter, we take a historical approach and follow the early development of quantum theory. Much of our understanding of quantum phenomena has come via spectroscopy—the study of the interaction of light with matter. Thus, we start with a description of the properties of waves and the wave theory of light.

10.1 Wave Properties of Light

Experiments by Newton in the seventeenth century demonstrated that white light may be separated into its component colors by the use of a glass prism and that the components can be recombined into white light by using an inverted prism. Later experiments involving interference effects demonstrated the wave nature of light. The wave picture of light became dominant following the theoretical work of Maxwell, so we begin with a discussion of waves.

The simplest one-dimensional waves are sinusoidal waves of the form

$$A = A_0 \sin(\nu t + \phi) \tag{10.1}$$

where A_0 is the amplitude of the wave, ν is the frequency (in unit of s^{-1} or hertz, Hz), t is time, and ϕ is the phase (Figure 10.1). The phase is an angle between 0° and 360° (0 and 2π). It describes the fraction of the wave cycle that has elapsed relative to the origin. The frequencies of visible light waves are rather large (on the order of 10^{14} s^{-1}) and thus we often use the wavelength λ, in units of nanometers, to describe a light

* Cropper, H. W. *The Quantum Physicists*, Oxford University Press, New York (1970), p. 7. Used by permission.

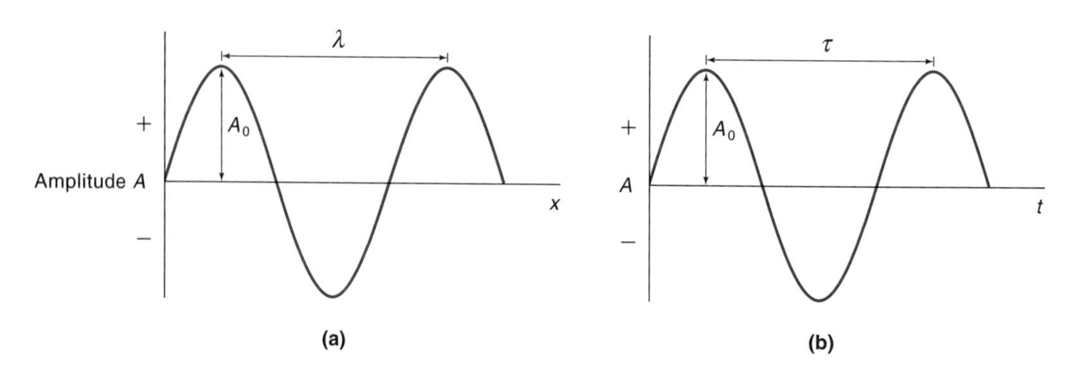

Figure 10.1
A sinusoidal wave with amplitude A_0 in the (a) spatial domain where λ is the wavelength and (b) time domain where τ is the period of oscillation.

wave. The wavelength is the distance between successive maxima (or minima) in the wave. The period τ of the wave is the time between maxima, and is the inverse of the frequency:

$$\tau = \frac{1}{\nu} \tag{10.2}$$

Because a wave travels a distance λ in time τ, the speed of the wave u is

We use u for speed (or the vector velocity) to distinguish it from ν (Greek nu).

$$\begin{aligned} u &= \frac{\lambda}{\tau} \\ &= \lambda\nu \end{aligned} \tag{10.3}$$

The speed of light c depends on the medium through which it travels, but for most purposes it can be taken as 3.00×10^8 m s^{-1}.

When two light waves combine, it is their amplitudes that add or subtract. The French mathematician Jean Baptiste Fourier (1768–1830) showed that any waveform may be written as the summation of sine waves. When we sum sine waves it is essential to consider the relative phases (or *phase difference*) of the waves; for example, do the peaks and troughs of the waves line up, or are they out of synchronization? The sum of two waves of the same amplitude and frequency illustrates the phenomena of constructive and destructive interference (Figure 10.2). Noise-reduction headphones for audio systems actively generate sound waves that are out of phase with external noise waves to minimize, for example, the tiring hum from an aircraft engine.

The phenomenon of interference is a convincing demonstration of the wave nature of light, first carried out by the English physicist Thomas Young (1773–1829) in the famous two-slit experiment (Figure 10.3). A monochromatic (i.e., one wavelength), coherent light source (i.e., the waves are all in phase) is used to illuminate two narrow slits, S_1 and S_2. The light source is an ordinary light bulb with a color filter and a small opening to select one color and one direction, respectively. The light waves from these slits interfere in varying amounts because of the phase difference created by the difference in distance traveled by waves from the two slits. The result of this interference is a pattern of continuously varying light and dark bands seen on a screen placed perpendicular to the traveling light.

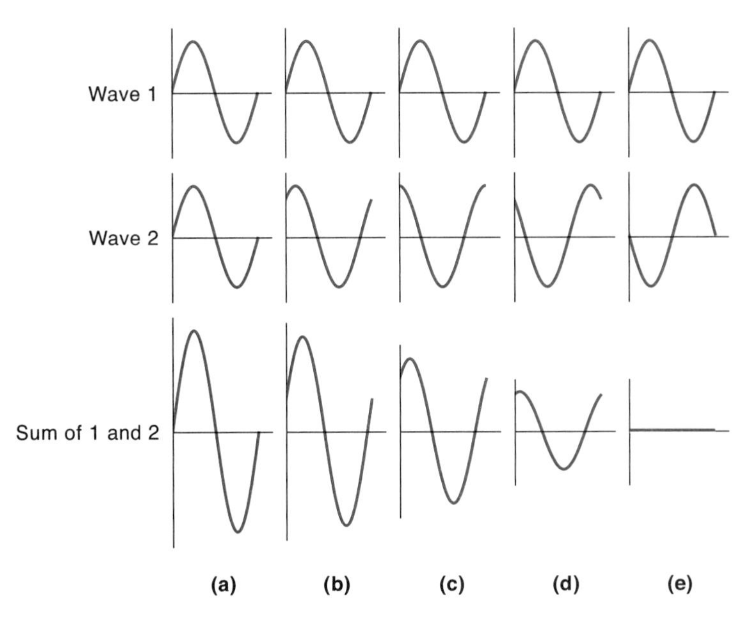

Figure 10.2
Interference between two waves of equal amplitude and wavelength: (a) two waves of the same phase show complete constructive interference; (b)–(d) waves partially out of phase; and (e) two out-of-phase waves show complete destructive interference.

In 1873 Maxwell showed that visible light is just one form of electromagnetic radiation, which has an electric field component and a magnetic field component (Figure 10.4). The complete spectrum of electromagnetic radiation ranges from radio waves to gamma rays (Figure 10.5). As we shall see in later chapters, different regions of the electromagnetic spectrum may be used to probe different chemical and physical properties of atoms and molecules.

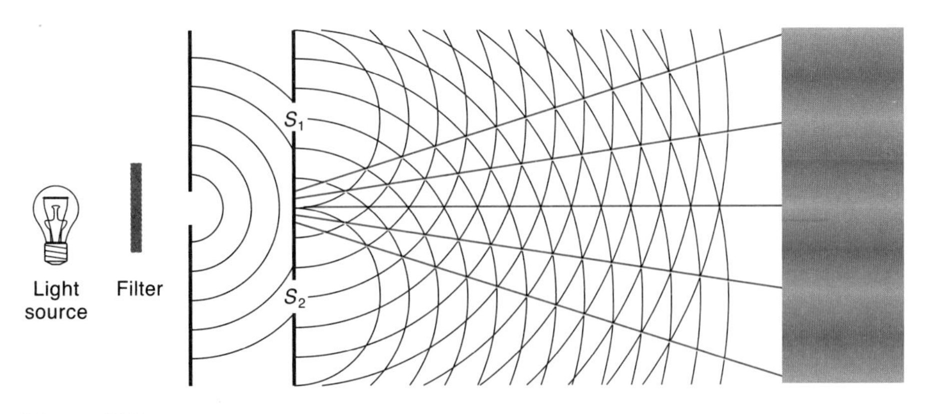

Figure 10.3
Young's two-slit experiment illustrating the interference phenomenon. The pattern on the screen consists of alternating bright and dark bands. The bright bands decrease in intensity with distance from the light source. The concentric circles represent wave fronts that differ from their neighbors by one wavelength.

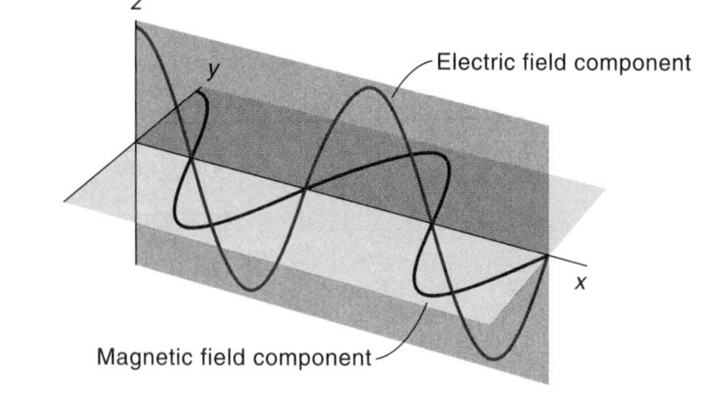

Figure 10.4
Electric and magnetic field components of an electromagnetic wave. The wave travels along the x axis. This electromagnetic wave is linearly polarized such that the electric field component lies in the x–z plane and the magnetic field component lies in the x–y plane.

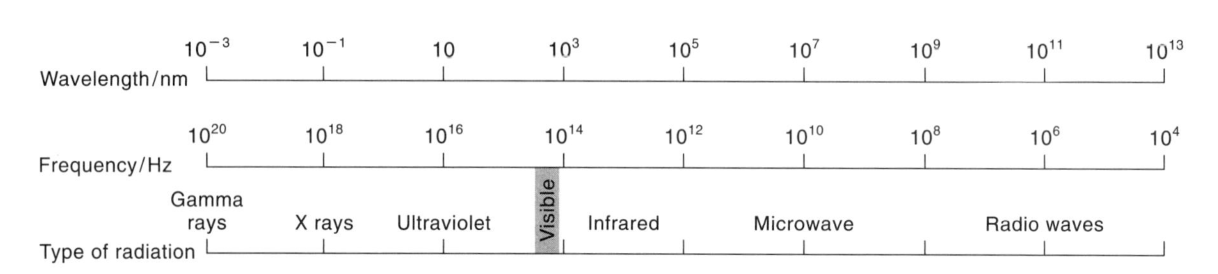

Figure 10.5
Types of electromagnetic radiation. The range of visible light extends from a wavelength of 400 nm (violet) to 750 nm (red).

10.2 Blackbody Radiation and Planck's Quantum Theory

In the 1800s it was recognized that the color of some substances depends on their temperature. Room-temperature (~300 K) objects such as iron, copper, silver, and gold have characteristic appearances, but they glow red when they are heated to ~1000 K. As they are heated to even higher temperatures, the color of light emitted shifts from red towards blue. The tungsten filament of an incandescent light bulb appears white when heated to ~3000 K.

There was considerable effort by the scientific community to provide a theoretical understanding of this phenomenon. In 1859 the German physicist Gustov Kirchoff (1824–1887) described a model in which an ideal substance absorbs all colors of light that strike it. Such a substance would reflect no light, and thus would appear black. For this reason, it is called a *blackbody.* A very good approximation of a blackbody is a pinhole in an empty container. Whatever light strikes the hole is lost to the interior of the container because it is eventually absorbed by the walls after many reflections. Because a blackbody is in thermal equilibrium with its surroundings, it is also a perfect emitter of radiation. Figure 10.6 shows the spectral radiant energy density ρ_λ (in units of J m^{-4}) of a blackbody at different temperatures versus wavelength. These plots are called blackbody radiation curves.

Using classical physics, many scientists tried to explain the distribution of radiation as a function of temperature. The British physicist Lord Rayleigh (1842–1919), with help from the British mathematician and physicist Sir James Hopwood Jeans (1877–1946), assumed that the light emitted by a blackbody came from a collection of

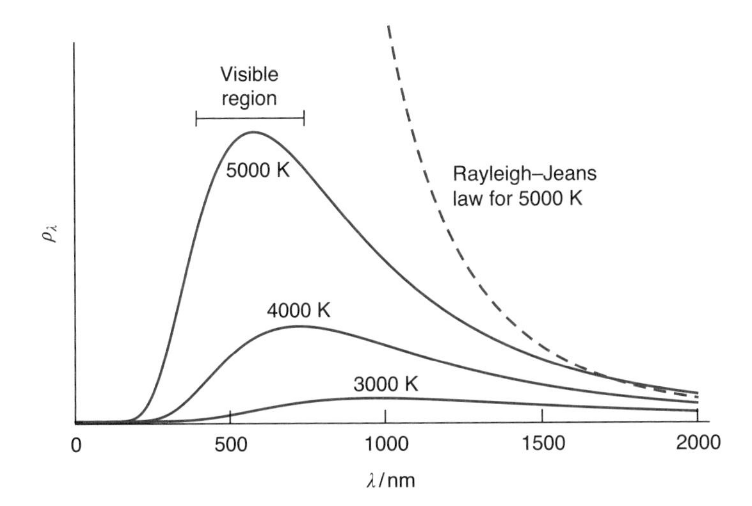

Figure 10.6
Emission spectrum from a blackbody radiator. As can be seen, the Rayleigh–Jeans law fits the experimental curve well in the long wavelength (infrared) region but fails completely in the short wavelength region, whereas Planck's law predicts the curves perfectly over all wavelengths at all temperatures.

oscillators (atoms or molecules) of all frequencies and proposed the following relationship between spectral radiant energy density and wavelength:

$$\rho_\lambda = \frac{8\pi k_B T}{\lambda^4} \tag{10.4}$$

where k_B is the Boltzman constant and T is the absolute temperature. Equation 10.4 is known as the Rayleigh–Jeans law. It provided excellent agreement with experiments in the infrared region, but poorer agreement in the visible, and predicts radiant energy densities approaching infinity in the ultraviolet.

The failure of classical theories to explain short-wavelength blackbody radiation has been termed the "ultraviolet catastrophe."

To resolve this failure of the classical theory, the German physicist Max Planck (1858–1957) came up with an assumption that departed drastically from classical physics. He proposed that the energy emitted by the oscillators could not have any arbitrary value; instead, the energy could be emitted only in small, discrete amounts that he called *quanta*. Planck then developed an equation that fit the experimental data remarkably well at *all* wavelengths.

$$\rho_\lambda = \frac{8\pi hc}{\lambda^5} \frac{1}{e^{\frac{hc}{\lambda k_B T}} - 1} \tag{10.5}$$

Equation 10.5 is known as the Planck distribution law* and h is called Planck's constant; it has the value 6.626×10^{-34} J s. In the classical view, energy could be varied continuously. In Planck's model, the energy emitted by an oscillator in a blackbody could take on only certain discrete values:

$$E = nh\nu \qquad n = 0, 1, 2, \ldots \tag{10.6}$$

That is, energy (E) is quantized in integral multiples of the frequency of oscillation. This was a revolutionary result. We can no longer assume that energy could take on any arbitrary value. By permitting only discrete values of energy, Planck was able

* For an interesting discussion of the Planck radiation law, see T. A. Lehman, *J. Chem. Educ.* **49**, 832 (1972).

to explain an experiment that had been challenging theoreticians for decades. His remarkable insight laid the foundation of quantum theory.

10.3 The Photoelectric Effect

The photoelectric effect was another experimental observation that could not be understood using classical theories. When light of a certain frequency shines on a clean metal surface in a vacuum, electrons are ejected from the metal. Experimentally it is found that (1) the number of electrons (called photoelectrons) ejected is proportional to the intensity of light; (2) the kinetic energy of the ejected electrons is proportional to the frequency of the light; and (3) no electrons can be ejected if the frequency of the light is lower than a certain value, called the *threshold frequency*, no matter how intense the light source. Figure 10.7 shows an apparatus for studying the photoelectric effect.

In the 1920s, the American physical chemist Gilbert Newton Lewis (for whom electron dot structures are named) suggested the term "photon" to be in line with electron, proton, etc.

According to the wave theory of light, the energy of radiation is proportional to the square of the amplitude. Thus, the energy is related to the intensity, not the frequency, of radiation. This seems to contradict point 2 above. The German-American physicist Albert Einstein (1879–1955) was able to explain the photoelectric effect in 1905 by assuming that light consisted of particles, called photons. The energy of each photon is given by

$$E_{\text{photon}} = h\nu \tag{10.7}$$

where h is Planck's constant and ν is the frequency of light. Thus, a single electron is ejected when a photon of sufficient energy strikes the metal surface. According to the law of conservation of energy, the energy input is equal to the energy output. If ν is above the threshold frequency, the following relationship must hold:

$$h\nu = \Phi + \tfrac{1}{2}m_e u^2 \tag{10.8}$$

where m_e is the electron mass and u is the speed of the ejected electron. The symbol Φ is called the *work function*. It measures how tightly the electrons are held in the surface of the metal. The work function explains the threshold behavior observed in the photoelectric effect. If the energy of a photon is less than the work function, then no electrons are emitted from the surface. At the threshold frequency, electrons are ejected with no kinetic energy. Above the threshold frequency, the kinetic energy of the ejected electrons is proportional to the light frequency. The intensity of light corresponds to the number of photons—the more intense the light, the greater the number of photons. At and above the threshold frequency, one photon produces one photoelectron; hence, more intense light produces more electrons and a greater photocurrent.

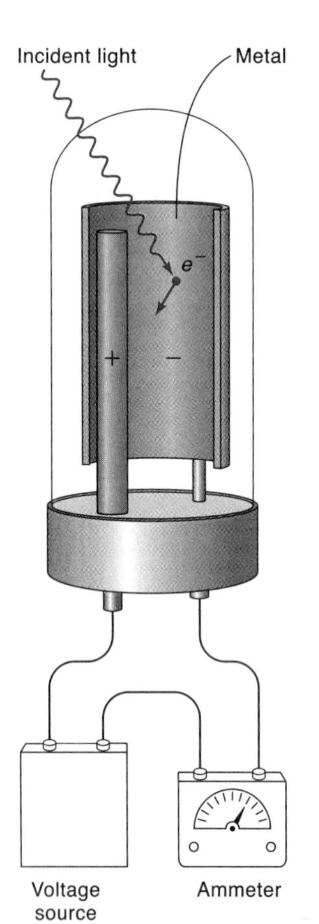

Figure 10.7
A phototube is a practical device for detecting photons and for studying the photoelectric effect. Light of a certain frequency falls on a clean metal surface (the *photocathode*) in a vacuum tube. Ejected electrons are attracted to the positive electrode (the *anode*) and the electric current is measured with an ammeter. Not shown is the grid used in a photoelectric effect experiment to apply a retarding potential that enables one to measure the kinetic energy of the ejected electrons.

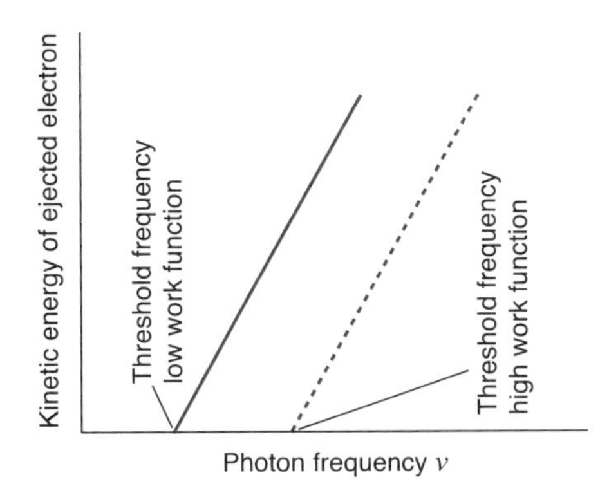

Figure 10.8
Plot of the kinetic energy of the ejected electrons from two different metals versus the frequency of incident light. The x intercept of each line corresponds to the threshold frequency, which is proportional to the work function of the metal.

Using Equation 10.8, we see that a plot of the kinetic energy of the ejected electron as a function of the frequency of incident radiation has an x-axis intercept related the work function (Figure 10.8). Indeed, this is a convenient method for measuring the work function of a metal surface. According to Einstein's formula, the slope of such a plot is equal to h, Planck's constant. Thus, photoelectric experiments provide an independent determination of this fundamental constant.

The photoelectric effect finds practical use in *photodetectors*, devices such as camera light meters that provide an electrical response to light input. Most photodetectors are sensitive only to UV and visible light. Red-sensitive photodetectors use metals or alloys with low work functions.

EXAMPLE 10.1

A photomultiplier tube (PMT) is a detector of electromagnetic radiation based on the photoelectric effect. First, photons of a certain wavelength strike a photocathode. The ejected electrons are then amplified with an electron multiplier and measured with an ammeter. For a so-called *solar-blind* PMT, calculate the maximum wavelength detectable assuming a photocathode with a work function of 4.0 eV is used. In what region of the electromagnetic spectrum does this wavelength lie? (The conversion factor is 1 $e\text{V} = 1.602 \times 10^{-19}$ J.)

ANSWER

To calculate the maximum detectable wavelength, we calculate the minimum detectable photon frequency using Equation 10.8 and zero kinetic energy for the photoelectron.

$$\nu = \frac{\Phi}{h}$$

$$= \frac{(4.0\ e\text{V} \times 1.602 \times 10^{-19}\ \text{J}\ e\text{V}^{-1})}{6.626 \times 10^{-34}\ \text{J s}}$$

$$= 9.67 \times 10^{14}\ \text{s}^{-1}$$

Then we convert frequency to wavelength.

$$\begin{aligned}\lambda &= \frac{c}{\nu} \\ &= \frac{(3.00 \times 10^8 \text{ m s}^{-1})(10^9 \text{ nm m}^{-1})}{(9.67 \times 10^{14} \text{ s}^{-1})} \\ &= 310 \text{ nm}\end{aligned}$$

This wavelength is in the ultraviolet region of the electromagnetic spectrum. This solar-blind detector will not "see" visible light.

By answering one question about light, Equation 10.8 poses another: What is the nature of light? On the one hand, the wave properties of light have been proven beyond doubt. On the other hand, the photoelectric effect can be explained only in terms of particulate photons. Can light be both wavelike and particlelike? This idea was strange and unfamiliar at the time the quantum theory was postulated, but scientists were beginning to realize that microscopic particles behave very differently from macroscopic objects.

10.4 Bohr's Theory of the Hydrogen Emission Spectrum

There were several other experimental observations at the start of the twentieth century that could only be explained by the quantum theory. We next consider atomic emission spectra, because they provide a straightforward starting point for quantum mechanics.

It had long been known that when atoms are subjected to high temperatures or an electrical discharge, they emit electromagnetic radiation that has characteristic frequencies. Figure 10.9 shows the arrangement for studying the emission spectrum of atomic hydrogen, which consists of a series of sharp, well-defined lines. Different atoms give rise to different sets of frequencies. Although the origin of these lines was not understood, this phenomenon was used to identify elements in unknown samples or in distant stars by matching their spectra with those of known elements.

Based on experimental data, the Swedish physicist Johannes Rydberg (1854–1919) formulated the following equation, which fits all the observed lines in the hydrogen emission spectra:

$$\tilde{\nu} = \frac{1}{\lambda} = \tilde{R}_{\text{H}}\left(\frac{1}{n_{\text{f}}^2} - \frac{1}{n_{\text{i}}^2}\right) \tag{10.9}$$

Equation 10.9 is known as the Rydberg formula, where $\tilde{\nu}$ is the wavenumber (number of waves per centimeter or per meter; it is a common unit in spectroscopy), $\tilde{R}_{\text{H}}$ is the

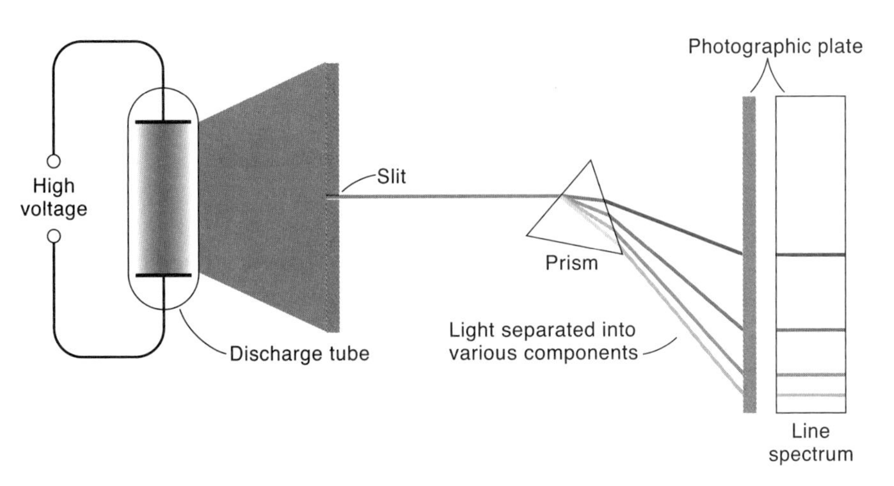

Figure 10.9
Experimental arrangement for studying the emission spectra of atoms and molecules. The gas (hydrogen) under study is placed in a discharge tube containing two electrodes. As electrons flow from the cathode to the anode, they collide with the H_2 molecules, which then dissociate into H atoms. The H atoms are formed in an excited state and quickly decay to the ground state with emission of light. The emitted light is spread into various components by a prism. Each component color shows up at a definite position according to its wavelength and forms an image of the slit on the screen (or photographic plate). The color images of the slit are called spectral lines.

Rydberg constant (109,737 cm^{-1}), and n_f and n_i are integers ($n_i > n_f$). The emission lines could be grouped according to particular values of n_f. Table 10.1 lists five series in the emission spectrum of hydrogen, which are named for their discoverers.

The structure of atoms was fairly well understood in the beginning of the twentieth century due to the work of the British physicist Joseph John Thomson (1856–1940), the New Zealand physicist Ernest Rutherford (1871–1937), and others. In an experiment in which he bombarded gold foil with α particles, Rutherford discovered that an atom consists of a nucleus made up of positively charged particles called protons. Neutral particles were postulated for nuclear stability, and the neutron was later

Table 10.1
Series in the Emission Spectrum of Atomic Hydrogen

Series	n_f	n_i	Region
Lyman	1	2, 3, …	UV
Balmer	2	3, 4, …	Visible, UV
Paschen	3	4, 5, …	IR
Brackett	4	5, 6, …	IR
Pfund	5	6, 7, …	IR

discovered by the British physicist James Chadwick (1891–1972). Because atoms are electrically neutral species, there must be an equal number of negatively charged particles, called electrons, as there are protons in each atom. Electrons were believed to be outside the nucleus, whirling around it in circular orbits at high velocities. Although this model was appealing because it resembled the motions of the planets around the sun, it had a serious problem. The laws of classical physics predict that such an electron would quickly lose energy and spiral into the nucleus, emitting electromagnetic radiation. Using Planck's quantum hypothesis and the notion that light consists of photons, the Danish physicist Neils Bohr (1885–1962) in 1913 presented a new model of the hydrogen atom that accounts for its emission spectrum.

Bohr's starting point was the same as the standard picture—electrons in atoms move in circular orbits of radius r about the nucleus. The force (F) holding the electron in a circular orbit is provided by the coulombic force of attraction between the proton and the electron (Coulomb's law):

$$F = \frac{Ze^2}{4\pi\varepsilon_0 r^2} \tag{10.10}$$

where Z is the atomic number (the number of protons in the nucleus)*, e the electronic charge of an electron, ε_0 the permittivity of free space (see Appendix 7.1), and r the radius of the orbit. The coulombic force is balanced by the centrifugal force:

$$F = \frac{m_e u^2}{r} \tag{10.11}$$

where m_e is the mass of the electron and u the instantaneous speed; that is any instant, the electron can be thought of as moving tangentially to the circular orbit. Equating the above two equations, we obtain

$$\frac{Ze^2}{4\pi\varepsilon_0 r^2} = \frac{m_e u^2}{r} \tag{10.12}$$

The total energy, E, of the electron can be expressed as the sum of the kinetic energy and the potential energy, where

$$E = \tfrac{1}{2} m_e u^2 - \frac{Ze^2}{4\pi\varepsilon_0 r} \tag{10.13}$$

The negative sign in front of the potential energy term indicates that the interaction between the electron and the nucleus is attractive. From Equation 10.12 we have

$$m_e u^2 = \frac{Ze^2}{4\pi\varepsilon_0 r} \tag{10.14}$$

* We include the atomic number here so the final result can also be applied to hydrogenlike ions (one-electron systems) such as He^+ and Li^{2+}.

Substituting Equation 10.14 into 10.13 gives

$$E = \tfrac{1}{2}m_e u^2 - m_e u^2$$
$$= -\tfrac{1}{2}m_e u^2 \qquad (10.15)$$

At this point, Bohr imposed a restriction, based on quantum theory, that the angular momentum (see Appendix A) of the electron ($m_e ur$) is quantized; that is, it can have only certain values, given by

$$m_e ur = n\frac{h}{2\pi} \qquad n = 1, 2, 3, \ldots$$
$$= n\hbar \qquad (10.16)$$

The n is a quantum number.

where $\hbar$ (h bar) is $h/2\pi$. (As we shall see, this symbol appears in many equations in quantum mechanics.) Dividing Equation 10.14 by Equation 10.16 we obtain

$$u = \frac{Ze^2}{2nh\varepsilon_0} \qquad (10.17)$$

Substituting Equation 10.17 into Equation 10.15 gives

$$E_n = -\frac{m_e Z^2 e^4}{8h^2\varepsilon_0^2}\frac{1}{n^2} \qquad n = 1, 2, 3, \ldots \qquad (10.18)$$

Note that we have added a subscript n to E in Equation 10.18 because each value of n (1, 2, 3, ...) gives a different value for E. The negative sign in this equation means that the allowed energies of the electron are *less* than the case in which the electron and the proton are infinitely separated, which is arbitrarily assigned to be zero. The more negative E_n is, the stronger the attraction between the electron and the proton. Thus, the most stable state is the one given by $n = 1$, which is called the *ground state*.

We can also derive an expression for the radius of the orbit as follows. From Equations 10.16 and 10.17,

$$r_n = \frac{n\hbar}{m_e u}$$
$$= \frac{n\hbar}{m_e} \times \frac{2nh\varepsilon_0}{Ze^2}$$
$$= \frac{n^2h^2\varepsilon_0}{Z\pi m_e e^2} \qquad (10.19)$$

where r_n is the radius of the nth orbit. Because the energies of the electron are quantized, we would expect that only certain orbits are available. Equation 10.19 confirms our expectation because the values of r_n are restricted by n. Furthermore, it predicts that the size of the orbit increases with n^2.

EXAMPLE 10.2

Calculate the radius of the smallest orbit of the hydrogen atom, known as the *Bohr radius*.

ANSWER

From Equation 10.19 and using the constants

$$\varepsilon_0 = 8.8542 \times 10^{-12}\ \mathrm{C^2\ N^{-1}\ m^{-2}} \qquad h = 6.626 \times 10^{-34}\ \mathrm{J\ s}$$

$$m_e = 9.109 \times 10^{-31}\ \mathrm{kg} \qquad e = 1.602 \times 10^{-19}\ \mathrm{C}$$

we write, for $n = 1$,

$$r = \frac{(1)^2(6.626 \times 10^{-34}\ \mathrm{J\ s})^2(8.8542 \times 10^{-12}\ \mathrm{C^2\ N^{-1}\ m^{-2}})}{(1)\pi(9.109 \times 10^{-31}\ \mathrm{kg})(1.602 \times 10^{-19}\ \mathrm{C})^2}$$

$$= 5.29 \times 10^{-11}\ \mathrm{m}$$

$$r = 0.529\ \text{Å}$$

where $1\ \text{Å} = 1 \times 10^{-10}$ m.

COMMENT

Although angstrom (Å) is not an SI unit, it is still used to describe atomic and molecular dimensions, because bond lengths are typically on the order of 1 Å.

Equation 10.18 provides a basis for analyzing the emission spectra of atomic hydrogen. Within Bohr's framework, when the electron undergoes a transition from a higher energy level to a lower one, a photon is emitted. The *resonance condition* for an electronic transition is given by

$$\Delta E = E_\mathrm{f} - E_\mathrm{i} = h\nu \tag{10.20}$$

where E_f and E_i are the energies of the final and initial levels involved in the transition and $h\nu$ is the energy of the emitted photon. Exactly the opposite happens in an absorption process (Figure 10.10). Applying Equation 10.18 to an emission process, in which an electron drops from a higher to a lower level, we write

$$\Delta E = E_\mathrm{f} - E_\mathrm{i} = \left(\frac{m_e Z^2 e^4}{8h^2\varepsilon_0^2}\right)\left(\frac{1}{n_\mathrm{i}^2} - \frac{1}{n_\mathrm{f}^2}\right) \tag{10.21}$$

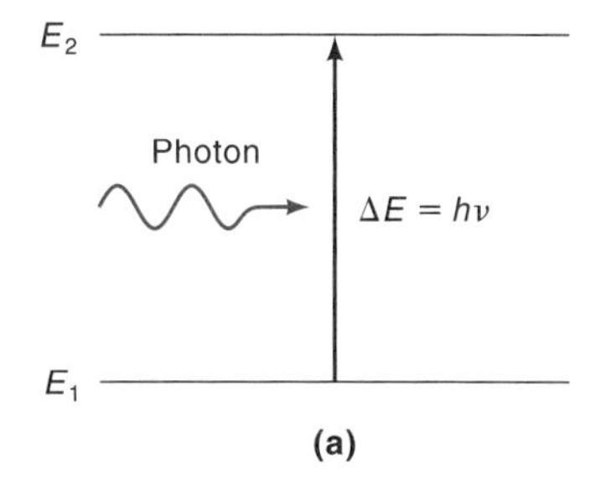

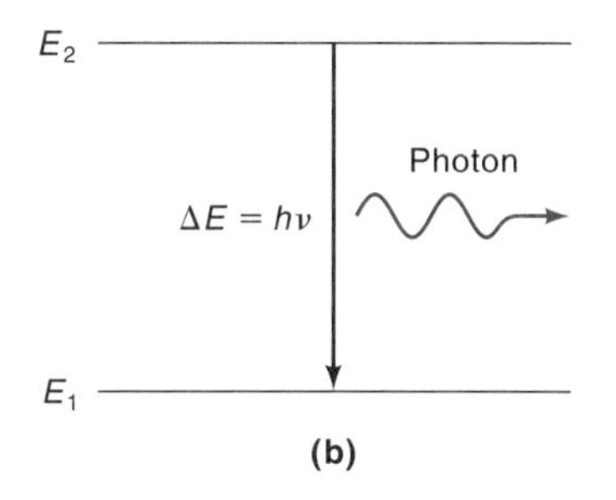

Figure 10.10
Interaction of electromagnetic radiation with energy levels in atoms and molecules. (a) Absorption; (b) emission. In each case, the energy of the photon ($h\nu$) is equal to ΔE, the energy difference between the two levels.

and the corresponding wavenumber is given by

$$\tilde{\nu} = \frac{1}{\lambda} = \frac{\nu}{c} = \frac{\Delta E}{hc} = \left(\frac{m_e Z^2 e^4}{8ch^3\varepsilon_0^2}\right)\left(\frac{1}{n_i^2} - \frac{1}{n_f^2}\right)$$

$$= \tilde{R}_H\left(\frac{1}{n_i^2} - \frac{1}{n_f^2}\right) \tag{10.22}$$

where the Rydberg constant (see Problem 10.13) is (for $Z = 1$)

$$\tilde{R}_H = \frac{m_e e^4}{8ch^3\varepsilon_0^2} = 109{,}737.31568539 \text{ cm}^{-1} \tag{10.23}$$

We shall use 109,737 cm^{-1} for $\tilde{R}_H$ in calculations. A comment about the signs for ΔE and $\tilde{\nu}$ in Equations 10.21 and 10.22 is in order. In absorption, $n_f > n_i$, so both ΔE and $\tilde{\nu}$ are positive. In emission, $n_f < n_i$, so ΔE is a negative value, which is consistent with the fact that energy is given off by the system to the surroundings. However, $\tilde{\nu}$ also becomes a negative value, which has no physical meaning. To ensure that the calculated value of $\tilde{\nu}$ will always be positive, regardless of whether the transition is an absorption or emission, we can take the *absolute value* (i.e., the magnitude but not the sign) of $[(1/n_i^2) - (1/n_f^2)]$.

Figure 10.11 shows the energy-level diagram of the hydrogen atom and the various emissions that give rise to some of the spectral series listed in Table 10.1.

EXAMPLE 10.3

Calculate the wavelength in nanometers of the $n = 4 \rightarrow 2$ transition in the hydrogen atom.

ANSWER

This is an emission process. Because $n_f = 2$, this line belongs to the Balmer series. We calculate the absolute value of $\tilde{\nu}$ from Equation 10.22:

$$\tilde{\nu} = (109{,}737 \text{ cm}^{-1})\left|\left(\frac{1}{4^2} - \frac{1}{2^2}\right)\right|$$

$$= 2.058 \times 10^4 \text{ cm}^{-1}$$

The vertical lines denote absolute values.

Therefore,

$$\lambda = \frac{1}{\tilde{\nu}} = \frac{1}{2.058 \times 10^4 \text{ cm}^{-1}}$$

$$= 4.86 \times 10^{-5} \text{ cm}$$

$$= 486 \text{ nm}$$

COMMENT

Four spectral lines in the Balmer series are in the visible region, including this case. A wavelength of 486 nm is perceived as a blue-green color.

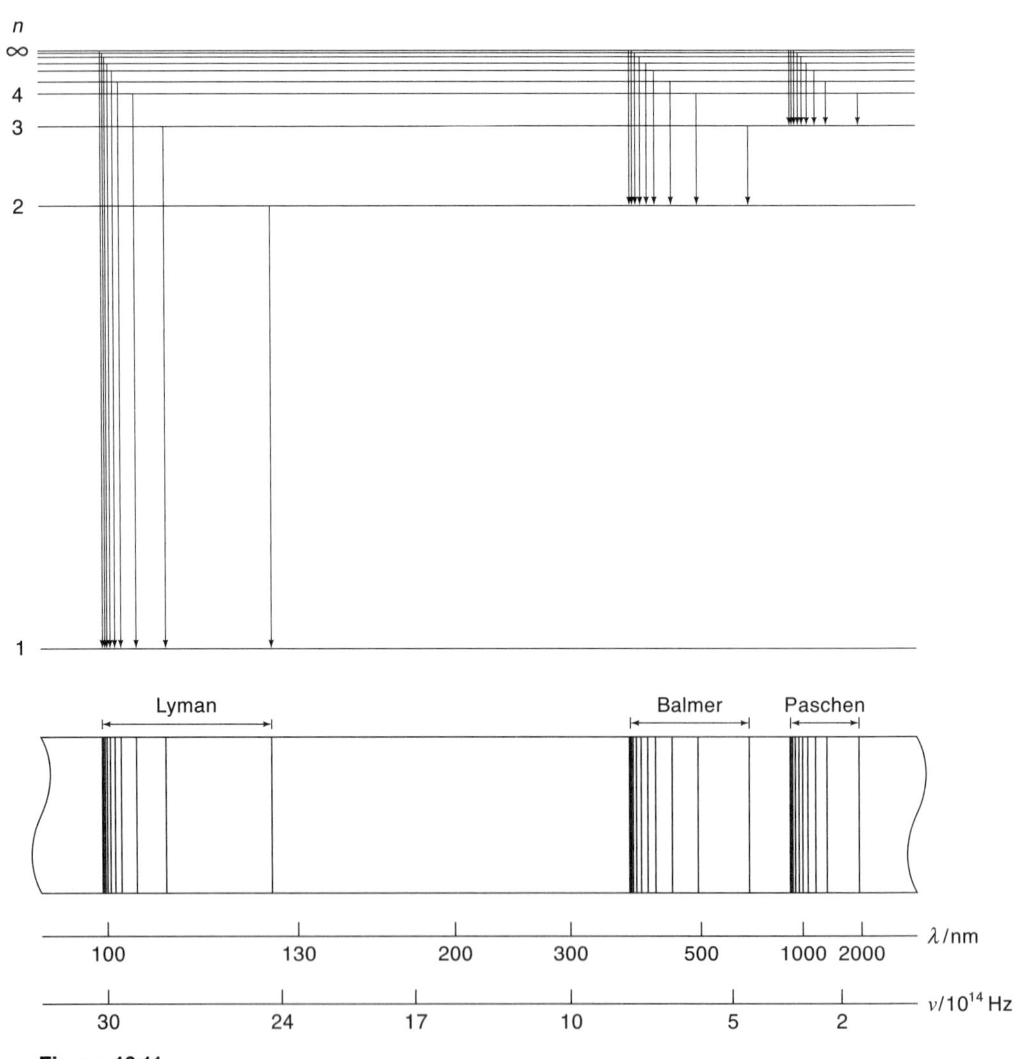

Figure 10.11
Energy levels and some of the hydrogen emission spectra series. [Adapted from McQuarrie, D. A., and J. D. Simon, *Physical Chemistry*, University Science Books, Sausalito, CA (1997)].

10.5 de Broglie's Postulate

Physicists were both mystified and intrigued by Bohr's theory. They questioned why the energies of the hydrogen electron would be quantized. Or, phrasing the question more concretely, why is the electron in a Bohr atom restricted to orbiting the nucleus at certain fixed distances? For a decade, no one—not even Bohr himself—had a logical explanation. In 1924, the French physicist Louis de Broglie (1892–1977) provided the answer.

de Broglie was from a noble family and he held the title of Prince.

de Broglie deduced the connection between particle and wave properties from the Einstein–Planck expression for the energy of an electromagnetic wave and the classical result for the momentum of such a wave. The two expressions are

$$E = h\nu$$

$$p = \frac{E}{c} \tag{10.24}$$

where p is the momentum, and c is the velocity of light. If we replace E by $h\nu = hc/\lambda$, we arrive at the de Broglie relation:

According to Einstein's special theory of relativity, photons have a zero rest mass. However, they do possess mass and hence momentum while in motion.

$$p = \frac{h}{\lambda}$$

or

$$\lambda = \frac{h}{p} = \frac{h}{mu} \tag{10.25}$$

Equation 10.25 says that any particle of mass m moving with velocity u will possess wavelike properties characterized by wavelength λ.

The experimental confirmation of Equation 10.25 was provided by the American physicists Clinton Davisson (1881–1958) and Lester Germer (1896–1972) in 1927, and the British physicist G. P. Thomson (1892–1975) in 1928. When Thomson bombarded a thin sheet of aluminum foil with electrons, the resulting pattern of concentric rings produced on a screen resembled the pattern made by X rays, which were known to be waves. Figure 10.12 shows the diffraction patterns of X rays and electrons arising from aluminum foil.

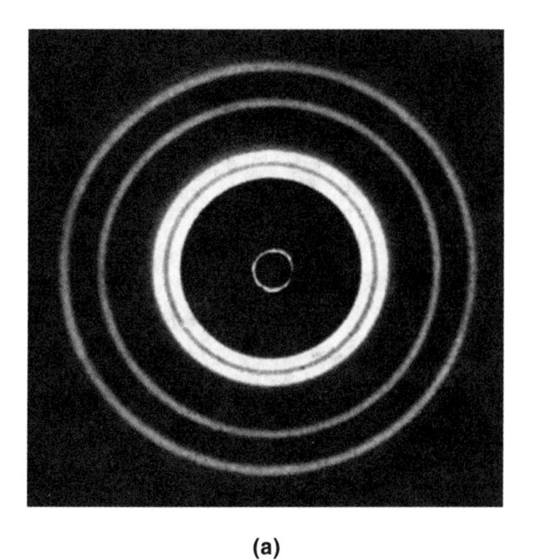

(a)

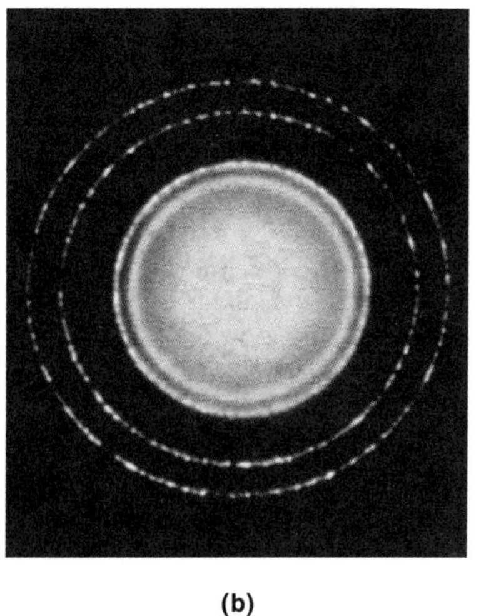

(b)

Figure 10.12
(a) X-ray diffraction pattern of aluminum foil. (b) Electron diffraction pattern of aluminum foil. The similarity of these two patterns shows that electrons can behave like X rays and display wavelike properties. Copyright 2013 Education Development Center, Inc. Reprinted with permission, all other rights reserved.

EXAMPLE 10.4

The fastest serves in tennis are about 150 mph, or 68 m s^{-1}. Calculate the wavelength associated with a 6.0×10^{-2} kg tennis ball traveling at this speed. Repeat the calculation for an electron traveling at the same speed.

ANSWER

Using Equation 10.25, we write

$$\lambda = \frac{6.626 \times 10^{-34}\ \text{J s}}{(6.0 \times 10^{-2}\ \text{kg})(68\ \text{m s}^{-1})}\left(\frac{1\ \text{kg m}^2\ \text{s}^{-2}}{1\ \text{J}}\right)$$

Therefore,

$$\lambda = 1.6 \times 10^{-34}\ \text{m}$$

This is an exceedingly small wavelength because the size of an atom itself is on the order of 1×10^{-10} m. For this reason, the wave properties of a tennis ball cannot be detected by any existing measuring device.

For the electron we have

$$\lambda = \frac{6.626 \times 10^{-34}\ \text{J s}}{(9.109 \times 10^{-31}\ \text{kg})(68\ \text{m s}^{-1})}\left(\frac{1\ \text{kg m}^2\ \text{s}^{-2}}{1\ \text{J}}\right)$$

$$= 1.1 \times 10^{-5}\ \text{m} = 1.1 \times 10^{4}\ \text{nm}$$

$$= 11\ \mu\text{m}$$

which corresponds to light with a wavelength in the infrared region.

COMMENT

This example illustrates that the de Broglie equation is important only for submicroscopic objects such as electrons, atoms, and molecules. We are not able to observe the wave properties of macroscopic objects.

According to de Broglie, an electron bound to the nucleus behaves like a *standing wave*. Standing waves can be generated by plucking, say, a guitar string. The waves are described as standing or stationary, because they do not travel along the string (Figure 10.13). Some points on the string, called *nodes*, do not move at all; that is, the amplitude of the wave at these points is zero. The greater the frequency of vibration, the shorter the wavelength of the standing wave and the greater the number of nodes. As Figure 10.13 shows, there can be only certain wavelengths in any of the allowed motions of the string. de Broglie argued that if an electron does behave like a standing wave in the hydrogen atom, the length of the wave must fit the circumference of the orbit exactly (Figure 10.14). Otherwise the wave would partially cancel itself on each

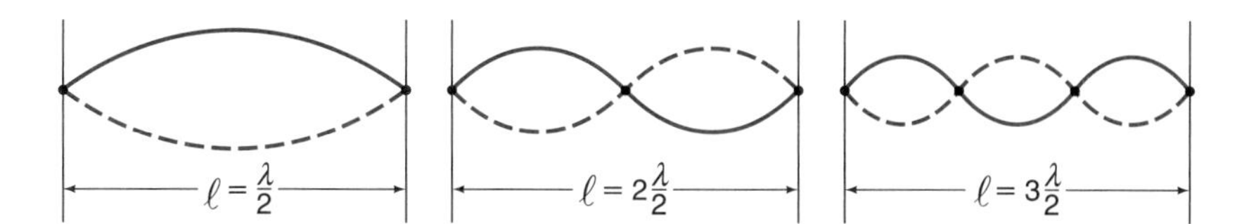

Figure 10.13
A representation of the standing waves generated by plucking a guitar string. The length of the string, ℓ, must be equal to an integral number of half wavelengths ($n\lambda/2$).

successive orbit. Eventually, the amplitude of the wave would be reduced to zero, and the wave would not exist.

The relation between the circumference of an allowed orbit ($2\pi r$) and the wavelength of the electron (λ) is given by

$$2\pi r = n\lambda \qquad n = 1, 2, 3, \ldots$$

Using the expression for λ in Equation 10.25, we get

$$2\pi r = n\frac{h}{m_e u}$$

Upon rearrangement,

$$m_e ur = n\hbar$$

Recall that $\hbar$ is $h/2\pi$.

which is identical to Equation 10.16. Thus, de Broglie's postulate leads to quantized angular momentum and to quantized energy states of the hydrogen atom.

One practical application of the wavelike behavior of electrons is in the use of the electron microscope. Human eyes are sensitive to light of wavelengths from about 400 nm to 700 nm. The ability to see details of small structures is limited by the resolving power, or resolution, of our optical systems. Resolution refers to the minimum distance at which objects can be distinguished as separate entities. Any two objects separated by less than that distance will blur together into a single object. The

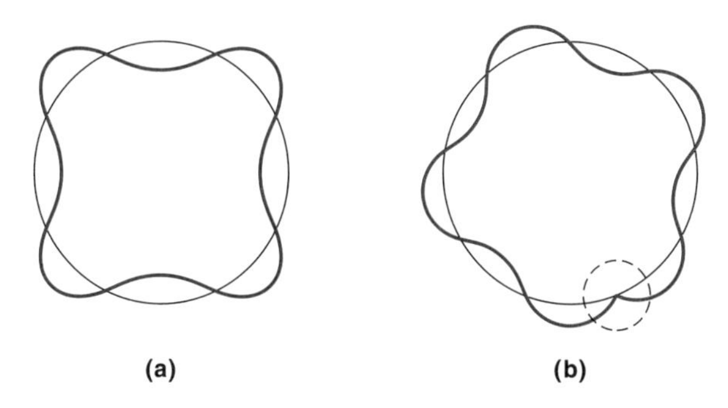

Figure 10.14
(a) The circumference of the orbit is equal to an integral number (4, in this case) of wavelengths. This is an allowed orbit. (b) The circumference of the orbit is not equal to an integral number of wavelengths. As a result, the electron wave does not constructively interfere with itself. This is a forbidden orbit.

Human hair is ~0.1 mm in diameter.

lower limit of resolution of the unaided human eye is approximately 0.1 mm, below which we cannot see individual objects. On the other hand, the lower limit of resolution of a light microscope is approximately 200 nm, or 0.2 μm. This means that with the aid of a light microscope, we can see objects that are the size of about one-half the wavelength of violet light (400 nm) but not smaller. Greater resolution is possible with an electron microscope because a selected beam of electrons has properties that correspond to wavelengths 100,000 times shorter than visible light. When a beam of electrons is directed through an accelerating electrostatic field (two parallel plates with a potential difference of V volts), the potential energy gained by each electron, eV, can be equated to its kinetic energy as follows:

$$eV = \tfrac{1}{2}m_e u^2$$

or

$$u = \sqrt{\frac{2eV}{m_e}}$$

where e is the electronic charge. Using the above expression for velocity in Equation 10.25, we obtain

$$\lambda = \frac{h}{\sqrt{2m_e eV}} \tag{10.26}$$

EXAMPLE 10.5

What is the wavelength of an electron when it is accelerated by 1.00×10^3 V?

ANSWER

From Equation 10.26, we write

$$\lambda = \frac{6.626 \times 10^{-34} \text{ J s}}{\sqrt{2(9.109 \times 10^{-31} \text{ kg})(1.602 \times 10^{-19} \text{ C})(1000 \text{ V})}}$$

Using the conversion factor 1 J = 1 C × 1 V, we find that

$$\lambda = 3.88 \times 10^{-11} \text{ m}$$
$$= 0.0388 \text{ nm}$$

Obtaining voltage in the kilo- or even megavolt range is relatively easy, so that very small wavelengths can be achieved. Thus, an electron microscope differs from a light microscope in that visible light is replaced by a beam of electrons. The much shorter wavelength produces better resolution. This technique enables us to "see" large molecules as well as heavy atoms. The major advantage of electron microscopy over X-ray diffraction is that electrons are charged particles, so they can be focused easily and thus imaged by electric and magnetic fields, which act as lenses. X rays are uncharged, so they cannot be focused this way; no condensing lenses are known for X rays.

10.6 The Heisenberg Uncertainty Principle

In 1927, the German physicist Werner Heisenberg (1901–1976) proposed a principle that has utmost importance in the philosophical groundwork for quantum mechanics. He deduced that when the uncertainties in the *simultaneous* measurements of momentum and position for a particle are multiplied together, the product is approximately equal to Planck's constant divided by 4π. Mathematically, this can be expressed as

$$\Delta x \Delta p \geq \frac{h}{4\pi} \tag{10.27}$$

where Δ means "uncertainty of." Thus, Δx is the uncertainty of position and Δp is the uncertainty of momentum. If the measured uncertainties of position and momentum are large, their product can be substantially greater than $h/4\pi$. The significance of Equation 10.27, which is the mathematical statement of the *Heisenberg uncertainty principle*, is that even in the most favorable conditions for measuring position and momentum, the lower limit of uncertainty is always given by $h/4\pi$.

Conceptually, we can see why the uncertainty principle should exist. Any measurement of a system must, by necessity, result in some disturbance on the system. Suppose that we want to determine the position of a quantum-mechanical object, say an electron. To locate the electron within a distance Δx, we might employ light with a wavelength on the order of $\lambda \approx \Delta x$. During the interaction (collision) between the photon and the electron, part of the photon's momentum ($p = h/\lambda$) will be transferred to the electron. Thus, the very act of trying to "see" the electron has changed its momentum. If we want to locate the electron more accurately, then we must use light of a shorter wavelength. Consequently, the photons of the light will possess greater momentum, resulting in a correspondingly larger change in the momentum of the electron. In essence, to make Δx as small as possible, the uncertainty in the momentum (Δp) will become correspondingly large at the same time. Similarly, if we design an experiment to determine the momentum of an electron as accurately as we can, then the uncertainty in its position will simultaneously become large. Keep in mind that this uncertainty is *not* the result of poor measurements or experimental techniques—it is a fundamental property of the act of measurement itself.

What about macroscopic objects? Because of their sheer size compared with that of quantum mechanical systems, the inaccuracies due to the interactions of observation in measuring the position and momentum of a baseball, for example, are completely negligible. Thus, we can accurately determine the position and momentum of a macroscopic object simultaneously. Planck's constant is such a small number that it becomes important only when we are dealing with particles on the atomic scale.

EXAMPLE 10.6

(a) In Example 10.2, we saw that the Bohr radius of the hydrogen atom is 0.529 Å, or 0.0529 nm. Assuming that we know the position of an electron in this orbit to an accuracy of 1% of the radius, calculate the uncertainty in the velocity of the electron. (b) A baseball (0.15 kg) thrown at 100 mph has a momentum of 6.7 kg m s^{-1}. If the uncertainty in measuring this momentum is 1.0×10^{-7} of the momentum, calculate the uncertainty in the baseball's position.

ANSWER

(a) The uncertainty in the electron's position is

$$\begin{aligned}\Delta x &= 0.01 \times 0.0529 \text{ nm} \\ &= 5.29 \times 10^{-4} \text{ nm} \\ &= 5.29 \times 10^{-13} \text{ m}\end{aligned}$$

From Equation 10.27,

$$\begin{aligned}\Delta p &= \frac{h}{4\pi \Delta x} \\ &= \frac{6.626 \times 10^{-34} \text{ J s}}{4\pi(5.29 \times 10^{-13} \text{ m})}\left(\frac{\text{kg m}^2 \text{ s}^{-2}}{\text{J}}\right) \\ &= 9.97 \times 10^{-23} \text{ kg m s}^{-1}\end{aligned}$$

Because $\Delta p = m_e \Delta u$, the uncertainty in speed is given by

$$\begin{aligned}\Delta u &= \frac{9.97 \times 10^{-23} \text{ kg m s}^{-1}}{9.1095 \times 10^{-31} \text{ kg}} \\ &= 1.1 \times 10^{8} \text{ m s}^{-1}\end{aligned}$$

Thus, the uncertainty in the electron's speed is of the same magnitude as the speed of light (3×10^8 m s^{-1}). At this level of uncertainty, we have virtually no idea of the speed of the electron.

(b) The uncertainty in the position of the baseball is

$$\begin{aligned}\Delta x &= \frac{h}{4\pi \Delta p} \\ &= \frac{6.626 \times 10^{-34} \text{ J s}}{4\pi \times 1.0 \times 10^{-7} \times 6.7 \text{ kg m s}^{-1}}\left(\frac{\text{kg m}^2 \text{ s}^{-2}}{\text{J}}\right) \\ &= 7.9 \times 10^{-29} \text{ m}\end{aligned}$$

This is such a small number as to be of no consequence.

COMMENT

The uncertainty principle is negligible in the world of macroscopic objects but is very important for objects with small masses, such as the electron. Note that we used the equal sign rather than the "greater than" sign in Equation 10.27 to get the minimum value of the uncertainty.

Finally, we note that the Heisenberg uncertainty principle can be expressed also in terms of energy and time, as follows. Because

$$\begin{aligned}\text{momentum} &= \text{mass} \times \text{speed}\\ &= \text{mass} \times \frac{\text{speed}}{\text{time}} \times \text{time}\\ &= \text{force} \times \text{time}\end{aligned}$$

Thus,

$$\begin{aligned}\text{momentum} \times \text{distance} &= \text{force} \times \text{distance} \times \text{time}\\ &= \text{energy} \times \text{time}\end{aligned}$$

or

$$\Delta x \Delta p = \Delta E \Delta t$$

where ΔE is the uncertainty in energy when the system is in a certain state, and Δt is the time interval during which the system is in the state. Equation 10.27 can now be written as

$$\Delta E \Delta t \geq \frac{h}{4\pi} \tag{10.28}$$

Thus, we cannot measure the (kinetic) energy of a particle with absolute precision (i.e., to have $\Delta E = 0$) in a finite span of time. Equation 10.28 is particularly useful for estimating spectral line widths (see Section 11.1). In quantum-mechanical language, momentum and position form a *conjugate pair*, as do energy and time. We shall return to this point in Chapter 11.

10.7 Postulates of Quantum Mechanics

Bohr's theory of the hydrogen atom was one of the early triumphs of the quantum theory. However, it was soon found to be inadequate, because it could not account for the emission spectra of many-electron atoms (atoms containing two or more electrons) or for the behavior of atoms in a magnetic field. Furthermore, the notion that the electron is circling the nucleus in a well-defined orbit is inconsistent with the uncertainty

In quantum chemistry, "many" means two or more electrons.

principle. There was a need for a general equation for submicroscopic systems—one comparable to Newton's equation for macroscopic bodies. In 1926 the Austrian physicist Erwin Schrödinger (1887–1961), motivated by de Broglie's wave ideas, furnished the necessary equation, which we shall consider in detail in Section 10.8. Here, we shall use the postulate approach to the development of quantum mechanics. Like Newton's laws and the laws of thermodynamics, the postulates cannot be derived from a first principle, but they have the advantage of simplicity and they predict results that are consistent with experimental observations.

Postulate One: The state of a system is completely described by a function $\Psi(\tau,t)$. This function is called the state function or the *wave function*. It depends on the spatial coordinates, τ, and on time, t. (Here, we use τ to represent general spatial coordinates.) For a particle in one dimension, we typically use the spatial coordinate x, and the wave function $\Psi(x,t)$. For a particle in three dimensions, we often use the Cartesian coordinates (x, y, and z) and the wave function is a function of four variables, $\Psi(x,y,z,t)$. For the moment, we shall ignore the time dependence of the wave function, and focus on the time-independent wave function, $\psi(\tau)$, which describes a *stationary state*.

The wave function ψ may be a real number or a *complex number*. Complex numbers contain both real and *imaginary* parts. Numbers containing i (where $i = \sqrt{-1}$) are referred to as imaginary numbers. For complex numbers, we write ψ^* as the *complex conjugate* of ψ. The complex conjugate is found by changing i to $-i$ everywhere in ψ. For example, let a and b be real constants. If ψ is a complex function given by $a + ib$, then $\psi^* = a - ib$ and $\psi^*\psi = (a + ib)(a - ib) = a^2 + b^2$. The product, $\psi^*\psi$, which is known as *the squared magnitude* of ψ, is always positive and real. The squared magnitude may also be written as $|\psi|^2$. In the case where ψ is a real function (i.e., if it does not contain i), then $\psi^* = \psi$ and $|\psi|^2 = \psi^2$.

The utility of the wave function is that it tells us everything we need to know, in a quantum mechanical sense. The wave function itself has no physical significance. In 1926, however, the German physicist Max Born (1882–1970) suggested that $|\psi|^2$ can be interpreted as a probability distribution or *probability density* for the system. If $\psi(x)$ is the wave function describing a particle, then $|\psi(x)|^2\,dx$ is the probability of finding the particle between x and $x + dx$. The idea of relating $|\psi|^2$ to probability density stemmed from a wave theory analogy. According to wave theory, the intensity of light is proportional to the squared magnitude of the amplitude of the wave, or A^2. The most likely place to find a photon is where the intensity is greatest, that is, where the value of A^2 is greatest. A similar argument associates a large magnitude of $|\psi|^2$ with higher likelihood of finding a particle, say an electron, in regions near the nucleus.

More generally, we write the probability as $|\psi|^2\,d\tau$ where $d\tau$ is an infinitesimally small volume element. In one dimension, $d\tau = dx$. In three-dimensional Cartesian coordinates, $d\tau = dxdydz$. The general notation $d\tau$ is used for convenience because it is not limited to a particular coordinate system.

Interpreting $|\psi|^2$ or $\psi^*\psi$ as a probability density imposes some restrictions on which types of functions are acceptable for a wave function ψ. To calculate the probability (P) of a particle lying in the region between 1 and 2, we integrate the probability density as follows:

$$P = \int_1^2 \psi^*\psi\,d\tau$$

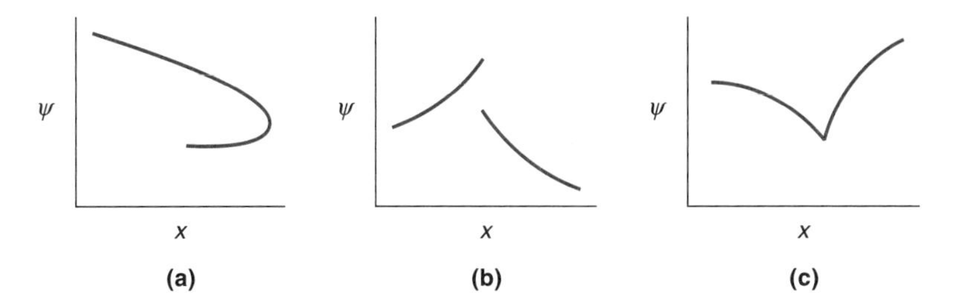

Figure 10.15
Unacceptable wave functions. (a) The function is not single valued. (b) The function is not continuous. (c) The slope of the function is discontinuous. There is a break in $d\psi/dx$.

Because the particle must lie somewhere within this region, the probability of finding the particle over the entire region must be equal to unity. Thus,

$$\int_{\text{all space}} \psi^* \psi \, d\tau = 1 \tag{10.29}$$

Equation 10.29 is called the *normalization condition.*

The types of functions that may be integrated over all space to produce a finite number must be finite and single-valued at any point, and "well behaved." Mathematically, *well behaved* means continuous and *quadratically integrable.* The latter condition implies that both ψ and $d\psi/d\tau$ are continuous. Figure 10.15 shows some examples of unacceptable wave functions.

Postulate Two: For every observable in classical mechanics, there is a corresponding linear, Hermitian operator in quantum mechanics.

Operators are simply mathematical tools that operate on functions. A caret over the mathematical symbol is used to designate an operator. We shall use functions of the variable x, $f(x)$, and $g(x)$ in the examples below. By convention, operators "operate" on a function (f) that is written to their right and produce a new function (g),

$$\hat{A}f(x) = g(x)$$

where $\hat{A}$ is an operator. For example, if $f(x) = x^2$ and $\hat{A}$ is d/dx, then

$$\hat{A}f(x) = \frac{d(x^2)}{dx} = 2x = g(x)$$

A *linear* operator satisfies the relation

$$\hat{A}[c_1 f(x) + c_2 g(x)] = c_1 \hat{A}f(x) + c_2 \hat{A}g(x) \tag{10.30}$$

where c_1 and c_2 are constants.

There are some special properties associated with Hermitian operators that make them useful in quantum chemistry. Hermitian operators satisfy an eigenfunction–eigenvalue equation

$$\hat{A}f(x) = af(x) \tag{10.31}$$

in which the function f (the eigenfunction) is the same on both sides of the equation, and a (the eigenvalue) is a *real* number. To use a specific example, consider the operator $\hat{A} = d/dx$ and the function $f(x) = e^{ax}$ so that

$$\hat{A}f(x) = \frac{d}{dx}(e^{ax}) = ae^{ax} = af(x)$$

Thus, this is an eigenfunction–eigenvalue relationship and the eigenvalue is a.

EXAMPLE 10.7

Do the following operators and functions satisfy an eigenfunction–eigenvalue relationship? If so, what is the eigenvalue?

	Operator	Function
(a)	$\times\, C$ (a constant)	$f(x)$
(b)	$\frac{d}{dx}$	e^{ax^2}
(c)	$\frac{d^2}{dx^2}$	$\sin ax$

ANSWER

(a) Multiplying the function by C, we obtain $Cf(x)$. Therefore, $f(x)$ is an eigenfunction of this operator, and the eigenvalue is C.

(b) $\frac{d}{dx}(e^{ax^2}) = 2ax\,e^{ax^2}$

Because the function is different on the right-hand side, this is not an eigenfunction–eigenvalue relationship.

(c) $\frac{d^2}{dx^2}(\sin ax) = -a^2 \sin ax$

Because we get the original function back, this is an eigenfunction–eigenvalue relationship. The eigenvalue is $-a^2$.

Note that like the wave function ψ, a Hermitian operator itself can also be complex. Technically, a Hermitian operator must satisfy the condition

$$\int_{\text{all space}} \psi_i^* \hat{A} \psi_j \, d\tau = \int_{\text{all space}} \psi_j \hat{A}^* \psi_i^* \, d\tau \qquad (10.32)$$

where the order of operations is significant—operators operate on the function to their right. The complex conjugate of an operator ($\hat{A}^*$) is obtained by replacing i with $-i$ in the operator ($\hat{A}$).

A second useful property of Hermitian operators is that their eigenfunctions are *orthonormal*; that is, they integrate over all space to either 0 or 1.

$$\int_{\text{all space}} \psi_i^* \psi_j \, d\tau = 0 \quad \text{when } i \neq j$$

$$\int_{\text{all space}} \psi_i^* \psi_j \, d\tau = 1 \quad \text{when } i = j$$

The following *orthonormalization condition* is a shorthand notation using the Krönecker delta symbol, δ_{ij}, which takes on the value of zero when $i \neq j$, and unity when $i = j$.

$$\int_{\text{all space}} \psi_i^* \psi_j \, d\tau = \delta_{ij} \tag{10.33}$$

Postulate Three: The only observable values, a_n, for a system in state ψ_n satisfy an eigenfunction–eigenvalue relationship $\hat{A}\psi_n = a_n\psi_n$.

The subscript n indicates a whole family of solutions, where each member is described by a quantum number, n. In the case of the hydrogen atom, each eigenfunction ψ_n describes a particular atomic orbital and the operator gives a unique eigenvalue a_n. This eigenvalue is the corresponding energy of that orbital. An *observable* is a quantity that can be determined by a single measurement of the system. We also see why it is relevant that operators corresponding to observables be Hermitian because their eigenvalues, which are observables, are *real* numbers.

Fundamentally, everything we can measure is quantized. In some cases, the separation between quantized values may be too small to be observable, but the underlying quantization concept remains. This postulate also suggests that there are values that *cannot* be the outcome of a particular measurement. Consider a six-sided die as a macroscopic analogy. When we roll the die, the result will be one of the six sides showing upright. Thus, a face value of 1, 2, 3, 4, 5, or 6 is the eigenvalue of the die. We cannot roll a 2.3 or a 2.7182 or a 3.14159. The only observables are the integers 1 to 6.

Postulate Four: For a state described by the wave function ψ_n, the average, or *expectation value*, for a series of measurements corresponding to the operator $\hat{A}$ is

$$\langle a \rangle = \frac{\int \psi_n^* \hat{A} \psi_n \, d\tau}{\int \psi_n^* \psi_n \, d\tau} \tag{10.34}$$

where $\langle a \rangle$ is the notation for the expectation value of a.

This postulate holds true whether or not ψ_n is an eigenfunction of the operator. Continuing the dice–rolling analogy, if the six-sided die is a fair one, then each of the integers 1 to 6 is an equally likely outcome. What result do we anticipate if we roll the same die over and over? If we roll the die a large number of times, the average roll produces a value of 3.5 $[\text{given by } (1 + 2 + 3 + 4 + 5 + 6)/6]$, which cannot be observed on an individual roll.

What happens when we roll a single die and hide the result under a cup? What number is the die showing before we reveal the answer by lifting the cup? Our expectation value is 3.5, but we know that the die must be showing an integer. How should we write a wave function to describe the hidden die? In the language of quantum mechanics, the die is in a *superposition* of states; that is, it is in all six states at once. Hence,

$$\begin{aligned}\Psi_{\text{hidden die}} &= \frac{1}{\sqrt{6}}(\psi_1 + \psi_2 + \psi_3 + \psi_4 + \psi_5 + \psi_6) \\ &= \frac{1}{\sqrt{6}}\sum_{n=1}^{6}\psi_n\end{aligned}$$

where ψ_n is the state in which n is showing face up. The factor of $1/\sqrt{6}$ is simply a normalization constant that is included to ensure that the probability of finding the die is equal to one. This is a paradox of quantum mechanics that was famously described by Schrödinger. In Schrödinger's thought experiment, a cat is placed in a closed chamber, which also contains a sealed vial containing a deadly poisonous gas (HCN) and a sample of radioactive material. If even a single nucleus has decayed, the emitted α particle will trigger a mechanism to break the vial and the subsequent release of the gas would kill the cat. However, without opening the chamber we have no way of knowing if such an event has occurred and hence we cannot know whether the cat is dead or alive. Thus, in quantum mechanical language, the cat is *both* alive and dead or it is in a superposition of states given by

$$\Psi_{\text{cat}} = \frac{1}{\sqrt{2}}(\psi_{\text{dead}} + \psi_{\text{alive}})$$

The Copenhagen interpretation is an interpretation of quantum mechanics formulated by Bohr and Heisenberg in Copenhagen around 1927.

where $1/\sqrt{2}$ is the normalization constant. According to the Copenhagen interpretation, the wave function describing the cat is in a superposition of states only when it is hidden and the wave function *collapses* to one of the two eigenfunctions (dead or alive) when the measurement is made; that is, when we open the chamber to examine the cat. (Keep in mind that Schrödinger's cat is only a model and not a true superposition of states because a real cat cannot convert from dead to alive and back. But Schrödinger's cat has even entered popular culture and the thought experiment serves to set up a discussion of the interpretation of a superposition state.)

Postulate Five: The wave function of a system varies in time according to the time-dependent Schrödinger equation.

$$\hat{H}\Psi_n(\tau,t) = i\hbar\frac{\partial\Psi_n(\tau,t)}{\partial t} \tag{10.35}$$

where $\hat{H}$ is the Hamiltonian operator [after the Irish mathematician William Hamilton (1805–1865)], which we shall consider in more detail in the next section. Through the use of this postulate, we may write the time-dependent wave function as the product of the time-independent wave function ψ and a time-dependent function

$$\Psi_n(\tau,t) = \psi_n(\tau)e^{-iE_nt/\hbar} \tag{10.36}$$

where E_n is the energy of the system in the nth state. The separation of the spatial (τ) and time (t) variables makes mathematical manipulations more straightforward. For example, if the system is in an eigenstate $\Psi_n(\tau,t)$, then the probability density is independent of time.

$$\Psi_n^*(\tau,t)\Psi_n(\tau,t)d\tau = \psi_n^*(\tau)e^{iE_nt/\hbar}\psi_n(\tau)e^{-iE_nt/\hbar}d\tau$$
$$= \psi_n^*(\tau)\psi_n(\tau)d\tau$$

Time-dependent quantum mechanics can be used to describe transition probabilities between stationary states. Spectroscopic selection rules may be derived using time-dependent quantum mechanics. In this text, we shall primarily state the results of the application of time-dependent quantum mechanics without detailed derivation.

10.8 The Schrödinger Wave Equation

From the postulates, we see the importance of knowing the wave function for a particular system. Schrödinger first developed the wave mechanics approach to quantum mechanics. The time-independent Schrödinger wave equation, or just the Schrödinger equation, looks deceptively simple:

$$\hat{H}\psi = E\psi \tag{10.37}$$

where $\hat{H}$ is the Hamiltonian operator, ψ is the wave function, and E is the energy of the system. To solve for the wave function, we "merely" solve the eigenfunction–eigenvalue problem for the Hamiltonian operator. Unfortunately, for most chemical systems there is no analytical solution for the Schrödinger equation. There are, however, several model systems where analytical solutions are straightforward. We shall consider these systems because they illustrate the principles involved and because they form the basis for approximate solutions to more complex situations.

The Hamiltonian operator $\hat{H}$ is the total energy operator for the system. In classical mechanics, the energy E is given by the sum of kinetic energy (K) and potential energy (V):

$$E = K + V \tag{10.38}$$

The Hamiltonian operator is similarly written as the sum of the kinetic and potential energy operators:

$$\hat{H} = \hat{K} + \hat{V} \tag{10.39}$$

Table 10.2 lists the most essential quantum mechanical operators for energy, momentum, position, and so on. We shall refer to these as needed.

We start by considering a single particle of mass m moving in one dimension. For convenience, we confine the particle to the x axis, so this is a *particle on a line* system. The particle's energy E is given by the sum of its kinetic and potential energies:

$$E = \tfrac{1}{2}mu^2 + V(x)$$
$$= \frac{p^2}{2m} + V(x)$$

Table 10.2
Some Observables from Classical Mechanics and their Corresponding Quantum-Mechanical Operators

Observable		Operator	
Name	Symbol	Symbol	Operation[a]
Position	x	$\hat{X}$	Multiply by x
Momentum	p_x	$\hat{P}_x$	$-i\hbar\frac{\partial}{\partial x}$
Kinetic energy	K_x	$\hat{K}_x$	$-\frac{\hbar^2}{2m}\frac{\partial^2}{\partial x^2}$
	K	$\hat{K}$	$-\frac{\hbar^2}{2m}\nabla^2$
Potential energy	$V(x)$	$\hat{V}_x$	Multiply by $V(x)$
	$V(x, y, z)$	$\hat{V}$	Multiply by $V(x, y, z)$
Total energy	E	$\hat{H}$	$-\frac{\hbar^2}{2m}\nabla^2 + V(x, y, z)$
Angular momentum along the z direction	L_z	$\hat{L}_z$	$-i\hbar\frac{\partial}{\partial \phi} = -i\hbar\left(x\frac{\partial}{\partial y} - y\frac{\partial}{\partial x}\right)$

[a] In Cartesian coordinates, $\nabla^2 = \left(\frac{\partial^2}{\partial x^2} + \frac{\partial^2}{\partial y^2} + \frac{\partial^2}{\partial z^2}\right)$

where $V(x)$ is the potential energy function, u is the speed of the particle, and p is its momentum. The total energy E is a constant, whereas the kinetic and potential energies are functions of the position, x. From Table 10.2, the Hamiltonian operator is

$$\hat{H} = \frac{-\hbar^2}{2m}\frac{d^2}{dx^2} + V(x)$$

and the Schrödinger equation for this particle on a line is

$$\frac{-\hbar^2}{2m}\frac{d^2\psi(x)}{dx^2} + V(x)\psi(x) = E\psi(x)$$

For a particle in *free space* (meaning that there are no electric, magnetic, or gravitational fields present), there is constant potential energy V everywhere. Because force is $-dV(x)/dx$, it follows that $V(x) =$ constant. If we set the constant to be zero, then the above equation is simplified to

$$\frac{-\hbar^2}{2m}\frac{d^2\psi(x)}{dx^2} = E\psi(x) \qquad -\infty < x < \infty \tag{10.40}$$

This is a second-order differential equation that we may solve simply by inspection. The sine, cosine, and exponential functions are good candidates for solutions to this equation. Let us check two trial solutions to this problem:

$$\psi(x) = Ae^{+ikx} \tag{10.41a}$$

and

$$\psi(x) = Ae^{-ikx} \tag{10.41b}$$

where A and k are constants. We solve for the energy of the first wave function by substituting Equation 10.41a into Equation 10.40:

$$\begin{aligned}\frac{-\hbar^2}{2m}\frac{d^2\psi(x)}{dx^2} &= \frac{-\hbar^2}{2m}\frac{d^2Ae^{ikx}}{dx^2}\\ &= (ik)^2\frac{-\hbar^2}{2m}(Ae^{ikx})\\ &= \frac{k^2\hbar^2}{2m}\psi(x)\end{aligned}$$

Our trial wave function is a valid eigenfunction of the Hamiltonian operator, with the eigenvalue E, where

$$E = \frac{k^2\hbar^2}{2m} \tag{10.42}$$

This is an interesting result because there is no quantization of the energy. The constant k may take on any value and thus any energy is allowed. If we perform the same calculation for the second wave function (Equation 10.41b), we get the same result.

EXAMPLE 10.8

From Table 10.2 we see that the momentum operator for a one-dimensional particle on a line is

$$\hat{P}_x = -i\hbar\frac{d}{dx}$$

Calculate the value of the momentum p_x for the particle described by the wave functions in Equation 10.41.

ANSWER

We test to see if the wave function given in Equation 10.41a is an eigenfunction of the momentum operator.

$$\hat{P}_x\psi(x) = -i\hbar\frac{d\psi(x)}{dx}$$

$$= -i\hbar\frac{d(Ae^{+ikx})}{dx}$$

$$= \hbar k(Ae^{+ikx})$$

$$= \hbar k\psi(x)$$

Thus, this is an eigenfunction–eigenvalue relationship, and the eigenvalue is $\hbar k$. When performing the same calculation on the second wave function (Equation 10.41b), we find the eigenvalue to be $-\hbar k$. The momenta of particles described by these states are $\hbar k$ and $-\hbar k$.

COMMENT

Our wave functions describe two possibilities for the particles: one moving to the right, and one moving to the left, as indicated by the + and − signs. An individual particle is free to move in either direction. For a large collection of particles, therefore, we expect no preference for moving to the left versus moving to the right, so the average momentum is expected to be zero; that is, $\langle P_x \rangle = 0$.

10.9 Particle in a One-Dimensional Box

We have seen how the Schrödinger equation may be used to solve the energy of a particle moving freely along a line. Now let us consider what happens when we restrict the movement of the particle to only a portion of the line. This problem is usually referred to as a *particle in a one-dimensional box*. This is a very simple model, but one that provides insight into more complicated systems. The merit of the particle-in-a-box system is that the solution is straightforward and the results illustrate most of the fundamental concepts of quantum mechanics. As we shall see, studying the particle-in-a-box problem helps us understand many chemical and biological systems.

Consider a particle of mass m confined to a one-dimensional box of length L. We again assume that the particle has zero potential energy inside the box (or on the line segment); that is, $V = 0$. The particle has only kinetic energy. At each end of the box is a wall of infinite potential energy, so there is no probability of finding the particle at the walls or outside the box. For simplicity, we chose the line segment to start at the origin, so x is restricted by $0 \le x \le L$ (Figure 10.16). The Schrödinger equation is similar to that for the free particle (see Equation 10.40), with the difference being that the value of x is constrained by the size of the box.

$$\frac{-\hbar^2}{2m}\frac{d^2\psi(x)}{dx^2} = E\psi(x) \qquad 0 \le x \le L \tag{10.43}$$

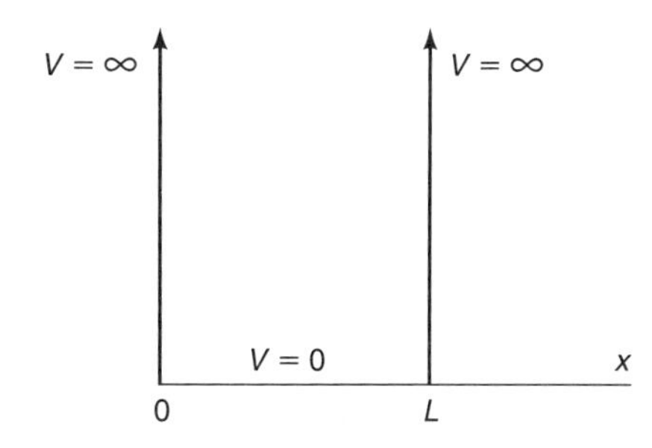

Figure 10.16
A one-dimensional box with infinite potential barriers.

Because the particle is constrained to the line segment, there is no probability that it will be found elsewhere on the line and thus we have the trivial solution

$$\psi(x) = 0 \qquad \text{for } x < 0 \text{ and for } x > L$$

As with all quantum mechanical problems, we are interested in knowing the values of E and ψ that the particle can possess. In the free particle problem, we found that the exponential function satisfied the Schrödinger equation. For the particle-in-a-box Schrödinger equation, let us try a slightly different trial wave function,

$$\psi(x) = A \sin(kx) + B \cos(kx) \tag{10.44}$$

The trigonometric and exponential functions are related as follows:
$\cos(x) + i\sin(x) = e^{ix}$ and
$\cos(x) - i\sin(x) = e^{-ix}$

where A, B, and k are constants to be determined. The solution to the Schrödinger equation for the particle in a box is similar to that for a free particle, but now we have the *boundary conditions* imposed by the infinite potential outside the box. When the potential is infinite, the wave function must have zero value. Because the wave function must be continuous, it similarly must have zero value at the edges of the box, that is, at the ends of the line segment. Thus,

$$\psi(x) = 0 \qquad \text{when } x \leq 0$$

and

$$\psi(x) = 0 \qquad \text{when } x \geq L$$

For $x = 0$, we have $\sin(0) = 0$ and $\cos(0) = 1$; therefore, the constant B in Equation 10.44 must be zero and the trial wave function is simplified to

$$\psi(x) = A \sin(kx) \tag{10.45}$$

Next, we solve for the constant k. We differentiate ψ with respect to x to get

$$\frac{d\psi(x)}{dx} = kA \cos(kx)$$

and differentiate a second time to get

$$\begin{aligned} \frac{d^2\psi(x)}{dx^2} &= -k^2 A \sin(kx) \\ &= -k^2 \psi(x) \end{aligned} \tag{10.46}$$

From Equations 10.43 and 10.46, we find that

$$k^2 = \frac{2mE}{\hbar^2}$$

or

$$k = \left(\frac{2mE}{\hbar^2}\right)^{1/2} \tag{10.47}$$

Substituting Equation 10.47 into Equation 10.45 gives

$$\psi(x) = A \sin\left[\left(\frac{2mE}{\hbar^2}\right)^{1/2} x\right] \tag{10.48}$$

So far, we have solved for the constants k and B, and only the constant A remains to be found. Mathematically, an infinite number of solutions can satisfy Equation 10.48 because A may take on any value. Physically, however, $\psi(x)$ must satisfy the boundary conditions noted above. We use the fact that when $x = L$,

$$\psi(L) = 0$$

and apply this condition to Equation 10.48:

$$0 = A \sin\left[\left(\frac{2mE}{\hbar^2}\right)^{1/2} L\right] \tag{10.49}$$

We neglect the trivial solution of $A = 0$, because this leads to a nonphysical solution of $\psi(x) = 0$ everywhere; in other words, there would be no probability of finding the particle in the box. Noting that

$$\sin \pi = \sin 2\pi = \sin 3\pi = \cdots = 0$$

we find that the general solution is thus

$$\left[\left(\frac{2mE}{\hbar^2}\right)^{1/2} L\right] = n\pi \qquad \text{where } n = 1, 2, 3, \ldots$$

Therefore,

$$\begin{aligned} E_n &= \frac{1}{2m}\left(\frac{n\pi\hbar}{L}\right)^2 \\ &= \frac{n^2h^2}{8mL^2} \qquad n = 1, 2, 3, \ldots \end{aligned} \tag{10.50}$$

where E_n is the energy for the nth level.

The corresponding wave functions are given by

$$\psi_n(x) = A \sin\left(\frac{n\pi x}{L}\right) \qquad n = 1, 2, 3, \ldots \tag{10.51}$$

where the subscript n is again used to indicate that there is a family of solutions.

Next we determine the normalization constant A. We start with the knowledge that because the particle must remain inside the box, the total probability of finding the particle between $x = 0$ and $x = L$ must be unity. Thus, the *normalization condition* is

$$\int_0^L \psi^*(x)\psi(x)dx = 1 \tag{10.52}$$

where $\psi^*(x)\psi(x)$ is the probability density and $\psi^*(x)\psi(x)dx$ is the probability of finding the particle between x and $x + dx$. Substituting the wave function (Equation 10.51) into the normalization condition (Equation 10.52), we have

$$A^2\int_0^L \sin^2\left(\frac{n\pi x}{L}\right)dx = 1 \tag{10.53}$$

The definite integral above gives us the *normalization constant A*.†

$$A^2\left(\frac{L}{2}\right) = 1$$

or

$$A = \left(\frac{2}{L}\right)^{1/2} \tag{10.54}$$

Finally, we have the *normalized* wave functions for the particle in a one-dimensional box:

$$\psi_n(x) = \left(\frac{2}{L}\right)^{1/2}\sin\left(\frac{n\pi x}{L}\right) \qquad n = 1, 2, 3, \ldots \tag{10.55}$$

Figure 10.17 shows the plots of the allowed energy levels, E_n, as well as ψ_n and ψ_n^2. Several important conclusions can be drawn from this model.

1. The (kinetic) energy of the particle is quantized according to Equation 10.50.
2. The lowest energy level is not zero but is equal to $h^2/8mL^2$. This *zero-point energy* can be accounted for by the Heisenberg uncertainty principle. If the particle could possess zero kinetic energy, then its speed would also be zero; consequently, there would be no uncertainty in determining its momentum. According to Equation 10.27, Δx would then be infinite. If the box is of finite size, however, the uncertainty in determining the particle's position cannot exceed L; therefore, zero kinetic energy would violate the Heisenberg uncertainty principle. Keep in mind that the zero-point energy means that the particle can never be at rest because its lowest kinetic energy is not zero.
3. The wave behavior of the particle is described by ψ_n (Equation 10.55), but the probability is described by ψ_n^2, which is always positive. (In fact, the wave functions look just like the standing waves set up in a vibrating string shown in Figure 10.13.) For $n = 1$, the maximum probability is at $x = L/2$; for

† This definite integral is evaluated by using the mathematical relation

$$\int \sin^2(ax)dx = \frac{x}{2} - \frac{\sin(2ax)}{4a}$$

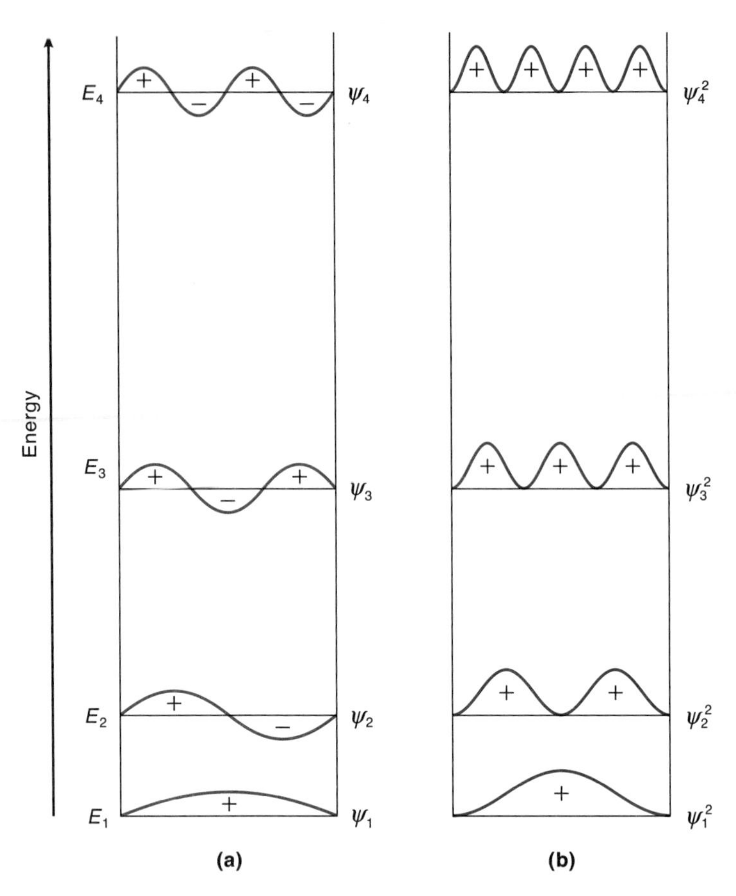

Figure 10.17
Plots of (a) ψ and (b) ψ^2 for the first four energy levels of a particle in a one-dimensional box. The sign of ψ and ψ^2 (which are dimensionless functions) is represented by + and − in the figure. The horizontal lines represent allowed energy levels; these have positive values with dimensions of energy (joules, for example).

More accurately, a node is a zero *crossing* of the wave function; the ends of the box where $x = 0$ and $x = L$ are not nodes.

$n = 2$, the maxima occur at $x = L/4$ and $x = 3L/4$, and the probability is zero at $x = L/2$. The point at which ψ (and hence ψ^2) is zero is called a *node*. The *nodal theorem* states that the ground state for a system is nodeless, and that the energy increases with an increasing number of nodes.

4. In classical mechanics, the probability of finding the particle is the same at all points along the box, regardless of its kinetic energy. The value of the probability density is a constant equal to $1/L$. The quantum mechanical solution shows that there cannot be a uniform probability density because this would result in a discontinuity in both ψ and ψ^2 at the ends of the box. If we let the value of the quantum number n become very large, we see that the probability density function begins to look uniform (Figure 10.18). This is an illustration of the *correspondence principle*, in which quantum mechanical and classical results agree in the limit of large quantum numbers. The large-quantum-number limit is known as the *classical limit.*
5. The energy of the system is inversely proportional to the mass of the particle (Equation 10.50). For macroscopic objects, m is very large, so the difference between successive energy levels is exceedingly small. This means that the

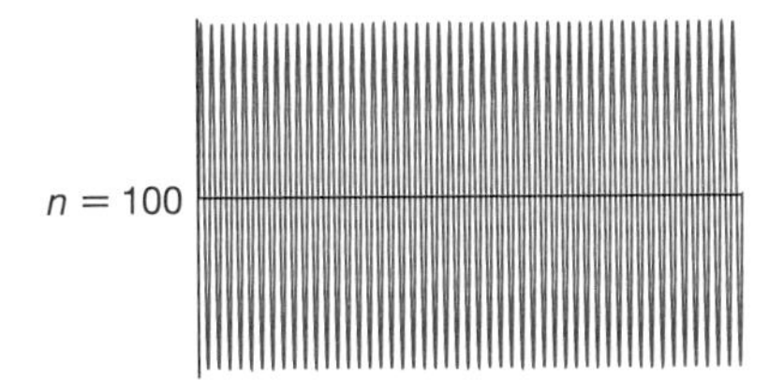

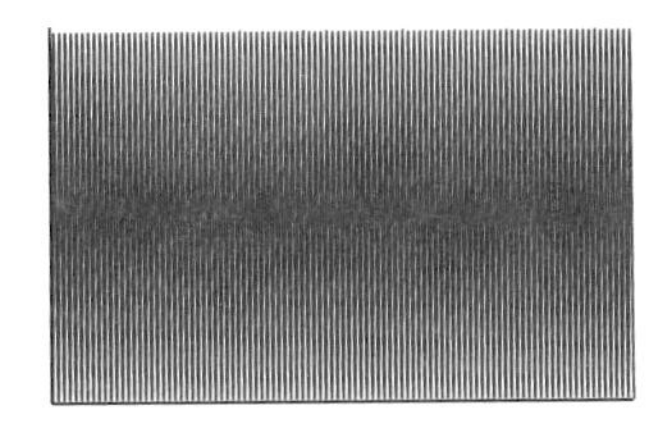

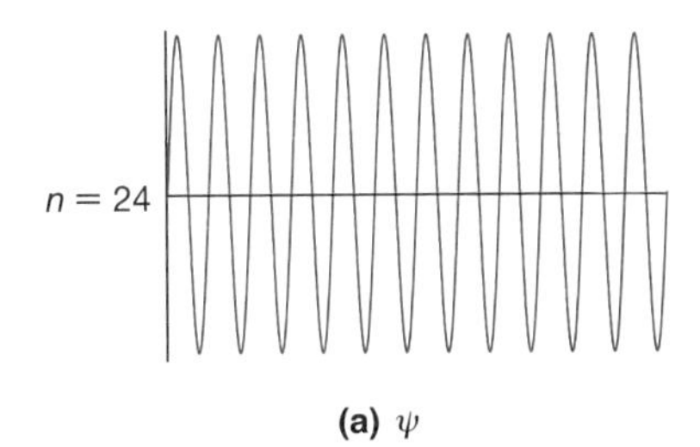

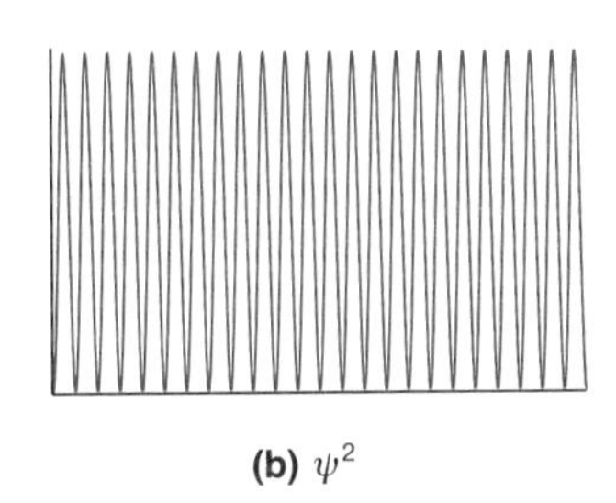

(a) ψ **(b)** ψ^2

Figure 10.18
Plots of (a) ψ and (b) ψ^2 for high energy levels of a particle in a one-dimensional box. As the quantum number becomes very large, the probability density, ψ^2, starts to resemble the uniform distribution expected for a classical (macroscopic) particle.

energy of the system is effectively not quantized; instead, it can essentially vary continuously. The inverse dependence of energy on L^2 means that if we confine a molecule in a container of macroscopic dimensions, then its energy will also effectively vary continuously rather than in a quantized fashion. We encountered this result earlier in the derivation of the translational kinetic energy of gases in Chapter 2. In summary, when we are dealing with systems of macroscopic magnitude, the quantum mechanical effects are no longer apparent and the systems are well described by classical mechanics.

EXAMPLE 10.9

An electron is placed in a one-dimensional box having a length of 0.10 nm (about the size of an atom). (a) Calculate the energy difference between the $n = 2$ and $n = 1$ states of the electron. (b) Repeat the calculation in (a) for an N_2 molecule in a container whose length is 10 cm. (c) Calculate the probability of finding the electron in (a) between $x = 0$ and $x = L/2$ in the $n = 1$ state.

ANSWER

(a) From Equation 10.50 we write the energy difference between the $n = 1$ and $n = 2$ states ΔE as

$$\begin{aligned}\Delta E &= E_2 - E_1\\ &= \frac{2^2h^2}{8mL^2} - \frac{1^2h^2}{8mL^2}\\ &= \frac{(4-1)(6.626 \times 10^{-34}\ \text{J s})^2}{8(9.109 \times 10^{-31}\ \text{kg})\left[(0.10\ \text{nm})\left(\frac{10^{-9}\ \text{m}}{\text{nm}}\right)\right]^2}\left(\frac{\text{kg m}^2\ \text{s}^{-2}}{\text{J}}\right)\\ &= 1.8 \times 10^{-17}\ \text{J}\end{aligned}$$

This energy difference is of comparable magnitude to the difference between the $n = 1$ and $n = 2$ state for the hydrogen atom (see Equation 10.28).

(b) The mass of an N_2 molecule is 4.65×10^{-26} kg, so we write

$$\Delta E = E_2 - E_1$$

$$= \frac{(4 - 1)(6.626 \times 10^{-34}\ \text{J s})^2}{8(4.65 \times 10^{-26}\ \text{kg})\left[(10\ \text{cm})\left(\frac{10^{-2}\ \text{m}}{\text{cm}}\right)\right]^2}\left(\frac{\text{kg m}^2\ \text{s}^{-2}}{\text{J}}\right)$$

$$= 3.5 \times 10^{-40}\ \text{J}$$

This energy difference is some 23 orders of magnitude less than that in (a), meaning that the translational energy of the N_2 molecule varies essentially continuously. This result supports our earlier statement that when molecules are confined to macroscopic systems, their translational motions are well described by classical mechanics.

(c) The probability (P) that the electron will be found between $x = 0$ and $x = L/2$ is

$$P = \int_0^{L/2} \psi^* \psi dx = \int_0^{L/2} \psi^2 dx$$

Using the normalized wave function in Equation 10.55 and setting $n = 1$, we find that

$$P = \frac{2}{L}\int_0^{L/2} \sin^2\left(\frac{x\pi}{L}\right)dx$$

which is readily solved with a table of integrals (see footnote on p. 415):

$$P = \frac{2}{L}\left[\frac{x}{2} - \frac{\sin(2\pi x/L)}{(4\pi/L)}\right]_0^{L/2}$$

$$= \frac{1}{2}$$

This is the expected result, either classically or quantum mechanically.

The problem of a particle in a one-dimensional box shows us that when a submicroscopic particle is in a *bound state*, that is, when potential barriers restrict its movement, its energy values are predicted to be quantized. This is precisely the case for electrons in atoms. Indeed, we can predict a number of atomic properties by using a particle in a three-dimensional box as a model. For example, the energies of an electron in a hydrogen atom must be quantized simply due to confinement. We shall discuss this and related systems shortly.

Electronic Spectra of Polyenes

One application of the particle in a one-dimensional box model is in the analysis of electronic spectra of polyenes. Polyenes are important conjugated π systems (with

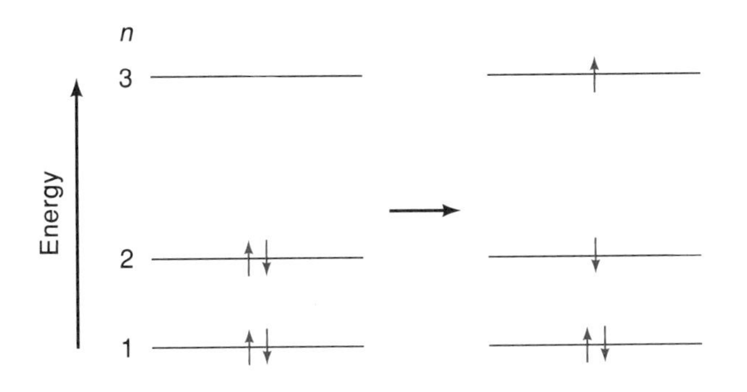

Figure 10.19
Energy levels for π orbitals in butadiene. The lowest energy electronic transition is from the highest filled level to the lowest unfilled level.

alternating C−C and C=C bonds) that play a role in photosynthesis and vision. Consider the simplest polyene, butadiene,

$$H_2C{=}\overset{H}{C}{-}\overset{H}{C}{=}CH_2$$

which contains four π electrons. Although butadiene, like all other polyenes, is not linear in shape, we assume that the π electrons move along the molecule like particles moving in a one-dimensional box. The potential energy along the chain is constant but rises sharply at the ends. Thus, the energies of the π electrons are quantized. This assumption, called the *free-electron model*, enables us to calculate the difference between energy levels and predict wavelengths associated with electronic transitions.

Butadiene has four π electrons, which occupy the two lowest-energy π orbitals when the molecule is in its electronic ground state (Figure 10.19). According to the *Pauli exclusion principle* (see Section 12.8), the electrons in each energy level have opposite spins. The electronic transition we are interested in is the one from the highest-filled level to the lowest-unfilled level (because it is the one that is usually measured experimentally); that is, the $n = 2$ to $n = 3$ transition. From Equation 10.50 we can derive a general expression for the wavelength for such a transition as follows. The number of filled energy levels is $N/2$, where N is the number of carbon atoms. This number ($N/2$) is also equal to the quantum number of the highest occupied level. The transition, then, is from the $N/2$ level to the $N/2 + 1$ level, and the energy difference is

$$\begin{aligned}\Delta E &= \frac{[(N/2) + 1]^2 h^2}{8m_e L^2} - \frac{(N/2)^2 h^2}{8m_e L^2} \\ &= \left[\left(\frac{N}{2} + 1\right)^2 - \left(\frac{N}{2}\right)^2\right]\frac{h^2}{8m_e L^2} \\ &= (N + 1)\frac{h^2}{8m_e L^2} \qquad (10.56)\end{aligned}$$

Using the relations $c = \lambda \nu$ and $\Delta E = h\nu$, we arrive at the following expression for the wavelength:

$$\lambda = \frac{hc}{\Delta E} = \frac{8m_e L^2 c}{h(N + 1)} \qquad (10.57)$$

For butadiene we have $N = 4$. To calculate L, the length of the molecule, we use bond lengths of 154 pm (1.54 Å) for C−C bonds and 135 pm for C=C bonds, plus the distance equal to a C atom radius at each end (77 pm). Thus, thc length of the molecule is (2 × 135 pm) + 154 pm + (2 × 77 pm) = 578 pm, or 5.78×10^{-10} m, so that

$$\lambda = \frac{8(9.109 \times 10^{-31}\ \text{kg})(5.78 \times 10^{-10}\ \text{m})^2(3.00 \times 10^8\ \text{m s}^{-1})}{(6.626 \times 10^{-34}\ \text{J s})(4 + 1)}$$

$$= 2.20 \times 10^{-7}\ \text{m} = 220\ \text{nm}$$

The experimentally measured wavelength is 217 nm. Considering the simplicity of the model, the agreement is remarkably good.

10.10 Particle in a Two-Dimensional Box

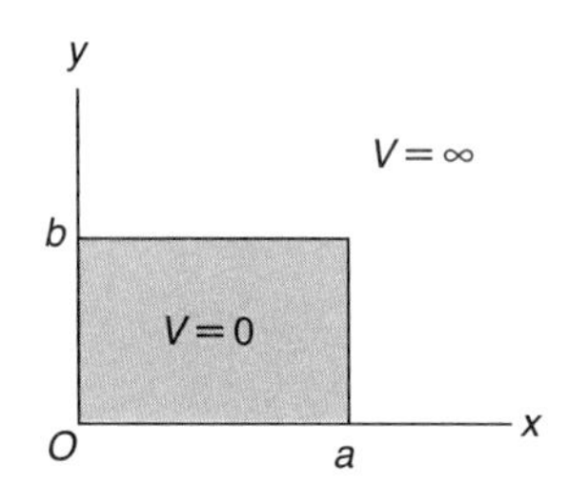

Figure 10.20
Particle in a two-dimensional box.

We now extend our discussion of a particle in a bound state to a two-dimensional system. In the model for a particle in a two-dimensional box, the particle is free to move about in two independent Cartesian coordinates, x and y. The potential energy is zero when the particle is in the box and infinite elsewhere (Figure 10.20). Our goal is to once again find the wave functions and the corresponding energies of the particle.

We start with the time-independent Schrödinger equation,

$$\hat{H}\psi(x, y) = E\psi(x, y) \qquad 0 \le x \le a,\ 0 \le y \le b$$

where a and b are the dimensions of the box. Because this is a zero-potential model, the potential $V = 0$. The Schrödinger equation, then, is expressed only in the kinetic energy term:

$$\hat{K}\psi(x, y) = E\psi(x, y)$$

$$-\frac{\hbar^2}{2m}\left[\frac{\partial^2}{\partial x^2} + \frac{\partial^2}{\partial y^2}\right]\psi(x, y) = E\psi(x, y) \tag{10.58}$$

With the two variables x and y, this is a more complicated situation than we have seen previously. To solve Equation 10.58, we assume that the variables x and y are independent of each other so that we may accomplish a *separation of variables.*

$$\psi(x, y) = X(x)Y(y) \tag{10.59}$$

The function $X(x)$ depends only on the x coordinate and the function $Y(y)$ depends only on the y coordinate. We substitute the wave function (Equation 10.59) into the Schrödinger equation (Equation 10.58),

$$-\frac{\hbar^2}{2m}\left[\frac{\partial^2 X(x)Y(y)}{\partial x^2} + \frac{\partial^2 X(x)Y(y)}{\partial y^2}\right] = EX(x)Y(y)$$

and then divide each term by the wave function $X(x)Y(y)$. Rearranging gives

$$\left[\frac{1}{X(x)}\frac{\partial^2 X(x)}{\partial x^2} + \frac{1}{Y(y)}\frac{\partial^2 Y(y)}{\partial y^2}\right] = -\frac{2mE}{\hbar^2} \tag{10.60}$$

We can write all of the terms that depend on x on one side of the equation and all the terms that depend on y on the other. Note that X is a function of the variable x, and that it does not depend on the variable y. We say that the function X is a constant with respect to the variable y. Similarly, recognizing that the function $Y(y)$ is a constant with respect to x, we rewrite Equation 10.60 as

$$\frac{1}{X(x)}\frac{\partial^2 X(x)}{\partial x^2} = -\frac{2mE}{\hbar^2} - \frac{1}{Y(y)}\frac{\partial^2 Y(y)}{\partial y^2}$$

We have separated the variables x and y so that they are on opposite sides of the equation. We now solve two independent one-dimensional differential equations. Because Y and y are constants with respect to x, we write

$$\frac{1}{X(x)}\frac{\partial^2 X(x)}{\partial x^2} = \text{constant} \tag{10.61}$$

which is equivalent to Equation 10.40. Using the results for the particle in a one-dimensional box, we may write the solutions directly as

$$X_{n_x}(x) = \left(\frac{2}{a}\right)^{1/2} \sin\left(\frac{n_x \pi x}{a}\right) \qquad n_x = 1, 2, 3, \ldots \tag{10.62}$$

and

$$E_{n_x} = \frac{n_x^2 h^2}{8ma^2} \tag{10.63}$$

There is a family of solutions and each of them is identified with a quantum number, n_x, where the subscript denotes the x dimension. The treatment of the function $Y(y)$ is the same. Thus,

$$Y_{n_y}(y) = \left(\frac{2}{b}\right)^{1/2} \sin\left(\frac{n_y \pi y}{b}\right) \qquad n_y = 1, 2, 3, \ldots \tag{10.64}$$

and

$$E_{n_y} = \frac{n_y^2 h^2}{8mb^2} \tag{10.65}$$

And when we combine our one-dimensional wave functions (Equations 10.62 and 10.64), we obtain the two-dimensional wave functions for a particle confined to a plane.

$$\psi_{n_x,n_y}(x,y) = \left(\frac{4}{ab}\right)^{1/2} \sin\left(\frac{n_x \pi x}{a}\right) \sin\left(\frac{n_y \pi y}{b}\right) \tag{10.66}$$

The total energy is the sum of the energies for the one-dimensional parts:

$$E_{n_x,n_y} = \frac{n_x^2 h^2}{8ma^2} + \frac{n_y^2 h^2}{8mb^2} \tag{10.67}$$

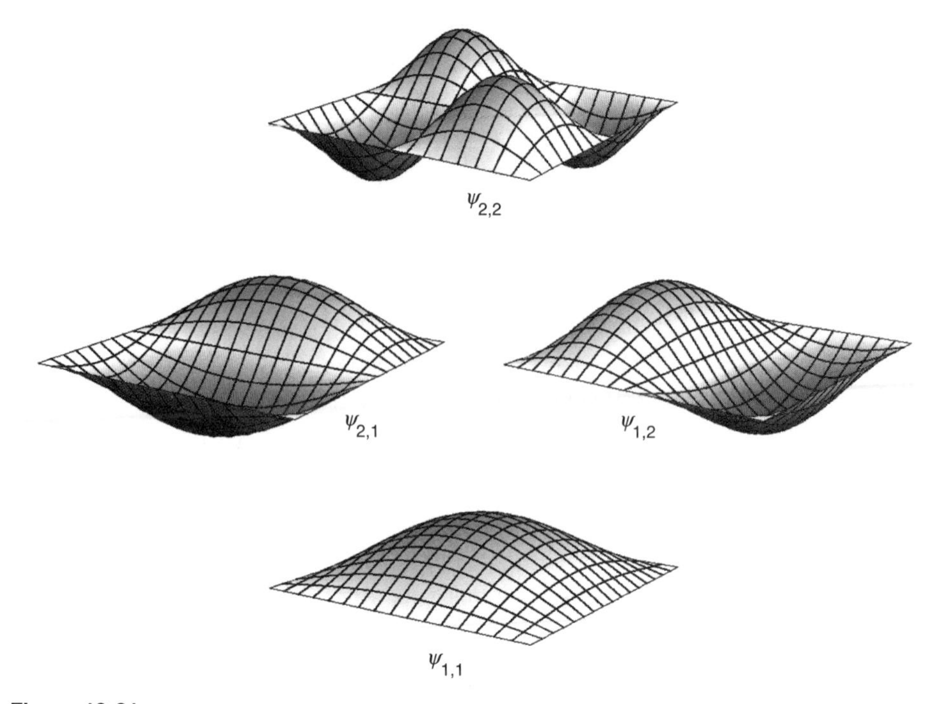

Figure 10.21
Some wave functions for a particle in a two-dimensional box with $a = b$.

Next we consider the shape of the box (square versus rectangle), which can affect both the wave functions and the energies of the particle. If the box is a square, that is, if $a = b$, the wave functions and energies are given by

$$\psi_{n_x,n_y}(x,y) = \frac{2}{a}\sin\left(\frac{n_x\pi x}{a}\right)\sin\left(\frac{n_y\pi y}{b}\right)$$

and

$$E_{n_x,n_y} = \frac{(n_x^2 + n_y^2)h^2}{8ma^2}$$

As with the particle in a one-dimensional box, the ground-state wave function $\psi_{1,1}(x,y)$, is nodeless and the first excited-state wave function has a single node (Figure 10.21). Because this is a square potential, the x and y coordinates are related by symmetry and the first excited state is doubly *degenerate*. States that are degenerate have the same energy. The wave function $\psi_{2,1}(x,y)$, which has a node in the x coordinate, has the same energy as the wave function $\psi_{1,2}(x,y)$, which has a node in the y coordinate. The probability density (Figure 10.22) for the ground state shows that the particle is most likely to be in the very middle of the box, and less likely to be near the edges. In the first excited state, however, the particle is likely to be found on one side or the other of the box, but not in the middle. In the limit of very large quantum numbers of n_x and n_y, the probability density function approaches the classical limit of uniform probability throughout the box.

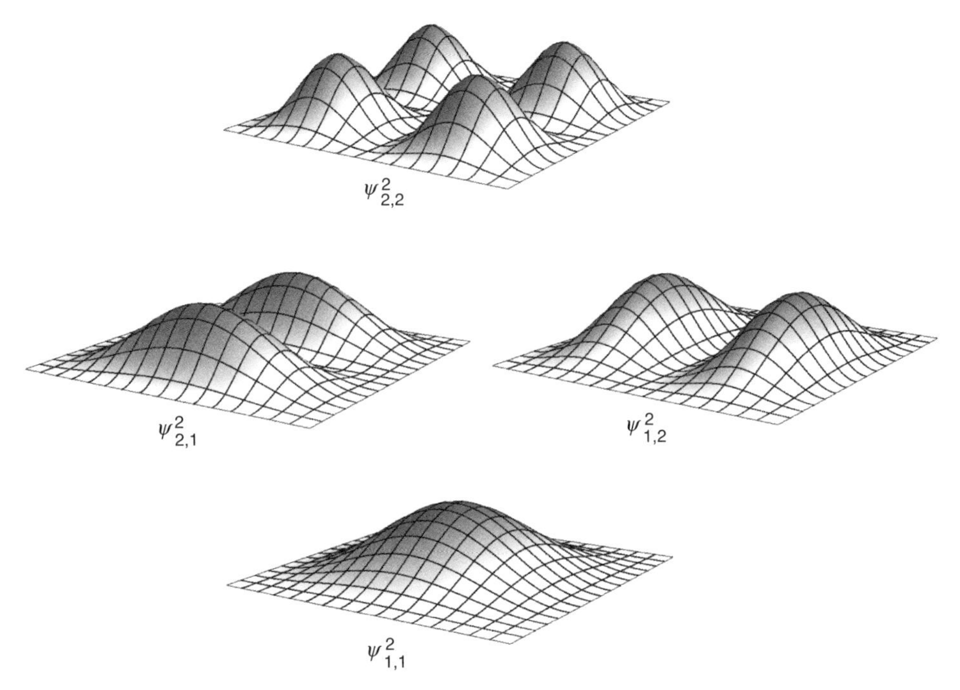

Figure 10.22
Probability density for a particle in a two-dimensional box with $a = b$.

The symbol for degeneracy is g.

The pattern of degeneracies in the energy levels depends on the relative sizes of the two sides of the box, a and b (Figure 10.23). For the square box ($a = b$), every state with $n_x = n_y$ is singly degenerate ($g = 1$), and for every state with $n_x \neq n_y$ there is a second state of equal energy with the two quantum numbers switched ($g = 2$). When the two-dimensional box is a rectangle ($a \neq b$), there are no degeneracies and the wave functions $\psi_{2,1}(x,y)$ and $\psi_{1,2}(x,y)$ have different energies, but there may be *accidental* degeneracies if a is an integral multiple of b. Extending our discussion to molecular systems, we might expect that high-symmetry molecules like benzene may possess degenerate electronic energy levels, whereas toluene or naphthalene would not. This is indeed the case.

There are some broad generalizations that are illustrated by the particle in one- and two-dimensional box problems. As stated earlier, there is one quantum number for each dimension. Without going through the derivation, the Schrödinger equation for the particle in a three-dimensional box has a solution whose energies and wave functions depend of three quantum numbers (we might call them n_x, n_y, and n_z). This turns out to be a general principle—there is one quantum number for each *degree of freedom* that describes the particle's motion (see Section 2.9).

A second generalization is that when we separate variables mathematically, the wave function is written as the product of single-variable wave functions, whereas the energy is written as the sum of the one-coordinate energies. Thus, for a particle in a three-dimensional box, we anticipate a family of solutions,

$$\psi_{n_x,n_y,n_z}(x, y, z) = X(x)Y(y)Z(z)$$

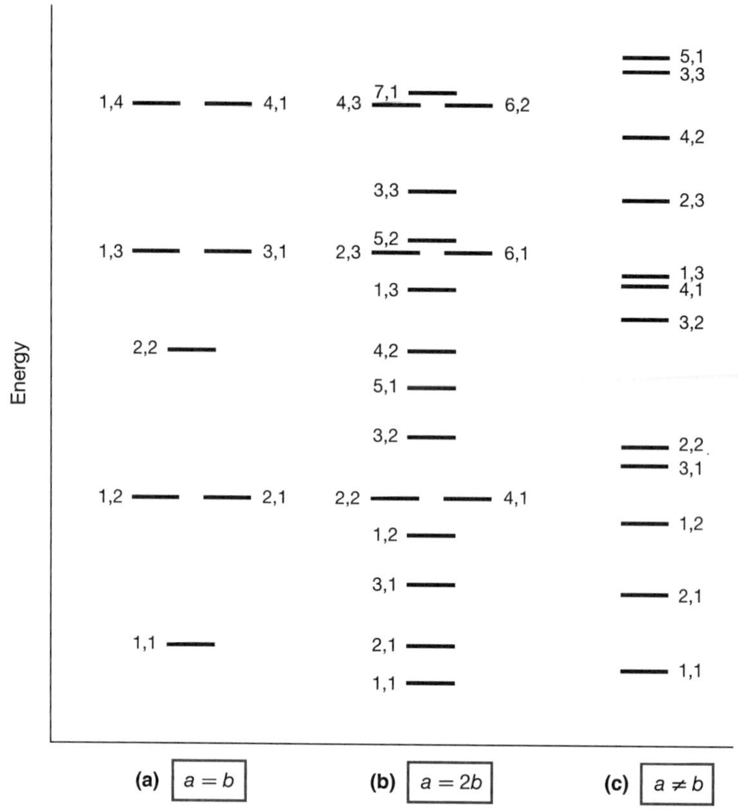

Figure 10.23
(a) Energy levels (in arbitrary units) for a particle in a two-dimensional box with $a = b$. We see that the wave functions $\psi_{1,2}$ and $\psi_{2,1}$ are doubly degenerate. This follows from the fact that because the box is square, we can convert $\psi_{1,2}$ into $\psi_{2,1}$ simply by rotating through 90°. The same holds for higher-order wave functions. (b) Energy levels for a particle in a two-dimensional box with $a = 2b$. Certain wave functions such as $\psi_{2,2}$ and $\psi_{4,1}$ are doubly degenerate. (c) Energy levels for a particle in a two-dimensional box where a is not an integral multiple of b. Normally no degeneracy exists in this case.

and

$$E_{n_x,n_y,n_z} = E_{n_x} + E_{n_y} + E_{n_z}$$

where the n's are independent quantum numbers:

$$n_x = 1, 2, 3, \ldots, \qquad n_y = 1, 2, 3, \ldots \qquad n_z = 1, 2, 3, \ldots$$

The spatial coordinates are each confined as follows:

$$0 \le x \le a \qquad 0 \le y \le b \qquad 0 \le z \le c$$

Applying our results to the hydrogen atom, which is a three-dimensional bound-state problem, we predict that the electron's energies are quantized and that its wave functions depend on three quantum numbers.

10.11 Particle on a Ring

Finally we consider a somewhat different bound state situation in which a particle is confined to a ring, known as the *particle-on-a-ring* model. In this framework, the particle is constrained to lie on a ring in a two-dimensional plane at a fixed distance from the center of the ring (Figure 10.24). The potential energy of the particle is zero on the ring, and infinite elsewhere. Like the particle-in-a-box model, this is a zero-potential model.

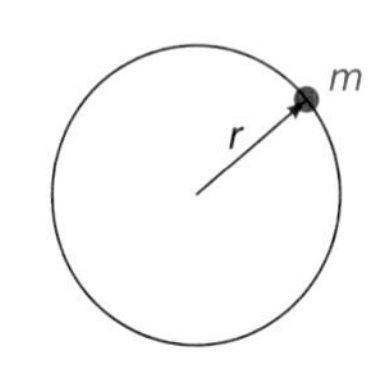

Figure 10.24
The particle-on-a-ring problem.

We start by placing the ring in the x–y plane. As with the particle in a two-dimensional box, the Schrödinger equation can be written in terms only of its kinetic energy operator (see Equation 10.58):

$$-\frac{\hbar^2}{2m}\left[\frac{\partial^2}{\partial x^2}+\frac{\partial^2}{\partial y^2}\right]\psi(x, y) = E\psi(x, y)$$

Solving the particle-on-a-ring problem may be simplified by a *transformation of variables*; that is, we replace the position in the x–y plane with a polar coordinate system in terms of a radius r and an angle ϕ (Figure 10.25). The advantage of using polar coordinates is that because r is a constant, the wave function depends only on a single variable—the angle ϕ; that is,

$$\psi = \psi(\phi)$$

In polar coordinates, the kinetic energy operator is also simplified,

$$\hat{K} = \frac{-\hbar^2}{2I}\frac{d^2}{d\phi^2} \tag{10.68}$$

where I, the *moment of inertia*, is given by

Moment of inertia is the angular equivalent of mass.

$$I = mr^2 \tag{10.69}$$

The Schrödinger equation can now be written as

$$\frac{-\hbar^2}{2I}\frac{d^2\psi(\phi)}{d\phi^2} = E\psi(\phi) \tag{10.70}$$

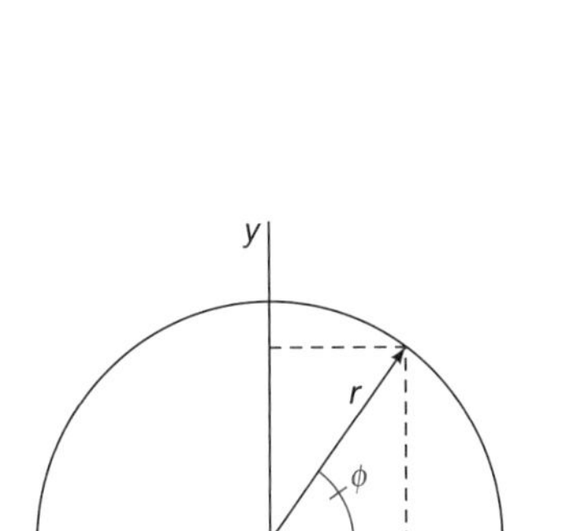

Figure 10.25
Polar coordinates in two dimensions. A point in the plane is described by a radius, $r = \sqrt{x^2 + y^2}$, and an angle, $\phi = \arctan(y/x)$. Note that $x = r\cos\phi$, and $y = r\sin\phi$.

As in previous cases, the solution to this equation is restricted by boundary conditions. First, the wave function must be zero everywhere except when the particle is on the ring. Second, the wave function must have the same value and phase when it comes completely around the circle. If this were not the case, then the wave function would destructively interfere with itself and the wave function would have a zero value. The boundary condition for constructive interference is

Equation 10.71 satisfies the condition that ψ is single-valued.

$$\psi(\phi) = \psi(\phi + 2\pi) \tag{10.71}$$

where the angle ϕ is expressed in radians (see Appendix A). The familiar trigonometric functions, sine and cosine, satisfy the boundary condition and the Schrödinger equation (Equation 10.70), but there is a more general solution. From the Euler relationship,

$$e^{\pm i\phi} = \cos\phi \pm i\sin\phi \tag{10.72}$$

The solution to the Schrödinger equation takes the form

$$\psi_m(\phi) = Ae^{im\phi} \qquad m = 0, \pm1, \pm2, \dots \tag{10.73}$$

where A is the normalization constant and m is a quantum number. Once again, imposing a boundary condition has resulted in a quantization of the energy. The solution to the differential equation is a family of wave functions, where each one is identified with a particular quantum number m. This quantum number differs from the the one we obtained for the particle-in-a-box problem in that m may take on zero and negative integer values (not just positive integers).

Our solution is not yet complete, however, because we still have to solve for A. The probability of finding the particle somewhere on the ring must be equal to unity; that is,

$$\int_{\text{all space}} \psi^*\psi d\tau = 1 \tag{10.74}$$

For the particle-on-a-ring problem, $d\tau = d\phi$, and the normalization condition is

$$\int_0^{2\pi} \psi_m(\phi)^*\psi_m(\phi)d\phi = 1 \tag{10.75}$$

To solve for A, we substitute Equation 10.73 into Equation 10.75:

$$\int_0^{2\pi} Ae^{-im\phi}Ae^{im\phi}d\phi = 1$$

$$A^2\int_0^{2\pi} d\phi = 1$$

$$A^2\,(2\pi) = 1$$

Thus, the normalization constant is given by

$$A = (2\pi)^{-1/2}$$

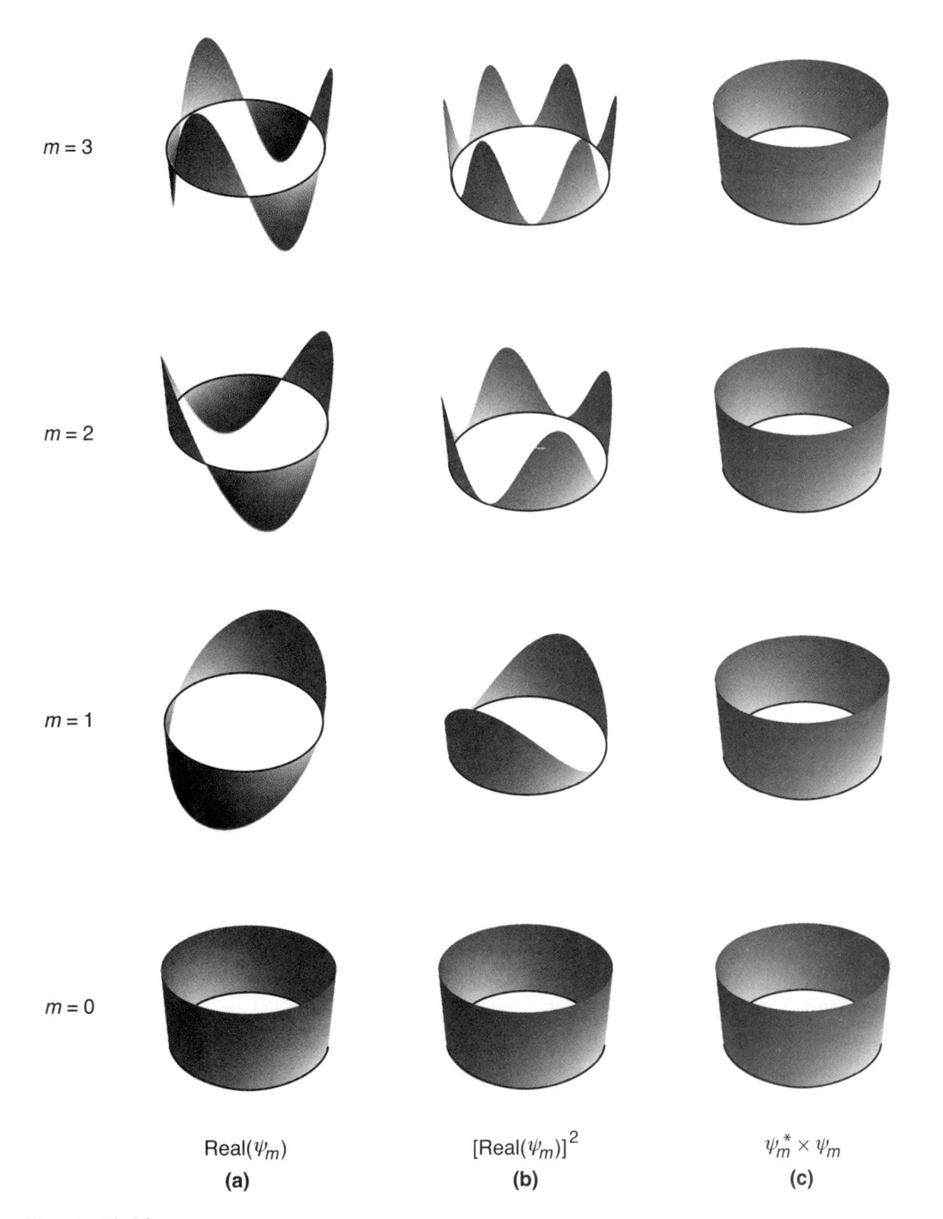

Figure 10.26
(a) Real part of the wave functions for the particle on a ring. (b) Square of the real part of wave functions. (c) Probability density given by the square magnitude of the wave function, that is, the complex conjugate of the wave function times the wave function.

and the wave functions are

$$\psi_m(\phi) = (2\pi)^{-1/2} e^{im\phi} \qquad m = 0, \pm 1, \pm 2, \ldots \tag{10.76}$$

It is difficult to plot complex wave functions, so we often simply plot the real part of the wave function. Consistent with the nodal theorem, the *real* part ground-state wave function is nodeless—it has a constant value everywhere on the ring. The *real* part of the first excited state wave function has a single node, and so on (Figure 10.26a). The square of the *real* part of the wave function (Figure 10.26b) also shows an increasing number of nodes with increasing quantum number. The probability density function is given by the square magnitude of the wave function, that is, the complex conjugate of the wave function times the wave function. As Figure 10.26c shows, this function is uniform for all quantum states.

All we need now are the energies, which we solve by substituting the wave function (Equation 10.76) into the Schrödinger equation (Equation 10.70) and differentiating with respect to ϕ twice:

$$\frac{-\hbar^2}{2I}\frac{d^2}{d\phi^2}[(2\pi)^{-1/2}e^{im\phi}] = E[(2\pi)^{-1/2}e^{im\phi}]$$

$$\frac{-\hbar^2}{2I}\frac{d}{d\phi}[im(2\pi)^{-1/2}e^{im\phi}] = E[(2\pi)^{-1/2}e^{im\phi}]$$

This is a good check that our wave function is indeed an eigenfunction.

$$\frac{\hbar^2 m^2}{2I}[(2\pi)^{-1/2}e^{im\phi}] = E[(2\pi)^{-1/2}e^{im\phi}]$$

We divide both sides of the equation by $\psi_m(\phi)$ to solve for the energy:

$$E_m = \frac{\hbar^2}{2I}m^2 \tag{10.77}$$

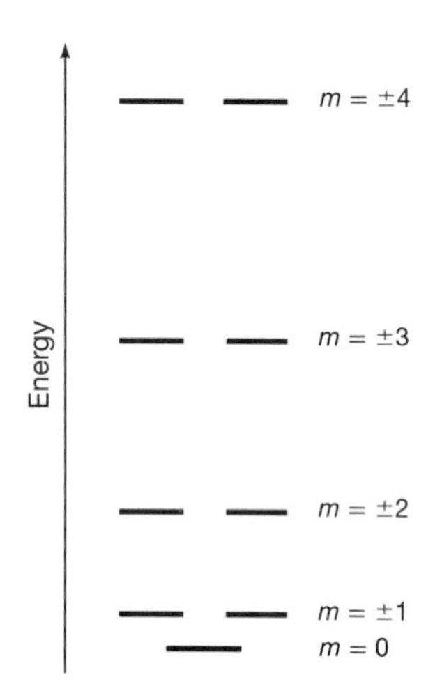

Figure 10.27
Energy level diagram for a particle on a ring.

Here we have an interesting situation—the ground state is singly degenerate ($g = 1$), and all of the excited states are doubly degenerate ($g = 2$), because with the exception of $m = 0$, the energy of each state with a positive m value is the same as that for a negative m value. The energy levels increase as m^2, but the energy level spacings ($\Delta E = E_{m+1} - E_m$) are proportional to m (Figure 10.27).

The particle-on-a-ring model differs from the particle-in-a-box model in that the integer quantum number may take on a negative or zero value. A simple physical interpretation of the positive and negative values of m is that they correspond to the particle moving in clockwise and counterclockwise direction around the ring. In the absence of other influences, which sign corresponds to which direction is arbitrary but the relative signs are not—they are important for the consideration of constructive and destructive interference. That the quantum number may take a value of zero may seem unusual. As we saw earlier for the particle-in-a-box problem, a quantum number of zero would lead to zero energy, which is a violation of the Heisenberg uncertainty principle. For the particle-on-a-ring system, if the quantum number $m = 0$ then the speed of the particle is zero, but there is no knowledge of the location of the particle because the angular position ϕ could take on any value. Therefore, the Heisenberg uncertainty principle is not violated.

10.12 Quantum Mechanical Tunneling

What would happen if the potential walls surrounding the particle in the one-dimensional box were not infinitely high? The particle would escape when its kinetic energy became equal to or greater than the potential energy of the barrier. What is more surprising, however, is the fact that we might find the particle outside the box even if its kinetic energy is insufficient to reach the top of the barrier! This phenomenon,

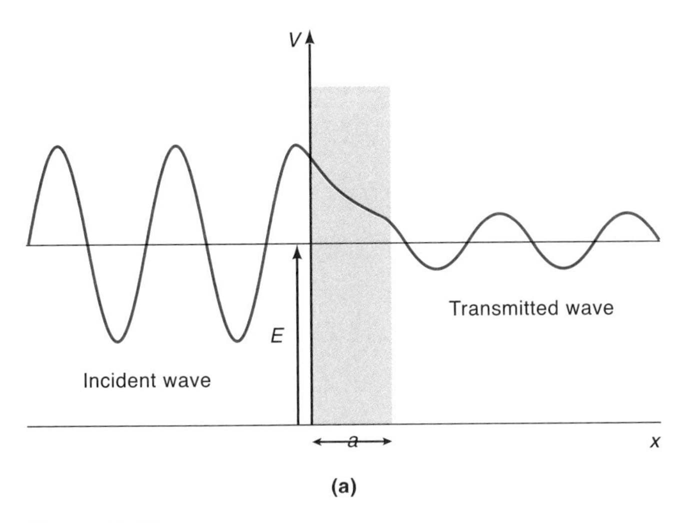

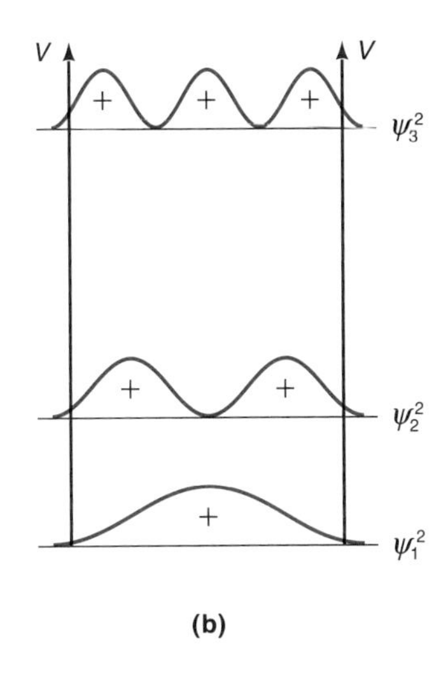

Figure 10.28
(a) Schematic diagram showing tunneling through a finite potential barrier. The particle is moving from left to right. Most of the incident wave of the particle is reflected back. A small part of the wave penetrates the barrier, emerging on the other side with diminished amplitude. (b) Plots of ψ^2 for a particle in a one-dimensional box with finite potential barriers. Note that in contrast to Figure 10.17b, these curves extend outside the box. There is some probability that the particle will be found inside the walls of the box.

known as *quantum mechanical tunneling*, has no analog in classical physics. It arises as a consequence of the wave nature of particles. Quantum mechanical tunneling has many profound consequences in the physical and biological sciences.*

The phenomenon of quantum mechanical tunneling was introduced by the Russian-American physicist George Gamow (1904–1968), among others, in 1928 to explain α decay. This is a process wherein a nucleus spontaneously decays by emitting an α particle, which is a helium nucleus (He^{2+}); for example,

$$^{238}_{92}\text{U} \longrightarrow {}^{234}_{90}\text{Th} + \alpha \qquad t_{1/2} = 4.51 \times 10^9 \text{ yr}$$

The dilemma facing physicists was the following: for U-238 decay, the measured (kinetic) energy of the emitted α particle is about 4 MeV, whereas the coulombic barrier is on the order of 250 MeV. (Imagine the α particle being at the center of the nucleus. It is surrounded by other protons and therefore behaves like a particle trapped in a box. The potential barriers are the result of electrostatic repulsion due to other protons present. The barrier height can be calculated from the radius of the nucleus and the atomic number.) The question naturally arose as to how the α particle can overcome the barrier and leave the nucleus. Gamow suggested that the α particle, being a quantum mechanical object, has wavelike properties that enable it to penetrate a potential barrier (Figure 10.28). This explanation turned out to be correct. In general, for finite potential barriers, there is always some probability of finding the particle outside of the box.

In nuclear physics and nuclear chemistry, the common unit for energy is the *e*V or Me*V* (1×10^6 *eV*), where 1 *eV* = 1.602×10^{-19} J.

* See W. T. Scott, *J. Chem. Educ.* **48**, 524 (1971), for an interesting illustration of quantum-mechanical tunneling.

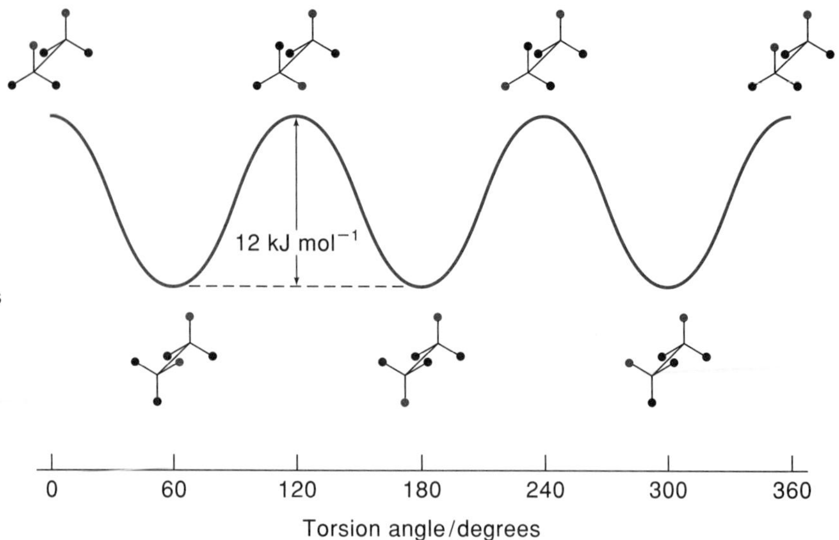

Figure 10.29
Potential energy diagram for internal rotation in ethane. Two of the H atoms are shown in red to indicate more clearly the process of rotation about the C—C bond.

The probability (P) of a particle of mass m tunneling through the barrier is proportional to the quantity*

$$P \propto \exp\left\{-\frac{4\pi a}{h}[2m(V-E)]^{1/2}\right\} \qquad V > E \tag{10.78}$$

where exp means exponential, V is the potential barrier, E is the energy of the particle, and a is the thickness of the barrier. Unless $V = \infty$ or $a = \infty$, there is always a probability that the particle will escape, although P may be a very small number. This is certainly the case for the α decay of U-238, which has an exceedingly large half-life. Physically, a very small P means that the α particles have to collide with the barrier many, many times before one of them can escape the nucleus. As Equation 10.78 shows, quantum mechanical tunneling is most likely to take place (for comparable V and a values) with light particles such as electrons, protons, and hydrogen atoms.

The energy profile for a chemical reaction is usually described in terms of reactant molecules acquiring sufficient energy to get over the activation energy barrier to form products (see Figure 15.11). However, in a number of cases (e.g., some electron exchange processes) reactions are known to proceed even when an insufficient amount of energy is available. Such results have been attributed to quantum mechanical tunneling. As another example, consider the internal rotation in ethane. Rotation about the C—C bond in ethane, while very rapid, is not completely free. The potential energy varies sinusoidally with the conformation of ethane (Figure 10.29). Spectroscopic studies of compounds containing a methyl group demonstrate that, at sufficiently low temperatures, when the molecules do not possess enough kinetic energy to overcome the potential energy barrier, rotation about the C—C bond still occurs. This phenomenon is also attributed to quantum mechanical tunneling.

* For the derivation, see F. L. Pilar, *Elementary Quantum Chemistry*, 2nd ed., McGraw-Hill Book Company, New York, 1990.

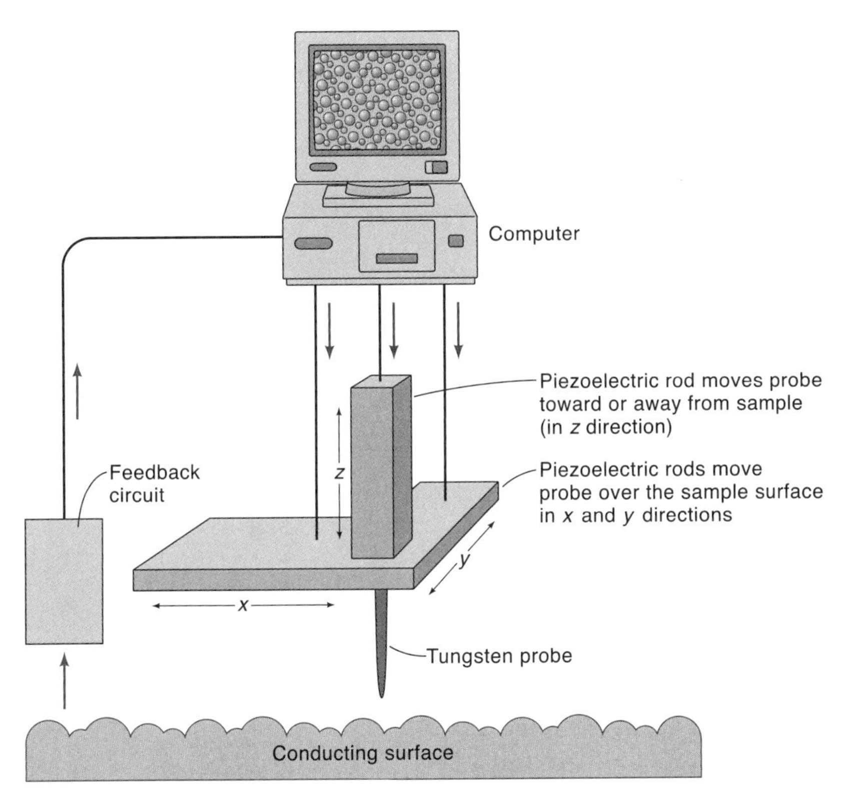

Figure 10.30
The scanning tunneling microscope. A tunneling current flows between the probe and the sample when there is a small voltage between them. A feedback circuit, which provides this voltage, senses the current and varies the voltage on a piezoelectric stepping device (the z drive) to keep the distance constant between the probe and the sample (a conducting surface). A computer provides voltages to the x drive and y drive to move the probe over the surface.

Scanning Tunneling Microscopy

A practical application of the quantum mechanical tunneling effect is the scanning tunneling microscope (STM), shown schematically in Figure 10.30. The STM consists of a tungsten metal needle with a very fine point, the source of the tunneling electrons. For atomic resolution, the tip is a single atom of tungsten! A voltage is maintained between the needle and the surface of the sample to induce electrons to tunnel through space to the sample. As the needle moves over the sample, at a distance of a few atomic diameters from the surface, the tunneling current, I, is measured. This current decreases exponentially with increasing distance s from the sample,

$$I \propto e^{-s} \tag{10.79}$$

and thus is very sensitive to height. By using an electronic feedback loop, the vertical position of the tip can be adjusted to a constant distance from the surface. The extent of these adjustments, which profile the sample, is recorded and displayed as a three-dimensional false-colored image. The STM is a useful tool in chemical, biological, and materials science research, but it requires a conducting surface. The Atomic Force Microscope (AFM) provides similar surface profile information for nonconducting surfaces.

Key Equations

$u = \lambda v$	(Speed of a wave)	(10.3)
$E = nhv \quad n = 0, 1, 2, \ldots$	(Planck's quantum theory)	(10.6)
$hv = \Phi + \frac{1}{2}m_e u^2$	(Photoelectric effect)	(10.8)
$E_n = -\frac{m_e Z^2 e^4}{8h^2\varepsilon_0^2}\frac{1}{n^2}$	(Energies of electron in H atom)	(10.18)
$\Delta E = E_\mathrm{f} - E_\mathrm{i} = hv$	(Resonance condition)	(10.20)
$\tilde{v} = \tilde{R}_\mathrm{H}\left(\frac{1}{n_\mathrm{i}^2} - \frac{1}{n_\mathrm{f}^2}\right)$	(Wave numbers of H atom electronic transitions)	(10.22)
$\lambda = \frac{h}{mu} = \frac{h}{p}$	(de Broglie wavelength for particles)	(10.25)
$\Delta x \Delta p \geq \frac{h}{4\pi}$	(Heisenberg uncertainty principle)	(10.27)
$\Delta E \Delta t \geq \frac{h}{4\pi}$	(Heisenberg uncertainty principle)	(10.28)
$\int_{\text{all space}} \psi_i^* \psi_j d\tau = \delta_{ij}$	(Orthonormalization condition)	(10.33)
$\hat{H}\psi = E\psi$	(Schrödinger wave equation)	(10.37)
$E_n = \frac{n^2 h^2}{8mL^2}$	(Energies of a particle in a one-dimensional box)	(10.50)
$\psi_n(x) = \left(\frac{2}{L}\right)^{1/2} \sin\left(\frac{n\pi x}{L}\right)$	(Wave function for particle in a one-dimensional box)	(10.55)
$E_{n_x,n_y} = \frac{n_x^2 h^2}{8ma^2} + \frac{n_y^2 h^2}{8mb^2}$	(Energies of a particle in a two-dimensional box)	(10.67)
$I = mr^2$	(Moment of inertia)	(10.69)
$E_m = \frac{\hbar^2}{2I}m^2$	(Energies of a particle on a ring)	(10.77)

APPENDIX 10.1

The Bracket Notation in Quantum Mechanics

The notations we used in this chapter and indeed in most physical chemistry texts are based on calculus (integral) formalisms. In the 1930s, the English physicist Paul Dirac (1902–1984) introduced what is known as the bracket notation for dealing with atomic structure, chemical bonding, and spectroscopy. In this notation, the wave function of a system is given by a quantity called *ket*, $|\psi\rangle$. Thus, instead of writing the eigenfunction–eigenvalue expression as

$$\hat{A}\psi = a\psi$$

we write

$$\hat{A}|\psi\rangle = a|\psi\rangle$$

Dirac used the *bra* notation to represent the complex conjugate ψ^* as $\langle\psi|$ and the * sign is not shown in the bra part. We can now write the orthonormalization condition (see Equation 10.33) as

$$\langle\psi_i|\psi_j\rangle = \delta_{ij}$$

where the integration over all space is understood. Likewise, the expectation value (see Equation 10.34) can be represented by

$$\langle a\rangle = \frac{\langle\psi_n|\hat{A}|\psi_n\rangle}{\langle\psi_n|\psi_n\rangle}$$

Note that the $\langle\ \rangle$ signs in $\langle a\rangle$ denote the average value of a resulting from measurements, and are not part of the bracket notation.

As you can see, the Dirac bracket notation is more compact and succinct because it avoids using cumbersome integral signs.

Suggestions for Further Reading

BOOKS

Atkins, P. W., *Quanta: A Handbook of Concepts*, Oxford University Press, New York, 1991.
Bell, R. P., *The Tunnel Effect in Chemistry*, Chapman and Hall, London, 1980.
Cropper, W. H., *The Quantum Physicists*, Oxford University Press, New York, 1970.
DeVault, D., *Quantum-Mechanical Tunneling in Biological Systems*, 2nd ed., Cambridge University Press, New York, 1984.
Feynman, R. P., R. B. Leighton, and M. Sands, *The Feynman Lectures on Physics*, Volumes I, II, and III, Addison-Wesley, Reading, MA, 1963.
Herzberg, G., *Atomic Spectra and Atomic Structure*, Dover Publications, New York, 1944.
Hochstrasser, R. M., *Behavior of Electrons in Atoms*, W. A. Benjamin, Menlo Park, CA, 1964.
Karplus, M., and R. N. Porter, *Atoms and Molecules: An Introduction for Students of Physical Chemistry*, W. A. Benjamin, New York, 1970.
Pilar, F. L., *Elementary Quantum Chemistry*, 2nd ed., McGraw-Hill Book Company, New York, 1990.
Ratner, M. A., and G. C. Schatz, *Introduction to Quantum Mechanics in Chemistry*, Prentice-Hall, Upper Saddle River, NJ, 2001.

ARTICLES

Quantum Theory
"The Limits of Measurement," R. Furth, *Sci. Am.* July 1950.
"The Quantum Theory," K. K. Darrow, *Sci. Am.* March 1952.
"What Is Matter?" E. Schrödinger, *Sci. Am.* September 1953.
"The Principle of Uncertainty," G. Gamow, *Sci. Am.* January 1958.
"The Bohr Atomic Model: Niels Bohr," A. B. Garrett, *J. Chem. Educ.* **39**, 534 (1962).
"Quantum Theory: Max Planck," A. B. Garrett, *J. Chem. Educ.* **40**, 262 (1963).
"Demonstration of the Uncertainty Principle," W. Laurita, *J. Chem. Educ.* **45**, 461 (1968).
"Particles, Waves, and the Interpretation of Quantum Mechanics," N. D. Christoudouleas, *J. Chem. Educ.* **52**, 573 (1975).
"The Mass of the Photon," A. S. Goldhaber and M. M. Nieto, *Sci. Am.* May 1976.
"The Spectrum of Atomic Hydrogen," T. W. Hänsch, A. L. Schawlow, and G. W. Series, *Sci. Am.* March 1979.
"Centrifugal Force and the Bohr Model of the Hydrogen Atom," B. L. Haendler, *J. Chem. Educ.* **58**, 719 (1981).
"Does Quantum Mechanics Apply to One or Many Particles?" F. Castano, L. Lain, M. N. Sanchez Rayo, and A. Torre, *J. Chem. Educ.* **60**, 377 (1983).
"Illustrating the Heisenberg Uncertainty Principle," G. D. Peckham, *J. Chem. Educ.* **61**, 868 (1984).
"Dice Throwing as an Analogy for Teaching Quantum Mechanics," B. de Barros Neto, *J. Chem. Educ.* **61**, 1044 (1984).
"Perspectives on the Uncertainty Principle and Quantum Reality," L. S. Bartell, *J. Chem. Educ.* **62**, 192 (1985).
"On Introducing the Uncertainty Principle," G. M. Muha and D. W. Muha, *J. Chem. Educ.* **63**, 525 (1986).
"Exercises in Quantum Mechanics," F. Rioux, *J. Chem. Educ.* **64**, 789 (1987).
"Heisenberg, Uncertainty, and the Quantum Revolution," D. C. Cassidy, *Sci. Am.* May 1992.

"On a Relation Between the Heisenberg and de Broglie Principles," O. G. Ludwig, *J. Chem. Educ.* **70**, 28 (1993).
"The Duality in Matter and Light," B.-G. Englert, M. O. Scully, and H. Walther, *Sci. Am.* December 1994.
"Using Natural and Artificial Light Sources to Illustrate Quantum Mechanical Concepts," G. A. Rechtsteiner and J. A. Ganske, *Chem. Educator* [Online] **3**, 1430 (1998) DOI 10.1333/s00897980230a.
"Introducing the Uncertainty Principle Using Diffraction of Light Waves," P. L. Muiño, *J. Chem. Educ.* **77**, 1025 (2000).
"The Dirac (Bracket) Notation in the Undergraduate Physical Chemistry Curriculum: A Pictorial Introduction," F. A. Khan and J. E. Hansen, *Chem. Educator* [Online] **5**, 113 (2000) DOI 10.1333/s00897000379a.
"The Planck Radiation Law: Exercises Using the Cosmic Background Radiation Data," S. Bluestone, *J. Chem. Educ.* **78**, 215 (2001).
"Using Optical Transforms to Teach Quantum Mechanics," F. Rioux and B. J. Johnson, *Chem. Educator* [Online] **9**, 12 (2004) DOI 10.1333/s00897040748a.
"Bohr Model Calculations for Atoms and Ions," F. Rioux, *Chem. Educator* [Online] **12**, 250 (2007) DOI 10.1333/s008970720061a.
"Determining Planck's Constant Using the Vernier LabQuest Interface and Power Amplifier," T. Thanel and M. Morgan, *Chem. Educator* [Online] **16**, 62 (2011) DOI 10.1333/s00897112347a.
"Comparing the Spectral Temperature of Incandescent and Compact Fluorescent Light Bulbs," R. C. Dudek, N. T. Anderson, and J. M. Donnelly, *Chem. Educator* [Online] **16**, 76 (2011) DOI 10.1333/s00897112348a.

Particle in a Box and Particle on a Ring

"A Particle in a Chemical Box," K. M. Jinks, *J. Chem. Educ.* **52**, 312 (1975).
"On the Momentum of a Particle in a Box," G. M. Muha, *J. Chem. Educ.* **63**, 761 (1986).
"How Do Electrons Get Across Nodes?" P. G. Nelson, J. *Chem. Educ.* **67**, 643 (1990).
"The Two-Dimensional Particle in a Box," G. L. Breneman, *J. Chem. Educ.* **67**, 866 (1990).
"Determination of Carbon-Carbon Bond Length From the Absorption Spectra of Cyanine Dyes," R. S. Moog, *J. Chem. Educ.* **68**, 506 (1991).
"More About the Particle-in-a Box System: The Confinement of Matter and the Wave-Particle Dualism," K. Volkamer and M. W. Lerom, *J. Chem. Educ.* **69**, 100 (1992).
"Visible Spectra of Conjugated Dyes: Integrating Quantum Chemical Concepts with Experimental Data," G. M. Shalhoub, *J. Chem. Educ.*, **69**, 1317 (1992).
"An Alternative Derivation of the Energy Levels of the 'Particle on a Ring' System," A. Vincent, *J. Chem. Educ.* **73**, 1001 (1996).
"Evaluating Experiment with Computation in Physical Chemistry: The Particle-in-a-Box Model with Cyanine Dyes," J. R. Bocarsly and C. W. David, *Chem. Educator* [Online] **2**, 1 (1997) DOI 10.1333/s00897970135a.
"Alternative Compounds for the Particle in a Box Experiment," B. D. Anderson, *J. Chem. Educ.* **74**, 985 (1997).
"Localization and Spread of the Particle in a Box," J. J. C. Mulder, *Chem. Educator* [Online] **7**, 71 (2002) DOI 10.1333/s00897020545a.
"Semiconductor Nanocrystals: A Powerful Visual Aid for Introducing the Particle in a Box," T. Kippeny, L. A. Swafford, and S. J. Rosenthal, *J. Chem. Educ.* **79**, 1094 (2002).
"Residual or Zero-Point Energy in Quantum Systems: Another View for Two Well-Known Cases," F. Enriquez and J. J. Quirante, *Chem. Educator* [Online] **8**, 238 (2003) DOI 10.1333/s00897030697a.

Quantum Mechanical Tunneling

"Quantum Chemical Reactions in the Deep Cold," V. I. Goldanskii, *Sci. Am.* February 1980.

"The Scanning Tunneling Microscope," G. Binnig and H. Rohrer, *Sci. Am.* August 1985.

"Electron Transfer in Biology," R. J. P. Williams, *Molec. Phys.* **68**, 1 (1989).

"Electron-Tunneling Pathways in Proteins," D. Beratan, J. N. Onuchic, J. R. Winkler, and H. B. Gray, *Science* **258**, 1740 (1992).

"Scanning Tunneling Microscopy," C. M. Lieber, *Chem. & Eng. News*, April 18, 1994.

"Quantum Mechanical Tunneling Through Barriers: A Spreadsheet Approach," A. Cedillo, *J. Chem. Educ.* **77**, 528 (2000).

"Localization and Spread of the Particle in a Box," J. J. C. Mulder, *Chem. Educ.* **7**, 1 (2002).

"Atomic Scale Imaging: A Hands-On Scanning Probe Microscopy Laboratory for Undergraduates," C.-J. Zhong, L. Han, M. M. Maye, J. Luo, N. N. Kariuki, W. E. Jones, Jr., *J. Chem. Educ.* **80**, 194 (2003).

"Potential Barriers and Tunneling," M. Ellison, *J. Chem. Educ.* **81**, 608 (2004).

"Electron Tunneling, a Quantum Probe for the Quantum World of Nanotechnology," K. W. Hipps and L. Scudiero, *J. Chem. Educ.* **82**, 704 (2005).

"The Particle Inside a Ring: A Two-Dimensional Quantum Problem Visualized by Scanning Tunneling Microscopy," M. D. Ellison, *J. Chem. Educ.* **85**, 1282 (2008).

"Thinking Outside the (Particle in a) Box: Tunneling, Uncertainty and Dimensional Analysis," N. C. Blank, K. Clemons, R. Crowdis, C. Estridge, M. Foster, S. Gash, B. Gish, B. Gollihue, C. Henzman, D. Hernandez, T. Ijaz, A. Ivey, J. Jones, A. Loveless, S. Roberts, T. Sauley, E. Velasco, C. Wilson, M. M. Blackburn, H. E. Montgomery, Jr., *Chem. Educator* [Online] **15**, 134 (2010) DOI 10.1007/s00897102266a.

Problems

Quantum Theory

10.1 Calculate the energy associated with a quantum (photon) of light with a wavelength of 500 nm.

10.2 The threshold frequency for dislodging an electron from a zinc metal surface is 8.54×10^{14} Hz. Calculate the minimum amount of energy required to remove an electron from the metal.

10.3 Calculate radii for the Bohr orbits with $n = 2$ and 3 for atomic hydrogen.

10.4 Calculate the frequency and wavelength associated with the transition from the $n = 5$ to the $n = 3$ level in atomic hydrogen.

10.5 What are the wavelengths associated with **(a)** an electron moving at 1.50×10^{8} cm s^{-1}, and **(b)** a 60-g tennis ball moving at 1500 cm s^{-1}?

10.6 A photoelectric experiment was performed by separately shining a laser at 450 nm (blue light) and a laser at 560 nm (yellow light) on a clean metal surface and measuring the number and kinetic energy of the ejected electrons. Which light would generate more electrons? Which light would eject electrons with greater kinetic energy? Assume that the same number of photons is delivered to the metal surface by each laser and that the frequencies of the laser lights exceed the threshold frequency.

10.7 Explain how scientists are able to estimate the temperature on the surface of the sun. (*Hint*: Treat solar radiation like radiation from a blackbody.)

10.8 In a photoelectric experiment, a student uses a light source whose frequency is greater than that needed to eject electrons from a certain metal. However, after continuously shining the light on the same area of the metal for a long period of time, the student notices that the maximum kinetic energy of ejected electrons begins to decrease, even though the frequency of the light is held constant. How would you account for this behavior?

10.9 A proton is accelerated through a potential difference of 3.0×10^{6} V, starting from rest. Calculate its final wavelength.

10.10 Suppose that the uncertainty in determining the position of an electron circling an atom in an orbit is 0.4 Å. What is the uncertainty in its velocity?

10.11 A person weighing 77 kg jogs at 1.5 m s^{-1}. **(a)** Calculate the momentum and wavelength of this person. **(b)** What is the uncertainty in determining his position at any given instant if we can measure his momentum to $\pm 0.05\%$? **(c)** Predict the changes that would take place in this problem if the Planck constant were 1 J s.

10.12 The diffraction phenomenon can be observed whenever the wavelength is comparable in magnitude to the size of the slit opening. To be "diffracted," how fast must a person weighing 84 kg move through a door 1 m in width?

10.13 **(a)** Show that for the hydrogen atom the first term on the right-hand side of Equation 10.18 is 2.18. $\times 10^{-18}$ J. **(b)** Use Equation 10.23 to calculate the Rydberg constant in cm^{-1}. Use the constants on the inside of back cover. (*Hint*: To obtain the full six significant figures used in the text for $\tilde{R}_H$, you would need at least seven significant figures in all your constants. In fact, $\tilde{R}_H$ is known to 14 significant figures.)

10.14 Spectral lines of the Lyman and Balmer series do not overlap. Verify this statement by calculating the longest wavelength associated with the Lyman series and the shortest wavelength associated with the Balmer series (in nm).

10.15 The He^+ ion contains only one electron and is therefore a hydrogenlike ion. Calculate the wavelengths, in increasing order, of the first four transitions in the Balmer series of the He^+ ion. Compare these wavelengths with the same transitions in a H atom. Comment on the differences. (The Rydberg constant for He^+ is 8.72×10^{-18} J.)

10.16 An electron in an excited state in a hydrogen atom can return to the ground state in two different ways: **(a)** via a direct transition in which a photon of wavelength λ_1 is emitted and **(b)** via an intermediate excited state reached by the emission of a photon of wavelength λ_2. This intermediate excited state then decays to the ground state by emitting another photon of wavelength λ_3. Derive an equation that relates λ_1 to λ_2 and λ_3.

10.17 The retina of a human eye can detect light when radiant energy incident on it is at least 4.0×10^{-17} J. For light of 600-nm wavelength, to how many photons does this correspond?

10.18 A 368-g sample of water absorbs infrared radiation at 1.06×10^4 nm from a carbon dioxide laser. Suppose all the absorbed radiation is converted to heat. Calculate the number of photons at this wavelength required to raise the temperature of the water by 5.00°C.

10.19 Ozone (O_3) in the stratosphere absorbs the harmful radiation from the sun by undergoing decomposition: $O_3 \rightarrow O + O_2$. **(a)** Referring to Appendix B, calculate the $\Delta_r H°$ for this process. **(b)** Calculate the maximum wavelength of photons (in nm) that possess this energy to bring about the photochemical decomposition of ozone.

10.20 Scientists have found interstellar hydrogen atoms with quantum number n in the hundreds. Calculate the wavelength of light emitted when a hydrogen atom undergoes a transition from $n = 236$ to $n = 235$. In what region of the electromagnetic spectrum does this wavelength fall?

10.21 A student records an emission spectrum of hydrogen and notices that there is one spectral line in the Balmer series that cannot be accounted for by the Bohr theory. Assuming that the gas sample is pure, suggest a species that might be responsible for this line.

10.22 In the mid-nineteenth century, physicists studying the solar emission spectrum (a continuum) noticed a set of dark lines that did not match any of the emission lines (bright lines) on Earth. They concluded that the lines came from a yet unknown element. Later this element was identified as helium. **(a)** What is the origin of the dark lines? How were these lines correlated with the emission lines of helium? **(b)** Why was helium so difficult to detect in Earth's atmosphere? **(c)** Where is the most likely place to detect helium on Earth?

10.23 How many photons at 660 nm must be absorbed to melt 5.0×10^2 g of ice? On average, how many H_2O molecules does one photon convert from ice to liquid water? (*Hint*: It takes 334 J to melt 1 g of ice at 0°C.)

Postulates of Quantum Mechanics

10.24 An alternative to the Copenhagen interpretation of the wave function is the so-called "many worlds" interpretation. Both interpretations are consistent with observation. Do some research on each of these interpretations and decide for yourself which one you prefer.

10.25 Which of the following functions are eigenfunctions of the operator $\frac{d^2}{dx^2}$?

(a) $f(x) = x^3$, **(b)** $f(x) = e^{6x}$, **(c)** $f(x) = \ln x$, **(d)** $f(x) = \sin x$, **(e)** $f(x) = e^{-ix}$, **(f)** $f(x) =$ constant.

10.26 Given the function $f(x)$, which of the following operators are Hermitian?

(a) $\frac{d^2}{dx^2}$, **(b)** $\frac{d}{dx}$, **(c)** the identity operator, **(d)** multiply by a real constant, **(e)** multiply by x.

10.27 For each of the following functions, identify whether or not it would be an acceptable wave function over the interval indicated, and for those that are not acceptable, explain why not.
(a) $f(x) = e^{-ix}\ [0, \infty]$, **(b)** $f(x) = ae^{-x^2}\ [-\infty, \infty]$, **(c)** $f(x) = a \sin(x)\ [0, 2\pi]$, **(d)** $f(x) = a \sin(x)\ [0, \frac{\pi}{2}]$, **(e)** $f(x) = \frac{1}{x}\ [0, \infty]$.

10.28 Which of the following pairs of functions are orthogonal over the interval indicated?
(a) $f(x) = \sin(x)$ and $g(x) = \cos(x)\ [-\infty, \infty]$, **(b)** $f(x) = \sin(x)$ and $g(x) = \cos(x)\ [0, 2\pi]$, **(c)** $f(x) = e^{i\pi x}$ and $g(x) = e^{i2\pi x}\ [-1, 1]$, **(d)** $f(x) = e^{i\pi x}$ and $g(x) = e^{-i\pi x}\ [-1, 1]$, **(e)** $f(x) = ix$ and $g(x) = -ix\ [-1, 1]$.

10.29 The following functions are eigenfunctions of the operator $\frac{d^2}{dx^2}$. For each of the following eigenfunctions list the eigenvalue.
(a) $f(x) = e^{ax}$, **(b)** $f(x) = \cos(\omega x)$, **(c)** $f(x) = a \sin(\omega x)$, **(d)** $f(x) = ae^{x}$.

10.30 Which of the following functions are normalized over the interval indicated?

(a) $f(x) = \frac{1}{\sqrt{a}}\ [0, a]$

(b) $f(x) = x\ [0, 1]$

(c) $f(x) = \sqrt{\frac{2}{a}} \sin\frac{5\pi x}{a}\ [0, a]$

(d) $f(x) = \frac{1}{\sqrt{\pi}} e^{-ax^2}\ [-\infty, \infty]$

(e) $f(x) = \left(\frac{a}{\pi}\right)^{1/4} e^{-ax^2/2}\ [-\infty, \infty]$

Particle in a One-Dimensional Box

10.31 Show that Equation 10.50 is dimensionally correct.

10.32 According to Equation 10.50, the energy is inversely proportional to the square of the length of the box. How would you account for this dependence in terms of the Heisenberg uncertainty principle?

10.33 What is the probability of locating a particle in a one-dimensional box between $L/4$ and $3L/4$, where L is the length of the box? Assume the particle to be in the lowest energy level.

10.34 Derive Equation 10.50 using de Broglie's relation. (*Hint*: First you must express the wavelength of the particle in the nth level in terms of the length of the box.)

10.35 An important property of the wave functions of the particle in a one-dimensional box is that they are orthogonal; that is,

$$\int_{\text{all space}} \psi_n^* \psi_m d\tau = 0 \quad m \neq n$$

or specifically for the particle in a one-dimensional box,

$$\int_0^L \psi_n \psi_m dx = 0 \quad m \neq n$$

Prove this statement using ψ_1 and ψ_2 and Equation 10.55.

10.36 Use Equation 10.57 to calculate the wavelength of the electronic transition in polyenes for $N = 6$, 8, and 10. Comment on the variation of λ with L, the length of the molecule.

10.37 Based on the particle in a one-dimensional box approximation for polyenes, suggest where along the line segment the $n = 1$ to $n = 2$ electronic transition would most likely take place. Explain your choice.

10.38 As stated in the chapter, the probability of locating a particle in a one-dimensional box is given by $\int \psi^* \psi \, dx$. Over a small distance, the probability can be calculated without integration as $\psi^* \psi \Delta x$. Consider an electron with $n = 1$ in a box of length 2.000 nm. Calculate the probability of locating the electron **(a)** between 0.500 nm and 0.502 nm and **(b)** between 0.999 nm and 1.001 nm. Comment on your results and on the validity of the approximation.

Particle in a Two-Dimensional Box

10.39 How would the energy levels differ for a particle in a two-dimensional box of size $a \times b$ when $a = 2b$ versus when $a > 2b$?

10.40 Under what conditions will the energy levels of a particle in a two-dimensional box show degeneracies?

Particle on a Ring

10.41 Explain in your own words the physical meaning of the quantum number $m = 0$ for a particle on a ring. How can $m = 0$ be interpreted using the Heisenberg uncertainty principle?

10.42 Calculate the angular momentum of an electron on a ring of diameter 10 pm with quantum number $m = 3$. In what direction does the angular momentum vector point?

10.43 Calculate the speed of an electron moving in a 132-pm diameter ring (about the size of a benzene molecule) with quantum number $m = 0$, 1, and 2.

10.44 To calculate the energy of core electrons in heavy atoms, it is necessary to use Einstein's theory of relativity because the electrons are moving at nearly the speed of light c. As a model illustration, calculate the (nonrelativistic) speed of an electron moving in a 2.0-pm-diameter ring with quantum number $m = 0$, 1, and 2. In each case, how does the speed of the electron compare with c? (Note that 2.0 pm is the order of the size of a 1s atomic orbital for a gold atom, and that the color of gold cannot be explained accurately without a relativistic theory. Although larger quantum numbers would predict speeds greater than c, actual electron speeds never exceed c.)

Additional Problems

10.45 Calculate the kinetic energy in units of electron volts for an electron emitted from a surface with a work function of 1.5 *e*V that has been illuminated with a 632.8-nm (red) helium neon laser, such as those used in a supermarket scanner. Perform the same calculation for a 543.5-nm (green) helium–neon laser.

10.46 Tungsten has a work function of 4.55 *e*V. Calculate the longest wavelength that may be used to eject a photoelectron from a clean tungsten surface in a vacuum.

10.47 When two atoms collide, some of their kinetic energy may be converted into electronic energy in one or both atoms. If the average kinetic energy ($\frac{3}{2}k_BT$) is about equal to the energy for some allowed electronic transition, an appreciable number of atoms can absorb enough energy through an inelastic collision to be raised to an excited electronic state. **(a)** Calculate the average kinetic energy per atom in a gas sample at 298 K. **(b)** Calculate the energy difference between the $n = 1$ and $n = 2$ levels in atomic hydrogen. **(c)** At what temperature is it possible to excite a hydrogen atom from the $n = 1$ level to $n = 2$ level by an average collision?

10.48 Photodissociation of water,

$$H_2O(g) \xrightarrow{h\nu} H_2(g) + \tfrac{1}{2}O_2(g)$$

has been suggested as a source of molecular hydrogen. The $\Delta_rH°$ for the reaction, calculated from thermochemical data, is 285.8 kJ mol^{-1} of water decomposed. Calculate the maximum wavelength (in nm) that would provide the necessary energy. In principle, is it feasible to use sunlight as a source of energy for this process?

10.49 Based on the discussion of decay and quantum mechanical tunneling, suggest a relation between the energy of emitted α particles and the half-life for the radioactive decay.

10.50 Only a fraction of the electrical energy supplied to a tungsten–filament light bulb is converted to visible light. The rest of the energy shows up as infrared radiation. A 75-W light bulb converts 15.0 percent of the energy supplied to it into visible light. If the wavelength is 550 nm, then how many photons are emitted by the light bulb per second? (1 W = 1 J s^{-1}.)

10.51 An electron in a hydrogen atom is excited from the ground state to the $n = 4$ state. Comment on the correctness of the following statements (true or false). **(a)** The $n = 4$ state is the first excited state. **(b)** It takes more energy to ionize (remove) the electron from $n = 4$ than from the ground state. **(c)** The electron is farther from the nucleus (on average) in $n = 4$ than in the ground state. **(d)** The wavelength of light emitted when the electron drops from $n = 4$ to $n = 1$ is longer than that from $n = 4$ to $n = 2$. **(e)** The wavelength the atom absorbs in going from $n = 1$ to $n = 4$ is the same as that emitted as it goes from $n = 4$ to $n = 1$.

10.52 The ionization energy of a certain element is 412 kJ mol^{-1}. When the atoms of this element are in the first excited state, however, the ionization energy is only 126 kJ mol^{-1}. Based on this information, calculate the wavelength of light emitted in a transition from the first excited state to the ground state.

10.53 The UV light that is responsible for sun tanning falls in the 320–400 nm region. Calculate the total energy (in joules) absorbed by a person exposed to this radiation for 2.0 hours, given that there are 2.0×10^{16} photons hitting Earth's surface per square centimeter per second over the range 320 nm to 400 nm and that the exposed body area is 0.45 m^2. Assume that only half of the radiation is absorbed and the other half is reflected by the body. (*Hint*: Use an average wavelength of 360 nm to calculate the energy of a photon.)

10.54 In 1996 physicists created an anti-atom of hydrogen. In such an atom, which is the antimatter equivalent of an ordinary atom, the electrical charges of all the component particles are reversed. Thus, the nucleus of an anti-atom is made of an anti-proton, which has the same mass as a proton but bears a negative charge, while the electron is replaced by an anti electron (also called positron) with the same mass as an electron, but bearing a positive charge. Would you expect the energy levels, emission spectra, and atomic orbitals of an anti-hydrogen atom to be different from those of a hydrogen atom? What would happen if an anti-atom of hydrogen collided with a hydrogen atom?

10.55 A student carried out a photoelectric experiment by shining visible light on a a clean piece of cesium metal. She determined the kinetic energy of ejected electrons by applying a retarding voltage such that the current due to the electrons reads exactly zero. The condition is reached when $e\text{V} = \frac{1}{2}m_e u^2$ where e is electric charge and V is the retarding potential. Her results are shown below:

λ/nm	405	435.8	480	520	577.7	650
V/volt	1.475	1.268	1.027	0.886	0.667	0.381

Rearrange Equation 10.3 to read

$$\nu = \frac{\Phi}{h} + \frac{e}{h}\text{V}$$

Determine both h and Φ graphically.

10.56 Use Equation 2.7 to calculate the de Broglie wavelength of a N_2 molecule at 300 K.

10.57 Alveoli are tiny sacs of air in the lungs. Their average diameter is 5.0×10^{-5} m. Calculate the minimum uncertainty in the velocity of an oxygen molecule (5.3×10^{-26} kg) trapped within a sac. (*Hint*: The maximum uncertainty in the position of the molecule is given by the diameter of the sac.)

10.58 The sun is surrounded by a white sphere of gaseous material called the corona, which becomes visible during a total eclipse of the sun. The temperature of the corona is in the millions of degrees Celsius, high enough to break up molecules and remove some or all of the electrons from atoms. One way astronomers have been able to estimate the temperature of the corona is by studying the emission lines of ions of certain elements. For example, the emission spectrum of Fe^{14+} ions has been recorded and analyzed. Knowing that it takes 3.5×10^4 kJ mol^{-1} to convert Fe^{13+} to Fe^{14+}, estimate the temperature of the sun's corona. (*Hint*: The average kinetic energy of one mole of a gas is $\frac{3}{2}RT$.)

10.59 Based on periodic trends, which would make a more red-sensitive (meaning, able to detect lower-energy photons) photocathode, one made of zinc or one made of vanadium?

10.60 A clean metal surface in a vacuum is illuminated with 1 W of orange light and 1 W of violet light. Assume that the photon energy from each source exceeds the work function of the metal. **(a)** The number of photons emitted from the surface illuminated by orange light is (greater than, less than, equal to) the number emitted from the surface illuminated by violet light. **(b)** The kinetic energy of photons emitted from the surface illuminated by orange light is (greater than, less than, equal to) the kinetic energy of electrons emitted from the surface illuminated by violet light.

10.61 Calculate the percentage of the electromagnetic radiation from the sun that lies in the visible region of the electromagnetic spectrum, assuming that the sun is a blackbody radiator of temperature 5800 K and that electromagnetic radiation in the X–ray and gamma–ray regions of the electromagnetic spectrum may be neglected. (*Hint*: Assume that the visible region of the electromagnetic spectrum lies between 400 and 750 nm, and perform an integration. Software such as Mathematica, Maple, or Mathcad would be useful in solving this problem.)

10.62 The 2006 Nobel Prize in Physics was awarded to John C. Mather and George F. Smoot "for their discovery of the blackbody form and anisotropy of the cosmic background radiation." Understanding subtle variations in the cosmic background is an active area of research in cosmology, because it has implications for the beginnings of the known universe. The background radiation from the "empty" part of space is well modeled by a 2.7 K blackbody. What is the wavelength of maximum emission for this radiation? In what region of the electromagnetic spectrum does it lie?

10.63 Consider a particle in the one-dimensional box shown below.

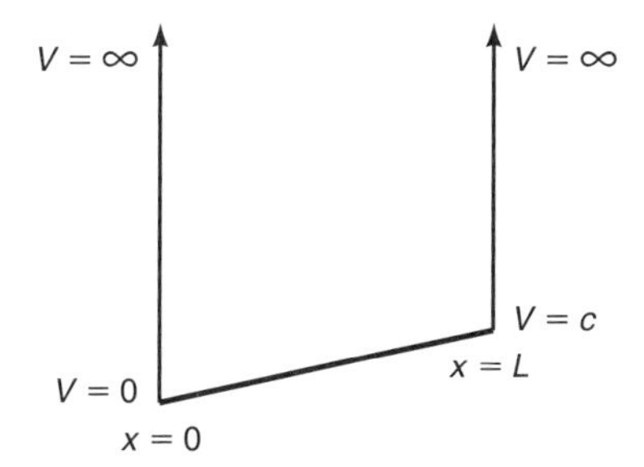

The potential energy of the particle is not zero everywhere in the box, but varies according to its position, x. (**a**) Write the Schrödinger wave equation for this system. (**b**) ψ_n will not be symmetrical about the center of the box, regardless of the value of n. Assume that the wave function for $n = 10$ in a one-dimensional box, where $V = 0$ throughout the box, looks like the following:

then sketch the corresponding wave function for the above box. Give a physical interpretation for your sketch. (*Hint*: Neither the amplitude nor the wavelength of the wave will be constant along the box.)

10.64 Consider a particle in a one-dimensional box in the ground state ψ_1 and the first excited state ψ_2 described by the wave functions listed below. For each wave function, calculate the expectation value of the position $\langle x \rangle$, the expectation value of the position squared $\langle x^2 \rangle$, the expectation value of the momentum $\langle p \rangle$, and the expectation value of the momentum squared $\langle p^2 \rangle$.

(**a**) $\psi_1(x) = \sqrt{\frac{2}{a}} \sin\frac{\pi x}{a}$ (**b**) $\psi_2(x) = \sqrt{\frac{2}{a}} \sin\frac{2\pi x}{a}$ $0 \leq x \leq a$

10.65 The standard deviation of a quantity, σ_x, may be used as a representation of its uncertainty (Δx). The standard deviation is the square root of the variance and may be calculated using the following formula:

$$\sigma_x = (\langle x^2 \rangle - \langle x \rangle^2)^{1/2}$$

Use the results of Problem 10.64 to calculate the uncertainty in the position and momentum in the ground and first excited states of the particle in a one-dimensional box and comment of the agreement with the Heisenberg uncertainty principle.

10.66 The following figure represents the emission spectrum of a hydrogenlike ion in the gas phase. All the lines result from the electronic transitions from the excited states to the $n = 2$ state.

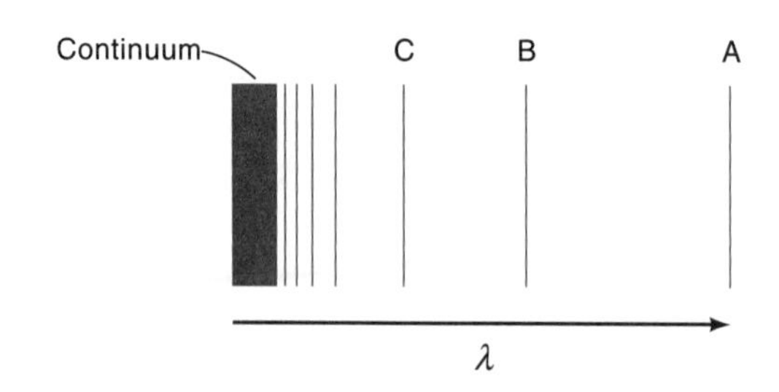

(**a**) What electronic transitions correspond to lines B and C? (**b**) If the wavelength of line C is 27.1 nm, calculate the wavelengths of lines A and B. (**c**) Calculate the energy needed to remove the electron from the ion in the $n = 4$ state. (**d**) What is the physical significance of the continuum? (*Hint*: Use the Rydberg constant for hydrogen in your calculation.)

10.67 In the beginning of the twentieth century, some scientists thought that a nucleus may contain both electrons and protons. Use the Heisenberg uncertainty principle to show that an electron cannot be confined within a nucleus. Repeat the calculation for a proton. Comment on your results.

10.68 The benzene molecule has a hexagonal symmetry, which may be approximated as a circular ring. Thus, we can apply the particle-on-a-ring model to predict the electronic structure of benzene. (**a**) From the C−C bond length in benzene, we can show that the radius of the ring is 132 pm. Starting with Equation 10.77, derive an expression for the transition wavenumbers (in cm^{-1} units) of this system. (**b**) Calculate the transition wavenumbers from the $m = 1$ state to the $m = 2$ state (or equivalently, from the $m = -1$ state to the $m = -2$ state). (**c**) The experimentally measured value for this transition is 37,900 cm^{-1}. Comment on the discrepancy between the predicted and measured values.

10.69 The following diagram shows the solar radiation, which may be approximated as a blackbody. According to Wien's law, the wavelength of maximum intensity in blackbody radiation is given by

$$\lambda_{max} = \frac{b}{T}$$

where b is a constant (2.898×10^6 nm K) and T is the temperature of the radiating body in kelvins. (**a**) Estimate the temperature at the surface of the sun. (**b**) What does this curve reveal about two consequences of great biological significance on Earth?

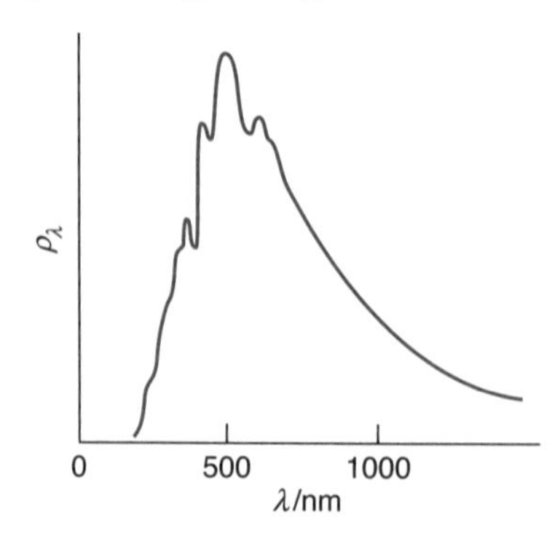

10.70 Shown below are contour plots of several wave functions for a particle in a two-dimensional box of dimension $a \times 2a$. Rank these in order of increasing energy and identify which (if any) might be degenerate.

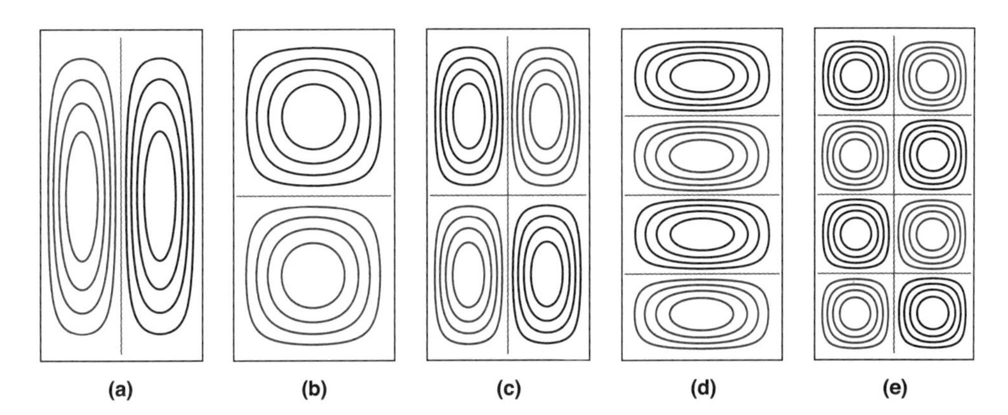

10.71 (**a**) An electron in the ground state of the hydrogen atom moves at an average speed of 5×10^6 m s^{-1}. If the speed is known to an uncertainty of 1 percent, what is the uncertainty in knowing its position? Given that the radius of the hydrogen atom in the ground state is 5.29×10^{-11} m, comment on your result. The mass of an electron is 9.1094×10^{-31} kg. (**b**) A 3.2-g Ping-Pong ball moving at 50 mph has a momentum of 0.073 kg m s^{-1}. If the uncertainty in measuring the momentum is 1.0×10^{-7} of the momentum, calculate the uncertainty in the Ping-Pong ball's position.

10.72 One wavelength in the atomic hydrogen emission spectrum is 1280 nm. What are the initial and final states of the transition responsible for this emission?

10.73 Owls have good night vision because their eyes can detect a light intensity as low as 5.0×10^{-13} W m^{-2}. Calculate the number of photons per second that an owl's eye can detect if its pupil has a diameter of 9.0 mm and the light has a wavelength of 500 nm. (1 W = 1 J s^{-1}.)

10.74 A student placed a large unwrapped chocolate bar in a microwave oven without a rotating glass plate. After turning on the oven for less than a minute, she noticed there were evenly spaced dents (due to heating) about 6 cm apart. Based on her observations, calculate the speed of light, given that the microwave frequency is 2.45 GHz.

10.75 The minimum uncertainty in the position of a certain moving particle is equal to its de Broglie wavelength. If the speed of the particle is 1.2×10^5 m s^{-1}, what is the minimum uncertainty in its speed?

10.76 In a photoelectric experiment, a student found that a wavelength of 351 nm is needed to just dislodge electrons from a zinc metal surface. Calculate the speed of an ejected electron when she employed light with a wavelength of 313 nm.

10.77 Calculate the mass of a proton moving at 99.5 percent the speed of light in the Large Hadron Collider. According to Einstein's special theory of relativity, the mass of a moving particle is related to its rest mass by

$$m_{\text{moving}} = m_{\text{rest}} / \sqrt{1 - (u/c)^2}$$

where u is the speed of the particle. (The photon has a rest mass of zero.) Also express your answer as a percentage of the rest mass of a proton.

CHAPTER 11

Applications of Quantum Mechanics to Spectroscopy

Nothing is more revealing than movement.
—Martha Graham*

We saw in Chapter 10 that solving the Schrödinger equation provided analytic solutions for the wave functions of the particle-in-a-box model together with the corresponding energy levels. In this chapter we consider the quantum mechanical treatment of the rigid rotor and the harmonic oscillator, which are useful models for molecular rotation and molecular vibration, respectively and their applications to two branches of spectroscopy: microwave spectroscopy and infrared spectroscopy. We also introduce Raman spectroscopy in this chapter. Many physical chemistry courses include a laboratory component, and spectroscopy experiments serve as an important way to illustrate the principles of quantum mechanics. But first we shall become acquainted with the general vocabulary of spectroscopy.

11.1 Vocabulary of Spectroscopy

Here, we introduce some of the terms that are common in spectroscopy.

Absorption and Emission

There are two primary categories of spectroscopy: absorption and emission. The fundamental equation for both absorption and emission is

$$\begin{aligned}\Delta E &= E_2 - E_1 \\ &= h\nu \end{aligned} \tag{11.1}$$

where E_1 and E_2 are the energies of the two quantized energy levels involved in a transition (see Figure 10.10), h is Planck's constant, and ν is the frequency of light. For a spectroscopic transition to occur, the energy of a photon must match the energy difference between the two atomic or molecular energy levels. Microwave spectroscopy, infrared spectroscopy, electronic spectroscopy, nuclear magnetic resonance, and electron spin resonance are usually studied in the absorption mode. Fluorescence and phosphorescence are emission processes. The word *laser* (*l*ight

* Martha Graham (1894–1991) was an American dancer and choreographer.

*a*mplification by *s*timulated *e*mission of *r*adiation) is an acronym for a special type of emission called stimulated emission, discussed in Section 14.4.

Units

The position of a spectral line in a spectrum corresponds to the difference in energy between two levels involved in a transition. This position can be reported in several different units that are readily interchangeable.

1. *Wavelength.* Wavelength (λ) can be given in meters (m), but for ultraviolet–visible (UV–vis) spectroscopy, nanometers (nm) are typically used.

$$1 \text{ nm} = 1 \times 10^{-9} \text{ m}$$

2. *Frequency.* Frequency (ν) is given by s^{-1} or Hz, the number of waves per second.
3. *Wavenumber.* Wavenumber ($\tilde{\nu}$) is the number of waves per unit length,

$$\tilde{\nu} = \frac{1}{\lambda} = \frac{\nu}{c} \tag{11.2}$$

where c is the speed of light. Wavenumber is usually given in terms of reciprocal centimeters (cm^{-1}). Variables expressed in wavenumbers are often written with a tilde.

4. *Energy.* Spectral lines are also reported as energies. In your reading, be careful to distinguish whether a transition energy is expressed per transition

$$E = h\nu = \frac{hc}{\lambda} = hc\tilde{\nu} \tag{11.3a}$$

or per mole of transitions

$$E = N_A h\nu = \frac{N_A hc}{\lambda} = N_A hc\tilde{\nu} \tag{11.3b}$$

where N_A is the Avogadro constant. Note that energy is directly proportional to wavenumber. Energy is also directly proportional to frequency, but unlike wavenumbers, frequencies are typically too large for practical use. For example, the characteristic red light from a helium–neon laser (which finds application in supermarket barcode scanners) may be equivalently expressed as

We use the symbol "≙" to mean "is equivalent to".

$$633 \text{ nm} \triangleq 4.74 \times 10^{14} \text{ Hz} \triangleq 1.58 \times 10^{4} \text{ cm}^{-1} \triangleq 3.14 \times 10^{-19} \text{ J} \triangleq 189 \text{ kJ mol}^{-1}$$

Regions of the Spectrum

Spectroscopic techniques utilize almost the entire electromagnetic spectrum to analyze atomic and molecular properties (Figure 11.1). The spectroscopic techniques that are most commonly used to study chemical and biological systems are infrared, visible and UV, nuclear magnetic resonance, and fluorescence. However, to gain a

	γ ray	X ray	Ultraviolet	Visible	Infrared	Microwave	Radio frequency
Phenomenon observed	Nuclear transitions	Inner electronic transitions $\sigma^* \leftarrow \sigma$	Outer electronic transitions $\pi^* \leftarrow \pi$, $\pi^* \leftarrow n$		Molecular vibration	Molecular rotation, electron spin resonance	Nuclear magnetic resonance
Type of spectroscopy	Mössbauer	UV	UV, Visible		IR, Raman	Microwave, ESR	NMR

Wavelength/nm	0.0003	0.03	10	30	400	800	1000	3×10^5	3×10^7	3×10^{11}	3×10^{13}
Frequency/Hz	1×10^{21}	1×10^{19}	3×10^{16}	1×10^{16}	8×10^{14}	4×10^{14}	3×10^{14}	1×10^{12}	1×10^{10}	1×10^6	1×10^4
Wavenumber/cm^{-1}	3×10^{10}	3×10^8	1×10^6	3×10^5	3×10^4	1.3×10^4	1×10^4	33	3	3×10^{-5}	3×10^{-7}
Energy/kJ mol^{-1}	4×10^8	4×10^6	1.2×10^4	4×10^3	330	170	125	0.4	4×10^{-3}	4×10^{-7}	4×10^{-9}

Figure 11.1
Types of spectroscopy. Mössbauer spectroscopy is not discussed in this text.

better overview, we also discuss microwave spectroscopy, electron spin resonance, and phosphorescence in this book.

Linewidth

Every spectral line has a finite, nonzero width, which is usually defined as the full width at half-height of the peak. If the two states involved in a spectroscopic transition have precisely well-defined energies, then their energy difference must also be a precisely measurable quantity. In this case we would observe a line of infinitesimal width. In reality, however, a number of phenomena cause every spectral line to have a definite width (Figure 11.2). We discuss the three most basic mechanisms below.

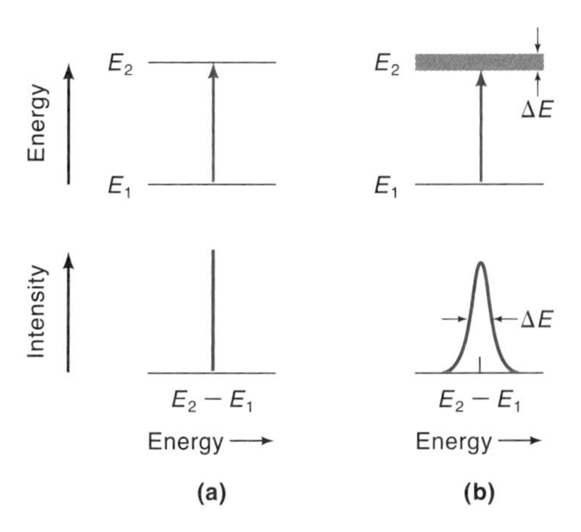

Figure 11.2
(a) A hypothetical absorption line having insignificant width. (b) An actual absorption line having a width ΔE at half height. The lifetime of the ground state is very long so that its energy is well defined.

Natural Linewidth. The so-called *natural linewidth* of a spectral line is a consequence of the Heisenberg uncertainty principle, which can be expressed as follows (see Equation 10.28)

$$\Delta E \Delta t \geq \frac{h}{4\pi} \tag{11.4}$$

This equation relates the uncertainty in the energy (one way of expressing the linewidth) and the lifetime of the system in a particular state. It says that the longer one takes to measure the energy of a system in that state (the larger Δt is), the more accurately this natural linewidth can be determined (the smaller ΔE is). For a transition between two states, we must consider the lifetime of each of the states. In practice, we usually only need to consider just one of the states. In an absorption process, for example, the final state (the excited state) has a finite lifetime t_{ex}; therefore, the duration in determining the lifetime, Δt, cannot exceed t_{ex}. Consequently, the uncertainty in the energy of that state is given by

$$\Delta E \geq \frac{h}{4\pi \Delta t} = \frac{h}{4\pi t_{ex}}$$

Because $E = h\nu$, we have $\Delta E = h\Delta\nu$ or

$$\Delta\nu = \frac{1}{4\pi \Delta t} = \frac{1}{4\pi t_{ex}} \tag{11.5}$$

We use an equals sign in Equation 11.5 to find the minimum value of $\Delta\nu$. The uncertainty in energy of a state (typically expressed in Hz) shows up as the width of the spectral line. This width is called the natural linewidth because it is inherent in the system and cannot be influenced (decreased) by external parameters such as temperature or concentration. Natural linewidths depend strongly on the type of transition. For transitions between rotational energy levels, for example, a typical excited-state lifetime is about 10^3 s, which translates into a natural linewidth of about 8×10^{-5} Hz (on a transition frequency of about 10^{10} Hz). On the other hand, an electronic excited state has a lifetime on the order of 10^{-8} s, resulting in an uncertainty in frequency of about 8×10^6 Hz! For an electronic transition frequency of 8×10^{14} Hz, this is only 1 part in 10^8.

The Doppler Effect. Experimentally, the widths of spectral lines are invariably much greater than those predicted solely by the lifetimes of excited states. Therefore, other mechanisms must also be responsible for broadening the lines. The Doppler broadening of spectral lines is an interesting result of the *Doppler effect* [after the Austrian physicist Christian Doppler (1803–1853)]. When radiation is emitted, its frequency depends on the speed of the atom or molecule relative to the detector.* For the same reason, the whistle of a railroad train traveling towards you seems to have a frequency higher than it really is, and when the train is moving away from you, the whistle sounds lower in frequency. The equation describing the Doppler effect is

$$\nu = \nu_0\left(1 \pm \frac{u}{c}\right) \tag{11.6}$$

* The same conclusion is obtained for an absorption process; that is, the Doppler effect influences (increases) the width of both emission and absorption spectral lines.

where ν_0 is the frequency of the emitting molecule, ν is the frequency registered by the detector, u is the average speed of the molecules in the sample, and c is the speed of light. The $\pm$ signs indicate that some molecules are moving toward the detector (+) while others are moving away from it (−).

It is possible to estimate how much broader an absorption line is as a result of the Doppler effect. For N_2 molecules at 300 K, we can use Equation 2.7 to calculate the root-mean-square speed, 517 m s^{-1}. Substituting this value into Equation 11.6, we get

$$\nu = \nu_0\left(1 \pm \frac{517 \text{ m s}^{-1}}{3.00 \times 10^8 \text{ m s}^{-1}}\right)$$
$$= \nu_0(1 \pm 1.72 \times 10^{-6})$$

Using a typical electronic transition frequency of 1×10^{15} Hz for N_2 (ν_0), we find a total frequency shift (plus and minus) of about 2×10^9 Hz, which is about 400 times that of the natural linewidth. Linewidth broadening due to the Doppler effect increases with temperature, because there is a larger spread in molecular speeds. To minimize this effect, spectra should be obtained from cold samples if possible.

The Pressure Effect. Another influence on the width of a spectral line is called pressure, or collisional, broadening. Molecular collisions can deactivate excited states, thereby shortening their lifetimes. If t is the mean time between collisions and each collision results in a transition between two states, then according to the Heisenberg uncertainty principle, there is a broadening $\Delta\nu$, given by $1/(4\pi t)$. Recall from Section 2.5 that $t = 1/Z_1$, where Z_1 is the collision frequency. Because Z_1 is directly proportional to pressure, it follows that an increase in pressure will lead to a broader spectral line. Figure 11.3 shows the electronic absorption spectra of benzene in the vapor and in solution. Because the collision frequency is greater in solution, the spectral lines are much broader. Collisional broadening can be minimized by recording spectra of molecules in the vapor state (if possible) at low pressures.

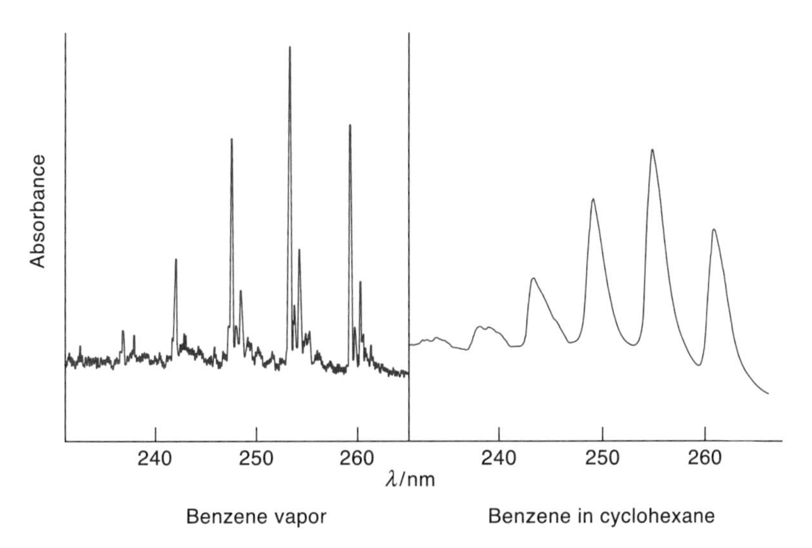

Figure 11.3
Electronic absorption spectrum of benzene. Left: vapor; right: in cyclohexane. Cyclohexane has negligible absorption over this wavelength range. (By permission of Varian Associates, Palo Alto, CA.)

Other Linebroadening Mechanisms. Rate processes, such as dissociation, rotation, and electron and proton transfer reactions, can also cause line broadening. We shall see some examples of these effects in later chapters. A transition may also appear to be broadened due to the presence of multiple transitions in the same spectral range. For example, a polyatomic molecule such as benzene (see Figure 11.3) does not exhibit merely a pure electronic transition in the ultraviolet region of the electromagnetic spectrum, but rather an electronic transition with simultaneous vibrational and rotational transitions (a *rovibronic* transition). Similarly, an infrared spectrum might appear to have broad lines due to the lack of resolution of rotational transitions appearing simultaneously with the vibrational transition (a *rovibrational* transition).

Resolution

Related to linewidth is the separation of one spectral line from another, called *resolution* (Figure 11.4). In all spectroscopic techniques, the *resolving power* (R) of an instrument is a measure of the ability of a spectrometer to distinguish closely spaced lines from one another. An instrument of high resolving power will show two closely spaced lines separately, whereas a low-resolution instrument will be unable to distinguish them. If $\Delta\lambda$ is the wavelength separation of the closest lines that can be seen to be two lines, the resolving power is given by

$$R = \frac{\lambda}{\Delta\lambda} \tag{11.7a}$$

where λ is the average wavelength of the two spectral lines. An analogous equation, expressed in terms of frequencies, is

$$R = \frac{\nu}{\Delta\nu} \tag{11.7b}$$

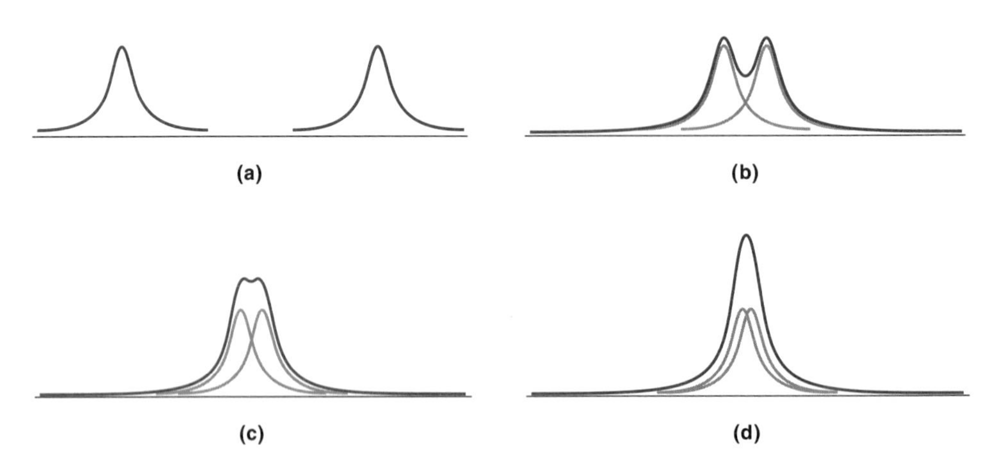

Figure 11.4
(a) Two well-resolved lines. (b)–(d) Two overlapping lines. From (b) through (d), the observed line shape is the sum of the two overlapping lines.

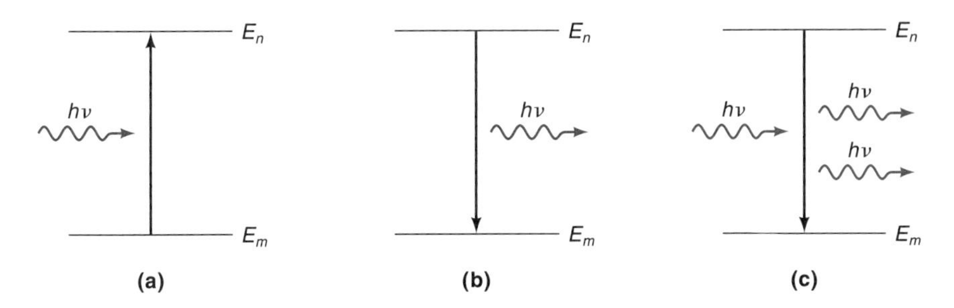

Figure 11.5
(a) Stimulated absorption. (b) Spontaneous emission. (c) Stimulated emission. The energy of the incoming or emitted photon is $h\nu$.

Intensity

A number of factors affect the intensity of an absorption line, which is related to the number of molecules participating in the spectroscopic transition. Here, we shall discuss the treatment presented by Einstein in 1917. Consider a two-state system separated by $\Delta E = E_n - E_m$. When molecules are exposed to radiation with frequency ν such that $\Delta E = h\nu$, they undergo a transition from the lower state m to the higher state n. The rate of transition, N_{mn}, to the higher level is proportional to the number of molecules, N_m, in the lower state and also to the *spectral radiant energy density* ρ_ν (in units J s m^{-3}) at this frequency. Thus,

$$\begin{aligned} N_{mn} &\propto N_m\rho_\nu \\ &= B_{mn}N_m\rho_\nu \end{aligned} \tag{11.8}$$

where B_{mn} is the *Einstein coefficient of stimulated absorption* (in units J^{-1} m^3 s^{-2}). Spectral radiant energy density is a measure of the amount of light available at the appropriate frequency to induce a transition. Einstein realized that radiation can also induce molecules in the excited state to undergo a transition to the lower state. The rate of this stimulated emission, N_{nm}, is

$$N_{nm} = B_{nm}N_n\rho_\nu \tag{11.9}$$

where B_{nm} is the *Einstein coefficient of stimulated emission* and N_n is the number of molecules in the excited state. When there is more population in the excited state n than the lower state m, then shining light of frequency ν will induce the molecules to undergo a transition from the excited state to the ground state. Only radiation of the *same* frequency as the transition can stimulate the shift from the excited state to the lower state. The two coefficients B_{mn} and B_{nm} are equal. In addition, molecules in an excited state can lose energy by spontaneous emission at a rate that is independent of the radiation frequency. This rate is given by $A_{nm}N_n$, where A_{nm} is the *Einstein coefficient of spontaneous emission* in units s^{-1}. These three situations are summarized in Figure 11.5. At equilibrium, the number of molecules going from the m to n state is equal to the number going from the n to m state, so

$$N_mB_{mn}\rho_\nu = N_nB_{nm}\rho_\nu + N_nA_{nm}$$

or

$$\rho_\nu = \frac{A_{nm}}{B_{nm}}\left(\frac{N_n}{N_m - N_n}\right)$$

$$= \frac{A_{nm}}{B_{nm}}\left(\frac{1}{\frac{N_m}{N_n} - 1}\right) \tag{11.10}$$

because $B_{mn} = B_{nm}$. The ratio N_n/N_m is given by the Boltzmann distribution law (see Equation 2.33):

$$\frac{N_n}{N_m} = e^{-h\nu/k_\mathrm{B}T} \tag{11.11}$$

and the inverse is

$$\frac{N_m}{N_n} = e^{h\nu/k_\mathrm{B}T} \tag{11.12}$$

Substituting Equation 11.12 into Equation 11.10, we obtain

$$\rho_\nu = \frac{A_{nm}}{B_{nm}}\left(\frac{1}{e^{h\nu/k_\mathrm{B}T} - 1}\right) \tag{11.13}$$

The spectral radiation energy density per unit frequency was shown by Planck to be

$$\rho_\nu = \frac{8\pi h\nu^3}{c^3}\left(\frac{1}{e^{h\nu/k_\mathrm{B}T} - 1}\right) \tag{11.14}$$

Substituting Equation 11.14 into Equation 11.13 gives the relationship between the Einstein A and B coefficients:

$$A_{nm} = B_{nm}\frac{8\pi h\nu^3}{c^3} \tag{11.15}$$

Note the dependence of A_{nm} on frequency. In electronic spectroscopy, ν is a large number, so the probability of spontaneous emission is usually higher than it is for stimulated emission. This explains the short lifetime of the excited electronic states mentioned earlier. When the frequencies are much smaller, as in microwave or magnetic resonance spectroscopies, the stimulated emission predominates. We shall return to the phenomenon of stimulated emission in Chapter 14.

Keep in mind that any type of spectrum (absorption or emission) is actually a superposition of a very large number of transitions from individual molecules. Most spectrophotometers are not designed to detect the energy absorbed or emitted by a single molecule. Furthermore, the interaction between a photon of electromagnetic radiation and a molecule can give rise to only one transition and hence one line. Any spectrum containing more than one line, as most do, is actually the statistical sum of all the transitions.

Specifically designed instruments can detect single photons of sufficient energy. Detecting photons from a single molecule is a greater challenge.

Selection Rules

Transitions do not take place between any two levels in an atom or molecule just because the frequency of radiation is appropriate to the resonance condition ($\Delta E = h\nu$). It is found experimentally that observed transitions obey certain *selection rules.* These selection rules can be justified using time-dependent quantum mechanics. Transitions, then, are classified as *allowed* (having a high probability) or *forbidden* (having a low probability), depending on how they occur according to selection rules. Formally, forbidden transitions have zero probability, under certain assumptions. Experimentally, the difference in probability for allowed and forbidden transitions might be a factor of a million or more. We predict two types of transitions as being forbidden: spin-forbidden transitions and symmetry-forbidden transitions.

Spin-Forbidden Transitions. Spin-forbidden transitions involve a change in *spin multiplicity.* The spin multiplicity is given by $(2S + 1)$, where S is the spin quantum number of the system (Table 11.1). The numerical value of S gives us the number of different ways in which the unpaired spins can line up in an external magnetic field. The selection rule is that the spin multiplicity must not change in a transition; that is, we must have $\Delta S = 0$. Normally, for example, a transition from a singlet state to a triplet state, or vice versa, is forbidden.

Symmetry-Forbidden Transitions. A quantitative measure for the intensity of a transition is provided by the *transition dipole moment* μ_{ij}, given by

$$\mu_{ij} = \int_{\text{all space}} \psi_i \hat{\mu} \psi_j d\tau \tag{11.16}$$

where ψ_i and ψ_j are the wave functions for the ith and jth state and $\hat{\mu}$ is the transition dipole moment operator connecting these two states. The integration is taken over all spatial coordinates and $d\tau$ represents the infinitesimal volume element ($d\tau = dxdydz$).

Table 11.1
Spin Multiplicity of Atoms and Molecules

Number of unpaired electrons	Electron spin S	$2S + 1$	Multiplicity
0	0	1	Singlet
1	$\frac{1}{2}$	2	Doublet
2	1	3	Triplet
3	$\frac{3}{2}$	4	Quartet
⋮	⋮	⋮	⋮

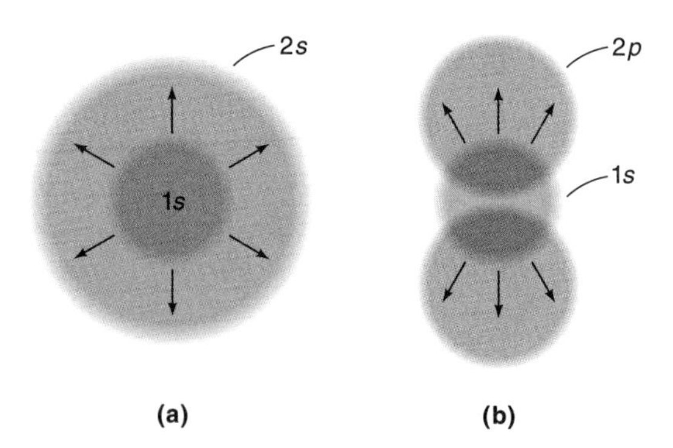

Figure 11.6
(a) There is no dipole moment associated with a $2s \leftarrow 1s$ transition because the electric charge migrates spherically. Consequently, this is a symmetry-forbidden transition.
(b) During a $2p \leftarrow 1s$ transition, there is a dipole associated with the charge migration. This is an allowed transition.

For an allowed transition, $\psi_i \hat{\mu} \psi_j$ must be an even function.* Because $\hat{\mu}$ depends only on the first power of the coordinates and is therefore odd, and ψ_i and ψ_j must have different symmetry with respect to each other (even–odd or odd–even) so that the product will be even.

By spectroscopic convention, the higher energy state is listed on the left, and the arrow indicates absorption (pointing to the left) or emission (pointing to the right). This convention is, however, frequently ignored.

To gain insight into the transition dipole moment, consider the electronic transition in the hydrogen atom. The physical significance of the dipole moment vector is that it denotes the electron charge migration during the transition. Figure 11.6 shows the charge migrations accompanying the transitions from hydrogen $1s$ to $2s$ and $2p$ states. From the figure, we can see the $2s \leftarrow 1s$ transition is forbidden, because the charge redistribution remains spherically symmetrical, while the $2p \leftarrow 1s$ transition is allowed, because the charge redistribution is dipolar. We can generalize these observations with the selection rule $\Delta\ell = \pm 1$, where ℓ is the angular momentum quantum number. Another way of viewing the selection rule is to consider conservation of angular momentum. Each photon has a single unit of angular momentum, so a transition that doesn't change angular momentum ($\Delta\ell = 0$, such as atomic hydrogen $2s \leftarrow 1s$) is forbidden in a one-photon process. Similarly a hydrogen atom $3d \leftarrow 1s$ transition requires a change in two units of angular momentum ($\ell = 2 \leftarrow \ell = 0$), so it also is a forbidden one-photon process.

Mechanisms too complex to list here cause various degrees of breakdown in the selection rules. Consequently, transitions predicted as forbidden may appear in a spectrum as weak lines.

Signal-to-Noise Ratio

Because of the manner in which the signals are detected, a recorded spectrum always contains random fluctuations of electronic signals called *noise*. The sensitivity of detection of any signal due to the sample being studied depends on how easily we can distinguish the signal from the noise. A large signal is not itself easy to detect if it is

* An even function $f(x)$ has the property that it is unchanged when we reverse the sign of x, that is, $f(x) = f(-x)$. The opposite holds true for an odd function, where $f(x) = -f(-x)$. Thus, x^2 is an even function and x^3 is an odd function.

accompanied by a similarly large noise. The practical quantity we wish to maximize is the signal-to-noise ratio, a dimensionless number. An effective way to increase the signal-to-noise ratio is by signal averaging, that is, by repeatedly recording the same spectrum and adding the signals. Theoretically, if we signal average a spectrum N times, the intensity will increase by a factor N, and the noise will increase by a factor $\sqrt{N}$. Thus, the signal-to-noise ratio will increase by $(N/\sqrt{N})$ or $\sqrt{N}$ times, so that scanning the same spectrum 10 times will enhance the signal-to-noise ratio by a factor of $\sqrt{10}$ or 3.2.

The Beer–Lambert Law

A useful equation for studying the quantitative aspects of absorption is the Beer–Lambert law. Consider the passage of a monochromatic beam of radiation (radiation of one wavelength) through a homogeneous medium, such as a solution. Let I_0 and I be the *intensity* of the incident and transmitted light, and I_x the intensity of the light at distance x (Figure 11.7). The incremental decrease in intensity, $-dI_x$, is proportional to $I_x dx$; that is,

The intensity of light is determined by the number of photons $cm^{-2}\ s^{-1}$.

$$
\begin{aligned}
-dI_x &\propto I_x dx \\
&= kI_x dx
\end{aligned} \tag{11.17}
$$

where I_x is the intensity at position x, and k is a proportionality constant whose value depends on the nature of the absorbing medium. Rearranging Equation 11.17, we obtain

$$\frac{dI_x}{I_x} = -kdx$$

Upon integration,

$$\ln I_x = -kx + C \tag{11.18}$$

where C is a constant of integration. At $x = 0$, $I_x = I_0$, so that $C = \ln I_0$. If we consider the entire length of the absorbing medium, I_x can be replaced by I (which is the intensity of the emerging light) and x by the pathlength b. Hence,

$$\ln I = -kb + \ln I_0$$

$$-\ln \frac{I}{I_0} = kb$$

or

$$-\log \frac{I}{I_0} = k'b \tag{11.19}$$

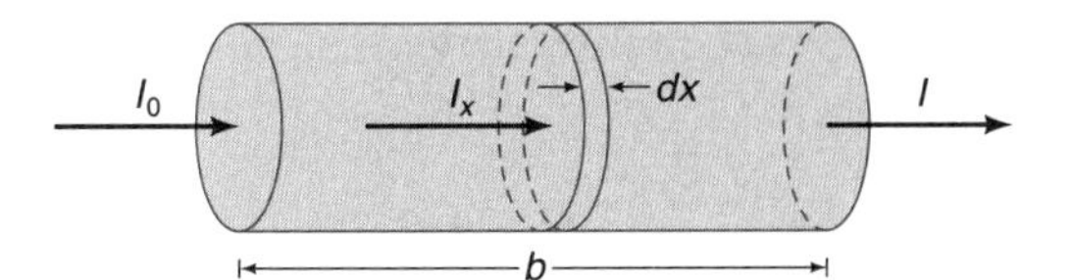

Figure 11.7
Absorption of light by a uniform medium of pathlength *b*.

where $k = 2.303k'$. The ratio I/I_0 is called the *transmittance* (T); it measures the amount of light transmitted after passing through the medium. Equation 11.19 can be expressed in a more convenient form as follows:

$$-\log T = \varepsilon bc$$

Alternatively,

$$A = \varepsilon bc \tag{11.20}$$

where A is the *absorbance*, ε is the *molar absorptivity* (also called *molar extinction coefficient*), b is the *pathlength* (cm), and c is the concentration in mol L^{-1}.

Beer's law mnemonic: The deeper the glass the darker the brew the less the amount of light that gets through.

Equation 11.20 is known as the Beer–Lambert law, after the German astronomer Wilhelm Beer (1797–1850) and the German mathematician Johann Heinrich Lambert (1728–1777). It is also known as Beer's law, or the Beer–Lambert–Bouger law after the French mathematician and astronomer Pierre Bouger (1698–1758). We see that the absorbance is equal to the negative log base 10 of transmittance. Thus, the smaller the transmittance (and hence $-\log T$ would be a larger number), the greater the absorption of light. The absorbance is a dimensionless quantity, so ε has the units L mol^{-1} cm^{-1}. In principle, A can have any positive value or zero; for conventional UV–visible spectrophotometers, it is practical to measure A between 0.001 and 3. For routine quantitative measurements, A is ideally kept between 0.1 and 0.9 by varying the concentration or pathlength of the sample.

The Beer–Lambert law forms the quantitative basis for all types of absorption spectroscopy. It holds as long as there is no interaction between solute molecules, such as dimerization or ion-pair formation. An interesting application of Equation 11.20 is the pulse oximeter, a device used to monitor the extent of oxygenation of hemoglobin molecules of patients without drawing blood. Light is directed through a fingertip and the absorbance of the arterial blood at different wavelengths is measured at different times. Using the known molar absorptivities of oxyhemoglobin and deoxyhemoglobin, the instrument determines the fraction of oxyhemoglobin in the blood.

11.2 Microwave Spectroscopy

In this section we shall study the rotational motion of molecules. For simplicity we focus our discussion mainly on diatomic molecules.

The Rigid Rotor Model

An equivalent name for the rigid rotor model is the particle-on-a-sphere model.

As a good approximation, we can treat a diatomic molecule as a rigid rotor, which consists of two point masses m_1 and m_2 separated by a fixed distance r (Figure 11.8a). The molecule is free to tumble in all directions. To consider the pure rotation, we find the *center of mass*, which is the point about which rotation occurs. Mass m_1 is at a distance r_1 from the center of mass, whereas m_2 is at a distance r_2. For this diatomic molecule, therefore,

$$r = r_1 + r_2$$

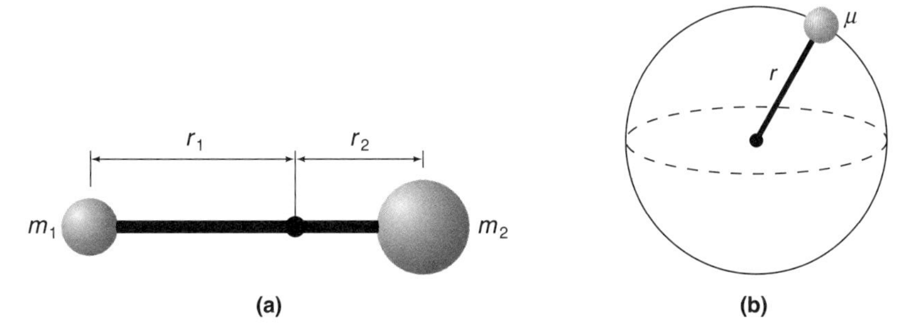

Figure 11.8
The rigid rotor model. (a) Two point masses separated by a fixed distance r. The masses may freely rotate about the center of mass, which is located at distance r_1 from mass m_1 along r. (b) The equivalent system to part (a). The reduced mass μ rotates at the fixed distance r about a point.

The location of the center of mass is found just as one would find the fulcrum of a see-saw, by setting the rotational masses equal to each other.

$$m_1 r_1 = m_2 r_2$$

If r, m_1, and m_2 are known, then r_1 and r_2 may be found algebraically.

$$r_1 = \frac{(m_2/m_1)r}{1 + (m_2/m_1)}$$

$$r_2 = \frac{(m_1/m_2)r}{1 + (m_1/m_2)}$$

The kinetic energy of the system is the rotational kinetic energy (Table 11.2), which depends on the *moment of inertia*, I.

$$\begin{aligned} I &= \textstyle\sum_i m_i r_i^2 \\ &= \mu r^2 \end{aligned} \tag{11.21}$$

Table 11.2
Linear and Rotating Systems

Linear				Angular			
Mass	m		kg	Moment of inertia	I	mr^2	kg m^2
Speed	u		m s^{-1}	Angular speed	u_{rot}	u/r	cycle s^{-1}
Momentum	p	mu	kg m s^{-1}	Angular momentum	ℓ	Iu_{rot}	kg m^2 s^{-1}
Kinetic E	K	$\frac{mu^2}{2}$	kg m^2 s^{-2}	Rotational kinetic E	K	$\frac{Iu_{rot}^2}{2}$	kg m^2 s^{-2}

where μ is the reduced mass defined by

$$\frac{1}{\mu} = \frac{1}{m_1} + \frac{1}{m_2} \tag{11.22}$$

or

$$\mu = \frac{m_1 m_2}{m_1 + m_2} \tag{11.23}$$

The kinetic energy of a rotating system is

$$K = \frac{1}{2} I u_{\text{rot}}^2 \tag{11.24}$$

where u_{rot} is the angular speed (in units of cycles per second).

Next we write the Schrödinger equation for this rotating system, and solve for the energy and the wave function.

$$\hat{H}\psi = E\psi$$

$$[\hat{K} + \hat{V}]\psi = E\psi$$

If we assume that the rotational motion is unrestricted, then the potential energy term is zero and we need only to consider the kinetic energy operator for the system.

$$\hat{K}\psi = E\psi$$

$$-\frac{\hbar^2}{2I}\left(\frac{\partial^2\psi}{\partial x^2} + \frac{\partial^2\psi}{\partial y^2} + \frac{\partial^2\psi}{\partial z^2}\right) = E\psi \tag{11.25}$$

which looks similar to the Schrödinger equation for a particle in a three-dimensional box, except that we use the moment of inertia rather than the mass. The term in parentheses in Equation 11.25 arises frequently and is called the Laplacian operator (here operating on the wave function ψ). The Laplacian operator has the symbol ∇^2 (del-squared). In Cartesian coordinates,

$$\nabla^2 = \frac{\partial^2}{\partial x^2} + \frac{\partial^2}{\partial y^2} + \frac{\partial^2}{\partial z^2} \tag{11.26}$$

Equation 11.22 can now be written as

$$-\frac{\hbar^2}{2I}\nabla^2\psi = E\psi \tag{11.27}$$

The solution to Equation 11.27 is rather complicated. We shall simply illustrate the main steps and focus on the results. Equation 11.27 has three variables (x, y, and z), but they are not independent of each other. We may view the rigid rotor model as equivalent to a point with reduced mass μ moving freely on the surface of a sphere of radius r (see Figure 11.8b). In this equivalent view, only two variables (analogous to a longitude and a latitude) are needed to describe the location of the point mass, and thus the orientation of the rigid rotor. First we transform the variables to a spherical

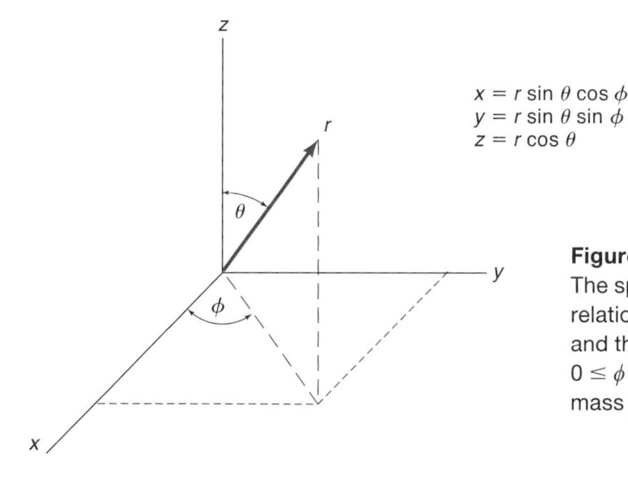

Figure 11.9
The spherical polar coordinate system and the relation between the spherical polar coordinates and the Cartesian coordinates ($0 \leq r \leq \infty$; $0 \leq \theta \leq \pi$; $0 \leq \phi \leq 2\pi$). For the rigid rotor, a point of reduced mass μ lies at a fixed distance r from the origin.

polar coordinate system, because this enables us to separate our fixed distance r from the two angles that describe the orientation of the rotor (Figure 11.9). Following the convention used by chemists, the angle θ is known as the co-latitude; it describes the angle from the z axis (i.e., from the North Pole). The angle ϕ is known as the azimuth; it describes the longitude with respect to the x axis. In this new coordinate system, the Laplacian operator becomes

$$\nabla^2 = \frac{1}{r^2}\frac{\partial}{\partial r}\left(r^2\frac{\partial}{\partial r}\right) + \frac{1}{r^2 \sin\theta}\frac{\partial}{\partial\theta}\left(\sin\theta\frac{\partial}{\partial\theta}\right) + \frac{1}{r^2\sin^2\theta}\left(\frac{\partial^2}{\partial\phi^2}\right) \qquad (11.28)$$

We define a new parameter

$$\beta = \frac{2IE}{\hbar^2} \qquad (11.29)$$

Combining Equations 11.27, 11.28, and 11.29, we obtain a simplified equation

$$\sin\theta\frac{\partial}{\partial\theta}\left(\sin\theta\frac{\partial\psi}{\partial\theta}\right) + \left(\frac{\partial^2\psi}{\partial\phi^2}\right) + (\beta\sin^2\theta)\psi = 0 \qquad (11.30)$$

The term in Equation 11.28 involving partial derivatives with respect to r drops out because r is a constant. Now we have a differential equation that depends on two variables, θ and ϕ. As we did for the particle in a two-dimensional box (see Section 10.11), we solve this differential equation by assuming that the variables are independent and therefore separable.

$$\psi(\theta,\phi) = \Theta(\theta)\Phi(\phi) \qquad (11.31)$$

The overall wave function $\psi(\theta,\phi)$ is assumed to be the product of two new functions Θ and Φ, each of which is a function of only a single variable. When we substitute Equation 11.31 into Equation 11.30, we see that indeed $\psi(\theta,\phi)$ is separable. A bit of algebraic manipulation gives

$$\left[\frac{\sin\theta}{\Theta(\theta)}\frac{d}{d\theta}\left(\sin\theta\frac{d\Theta(\theta)}{d\theta}\right) + \beta\sin^2\theta\right] + \left[\frac{1}{\Phi(\phi)}\frac{d^2\Phi(\phi)}{d\phi^2}\right] = 0 \qquad (11.32)$$

Table 11.3
The First Few Spherical Harmonic Functions

ℓ	m_ℓ	$Y_\ell^{m_\ell}(\theta,\phi)$[a]
0	0	$\left(\frac{1}{4\pi}\right)^{1/2}$
1	0	$\left(\frac{3}{4\pi}\right)^{1/2}\cos\theta$
1	±1	$\mp\left(\frac{3}{8\pi}\right)^{1/2}(\sin\theta)e^{\pm i\phi}$
2	0	$\left(\frac{5}{16\pi}\right)^{1/2}(3\cos^2\theta - 1)$
2	±1	$\mp\left(\frac{15}{8\pi}\right)^{1/2}(\sin\theta)(\cos\theta)e^{\pm i\phi}$
2	±2	$\left(\frac{15}{32\pi}\right)^{1/2}(\sin^2\theta)e^{\pm 2i\phi}$

[a]The negative signs in $Y_1^1(\theta, \phi)$ and $Y_2^1(\theta, \phi)$ are simply a convention.

where the term in the first square brackets depends only on θ, and the term in the second square brackets depends only on ϕ.

We may now solve separately the two parts to Equation 11.32. The part in the second square brackets may look familiar—it is equivalent to the particle on a ring, which we solved in Section 10.10. The solution is of the form

$$\Phi(\phi) = Ae^{im\phi} \qquad m = 0, \pm1, \pm2, \ldots \tag{11.33}$$

where A is a normalization constant and m is a quantum number. The part of Equation 11.32 in the first square brackets is much more difficult to solve. Fortunately, mathematicians had solved this type of differential equation prior to the development of quantum mechanics. There is a whole family of quantized solutions; these are called the associated Legendre polynomials after the French mathematician Adrien-Marie Legendre (1752–1853). But our goal is to find $\psi(\theta,\phi)$, which is the product of a Legendre polynomial and a particle-on-a-ring wave function (Equation 11.33). The full solutions to the rigid rotor wave function are a quantized family of functions known as the spherical harmonics, $Y_\ell^{m_\ell}(\theta,\phi)$.

The subscript in m_ℓ shows the connection of m with ℓ.

$$\psi_{\ell,m_\ell}(\theta,\phi) = Y_\ell^{m_\ell}(\theta,\phi) \quad \ell = 0, 1, 2, \ldots \quad m_\ell = 0, \pm1, \pm2, \ldots, \pm\ell \tag{11.34}$$

There are two variables (θ and ϕ) that describe the rigid rotor; hence, there are two quantum numbers, ℓ and m_ℓ. Also note that this system has two degrees of freedom, as discussed in Section 10.10. An examination of the first few spherical harmonic wave functions (Table 11.3 and Figure 11.10) reveals a few trends that hold for the whole

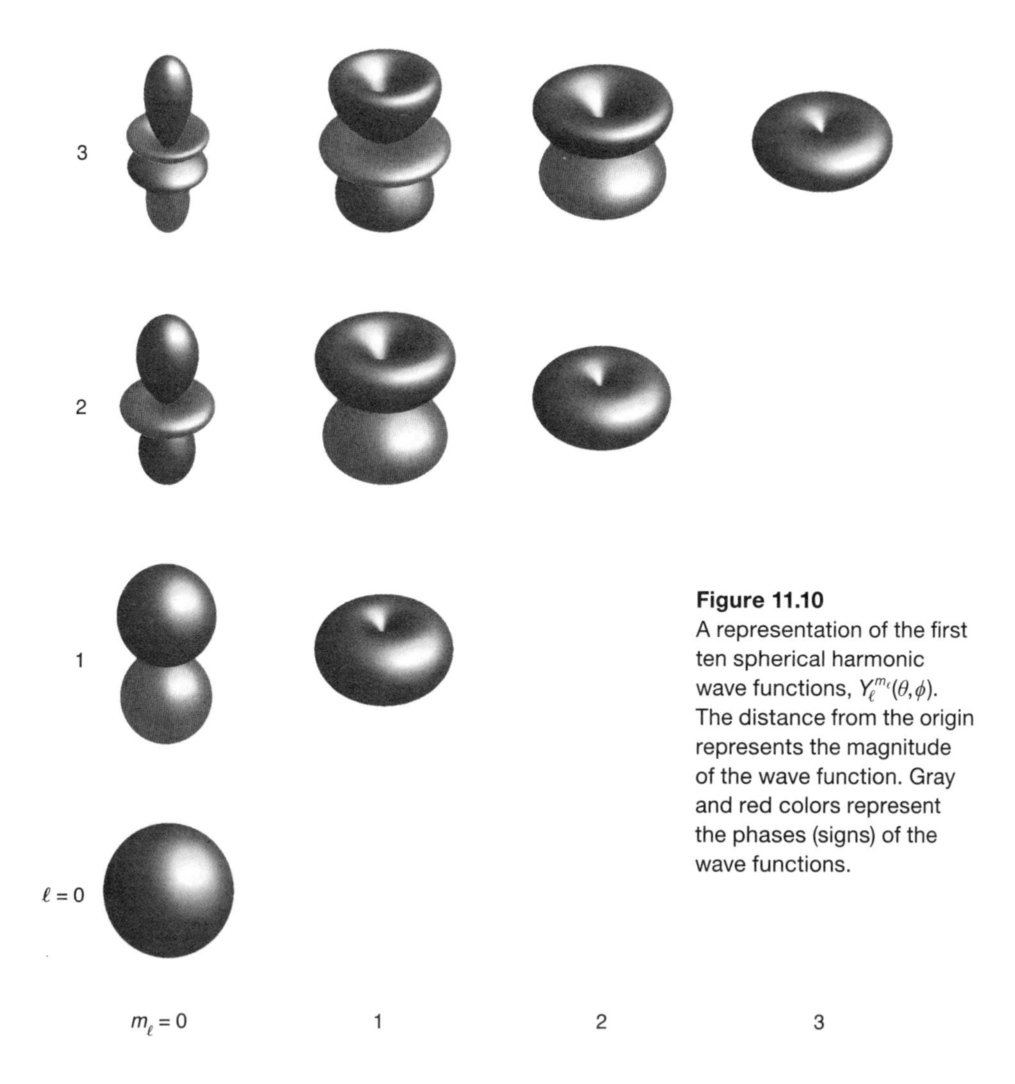

Figure 11.10
A representation of the first ten spherical harmonic wave functions, $Y_\ell^{m_\ell}(\theta,\phi)$. The distance from the origin represents the magnitude of the wave function. Gray and red colors represent the phases (signs) of the wave functions.

family of functions—the number of nodes is given by the quantum number ℓ and the orientation of the node(s) is (are) given by the quantum number m_ℓ. For the rigid rotor, the shape of each node is a circle.

Rigid Rotor Energy Levels

Now that we have solved for the rigid rotor wave functions, $\psi_{\ell,m_\ell}(\theta,\phi)$, we can find the energy E_ℓ associated with each wave function using the Schrödinger equation:

$$\hat{H}\psi_{\ell,m_\ell}(\theta,\phi) = E_\ell\psi_{\ell,m_\ell}(\theta,\phi) \quad \ell = 0, 1, 2, \ldots \quad m_\ell = 0, \pm1, \pm2, \ldots, \pm\ell \quad (11.35)$$

We have placed a subscript ℓ on the energy because the energy depends on the quantum number ℓ (but not on the quantum number m_ℓ). The result is

$$E_\ell = \ell(\ell+1)\frac{\hbar^2}{2I} \qquad \ell = 0, 1, 2, \ldots \quad (11.36)$$

From Equation 11.29, $\beta = \ell(\ell + 1)$.

Within an energy level, each energy state is described by a unique quantum number m_ℓ. The quantum number m_ℓ takes on values ranging from $-\ell$ to ℓ, for a total of

$(2\ell + 1)$ different values. At each energy level E_ℓ there are $(2\ell + 1)$ energy states; thus, the degeneracy g of an energy level equals $(2\ell + 1)$.

The quantum number ℓ is related to the total energy of the system. Because there is no potential energy associated with the rigid rotor, ℓ describes the (rotational) kinetic energy. Thus, ℓ is called the *angular momentum quantum number*. To determine the magnitude of the angular momentum (in a sense, how fast the rotor is spinning), we use the angular momentum squared operator, $\hat{L}^2$. The rigid rotor wave functions are eigenfunctions of the angular momentum squared operator:

$$\hat{L}^2\psi_{\ell,m_\ell}(\theta,\phi) = \hbar^2\ell(\ell + 1)\psi_{\ell,m_\ell}(\theta,\phi) \tag{11.37}$$

Thus, the square magnitude of the angular momentum is

$$L^2 = \hbar^2\ell(\ell + 1) \tag{11.38}$$

and the magnitude of angular momentum is

$$L = \hbar\sqrt{\ell(\ell + 1)} \tag{11.39}$$

We would like to know in which direction the angular momentum vector is pointing, that is, about which direction the rigid rotor is rotating. However, the Heisenberg uncertainty principle limits our knowledge of this quantity. We cannot measure the direction precisely because this would give zero uncertainty in the angular momentum. We can only precisely measure one of the three Cartesian components of the angular momentum. By convention, we choose the z-axis component to be the one we measure. We choose the z component for a simple reason—the rigid rotor wave functions are eigenfunctions of the z component of the angular momentum operator $\hat{L}_z$. In Cartesian coordinates, the z component of the angular momentum operator looks complicated (see Table 10.2), but in spherical polar coordinates it is much simpler.

$$\hat{L}_z = -i\hbar\frac{\partial}{\partial\phi} \tag{11.40}$$

The rigid rotor wave functions satisfy an eigenfunction–eigenvalue relationship,

$$\hat{L}_z\psi_{\ell,m_\ell}(\theta,\phi) = \hbar m_\ell\psi_{\ell,m_\ell}(\theta,\phi) \tag{11.41}$$

where the eigenvalues are the allowed z components of angular momentum:

$$L_z = \hbar m_\ell \tag{11.42}$$

Now we can provide a more physical interpretation for the rigid rotor quantum numbers ℓ and m_ℓ. The total energy of the system, the kinetic energy of the system, the total angular momentum, and the speed of rotation are described by ℓ. The direction of rotation is described by m_ℓ.

Microwave Spectra

Microwave spectra arise as a result of the transitions among rotational energy levels. These transitions are generally well described using the rigid rotor model. In common spectroscopic notation, the quantum number ℓ is replaced with the rotational quantum number J. Thus rotational energies are given by (see Equation 11.36)

$$E_J = \frac{\hbar^2}{2I} J(J+1) \qquad J = 0, 1, 2, \ldots$$
$$= BhJ(J+1) \tag{11.43}$$

where B is the *rotational constant*, given by $\hbar/4\pi I$. The rotational constant has units of Hz or s^{-1}. Expressed in units of cm^{-1}, the wavenumber version is given a tilde,

$$\tilde{B} = \frac{B}{c} \tag{11.44}$$

where c is speed of light.

A transition from a lower energy level to a higher one can be induced by irradiating a sample of molecules with the appropriate microwave frequency. Not all the transitions are allowed because of the selection rule $\Delta J = \pm 1$. This selection rule is derived using time-dependent quantum mechanics, but we can rationalize it using the following argument. A photon carries a single unit of angular momentum, and angular momentum, like energy, is conserved. Thus, for the absorption of a single photon, the molecule can only increase its angular momentum by one unit: $\Delta J = +1$. On emission of a photon, $\Delta J = -1$. From Equation 11.36, the energy change, ΔE, for a transition from $J = 0$ to the $J = 1$ level is given by

$$\Delta E_{rot} = E_1 - E_0$$
$$= 2Bh$$

For the $J = 2 \leftarrow 1$ transition, ΔE is given by

$$\Delta E_{rot} = Bh[2(2+1) - 1(1+1)] = 4Bh$$

and so on. We can generalize the results by letting J' and J'' be the rotational quantum numbers for the upper and lower levels, respectively. The energy difference, then, is given by

$$\Delta E_{rot} = BJ'(J'+1)h - BJ''(J''+1)h$$
$$= Bh[J'(J'+1) - J''(J''+1)]$$

Because $J' - J'' = 1$, the above equation becomes

$$\Delta E_{rot} = 2BhJ' \qquad J' = 1, 2, 3, \ldots \tag{11.45}$$

Thus, absorptions $J = 1 \leftarrow 0$, $2 \leftarrow 1$, $3 \leftarrow 2$, … have energy differences of $2Bh$, $4Bh$, $6Bh$, …, and a rotational spectrum with a set of equally spaced lines with separations of $2Bh$ is obtained (Figure 11.11).

Under high resolution, the spacing between adjacent lines in a rotational spectrum is found to decrease with increasing J. At higher energy levels, a molecule rotates faster, so the internuclear bond is stretched somewhat by the centrifugal force. An increase in bond length r increases the moment of inertia I, causing E_{rot} to decrease

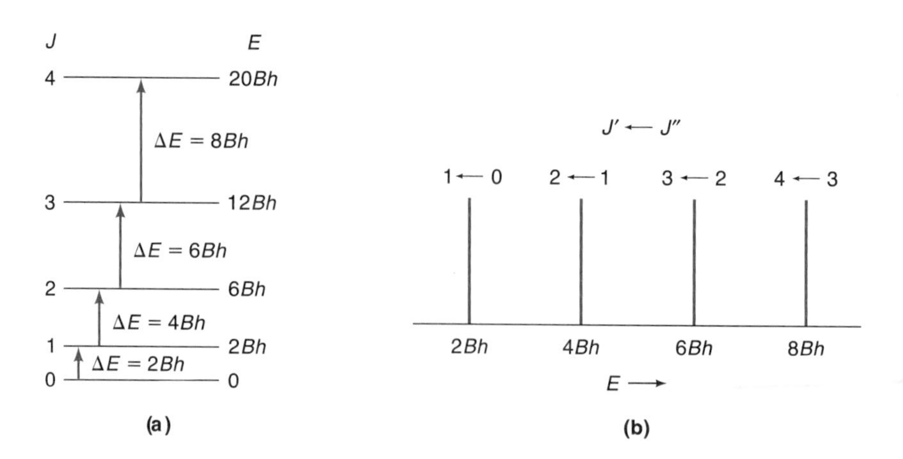

Figure 11.11
(a) Allowed resonance conditions for microwave transitions for a diatomic molecule. The degeneracies of the energy levels are not shown here. (b) Equally spaced rotational lines in a microwave absorption spectrum. In practice, the lines are not of equal intensity.

(see Equation 11.43). For better accuracy, this effect can be corrected by adding a term to Equation 11.43 as follows:

$$E_J = BhJ(J + 1) - Dh[J(J + 1)]^2 \tag{11.46}$$

where D, the *centrifugal constant*, is about 1000 times smaller than the rotational constant B. Thus, the second term can be neglected unless J is a large number.

Microwave spectroscopy is an important tool for determining molecular geometry. From the separation between adjacent lines, we obtain the rotational constant B and hence the moment of inertia and the internuclear distance r.

EXAMPLE 11.1

The microwave spectrum of carbon monoxide consists of a series of lines separated by 1.15×10^{11} Hz. Calculate the bond length of $^{12}C^{16}O$.

ANSWER

The spacing between successive lines (Figure 11.12) is $2Bh$, which is equal to ΔE_{rot}. Therefore, the difference in frequencies, $\Delta\nu$, is given by

$$\Delta\nu = \frac{\Delta E_{rot}}{h} = 2B = 2\left(\frac{\hbar}{4\pi I}\right)$$

Solving for the moment of inertia, I, we get

$$I = \frac{\hbar}{2\pi\Delta\nu}$$

$$= \frac{(1.055 \times 10^{-34}\ \text{J s})}{2\pi(1.15 \times 10^{11}\ \text{s}^{-1})}\left(\frac{\text{kg m}^2\ \text{s}^{-2}}{\text{J}}\right)$$

$$= 1.46 \times 10^{-46}\ \text{kg m}^2$$

From Equation 11.21,

$$I = \mu r^2 = \frac{m_1 m_2}{m_1 + m_2} r^2$$

so by algebraic manipulation

$$r = \left[\frac{I(m_1 + m_2)}{m_1 m_2}\right]^{1/2}$$

$$= \left\{\frac{(1.46 \times 10^{-46}\ \text{kg m}^2)[(0.0120 + 0.0160)\ \text{kg mol}^{-1}](6.022 \times 10^{23}\ \text{mol}^{-1})}{(0.0120\ \text{kg mol}^{-1})(0.0160\ \text{kg mol}^{-1})}\right\}^{1/2}$$

$$= 1.13 \times 10^{-10}\ \text{m}$$

$$= 1.13\ \text{Å}$$

COMMENT

For precise work, we include the mass of the individual isotopes (here ^{12}C and ^{16}O) rather than the average atomic mass. Many bond lengths have been determined to five or more digit precision by microwave spectroscopy.

The analysis of the microwave spectra of polyatomic molecules can be quite complex and is left for more advanced texts. Note that three-dimensional molecules (and non-linear molecules) have three rotational degrees of freedom, and thus three rotational quantum numbers and three rotational constants (A, B, C). One might visualize the three rotational degrees of freedom as rotation about the x, y, and z coordinate axes. Here, we shall briefly discuss carbonyl sulfide (OCS), a linear triatomic molecule. Although OCS has two rotational degrees of freedom, they are equivalent, so OCS rotations are described by a single rotational quantum number. Like CO, OCS gives a series of equally spaced lines in its microwave spectrum. However, because there are two bond lengths, the moment of inertia is expressed in terms of two unknowns, r_{CO} and r_{CS}. We can overcome this difficulty by making the (reasonable) assumption that isotopic substitution does not alter bond lengths. Recording the microwave spectra of $^{16}O^{12}C^{32}S$ and $^{16}O^{12}C^{34}S$ yields two moments of inertia corresponding to these species and solving a system of two equations and two unknowns provides the two bond lengths.

Rotation about the bond axis of a linear molecule does not move the position of any atoms and thus is not considered a rotational degree of freedom.

Nonpolar molecules (e.g., homonuclear diatomic molecules like N_2 and O_2) do not absorb radiation in the microwave region and are said to be *microwave inactive*. To see why a polar molecule like CO behaves differently, let us consider the interaction between a dipole and an oscillating electric field from the electromagnetic wave (Figure 11.12). In Figure 11.12a, the negative end of the dipole follows the propagation of the wave (the positive region) and rotates in a clockwise direction. If, after

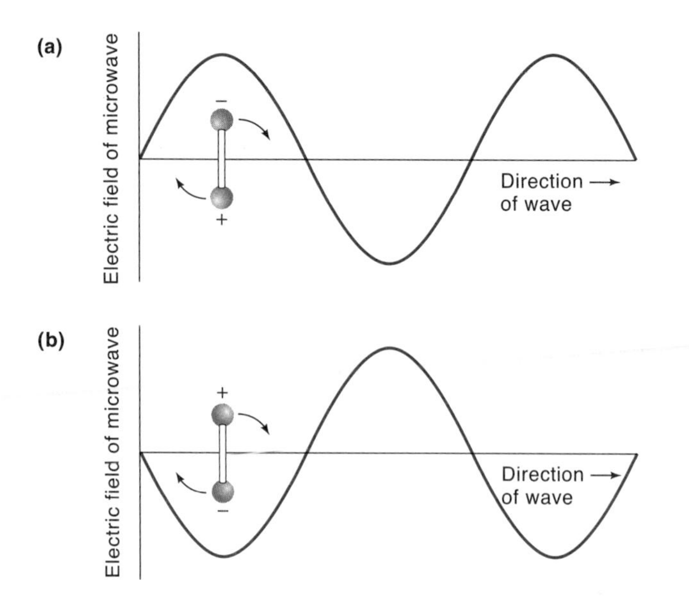

Figure 11.12
Interaction between the electric field component of microwave radiation and an electric dipole. (a) The negative end of the dipole follows the propagation of the wave (the positive region) and rotates in a clockwise direction. (b) If, after the molecule has rotated to the positive region, the radiation has also moved along to its next half cycle, the positive end of the dipole will move into the negative region of the wave, and the negative end will be pushed up. Thus, the molecule will rotate faster. No such interaction can occur with nonpolar molecules.

the molecule has rotated by 180°, the wave has also moved along to its next cycle (Figure 11.12b), the positive end of the dipole will move into the negative region of the wave while the negative end will be pushed up. Unless the frequency of the radiation is equal to that of the molecular rotation, the dipole cannot absorb energy from the radiation to rotate faster. This classical wave-nature description supplements the quantum mechanical picture for a transition from a lower rotational energy level to a higher one. It also explains why nonpolar molecules are microwave inactive, because they cannot interact with the electric field component of the radiation.

Next, we turn our attention to the intensity of spectral lines. As mentioned earlier, the intensity of a line in absorption spectroscopy is based on the Boltzmann distribution. In rotational spectra, another factor must also be included; that is, the degeneracy of the rotational level. For the rotational level corresponding to the rotational quantum number J, there are $(2J + 1)$ states. In the absence of an external electric field, this degeneracy must be incorporated into the Boltzmann equation. Thus, the ratio of the number of molecules in the $J = J$ level (N_J) to the $J = 0$ level (N_0) is given by

$$\frac{N_J}{N_0} = (2J + 1)e^{-\Delta E/k_B T} = (2J + 1)e^{-(E_J - E_0)/k_B T}$$
$$= (2J + 1)e^{-BhJ(J+1)/k_B T}$$

Figure 11.13 shows a plot of N_J/N_0 versus J for CO at 300 K. Initially, when J is small, the exponential term is close to unity, so N_J/N_0 is actually greater than 1. As J increases, the $(E_J - E_0)$ term becomes important, and N_J/N_0 begins to decrease. The

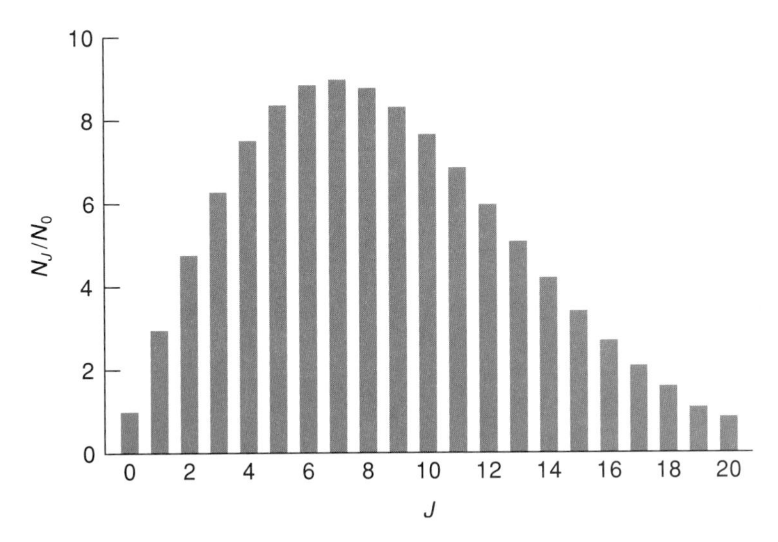

Figure 11.13
Plot of population ratio (N_J/N_0) versus rotational quantum number (J) for the first 20 states of CO at 300 K.

intensity of the lines in the microwave absorption spectrum of CO follows closely the pattern in Figure 11.13.

Finally, microwave spectroscopy generally applies only to molecules in the gas phase. In solution, the frequency of molecular collision is much greater than the frequency of molecular rotation. Consequently, molecules cannot execute full rotational motion, and hence no microwave spectra can be obtained.

11.3 Infrared Spectroscopy

Having discussed microwave spectroscopy, we now turn our attention to infrared spectroscopy, which deals with the vibrational motion of molecules. First, we need to develop another simple quantum mechanical model that has an analytical solution—the harmonic oscillator.

The Harmonic Oscillator

It is useful to start with a system that behaves according to classical, that is, Newtonian–mechanics. An object of mass m is attached to a spring (Figure 11.14). According to Hooke's law (after the English natural philosopher and physicist, Robert Hooke, 1635–1703), the force F acting on the object is proportional to the displacement x from the equilibrium position,

$$\begin{aligned} F &\propto -x \\ &= -kx \end{aligned} \tag{11.47}$$

where k is the *force constant*, a characteristic of the stiffness of the spring, measured in N m^{-1}. The negative sign means that the force acts in the direction opposite to x; if

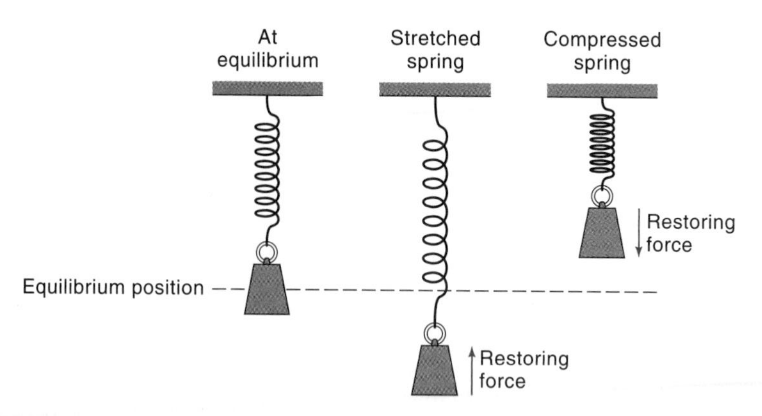

Figure 11.14
In a harmonic oscillator, a weight attached to a spring exhibits simple harmonic motion when it is pulled downward and then released. The ideal spring obeys Hooke's law, $F = -kx$, and gravity is neglected.

x is positive (i.e., if the spring is stretched), then F is negative, meaning that there is a restoring force pulling the object upward. The reverse holds true for a compressed spring. If we pull the object downward and let go, it will undergo a periodic vibrational motion known as *simple harmonic motion*, for which the displacement x at any time t is given by a sinusoidal function,

$$x = A \sin 2\pi\nu t \tag{11.48}$$

where A is the amplitude of vibration and ν is a constant equal to the frequency of vibration expressed in cycles per second (Hz):

$$\nu = \frac{1}{2\pi}\sqrt{\frac{k}{m}} \tag{11.49}$$

Because every term on the right-hand side of Equation 11.49 is a constant, this equation says that there is one characteristic, or natural, frequency of vibration. During the complete period of a vibration, the kinetic energy of the particle is being converted to the potential energy of the spring, and vice versa. The potential energy V of the system is given by

$$V = \frac{1}{2}kx^2 \tag{11.50}$$

To apply the simple harmonic motion to the vibration of molecules, we substitute the reduced mass μ for the mass m in Equation 11.49,

$$\nu = \frac{1}{2\pi}\sqrt{\frac{k}{\mu}} \tag{11.51}$$

where ν is now the fundamental frequency of vibration of the molecule. If a diatomic molecule behaves as a harmonic oscillator, then its potential energy is given by

$$V = \frac{1}{2}k(r - r_e)^2 \tag{11.52}$$

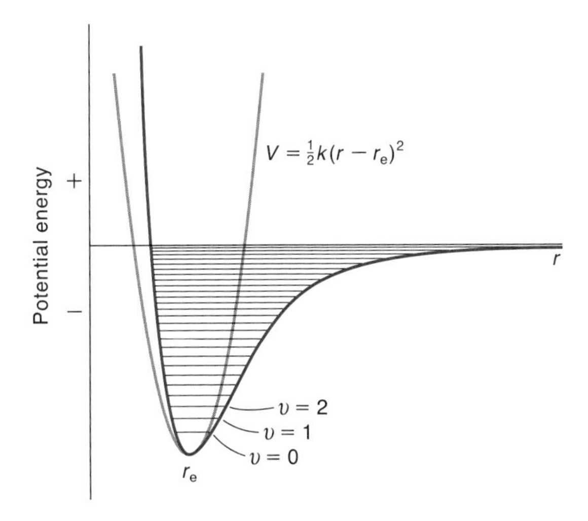

Figure 11.15
Potential energy curve for a diatomic molecule. The symmetric curve is given by Equation 11.52; the other curve represents the actual potential energy of the molecule. The vertical distance from $\upsilon = 0$ to the r axis is the bond dissociation energy of the molecule. Horizontal lines represent vibrational energy levels.

where r is the distance between the two atoms and r_e is the equilibrium bond distance. Thus, we have the displacement x written as

$$x = r - r_e$$

A plot of V versus x yields a parabolic curve.

Figure 11.15 compares the potential energy curve based on Equation 11.52 with that of a real molecule, such as carbon monoxide, CO. At small displacements from r_e, the two curves match quite well, and the oscillation can be described as simple harmonic motion. As r increases due to more energetic vibrations, however, appreciable deviation occurs. The character of the vibrations now is said to be *anharmonic* because the stretching is not equivalent to the compression. We shall return to this point later.

Quantum–Mechanical Solution to the Harmonic Oscillator

To solve the energy and wave function of the quantum mechanical system, we follow a familiar path. We start by writing the Schrödinger equation for the one-dimensional harmonic oscillator,

$$\hat{H}\psi(x) = E\psi(x)$$

$$(\hat{K} + \hat{V})\psi(x) = E\psi(x)$$

and substituting the kinetic and potential energy operators,

$$\left(-\frac{\hbar^2}{2\mu}\frac{d^2}{dx^2} + \frac{1}{2}kx^2\right)\psi(x) = E\psi(x) \tag{11.53}$$

The solution to Equation 11.53 shows that the energy of the harmonic oscillator is quantized,

$$E_\upsilon = \left(\upsilon + \frac{1}{2}\right)h\nu \qquad \upsilon = 0, 1, 2, \ldots \tag{11.54}$$

Distinguish the *vibrational quantum number* υ (vee) from the *vibrational frequency* ν (nu).

where υ is the vibrational quantum number. Like the particle-in-a-box model, but unlike the particle-on-a-ring model, the lowest vibrational energy ($\upsilon = 0$) is not zero. The ground-state energy of the harmonic oscillator is equal to $h\nu/2$. This means that a molecule will execute vibrational motion even at the absolute zero of temperature. This *zero-point* energy is in accord with the Heisenberg uncertainty principle. If a molecule did not vibrate, then its energy and the momentum associated with the motion would be zero. In this case, the uncertainty in the momentum would also be zero, which means the uncertainty in the position (in locating the atoms) would be infinite. But the atoms are separated by a finite distance, so the uncertainty should be comparable to the bond distance and not infinite. A classical harmonic oscillator has a motionless ground state with momentum $p = 0$, and thus $\Delta p = 0$, which is inconsistent with quantum mechanics.

For the quantum-mechanical harmonic oscillator, the energy levels are evenly spaced and separated by energy $h\nu$ (Figure 11.16). The resonance condition is

$$\Delta E = h\nu$$

Thus, to promote a quantum harmonic oscillator from the ground vibrational state to the first excited vibrational state, we have

$$\begin{aligned}\Delta E &= E_1 - E_0 \\ &= \hbar\left(\frac{k}{\mu}\right)^{1/2}\end{aligned} \tag{11.55}$$

The transition wavenumber is

$$\tilde{\nu} = \frac{1}{2\pi c}\left(\frac{k}{\mu}\right)^{1/2} \tag{11.56}$$

For real molecules, $\tilde{\nu}$ typically lies in the range 400 to 3000 cm^{-1}, which is in the infrared region of the electromagnetic spectrum.

The wave functions corresponding to E_υ for a harmonic oscillator are given by

$$\psi_\upsilon(x) = N_\upsilon H_\upsilon(\alpha^{1/2}x)e^{-\alpha x^2/2} \qquad \upsilon = 0, 1, 2, \ldots \tag{11.57}$$

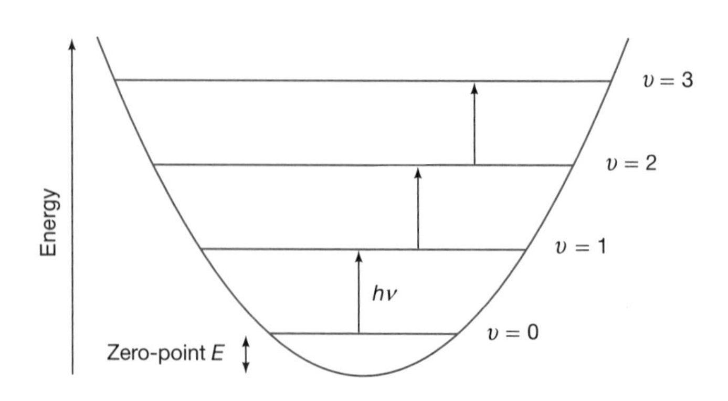

Figure 11.16
The energy levels of a quantum-mechanical harmonic oscillator. The energy levels are each separated by $h\nu$, and the zero-point energy is $h\nu/2$. The symmetric curve is given by Equation 11.52.

Table 11.4
The First Few Hermite Polynomial Functions[a]

$H_0(\xi) = 1$
$H_1(\xi) = 2\xi$
$H_2(\xi) = 4\xi^2 - 2$
$H_3(\xi) = 8\xi^3 - 12\xi$

[a] The variable ξ is equal to $\alpha^{1/2}x$. ξ is also equal to $(k\mu)^{1/4}\hbar^{-1/2}x$.

where

$$\alpha = \left(\frac{k\mu}{\hbar^2}\right)^{1/2} \tag{11.58}$$

The normalization constant N_υ is

$$N_\upsilon = \frac{1}{(2^\upsilon \upsilon!)^{1/2}}\left(\frac{\alpha}{\pi}\right)^{1/4} \tag{11.59}$$

where $\upsilon! = 1 \times 2 \times 3 \times \cdots \times \upsilon$. The term $H_\upsilon(\alpha^{1/2}x)$ represents the Hermite polynomials, a family of functions (Table 11.4) that gives the harmonic oscillator wave functions their characteristic shape (Figure 11.17). Note that $H_\upsilon(\alpha^{1/2}x)$ is a υth-order polynomial in $\alpha^{1/2}x$, so there are υ nodes in the wave function ψ_υ. The ground state wave function for the harmonic oscillator ($\upsilon = 0$) is nodeless and has the form of a Gaussian function:

Recall that $0! = 1$.

$$\psi_0(x) = \left(\frac{\alpha}{\pi}\right)^{1/4} e^{-\alpha x^2/2} \tag{11.60}$$

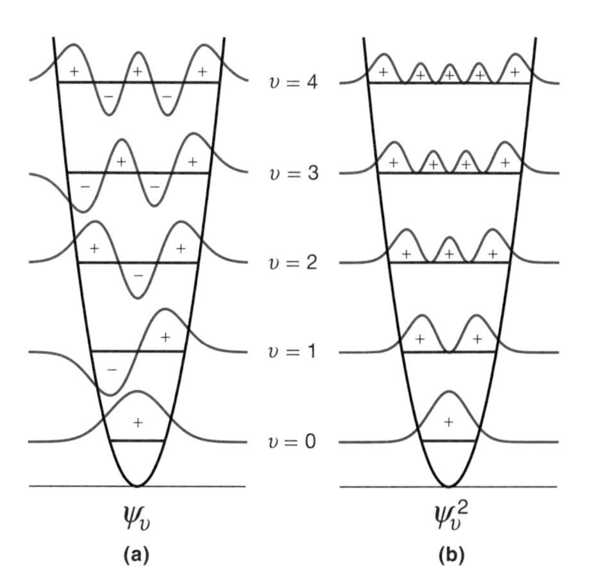

Figure 11.17
(a) The wave functions for the first few levels of the harmonic oscillator. (b) The corresponding probability densities (square of the wave functions).

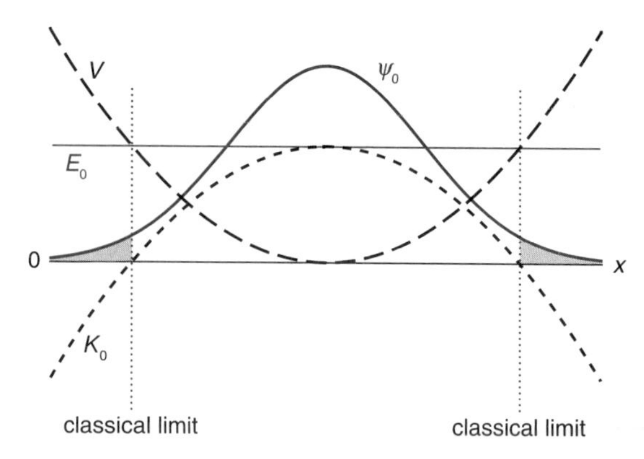

Figure 11.18
Tunneling in the quantum harmonic oscillator. The red horizontal line represents the zero-point energy $\left(\frac{1}{2}h\nu\right)$ and the shaded region is the classically forbidden region in which $K_0 < 0$.

Tunneling and the Harmonic Oscillator Wave Functions

For the quantum harmonic oscillator, the wave function (and hence the probability density function) has a nonzero value outside the points where the energy level crosses the potential energy curve. Consider the ground state $\psi_0(x)$ (Figure 11.18). The horizontal line at $h\nu/2$ is the total energy E_0, which is a constant independent of the position x. The potential energy (V) is a function of position, $V = kx^2$, and the kinetic energy (K_0) is the total energy minus the potential energy, $K_0 = E_0 - V$. The *classical turning points* occur when the kinetic energy is zero—these are the points when the mass on the spring is at the two extreme positions and all of the energy of the system is potential energy. The quantum mechanical oscillator, however, has probability density beyond the classical turning points, in a region with "negative kinetic energy." This is the quantum mechanical tunneling phenomenon we saw in Section 10.12. There is no direct indication of quantum mechanical tunneling from an IR spectrum (as is the case of α-particle decay based on the particle-in-a-box model). The fact that the experimental potential energy curve deviates from the simple harmonic oscillator model, however, is at least partially the result of tunneling, which alters (increases) the bond length. (The other contributing factor to the asymmetry of the curve is anharmonicity, to be discussed shortly.)

Another way in which the quantum harmonic oscillator differs from the classical harmonic oscillator is in the location of the most probable position. In the ground-state quantum harmonic oscillator, the most probable position is the equilibrium position; the maximum in the wave function and the wave function squared (the probability density function) lies at $x = 0$ (see Figure 11.17). In contrast, the classical harmonic oscillator is *least* likely to be found at the equilibrium position and most likely to be found near the classical turning points. The classical oscillator is moving the fastest when $x = 0$ and moving the slowest at the turning points. Thus, the turning points correspond to the most probable positions. The situation changes for excited-state quantum harmonic oscillators. At larger values of the vibrational quantum number υ, the probability density function starts to resemble the classical probability density function (Figure 11.19), and the oscillator is more likely to be found away from equilibrium ($x = 0$) and near the classical turning points.

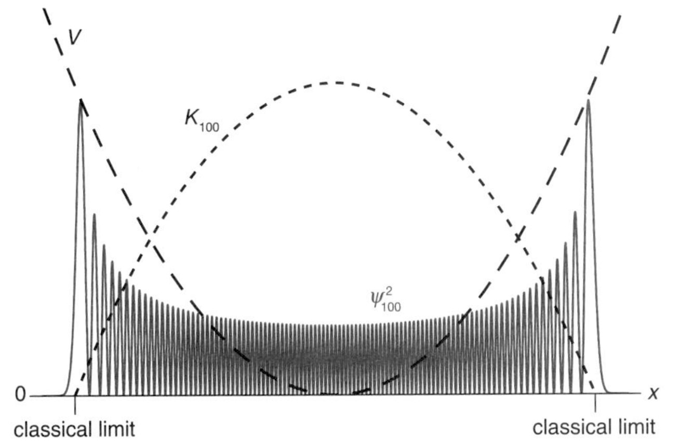

Figure 11.19
The probability density plot for the υ = 100th state of the quantum harmonic oscillator.

IR Spectra

We have shown how the harmonic oscillator model may be used to predict the fundamental vibrational frequencies of molecules, and how these transitions lie in the infrared region of the electromagnetic spectrum. Thus, the harmonic oscillator model finds great utility in interpreting infrared spectra. As the following example shows, the fundamental frequency determined from an IR spectrum provides information about the stiffness (which is related to the strength) of chemical bonds.

EXAMPLE 11.2

The fundamental frequency of vibration for the $H^{35}Cl$ molecule is 2886 cm^{-1}. Calculate the force constant of the molecule.

ANSWER

First, we calculate the reduced mass of the molecule, using ^{35}Cl:

$$\mu = \frac{m_{\text{H}} m_{\text{Cl}}}{m_{\text{H}} + m_{\text{Cl}}}$$

$$= \frac{(0.001008 \text{ kg mol}^{-1})(0.03497 \text{ kg mol}^{-1})}{(0.001008 + 0.03497)\text{kg mol}^{-1}(6.022 \times 10^{23} \text{ mol}^{-1})}$$

$$= 1.627 \times 10^{-27} \text{ kg}$$

Rearranging Equation 11.51 and noting that $\nu = c\tilde{\nu}$ we write

$$k = 4\pi^2 c^2 \tilde{\nu}^2 \mu$$

$$= 4\pi^2 (2.998 \times 10^{10} \text{ cm s}^{-1})^2 (2886 \text{ cm}^{-1})^2 (1.627 \times 10^{-27} \text{ kg})$$

$$= 480.8 \text{ kg s}^{-2}$$

Note that 1 N = 1 kg m s^{-2}, so

$$k = 480.8 \text{ N m}^{-1}$$

COMMENT

The force constant measures how much force is needed to stretch a bond by unit length (per meter or per angstrom). As expected, triple bonds have larger force constants than double bonds, which have larger values than single bonds. For example, k is approximately 450 N m^{-1} for a C−C bond, 930 N m^{-1} for a C=C bond, and 1600 N m^{-1} for a C≡C bond. A spring with k = 500 N m^{-1} would be deflected about 1 mm by the weight of a 50 kg person; this stiffness is the order of what we might expect from the springs on a full-sized pick-up truck.

Molecules do not behave exactly like harmonic oscillators. For example, as the bond length r increases past the equilibrium bond distance, the chemical bond weakens, and eventually dissociation takes place. For the simple harmonic oscillator, r could take on negative values, whereas r must be a positive number for real molecules. A more realistic description of molecular vibration is presented by the asymmetric potential energy curve in Figure 11.15. Each horizontal line represents a vibrational level. The spacing between successive levels decreases with increasing υ, due to the anharmonic character of the vibration. As a correction, we rewrite Equation 11.54 as

Note that x_e is a constant, not a coordinate.

$$E_\upsilon = \left(\upsilon + \frac{1}{2}\right)h\nu - x_e\left(\upsilon + \frac{1}{2}\right)^2 h\nu \tag{11.61}$$

where x_e is the *anharmonicity constant.* As in the case of the centrifugal distortion of rotation, the second term in Equation 11.61 can often be ignored because x_e typically lies in the range of 0.002 to 0.02 for diatomic molecules. Inclusion of the anharmonic term becomes more important as the vibrational quantum number υ increases.

The *selection rule* for transitions between vibrational energy levels is $\Delta\upsilon = \pm 1$. This means that transitions are only allowed to occur when the vibrational quantum number changes by one unit. This selection rule, like others previously discussed, may be derived from time-dependent quantum mechanics. Because the spacing between energy levels is large, most of the molecules reside in the ground level at room temperature. Therefore, the absorption of IR radiation almost always involves the $\upsilon = 1 \leftarrow 0$ transition (called *the fundamental band*). If the molecule behaves as a harmonic oscillator, then the $\upsilon = 2 \leftarrow 1$ transition, called a *hot band* (because its intensity increases with increasing temperature), will occur at the same frequency as the fundamental band. However, if there is appreciable anharmonicity, then the hot band can be distinguished from the fundamental band by its slightly lower frequency in the spectrum. Another consequence of anharmonicity is the breakdown of the selection rule, so that it is possible to have $\upsilon = 2 \leftarrow 0, 3 \leftarrow 0, \ldots$transitions, which are called *overtones*. The first overtone ($\upsilon = 2 \leftarrow 0$) band does *not* appear at exactly twice the frequency of the fundamental band. As stated in Example 11.2, the fundamental band of HCl occurs at 2886 cm^{-1}, whereas the first overtone is observed at 5668 cm^{-1}, which is somewhat less than 2×2886 cm^{-1}, or 5772 cm^{-1}. The difference is due to anharmonicity.

In order for a particular vibration to absorb IR radiation, or to be IR *active*, we must have

$$\frac{d\mu}{dx} \neq 0 \tag{11.62}$$

That is, the electric dipole moment μ must change with bond displacement x during a vibration. More generally, the electric dipole moment must change with molecular motion (such as a stretching or bending or wagging) during a vibration. Using a more intuitive wave-nature picture of radiation, we see that in order for a vibrational mode to be excited, the frequency of the oscillating electric field of the IR radiation must match that of the bond vibration. Keep in mind that when energy is absorbed by the molecule, the frequency of the bond vibration does not change. It is the amplitude that increases.* The requirement that the dipole moment must change with bond distance during vibration rules out IR activity in all homonuclear diatomic molecules.

A diatomic molecule has only one vibrational degree of freedom and so only one fundamental frequency of vibration. A nonlinear polyatomic molecule containing N atoms has $(3N - 6)$ vibrational degrees of freedom (see Section 2.9). Therefore, H_2O and SO_2 each has $(3 \times 3 - 6)$ or three different modes of vibration. The apparently complex vibrational motions of such molecules can be analyzed in terms of these three fundamental frequencies (Figure 11.20). We expect a total of three lines from this scheme if we consider only the $\upsilon = 1 \leftarrow 0$ transition for each vibrational mode. Each of these vibrational modes is termed a *normal mode*. Any vibrational motion of the molecule may be written as a linear combination of these normal modes.

As another example, let us consider a linear molecule, CO_2, which has four vibrational degrees of freedom ($3N - 5 = 4$) (Figure 11.21). Although CO_2 does not possess a permanent electric dipole moment, three of the four vibrations are IR active because there is a change in dipole moment with respect to bond displacement. Two of the four vibrations are *degenerate* because they have the same frequency and thus the same energy. In an IR absorption spectrum, CO_2 has two vibrational bands corresponding to an asymmetric stretch (2350 cm^{-1}) and a bending motion ($\sim$667 cm^{-1}). If we are not careful to account for the IR absorption of ambient air (by background subtraction or by purging with nitrogen or argon), then absorption due to CO_2 is seen in many IR spectra. In a polyatomic molecule, it is also possible to simultaneously excite two different vibrational modes. These *combination bands* appear in IR spectra at energies corresponding to the sum of the energies of the component vibrations because energy is conserved. Combination band absorption is much less probable, and thus combination bands are much less intense than fundamental absorptions.

* An analogy of this effect is the situation of pushing someone on a playground swing. You must push "in phase" with the oscillation of the swing, and your energy goes into increasing the amplitude, not the frequency, of the swinging motion.

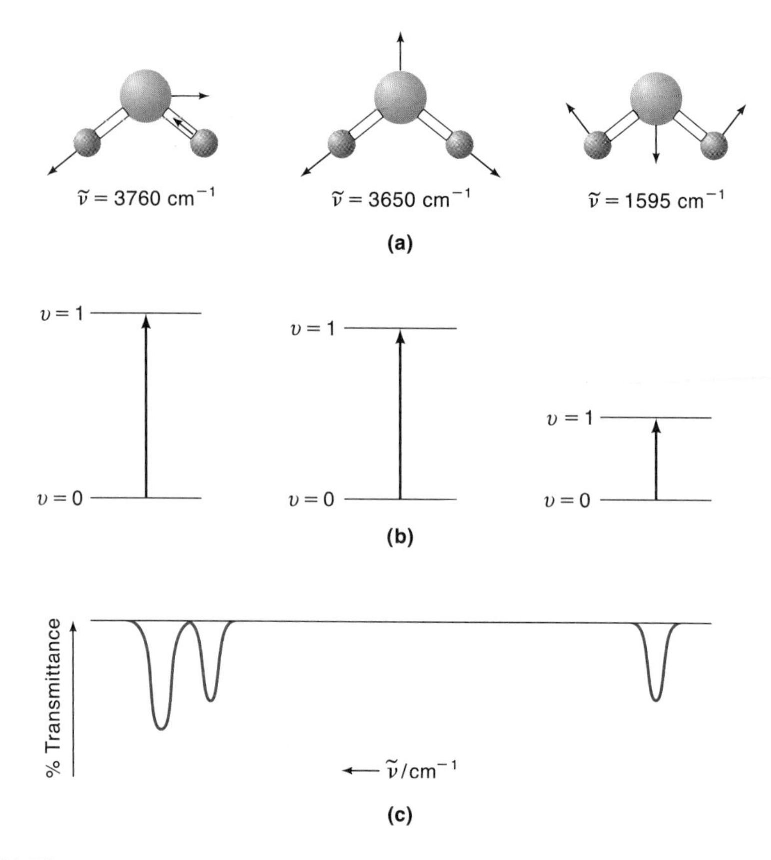

Figure 11.20
(a) The three normal vibrational modes of H_2O, all of which are infrared active. Note that in each case, the center of mass of the molecule remains unchanged during the course of a vibration. (b) Energy levels. (c) IR transmission spectrum.

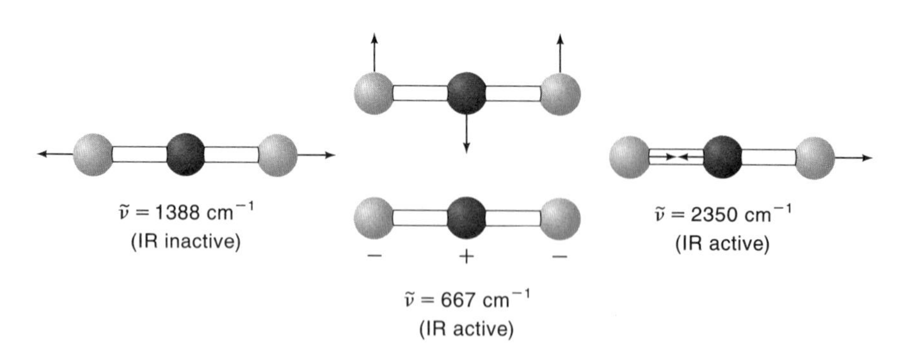

Figure 11.21
(a) The four normal vibrational modes of CO_2. The middle two vibrations have the same frequency (the same energy) and are described as degenerate. The "+" and "−" signs indicate movement of the atoms into and out of the plane of the paper. The frequency of the IR inactive mode is determined by another spectroscopic technique, called Raman scattering.

Simultaneous Vibrational and Rotational Transitions

Associated with any given vibrational state v is a set of rotational levels. Thus, a molecule is constantly executing both rotational and vibrational motions. The rotational energy levels shown in Figure 11.11 are assumed to be associated with the $v = 0$ level. Generally, then, a $v = 1 \leftarrow 0$ transition is accompanied by a simultaneous transition between two rotational levels associated with the lower and upper vibrational states. The selection rule $\Delta J = \pm 1$ still holds in this case. In solution, however, collisions between molecules hinder molecular rotation. The vibrations of molecules, on the other hand, are little affected by their neighbors because vibrational frequencies are greater than the frequency of collision. The situation is different for molecules in the gas phase, where simultaneous rotational and vibrational energy changes can occur. For example, it is possible to obtain a high-resolution spectrum for diatomic molecules in the gas phase because the rotational fine structure can be observed (Figure 11.22).

Starting with Equations 11.43 and 11.54, we can write the energy difference for a *simultaneous* transition between vibrational and rotational levels as

$$\Delta E = \left(v' + \frac{1}{2}\right)h\nu + BJ'(J' + 1)h - \left(v'' + \frac{1}{2}\right)h\nu - BJ''(J'' + 1)h \tag{11.63}$$

where v' and v'' represent the higher and lower vibrational states, and J' and J'' the rotational levels in the v' and v'' states, respectively. For the $v = 1 \leftarrow 0$ transition, that is, $v' = 1$ and $v'' = 0$, we obtain

$$\Delta E = h\nu + Bh[J'(J' + 1) - J''(J'' + 1)] \tag{11.64}$$

Or, expressed in terms of wavenumbers, it is

$$\tilde{\nu}_{\text{obs}} = \tilde{\nu} + \tilde{B}[J'(J' + 1) - J''(J'' + 1)] \tag{11.65}$$

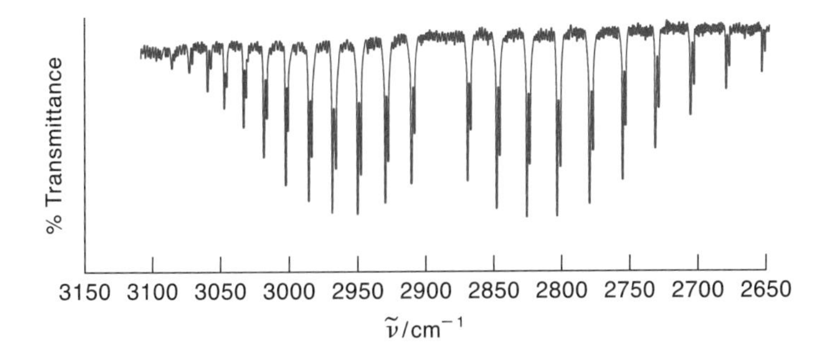

Figure 11.22
Infrared absorption spectrum of HCl(g). This region of the spectrum corresponds to the fundamental vibrational transition ($v = 1 \leftarrow 0$) recorded at sufficiently high resolution to see individual rotational transitions as well. For each rotational transition there are two lines, corresponding to the two isotopomers of hydrogen chloride, $H^{35}Cl$ and $H^{37}Cl$. [From J. L. Hollenberg, *J. Chem. Educ.* **47**, 2 (1970).]

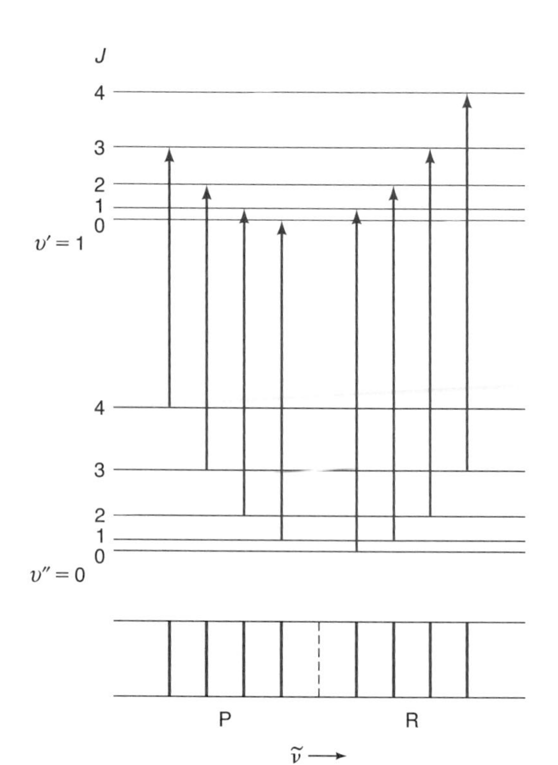

Figure 11.23
Simultaneous rotational energy level transitions accompanying the $\upsilon = 1 \leftarrow 0$ transition for a diatomic molecule.

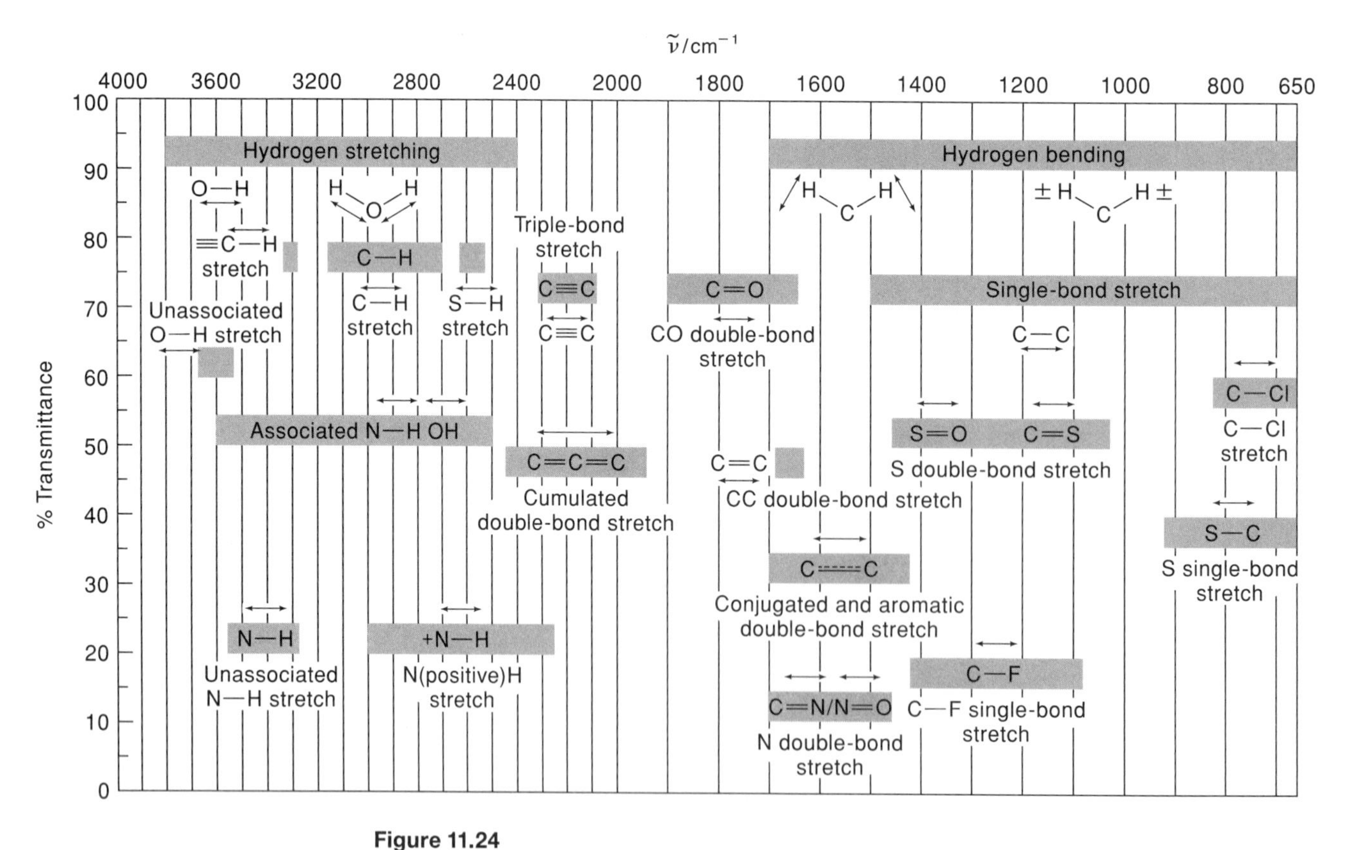

Figure 11.24
Group-frequency correlation chart for some common functional groups. (By permission of the Perkin-Elmer Corporation, Norwalk, CT.)

In many cases, an IR spectrum can be divided into two branches, called P and R, according to the following conditions (Figure 11.23):

$$\text{P branch:}\quad J' = J'' - 1 \quad \tilde{\nu}_{\text{obs}} = \tilde{\nu} - 2\tilde{B}J'' \quad J'' = 1, 2, 3, \ldots \tag{11.66}$$

$$\text{R branch:}\quad J' = J'' + 1 \quad \tilde{\nu}_{\text{obs}} = \tilde{\nu} + 2\tilde{B}(J'' + 1) \quad J'' = 0, 1, 2, \ldots \tag{11.67}$$

A mnemonic: P for *Poor* or decreasing J, and R for *Rich* or increasing J.

In some situations for polyatomic molecules, it is possible to observe transitions in which $\Delta\upsilon = 1$ and $\Delta J = 0$. This results in a Q branch, which lies between the P and R branches in an IR spectrum.

$$\text{Q branch:}\quad J' = J'' \quad \tilde{\nu}_{\text{obs}} = \tilde{\nu} \quad J'' = 0, 1, 2, \ldots \tag{11.68}$$

Because the transition energy is approximately the same for all values of J, Q branches appear in IR spectra as narrow peaks relative to the broad P and R branches. A more detailed interpretation of IR spectra recognizes that the rotational constant $\tilde{B}$ varies with vibrational energy level. In general, the rotational constant increases with increasing vibration ($\tilde{B}' > \tilde{B}''$), so the Q branch lines with different J'' values do not lie at exactly the same wavenumbers.

Infrared spectroscopy is a highly useful technique for chemical analysis. The complexity of molecular vibration virtually assures that any two different molecules cannot produce identical IR spectra. Matching the IR spectrum of an unknown with that of a standard compound, a procedure known as *fingerprinting*, is an unequivocal method of identification. So far, over 300,000 reference spectra have been recorded and stored for fingerprinting. The details of an IR spectrum reveal much useful information about the structure and bonding of the molecule. A group-frequency correlation chart for some organic functional groups is given in Figure 11.24. Figure 11.25 shows the IR spectrum of a relatively simple molecule, 2-propenitrile (CH_2=CHCN), and the assignment of its major peaks.

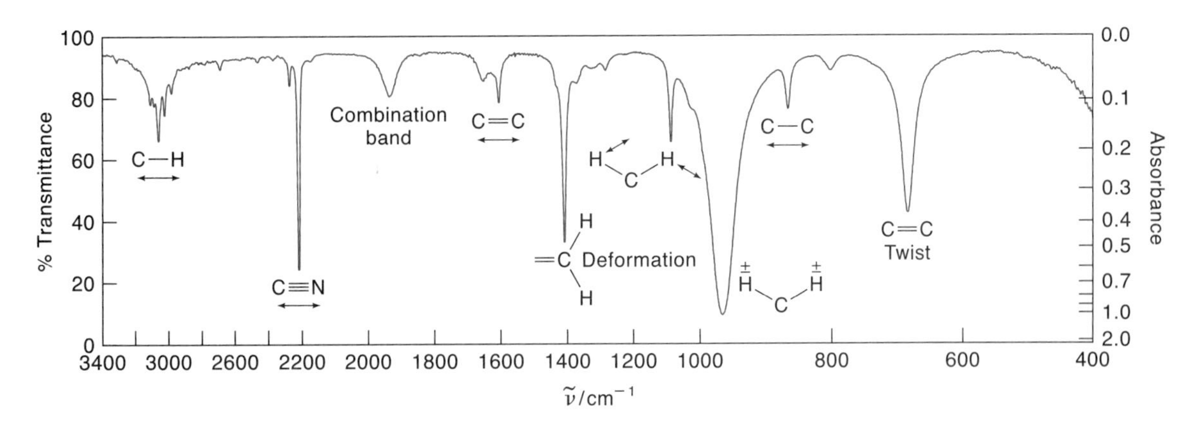

Figure 11.25
Infrared absorption spectrum of 2-propenenitrile, CH_2=CHCN.

11.4 Symmetry and Group Theory*

Molecular symmetry derives from the shape, or three-dimensional arrangement, of atoms in a molecule. We introduce molecular symmetry at this point because it is useful in describing molecular vibrational modes and in determining the selection rules for vibrational transitions.

Symmetry Elements

The planar water molecule provides a good illustration of symmetry. When the water molecule is rotated about an imaginary axis through the oxygen atom by 180° in either direction, as indicated in Figure 11.26, its new configuration is indistinguishable from the original one. Rotating a molecule in this fashion is a *symmetry operation*. The axis about which the water molecule is rotated is called a *symmetry element*. A symmetry element is a geometric entity such as a point, a plane, or a line, with respect to which a symmetry operation is performed. One symmetry element, called the *identity element*, E, does not involve a symmetry operation and is therefore possessed by all molecules. We summarize the elements of symmetry below.

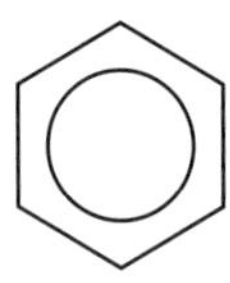

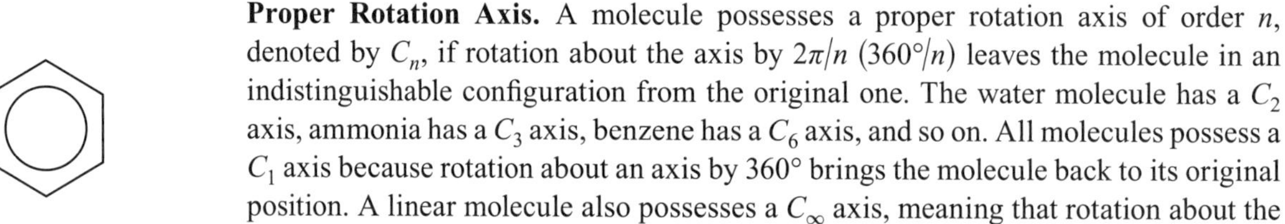

Proper Rotation Axis. A molecule possesses a proper rotation axis of order n, denoted by C_n, if rotation about the axis by $2\pi/n$ $(360°/n)$ leaves the molecule in an indistinguishable configuration from the original one. The water molecule has a C_2 axis, ammonia has a C_3 axis, benzene has a C_6 axis, and so on. All molecules possess a C_1 axis because rotation about an axis by 360° brings the molecule back to its original position. A linear molecule also possesses a C_∞ axis, meaning that rotation about the internuclear axis by any amount would leave the molecule unchanged.

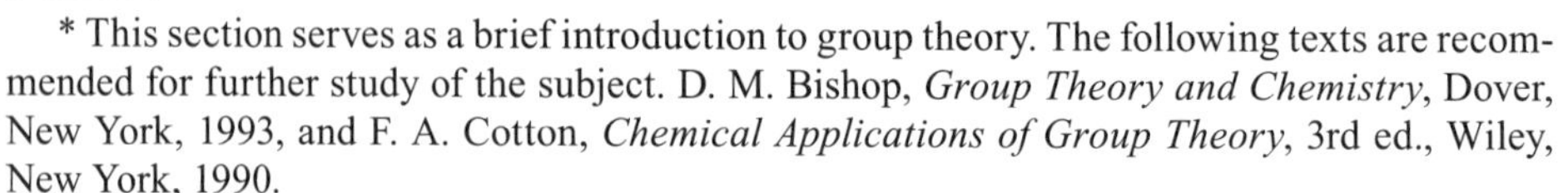

* This section serves as a brief introduction to group theory. The following texts are recommended for further study of the subject. D. M. Bishop, *Group Theory and Chemistry*, Dover, New York, 1993, and F. A. Cotton, *Chemical Applications of Group Theory*, 3rd ed., Wiley, New York, 1990.

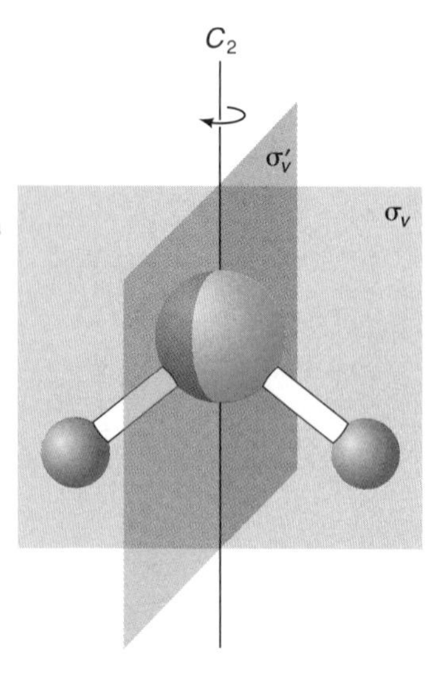

Figure 11.26
Symmetry of a water molecule. The molecule has two planes of symmetry (labeled σ_v and σ_v'). One plane contains all three atoms; the other is perpendicular to this plane. Rotating the molecule 180° about the axis labeled C_2 results in a configuration that is indistinguishable from the original one.

Plane of Symmetry. A molecule possesses a *plane of symmetry* (σ) if a reflection through the plane leaves the molecule in an indistinguishable position from the original one. Thus, the water molecule has two planes of symmetry (see Figure 11.26). A planar molecule has at least one plane of symmetry, which is the plane containing the atoms. Planes of symmetry are given a subscript to designate their orientation with respect to a molecule. Those that include the proper rotation axis of highest symmetry are called vertical, σ_v, and those that are perpendicular to the axis of highest symmetry are called horizontal, σ_h. Both of the water molecule planes of symmetry are designated σ_v. The planar molecule boron trifluoride contains three σ_v planes of symmetry; each σ_v plane of symmetry contains a B−F bond and the C_3 proper rotation axis. It also has a single σ_h plane of symmetry that is the plane of the molecule.

The Greek letter sigma is equivalent to the letter *s*, which starts the German word "Spiegel" (mirror).

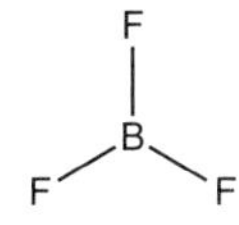

Center of Symmetry. If the coordinates (x, y, z) of every atom in a molecule are changed into $(-x, -y, -z)$ and the molecule's configuration is indistinguishable from the original one, then the point of origin, that is, $(0, 0, 0)$, is the *center of symmetry* (i), where i stands for *inversion*. The carbon dioxide and benzene molecules each have a center of symmetry, whereas the water and boron trifluoride molecules do not. A molecule can have only one center of symmetry.

Improper Rotation Axis. The *improper rotation axis* (S_n) is more complicated than the preceding three symmetry elements because it involves two operations. A molecule possesses an improper rotation axis of order n if rotation about the axis by $2\pi/n$ followed by reflection in a plane perpendicular to the axis leaves the molecule in an indistinguishable configuration from the original one. Figure 11.27 shows that methane has an S_4 axis.

Molecular Symmetry and Dipole Moment

We can draw some interesting and useful conclusions from molecular symmetry. For example, we can deduce the presence of a permanent electric dipole moment from a molecule's symmetry elements. Because dipole moment is a vector quantity and because it remains unchanged by any symmetry operation, it must lie along a symmetry element. Therefore, a molecule cannot have a dipole moment if it possesses a center of symmetry, because a vector cannot be a point. Ethylene and acetylene, for example, each possess a center of symmetry and are nonpolar. Similarly, a molecule

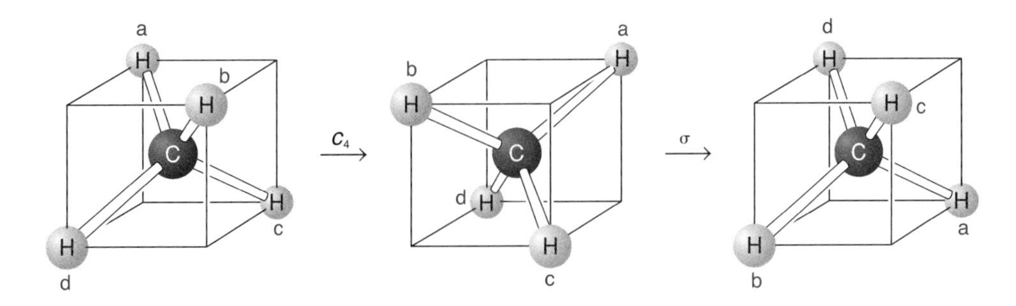

Figure 11.27
A methane molecule has an S_4 axis. Rotation about the molecular axis by 90° followed by reflection in a plane perpendicular to the axis brings the molecule back to a position that is indistinguishable from the original one.

cannot have a dipole moment if it possesses more than one C_n axis (where $n \geq 2$) because a vector cannot lie along two different axes. Boron trifluoride, for example, has both C_2 and C_3 axes, but has no center of symmetry and is nonpolar.

Point Groups

Molecules may be described by the symmetry elements that they contain. For example, the water molecule contains the symmetry elements of a C_2 axis, two planes of symmetry (each of which contains the C_2 axis), and the identity E. Because it contains these four symmetry elements, water belongs to the point group C_{2v}. A *point group* is a mathematical structure that may be identified by the symmetry elements that it contains. Point groups find utility in chemistry for concisely describing the symmetry features of a molecule, but also for describing electronic structure and bonding, and for predicting spectroscopic selection rules. The branch of mathematics known as group theory may be used to predict, for example, whether a vibrational mode is IR active, that is, whether the vibrational transition is allowed. The details of the mathematics are contained in more specialized texts. Here we present a qualitative description to illustrate the general procedure.

Character Tables

To describe the symmetry of translational, rotational, and vibrational motions of a molecule, we use *representations* of that motion. The simplest examples are the so-called *irreducible representations*, which appear in a *character table*. We use the water molecule, which is described by the C_{2v} point group, as an example. For each molecular motion, we consider how that motion is changed by each symmetry element in the group. The molecule has three atoms, so the total number of coordinates needed to define the system is 3 × 3 or 9, which is also the number of degrees of freedom.

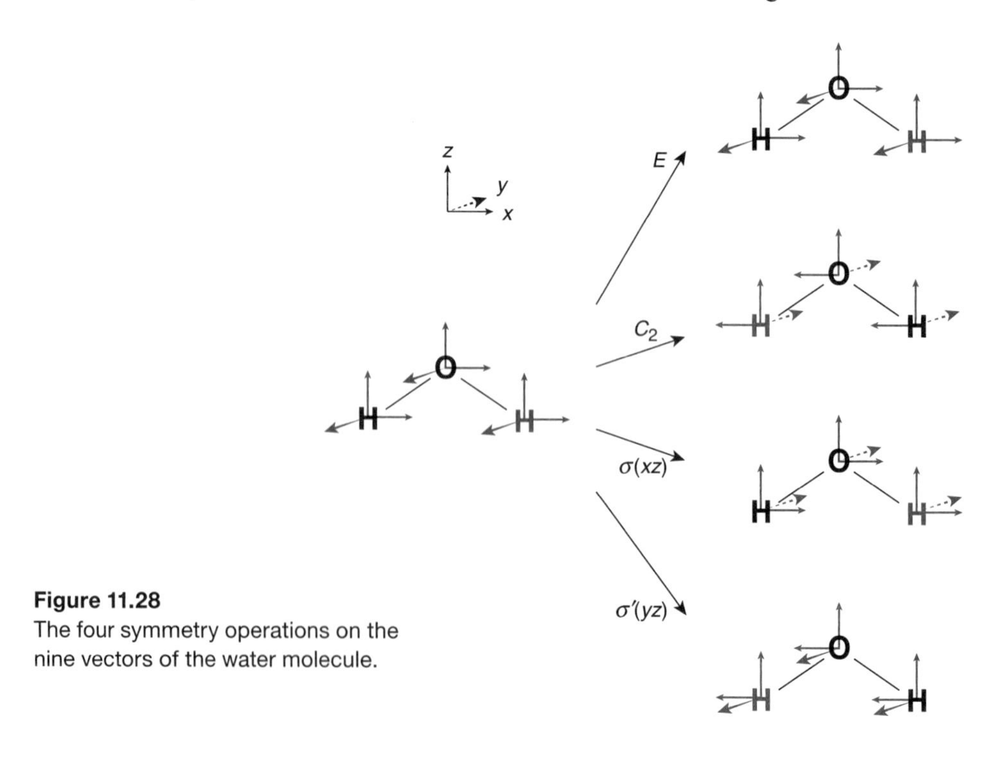

Figure 11.28
The four symmetry operations on the nine vectors of the water molecule.

Table 11.5
The Character Table for the C_{2v} Point Group

C_{2v}	E	C_2	$\sigma(xz)$	$\sigma'(yz)$		
A_1	1	1	1	1	z	x^2, y^2, z^2
A_2	1	1	−1	−1		xy
B_1	1	−1	1	−1	x	xz
B_2	1	−1	−1	1	y	yz

Figure 11.28 shows that each atom is labeled with three vectors in the x, y, and z directions. We classify the motion as symmetric or antisymmetric under that symmetry operation and designate the *character* as +1 for symmetric and −1 for antisymmetric. Consider translation of the water molecule. If we translate the water molecule in the x direction, then the identity operator would not change this motion, so it has the character 1. The rotation about the C_2 axis would convert motion in the x direction to motion in the $-x$ direction, so the character is −1. Reflection in the plane that includes all three atoms (the x–z plane) would not change translation in the x direction, so the character is 1. Finally, reflection in the second mirror plane (the y–z plane) would convert motion in the x direction to motion in the $-x$ direction, so the character is −1. These four characters, 1, −1, 1, −1, constitute an irreducible representation. We write each irreducible representation as a row in a character table. In the C_{2v} point group, translation in the x direction has the irreducible representation named B_1, which is the third row of the character table (Table 11.5). At the right-hand side of the character table, there is an x to indicate which irreducible representation (which row) corresponds to translation in the x direction. If we repeat the exercise for translation of water in the z direction, we find that the translation is unchanged by each of the four symmetry operations, and thus the irreducible representation is 1, 1, 1, 1, and is shown in the first row of the character table, which is named A_1. We could continue this type of exercise to complete the C_{2v} character table.

How do we determine the selection rules for infrared spectroscopy? For a vibrational transition to be allowed, the vibrational mode of the molecule must have the same irreducible representation as translation along the x, y, or z axis. (This selection rule arises from a symmetry argument in calculating the integral in the transition dipole moment.) Water has three vibrational modes: bending, symmetric stretching, and asymmetric stretching (see Figure 11.19). The bending vibrational mode of a water molecule is unchanged by any of the four symmetry elements of the C_{2v} point group, so the bending vibration has the irreducible representation A_1. Translation along the z axis also has the irreducible representation A_1, so the bending motion is IR active and we are able to detect the bending of a water molecule using IR absorption spectroscopy. The symmetric stretch also transforms in the same way, so it too is IR active. Finally, the asymmetric stretch of the water molecule is described by the irreducible representation B_1, which is the same as translation along the x axis, so this vibrational mode is also IR active. Thus, for the water molecule example, all three vibrational modes may be probed by IR absorption spectroscopy. If a molecule in the C_{2v} point group has a vibrational mode that transforms as the A_2 irreducible representation, then this mode could *not* be detected using IR absorption spectroscopy. In the C_{2v} point group, A_2 vibrational modes are IR inactive because they do not translate along any of the three axes; the pure vibrational transitions are thus forbidden. This A_2 mode may be detected, however, using other techniques, such as Raman scattering, which is described in the next section.

11.5 Raman Spectroscopy

Vibrational transitions in homonuclear diatomic molecules, for example, are IR forbidden because there is no change in the molecular dipole moment with vibration. How, then, can we measure the vibrational frequency, and thus the force constant, of a simple molecule such as N_2 or O_2? One experimental technique available is Raman spectroscopy. Raman spectroscopy is a light *scattering* technique as opposed to a light absorption technique. This technique was first discovered in 1928 by the Indian physicist Chandrasekhara Venkata Raman (1888–1970), who used sunlight as the source and the human eye as a detector. It has become more practical with the development of laser light sources and sensitive detectors.

Raman was the first Asian to win a Nobel Prize in science.

In absorption spectroscopy, we are concerned with the light that is resonant with energy spacings within the molecule. Light scattering techniques, such as Raman spectroscopy, employ nonresonant light that is not absorbed. When nonresonant light interacts with a molecule, a small fraction of the incident light may be transiently absorbed to a *virtual state*, and then quickly re-emitted or scattered. (A virtual state may be thought of as an extremely short-lived excited state.) Most of the scattered light is of the same energy as the incident light; that is, it is *elastically* scattered and is called Rayleigh scattering (Figure 11.29). The intensity of Rayleigh scattering is inversely proportional to the wavelength raised to the fourth power. This dependence favors the shorter-wavelength end of the visible spectrum. (Thus, the sky appears blue as sunlight is scattered by the air molecules.) A small amount of the light, however, will be *inelastically* scattered, which gives rise to the *Raman effect*. The most common application of Raman scattering is in vibrational Raman spectroscopy, although rotational Raman spectroscopy is also a useful tool.

When a beam of light shines on an assembly of molecules, the photons of energy $h\nu$ (assuming monochromatic radiation) collide with the molecules and one of two things can happen.* If the collision is elastic, then the deflected photons, that is, the scattered radiation, will have the same energy as the incident photons. On the other hand, if the collision is inelastic, then the deflected photons will have either higher or lower energy than the incident photons. Conservation of energy requires that

$$h\nu + E = h\nu' + E' \tag{11.69}$$

where E represents the rotational, vibrational, and electronic energy of the molecule before collision and E' represents the same values after collision. Rearranging Equation 11.69, we get

$$\frac{E' - E}{h} = \nu - \nu' \tag{11.70}$$

* We assume no photochemical reactions would occur.

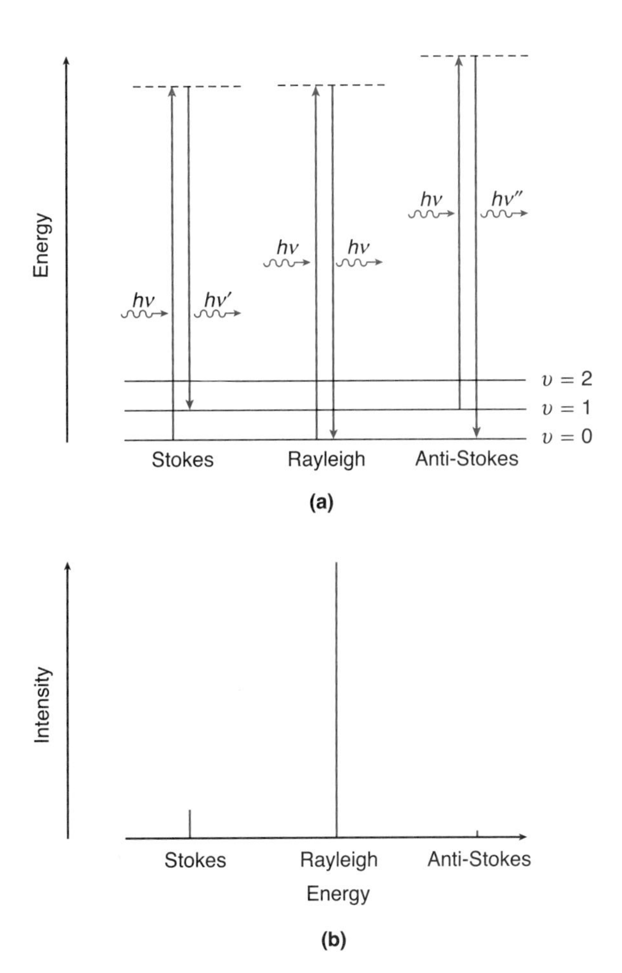

Figure 11.29
Light-scattering processes. (a) Energy level changes in Rayleigh scattering, Stokes Raman scattering, and anti-Stokes Raman scattering. The dashed lines denote the virtual state. (b) Relative intensities of the three types of scattering (not drawn to scale). Frequency increases from left to right.

The scattered radiation is classified as

$E < E' \ (\nu > \nu')$	Stokes Raman scattering
$E = E \ (\nu = \nu)$	Rayleigh scattering
$E > E'' \ (\nu < \nu'')$	Anti-Stokes Raman scattering

Thus, in the Raman scattering case, energy is either given to or taken away from the molecule as a result of the interactions. Figure 11.29a shows the energy level diagram for these interactions.

When a molecule is promoted from the ground vibration state to a higher, unstable vibrational state by the incident radiation, it can either return to the original state or to a different vibrational state. The former gives rise to Rayleigh scattering and the latter to Raman scattering which, in this case, gives rise to the *Stokes* lines (after the British physicist Sir George Gabriel Stokes, 1819–1903). If the molecule is initially in the first excited vibrational state, it can be promoted to a higher, unstable state and then return to the ground state. This is also Raman scattering, which gives rise to *anti-Stokes lines*. The intensity of the Rayleigh line is much stronger than that of the Stokes lines, which in turn is much stronger than the anti-Stokes lines (Figure 11.29b). We also have changes in the rotational energy levels in addition to the vibrational energy levels.

Raman spectroscopy is similar to infrared spectroscopy in that it may be used to probe molecular vibrations. Raman spectroscopy, however, obeys different selection rules. For a transition to be Raman allowed, the *polarizability* of the molecule must change with vibration. Polarizability is a measure of how easy it is to move electrons about a molecule. In a neutral nonpolar species such as the helium atom, the electron charge density is spherically symmetrical. When placed in an electric field (E), the electrostatic interaction will cause a redistribution of the charge density. The atom will then acquire an induced dipole moment μ_{ind} given by

$$\mu_{\text{ind}} \propto E$$

$$= \alpha E \tag{11.71}$$

where α is the polarizability of helium. (The same effect applies to polar molecules.) More loosely held electrons, such as those in molecular iodine, are more polarizable than tightly held electrons, such as those in molecular fluorine (Figure 11.30). The vibrations of homonuclear diatomic molecules change the charge distribution from longer and thinner to shorter and fatter, so the polarizability changes, and the vibrational transition may be observed using Raman spectroscopy.

Here, the tensor describes how one vector is transformed to a second vector, analogous to how a function transforms one scalar to a second scalar.

Determining by inspection whether or not polarizability changes with vibration can be a challenging process for all but the smallest molecules. Fortunately, group theory simplifies the process considerably. Vibrational modes that are Raman allowed have the same irreducible representation (row on a character table) as a *polarizability tensor*, α. Again referring to the character table for the C_{2v} point group (see Table 11.5), we see that the polarizability tensors denoted by two Cartesian coordinates xy, xz, x^2, etc., are placed in the far right column. Vibrational modes that have the same irreducible representation as a polarizability tensor are *Raman active*—that is, they may be

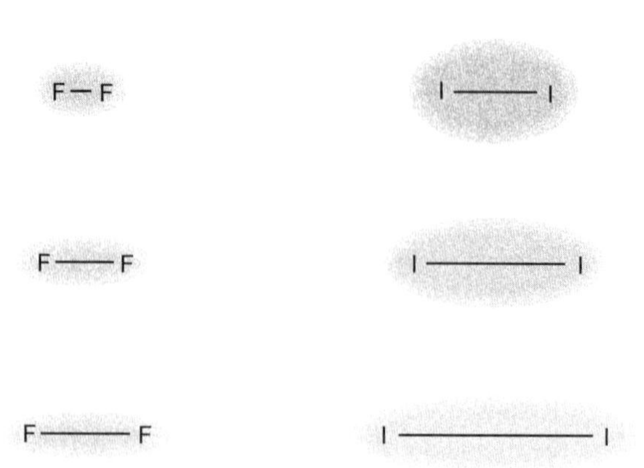

Figure 11.30
Cartoon of the changes in the polarizability of F_2 and I_2 electron clouds with vibration.

detected using Raman scattering. So, if we have a character table, then the real work is to identify the irreducible representation of vibrational mode. For the water molecule, the bending motion is described by the A_1 irreducible representation. This motion is Raman active because A_1 also describes the polarizability tensors listed in the character table as x^2, y^2, and z^2. In fact, all vibrations that are IR active (the A_1, B_1, and B_2 irreducible representations) are also Raman active. Molecules within the C_{2v} point group may have vibrations that are Raman active but IR inactive (the A_2 irreducible representation), but there are no vibrational modes that are both IR and Raman inactive. Similarly, we can show that homonuclear diatomic molecules such as N_2 and O_2 are Raman active because their vibrations correspond to the irreducible representation described by the polarizabilty tensors in the character table for such molecules.

There is a general selection rule that molecules with a center of symmetry (i) have vibrational modes that are either infrared or Raman allowed, but not both. For example, IR and Raman spectroscopies may be used to distinguish between *ortho*-difluorobenzene and *para*-difluorobenzene:

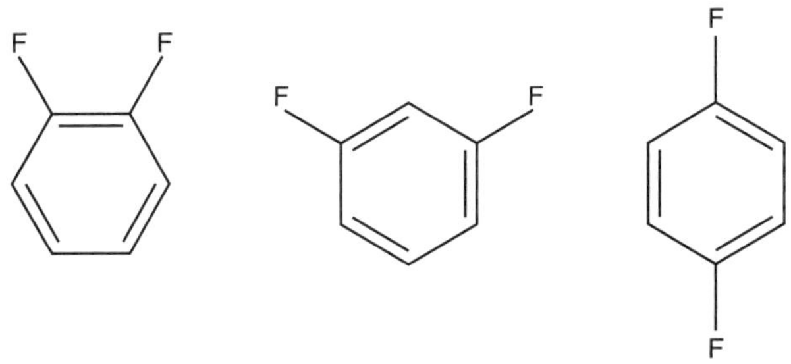

Because *para*-difluorobenzene has a center of symmetry, it does not have a vibrational mode that is both IR and Raman active. On the other hand, *ortho*-difluorobenzene is in the C_{2v} point group; it has no center of symmetry and has numerous vibrational modes that are both IR and Raman active. On the basis of group theory, however, we cannot distinguish *ortho*- from *meta*-difluorobenzene because they both belong to the C_{2v} point group.

Raman spectroscopy is less commonly employed than IR spectroscopy because it is generally more expensive to implement and less useful as an analytical tool. Raman spectroscopy does, however, have several advantages over infrared spectroscopy. Because water has a very large and broad infrared absorption cross section and a relatively weak Raman scattering cross section, Raman spectroscopy is a useful probe of aqueous environments. Note, however, that all three vibrational modes of water (bend, symmetric stretch, asymmetric stretch) are both IR and Raman active. Group theory tells us whether a vibrational transition is active (allowed) or inactive (forbidden), but does not otherwise describe the relative probability of a transition. Because they employ shorter wavelength visible light rather than IR radiation, Raman spatial imaging techniques have higher resolution than IR and thus may be used to probe biological cells. However, fluorescence often interferes with Raman spectra but it is not a problem with IR.

Rotational Raman Spectra

With a high-resolution (laser) excitation source and a high-resolution spectrophotometer, it is possible to record Raman spectra that resolve rotational transitions. The resolution of individual rotational transitions is only possible for small molecules in the gas phase. The energy difference in a Raman rotational transition is

$\Delta J = +2$ is Stokes and $\Delta J = -2$ is anti-Stokes rotational scattering.

$$\begin{aligned}\Delta E_{\text{rot}} &= BJ'(J'+1)h - BJ''(J''+1)h \\ &= B(4J+6)h \quad \text{for Stokes} \\ &= -B(4J+2)h \quad \text{for anti-Stokes}\end{aligned} \tag{11.72}$$

where J denotes the starting rotational quantum number. Note that unlike microwave spectra, the selection rule here is $\Delta J = \pm 2$. The reason is that one photon carries one unit of angular momentum, but the Raman effect is a two-photon process. Thus, the molecule must lose or gain two quanta of angular momentum per transition. Rotational Raman spectra provide essentially the same information as microwave spectra in that the spacings of the observed Raman signals can be used to find the rotational constant, the moment of inertia, and detailed information about molecular geometry. Rotational Raman spectroscopy relies on a change in polarizability, and may be applied to molecules without a permanent dipole moment. Homonuclear diatomic molecules such as O_2 and N_2 do not absorb microwave radiation, and thus may not be probed using pure rotational (absorption) spectroscopy. However, rotational Raman spectroscopy may be employed to determine their bond lengths.

Key Equations

$$\tilde{\nu} = \frac{1}{\lambda} = \frac{\nu}{c}$$ (Wavenumber) (11.2)

$$\mu_{ij} = \int \psi_i \hat{\mu} \psi_j d\tau$$ (Transition dipole moment) (11.16)

$$A = \varepsilon bc$$ (Beer–Lambert law) (11.20)

$$\frac{1}{\mu} = \frac{1}{m_1} + \frac{1}{m_2}$$ (Reduced mass) (11.22)

$$E_J = \frac{\hbar^2}{2I} J(J+1)$$ (Quantized rotational energies) (11.43)

$$\nu = \frac{1}{2\pi}\sqrt{\frac{k}{\mu}}$$ (Fundamental vibrational frequency) (11.51)

$$E_\upsilon = \left(\upsilon + \frac{1}{2}\right)h\nu$$ (Quantized vibrational energies) (11.54)

$$\Delta E = \left(\upsilon' + \frac{1}{2}\right)h\nu + BJ'(J'+1)h - \left(\upsilon'' + \frac{1}{2}\right)h\nu - BJ''(J''+1)h$$

(Simultaneous transitions between vibrational and rotational levels) (11.63)

APPENDIX 11.1

Fourier-Transform Infrared Spectroscopy

In recent years, the application of Fourier transform to data acquisition techniques has had a profound impact on spectroscopy over a wide range of the electromagnetic spectrum, including NMR, microwave, IR, and UV–visible. Most commercial NMR and IR spectrometers, for example, are now Fourier-transform instruments. Here, we discuss Fourier-transform IR (FT-IR), a method commonly used in organic chemistry for functional group identifications.

But before we discuss FT-IR, we need an idea as to how conventional IR spectrometers work. These spectrometers are of the *dispersive* type; that is, they separate the individual frequencies of radiation emitted from the IR source (a glowing blackbody) by means of a prism or diffraction grating. An IR prism works much the same way as a glass prism, which separates visible light into its colors. A grating is a more modern dispersive element that separates the frequencies in IR radiation with better resolution. The detector measures the amount of energy at each frequency that has passed through the sample. The resulting spectrum is a plot of either absorbance or percent transmittance versus frequency, wavelength, or wavenumber. Although this technique has been used reliably for many years, it has some serious limitations. First, because dispersive instruments measure frequencies individually, the recording of a single spectrum may take minutes. This length of time can be a problem when one has to study hundreds of samples in, say, environmental analysis. Second, such spectrometers have fairly low sensitivity. Finally, the instruments have many moving parts and are prone to mechanical breakdown.

A FT-IR spectrometer provides better sensitivity, speed, and accuracy of wavelength measurement than dispersive IR instruments. It measures all of the IR frequencies *simultaneously* with the aid of an optical device called an *interferometer*. No prism or grating is needed. The interferometer produces a unique type of signal that has all of the IR frequencies encoded into it. This signal can be measured very quickly, usually in one second or so; therefore, signal averaging with FT-IR can be carried out conveniently.

Figure 11.31 is a schematic diagram of a FT-IR spectrometer. IR radiation of all frequencies from the source (a) strikes a *beam splitter*, a flat, partially transmitting plate that splits the incoming beam into two parts, beams b and c. Beam b reflects off a flat mirror that is fixed in space, producing beam d. Beam c reflects off a flat moving mirror, producing beam e. Beams d and e meet back at the beam splitter, which then transmits part of beam d and reflects part of beam e to give the combined beam f. Beam f passes through the sample, and the emerging beam g strikes the detector.

Because the path that one beam travels (abd) is fixed in length and the other (ace) is constantly changing as its mirror moves, the signal that exits the interferometer is

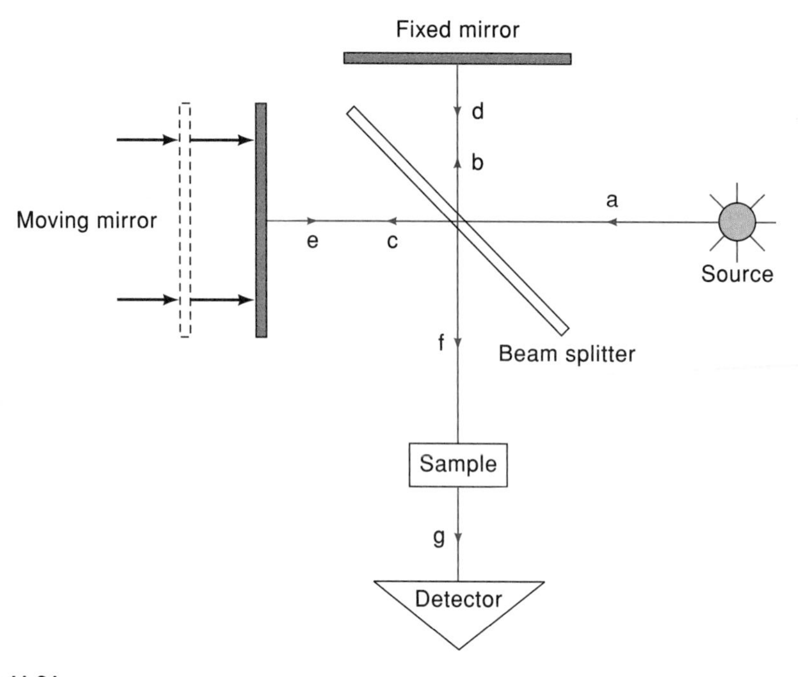

Figure 11.31
A FT-IR spectrometer. The distance traversed by the moving mirror (a few millimeters) is accurately monitored by a helium–neon laser (not shown).

the product of these two beams "interfering" with each other.* The resulting signal, called an *interferogram* (Figure 11.32a), has the unique property that every data point (a function of the moving mirror position) in the signal carries information about every IR frequency that comes from the source. Therefore, after the interferogram has passed through the sample, all frequencies are measured simultaneously. This is the reason a FT-IR spectrometer can record a spectrum so much more quickly than a conventional IR spectrometer.

The mathematical details of converting an interferogram into a frequency spectrum (a plot of the intensity at each frequency), a task carried out by Fourier transformation, are beyond the scope of this book. The following two equations relate the radiation intensity falling on the detector, $I(\delta)$, where δ is the difference in pathlength traveled by beams c and e as compared with beams b and d, to the spectral density, $B(\tilde{\nu})$, which is the intensity of radiation reaching the detector as a function of wavenumber:

$$I(\delta) = \int_{-\infty}^{+\infty} B(\tilde{\nu}) \cos(2\pi\tilde{\nu}\delta) d\tilde{\nu} \tag{1}$$

and

$$B(\tilde{\nu}) = \int_{-\infty}^{+\infty} I(\delta) \cos(2\pi\tilde{\nu}\delta) d\delta \tag{2}$$

* A visible light analog of the interference phenomenon is the spectrum of color reflected from a soap bubble or from a thin film of gasoline on a wet street.

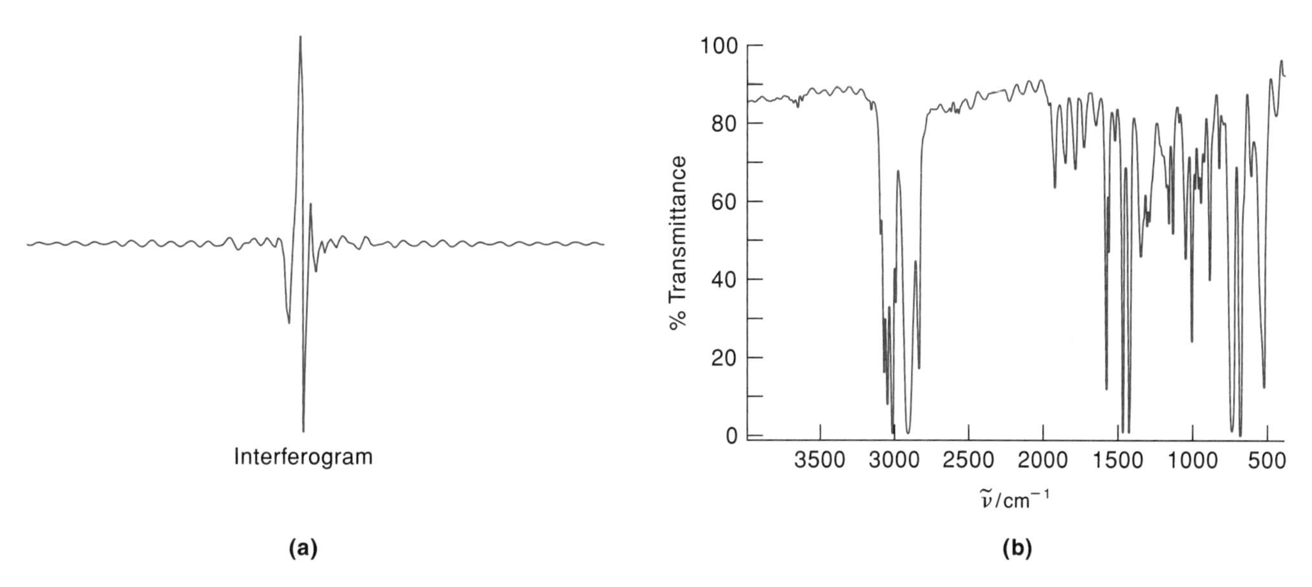

Figure 11.32
(a) The interferogram of polystyrene on a film. The plot shows intensity versus distance. Note the strong centerburst, which occurs at $\delta = 0$. (b) The Fourier transform of the interferogram, which is a plot of percent transmittance versus wavenumber. (Courtesy Nicolet Instruments.)

These two equations are interconvertible and are known as a Fourier-transform pair. In practice, the measured interferogram signal is digitized and sent to a computer where the Fourier transformation takes place according to Equation 2. The resulting IR spectrum is usually displayed as a plot of percent transmittance versus wavenumber (Figure 11.32b).

Suggestions for Further Reading

BOOKS

Barrow, G. M., *Introduction to Molecular Spectroscopy*, McGraw-Hill, New York, 1962.
Dunford, H. B., *Elements of Diatomic Molecula Spectra*, Addison-Wesley, Reading, MA, 1968.
Harris, D. C., and M. D. Bertolucci, *Symmetry and Spectroscopy: An Introduction to Vibrational and Electronic Spectroscopy*, Dover, New York, 1989.
Heilbronner, E., and J. D. Dunitz, *Reflections on Symmetry*, VCH Publishers, New York, 1993.
Jaffe, H. H., and M. Orchin, *Symmetry in Chemistry*, Dover, New York, 2002.
Lide, D. R., *CRC Handbook of Chemistry and Physics*, 94th ed., Taylor-Francis/CRC Press, Boca Raton, FL, 2013. New editions of this valuable reference book appear regularly.
McQuarric, D. A., *Quantum Chemistry*, 2nd ed., University Science Books, Sausalito, CA, 2008.
Ratner, M. A., and G. C. Schatz, *Introduction to Quantum Mechanics in Chemistry*, Prentice-Hall, Upper Saddle River, NJ, 2001.
Steinfeld, J. I., *Molecules and Radiation: An Introduction to Modern Molecular Spectroscopy*, 2nd ed., MIT Press, Cambridge, MA, 1985.
Walton, P. H., *Beginning Group Theory for Chemistry*, Oxford University Press, New York, 1998.

ARTICLES

General Spectroscopy

"Demonstration of the Doppler Effect," D. C. Luehrs and J. M. Luehrs, *J. Chem. Educ.* **52**, 567 (1975).
"Einstein Coefficients, Cross Sections, *f* Values, Dipole Moments, and all that," R. C. Hilborn, *Am. J. Phys.* **50**, 982 (1981).
"The Early History of Spectroscopy," N. C. Thomas, *J. Chem. Educ.* **68**, 631 (1991).
"The Beer-Lambert Law Revisited," P. Lykos, *J. Chem. Educ.* **69**, 730 (1992).
"Why Do Spectral Lines Have A Linewidth?," V. B. E. Thomsen, *J. Chem. Educ.* **72**, 616 (1995).
"A Unified Approach to Absorption Spectroscopy at the Undergraduate Level," R. S. Macomber, *J. Chem. Educ.* **74**, 65 (1997).

Microwave Spectroscopy and the Rigid Rotor

"Elements of the Quantum Theory: V. The Rigid Rotator," S. Dushman, *J. Chem. Educ.* **12**, 436 (1935).
"An Approximate Wave Mechanical Treatment of the Harmonic Oscillator and Rigid Rotator," R. Rich, *J. Chem. Educ.* **40**, 365 (1963).
"Microwave Absorption Spectroscopy," G. W. Ewing, *J. Chem. Educ.* **43**, A683 (1966).
"Rationalization of the $\Delta J = \pm 1$ Selection Rule for Rotational Transitions," C. T. Moynihan, *J. Chem. Educ.* **46**, 431 (1969).
"Microwave Spectroscopy in the Undergraduate Laboratory," R. H. Schwendeman, H. N. Volltrauer, V. W. Laurie, and E. C. Thomas, *J. Chem. Educ.* **47**, 526 (1970).
"On the Solution of the Quantum Rigid Rotor," C. Pye, *J. Chem. Educ.* **83**, 460 (2006).
"Dinner and a Show," M. Fischetti, *Sci. Am.* November 2008.

Infrared and Raman Spectroscopy and the Harmonic Oscillator

"An Elementary Approach to the Wave-Mechanical Harmonic Oscillator," C. J. H. Schutte, *J. Chem. Educ.* **45**, 567 (1968).

"Resonance Raman Spectroscopy," D. P. Strommen and K. Nakamoto, *J. Chem. Educ.* **54**, 474 (1977).
"Integral of ξ^v Over Harmonic Oscillator Wave Functions," B. D. Joshi, S. E. LaGrou, and D. W. Spooner, *J. Chem. Educ.* **58**, 39 (1981).
"Demonstration of Maxwell Distribution Law of Velocity by Spectral Line Shape Analysis," C. L. Berg and R. Chang, *Am. J. Phys.* **52**, 80 (1984).
"FTIR Rotational Spectroscopy," R. Woods and G. Henderson, *J. Chem. Educ.* **64**, 921 (1987).
"From Quantum Mechanical Oscillators to Classical Ones Through Maximization of Entropy," B. Boulil and O. Henri-Rousseau, *J. Chem. Educ.* **66**, 467 (1989).
"Introduction to a Quantum Mechanical Harmonic Oscillator Using a Modified Particle-in-a-Box Problem," H. F. Blanck, *J. Chem. Educ.* **69**, 98 (1992).
"The Fundamental Rotational-Vibrational Band of CO and NO: Teaching the Theory of Diatomic Molecules," H. H. R. Schor and E. L. Teixeira, *J. Chem. Educ.* **71**, 771 (1994).
"Vibrational Spectroscopy: An Integrated Experiment," B. A. DeGraff, T. C. DeVore, and D. Sauder, *Chem. Educator* [Online] **1**, 6 (1997) DOI 10.1333/s00897970069a.
"Extending the Diatomic FTIR Experiment: A Computational Exercise to Calculate Potential Energy Curves," O. Sorkhabi, W. M. Jackson, and I. Daizadeh, *J. Chem. Educ.* **75**, 238 (1998).
"Raman Spectroscopy of Symmetric Oxyanions," M. G. Comstock and J. A. Gray, *J. Chem. Educ.* **76**, 1272 (1999).
"Educational Applications of IR and Raman Spectroscopy: A Comparison of Experiment and Theory," B. L. McClain, S. M. Clark, R. L. Gabriel, and D. Ben-Amotz, *J. Chem. Educ.* **77**, 654 (2000).
"The Raman Effect: A Large-Scale Lecture Demonstration," G. C. Weaver and R. W. Schwenz, *Chem. Educator* [Online] **6**, 164 (2004) DOI 10.1007/s00897010476a.
"The Effect of Anharmonicity on Diatomic Vibration: A Spreadsheet Simulation," K. F. Lim, *J. Chem. Educ.* **82**, 1263 (2005).
"Observation and Analysis of N_2O Rotation-Vibration Spectra," M. S. Bryant, S. W. Reeve, and W. A. Burns, *J. Chem. Educ.* **85**, 121 (2008).
"Infrared and Raman Spectroscopy: A Discovery-Based Activity for the General Chemistry Curriculum," K. L. Borgsmiller, D. J. O'Connell, K. M. Klauenberg, P. M. Wilson, and C. J. Stromberg, *J. Chem. Educ.* **89**, 365 (2012).

Symmetry and Group Theory

"An Introduction to Molecular Symmetry and Symmetry Point Groups," M. Zeldin, *J. Chem. Educ.* **43**, 17 (1966).
"An Introduction to Group Theory for Chemists," J. E. White, *J. Chem. Educ.* **44**, 128 (1967).
"Symmetry," C. A. Coulson, *Chem. Brit.* **4**, 113 (1968).
"Symmetry, Point Groups, and Character Tables. Three Parts," M. Orchin and H. H. Jaffe, *J. Chem. Educ.* **47**, 246 (1970).
"The Use of Group Theory to Determine Molecular Geometry from IR Spectra," C. H. Thomas, *J. Chem. Educ.* **51**, 91 (1974).
"Group Theory: From Common Objects to Molecules," M. Herman and J. Lievin, *J. Chem. Educ.* **54**, 596 (1977).
"Group Theory Calculations of Molecular Vibrations Using Spreadsheets," S. M. Condren, *J. Chem. Educ.* **71**, 487 (1994).
"Applications of Group Theory: Infrared and Raman Spectra of the Isomers of 1,2-Dichloroethylene," N. C. Craig and N. N. Lacuesta, *J. Chem. Educ.* **81**, 1199 (2004).

Problems

Vocabulary of Spectroscopy

11.1 Convert 15,000 cm^{-1} into wavelength (nm) and frequency.

11.2 Convert 450 nm into wavenumber (cm^{-1}) and frequency.

11.3 Convert the following percent transmittance to absorbance: **(a)** 100%, **(b)** 50%, and **(c)** 0%.

11.4 Convert the following absorbance to percent transmittance: **(a)** 0.0, **(b)** 0.12, and **(c)** 4.6.

11.5 The molar absorptivity of a solute at 664 nm is 895 L mol^{-1} cm^{-1}. When light at that wavelength is passed through a 2.0-cm cell containing a solution of the solute, 74.6% of the light is absorbed. Calculate the concentration of the solution.

11.6 The absorption of radiation energy by a molecule results in the formation of an excited molecule. It would seem that given enough time all of the molecules in a sample would have been excited and no more absorption would occur. Yet in practice we find that the absorbance of a sample at any wavelength remains unchanged with time. Why?

11.7 The mean lifetime of an electronically excited molecule is 1.0×10^{-8} s. If the emission of the radiation occurs at 610 nm, what are the uncertainties in frequency ($\Delta\nu$) and wavelength ($\Delta\lambda$)?

11.8 The frequency of molecular collision in the liquid phase is about 1×10^{13} s^{-1}. Ignoring all other mechanisms contributing to linewidth, calculate the width (in Hz) of vibrational transitions if **(a)** every collision is effective in deactivating the molecule vibrationally, and **(b)** that one collision in 40 is effective.

11.9 Analysis of lines, broadened by the Doppler effect shows that the width at half-height, $\Delta\lambda$, is given by

$$\Delta\lambda = 2\left(\frac{\lambda}{c}\right)\left(\frac{k_B T}{m}\right)^{1/2}$$

where c is the speed of light, T is the temperature (in kelvins), and m is the mass of the species involved in the transition. The corona of the sun emits a spectral line at about 677 nm due to the presence of an ionized ^{57}Fe atom (molar mass: 0.0569 kg mol^{-1}). If the line has a width of 0.053 nm, then what is the temperature of the corona?

11.10 The familiar yellow D line of sodium is actually a doublet at 589.0 nm and 589.6 nm. Calculate the difference in energy (in J) between these two lines. Convert your answer to cm^{-1} units.

11.11 The resolution of visible and UV spectra can usually be improved by recording the spectra at low temperatures. Why does this procedure work?

11.12 Assuming that the width of a spectral line to be the sole result of lifetime broadening, estimate the lifetime of a state that gives rise to a line of width **(a)** 1.0 cm^{-1}, **(b)** 0.50 Hz.

11.13 What is the molar absorptivity of a solute that absorbs 86 percent of a certain wavelength of light when the beam is passed through a 1.0-cm cell containing a 0.16 *M* solution?

11.14 The molar absorptivity of a benzene solution of an organic compound is 1.3×10^2 L mol^{-1} cm^{-1} at 422 nm. Calculate the percentage reduction in light intensity when light of that wavelength passes through a 1.0-cm cell containing a solution of concentration 0.0033 *M*.

11.15 A single NMR scan of a dilute sample exhibits a signal-to-noise (S/N) ratio of 1.8. If each scan takes 8.0 minutes, calculate the minimum time required to generate a spectrum with a S/N ratio of 20.

11.16 What is the wavelength of a photon that has four times the energy of a photon whose wavelength is 1064 nm?

11.17 For each of the following electromagnetic radiation sources, tabulate the frequency (Hz), the wavenumber (cm^{-1}), the energy (eV), and the region of the electromagnetic spectrum (IR, vis, etc.). **(a)** Frequency-doubled Nd:YAG laser at 532 nm, **(b)** CO_2 laser at 10.6 μm, **(c)** klystron at 1.23 cm, **(d)** Cu K_α emission at 154.1 pm, **(e)** an oscillator used in NMR at 1.5 m.

Microwave Spectroscopy and the Rigid Rotor

11.18 Which of the following molecules are microwave active: **(a)** C_2H_2, **(b)** CH_3Cl, **(c)** C_6H_6, **(d)** CO_2, **(e)** H_2O, **(f)** HCN.

11.19 What is the degeneracy of the rotational energy level with $J = 7$ for a diatomic rigid rotor?

11.20 The $J = 4 \leftarrow 3$ transition for a diatomic molecule occurs at 0.50 cm^{-1}. What is the wavenumber for the $J = 7 \leftarrow 6$ transition for this molecule? Assume a rigid rotor.

11.21 Derive an expression for the value of J corresponding to the most populous rotational energy level of a rigid diatomic rotor at temperature T. Evaluate the expression for HCl ($\tilde{B} = 10.59\ cm^{-1}$) at 25°C. (*Hint*: Remember to include the degeneracy.)

11.22 The equilibrium bond length in nitric oxide ($^{14}N^{16}O$) is 1.15 Å. Calculate **(a)** the moment of inertia of NO, and **(b)** the energy for the $J = 1 \leftarrow 0$ transition. **(c)** How many times does the molecule rotate per second in the $J = 1$ level?

11.23 Will the molecule HD ($^1H^2H$) have an absorption-line in the microwave region? Explain.

11.24 For a given diatomic molecule, a single absorption line is observed in the microwave region of the electromagnetic spectrum. Is this sufficient information to determine the bond length of the molecule? Explain.

11.25 The pure rotational spectrum of the molecule HF has a series of absorption lines separated by 41.878 cm^{-1}. Calculate the bond length for this molecule.

11.26 In Example 11.1, we calculated the interatomic distance for CO. A more precise value of the separation between absorption lines is 1.15270×10^{11} Hz for isotopically pure $^{12}C^{16}O$. Assume that the bond length does not change with isotopic substitution and predict the separation between absorption lines for $^{13}C^{16}O$. Repeat the calculation for $^{12}C^{18}O$. Use the relative atomic masses: $^{12}C = 12.000000$, $^{13}C = 13.003355$, $^{16}O = 15.994915$, and $^{18}O = 17.999160$.

Infrared Spectroscopy and the Harmonic Oscillator

11.27 Which of the following molecules are IR active: **(a)** N_2, **(b)** HBr, **(c)** CH_4, **(d)** HD, **(e)** H_2O_2, **(f)** NO?

11.28 Give the number of normal vibrational modes of **(a)** O_3, **(b)** C_2H_2, **(c)** CBr_4, **(d)** C_6H_6.

11.29 Draw a normal vibrational mode of the BF_3 molecule that is IR inactive.

11.30 An object of mass 500 g suspended from the end of a rubber band has a vibrational frequency of 4.2 Hz. Calculate the force constant of the rubber band.

11.31 The fundamental frequency of vibration for carbon monoxide is 2143.3 cm^{-1}. Calculate the force constant of the carbon-oxygen bond.

11.32 If molecules did not possess zero-point energy, would they be able to undergo the $\upsilon = 1 \leftarrow 0$ transition?

11.33 Under what conditions can one observe a hot band in an IR spectrum?

11.34 Show all the fundamental vibration modes of **(a)** carbon disulfide (CS_2) and **(b)** carbonyl sulfide (OCS) and indicate which ones are IR active.

11.35 Calculate the number of vibrational degrees of freedom of the hemoglobin molecule which contains 9272 atoms.

11.36 Which of the following molecules has the highest fundamental frequency of vibration? H_2, D_2, HD. (D = ^2H)

11.37 The fundamental wavenumber of vibration for $D^{35}Cl$ is given by $\tilde{\nu} = 2081.0\ cm^{-1}$. Calculate the force constant k and compare this value with the force constant obtained for $H^{35}Cl$ in Example 11.2. Comment on your result.

11.38 The molecule $H^{79}Br$ has a force constant of 405.7 N m^{-1}. At what wavenumber will $H^{79}Br$ absorb in the infrared?

11.39 For a given diatomic molecule, a single absorption peak is observed in the infrared region of the electromagnetic spectrum. Is this sufficient information to determine the bond length of the molecule? Is this sufficient information to determine the force constant of the molecule? Explain.

11.40 Consider the 2-propenenitrile molecule whose IR spectrum is shown in Figure 11.25. Which of the following types of molecular motion has the largest number of energy levels appreciably occupied at 300 K? Electronic, C—H stretching vibration, C=C stretching vibration, HCH bending motion, or rotational motion.

Raman Spectroscopy

11.41 For the molecule SO_2, identify whether each of the following vibrational modes is Raman active or forbidden: **(a)** bend, **(b)** symmetric stretch, **(c)** asymmetric stretch.

11.42 For the molecule BeH_2, identify whether each of the following vibrational modes is Raman active or forbidden: **(a)** bend, **(b)** symmetric stretch, **(c)** asymmetric stretch. Compare and contrast your results with those from the previous problem.

11.43 Given that the Raman scattering cross section is proportional to the change in polarizability with vibration, predict the trend in the amplitude of the Raman scattering for the series of homonuclear diatomic molecules F_2, Br_2, Cl_2, and I_2. Next predict the trend in Raman scattering cross section for the series C_2, N_2, O_2, and F_2.

11.44 Explain why Stokes Raman scattered light is so much more intense than anti-Stokes Raman scattered light.

11.45 In a process known as Raman shifting, Raman scattering may be used to extend the spectrum of light available from a laser source. Focusing a laser in a gas cell filled with high pressure $H_2(g)$ will produce light that is both Stokes and anti-Stokes shifted by integer multiples of the vibrational frequency of $H_2(\tilde{\nu} = 4155\ \text{cm}^{-1})$. For a Nd:YAG laser operating at 532 nm, calculate the wavelengths of the first three Stokes and anti-Stokes lines. For those that lie in the visible region of the electromagnetic spectrum, what color are they?

11.46 A 633-nm HeNe laser is used to irradiate a liquid sample of benzene, which has a prominent vibrational mode of 992 cm^{-1}. Describe the absolute wavelengths and the relative strengths of the three types of scattered light: Rayleigh, Stokes Raman, and anti-Stokes Raman.

11.47 A mercury lamp, which emits several discrete wavelengths of light, finds many uses in spectroscopy. Mercury lines are employed as wavelength standards, as UV sources, and historically as excitation sources for Raman scattering. The 435.83-nm light from a mercury lamp is used to irradiate a pure liquid CCl_4 sample, and scattered light is observed at 447.57, 442.19, and 440.05 nm (among others). What CCl_4 vibrational wavenumbers correspond to the scattered light observed here?

11.48 Why is Raman spectroscopy performed with monochromatic excitation light, and not polychromatic or blackbody radiation (such as a tungsten–filament lightbulb)?

Additional Problems

11.49 A typical rotational transition wavenumber is on the order of 1 cm^{-1} and a typical vibrational transition wavenumber is on the order of 1000 cm^{-1}. Calculate the energy (in kJ mol^{-1}) for typical rotational and vibrational transitions. Compare the period of rotation (the time required for one revolution) with the period of vibration.

11.50 This problem deals with the amplitude of molecular vibration of a diatomic molecule in its ground vibrational state. **(a)** When the molecule is stretched by an extent x from the equilibrium position, the increase in the potential energy is given by the integral

$$\int_0^x kx\,dx$$

where k is the force constant. Evaluate this integral. **(b)** Calculate the amplitude of vibration by equating the potential energy with the vibrational energy in the ground state. Use x_{max} to represent the maximum displacement. **(c)** Given that the force constant for $H^{35}Cl$ is $4.84 \times 10^2\ \text{N m}^{-1}$, calculate the amplitude of vibration in the $\upsilon = 0$ state. **(d)** What is the percent of the amplitude compared to the bond length (1.27 Å)? **(e)** Repeat the calculations in **(c)** and **(d)** for carbon monoxide, given that the force constant is $1.85 \times 10^3\ \text{N m}^{-1}$ and the bond length is 1.13 Å. (^{35}Cl = 34.97 amu.)

11.51 The IR spectrum of the carbon monoxide-hemoglobin complex shows a peak at about 1950 cm^{-1}, which is due to the carbonyl stretching frequency. **(a)** Compare this value with the fundamental frequency of free CO, which is 2143.3 cm^{-1}. Comment on the difference. **(b)** Convert this frequency to kJ mol^{-1}. **(c)** What conclusion can you draw from the fact that there is only one band present?

11.52 For predicting the electronic absorption spectrum of the C_{60} molecule [the fullerene known as a "Buckyball", named after American inventor R. Buckminster Fuller (1895–1983)], which of the following models would you expect to provide the best prediction of the first three transitions, and why? Particle-on-a-ring, rigid rotor, harmonic oscillator, particle-in-a-box.

11.53 Microwave spectroscopy has been used to determine the equilibrium geometry of van der Waals complexes such as the water dimer. Explain how microwave spectroscopy could distinguish between various possible geometries of water dimers. Draw at least three possible geometries. Speculate which of your geometries is most plausible, and why.

11.54 Explain in your own words why molecules vibrate, even at absolute zero of temperature.

11.55 The lanthanide elements may form diatomic hydrides and diatomic oxides in the gas phase under laboratory conditions. Compare and contrast the appearance of the infrared and microwave absorption spectra of diatomic lanthanide hydrides with diatomic lanthanide oxides.

11.56 Is the rotational constant B' for an excited vibrational state greater than, less than, or equal to the rotational constant B'' for the ground vibrational state? Explain your reasoning.

11.57 Is the vibrational frequency of a diatomic molecule in an excited vibrational state ($v' = 1$) greater than, less than, or equal to the vibrational frequency of the molecule in its ground vibrational state ($v'' = 0$)? Explain.

11.58 Does the vibrational frequency depend on which rotational state the molecule is in? Explain.

11.59 For a small organic molecule, how many more normal vibrational modes are there when a methyl group ($-CH_3$) is substituted for a hydrogen atom ($-H$)? How many more rotational modes?

11.60 Identify each of the following functions as even, odd, or neither even nor odd. **(a)** $f(x) =$ constant, **(b)** $f(x) = -x$, **(c)** $f(x) = \sin(x)$, **(d)** $f(x) = \cos(x)$, **(e)** $f(x) = \sin(x)\cos(x)$, **(f)** $f(x) = 3\cos^2(x) - 1$, **(g)** $f(x) = \sin^2(x)e^{2ix}$.

11.61 Terahertz spectroscopy is a relatively new area of research because the technology for the production and detection of terahertz radiation has been developed only recently. Terahertz radiation finds application in imaging because many objects are transparent to terahertz radiation. Terahertz radiation can, for example, be used to "see through" clothing in detecting weapons, and to "see through" cereal boxes to image the raisins in raisin bran. **(a)** Calculate the wavelength and the wavenumber for a "T ray" with frequency of 1 THz (1×10^{12} s^{-1}). **(b)** In what region of the electromagnetic spectrum does this lie with respect to microwave, IR, visible, and UV radiation? **(c)** What sort of molecular motions do you think might be excited by terahertz radiation?

11.62 For each of the following molecules, identify how many unique rotational quantum numbers are needed to model the rotational behavior of the molecule. **(a)** NO, **(b)** CH_4, **(c)** CH_3Cl, **(d)** CO_2, **(e)** SO_2, **(f)** OCS, **(g)** SF_6, **(h)** SF_4.

11.63 In addition to microwave absorption spectroscopy, rotational energy levels may also be probed by a light scattering technique known as rotational Raman spectroscopy. With the knowledge that rotational Raman involves the interaction of rotational energy levels with two photons and that each photon carries one unit of angular momentum, predict the selection rules for rotational Raman spectroscopy. (*Hint*: Your answer should limit the allowed changes in the quantum number J.)

11.64 For the following series of isotopomers of the hydrogen molecule, rank them in order of increasing zero–point energy. Then rank them in order of increasing dissociation energy. Assume that all of the isotopomers have the same potential energy surface. H_2, HD, D_2, HT, T_2, DT. (Note H = ^{1}H, D = ^{2}H, and T = ^{3}H)

11.65 For a linear molecule with the molecular formula XY_2, explain how vibrational spectroscopy can be used to distinguish between the two possible structures, X–Y–Y and Y–X–Y.

11.66 Isotopic substitution can be used to "tag" vibrational modes in small molecules because the isotopic substitution will result in a shift in the frequency of vibrational modes. In general, which will result in a larger frequency shift, a substitution of ^{2}H for ^{1}H or a substitution of ^{13}C for ^{12}C? Why?

11.67 When the rigid rotor is in its ground state, it has zero kinetic energy. How can this be interpreted using the Heisenberg uncertainty principle?

11.68 Explain how Raman and IR spectroscopies might be used to distinguish the *ortho-*, *meta-*, and *para-* isomers of dichlorobenzene.

11.69 Coherent Anti-Stokes Raman Scattering (CARS) is an experimental technique that has been used to probe the temperature of gases in internal combustion engines (cars). Do a literature search concerning CARS, and then identify the advantages that this technique might have over conventional Raman spectroscopy and over infrared spectroscopy.

11.70 A rigid rotor is described by the wave function

$$\psi(\theta, \phi) = \left(\frac{15}{8\pi}\right)^{1/2} \sin\theta \cos\theta \, e^{-i\phi}$$

What are the quantum numbers ℓ and m_ℓ that correspond to this wave function? What is the energy, the magnitude of angular momentum, and the z-component of angular momentum for a rigid rotor described by this wave function? Express your answers in terms of $\hbar$ and I (moment of inertia).

11.71 Is it possible to have a vibrational mode that is neither infrared nor Raman active? (*Hint*: For a molecule containing such a vibrational mode, consider the general appearance of its character table. What would be missing?)

11.72 Apply dimensional analysis to determine the units of ρ_λ and ρ_ν. (*Hint*: See Equations 10.5 and 11.14.)

CHAPTER 12

Electronic Structure of Atoms

The underlying physical laws necessary for the mathematical theory of a large part of physics and the whole of chemistry are thus completely known, and the difficulty is only that the application of these laws leads to equations much too complicated to be soluble.

—P. A. M. Dirac, *Proc. Roy. Soc.*, **A123**, 714 (1929).

We now return to one of the experiments that puzzled classical physicists circa 1900, that of atomic emission. The hydrogen atom emission spectrum was explained by the Bohr theory (Section 10.5), but the Bohr theory could not explain the behavior of atoms or ions containing more than a single electron. In this chapter, we consider hydrogen atomic structure using the Schrödinger equation. For multi-electron atoms, there is no exact analytical solution, so we present techniques that may be used to find approximate solutions.

12.1 The Hydrogen Atom

The simplest atomic system, the hydrogen atom, consists of one electron and one proton. As with other simple systems, we use the Schrödinger equation to find analytical solutions for the energy and wave function of the hydrogen atom. This is a three-dimensional problem, so the wave function ψ for the electron depends on the x, y, and z coordinates. We consider the nucleus to be fixed in space, so the kinetic energy of the system depends only on the motion to the electron.

We write this problem in the frame of reference of a fixed nucleus. A more accurate treatment uses the center of mass as the frame of reference.

The kinetic-energy operator is the same as we used in the particle-in-a-box, harmonic oscillator, and rigid rotor systems. The potential energy is the Coulombic interaction between the electron and the nucleus, given by $-e^2/4\pi\varepsilon_0 r$, where e is the electron charge, r is the distance between the electron and the nucleus, and ε_0 is the permittivity of free space. The time-independent Schrödinger wave equation is given by

$$\hat{H}\psi = E\psi$$

$$(\hat{K} + \hat{V})\psi = E\psi$$

$$\frac{-\hbar^2}{2m_e}\nabla^2\psi - \frac{e^2\psi}{4\pi\varepsilon_0 r} = E\psi \tag{12.1}$$

where m_e is the mass of the electron. Because the attraction has spherical symmetry (i.e., it depends only on r), this system is known as a *central force problem*. A Cartesian coordinate system (x, y, z) is inconvenient, because the solution to Equation 12.1 does not separate these variables. So, as we did with the rigid rotor Schrödinger equation, we express Equation 12.1 in terms of (r, θ, ϕ) *spherical polar coordinates* (See Figure 11.9). The transformation to spherical polar coordinates is a long but straightforward process. With this transformation, we can rewrite Equation 12.1 as

$$\frac{-\hbar^2}{2m_e}\left[\frac{1}{r^2}\frac{\partial}{\partial r}\left(r^2\frac{\partial}{\partial r}\right) + \frac{1}{r^2 \sin\theta}\frac{\partial}{\partial\theta}\left(\sin\theta\frac{\partial}{\partial\theta}\right) + \frac{1}{r^2\sin^2\theta}\left(\frac{\partial^2}{\partial\phi^2}\right)\right]\psi - \frac{e^2\psi}{4\pi\varepsilon_0 r} = E\psi \quad (12.2)$$

Fortunately, this multivariable equation has already been solved, so we need be concerned only with the results. As we did with the two-dimensional particle-in-a-box problem (see Section 10.10), we assume that the variables may be separated and write the total wave function as the product of three single-coordinate functions

$$\psi(r,\theta,\phi) = R(r)\Theta(\theta)\Phi(\phi) \quad (12.3)$$

Thus, ψ is given as a product of two independent functions, the *radial part*, $R(r)$, and the *angular part*, $\Theta(\theta)\Phi(\phi)$, of the wave function, respectively.

The solution of Equation 12.2 is nontrivial.* It should come as no surprise, however, that the solution yields a family of quantized wave functions that are described by integer quantum numbers. The angular part of the wave function for the hydrogen atom takes the same form as the rigid rotor; it is a collection of functions known as the spherical harmonics (see Section 11.2),

$$\Theta(\theta)\Phi(\phi) = Y_\ell^{m_\ell}(\theta,\phi) \quad (12.4)$$

where $\ell = 0, 1, 2, \ldots$ and $m_\ell = 0, \pm1, \pm2, \ldots, \pm\ell$. As in the case of the rigid rotor, we have added a subscript ℓ to the quantum number m. A complete solution to the Schrödinger equation for the hydrogen atom consists of a family of functions described by three quantum numbers, n, ℓ, and m_ℓ.

$$\psi_{n,\ell,m_\ell}(r,\theta,\phi) = \underbrace{R_{n\ell}(r)}_{\text{Radial part}}\underbrace{Y_\ell^{m_\ell}(\theta,\phi)}_{\text{Angular part}} \quad (12.5)$$

The radial part of the wave function is described in the next section. Because ψ is a function of three spatial coordinates, we should expect three quantum numbers for the time-independent solutions. The quantum numbers describe the hydrogen atom wave functions, which are called orbitals. These quantum numbers may take on the following values:

$$n = 1, 2, 3, \ldots \quad (12.6a)$$

$$\ell = 0, 1, 2, \ldots, (n - 1) \quad (12.6b)$$

$$m_\ell = 0, \pm1, \pm2, \ldots, \pm\ell \quad (12.6c)$$

---

* See the physical chemistry texts listed on p. 548 for a detailed discussion of the solution of Equation 12.2.

The *principal quantum number*, n, determines the size of the wave function and the energy of the electron. The *azimuthal quantum number* or the *orbital angular momentum quantum number*, ℓ, determines the shape of the wave function. Finally, the *magnetic quantum number*, m_ℓ, which gives the z component of the angular momentum, determines the orientation of the wave function in space. We shall soon see in more detail how these quantum numbers are used to describe the electron in hydrogen atom.

All the orbitals having a given value of n form a single *shell*. The shells are referred to by capital letters:

n:	1	2	3	4	5	6	...
Name of shell:	K	L	M	N	O	P	...

The orbitals with the same value of n but different values of ℓ form the *subshells* of a given shell. These subshells are generally designated by the small letters $s, p, d, \ldots$, as follows:

ℓ:	0	1	2	3	4	5	...
Name of subshell:	s	p	d	f	g	h	...

Thus, if $n = 2$ and $\ell = 1$, we have a $2p$ subshell, and its three orbitals (corresponding to $m_\ell = +1, 0$, and -1) are called the $2p$ orbitals. The unusual sequence of letters (s, p, d, and f) has a historical origin. Physicists who studied atomic emission spectra tried to correlate the observed spectral lines with the particular energy states involved in the transitions. They noted that some of the lines were *s*harp; some were rather spread out, or *d*iffuse; and some were very strong and hence referred to as *p*rincipal lines. Subsequently, the initial letters of each characteristic were assigned to those energy states. However, starting with the letter *f* (for *f*undamental), the orbital designations follow alphabetical order.

12.2 The Radial Distribution Function

For a solution to the radial portion of the Schrödinger equation $R_{n\ell}(r)$ for the hydrogen atom, we again rely on the prior work of mathematicians. There is a whole family of solutions that depend on the associated Laguerre polynomials $L_{n+\ell}^{2\ell+1}(x)$ (after the French mathematician Edmond Nicolas Laguerre, 1834–1886):

$$R_{n\ell}(r) = \underbrace{N_{n\ell}}_{\text{Normalization constant}} \times \underbrace{r^\ell e^{-r/na_0}}_{\text{Exponential part}} \times \underbrace{L_{n+\ell}^{2\ell+1}\left(\frac{2r}{na_0}\right)}_{\text{Associated Laguerre polynomial}} \tag{12.7}$$

Recall that a_0 is the Bohr radius (0.529 Å).

An examination of the first few associated Laguerre polynomials shows that they are polynomials in x (Table 12.1). In the radial part of the wave function $R_{n\ell}(r)$, we set x equal to $2r/na_0$, where a_0 is the Bohr radius and n is the principal quantum number.

Table 12.1
The First Few Associated Laguerre Polynomials[a]

n	ℓ	Associated Laguerre polynomial, $L_{n+\ell}^{2\ell+1}(x)$
1	0	$L_1^1(x) = -1$
2	0	$L_2^1(x) = -2(2 - x)$
2	1	$L_3^3(x) = -6$
3	0	$L_3^1(x) = -6(3 - 3x + \frac{1}{2}x^2)$
3	1	$L_4^3(x) = -24(4 - x)$
3	2	$L_5^5(x) = -120$
		$L_n^k(x) = \sum_{m=0}^{n}(-1)^m \frac{(n + k)!}{(n - m)!(k + m)!m!}x^m$

[a] When the Laguerre polynomials are used in hydrogen atom wave functions, $x = 2r/na_0$, where a_0 is the Bohr radius. [After D. A. McQuarrie, and J. D. Simon, *Physical Chemistry: A Molecular Approach*, University Science Books, Sausalito, CA (1997).]

The radial portion of the wave function also depends on the quantum number ℓ, but not on the quantum number m_ℓ.

Table 12.2 lists the radial and angular wave functions for $n = 1, 2,$ and 3. Having considered the angular part when studying the rigid rotor (Section 11.2), let us now look at the radial part of the hydrogen atom wave function in some detail. We would like to know the location of the electron with respect to the nucleus, so we plot the square of the radial part of the wave function (Figure 12.1). The term $R(r)^2dr$ gives the probability of finding the electron between r and $r + dr$ along a certain direction away from the nucleus. But this isn't quite what we wanted, because for the hydrogen 1*s* orbital it gives a maximum at the nucleus ($r = 0$). A more informative plot should give the total probability of finding the electron in a volume element between r and $r + dr$ in all directions. To do this, we consider two concentric spheres of radii r and $r + dr$ (Figure 12.2). The volume between these two spheres is $4\pi r^2 dr$* and the probability of finding the electron within this spherical shell is $4\pi r^2 R(r)^2 dr$. The function $4\pi r^2 R(r)^2$ is called the *radial distribution function*. The radial distribution function for 1*s*, 2*s*, 2*p*, 3*s*, 3*p*, and 3*d* orbitals and indeed for all higher orbitals has zero value at the nucleus (Figure 12.3). It is noteworthy that for the 1*s* orbital, the maximum value of the radial distribution function occurs at

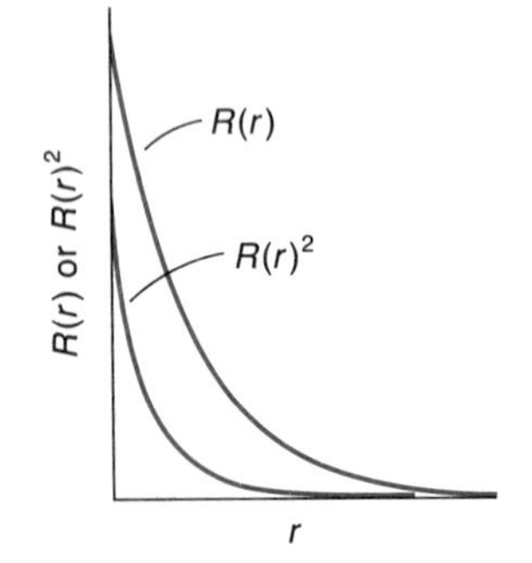

Figure 12.1
Plots of $R(r)$ and $R(r)^2$ versus r for the hydrogen 1s orbital.

* This function is obtained by taking the difference between the two volumes: $(4\pi/3)(r + dr)^3 - (4\pi/3)r^3$ and neglecting the $(dr)^2$ and $(dr)^3$ terms as trivially small in comparison to the linear term dr.

Table 12.2
Complex Hydrogen Atom Wave Functions for $n = 1, 2,$ and 3

n	ℓ	m_ℓ	ψ_{n,ℓ,m_ℓ}	$R_{n\ell}(r)$[a]	$\Theta_{\ell,m_\ell}(\theta)$	$\Phi_{m_\ell}(\phi)$[b]
1	0	0	$1s$	$\frac{2}{\sqrt{a_0^3}}e^{-\rho}$	$\frac{1}{\sqrt{2}}$	$\frac{1}{\sqrt{2\pi}}$
2	0	0	$2s$	$\frac{1}{\sqrt{2a_0^3}}\left(1-\frac{\rho}{2}\right)e^{-\rho/2}$	$\frac{1}{\sqrt{2}}$	$\frac{1}{\sqrt{2\pi}}$
2	1	0	$2p_0$	$\frac{1}{\sqrt{24a_0^3}}\rho e^{-\rho/2}$	$\sqrt{\frac{3}{2}}\cos\theta$	$\frac{1}{\sqrt{2\pi}}$
2	1	±1	$2p_{\pm1}$	$\frac{1}{\sqrt{24a_0^3}}\rho e^{-\rho/2}$	$\sqrt{\frac{3}{4}}\sin\theta$	$\frac{1}{\sqrt{2\pi}}e^{\pm i\phi}$
3	0	0	$3s$	$\frac{2}{\sqrt{27a_0^3}}\left(1-\frac{2}{3}\rho+\frac{2}{27}\rho^2\right)e^{-\rho/3}$	$\frac{1}{\sqrt{2}}$	$\frac{1}{\sqrt{2\pi}}$
3	1	0	$3p_0$	$\frac{8}{27\sqrt{6a_0^3}}\rho\left(1-\frac{\rho}{6}\right)e^{-\rho/3}$	$\sqrt{\frac{3}{2}}\cos\theta$	$\frac{1}{\sqrt{2\pi}}$
3	1	±1	$3p_{\pm1}$	$\frac{8}{27\sqrt{6a_0^3}}\rho\left(1-\frac{\rho}{6}\right)e^{-\rho/3}$	$\sqrt{\frac{3}{4}}\sin\theta$	$\frac{1}{\sqrt{2\pi}}e^{\pm i\phi}$
3	2	0	$3d_0$	$\frac{4}{81\sqrt{30a_0^3}}\rho^2e^{-\rho/3}$	$\sqrt{\frac{5}{8}}(3\cos^2\theta-1)$	$\frac{1}{\sqrt{2\pi}}$
3	2	±1	$3d_{\pm1}$	$\frac{4}{81\sqrt{30a_0^3}}\rho^2e^{-\rho/3}$	$\frac{\sqrt{15}}{2}\sin\theta\cos\theta$	$\frac{1}{\sqrt{2\pi}}e^{\pm i\phi}$
3	2	±2	$3d_{\pm2}$	$\frac{4}{81\sqrt{30a_0^3}}\rho^2e^{-\rho/3}$	$\frac{\sqrt{15}}{4}\sin^2\theta$	$\frac{1}{\sqrt{2\pi}}e^{\pm 2i\phi}$

[a] The variable $\rho = r/a_0$, where a_0 is the Bohr radius.
[b] The symbol i denotes the imaginary number, $\sqrt{-1}$.

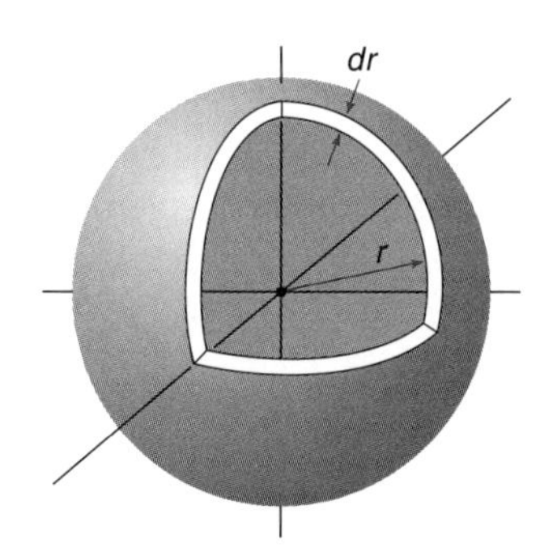

Figure 12.2
The radial distribution function gives the total probability of finding an electron in a spherical shell of thickness dr at a distance r from the nucleus. Note that the volume of the shell is proportional to r^2 and is zero when $r = 0$ (at the nucleus).

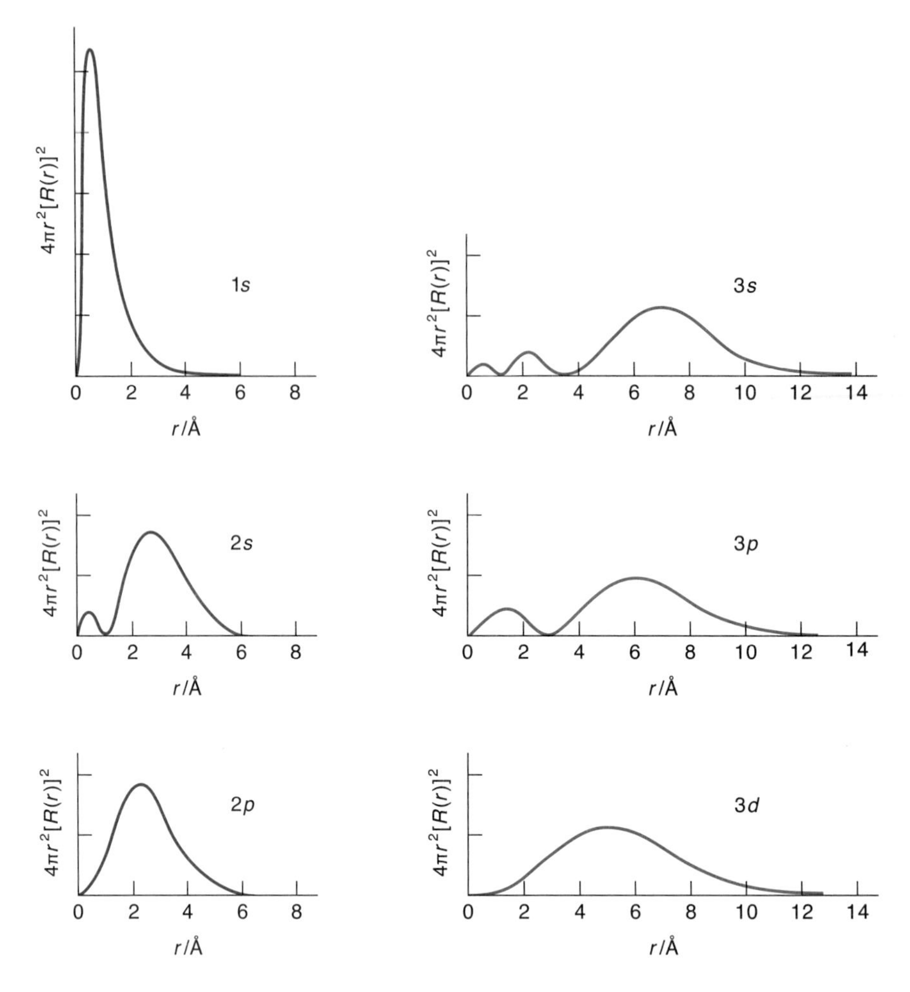

Figure 12.3
Radial distribution functions for the hydrogen 1*s*, 2*s*, 2*p*, 3*s*, 3*p*, and 3*d* orbitals. Note that the maximum for the 1*s* orbital occurs at 0.529 Å (52.9 pm), which is the Bohr radius.

0.529 Å (5.29 pm), which is the Bohr radius. From these plots we see that an electron in any particular orbital does not have a well-defined position; therefore, it is more convenient to use the term *electron density* to describe the probability of locating the electron. Mathematically, the probability vanishes only when r approaches infinity. Physically, however, we need only consider each orbital over a relatively small distance (a few angstroms) because the function decreases exponentially rapidly with increasing r.

The 2*s* orbital has two maxima. In this case, we can imagine two concentric spheres with a radial node somewhere in between. The shape of the node is spherical with radius $r = 2a_0$. The 3*s* orbital has two radial nodes whose radii may be determined by setting the radial portion of the wave function equal to zero.

EXAMPLE 12.1

Find the location of the radial nodes in a $3s$ orbital.

ANSWER

A node occurs when the value of the wave function is zero. Set the radial portion of the wave function equal to zero (see Table 12.2).

$$0 = R_{3s}(r)$$

$$0 = \frac{2}{\sqrt{27a_0^3}}\left(1 - \frac{2}{3}\rho + \frac{2}{27}\rho^2\right)e^{-\rho/3}$$

$$0 = \left(1 - \frac{2}{3}\rho + \frac{2}{27}\rho^2\right)$$

Solve for the roots of this polynomial in ρ using the quadratic formula:

$$\rho = \frac{-b \pm \sqrt{b^2 - 4ac}}{2a}$$

$$\rho = \frac{-\left(-\frac{2}{3}\right) \pm \sqrt{\left(-\frac{2}{3}\right)^2 - 4\left(\frac{2}{27}\right)}}{2\left(\frac{2}{27}\right)}$$

$$\rho = \frac{9 \pm \sqrt{27}}{2}$$

$$\rho = 1.90 \text{ and } 7.10$$

Because $\rho = r/a_0$, there are radial nodes at $r = 1\ 90a_0$ and $7.10a_0$. These correspond to spherical nodes with radius $r = 1.01$ Å and 3.75 Å, respectively.

COMMENT

Nodes are only located where the wave function crosses zero, so there is neither a node at $r = 0$ (at the nucleus) nor at $r = \infty$ (Figure 12.3). Here we have found the radial nodes; in general there may also be angular nodes.

The radial distribution function plots for the p and d orbitals are more complex in form, but they can be interpreted in a similar manner. The location of the radial nodes may be found using the technique illustrated in Example 12.1. The number of radial nodes may be generalized by considering the properties of the Laguerre polynomials and confirmed by examining the radial distribution plots. The number of

Table 12.3
Nodes for the Hydrogenlike Wave Functions

n	ℓ	Radial nodes	Angular nodes	Total nodes
1	0	0	0	0
2	0	1	0	1
2	1	0	1	1
3	0	2	0	2
3	1	1	1	2
3	2	0	2	2
n	ℓ	$(n - \ell - 1)$	ℓ	$(n - 1)$

radial nodes is given by $(n - \ell - 1)$. The total number of nodes is $(n - 1)$; therefore, the number of angular nodes is given by ℓ (Table 12.3). That the total number of nodes depends only on the principal quantum number n is consistent with the fact that the energy of a hydrogen atom (in the absence of an external field) depends only on n (see Equation 10.18).

12.3 Hydrogen Atomic Orbitals

The product of the radial and angular parts gives a complete wave function for the hydrogen atom (Equation 12.5)—the size of the orbital is determined by the former and the shape by the latter. An orbital can be represented in several ways, as Figure 12.4 shows. The boundary-surface representation is the simplest to use, although it is also the least informative. Hydrogen atom wave functions extend to infinity, so boundary surfaces are chosen to contain an arbitrary fraction of the total probability density, typically 90 percent. The contour-surface and electron-density representations provide a more detailed description, but they take more effort to draw. In most graphical representations, a contour or a surface is chosen such that the value of the wave function, and thus the value of the probability density, is a constant.

$$\psi(r,\theta,\phi) = \text{constant}$$

The task is to display four variables (three coordinates and the value of the wave function or the probability density function) on a flat piece of paper. The situation becomes even more challenging when we want to display the phase of the wave function as well. Computer graphics and false-colored representations can help with these visualization challenges.

The shapes of the orbitals are determined by the angular part of the wave functions, which are also listed in Table 12.4. The s orbital angular distribution functions are constants; therefore, s orbitals are spherically symmetric. The p orbitals, on the other hand, are more difficult to represent because they not only depend on θ and ϕ,

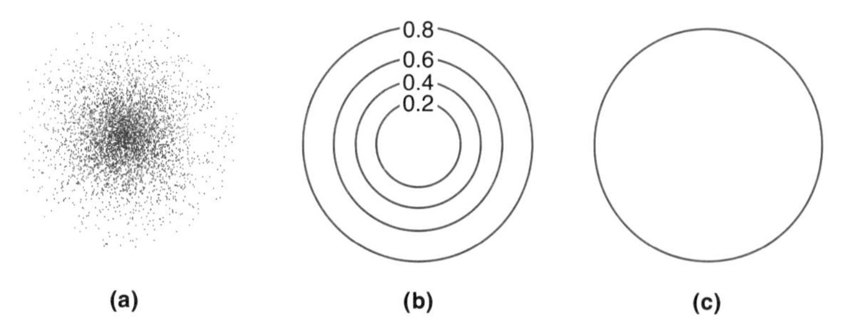

Figure 12.4
Representations of the hydrogen 1*s* orbital. (a) Charge cloud, (b) contour surfaces (the numbers represent the relative charge densities), and (c) boundary surface.

but are complex as well. From Table 12.2, the angular parts of the $\ell = 1$ states are given by

$$p_1 = \sqrt{\frac{3}{8\pi}} \sin\theta e^{i\phi} \tag{12.8a}$$

$$p_0 = \sqrt{\frac{3}{4\pi}} \cos\theta \tag{12.8b}$$

$$p_{-1} = \sqrt{\frac{3}{8\pi}} \sin\theta e^{-i\phi} \tag{12.8c}$$

The probability densities associated with p_1 and p_{-1} are the same:

$$|p_1|^2 = \frac{3}{8\pi}\sin^2\theta \quad \text{and} \quad |p_{-1}|^2 = \frac{3}{8\pi}\sin^2\theta$$

Because p_1 and p_{-1} correspond to the same energy, it follows that any linear combination of p_1 and p_{-1} is also a function that yields the same energy. Thus, we can generate the *real* orbitals by writing

$$p_x = \frac{1}{\sqrt{2}}(p_1 + p_{-1}) = \sqrt{\frac{3}{4\pi}} \sin\theta \cos\phi \tag{12.9a}$$

Recall the Euler relations: $e^{\pm ix} = \cos x \pm i \sin x$

$$p_y = \frac{1}{i\sqrt{2}}(p_1 - p_{-1}) = \sqrt{\frac{3}{4\pi}} \sin\theta \sin\phi \tag{12.9b}$$

$$p_z = p_0 = \sqrt{\frac{3}{4\pi}} \cos\theta \tag{12.9c}$$

Note that p_0 is not a complex orbital and is equal to p_z. Figure 12.5 shows the boundary-surface diagrams of the three *p* orbitals along the *x*, *y*, and *z* coordinates. Each orbital consists of two adjacent lobes in a dumbbell-like configuration. Furthermore, the *phase* of the wave function is positive in one region and negative in the other, with

Table 12.4
Real Hydrogen Atom Wave Functions for n = 1, 2, and 3

n	ℓ	m_ℓ	ψ_{n,ℓ,m_ℓ}	$R_{n\ell}(r)$[a]	$\Theta_{\ell,m_\ell}(\theta)$	$\Phi_{m_\ell}(\phi)$
1	0	0	$1s$	$\frac{2}{\sqrt{a_0^3}}e^{-\rho}$	$\frac{1}{\sqrt{2}}$	$\frac{1}{\sqrt{2\pi}}$
2	0	0	$2s$	$\frac{1}{\sqrt{2a_0^3}}\left(1-\frac{\rho}{2}\right)e^{-\rho/2}$	$\frac{1}{\sqrt{2}}$	$\frac{1}{\sqrt{2\pi}}$
2	1	0	$2p_z$	$\frac{1}{\sqrt{24a_0^3}}\rho e^{-\rho/2}$	$\sqrt{\frac{3}{2}}\cos\theta$	$\frac{1}{\sqrt{2\pi}}$
2	1	±1	$2p_x$	$\frac{1}{\sqrt{24a_0^3}}\rho e^{-\rho/2}$	$\sqrt{\frac{3}{4}}\sin\theta$	$\frac{1}{\sqrt{\pi}}\cos\phi$
			$2p_y$	$\frac{1}{\sqrt{24a_0^3}}\rho e^{-\rho/2}$	$\sqrt{\frac{3}{4}}\sin\theta$	$\frac{1}{\sqrt{\pi}}\sin\phi$
3	0	0	$3s$	$\frac{2}{\sqrt{27a_0^3}}\left(1-\frac{2}{3}\rho+\frac{2}{27}\rho^2\right)e^{-\rho/3}$	$\frac{1}{\sqrt{2}}$	$\frac{1}{\sqrt{2\pi}}$
3	1	0	$3p_z$	$\frac{8}{27\sqrt{6a_0^3}}\rho\left(1-\frac{\rho}{6}\right)e^{-\rho/3}$	$\sqrt{\frac{3}{2}}\cos\theta$	$\frac{1}{\sqrt{2\pi}}$
3	1	±1	$3p_x$	$\frac{8}{27\sqrt{6a_0^3}}\rho\left(1-\frac{\rho}{6}\right)e^{-\rho/3}$	$\sqrt{\frac{3}{4}}\sin\theta$	$\frac{1}{\sqrt{\pi}}\cos\phi$
			$3p_y$	$\frac{8}{27\sqrt{6a_0^3}}\rho\left(1-\frac{\rho}{6}\right)e^{-\rho/3}$	$\sqrt{\frac{3}{4}}\sin\theta$	$\frac{1}{\sqrt{\pi}}\sin\phi$
3	2	0	$3d_{z^2}$	$\frac{4}{81\sqrt{30a_0^3}}\rho^2e^{-\rho/3}$	$\sqrt{\frac{5}{8}}(3\cos^2\theta-1)$	$\frac{1}{\sqrt{2\pi}}$
3	2	±1	$3d_{xz}$	$\frac{4}{81\sqrt{30a_0^3}}\rho^2e^{-\rho/3}$	$\frac{\sqrt{15}}{2}\sin\theta\cos\theta$	$\frac{1}{\sqrt{\pi}}\cos\phi$
			$3d_{yz}$	$\frac{4}{81\sqrt{30a_0^3}}\rho^2e^{-\rho/3}$	$\frac{\sqrt{15}}{2}\sin\theta\cos\theta$	$\frac{1}{\sqrt{\pi}}\sin\phi$
3	2	±2	$3d_{x^2-y^2}$	$\frac{4}{81\sqrt{30a_0^3}}\rho^2e^{-\rho/3}$	$\frac{\sqrt{15}}{4}\sin^2\theta$	$\frac{1}{\sqrt{\pi}}\cos2\phi$
			$3d_{xy}$	$\frac{4}{81\sqrt{30a_0^3}}\rho^2e^{-\rho/3}$	$\frac{\sqrt{15}}{4}\sin^2\theta$	$\frac{1}{\sqrt{\pi}}\sin2\phi$

[a] The variable $\rho = r/a_0$, where a_0 is the Bohr radius.

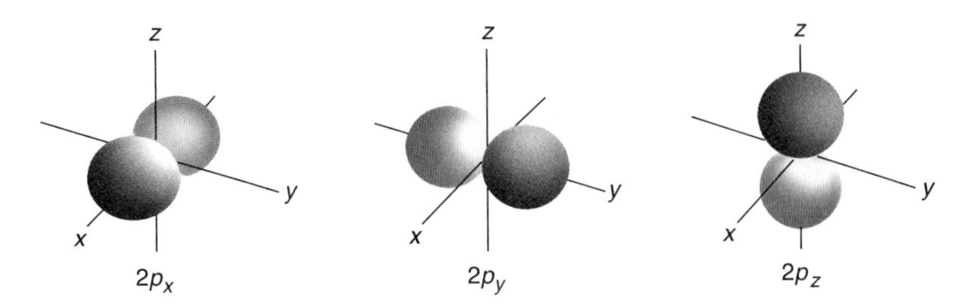

Figure 12.5
Plots of the angular parts of the real representation of the hydrogen wave functions for $\ell = 1$. These are the $2p_x$, $2p_y$, and $2p_z$ orbitals. The gray and red colors represent the phases (+) and (–) of the wave functions, respectively.

a nodal plane in between. The value of the wave function is zero in this plane, which also contains the nucleus. These three p orbitals are entirely equivalent except for their orientations. Thus, all three p orbitals have the same energy and are said to be *degenerate*.* There is no physical significance in the signs by themselves; the positive and negative lobes of a $2p$ orbital have the same charge distribution. The physically meaningful quantity is the probability of finding the electron given by the square magnitude of the wave function, which is always positive. The signs will be useful, however, when we consider the interaction between orbitals that occurs, for example, in chemical bond formation.

The five d orbitals are shown in Figure 12.6. These orbitals are energetically equivalent, but differ in their spatial orientations. Just as with the p orbitals, the d orbitals may be written as real or complex (Tables 12.2 and 12.4). The real orbitals are useful for describing directional chemical bonds.

* In quantum mechanics, the terms "state" and "energy level" have different meanings. A (stationary) state is specified by a particular wave function ψ. Each different ψ is a different state. An energy level is specified by giving the value of the energy. Each different value of E is a different energy level. For example, the three different states giving rise to the three $2p$ orbitals belong to the same energy level.

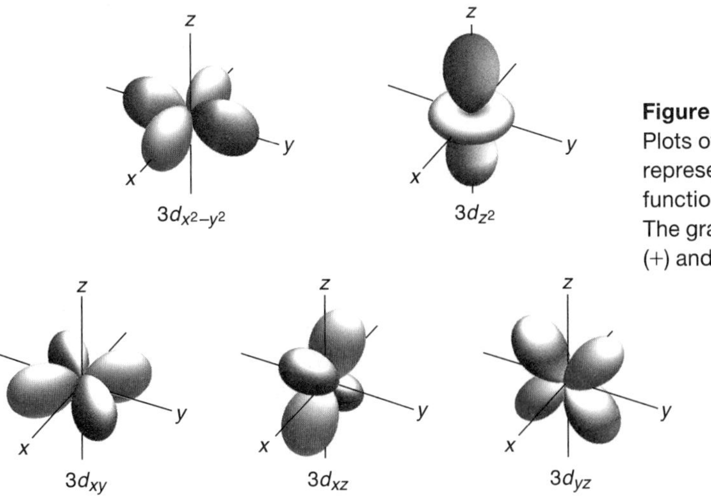

Figure 12.6
Plots of the angular parts of the real representation of the hydrogen wave functions for $\ell = 2$. These are the d orbitals. The gray and red colors represent the phases (+) and (–) of the wave functions, respectively.

12.4 Hydrogen Atom Energy Levels

Now that we have solutions for the hydrogen atom wave functions (the orbitals), we may find the corresponding energies using the Schrödinger equation. The wave functions are eigenfunctions of the Hamiltonian operator and the energies are the eigenvalues.

$$\hat{H}\psi_{n,\ell,m_\ell} = E_n\psi_{n,\ell,m_\ell}$$

The energy of the electron in a hydrogen atom is given by

$$E_n = -\frac{m_e e^4}{8h^2\varepsilon_0^2}\frac{1}{n^2} \qquad n = 1, 2, 3, \ldots \tag{12.10}$$

which is identical to Equation 10.18 if we set $Z = 1$. Thus, for an isolated hydrogen atom in the absence of an electric or magnetic field, all orbitals within a given shell have the same energy. For a hydrogen atom, then, $2s$, $2p_x$, $2p_y$, $2p_z$ (or $2s$, $2p_{-1}$, $2p_1$, and $2p_0$) orbitals have *exactly* the same energy; that is, they are degenerate. Based on the allowed quantum numbers (Equation 12.6) we can derive the degeneracy g_n of a particular shell, which gives the number of states with the same energy for a particular quantum number n, where

$$g_n = n^2 \tag{12.11}$$

EXAMPLE 12.2

Name the hydrogen atom orbitals for $n = 3$, and calculate the degeneracy.

ANSWER

The degeneracy is

$$g_n = n^2 = 3^2 = 9$$

For $n = 3$, we can have $\ell = 0$, 1, and 2. For each value of ℓ we can have

$$m_\ell = 0, \pm 1, \ldots, \pm\ell.$$

So the nine orbitals are

$$3s, 3p_0, 3p_{\pm 1}, 3d_0, 3d_{\pm 1}, 3d_{\pm 2}$$

COMMENT

All nine of these atomic orbitals have the same energy in an isolated hydrogen atom. The answer above lists the complex orbitals, which are one set of solutions to the Schrödinger equation. Alternatively, one might also list the nine real orbitals as

$$3s, 3p_x, 3p_y, 3p_z, 3d_{xy}, 3d_{xz}, 3d_{yz}, 3d_{z^2}, 3d_{x^2-y^2}$$

As we shall see, when a one-electron atom or ion is placed in an external electric or magnetic field, or when a second electron is added, the degeneracy of the subshells is lifted. For example, in many-electron atoms, $2s$ orbitals have lower energies than $2p$ orbitals.

As with the harmonic oscillator and the rigid rotor, a series of selection rules for transitions between hydrogen atom energy levels may be derived using the time-dependent Schrödinger equation. The selection rules are

$$\Delta\ell = \pm 1 \tag{12.12}$$

$$\Delta m_\ell = 0, \pm 1 \tag{12.13}$$

and there are no restrictions on the quantum number n. Thus, in the emission process, a $2s \longrightarrow 1s$ transition is not allowed because there is no change in the quantum number ℓ, but a $2p \longrightarrow 1s$ transition can take place because $\Delta\ell = -1$. The selection rule $\Delta\ell = \pm 1$ may once again be understood by applying the conservation of angular momentum. A photon carries a single unit of angular momentum, so the hydrogen atom must gain a unit of angular momentum when a photon is absorbed ($\Delta\ell = 1$) and lose a unit of angular momentum when a photon is emitted ($\Delta\ell = -1$).

12.5 Spin Angular Momentum

The solution of the Schrödinger equation for the hydrogen atom provides us with three quantum numbers for an electron. To completely describe an electron, however, we need a fourth quantum number. It is convenient to view an electron as a particle and to consider that each electron spins about its own axis in either a clockwise or counterclockwise direction (Figure 12.7). The spinning motion of a charged particle generates a magnetic field, so each electron behaves like a small magnet. In quantum mechanics, we say that the electron has a spin, S, of $\frac{1}{2}$ and spin quantum number, $m_s = \pm\frac{1}{2}$. The value of m_s gives the orientation of the magnetic dipole moment of the electron, which is a vector quantity showing the positions of the north and south poles of the magnet. Thus, m_s is analogous to m_ℓ, which determines the orientation of the orbitals in space. When electrons of spin $m_s = \pm\frac{1}{2}$ are drawn in energy level diagrams, they are usually represented by upward and downward pointing arrows.

Electrons spinning on their own axes is a convenient model, but it is not physically accurate. Spin angular momentum is a purely quantum mechanical phenomenon.

A complete description of the wave function for an electron must therefore include the spin portion as well as the spatial portion. To a good approximation, the spin and spatial portions of the wave function may be separated. We shall not consider the spin angular momentum in detail, other than to note that each spatial orbital may contain two electrons; one spin up and one spin down.

Experimental evidence for spin angular momentum was provided by the German physicists Otto Stern (1888–1969) and Walther Gerlach (1889–1979). In their important

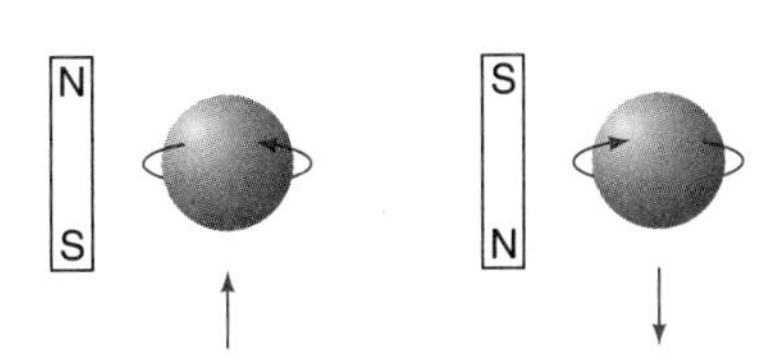

Figure 12.7
An analogy for electron spin. The up and down arrows are the symbols commonly employed to denote electron spin direction. The magnetic fields generated by the spinning motion are equivalent to those of two bar magnets.

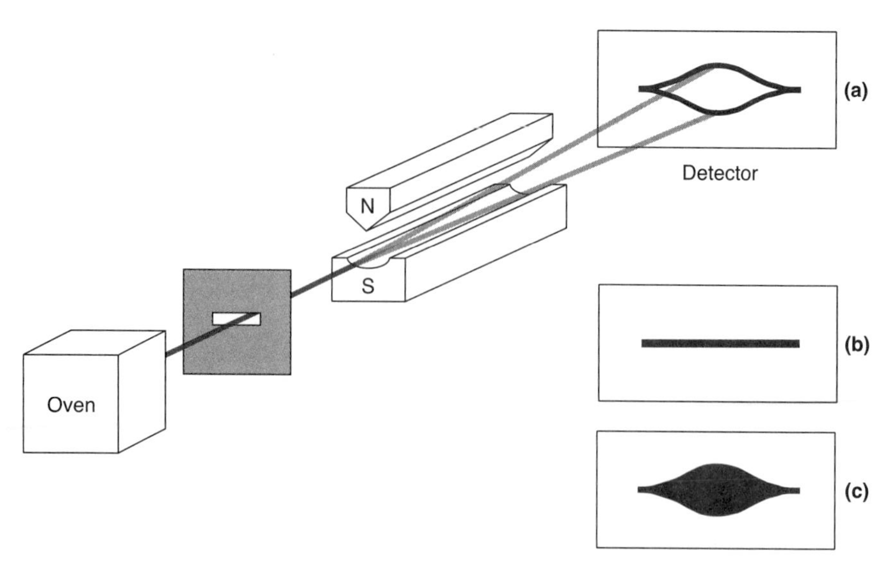

Figure 12.8
A schematic diagram of the Stern–Gerlach experiment. The entire apparatus is contained in a vacuum chamber. A beam of silver atoms is generated in an oven, and is deposited on a glass slide after traveling through an inhomogeneous magnetic field. (a) A representation of the image produced on a glass plate in the actual experiment. (b) How the image appears in the absence of a magnetic field. (c) How the image might appear if the angular momentum of the silver atoms was not quantized.

A scalar dot product is a real number (scalar) that is the product of two vectors.

experiment, Stern and Gerlach generated a beam of silver atoms (containing an unpaired electron) by heating silver to 1000°C in an oven and allowing the atoms to pass through a narrow slit (Figure 12.8). The atomic beam then traveled through an inhomogeneous magnetic field and deposited on a glass plate. The potential energy V due to the interaction between the magnetic dipole μ_z and the magnetic field B_z is given by the dot product,

$$V = -\mu_z \cdot B_z \tag{12.14}$$

where z is the direction of the magnetic field and μ_z is the z component of the magnetic moment associated with the unpaired electron. The silver atoms experience a displacement force F_z given by

$$\frac{\partial V}{\partial z} = F_z$$

$$F_z = -\mu_z \frac{\partial B_z}{\partial z} \tag{12.15}$$

The negative sign indicates that the force is attractive. If a uniform magnetic field had been used (i.e., $\partial B_z/\partial z = 0$), then the silver atoms would experience zero force, and they would not be displaced. Classical theories would predict that the deposited silver atoms would have their magnetic moments oriented at all possible angles with respect to the z axis and thus be deposited symmetrically and continuously about the beam axis. What Stern and Gerlach observed was a splitting of the atomic beam—two distinct lines rather than a continuous distribution (see Figure 12.8). They observed a quantized result. At the time of the experiment, the interpretation was that the magnetic moment of the silver atom was due to the orbital angular momentum (ℓ). Later theoretical interpretations showed that the orbital angular momentum of the silver atom is zero (because the unpaired electron is in an s orbital, so $l = 0$), and that the magnetic moment is due to quantized spin angular momentum ($S = \frac{1}{2}$) of the unpaired electron in a silver atom. The two lines observed correspond to the magnetic moment of the electron being oriented with or opposed to ($m_s = \pm\frac{1}{2}$) the external field.

12.6 The Helium Atom

Having examined the analytical solutions to the Schrödinger equation for the hydrogen atom, we now turn our attention to more complicated atoms, such as helium. The helium atom is an example of the so-called *three-body problem*; it has no exact analytic solution. Indeed, no atomic system with more than one electron has an exact analytic solution. Fortunately, quantum mechanics provides several approaches by which we can arrive at approximations that are in excellent agreement with experiment. To outline these approaches, we shall use helium as an example for all many-electron atoms.

Helium has a nucleus with charge $Z = 2$ and two electrons located at distances r_1 and r_2 from the nucleus (Figure 12.9). The vector r_{12} gives the distance between the two electrons. We can set up the Schrödinger equation to see where the difficulty lies. We start by writing the kinetic and potential energy terms. Just as we did with the hydrogen atom, we neglect the motion of the nucleus and consider the problem in the frame of reference of the fixed nucleus.

$$\hat{H}\psi = E\psi$$

$$(\hat{K} + \hat{V})\psi = E\psi$$

$$\underbrace{-\frac{h^2}{2m_e}\nabla_1^2\psi - \frac{h^2}{2m_e}\nabla_2^2\psi}_{\text{Electron kinetic energy terms}} \underbrace{- \frac{Ze^2\psi}{4\pi\varepsilon_0 r_1} - \frac{Ze^2\psi}{4\pi\varepsilon_0 r_2} + \frac{e^2\psi}{4\pi\varepsilon_0 r_{12}}}_{\text{Coulombic potential energy terms}} = E\psi \qquad (12.16)$$

where the subscripts 1 and 2 refer to the two electrons. The first two terms in Equation 12.16 are the kinetic energy terms for the two electrons; the third and fourth terms

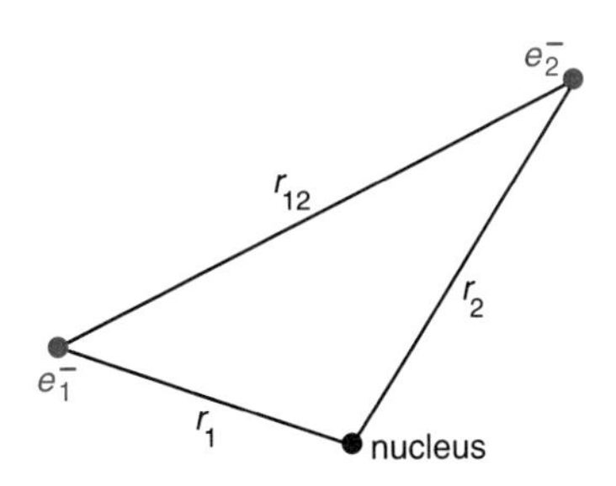

Figure 12.9
The coordinate system used for the helium atom. The two electrons are arbitrarily labeled numbers 1 and 2.

Table 12.5
Atomic Units and their SI Equivalents

Property	Atomic unit	SI equivalent
Mass	m_e, mass of an electron	9.1094×10^{-34} kg
Charge	e, proton charge	1.6022×10^{-19} C
Distance	a_0, Bohr radius	5.2918×10^{-11} m
Energy	$E_h = \frac{m_e e^4}{4\varepsilon_0^2 h^2}$, hartree	4.3597×10^{-18} J
Angular momentum	$\hbar$, Planck constant$/2\pi$	1.0546×10^{-34} J s
Permittivity	$4\pi\varepsilon_0$	1.1127×10^{-10} C^2 J^{-1} m^{-1}

Note that using atomic units means you can no longer use units and dimensional analysis to check your calculations.

The unit *hartree* is named after the British physicist Douglas Rayner Hartree (1897–1958).

are potential energy terms for the electron–nuclear attraction; and the fifth term is the potential energy due to electron–electron repulsion.

At this point, it is convenient to introduce a new system of notation. We shall write everything in atomic units (Table 12.5) such that Equation 12.16 becomes greatly simplified. In atomic units, quantities such as the mass and charge of the electron as well as $\hbar$ and $4\pi\varepsilon_0$ are all equal to one (i.e., one atomic unit). The atomic unit for energy, called a *hartree*, is twice the amount of energy required to separate a proton and an electron that are one Bohr radius (one atomic unit of distance, 0.529 Å) from each other. Thus, one hartree (E_h) is equal to twice the negative ground-state energy of a H atom (E_H); that is, $E_h = -2E_H$ or $E_h = -2 \times 1312.75$ kJ mol^{-1} = 2625.5 kJ mol^{-1}.* The Schrödinger equation now becomes

$$-\frac{1}{2}\nabla_1^2\psi - \frac{1}{2}\nabla_2^2\psi - \frac{Z\psi}{r_1} - \frac{Z\psi}{r_2} + \frac{\psi}{r_{12}} = E\psi \tag{12.17}$$

Now we recognize that the first and third terms are just that of a one-electron hydrogenlike system, with nuclear charge $Z = 2$ in the case of helium. Likewise, the second and fourth terms of Equation 12.17 also describe a hydrogenlike atom. Thus, we write the helium atom Schrödinger equation as

$$\hat{H}_1\psi + \hat{H}_2\psi + \frac{\psi}{r_{12}} = E\psi \tag{12.18}$$

where the Hamiltonian operators, $\hat{H}_1$ and $\hat{H}_2$, are those for hydrogenlike atoms. We know how to solve these one-electron Hamiltonians ($\hat{H}_1$ and $\hat{H}_2$), but there is no way to

* The E_H value used here is based on the Rydberg constant using the reduced mass. A slightly different value for the Rydberg constant is obtained if we consider the proton to be fixed at the center of revolution (see Problem 12.38).

avoid the electron–electron repulsion term, $1/r_{12}$. Consequently, we are unable to separate the variables in many-electron Schrödinger equations as we did for the hydrogen atom. Therefore, we must resort to approximations.

A useful approach is the *orbital approximation*. In the orbital approximation, we write the total wave function ψ of a system as the product of one-electron wave functions, ϕ. So, for the helium atom with two electrons, we write

$$\psi(\tau_1, \tau_2) = \phi_1(\tau_1)\phi_2(\tau_2) \tag{12.19}$$

where τ represents the three-dimensional spatial coordinates of an electron. Using spherical polar coordinates r, θ, and ϕ, Equation 12.19 becomes

$$\psi(r_1,\theta_1,\phi_1,r_2,\theta_2,\phi_2) = \phi_1(r_1,\theta_1,\phi_1)\phi_2(r_2,\theta_2,\phi_2) \tag{12.20}$$

Note that in Equation 12.20 we use the same symbol (ϕ) to represent both polar coordinate and wave function.

where the one-electron wave functions ϕ_1 and ϕ_2 are hydrogenlike wave functions.

The separation of the wave functions enables us to apply the *independent-electron approximation*, which is a crude but useful approach for comparison purposes. Here, we neglect electron–electron repulsion completely by omitting the $1/r_{12}$ term. For the helium atom, each electron behaves independently like a hydrogen-atom electron with $Z = 2$. The helium orbitals look like hydrogen-atom orbitals only smaller in the coordinate r because of greater nuclear attraction. In the independent electron approximation, the energy of a helium atom is equal to twice the energy of a helium cation.

$$E_{\text{He}} = E_{\text{He}^+} + E_{\text{He}^+} = -4E_{\text{h}} \tag{12.21}$$

Note that because He^+ is a hydrogenlike ion, we can calculate its energy. The experimental value (from ionization energy measurements) for the helium atom energy is −2.9033 hartree, so our crude approximation of −4 hartree is "only" 38% off. This isn't bad for such a simplistic model, but a more accurate approximation is needed to explain experimental observations.

In addition to ignoring electron–electron repulsion, the approach described above is deficient in other respects. For example, we also need to consider electron spin and the Pauli exclusion principle, which help to identify which of the possible electron configurations are allowed and which are forbidden. This is discussed in the next section.

12.7 Pauli Exclusion Principle

Only two electrons with paired spins may occupy an orbital, that is, one spin up and one spin down. This is a familiar consequence of the Pauli exclusion principle [named after the Austrian physicist Wolfgang Pauli (1900–1958)]. When placing electrons in orbitals, each orbital may hold only two electrons. A slightly broader result of the Pauli exclusion principle is that no two electrons may have the same set of four quantum numbers (n, ℓ, m_ℓ, and m_s). Within an atom, each electron is identified by its own label—a unique set of four quantum numbers. But this principle can be stated even more generally. Let us return to the case of helium to illustrate a broader definition of the Pauli exclusion principle.

The ground electronic state of helium has two electrons, each in a 1*s* hydrogenlike orbital. If we arbitrarily call these electrons 1 and 2, we can write five possibilities for how these electrons may be arranged with electron spin up (called an α electron) and spin down (called a β electron).

1. $\alpha(1)\beta(2)$
2. $\alpha(1)\alpha(2)$
3. $\beta(1)\beta(2)$
4. $[\alpha(1)\beta(2) + \alpha(2)\beta(1)]$
5. $[\alpha(1)\beta(2) - \alpha(2)\beta(1)]$

Because electrons are indistinguishable, possibility 1 is not allowed. Recall that we assigned the numbers 1 and 2 arbitrarily; there is no way that we can really identify individual electrons. Possibilities 2 and 3 don't distinguish electrons, but they would give both electrons the exact same set of quantum numbers—a situation that violates the Pauli exclusive principle. The fourth and fifth possibilities are *superposition states* (see Section 10.7), and at first glance they both look like good candidates. They don't distinguish electrons, and they also don't give the same two quantum numbers to the electrons. Only the fifth possibility is observed in nature, however, because it agrees with the broader formulation of the Pauli exclusion principle, which states that an acceptable electronic wave function must be *antisymmetric* (i.e., changes signs) with respect to the exchange of a pair of electrons. Thus, if we were to exchange electrons 1 and 2, the fifth wave function becomes $[\alpha(2)\beta(1) - \alpha(1)\beta(2)]$, or $-[\alpha(1)\beta(2) - \alpha(2)\beta(1)]$, which has the opposite sign to the original wave function. On the other hand, if we exchange electrons 1 and 2 in the fourth wave function, we get the same wave function back. In the language of quantum mechanics, we say that wave function number 4 is *symmetric* with respect to exchange. A statement of the Pauli exclusion principle is sometimes taken as a postulate of quantum mechanics (see Section 10.7); all electronic wave functions must be antisymmetric with respect to the exchange of *any* two electrons.

Writing wave functions that are properly antisymmetric could be a tedious process when there are many electrons present. Fortunately, the American physicist/chemist John Clarke Slater (1900–1976) developed a simplified procedure for writing wave functions that ensures they obey the Pauli exclusion principle. The procedure uses the mathematical concept of a *determinant*, and the resulting wave functions are known as *Slater determinants.* A determinant of nth order is a square $n \times n$ array of numbers symbolically enclosed by vertical lines. For example, a 2×2 determinant is given by

$$\begin{vmatrix} a_1 & b_1 \\ a_2 & b_2 \end{vmatrix} = a_1 b_2 - b_1 a_2$$

and a 3×3 determinant takes the form

$$\begin{vmatrix} a_1 & b_1 & c_1 \\ a_2 & b_2 & c_2 \\ a_3 & b_3 & c_3 \end{vmatrix} = a_1\begin{vmatrix} b_2 & c_2 \\ b_3 & c_3 \end{vmatrix} - b_1\begin{vmatrix} a_2 & c_2 \\ a_3 & c_3 \end{vmatrix} + c_1\begin{vmatrix} a_2 & b_2 \\ a_3 & b_3 \end{vmatrix}$$

$$= a_1b_2c_3 - a_1b_3c_2 - b_1a_2c_3 + b_1a_3c_2 + c_1a_2b_3 - c_1a_3b_2$$

For an N-electron system, the Slater determinant incorporates different one-electron wave functions in each of N columns and different electron numbers in each of N rows. To ensure that the wave function is normalized, the determinant is multiplied by a normalization constant, $1/\sqrt{N!}$. The wave function including both the spin and spatial parts for the helium atom (with $N = 2$) is written using the Slater determinant

$$\psi_{\text{He}} = \frac{1}{\sqrt{2!}}\begin{vmatrix} 1s\alpha(1) & 1s\beta(1) \\ 1s\alpha(2) & 1s\beta(2) \end{vmatrix} \tag{12.22}$$

If we expand the Slater determinant (Equation 12.22) and evaluate the factorial 2!, we find

$$\psi_{\text{He}} = \frac{1}{\sqrt{2}}[1s\alpha(1)1s\beta(2) - 1s\alpha(2)1s\beta(1)] \tag{12.23}$$

This wave function has the same form as the fifth possibility shown above and it is normalized. Besides its compact form, the Slater determinant also incorporates the Pauli exclusion principle. If we interchange the two electrons, which is the same as swapping the two rows, then Equation 12.22 becomes

$$\psi'_{\text{He}} = \frac{1}{\sqrt{2!}}\begin{vmatrix} 1s\alpha(2) & 1s\beta(2) \\ 1s\alpha(1) & 1s\beta(1) \end{vmatrix}$$

On expansion, we get

$$\psi'_{\text{He}} = \frac{1}{\sqrt{2}}[1s\alpha(2)1s\beta(1) - 1s\alpha(1)1s\beta(2)]$$

$$= -\frac{1}{\sqrt{2}}[1s\alpha(1)1s\beta(2) - 1s\alpha(2)1s\beta(1)] = -\psi_{\text{He}} \tag{12.24}$$

We see that this wave function is equivalent to multiplying the wave function in Equation 12.23 by -1. Thus, the wave function is antisymmetric to the exchange of electrons.

EXAMPLE 12.3

Write a properly antisymmetrized wave function for a ground-state lithium atom, including both the spatial part and the spin part.

ANSWER

If we place each electron in the three lowest energy atomic orbitals,

$$1s\alpha(1)1s\beta(2)2s\alpha(3)$$

we can quickly recognize that it will be challenging to write a wave function that is antisymmetric with respect to exchange. Writing the Slater determinant provides a straightforward formulation of the appropriate wave function:

$$\psi_{\text{Li}} = \frac{1}{\sqrt{3!}}\begin{vmatrix} 1s\alpha(1) & 1s\beta(1) & 2s\alpha(1) \\ 1s\alpha(2) & 1s\beta(2) & 2s\alpha(2) \\ 1s\alpha(3) & 1s\beta(3) & 2s\alpha(3) \end{vmatrix}$$

If we multiply out the determinant, the expression is quite long, so we can see why in practice we don't bother to write properly antisymmetrized wave functions.

$$\psi_{\text{Li}} = \frac{1}{\sqrt{6}}[1s\alpha(1)1s\beta(2)2s\alpha(3) - 1s\alpha(1)1s\beta(3)2s\alpha(2) - 1s\alpha(2)1s\beta(1)2s\alpha(3) + 1s\alpha(3)1s\beta(1)2s\alpha(2) + 1s\alpha(2)1s\beta(3)2s\alpha(1) - 1s\alpha(3)1s\beta(2)2s\alpha(1)]$$

COMMENT

As mentioned above, swapping any two rows in a determinant is equivalent to multiplying the determinant by -1. Try it!

The Pauli exclusion principle applies to more than just electrons; it applies to any wave function of fermions. *Fermions* are particles with half-integer spin ($\frac{1}{2}, \frac{3}{2}, \frac{5}{2}, \ldots$). Like electrons, protons and neutrons are also fermions. Particles with integer spin (0, 1, 2, ...) are called *bosons*. Examples of bosons are photons, helium atoms, and hydrogen atoms. Fermions are "excluded" from occupying the same place, as noted by the Pauli exclusion principle, whereas bosons have no such limitations. As a result, many photons with the same wavelength and phase may occupy the same space at the same time. This situation is realized in laser light, among the brightest light sources known. In the 1920s, the Indian physicist Satyendra Nath Bose (1894–1974) and Einstein had predicted that matter made of bosons could, at very low temperatures, merge into a single entity and a new form of matter. It wasn't until 1995 that Colorado scientists Eric Cornell (1961–) and Carl Wieman (1951–) were able to realize this new state of matter, called a *Bose–Einstein condensate*, in a collection of thousands of ^{87}Rb atoms. For this work, they shared the 2001 Nobel Prize in Physics with Wolfgang Ketterle (1957–).

12.8 Aufbau Principle

For the hydrogen atom and hydrogenlike ions, the energy of an electron depends only on the principal quantum number n (see Equation 12.10). Therefore, the orbitals can be arranged in order of increasing energy (decreasing stability) as follows:

$$1s < 2s = 2p < 3s = 3p = 3d < 4s = 4p = 4d < 5s \cdots$$

For many-electron atoms, however, the electron's energy depends on both n and ℓ, so the order of increasing energy is given by

$$1s < 2s < 2p < 3s < 3p < 4s < 3d < 4p < 5s < 4d < \cdots$$

Figure 12.10 shows the order in which atomic subshells are filled in a many-electron atom. The difference between a hydrogen atom and a many-electron atom can be qualitatively explained as follows. Considering the $2s$ and $2p$ orbitals, we see from Figure 12.3 that even though the most probable location of a $2p$ electron is closer to the nucleus than that of a $2s$ electron, the electron density close to the nucleus is actually greater for a $2s$ electron. Put another way, we say that an s electron is more *penetrating* than a p electron. Thus, the $2p$ electron is more *shielded* from the nucleus by the $2s$ electron than the other way around. Consequently, the energy of the $2p$ electron is higher than that of a $2s$ electron because the $2s$ electron shields it somewhat from the full attractive force of the nucleus. In general, for the same value of n, the penetrating power decreases as follows:

$$s > p > d > f > \cdots \tag{12.25}$$

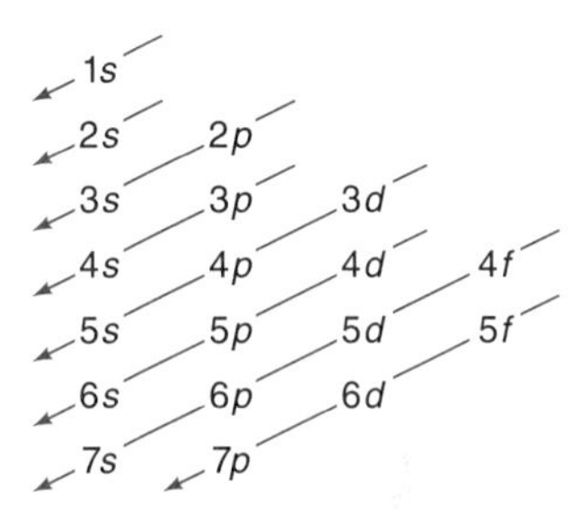

Figure 12.10
A mnemonic device for the order in which electrons fill orbitals in many-electron atoms.

The result of shielding means that each electron experiences a different *effective nuclear charge* ζ (Greek zeta), given by

$$\zeta = Z - \sigma \tag{12.26}$$

where Z is the nuclear charge of the atom and σ is called the shielding constant. For carbon ($Z = 6$), the effective nuclear charge for the $1s$ electrons is 5.7; for the $2s$ electrons it is 3.2; for the $2p$ electrons it is 3.1. In a hydrogen atom or a hydrogenlike ion there is only one electron; hence, no shielding occurs. The general patterns are that the effective charges increase from left to right across a period and from top to bottom in a group.

When the electron in the hydrogen atom is in the lowest possible energy level (the ground state), the electron configuration is $1s^1$, meaning that there is one electron in the $1s$ orbital. The four quantum numbers (n, ℓ, m_ℓ, and m_s) of the electron can be either $(1, 0, 0, +\frac{1}{2})$ or $(1, 0, 0, -\frac{1}{2})$. In the absence of a magnetic field, the energy of the electron is the same whether $m_s = +\frac{1}{2}$ or $-\frac{1}{2}$. Helium has two electrons, so its ground-state electron configuration is $1s^2$. Helium atoms are *diamagnetic*, which means that the net magnetic field generated is zero. (Diamagnetic substances are slightly repelled by a magnet.) The two electrons must have opposite spins; one electron must have $m_s = +\frac{1}{2}$ and the other $m_s = -\frac{1}{2}$. The other three quantum numbers are the same. According to the Pauli exclusion principle, the third electron in a lithium atom may not enter the $1s$ orbital, so we place it in the $2s$ orbital and the electron configuration of lithium is $1s^2 2s^1$.

Note that the electrons in the outermost shell of an atom are called the *valence electrons* because they are largely responsible for the chemical bonds that the atom forms. Thus, hydrogen has one valence electron in the 1*s* orbital and lithium has one valence electron in the 2*s* orbital.

Hund's Rules

As procedure of filling orbitals continues across the second row of the periodic table, the carbon atom ($1s^2 2s^2 2p^2$) makes possible different options for adding the two valence electrons to the three *p* orbitals. To determine which arrangement of the electrons yields the lowest energy, we refer to Hund's rules [after the German physicist Frederick Hund (1896–1997)]. The rules are:

1. The arrangement that gives rise to the maximum value of spin angular momentum S is the most stable. Here S is given by

$$S = \sum_i m_{s,i}$$

   where $m_{s,i}$ is the spin quantum number of the ith electron. For maximum S, the electrons must have the same m_s value ($+\frac{1}{2}$ or $-\frac{1}{2}$). For closed shells, S is zero because all the electrons are paired.

2. For a given value of S, the level with the maximum value of orbital angular momentum L is the most stable. Here L is given by

$$L = \sum_i m_{\ell,i}$$

   where $m_{\ell,i}$ is the magnetic quantum number of the ith electron.

3. For levels with the same S and L values, the one with the lowest energy depends on the extent to which the subshell is filled.
   a. If the subshell is less than half-filled, the state with the smallest value of J, which is the total angular momentum quantum number (given by $L + S$), is the most stable.
   b. If the subshell is more than half-filled, the state with the largest value of J is the most stable.

To understand Rule 1, maximum S value for two electrons implies parallel spin ($S = 1$). Consequently, the two electrons must occupy separate orbitals according to the Pauli exclusion principle. This arrangement minimizes electrostatic repulsion. Rule 2 also has an electrostatic origin, while Rule 3 is a consequence of the spin-orbit (magnetic) interaction. For a ground-state carbon atom, the three possibilities are

↑	↑	
$2p_1$	$2p_0$	$2p_{-1}$

(a)

↑		↑
$2p_1$	$2p_0$	$2p_{-1}$

(b)

	↑	↑
$2p_1$	$2p_0$	$2p_{-1}$

(c)

Each square in the so-called *orbital diagram* represents an orbital. According to Rule 1, we have $S = 1$ and the spin multiplicity is $(2S + 1)$ or 3, which is a triplet state. For the $2p$ subshell, $\ell = 1$ and the m_ℓ values are 1, 0, and −1. For the carbon atom, the maximum value of L is the maximum sum of the m_ℓ values of the two electrons, which is (1 + 0) or 1. Applying Rule 2, we see that the most stable ground state is (a). As expected, experimentally we find that the ground-state carbon atom is *paramagnetic*, containing two unpaired electrons. (A paramagnetic substance contains one or more unpaired spins and is attracted by a magnet.) Note that if we had represented the $2p$ orbitals as $2p_x$, $2p_y$, and $2p_z$,

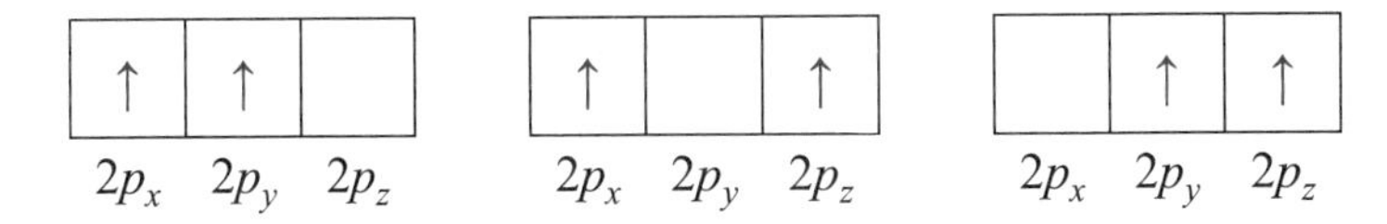

then all three arrangements are equivalent because the designations x, y, and z are arbitrary. In this representation, only Rule 1 applies.

The German word "Aufbau" means "building up."

The stepwise procedure of writing the electron configuration of elements is based on the *Aufbau principle*, which states that as protons are added one by one to the nucleus to build up the elements, electrons are similarly added to the atomic orbitals. Table 12.6 lists the ground-state electron configurations of elements from H ($Z = 1$) through Cn ($Z = 112$). The electron configurations of all elements except hydrogen and helium are represented by a *noble gas core*, which shows in brackets the noble gas element that most nearly precedes the element, followed by the symbol for the highest-filled subshells in the outermost shells. Notice that the electron configurations of the highest-filled subshells in the outermost shells for the elements sodium ($Z = 11$) through argon ($Z = 18$) follow a pattern similar to those of lithium ($Z = 3$) through neon ($Z = 10$). Most electron configurations may be read directly from the periodic table, once you recognize the periodic trends.

The $4s$ subshell is filled before the $3d$ subshell in a many-electron atom. Thus, the electron configuration of potassium ($Z = 19$) is $1s^2 2s^2 2p^6 3s^2 3p^6 4s^1$. Because $1s^2 2s^2 2p^6 3s^2 3p^6$ is the electron configuration of argon, we can simplify the electron configuration of potassium by writing [Ar]$4s^1$, where [Ar] denotes the *argon core*. Similarly, we can write the electron configuration of calcium ($Z = 20$) as [Ar]$4s^2$. The placement of the outermost electron in the $4s$ orbital (rather than in the $3d$ orbital) of potassium is strongly supported by experimental evidence. The following comparison also suggests that this is the correct configuration. The chemistry of potassium is very similar to that of lithium and sodium, the first two alkali metals. The outermost electron of both lithium and sodium is in an s orbital (there is no ambiguity in assigning their electron configurations); therefore, we expect the valence electron in potassium to occupy the $4s$ rather than the $3d$ orbital.

The elements from scandium ($Z = 21$) to copper ($Z = 29$) are transition metals. *Transition metals* either have incompletely filled d subshells or readily give rise to cations that have incompletely filled d subshells. Consider the first transition metal series, from scandium through copper. In this series, additional electrons are placed in the $3d$ orbitals, according to Hund's rule. However, there are two irregularities. The electron configuration of chromium ($Z = 24$) is [Ar]$4s^1 3d^5$ and not [Ar]$4s^2 3d^4$, as we might expect. A similar break in the pattern is observed for copper, whose electron configuration is [Ar]$4s^1 3d^{10}$ rather than [Ar]$4s^2 3d^9$. The full reason for these irregularities requires a quantum-mechanical description with no classical analogy. However, we may predict the ground-state electron configuration of most elements

Table 12.6
The Ground-State Electron Configurations of the Elements[a]

Atomic number	Symbol	Electron configuration	Atomic number	Symbol	Electron configuration	Atomic number	Symbol	Electron configuration
1	H	$1s^1$	38	Sr	[Kr]$5s^2$	75	Re	[Xe]$6s^24f^{14}5d^5$
2	He	$1s^2$	39	Y	[Kr]$5s^24d^1$	76	Os	[Xe]$6s^24f^{14}5d^6$
3	Li	[He]$2s^1$	40	Zr	[Kr]$5s^24d^2$	77	Ir	[Xe]$6s^24f^{14}5d^7$
4	Be	[He]$2s^2$	41	Nb	[Kr]$5s^14d^4$	78	Pt	[Xe]$6s^14f^{14}5d^9$
5	B	[He]$2s^22p^1$	42	Mo	[Kr]$5s^14d^5$	79	Au	[Xe]$6s^14f^{14}5d^{10}$
6	C	[He]$2s^22p^2$	43	Tc	[Kr]$5s^24d^5$	80	Hg	[Xe]$6s^24f^{14}5d^{10}$
7	N	[He]$2s^22p^3$	44	Ru	[Kr]$5s^14d^7$	81	Tl	[Xe]$6s^24f^{14}5d^{10}6p^1$
8	O	[He]$2s^22p^4$	45	Rh	[Kr]$5s^14d^8$	82	Pb	[Xe]$6s^24f^{14}5d^{10}6p^2$
9	F	[He]$2s^22p^5$	46	Pd	[Kr]$4d^{10}$	83	Bi	[Xe]$6s^24f^{14}5d^{10}6p^3$
10	Ne	[He]$2s^22p^6$	47	Ag	[Kr]$5s^14d^{10}$	84	Po	[Xe]$6s^24f^{14}5d^{10}6p^4$
11	Na	[Ne]$3s^1$	48	Cd	[Kr]$5s^24d^{10}$	85	At	[Xe]$6s^24f^{14}5d^{10}6p^5$
12	Mg	[Ne]$3s^2$	49	In	[Kr]$5s^24d^{10}5p^1$	86	Rn	[Xe]$6s^24f^{14}5d^{10}6p^6$
13	Al	[Ne]$3s^23p^1$	50	Sn	[Kr]$5s^24d^{10}5p^2$	87	Fr	[Rn]$7s^1$
14	Si	[Ne]$3s^23p^2$	51	Sb	[Kr]$5s^24d^{10}5p^3$	88	Ra	[Rn]$7s^2$
15	P	[Ne]$3s^23p^3$	52	Te	[Kr]$5s^24d^{10}5p^4$	89	Ac	[Rn]$7s^26d^1$
16	S	[Ne]$3s^23p^4$	53	I	[Kr]$5s^24d^{10}5p^5$	90	Th	[Rn]$7s^26d^2$
17	Cl	[Ne]$3s^23p^5$	54	Xe	[Kr]$5s^24d^{10}5p^6$	91	Pa	[Rn]$7s^25f^26d^1$
18	Ar	[Ne]$3s^23p^6$	55	Cs	[Xe]$6s^1$	92	U	[Rn]$7s^25f^36d^1$
19	K	[Ar]$4s^1$	56	Ba	[Xe]$6s^2$	93	Np	[Rn]$7s^25f^46d^1$
20	Ca	[Ar]$4s^2$	57	La	[Xe]$6s^25d^1$	94	Pu	[Rn]$7s^25f^6$
21	Sc	[Ar]$4s^23d^1$	58	Ce	[Xe]$6s^24f^15d^1$	95	Am	[Rn]$7s^25f^7$
22	Ti	[Ar]$4s^23d^2$	59	Pr	[Xe]$6s^24f^3$	96	Cm	[Rn]$7s^25f^76d^1$
23	V	[Ar]$4s^23d^3$	60	Nd	[Xe]$6s^24f^4$	97	Bk	[Rn]$7s^25f^9$
24	Cr	[Ar]$4s^13d^5$	61	Pm	[Xe]$6s^24f^5$	98	Cf	[Rn]$7s^25f^{10}$
25	Mn	[Ar]$4s^23d^5$	62	Sm	[Xe]$6s^24f^6$	99	Es	[Rn]$7s^25f^{11}$
26	Fe	[Ar]$4s^23d^6$	63	Eu	[Xe]$6s^24f^7$	100	Fm	[Rn]$7s^25f^{12}$
27	Co	[Ar]$4s^23d^7$	64	Gd	[Xe]$6s^24f^75d^1$	101	Md	[Rn]$7s^25f^{13}$
28	Ni	[Ar]$4s^23d^8$	65	Tb	[Xe]$6s^24f^9$	102	No	[Rn]$7s^25f^{14}$
29	Cu	[Ar]$4s^13d^{10}$	66	Dy	[Xe]$6s^24f^{10}$	103	Lr	[Rn]$7s^25f^{14}6d^1$
30	Zn	[Ar]$4s^23d^{10}$	67	Ho	[Xe]$6s^24f^{11}$	104	Rf	[Rn]$7s^25f^{14}6d^2$
31	Ga	[Ar]$4s^23d^{10}4p^1$	68	Er	[Xe]$6s^24f^{12}$	105	Db	[Rn]$7s^25f^{14}6d^3$
32	Ge	[Ar]$4s^23d^{10}4p^2$	69	Tm	[Xe]$6s^24f^{13}$	106	Sg	[Rn]$7s^25f^{14}6d^4$
33	As	[Ar]$4s^23d^{10}4p^3$	70	Yb	[Xe]$6s^24f^{14}$	107	Bh	[Rn]$7s^25f^{14}6d^5$
34	Se	[Ar]$4s^23d^{10}4p^4$	71	Lu	[Xe]$6s^24f^{14}5d^1$	108	Hs	[Rn]$7s^25f^{14}6d^6$
35	Br	[Ar]$4s^23d^{10}4p^5$	72	Hf	[Xe]$6s^24f^{14}5d^2$	109	Mt	[Rn]$7s^25f^{14}6d^7$
36	Kr	[Ar]$4s^23d^{10}4p^6$	73	Ta	[Xe]$6s^24f^{14}5d^3$	110	Ds	[Rn]$7s^25f^{14}6d^8$
37	Rb	[Kr]$5s^1$	74	W	[Xe]$6s^24f^{14}5d^4$	111	Rg	[Rn]$7s^25f^{14}6d^9$
						112	Cn	[Rn]$7s^25f^{14}6d^{10}$

[a] The symbol [He] is called the *helium core* and represents $1s^2$. [Ne] is called the *neon core* and represents [He]$2s^22p^6$. [Ar] the *argon core* and represents [Ne]$3s^23p^6$. [Kr] is called the *krypton core* and represents [Ar]$4s^23d^{10}4p^6$. [Xe] is called the *xenon core* and represents [Kr]$5s^24d^{10}5p^6$. [Rn] is called the *radon core* and represents [Xe]$6s^24f^{14}5d^{10}6p^6$.

by noting there to be a slightly greater stability associated with the exactly half-filled ($3d^5$) and completely filled ($3d^{10}$) subshells. According to Hund's Rule 1, the orbital diagram for Cr is

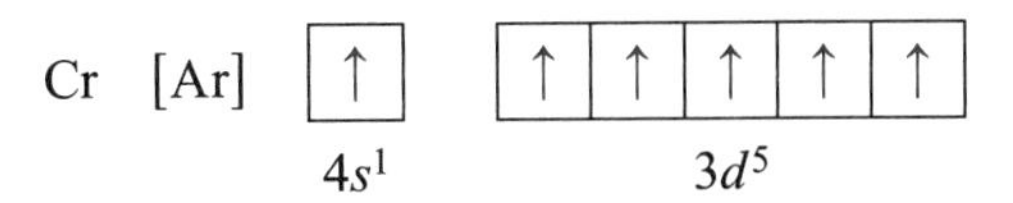

Having the d electrons in separate orbitals reduces the electrostatic repulsion. Thus, Cr has a total of six unpaired electrons. The orbital diagram for copper is

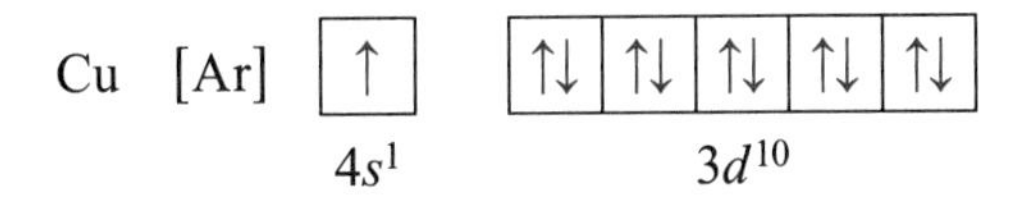

The qualitative explanation for this configuration is as follows. Electrons in the same subshell have equal energy but different spatial distributions; consequently, their shielding of one another is relatively slight. Therefore, the effective nuclear charge increases as the actual nuclear charge increases, so that a completely filled subshell (d^{10}) has a high stability and a spherical distribution of charge.

For elements zinc ($Z = 30$) through krypton ($Z = 36$), the $4s$ and $4p$ subshells fill in a straightforward manner. With rubidium ($Z = 37$), electrons begin to enter the $n = 5$ energy level. The electron configurations in the second transition metal series [yttrium ($Z = 39$) to silver ($Z = 47$)] are also irregular, but we will not discuss the details here.

The sixth period of the periodic table begins with cesium ($Z = 55$) and barium ($Z = 56$), whose electron configurations are $[Xe]6s^1$ and $[Xe]6s^2$, respectively. Next, we come to lanthanum ($Z = 57$). The energies of the $5d$ and $4f$ orbitals are very close; in fact, for lanthanum, $4f$ is slightly higher in energy than $5d$. Thus, lanthanum's electron configuration is $[Xe]6s^2 5d^1$ and not $[Xe]6s^2 4f^1$. Following lanthanum are the 14 elements known as the *lanthanides*, or *rare earth series* [cerium ($Z = 58$) to lutetium ($Z = 71$)]. The rare earth metals have incompletely filled $4f$ subshells or readily give rise to cations that have incompletely filled $4f$ subshells. In this series, the added electrons are placed in the $5d$ subshell of lutetium. Note that the electron configuration of gadolinium ($Z = 64$) is $[Xe]6s^2 4f^7 5d^1$ rather than $[Xe]6s^2 4f^8$. Like chromium, gadolinium gains extra stability by having a half-filled subshell ($4f^7$). The third transition metal series, including lanthanum and hafnium ($Z = 72$) and extending through gold ($Z = 79$), is characterized by the filling of the $5d$ orbitals. The $6s$ and $6p$ subshells are filled next, bringing us to radon ($Z = 86$). The next row of elements includes the *actinide series*, which starts at thorium ($Z = 90$) and has incompletely filled $5f$ subshells. Most of these elements are not found in nature but have been synthesized.

Finally, let us look at the procedure for writing electron configurations for ions. For cations, we first remove p valence electrons, then s valence electrons, and then as many d electrons as necessary to achieve the required charge. For example, the electron configuration of Mn is $[Ar]3d^5 4s^2$ so that the electron configuration of Mn^{2+} is $[Ar]3d^5$. The electron configurations of anions are derived by adding electrons to the atoms until the next noble gas core has been reached. Thus, for the oxide ion (O^{2-}) we add two electrons to $[He]2s^2 2p^4$, reaching $[He]2s^2 2p^6$, which is the same as the electron configuration for neon.

Periodic Variations in Atomic Properties

For most periodic trends, F ($Z = 9$) and Fr ($Z = 87$) lie at the extremes.

The periodic trends in electron configuration result in periodic trends in chemical and physical properties. Here we consider atomic radius, ionization energy, and electron affinity. The general periodic trends are such that as we move across a period from left to right, the metallic properties decrease; down a particular group, the metallic character increases. These trends do not apply to transition elements, which are all metallic and possess similar properties.

Atomic Radius. An atom does not have a definite size. Mathematically, the wave function of an atom extends to infinity. Therefore, we need to define it in a somewhat arbitrary manner. One way is to use the *covalent radius*, obtained from measurements of distances between the nuclei of atoms in molecules, as a measure of atomic size. Figure 12.11 shows a plot of atomic radii versus atomic number. Consider the second-period elements. From Li to Ne, atomic number increases and electrons are added to the 2*s* and 2*p* orbitals. Because electrons in the same subshells do not shield each other well, the effective nuclear charge ζ increases, concentrating the electron density, in which case the size of the atom decreases. Within a periodic group, the atomic radius increases with increasing atomic number. In the Group 1 alkali metals, for example, the outermost electron resides in the *ns* orbital. Because orbital size increases as the principal quantum number *n* increases, the outermost electron is farther away from the nucleus. As a result, the size of the atoms increases from Li to Cs, even though the effective nuclear charge also increases from Li to Cs.

Ionization Energy. *Ionization energy* is the minimum energy required to remove an electron from a gaseous atom in its ground state,

$$\text{energy} + \text{X}(g) \rightarrow \text{X}^+(g) + e^- \tag{12.27}$$

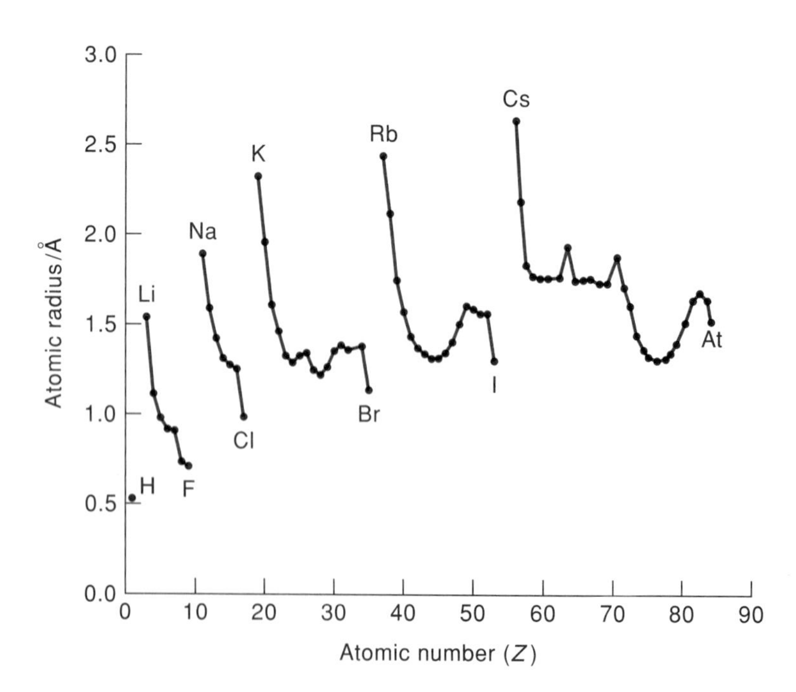

Figure 12.11
Plot of atomic radii versus atomic number (Z). Generally, atomic radii decrease from left to right in a row and increase down a column of the periodic table. (1 Å = 0.1 nm.)

Table 12.7
The Ionization Energies ($kJ\ mol^{-1}$) of the First 20 Elements

Z	Element	First	Second	Third	Fourth	Fifth	Sixth
1	H	1312					
2	He	2373	5251				
3	Li	520	7300	11815			
4	Be	899	1757	14850	21005		
5	B	801	2430	3660	25000	32820	
6	C	1086	2350	4620	6220	38000	47300
7	N	1400	2860	4580	7500	9400	53000
8	O	1314	3390	5300	7470	11000	13000
9	F	1680	3370	6050	8400	11000	15200
10	Ne	2080	3950	6120	9370	12200	15000
11	Na	495.9	4560	6900	9540	13400	16600
12	Mg	738.1	1450	7730	10500	13600	18000
13	Al	577.9	1820	2750	11600	14800	18400
14	Si	786.3	1580	3230	4360	16000	20000
15	P	1012	1904	2910	4960	6240	21000
16	S	999.5	2250	3360	4660	6990	8500
17	Cl	1251	2297	3820	5160	6540	9300
18	Ar	1521	2666	3900	5770	7240	8800
19	K	418.7	3052	4410	5900	8000	9600
20	Ca	589.5	1145	4900	6500	8100	11000

where X represents an atom of any element. Ionization is always an endothermic process. This measurement gives the first ionization energy. The process can be continued to give the second, third, etc., ionization energies as follows:

$$\text{energy} + X^{+}(g) \rightarrow X^{2+}(g) + e^{-}$$

$$\text{energy} + X^{2+}(g) \rightarrow X^{3+}(g) + e^{-}$$

For a given element, the third ionization energy is always greater than second ionization energy, which is greater, in turn, than first ionization energy—the greater the positive charge, the greater the amount of energy required to remove an electron. Table 12.7 lists ionization energies for the first 20 elements and Figure 12.12 plots the first ionization energy versus atomic number. The interpretation of the plot is similar to that for atomic radii in Figure 12.11. The effective nuclear charge increases across a period from left to right, so the ionization energy also increases because the outermost electron is held more tightly. Down a group, the outermost electron is placed

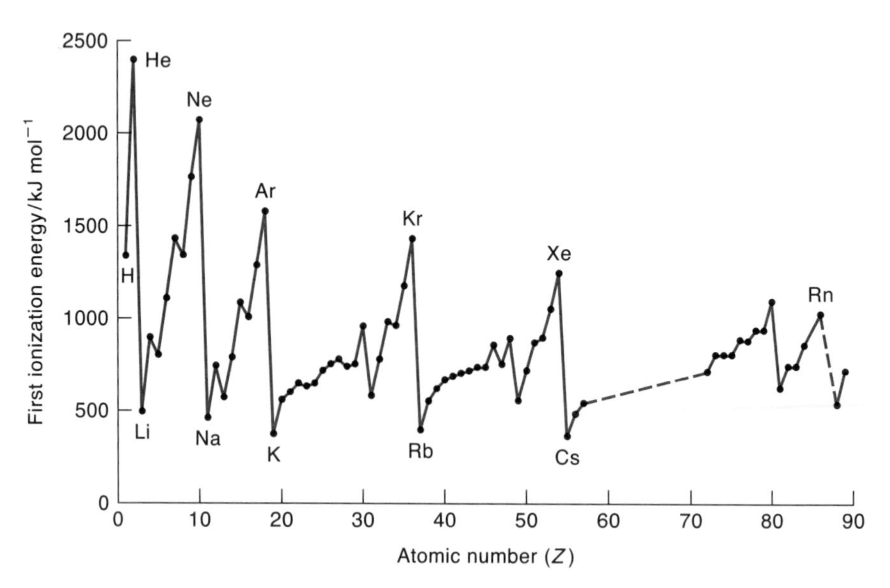

Figure 12.12
Plot of first ionization energy versus atomic number (Z).

in successive outer shells, where it is increasingly farther away from the nucleus. Consequently, it can be removed more easily than the outermost electron in the element above it, even though the effective nuclear charge increases from top to bottom.

Another way of defining the electron affinity of X is to equate it to the first ionization energy of the anion X^-.

Electron Affinity. *Electron affinity* is defined as the negative of the energy change that occurs when an electron is accepted by an atom in the gaseous state to form an anion:

$$X(g) + e^- \rightarrow X^-(g) \tag{12.28}$$

In contrast to ionization energies, electron affinities are difficult to measure because isolated anions bearing more that one negative charge are often unstable. Table 12.8 lists the electron affinities of a number of elements. Electron affinities may take on positive or nearly zero values. The more positive the electron affinity, the greater the tendency of the atom to accept an electron.

12.9 Variational Principle

We noted earlier there is no exact analytical solution to the Schrödinger equation for atoms or ions with more than one electron. This difficulty can be dealt with by applying a very powerful technique known as the *variational method*, which is based on the variational principle. Suppose we wish to calculate the ground-state energy of a system. We can write the Hamiltonian operator for the system without knowing the true wave function. We can make intelligent guesses and use some trial wave function for our calculation. According to the *variational principle*, the energy we calculate using any trial wave function cannot be below the true (experimentally determined)

Table 12.8
Electron Affinities (kJ mol^{-1}) of Some Representative Elements and the Noble Gases[a]

1A	2A	3A	4A	5A	6A	7A	8A
H							He
73							<0
Li	Be	B	C	N	O	F	Ne
60	≤0	27	122	≤0	141	328	<0
Na	Mg	Al	Si	P	S	Cl	Ar
53	≤0	44	134	72	200	349	<0
K	Ca	Ga	Ge	As	Se	Br	Kr
48	2.4	29	118	77	195	325	<0
Rb	Sr	In	Sn	Sb	Te	I	Xe
47	4.7	29	121	101	190	295	<0
Cs	Ba	Tl	Pb	Bi	Po	At	Rn
45	14	30	110	110	183	270	<0

[a] The electron affinities of the noble gases, Be, N, and Mg have not been determined experimentally but are believed to be close to zero or negative.

energy for the ground state of the system. Using the variational principle, we can put as many parameters (known as *variational parameters*) as we like in the trial wave function, but the energy we calculate (E_ϕ) will always be greater than, or in the most favorable case, equal to, the true ground-state energy (E_0). (See Appendix 12.1 for proof of Equation 12.29.)

$$E_\phi \geq E_0 \tag{12.29}$$

If the calculated energy is lower than the experimentally determined energy, then we are using the wrong Hamiltonian operator.

To illustrate the variational method, we return to the case of the helium atom. We make the orbital approximation and treat the two electrons independently; that is, the helium wave function is given as a product of two hydrogen wave functions,

$$\psi_{\mathrm{He}}(1,2) = \phi_{\mathrm{H}}(1)\phi_{\mathrm{H}}(2)$$

where each hydrogen wave function is given by

$$\phi_{\mathrm{H}} = N_{\mathrm{H}} e^{-Zr/a_0}$$

and N_{H} is the normalization constant. But rather than using hydrogen-atom wave functions, we modify the hydrogen 1*s* orbital to account for the shielding of the nucleus by the other electron. We replace the integer nuclear charge Z by the effective nuclear charge ζ (see Equation 12.26) and write

$$\phi_{\mathrm{He}} = N_{\mathrm{He}} e^{-\zeta r / a_0} \tag{12.30}$$

where N_{He} is the normalization constant. Thus, $\psi_{\mathrm{He}}(1,2) = \phi_{\mathrm{He}}(1)\phi_{\mathrm{He}}(2)$. We then treat ζ as a variational parameter and find the minimum energy. Because helium has $Z = 2$ and only a single electron shields the nucleus from the other electron, we anticipate

$$1 < \zeta < 2$$

will give the optimal results. The variational principle states that the best variational parameter is the one that gives the lowest energy. Here we calculate the energy in the usual way, using the postulates of quantum mechanics (see Section 10.7). The average or expectation value of the energy is given by

$$\begin{aligned} E_{\psi} &= \langle E \rangle \\ &= \frac{\int \psi_{\mathrm{He}}^{*} \hat{H} \psi_{\mathrm{He}} d\tau}{\int \psi_{\mathrm{He}}^{*} \psi_{\mathrm{He}} d\tau} \end{aligned} \tag{12.31}$$

Because this is a one-parameter problem, we can make a plot of energy versus our variational parameter (ζ) and find the minimum energy graphically (Figure 12.13). The mathematics in solving the integrals in Equation 12.31 are a bit lengthy, but straightforward. The result is

$$E_{\psi}(\zeta) = \underbrace{\frac{m_e e^4}{4\varepsilon_0^2 h^2}}_{\text{One hartree}} \left(\zeta^2 - \frac{27}{8}\zeta \right) \tag{12.32}$$

The term before the parentheses in the above equation happens to be equal to one atomic unit (or one hartree; see Table 12.5), so we can express the trial energy in atomic units as

$$E_{\psi}(\zeta) = \left(\zeta^2 - \frac{27}{8}\zeta \right) \tag{12.33}$$

In general, to find the best value of a variational parameter we take the derivative of the energy with respect to that parameter, set that quantity equal to zero, and solve for the value of the parameter. For complicated problems, we calculate the derivative

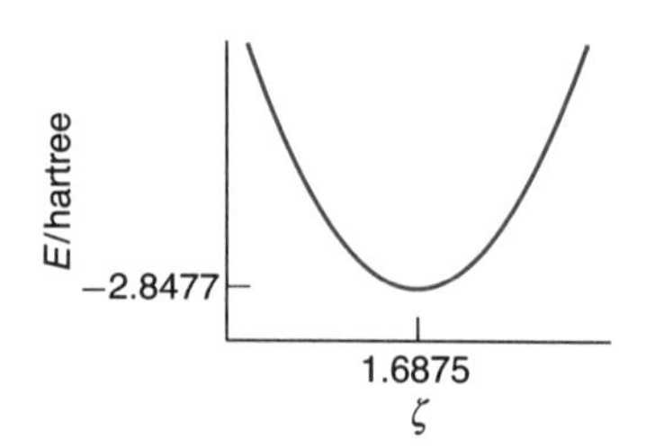

Figure 12.13
Plot of the calculated average energy of the helium versus the variational parameter ζ (the effective nuclear charge).

numerically; but, here we may readily find an analytical solution. In our helium example, the variational parameter is ζ, so we set

$$\frac{dE_\psi(\zeta)}{d\zeta} = 0 \tag{12.34}$$

From Equation 12.33, we find that

$$0 = \frac{d\left(\zeta^2 - \frac{27}{8}\zeta\right)}{d\zeta}$$

$$= 2\zeta - \frac{27}{8}$$

Note that the condition in Equation 12.34 only means that ζ is either a maximum or a minimum. In this case, however, we do find the minimum value for energy.

Therefore, $\zeta_{\text{min}} = 27/16 = 1.6875$, which is the value of ζ that yields the lowest energy.

Substituting ζ_{min} into Equation 12.33, we get

$$E_{\psi,\text{min}} = \left[\left(\frac{27}{16}\right)^2 - \frac{27}{8}\left(\frac{27}{16}\right)\right]$$

$$= -2.8477 \text{ hartree or } -7476 \text{ kJ mol}^{-1} \tag{12.35}$$

This trial helium atom energy lies above the experimental value of -2.9033 hartree (-7624 kJ mol^{-1}), which is the sum of the first and second ionization energies of He, but is considerably better than the independent electron approximation value of -4 hartree.

To improve the accuracy of the calculation, we can add a second variational parameter, and find values of both variational parameters that minimize the expectation value of the energy. A common approach for atoms, molecules, and ions is to use *Slater-type atomic orbitals* (STOs), which in essence treat both the principal quantum number n and ζ as variational parameters. An STO has the form

$$\psi = N r^{n-1} e^{\zeta r} Y_\ell^{m_\ell} \tag{12.36}$$

where N is a normalization constant and $Y_\ell^{m_\ell}$ are the spherical harmonic functions. Here, ℓ and m_ℓ are the usual integer quantum numbers, but n is a real-number variational parameter. To find the best solution for the helium atom we set both

$$\left[\frac{\partial E_\psi(\zeta,n)}{\partial \zeta}\right]_n = 0 \tag{12.37}$$

and

$$\left[\frac{\partial E_\psi(\zeta,n)}{\partial n}\right]_\zeta = 0 \tag{12.38}$$

and after some more mathematics we find the optimal (minimal energy) solution is $E_{\psi,\text{min}} = -2.8542$ hartree, when $n_{\text{min}} = 0.995$ and $\zeta_{\text{min}} = 1.6116$. This is an improvement over the single-parameter result, but requires considerably more computational effort. As with many areas of physical chemistry, we find that more parameterization leads to more accurate results, but involves more work and leads to less physical insight into the parameters. To obtain results that agree with spectroscopic measurements, computational chemists might perform calculations using two or three (or more) effective nuclear charges to describe a single orbital (so-called *double-zeta-* and *triple-zeta-*quality calculations).

EXAMPLE 12.4

Apply the variational principle to solve for the ground-state energy of a hydrogen atom using a trial wave function of the form,

$$\phi(r) = Ne^{-ar^2}$$

where N is a normalization constant, r is the distance from the nucleus, and a is a variational parameter to be optimized. Because we seek the spherical ground state, the terms containing θ and ϕ drop out and the Hamiltonian operator for the hydrogen atom (Equation 12.2) is simplified to

$$\hat{H} = \frac{-\hbar^2}{2m_e}\frac{1}{r^2}\frac{d}{dr}\left(r^2\frac{d}{dr}\right) - \frac{e^2}{4\pi\varepsilon_0 r}$$

and the spherical volume element becomes

$$d\tau = 4\pi r^2 dr$$

ANSWER

We write the expectation value of the trial energy as follows:

$$E_\phi = \frac{\int(\phi^*\hat{H}\phi)4\pi r^2 dr}{\int\phi^*\phi 4\pi r^2 dr}$$

$$= \frac{4\pi N^2\int_0^\infty e^{-ar^2}\left[\frac{-\hbar^2}{2m_e r^2}\frac{d}{dr}\left(r^2\frac{d}{dr}\right) - \frac{e^2}{4\pi\varepsilon_0 r}\right]e^{-ar^2}r^2 dr}{4\pi N^2\int_0^\infty e^{-ar^2}r^2 dr}$$

$$= \frac{\frac{-\hbar^2}{2m_e}\int_0^\infty(4a^2r^4 - 6ar^2)e^{-2ar^2}dr - \frac{e^2}{\varepsilon_0}\int_0^\infty re^{-2ar^2}dr}{\frac{1}{8a}\left(\frac{\pi}{2a}\right)^{1/2}}$$

$$= \frac{\frac{-\hbar^2}{2m_e}\left[-\frac{3}{8}\left(\frac{\pi}{2a}\right)^{1/2}\right] - \frac{e^2}{4\pi\varepsilon_0}\frac{1}{4a}}{\frac{1}{8a}\left(\frac{\pi}{2a}\right)^{1/2}}$$

$$= \frac{3\hbar^2 a}{2m_e} - \frac{e^2 a^{1/2}}{2^{1/2}\pi^{3/2}\varepsilon_0}$$

Now that we have an expression for the energy as a function of the variational parameter a, we find the minimum value of the trial energy by taking the derivative with respect to a and setting it equal to zero.

$$\frac{dE_\phi}{da} = \frac{3\hbar^2}{2m_e} - \frac{e^2}{(2\pi)^{3/2}\varepsilon_0(a_{\text{min}})^{1/2}} = 0$$

Solving for the value of the variational parameter that gives the minimum energy, we see that

$$a_{\text{min}} = \frac{m_e^2 e^4}{18\pi^3\varepsilon_0^2\hbar^4}$$

Substituting a_{min} into the energy expression, we find the optimum trial energy.

$$E_\phi = \frac{3\hbar^2}{2m_e}\left(\frac{m_e^2 e^4}{18\pi^3\varepsilon_0^2\hbar^4}\right) - \frac{e^2}{2^{1/2}\pi^{3/2}\varepsilon_0}\left(\frac{m_e^2 e^4}{18\pi^3\varepsilon_0^2\hbar^4}\right)^{1/2}$$

$$= -\left(\frac{1}{3\pi}\right)\frac{m_e e^4}{\varepsilon_0^2 h^2}$$

COMMENT

The true ground-state energy of the hydrogen atom (Equation 12.10) is

$$E_{\text{H}} = -\left(\frac{1}{8}\right)\frac{m_e e^4}{\varepsilon_0^2 h^2}$$

so our simple trial function is only 15% above the true energy. Note that the true ground-state wave function for the hydrogen atom is (see Table 12.2)

$$\psi_{1s}(r) = \frac{2}{\sqrt{a_0^3}} e^{-r/a_0}$$

where a_0 is the Bohr radius.

EXAMPLE 12.5

Apply the variational principle to estimate the ground-state energy of a particle of mass m in a one-dimensional box of length L using a trial wave function,

$$\phi(x) = Nx(L - x)$$

where N is a normalization constant.

ANSWER

The Hamiltonian operator for the particle in a one-dimensional box is (Equation 10.43)

$$\hat{H} = \frac{-\hbar^2}{2m}\frac{d^2}{dx^2}$$

We write the expectation value of the trial energy as follows:

$$\begin{aligned} E_\phi &= \frac{\int_0^L \phi^* \hat{H} \phi dx}{\int_0^L \phi^* \phi dx} \\ &= \frac{N^2 \frac{\hbar^2}{m} \int_0^L x(L-x) dx}{N^2 \int_0^L (xL - x^2)^2 dx} \\ &= \frac{\hbar^2}{m}\left(\frac{L^3/6}{L^5/30}\right) \\ &= \left(\frac{5}{4\pi^2}\right)\frac{h^2}{mL^2} \end{aligned}$$

COMMENT

The trial energy is 0.12665 $h^2m^{-1}L^{-2}$, whereas the true ground-state energy (see Equation 10.50) is $h^2/(8mL^2) = 0.12500\ h^2m^{-1}L^{-2}$, so this trial energy is only 1.321% higher. In this case, there were no variational parameters to optimize.

12.10 Hartree–Fock Self-Consistent-Field Method

As mentioned, solving the Schrödinger equation for helium and other many-electron atoms requires the use of approximations. A widely used approach is called the *Hartree–Fock self-consistent-field* (HF–SCF) *method*.* The HF–SCF method is a variational method for finding an approximate solution to the Schrödinger wave equation for a system with N electrons. It uses the orbital approximation in which electrons are described individually by wave functions that we call orbitals.

* Vladimir Fock (1898–1974) was a Russian chemist/physicist. He generalized the equations of Hartree by using wave functions that are antisymmetric with respect to the exchange of electrons (the Pauli exclusion principle).

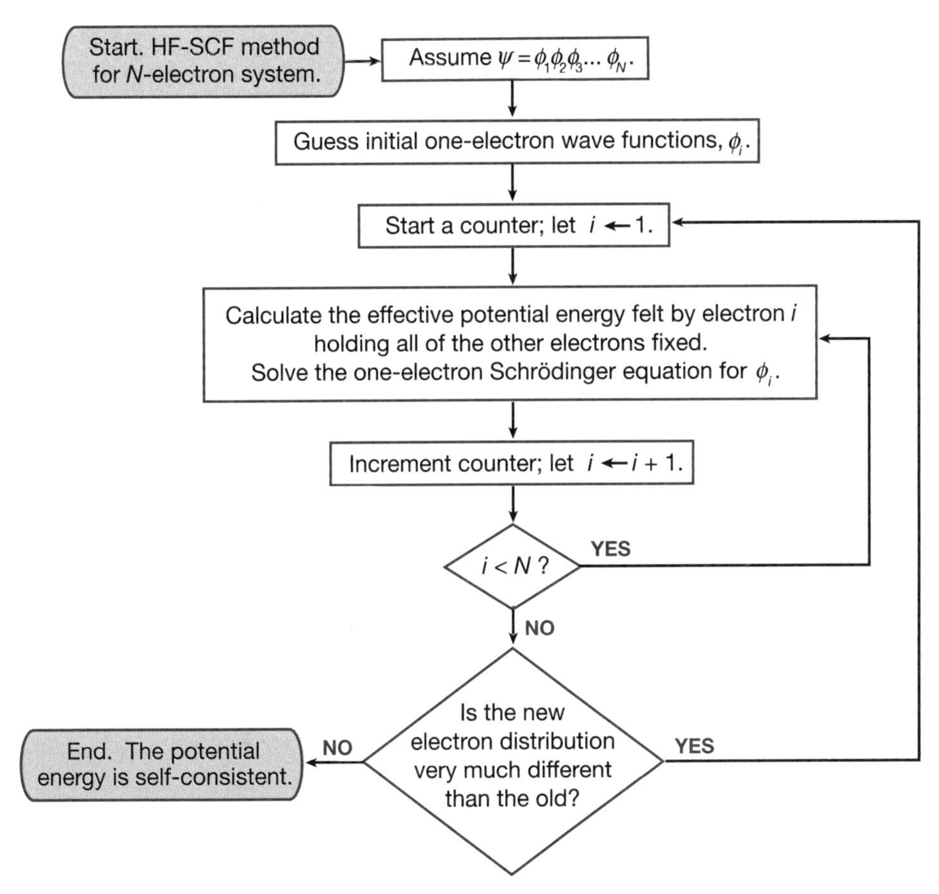

Figure 12.14
A flowchart representation of the SCF method for obtaining the wave functions of a many-electron system

In general, we start the HF–SCF method by writing the total wave function, ψ, as a product of N one-electron wave functions.

$$\psi = \phi_1\phi_2\phi_3\ldots\phi_\text{N}$$

A wave function is guessed for each electron except one. For example, we guess the wave functions for electrons 2, 3, 4, …, N to be ϕ_2, ϕ_3, ϕ_4, …, ϕ_N. We then solve the Schrödinger equation for electron 1, which is moving in a potential field generated by the nucleus and by the electrons in orbitals ϕ_2, ϕ_3, ϕ_4, …, ϕ_N. The repulsions between electron 1 and the rest of electrons are calculated at each point in space from the sum of the average electron densities around that point. This procedure gives us the wave function for electron 1, which we call ϕ_1'. Next, we do a similar calculation for electron 2 moving in a field of electrons described by the wave functions ϕ_1', ϕ_3, ϕ_4, …, ϕ_N. This step yields a new function ϕ_2' for electron 2. We repeat this procedure for the rest of electrons until we obtain a set of new wave functions ϕ_1', ϕ_2', ϕ_3', …, ϕ_N' for all the electrons. This process is then repeated as many times as necessary until we obtain a set of wave functions that is practically identical to the previous set. At that point, a self-consistent field is achieved and no more calculations are necessary. Figure 12.14 summarizes the procedure. The advent of high-speed computers has made it possible to calculate accurately the orbitals and energies of complex atoms. The results show that the orbitals of many-electron atoms are qualitatively similar to those of the hydrogen atom, and so we label them with the same quantum numbers used to describe the hydrogen atomic orbitals.

Let us now apply the Hartree–Fock method to the helium-atom electronic structure problem. We start once again with the orbital approximation (Equation 12.19),

$$\psi(\tau_1, \tau_2) = \phi(\tau_1)\phi(\tau_2)$$

where τ_1 and τ_2 represent the three-dimensional coordinates of electrons number 1 and 2, respectively. For the ground-state helium, both electrons occupy the same atomic orbital. In the Hartree–Fock approximation, the first electron "feels" an *effective potential* V^{eff} due to the charge or probability distribution of the second electron.

$$V_1^{\text{eff}}(\tau_1) = \int \phi^*(\tau_2)\frac{1}{r_{12}}\phi(\tau_2)d\tau_2 \tag{12.39}$$

where r_{12} is the distance between the two electrons and the integral is taken over all space. We can now write an effective one-electron Hamiltonian operator that includes the kinetic energy operator, the electron–nuclear Coulomb interaction, and this effective potential in atomic units,

$$\hat{H}_1^{\text{eff}}(\tau_1) = -\frac{1}{2}\nabla_1^2 - \frac{2}{r_1} + V_1^{\text{eff}}(\tau_1) \tag{12.40}$$

where the superscript "eff" has again been used to emphasize the effective, average nature of the operator. This effective Hamiltonian operator is used in the Schrödinger equation for electron 1:

$$\hat{H}_1^{\text{eff}}(\tau_1)\phi(\tau_1) = \varepsilon_1\phi(\tau_1) \tag{12.41}$$

where ε_1, is the orbital energy of electron 1. For helium, the Schrödinger equation for electron number 2 is the same. You may have noticed that there appears to be some circular logic in this presentation. To solve the Schrödinger equation (Equation 12.41) for the one-electron wave function ϕ, we need the effective Hamiltonian operator (Equation 12.40), which in turn needs the effective potential (Equation 12.39), which itself needs the wave function ϕ. Where do we begin? We start by making an educated guess for the functional form of ϕ. We use this initial guess to find the effective potential and the effective Hamiltonian operator. We then solve the Schrödinger equation to find a new ϕ. We use the new electron wave function to calculate a new effective potential and repeat the cycle until ϕ doesn't change very much between cycles. When this occurs, the solution is said to converge, and the field is said to be *self-consistent*. The one-electron wave functions ϕ obtained in this fashion are called the *Hartree–Fock orbitals*.

In computational chemistry, the orbitals are written as a linear combination of atomic orbitals. The collection of atomic orbitals used comprises the *basis set* for the calculation. When the basis set size is extrapolated to be infinitely large, we reach the *Hartree–Fock limit*, which is the best answer obtainable when we write the total wave

Table 12.9
Ground-State Energy of the Helium Atom

Method	Energy[a]/E_h	Ionization Energy[a]/E_h
Experimental value[b] (for ^{4}He)	−2.90357059	0.90357059
Complete neglect of $1/r$ (Section 12.6)	−4.000	2.000
Variational, with $\zeta = 1.6875$ (Section 12.9)	−2.8477	0.8477
Variational, with $\zeta = 1.61162$ and $n = 0.995$	−2.8542	0.8542
Variational, with 1078 parameters[c]	−2.903724375	0.903724375
Hartree–Fock[d] (Section 12.10)	−2.8617	0.8617
First-order perturbation theory (Section 12.11)	−2.7500	0.7500
Second-order perturbation theory	−2.9077	0.9077
Thirteenth-order perturbation theory[e]	−2.90372433	0.90372433

[a] Energies are in atomic units. $E_h = 1$ hartree. The theoretical energies are calculated using nonrelativistic, fixed-nucleus approximation methods.
[b] G. Herzberg, *Proc. Roy. Soc. (London)* **248**, 309 (1958). The values for ^{3}He are 4.78×10^{-5} hartree smaller. The experimental energy is the sum of the two ionization potentials.
[c] C. L. Pekeris, *Phys. Rev.* **115**, 1126 (1959).
[d] E. Clementi and C. Roetti, *At. Data Nucl. Data Tables* **14**, 177 (1974). This is the Hartree–Fock limit, with an extrapolation to an infinitely large basis set.
[e] C. W. Scheer and R. E. Knight, *Rev. Mod. Phys.* **35**, 426 (1963).

function as a product of one-electron wave functions. But the Hartree–Fock limit is still above the true (or exact, or spectroscopic) energy of the system (see Table 12.9). The reason is that the Hartree–Fock approach does not fully account for *electron correlation*. In the Hartree–Fock approximation, the electrons, are said to be uncorrelated because they "feel" each other only in an average or effective sense. The difference between the exact energy and the Hartree–Fock energy is known as the *correlation energy* (CE), where

$$\text{CE} = E_{\text{exact}} - E_{\text{HF}} \tag{12.42}$$

For helium, the correlation energy is 0.0419 hartree, or 1.44% of the exact energy. While this is a small percentage, 0.0419 hartree is 110 kJ mol^{-1}, the same order of magnitude as a chemical bond enthalpy.

The Hartree–Fock approach is generally more complicated than that shown for helium. With helium, both electrons occupy the same orbital, so we are able to separate the spatial and spin portions of the total wave function. For more complex systems, we need to write the complete wave function, including both spatial and spin portions (see Example 12.3). This means that the effective Hamiltonian operator is more complicated than in the case of helium. Hartree–Fock methods are among the most commonly used in computational chemistry because they provide reasonable agreement with experiment at an acceptable cost (in computer time and memory).

12.11 Perturbation Theory

Another approach for finding approximate solutions to the Schrödinger equation for systems where no exact analytical solution is possible is known as perturbation theory. Perturbation theory differs from the variational method in that the energies obtained from a perturbation theory calculation are not guaranteed to lie above the true ground-state energy of the system. Nevertheless, perturbation theory methods are quite useful because they provide reasonable quantitative results for many systems. Perturbation theory methods work well for systems that resemble a well-understood quantum mechanical model, such as a particle in a box, a harmonic oscillator, a rigid rotor, a hydrogen atom, and so on. For example, if there is a particle in a potential box with a small bump (or potential barrier) in the middle, then the system is modeled with a zero-potential box, and the bump is treated as a small *perturbation* to the system. To explain perturbation theory, we'll first give a general description of the method, and then apply the method to some familiar systems.

"Zero order" means "unperturbed."

The general strategy in perturbation theory is to choose a simple zero-order wave function $\psi^{(0)}$ that approximates the system of interest, and then to use the complete, correct Hamiltonian operator to calculate the expectation value of the energy using postulate four of quantum mechanics (see Section 10.7):

$$\langle E \rangle = \frac{\int \psi^{(0)*} \hat{H} \psi^{(0)} d\tau}{\int \psi^{(0)*} \psi^{(0)} d\tau} \tag{12.43}$$

The superscript (0) is used to denote the zero-order terms. But here, the wave functions are not eigenfunctions of the Hamiltonian operator because of the perturbation.

$$\hat{H}\psi^{(0)} \neq E\psi^{(0)}$$

The complete, correct Hamiltonian operator $\hat{H}$ is written as the sum of the zero-order Hamiltonian, plus a first-order perturbation term,

$$\hat{H} = \hat{H}^{(0)} + \hat{H}^{(1)} \tag{12.44}$$

where the zero-order Hamiltonian has the zero-order wave functions as its eigenfunctions:

$$\hat{H}^{(0)}\psi_n^{(0)} = E_n^{(0)}\psi_n^{(0)} \tag{12.45}$$

The superscript (1) is used on the terms involving a small perturbation; we call them first-order in anticipation that we might have second-order, third-order, and so on perturbation terms at a higher level of theory. In first-order perturbation theory, the expectation value of the energy is the sum of the zero-order energy plus a first-order perturbation term.

$$\langle E \rangle = E^{(0)} + E^{(1)} \tag{12.46}$$

As the name suggests, the perturbation energy $E^{(1)}$ is generally much smaller than the zero-order energy $E^{(0)}$. The first-order correction to the energy is calculated as in Equation 12.43:

$$E^{(1)} = \langle E^{(1)} \rangle = \frac{\int \psi^{(0)*} \hat{H}^{(1)} \psi^{(0)} d\tau}{\int \psi^{(0)*} \psi^{(0)} d\tau} \tag{12.47}$$

One of the keys to perturbation theory is the judicious choice of a zero-order Hamiltonian and associated zero-order wave functions. To make the calculations run smoothly, the zero-order system is usually chosen to be one that has an exact analytical solution. The simple quantum-mechanical systems that we have considered so far (particle in a box, particle on a ring, rigid rotor, harmonic oscillator, hydrogen atom) are commonly used.

Let us apply perturbation theory to the helium atom problem that we have been considering in this chapter. We chose the hydrogen atom Hamiltonian and the hydrogen atom wave functions as our zero-order starting point. We write the helium atom wave function as we have done previously, using the independent electron (or independent orbital) approach:

$$\Psi_{\text{He}}^{(0)}(\tau_1, \tau_2) = \psi_1^{(0)}(\tau_1)\psi_2^{(0)}(\tau_2) \tag{12.48}$$

where τ_1 and τ_2 are the spatial coordinates of electrons 1 and 2. We write the Hamiltonian operator using atomic units (see Equation 12.16):

$$\hat{H} = \hat{H}_{\text{exact}} = \underbrace{-\frac{1}{2}\nabla_1^2 - \frac{Z}{r_1}}_{\substack{\text{H-atom} \\ \text{Hamiltonian}}} \underbrace{-\frac{1}{2}\nabla_2^2 - \frac{Z}{r_2}}_{\substack{\text{H-atom} \\ \text{Hamiltonian}}} \underbrace{+\frac{1}{r_{12}}}_{\text{Perturbation}}$$

The terms have been ordered in such as way as to emphasize that the first four terms are merely the hydrogen-atom Hamiltonians (with $Z = 2$) for electrons 1 and 2, respectively.

$$\hat{H} = \hat{H}_1^{(0)} + \hat{H}_2^{(0)} + \frac{1}{r_{12}} \tag{12.49}$$

The $1/r_{12}$ term, which prevents us from obtaining an exact analytical solution, is now treated as a first-order perturbation term,

$$\hat{H} = \hat{H}_1^{(0)} + \hat{H}_2^{(0)} + \hat{H}_{12}^{(1)} \tag{12.50}$$

where the subscript "12" is added to emphasize that the perturbation operator depends on the coordinates of both electrons 1 and 2. Now we are ready to apply postulate four of quantum mechanics (see Equation 10.34) to obtain the expectation value of the energy. We substitute our zero-order wave function (Equation 12.48) into the energy expression (Equation 12.43) and obtain the following intimidating-looking equation:

$$\langle E \rangle = \frac{\iint \psi_1^{(0)*}(\tau_1)\psi_2^{(0)*}(\tau_2)[\hat{H}_1^{(0)} + \hat{H}_2^{(0)} + \hat{H}_{12}^{(1)}]\psi_1^{(0)}(\tau_1)\psi_2^{(0)}(\tau_2)d\tau_1 d\tau_2}{\iint \psi_1^{(0)*}(\tau_1)\psi_2^{(0)*}(\tau_2)\psi_1^{(0)}(\tau_1)\psi_2^{(0)}(\tau_2)d\tau_1 d\tau_2} \quad (12.51)$$

Our judicious choice of zero-order function, however, enables us to readily simplify this expression. The denominator is equal to one because we have chosen normalized wave functions. The numerator may be separated into three terms, based on the three parts of the Hamiltonian operator:

$$\begin{aligned}\langle E \rangle = &\int \psi_1^{(0)*}(\tau_1)\hat{H}_1^{(0)}\psi_1^{(0)}(\tau_1)d\tau_1 \int \psi_2^{(0)*}(\tau_2)\psi_2^{(0)}(\tau_2)d\tau_2 + \\ &\int \psi_2^{(0)*}(\tau_2)\hat{H}_2^{(0)}\psi_2^{(0)}(\tau_2)d\tau_2 \int \psi_1^{(0)*}(\tau_1)\psi_1^{(0)}(1)d\tau_1 + \\ &\iint \psi_1^{(0)*}(\tau_1)\psi_2^{(0)*}(\tau_2)\hat{H}_{12}^{(1)}\psi_1^{(0)}(\tau_1)\psi_2^{(0)}(\tau_2)d\tau_1 d\tau_2 \end{aligned} \quad (12.52)$$

In the first of the three terms, the first integral is the hydrogen atom problem, and the second integral is the normalization condition (equal to one). The same holds in the second term, so we have

$$\langle E \rangle = E_1^{(0)} + E_2^{(0)} + \iint \psi_1^{(0)*}(\tau_1)\psi_2^{(0)*}(\tau_2)\hat{H}_{12}^{(1)}\ \psi_1^{(0)}(\tau_1)\psi_2^{(0)}(\tau_2)d\tau_1 d\tau_2 \quad (12.53)$$

The remaining double integral is a challenging one to solve, but it does have an analytical solution.[†] The result in atomic units is

$$\langle E \rangle = E_1^{(0)} + E_2^{(0)} + \frac{5}{8}Z$$

where Z is the nuclear charge. For helium,

$$\langle E \rangle = -2 - 2 + \frac{5}{8}(2) = -2.75 \text{ hartree}$$

which is quite close to the experimental value (−2.9033 hartree).

[†] See, for example, M. Karplus and R. N. Porter, *Atoms and Molecules: An Introduction for Students of Physical Chemistry*, W. A. Benjamin, New York, 1970, pages 174–177.

EXAMPLE 12.6

Apply perturbation theory to solve for the ground-state energy of an oscillator governed by the following potential energy function:

$$V(x) = \frac{k}{2}x^2 + bx^4$$

where k and b are constants. Assume that b is sufficiently small to be treated as a perturbation.

ANSWER

We choose the harmonic oscillator as our unperturbed system and treat the quartic part (the part containing x^4) of the potential as the perturbation

$$\hat{H}^{(1)} = bx^4$$

The zero-order Hamiltonian operator is the same as for the harmonic oscillator (Equation 11.53), but we do not need to write this out. The zero-order energy is taken from the harmonic oscillator (Equation 11.54) with the quantum number $\upsilon = 0$,

$$E_0 = \left(\upsilon + \frac{1}{2}\right)h\nu = \frac{1}{2}h\nu$$

where ν is the fundamental frequency of the harmonic oscillator. For the zero-order wave function, we look up the ground-state wave function of the harmonic oscillator (Equation 11.57),

$$\psi_0(x) = \left(\frac{\alpha}{\pi}\right)^{1/4} e^{-\alpha x^2/2}$$

where

$$\alpha = \left(\frac{k\mu}{\hbar^2}\right)^{1/2}$$

and μ is the reduced mass. Now we can calculate the first-order perturbation energy:

$$E^{(1)} = \int \psi^{(0)*}\hat{H}^{(1)}\psi^{(0)}d\tau$$

$$= \int_{-\infty}^{\infty} bx^4\left(\frac{\alpha}{\pi}\right)^{1/2} e^{-\alpha x^2}dx$$

$$= 2b\left(\frac{\alpha}{\pi}\right)^{1/2}\int_0^{\infty} x^4 e^{-\alpha x^2}dx$$

Consulting the standard integral tables in, for example, the Handbook of Chemistry and Physics, we find

$$E^{(1)} = \frac{3b}{4\alpha^2}$$

$$= \frac{3b\hbar^2}{4k\mu}$$

$$= \frac{3bh^2\nu^2}{4k^2}$$

The total energy of the system is thus

$$E = E^{(0)} + E^{(1)}$$

$$= \frac{h\nu}{2} + \frac{3bh^2\nu^2}{4k^2}$$

COMMENT

You should try dimensional analysis to confirm that your answer has the correct units.

First-order perturbation theory does not do quite as well as the variational approach using the effective nuclear charge ζ as the sole variational parameter (see Equation 12.35). To improve the results of perturbation theory, we use higher-order perturbation theory. In general, our goal is to solve

$$\hat{H}\psi_n = E_n\psi_n$$

As with first-order perturbation theory, we select a zero-order Hamiltonian and associated eigenfunctions:

$$\hat{H}^{(0)}\psi_n^{(0)} = E_n^{(0)}\psi_n^{(0)}$$

We assume that the exact Hamiltonian is the zero-order Hamiltonian plus a perturbation, but now introduce a parameter λ, the value of which determines the magnitude of the perturbation:

$$\hat{H} = \hat{H}^{(0)} + \lambda\hat{H}^{(1)} \qquad (12.54)$$

Equation 12.54 can be extended to include higher-order terms. The generalized solution for the wave function is

$$\psi_n = \psi_n^{(0)} + \lambda\psi_n^{(1)} + \lambda^2\psi_n^{(2)} + \cdots \qquad (12.55)$$

and for the energy is

$$E_n = E_n^{(0)} + \lambda E_n^{(1)} + \lambda^2 E_n^{(2)} + \lambda^3 E_n^{(3)} + \cdots \qquad (12.56)$$

where (1), (2), (3), ... represent first-order, second-order, third-order, ... perturbations, respectively. These higher order wave functions $[\psi_n^{(2)}, \psi_n^{(3)}, \ldots]$ and energies $[E_n^{(2)}, E_n^{(3)}, \ldots]$ may be determined with just a first-order Hamiltonian operator $[\hat{H}^{(1)}]$. Commonly, we calculate the energy to one more order than the wave function. In our example with the helium atom, we had $\lambda = 1$ and found the first-order perturbation energy $E_n^{(1)}$ and the zero-order wave function $\psi_n^{(0)}$. The zero-order wave functions for helium are merely the hydrogen-atom wave functions with $Z = 2$.

As we increase the order of perturbation theory used (see Table 12.9), we increase both the accuracy of the calculation and the computational complexity. Increasing the order of perturbation theory leads to less and less physical significance of the perturbation terms. A common trade-off in computational chemistry is that of accuracy versus computational cost. We must decide what our ultimate goal is and when to stop calculating.

Perturbation theory is a significantly different approach than the variational methods described in the previous section. An advantage of perturbation theory over variational theory is that perturbation theory may be straightforwardly applied to excited states, whereas the variational method applies to the ground state. We might favor variational methods because they are guaranteed to give energies above the true energy. As stated earlier, unlike variational methods, the perturbation theory approach may yield energies below the true energy. As we increase the order of perturbation theory or the number of variational parameters, however, both perturbation theory and variational methods converge on the true energies and wave functions (Table 12.9). In this case, the difference between the thirteenth-order perturbation theory calculation, for example, and the experimental value is partly due to the approximations made in the perturbation theory calculation. Both the perturbation theory and variational method calculations neglect relativistic effects* and also used the fixed-nucleus approximation. Nuclear motion and relativistic effects are thought to account for most of the difference between the experimental and high-level theoretical values.

* Looking down the periodic table, the nuclear charge becomes larger. As a result, the speed of the inner electrons increases. For heavy elements, the electron speed can approach that of light. In these cases, corrections have to be made for the mass of the electron and other electronic properties such as the size of the orbitals.

Key Equations

$$\frac{-\hbar^2}{2m_e}\left[\frac{1}{r^2}\frac{\partial}{\partial r}\left(r^2\frac{\partial}{\partial r}\right) + \frac{1}{r^2\sin\theta}\frac{\partial}{\partial\theta}\left(\sin\theta\frac{\partial}{\partial\theta}\right) + \frac{1}{r^2\sin^2\theta}\left(\frac{\partial^2}{\partial\phi^2}\right)\right]\psi - \frac{e^2\psi}{4\pi\varepsilon_0 r} = E\psi$$

(Schrödinger wave equation for the hydrogen atom) (12.2)

$$\psi_{n,\ell,m_\ell}(r,\theta,\phi) = \underbrace{R_{n\ell}(r)}_{\text{Radial part}}\underbrace{Y_\ell^{m_\ell}(\theta,\phi)}_{\text{Angular part}}$$

(Separation of the hydrogen wave function) (12.5)

$$E_n = -\frac{m_e e^4}{8h^2\varepsilon_0^2}\frac{1}{n^2}$$

(Energy of the electron in a hydrogen atom) (12.10)

$$\underbrace{-\frac{\hbar^2}{2m_e}\nabla_1^2\psi - \frac{\hbar^2}{2m_e}\nabla_2^2\psi}_{\text{Electron kinetic energy terms}} \underbrace{- \frac{Ze^2\psi}{4\pi\varepsilon_0 r_1} - \frac{Ze^2\psi}{4\pi\varepsilon_0 r_2} + \frac{e^2\psi}{4\pi\varepsilon_0 r_{12}}}_{\text{Coulombic potential energy terms}} = E\psi$$

(Schrödinger wave equation for the helium atom) (12.16)

$$\zeta = Z - \sigma$$

(Effective nuclear charge) (12.26)

$$E_\phi \geq E_0$$

(Variation method) (12.29)

APPENDIX 12.1

Proof of the Variational Principle

Our goal is to find approximations for the wave function ψ_0 and the energy E_0 of the ground state of a system for which we may have no analytical solution. The variational principle tells us that $E_\phi \geq E_0$ (Equation 12.29); the energy of our trial wave function E_ϕ will always be greater than or equal to the true ground-state energy. In this appendix, we outline a proof of the variational principle.

Let ψ_n be the true (unknown, and perhaps unknowable) wave functions that describes the system of interest, so that

$$\hat{H}\psi_n = E_n\psi_n \qquad n = 0, 1, 2, \ldots \tag{1}$$

where $\hat{H}$ is the Hamiltonian operator for the system, and E_n are the true energies. Let ϕ be our trial wave function, which is an approximation to the wave function ψ_0 that describes the true ground state of the system. We can write our trial wave function as a linear combination of the true wave functions,

$$\phi = \textstyle\sum_n c_n \psi_n \tag{2}$$

where c_n are the coefficients, which may be real or imaginary:

$$0 \leq |c_n|^2 \leq 1 \tag{3}$$

We write the energy E_ϕ of our trial wave function just as we write the average value of the energy of a wave function using postulate 4 of quantum mechanics (see Section 10.7)

$$E_\phi = \frac{\int \phi^* \hat{H} \phi \, d\tau}{\int \phi^* \phi \, d\tau} \tag{4}$$

Next we substitute the linear combination of true wave functions that describes our trial wave function (Equation 2) into Equation 4:

$$E_\phi = \frac{\int \left(\sum_m c_m^* \psi_m^*\right) \hat{H} \left(\sum_n c_n \psi_n\right) d\tau}{\int \left(\sum_m c_m^* \psi_m^*\right)\left(\sum_n c_n \psi_n\right) d\tau} \tag{5}$$

Here we have changed the dummy variable from n to m in the complex conjugates merely to keep track of the fact that we are doing two summations in the numerator and two summations in the denominator. Applying the Hamiltonian operator

(Equation 1) in the numerator gives

$$E_\phi = \frac{\int\left(\sum_m c_m^* \psi_m^*\right)\left(\sum_n c_n E_n \psi_n\right) d\tau}{\int\left(\sum_m c_m^* \psi_m^*\right)\left(\sum_n c_n \psi_n\right) d\tau} \tag{6}$$

Distributing the summations, we find

$$E_\phi = \frac{\sum_m \sum_n \left[c_m^* c_n E_n \int \psi_m^* \psi_n d\tau\right]}{\sum_m \sum_n \left[c_m^* c_n \int \psi_m^* \psi_n d\tau\right]} \tag{7}$$

We simplify the integrals by recognizing that the true wave functions, ψ, are *orthonormal*, so

$$E_\phi = \frac{\sum_m \sum_n \left[c_m^* c_n E_n \delta_{mn}\right]}{\sum_m \sum_n \left[c_m^* c_n \delta_{mn}\right]} \tag{8}$$

where δ_{mn} is the Krönecker delta. Once again, we use the orthonormality condition to simplify the summations. The double summations are collections of terms that have nonzero values only when $m = n$, so we may write them as single summations:

$$E_\phi = \frac{\sum_n c_n^* c_n E_n}{\sum_n c_n^* c_n} \tag{9}$$

Next we subtract the true ground state energy, E_0, from each side:

$$E_\phi - E_0 = \frac{\sum_n c_n^* c_n (E_n - E_0)}{\sum_n c_n^* c_n} \tag{10}$$

Now we recognize that

$$c_n^* c_n \geq 0 \tag{11}$$

and by definition

$$E_n \geq E_0 \tag{12}$$

The denominator of Equation 9 is positive, and each term of the numerator is greater than or equal to zero, so the whole right-hand side of Equation 10 is zero or a positive number. Thus,

$$E_\phi \geq E_0$$

This is the important result of the variational principle (Equation 12.29). It does not matter what trial wave function we have chosen, it does not matter how many variational parameters we have included, and it does not matter if we have chosen the true wave function as our trial wave function; the energy of our trial wave function will always be greater than or equal to the true ground-state energy. If we find that our variational energy is less than the true energy, then we have used the wrong (or perhaps merely incomplete) Hamiltonian operator. If we choose an invalid wave function, such as a discontinous or non-normalized wave function, then we may also calculate the variation energy that is less than the true energy. Thus, we shall be careful in selecting our trial wave functions.

Suggestions for Further Reading

BOOKS

Feynman, R. P., R. B. Leighton, and M. Sands, *The Feynman Lectures on Physics*, Volumes I, II, and III, Addison-Wesley, Reading, MA, 1963.

Herzberg, G., *Atomic Spectra and Atomic Structure*, Dover Publications, New York, 1944.

Karplus, M., and R. N. Porter, *Atoms and Molecules: An Introduction for Students of Physical Chemistry*, W. A. Benjamin, New York, 1970.

Levine, I. N., *Quantum Chemistry*, 7th ed., Prentice-Hall, New York, 2013.

McQuarrie, D. A., *Quantum Chemistry*, 2nd ed., University Science Books, Sausalito, CA, 2008.

McQuarrie, D. A., and J. D. Simon, *Physical Chemistry: A Molecular Approach*, University Science Books, Sausalito, CA, 1997.

Pilar, F. L., *Elementary Quantum Chemistry*, 2nd ed., Dover Publications, New York, 2001.

Ratner, M. A., and G. C. Schatz, *Introduction to Quantum Mechanics in Chemistry*, Dover Publications, New York, 2002.

ARTICLES

Hydrogen Atom

"The Stability of the Hydrogen Atom," F. Rioux, *J. Chem. Educ.* **50**, 550 (1973).

"Radial Probability Density and Normalization in Hydrogenic Atoms," L. Lain, A. Toree, and J. M. Alvariño, *J. Chem. Educ.* **58**, 617 (1981).

"Why Doesn't the Electron Fall into the Nucleus?," F. P. Mason and R. W. Richardson, *J. Chem. Educ.* **60**, 40 (1983).

"Some Further Comments about the Stability of the Hydrogen Atom," P. Blaise, O. Henri-Rousseau, and N. Merad, *J. Chem. Educ.* **61**, 957 (1984).

"On the Problem of the Exact Shape of Orbitals," E. Peacock-López, *Chem. Educator* [Online] **8**, 96 (2003) DOI 10.1333/s00897030676a.

"Schrödinger Equation Solutions That Lead to the Solution for the Hydrogen Atom," P. F. Newhouse and K. C. McGill, *J. Chem. Educ.* **81**, 424 (2004).

"The Scales of Time, Length, Mass, Energy, and Other Fundamental Physical Quantities in the Atomic World and the Use of Atomic Units in Quantum Mechanical Calculations," B. K. Teo and W. K. Li, *J. Chem. Educ.* **88**, 921, (2011).

Atomic Structure

"The Exclusion Principle," G. Gamow, *Sci. Am.* July 1959.

"Atomic Orbitals," R. S. Berry, *J. Chem. Educ.* **43**, 283 (1966).

"The Five Equivalent *d* Orbitals," R. E. Powell, *J. Chem. Educ.* **45**, 45 (1968).

"Five Equivalent *d* Orbitals," L. Pauling and V. McClure, *J. Chem. Educ.* **47**, 15 (1970).

"The Pauli Principle and Electronic Repulsion in Helium," R. L. Snow and J. L. Bills, *J. Chem. Educ.* **51**, 585 (1974).

"4*s* is Always Above 3*d*! or, How to tell the Orbitals from the Wavefunctions," F. L. Pilar, *J. Chem. Educ.* **55**, 2 (1978).

"Highly Excited Atoms," D. Kleppner, M. G. Littman, and M. L. Zimmerman, *Sci. Am.* May 1981.

"Teaching the Shapes of the Hydrogenlike and Hybrid Atomic Orbitals," R. D. Allendoerfer, *J. Chem. Educ.* **67**, 37 (1990).

"Relative Energies of 3*d* and 4*s* Orbitals," P. G. Nelson, *Educ. Chem.* **29**, 84 (1992).
"Transition Metals and the Aufbau Principle," L. G. Vanquickenborne, K. Pierloot, and D. Devoghel, *J. Chem. Educ.* **71**, 469 (1994).
"Understanding Electron Spin," J. C. A. Boeyens, *J. Chem. Educ.* **72**, 412 (1995).
"Why the 4*s* is Occupied Before the 3*d*," M. Melrose and E. R. Scerri, *J. Chem. Educ.* **74**, 498 (1996).
"Ionization Energies of Atoms and Atomic Ions," P. F. Lang and B. C. Smith, *J. Chem. Educ.* **80**, 938 (2003).
"The Noble Gas Configuration - Not the Driving Force but the Rule of the Game in Chemistry," R. Schmid, *J. Chem. Educ.* **80**, 931 (2003).
"The Meaning of *d*-Orbital Labels," G. Ashkenazi, *J. Chem. Educ.* **82**, 323 (2005).
"How is an Orbital Defined?" D. Keeports, *Chem. Educator* [Online] **11**, 1 (2006) DOI 10.1333/s00897060992a.
"Hund's Multiplicity Rule Revisited," F. Rioux, *J. Chem. Educ.* **84**, 358 (2007).
"Hund's Rule in Two-Electron Atomic Systems," J. E. Harriman, *J. Chem. Educ.* **85**, 451 (2008).
"Applying Electron Exchange Symmetry Properties to Better Understand Hund's Rule," P. E. Fleming, *Chem. Educator* [Online] **13**, 141 (2008) DOI 10.1333/s00897082137a.
"Cartesian Approach to Atomic and Molecular Orbitals," C. W. David, *Chem. Educator* [Online] **13**, 270 (2008) DOI 10.1333/s00897082159a.
"The Shape of Atoms," D. Castelvecchi, *Sci. Am.* December 2009.
"Aspects of Quantum Mechanics Clarified by Lateral Thinking," Y. Liu, Y. Liu, and M. G. B. Drew, *Chem. Educator* [Online] **16**, 272 (2011) DOI 10.1007/s00897112390a.

Periodic Trends

"Periodic Contractions Among the Elements: or, On Being the Right Size," J. Mason, *J. Chem. Educ.* **65**, 17 (1988).
"The Periodicity of Electron Affinity," R. T. Meyers, *J. Chem. Educ.* **67**, 307 (1990).
"Ionization Energies Revisited," N. C. Pyper and M. Berry, *Educ. Chem.* **27**, 135 (1990).
"Electron Affinities of the Alkaline Earth Metals and the Sign Convention for Electron Affinity," J. C. Wheeler, *J. Chem. Educ.* **74**, 123 (1997).
"The Bose-Einstein Condensate," E. A. Cornell and C. E. Weiman, *Sci. Am.* March 1998.
"The Evolution of the Periodic System," E. R. Scerri, *Sci. Am.* September 1998.
"Two Particles in a Box," I. Noval, *J. Chem. Educ.* **78**, 395 (2001).
"Stern and Gerlach: How a Bad Cigar Helped Reorient Atomic Physics," B. Friedrich and D. Herschbach, *Physics Today*, December 2003.
"Screened Atomic Potential: A Simple Explanation of the Aufbau Model," W. Eck, S. Nordholm, and G. B. Bacskay, *Chem. Educator* [Online] **11**, 235 (2006) DOI 10.1333/s00897061050a.
"The Past and Future of the Periodic Table," E. R. Scerri, *Am. Scientist* January-February 2008.

Variational Principle

"A Simple Illustration of the SCF-LCAO-MO Method," R. L. Snow and J. L. Bills, *J. Chem. Educ.* **52**, 506 (1975).
"Atomic Variational Calculations: Hydrogen to Boron," F. Rioux, *Chem. Educator* [Online] **4**, 40 (1999) DOI 10.1333/s00897990292a.
"Variational Principle for a Particle in a Box," J. I. Casaubon and G. Doggett, *J. Chem. Educ.* **77**, 1221 (2000).
"Variational Methods Applied to the Particle in a Box," W. T. Gribbs, *J. Chem. Educ.* **78**, 1557 (2001).

Perturbation Theory

"The Perturbation MO Method. Quantum Mechanics on the Back of an Envelope," W. B. Smith, *J. Chem. Educ.* **48**, 749 (1971).

"Helium Revisited: An Introduction to Variational Perturbation Theory," H. E. Montgomery, Jr., *J. Chem. Educ.* **54**, 748 (1977).

"Applications of the Perturbational Molecular Orbital Method," F. Freeman, *J. Chem. Educ.* **55**, 26 (1978).

"Simple Perturbation Example for Quantum Chemistry," P. L. Goodfriend, *J. Chem. Educ.* **62**, 202 (1985).

"Using the Perturbed Harmonic Oscillator to Introduce Rayleigh-Schrödinger Perturbation Theory," K. Sohlberg and D. Shreiner, *J. Chem. Educ.* **68**, 203 (1991).

"Perturbation Theory for a Particle in a Box," H. E. Montgomery, Jr., and W. P. Crummett, *Chem. Educator* [Online] **10**, 169 (2005) DOI 10.1333/s00897050896a.

Problems

The Hydrogen Atom

12.1 For the following hydrogen atom orbitals, identify the quantum numbers n, ℓ, and m_ℓ: $3s$, $4d_{xy}$, $5p_z$, $6f_0$.

12.2 Sketch a $4s$ hydrogen atom orbital, including an approximate scale, and clearly identify the location of any nodes.

12.3 Sketch a $3p_0$ hydrogen atom orbital, including an approximate scale, and clearly identify the location of any nodes.

12.4 A hydrogen atom $5d$ orbital has the radial wave function

$$R_{52}(r) = \frac{1}{150\sqrt{70a_0^3}}(42 - 14\rho + \rho^2)\rho^2 e^{-\rho/2}$$

How many radial nodes does a $5d$ orbital have and at what radii do they occur?

12.5 For the principal quantum number $n = 6$, which hydrogen atom subshell does not contain any radial nodes? Give the value of the orbital angular momentum quantum number ℓ and the letter used to describe it.

12.6 Write the complex hydrogen atom wave functions, $2p_{-1}$, $2p_0$, and $2p_1$ in terms of the real wave functions $2p_x$, $2p_y$, and $2p_z$.

12.7 Write the real hydrogen atom wave functions, $3d_{z^2}$, $3d_{xy}$, $3d_{xz}$, $3d_{yz}$, and $3d_{x^2-y^2}$ in terms of the complex wave functions, $3d_{-2}$, $3d_{-1}$, $3d_0$, $3d_1$, and $3d_2$.

12.8 Rank the following hydrogen atom orbitals in order of increasing energy, and identify any degeneracies. Assume there are no external fields present. $6s$, $5s$, $4s$, $4p_z$, $4p_x$, $5d_{xy}$, $5d_{xz}$.

12.9 Obtain an expression for the most probable radius at which an electron will be found when it occupies the hydrogen $1s$ orbital. (*Hint*: Differentiate the $1s$ wave function in Table 12.2 with respect to r.)

12.10 Use the hydrogen $2s$ wave function given in Table 12.2 to calculate the value of r (other than $r = \infty$) at which this wave function becomes zero.

Electronic Configuration and Atomic Properties

12.11 Write the ground-state electron configurations of the following ions, which play important roles in biochemical processes in our bodies: **(a)** Na^+, **(b)** Mg^{2+}, **(c)** Cl^-, **(d)** K^+, **(e)** Ca^{2+}, **(f)** Fe^{2+}, **(g)** Cu^{2+}, **(h)** Zn^{2+}.

12.12 Explain, in terms of their electron configurations, why Fe^{2+} is more easily oxidized to Fe^{3+} than Mn^{2+} to Mn^{3+}.

12.13 Ionization energy is the energy required to remove a ground-state ($n = 1$) electron from an atom. It is usually expressed in units of kJ mol^{-1}. **(a)** Calculate the ionization energy for the hydrogen atom. **(b)** Repeat the calculation, assuming in this case that the electrons are removed from the $n = 2$ state.

12.14 The formula for calculating the energies of an electron in a hydrogenlike ion is given in Equation 10.18. This equation cannot be applied to many-electron atoms. One way to modify it for the more complex atoms is to replace Z with $(Z - \sigma)$, where Z is the atomic number and σ is a positive dimensionless quantity called the *shielding constant*. Consider the helium atom as an example. The physical significance of σ is that it represents the extent of shielding that the two 1s electrons exert on each other. Thus, the quantity $(Z - \sigma)$ is appropriately called the *effective nuclear charge*. Calculate the value of σ if the first ionization energy of helium is 3.94×10^{-18} J per atom. (In your calculations, ignore the minus sign in the given equation.)

12.15 Plasma is a state of matter consisting of positive gaseous ions and electrons. In the plasma state, a mercury atom could be stripped of its 80 electrons and therefore would exist as Hg^{80+}. Calculate the energy required for the last ionization step; that is,

$$Hg^{79+}(g) \rightarrow Hg^{80+}(g) + e^-$$

12.16 A technique called photoelectron spectroscopy (see Section 14.5) is used to measure the ionization energy of atoms. A sample is irradiated with UV light, and electrons are ejected from the valence shell. The kinetic energies of the ejected electrons are measured. Since the energy of the UV photon and the kinetic energy of the ejected electron are known, we can write

$$h\nu = IE + \tfrac{1}{2}m_e u^2$$

where ν is the frequency of the UV light, and m_e and u are the mass and speed of the electron, respectively. In one experiment the kinetic energy of the ejected electron from potassium is found to be 5.34×10^{-19} J using a UV source of wavelength 162 nm. Calculate the ionization energy of potassium. How can you be sure that this ionization energy corresponds to the electron in the valence shell (i.e., the most loosely held electron)?

12.17 The energy needed for the following process is 1.96×10^4 kJ mol^{-1}:

$$Li(g) \rightarrow Li^{3+}(g) + 3e^-$$

If the first ionization of lithium is 520 kJ mol^{-1}, calculate the second ionization of lithium, that is, the energy required for the process

$$Li^+(g) \rightarrow Li^{2+}(g) + e^-$$

12.18 Experimentally, the electron affinity of an element can be determined by using a laser light to ionize the anion of the element in the gas phase:

$$X^-(g) + h\nu \rightarrow X(g) + e^-$$

Referring to Table 12.8, calculate the photon wavelength (in nm) corresponding to the electron affinity for chlorine. In what region of the electromagnetic spectrum does this wavelength lie?

12.19 The standard enthalpy of atomization of an element is the energy required to convert one mole of an element in its most stable form at 25°C to one mole of monatomic gas. Given that the standard enthalpy of atomization for sodium is 108.4 kJ mol^{-1}, calculate the energy in kilojoules required to convert one mole of sodium metal at 25°C to one mole of gaseous Na^+ ions.

12.20 Explain why the electron affinity of nitrogen is approximately zero, while the elements on either side, carbon and oxygen, have substantial positive electron affinities.

12.21 Calculate the maximum wavelength of light (in nm) required to ionize a single sodium atom.

12.22 The first four ionization energies of an element are approximately 738 kJ mol^{-1}, 1450 kJ mol^{-1}, 7.7×10^3 kJ mol^{-1}, and 1.1×10^4 kJ mol^{-1}. To which periodic group does this element belong? Why?

Variational Principle

12.23 Explain in your own words the main principles of the variational method in quantum mechanics.

12.24 Use variational theory to calculate the energy of a particle in a one-dimensional box of length L with the following trial function,

$$\phi(x) = Nx^2(L^2 - x^2)$$

where N is a normalization constant to be determined. Recognizing that the true energy is $h^2/(8mL^2)$, calculate the percent error.

12.25 In using the variational method to calculate the energy of a system, do not forget that your answer may, in fact, be equal to the exact energy. Apply the variational method to the harmonic oscillator using the trial wave function,

$$\phi(x) = Ne^{-cx^2}$$

where N is a normalization constant, c is a variational parameter, and x is the displacement from equilibrium. First solve for the normalization constant, N. Next derive an expression for the energy in terms of the constant c, and find the minimum energy with respect to c. [*Hint*: You should find $N = (2c/\pi)^{1/4}$ and $E_\phi(c) = k/8c + c\hbar^2/2m$ where k is the force constant and m is the mass.]

12.26 Apply the variational method to the harmonic oscillator using the trial wave function

$$\phi(x) = Ne^{-c|x|}$$

where N is a normalization constant, c is a variational parameter, and $|x|$ is the absolute value of the displacement from equilibrium. Because of the discontinuity in the derivative of the trial function at $x = 0$, one would not expect this trial function to give a good solution. Nevertheless, it may be used as a trial wave function. Calculate the variational energy and compare your results to the exact solution. (*Hint*: First calculate $\langle E \rangle$ as a function of c, then take the derivative $d\langle E \rangle/dc$ to find the value of c which minimizes the energy.)

12.27 If we apply the variational method to the hydrogen atom, and choose a trial wave function of the form

$$\phi(r) = ce^{-ar} + de^{-br^2}$$

where a, b, c, and d are variational parameters, then we can calculate the minimum energy for the system. Without performing any calculations, identify the values of a, b, c, d, and $E_{\min}$ for the hydrogen atom.

Perturbation Theory

12.28 Perturbation theory considers small influences on systems with analytical solutions. For each of the following cases, identify the zero-order Hamiltonian operator $\hat{H}^{(0)}$, the perturbation Hamiltonian operator $\hat{H}^{(1)}$, the zero-order wave function $\psi^{(0)}$, and the zero-order energy $E^{(0)}$. It is not necessary to calculate solutions—just use a model quantum mechanical system (such as particle in a box, particle on a ring, rigid rotor, harmonic oscillator, or hydrogen atom) for zero-order items. **(a)** A lithium ion, Li^+. **(b)** A helium atom, He. **(c)** A particle on a line segment of length L, with the potential energy function,

$$V(x) = \infty \qquad x < 0, \text{ or } x > L$$

$$V(x) = bx \qquad 0 \le x \le L$$

where b is a constant. **(d)** An anharmonic oscillator with the potential energy function,

$$V(x) = ax^2 + bx^3 + cx^4$$

where a, b, and c are constants. **(e)** A Morse oscillator with the potential energy function,

$$V(x) = D(1 - e^{-\beta x})^2$$

where D and β are constants. (*Hint*: Use a Maclaurin series expansion:

$$e^{-x} = 1 - x + \frac{x^2}{2!} - \frac{x^3}{3!} + \ldots$$

and ignore x^4 and higher terms. The x^3 is the perturbation term.) **(f)** A rigid rotor in an electric field of strength E. The Hamiltonian operator that describes this system is

$$\hat{H} = -\frac{\hbar^2}{2I}\nabla^2 + \mu E \cos\theta$$

where I is the moment of inertia and μ is the dipole moment, and θ is the angle between the electric field and the dipole moment vector. **(g)** A hydrogen atom in a magnetic field of strength B_z in the z direction. The Hamiltonian operator that describes this system is

$$\hat{H} = -\frac{\hbar^2}{2m_e}\nabla^2 - \frac{e^2}{4\pi\varepsilon_0 r} - i\mu_B B_z\left(x\frac{\partial}{\partial y} - y\frac{\partial}{\partial x}\right)$$

where μ_B is the Bohr magneton (a constant), and the other symbols have their usual meaning.

12.29 For the quartic oscillator with the potential energy function,

$$V(x) = cx^4$$

where c is a constant, we might use perturbation theory to find the ground-state energy. Consider the oscillator from Example 12.6 as a comparison. When using first-order perturbation theory, how do the ground-state energies of these two oscillators differ?

12.30 Using a particle in a box as the zero-order system, apply perturbation theory to calculate the energy of a particle of mass m in a box of length a that has the step-potential energy function,

$$V(x) = 0 \qquad 0 \leq x \leq \frac{a}{2}$$

$$V(x) = c \qquad \frac{a}{2} \leq x \leq a$$

$$V(x) = \infty \qquad x < 0, \text{ or } x > a$$

where c is a constant.

12.31 Use perturbation theory to calculate the first-order energy correction term for a particle of mass m in a box of length a described by the slanting potential energy function,

$$V(x) = cx \qquad 0 \leq x \leq a$$

$$V(x) = \infty \qquad x < 0, \text{ or } x > a$$

where c is a constant.

12.32 Using first-order perturbation theory, calculate the ground-state energy of an anharmonic oscillator of reduced mass μ, with the potential energy function,

$$V(x) = \frac{k}{2}x^2 + bx$$

where k and b are constants. Assume b is sufficiently small that the bx term may be treated as a perturbation.

12.33 Using the harmonic oscillator as the zero-order system, calculate the first-order perturbation energy for the ground state of an oscillator that is governed by the potential energy function,

$$V(x) = \frac{k}{2}x^2 \qquad -a \leq x \leq a$$

$$V(x) = \infty \qquad x < -a, \text{ or } x > a$$

where k is a constant.

Additional Problems

12.34 In the hydrogen atom and hydrogenlike ions, the electronic energy levels depend only on the principal quantum number n, whereas in many-electron atoms the energy levels depend on both n and the orbital angular momentum quantum number ℓ. In your own words, explain this difference.

12.35 In the Stern–Gerlach experiment, what would happen if a uniform magnetic field were employed, rather than a nonuniform magnetic field?

12.36 The variational method generally applies to ground-state wave functions. It may be extended to excited states under the condition that the trial wave function is orthogonal to the exact ground-state wave function. Under these circumstances, the trial wave function gives an upper limit for the energy of the first excited state. Given that the wave function,

$$\phi(x) = N\left(x^3 - \frac{3}{2}Lx^2 + \frac{1}{2}L^2x\right)$$

where N is a normalization constant, satisfies these conditions for a particle in a box of length L, determine the formula for the variational energy of the lowest-energy excited state. Next, calculate the energy for an electron in a 0.80-nm box. Then calculate the exact energy for an electron in the first excited state in a 0.80-nm box. How good an approximation is this trial wave function?

12.37 Calculate the radius (in pm) of the He^+ ion in its ground ($n = 1$) state.

12.38 Using Equation 10.23, confirm the value of the Rydberg constant for the hydrogen (^1H) atom. Next, substitute the reduced mass μ for the mass of the electron m_e, and calculate the value of the Rydberg constant for the hydrogen (^1H) atom. Again using the reduced mass, calculate the Rydberg constant for the deuterium (^2H) atom. Compare and contrast your results. The mass of the deuterium atom nucleus is $3.34358320 \times 10^{-27}$ kg.

12.39 Unsöld's theorem states that when an atomic subshell of electrons is completely filled or half-filled, the distribution of those electrons is spherically symmetric. That is, for a given value of the quantum number ℓ, the sum of the electron probability densities is independent of the angles θ and ϕ; in other words,

$$\sum_{\ell,m_\ell} [\Theta_{\ell,m_\ell}(\theta)\Phi_{m_\ell}(\phi)]^2 = \text{constant}$$

Calculate this sum for the complex $2p$ atomic orbitals and show that it is independent of θ and ϕ. Repeat the calculation for the real $2p$ atomic orbitals.

12.40 The ionization energy of a certain element is 412 kJ mol^{-1}. When the atoms of the element are in the first excited electronic state, however, the ionization energy is only 126 k mol^{-1}. Based on this information, calculate the wavelength of light emitted in a transition from the first excited state to the ground state.

CHAPTER 13

Molecular Electronic Structure and the Chemical Bond

I believe the chemical bond is not so simple as some people seem to think.

—Robert S. Mulliken

Having developed exact solutions for the Schrödinger equation for some simple model systems (particle in a box, rigid rotor, harmonic oscillator) and for one real system (the hydrogen atom), we now turn our attention to the chemical bond. Understanding the nature of covalent bonding lies at the heart of chemistry. We start by considering the hydrogen molecular cation (H_2^+), then move on to the hydrogen molecule (H_2), homo- and heteronuclear diatomic molecules, and finally go to polyatomic systems.

13.1 The Hydrogen Molecular Cation

Consisting of two protons and a single electron, the hydrogen molecular cation (H_2^+) is the simplest covalently bonded species. Yet even this species defies description by the methods of classical physics. Coulomb's law suffices to describe ionic bonding, such as that found in compounds like NaCl, but quantum chemistry is needed to explain covalent bonding. As with the model systems described in Chapter 12, we would like to be able to write wave functions and calculate their energies for the H_2^+ system. Unfortunately, the hydrogen molecular cation has no such mathematical solution. This is an example of the "three-body problem"—one that has no exact analytical solution. All is not lost, however, because we can make some simplifying approximations that enable us to write mathematical descriptions, gain physical insight, and make reliable predictions.

The original three-body problem was to describe two planets orbiting the sun.

The approximation used for the H_2^+ cation is one that we will use extensively with molecular systems. The mass of an electron is nearly 2,000 times less than that of a proton. As a result, the electron moves much more rapidly than an atomic nucleus. The *Born–Oppenheimer approximation* [after Max Born and the American physicist J. Robert Oppenheimer (1904–1967)] recognizes this difference in speed, and treats the atomic nuclei as stationary. To solve for the energy levels of a molecule, then, we only need to solve the Schrödinger equation in this frame of reference. Solutions for the Schrödinger equation give the energy levels and the wave functions for the electron. For H_2^+, the positions of the nuclei are described by a single parameter, the bond distance R. Additionally, r_A and r_B are the distances of the electron from nucleus A and nucleus B, respectively (Figure 13.1). Using the Born–Oppenheimer approximation, we treat the variables R and r_A and r_B independently. Analogous to the particle in a

Figure 13.1
Coordinates for the hydrogen molecular cation.

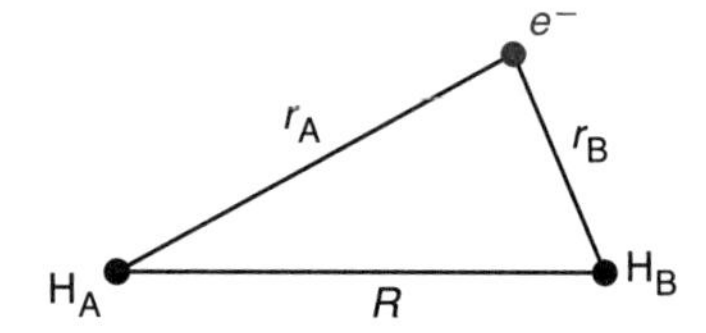

two-dimensional box problem, the overall wave function $\psi(r_A, r_B, R)$ can be written as a *product* of the electron and nuclear wave functions ϕ and χ, respectively:

$$\psi(r_A, r_B, R) = \phi(r_A, r_B)\chi(R) \tag{13.1}$$

The overall energy is the *sum* of the electronic and nuclear energies:

$$E = E_{elec} + E_{nuclei} \tag{13.2}$$

Once we have separated the variables, we fix the nuclei at a given bond distance, R, and solve a simplified Schrödinger equation for the wave function ϕ and energy E of the electron:

$$\hat{H}\phi(r_A, r_B) = E\phi(r_A, r_B) \tag{13.3}$$

The full Hamiltonian operator $\hat{H}$ consists of three kinetic energy and three potential energy terms:

$$\hat{H} = \underbrace{\hat{H}_{\text{kin A}} + \hat{H}_{\text{kin B}} + \hat{H}_{\text{kin }e^-}}_{\text{Kinetic}} + \underbrace{\hat{H}_{e^-\text{A}} + \hat{H}_{e^-\text{B}} + \hat{H}_{\text{AB}}}_{\text{Potential}}$$

Here, A and B denote the two nuclei (protons) and e^- refers to the electron. Using the Born–Oppenheimer approximation, we neglect the terms describing the kinetic energy of the nuclei $(\hat{H}_{\text{kin A}} + \hat{H}_{\text{kin B}})$ and write the electron kinetic energy term using the mass of the electron m_e, Planck's constant over $2\pi(\hbar)$, and the Laplacian operator ∇^2:

$$\hat{H}_{\text{kin }e^-} = \frac{-\hbar^2}{2m_e}\nabla^2$$

We use the general Z_A and Z_B so that the solution applies to any atomic nuclei. For H_2^+, $Z_A = Z_B = 1$.

The three potential energy terms are written using Coulomb's law, where the charges on nuclei A and B are Z_A and Z_B, respectively. Using atomic units, the Hamiltonian operator becomes

$$\hat{H} = -\frac{1}{2}\nabla^2 - \frac{Z_A}{r_A} - \frac{Z_B}{r_B} + \frac{Z_A Z_B}{R} \tag{13.4}$$

The Schrödinger equation is given by

$$-\frac{1}{2}\nabla^2\phi - \frac{Z_A}{r_A}\phi - \frac{Z_B}{r_B}\phi + \frac{Z_A Z_B}{R}\phi = E\phi \tag{13.5}$$

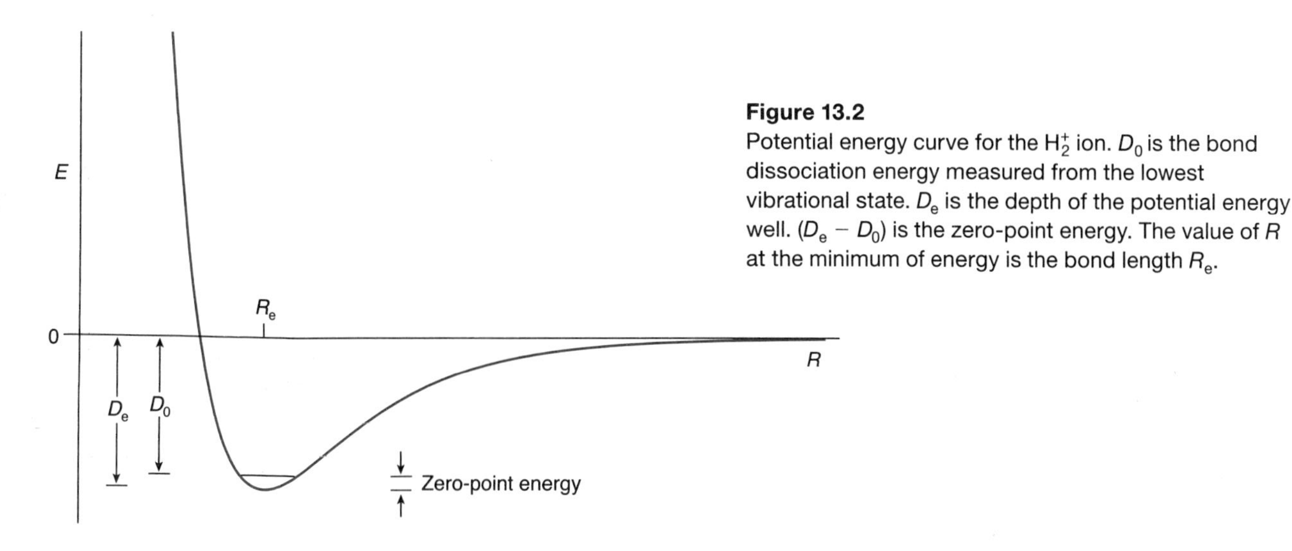

Figure 13.2
Potential energy curve for the H_2^+ ion. D_0 is the bond dissociation energy measured from the lowest vibrational state. D_e is the depth of the potential energy well. ($D_e - D_0$) is the zero-point energy. The value of R at the minimum of energy is the bond length R_e.

As mentioned earlier, the internuclear separation R is treated as a constant so that the electron position is the only variable. We have made the problem more straightforward, but Equation 13.5 is still a three-dimensional differential equation. To further simplify the problem, we resort to a mathematical technique that we used in solving the rigid rotator problem and transform the spatial coordinates to a new coordinate system. Because it contains *two* nuclei, the hydrogen molecular cation is not naturally represented in a Cartesian (x, y, z) coordinate system; an elliptical representation in which the nuclei lie at the foci of the ellipse is more appropriate in describing the system. The advantage to transforming to a new coordinate system is that we now have an analytical solution. As in the example of the particle in a two-dimensional box, the separation of variables enables us to independently solve two one-dimensional problems.

We omit the details of the solution to the Schrödinger equation for H_2^+ and show the results qualitatively. A plot of the energy as a function of nuclear geometry is known as a *potential-energy surface*. In this case, we have only a single variable (the bond length R) to describe the geometry, so we may construct a potential energy curve to describe the chemical bond (Figure 13.2). At each value of R, we calculate the energy of the system, $E(R)$. The internuclear separation, R, is only meaningful for positive values. We calibrate the potential energy by setting E to zero at infinite nuclear separation, where there is *no* chemical bonding. The value of the potential energy may be positive or negative. Negative potential energy means that the system is more stable than separated nuclei; it refers to the chemically *bound* situation.

From the potential energy curve we may derive two important values, the equilibrium bond distance and the bond energy. The *bond distance* is the value of R at the minimum in the potential energy curve. This value is also called the equilibrium bond distance, and is given the symbol R_e. Chemical bond lengths are typically on the order of 100 pm (100 pm $= 10^{-10}$ m $= 1$ Å). The *bond energy* or *bond strength* is the depth of the potential energy well (D_e); that is, the difference in energy between the equilibrium geometry and infinite atomic separation. For practical purposes, infinite separation is approximately 10 Å. Chemical bond energies are on the order of hundreds of kilojoules per mole.

D_e differs from the experimentally measured bond dissociation energy, D_0, due to zero-point energy (see Section 11.3). Here, $D_e = D_0 + h\nu/2$.

The details of the solutions to the Schrödinger equation provides some insight into covalent bonding. A common view is that bonding occurs due to coulombic attraction between the electron and the nuclei. In fact, the lowering of the kinetic energy of the electrons that occurs with bonding is more important. Using an analogy from the particle-in-a-box model, the H_2^+ molecule is bound at R near R_e, because the electron has a larger space in which to move (a larger box) and thus possesses lower energy. At large values of R, the electron becomes localized near a single proton, so the H_2^+ system is at higher energy.

When we use the simple choice that the H_2^+ wave function is merely the (normalized) sum of hydrogenlike 1*s* orbitals on the two nuclei, then the solution of the Schrödinger equation for H_2^+ provides moderate agreement with experimental values. The calculated bond length is 132 pm and the experimental bond length is 106 pm. The calculated bond energy is 170 kJ mol^{-1}, compared with the experimental bond energy of 269 kJ mol^{-1}. A more detailed theory, such as that described for the neutral hydrogen molecule in the next section, provides more accurate values for the hydrogen molecular cation bond distance and bond energy.

As with the hydrogen atom, solving the Schrödinger equation for H_2^+ provides a family of wave functions, each with a different energy. The solution shown in Figure 13.2 is the ground state, which has the lowest energy. The next lowest energy solution yields the first excited state, whose potential-energy curve is dissociative (Figure 13.3). By this we mean that the most stable internuclear separation is at infinity. If H_2^+ is excited to this state, it will rapidly dissociate to a hydrogen atom and a proton.

Let us consider the wave functions for the ground and first electronic excited states in a qualitative way to help us understand why one is bound and the other is dissociative. The ground-state wave function resembles two overlapping hydrogen 1*s* wave functions (Figure 13.4). Both the ground-state and excited-state orbitals are designated as σ(sigma) orbitals. Loosely speaking, σ orbitals have the symmetry of a cylinder that is localized along the internuclear axis. The ground-state σ orbital has electron density between the two nuclei. This electron density helps to shield the nuclear charges from each other, and thus lower the nuclear–nuclear repulsion. This is an example of a *bonding orbital*. The excited-state wave function, in contrast, has a nodal plane (where the wave function equals zero) that bisects the internuclear axis; it more closely resembles two isolated hydrogen 1*s* wave functions. This gives

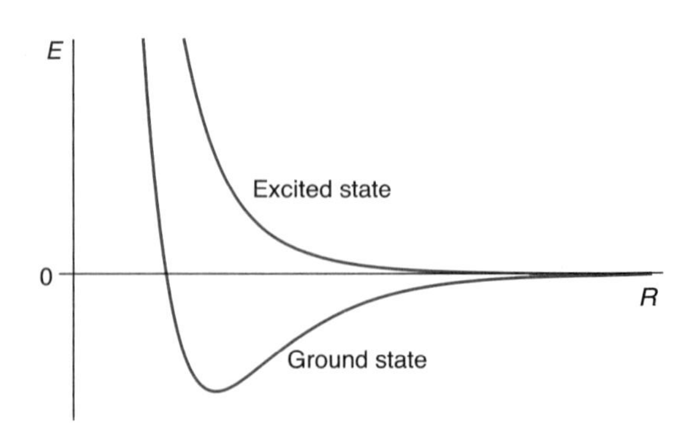

Figure 13.3
Potential energy curves for the two lowest energy states of H_2^+. The ground state is bound, while the excited state is dissociative.

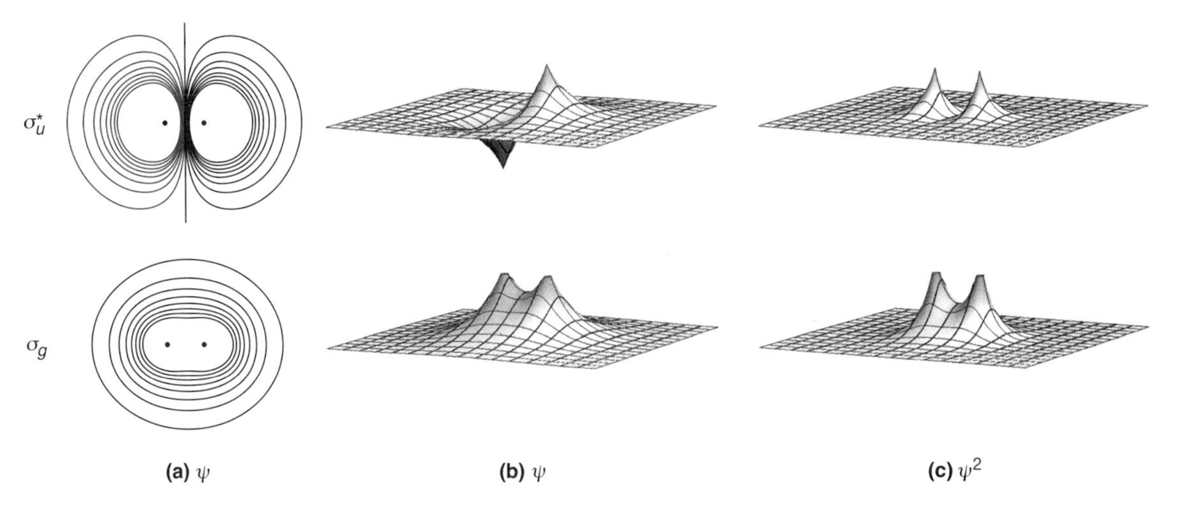

Figure 13.4
(a) Contour diagram of the wave functions, (b) wave functions, and (c) probability density for the H_2^+ ground (σ_g) and first excited states (σ_u^*).

rise to an *antibonding orbital* and is designated with an asterisk, σ^* (pronounced sigma star).

The ground-state H_2^+ orbital is given a subscript "g" (for "gerade") because the wave function is symmetric with respect to inversion through the bond center. Conversely, the excited state wave function is given a subscript "u" (for "ungerade") because here the wave function is antisymmetric with respect to inversion. If we set the origin to be the center of the internuclear axis, then the sign of an ungerade wave function at a point (x, y, z) is opposite that at the point $(-x, -y, -z)$. For gerade, $\psi_g(x, y, z) = \psi_g(-x, -y, -z)$; whereas for ungerade, $\psi_u(x, y, z) = -\psi_u(-x, -y, -z)$.

Gerade is German for even. *Ungerade* is odd.

13.2 The Hydrogen Molecule

The smallest and simplest of the stable neutral molecular systems, the hydrogen molecule (H_2), consists of two protons and a shared pair of electrons held together by a covalent bond (Figure 13.5). The Born–Oppenheimer approximation enabled us to determine analytical solutions for the energy and wave function of H_2^+, but with the introduction of a second electron to form H_2, this is no longer feasible. To solve the Schrödinger equation for H_2, we need to make additional approximations. We shall outline the steps along the way to obtain reasonable predictions for the energy and wave function of the molecular system.

The Hamiltonian operator for H_2 describes the interactions of the four subatomic particles (two protons and two electrons),

$$\hat{H} = \underbrace{\hat{H}_{\text{kin A}} + \hat{H}_{\text{kin B}} + \hat{H}_{\text{kin 1}} + \hat{H}_{\text{kin 2}}}_{\text{Kinetic energy terms}} + \underbrace{\hat{H}_{1\text{A}} + \hat{H}_{1\text{B}} + \hat{H}_{2\text{A}} + \hat{H}_{2\text{B}} + \hat{H}_{\text{AB}} + \hat{H}_{12}}_{\text{Potential energy terms}}$$

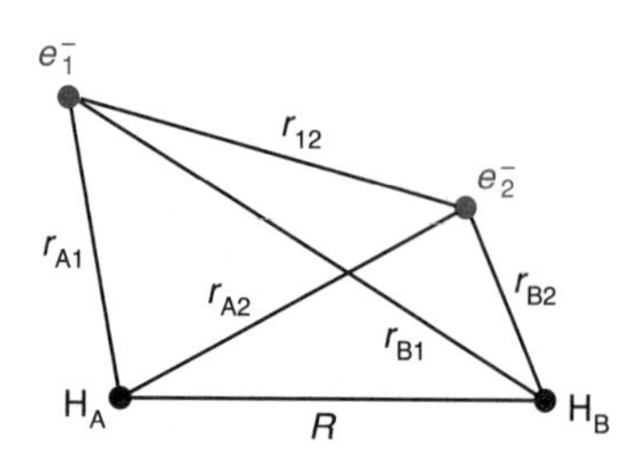

Figure 13.5
Coordinates for the hydrogen molecule.

where A and B refer to the atomic nuclei and the 1 and 2 refer to the electrons. Applying the Born–Oppenheimer approximation, we neglect the kinetic energy of the nuclei and write

$$\hat{H} = \hat{H}_{\text{kin 1}} + \hat{H}_{\text{kin 2}} + \hat{H}_{1\text{A}} + \hat{H}_{1\text{B}} + \hat{H}_{2\text{A}} + \hat{H}_{2\text{B}} + \hat{H}_{\text{AB}} + \hat{H}_{12}$$

Following the procedure we used for H_2^+, letting the nuclear charges $Z_A = Z_B = 1$, and using atomic units,

$$\hat{H} = -\frac{1}{2}\nabla_1^2 - \frac{1}{2}\nabla_2^2 - \frac{1}{r_{1\text{A}}} - \frac{1}{r_{1\text{B}}} - \frac{1}{r_{2\text{A}}} - \frac{1}{r_{2\text{B}}} + \frac{1}{R} + \frac{1}{r_{12}} \quad (13.6)$$

The Hamiltonian operator for H_2 (Equation 13.6) closely resembles that for H_2^+ (see Equation 13.4). Rewriting Equation 13.6 by adding and subtracting a $1/R$ term and rearranging

$$\hat{H} = \underbrace{-\frac{1}{2}\nabla_1^2 - \frac{1}{r_{1\text{A}}} - \frac{1}{r_{1\text{B}}} + \frac{1}{R}}_{\text{H}_2^+ \text{ for electron \#1}} \underbrace{-\frac{1}{2}\nabla_2^2 - \frac{1}{r_{2\text{A}}} - \frac{1}{r_{2\text{B}}} + \frac{1}{R}}_{\text{H}_2^+ \text{ for electron \#2}} - \frac{1}{R} + \frac{1}{r_{12}} \quad (13.7)$$

we see that the Hamiltonian operator for H_2 is twice the Hamiltonian operator for H_2^+ plus a $1/r_{12}$ term and minus an additional constant term $(1/R)$. The $1/R$ term represents the nuclear–nuclear repulsion; it is a constant parameter in the Born–Oppenheimer approximation. The electron–electron repulsion term $1/r_{12}$ is again the one that causes us difficulty. Because the $1/r_{12}$ term depends on the distance between the two electrons, we cannot separate it into parts that depend only on a single variable (a single electron coordinate). Consequently, there is no analytical solution for the Schrödinger equation for the H_2 molecule. We must make additional approximations to determine the wave functions and energies of the H_2 molecule. Two approaches that are commonly used are valence bond (VB) theory and molecular orbital (MO) theory. VB theory is, in a sense, a quantification of Lewis electron dot diagrams. Here chemical bonds are formed by the constructive overlap of two atomic orbitals and are localized between two atoms. In MO theory, bonds (and antibonds) are created by linear combination of atomic orbitals from all the atoms in a molecule. In VB theory, electrons retain their atomic orbital character, whereas in MO theory electrons are described by molecular orbitals. We shall see that both VB and MO approaches have their advantages and drawbacks. With additional approximations, they actually converge to give the *same* results for the hydrogen molecule in good agreement with experiment.

13.3 Valence Bond Approach

The valence bond approach might be considered a more quantitative description of Lewis electron dot diagrams. It is sometimes known as the Heitler–London approach, after the German-Irish physicist Walter Heitler (1904-1981) and the American physicist Fritz London (1900–1954). Linus Pauling and John Slater also made significant contributions to VB theory. The theory rests on a few simple assumptions. Covalent bonds are formed when a pair of valence (outer shell) electrons is localized between two atoms. Bonding orbitals are formed by the overlap of atomic orbitals. The most stable bonding occurs when each of the atoms has a complete shell of electrons.

For the hydrogen molecule, we can write three simple Lewis electron dot diagrams to describe electronic structure (Figure 13.6). Because hydrogen is a homonuclear diatomic molecule, we expect that the chemical bond is covalent in nature and that Figure 13.6a describes the interaction between the two atoms. To accurately describe the chemical bond in molecular hydrogen, we need a small contribution from the ionic bonding form (Figure 13.6b). However, VB theory neglects the ionic forms and considers only localized covalent bonds, which are the "sticks" in Lewis dot diagrams.

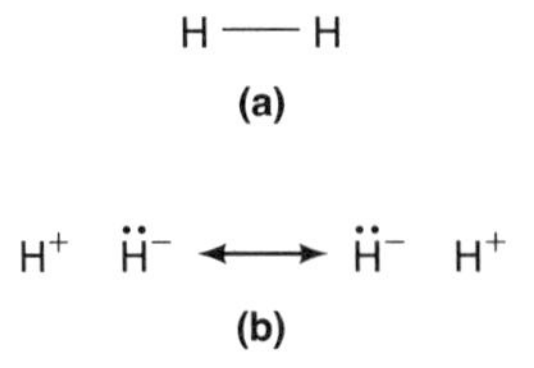

H — H

(a)

H^+ $\ddot{H}^-$ ⟷ $\ddot{H}^-$ H^+

(b)

Figure 13.6
Lewis electron dot diagrams for the hydrogen molecule. (a) One covalently bonded. (b) Two equivalent ionically bonded. In describing the actual electronic structure of H_2, the covalent description is substantially more important than the ionic.

The analytical result obtained for the H_2^+ cation suggests that the wave function for the electron resembles two overlapping hydrogen atom wave functions. We use this as a guide for the hydrogen molecule. As a first guess, we write

$$\psi_{\text{trial VB}}(1,2) = N 1s_A(1) 1s_B(2) \tag{13.8}$$

where N is a normalization constant, $1s_A$ and $1s_B$ indicate hydrogen $1s$ orbitals on nuclei A and B, respectively, and the 1 and 2 refer to the two electrons. This trial wave function is invalid, however. Because electrons are indistinguishable, it should not matter whether electron 1 or electron 2 is located on nucleus A. Another valence-bond-type wave function is

$$\psi_{\text{trial VB+}}(1,2) = N\big[\underbrace{1s_A(1)1s_B(2)}_{\psi_{\text{trial VB}}(1,2)} + \underbrace{1s_A(2)1s_B(1)}_{\psi_{\text{trial VB}}(2,1)}\big] \tag{13.9}$$

where VB+ indicates that the wave function is the *sum* of two equivalent wave functions and therefore does not distinguish between electrons. This wave function is still not a complete description of the hydrogen molecule, however, because it does not satisfy the Pauli principle (see Section 12.7). When we exchange the two electrons in Equation 13.9, the wave function remains the same:

$$\psi_{\text{trial VB+}}(1,2) = \psi_{\text{trial VB+}}(2,1)$$

That is, the wave function is symmetric with respect to the exchange. To satisfy the Pauli principle, the wave function must be antisymmetric with respect to exchange; that is, when you swap the two electrons (trade the 1 and the 2 in this case), you should get back the negative of the original wave function. To make a wave function that satisfies the Pauli principle, we also need to consider electron spin. As is customary, we use $\alpha(1)$ to represent a "spin up" electron number one, $\beta(2)$for a "spin down" electron number 2, and so forth. Thus, a proper valence bond wave function for the hydrogen molecule is

$$\psi_{VB+}(1,2) = N_+\underbrace{\left[1s_A(1)1s_B(2) + 1s_A(2)1s_B(1)\right]}_{\text{Space part}}\underbrace{\left[\alpha(1)\beta(2) - \beta(1)\alpha(2)\right]}_{\text{Spin part}} \quad (13.10)$$

where N_+ is a normalization constant. The wave function ψ_{VB+} now satisfies the Pauli principle as shown here:

$$\psi_{VB+}(1,2) = -\psi_{VB+}(2,1)$$

This somewhat long expression (Equation 13.10) signifies a covalent bond represented by a Lewis electron dot diagram (see Figure 13.6a). Two electrons are shared between two nuclei; one electron is spin up, and the other is spin down.

It is also possible to write wave functions for the hydrogen molecule in which the spatial portion of the wave function is written as a difference rather than a sum. When the spatial portion of the wave function is antisymmetric, the spin part must be symmetric, so that the overall wave function is antisymmetric with respect to exchange of the electrons. There are three ways to accomplish this:

$$\psi_{VB-1}(1,2) = N_-[s_A(1)s_B(2) - s_B(1)s_A(2)]\frac{1}{\sqrt{2}}[\alpha(1)\beta(2) + \beta(1)\alpha(2)] \quad (13.11)$$

$$\psi_{VB-2}(1,2) = N_-[s_A(1)s_B(2) - s_B(1)s_A(2)][\alpha(1)\alpha(2)] \quad (13.12)$$

$$\psi_{VB-3}(1,2) = N_-[s_A(1)s_B(2) - s_B(1)s_A(2)][\beta(1)\beta(2)] \quad (13.13)$$

where VB− indicates that the wave function is the *difference* of two equivalent wave functions and N_- is a normalization constant. We write the normalization constants N_+ and N_- in compact form as

$$N_\pm = (2 \pm 2S^2)^{-1/2} \quad (13.14)$$

to indicate that the plus sign is used for N_+ and the minus sign is used for N_-. The term S is called the *overlap integral*, which occurs frequently in molecular electronic structure calculations. It is given by

$$\begin{aligned} S &= \langle s_A | s_B \rangle \\ &= \int s_A^* s_B d\tau \end{aligned} \quad (13.15)$$

where the asterisk indicates the complex conjugate. The overlap integral S ranges in numerical value from one (in the limit when the two nuclei are in the same place) to zero (in the limit when the nuclei are infinitely far apart) (Figure 13.7). Physically, S may be interpreted as the overlap of the wave functions, but note that S also considers the relative phases of the wave functions.

The three ψ_{VB-} wave functions (Equations 13.11, 13.12, and 13.13) have the same spatial part (and different spin part); thus, they are degenerate, meaning they have the same energy in the absence of an external electric or magnetic field. To calculate

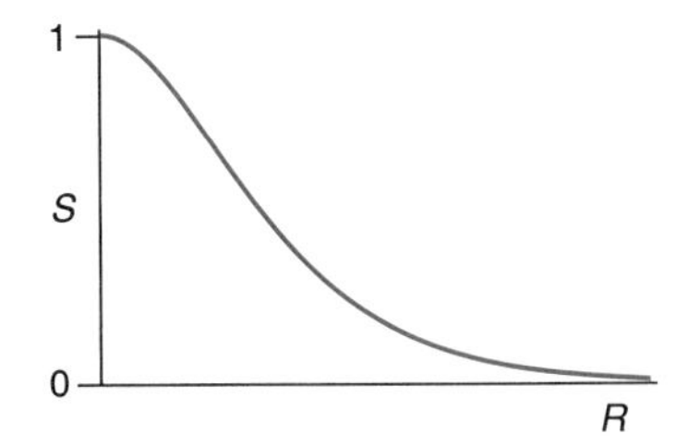

Figure 13.7
The overlap integral S decrease monotonically with internuclear separation R.

the energy of the wave functions ψ_{VB+} and ψ_{VB-}, we use the Hamiltonian operator and calculate the expectation value. We have noted that there is no analytical solution to the Schrödinger equation for the hydrogen molecule. To solve for the energy of H_2, we adopt a perturbation theory approach. We rearrange the terms in the H_2 Hamiltonian operator (see Equation 13.6), so that it is written as two hydrogen-atom Hamiltonian operators (see Equation 12.1), plus a perturbation term.

$$\hat{H} = \underbrace{-\frac{1}{2}\nabla_1^2 - \frac{1}{r_{1A}}}_{\text{H-atom } e^- \#1} \underbrace{-\frac{1}{2}\nabla_2^2 - \frac{1}{r_{2B}}}_{\text{H-atom } e^- \#2} \underbrace{-\frac{1}{r_{1B}} - \frac{1}{r_{2A}} - \frac{1}{R} + \frac{1}{r_{12}}}_{\text{Perturbation term}} \tag{13.16}$$

Under the Born–Oppenheimer approximation, the internuclear separation R is a fixed parameter, and we can solve for the energy of the system. Here again, we omit the details of the calculation and highlight the important results. In compact notation, we write the expectation value of the energy of the system as

$$\langle E_{\pm} \rangle = 2E_{1s} + \frac{J \pm K}{1 \pm S^2} \tag{13.17}$$

where J is the *Coulomb integral*, K is the *exchange integral* (discussed below), S is the overlap integral, and E_{1s} is the energy of a hydrogen atom. $E_{\pm}$ is a shorthand notation for the energy of the wave functions ψ_{VB+} (Equation 13.10) and ψ_{VB-} (any one of Equations 13.11, 13.12, and 13.13). If we repeat this calculation for a series of internuclear separations (R), we generate potential-energy curves as that shown in Figure 13.8. The ψ_{VB+} wave function is bound, with a well depth of 300 kJ mol^{-1}, whereas ψ_{VB-} is antibonding (note that the curve does not have a minimum). Experimentally, the curve is found to have a depth of 458 kJ mol^{-1}.

The J term represents the Coulomb integral, given by

$$J = \iint 1s_A(1)^2 \left[-\frac{1}{r_{1B}} - \frac{1}{r_{2A}} - \frac{1}{R} + \frac{1}{r_{12}} \right] 1s_B(2)^2 d\tau_1 d\tau_2 \tag{13.18}$$

Equation 13.18 is a double integral over the coordinates of electrons 1 and 2 ($d\tau_1$ and $d\tau_2$). The term $1s_A(1)^2$ represents the charge density of electron 1 localized in a hydrogen-atom $1s$ orbital centered on nucleus A, and $1s_B(2)^2$ applies to electron 2 on nucleus B. The Coulomb integral incorporates *all* of the Coulombic interactions between charged particles (electron–electron, nuclear–nuclear, and electron–nuclear). It is noteworthy that J accounts for only 10% of the chemical bonding in

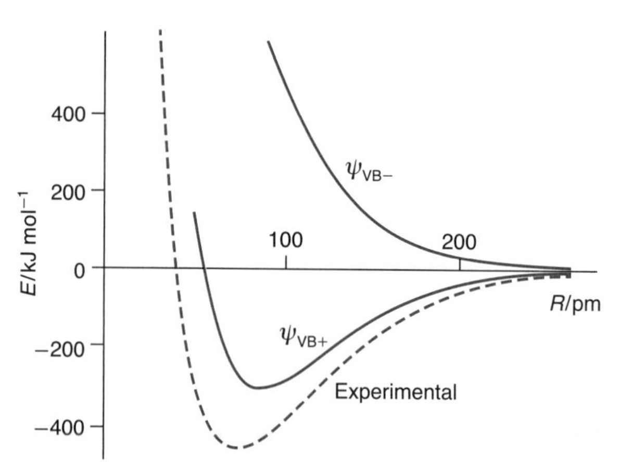

Figure 13.8
Experimental and theoretical potential energy curves for the hydrogen molecule.

molecular hydrogen. In general, the dominant term is the exchange integral K, given by

$$K = \iint 1s_{\rm A}(1) 1s_{\rm B}(1) \left[-\frac{1}{r_{1\rm B}} - \frac{1}{r_{2\rm A}} - \frac{1}{R} + \frac{1}{r_{12}} \right] 1s_{\rm A}(2) 1s_{\rm B}(2) d\tau_1 d\tau_2 \tag{13.19}$$

The exchange integral is also a double integral, but its physical interpretation is trickier because it describes a purely quantum mechanical phenomenon. In a sense, K accounts for the lowering of energy due to the spreading (and/or sharing) of the electron probability density. Analogous to the particle-in-a-box case (see Section 10.9), the system's energy is lowered because the electrons have more room to move about. The exchange integral is sometimes called the resonance integral, which originates from classical physics. It vanishes when the orbitals do not overlap. The overlap integral is always positive while the Coulomb and exchange integrals are both negative. That the potential energy curve for $\psi_{\rm VB-}$ is always positive illustrates that the exchange integral is of larger magnitude than the Coulomb integral.

$K < J < 0$ and $0 \le S \le 1$.

Examination of the hydrogen molecule VB wave functions $\psi_{\rm VB+}$ and $\psi_{\rm VB-}$ reveals, not surprisingly, that they closely resemble the ground-state and excited-state wave functions for the hydrogen molecular cation (Figure 13.9). The lower energy

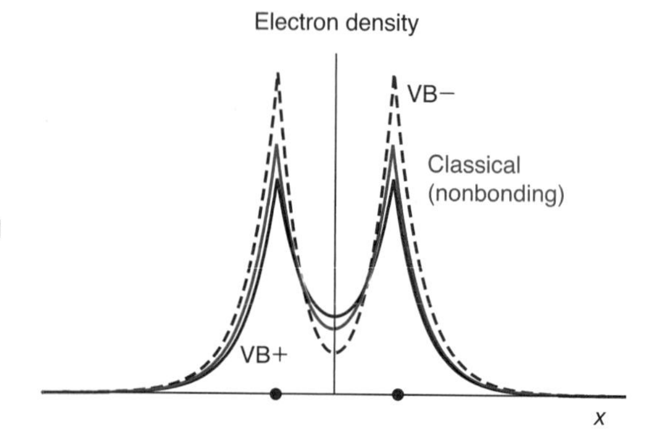

Figure 13.9
Electron density for H_2 calculated from the $\psi_{\rm VB+}$ wave function (solid black line), the $\psi_{\rm VB-}$ wave function (dashed black line), and the sum of two noninteracting hydrogen atom 1*s* wave functions (red line).

VB wave function ψ_{VB+} is a σ_g orbital, which is bonding, and the higher energy VB wave function ψ_{VB-} is an antibonding σ_u^* orbital. The ψ_{VB+} wave function has more electron density localized between the two nuclei than does the superposition of two overlapping hydrogen-atom 1*s* orbitals. This agrees with our qualitative view of electron density in a covalent bond.

VB theory is a valuable approach in that it is straightforward and it accounts for many important chemical concepts. It quantifies predictions from Lewis electron dot diagrams. VB theory may be improved, for instance, by incorporating ionic contributions (as suggested in Figure 13.6). Including ionic contributions using perturbation theory, for example, significantly improves the agreement between the predicted and experimental bond energy. Simple VB theory does not, however, always agree with experiment in predicting photoelectron spectra (see Section 14.5) and magnetic properties (such as paramagnetism, as we shall see in Section 13.5). In the next section, we consider an alternative approach, molecular orbital theory, which addresses several shortcomings of VB theory.

13.4 Molecular Orbital Approach

The molecular orbital (MO) approach to covalent bonds places pairs of electrons into molecular orbitals. Whereas in the VB approach valence electrons are localized between two adjacent nuclei, molecular orbitals may spread about the entire molecule. In this section we continue our consideration of the hydrogen molecule. The MO approach was developed mainly by Friedrich Hund and the American chemist Robert Mulliken (1896–1986).

The common procedure to construct MOs is called the *linear combinations of atomic orbitals* and hence the acronym LCAO-MO. For the hydrogen molecule, we use hydrogen 1*s* AOs. For other atoms, such as oxygen, we use hydrogenlike orbitals. The notation is the same as those used in VB theory: $1s_A(1)$ represents electron number 1 in a hydrogenlike 1*s* atomic orbital centered on nucleus A. Similarly, $1s_B(2)$ represents electron number 2 in a hydrogenlike 1*s* orbital centered on nucleus B. For the hydrogen molecule, there are two simple ways to make a MO—addition and subtraction of hydrogen 1*s* orbitals on A and B. For electron 1 we have

Linear here means that we take weighted sums of atomic orbitals that are raised to the *first* power.

$$\phi_g(1) = N_g[1s_A(1) + 1s_B(1)] \tag{13.20}$$

$$\phi_u(1) = N_u[1s_A(1) - 1s_B(1)] \tag{13.21}$$

where N_g and N_u are normalization constants. Similar equations are written for electron 2. For simplicity, we frequently omit the normalization constant. These ϕs are one-electron wave functions that are the LCAO-MOs.* As we might expect, the hydrogen molecule LCAO-MOs look like the ground-state and excited-state wave functions for the hydrogen molecular cation. The MO produced by the sum of AOs is a σ bonding orbital with gerade symmetry, so we label it ϕ_g. The MO formed from the

* In this presentation, we use a *minimal basis set*, meaning that only 1*s* orbitals are used to construct the molecular orbitals. For more accurate results, as is common in computational chemistry software, we could employ a larger basis set, incorporating contributions from additional atomic orbitals (such as larger 1*s*-like, 2*s*, and 2*p* atomic orbitals) into the molecular orbital.

difference of AOs is an ungerade antibonding sigma orbital, ϕ_u^*. The energies of the gerade and ungerade orbitals are

$$E_g = \frac{J' + K'}{1 + S} \qquad E_u = \frac{J' - K'}{1 - S} \tag{13.22}$$

where S is the overlap integral as described in the previous section. The Coulomb (J') and exchange (K') integrals are different from those in the VB approach, and thus are noted with primes. Here, the J' and K' integrals are single-electron integrals over the spatial coordinates of that electron.

$$J' = \int 1s_A(1)\left[-\frac{1}{r_{1B}} - \frac{1}{r_{2A}} - \frac{1}{R} + \frac{1}{r_{12}}\right]1s_A(1)d\tau \tag{13.23}$$

$$K' = \int 1s_A(1)\left[-\frac{1}{r_{1B}} - \frac{1}{r_{2A}} - \frac{1}{R} + \frac{1}{r_{12}}\right]1s_B(1)d\tau \tag{13.24}$$

These are similar to the two-electron Coulomb and exchange integrals J (Equation 13.18) and K (Equation 13.19) that we encountered in the VB method. The exchange integral, which is a quantum mechanical phenomenon, accounts for most of the bonding energy.

The ground-state wave function for the H_2 molecule, ignoring the spin part and the normalization constant, is given by

$$\begin{aligned}\psi_{MO}(1,2) &= \phi_g(1)\phi_g(2)\\ &= [1s_A(1) + 1s_B(1)][1s_A(2) + 1s_B(2)]\end{aligned}$$

We multiply the terms in brackets to obtain four product terms.

$$\psi_{MO}(1,2) = [1s_A(1)1s_B(2) + 1s_A(2)1s_B(1) + 1s_A(1)1s_A(2) + 1s_B(1)1s_B(2)] \tag{13.25}$$

Comparing these four terms with the two obtained with VB theory (see Equation 13.9),

$$\psi_{VB}(1,2) = [1s_A(1)1s_B(2) + 1s_A(2)1s_B(1)] \tag{13.26}$$

we see that the last two terms in Equation 13.25 are ionic terms in which both electrons reside on one nucleus. Thus, we may write an ionic wave function that describes the H_2 molecule as if it were H^-H^+ and H^+H^-; that is, a pure ionic bond.

$$\psi_{ionic}(1,2) = \underbrace{1s_A(1)1s_A(2)}_{H^-\,H^+} + \underbrace{1s_B(1)1s_B(2)}_{H^+\,H^-} \tag{13.27}$$

Furthermore, the MO wave function for the hydrogen molecule may be written as the sum of a VB description, in which electrons are equally shared, and an ionic description, in which electrons are completely transferred:

$$\psi_{MO} = \psi_{VB} + \psi_{ionic} \tag{13.28}$$

Note that to satisfy the Pauli principle for the MO, VB, and ionic wave functions (Equations 13.25, 13.26, and 13.27, respectively), we need to include the proper spin wave function, given by $[\alpha(1)\beta(2) - \beta(1)\alpha(2)]$. In a sense, the VB picture and the ionic

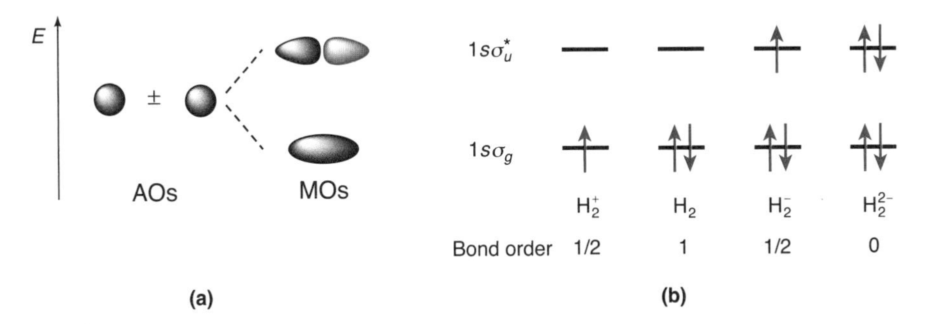

Figure 13.10
(a) The interaction between two hydrogen 1s AOs generates a bonding and an antibonding σ orbital. (b) Energy levels and bond orders of four diatomic hydrogen species.

picture are two extremes, and the true nature of the chemical bond lies somewhere in between. A chemical bond is neither 100 percent ionic, nor 100 percent covalent. The VB picture is zero percent ionic, and thus contains too little ionic character. The MO picture is 50 percent ionic, and thus contains too much ionic character. The fact that the VB treatment does a better job at predicting the energy of the hydrogen molecule than the MO treatment suggests that the H_2 bond has only a small ionic character. In the next section, we show how the percent ionic character of a bond might be estimated.

Using the Aufbau principle, we fill MOs with electrons just as we do with AOs. Here we illustrate the procedure with the family of diatomic hydrogen species H_2^+, H_2, H_2^-, and H_2^{2-} containing one, two, three, and four electrons, respectively (Figure 13.10). Combining two hydrogen 1*s* AOs produces a bonding MO (σ_g) and an antibonding MO (σ_u^*). We write their ground-state electron configurations as follows: H_2^+:$(\sigma_g 1s)^1$, H_2:$(\sigma_g 1s)^2$, H_2^-:$(\sigma_g 1s)^2(\sigma_u^* 1s)^1$, and H_2^{2-}:$(\sigma_g 1s)^2(\sigma_u^* 1s)^2$, where the asterisk denotes the antibonding MO and the superscript denotes the number of electrons in the MO. A qualitative measure of the bond strength is the *bond order* (BO), which is the number of electrons in bonding orbitals minus the number of electrons in antibonding orbitals, divided by 2.

$$\text{BO} = \frac{(\#\ \text{bonding}\ e^- - \#\ \text{antibonding}\ e^-)}{2} \tag{13.29}$$

Thus, molecular hydrogen is predicted to have a bond order of 1; that is, a single covalent bond. Double bonds have bond order 2 and triple bonds have bond order 3. Species such as H_2^+ and H_2^- with a bond order of $\frac{1}{2}$ are difficult to describe using VB theory. Species with a bond order of zero, such as H_2^{2-} and He_2, contain no covalent bond at all, and therefore are predicted not to exist. The helium dimer, He_2, has been observed in the laboratory at low temperatures. It is held together by the weakest of van der Waals forces, and is not considered to be a chemically bound species.

In addition to predicting the qualitative stability of a species, the bond order may be used to predict bond length, bond strength, and bond stiffness (or force constant). In general, bond strength and bond stiffness increase with increasing bond order, whereas bond length decreases with increasing bond order. Thus, double bonds are shorter, stronger, and stiffer than single bonds. Broadly speaking, double bonds have twice the bond strength of single bonds, but they are neither twice as stiff, nor half the length. In the following sections, we shall see some quantitative examples of bond strength and bond length as related to bond order.

13.5 Homonuclear and Heteronuclear Diatomic Molecules

In this section we extend the application of molecular orbital (MO) theory to diatomic molecules beyond hydrogen.

Homonuclear Diatomic Molecules

We again construct MOs using the LCAO-MO approach, starting with second-period homonuclear diatomic molecules (Li_2, Be_2, B_2, C_2, N_2, O_2, and F_2). In constructing MOs, we are guided by a few general principles. In general, AOs should have similar energy in order for them to be combined to form MOs. For homonuclear diatomic molecules, this suggests that only orbitals with the same principal quantum number n will be combined. The stabilization (lowering in energy) of the bonding MO relative to the component AOs is about the same as the destabilization of the antibonding MO. Another general principle is that AOs must have the proper symmetry and orientation to make MOs. Viewing orbitals as wave functions, we note that to combine two AOs to make bonding orbitals, they must have the proper phase for constructive interference.

In schematic depictions of AOs and MOs, we use gray and red colors throughout to represent the phases (+) and (−) of the wave functions, respectively.

So far we have discussed the formation of MOs from 1*s* orbitals. The situation is more involved with 2*p* orbitals because they can interact in two different ways. Consider the "head-on" approach of two 2*p* orbitals. By convention, we take the *z* axis to be the internuclear axis and make the plane of the paper the *x*–*z* plane. Figure 13.11a shows that the constructive and destructive interference between the two orbitals produce a $\sigma_g 2p_z$ bonding MO and a $\sigma_u^* 2p_z$ antibonding MO. Alternatively, these two 2*p* orbitals can approach each other sideways (i.e., aligned parallel to each other), Here, the constructive and destructive interferences produce a $\pi_u 2p_x$ bonding MO and a $\pi_g^* 2p_x$ antibonding MO (Figure 13.11b).

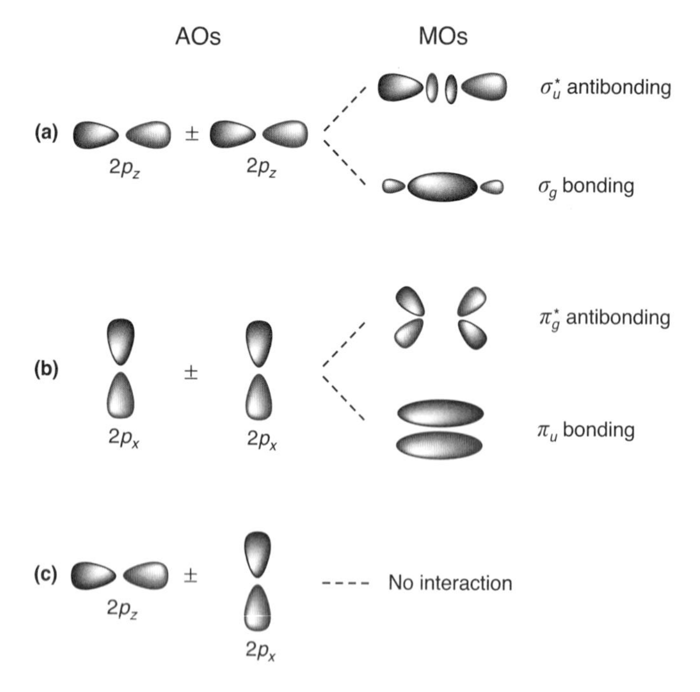

Figure 13.11
(a) The end-to-end interaction of two 2*p* AOs generates a bonding and an antibonding σ MO. (b) The sideways interaction of two 2*p* AOs generates a bonding and an antibonding π MO. (c) No MOs are formed from this interaction.

EXAMPLE 13.1

What types of MOs may be formed from a 2*s* AO from one atom and a 2*p* AO from a second atom?

ANSWER

In combining a 2*s* orbital from one atom with a 2*p* orbital from a second atom, the type of MO produced depends on the relative orientation and phase of the two AOs. If we take the *z* axis as the internuclear axis, then the "head-on" overlap shown in Figure 13.12a results in a bonding MO (a σ orbital) and an antibonding MO (a σ^* orbital). If the 2*s* and the 2*p* orbitals approach each other in the manner shown in Figure 13.12b, then no MOs will be formed.

COMMENT

With *s* AOs, we may either form σ MOs or no MO; we may not form π MOs.

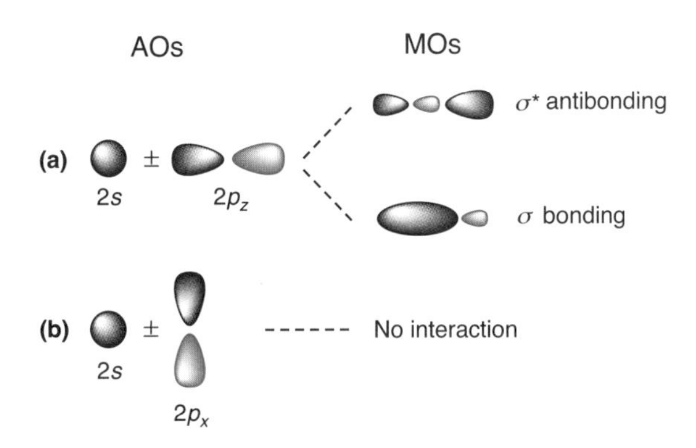

Figure 13.12
(a) A 2*s* AO may interact with a 2*p* AO to make a bonding MO and an antibonding σ MO.
(b) No MOs are formed from this interaction.

Because they lack a center of symmetry, the MOs formed from 2*s* and $2p_z$ AOs are neither gerade (g) nor ungerade (u).

The energetics of the homonuclear diatomic MOs may be understood with a few guiding principles. In general, we would expect that the nodal theorem applies; that is, MOs with fewer nodes would be expected to be of lower energy, and thus filled first by the Aufbau principle. We also anticipate that the energies of the AOs would roughly predict the energies of the corresponding MOs. We examine these general principles in the form of a *correlation diagram*, which shows the AOs that are used to form MOs (Figure 13.13). In the example of homonuclear diatomic molecules, $\sigma_g 1s$ orbitals are far lower in energy (more stable) than $\sigma_g 2s$ orbitals and are omitted in Figure 13.13.

The oxygen molecule illustrates some of the principles and successes of MO theory. A Lewis electron dot diagram and a VB treatment would suggest that molecular oxygen has a bond order of 2 and no unpaired electrons:

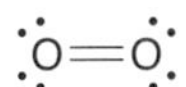

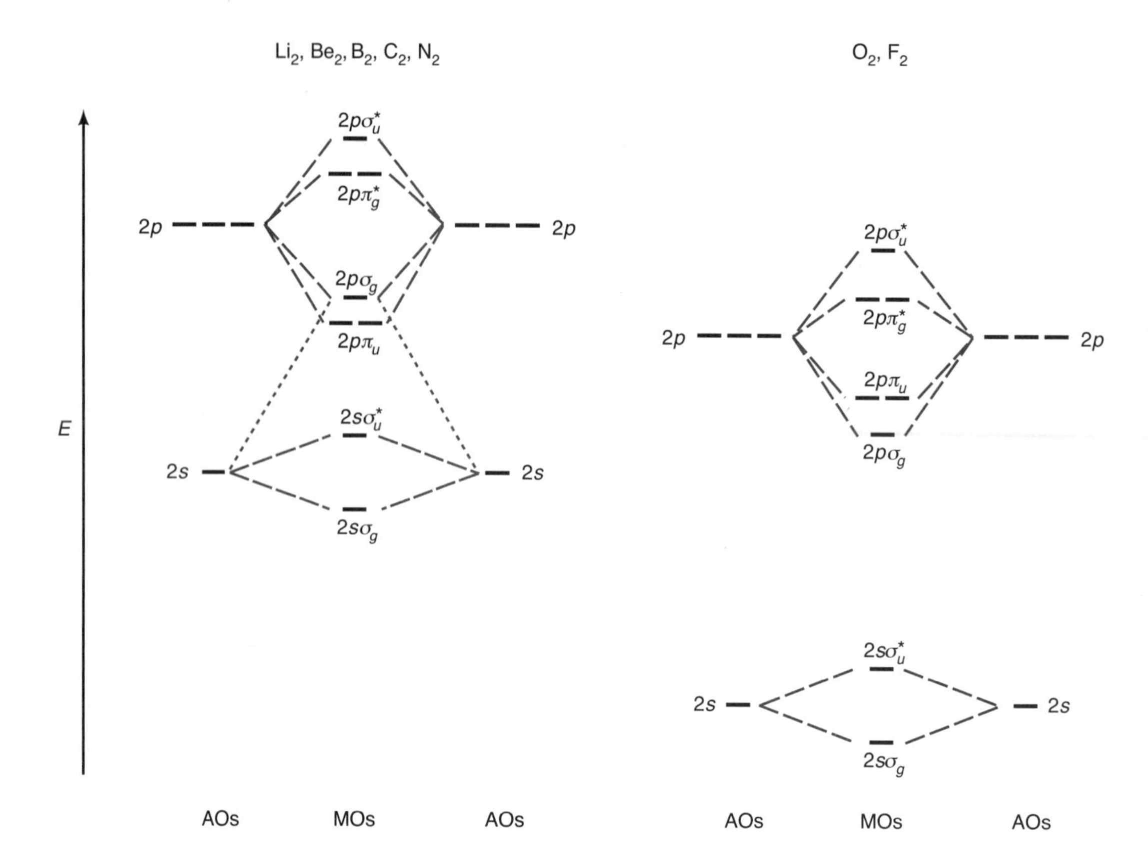

Figure 13.13
Schematic correlation diagram for the second period homonuclear diatomic molecules. The dashed red lines show which AOs are used to make MOs in the LCAO–MO approximation. (a) Li_2, Be_2, B_2, C_2, and N_2 have energy levels sufficiently close together such that the 2*s* AOs contribute to the $2p\sigma_g$ MO and raise its energy above the $2p\pi_u$ MO, as indicated by the dotted red lines. (b) For O_2 and F_2, the 2*s* and 2*p* AOs are sufficiently different in energy so that the 2*s* AOs do not significantly contribute to the MOs made from 2*p* AOs.

Molecules with no unpaired electrons have no net electron spin and are weakly repelled by a magnetic field; they are called *diamagnetic*. Experimentally, we find that molecular oxygen is attracted to an external magnetic field. Species that are attracted to an external magnetic field are called *paramagnetic*. Paramagnetism is a characteristic of molecules with one or more unpaired electrons. Examination of the MO theory treatment of O_2 (see Figure 13.13) shows that O_2 has two unpaired electrons, in agreement with experiment. These unpaired electrons are predicted to be in $\pi_g^* 2p$ MOs. The ground-state electron configuration is

$$(\sigma_g 1s)^2(\sigma_u^* 1s)^2(\sigma_g 2s)^2(\sigma_u^* 2s)^2(\sigma_g 2p_z)^2(\pi_u 2p_x)^2(\pi_u 2p_y)^2(\pi_g^* 2p_x)^1(\pi_g^* 2p_y)^1$$

The MO prediction is that O_2 has four net bonding electrons (10 bonding and 6 antibonding) for a bond order of 2. Thus, in terms of the bond order, MO theory agrees with VB theory, and this bond order is consistent with experimental bond lengths and bond strengths. Note that bond order can't be measured, but we can measure bond length and bond strength, which are roughly related to bond order (Table 13.1).

Table 13.1
Bond Orders, Bond Lengths, and Bond Energies for Diatomic Oxygen Species

Species	Bond order	Bond length/pm	Bond energy/kJ mol^{-1}	Magnetic properties
O_2^+	$\frac{5}{2}$	112	647.8	Paramagnetic
O_2	2	121	497.4	Paramagnetic
O_2^-	$\frac{3}{2}$	135	395.9	Paramagnetic
O_2^{2-}	1	167[a]	67.8	Diamagnetic

[a] The peroxide ion, O_2^{2-} is unstable in the gas phase; a calculation at a high level of theory [H. Nakatsuji and H. Nakai, *Chem. Phys. Letts.* **197**, 339 (1992)] predicts a metastable electronic state with a bond length and barrier to dissociation as shown in the table. All other values are experimental.

EXAMPLE 13.2

Using MO theory, compare the (a) bond order, (b) magnetic properties, and (c) bond length of N_2 and N_2^+.

ANSWER

We refer to Figure 13.13. N_2 has 14 electrons, so its electron configuration is

$$(\sigma_g 1s)^2(\sigma_u^* 1s)^2(\sigma_g 2s)^2(\sigma_u^* 2s)^2(\pi_u 2p_x)^2(\pi_u 2p_y)^2(\sigma_g 2p_z)^2$$

With one fewer electron, the electron configuration of N_2^+ is

$$(\sigma_g 1s)^2(\sigma_u^* 1s)^2(\sigma_g 2s)^2(\sigma_u^* 2s)^2(\pi_u 2p_x)^2(\pi_u 2p_y)^2(\sigma_g 2p_z)^1$$

(a) N_2 has 10 electrons in bonding MOs and four electrons in antibonding MOs, so the bond order is $(10 - 4)/2 = 3$. For N_2^+, the bond order is $(9 - 4)/2 = 2.5$.

(b) There are no unpaired electrons in N_2, so it is diamagnetic. N_2^+ has one unpaired electron and so is a paramagnetic species.

(c) We expect N_2^+ to have a longer bond length because it has a smaller bond order. Indeed, spectroscopic measurements show that the bond length of N_2 is 110 pm, compared with 112 pm for N_2^+.

Heteronuclear Diatomic Molecules

MO theory applies to heteronuclear diatomics much in the way it applies to homonuclear diatomics. The added twist is that the component AOs in heteronuclear diatomics are not of the same energy, and thus the relative energy stabilization is not as great. In general, the closer in energy the AOs, the greater the stability of the MOs relative to the AOs, subject to symmetry and phase considerations as shown above.

The CO Molecule. The Lewis structure of carbon monoxide is

$$^{-}:C\equiv O:^{+}$$

Note that the negative formal charge is on the less electronegative carbon atom. This is consistent with the fact that CO has a rather small dipole moment of about 0.1 D. Carbon monoxide is *isoelectronic* (i.e., has the same number of electrons) with N_2. Figure 13.14 shows the correlation diagram for CO. The electron configuration of CO is given by

$$(\sigma 1s)^2(\sigma^* 1s)^2(\sigma 2s)^2(\sigma^* 2s)^2(\pi 2p_x)^2(\pi 2p_y)^2(\sigma 2p_z)^2$$

The carbon-to-oxygen triple bond in CO is the strongest bond known. CO is a colorless, odorless, and toxic gas because of its high affinity for hemoglobin molecules.

The order of MOs is similar to N_2 rather than O_2. According to Equation 13.29, CO has a bond order of 3. Note that heteronuclear diatomic molecules lack a center of inversion; therefore, we do not use the gerade/ungerade notation for their MOs. When linear combinations of these two AOs are taken to form MOs, the resulting $\sigma 2s$ orbital is lower in energy than the oxygen $2s$ AO and the $\sigma^* 2s$ orbital is higher in energy than

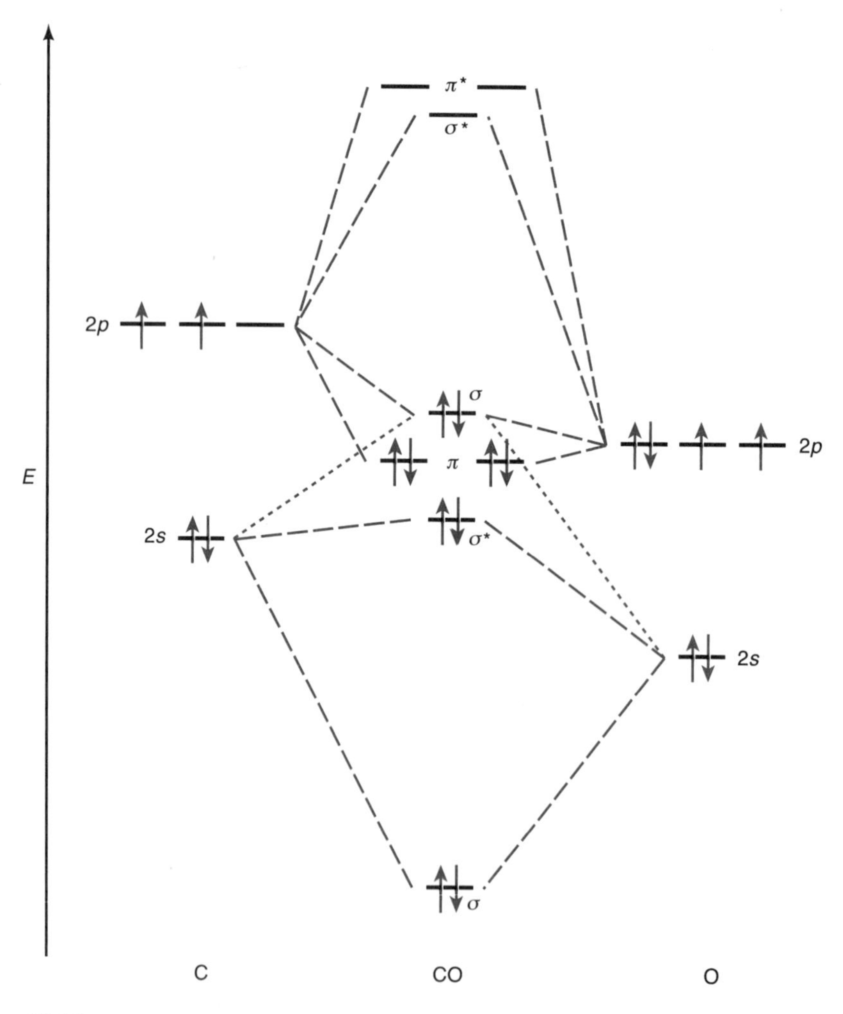

Figure 13.14
Correlation diagram for the heteronuclear diatomic molecule CO (carbon monoxide).

the carbon 2*s* AO. The carbon monoxide MOs also illustrate that the magnitude of the stabilization and destabilization when MOs are made from AOs need not be the same.

The HF Molecule. Hydrogen fluoride has a single bond and three lone pairs on the F atom:

H—F:

Figure 13.15 shows the correlation diagram for the HF molecule. The fluorine 1*s* and 2*s* atomic orbitals lie far below the hydrogen 1*s* orbital. Consequently, they are essentially unchanged from the F atom to the HF molecule. Although the energies of the fluorine 2*p* orbitals are close to that of the hydrogen 1*s* orbital, only the $2p_z$ orbital has the correct symmetry for bonding to occur. The resulting wave function for the σ bonding MO by the H1*s*-F$2p_z$ overlap is

$$\psi_\sigma = c_{\rm H}\phi_{{\rm H}1s} + c_{\rm F}\psi_{{\rm F}2p_z} \tag{13.30}$$

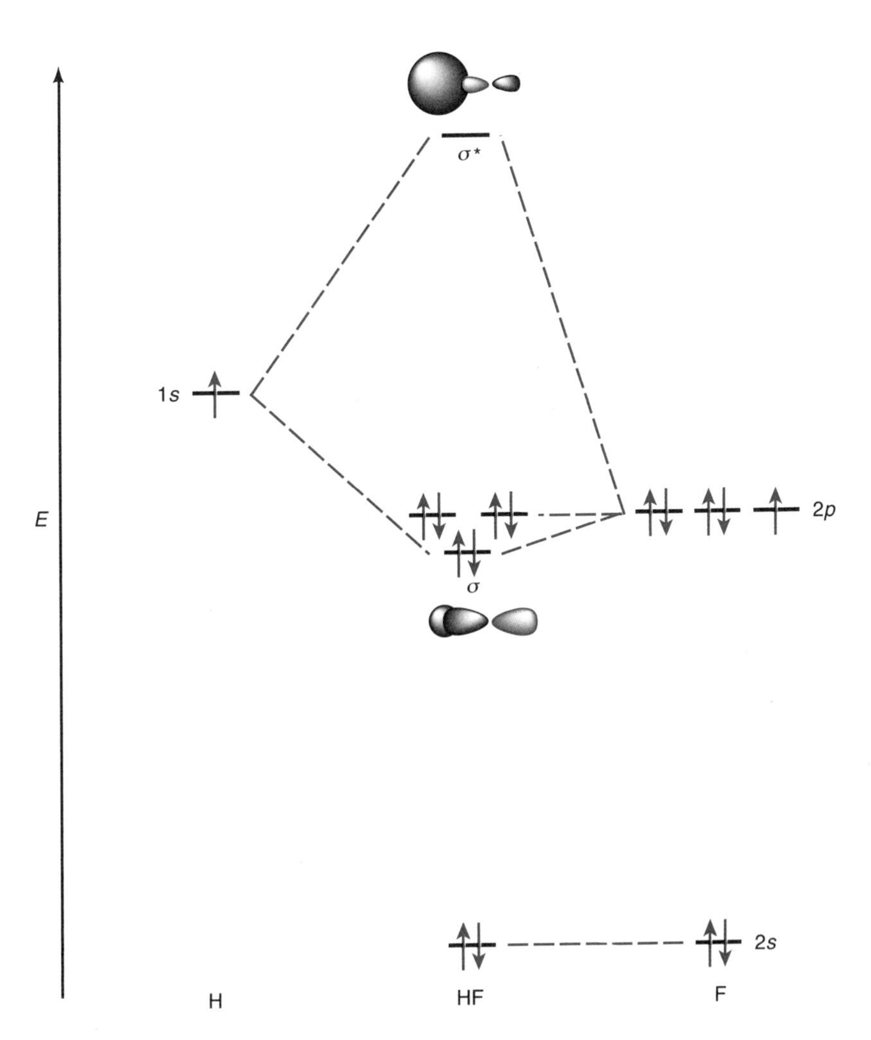

Figure 13.15
Correlation diagram for the heteronuclear diatomic molecule HF (hydrogen fluoride).

Here $c_F > c_H$, meaning that the electron density is more concentrated on the F atom. The F2p_x and F2p_y orbitals remain largely unchanged. Together with the F2s orbital, they constitute nonbonding MOs in which the three lone pairs reside.

Electronegativity, Polarity, and Dipole Moments

In a homonuclear diatomic molecule like H_2 or N_2, the electron density is evenly distributed over the entire molecule and equally shared by the two atoms. Such is not the case for heteronuclear diatomic molecules like CO or HF. The unequal distribution of electron density in these cases results directly from the difference in *electronegativity* (χ), which is the tendency of an atom to attract electrons in a molecule. There are several ways to compare the electronegativity of elements. Mulliken defined electronegativity as the average of the first ionization energy and the electron affinity:

$$\chi = \tfrac{1}{2}(E_{IE} + E_{EA}) \tag{13.31}$$

The most commonly used definition of electronegativity, however, was introduced by Linus Pauling in 1932. Pauling observed that the A–B average bond energy exceeds the mean of the A–A and B–B bond energies by an amount that increases with the increasing polarity of the A–B bond. The Pauling electronegativity scale defines the electronegativity difference between elements A and B as

$$|\chi_A - \chi_B| = 0.102\sqrt{D_{AB} - 0.5(D_{A_2} + D_{B_2})} \tag{13.32}$$

where D_{AB}, D_{A_2}, and D_{B_2} are the bond dissociation energies (kJ mol^{-1}) of molecules AB, A_2, and B_2, respectively. Because only electronegativity differences are obtained from Equation 13.32, one element must be assigned a specific electronegativity value, and then the values for the other elements can be calculated readily. By arbitrarily defining $\chi_H = 2.1$, Pauling developed the electronegativity scale shown in Figure 13.16. Note that electronegativity is not an experimentally measurable quantity, but it is a useful concept in describing chemical bonding.

In heteronuclear diatomic molecules, the electrons are not shared equally. Thus, the chemical bond is *polar* and the molecule posses a permanent electric *dipole moment*. Mathematically, when there is a separation R between the center of the positive charge $+Q$ and the center of the negative charge $-Q$, we define the *magnitude* of the dipole moment μ as

$$\mu = QR \tag{13.33}$$

The dipole moment vector points from the center of positive charge to the center of negative charge.

Physicists use the opposite sign convention from chemists; physicists define the dipole moment vector as pointing from the center of negative charge to the center of positive charge.

The dipole moment of a molecule may be measured experimentally and calculated theoretically. Units for dipole moments are named debye, after Peter Debye. One debye (D) is equal to 0.208 unit of electron charge (e) separated by one Ångstrom (100 pm) so that

$$1\ \text{D} = 3.336 \times 10^{-30}\ \text{C m}$$

1	2	3	4	5	6	7	8	9	10	11	12	13	14	15	16	17	18
H 2.1																	He
Li 1.0	Be 1.5											B 2.0	C 2.5	N 3.0	O 3.5	F 4.0	Ne
Na 0.9	Mg 1.2											Al 1.5	Si 1.8	P 2.1	S 2.5	Cl 3.0	Ar
K 0.8	Ca 1.0	Sc 1.3	Ti 1.5	V 1.6	Cr 1.6	Mn 1.5	Fe 1.8	Co 1.9	Ni 1.9	Cu 1.9	Zn 1.6	Ga 1.6	Ge 1.8	As 2.0	Se 2.4	Br 2.8	Kr 3.0
Rb 0.8	Sr 1.0	Y 1.2	Zr 1.4	Nb 1.6	Mo 1.8	Tc 1.9	Ru 2.2	Rh 2.2	Pd 2.2	Ag 1.9	Cd 1.7	In 1.7	Sn 1.8	Sb 1.9	Te 2.1	I 2.5	Xe 2.6
Cs 0.7	Ba 0.9	La–Lu 1.0–1.2	Hf 1.3	Ta 1.5	W 1.7	Re 1.9	Os 2.2	Ir 2.2	Pt 2.2	Au 2.4	Hg 1.9	Tl 1.8	Pb 1.9	Bi 1.9	Po 2.0	At 2.2	Rn 2.0
Fr 0.7	Ra 0.9	Ac–No 1.1–1.4	Rf	Db	Sg	Bh	Hs	Mt	Ds	Rg	Cn		Fl		Lv		

Figure 13.16
Pauling's electronegativity scale for the elements.

For polyatomic molecules, the net dipole moment is calculated as the vector sum of the individual *bond moments*. Bond moments are the dipole moment between two bonded atoms within a molecule.

Experimental measurement of a dipole moment may be used to calculate the percent ionic character of a covalent bond. Let us return to our example of the heteronuclear diatomic molecule hydrogen fluoride. Experimentally, hydrogen fluoride has a bond length of 91.7 pm and a dipole moment of 1.92 D. If hydrogen fluoride formed a purely covalent bond, it would have equally shared valence electrons and zero dipole moment. On the other hand, if hydrogen fluoride formed a purely ionic bond between spherical ions H^+ and F^-, then the magnitude of each charge would be the fundamental unit of charge e (1.602×10^{-19} C). If we take the charge separation to be the experimental internuclear separation (91.7 pm), then this hypothetical H^+F^- would have a dipole moment of

$$
\begin{aligned}
\mu &= QR \\
&= (1.602 \times 10^{-19}\ \mathrm{C})(91.7\ \mathrm{pm})\left(\frac{10^{-12}\ \mathrm{m}}{\mathrm{pm}}\right)\left(\frac{1\ \mathrm{D}}{3.336 \times 10^{-30}\ \mathrm{C\ m}}\right) \\
&= 4.40\ \mathrm{D}
\end{aligned}
$$

The percent ionic character is defined as

$$
\begin{aligned}
\text{ionic character} &= \frac{\mu_{\text{exptl}}}{\mu_{\text{ionic}}} \times 100\% \\
&= \frac{\mu_{\text{exptl}}}{eR} \times 100\% \qquad (13.34)
\end{aligned}
$$

For hydrogen fluoride, we find

$$\text{ionic character} = \frac{1.92\ \text{D}}{4.40\ \text{D}} \times 100\%$$

$$= 43.6\%$$

The percent ionic character here is consistent with the large difference between the electronegativities of H and F.

EXAMPLE 13.3

Gas-phase lithium fluoride has a dipole moment of 6.28 D and a bond length of 153 pm. Calculate the percent ionic character and comment on your results.

ANSWER

Use Equation 13.35:

$$\text{ionic character} = \frac{\mu_{\text{exptl}}}{\mu_{\text{ionic}}} \times 100\%$$

$$= \frac{6.28\ \text{D}}{(1.602 \times 10^{-19}\ \text{C})(153\ \text{pm})} \times 100\% \times \frac{3.336 \times 10^{-30}\ \text{C m}}{1\ \text{D}} \times \frac{10^{12}\ \text{pm}}{\text{m}}$$

$$= 85.5\ \%$$

The lithium fluoride molecule is not purely ionic, but has some covalent character.

13.6 Polyatomic Molecules

The study of polyatomic molecules requires that we be able to account for both the bonding scheme and geometry. We shall first discuss molecular geometry in terms of a simple and effective approach and then deal with bonding in these molecules.

Molecular Geometry

There is a simple procedure that enables us to predict with considerable success the overall geometry of a molecule if we know the number of electrons surrounding a central atom in its Lewis structure. The basis of this approach is the assumption that electron pairs in the valence shells of an atom repel each other. In a polyatomic molecule, where there are two or more bonds between the central atom and the surrounding atoms, the repulsion between electrons in different bonding pairs causes them to remain as far apart as possible. The geometry that the molecule ultimately assumes (as defined by the positions of all the atoms) minimizes the repulsion. This approach to the study of molecular geometry is called the *valence-shell electron-pair repulsion (VSEPR) model.*

Valence shells are the outermost electron occupied shell of an atom; they hold electrons that are usually involved in bonding.

Two general rules govern the use of the VSEPR model:

1. As far as electron repulsion is concerned, double bonds and triple bonds can be treated like single bonds.
2. If the central atom possesses lone pair(s) in addition to bonding pairs, the repulsion decreases in the following order:

lone-pair vs. lone-pair repulsion	>	lone-pair vs. bonding-pair repulsion	>	bonding-pair vs. bonding-pair repulsion

because lone pairs occupy more space and hence experience a greater repulsion from neighboring lone pairs and bonding pairs.

Table 13.2 shows the geometries of a number of molecules. Methane (CH_4), with four covalent bonds, has a tetrahedral geometry with a bond angle of 109.5° (see Problem 13.49). Ammonia (NH_3) has three covalent bonds plus a lone pair. The four electron pairs have a tetrahedral arrangement and the molecule has a trigonal pyramidal geometry. Because of the greater lone-pair versus bonding-pair repulsion, the bond angle is reduced to 107.3°. The sulfur tetrafluoride (SF_4) molecule has five electron pairs with a trigonal bipyramidal arrangement. Because one of the electron pairs is a lone pair, the molecule must have one of the following geometries:

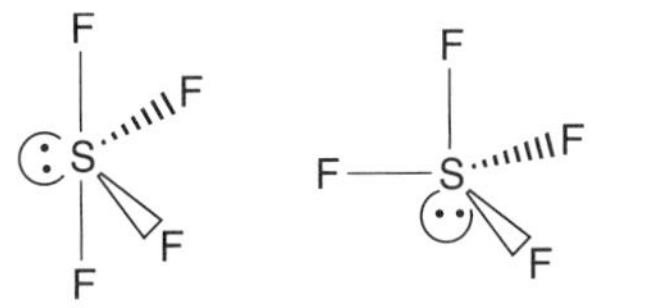

(a) (b)

In (a) the lone pair occupies an equatorial position, and in (b) it occupies an axial position. The axial position has three neighboring pairs at 90° and one at 180°, while the equatorial position has two neighboring pairs at 90° and two more at 120°. The repulsion is smaller for (a), and indeed (a) is the structure observed experimentally. This geometry is described as "seesaw."

Hybridization of Atomic Orbitals

The VSEPR model, while useful, is based on Lewis structures, which explain neither why nor how chemical bonds are formed. Here we treat both chemical bond formation and geometry in terms of the hybridization of atomic orbitals, which is based on VB theory. The concept of hybridization is not used in contemporary computational chemistry, but it is a convenient model for understanding chemical bonding from Lewis structures. While MO theory is favored for describing inorganic complexes, hybridization still plays a role in describing organic molecules. We shall focus on three of the simplest organic molecules: methane, ethylene, and acetylene.

Linus Pauling introduced the term "hybridization" in 1931.

Methane (CH_4). Methane is the simplest hydrocarbon. Physical and chemical studies show that all four C–H bonds are identical in length and strength, and the molecule has tetrahedral symmetry. The HCH bond angle is 109.5°. How can we explain carbon's tetravalency? A carbon atom in its ground state would not form four single

Table 13.2
Molecular Geometries as Predicted by VSEPR Theory

Notation[a]	Groups of e^-	Lone pairs	Hybridization of central atom	Molecular geometry	Example molecule	Molecular structure
AB_2	2	0	*sp*	Linear	BeH_2	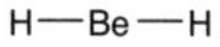 H—Be—H
AB_3	3	0	sp^2	Trigonal planar	BF_3	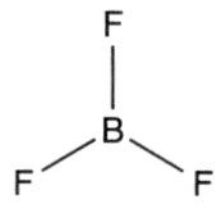
AB_2E	3	1	sp^2	Bent, ~120°	SO_2	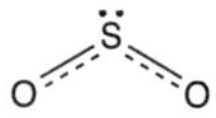
AB_4	4	0	sp^3	Tetrahedral	CH_4	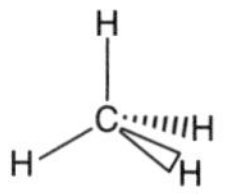
AB_3E	4	1	sp^3	Trigonal pyramidal	NH_3	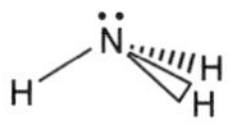
AB_2E_2	4	2	sp^3	Bent, ~109.5°	H_2O	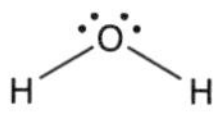
AB_5	5	0	sp^3d	Trigonal pyramidal	PF_5	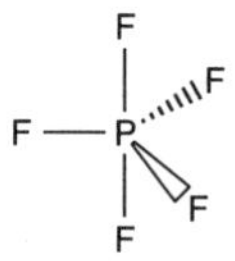
AB_4E	5	1	sp^3d	See-saw	SF_4	
AB_4E_2	5	3	sp^3d	Linear	XeF_2	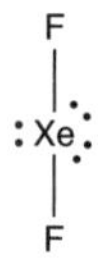
AB_6	6	0	sp^3d^2	Octahedral	SF_6	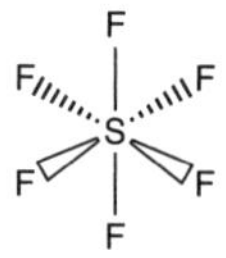
AB_5E	6	1	sp^3d^2	Square pyramidal	ClF_5	
AB_4E_2	6	2	sp^3d^2	Square planar	XeF_4	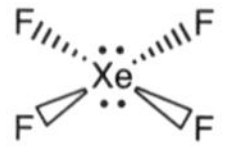

[a] "A" represents the central atom, "B" a bond pair, and "E" a lone pair of electrons.

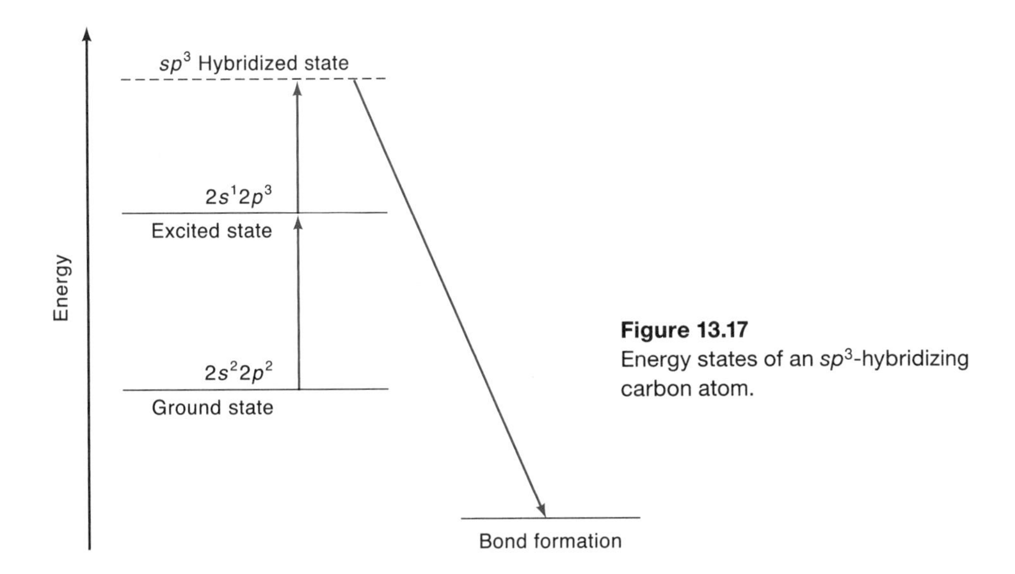

Figure 13.17
Energy states of an sp^3-hybridizing carbon atom.

bonds. Promoting a $2s$ electron into the empty $2p$ orbital would result in four unpaired electrons ($2s^12p^3$), which could form four C—H bonds. If this were the case, methane would contain three C—H bonds of one type and a fourth C—H bond of a different type. This configuration is contrary to experimental evidence. The fact that all four bonds are identical suggests that the bonding atomic orbitals of carbon are all equivalent, meaning that the s and p orbitals are mixed, so that hybridized, or *hybrid*, orbitals are formed. Because there are one s and three p orbitals, this process is called sp^3 hybridization.

Figure 13.17 shows the energy changes that occur in the process of hybridization. The excited state labeled $2s^12p^3$ is real and can be detected spectroscopically. The valence state, that is, the state in which the four equivalent hybrid orbitals are formed, is not real in the sense that it does not exist for an isolated carbon atom, but for us to imagine such a state just before the formation of the methane molecule is convenient. As Figure 13.17 shows, extra energy is needed to reach this state or to hybridize the atomic orbitals, but the investment is more than compensated by the release of energy that results from bond formation.

Mathematically, the mixing of atomic orbitals to form four hybrid orbitals ϕ_1, ϕ_2, ϕ_3, and ϕ_4, can be represented by*

$$\phi_1 = \tfrac{1}{2}(2s + 2p_x + 2p_y + 2p_z)$$

$$\phi_2 = \tfrac{1}{2}(2s - 2p_x - 2p_y + 2p_z)$$

$$\phi_3 = \tfrac{1}{2}(2s + 2p_x - 2p_y - 2p_z)$$

$$\phi_4 = \tfrac{1}{2}(2s - 2p_x + 2p_y - 2p_z) \tag{13.35}$$

* Do not confuse the linear combination of atomic orbitals here with the LCAO-MO procedure discussed earlier. Because all the atomic orbitals originate from the *same* atom, the resulting orbitals ($\phi_1, \phi_2, \ldots$) are still atomic orbitals, although they are no longer "pure" s and p orbitals. No MOs are generated at this stage.

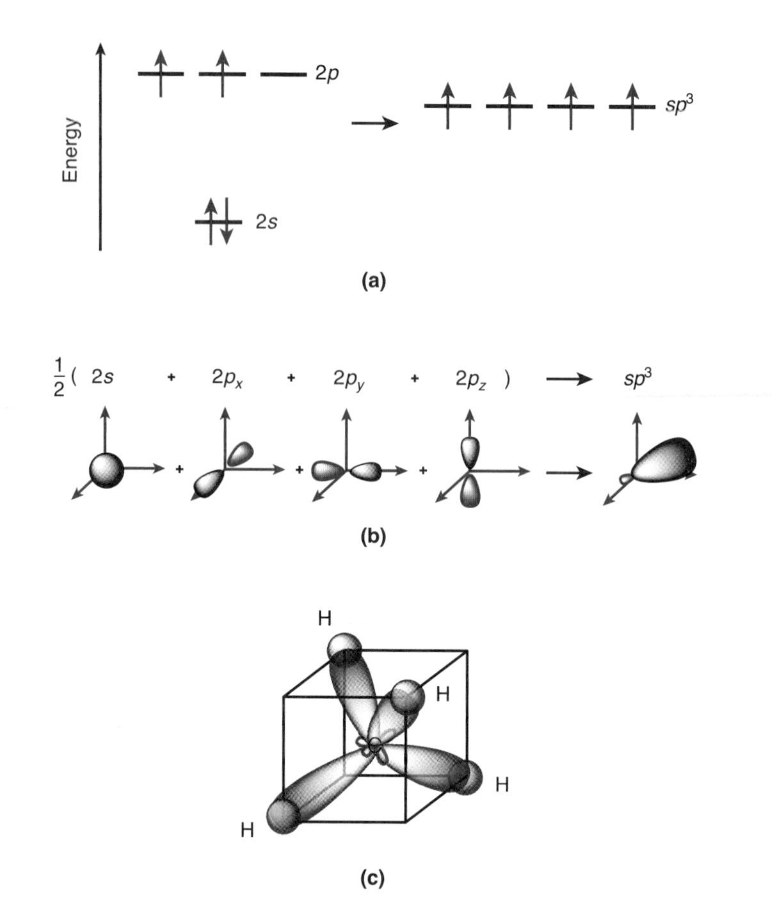

Figure 13.18
(a) Arrangement of carbon 2*s* and 2*p* electrons in sp^3 hybridization. (b) Formation of an sp^3 orbital according to Equation 13.35. (c) The formation of the four C–H bonds between the sp^3 hybrid orbitals and the hydrogen 1*s* orbitals.

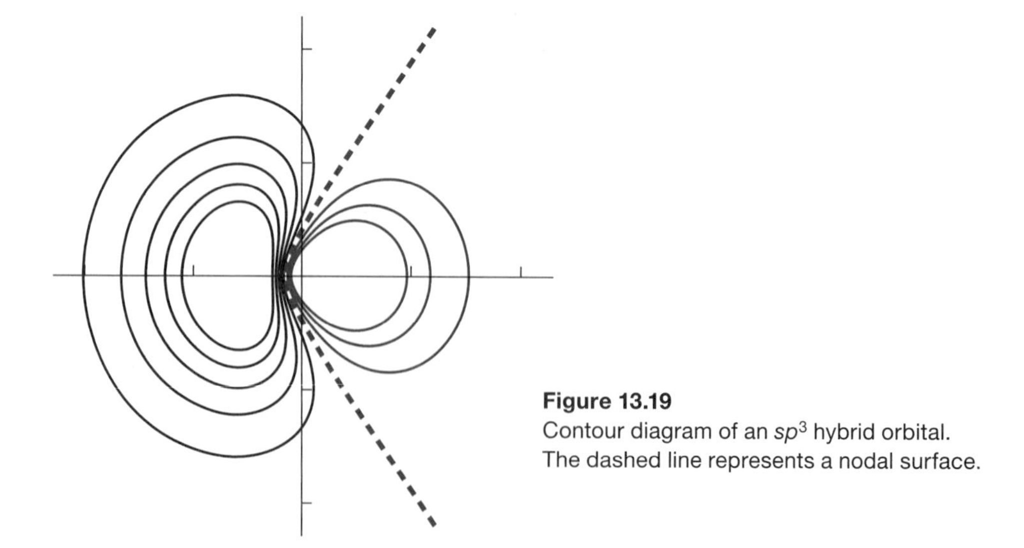

Figure 13.19
Contour diagram of an sp^3 hybrid orbital. The dashed line represents a nodal surface.

where $2s$, $2p_x$, $2p_y$, and $2p_z$ represent the carbon atomic orbitals and $\frac{1}{2}$ is the normalization constant. Note that the combinations of s and p character to each of the orbitals are $\frac{1}{4}$ and $\frac{3}{4}$, respectively. Each sp^3 hybrid orbital has the shape shown in Figure 13.18; its direction is determined by the relative signs in Equation 13.35. A C−H σ bond can then be formed by the overlap between a carbon sp^3 hybrid orbital and a hydrogen $1s$ orbital. Figure 13.19 shows the cross section of a computer-generated sp^3 hybrid orbital.

Ethylene (C_2H_4). Ethylene is a planar molecule; the HCH angle is about 120°. In contrast to methane, each carbon atom is bonded to only three atoms. Both the geometry and bonding can be understood if we assume that each carbon atom is sp^2 hybridized. As Figure 13.20a shows, mixing the s orbital with only two p orbitals (say, $2p_x$ and $2p_y$) produces three sp^2 hybrid orbitals (which all lie in the xy plane), plus a pure p orbital (the $2p_z$ orbital). These three hybrid orbitals are then used to form two σ bonds with the two hydrogen atoms and a σ bond with the other carbon atom (Figure 13.20b). The $2p_z$ orbitals on the two carbon atoms can also overlap sideways to form a π bond (Figure 13.20c). These three hybrid orbitals are represented by

$$\phi_1 = \frac{1}{\sqrt{3}}2s + \sqrt{\frac{2}{3}}2p_x$$

$$\phi_2 = \frac{1}{\sqrt{3}}2s - \frac{1}{\sqrt{6}}2p_x + \frac{1}{\sqrt{2}}2p_y$$

$$\phi_3 = \frac{1}{\sqrt{3}}2s - \frac{1}{\sqrt{6}}2p_x - \frac{1}{\sqrt{2}}2p_y \tag{13.36}$$

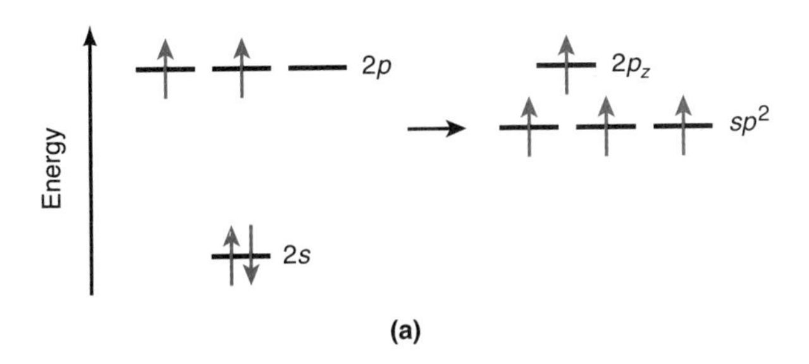

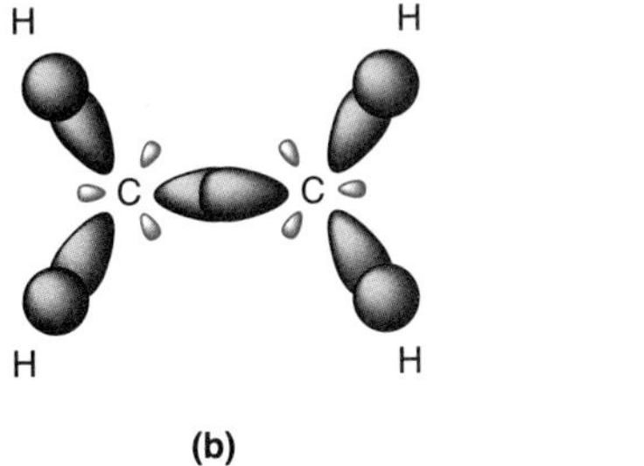

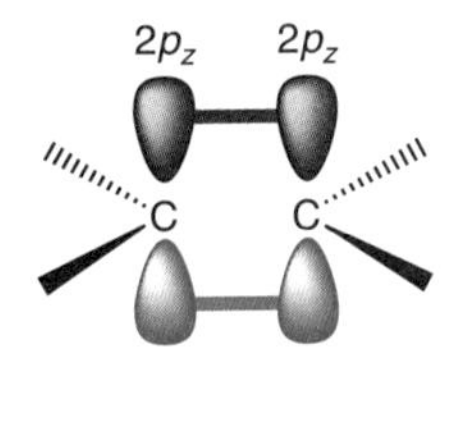

Figure 13.20
(a) Arrangement of carbon $2s$ and $2p$ electrons in sp^2 hybridization. (b) The σ bonds between carbon and hydrogen and between the two carbon atoms in ethylene. (c) The π bond formed by the sideways overlap of the two $2p_z$ orbitals.

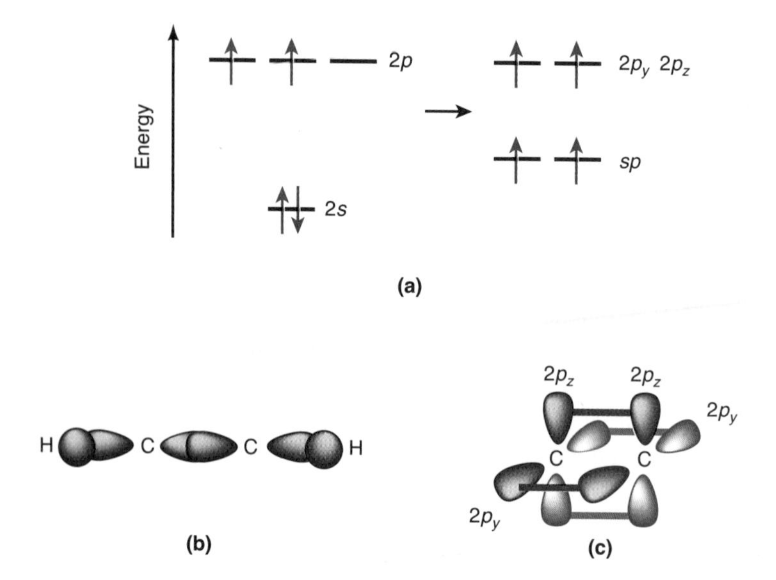

Figure 13.21
(a) Arrangement of carbon 2*s* and 2*p* electrons in *sp* hybridization. (b) The σ bonds between carbon and hydrogen and between the two carbon atoms in acetylene. (c) The π bonds between the two $2p_x$ and between the two $2p_y$ orbitals.

Acetylene (C_2H_2). Acetylene is a linear molecule. From Figure 13.21a we see that by mixing the 2*s* orbital with only one *p* orbital (the $2p_x$ orbital), we obtain two *sp* hybrid orbitals plus two pure *p* orbitals (the $2p_y$ and $2p_z$ orbitals). Consequently, each carbon atom forms two σ bonds (one with the hydrogen atom and one with the other carbon atom) and two π bonds (both with the other carbon atom), as shown in Figures 13.21b and 13.21c. The two hybrid orbitals are

$$\phi_1 = \sqrt{\frac{1}{2}}(2s + 2p_x)$$

$$\phi_2 = \sqrt{\frac{1}{2}}(2s - 2p_x) \quad (13.37)$$

Again, $\sqrt{1/2}$ is the normalization constant.

The concept of hybridization applies equally well to elements other than carbon. In ammonia, for example, each N—H bond points to the apex of a slightly distorted tetrahedron; the angle between any two N—H bonds is 107.3°. Because the electron configuration of nitrogen is $1s^22s^22p^3$, the bonding in NH_3 might be explained by assuming that the three *p* orbitals and the hydrogen 1*s* orbitals overlap. But if this were the case, the bond angles would be 90°, because the *p* orbitals are mutually perpendicular. The assumption that nitrogen is sp^3 hybridized is more valid. One of the hybrid orbitals contains the lone pair. Repulsion between the lone-pair electrons and those in the bonding orbitals results in a 107.3° bond angle, which is slightly less than the tetrahedral angle (109.5°). The lone pair is also responsible for the basicity of ammonia. When ammonia dissolves in water, it readily accepts a proton to form the NH_4^+ ion, which possesses perfect tetrahedral symmetry. Although we have looked at

Table 13.3
Covalent Radii for Atoms/Å[a]

	Single-bond radius	Double-bond radius	Triple-bond radius
H	0.37		
C	0.772	0.667	0.603
N	0.70	0.62	0.55
O	0.66	0.62	
F	0.64		
Si	1.17	1.07	1.00
P	1.10	1.00	0.93
S	1.04	0.94	0.87
Cl	0.99	0.89	
Br	1.14	1.04	
I	1.33	1.23	

[a] Reprinted from Pauling, L. *The Nature of the Chemical Bond*, 3rd ed., p. 224. Copyright 1939 and 1940, 3rd ed. copyright 1960 by Cornell University. Used by permission of Cornell University Press.

hybridization only in terms of the *s* and *p* orbitals, the participation of *d* orbitals is also possible for the third-period elements and beyond (See Table 13.2).

Finally, note that the distance between atoms A and B forming a covalent bond remains fairly constant in molecules. The bond length can be expressed as a sum $r_A + r_B$, where r_A and r_B are the *covalent radii* of A and B. For example, the C−C bond distance in many compounds is approximately 1.54 Å, so that the covalent radius for a carbon single bond is taken to be $1.54/2 = 0.77$ Å. Now the C−Cl distance in many compounds is approximately 1.76 Å; therefore, it follows that the covalent radius for Cl is $(1.76 - 0.77)$ Å $= 0.99$ Å. Using this procedure, we can find the covalent radius of many atoms. The covalent radii of several atoms are listed in Table 13.3.

13.7 Resonance and Electron Delocalization

The advantage of the orbital hybridization is that, in addition to explaining the geometry of molecules, it enables us to continue thinking of a chemical bond as the pairing of electrons. The properties of a molecule, however, cannot always be completely represented by a single structure. A case in point is the carbonate ion:

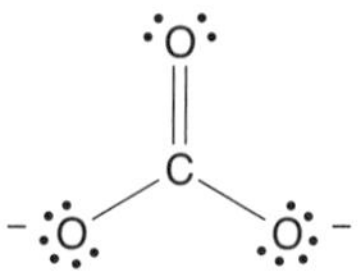

The ion has a planar structure and the OCO bond angle is 120°, which can be readily explained if we assume that the carbon is sp^2 hybridized. But because experimental studies indicate all three carbon–oxygen bonds have equal length and equal strength, the structure shown above is inadequate to describe the ion. The position of the C=O double bond is chosen arbitrarily, so we must consider the following three structures together:

A B C

The double-headed arrow indicates that we are looking at *resonance structures* of the carbonate ion. The term *resonance* means the use of two or more Lewis structures for a molecule (or ion) that cannot be described fully with a single Lewis structure. The point is that none of the individual resonance structures accurately represents the carbonate ion; the ion is best represented by the superposition of all three resonance structures. Clearly, then, the character of each carbon–oxygen bond is somewhere between a single bond and a double bond, in accord with experimental observations. No evidence whatsoever indicates that these three structures actually oscillate back and forth. In fact, each of the three resonance structures is of a nonexistent ion. The model above merely enables us to solve the dilemma of trying to explain properties such as bond length and bond strength. In VB theory, we can write the wave functions for the carbonate ion as

$$\psi = c_A\psi_A + c_B\psi_B + c_C\psi_C \tag{13.38}$$

where ψ_A, ψ_B, and ψ_C are the wave functions for the three individual resonance structures and c_A, c_B, and c_C are all equal.

The concept of resonance is applied most often to aromatic hydrocarbons. In 1865, the German chemist August Kekulé (1829–1896) first proposed the ring structure for benzene. Since then, considerable progress has been made in the study of these molecules. The measured carbon–carbon distance in benzene is 1.40 Å, which lies between the single C−C bond (1.54 Å) and the double C=C bond (1.33 Å). It is more realistic to describe the resonance between two Kekulé structures as follows:

The planar hexagonal shape of benzene was definitively confirmed by the Irish crystallographer Kathleen Lonsdale (1903–1971) using neutron and X-ray diffraction.

In VB theory, we view the electron distribution to be a superposition of the two resonance structures. The single bonds are σ bonds, and each double bond consists of a σ bond plus a π bond. In the superposition state, the carbon–carbon bond order is 1.5, which is the average of the two resonance structures. The bond between adjacent carbon atoms may thus be described by the average of the two resonance structures, having $\frac{2}{3}$ σ character and $\frac{1}{3}$ π character. There are three groups of electrons surrounding each carbon atom, and therefore each carbon atom is sp^2 hybridized. The

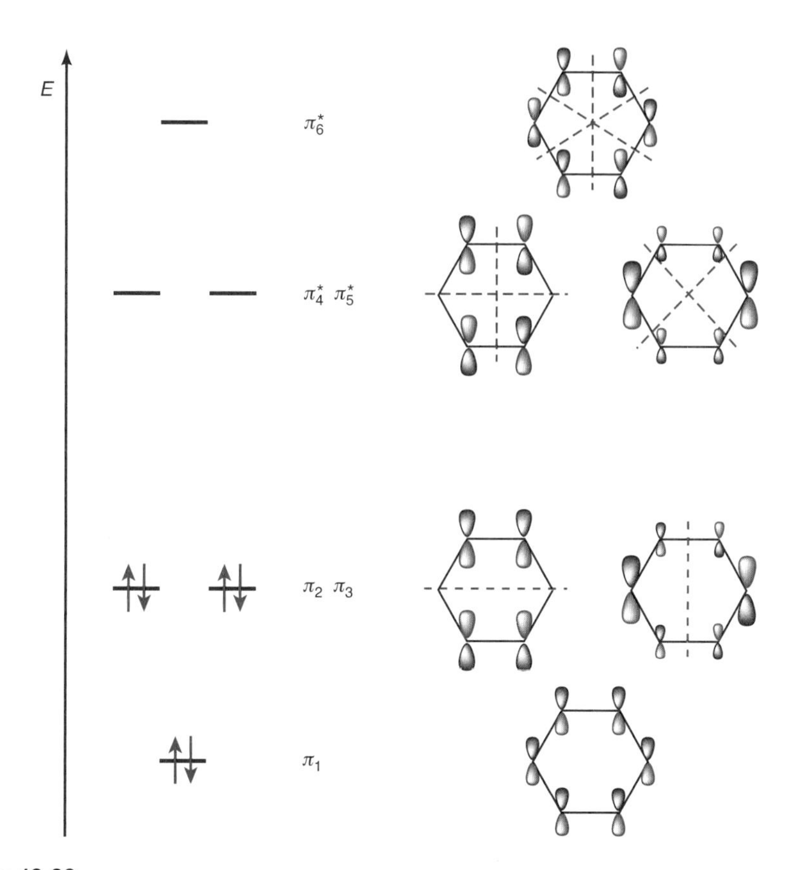

Figure 13.22
Energy-level diagram for the three bonding and three antibonding π MOs in benzene, together with the π electron wave functions. The size of the lobe is proportional to the magnitude of the coefficient of the AO in the MO.

carbon–hydrogen bond order is 1 (a single σ bond) and the predicted symmetry is hexagonal, in agreement with experiment. Just as when we mix AOs to form MOs that are lower in energy, when we invoke the concept of resonance there is a *resonance stabilization* of the resulting electronic structure. Benzene is more stable than we would predict based on a single Lewis structure.

MO theory takes a different approach to explaining the properties discussed above. It uses six carbon $2p$ AOs to form six MOs, of which three are π orbitals and three are π^* orbitals (Figure 13.22). The ordering of the MOs by energy is consistent with the nodal theorem. Each of the six MOs has a nodal plane that is the plane of the molecule. The lowest energy π orbital has no other nodes, and is bonding between all neighboring carbon atoms. There are two additional degenerate bonding orbitals, each of which contains a single nodal plane that is perpendicular to the plane of the molecule. Although the perpendicular node might suggest some antibonding character to these orbitals, they are net bonding orbitals. Continuing up in energy, the next two degenerate orbitals have two nodal planes that are perpendicular to the molecular plane. Although there is still some bonding character to these MOs, they are in fact net antibonding orbitals. At the top on the energy ladder, we find a single MO with three planar nodes perpendicular to the molecular plane, in addition to a node in the molecular plane. This is a purely antibonding orbital—the nodes separate each carbon atom from the others. Within the framework of MO theory, the benzene molecule is often represented as

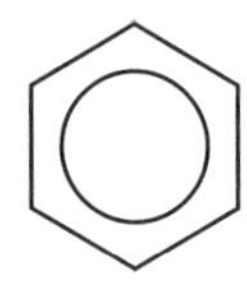

in which the circle indicates that the π bonds are not confined to individual pairs of atoms; rather, the π electron densities are evenly distributed throughout the molecule. This description also accounts for the equal carbon–carbon bond lengths and equal bond strength in benzene.

EXAMPLE 13.4

Using VB theory, describe the geometry of the ozone molecule (O_3), the hybridization of the central atom, and the bond order.

ANSWER

Ozone has 18 valence electrons, six from each of the three O atoms. We can draw two equivalent resonance structures that satisfy the octet rule (Figure 13.23a). Each resonance structure has an O—O single bond and an O=O double bond; thus, the net oxygen–oxygen bond order is 1.5. There are three groups of electrons surrounding the central oxygen atom, so it is sp^2 hybridized and the geometry of the molecule is bent at about 120°.

COMMENT

Recall that bond order is not an experimentally measurable quantity. A bond order of 1.5 implies that the bond length and bond strength are intermediate between those for typical single and double bonds.

Figure 13.23
Ozone (O_3) electronic structure. (a) The Lewis electron–dot diagram has two equivalent resonance structures. (b) The three AOs combine to form three MOs. The size of the lobe is proportional to the coefficient of the AO in the MO.

EXAMPLE 13.5

Sketch the π MOs for the ozone molecule (O_3) and rank them by increasing energy.

ANSWER

We saw in Example 13.4 that ozone is a bent molecule and that it contains a single π bond that is delocalized over all three atoms. In simple MO theory, each of the three atoms contributes a $2p$ AO to form MOs, so we should have three MOs. Like the AOs, each MO contains a nodal plane that is the plane of the molecule. By the nodal theorem, we predict that the lowest-energy π MO will not contain any additional nodes. The other MOs contain one and two nodal planes that are perpendicular to the plane of the molecule, and the MO energy increases with the number of nodes (Figure 13.23b). Of these three MOs, the lowest energy is a π bonding orbital, and the highest-energy orbital is a π^* (antibonding) MO. The third orbital is neither bonding nor antibonding, so it is intermediate in energy between the other two. This is a "n" orbital, for nonbonding, which is analogous to a lone pair in the VB picture.

13.8 Hückel Molecular Orbital Theory

Up to this point, we have relied on qualitative arguments to describe molecular orbitals and energies. In this section we introduce approximations to quantitatively describe conjugated π systems, an important class of organic molecules. A simple way to describe quantitatively planar systems with π bonds is to use Hückel molecular orbital (HMO) theory. HMO theory makes some gross assumptions, yet it provides good descriptions of conjugated π systems. In higher-level molecular electronic structure calculations, HMO theory (or extended HMO theory) is often employed to make initial guesses for the electronic structure of molecules.

As formulated by Erich Hückel (see Section 7.5), HMO theory places electrons in molecular orbitals, just as have other theories we have considered. HMO theory divides valence electrons into two types: the σ network and the π network, then completely ignores the σ network. This assumption is justified on the grounds that it is the π electrons that are responsible for the chemistry and UV–vis electronic spectroscopy in molecules with conjugated π systems. HMO theory uses the Schrödinger equation to find the energies and the wave functions of the π electrons. We start by writing molecular orbitals ψ as linear combinations of atomic orbitals ϕ:

$$\psi = \sum_i c_i \phi_i \tag{13.39}$$

where c_i is the coefficient of the $2p$ atomic orbital (ϕ_i) on atom i that is perpendicular to the σ framework. In general, the coefficients c may be complex numbers, but for the MO to be normalized we have

$$0 \le c_i^* c_i \le 1 \tag{13.40}$$

Ethylene (C_2H_4)

We start with a HMO treatment of ethylene and then examine butadiene, one of the simplest conjugated π systems. For ethylene, we have two carbon atoms and hence two $2p$ AOs (ϕ_1 and ϕ_2) so that

$$\psi = c_1\phi_1 + c_2\phi_2 \tag{13.41}$$

Next, we find the energy by the expectation value of the Hamiltonian operator:

$$E = \frac{\langle\psi|\hat{H}|\psi\rangle}{\langle\psi|\psi\rangle} \tag{13.42}$$

Substituting Equation 13.41 for ψ in Equation 13.42, we obtain

$$E = \frac{c_1^2\langle\phi_1|\hat{H}|\phi_1\rangle + c_2^2\langle\phi_2|\hat{H}|\phi_2\rangle + 2c_1c_2\langle\phi_1|\hat{H}|\phi_2\rangle}{c_1^2 + c_2^2} \tag{13.43}$$

There are only two terms in the denominator because the AOs, ϕ_i, are orthonormal in the Hückel model; that is, they overlap perfectly with themselves, but not at all with any other $2p$ orbital. The overlap integral, S, is equal to zero or one. We represent this with the Krönecker delta symbol, δ_{ij}:

Recall $\delta_{i=j} = 1$ and $\delta_{i\neq j} = 0$.

$$\begin{aligned} S_{ij} &= \langle\phi_i|\phi_i\rangle \\ &= \delta_{ij} \end{aligned} \tag{13.44}$$

Thus, $S_{11} = S_{22} = 1$ and $S_{12} = S_{21} = 0$. To simplify Equation 13.43, we introduce the conventional notation for integrals,

$$H_{11} = \langle\phi_1|\hat{H}|\phi_1\rangle$$
$$H_{22} = \langle\phi_2|\hat{H}|\phi_2\rangle$$
$$H_{12} = \langle\phi_1|\hat{H}|\phi_2\rangle$$
$$H_{21} = \langle\phi_2|\hat{H}|\phi_1\rangle$$

where H_{11} and H_{22} are the Coulomb integrals and H_{12} and H_{21} are the exchange or resonance integrals (see p. 568). Equation 13.43 now becomes

$$E = \frac{c_1^2H_{11} + c_2^2H_{22} + 2c_1c_2H_{12}}{c_1^2 + c_2^2} \tag{13.45}$$

Rearranging, we find

$$c_1^2(H_{11} - E) + c_2^2(H_{22} - E) + 2c_1c_2H_{12} = 0 \tag{13.46}$$

Next we apply the variational principle. We recognize that the ground-state wave function will be the lowest energy with respect to variation in the coefficients c_i. The coefficients are the variational parameters in this problem. For each variational parameter, we take the partial derivative of the energy expression with respect to that parameter, and set the result to zero. For the coefficient c_1 we have the expression,

$$\left(\frac{\partial E}{\partial c_1}\right)_{c_2} = 0$$

which simplifies to

$$2c_1(H_{11} - E) + 2c_2H_{12} = 0$$

or

$$c_1(H_{11} - E) + c_2H_{12} = 0 \tag{13.47}$$

Similarly for c_2,

$$c_1H_{21} + c_2(H_{22} - E) = 0 \tag{13.48}$$

Note that we are now using the symbol E for the minimum energy rather than $\langle E \rangle$, because this represents the best possible value of the energy. Equations 13.47 and 13.48 are known as the *secular equations*. Here we have two equations with two unknowns (c_1 and c_2). They have nontrivial solutions only if the secular determinant of the coefficients is equal to zero:

The trivial solutions to the equations are $c_1 = 0$ and $c_2 = 0$. Determinants are discussed in Section 12.7.

$$\begin{vmatrix} H_{11} - E & H_{12} \\ H_{21} & H_{22} - E \end{vmatrix} = 0 \tag{13.49}$$

Now we introduce some additional assumptions of HMO theory. All carbon atoms are treated in the same way, so that all of the carbon-atom Coulomb integrals (H_{ii}) are equal. By convention, the value of this integral is denoted by α and

$$H_{ii} = \alpha \tag{13.50}$$

As mentioned earlier, α is the energy of an electron in the orbital i due to interaction with the nucleus and the other electrons. In applying HMO theory to hydrocarbon molecules, α is approximately the energy of a $2p$ orbital in a carbon atom. For heteroatoms, α takes on a different value.

Heteroatoms are those other than carbon or hydrogen.

Another Hückel assumption is that the resonance integral (H_{ij}) between p orbitals on nearest neighbors is the same value (conventionally called β):

$$H_{ij} = \beta \qquad \text{for } i \text{ "next to" } j \tag{13.51}$$

The resonance integral for non-neighbors is zero:

$$H_{ij} = 0 \qquad \text{for } i \text{ "not next to" } j \tag{13.52}$$

For ethylene, we have two neighboring carbon atoms, and Equation 13.49 becomes

$$\begin{vmatrix} \alpha - E & \beta \\ \beta & \alpha - E \end{vmatrix} = 0 \tag{13.53}$$

This is a small enough determinant that we can solve it "by hand," using a simple substitution. We divide each term in the determinant by β and let $x = (\alpha - E)/\beta$, so the determinant becomes

$$\begin{vmatrix} x & 1 \\ 1 & x \end{vmatrix} = 0 \tag{13.54}$$

Expanding, we obtain

$$x^2 - 1 = 0 \tag{13.55}$$

This quadratic equation has two solutions:

$$x = \pm 1 \tag{13.56}$$

Substitution of these solutions into the definition of x gives us the energies of the MOs (Figure 13.24):

$$E = \alpha \pm \beta \tag{13.57}$$

Both α and β are negative quantities, so the energy of the lowest–energy Hückel MO is

$$E_1 = \alpha + \beta$$

where the subscript is the orbital number. We assign the index number 1 to the lowest energy orbital and follow the convention that orbital energy increases with number. Under the HMO approximation, there is only one more π orbital (an antibonding orbital) with energy

$$E_2 = \alpha - \beta$$

The total π electronic energy is the sum of the π orbital energies for each occupied orbital,

$$E = \sum_{j=1}^{2} n_j E_j \tag{13.58}$$

where n_j is the number of electrons in orbital j ($n_j = 0$, 1, or 2). For ethylene, we place the two π electrons in the lower energy orbital and calculate the total π electronic energy:

$$\begin{aligned} E &= 2E_1 \\ &= 2\alpha + 2\beta \end{aligned} \tag{13.59}$$

HOMO = Highest Occupied Molecular Orbital; LUMO = Lowest Unoccupied Molecular Orbital.

An individual carbon atom has energy α; thus, two separated carbon atoms have energy 2α, and ethylene has a binding energy of 2β.

The parameter β can be measured spectroscopically. The energy difference between E_1 and E_2 is 2β. This transition involves the promotion of an electron from one orbital to the next (the HOMO to the LUMO) and is the lowest energy electronic transition. In general, π–π transitions such as this one can be measured using UV–vis absorption spectroscopy.

Our next step is to solve for the wave functions. The equations for c_1 and c_2 associated with Equations 13.47 and 13.48 are

$$c_1(\alpha - E) + c_2\beta = 0 \qquad \text{and} \qquad c_1\beta + c_2(\alpha - E) = 0$$

For the bonding π MO, $E = \alpha + \beta$, so that from either of the above equations we get $c_1 = c_2$ and

$$\psi_1 = c_1\phi_1 + c_1\phi_2$$

By imposing the condition that ψ_1 is normalized; that is, $\int \psi_1^*\psi_1 d\tau = 1$, we have $c_1^2(1 + 2S + 1) = 1$. Because in the HMO approximation the atomic orbitals ϕ_1 and ϕ_2 do not overlap, $S = 0$, so $c_1 = 1/\sqrt{2}$. Thus,

$$\psi_1 = \sqrt{\frac{1}{2}}\phi_1 + \sqrt{\frac{1}{2}}\phi_2 \tag{13.60}$$

For the antibonding π MO, $E = \alpha - \beta$ and $c_1 = -c_2$. Hence,

$$\psi_2 = \sqrt{\frac{1}{2}}\phi_1 - \sqrt{\frac{1}{2}}\phi_2 \tag{13.61}$$

Our results are consistent with the nodal theorem—the lowest-energy orbital has no nodes perpendicular to the plane of the carbon atoms, and the higher energy orbital has one node (Figure 13.24).

Once we have the coefficients, c_i, HMO theory enables us to calculate *charge densities*. The charge density, q_i, on a particular atom i is given by the summation over each MO j,

$$q_i = \sum_{j=1}^{2} n_j(c_{ji})^2 \tag{13.62}$$

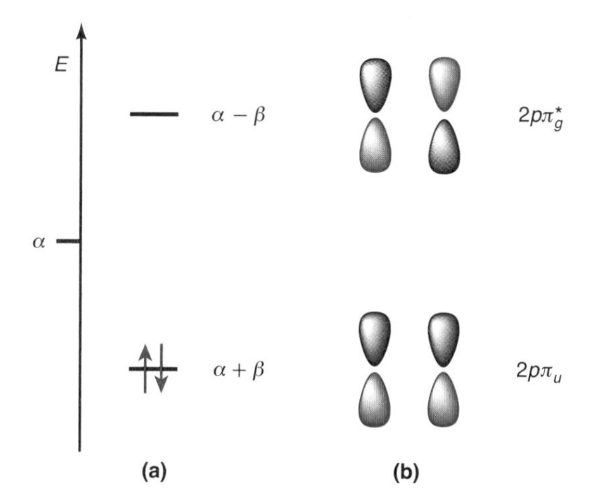

Figure 13.24
(a) Energy-level diagram for the π MOs of ethylene. (b) The wave functions for the bonding and antibonding π MOs.

where c_{ji} is the coefficient of the AO on atom i in the jth MO and n_j is the number of electrons in that MO. In the case of ethylene, $j = 1$ or 2, because we are considering only two molecular orbitals. Applying Equation 13.62 to atom number 1 in ethylene, we see that there are two electrons in the bonding π MO and none in the π^* MO. Therefore, the electron density is given by

$$\begin{aligned} q_1 &= n_1(c_{11})^2 + n_2(c_{21})^2 \\ &= 2\left(\frac{1}{\sqrt{2}}\right)^2 + 0\left(\frac{1}{\sqrt{2}}\right)^2 \\ &= 1 \end{aligned}$$

Performing the same calculation for the second carbon atom yields

$$\begin{aligned} q_2 &= n_1(c_{12})^2 + n_2(c_{22})^2 \\ &= 2\left(\frac{1}{\sqrt{2}}\right)^2 + 0\left(\frac{-1}{\sqrt{2}}\right)^2 \\ &= 1 \end{aligned}$$

That the π electron density is equally distributed between the two carbon atoms is consistent with the Lewis structure for ethylene.

HMO theory may also be used to calculate the bond order between two atoms. In Hückel MO theory, the π bond order P_{ik}^{π} between atoms i and k is given as the sum over all molecular orbitals j of the number of electrons n times the coefficients on the two carbon atoms:

$$P_{ik}^{\pi} = \sum_{j=1}^{2} n_j c_{ji} c_{jk} \tag{13.63}$$

For ethylene, there are two molecular orbitals to consider, so $j = 1$ or 2; and there are two carbon atoms, so $i = 1$ or 2 and $k = 1$ or 2. The π bond order (between carbon atoms 1 and 2) is

$$\begin{aligned} P_{12}^{\pi} &= n_1(c_{11})(c_{12}) + n_2(c_{21})(c_{22}) \\ &= 2\left(\frac{1}{\sqrt{2}}\right)\left(\frac{1}{\sqrt{2}}\right) + 0\left(\frac{1}{\sqrt{2}}\right)\left(\frac{-1}{\sqrt{2}}\right) \\ &= 1 \end{aligned}$$

The total bond order is simply the sum of the σ and π bond orders:

$$P_{ik}^{\text{total}} = P_{ik}^{\sigma} + P_{ik}^{\pi} \tag{13.64}$$

For ethylene

$$P_{12}^{\text{total}} = 1 + 1 = 2$$

Thus, simple HMO theory predicts that ethylene has a double bond between the two carbon atoms, and that the double bond consists of a σ bond and a π bond. This conclusion is consistent with VB theory.

Butadiene (C_4H_6)

The power of HMO theory is better illustrated when it is applied to systems with delocalized MOs, such as butadiene. We number the carbon atoms as follows

Butadiene exists in both cis and trans configurations. Here we treat it as a linear molecule.

$$\underset{1}{H_2C}=\underset{2}{\overset{H}{C}}-\underset{3}{\overset{H}{C}}=\underset{4}{CH_2}$$

The molecular orbital for the π electrons is written as a linear combination of the atomic $C2p_z$ orbitals:

$$\psi = c_1\phi_1 + c_2\phi_2 + c_3\phi_3 + c_4\phi_4 \tag{13.65}$$

As in the ethylene case, we apply the variation principle to find the optimum values for these coefficients by minimizing the energies. The resulting equations are

$$c_1(H_{11} - ES_{11}) + c_2(H_{12} - ES_{12}) + c_3(H_{13} - ES_{13}) + c_4(H_{14} - ES_{14}) = 0 \tag{13.66}$$

$$c_1(H_{21} - ES_{21}) + c_2(H_{22} - ES_{22}) + c_3(H_{23} - ES_{23}) + c_4(H_{24} - ES_{24}) = 0 \tag{13.67}$$

$$c_1(H_{31} - ES_{31}) + c_2(H_{32} - ES_{32}) + c_3(H_{33} - ES_{33}) + c_4(H_{34} - ES_{34}) = 0 \tag{13.68}$$

$$c_1(H_{41} - ES_{41}) + c_2(H_{42} - ES_{42}) + c_3(H_{43} - ES_{43}) + c_4(H_{44} - ES_{44}) = 0 \tag{13.69}$$

For nontrivial solutions, the following secular determinant is equal to zero:

$$\begin{vmatrix} H_{11} - ES_{11} & H_{12} - ES_{12} & H_{13} - ES_{13} & H_{14} - ES_{14} \\ H_{21} - ES_{21} & H_{22} - ES_{22} & H_{23} - ES_{23} & H_{24} - ES_{24} \\ H_{31} - ES_{31} & H_{32} - ES_{32} & H_{33} - ES_{33} & H_{34} - ES_{34} \\ H_{41} - ES_{41} & H_{42} - ES_{42} & H_{43} - ES_{43} & H_{44} - ES_{44} \end{vmatrix} = 0 \tag{13.70}$$

Defining the α and β terms as before and recalling that $S_{ij} = \delta_{ij}$, we write

$$\begin{vmatrix} \alpha - E & \beta & 0 & 0 \\ \beta & \alpha - E & \beta & 0 \\ 0 & \beta & \alpha - E & \beta \\ 0 & 0 & \beta & \alpha - E \end{vmatrix} = 0 \tag{13.71}$$

Next, we divide each element by β and let $x = (\alpha - E)/\beta$ to simplify the determinant and obtain

$$\begin{vmatrix} x & 1 & 0 & 0 \\ 1 & x & 1 & 0 \\ 0 & 1 & x & 1 \\ 0 & 0 & 1 & x \end{vmatrix} = 0 \tag{13.72}$$

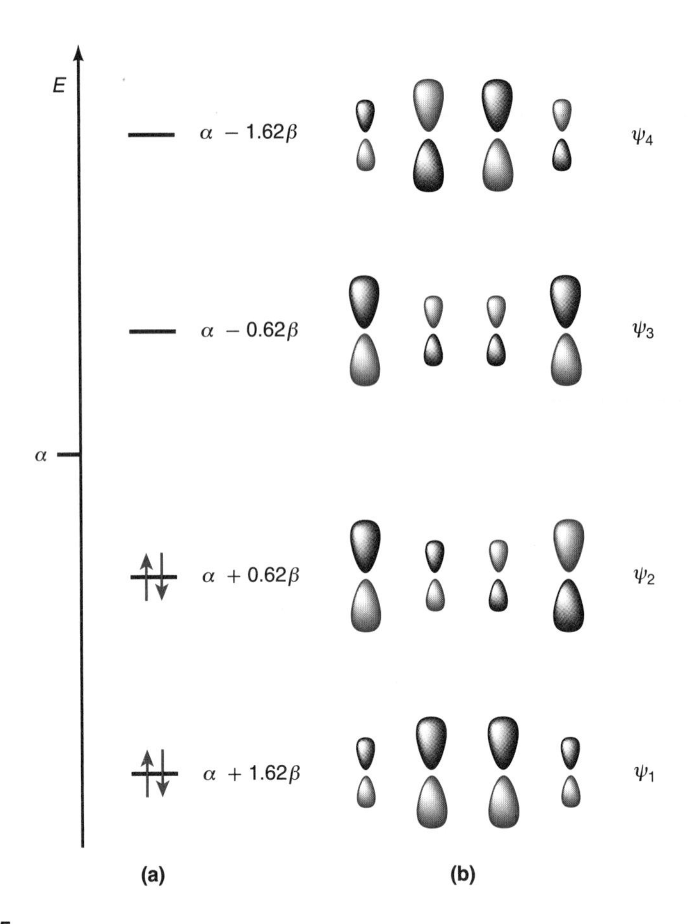

Figure 13.25
(a) Energy-level diagram for the π MOs of butadiene. (b) The wave functions for the bonding and antibonding π MOs. The size of the lobe is proportional to the coefficient of the AO in the MO.

This 4×4 determinant can be expanded to give

$$x^4 - 3x^2 + 1 = 0$$

This equation has four roots, which may be found by using the quadratic formula for x^2 and then taking the square root, giving

$$x = \pm 1.62 \quad \text{and} \quad x = \pm 0.62$$

In its ground electronic state, butadiene has four π electrons, which fill the two lowest-energy MOs (Figure 13.25). Thus, using Equation 13.58, the total π binding energy of butadiene is

$$\begin{aligned} E &= 2(\alpha + 1.62\beta) + 2(\alpha + 0.62\beta) \\ &= 4\alpha + 4.48\beta \end{aligned}$$

Recall that the total energy of an ethylene molecule is $(2\alpha + 2\beta)$. We can determine the difference between a butadiene molecule and two ethylene molecules by writing

$$E_{\text{butadiene}} - 2E_{\text{ethylene}} = (4\alpha + 4.48\beta) - 2(2\alpha + 2\beta)$$
$$= 0.48\beta$$

Thus, butadiene is more stable than two ethylene molecules by 0.48β, which is known as the *delocalization energy*, or the *resonance energy*. In VB theory, the π electrons are localized between pairs of carbon atoms. In the HMO picture, the π electrons are delocalized over the entire molecule. As we saw in the case of a particle in a box, when electrons have more space to spread out, there is a lowering of the energy of the system.

Our next step is to obtain the wave functions for the π MOs. Applying the Hückel approximations and again using $x = (\alpha - E)/\beta$, we simplify Equations 13.66–13.69 as follows:

$$c_1 x + c_2 = 0 \tag{13.73}$$

$$c_1 + c_2 x + c_3 = 0 \tag{13.74}$$

$$c_2 + c_3 x + c_4 = 0 \tag{13.75}$$

$$c_3 + c_4 x = 0 \tag{13.76}$$

From Equations 13.73 and 13.74, we get

$$c_2\left(x - \frac{1}{x}\right) + c_3 = 0$$

Focusing on the lowest π MO, we write $E_1 = \alpha + 1.62\beta$ or $x = -1.62$. From Equation 13.73 we have $c_2 = 1.62c_1$. Substituting the value of x in the above equation, we find that $c_2 = c_3$. Similarly, we can show that $c_1 = c_4$. Now, because the wave functions are all normalized, we can write

$$c_1^2 + c_2^2 + c_3^2 + c_4^2 = c_1^2 + (1.62c_1)^2 + (1.62c_1)^2 + c_1^2 = 1$$

Solving, we get $c_1 = 0.37$ and $c_2 = 0.60$. Thus, the wave function ψ_1 is given by

$$\psi_1 = 0.37\phi_1 + 0.60\phi_2 + 0.60\phi_3 + 0.37\phi_4 \tag{13.77}$$

Using a similar procedure, we write the wave functions for the other three π MOs as

$$\psi_2 = 0.60\phi_1 + 0.37\phi_2 - 0.37\phi_3 - 0.60\phi_4 \tag{13.78}$$

$$\psi_3 = 0.60\phi_1 - 0.37\phi_2 - 0.37\phi_3 + 0.60\phi_4 \tag{13.79}$$

$$\psi_4 = 0.37\phi_1 - 0.60\phi_2 + 0.60\phi_3 - 0.37\phi_4 \tag{13.80}$$

Knowing the coefficients enables us to calculate the π bond order using Equation 13.63:

$$P_{ik}^{\pi} = \sum_{j=1}^{4} n_j c_{ji} c_{jk}$$

Between carbon atoms 1 and 2 we write

$$\begin{aligned} P_{12}^{\pi} &= n_1(c_{11})(c_{12}) + n_2(c_{21})(c_{22}) + n_3(c_{31})(c_{32}) + n_4(c_{41})(c_{42}) \\ &= 2(0.37)(0.60) + 2(0.60)(0.37) + 0 + 0 \\ &= 0.89 \end{aligned}$$

By symmetry, we have $P_{34}^{\pi} = = 0.89$. In contrast, a simple Lewis structure predicts that the π bond order should be unity. The π bond order between carbon atoms 2 and 3 is given by

$$\begin{aligned} P_{23}^{\pi} &= n_1(c_{12})(c_{13}) + n_2(c_{22})(c_{23}) + n_3(c_{32})(c_{33}) + n_4(c_{42})(c_{43}) \\ &= 2(0.60)(0.60) + 2(0.37)(-0.37) + 0 + 0 \\ &= 0.22 \end{aligned}$$

Here, the HMO predicts some π bond character, whereas the Lewis structure shows only a σ bond and no π bond at all.

Cyclobutadiene (C_4H_4)

Finally, it is interesting to compare our butadiene results to cyclobutadiene (C_4H_4), a cyclic system also containing four C atoms:

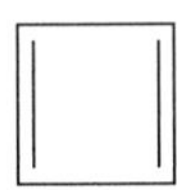

We write the determinant:

$$\begin{vmatrix} \alpha - E & \beta & 0 & \beta \\ \beta & \alpha - E & \beta & 0 \\ 0 & \beta & \alpha - E & \beta \\ \beta & 0 & \beta & \alpha - E \end{vmatrix} = 0 \tag{13.81}$$

Because each carbon atom has *two* nearest neighbors, there are additional β terms in the lower left and upper right of the determinant, in contrast to the linear butadiene case (Equation 13.71). As usual, we divide each term by β and let $x = (\alpha - E)/\beta$,

$$\begin{vmatrix} x & 1 & 0 & 1 \\ 1 & x & 1 & 0 \\ 0 & 1 & x & 1 \\ 1 & 0 & 1 & x \end{vmatrix} = 0 \tag{13.82}$$

which simplifies to

$$0 = x^4 - 4x^2$$

This quadratic equation has four roots,

$$x = 2, 0, 0, \text{ and } -2 \tag{13.83}$$

and thus the energies of the system are

$$E = (\alpha - 2\beta), \alpha, \alpha, \text{ and } (\alpha + 2\beta) \tag{13.84}$$

The ground electronic state of cyclobutadiene has energy

$$\begin{aligned} E &= 2E_1 + E_2 + E_3 \\ &= 2(\alpha + 2\beta) + \alpha + \alpha \\ &= 4\alpha + 4\beta \end{aligned} \tag{13.85}$$

But this is the same energy that HMO predicts for two ethylene molecules. Thus, the delocalization energy is zero, and cyclobutadiene is *not* resonance stabilized with respect to two ethylene molecules.

EXAMPLE 13.6

Set up (but don't bother to solve) the secular determinants using HMO theory for hexa-1,3,5-triene and benzene.

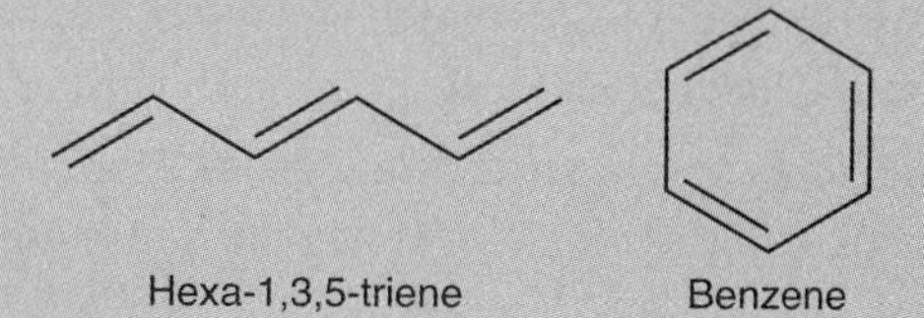

ANSWER

By analogy to the determinant for butadiene, we can write the determinant for hexa-1,3,5-triene as

$$\begin{vmatrix} \alpha - E & \beta & 0 & 0 & 0 & 0 \\ \beta & \alpha - E & \beta & 0 & 0 & 0 \\ 0 & \beta & \alpha - E & \beta & 0 & 0 \\ 0 & 0 & \beta & \alpha - E & \beta & 0 \\ 0 & 0 & 0 & \beta & \alpha - E & \beta \\ 0 & 0 & 0 & 0 & \beta & \alpha - E \end{vmatrix} = 0$$

Like cyclobutadiene, benzene is a cyclic system. In benzene, each carbon atom has two nearest neighbors, so we add a β term to the lower left and upper right corners.

$$\begin{vmatrix} \alpha - E & \beta & 0 & 0 & 0 & \beta \\ \beta & \alpha - E & \beta & 0 & 0 & 0 \\ 0 & \beta & \alpha - E & \beta & 0 & 0 \\ 0 & 0 & \beta & \alpha - E & \beta & 0 \\ 0 & 0 & 0 & \beta & \alpha - E & \beta \\ \beta & 0 & 0 & 0 & \beta & \alpha - E \end{vmatrix} = 0$$

COMMENT

A numerical solution to these determinants shows that benzene has resonance stabilization energy of 2β with respect to 1,3,5-hexatriene. With these few examples, you should be able to write the simple HMO determinants for most planar π systems. Calculating the energies may be nontrivial, but there are numerous computer programs that can simplify the algebra.

In summary, HMO theory for π systems rests on a few simple assumptions:

1. The σ and π electrons are separated, and only the π electrons are considered.
2. The π electrons are placed in molecular orbitals that are made of linear combinations of atomic orbitals.
3. The $2p$ atomic orbitals are orthonormal; that is, $S_{ij} = \delta_{ij}$.
4. All of the carbon atoms are treated equally, so that all Coulomb integrals are given the same value; that is, $H_{ii} = \alpha$ for all carbon atoms, i.
5. For carbon atoms that are nearest neighbors, the resonance integral is always the same value; that is, $H_{ij} = \beta$. When the atoms are not nearest neighbors, $H_{ij} = 0$.

With these assumptions, we can calculate atomic charge densities, bond orders, HOMO–LUMO energy gaps, and resonance stabilization energies for a large family of molecules.

13.9 Computational Chemistry Methods

The particle-in-a-box model and Hückel MO theory have been used successfully to describe the electronic structure of small molecules. These models are simple enough to calculate "by hand," but the level of accuracy may not be very high. To make more accurate predictions of molecular energetics or molecular geometry, we seek other theories and approximations. In this section we outline three types of computational methods that may be used to make quantitative (as well as qualitative) predictions:

molecular mechanics, empirical, and *ab initio* calculations. With increasing complexity of a computational method, its predictions become more accurate, but the goal is accomplished at the expense of more computer time and memory. Increasingly complex theories also tend to lose physical insight. For example, in perturbation theory calculations, a perturbation term or two may be correlated with a physical phenomenon such as an external magnetic field. But when there are over 10 perturbation terms or over 1000 variational parameters (as, for example, in the calculation of the ground-state helium energy), it is neither plausible nor practical to attribute a physical meaning to each term or parameter. We shall start with the simplest approach to computational chemistry.

Molecular Mechanics (Force Field) Methods

Molecular mechanics (MM), or force field (FF), methods are employed to find the minimum energy geometry of the electronic ground state of molecules. The energy of a molecule is calculated empirically as a function of the atomic coordinates, and the most stable configuration is the one that has minimum energy. MM calculations use classical physics to calculate the energy of a particular geometry, treating molecules as if they were made of balls and springs. The total energy, E_{FF}, is written as the sum of the energies of various interactions, and the molecular geometry is varied to find the minimum total energy:

$$E_{\text{FF}} = E_{\text{stretch}} + E_{\text{bend}} + E_{\text{torsion}} + E_{\text{van der Waals}} + E_{\text{elec}} \; (+ \cdots) \tag{13.81}$$

Each of the individual interaction energies is parameterized to fit a data set of known molecular interactions. Consider the first term, E_{stretch}, as an example. A harmonic oscillator could be employed, and only two parameters are needed—the spring constant, k, and the equilibrium bond length R_e. For each type of chemical bond, the parameters will be different. A C−Cl bond is shorter and "stiffer" (has a larger value of the spring constant, k) than a C−H bond, but as a first approximation we might treat all C−Cl bonds as following the same potential energy function. In a similar fashion, the bending motions, torsional motions, van der Waals interactions, Coulombic interactions, etc., are also fit to force-field parameters. MM calculations are very fast, so they may be applied to macromolecules of any size.

Empirical and Semi-Empirical Methods

Empirical and semi-empirical methods rely on parameters that have been derived from experiments to calculate molecular properties. While molecular mechanics methods may also be considered empirical methods, we shall distinguish MM from empirical methods that consider electronic structure. Empirical methods are characterized by the extent of parameterization or approximation employed. Examples of semi-empirical methods include Hückel MO (HMO) theory and extended Hückel theory (EHT). As we saw in the previous section, the HMO theory considers only pi orbitals; EHT considers valence sigma orbitals as well. Examples of parameterization from HMO and EHT include the use of the negative ionization energy of a carbon $2p$ orbital, rather than a mathematical calculation, to determine the integral α (see Equation 13.50). Semi-empirical calculations might be best for large molecules.

Another use of a semi-empirical method is to optimize a molecular geometry for use as the starting point in an *ab initio* calculation.

Ab Initio Methods

The Latin phrase *ab initio* translates to "from the beginning" or in the context of computational chemistry, "from first principles." *Ab initio* calculations have no empirical parameters and thus are potentially very accurate, but also very slow because of the large number of integrals that must be computed. Arbitrarily, we divide *ab initio* calculations into two groups, those based on the Hartree–Fock (HF) method (see Section 12.10), and the so-called "post-Hartree–Fock" methods. One of the factors that determines the accuracy (and the expense of computer time and space) of a HF calculation is the size of the *basis set* employed in the calculation. The basis set may be considered all of the atomic orbitals that are employed in the calculation. At the simplest computational level, a minimum basis set (MBS) is employed—one that has just enough atomic orbitals to form molecular orbitals for every electron. A minimum basis set for hydrocarbons might include a $1s$ orbital on each hydrogen atom and $1s$, $2s$, and $2p$ orbitals for each carbon atom. Larger basis sets might incorporate, for example, hydrogenlike $3s$, $3p$, and even $3d$ orbitals (basis functions) in describing a ground-state carbon atom in a molecule. It has been humorously noted that there are as many basis sets as there are computational chemists. While it is generally true that a larger basis set will increase the accuracy (and computational time) of a calculation, there is a diminishing return for effort. Ultimately, however, HF methods are limited in that they neglect electron–electron correlations. Even at the *Hartree–Fock limit* (an extrapolation to an infinitely large number of basis functions), calculated molecular energies are still greater than those measured experimentally.

Post Hartree–Fock Methods. There are several approaches to improve the accuracy of energy calculations beyond the Hartree–Fock limit and account for electron correlation (EC). EC is the term used to describe the energy that is not accounted for when applying the HF approximation. EC arises in part because the HF method is essentially a single-electron method that omits part of the energy of the system due to the motion of the electrons to avoid each other. In general, we can express EC as

$$\text{EC} = E_{\text{lowest for that basis set}} - E_{\text{HF for that basis set}} \tag{13.82}$$

One approach to account for EC is to use perturbation theory, as we did in Section 12.11. Møller–Plesset theory is a type of perturbation theory that has been widely employed. It is called MP2 when perturbation up to second order in energy is considered, MP3 for third order, and so on. Other approaches for handling electron correlation include coupled cluster (CC) and configuration interaction (CI) methods. CC methods also involve perturbation theory, whereas CI is a variational method (see Section 12.9) that includes a linear combination of excited states contributing to the ground state.

Density Functional Theory (DFT). One of the most popular approaches in computational chemistry is density functional theory (DFT) because it incorporates some electron correlation at considerably less cost than CC or CI techniques. DFT is often listed as an *ab initio* theory, although it does include some parameterization. DFT differs from other methods in that the total electron density is considered, rather than placing electrons in individual orbitals. In essence, if the electron density is known precisely, then we can in principle determine the total energy (and all other properties of the system) precisely. However, the mathematical function that relates electron density to the energy is not known, so it must be guessed rather than derived.

With all of the computational methods available, and with new methods being continually developed, how do we decide which method to employ? The trade-off that is always present is one of accuracy versus time. For small molecules and high-level theory, it is possible to make remarkably accurate predictions of molecular geometries, energies, and spectroscopic properties. For larger molecules, the computational times with *ab initio* methods can become prohibitively large. Fortunately, computational hardware and software continue to improve at a rapid pace, so the frontiers of computational methods are constantly being expanded. At present, an array of computational chemistry software packages are available with user-friendly interfaces, so computational chemistry is now accessible to interested scientists.

Key Equations

$$\psi_{\text{MO}}(1,2) = [1s_{\text{A}}(1)1s_{\text{B}}(2) + 1s_{\text{A}}(2)1s_{\text{B}}(1) + 1s_{\text{A}}(1)1s_{\text{A}}(2) + 1s_{\text{B}}(1)1s_{\text{B}}(2)]$$

(MO wave function for H_2) (13.25)

$$\psi_{\text{VB}}(1,2) = [1s_{\text{A}}(1)1s_{\text{B}}(2) + 1s_{\text{A}}(2)1s_{\text{B}}(1)]$$

(VB wave function for H_2) (13.26)

$$\psi_{\text{ionic}}(1,2) = \underbrace{1s_{\text{A}}(1)1s_{\text{A}}(2)}_{\text{H}^-\,\text{H}^+} + \underbrace{1s_{\text{B}}(1)1s_{\text{B}}(2)}_{\text{H}^+\,\text{H}^-}$$

(Wave function for H_2 ionic bond) (13.27)

$$\psi_{\text{MO}} = \psi_{\text{VB}} + \psi_{\text{ionic}}$$

(Relation between MO and VB treatment) (13.28)

$$\text{BO} = \frac{(\#\text{ bonding } e^- - \#\text{ antibonding } e^-)}{2}$$

(Definition of bond order) (13.29)

$$|\chi_{\text{A}} - \chi_{\text{B}}| = 0.102\sqrt{D_{\text{AB}} - 0.5(D_{\text{A}_2} + D_{\text{B}_2})}$$

(Pauling definition of electronegativity) (13.32)

Suggestions for Further Reading

BOOKS

Coulson, C. A., *Valence*, 3rd ed., Oxford University Press, New York, 1979. Also known as *Coulson's Valence*, R. McWeeny, Ed.

DeKock, R. L., and H. B. Gray, *Chemical Structure and Bonding*, University Science Books, Sausalito, CA, 1989.

Foresman, J. B., and A. Frisch, *Quantum Exploring Chemistry with Electronic Structure Methods*, 2nd ed., Gaussian, Inc., Pittsburgh, PA, 1996.

Hehre, W. J., L. Radom, P. von R. Schleyer, and J. Pople, *Ab Initio Molecular Orbital Theory*, John Wiley & Sons, New York, 1986.

Karplus, M., and R. N. Porter, *Atoms & Molecules: An Introduction For Students of Physical Chemistry*, Benjamin/Cummings, Menlo Park, CA, 1970.

Levine, I. N., *Quantum Chemistry*, 7th ed., Prentice-Hall, New York, 2013.

McQuarrie, D. A., *Quantum Chemistry*, 2nd ed., University Science Books, Sausalito, CA, 2008.

Pauling, L., *The Nature of the Chemical Bond*, 3rd ed., Cornell University Press, Ithaca, NY, 1960.

Pauling, L., and E. B. Wilson, Jr., *Introduction to Quantum Mechanics*, Dover Publications, New York, 1985. The original text is copyright 1935.

Szabo, A., and N. S. Ostlund, *Modern Quantum Chemistry*, Dover Publications, New York, 1996.

Weinhold, F., and C. R. Landis, *Discovering Chemistry with Natural Bond Orbitals*, John Wiley & Sons, Hoboken, New Jersey, 2012.

ARTICLES

"Quantum Theory of Molecules," M. Born and J. R. Oppenheimer, *Ann. Physik* **84**, 457 (1927).

"The Probability Equals Zero Problem in Quantum Mechanics," F. O. Ellison and C. A. Hollingsworth, *J. Chem. Educ.* **53**, 767 (1976).

"How do Electrons Get Across Nodes?," P. G. Nelson, *J. Chem. Educ.* **67**, 643 (1990).

"Spectroscopy and Quantum-Mechanics of the Hydrogen Molecular Cation—A Test of Molecular Quantum Mechanics," C. A. Leach and R. E. Moss, *Ann. Rev. Phys. Chem.* **46**, 55 (1995).

"Kinetic Energy and the Covalent Bond in H_2^+," F. Rioux, *Chem. Educator* [Online] **2**, 40 (1997) DOI 10.1333/s00897970153a.

"Quantum Chemistry Comes of Age," G. B. Kauffman and L. M. Kauffman, *Chem. Educator* [Online] **4**, 259 (2001) DOI 10.1007/s00897990337a.

"The Covalent Bond in H_2," F. Rioux, *Chem. Educator* [Online] **6**, 288 (2001) DOI 10.1007/s00897010509a.

"The Hydrogen Molecular Ion Revisited," J.-P. Grivet, *J. Chem. Educ.* **79**, 127 (2002).

"The Excited States of Molecular Oxygen," D. Tudela and V. Fernandez, *J. Chem. Educ.* **80**, 1381 (2003).

"A Conversation on VB vs MO Theory: A Never-Ending Rivalry?" R. Hoffman, S. Shaik, and P. C. Hiberty, *Acc. Chem. Res.* **36**, 750 (2003).

"Misconceptions in Sign Convention: Flipping the Electric Dipole Moment," J. W. Hovick and J. C. Poler, *J. Chem. Educ.* **82**, 889 (2005).

"The Old Quantum Theory for H_2^+: Some Chemical Implications," S. K. Knudson, *J. Chem. Educ.* **83**, 464 (2006).

"The Concept of Resonance," D. G. Truhlar, *J. Chem. Educ.* **84**, 781 (2007).

"On the Rule of *d* Orbital Hybridization in the Chemistry Curriculum," J. M. Galbraith, *J. Chem. Educ.* **84**, 783 (2007).

"The Weakest Link: Bonding Between Helium Atoms," L. L. Lohr and S. M. Blinder, *J. Chem. Educ.* **84**, 860 (2007).

"Simple Molecular Orbital Calculations for Diatomics: Oxygen and Carbon Monoxide," S. G. Lieb, *Chem. Educator* [Online] **13**, 333 (2008) DOI 10.1333/s00897082173a.

"Bond Order and Chemical Properties of BF, CO, and N_2," R. J. Martine, J. J. Bultema, M. N. Vander Wal, B. J. Burkhart, D. A. Vander Griend, and R. J. DeKock, *J. Chem. Educ.* **88**, 1094 (2011).

"Connections between Concepts Revealed by the Electronic Structure of Carbon Monoxide," Y. Liu, B. Liu, and M. G. B. Drew, *J. Chem. Educ.* **89**, 355 (2012).

Problems

Hydrogen Molecule and Molecular Ions

13.1 In solving the Schrödinger equation (Equation 13.5) for the hydrogen molecular cation, qualitatively describe how your answers differ or not for HD^+ and D_2^+ in comparison to H_2^+.

13.2 In solving the Schrodinger equation (Equation 13.5), qualitatively describe how the bond energies and bond lengths would vary amongst a series of one-electron cations: H_2^+, HHe^{2+}, He_2^{3+}, and LiH^{3+}.

13.3 To find an analytical solution to the Schrödinger equation for H_2^+ (Equation 13.5), we carry out a transformation of variables for the coordinate of the electron from Cartesian coordinates to elliptic coordinates, (λ, μ, ϕ), where $\lambda = (r_A + r_B)/R$, $\mu = (r_A - r_B)/R$, and ϕ is the angle about the internuclear axis. The vectors r_A and r_B describe the vectors from nuclei A and B to the electron, respectively, and R is the distance between the nuclei (see Figure 13.1). The limits on the coordinate ϕ are $0 \leq \phi \leq 2\pi$. What are the limits on values of λ and μ?

13.4 A sigma bonding orbital is taken as the sum of two hydrogen atom 1s orbitals,

$$\sigma(1) = N[s_A(1) + s_B(1)]$$

where "1" represents the coordinates of an electron numbered one. Assuming that the hydrogen atomic orbitals are normalized, calculate the normalization constant, N, for this sigma orbital. Express your answer in terms of the overlap integral, S, defined in Equation 13.15.

13.5 A sigma antibonding MO is taken as the difference of two hydrogen 1s AOs,

$$\sigma^*(1) = N^*[s_A(1) - s_B(1)]$$

where "1" represents the coordinates of an electron that is arbitrarily numbered one. Assuming that the hydrogen atomic orbitals are normalized, calculate the normalization constant, N^*, for this antibonding sigma orbital. Express your answer in terms of the overlap integral, S.

13.6 Sigma bonding and antibonding orbitals (σ and σ^*) are defined in Problems 13.4 and 13.5. Show that these two orbitals are orthogonal.

13.7 One way to describe chemical bonds is by their percentage ionic character. The extreme cases are (1) ionic bonding, with complete transfer of an electron and 100% ionic character and (2) covalent bonding, with perfect sharing of electron pairs and 0% ionic character. In this picture, what percent ionic character does a H_2 molecular orbital (Equation 13.25) contain?

13.8 The valence bond and molecular orbital pictures are two ways of describing a chemical bond. In a sense, the valence bond approach contains too little ionic character and the molecular orbital approach contains too much ionic character. One technique that might be applied is that of perturbation theory, in which ionic bonding is treated as a perturbation of valence bonding. Outline how to solve for the energetics of a chemical bonding system using perturbation theory.

13.9 An antibonding orbital contains a nodal plane that bisects the internuclear axis. This means that there is a plane where there is zero probability of finding an electron. Describe in your own words how an electron in an antibonding orbital crosses this nodal plane. [For an extensive discussion of this problem, see the series of articles in *J. Chem. Educ.* **53**, 767 (1976); **67**, 643 (1990); and **70**, 345 & 346 (1993).]

13.10 Explain why the overlap integral S (Equation 13.15) is never zero for two hydrogen atom $1s$ orbitals.

13.11 Under what conditions would the value of the overlap integral S (Equation 13.15) be equal to exactly zero?

13.12 The four–electron species H_2^{2-} is not covalently bound in its ground electronic state, but has a bond order of one in a low-lying excited electronic state. Explain this phenomenon by drawing a molecular orbital diagram constructed from hydrogen $1s$ and $2s$ orbitals, and assigning electrons to molecular orbitals.

13.13 For each of the following excited state electronic configurations of the hydrogen molecule, predict the probability of homolytic (to form two H atoms) versus heterolytic (to form H^+ and H^-) bond cleavage, **(a)** $1s\sigma_g 1s\sigma_u^*$, **(b)** $(1s\sigma_u^*)^2$, **(c)** $1s\sigma_g 2s\sigma_g$, **(d)** $(2s\sigma_g)^2$.

13.14 Explain the difference between the (often interchanged) terms "molecular orbital" and "molecular wave function."

Homonuclear and Heteronuclear Diatomic Molecules

13.15 Lithium dimers are a starship fuel of science fiction. **(a)** Using $1s$ and $2s$ hydrogenlike AOs, construct a MO diagram for Li_2. Label each orbital as σ or π, bonding or antibonding, and gerade or ungerade. **(b)** Using your MO diagram, is Li_2 paramagnetic or diamagnetic? **(c)** Does the bond order predicted by MO theory agree with VB theory from a Lewis electron dot diagram?

13.16 Diatomic neon (Ne_2) has a dissociative ground state and simple molecular orbital theory predicts bond order zero. Excited states of diatomic neon may have nonzero bond order. Sketch the molecular orbital diagram for diatomic neon, and place electrons in orbitals to produce an excited state with a nonzero bond order. Identify the bond order of your excited state.

13.17 For the series of diatomic nitrogen species in their ground electronic state—N_2^+, N_2, N_2^-—identify the bond order and whether the species is paramagnetic or diamagnetic. Which species would have the longest bond length? Which would have the greatest bond strength? Use molecular orbital diagrams to support your conclusions.

13.18 For the homonuclear diatomic ions in their ground electronic state—Be_2^+, B_2^+, C_2^+—identify the bond order and whether the species is paramagnetic or diamagnetic. Which species would have the longest bond length? Which would have the greatest bond strength? Use molecular orbital diagrams to support your conclusions.

13.19 Write the Hamiltonian operator for the LiH molecule. Which of the terms in the Hamiltonian operator may be neglected when using the Born–Oppenheimer approximation?

13.20 Among the simplest heteronuclear diatomic species is HeH^+. Sketch a correlation diagram for the molecular orbitals of this species. What is the predicted bond order for the ground electronic state? If this species is thermally dissociated, remaining on the ground state potential energy surface, which atomic species would be formed, He and H^+ or He^+ and H? Next draw the electron configuration for when an electron is promoted from the HOMO (highest-occupied molecular orbital) to the LUMO (lowest-unoccupied molecular orbital). Is this excited electronic state bound or dissociative? If HeH^+ dissociates from this excited state, will it form He and H^+ or He^+ and H?

13.21 Table 13.1 shows the bond order, bond length, and bond energy for a series of diatomic oxygen species. Add a row to this table for the species O_2^{2+} and fill in the row with your predictions based on molecular orbital theory.

13.22 First consider a homonuclear diatomic molecule, then consider a heteronuclear diatomic molecule. For each molecule, is it possible to form a molecular orbital from the combination of a $1s$ atomic orbital and a $2s$ atomic orbital? If so, would it be σ or π or neither? g(gerade) or u(ungerade) or neither? Bonding, antibonding or nonbonding?

13.23 Gas-phase potassium bromide has one of the largest dipole moments of any diatomic, 10.5 debye. Using rovibrational spectroscopy, the potassium bromide bond distance is measured to be 282 pm. Calculate the percent ionic character of the bond.

13.24 Quantitatively estimate the largest value of the dipole moment that could be obtained for a diatomic species. Explicitly state any assumptions that you make. Using the scientific literature, look up any information that you might need to make your estimate more accurate.

13.25 The diatomic molecule lithium hydride has drawn attention from the scientific community because of the use of lithium hydride battery technology in cell phones and electric automobiles. The dipole moment of LiH is 6.00 debye. The dipole moment of HF is 1.92 debye. Account for the difference between these two molecules.

13.26 Based on the following table, describe the periodic trend in percent ionic character of these bonds. Which neutral diatomic species has the most covalent bond? Does this agree with your expectation based on electronegativity differences?

Species	μ/debye	eR/debye	R/pm
BH	1.733	5.936	123.6
CH	1.570	5.398	112.4
NH	1.627	4.985	103.8
OH	1.780	4.661	97.05
FH	1.942	4.405	91.71

Polyatomic Molecules

13.27 Lithium trimers are the star-destroying material of science fiction. Consider the series of triatomic species Li_3, Li_3^+, and Li_3^-. For each species, draw a Lewis electron dot diagram, predict the geometry, identify the Li–Li bond order, and identify the hybridization of the central lithium atom.

13.28 Consider the series of hydride molecules LiH, BeH_2, BH_3, CH_4, NH_3, H_2O, and HF. For each molecule, would the covalent single bond to the hydrogen atom be better described using simple VB or MO theory? For this series, identify molecular properties that might best be described by VB theory. Recall that VB theory underestimates and MO theory overestimates the percent ionic character in a covalent bond. Is there a periodic trend in ionic character?

13.29 Xenon octafluoride ion, XeF_8^{2-}, consists of a central xenon atom surrounded by eight fluorine atoms. (Note: This structure is not covered in Table 13.2.) Use VSEPR theory to predict the geometry of this ion. Are all of the fluorine atoms equivalent? In other words, do all of the Xe–F bonds have the same length and same bond energy? In the valence bond view of this ion, what must be the hybridization of the central xenon atom?

13.30 Table 13.2 lists most of the more common situations where VSEPR theory is applied to small molecules. Omitted from this table is the situation where four groups of electrons surround a central atom and three of these groups of electrons are lone pairs. **(a)** What might the hybridization of the central atom be under these circumstances? **(b)** What would the molecular geometry be? **(c)** Give an example of a molecule that might be described by this electronic structure. **(d)** Discuss the advantages and disadvantages of applying VSEPR theory in this situation.

13.31 For the molecular geometries considered in Table 13.2, what is the smallest bond angle predicted? What is the largest bond angle? Consider only the terminal atom–central atom–terminal atom (B–A–B) bond angles.

13.32 An extension of Table 13.2 is a molecule or ion in which six groups of electrons surround a central atom and three of these groups of electrons are lone pairs. **(a)** What might the hybridization of the central atom be under these circumstances? **(b)** What would the molecular geometry be?

Hückel Molecular Orbital Theory

13.33 How would isotopic substitution influence the results of Hückel MO theory when applied to hydrocarbon molecules?

13.34 How would the substitution of a fluorine atom for a single hydrogen atom influence the results of a simple Hückel MO theory calculation on an aromatic hydrocarbon molecule? What if all hydrogen atoms were replaced with fluorine atoms?

13.35 Using simple Huckel MO theory, calculate the HOMO–LUMO energy in terms of the Hückel parameters α and β for the series of benzene species $C_6H_6^{2+}$, $C_6H_6^{+}$, C_6H_6, $C_6H_6^{-}$, and $C_6H_6^{2-}$.

13.36 Use Hückel MO theory to compare two ethylene molecules (two C_2H_4) with butadiene (C_4H_6) and with cyclobutadiene (C_4H_4). Express the energies of these three systems in terms of the Hückel parameters α and β. In the simple Hückel approximation, which system has the lowest π energy?

13.37 Explain in your own words how one might measure experimentally the value of the Hückel MO theory term β.

13.38 Set up (but do not solve by hand) the Hückel determinants for the following series of C_8H_8 molecules. Predict which of these molecules will be aromatic.

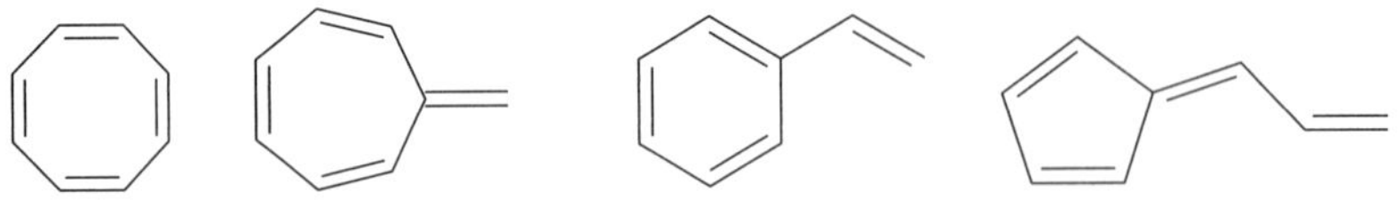

According to Hückel's rule, aromatic molecules have $(4n + 2)$ delocalized π electrons in a planar ring or rings, where $n = 1, 2, 3, \ldots$.

13.39 Does simple Hückel MO theory distinguish between the *Z* (cis-like) and *E* (trans-like) forms of 1,3,5-hexatriene shown here? Why or why not?

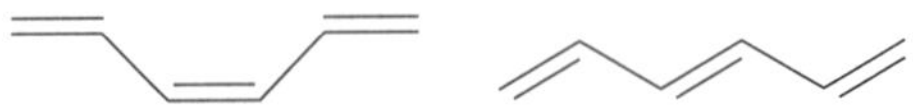

Z *E*

13.40 Obtain a computer program that is able to perform simple Hückel MO theory electronic structure calculations (there are many such programs available). Perform calculations on the bicyclic $C_{10}H_8$ constitutional isomers naphthalene and azulene shown here and comment on the difference. Are each of these molecules aromatic?

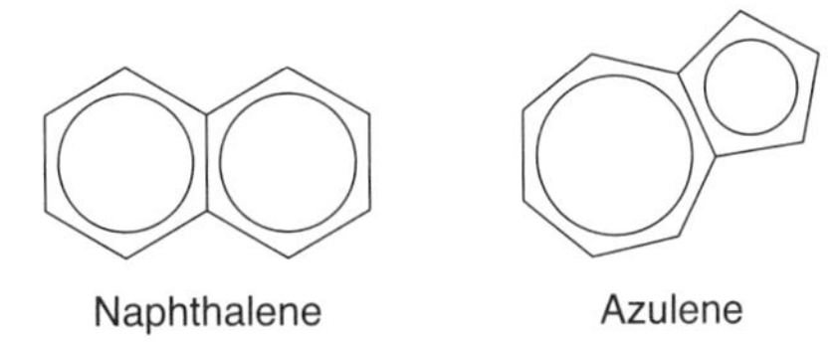

13.41 Obtain a computer program that is able to perform simple Hückel MO theory electronic structure calculations. Perform calculations on the series of aromatic hydrocarbons benzene, naphthalene, anthracene, and tetracene, and predict the trend in the HOMO–LUMO energy. Compare and contrast your results with those obtained from the particle in a 2-dimensional box model using a $d \times d$ square for benzene, a $2d \times d$ rectangle for naphthalene, a $3d \times d$ rectangle for anthracene and a $4d \times d$ rectangle for tetracene (see Section 10.10). Use $d = 278$ pm for the dimension of a benzene square.

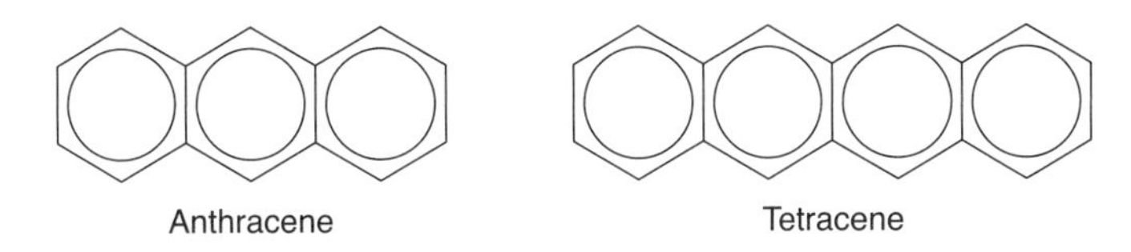

Additional Problems

13.42 Noting the appearance of the $\sigma_g 1s$ and $\sigma_g 2s$ orbitals for the H_2^+ molecular ion, sketch the appearance of the $\sigma_g 3s$ molecular orbital. How many nodes does the $\sigma_g 3s$ orbital contain? Describe their shape.

13.43 Noting the appearance of the $\pi_u 2p$ orbital for the H_2^+ molecular ion, sketch the appearance of the $\pi_u 3p$ molecular orbital. How many nodes does the $\pi_u 3p$ orbital contain? Describe the shape of each node in the $\pi_u 3p$ orbital.

13.44 For which of the following chemical species is it possible to find an analytical solution to the time-independent electronic Schrödinger equation using the Born–Oppenheimer approximation? Explain. ${}^2_1H$, ${}^2_1H_2^+$, H_3^{2+}, ${}^3_2He$, ${}^4_2He^+$, Li_2^{5+}, C_2, N^{6+}, an α particle (${}^4_2He^{2+}$).

13.45 To explain covalent bonding in the hydrogen fluoride molecule, HF(g), the $1s$ orbital on the hydrogen atom overlaps with an orbital on the fluorine atom. For the fluorine atom, there are several possibilities: **(a)** the $2p_z$ atomic orbital, **(b)** an sp hybrid orbital, **(c)** an sp^2 hybrid orbital, or **(d)** an sp^3 hybrid orbital. Explain which of these possibilities is most reasonable and which is least reasonable, and why. Suggest experimental measurements that might be done to identify which of these bonding models best describes the real molecule.

13.46 The resonance concept is sometimes described by analogy to a mule, which is a cross between a horse and a donkey. Compare this analogy with the description of a rhinoceros as a cross between a griffin and a unicorn. Which description is more appropriate? Explain.

13.47 Which of the following molecules has the shortest nitrogen-to-nitrogen bond: N_2H_4, N_2O, N_2, N_2O_4?

13.48 A single bond is almost always a sigma bond, and a double bond is almost always made up of a sigma bond and a pi bond. There are very few exceptions to this rule. Show that the B_2 and C_2 molecules are examples of the exceptions.

13.49 The molecular geometries discussed in Table 13.2 all lend themselves to fairly straightforward elucidation of bond angles. The exception is the tetrahedron, because its bond angles are hard to visualize. Consider the CCl_4 molecule, which has a tetrahedral geometry and is nonpolar. By equating the bond dipole moment of a particular C—Cl bond to the resultant bond dipole moments of the other three C—Cl bonds in opposite directions, show that the bond angles are all equal to 109.5°. (*Hint*: Place Cl atoms at the corners of a cube, as in Figure 13.18c.)

13.50 The Lewis structure for O_2 is

$$\ddot{\underset{..}{O}}=\ddot{\underset{..}{O}}$$

Use the molecular orbital theory to show that the structure actually corresponds to an excited state of the oxygen molecule.

13.51 Set up the Hückel secular determinant (but do not solve) for pyridine (C_6H_5N). Because N is a heteroatom, use the following symbols for the Coulomb integrals (α) and resonance integrals (β) for C and N: α_C, β_{CC}, α_N, β_{CN}. Number the N atom as atom 1.

13.52 Does the molecule HBrC=C=CBrH have a dipole moment? Explain.

CHAPTER 14

Electronic Spectroscopy and Magnetic Resonance Spectroscopy

WARNING: Do not look into laser with remaining eye!

—Laser safety sign

Spectroscopy is the study of the interaction between electromagnetic radiation and matter. Detailed information about structure and bonding, as well as about various intra- and intermolecular processes, can be obtained from the analysis of atomic and molecular spectra. In Chapter 11 we discussed microwave, infrared, and Raman spectroscopies, which deal with molecular rotation and vibration. In this chapter we shall focus on two other major branches of spectroscopy: electronic spectroscopy and magnetic resonance spectroscopy—nuclear magnetic resonance and electron spin resonance. In discussing electronic spectroscopy, we cover absorption, emission, and photoelectron spectroscopies. Lasers and their applications are discussed in more detail.

14.1 Molecular Electronic Spectroscopy

The absorption of radiation in the UV and visible regions involves electronic transitions and gives rise to electronic spectra. There is a major difference in the appearance of the electronic spectra of diatomic molecules and polyatomic molecules. Diatomic molecules are described by a single rotational degree of freedom and a single vibrational degree of freedom, whereas polyatomic molecules have three rotational degrees of freedom and $\sim 3N$ vibrational degrees of freedom, where N is the number of atoms in the molecule. Electronic transitions may simultaneously involve vibrational and rotational transitions. Thus, polyatomic molecules tend to have considerably more complex electronic spectra. For simplicity, we shall first discuss diatomic molecules.

Degrees of freedom are discussed in Section 2.9.

Figure 14.1 shows potential energy curves for the ground state and an excited state of a diatomic molecule and their respective vibrational energy levels υ'' and υ'. According to the Boltzmann distribution (see Equation 2.33), $h\nu > k_{\mathrm{B}}T$ at room temperature, so that in most cases practically all the transitions originate in the ground vibrational state. Two features of electronic transitions are worth noting. First, the selection rule $\Delta\upsilon = \pm 1$, which holds for vibrational transitions within a given electronic state, is not valid here. In electronic transitions, $\Delta\upsilon$ can assume any value (see Figure 14.1). Second, we can use the Franck–Condon principle [named after the German physicist James Franck (1882–1964) and the American physicist Edward Uhler Condon (1902–1974)] to predict the relative intensities of the bands. This principle states that because it takes much more time for a molecule to execute a vibration

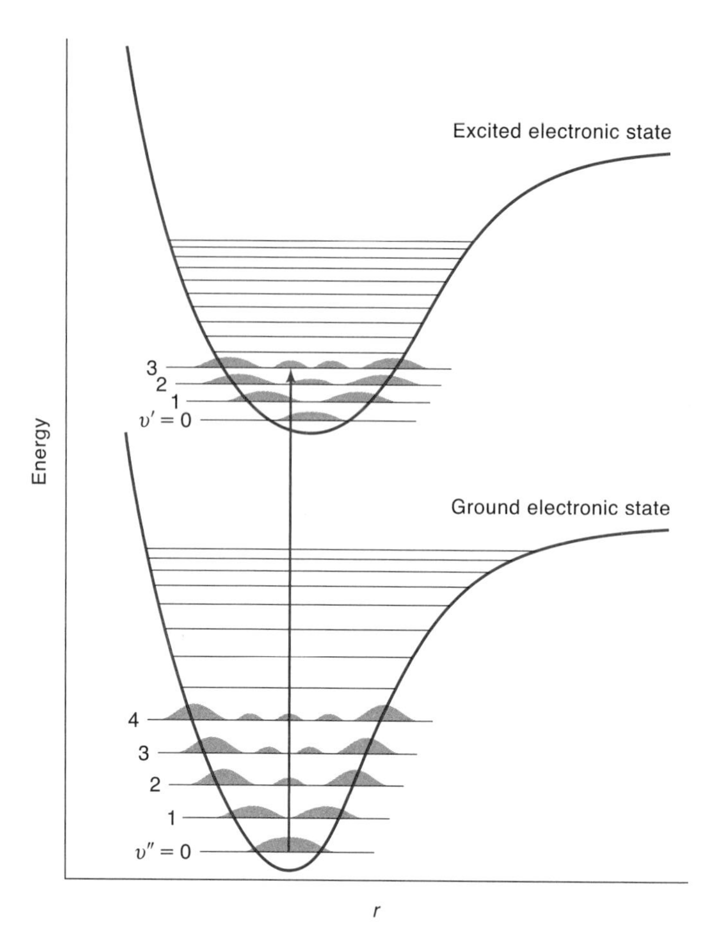

Figure 14.1
Diagram showing the most probable electronic transition of a diatomic molecule. A transition is favored (i.e., it will give rise to a strong line) when the internuclear distance is such that the transition connects probable states of the molecule. In other words, the transition starts in the ground state at some point along r where the vibrational probability density function (ψ^2) is large, and ends in the excited state at some point along r where the value of ψ^2 is also appreciable.

(about 10^{-12} s) than it does for an electronic transition (about 10^{-15} s), the nuclei do not appreciably alter their positions during an electronic transition (see Appendix 14.1). Therefore, the most probable (most intense) transitions are those for which the internuclear distance remains unchanged. Thus, on a potential energy diagram such as Figure 14.1, electronic transitions are represented as vertical lines. The most probable transitions also have significant overlap of the wave functions for the two states.

Homonuclear diatomic molecules, lacking a permanent dipole moment, exhibit neither pure rotational (microwave) nor rovibrational (IR) spectra. However, these general selection rules do not hold for the simultaneous rotational and vibrational transitions in electronic spectra (UV–visible), so it is possible to determine the geometries of these molecules. In the gas phase, the electronic spectra of diatomic molecules under high resolution show both vibrational bands and rotational fine structure. Although these spectra are quite complex, consisting of hundreds or even thousands of lines, for a number of molecules they have been assigned to specific vibrational and rotational transitions. Such an analysis enables us to determine bond lengths of homonuclear diatomic molecules such as N_2 and I_2. Bond lengths for heteronuclear diatomic molecules such as HCl and NO may be determined from their pure rotational or rovibrational spectra (see Sections 11.2 and 11.3).

The situation is quite different for polyatomic molecules. Large moments of inertia and spectral congestion can make it difficult to resolve the rotational fine structure of these molecules. In solution, polyatomic (and diatomic) molecules usually produce electronic spectra with broad, unresolved bands. We shall briefly discuss the electronic spectra of organic molecules and molecules that undergo charge-transfer interactions.

Bond lengths of homonuclear diatomic molecules can also be obtained from rotational Raman spectra.

Organic Molecules

In saturated organic molecules, such as the alkanes, electronic transitions are of the $\sigma^* \leftarrow \sigma$ type, meaning that an electron is promoted from a σ bonding MO to an unoccupied σ antibonding MO. Aromatic molecules and compounds containing C=C and C=O groups also have $\pi^* \leftarrow \pi$, $\pi^* \leftarrow \sigma$, and $\pi^* \leftarrow n$ transitions, where n denotes a nonbonding (lone pair) orbital. Typically, the $\pi^* \leftarrow \sigma$ and $\pi^* \leftarrow n$ transitions are weaker because they are symmetry forbidden. Analogous to the group vibrational frequencies in IR, electronic spectra can often be characterized by the absorption of special groups of atoms, called *chromophores*. Table 14.1 lists absorption wavelengths of some common chromophores. The specific wavelength of maximum

Table 14.1
Some Common Chromophores and Their Approximate Wavelengths of Maximum Absorbance

Chromophore	λ_{max}/nm
>C=C<	190
>C=C—C=C<	210
—C₆H₅ (phenyl)	190 260
>C=O	190 280
—C≡N	160
—COOH	200
—N=N—	350
$—NO_2$	270

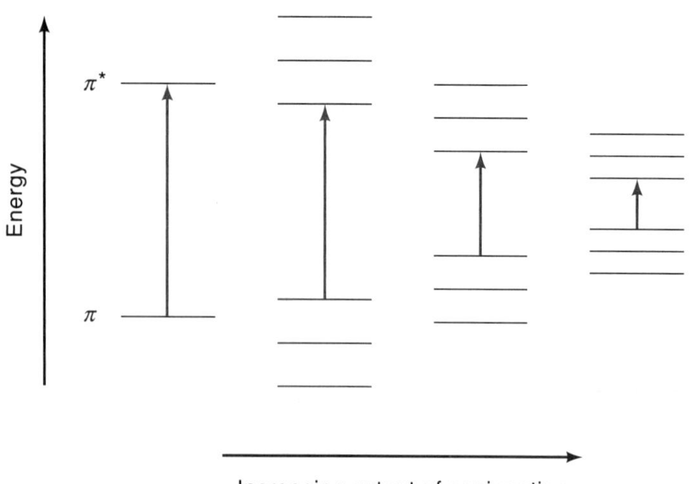

Figure 14.2
Effect of increased conjugation on the $\pi^* \leftarrow \pi$ transitions in polyenes. The decrease in the energy gap can be accounted for by using the particle in a one-dimensional box model.

absorbance for these chromophores depends not only on the compound involved but also on the environment, because it varies with changes in variables such as solvent and temperature.

In Section 10.9 we used the free-electron model to analyze the electronic spectrum of butadiene. Butadiene is a simple example from a series of related molecules called polyenes that contain conjugated single and double bonds. Figure 14.2 shows the effect of increasing conjugation on the $\pi^* \leftarrow \pi$ transitions in diphenylpolyenes, $C_6H_5{-}(CH{=}CH)_n{-}C_6H_5$. For $n = 1$ and 2, the compounds are colorless, because absorption occurs in the UV range. As n increases, there is a gradual shift toward the visible region, and the color of the compound changes from pale yellow for $n = 3$ to greenish black for $n = 15$. Recall that the color of a compound is the complement of the color that is absorbed; thus, a molecule that appears green absorbs in the red region of the electromagnetic spectrum (Table 14.2). The artist's color wheel is a useful mnemonic device for identifying complementary colors.

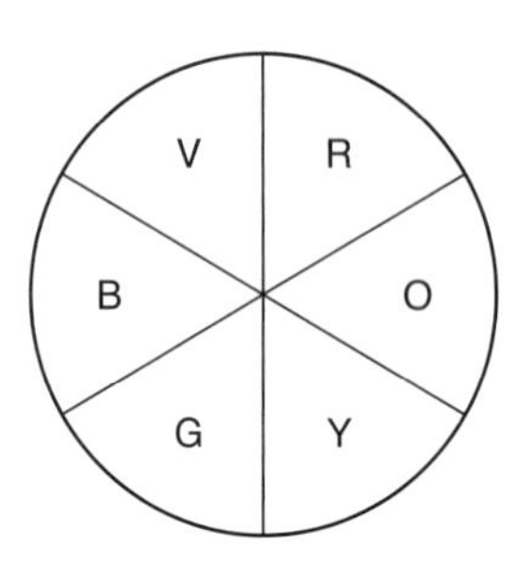

The artist's color wheel

The electronic spectra of most amino acids arise from the $\sigma^* \leftarrow \sigma$ transitions that occur in the far UV, below 230 nm. The exceptions are phenylalanine, tryptophan, and tyrosine, all of which contain the phenyl ($-C_6H_5$) chromophore and absorb strongly above 250 nm (Figure 14.3). Absorbance at 280 nm, due mainly to the tryptophan and tyrosine residues, is useful in measuring the concentration of protein solutions.

To learn about the optical properties of DNA and RNA, researchers have studied the electronic spectra of purines (adenine and guanine) and pyrimidines (cytosine, thymine, and uracil), shown in Figure 14.4 on p. 616. The concentration of nucleic acid solutions is determined by measuring the absorbance at 260 nm.

Both DNA and RNA exhibit an interesting phenomenon called *hypochromism*. In general, the molar absorptivity of intact DNA is some 20 to 40 percent lower than we would expect it to be, given the number of nucleotides present. For example, molar absorptivity of calf thymus DNA at 260 nm increases from about 6500 to 9500 L mol^{-1} cm^{-1} when the polymer undergoes thermal denaturation. Although the

Table 14.2
Wavelengths of Visible Light, Their Corresponding Color, and Their Complementary Color[a]

Wavelength region/nm	Color	Complementary color
400–435	violet	yellow-green
435–480	blue	yellow
480–490	green-blue	orange
490–500	blue-green	red
500–560	green	purple
560–580	yellow-green	violet
580–595	yellow	blue
595–650	orange	green-blue
650–750	red	blue-green

[a] The colors of the rainbow, red-orange-yellow-green-blue-(indigo)-violet, may be remembered in order of increasing photon energy by the mnemonic name ROY G. BIV. Complementary colors lie opposite each other on the artist's color wheel.

theory of hypochromism is beyond the scope of this text, the phenomenon is attributed to Coulombic interactions between electric dipoles induced by light absorption in the base pairs. This interaction depends on the orientation of the dipoles relative to one another. In a random orientation, there would be little or no interaction and no effect on the absorption spectrum. In the native state, the dipoles are stacked parallel on top of one another, leading to a decrease the absorbance. This

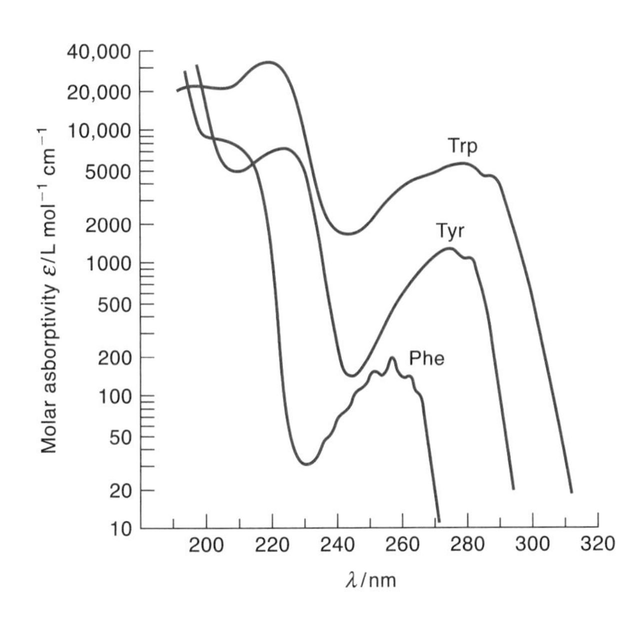

Figure 14.3
UV spectra of phenylalanine (Phe), trytophan (Trp), and tyrosine (Tyr). [From D. C. Neckers, *J. Chem. Educ.* **50**, 164 (1973).]

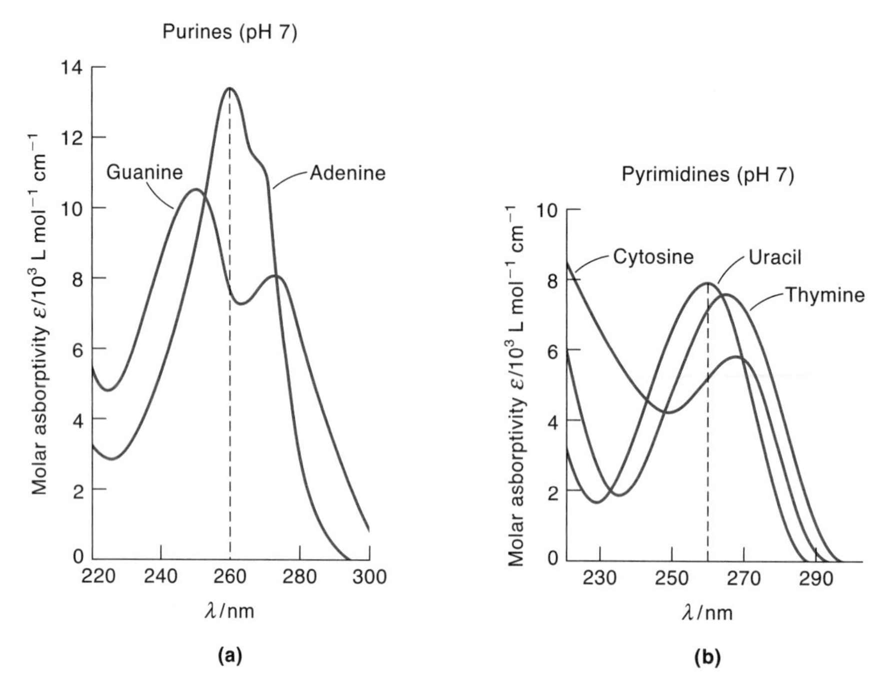

Figure 14.4
UV spectra of purines and pyrimidines. (From A. L. Lehninger, *Biochemistry*, 2nd ed., Worth Publishers, New York, 1975.)

property has been successfully employed to monitor the helix–coil transition in DNA. Figure 14.5 shows the *melting curve* of a DNA solution. The term *melting* here refers to the unwinding of the double-helical structure. The melting temperature T_m corresponds to the inflection point of the curve; its value depends on the base-pair composition of the DNA.

Charge-Transfer Interactions

A special type of electronic spectrum arises from *charge-transfer* interaction between a pair of molecules. When tetracyanoethylene $[(CN)_2C{=}C(CN)_2]$, an electron acceptor, dissolves in carbon tetrachloride, the resulting solution is colorless. The reason is that the $\pi^* \leftarrow \pi$ transition occurs in the UV region. Upon the addition to the solution of a small amount of an electron donor (an aromatic hydrocarbon such as benzene or toluene), the

Tetracyanoethylene
TCNE

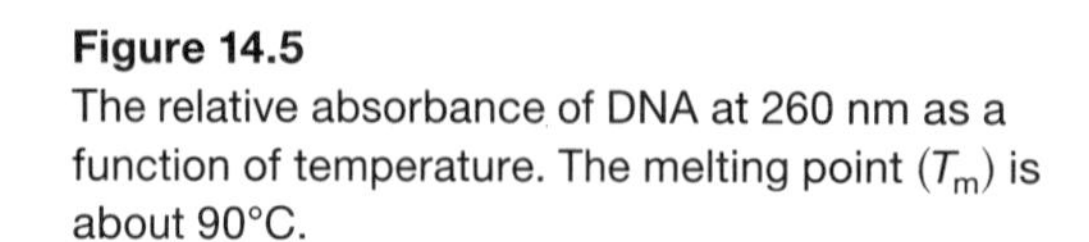

Figure 14.5
The relative absorbance of DNA at 260 nm as a function of temperature. The melting point (T_m) is about 90°C.

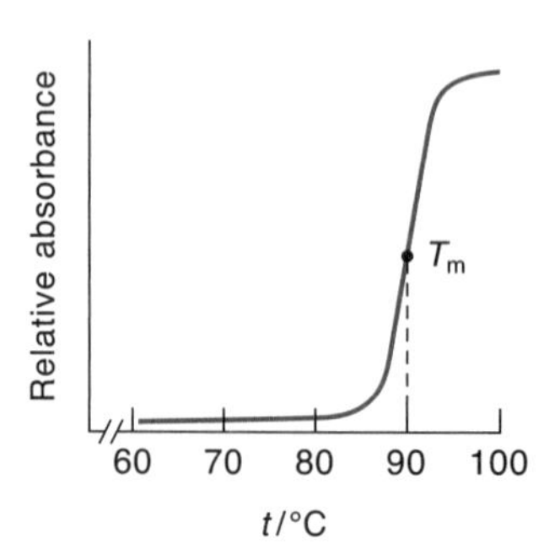

solution immediately turns yellow (Figure 14.6). Many similar reactions, including that between iodine and benzene, have been observed. In 1952 Robert Mulliken proposed the following scheme to explain the spectra of charge-transfer complexes:

$$\mathrm{D} + \mathrm{A} \rightleftharpoons \underset{\text{ground state}}{[(\mathrm{D}, \mathrm{A})]} \xrightarrow{h\nu} \underset{\text{excited state}}{[(\mathrm{D}^+, \mathrm{A}^-)]^*} \tag{14.1}$$

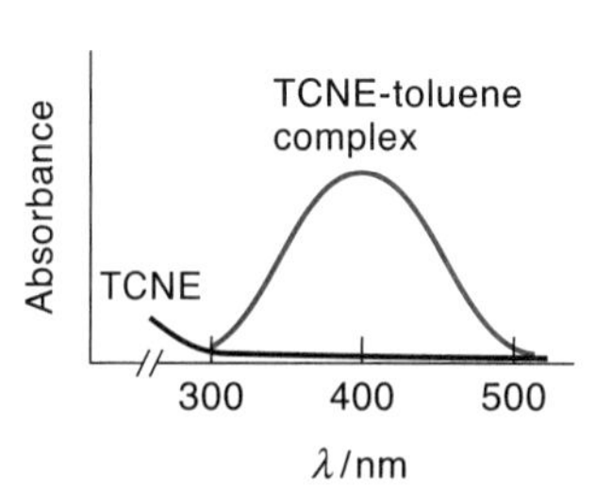

Figure 14.6
Visible absorption spectrum of the tetracyanoethylene–toluene charge-transfer complex in carbon tetrachloride. (TCNE = tetracyanoethylene.)

where D and A are the donor and acceptor molecules, and (D, A) and (D^+, A^-) represent covalent and ionic resonance structures of the charge-transfer complex, respectively. In the ground state, van der Waals forces hold the molecules together, and there is little, if any, actual transfer of charge from D to A. However, when the complex is excited at a suitable wavelength, a large charge transfer takes place, and the ionic structure makes a major contribution to the excited state. If the exciting wavelength falls in the visible region, the solution will appear colored. There is an interesting difference between this electronic transition and normal absorption. In this case, an electron is excited from a lower level (bonding molecular orbital) in the donor molecule to a higher level (antibonding molecular orbital) in the acceptor molecule. The tendency for charge-transfer formation generally depends on the ionization energy of the donor and the electron affinity of the acceptor. Many transition-metal complexes also produce charge-transfer spectra; these are due to intramolecular charge-transfer processes. There the absorption process is accompanied by the transfer of an electron from a ligand to the metal, or from the metal to a ligand. These charge-transfer transitions often give rise to intense bands, but they can be distinguished from the *d–d* transitions because metal–ligand charge-transfer transitions fall in the far UV region, whereas most *d–d* transitions take place in the visible region (hence the color of these complex ions).

Charge-transfer absorptions tend to have large molar absorptivities, and thus intense bands, because there is a large change in dipole moment with transition.

Application of the Beer–Lambert Law

UV–visible spectroscopy is not as reliable for compound identification as IR or NMR (see Chapters 11 and Section 14.6), because an electronic spectrum generally does not possess the fine details of an IR or NMR spectrum. Electronic spectroscopy is, however, a useful tool in quantitative analysis. The concentration of a solution can be determined readily using the Beer–Lambert law (see Section 11.1) by measuring the absorbance (if the molar absorptivity is known). A commonly encountered case is the analysis of a solution containing two species X and Y whose absorption bands overlap. The absorbance A at a wavelength λ is additive, so that, from the Beer–Lambert law, we have

$$\begin{aligned} A_\lambda = A_\lambda^{\mathrm{X}} + A_\lambda^{\mathrm{Y}} &= \varepsilon_\lambda^{\mathrm{X}} b[\mathrm{X}] + \varepsilon_\lambda^{\mathrm{Y}} b[\mathrm{Y}] \\ &= b(\varepsilon_\lambda^{\mathrm{X}}[\mathrm{X}] + \varepsilon_\lambda^{\mathrm{Y}}[\mathrm{Y}]) \end{aligned} \tag{14.2}$$

where ε is the molar absorptivity and b is the pathlength. If the molar absorptivities of X and Y are known at two different wavelengths, λ_1 and λ_2, then the absorbances are given by

In common laboratory cuvettes, $b = 1.00$ cm.

$$A_1 = b(\varepsilon_1^{\mathrm{X}}[\mathrm{X}] + \varepsilon_1^{\mathrm{Y}}[\mathrm{Y}]) \tag{14.3a}$$

and

$$A_2 = b(\varepsilon_2^{\mathrm{X}}[\mathrm{X}] + \varepsilon_2^{\mathrm{Y}}[\mathrm{Y}]) \tag{14.3b}$$

From Equations 14.3a and 14.3b, we can solve for [X] and [Y] as follows:

$$[\mathrm{X}] = \frac{1}{b}\frac{\varepsilon_2^{\mathrm{Y}}A_1 - \varepsilon_1^{\mathrm{Y}}A_2}{\varepsilon_1^{\mathrm{X}}\varepsilon_2^{\mathrm{Y}} - \varepsilon_2^{\mathrm{X}}\varepsilon_1^{\mathrm{Y}}} \tag{14.4a}$$

and

$$[\mathrm{Y}] = \frac{1}{b}\frac{\varepsilon_1^{\mathrm{X}}A_2 - \varepsilon_2^{\mathrm{X}}A_1}{\varepsilon_1^{\mathrm{X}}\varepsilon_2^{\mathrm{Y}} - \varepsilon_2^{\mathrm{X}}\varepsilon_1^{\mathrm{Y}}} \tag{14.4b}$$

Now consider a situation in which at some wavelength within the overlapped region the molar absorptivities of the two species are equal. If the sum of the molar concentrations of these two compounds in solution is held constant, then there will be *no* change in absorbance at this wavelength as the ratio of these two compounds is varied. This invariant point is called the *isosbestic point.* At the isosbestic point, we can write

$$A_{\mathrm{iso}} = \varepsilon_{\mathrm{iso}}b[\mathrm{X}] + \varepsilon_{\mathrm{iso}}b[\mathrm{Y}] \tag{14.5}$$

or

$$[\mathrm{X}] + [\mathrm{Y}] = \frac{A_{\mathrm{iso}}}{\varepsilon_{\mathrm{iso}}b} \tag{14.6}$$

Thus, measurements at the isosbestic wavelength, plus one other wavelength where the molar absorptivities of the two compounds differ, will enable us to determine both [X] and [Y].

The existence of one or more isosbestic points in a system can be an indication of chemical equilibrium between two compounds. Figure 14.7 shows the absorption

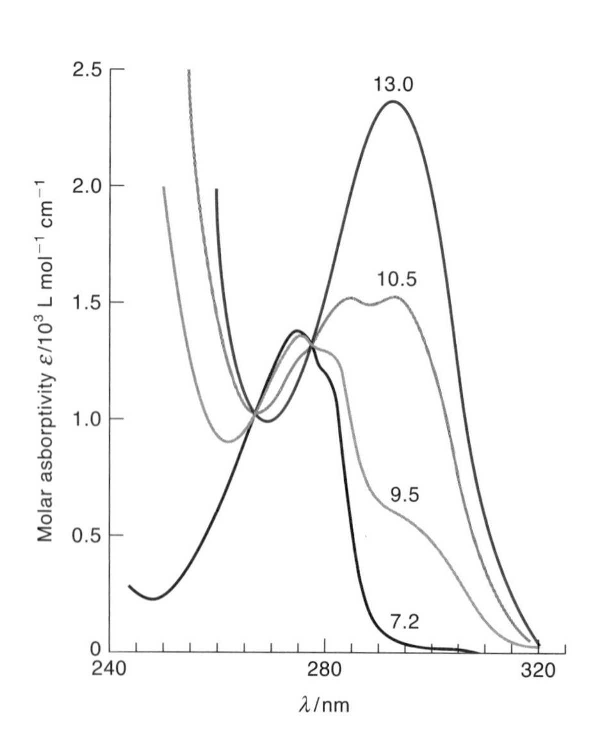

Figure 14.7
Absorption spectrum of tyrosine at pH values indicated. Note the isosbestic points at 267 nm and 277.5 nm. [From D. Schugar, *Biochem. J.* **52**, 142 (1952).]

spectra of the amino acid tyrosine at various pH values. There are actually two isosbestic points, at 267 nm and 277 nm, due to the following equilibrium process:

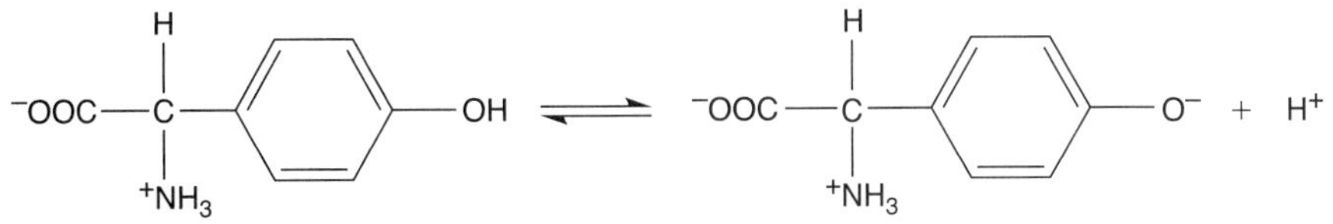

$$^{-}OOC-CH(NH_3^{+})-C_6H_4-OH \rightleftharpoons {}^{-}OOC-CH(NH_3^{+})-C_6H_4-O^{-} + H^{+}$$

Analysis shows that the pK_a the hydroxyl group, which is related to the acid dissociation constant, is 10.1. If we wanted to measure the total concentration of tyrosine (the sum of the concentrations of the protonated and deprotonated species), then we could monitor the absorbance at either 267 nm or 277 nm using Equation 14.6.

14.2 Fluorescence and Phosphorescence

In electronic absorption spectroscopy, molecules are promoted to an excited electronic state by absorption of UV–visible radiation. When a molecule is in an excited electronic state, it may react (via dissociation or rearrangement) to form a new chemical species, or return to the ground electronic state. There are two types of pathways for returning to the ground state, radiationless and radiative. Radiationless pathways typically involve collisional energy transfer to other molecules such as solvent molecules. Radiative pathways involve the release of a photon of energy $h\nu$. The liberated photon shows up as luminescent emission. In luminescence, there are two paths for energy depletion in the excited molecule—fluorescence and phosphorescence.

Fluorescence

Fluorescence is the emission of radiation resulting from the transition of an electron from an excited electronic state to a ground state without any change in spin multiplicity. Because electrons in molecules are paired according to the Pauli exclusion principle and because most molecules have no net electron spin, the initial absorption is from the ground singlet state S_0 to the first excited singlet state S_1 (or some higher singlet level). At first, fluorescence might appear to be exactly the reverse of the absorption process. This is true for isolated atoms, but a comparison of the absorption and emission spectra of molecules shows them not to be superimposable. Instead, they usually form the mirror image of each other, with the emission spectrum displaced toward the longer wavelengths (Figure 14.8). Because the time required for vibrational energy transfer (about 10^{-13} s) is much shorter than the decay or mean lifetime (about 10^{-9} s) of the fluorescent state, most of the excess vibrational energy dissipates to the surroundings as heat, and the electronically excited molecules will decay from their ground vibrational levels.

We define the *quantum yield* of fluorescence emission Φ_F as the ratio of photons emitted through fluorescence to the total number of photons originally absorbed. The maximum value of Φ_F is 1, although it can be appreciably smaller than 1 if there are other processes present to deactivate the excited molecules, as is usually the case. An apparent reading of $\Phi_F > 1$ may be an indicator of a chemical chain

The molecule fluorescein, which emits green light (~520 nm), has $\Phi_F = 0.95$.

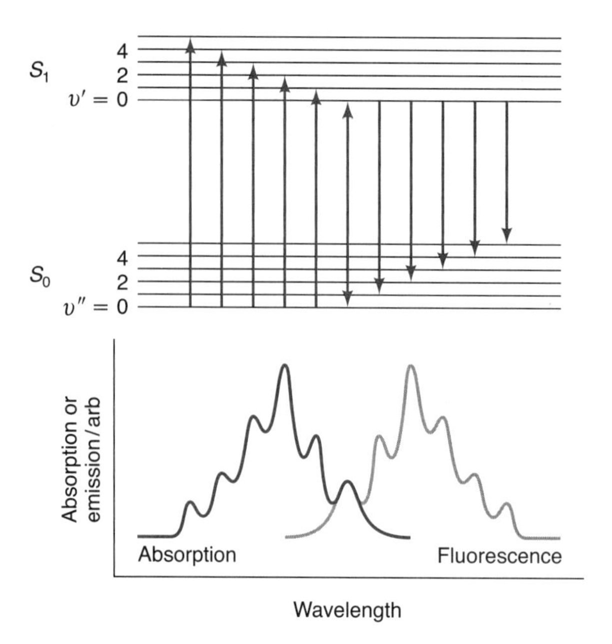

Figure 14.8
Relation between absorption and fluorescence. (Adapted from D. A. McQuarrie and J. D. Simon, *Physical Chemistry*, University Science Books, Sausalito, CA, 1997.)

reaction (see p. 690). The intensity of radiation emitted after the exciting light is turned off is given by

$$I = I_0 e^{-t/\tau} \tag{14.7}$$

where I is the intensity at time t, I_0 is the intensity at $t = 0$, and τ is the mean lifetime of the fluorescent state. The mean lifetime is equal to the time it takes for the original intensity to decrease to $1/e$, or 0.368, of its original value. Thus, when $t = \tau$, we observe $I = I_0/e$. The decay of fluorescence obeys first-order kinetics (see Equation 15.7) and the rate constant k for the decay is given by $1/\tau$.

Liquid Scintillation Counting. The fluorescence technique, besides yielding information about the electronic structure of molecules in excited states, is also useful in chemical and biochemical analysis. For example, it is employed in *liquid scintillation counting*, a common method of assaying radioactive tracer-labeled compounds containing ^3H, ^{14}C, ^{32}P, and ^{35}S. Scintillators are compounds that can be excited either in the solid state or in solution. The intensity of the subsequent fluorescence of these compounds is related to the amount of the exciting source present. The general procedure in liquid scintillation counting is to first dissolve the scintillator (called a fluor) in a solvent (toluene or dioxane, depending on the nature of the sample to be studied), giving what is referred to as a "cocktail." Then the radioactive sample is added to the "cocktail" and the following sequence of events takes place: (1) The solvent molecules are excited by bombardment with β particles emitted from the radioactive nuclei; (2) the excited solvent molecules transfer the energy to the scintillator; (3) the fluorescence of the scintillator molecules is measured; and (4) the amount of radioactive nuclei present in the original sample is determined from previously calibrated fluorescence versus concentration measurements.

The fluor, characterized by a large Φ_F value, is excited by energy from the solvent (say, toluene), received via a nonradiative mechanism such as collision. This singlet–singlet energy transfer is represented by

$$D(S_1) + A(S_0) \rightarrow D(S_0) + A(S_1) \tag{14.8}$$

Here the donor molecule (D) is the excited toluene and the acceptor molecule (A) is the fluor. If the wavelengths of the photons emitted by the fluor are not in the region of highest sensitivity of the detector, a second fluor is added. The secondary fluor absorbs the photons emitted by the primary fluor and re-emits them as fluorescence at a longer wavelength, which is better suited for the detector. The most commonly used primary fluor is 2,5-diphenyloxazole (PPO) and the most commonly used secondary fluor is 1,4-bis-2-(5-phenyloxazole)benzene (POPOP):

PPO POPOP

Another mechanism for deactivation of excited molecules is known by its acronym FRET [Förster resonance energy transfer, after the German physicist Theodor Förster (1910–1974)]. FRET is responsible for long-range intermolecular transfer of energy and it depends on the overlap between the emission band of the donor molecule and the absorption band of the acceptor molecule. The excited donor interacts with the ground-state acceptor through a dipole–dipole mechanism. FRET is highly sensitive to the distance between the donor and acceptor molecules and, thus, is a useful probe of molecular dynamics, such as protein conformational changes.

In this context, "long-range" means 1 to 10 nm.

Phosphorescence

Phosphorescence offers a different path for the return of an excited molecule to the ground electronic state with the emission of light. Phosphorescence can be distinguished readily from fluorescence by two characteristics. First, phosphorescence has a much longer decay period than fluorescence, about 10^{-3} to several seconds. Second, a molecule in the phosphorescent state is paramagnetic, containing two unpaired electrons; that is, it is in a triplet state. The relation between the excited singlet and triple electronic states is conveniently illustrated by a *Jablonski diagram* [after the Polish physicist Alexander Jablonski (1898–1980)] (Figure 14.9). Initially, an electron is promoted from the singlet ground state S_0 to the lowest-energy excited singlet state S_1. The promotion is followed by a process called the *radiationless transition*, in which the electron flips its spin and drops from S_1 to T_1, the lowest triplet state, without the emission of light. This radiationless transition is known as *intersystem crossing* (ISC) because the molecule has moved from one system, the singlet manifold, to another system, the triplet manifold. In the end, a radiative transition from T_1 to S_0 occurs. This radiative step is called phosphorescence. Because the transition involves a change in spin multiplicity (from triplet to singlet), it is spin forbidden and therefore has a low probability, accounting for the long lifetimes observed. The excited state (T_1) is easily deactivated

Many "glow-in-the-dark" toys are based on the yellow-green phosphorescence of copper-doped zinc sulfide.

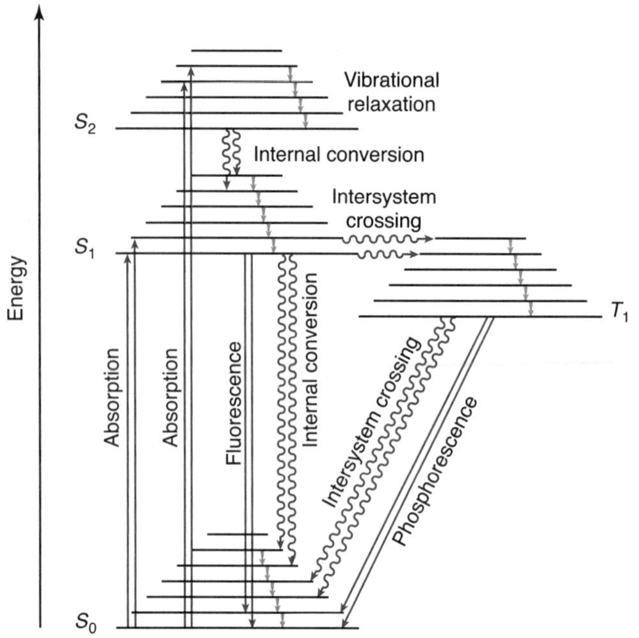

Figure 14.9
Jablonski diagram showing absorption, fluorescence, phosphorescence, internal conversion, and intersystem crossing. The wavy lines indicate radiationless transitions; the closely spaced lines represent the vibrational levels. Singlet states are labeled S and a triplet state is labeled T.

by collision due to its long lifetime; therefore, phosphorescence, unlike fluorescence, cannot be studied in a liquid phase. Phosphorescence is best studied when the sample is in a frozen clear glass at or below the temperature of liquid nitrogen (77 K).

14.3 Lasers

Laser is an acronym for "light amplification by stimulated emission of radiation." It is a special type of emission involving either atoms or molecules (or their ions). Laser emission occurs when a species in an excited state is stimulated to fall to a lower energy state by light with photon energy equal to the difference between the two energy levels. For lasing to occur, there must be more species in the excited state than in the lower energy state. This situation, called a *population inversion*, is not trivial to achieve. Consider first a two-level system. Suppose that N molecules are irradiated by light at spectral radiant energy density ρ_ν. In Section 11.1 we saw that the rate of stimulated absorption is given by $B_{mn}\rho_\nu N(1-x)$, and the rate of stimulated emission is $B_{nm}\rho_\nu Nx$, where x is the fraction of the molecules in the excited state. In addition, an excited molecule undergoes spontaneous emission, the rate of which is $A_{nm}Nx$. A_{nm} and B_{nm} are the Einstein coefficients. At equilibrium, the rates of absorption and emission are equal so that

$$B_{mn}\rho_\nu N(1-x) = B_{mn}\rho_\nu Nx + A_{mn}Nx \tag{14.9}$$

or

$$x = \frac{B_{mn}\rho_\nu}{2B_{mn}\rho_\nu + A_{nm}} \tag{14.10}$$

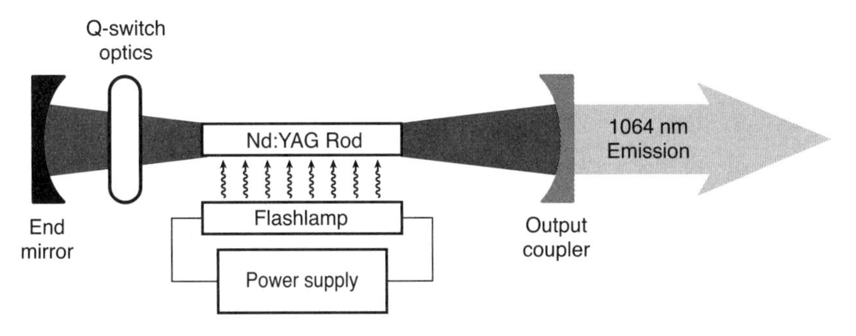

Figure 14.10
Schematic diagram of a Nd:YAG laser.

(Recall that $B_{mn} = B_{nm}$.) It follows, therefore, that the maximum value of x is 0.5, reached only in the limit as ρ_v approaches infinity.

The above discussion means that by an absorption process the population of the excited state can never exceed the population of the ground state in a two-level system. But if we can somehow populate the upper level without using the normal radiation process, we might be able to bring about a population inversion, causing x to exceed 0.5. In such a case, intense emission could be induced by irradiating the system with photons of the appropriate frequency. In fact, this objective can be reached by using a three- or four-level system, on which most laser actions are based.

An example of a widely used four-level system is provided by the Nd:YAG laser. A solid rod made of yttrium aluminum garnet ($Y_3Al_5O_{12}$) is doped with Nd^{3+} ions, which replace some of the Al^{3+} ions in the crystal lattice. A schematic diagram of a Nd:YAG laser shows elements common to all lasers: an excitation source (the flash lamp), a *gain medium* (the Nd-doped YAG rod), and an optical resonator (the pair of mirrors) (Figure 14.10). This laser also contains a so-called *Q*-switch,* which serves to generate laser pulses that are nanoseconds long. In general, the gain medium consists of atoms, molecules, ions, or solids that store energy until it is released by stimulated emission. Although Nd:YAG lasers are commonly called "YAG" lasers, it is the energy levels of the Nd^{3+} ion that comprise the four-level system for lasing action (Figure 14.11). At first, the laser system is excited (by discharging a flashlamp, for example) by short, intense visible or near-IR irradiation, known as *optical pumping*, which causes transitions from the E_0 to E_3 levels of Nd^{3+}. The excited states then decay to the E_2 state by radiationless transitions. The fluorescence lifetime of the E_2 state is relatively long—about 230 μs at room temperature—because the $E_2 \rightarrow E_0$ transition is spin forbidden. If the pumping is effective, the population in the E_2 state will exceed that in the E_1 state, and a laser transition can be effected by stimulating the transition with IR photons of wavelength 1064 nm. The first 1064-nm photons are generated by spontaneous emission, but just a few of these photons will rapidly be amplified to form laser radiation.

---

* If a shutter is placed within the laser cavity, a sizable population inversion can be established in the gain medium without any appreciable amount of stimulated emission taking place. The shutter is then opened so that the cavity can act as a resonator. The energy stored in the gain medium can be released in a single pulse of intense laser light. The term *Q-switching* means that at first we reduce and then quickly increase the quality, or *Q*-factor, of the laser.

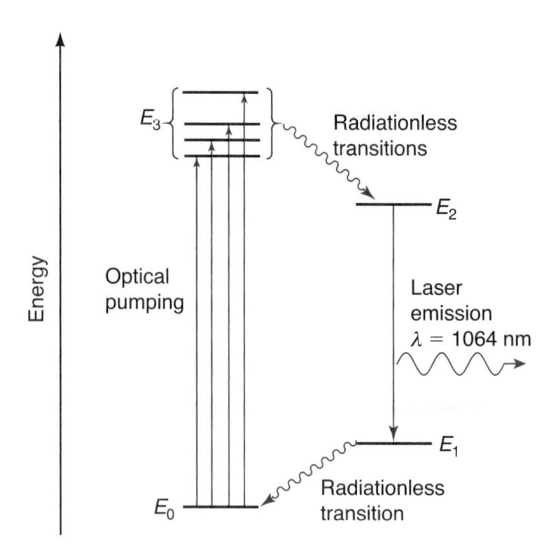

Figure 14.11
Energy-level diagram for a Nd^{3+} ion used in a Nd:YAG laser.

Red laser pointers and lasers for reading CDs and DVDs are based on solid-state semiconductor diodes.

A wide variety of practical lasers are now commercially available. Operating in solid, liquid, and gaseous media, lasers have been constructed that emit radiation ranging from infrared to ultraviolet and X ray (although any single laser emits in a narrow region of the electromagnetic spectrum). We will not comprehensively survey the types of lasers here, except to point out that there are several mechanisms for achieving population inversions. In a helium–neon laser (an example of an atomic gas laser), He atoms are first excited by collisions with electrons to the higher electronic

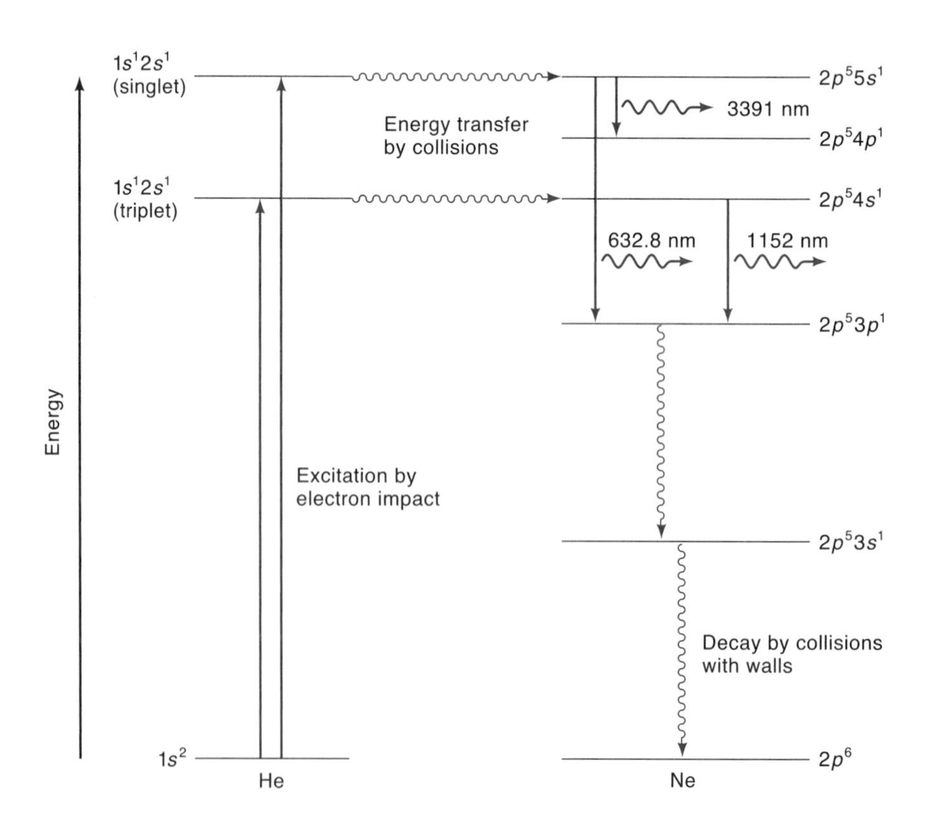

Figure 14.12
Energy-level diagram for a helium–neon laser. With the appropriate choice of mirrors, a helium neon laser may lase at 3391 nm, 1152 nm, or the characteristic red 632.8 nm.

states, which are then deactivated by collisions with the Ne atoms (Figure 14.12). Ne has an excited electronic energy state ($2p^5 5s^1$) that is energetically close to that of excited helium($1s^1 2s^1$), so this energy transfer process is favorable. The populations in the upper excited states in Ne build up to exceed those of the lower excited states, so laser transitions can occur.

Table 14.3 summarizes the properties of some selected lasers. Lasers are operated in one of two different modes: continuous wave (cw) or pulsed. As the names suggest, in the cw mode the laser light is emitted continuously, whereas in the pulsed mode the light comes out in pulses, some of which may be as short as 1×10^{-14} s, or 10 fs (1 femtosecond = 10^{-15} s). As of 2012, the fastest developed laser produced pulse lengths of only 67 as (1 as = 1 attosecond = 10^{-18} s). With such a short pulse length, the Heisenberg uncertainty principle tells us that the uncertainty in the wavelength (and thus the linewidth) will be correspondingly large. For lasers in general, the mode of laser action depends on the system, the method of pumping, and the design of the apparatus. For example, if the rate of pumping is less than the decay rate from the upper laser level, then a population inversion cannot be sustained, and pulsed operation must be used (with a pulse duration governed by the decay kinetics). A laser can operate continuously if the heat it generates is easily dissipated and population inversion can be sustained. Otherwise, it must be operated in a pulsed mode.

Table 14.3
Some Common Fixed-Wavelength Laser Systems

Laser	Emitted wavelength(s)/nm	Mode
$F_2(g)$	157	pulsed
$ArF(g)$	193	pulsed
$XeCl(g)$	308	pulsed
$N_2(g)$	337.1	pulsed
$Ar^+(g)$	457 488 514.5	cw
Ruby[a]	694.3	pulsed
He–Ne(g)	623.8 1152 3391	cw
Nd:YAG[b]	1064	cw/pulsed
$CO_2(g)$[c]	10600	cw/pulsed

[a] Ruby is corundum (Al_2O_3) that has been doped with chromium. It is Cr^{3+} that lases.

[b] This laser system is made of neodymium ions (Nd^{3+}) trapped in *y*ttrium *a*luminum garnet crystal ($Y_3Al_5O_{12}$).

[c] The CO_2 laser is based on vibrational energy transitions. All of the other lasers in this table are based on electronic transitions.

Properties of Laser Light

A laser beam is characterized by several properties: high intensity, high coherence, high monochromaticity, and low divergence. We shall briefly discuss these properties and applications based on them.

Intensity. Laser light has the highest intensity of any light on Earth. As an example, consider a Q-switched Nd:YAG laser that produces 7.0×10^{15} photons at 1064.1 nm during a pulse lasting 150 ps (1 ps = 10^{-12} s). Because $E = h\nu = hc/\lambda$, the total energy output per pulse is given by

$$\begin{aligned} E &= \left(\frac{hc}{\lambda}\text{ photon}^{-1}\right)(7.0 \times 10^{15}\text{ photons}) \\ &= \frac{(6.626 \times 10^{-34}\text{ J s})(3.00 \times 10^{8}\text{ m s}^{-1})(7.0 \times 10^{15})}{1064.1 \times 10^{-9}\text{ m}} \\ &= 1.3 \times 10^{-3}\text{ J} \end{aligned}$$

Now 1.3 millijoule may not seem like much, but it is generated in an extremely short period of time. We can calculate the peak power associated with such a laser beam as follows:

$$\begin{aligned} \text{power} &= \frac{\text{energy}}{\text{time}} \\ &= \frac{1.3 \times 10^{-3}\text{ J}}{150 \times 10^{-12}\text{ s}} \\ &= 8.7 \times 10^{6}\text{ J s}^{-1} \end{aligned}$$

The unit of power is the watt, denoted by W, where 1 W = 1 J s^{-1}. Thus, 8.7 megawatts is the power output during the pulse of laser action. When such a laser beam is focused on a small target of, say, area 0.01 cm^2, the *power flux density* is given by

$$\begin{aligned} \text{power flux density} &= \frac{\text{power}}{\text{area}} \\ &= \frac{8.7 \times 10^{6}\text{ W}}{0.01\text{ cm}^2} \\ &= 8.7 \times 10^{8}\text{ W cm}^{-2} \end{aligned}$$

This is nearly a gigawatt per square centimeter of instantaneous power flux density, which is sufficient to drill a hole in almost anything. Shorter pulses and smaller areas provide even higher power flux densities.

Intense laser beams have been used to cut and weld metals and even to produce nuclear fusion. Medically, lasers are used in surgery. For example, a pulsed argon ion laser is employed to "spot-weld" a detached retina back onto its support (the choroid). This procedure has some advantages over the traditional treatment in that it is noninvasive and does not require the administration of anesthetics. A pulsed

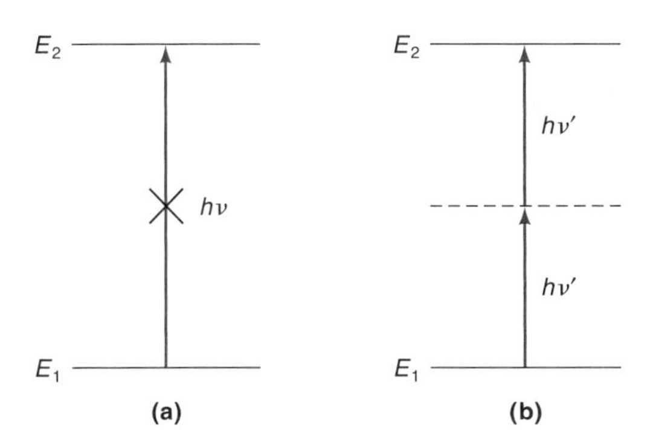

Figure 14.13
(a) A forbidden transition in a one-photon absorption process. (b) The same transition becomes allowed in a two-photon absorption process. Note that $\nu = 2\nu'$.

ultraviolet argon–fluoride *excimer* (*exc*ited d*imer*) laser is used to cut the cornea of a human eye in the LASIK (laser-assisted *in situ* keratomileusis) procedure for correcting vision.

The high intensity of a laser beam also facilitates *multiphoton absorption*, a process in which a molecule absorbs two or more photons during a spectroscopic transition. In a conventional spectroscopic measurement, an atom or molecule absorbs a photon of the same energy as that separating the ground state and the excited state. This is the normal one-photon process. However, when a system is irradiated with a high-power laser beam of frequency ν', it may undergo a two-photon process in which the system attains the excited state by absorbing two photons in a concerted fashion (Figure 14.13). A two-photon process requires a very intense laser beam because the two photons must pass essentially simultaneously through the region of space occupied by one molecule. An interesting aspect of multiphoton spectroscopy is that the selection rules are different, so that transitions that are strictly forbidden for one-photon absorption can now occur in a two-photon process, and vice versa.* For example, in the hydrogen atom, the forbidden $2s \leftarrow 1s$ and $3s \leftarrow 1s$ transitions are allowed in two-photon processes. In addition, transitions that take place in the UV region can be probed using two photons of visible laser light because of the additivity of frequencies.

Coherence. By *coherent* we mean that the photons in a laser are emitted *in phase* with one another and are moving in exactly the same direction. The high degree of coherence arises from the fact that the stimulated emission synchronizes the radiation of the individual molecules, so that the photon emitted from one molecule stimulates another molecule to emit a photon of the same wavelength that is exactly in phase with the first photon and so on. One application of laser coherence is *holography*, a technique for producing three-dimensional images. Coherent laser light is reflected off a subject and is combined with light that has not been reflected to produce an interference pattern. This process produces a *hologram*, which contains information

* As mentioned in Section 11.1, for an allowed transition ψ_i and ψ_j must have different symmetries with respect to each other. For the two-photon process we can imagine the presence of an intermediate state between the initial and final states, that is, the transition takes place in two steps. Thus, the initial and final states must have the *same* symmetry (even-even or odd-odd). Noting that each photon has one unit of angular momentum, we can also rationalize how the atom or molecule must therefore change by 2 or 0 units of angular momentum.

not only on the intensity (as in a conventional two-dimensional photograph), but also on the phase of light reflected from the subject. Subsequent illumination of the hologram reconstructs a three-dimensional image. Holography has been used to record the three-dimensional structure of, for example, art objects. One of the motivations for developing an X-ray laser is the desire to produce a hologram of the contents of a living cell with high spatial resolution.

Monochromaticity. Laser light is highly monochromatic (having the same wavelength) because all the photons are emitted as a result of a transition between the same two atomic or molecular energy levels. Therefore, they possess the same frequency and wavelength. In the Nd:YAG laser, for example, the emitted light is centered at 1064 nm with a width less than 0.5 nm. Although narrow linewidths can be obtained also with ordinary light (from, say, an incandescent light bulb) and a *monochromator* (a prism or diffraction grating which separates light of different wavelengths), the intensity of a laser beam at a particular wavelength can be six orders of magnitude or more greater than that from a conventional source.

The highly monochromatic nature of laser light makes it possible to induce and identify specific transitions between electronic, vibrational, and even rotational energy levels for many molecules and therefore to obtain well-resolved absorption spectra. However, the laser systems discussed so far are all fixed-frequency lasers, which provide emission of light at one or a few discrete wavelengths. Thus, they are unsuitable for the usual absorption methods, which require scanning over a continuous range of wavelengths. Organic dye lasers are appropriate for this application because they are *tunable*, that is, they provide wavelengths over a continuous range. One of the most widely used organic dyes is Rhodamine 6G, a molecule that has many modes of vibration.

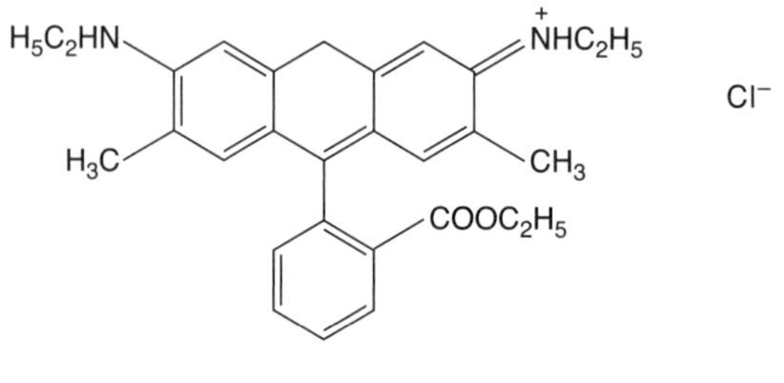

H_5C_2HN $\overset{+}{N}HC_2H_5$ Cl^-
H_3C CH_3
$COOC_2H_5$

The electronic spectrum of Rhodamine 6G in solution (Figure 14.14) shows a broad peak due to strong molecular interaction in the liquid state. Collisions with solvent molecules broaden the vibrational structure of the transitions into unresolved bands. Consequently, the fluorescence of the dye, which occurs at a longer wavelength, also appears as a broad peak. By *pumping* (i.e., electronically exciting) the solution with a laser, it is possible to bring about a population inversion in Rhodamine 6G and subsequent laser action. By using a wavelength-tuning element such as diffraction grating as part of the optical resonator, it is possible to vary the output wavelength of a dye laser. For example, a dye laser using Rhodamine 6G in methanol solution as the gain medium is continuously tunable over the range 570–660 nm. Using different organic dyes, spectroscopists have expanded the tunable range of dye lasers to between 310 and 1200 nm. This technique has greatly broadened the scope of high-resolution spectroscopy. Tunable solid-state gain media, such as titanium-doped sapphire ($Ti:Al_2O_3$, often

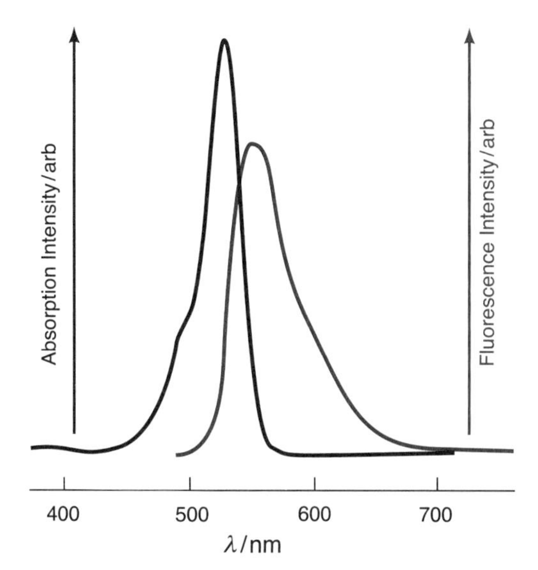

Figure 14.14
Absorption and fluorescence spectra of the laser dye Rhodamine 6G in alcohol solution. Note the wavelength of maximum absorbance (λ_{max}) is close to 532 nm, the wavelength of a frequency-doubled Nd:YAG laser, which is often used to excite the dye. The range of emission wavelengths loosely corresponds to the range of lasing possible.

written as "Ti:sapph"), are convenient alternatives when available. Ti:sapph lasers are tunable in the range 650 to 1100 nm and may be operated in cw or pulsed mode.

Spatial collimation. A property of many lasers is that they have very low spatial divergence; that is, they are highly spatially collimated. Unlike an incandescent light bulb, where the intensity decreases with distance, laser light tends to remain in a small tight-waisted beam. Taking advantage of this collimation, a pulsed laser can be aimed at the moon, where a retroflector placed by astronauts reflects the laser back to Earth. Timing of the round-trip time for the laser pulse is used to measure the distance between Earth and its moon to a precision of about 1 cm. The size of the laser spot on the moon is only a few km in diameter, which is a divergence of less than 5×10^{-4} degrees.

14.4 Applications of Laser Spectroscopy

Since their development in 1960, lasers have been used for a wide variety of consumer, industrial, medical, and scientific applications. In this section, we illustrate a few spectroscopic applications that take advantage of the unique properties of laser light.

Laser-Induced Fluorescence

The high intensity, high monochromaticity, and tunability of laser light enable us to excite atoms and molecules and monitor the subsequent fluorescence of these species. The advantage of *laser-induced fluorescence* (LIF) over fluorescence produced by a conventional light source is that LIF is both highly sensitive and highly selective. In elemental analysis, for example, a sample solution is atomized (decomposed into atomic species) in a furnace or a flame and then irradiated with a laser beam to induce atomic fluorescence. By this procedure it is possible to detect concentrations (of the original solution) in the 10^{-11} g mL^{-1} range. The high monochromaticity and narrow linewidth of a laser source enable us to excite molecules to specific vibrational levels (and in some cases, specific rotational levels) in excited electronic states and to observe the subsequent fluorescence. This method yields valuable information about the electronic structure of excited states and is particularly useful for studying small transient species, such as the free radicals involved in atmospheric chemistry and in flames.

Detection of fluorescence at right angles to the laser beam allows for very high spatial resolution. In the simplest form, the detection volume is the intersection of a laser beam and the viewing region of the fluorescence collection optics. Spatial resolution of a cubic micrometer is possible. When laser light is imaged into a sheet using a cylindrical lens, a fluorescence image may be collected using an array detector, such as a digital camera. This approach of *planar LIF* has been used to image the distribution of OH radicals in flames, and thus identify the flame front, where most of the chemistry takes place.

Ultrafast Spectroscopy

Prior to the development of the laser, there were few methods for measuring directly the time scale for very fast chemical processes. With the development of increasingly fast laser technology it is now possible to measure directly the time scale of the most fundamental of chemical processes, including bond breaking and intramolecular energy redistribution. The microsecond flash photography methods of the American electrical engineer Harold Edgerton (1903–1990) captured bullets in flight and droplets of milk splashes by "freezing" the motion of macroscopic objects. To "freeze" the motion of molecules requires a flash that is another five orders of magnitude faster. Consider the time scale needed to observe fundamental motions in molecules. A typical C–H stretching motion, which occurs at a wavenumber of 3000 cm^{-1}, corresponds to a vibrational period of about 10^{-14} s or 10 fs (femtoseconds). Direct bond dissociation can be considered half a bond stretching period, and thus might occur on a femtosecond time scale. Similarly, a rotational motion at a wavenumber of 1 cm^{-1} corresponds to a 33 ps (picosecond) period. The time scales of pico- and femtoseconds are commonly referred to as "ultrafast," and the development of ultrafast laser technology since the 1970s and 1980s has changed the nature of chemical questions that may be asked. The Egyptian-American chemist Ahmed Zewail (1946–) won the 1999 Nobel Prize in Chemistry for his work in *femtochemistry*—the study of ultrafast reaction kinetics and dynamics. A typical ultrafast experiment employs two laser pulses: a *pump* pulse, which is used to initiate a chemical reaction or prepare an excited state of a molecule, followed by a *probe* pulse, which measures what happens to the system in time. Synchronizing two separate ultrafast lasers in time is impractical, so the pump and probe pulses usually are generated from a common source. The time delay between the pump and probe pulse is typically controlled by translating a mirror and varying the path length that the probe pulse travels. Because the speed of light is 3×10^8 m s^{-1}, a delay of 1 ps correspond to a distance of 0.3 mm, or about three times the width of a typical human hair.

Light travels approximately 1 foot in 1 nanosecond (1 ns = 10^{-9} s).

Experiments such as those performed by Zewail and his colleagues provide the opportunity to examine chemical reactions in unprecedented detail. With ultrafast spectroscopy, we can follow the course of the chemical reaction from reactants through the transition state, and into the formation of products. The first demonstration of this technique was for the gas-phase photofragmentation reaction of iodine cyanide:

$$\mathrm{ICN} \xrightarrow{\lambda_1} \mathrm{ICN}^* \longrightarrow \mathrm{I} + \mathrm{CN} \tag{14.11}$$

The reaction is initiated by a femtosecond laser pulse, λ_1, which promotes or *pumps* the molecule to an excited electronic state (indicated with an asterisk). This excited state immediately starts to dissociate into two photofragments, an iodine atom and a CN radical. The dissociation process is *probed* using laser-induced fluorescence with a second femtosecond laser pulse, λ_2 (or λ_3), which promotes ICN* to an even higher-energy state ICN** (labeled with two asterisks). ICN** emits fluorescent light λ_{fl} on returning to the state ICN*.

$$\mathrm{ICN}^* \xrightarrow{\lambda_2\,(\mathrm{or}\ \lambda_3)} \mathrm{ICN}^{**} \longrightarrow \mathrm{ICN}^* + \lambda_{\mathrm{fl}} \tag{14.12}$$

The resulting fluorescence λ_{fl} is monitored as a function of delay time between the pump and probe laser pulses. Using λ_2 and λ_3 to probe short and long I–CN bond lengths, respectively, the data show that the bond cleavage occurs on a time scale of about 200 fs (Figure 14.15).

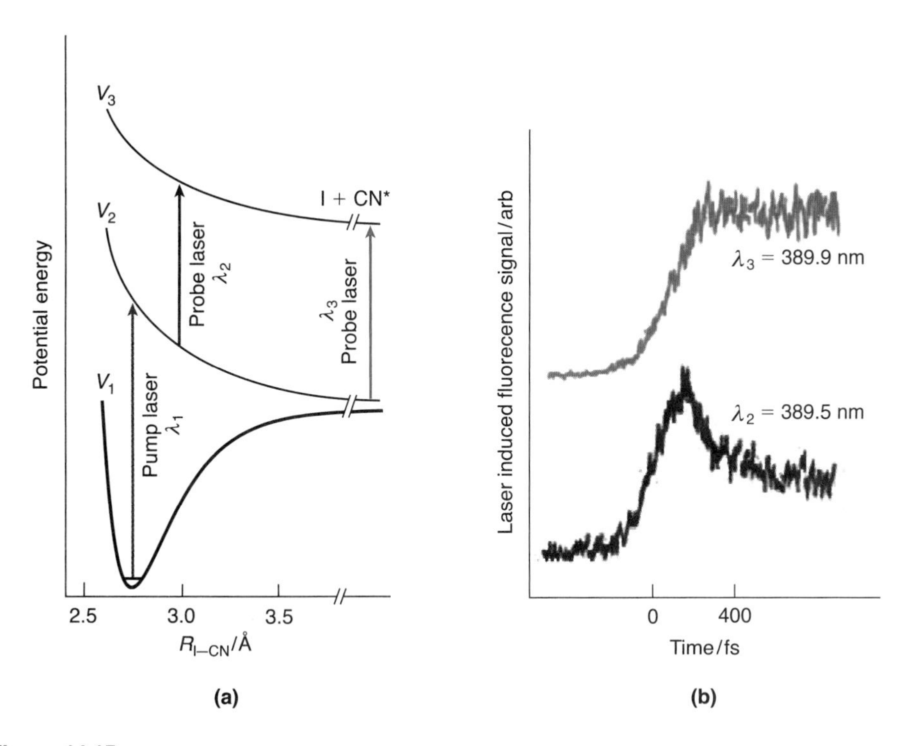

Figure 14.15
(a) Potential energy surfaces for the iodine cyanide (ICN) molecular system. An ultrafast laser pulse, λ_1, excites ICN from the ground state surface V_1, where the I—CN separation is 2.75 Å, to a dissociative excited state V_2. A short time later, a second ultrafast laser pulse, λ_2 or λ_3, probes the excited ICN by promoting it to a higher energy surface V_3. The probe lasers are tuned such that λ_2 is absorbed when the I—CN separation is about 3 Å, and λ_3 is absorbed when the I—CN separation is more than 6 Å (which is effectively dissociated to I and CN radicals). (b) Transient laser induced fluorescence signals excited by λ_2 and λ_3 show that the I—CN dissociation takes place in 205 ± 30 femtoseconds. (*Ahmed Zewail Collections, California Institute of Technology.*)

Femtochemistry has been applied to unravel the mechanisms of many chemical reactions and biological processes such as photosynthesis and vision. It has added a new dimension to chemical kinetics.

Single-Molecule Spectroscopy

In 1959, the American physicist Richard Feynman (1918–1988) gave a forward-looking speech in which he noted that there was "plenty of room at the bottom" and that devices could keep getting smaller and smaller for quite some time before we reached any physical limits. His talk was one of the motivating moments for the new field of nanotechnology. Part of that vision has now been realized. Using laser-induced fluorescence, the ultimate in detection limit has been achieved in *single molecule spectroscopy.* As a result of advances in optical and detector technology, it is straightforward to detect molecules one at a time. One such apparatus employs a high-powered optical microscope to both deliver a laser beam to a small volume and collect the resulting fluorescence (Figure 14.16). Single molecule spectroscopy relies on two general principles to achieve its sensitivity. First, it requires that there is only one molecule in the volume that is probed. Second, that the signal-to-noise ratio is sufficient in the detection system. Having a single molecule in view is assured by dilute samples and the microscopic probe for a sample volume on the order of 1 μm^3. Detection sensitivity and signal-to-noise ratio are obtained by using detectors with good quantum efficiency (to respond to a single photon) and by a careful choice of sample molecule and detection technique. Many single-molecule spectroscopy techniques have employed laser-induced fluorescence of laser dyes. Excitation of Rhodamine 6G with green light, for example, produces red fluorescence with a quantum yield of 0.95 (see Figure 14.14). Because the color of the fluorescence light is significantly different from that of the excitation light, the excitation light may be readily blocked with a filter. Thus, there is very little scattered light that reaches the detector.

One molecule in 1 μm^3 corresponds to about 1 nanomolar concentration.

Single-molecule spectroscopy provides information unavailable in conventional techniques because it removes ensemble averaging. Rather than reporting the mean of what is happening in a bulk sample, single-molecule spectroscopy shows the actual distribution (a histogram) of the values being measured. The single molecule is a reporter on the local *nanoenvironment* which is particularly useful for inhomogeneous systems.

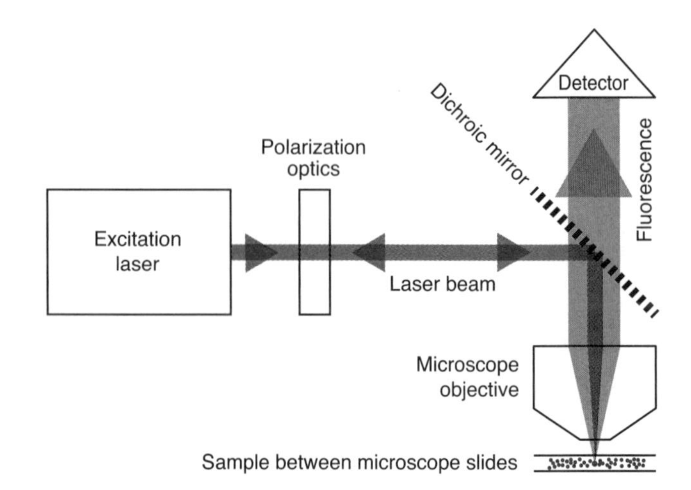

Figure 14.16
Apparatus for single-molecule spectroscopy. The polarization optics are used for shuttering the laser on and off and for preventing the laser from shinning back on itself. The dichroic mirror reflects the laser light but transmits the fluorescence. [Adapted from W. E. Moerner, *J. Phys. Chem. B* **106**, 910 (2002), Figure 14b.]

Using single-molecule spectroscopy, we can determine if a bulk sample is characterized by uniform composition, or by local domains. By analyzing the details of the fluorescence emission, it is possible to measure the diffusion constants, orientational or conformational changes, intersystem crossing, reaction kinetics, or lifetime of the molecule.

14.5 Photoelectron Spectroscopy

The electronic spectroscopies discussed so far provide information about readily accessible molecular electronic states, that is, states that are accessible by allowed electronic transitions with energies typically in the ultraviolet and visible (200–700 nm). It is certainly possible to probe electronic states using vacuum ultraviolet spectroscopy, with wavelengths in the range 100–200 nm, but it is an experimentally inconvenient region of the electromagnetic spectrum because air (especially molecular oxygen) absorbs extensively in this region. A sensitive technique for probing molecular electronic structure of lower lying orbitals is that of photoelectron spectroscopy. Photoelectron spectroscopy is based on the photoelectric effect, which is described in Section 10.3.

We distinguish between ultraviolet photoelectron spectroscopy (UPS), which probes valence orbitals, and X-ray photoelectron spectroscopy (XPS), which probes both core and valence orbitals. When XPS is applied to the analysis of the elemental composition of surfaces, the acronym ESCA (electron spectroscopy for chemical analysis) is often used.

In a photoelectron spectroscopy experiment, a gaseous sample is illuminated with light of a particular frequency ν and the kinetic energy of the ejected electron is measured (Figure 14.17). The fundamental equation describing photoelectron spectroscopy is essentially the same as that for the photoelectric effect, and is based on the

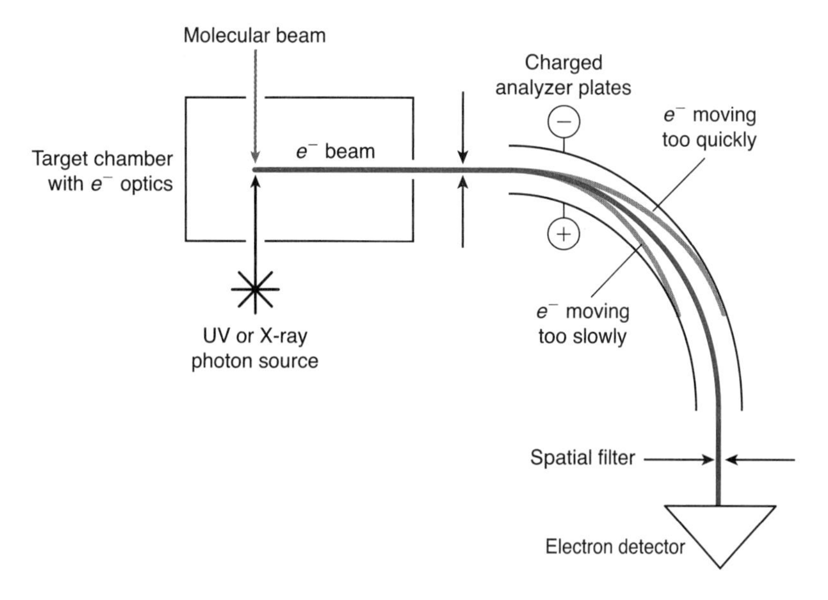

Figure 14.17
Schematic diagram of an ultraviolet photoelectron spectrometer. The instrumentation is in a vacuum so that electrons will not collide with gas molecules. Electron kinetic energy is measured by varying the voltage across the curved parallel plates. For X-ray photoelectron spectroscopy, the He(I) source is replaced with an X ray source.

Table 14.4
Photon Sources for Photoelectron Spectroscopy

Photon source	Energy/eV	Wavelength/nm	Type	Electron shell probed
He (I)	21.22	58.46	UPS	valence
He (II)	42.44	29.23	UPS	valence
Mg K_α	1253.6	0.989	XPS	core
Al K_α	1486.6	0.83386	XPS	core
Cr K_α	5409.	0.2292	XPS	core
Cu K_α	8040.	0.1542	XPS	core
Synchrotron	tunable	tunable	UPS & XPS	valence & core

conservation of energy. The energy of the impinging photon is equal to the binding energy of the electron plus the energy of the ejected electron,

$$E_{\text{photon}} = E_{\text{ionization, electron}} + E_{\text{kinetic, electron}}$$
$$h\nu_{\text{photon}} = IE + \tfrac{1}{2}m_e u^2 \qquad (14.13)$$

where IE denotes ionization energy. UPS employs vacuum ultraviolet photons with energies in the range 10-45 eV to probe electrons in valence orbitals (Table 14.4). XPS uses soft X rays with photon energies in the range 200–2000 eV to probe core electrons in addition to valence electrons. The photoelectron equation shows that for a fixed excitation wavelength, a lower energy photoelectron corresponds to a more tightly held electron, and vice versa. Correspondingly, fast-moving photoelectrons correspond to loosely bonded electrons. According to Koopmans' theorem [after the Dutch polymath Tjaling Charles Koopmans, (1910–1985)], the ionization energy for a particular electron is approximately equal to the negative of the energy of the orbital from which it originated,

Koopmans shared the 1975 Nobel Prize in Economics.

$$IE_j \cong -\varepsilon_j \qquad (14.14)$$

where the subscript j denotes the jth orbital. The approximation arises from the neglect of relaxation of the remaining electrons when an electron is removed. This relaxation occurs to minimize the energy of the resulting ion. Because of differing electron-electron repulsion energies, orbital energies in neutral and ion species are not the same. In the Hartree–Fock approximation (see Section 12.10), the ionization energy is also equal to the negative of the orbital energy. Thus, we commonly apply Koopmans' theorem to photoelectron spectroscopy, and relate each peak in the spectrum to an orbital energy.

For molecular photoelectron spectra, we also have to account for the internal energies of the parent molecule and its cation. Typically, we employ a room temperature or colder molecular source, so that the parent molecules contain a negligible

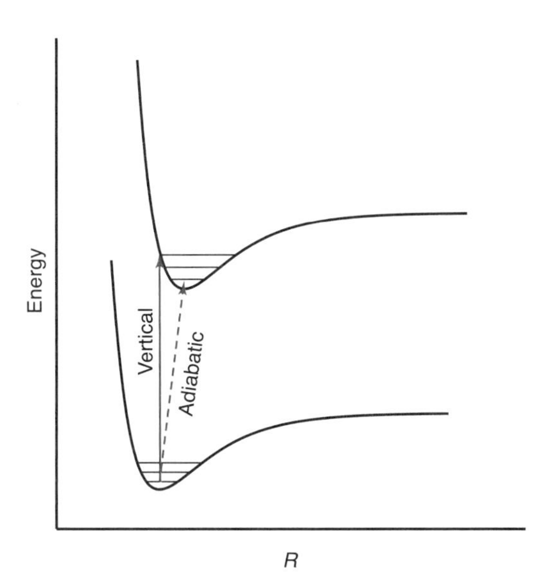

Figure 14.18
Potential-energy diagram showing vertical (1) and adiabatic (2) transitions.

amount of vibrational excitation. Experimentally, most photoelectron spectra do not contain sufficient resolution to observe rotational structure. The relevant energy equation becomes

$$h\nu_{\text{photon}} = IE + E_{\text{vib}} + E_{\text{rot}} + \tfrac{1}{2}m_e u^2 \tag{14.15}$$

where E_{vib} and E_{rot} are the vibrational and rotational energies, respectively, of the cation. We distinguish the adiabatic ionization energy from the vertical ionization energy, *IE* (Figure 14.18). Adiabatic ionization energy describes the minimum energy required to ionize a molecule, but necessitates a change in geometry between the neutral and ionic species. Vertical ionization energies are measured in photoelectron spectroscopy, and are consistent with the Born–Oppenheimer approximation (see Section 13.1). The ultraviolet photoelectron spectrum of molecular hydrogen contains a series of closely spaced peaks (Figure 14.19). The separation between these peaks corresponds to the separation in energy of the vibrational energy levels in the hydrogen molecular cation.

Gas-phase photoelectron spectra provide information about molecular orbitals. Consider the water molecule as an example. Water has 10 electrons, with two electrons in the oxygen atom 1*s* core orbital, four electrons in two OH bonding orbitals, and four electrons in two lone pair nonbonding orbitals. We might thus expect three peaks in a photoelectron spectrum, corresponding to the core, bonding, and lone pairs orbitals. Experimentally, however, five peaks are observed (Figure 14.20), with one peak corresponding to each of the orbitals. A molecular orbital picture for the bonding in the water molecule provides a clearer interpretation than the valence bond (Lewis electron dot diagram) picture for the bonding in molecular water (Figure 14.21). Rather than two equivalent valence bonds, the covalent bonding is better described as three sigma molecular orbitals and a nonbonding orbital.

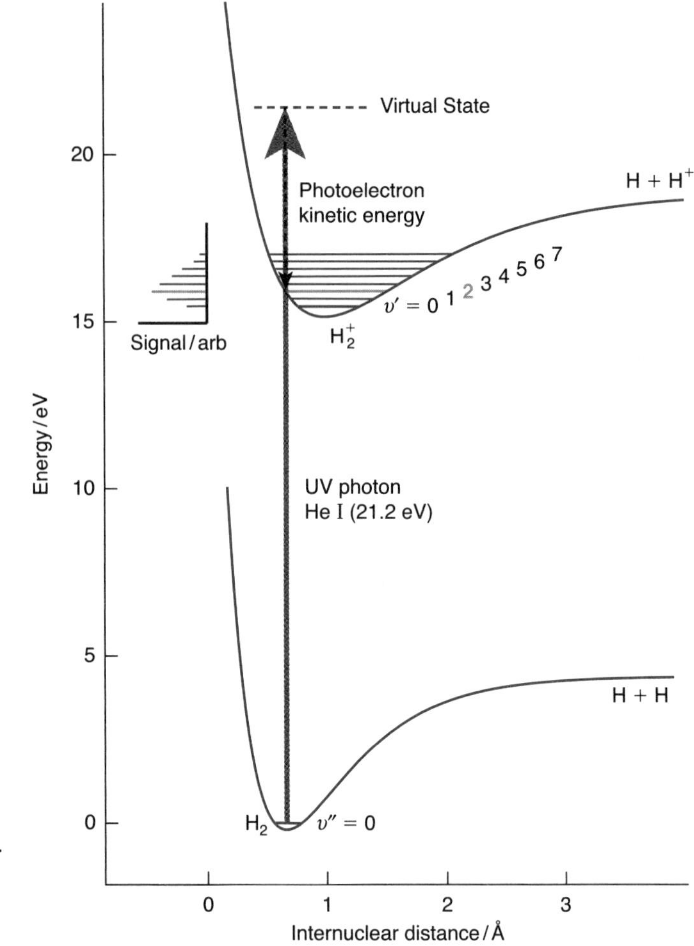

Figure 14.19
Ultraviolet photoelectron spectrum of molecular hydrogen.

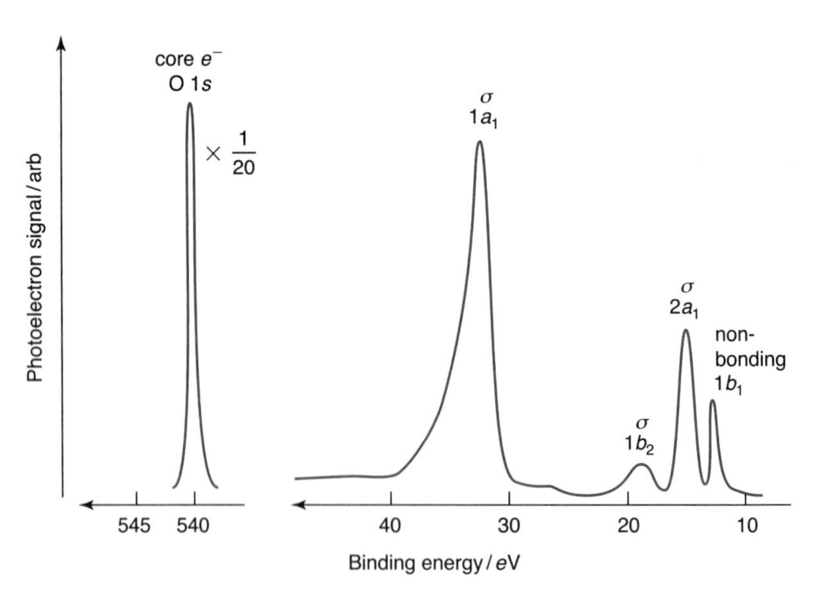

Figure 14.20
X-ray photoelectron spectrum of water.

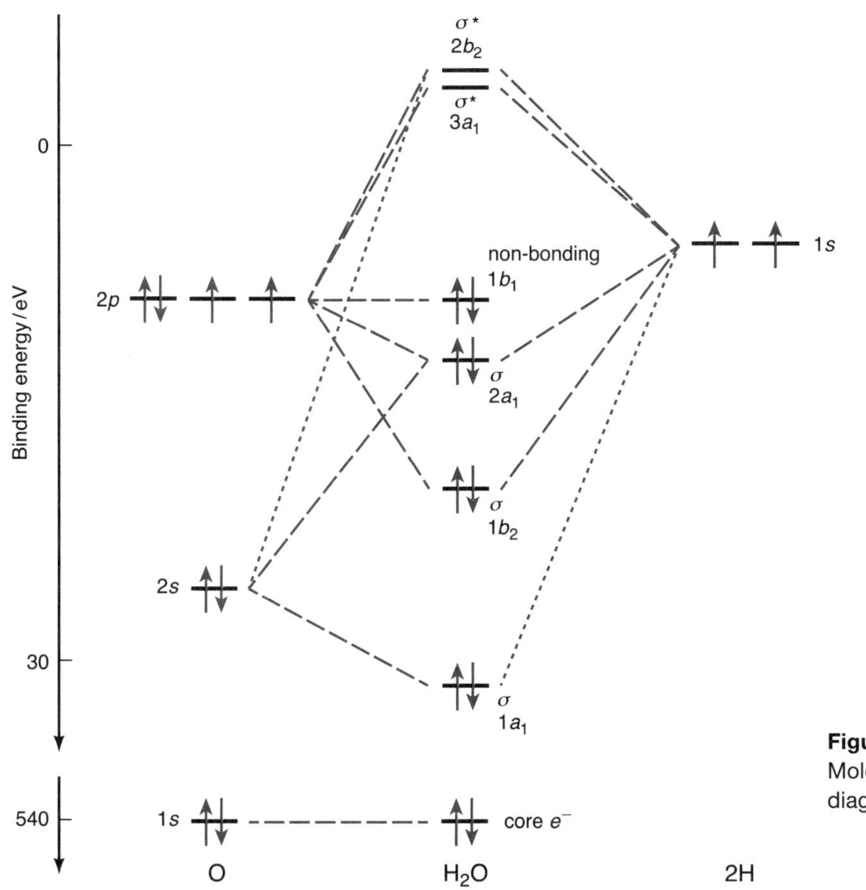

Figure 14.21
Molecular orbital energy level diagram of water.

14.6 Nuclear Magnetic Resonance Spectroscopy

Like electrons, some nuclei possess spin and therefore a magnetic moment associated with them. The nuclear spin, I, may have one of the following values:

$$I = 0, \tfrac{1}{2}, 1, \tfrac{3}{2}, 2, \ldots$$

A value of zero means the nucleus has no spin. Table 14.5 gives rules for determining nuclear spin on the basis of atomic number and the number of neutrons present. For a given value of I, there are $(2I + 1)$ values of m_I, the *nuclear spin quantum number*, which denotes the spin orientations. The corresponding energy levels are all degenerate in the absence of a magnetic field. Consider the proton (^{1}H), for which $I = \frac{1}{2}$. The two m_I values are $+\frac{1}{2}$ and $-\frac{1}{2}$, respectively. When an external magnetic field is applied, the degeneracy is removed. The energy of a given nuclear spin state, E_{m_I}, is directly proportional to the value of m_I and the magnetic field strength B_0,

$$E_{m_I} = -m_I B_0 \gamma \hbar \tag{14.16}$$

where γ is the *gyromagnetic* ratio (also called the *magnetogyric ratio*), which is a constant characteristic of the nucleus being studied. The minus sign follows the convention

Table 14.5
Rules for Predicting Nuclear Spin

Number of protons (Z)	Number of neutrons[a]	Nuclear spin (I)
Even	Even	0
Even	Odd	$\frac{1}{2}$ or $\frac{3}{2}$ or $\frac{5}{2}$ ⋯
Odd	Even	$\frac{1}{2}$ or $\frac{3}{2}$ or $\frac{5}{2}$ ⋯
Odd	Odd	1 or 2 or 3 ⋯

[a] In the only case in which the nucleus has no neutrons, that is, the ^{1}H isotope, "0" is treated as an even number, and $I = \frac{1}{2}$.

that a positive m_I value corresponds to a lower (negative) energy than that of its negative counterpart. Figure 14.22 shows the splitting of the nuclear spin energy levels relative to magnetic field strength for $I = \frac{1}{2}$. The energy difference, ΔE, is given by

$$\begin{aligned}\Delta E &= E_{-1/2} - E_{+1/2} \\ &= -\left[\left(-\tfrac{1}{2}\right) - \left(+\tfrac{1}{2}\right)\right] B_0 \gamma \hbar \\ &= B_0 \gamma \hbar \end{aligned} \tag{14.17}$$

A nuclear magnetic resonance (NMR), that is, a transition from the $m_I = +\frac{1}{2}$ level to the $m_I = -\frac{1}{2}$ level, can be observed by varying either the frequency, ν, of the applied radiation (given by $\Delta E/h$ or $B_0\gamma/2\pi$) or the intensity of the magnetic field, B_0, until

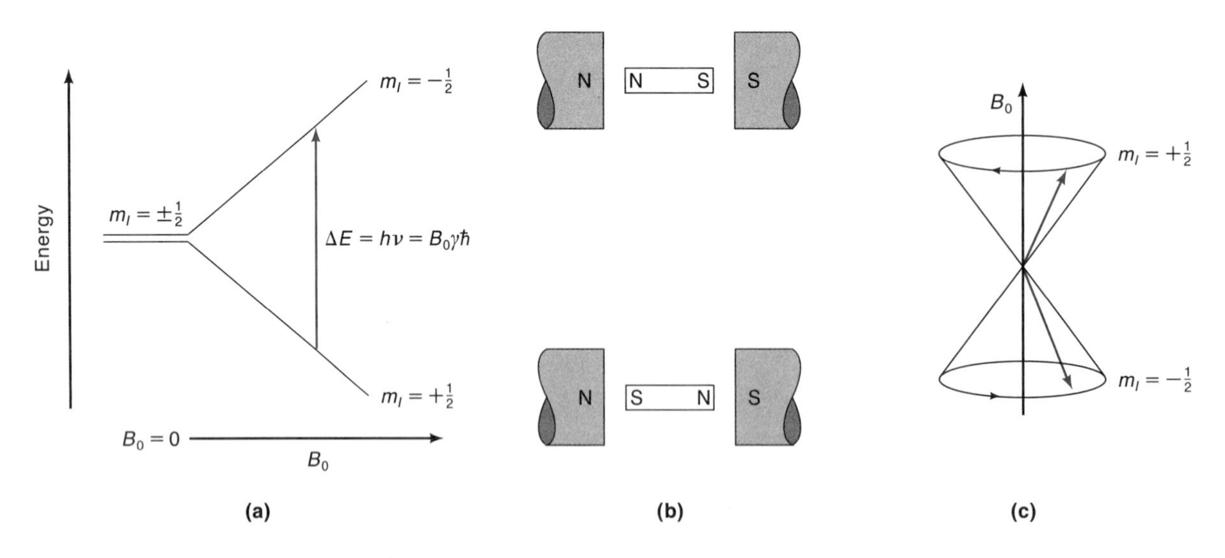

Figure 14.22
(a) Difference in nuclear spin energy levels in an external magnetic field B_0. (b) Classical description of the alignment of the nuclear spins parallel and antiparallel to the external magnetic field. (c) The precession of the nuclear spins at their Larmor frequency.

the resonance condition ($\Delta E = h\nu$) is satisfied.* The selection rule for nuclear spin energy-level transitions is $\Delta m_I = \pm 1$.

The two magnetic moments associated with $m_I = \pm\frac{1}{2}$ are not aligned statically with or against the external magnetic field; rather, they wobble around (like a spinning top), or precess about, the axis of the applied field (Figure 14.22c). The frequency of this precession, called the *Larmor frequency* (ω), is given by

$$\omega = B_0\gamma \tag{14.18}$$

The Larmor frequency is given in radians per second (rad s^{-1}), but it can be converted into linear frequency, ν, as follows (see Appendix A):

$$\nu_{\text{precession}} = \frac{\omega}{2\pi} = \frac{B_0\gamma}{2\pi} \tag{14.19}$$

This precession frequency is independent of m_I, so all spin orientations of a given nucleus precess at this frequency in a magnetic field. Note that Equation 14.19 looks the same as the frequency for observing nuclear magnetic resonance mentioned above. The reason is that the frequency of the applied radiation must be equal to the Larmor frequency for resonance to occur.

The strength of a magnetic field is measured in *tesla* (T), after the Serbian engineer and inventor Nikola Tesla (1856–1943), where

A tesla coil, which generates a high-voltage, low-current electrical discharge, finds use for detecting leaks in glass vacuum manifolds.

$$1 \text{ T} = 10^4 \text{ gauss}$$

The gyromagnetic ratio has the units T^{-1} s^{-1}. Table 14.6 lists the gyromagnetic ratios, NMR frequencies (in a 4.7-T field), and natural abundances of several isotopes.

* Note that unlike microwave, IR, and electronic spectroscopies, NMR spectroscopy involves the interaction between the magnetic-field component of electromagnetic radiation and the magnetic moment of the nuclei.

Table 14.6
Gyromagnetic Ratios, NMR Frequencies (in a 4.7-T Field), and Natural Abundances of Isotopes

Isotope	I	$\gamma/10^7$ T^{-1} s^{-1}	ν/MHz	Natural abundance (%)
^{1}H	$\frac{1}{2}$	26.75	200	99.985
^{2}H	1	4.11	30.7	0.015
^{13}C	$\frac{1}{2}$	6.73	50.3	1.108
^{14}N	1	1.93	14.5	99.63
^{15}N	$\frac{1}{2}$	2.71	20.3	0.37
^{17}O	$\frac{5}{2}$	3.63	27.2	0.037
^{19}F	$\frac{1}{2}$	25.17	188.3	100
^{31}P	$\frac{1}{2}$	10.83	81.1	100
^{33}S	$\frac{3}{2}$	2.05	15.3	0.76

These frequencies are in the radiofrequency region. For a given nucleus, the larger the γ, the easier it is to detect the corresponding NMR signal. Thus, the most readily studied nuclei are ^{1}H, ^{19}F, and ^{31}P. With modern instrumentation (see p. 649), however, the NMR of the ^{13}C nucleus, which has a small γ value and very low natural abundance but is of great importance in organic chemistry and biochemistry, can be studied straightforwardly.

EXAMPLE 14.1

Calculate the magnetic field, B_0, that corresponds to a precession frequency of 400 MHz for ^{1}H.

ANSWER

From Equation 14.19 and Table 14.6,

$$B_0 = \frac{2\pi v}{\gamma}$$

$$= \frac{2\pi(400 \times 10^6\ \text{s}^{-1})}{26.75 \times 10^7\ \text{T}^{-1}\ \text{s}^{-1}}$$

$$= 9.40\ \text{T}$$

COMMENT

NMR studies of nuclei with frequencies at or greater than 200 MHz require a strong magnetic field; therefore, superconducting magnets must be used for these experiments.

The Boltzmann Distribution

Because NMR is a branch of absorption spectroscopy, its sensitivity is governed by the Boltzmann distribution. Consider a sample of ^{1}H nuclei in a magnetic field measuring 9.40 T at 300 K. From Equation 14.17, we have

$$\Delta E = \frac{(9.40\ \text{T})(26.75 \times 10^7\ \text{T}^{-1}\ \text{s}^{-1})(6.626 \times 10^{-34}\ \text{J s})}{2\pi}$$

$$= 2.65 \times 10^{-25}\ \text{J}$$

and $k_B T = 4.14 \times 10^{-21}$ J. Thus, the ratio of the number of nuclear spins in the upper energy level to that in the lower energy level is

$$\frac{N_{-1/2}}{N_{+1/2}} = e^{-\Delta E/k_B T}$$

$$= \exp\left(\frac{-2.65 \times 10^{-25}\ \text{J}}{4.14 \times 10^{-21}\ \text{J}}\right)$$

$$= 0.99994$$

This number is very close to one, meaning that the two levels are almost equally populated.* This distribution is the result of strong thermal motion in the sample, which overwhelms the tendency to orient the spins in a magnetic field. Nevertheless, even the slight excess of spins in the lower level is sufficient to give rise to detectable NMR signals.

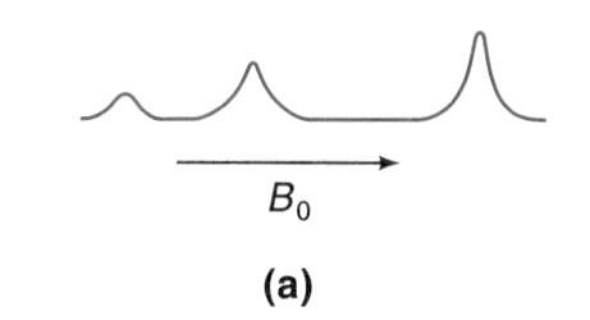

Chemical Shifts

The discussion so far may seem to suggest that all protons resonate at the same frequency, but this is not the case. In reality, at a particular magnetic field strength, the resonance frequency for any given ^1H nucleus depends on its position in the molecule under study. This effect, called the *chemical shift*, is what makes NMR spectroscopy so useful.

Figure 14.23a shows the proton NMR spectrum of ethanol (CH_3CH_2OH). The three peaks of relative areas 1:2:3 correspond to the hydroxyl, methylene, and methyl protons, respectively. The fact that three separate peaks are observed means that the local magnetic field, B, present at each type of nucleus is different from the external magnetic field, B_0. These magnetic fields are related according to

$$B = B_0(1 - \sigma) \tag{14.20}$$

where σ, a dimensionless constant, is called the *screening* or *shielding constant*. As a result of this shielding, the resonance frequency for a given nucleus becomes

$$\nu = \frac{B_0(1 - \sigma)\gamma}{2\pi} \tag{14.21}$$

Thus, the resonance frequency for a nucleus in an atom (or molecule) is lower than that for a bare nucleus, that is, for an isolated proton. In general, σ is a small number (about 10^{-5} for protons), whose magnitude depends on the electronic structure around the nucleus in question. The modified resonance condition is shown in Figure 14.24.

In general, we are interested not in the absolute shifts of the NMR peaks from that expected for a bare proton but in the relative positions of the peaks. Thus, common

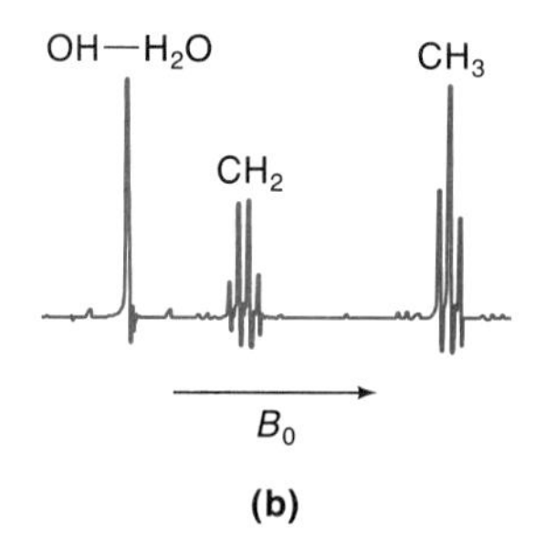

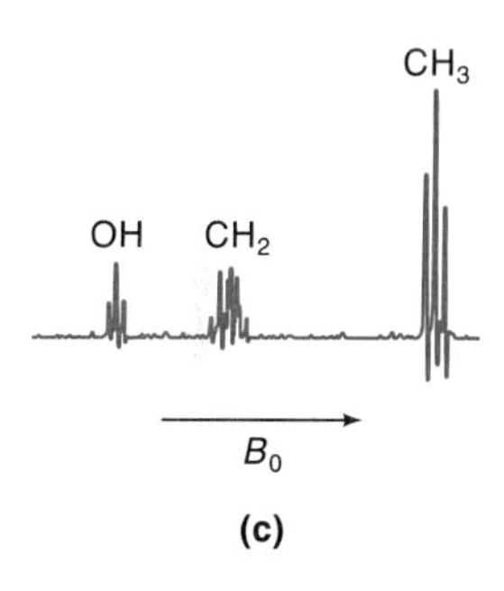

Figure 14.23
(a) Low-resolution and (b) high-resolution proton NMR spectrum of ethanol. (c) NMR spectrum of anhydrous pure ethanol. [Parts (b) and (c) from G. Glaros and N. H. Cromwell, *J. Chem. Educ.* **48**, 202 (1971).]

* This ratio is *much* smaller (favoring the absorption process) for IR and electronic spectroscopy because of the appreciably larger separation in energy levels.

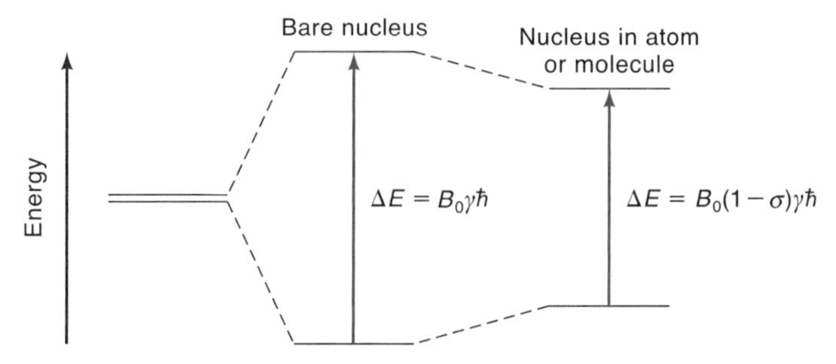

Figure 14.24
Effect of electron shielding on the nuclear magnetic resonance condition. Diagram not drawn to scale.

practice is to define the chemical shift by the difference in resonance frequencies between a nucleus of interest (ν) and a reference nucleus (ν_{ref}) in terms of the chemical shift parameter (δ), where

$$\delta = \frac{\nu - \nu_{ref}}{\nu_{spec}} \times 10^6 \text{ ppm} \tag{14.22}$$

where ν_{spec} is the spectrometer frequency. Because the difference between ν and ν_{ref} is typically on the order of hundreds of hertz and ν_{spec} is typically hundreds of megahertz, the ratio is multiplied by 10^6 to make δ a convenient number with which to work. For this reason, chemical shifts are expressed in units of ppm (parts per million). Note that the frequency difference ($\nu - \nu_{ref}$) is divided by ν_{spec}. This means that δ is *independent* of the magnetic field used to measure it. The reference proton compound chosen for most organic systems is tetramethylsilane (TMS), $(CH_3)_4Si$, because it has the following advantages: (1) it contains 12 protons of the same type, so only a small amount is needed as an internal reference, (2) it is chemically inert, and (3) its protons have a smaller resonance frequency than that observed for most other protons, so their chemical shifts can be assigned positive values. Figure 14.25 shows chemical shifts for various types of protons relative to TMS in ppm.

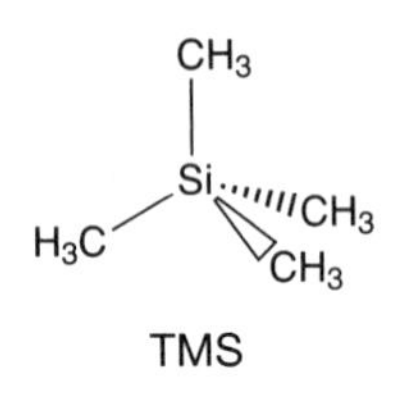

TMS

Conventionally, NMR spectra are plotted with ν (and δ) increasing from right to left. Thus, more heavily shielded protons (larger σ, smaller ν, smaller δ) appear toward the right side of the spectrum. Sometimes chemists refer to chemical shifts as "upfield" or "downfield," meaning "more shielded" or "less shielded," respectively. Chemical shifts can be readily converted back to the frequencies separating the sample and reference peaks using Equation 14.22. For example, the chemical shift of benzene is about 7.30 ppm so that, if the spectrometer operates at 200 MHz frequency,

$$\begin{aligned}\nu_{benzene} - \nu_{TMS} &= \delta \times \nu_{spec} \\ &= (7.30 \times 10^{-6})(200 \times 10^6 \text{ Hz}) \\ &= 1.46 \times 10^3 \text{ Hz}\end{aligned}$$

You can see that if the signals were recorded with a 400-MHz spectrometer, the difference in frequencies would be 2.92×10^3 Hz. So, the separation between peaks in a given NMR spectrum is directly proportional to the spectrometer frequency (or the magnetic field), but δ is independent of the frequency. For this reason, high-field NMR (approaching 1000 MHz as of 2013) is becoming increasingly popular in the study of protein solutions, for which separation of the many overlapping peaks observed is critical to meaningful analysis.

Spin–Spin Coupling

At high resolution, the NMR spectrum of ethanol is as shown in Figure 14.23b. The $-CH_2$ and $-CH_3$ peaks actually consist of four and three lines, respectively, with relative intensities of 1:3:3:1 and 1:2:1. The spacing between each group of lines is *independent* of the spectrometer frequency. Therefore, it cannot be a chemical-shift effect as discussed above. How can we explain this observation? Each nucleus with $I \neq 0$ has a nuclear magnetic moment, and the magnetic field generated by this nucleus can affect the magnetic field experienced by a neighboring nucleus, thereby slightly changing the frequency at which the neighboring nucleus will undergo NMR

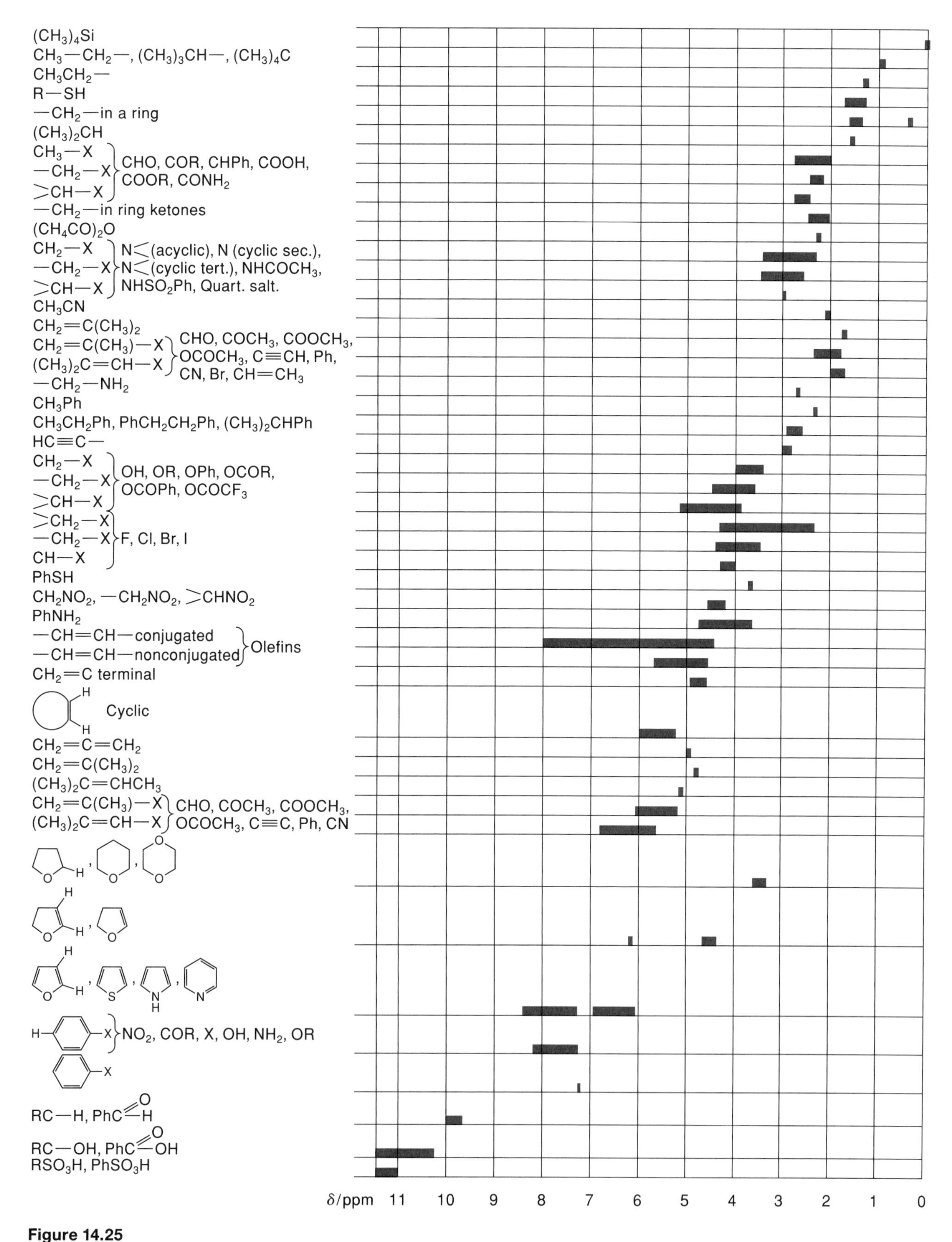

Figure 14.25
Chemical shifts for various organic compounds. TMS has a chemical shift of zero, because it is the reference compound. [From E. Mohacsi, *J. Chem. Educ.* **41**, 38 (1964).]

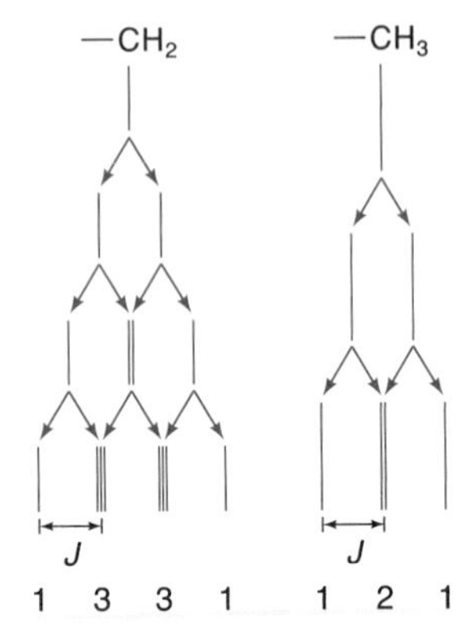

Figure 14.26
Spin–spin splitting between $-CH_2$ and $-CH_3$ groups in ethanol that gives rise to the triplet and quartet in the NMR spectrum. The coupling constant (J) is the same in both cases.

absorption. In the liquid or gas phase, where rapid molecular rotation occurs, the direct nuclear spin–spin interaction, called dipole interaction, averages to zero.

There is, however, an additional, indirect interaction between the nuclear spins that is transmitted through the bonding electrons. This interaction is unaffected by molecular rotation and causes splitting of the NMR peaks. In ethanol, with two possible orientations for each nuclear spin in the methylene group, the methyl peak is split into two lines by the magnetic field generated by the first methylene proton. Each of these two lines is then further split by the second methylene proton, and a total of four lines result. We see only three lines for $-CH_3$ because two of the lines fall on top of each other, giving rise to the observed intensity pattern of 1:2:1. Similarly, four lines are obtained for the $-CH_2$ group due to splitting by the methyl protons (Figure 14.26). The separation between the lines in each group gives the *spin–spin* coupling constant (J), whose magnitude is determined by the extent of this magnetic interaction. The following points are worth noting about spin–spin coupling.

1. Nuclei must be magnetically nonequivalent to produce spin–spin coupling. For example, the protons in the methyl group in ethanol are magnetically equivalent, so they do not interact with each other. They cause the splitting of the methylene peak only because the methylene protons are magnetically nonequivalent to the methyl protons.
2. Spin-spin coupling is observed only for two nuclei separated by no more than three bonds.
3. For ^1H (or any nucleus with $I = \frac{1}{2}$), the splitting of a line by a group of n equivalent protons is governed by the $(n + 1)$ rule, and the intensities are given by the *binomial distribution* (Table 14.7). The binomial distribution satisfactorily explains the NMR splitting patterns in ethanol and other hydrogen-containing compounds.

NMR and Rate Processes

To finish our discussion of the ethanol spectrum, we must account for the absence of spin–spin interaction between the methylene and hydroxyl groups. Actually, in

Table 14.7
The Coefficients of the Binomial Distribution[a]

n	Intensity ratio	Multiplicity
0	1	singlet
1	1 1	doublet
2	1 2 1	triplet
3	1 3 3 1	quartet
4	1 4 6 4 1	quintet
5	1 5 10 10 5 1	sextet

[a] The coefficients are generated by the equation $(1 + x)^n$. The intensity ratio starts with the number 1, and is followed by the coefficients of x, x^2, ...

pure ethanol, the hydroxyl peak is indeed split into a 1:2:1 triplet by the methylene group, and each of the four lines in the methylene group is further split into a doublet of equal intensity by the hydroxyl proton (Figure 14.23c). There is no observable splitting between the $-OH$ and $-CH_3$ groups because these protons are separated by more than three bonds. In the presence of a small amount of water, a rapid proton-exchange reaction between the $-OH$ group and H_2O and between C_2H_5OH and protonated C_2H_5OH effectively eliminates the spin–spin interaction between the $-OH$ and $-CH_2$ groups:

$$C_2H_5-\overset{\mathsf{H}}{O} + H-\overset{H}{\underset{\oplus}{O}}-H \rightleftharpoons C_2H_5-\overset{H}{\underset{\oplus}{O}}-\mathsf{H} + \overset{H}{O}-H$$

$$C_2H_5-\overset{H}{\underset{\oplus}{O}}-\mathsf{H} + \overset{H}{O}-H \rightleftharpoons C_2H_5-\overset{H}{O} + \mathsf{H}-\overset{H}{\underset{\oplus}{O}}-H$$

$$C_2H_5-\overset{H}{\underset{\oplus}{O}}-\mathsf{H} + \overset{H}{O}-C_2H_5 \rightleftharpoons C_2H_5-\overset{H}{O} + \mathsf{H}-\overset{H}{\underset{\oplus}{O}}-C_2H_5$$

We use H to show the proton involved in the exchange reaction.

In fact, NMR spectroscopy is convenient for studying the rates of proton-exchange reactions and of many other chemical processes such as rotation about a chemical bond and ring inversion. Consider, for example, the conformational change, or "ring inversion," that occurs in cyclohexane:

H_A H_B $\rightleftharpoons$ H_A H_B

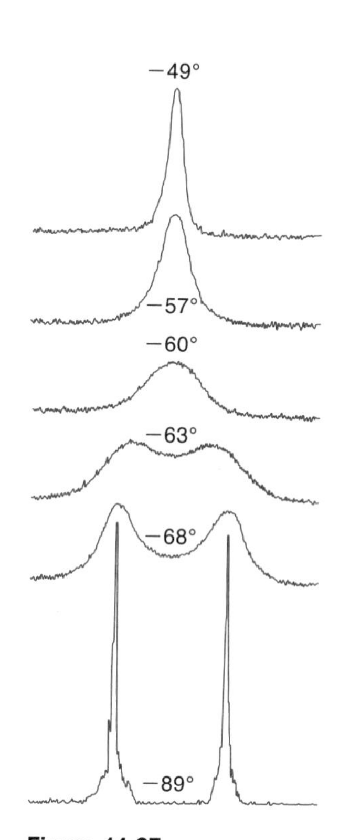

Figure 14.27
Proton NMR spectra of deuterated cyclohexane ($C_6D_{11}H$) at various temperatures (degrees Celsius). The magnetic field increases from left to right. (From F. A. Bovey, *Nuclear Magnetic Resonance Spectroscopy*, Academic Press, Inc., New York, 1969.)

The NMR spectrum of cyclohexane is rather complex because of spin–spin interactions. By using a deuterated compound, $C_6D_{11}H$, and applying a procedure known as *spin decoupling*, however, we eliminate these interactions, which leaves only two lines for observation, one representing the proton in the axial position and the other corresponding to the equatorial position (Figure 14.27). At −89°C, the ring-inversion rate is very slow so that two lines are observed corresponding to a state in which only half of the protons in the sample are in the axial position and half are in the equatorial position. Warming the sample causes the peaks to broaden. At −60°C, the peaks merge into a single line, which sharpens as temperature is increased further. This chemical exchange (between the axial and equatorial positions) can be understood in terms of the Heisenberg uncertainty principle (see Section 10.6):

$$\Delta E = \frac{h}{4\pi\tau}$$

or

$$\Delta\nu = \frac{1}{4\pi\tau}$$

where τ is the mean lifetime of the proton in a particular magnetic environment, and $\Delta\nu$ is the width of an NMR line. The exchange process causes the lifetime to shorten, leading to a large $\Delta\nu$ and hence a broadening of the line that exceeds its natural line width (lifetime broadening). The ring-inversion rate increases rapidly with temperature. When the exchange rate, $1/\tau$, is large compared with the frequency difference between the two lines, the spectrum collapses into a single line. At a still higher rate of inversion, the two protons change sites so rapidly that the system behaves as though there were only one type of proton present. The spectrum observed is in the so-called exchange-narrowing region (the peak recorded at −49°C). From an analysis of the change in linewidth with temperature, the energy of activation for the cyclohexane ring inversion is found to be 42 kJ mol^{-1}.

NMR of Nuclei Other Than ^{1}H

Proton magnetic resonance is the most frequently encountered form of NMR, but other nuclei are also important to investigations of chemical and biological systems. As a NMR–active nucleus, ^{13}C is second in popularity only to ^{1}H, thanks to the development of Fourier-transform spectroscopy. The spin–spin coupling discussed for protons also occurs between the ^{13}C nucleus and any protons attached to it. Thus, the ^{13}C NMR spectrum of a methylene group is observed as a triplet due to the interaction of ^{13}C with the two protons. The low natural abundance of the ^{13}C isotope does have an advantage: we do not observe ^{13}C—^{13}C splitting because the probability of two ^{13}C atoms being bonded to each other is very small. In practice, ^{13}C NMR spectra are *proton-decoupled*, a procedure that eliminates all ^{13}C—^{1}H splitting so that the spectra are more easily analyzed.* Moreover, the chemical shifts of ^{13}C NMR span a range about 250 ppm in magnitude, which is an order of magnitude larger than the range for protons.

In addition to ^{13}C, the isotopes ^{15}N, ^{19}F, and ^{31}P are important in NMR spectroscopy because these elements are found in many chemical and biological compounds. As an interesting example, note that in the 1990s chemists detected spin–spin coupling between two nitrogen-15 nuclei taking part in hydrogen bonding, that is, N—H···N. This finding is significant because it provides strong evidence that this hydrogen bond, and indeed all hydrogen bonds in general, possess some covalent character. We saw earlier that spin–spin coupling is transmitted through bonding electrons; therefore, such an interaction could not arise if hydrogen bonds were purely electrostatic attractions. Thus, there must be some overlap of the wave functions between the proton donor group (N—H) and the acceptor atom (N).

Figure 14.28 shows the ^{1}H, ^{13}C, and ^{31}P NMR spectra of a small important biomolecule, adenosine-5′-triphosphate (ATP).

* Decoupling is achieved by irradiating the sample at the ^{1}H resonance frequency while the ^{13}C spectrum is being recorded. The ^{13}C NMR spectrum of a C—H group shows a doublet (assuming no other magnetic nuclei are present). When the power at the decoupling frequency is large enough, the rate of change of the ^{1}H spin orientations becomes much greater than the coupling constant, and the doublet collapses into a single peak.

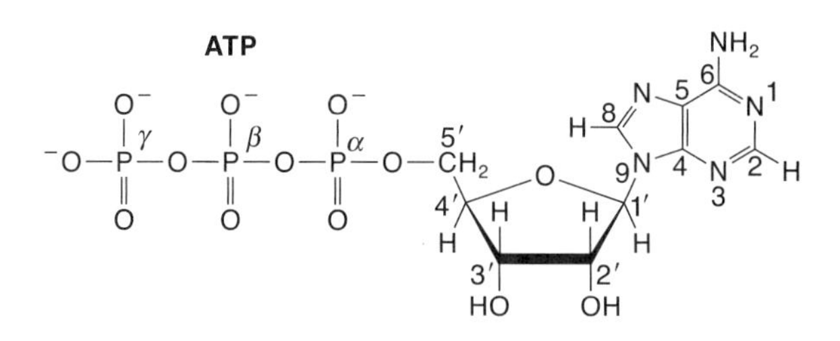

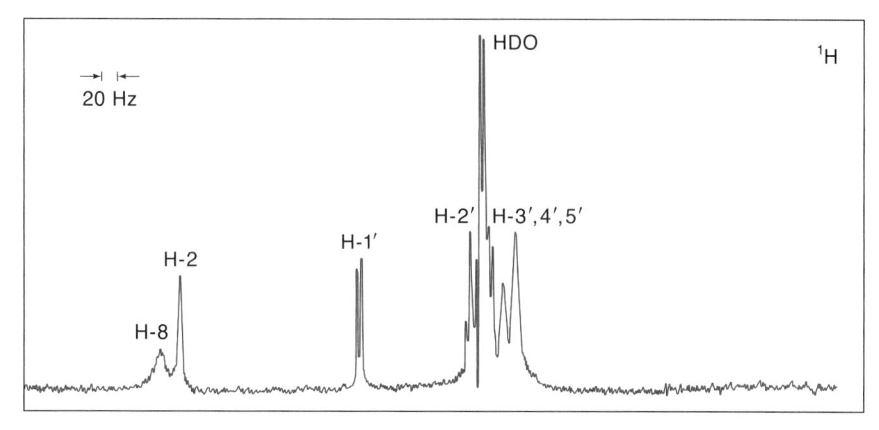

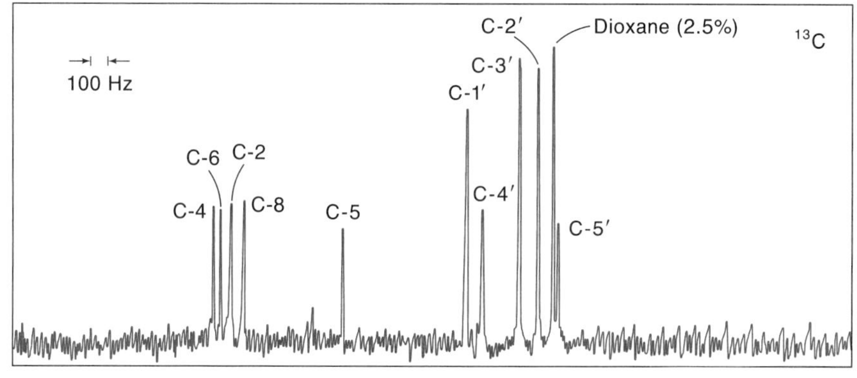

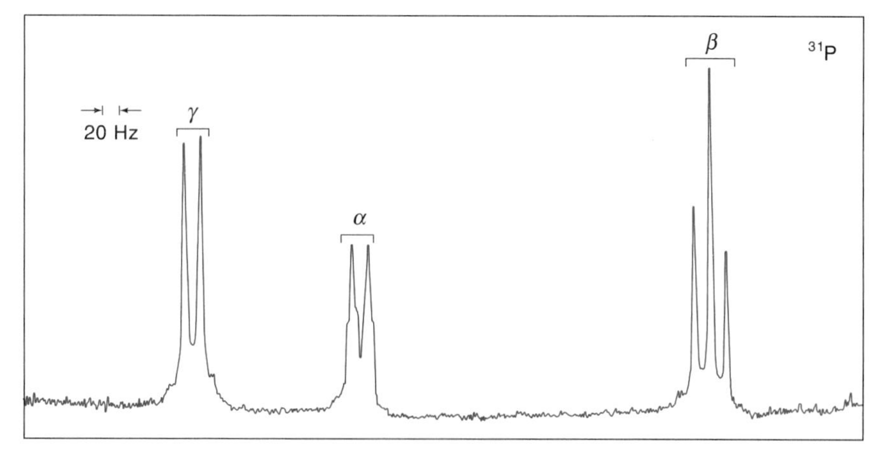

Figure 14.28
^{1}H, ^{13}C, and ^{31}P NMR spectra of adenosine-5′-triphosphate (ATP). Note that the ^{13}C spectrum is proton-decoupled so that only the chemical shifts of different types of carbon atoms are shown. (By permission of Varian Associates, Palo Alto, CA.)

Solid-State NMR

The difference between liquid- and solid-phase NMR spectra lies in the linewidth. A main contributor to the width of lines in an NMR spectrum is the magnetic dipole–dipole interaction between two nearby nuclei, which can increase or decrease the resonance frequency and hence cause the lines to broaden. Theory shows that the frequency shift resulting from the dipolar interaction, $\Delta\nu$, is given by

$$\Delta\nu \propto \frac{\mu_i \mu_j}{r_{ij}^3}(3\cos^2\theta_{ij} - 1) \tag{14.23}$$

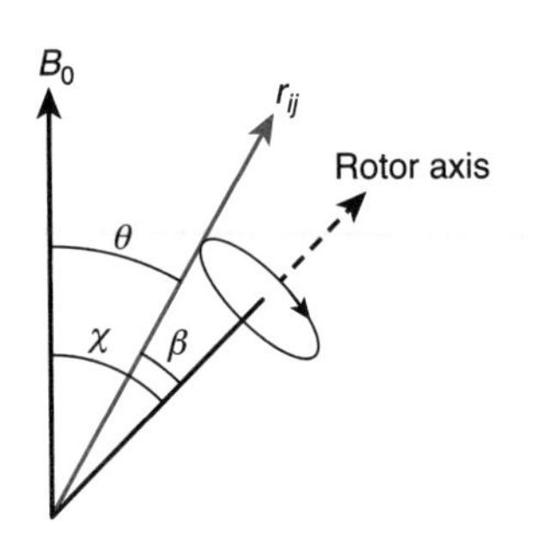

Figure 14.29
The angle relationships in a magic angle spinning NMR experiment.

where μ_i and μ_j are the magnetic dipoles of nuclei i and j, r_{ij} is the distance between the nuclei, and θ_{ij} is the angle between the external magnetic field and the line joining the two nuclei. The rapid molecular tumbling motion in a liquid averages this interaction to zero, resulting in well-resolved spectra. The situation is different in a solid, however. Here, the molecules are locked in position, so no such motional averaging can occur. As a result, linewidths on the order of 10^4 Hz (compared to 1 Hz or less for liquids) are not uncommon for solids.

In a solid-state NMR experiment, a rotating sample tube containing the solid substance is oriented at an angle χ from the magnetic field B_0 (Figure 14.29). Analysis shows that the time average of the term $(3\cos^2\theta_{ij} - 1)$ is given by

$$\langle 3\cos^2\theta_{ij} - 1\rangle = (3\cos^2\chi - 1)\left(\frac{3\cos^2\beta_{ij} - 1}{2}\right) \tag{14.24}$$

where β_{ij} is the angle between the vector connecting the dipoles i and j and the rotation axis. We see that when $\chi = 54.74°$, $3\cos^2\chi - 1 = 0$. If the sample tube is rotating rapidly, then all pairs of coupled dipoles have the same value of χ, even though they have different values of β_{ij}. In this way we can effectively reduce the dipolar broadening to nearly zero. This technique, called magic angle spinning (MAS), requires that the frequency of spinning must be greater than the width of the spectrum. This can now be accomplished using gas-driven spinners. Figure 14.30 compares the ^{31}P NMR spectrum of ammonium dihydrogen phosphate ($NH_4H_2PO_4$) with and without MAS.

Dental drills were used in the early MAS experiments.

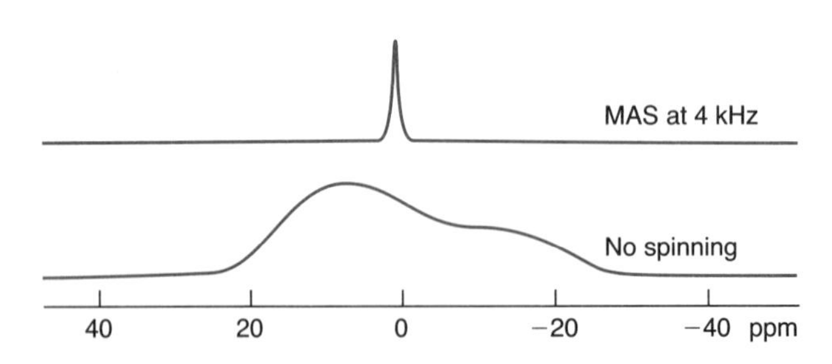

Figure 14.30
Proton-decoupled ^{31}P NMR spectra of ammonium dihydrogen phosphate ($NH_4H_2PO_4$). Bottom: The powder spectrum. Top: Magic angle spinning at 4 kHz. The asymmetric appearance of the spectrum with no spinning is due to chemical shift anisotropy, or the spread in the chemical shift of chemically equivalent nuclei as a result of different chemical environment. (Used with permission from Glenn A. Facey, "University of Ottawa NMR Facility BLOG," http://www.u-of-o-nmr-facility.blogspot.ca, accessed 1-Apr-2013.)

Fourier-Transform NMR

In recent years, the technique of Fourier transforms has had a profound impact on several branches of spectroscopy. In Chapter 11 we became acquainted with FT-IR (p. 491). Here we shall discuss Fourier-transform NMR (FT-NMR).

In principle, NMR can be observed either by keeping the external magnetic field constant and varying the radio frequency (rf) of the applied radiation or by keeping the radio frequency constant and sweeping the magnetic field until the resonance condition is satisfied. Technically it is simpler to maintain a constant radio frequency and vary the magnetic field. In either case, however, the spectrometer operates in the cw (continuous wave) mode because radiation is supplied continuously to the sample. As a result, it takes minutes to record an NMR spectrum with such an instrument. On the other hand, FT-NMR uses pulsed rf radiation and is much faster.

To understand how FT-NMR works, consider a sample consisting of many identical spins of $I = \frac{1}{2}$. When placed in a strong external magnetic field B_0, the nuclei align themselves parallel or antiparallel to the applied field, with a very slight excess in parallel alignment, which corresponds to the lower energy level. The net magnetization, M, then precesses around B_0 (the z axis) at the Larmor frequency (Figure 14.31). In a pulsed NMR experiment, a single, short, intense burst of magnetic field of strength B_1 along the x axis is applied to the sample. As a result, the net magnetization is rotated through an angle α given by

$$\alpha = \gamma B_1 t_\text{p} \tag{14.25}$$

where γ is the gyromagnetic ratio and t_p is the duration of the applied pulse, which is on the order of microseconds. The appropriate choice of t_p causes the magnetization to rotate from the z axis to the y axis. (This is called a 90° pulse because $\alpha = 90°$.) The NMR signal is measured by a detecting coil along the y axis. Immediately after the pulse (when B_1 is turned off), the magnetization vector begins to rotate in the xy plane at the Larmor frequency. Relaxation mechanisms subsequently cause the magnetization to decrease along the y axis, and eventually, at thermal equilibrium, it is restored along the z axis. The decay in NMR signal as a function of time is called the *free induction decay* (FID). Finally, by applying a Fourier transformation (see below), the FID is converted to an absorption peak.

Figure 14.31d applies to the situation in which the nuclei are identical and the rf radiation (B_1) is chosen to match the Larmor frequency. More often than not, we study nuclei that differ in Larmor frequency as a result of chemical shifts and spin–spin coupling. In those cases, different groups of nuclei precess at different frequencies, and interference effects can occur so that the FID will have a much more complex appearance.

An essential feature of pulsed NMR is that we can excite simultaneously nuclei with different chemical shifts even if we apply an rf pulse at a single frequency. Consider a pulse of 10 μs (1 μs = 10^{-6} s) duration. From the Heisenberg uncertainty principle (see Equation 10.28),

$$\Delta E \Delta t = \frac{h}{4\pi}$$

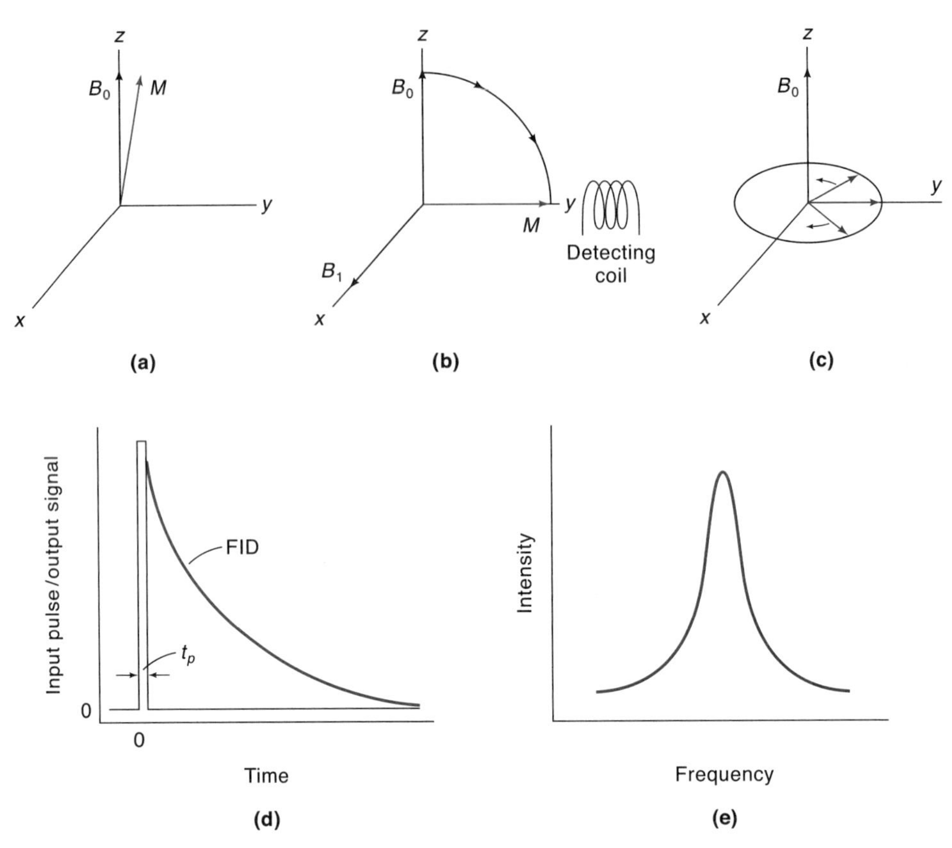

Figure 14.31
(a) Precession of the net magnetization M of a collection of spins about an external magnetic field B_0 along the z axis. (b) A 90° pulse (by a rf field B_1 along the x axis) flips the magnetization vector along the y axis, where the detecting coil is located. (c) Right after the pulse, the spins begin to precess in the xy plane. As the precession continues, the magnetization vector gives a signal with alternating maxima and minima. (d) The time sequence between the 90° pulse and the subsequent free induction decay. (e) Fourier transform of the FID $[f(t)]$ produces an intensity versus frequency spectrum.

Because $\Delta E = h\Delta\nu$, we have

$$\begin{aligned}\Delta\nu &= \frac{1}{4\pi\Delta t} \\ &= \frac{1}{4\pi(10 \times 10^{-6}\ \mathrm{s})} \\ &\approx 8 \times 10^3\ \mathrm{s}^{-1} \\ &= 8\ \mathrm{kHz}\end{aligned}$$

This frequency range is broad enough to cover most proton chemical shifts. Because of the difficulty in interpreting $f(t)$, the function that describes the variation in intensity of the signal with time associated with the FID, we must convert it to the more recognizable form $I(\nu)$, which is a function that describes the variation in intensity

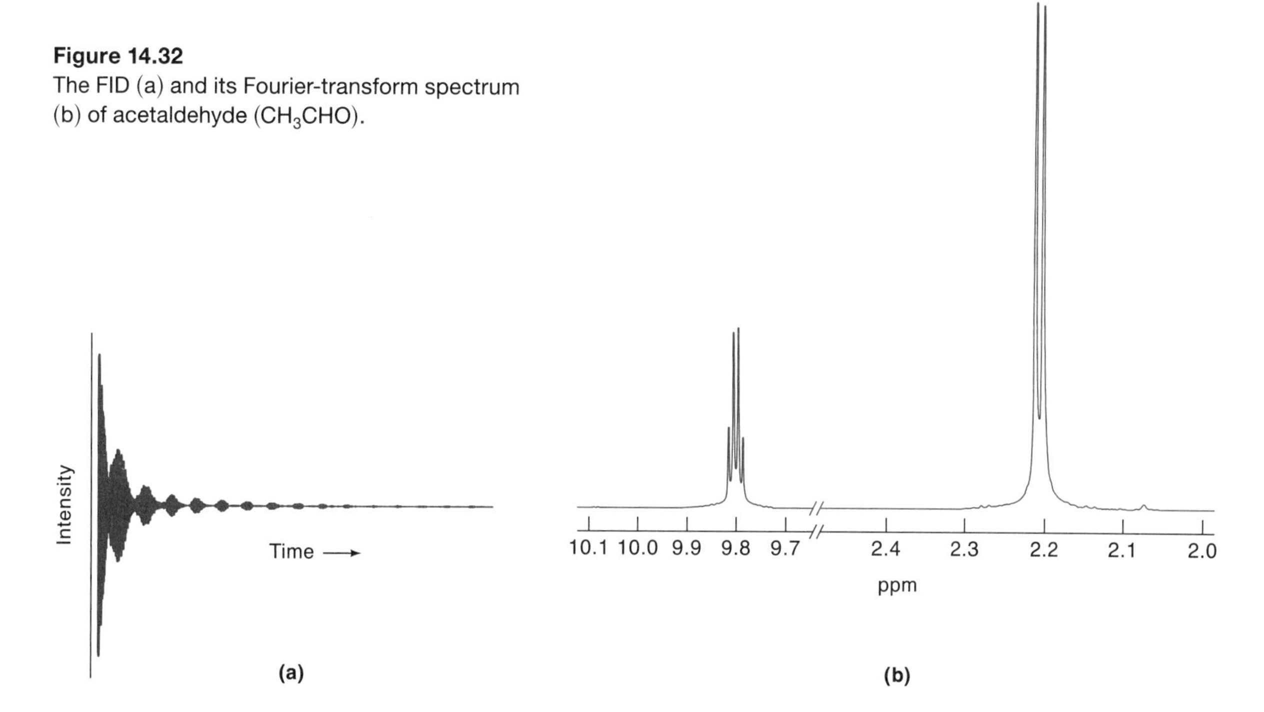

Figure 14.32
The FID (a) and its Fourier-transform spectrum (b) of acetaldehyde (CH_3CHO).

with frequency. These two spectral functions are related by the Fourier transformation as follows:

$$f(t) = \int_{-\infty}^{+\infty} I(v)\cos(vt)dv \quad (14.26)$$

$$I(v) = \int_{-\infty}^{+\infty} f(t)\cos(vt)dt \quad (14.27)$$

Figure 14.32a shows the FID for acetaldehyde (CH_3CHO). The FID curve is so complex because it arises from the precession of a magnetization vector that is composed of six components (from the six peaks), each of which processes with a characteristic frequency. By performing the Fourier transformation using Equation 14.27, we obtain the conventional NMR spectrum of acetaldehyde in which the intensities of the peaks are plotted against frequency (Figure 14.32b).

Because of the rapidity with which data can be collected and processed, FT-NMR enables us to record hundreds or thousands of similar spectra over a relatively short time period for signal averaging. In addition, the advances in instrumentation in recent years have made NMR one of the most powerful and versatile spectroscopic techniques.

Magnetic Resonance Imaging (MRI)

MRI is a noninvasive technique for obtaining cross-sectional pictures through the human body without exposing the patient to ionizing radiation, as in X-ray computerized tomography, or CT scanning. Figure 14.33 illustrates the basic idea behind MRI.

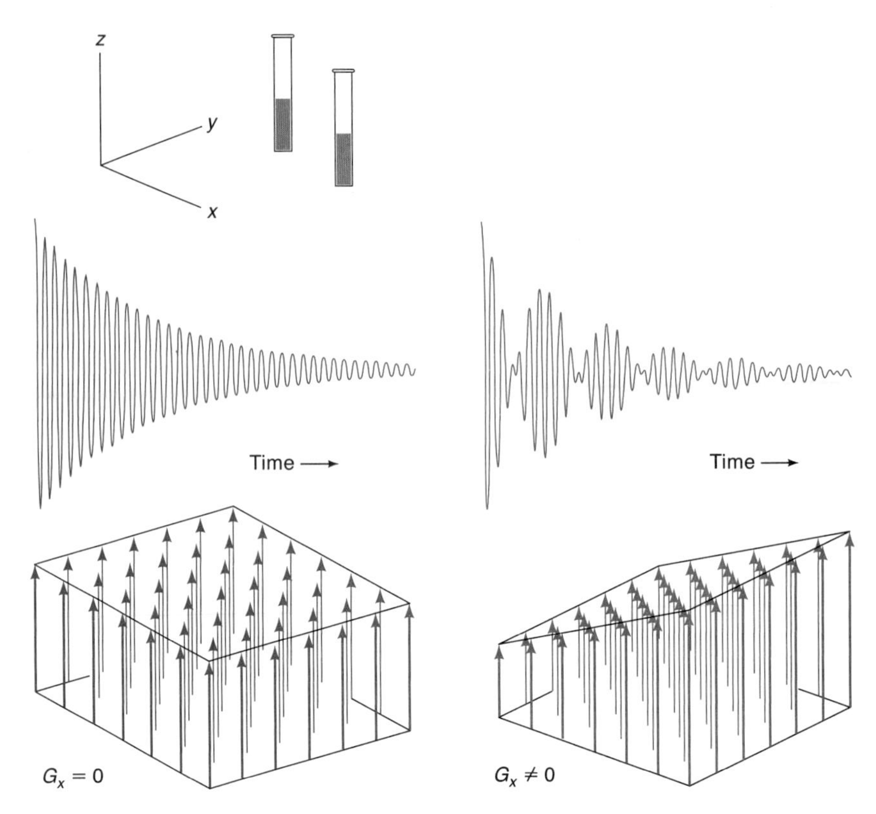

Figure 14.33
Top: Two small water-filled cylinders. Left: In the absence of a gradient, the B_0 field (along the z axis) is the same at all points in the cylinders as depicted by the set of vectors. Application of a short rf pulse yields an FID, which gives rise to a single line. Right: Here, a gradient field G_x is applied along the x axis. The water molecules in the two cylinders are now exposed to different magnetic fields. Consequently, the Fourier transformation of the FID will give rise to two signals.

Two water samples are separated in space (as they might be for water in two different regions of the body). The normal NMR spectrum of water shows a single peak, because the two protons are magnetically equivalent. Suppose that, in addition to the normal magnetic field B_0, a gradient magnetic field, G_x, is also applied along the x direction. Because the gradient field varies with distance, the field at the left cylinder will be slightly different from that at the second cylinder. After a pulsed NMR experiment, the Fourier transform of the FID will now show two peaks instead of one. To make a three-dimensional image, however, we need to apply gradient fields along the x, y, and z axis. Furthermore, the signals are shown as intensity versus distance rather than intensity versus frequency. A medical application of MRI is shown in Figure 14.34.

14.7 Electron Spin Resonance Spectroscopy

Electron spin resonance (ESR), also called electron paramagnetic resonance (EPR), is very similar to NMR in theory. The electron has a spin of $S = \frac{1}{2}$. The spinning motion

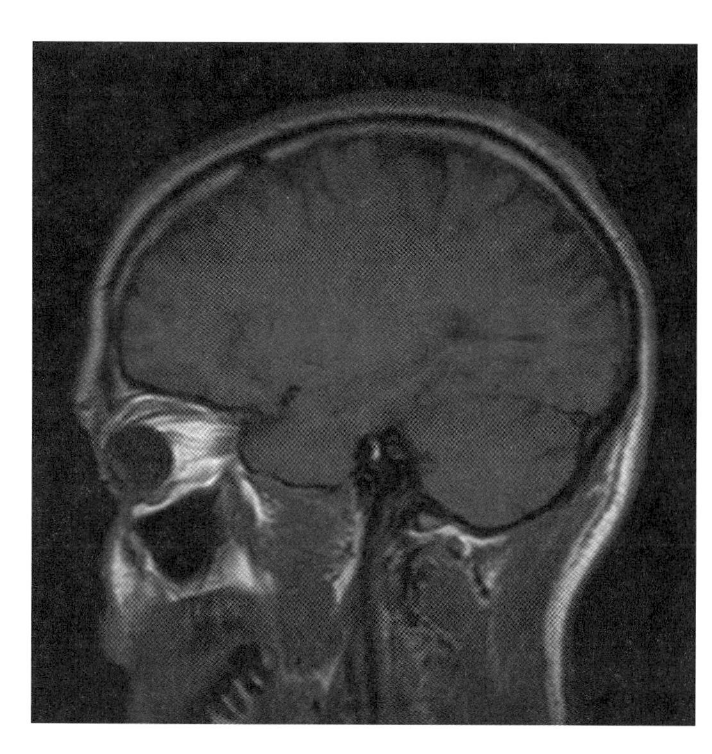

Figure 14.34
A magnetic resonance image of a slice of a human head.

of the electron generates a magnetic field, and the orientation of the electron magnetic moment in ah external magnetic field (B_0) is characterized by the electron spin quantum numbers, $m_s = \pm\frac{1}{2}$. The resonance condition is given by

$$\Delta E = h\nu = g\mu_B B_0 \tag{14.28}$$

where g is a dimensionless constant called the Landé g factor, which, for a free electron, is equal to 2.0023*; μ_B is the Bohr magneton, given by $eh/2\pi m_e c$, where e and m_e are the electronic charge and mass, respectively; and c is the speed of light (Figure 14.35).

Because the magnetic moment for an electron is about 600 times greater than that for a proton, ESR measurements are usually carried out in a magnetic field of about 0.34 T at a frequency of 9.5×10^9 Hz, or 9.5 GHz, which falls in the microwave region. Most spectrometers are designed to present the ESR transitions as the first derivative of the absorption lines (Figure 14.36).

Although isolated electrons, or electrons trapped in a matrix, undergo only one transition and therefore only one line is observed, the ESR spectrum of the hydrogen atom consists of two lines of equal intensity, as shown in Figure 14.37b. This *hyperfine splitting* results from the magnetic interaction between the unpaired electron and the nucleus, which is analogous to the spin–spin interaction discussed earlier for NMR. Only two transitions are allowed, however, because of the selection rules $\Delta m_s = \pm 1$ and $\Delta m_I = 0$. One interpretation of the selection rules is that the motion of a nucleus is much slower than that of an electron, so that during the time it takes for an electron to change its orientation, the nuclear spin has no time to reorient. The separation

* The g factor is the ratio of the electron's magnetic moment to its spin angular momentum.

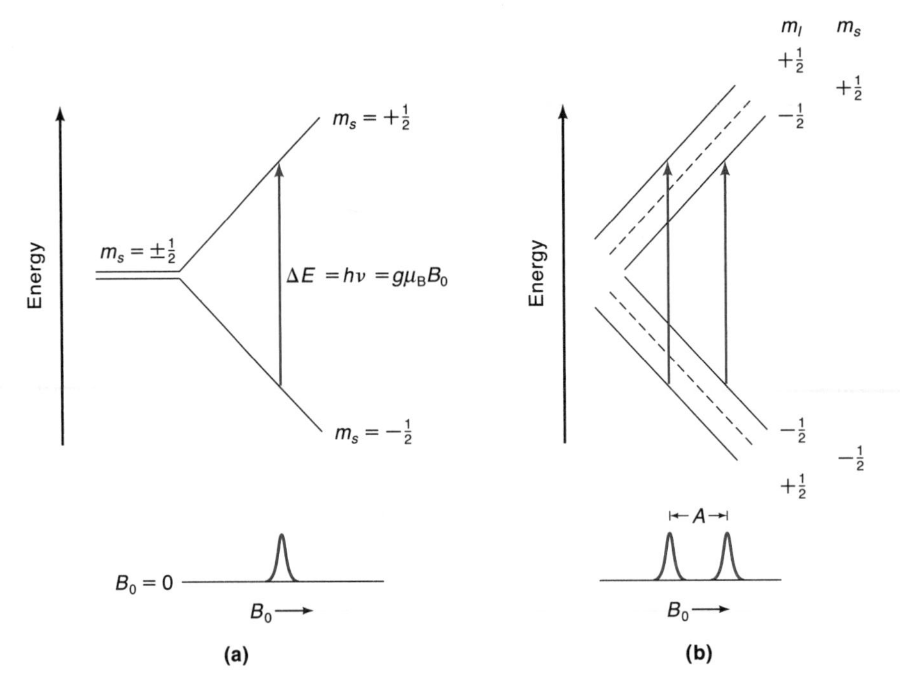

Figure 14.35
(a) Resonance condition for an electron. (b) Resonance conditions for the electron in a hydrogen atom.

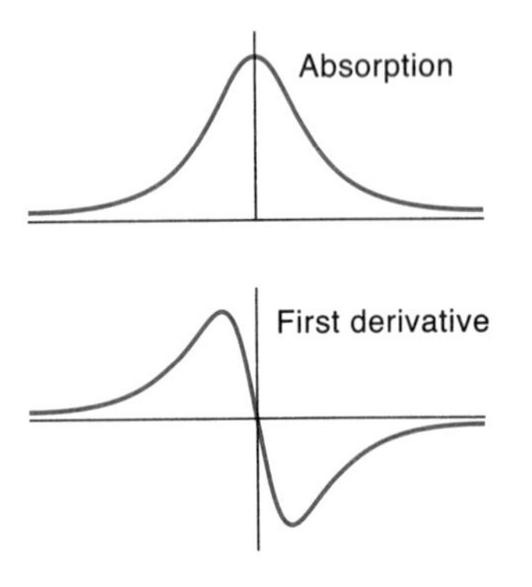

Figure 14.36
Relation between an absorption line and its first derivative.

between these two lines is the *hyperfine splitting constant* (A). In general, the number of hyperfine lines can be predicted by the quantity $(2nI + 1)$, where n is the number of equivalent nuclei, and I is the nuclear spin. As in NMR, the relative intensity of the lines arising from proton hyperfine splitting is given by the binomial distribution.

In most molecules, electrons regularly are paired with electrons that have opposite spins, as required by the Pauli exclusion principle; hence, ESR experiments cannot be performed on them. A few molecules, including NO, NO_2, ClO_2, and O_2, do contain one or more unpaired electrons in their electronic ground states. The ESR spectra for these molecules have been studied. It is also possible to reduce diamagnetic molecules by chemical or electrochemical means, converting them to anion radicals. For example, when benzene and naphthalene are dissolved in an inert organic solvent, such as tetrahydrofuran, and are treated with potassium metal in the absence of oxygen and water, the benzene and naphthalene anion radicals are generated (Figure 14.37a and b).

$$C_6H_6 + K \rightarrow C_6H_6^-K^+$$

$$C_{10}H_8 + K \rightarrow C_{10}H_8^-K^+$$

An important class of stable, neutral radicals are the nitroxides. In these molecules, the unpaired electron is localized on the nitrogen and oxygen atoms. An example is the di-*tert*-butyl nitroxide radical:

$$(CH_3)_3C-\underset{\underset{O\cdot}{|}}{N}-C(CH_3)_3$$

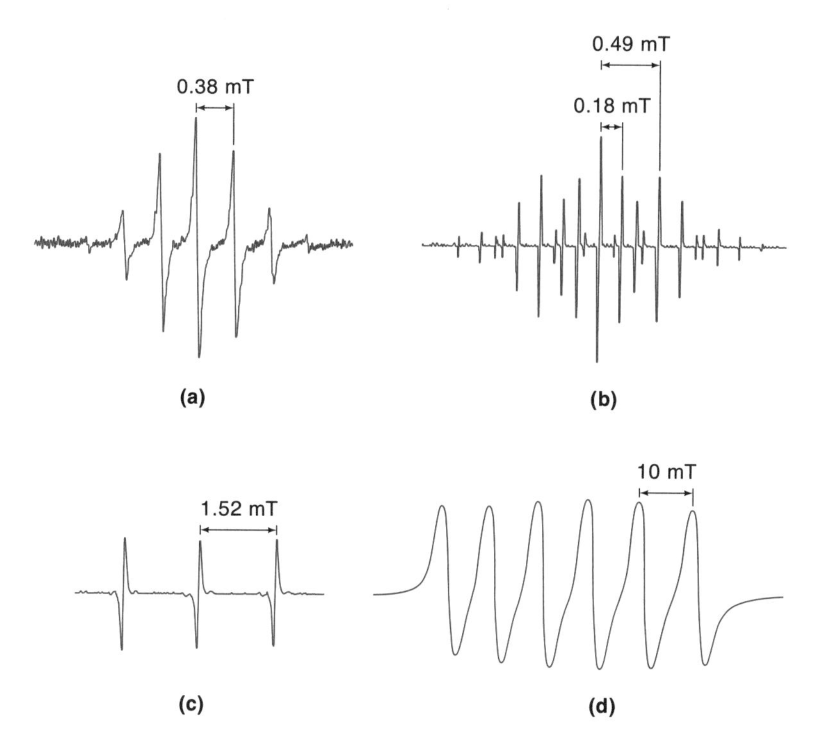

Figure 14.37
ESR spectra: (a) benzene anion radical; (b) naphthalene anion radical; (c) di-*tert* butyl nitroxide radical; and (d) Mn^{2+} in water. Note that there are two types of protons in naphthalene and therefore two different coupling constants.

Because ^{16}O has no magnetic moment ($I = 0$), the hyperfine splittings are due solely to the ^{14}N nucleus; a total of three lines of equal intensity are observed (Figure 14.37c).* Because of their stability and the simplicity of their ESR spectra, nitroxides have been used extensively as spin labels to probe the structure and dynamics of proteins.

Many transition-metal ions contain unpaired *d* electrons and are particularly suited for ESR studies.† Of particular interest are the Cu^{2+}, Co^{2+}, Fe^{3+}, Ni^{3+}, and Mn^{2+} ions because they occur in biological systems. The ESR spectrum of Mn^{2+} ($I = \frac{5}{2}$) gives rise to six equally spaced lines (Figure 14.37d). The Cu^{2+} ion is specially amenable to ESR investigation because it has only one unpaired electron. Due to hyperfine interaction (both ^{63}Cu and ^{65}Cu have nuclear spin $I = \frac{3}{2}$), its spectrum gives rise to a four-line pattern.

---

* The nuclear spin of ^{14}N is 1, according to the $(2nI + 1)$ rule, we have $2 \times 1 \times 1 + 1 = 3$.

† According to a theorem by the Dutch physicist Hendrik Kramers (1894–1952), the most suitable ions for ESR study are those containing an *odd* number of electrons.

Key Equations

$I = I_0 e^{-t/\tau}$	(Fluorescence intensity decay)	(14.7)
$h\nu_{\text{photon}} = \frac{1}{2}m_e u^2 + IE$	(Photoelectron spectroscopy)	(14.13)
$\Delta E = B_0 \gamma \hbar$	(Energy change in NMR)	(14.17)
$\delta = \dfrac{\nu - \nu_{\text{ref}}}{\nu_{\text{spec}}} \times 10^6 \text{ ppm}$	(Chemical shift in ppm)	(14.22)
$\Delta\nu \propto \dfrac{\mu_i \mu_j}{r_{ij}^3}(3\cos^2\theta_{ij} - 1)$	(Dipolar interaction)	(14.23)
$\Delta E = h\nu = g\mu_{\text{B}} B_0$	(Energy change in ESR)	(14.28)

APPENDIX 14.1

The Franck–Condon Principle

The Franck–Condon principle states that because the time required for a molecule to execute a vibration ($\sim 10^{-12}$ s) is much longer than that required for an electronic transition ($\sim 10^{-15}$ s), the nuclei do not move appreciably during an electronic transition. This means that the internuclear distance R does not change during a transition, and thus the intensity of the absorption bands can be predicted from vertical lines on a potential energy surface. As mentioned in Chapter 11, the strength of a spectroscopic transition is governed by the transition dipole moment μ_{12} that connects states described by the wave functions ψ_1 and ψ_2.

$$\mu_{12} = \int \psi_1 \hat{\mu} \psi_2 d\tau \tag{1}$$

where $\hat{\mu}$ is the transition dipole moment operator. We assume that the wave function is separable into electronic, vibrational, and rotational parts

$$\psi = \psi_e \psi_{vib} \psi_{rot} \tag{2}$$

and then we neglect the rotational contribution to the transition dipole moment.* Setting aside consideration of rotational motion is reasonable because many practical molecular electronic spectroscopy experiments do not resolve individual rotational transitions. Using primes to indicate the upper energy state and double primes to indicate the lower energy state, we thus write the transition dipole moment

$$\mu_{12} = \int \psi'_e \psi'_{vib} \hat{\mu} \psi''_e \psi''_{vib} d\tau \tag{3}$$

We write the overall dipole moment operator $\hat{\mu}$ as the sum of the electronic and nuclear dipole moment operators

$$\hat{\mu} = \hat{\mu}_e + \hat{\mu}_n \tag{4}$$

and the finite volume element $d\tau$ as the product of its electronic and nuclear components

$$d\tau = d\tau_e d\tau_n \tag{5}$$

* For completeness, the rotational-state dependence of the transitional dipole moment is governed by the Honl–London factor $\int \psi'_{rot} \psi''_{rot} d\tau_n$ which is analogous to the Franck–Condon factor.

With these substitutions, we expand equation (3),

$$\mu_{12} = \underbrace{\int \psi_e' \hat{\mu}_e \psi_e'' d\tau_e}_{\text{Constant, for a given electronic transition}} \underbrace{\int \psi_{vib}' \psi_{vib}'' d\tau_n}_{\text{The Franck–Condon factor}} + \underbrace{\int \psi_e' \psi_e'' d\tau_e}_{\text{Zero, because the wave functions are orthogonal}} \underbrace{\int \psi_{vib}' \hat{\mu}_n \psi_{vib}'' d\tau_n}_{\text{A number}} \tag{6}$$

which gives us the important result

The integral in Equation 7 is called the Franck–Condon factor.

$$\mu_{12} \propto \int \psi_{vib}' \psi_{vib}'' d\tau_n \tag{7}$$

For a spectroscopic transition from a given electronic state to a second electronic state, the strength of the transition (as given by the transition dipole moment) depends on the overlap of the vibrational wave functions (see Figure 14.1).

APPENDIX 14.2

A Comparison of FT-IR and FT-NMR

In Chapter 11 (p. 491) and this chapter (p. 649) we discussed FT-IR and FT-NMR. Because of the large difference in energy between IR and NMR, it is instructive to compare the operations of these two techniques.*

We have seen that an FT-NMR spectrometer records $f(t)$, which is in the time domain. Recalling the Heisenberg uncertainty principle (see Equation 10.28), we write

$$\Delta E \Delta t = \frac{h}{4\pi} \tag{1}$$

$$h\Delta \nu \Delta t = \frac{h}{4\pi}$$

or

$$\Delta \nu \Delta t = \frac{1}{4\pi} \tag{2}$$

This means that the information contained in $f(t)$ can be Fourier-transformed to $I(\nu)$, which is in the frequency domain. A FT-IR spectrometer, on the other hand, measures $I(\delta)$, which is a function of the position variable. The Fourier transform of $I(\delta)$ gives a more convenient IR spectrum, $B(\tilde{\nu})$, as a function of the wavenumber, $\tilde{\nu}$. Again, we can use the Heisenberg uncertainty principle to show that distance and wavenumber are also conjugate variables:

$$\Delta x \Delta p = \frac{h}{4\pi} \tag{3}$$

Using the de Broglie relation ($\lambda = h/p$), we write

$$p = \frac{h}{\lambda} = h\tilde{\nu}$$

and

$$\Delta p = h\Delta \tilde{\nu}$$

* The discussion here follows closely the excellent account by M. K. Ahn, *J. Chem. Educ.* **66**, 802 (1989).

so that

$$\Delta x \Delta \tilde{\nu} = \frac{1}{4\pi} \tag{4}$$

Thus, Fourier transforms always take place between conjugate variables.

In principle, a spectrum can be recorded in any domain defined by one of the four variables (frequency, time, distance, and wavenumber). The choice of the spectral function to be detected first in a given spectrometer depends on the design of the particular instrument. In a cw NMR spectrometer, we record $I(\nu)$ in the frequency domain, which is what we want, but the process is slow. Using pulsed NMR, we generate a FID that decays exponentially as follows (see Figure 14.31d):

$$f(t) = f(0)e^{-t/T_2} \tag{5}$$

where $f(0)$ is the intensity of the signal at $t = 0$. The quantity T_2 is called the *spin–spin relaxation time*, a characteristic time that measures how quickly the magnetization (from a collection of spins) along the y axis (after the 90° pulse) spreads out in the xy plane. The uncertainty in frequency of an NMR line, $\Delta\nu$, is identified with the line width at half-height and the uncertainty in time is identified with T_2. If an NMR line has a width of 10 Hz (a fairly broad line), the uncertainty relation yields $T_2 = 8$ ms (1 ms $= 10^{-3}$ s). If we assume that a 10-fold exponential decay brings $f(t)$ to the baseline (i.e., for the signal to vanish), then the full spectrum is collected in 10×8 ms, or 0.08 s. This is a long enough time period for the spectrometer to record all the pertinent data.

In contrast, FT-IR measures transitions among energy levels on a much greater scale. Consider an IR line of width $\Delta\tilde{\nu} = 10\ \text{cm}^{-1}$. Using the relation $\Delta\nu = c\Delta\tilde{\nu}$, we obtain 3×10^{11} Hz for the line width. This uncertainty in frequency gives an uncertainty in time of about 3×10^{-13} s, which is much too short a time in which to collect the data. Instead, the spectrum, $I(\delta)$, is recorded first as a function of the position variable. The Fourier transform of $I(\delta)$ then gives the more conventional IR spectrum, $B(\tilde{\nu})$, as a function of the wavenumber, $\tilde{\nu}$.

Suggestions for Further Reading

BOOKS

General Molecular Electronic Spectroscopy

Barrow, G. M., *Molecular Spectroscopy*, McGraw-Hill, New York, 1962.

Hollas, J. M., *Modern Spectroscopy*, 4th ed., John Wiley & Sons, New York, 2004.

Lakowicz, J. R., *Principles of Fluorescence Spectroscopy*, Springer, New York, 2006.

Skoog, D. A., F. J. Holler, and S. R. Crouch, *Principles of Instrumental Analysis*, 6th ed., Thomson Brooks/Cole, Belmont, CA, 2006.

Steinfeld, J. I., *Molecules and Radiation*, 2nd ed., MIT Press, Cambridge, MA, 1996.

Willard, H. H., L. L. Merritt, Jr., J. A. Dean, and F. A. Settle, Jr., *Instrumental Methods of Analysis*, 7th ed., Wadsworth Publishing Co., Belmont, CA, 1988.

Lasers and Laser Spectroscopy

Andrews, D. L., *Lasers in Chemistry*, 3rd ed., Springer-Verlag, New York, 1997.

Demtroder, W., *Laser Spectroscopy: Basic Concepts and Instrumentation*, 3rd ed., Springer-Verlag, New York, 2002.

Zare, R. N., *Laser: Experiments for Beginners*, University Science Books, Sausalito, CA, 1995.

Photoelectron Spectroscopy

Baker, A. D., and D. Betteridge, *Photoelectron Spectroscopy: Chemical and Analytical Aspects*, Pergamon Press, New York, 1972.

Ellis, A. M., M. Feher, and T. G. Wright, *Electronic and Photoelectron Spectroscopy: Fundamentals and Case Studies*, Cambridge University Press, Cambridge, UK, 2005.

Hüfner, S., *Photoelectron Spectroscopy: Principles and Applications*, 3rd ed., Springer-Verlag, New York, 2003.

Magnetic Resonance Spectroscopy

Carrington, A., and A. D. McLachlan, *Introduction to Magnetic Resonance*, Harper & Row Publishers, New York, 1967.

Freeman, R., *Magnetic Resonance in Chemistry and Medicine*, Oxford University Press, New York, 2003.

Hore, P. J., *Nuclear Magnetic Resonance*, Oxford University Press, New York, 1995.

Macomber, R. S., *A Complete Introduction to Modern NMR Spectroscopy*, John Wiley & Sons, New York, 1998.

Roberts, J. D., *ABCs of FT-NMR*, University Science Books, Sausalito, CA, 2000.

Wertz, J. E., and J. R. Bolton, *Electron Spin Resonance: Elementary Theory and Practical Applications*, Wiley-Interscience, New York, 2007.

ARTICLES

General

"The Ultraviolet Spectra of Aromatic Molecules," P. E. Stevenson, *J. Chem. Educ.* **41**, 234 (1964).

"The Fates of Electronic Excitation Energy," H. H. Jaffé and A, L. Miller, *J. Chem. Educ.* **43**, 469 (1966).

"Applications of Absorption Spectroscopy in Biochemistry," G. R. Penzer, *J. Chem. Educ.* **45**, 692 (1968).

"Light," G. Feinberg, *Sci. Am.* September 1968.

"How Light Interacts With Matter," V. F. Weisskopf, *Sci. Am.* September 1968.

"The Triplet State," N. J. Turro, *J. Chem. Educ.* **46**, 2 (1969).
"Liquid Scintillation Counting," W. Yang and E. K. C. Lee, *J. Chem. Educ.* **46**, 277 (1969).
"Energy States of Molecules," J. L. Hollenberg, *J. Chem. Educ.* **47**, 2 (1970).
"The Chemical Origin of Color," M. V. Orna, *J. Chem. Educ.* **55**, 478 (1978).
"The Causes of Color," K. Nassau, *Sci. Am.* October 1980.
"A Time Scale for Fast Events," D. Onwood, *J. Chem. Educ.* **63**, 680 (1986).
"Band Breadth of Electronic Transitions and the Particle-in-a-Box Model," L.-F. Olsson, *J. Chem. Educ.* **63**, 756 (1986).
"Radiationless Relaxation and Red Wine," H. D. Burrows and A. C. Cardoso, *J. Chem. Educ.* **64**, 995 (1987).
"The Fourier Transform," R. N. Bracewell, *Sci. Am.* June 1989.
"Using Fourier Transform to Understand Spectral Line Shape," E. Grunwald, J. Herzog, and C. Steel, *J. Chem. Educ.* **72**, 210 (1995).
"Experiments with Glow-in-the-Dark Toys: Kinetics of Doped ZnS Phosphorescence," G. C. Lisensky, M. N. Patel, and M. L. Reich, *J. Chem. Educ.* **73**, 1048 (1996).
"A Unified Approach to Absorption Spectroscopy at the Undergraduate Level," R. S. Macomber, *J. Chem. Educ.* **74**, 65 (1997).
"Turning on the Light: Lessons from Luminescence," P. B. O'Hara, C. Engelson, and W. St. Peter, *J. Chem. Educ.* **82**, 49 (2005).
"Demonstrations for Fluorescence and Phosphorescence," D. P. Richardson and R. Chang, *Chem. Educator* [Online] **12**, 279 (2007) DOI 10.1333/s00897072049a.

Lasers
"Laser Chemistry," D. L. Rousseau, *J. Chem. Educ.* **43**, 566 (1966).
"Advances in Holography," K. S. Pennington, *Sci. Am.* February 1968.
"Laser Light," A. L. Schawlow, *Sci. Am.* September 1968.
"Applications of Laser Light," D. R. Harriott, *Sci. Am.* September 1968.
"Organic Lasers," P. Sorokin, *Sci. Am.* February 1969.
"Laser Spectroscopy," M. S. Feld and V. S. Letokhov, *Sci. Am.* December 1973.
"Applications of Lasers to Chemical Research," S. R. Leone, *J. Chem. Educ.* **53**, 13 (1976).
"Laser Chemistry," A. M. Ronn, *Sci. Am.* May 1979.
"Laser—An Introduction," W. F. Coleman, *J. Chem. Educ.*, **59**, 441 (1982).
"Lasers: A Valuable Tool for Chemists," E. W. Findsen and M. R. Ondrias, *J. Chem. Educ.* **63**, 479 (1986).
"Detecting Individual Atoms and Molecules With Lasers," V. S. Letokhov, *Sci. Am.* September 1988.
"The Birth of Molecules," A. H. Zewail, *Sci. Am.* December 1990.
"Laser Surgery," M. W. Berns, *Sci. Am.* June 1991.
"Diode Lasers," M. G. D. Baumann, J. C. Wright, A. B. Ellis, T. Kuech, and G. C. Lisensky, *J. Chem. Educ.* **69**, 89 (1992).
"Laser Control of Chemical Reactions," P. Brumer and M. Shapiro, *Sci. Am.* March 1995.
"Using Lasers to Demonstrate the Concept of Polarizability," G. R. van Hecke, K. K. Karukstis, and J. M. Underhill, *Chem. Educator* [Online] **2**, 5 (1997) DOI 10.1333/s00897970147a.
"Innovative Laser Techniques in Chemical Kinetics," L. J. Kovalenko and S. R. Leone, *J. Chem. Educ.* **65**, 681 (1998).

Laser Spectroscopy
"Laser-Induced Fluorescence in Spectroscopy, Dynamics, and Diagnostics," D. R. Crosley, *J. Chem. Educ.* **59**, 446 (1982).
"Optical Studies of Single Molecules at Room Temperature," X. S. Xie and J. K. Trautman, *Ann. Rev. Phys. Chem.* **49**, 441 (1998).

"Ultrashort-Pulse Lasers: Big Payoffs in a Flash," J.-M. Hopkins and W. Sibbett, *Sci. Am.* September 2000.
"Freezing Atoms in Motion: Principles of Femtochemistry and Demonstration by Laser Stroboscopy," J. S. Baskin and A. H. Zewail, *J. Chem. Educ.* **78**, 737 (2001).
"Blue Diode Lasers: New Opportunities in Chemical Education," J. E. Whitten, *J. Chem. Educ.* **78**, 1096 (2001).
"A Dozen Years of Single-Molecule Spectroscopy in Physics, Chemistry, and Biophysics," W. E. Moerner, *J. Phys. Chem. B* **106**, 910 (2002).
"Fluorescence Microscopy of Single Molecules," J. Zimmerman, A. van Dorp, and A. Renn, *J. Chem. Educ.* **81**, 553 (2004).

Photoelectron Spectroscopy
"Photoelectron Spectroscopy," T. L. James, *J. Chem. Educ.* **48**, 712 (1971).
"Photoelectron Spectra. An Experimental Approach to Teaching Molecular Orbital Models," H. Bock and P. D. Mollere, *J. Chem. Educ.* **51**, 506 (1974).
"Multiplets in Atoms and Ions Displayed by Photoelectron Spectroscopy," S. Suzer, *J. Chem. Educ.* **59**, 814 (1982).
"Hückel Theory and Photoelectron Spectroscopy," E. I. von Nagy-Felsobuki, *J. Chem. Educ.* **66**, 821 (1989).
"Why Equivalent Bonds Appear as Distinct Peaks in Photoelectron Spectra," J. Simons, *J. Chem. Educ.* **69**, 522 (1992).

Nuclear Magnetic Resonance
"NMR Imaging in Medicine," I. L. Pykett, *Sci. Am.* May 1982.
"The NMR Time Scale," R. G. Bryant, *J. Chem. Educ.* **60**, 933 (1983).
"Atomic Memory," R. G. Brewer and E. L. Hahn, *Sci. Am.* December 1984.
"Sensitivity Enhancement by Signal Averaging in Pulsed/Fourier Transform NMR Spectroscopy," D. L. Rabenstein, *J. Chem. Educ.* **61**, 909 (1984).
"A Primer on Fourier Transform NMR," R. S. Macomber, *J. Chem. Educ.* **62**, 213 (1985).
"Fourier Transforms for Chemists. Part 1. Introduction to the Fourier Transform," L. Glasser, *J. Chem. Educ.* **64**, A228 (1987).
"Fourier Transforms for Chemists. Part 2. Fourier Transforms in Chemistry and Spectroscopy," L. Glasser, *J. Chem. Educ.* **64**, A260 (1987).
"Fourier Transforms for Chemists. Part 3. Fourier Transforms in Data Treatment," L. Glasser, *J. Chem. Educ.* **64**, A306 (1987).
"A Step-by-Step Picture of Pulsed (Time-Domain) NMR," L. J. Schwartz, *J. Chem. Educ.* **65**, 959 (1988).
"Spin-Lattice Relaxation Times in ^{1}H NMR Spectroscopy," D. J. Wink, *J. Chem. Educ.* **66**, 810 (1989).
"A Demonstration of Imaging on an NMR Spectrometer," L. A. Hull, *J. Chem. Educ.* **67**, 782 (1990).
"Gas-Phase NMR Spectroscopy," C. Suarez, *Chem. Educator* [Online] **3**, 1 (1998) DOI 10.1007/s00897980202a.
"A Solid-State NMR Experiment: Analysis of Local Structural Environments in Phosphate Glasses," S. E. Anderson, D. Saiki, H. Eckert, and K. Meise-Gresch, *J. Chem. Educ.* **81**, 1034 (2004).
"Using an NMR Spectrometer To Do Magnetic Resonance Imaging," W. E. Steinmetz and M. C. Maher, *J. Chem. Educ.* **84**, 1830 (2007).
"NMR Spectroscopy and Its Value: A Primer," S. Veeraraghavan, *J. Chem. Educ.* **85**, 537 (2008).

Electron Spin Resonance
"ESR Study of Organic Electron Transfer Reactions," R. Chang, *J. Chem. Educ.* **47**, 563 (1970).

"Phosphorescence and Electron Paramagnetic Resonance Study of a Photoexcited Triplet State: Advanced Undergraduate Experiment in Molecular Physics," R. Chang and W. R. Moomaw, *Am. J. Phys.* **44**, 455 (1976).
"Structural Information from Liquid- and Solid-Phase ESR: The Example of Triphenyl-3 Substituted Radical $(C_6H_5)_3A\bullet$ of Group IVB Elements," M. Geoffroy and J. H. Hammons, *J. Chem. Educ.* **58**, 389 (1981).
"EPR Studies of Spin-Spin Exchange Processes: A Physical Chemistry Experiment," M. P. Eastman, *J. Chem. Educ.* **59**, 677 (1982).
"ESR Studies and HMO Calculations on Benzosemiquinone Radical Anions," R. Beck and J. W. Nibler, *J. Chem. Educ.* **66**, 263 (1989).
"An EPR Experiment for the Undergraduate Physical Chemistry Laboratory," R. A. Butera and D. H. Waldeck, *J. Chem. Educ.* **77**, 1489 (2000).
"Use of EPR Spectroscopy in Elucidating Electronic Structures of Paramagnetic Transition Metal Complexes," P. Basu, *J. Chem. Educ.* **78**, 666 (2001).

Problems

Molecular Electronic Spectroscopy, Fluorescence, and Phosphorescence

14.1 Rank the following molecules in order of increasing wavelength of maximum absorbance, λ_{max}, and explain your reasoning.

$$CH_2{=}CHC(O)CH_3,\ CH_3CH_2CH_2CH_3,\ CH_3CH{=}CHCH_3,\ CH_3CH_2C(O)CH_3$$

14.2 Using the Franck–Condon principle, qualitatively explain how the electronic spectra originating from the following two potential-energy diagrams will differ.

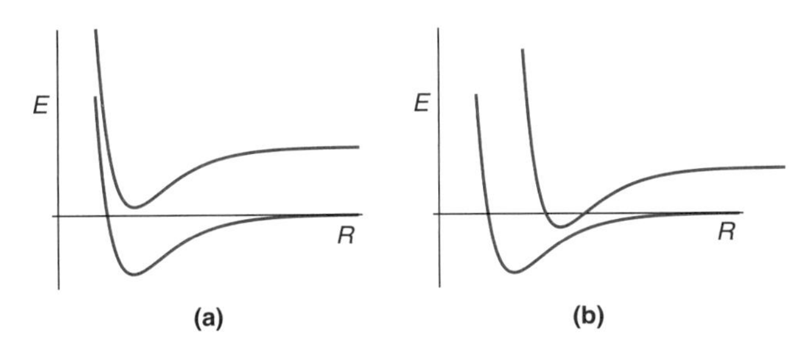

14.3 What is indicated by the *absence* of an isosbestic point?

14.4 Rank the following in order of increasing rate: absorption, fluoresecence, phosphorescence, vibrational energy transfer, internal conversion, intersystem crossing. (Some of these may occur at approximately the same rate.)

14.5 An aqueous solution contains two species A and B. The absorbance at 300 nm is 0.372 and at 250 nm is 0.478. The molar absorptivities of A and B are as follows:

$$\text{A: } \varepsilon_{300} = 3.22 \times 10^4 \text{ L mol}^{-1}\text{ cm}^{-1}$$

$$\varepsilon_{250} = 4.05 \times 10^4 \text{ L mol}^{-1}\text{ cm}^{-1}$$

$$\text{B: } \varepsilon_{300} = 2.86 \times 10^4 \text{ L mol}^{-1}\text{ cm}^{-1}$$

$$\varepsilon_{250} = 3.76 \times 10^4 \text{ L mol}^{-1}\text{ cm}^{-1}$$

If the pathlength of the cell is 1.00 cm, calculate the concentrations of A and B in mol L^{-1}.

14.6 Explain why fluorescence is generally a more sensitive detection technique than absorption or phosphorescence.

14.7 How are absorptions due to $\pi^* \leftarrow \pi$, $\pi^* \leftarrow n$, and charge transfer processes distinguished?

14.8 List some important differences between fluorescence and phosphorescence.

14.9 The lowest triplet state in naphthalene ($C_{10}H_8$) is about 11,000 cm^{-1} below the lowest excited singlet electronic level at 77 K. Calculate the ratio of populations in these two states at equilibrium. [*Hint*: The Boltzmann equation is given by $N_2/N_1 = (g_2/g_1)\exp(-\Delta E/k_BT)$, where g_1 and g_2 are the degeneracies for levels 1 and 2.]

14.10 The luminescent first-order decay of a certain organic molecule yields the following data:

t/s	0	1	2	3	4	5	10
I	100	43.5	18.9	8.2	3.6	1.6	0.02

where I is the relative intensity. Calculate the mean lifetime τ for the process. Is it fluorescence or phosphorescence?

14.11 Give a qualitative explanation as to why PPO absorbs light at a shorter wavelength than that for POPOP (see p. 621).

14.12 The fluorescence of a protein is due to tryptophan, tyrosine, and phenylalanine (assuming that the protein does not contain a prosthetic group that is fluorescent). Iodide ions are known to quench the fluorescence of tryptophan. If a protein is known to contain only one tryptophan residue and iodide fails to quench its fluorescence, what can you conclude about the location of the tryptophan residue?

14.13 How many unpaired electrons are in a molecule in a quartet state?

14.14 Molecules with a net electron spin of $S = \frac{1}{2}$ have a ground electronic state called a doublet state because the multiplicity $2S + 1 = 2$. For molecules with a doublet ground state, what is the multiplicity of the electronic state from which phosphorescence originates? (Note: Multiplicity = singlet, doublet, triplet, quartet, ...)

14.15 Some molecules, such as molecular oxygen, have a triplet ground electronic state. For molecules with triplet ground states, identify the multiplicity of the excited states from which fluorescence originates.

Lasers and Laser Spectroscopy

14.16 Name four characteristic properties of lasers.

14.17 Explain why we cannot produce laser light with only two stable energy levels.

14.18 Consider a three-level laser system. The wavelength for an absorption from level A to level C is 466 nm and that the wavelength for a transition between levels B and C is 752 nm. What is the wavelength for a transition between levels A and B?

14.19 What is the difference between a one-photon and a multiphoton process? Why does the use of lasers make it favorable to observe, say, a two-photon process?

14.20 The Raman effect was first discovered prior to the invention of the laser. Why is essentially all contemporary Raman spectroscopy performed using a laser? (Section 11.5 describes Raman spectroscopy.)

14.21 Which of the following lasers are tunable? **(a)** Nd:YAG; **(b)** diode; **(c)** dye; and **(d)** He–Ne.

14.22 To extend the wavelength range of lasers, frequency doubling, tripling, and quadrupling may be used when suitable crystalline materials are available for the conversion process. A Nd:YAG laser, operating at 1064 nm is frequency doubled to make 532 nm (green) light. Calculate the wavelengths obtained by frequency tripling and quadrupling a Nd:YAG laser. In what region of the electromagnetic spectrum do these lie?

14.23 A Nd:YAG-pumped dye laser has a pulse duration of 6 ns. Calculate the length (in meters) of such a laser pulse.

14.24 A laser operating at 10 Hz produces pulses with a wavelength of 600 nm and a pulse duration of 8 ns. The average power of the laser is measured to be 400 mW. Calculate the energy per laser pulse, the peak power of a laser pulse, and the number of photons per laser pulse.

14.25 An ultrafast laser produces pulses of 80 fs duration at a rate of 2 kHz and a total average power of 100 mW. For an individual laser pulse, calculate **(a)** the length in meters, **(b)** the energy in joules, and **(c)** the peak power in watts.

Photoelectron Spectroscopy

14.26 Compare and contrast ultraviolet photoelectron spectroscopy (UPS) with X-ray photoelectron spectroscopy (XPS).

14.27 The excitation wavelength is decreased when measuring a photoelectron spectrum. Will each of the following increase, decrease, or remain the same? **(a)** The speed of the ejected photoelectron; **(b)** the kinetic energy of the ejected photoelectron; **(c)** the bonding energy of the ejected photoelectron; and **(d)** the vibrational spacing present in the photoelectron spectrum.

14.28 Explain in your own words why methane (CH_4) exhibits two (only) different peaks corresponding to C—H bonds in its photoelectron spectrum.

14.29 Why might certain core orbitals NOT appear in a photoelectron spectrum?

14.30 A high-level molecular electronic structure calculation predicts oxygen valence molecular orbitals with binding energies of 0.5389, 0.60354, and 0.7171 hartree. With a 58.46-nm photon source, what will the corresponding electron kinetic energies (in *eV*) be in a photoelectron spectrum? (1 hartree = 27.211 *eV*)

14.31 In the high-resolution photoelectron spectrum of molecular hydrogen, peaks are observed with a separation of 0.515 *eV*. This separation corresponds to the vibrational energy level spacing in the molecular hydrogen cation (H_2^+). **(a)** Based on this spectrum, what is the vibrational wavenumber of the molecular hydrogen cation? **(b)** Could this vibrational wavenumber be obtained with FT-IR spectroscopy? **(c)** Could photoelectron spectroscopy be used to measure the rotational constant, B, for the molecular hydrogen cation? Explain your answers.

14.32 One way to measure the kinetic energy of ejected photoelectrons is to measure their stopping voltage, that is, the voltage required to stop the transmission of photoelectrons and thus result in zero current. The stopping voltages for the photoelectrons from a particular sample are 952.9, 1083.9, and 1201.2 *eV* when using an aluminum K_α X-ray source. What are the binding energies of the corresponding molecular orbitals? Based on these binding energies, from what type of atomic or molecular orbital do you anticipate that these photoelectrons originate, core or valence?

14.33 Based on the x-axis labels in Figure 14.20, what photon source(s) might be used to acquire the photoelectron spectrum?

Magnetic Resonance Spectroscopy

14.34 The NMR signal of a compound is found to be 240 Hz downfield from the TMS peak using a spectrometer operating at 60 MHz. Calculate its chemical shift in ppm relative to TMS.

14.35 What is the field strength (in tesla) needed to generate a ^1H frequency of 600 MHz?

14.36 Suppose the NMR spectrum of acetaldehyde (see Figure 14.32) is recorded at 200 MHz and 400 MHz. State whether each of the following quantities remains unchanged or is different from 200 MHz to 400 MHz: **(a)** sensitivity of detection, **(b)** $|\delta_{CH_3} - \delta_H|$, **(c)** $|\nu_{CH_3} - \nu_H|$, and **(d)** J.

14.37 For an applied field of 9.4 T (used in a 400 MHz NMR spectrometer), calculate the difference in frequencies for two protons whose δ values differ by 2.5 ppm.

14.38 For each of the following molecules, state how many proton NMR peaks occur, and whether each peak is a singlet, doublet, triplet, etc.: **(a)** CH_3OCH_3, **(b)** $C_2H_5OC_2H_5$, **(c)** C_2H_6, **(d)** CH_3F, and **(e)** $CH_3COOC_2H_5$.

14.39 Sketch the NMR spectrum of isobutyl alcohol $[(CH_3)_2CHCH_2OH]$, given the following chemical shift data: $-CH_3$: 0.89 ppm, $-CH$: 1.67 ppm, $-CH_2$: 3.27 ppm, $-OH$: 4.50 ppm.

14.40 The toluene proton NMR spectrum, consisting of two peaks due to the methyl and aromatic protons, has been recorded at 60 MHz and 1.41 T. **(a)** What would the magnetic field be at 300 MHz? **(b)** At 60 MHz, the resonance frequencies are methyl: 140 Hz; aromatic: 430 Hz. What would the frequencies be using a 300 MHz spectrometer? **(c)** Calculate the chemical shifts (δ) of the two signals, using both the 60 MHz and 300 MHz data. **(d)** What are the positions of the methyl and aromatic signals in Hz at 300 MHz?

14.41 The methyl radical has a planar geometry. How many lines would you observe in the ESR spectrum of $\bullet CH_3$? Of $\bullet CD_3$?

14.42 Account for the number of lines observed in the ESR spectra of benzene and naphthalene anion radicals shown in Figure 14.37. How would you use isotopic substitution to assign the two hyperfine splitting constants in naphthalene?

14.43 One way to study membrane structure is to use a spin label, such as a nitroxide radical having the following structure,

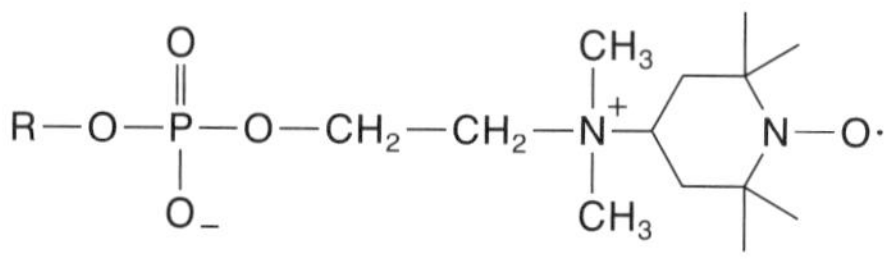

$$R-O-\overset{\overset{O}{\|}}{\underset{\underset{O^-}{|}}{P}}-O-CH_2-CH_2-\overset{\overset{CH_3}{|}}{\underset{\underset{CH_3}{|}}{N^+}}-\langle\ \ \rangle N-O\cdot$$

where the R group represents the hydrophobic tail part of the phosphatidic acid derivative. Like the di-*tert*-butyl nitroxide, the ESR spectrum of this spin label shows three lines of equal intensities. The ESR signals disappear rapidly when the nitroxide comes in contact with a reducing agent such as ascorbate. In one experiment, these spin-label molecules were incorporated in the biological membrane lipid bilayer structure at about 5% concentration. The amplitude of the nitroxide ESR signals decreased to 35% of the initial value within a few minutes of the addition of ascorbate. The amplitude of the residual spectrum decayed exponentially with a half-life of about 7 hr. Explain these observations.

14.44 Both NMR and ESR spectroscopy differ from other branches of spectroscopy discussed in this chapter in one important aspect. Explain.

Additional Problems

14.45 Calculate the probe volume of a 10^{-10} M solution needed to contain a single molecule.

14.46 A laser dye is dissolved in a polymer to a concentration of 5.5×10^{-9} *M*. A 1.0-μm-thick layer of the polymer is deposited on a microscope slide. Calculate the radius of a probe laser needed to contain a single laser dye molecule.

14.47 The typical energy differences for transitions in microwave, IR, and electronic spectroscopies are 5×10^{-22} J, 0.5×10^{-19} J, and 1×10^{-18} J, respectively. Calculate the ratio of number of molecules in the two adjacent energy levels (e.g., the ground level and the first excited level) at 300 K in each case.

14.48 The molar absorptivity of a solute at 664 nm is 895 L mol^{-1} cm^{-1}. When light at that wavelength is passed through a 2.0-cm cell containing a solution of the solute, 74.6% of the light is absorbed. Calculate the concentration of the solution.

14.49 The NMR spectrum of *N,N*–dimethylformamide shows two methyl peaks at 25°C. When heated to 130°C, there is only one peak due to the methyl protons. Explain.

```
O       CH3
 \\     /
  C—N
 /     \
H       CH3
```

14.50 The frequency of molecular collision in the liquid phase is about 1×10^{13} s^{-1}. Ignoring all other mechanisms contributing to linewidth, calculate the width (in Hz) of vibrational transitions if **(a)** every collision is effective in deactivating the molecule vibrationally and **(b)** one collision in 40 is effective.

14.51 Consider the IR spectrum of 2–propenenitrile molecule shown in Figure 11.25. Which of the following types of energy has the largest number of energy levels appreciably occupied at 300 K? **(a)** Electronic, **(b)** C—H stretching vibration, **(c)** C=C stretching vibration, **(d)** H—C—H bending motion, and **(e)** rotational.

14.52 The molecule fluorescein is a red solid known as the color additive "D&C Yellow no. 7" because it appears yellow in aqueous solution. When excited, fluorescein emits green light (~520 nm). Draw a schematic energy level diagram that explains these colors.

14.53 Analyze the ^{31}P NMR spectrum of ATP shown in Figure 14.28.

14.54 The Beer-Lambert law predicts a linear dependence of absorbance on concentration. The law often breaks down at very high concentrations. Why?

14.55 The wavelengths of absorption of chromophores in electronic spectra are often influenced by the solvent. For example, polar solvents stabilize the ground state of $\pi^* \leftarrow n$ transitions more than the excited state. On the other hand, for $\pi^* \leftarrow \pi$ transitions, the excited state is more stabilized. Sketch diagrams to show the changes in energy levels involved in the electronic transition when a chromophore changes from a nonpolar to a polar solvent environment and predict the shift in wavelength in each case.

14.56 **(a)** Calculate the energy difference between the two spin states of ^{1}H and of ^{13}C in a magnetic field of 4.7 T. **(b)** What is the precession frequency of a ^{1}H nucleus at this magnetic field? Of a ^{13}C nucleus? **(c)** At what magnetic field do protons precess at a frequency of 500 MHz?

14.57 The absorbance at 260 nm of a DNA solution is 0.120 at 25°C and below (all double helices) and 0.142 at 90°C and above (totally denatured). Calculate the fraction of the double helix remaining at 70°C if the absorbance is 0.131.

14.58 Oxygen is an effective quencher of fluorescence because it is a triplet in its electronic ground state. The unpaired spins of O_2 can induce the excited state of the fluorescent molecule to undergo intersystem crossing from the singlet state to the triplet statc, that is, $S_1 \rightarrow T_1$ (see Figure 14.9). **(a)** How would you verify the mechanism experimentally? **(b)** Assume that the quenching rate constant (i.e., the rate constant for the collision between O_2 and fluorescent molecules) to be $1 \times 10^{10}\ M^{-1}\ \mathrm{s}^{-1}$ at 25°C. How many collisions s^{-1} on average does each fluorescent molecule in solution experience? The concentrations are: $[O_2] = 3.4 \times 10^{-4}\ M$ and $[\mathrm{F}] = 0.50\ M$, where F is the fluorescent molecule. **(c)** The fluorescence lifetime of pyrene, a molecule that is often used to probe biological systems, is 500 ns, while that of tryptophan is about 5 ns. Explain why under normal atmospheric conditions O_2 can interfere only with the fluorescence of pyrene but not that of tryptophan. **(d)** A quantitative relationship of fluorescence quenching is the Stern–Volmer equation,

$$\frac{I_0}{I} = 1 + k_\mathrm{Q}\tau_0[\mathrm{Q}]$$

where I_0 and I are the fluorescence intensities in the absence and presence of the quencher, k_Q is the quenching rate constant, τ_0 is the mean lifetime of the fluorescent state in the absence of the quencher, and $[\mathrm{Q}]$ is the concentration of the quencher. Use this equation to support your conclusion in **(c)**. **(e)** What air pressure is needed to get 50% quenching of tryptophan in solution?

CHAPTER 15

Chemical Kinetics

The rate of chemical reactions is a very complicated subject. This statement is to be interpreted as a challenge to enthusiastic and vigorous chemists; it is not to be interpreted as a sad sigh of defeat.

—Harold S. Johnston*

The aims of studying chemical kinetics are to determine experimentally the rate of a reaction and its dependence on parameters such as concentration, temperature, and catalysts, and to understand the mechanism of a reaction, that is, the number of steps involved and the nature of intermediates formed.

The subject of chemical kinetics is conceptually easier to understand than some other topics in physical chemistry, such as thermodynamics and quantum mechanics, although rigorous theoretical treatment of the energetics involved is possible only for very simple systems in the gas phase. Nevertheless, the macroscopic, empirical approach to the subject can provide much useful information.

In this chapter, we discuss general topics in chemical kinetics and consider some important examples, including fast reactions and enzyme kinetics.

15.1 Reaction Rate

The rate of a reaction is expressed as the change in reactant concentration with time. Consider the stoichiometrically simple reaction

$$\mathrm{R} \rightarrow \mathrm{P}$$

Let the concentrations (in mol L^{-1}) of R at times t_1 and t_2 ($t_2 > t_1$) be $[\mathrm{R}]_1$ and $[\mathrm{R}]_2$. The rate of the reaction over the time interval $(t_2 - t_1)$ is given by

$$\frac{[\mathrm{R}]_2 - [\mathrm{R}]_1}{t_2 - t_1} = \frac{\Delta[\mathrm{R}]}{\Delta t}$$

Because $[\mathrm{R}]_2 < [\mathrm{R}]_1$, we introduce a minus sign so that the rate will be a positive quantity:

$$\text{rate} = -\frac{\Delta[\mathrm{R}]}{\Delta t}$$

* Johnston, H. S., *Gas Phase Reaction Rate Theory*, The Ronald Press, New York, 1966. Used by permission.

The rate can be expressed also in terms of the appearance of a product

$$\text{rate} = \frac{[\text{P}]_2 - [\text{P}]_1}{t_2 - t_1} = \frac{\Delta[\text{P}]}{\Delta t}$$

In this case, we have $[\text{P}]_2 > [\text{P}]_1$. In practice, we find that the quantity of interest is not the rate over a certain time interval (because this is only an average quantity whose value depends on the particular value of Δt); rather, we are interested in the instantaneous rate. In the language of calculus, as Δt becomes smaller and eventually approaches zero, the rate of the foregoing reaction at a specific time t is given by

$$\text{rate} = -\frac{d[\text{R}]}{dt} = \frac{d[\text{P}]}{dt}$$

The units of reaction rates are usually $M\ \text{s}^{-1}$ or $M\ \text{min}^{-1}$.

For stoichiometrically more complicated reactions, the rate must be expressed in an unambiguous manner. Suppose that the reaction of interest is

$$2\text{R} \longrightarrow \text{P}$$

The ratios $-d[\text{R}]/dt$ and $d[\text{P}]/dt$ still express the rate of change of the reactant and the product, respectively, but they are no longer equal to each other because the reactant is disappearing twice as fast as the product is appearing. For this reason, we write the rate of this reaction as

$$\text{rate} = -\frac{1}{2}\frac{d[\text{R}]}{dt} = \frac{d[\text{P}]}{dt}$$

In general, for the reaction

$$a\text{A} + b\text{B} \longrightarrow c\text{C} + d\text{D}$$

the rate is given by

$$\text{rate} = -\frac{1}{a}\frac{d[\text{A}]}{dt} = -\frac{1}{b}\frac{d[\text{B}]}{dt} = \frac{1}{c}\frac{d[\text{C}]}{dt} = \frac{1}{d}\frac{d[\text{D}]}{dt} \tag{15.1}$$

where the expressions in brackets refer to the concentrations of the reactants and products at time t after the start of the reaction.

15.2 Reaction Order

The relationship between the rate of a chemical reaction and the concentrations of the reactants is a complex one that must be determined experimentally. Referring to the general equation above, however, we find that usually (but by no means always) the reaction rate can be expressed as

$$\begin{aligned}\text{rate} &\propto [\text{A}]^x[\text{B}]^y \\ &= k[\text{A}]^x[\text{B}]^y \end{aligned} \tag{15.2}$$

This equation, known as the *rate law*, tells us that the rate of a reaction is not constant; its value at any time, t, is proportional to the concentrations of A and B raised to some powers. The proportionality constant, k, is called the *rate constant*. The rate law is defined in terms of the reactant concentrations, but the rate constant for a given reaction does not depend on the concentrations of the reactants. The rate constant is affected only by temperature, as we shall see later.

Expressing the rate of a reaction as shown in Equation 15.2 enables us to define the *order of a reaction*. We say that the reaction is x order with respect to A and y order with respect to B. Thus, the reaction has an overall order of $(x + y)$. It is important to understand that, in general, there is no connection between the order of a reactant in the rate expression and its stoichiometric coefficient in the balanced chemical equation. For example, the rate of the reaction

This must be the case because a chemical equation can be balanced in many different ways.

$$2N_2O_5(g) \rightarrow 4NO_2(g) + O_2(g)$$

is given by

$$\text{rate} = k[N_2O_5]$$

The reaction is first order in N_2O_5—not second order as we might have inferred from the balanced equation.

The order of a reaction specifies the empirical dependence of the rate on concentrations. It may be zero, an integer, a negative integer, or even a noninteger. We can use the rate law to determine the concentrations of reactants at any time during the course of a reaction. To do so, we need to integrate the rate law expressions. For simplicity, we shall focus only on reactions that have positive integral orders.

Zero-Order Reactions

The rate law for a zero-order reaction of the type

$$A \rightarrow \text{product}$$

is given by

$$\text{rate} = -\frac{d[A]}{dt} = k[A]^0 = k \qquad (15.3)$$

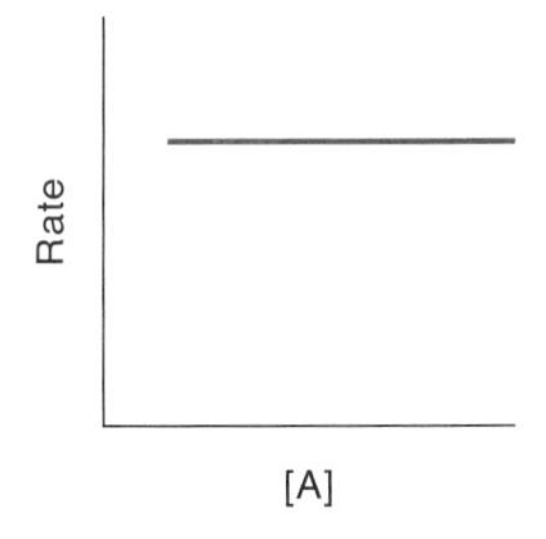

Figure 15.1
Zero-order reaction. The rate is independent of concentration.

The quantity k (M s^{-1}) is the zero-order rate constant. As you can see, the rate of the reaction is independent of the reactant concentration (Figure 15.1). Rearranging Equation 15.3, we obtain

$$d[A] = -kdt$$

Integration between $t = 0$ and $t = t$ at concentrations $[A]_0$ and $[A]$ gives

$$\int_{[A]_0}^{[A]} d[A] = [A] - [A]_0 = -\int_0^t kdt = -kt$$

or

$$[A] = [A]_0 - kt \qquad (15.4)$$

Note that Equation 15.4 gives the time dependence of [A], but cannot be the full description of the factors affecting the rate. To illustrate this point, consider the decomposition of gaseous ammonia on a tungsten surfacc:

$$NH_3(g) \rightarrow \tfrac{1}{2}N_2(g) + \tfrac{3}{2}H_2(g)$$

Under certain conditions, this reaction obeys zero-order rate law. Such a zero-order reaction can occur if the rate is limited, for example, by the concentration of a catalyst. The rate of the reaction is given by

$$\text{rate} = k'\theta A$$

where k' is a constant, θ is the fraction of metal surface covered by the adsorbed ammonia molecules, and A is the total catalyst surface area. If the pressure of ammonia is large enough, $\theta = 1$, and the reaction is zero order in ammonia. At sufficiently low pressures, however, θ is proportional to $[NH_3]$ in the gas phase and the reaction becomes first order in ammonia. Note that the rate will also depend on the amount of catalyst, that is, on the area A.

First-Order Reactions

A first-order reaction is one in which the rate of the reaction depends only on the concentration of the reactant raised to the first power,

$$\text{rate} = -\frac{d[A]}{dt} = k[A] \tag{15.5}$$

where k (s^{-1}) is the first-order rate constant. Rearranging Equation 15.5, we get

$$-\frac{d[A]}{[A]} = kdt$$

Integrating between $t = 0$ and $t = t$ at concentrations $[A]_0$ and $[A]$, we obtain

$$\int_{[A]_0}^{[A]} \frac{d[A]}{[A]} = -\int_0^t kdt$$

$$\ln\frac{[A]}{[A]_0} = -kt \tag{15.6}$$

or

$$[A] = [A]_0 e^{-kt} \tag{15.7}$$

A plot of $\ln([A]/[A]_0)$ versus t gives a straight line whose slope, which is negative, is given by $-k$ (Figure 15.2a). Equation 15.7 shows that in first-order reactions, the decrease in reactant concentration with time is exponential (Figure 15.2b).

Radioactive decays fit first-order kinetics. An example is

$$^{222}_{86}Rn \rightarrow {}^{218}_{84}Po + \alpha$$

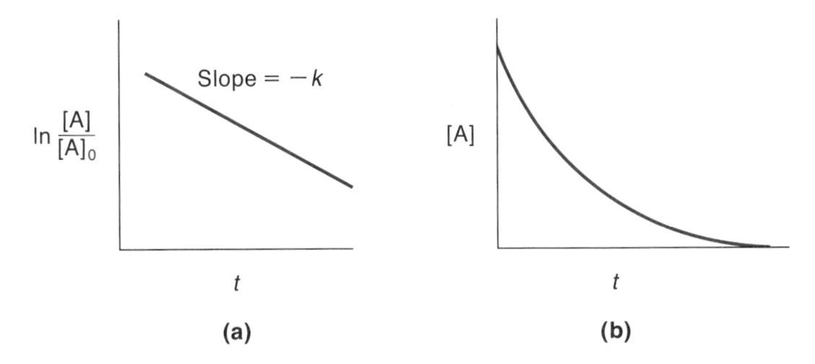

Figure 15.2
First-order reaction. (a) Plot based on Equation 15.6 with a slope of $-k$. (b) Exponential decay of [A] with time according to Equation 15.7.

where α represents the helium nucleus (He^{2+}). The thermal decomposition of N_2O_5 mentioned earlier is first order in N_2O_5. Another example is the rearrangement of methyl isonitrile to acetonitrile:

$$CH_3NC(g) \rightarrow CH_3CN(g)$$

Half-Life of a Reaction. A measure of considerable practical importance in kinetic studies is the *half-life* ($t_{1/2}$) of a reaction. The half-life of a reaction is defined as the time it takes for the concentration of the reactant to decrease by half of its original value. For example, in a first-order reaction, as $[A] = [A]_0/2$, $t = t_{1/2}$ and Equation 15.6 becomes

$$\ln \frac{[A]_0/2}{[A]_0} = -kt_{1/2}$$

or

$$t_{1/2} = \frac{\ln 2}{k} = \frac{0.693}{k} \qquad (15.8)$$

Thus, the half-life of a first-order reaction is *independent* of the initial concentration (Figure 15.3). For A to decrease from 1 M to 0.5 M takes just as much time as it does

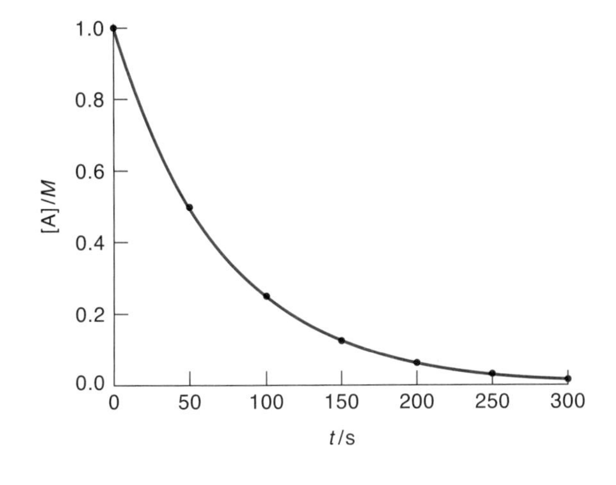

Figure 15.3
The half-lives of a first-order reaction (A → product). The initial concentration is arbitrarily set at 1 M, and A reacts with a constant half-life of 50 s.

Table 15.1
Half-lives of Some Common Radioisotopes

Isotope	Decay process	$t_{1/2}$
${}^{3}_{1}H$	${}^{3}_{1}H \rightarrow {}^{3}_{2}He + {}^{0}_{-1}\beta$	12.3 yr
${}^{14}_{6}C$	${}^{14}_{6}C \rightarrow {}^{14}_{7}N + {}^{0}_{-1}\beta$	5.73×10^{3} yr
${}^{24}_{11}Na$	${}^{24}_{11}Na \rightarrow {}^{24}_{12}Mg + {}^{0}_{-1}\beta$	15 h
${}^{32}_{15}P$	${}^{32}_{15}P \rightarrow {}^{32}_{16}S + {}^{0}_{-1}\beta$	14.3 d
${}^{35}_{16}S$	${}^{35}_{16}S \rightarrow {}^{35}_{17}Cl + {}^{0}_{-1}\beta$	88 d
${}^{60}_{27}Co$	Emission of γ rays	5.26 yr
${}^{99m}_{43}Tc$ [a]	Emission of γ rays	6.0 h
${}^{131}_{53}I$	${}^{131}_{53}I \rightarrow {}^{131}_{54}Xe + {}^{0}_{-1}\beta$	8.05 d

[a] The superscript m denotes the metastable excited nuclear energy state.

for A to decrease from 0.1 *M* to 0.05 *M*. Table 15.1 lists the half-lives of radioactive isotopes that are used extensively in biochemical research and medicine.

In contrast to first-order reactions, the half-lives of other types of reaction all depend on the initial concentration. In general, we can show that (see Appendix 15.1 on p. 724)

$$t_{1/2} \propto \frac{1}{[A]_0^{n-1}} \tag{15.9}$$

where n is the order of the reaction.

EXAMPLE 15.1

The thermal decomposition of 2,2′-azobisisobutyronitrile (AIBN)

$$N{\equiv}C{-}C(CH_3)_2{-}N{=}N{-}C(CH_3)_2{-}C{\equiv}N \xrightarrow{\Delta} 2N{\equiv}C{-}\dot{C}(CH_3)_2 + N_2$$

has been studied in an inert organic solvent at room temperature. The progress of the reaction can be monitored by the optical absorption of AIBN at 350 nm. The following data are obtained:

t/s	A
0	1.50
2000	1.26
4000	1.07
6000	0.92
8000	0.81
10,000	0.72
12,000	0.65
∞	0.40

where A is the absorbance. Assume that the reaction is first order in AIBN, and calculate the rate constant.

ANSWER

From Equation 15.6, we have

$$\ln \frac{[\text{AIBN}]}{[\text{AIBN}]_0} = -kt$$

The difference in absorbance at $t = 0$ and at $t = \infty$, $(A_0 - A_\infty)$, is proportional to the initial concentration of AIBN in the solution. Similarly, the difference $(A_t - A_\infty)$, where A_t is the absorbance of AIBN at time t, is proportional to the instantaneous concentration [AIBN].* The rate equation can now be expressed as

$$\ln \frac{A_t - A_\infty}{A_0 - A_\infty} = -kt$$

Because $A_0 = 1.50$ and $A_\infty = 0.40$, we have

t/s	$\ln \frac{A_t - A_\infty}{A_0 - A_\infty}$
2000	−0.246
4000	−0.496
6000	−0.749
8000	−0.987
10,000	−1.24
12,000	−1.48

The first-order rate constant can be obtained by plotting the natural logarithmic term versus t, as shown in Figure 15.4. It is given by the slope of the straight line, which is $1.24 \times 10^{-4}\ \text{s}^{-1}$.

* These statements hold true if little or no AIBN remains unreacted as t approaches infinity and the absorbance of products does not interfere with that of AIBN at 350 nm.

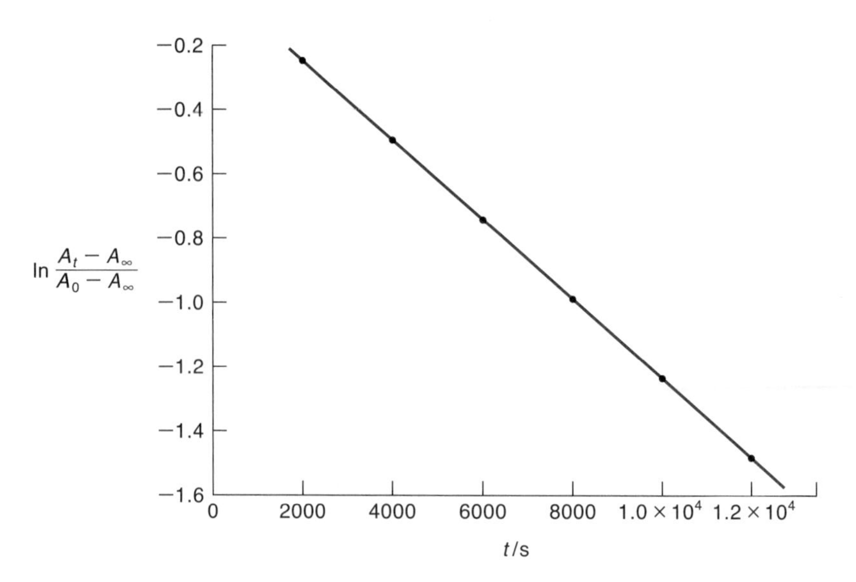

Figure 15.4
AIBN decomposition. The fit of the linear equation is $y = -0.000124x - 0.00167$. Therefore, the first-order rate constant, which is equal to the negative slope, is 1.24×10^{-4} s^{-1}.

Second-Order Reactions

We consider two types of second-order reactions here. In one type, there is just one reactant. The second type involves two different reactants. The first type is represented by the general reaction

$$\text{A} \rightarrow \text{products}$$

and this rate is

$$\text{rate} = -\frac{d[\text{A}]}{dt} = k[\text{A}]^2 \tag{15.10}$$

That is, the rate is proportional to the concentration of A raised to the second power, and k (M^{-1} s^{-1}) is the second-order rate constant. Separating the variables and integrating, we obtain

$$\int_{[\text{A}]_0}^{[\text{A}]} \frac{d[\text{A}]}{[\text{A}]^2} = -\int_0^t k dt$$

$$\frac{1}{[\text{A}]} - \frac{1}{[\text{A}]_0} = kt$$

or

$$\frac{1}{[\text{A}]} = kt + \frac{1}{[\text{A}]_0} \tag{15.11}$$

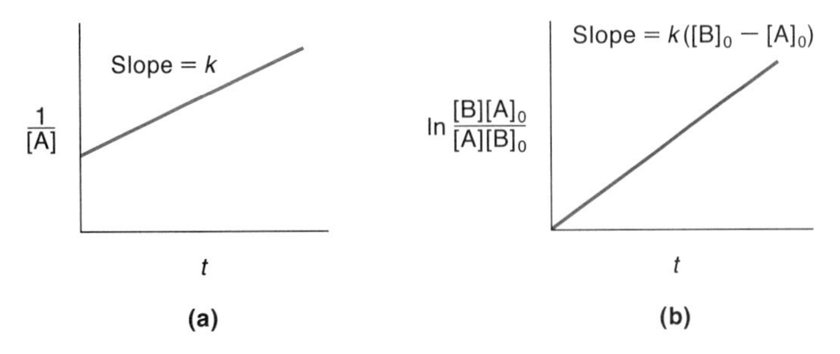

Figure 15.5
Second-order reactions. (a) Plot based on Equation 15.11. (b) Plot based on Equation 15.14.

where $[\mathrm{A}]_0$ is the initial concentration. Thus, a plot of $1/[\mathrm{A}]$ versus t gives a straight line with a slope equal to k (Figure 15.5a). To derive the half-life of this type of second-order reaction, we set $[\mathrm{A}] = [\mathrm{A}]_0/2$ in Equation 15.11 and write

$$\frac{1}{[\mathrm{A}]_0/2} = kt_{1/2} + \frac{1}{[\mathrm{A}]_0}$$

or

$$t_{1/2} = \frac{1}{k[\mathrm{A}]_0} \tag{15.12}$$

As mentioned earlier, except for first-order reactions, the half-lives of all other reactions are concentration dependent.

The second type of second-order reaction is represented by

$$\mathrm{A} + \mathrm{B} \rightarrow \text{products}$$

and

$$\text{rate} = -\frac{d[\mathrm{A}]}{dt} = -\frac{d[\mathrm{B}]}{dt} = k[\mathrm{A}][\mathrm{B}] \tag{15.13}$$

This reaction is first order in A, first order in B, and second order overall. Let

$$[\mathrm{A}] = [\mathrm{A}]_0 - x$$

$$[\mathrm{B}] = [\mathrm{B}]_0 - x$$

where x (in mol L^{-1}) is the amount of A and B consumed in time t. From Equation 15.13,

$$-\frac{d[\mathrm{A}]}{dt} = -\frac{d([\mathrm{A}]_0 - x)}{dt} = \frac{dx}{dt} = k[\mathrm{A}][\mathrm{B}]$$

$$= k([\mathrm{A}]_0 - x)([\mathrm{B}]_0 - x)$$

Rearranging, we obtain

$$\frac{dx}{([A]_0 - x)([B]_0 - x)} = kdt$$

By the somewhat tedious but straightforward method of integration by partial functions, we can obtain the final result:

$$\frac{1}{[B]_0 - [A]_0} \ln \frac{([B]_0 - x)[A]_0}{([A]_0 - x)[B]_0} = kt$$

or

$$\frac{1}{[B]_0 - [A]_0} \ln \frac{[B][A]_0}{[A][B]_0} = kt \tag{15.14}$$

Equation 15.14 was derived by assuming that $[A]_0 < [B]_0$. If $[A]_0 = [B]_0$, the solution is the same as that in Equation 15.11. Note that Equation 15.11 cannot be obtained from Equation 15.14 by setting $[A]_0 = [B]_0$. A plot of Equation 15.14 is shown in Figure 15.5b.

Below are a few examples of second-order reactions:

$$CH_3CHO(g) \rightarrow CH_4(g) + CO(g)$$

$$2NO_2(g) \rightarrow 2NO(g) + O_2(g)$$

$$C_2H_5Br(aq) + OH^-(aq) \rightarrow C_2H_5OH(aq) + Br^-(aq)$$

Pseudo-First-Order Reactions. An interesting special case of second-order reactions occurs when one of the reactants is present in great excess. An example is the hydrolysis of acetyl chloride:

$$CH_3COCl(aq) + H_2O(l) \rightarrow CH_3COOH(aq) + HCl(aq)$$

Because the concentration of water in the acetyl chloride solution is quite high (about 55.5 *M*, the concentration of pure water) and the concentration of acetyl chloride is of the order of 1 *M* or less, the amount of water consumed is negligible compared with the amount of water originally present. Thus, we can express the rate as

$$\frac{d[CH_3COCl]}{dt} = k'[CH_3COCl][H_2O]$$

$$= k[CH_3COCl] \tag{15.15}$$

where $k = k'[H_2O]$. The reaction therefore appears to follow first-order kinetics and is called a *pseudo-first-order reaction.* (To measure k', the second-order rate constant, we need to measure k for many different initial concentrations of H_2O, then a plot of k versus $[H_2O]$ will yield a straight line with a slope equal to k'.)

Table 15.2
Summary of Rate Equations for A → Products

Order	Differential form	Integrated form	Half-life	Units of the rate constant
0	$-\frac{d[A]}{dt} = k$	$[A]_0 - [A] = kt$	$\frac{[A]_0}{2k}$	$M\ s^{-1}$
1	$-\frac{d[A]}{dt} = k[A]$	$[A] = [A]_0 e^{-kt}$	$\frac{\ln 2}{k}$	s^{-1}
2	$-\frac{d[A]}{dt} = k[A]^2$	$\frac{1}{[A]} - \frac{1}{[A]_0} = kt$	$\frac{1}{[A]_0 k}$	$M^{-1}\ s^{-1}$
2[a]	$-\frac{d[A]}{dt} = k[A][B]$	$\frac{1}{[B]_0 - [A]_0} \ln \frac{[B][A]_0}{[A][B]_0} = kt$	—	$M^{-1}\ s^{-1}$

[a] For A + B → products.

Table 15.2 summarizes the rate laws and half-life expressions for zero-, first-, and second-order reactions. Third-order reactions are known but are uncommon, and so we shall not discuss them.

Determination of Reaction Order

In the study of chemical kinetics, one of the first tasks is to determine the order of the reaction. Several methods are available for determining the order of a reaction, and we shall briefly discuss four common approaches.

1. Integration Method. An obvious procedure is to measure the concentration of the reactant(s) at various time intervals of a reaction and to substitute the data into the equations listed in Table 15.2. The equation giving the most constant value of the rate constant for a series of time intervals is the one that corresponds best to the correct order of the reaction. In practice, this method is not precise enough to do more than to distinguish between, say, first- and second-order reactions.

2. Differential Method. This method was developed by van't Hoff in 1884. Because the rate v of an nth-order reaction is proportional to the nth power of the concentration of the reactant, we write

$$v = k[A]^n \tag{15.16}$$

Taking common logarithms of both sides, we obtain

$$\log v = n \log[A] + \log k \tag{15.17}$$

Thus, by measuring v at several different concentrations of A, we can obtain the value of n from a plot of $\log v$ versus $\log[A]$. A satisfactory procedure is to measure the

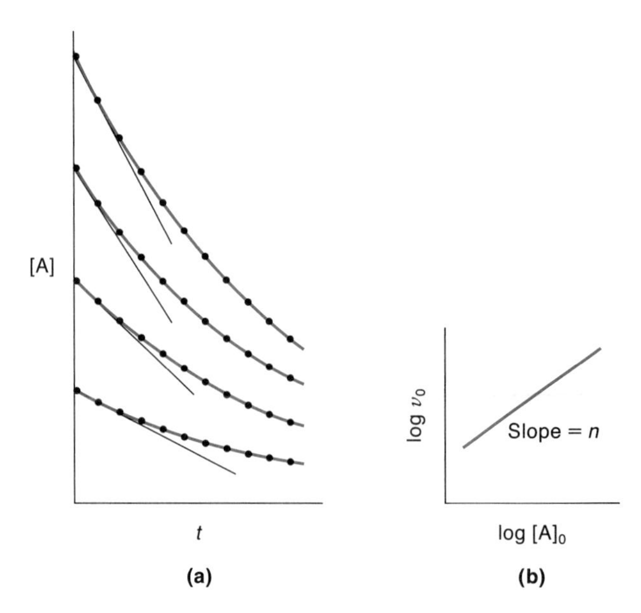

Figure 15.6
(a) Measurement of the initial rates, v_0, of a reaction at different concentrations. (b) Plot of log v_0 versus log [A].

initial rates (v_0) of the reaction for several different starting concentrations of A, as shown in Figure 15.6. The advantages of using initial rates are (1) that it avoids possible complications due to the presence of products that might affect the order of the reaction and (2) that the reactant concentrations are known most accurately at this time.

3. Half-Life Method. Another simple method of determining reaction order is to find the dependence of the half-life of a reaction on the initial concentration, again using the equations in Table 15.2 or Equation 15.9. Thus, measuring the half-life of a reaction will help us determine the order of the reaction. This procedure is particularly useful for first-order reactions because their half-lives are independent of concentration.

4. Isolation Method. If a reaction involves more than one type of reactant, we can keep the concentrations of all but one reactant constant and measure the rate as a function of its concentration. Any change in rate must be due to that reactant alone. Once we have determined the order with respect to this reactant, we repeat the procedure for the second reactant, and so on. In this way, we can obtain the overall order of the reaction.

The four methods described above apply only in ideal cases. In practice, determining reaction order can be very difficult because of uncertainty in concentration measurements (e.g., when there are small concentration changes in initial rate determinations) as well as the complexity of the reactions (e.g., when reactions are reversible and reactions occur between reactants and products). To a certain extent, the procedure is one of trial and error. The use of computers has significantly facilitated the analysis of kinetic data.

Once the order of the reaction has been determined, the rate constant at a particular temperature can be calculated from the ratio of the rate and the concentrations of the reactants, each raised to the power of its order. Knowing the order and rate constant then enables us to write the rate law for the reaction.

15.3 Molecularity of a Reaction

Knowing the order of a reaction is but one step toward a detailed understanding of how a reaction occurs. A reaction seldom takes place in the manner suggested by a balanced chemical equation. Typically, the overall reaction is the sum of several steps; the sequence of steps by which a reaction occurs is called the *mechanism* of the reaction. To know the mechanism of a reaction is to know how molecules approach one another during a collision, how chemical bonds are broken and formed, how charges are transferred, and so on when the reactant molecules are in close proximity. The mechanism proposed for a given reaction must account for the overall stoichiometry, the rate law, and other known facts. Consider the decomposition of hydrogen peroxide:

$$2H_2O_2(aq) \rightarrow 2H_2O(l) + O_2(g)$$

When this reaction is catalyzed by iodide ions, the rate law is found to be

$$\text{rate} = k[H_2O_2][I^-]$$

Thus, the reaction is first order with respect to both H_2O_2 and I^-.

Whereas the word *order* reflects the overall change in going from the reactants to products, the *molecularity* of a reaction refers to a single, definite kinetic process that may be only one step in the overall reaction. For example, evidence suggests that the decomposition of hydrogen peroxide takes place in two steps, as follows:

$$\text{Step (1)} \qquad H_2O_2 + I^- \xrightarrow{k_1} H_2O + IO^-$$

$$\text{Step (2)} \qquad H_2O_2 + IO^- \xrightarrow{k_2} H_2O + O_2 + I^-$$

Each of these so-called *elementary steps* describes what actually happens at the molecular level. How do we account for the observed rate dependence in terms of these two steps? We simply assume that the rate for the first step is much slower than that for the second step, that is, $k_1 \ll k_2$. The overall rate of decomposition, then, is completely controlled by the rate of the first step, which is aptly called the *rate-determining* step, and we have: rate $= k_1[H_2O_2][I^-]$, where $k_1 = k$. Note that the sum of steps 1 and 2 gives us the overall reaction, because the species IO^- cancels out. Such a species is called an *intermediate* because it appears in the mechanism of the reaction but is not in the overall balanced chemical equation. An intermediate is always formed in an early elementary step and consumed in a later elementary step. On the other hand, a catalyst (I^- in this case) appears as a reactant in the initial elementary step. It invariably forms an intermediate and is regenerated at the end of the reaction.

The preceding discussion shows that our insight into a reaction comes from an understanding of molecularity, not of order. Once we know the mechanism and the rate-determining step, we can write the rate law for the reaction, which must agree with the experimentally determined rate law. Although most reactions are kinetically complex, the mechanisms for some of them are sufficiently well understood to be discussed in molecular terms. In general, however, proving the uniqueness of a mechanism is very difficult or impossible, especially for a complex reaction.

As in a court of law, we ask only for proof beyond a reasonable doubt.

We shall now examine the three different types of molecularity. Unlike the order of a reaction, molecularity cannot be zero or a noninteger.

Unimolecular Reactions

Reactions such as cis-trans isomerization, thermal decomposition, ring opening, and racemization are usually *unimolecular*, that is, they involve only one reactant molecule in the elementary step. For example, the following gas-phase elementary steps are unimolecular:

$$N_2O_4(g) \longrightarrow 2NO_2(g)$$

$$\underset{\text{Cyclopropane}}{\text{(CH}_2\text{)}_3} \longrightarrow \underset{\text{Propene}}{CH_3CH{=}CH_2}$$

Cyclopropane Propene

Unimolecular reactions often follow a first-order rate law. Because these reactions presumably occur as the result of a binary collision through which the reactant molecules acquire the necessary energy to change forms, we would expect them to be bimolecular processes and hence second-order reactions. How do we account for the discrepancy between the predicted and observed rate laws? To answer this question, let us consider the treatment put forward by the British chemist Frederick Alexander Lindemann (1886–1957) in 1922.† Every now and then a reactant molecule, A, collides with another A molecule, and one becomes energetically excited at the expense of the other:

$$A + A \xrightarrow{k_1} A + A^*$$

where the asterisk denotes the activated molecule. The activated molecule can form the desired product according to the elementary step

$$A^* \xrightarrow{k_2} \text{product}$$

Another process that may also be going on is the deactivation of the A* molecule:

$$A^* + A \xrightarrow{k_{-1}} A + A$$

The rate of product formation is given by

$$\frac{d[\text{product}]}{dt} = k_2[A^*] \tag{15.18}$$

All that remains for us to do is to derive an expression for $[A^*]$. Because A^* is an energetically excited species, it has little stability and a short lifetime. Its concentration in the gas phase is not only low but probably fairly constant as well. Using this assumption, we can apply the *steady-state approximation* as follows.‡ The rate of change of $[A^*]$ is given by the steps leading to the formation of A^* minus the steps leading to

† A similar treatment was proposed independently and almost simultaneously by the Danish chemist Jens Anton Christiansen (1888–1969).

‡ Note that the steady-state approximation does not always apply to intermediates. Its use must be justified by either experimental evidence or theoretical considerations.

the removal of A^*. According to the steady-state approximation, however, this rate of change must be zero. Mathematically, we have

$$\frac{d[A^*]}{dt} = 0 = k_1[A]^2 - k_{-1}[A][A^*] - k_2[A^*] \tag{15.19}$$

Solving for $[A^*]$, we obtain

$$[A^*] = \frac{k_1[A]^2}{k_2 + k_{-1}[A]} \tag{15.20}$$

The rate of product formation is now given by

$$\frac{d[\text{product}]}{dt} = k_2[A^*] = \frac{k_1k_2[A]^2}{k_2 + k_{-1}[A]} \tag{15.21}$$

Two important limiting cases may be applied to this equation. At higher pressures ($\geq$ 1 atm), most A^* molecules will be deactivated instead of forming product, and we have

$$k_{-1}[A][A^*] \gg k_2[A^*]$$

or

$$k_{-1}[A] \gg k_2$$

The rate in this case is given by

$$\frac{d[\text{product}]}{dt} = \frac{k_1k_2}{k_{-1}}[A] \tag{15.22}$$

and the reaction is first order in A. On the other hand, if the reaction is run at low pressures ($<$ 0.01 atm) so that most A^* molecules form the product instead of being deactivated, the following inequality will hold:

$$k_{-1}[A][A^*] \ll k_2[A^*]$$

or

$$k_{-1}[A] \ll k_2$$

The rate now becomes

$$\frac{d[\text{product}]}{dt} = k_1[A]^2 \tag{15.23}$$

which is second order in A.

Lindemann's theory has been tested for a number of systems and is found to be essentially correct. The analysis for the intermediate case (i.e., $k_{-1}[A][A^*] \approx k_2[A^*]$) is more complex and will not be discussed here.

Bimolecular Reactions

Any elementary step that involves two reactant molecules is a *bimolecular reaction*. Some of the examples in the gas phase are

$$H + H_2 \rightarrow H_2 + H$$

$$NO_2 + CO \rightarrow NO + CO_2$$

$$2NOCl \rightarrow 2NO + Cl_2$$

In the solution phase we have

$$2CH_3COOH \rightarrow (CH_3COOH)_2 \text{ (in nonpolar solvents)}$$

$$Fe^{2+} + Fe^{3+} \rightarrow Fe^{3+} + Fe^{2+}$$

Termolecular Reactions

Finally, an elementary step that involves the simultaneous encounter of three reactant molecules is called a *termolecular reaction*. The probability of a three-body collision is usually quite small and only a few such reactions are known. Interestingly, they all involve nitric oxide as one of the reactants:

$$2NO(g) + X_2(g) \rightarrow 2NOX(g)$$

where X = Cl, Br, or I. Another type of termolecular "reaction" involves atomic recombinations in the gas phase; for example,

$$H + H + M \rightarrow H_2 + M$$

$$I + I + M \rightarrow I_2 + M$$

where M is usually some inert gas such as N_2 or Ar. When atoms combine to form diatomic molecules, they possess an excess of kinetic energy, which is converted to vibrational motion, resulting in bond dissociation. Through three-body collisions, the M species can take away some of this excess energy to prevent the break-up of the diatomic molecules.

No elementary steps with a molecularity greater than three are known.

15.4 More Complex Reactions

All the reactions discussed so far are simple in the sense that only one reaction is taking place in each case. Unfortunately, this condition is often not satisfied in actual practice. Three examples of more complex reactions will now be discussed.

Reversible Reactions

Most reactions are reversible to a certain extent, and we must consider both the forward and reverse rates. For the reversible reaction that proceeds by two elementary steps,

$$A \underset{k_{-1}}{\overset{k_1}{\rightleftharpoons}} B$$

we represent the net rate of change in [A] as

$$\frac{d[\mathrm{A}]}{dt} = -k_1[\mathrm{A}] + k_{-1}[\mathrm{B}] \tag{15.24}$$

At equilibrium, there is no net change in the concentration of A with time, that is, $d[\mathrm{A}]/dt = 0$, so that

$$k_1[\mathrm{A}] = k_{-1}[\mathrm{B}] \tag{15.25}$$

This expression leads to

$$\frac{[\mathrm{B}]}{[\mathrm{A}]} = \frac{k_1}{k_{-1}} = K \tag{15.26}$$

where K is the equilibrium constant.

The discussion of the relationship between reaction rates and equilibria is rooted in a principle of great importance in chemical kinetics. The *principle of microscopic reversibility* states that at equilibrium, the rates of the forward and reverse processes are equal for every elementary reaction occurring.* It means that the process A → B is exactly balanced by B → A so that equilibrium cannot be maintained by a cyclic process in which the forward reaction is A → B and the reverse reaction is B → C → A:

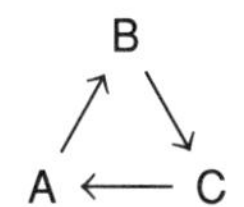

Instead, for every elementary reaction we must write a reverse reaction as follows:

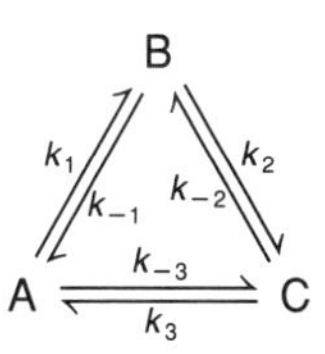

such that

$$k_1[\mathrm{A}] = k_{-1}[\mathrm{B}]$$

$$k_2[\mathrm{B}] = k_{-2}[\mathrm{C}]$$

$$k_3[\mathrm{C}] = k_{-3}[\mathrm{A}]$$

These rate constants are not all independent. By simple algebraic manipulation, we can show that $k_1k_2k_3 = k_{-1}k_{-2}k_{-3}$ (see Problem 15.65). The usefulness of the principle of microscopic reversibility is that it tells us that the reaction pathway for the reverse of

* The principle of microscopic reversibility is a consequence of the fact that the fundamental equations for the microscopic dynamics of a system (i.e., Newton's laws or the Schrödinger equation) have the same form when time t is replaced by $-t$ and when the signs of all velocities are also reversed. See B. H. Mahan, *J. Chem. Educ.* **52**, 299 (1975).

a reaction at equilibrium is the exact opposite of the pathway for the forward reaction. Therefore, the transition states* for the forward and reverse reactions are identical.

Consider the base-catalyzed esterification between acetic acid and ethanol,

O
CH_3—C
OH
CH_3CH_2—O
H
B̈
⟶
O⁻
CH_3—C
OH
O
CH_3CH_2
H
B^+
⟶
CH_3
C=O
CH_3CH_2O + OH^-

where B is a base (e.g., OH^-). The species formed in the first step is a tetrahedral intermediate. Now, according to the principle of microscopic reversibility, the reverse reaction, that is, the hydrolysis of ethyl acetate, must involve the acid-catalyzed expulsion of ethoxide ion from the same tetrahedral intermediate:

CH_3
C=O
CH_3CH_2O
OH^-
⟶
O⁻
CH_3—C
OH
O
CH_3CH_2
H
B^+
⟶
O
CH_3—C
OH
+ CH_3CH_2OH + B̈

Thus, when the likelihood of a certain mechanism is being considered, we can always turn to the principle for guidance. If the reverse mechanism looks implausible, then chances are that the proposed mechanism is wrong and we must search for another mechanism.

Consecutive Reactions

A consecutive reaction is one in which the product from the first step becomes the reactant for the second step, and so on. The thermal decomposition of acetone in the gas phase is an example:

$$CH_3COCH_3 \rightarrow CH_2{=}CO + CH_4$$

$$CH_2{=}CO \rightarrow CO + \tfrac{1}{2}C_2H_4$$

Many nuclear decays are also consecutive reactions. For example, upon the capture of a neutron, a uranium-238 isotope is converted to a uranium-239 isotope, which then decays as follows:

$$^{239}_{92}U \rightarrow\ ^{239}_{93}Np +\ ^{0}_{-1}\beta$$

$$^{239}_{93}Np \rightarrow\ ^{239}_{94}Pu +\ ^{0}_{-1}\beta$$

* The transition state of a reaction is the complex formed between the reactants and products along the reaction coordinate (discussed further in Section 15.7).

For a two-step consecutive reaction, we have

$$\mathrm{A} \xrightarrow{k_1} \mathrm{B} \xrightarrow{k_2} \mathrm{C}$$

Because each step is first order, the rate law equations are

$$\frac{d[\mathrm{A}]}{dt} = -k_1[\mathrm{A}] \tag{15.27}$$

$$\frac{d[\mathrm{B}]}{dt} = k_1[\mathrm{A}] - k_2[\mathrm{B}] \tag{15.28}$$

$$\frac{d[\mathrm{C}]}{dt} = k_2[\mathrm{B}] \tag{15.29}$$

We assume that initially only A is present and its concentration is $[\mathrm{A}]_0$ so that

$$[\mathrm{A}] = [\mathrm{A}]_0 e^{-k_1 t} \tag{15.30}$$

The rate equation for the intermediate B is quite complex and will not be fully discussed here. The treatment can be simplified, however, by applying the steady-state approximation to B, that is, by assuming that the concentration of B remains constant over a certain time period so that we can write

$$\frac{d[\mathrm{B}]}{dt} = 0 = k_1[\mathrm{A}] - k_2[\mathrm{B}] \tag{15.31}$$

or

$$[\mathrm{B}] = \frac{k_1}{k_2}[\mathrm{A}] = \frac{k_1}{k_2}[\mathrm{A}]_0 e^{-k_1 t} \tag{15.32}$$

Equation 15.32 holds if $k_2 \gg k_1$. Under this condition, B molecules are converted to C as soon as they are formed, so $[\mathrm{B}]$ is kept constant and low compared to $[\mathrm{A}]$.

To get an expression for $[\mathrm{C}]$, we note that at any instant we have $[\mathrm{A}]_0 = [\mathrm{A}] + [\mathrm{B}] + [\mathrm{C}]$. Therefore, from Equations 15.30 and 15.32 we obtain

$$\begin{aligned} [\mathrm{C}] &= [\mathrm{A}]_0 - [\mathrm{A}] - [\mathrm{B}] \\ &= [\mathrm{A}]_0\left(1 - e^{-k_1 t} - \frac{k_1}{k_2} e^{-k_1 t}\right) \\ &= [\mathrm{A}]_0(1 - e^{-k_1 t}) \end{aligned} \tag{15.33}$$

The $(k_1/k_2)\exp(-k_1 t)$ term is eliminated because it is much smaller than 1.

Figure 15.7 shows plots of $[\mathrm{A}]$, $[\mathrm{B}]$, and $[\mathrm{C}]$ with time for different rate constant ratios. In all cases, $[\mathrm{A}]$ falls steadily from $[\mathrm{A}]_0$ to zero while $[\mathrm{C}]$ rises from zero to $[\mathrm{A}]_0$. The concentration of B rises from zero to a maximum and then falls back to zero. Note that as k_2 becomes larger than k_1, the steady-state approximation becomes valid over the time period when $[\mathrm{B}]$ remains constant (Figure 15.7c).

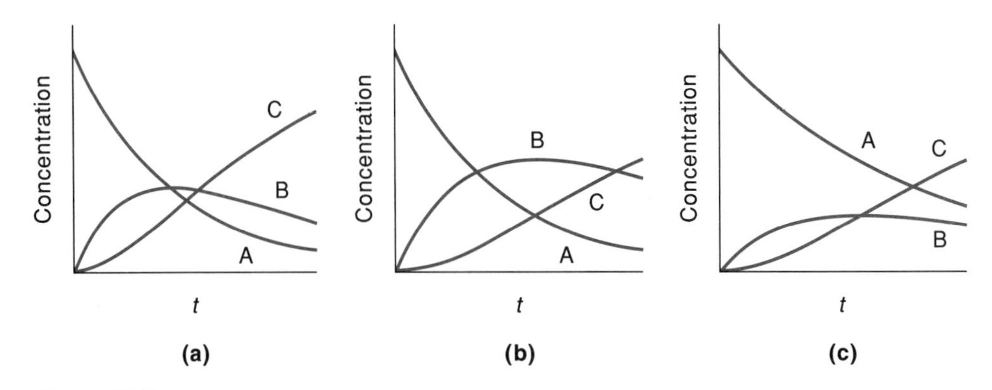

Figure 15.7
Variation in the concentrations of A, B, and C with time for a consecutive reaction $A \xrightarrow{k_1} B \xrightarrow{k_2} C$. (a) $k_1 = k_2$; (b) $k_1 = 2k_2$; (c) $k_1 = 0.5k_2$.

A more complicated but common consecutive reaction is

$$A + B \underset{k_{-1}}{\overset{k_1}{\rightleftharpoons}} C \xrightarrow{k_2} P$$

where P denotes product. This scheme involves a *pre-equilibrium*, in which an intermediate is in equilibrium with the reactants. A pre-equilibrium arises when the rates of formation of the intermediate and of its decay back into reactants are much faster than its rate of formation of products, that is, when $k_{-1} \gg k_2$. Because A, B, and C are assumed to be in equilibrium, we can write

$$K = \frac{[C]}{[A][B]} = \frac{k_1}{k_{-1}}$$

and the rate of formation of P is given by

$$\frac{d[P]}{dt} = k_2[C] = k_2K[A][B]$$

$$= \frac{k_1k_2}{k_{-1}}[A][B] \qquad (15.34)$$

Chain Reactions

One of the best-known gas-phase chain reactions involves the formation of hydrogen bromide from molecular hydrogen and bromine between 230°C and 300°C:

$$H_2(g) + Br_2(g) \rightarrow 2HBr(g)$$

The complexity of this reaction is indicated by the rate equation,

$$\frac{d[HBr]}{dt} = \frac{\alpha[H_2][Br_2]^{1/2}}{1 + \beta[HBr]/[Br_2]} \qquad (15.35)$$

where α and β are some constants. Thus, the reaction does not have an integral reaction order. It has taken many experiments and considerable chemical intuition to come up with Equation 15.35. We assume that a chain of reactions proceeds as follows:

$$Br_2 \xrightarrow{k_1} 2Br \qquad \text{chain initiation}$$

$$Br + H_2 \xrightarrow{k_2} HBr + H \qquad \text{chain propagation}$$

$$H + Br_2 \xrightarrow{k_3} HBr + Br \qquad \text{chain propagation}$$

$$H + HBr \xrightarrow{k_4} H_2 + Br \qquad \text{chain inhibition}$$

$$Br + Br \xrightarrow{k_5} Br_2 \qquad \text{chain termination}$$

The following reactions play only a minor role in determining the rate:

$$H_2 \rightarrow 2H \qquad \text{chain initiation}$$

$$Br + HBr \rightarrow Br_2 + H \qquad \text{chain inhibition}$$

$$H + H \rightarrow H_2 \qquad \text{chain termination}$$

$$H + Br \rightarrow HBr \qquad \text{chain termination}$$

For this reason, they are not included in the kinetic analysis. By applying the steady-state approximation to the intermediates H and Br, we can derive Equation 15.35 using the first five elementary steps (see Problem 15.20).

15.5 The Effect of Temperature on Reaction Rate

Figure 15.8 shows four types of temperature dependence for reaction rate constants. Type (a) represents normal reactions whose rates increase with increasing temperature. Type (b) shows a rate that initially increases with temperature, reaches a maximum, and finally decreases with further temperature rise. Type (c) shows a steady decrease of rate with temperature. The behavior outlined in (b) and (c) may be surprising, because we might expect the rate of a reaction to depend on two quantities: the number of collisions per second and the fraction of collisions that activate molecules

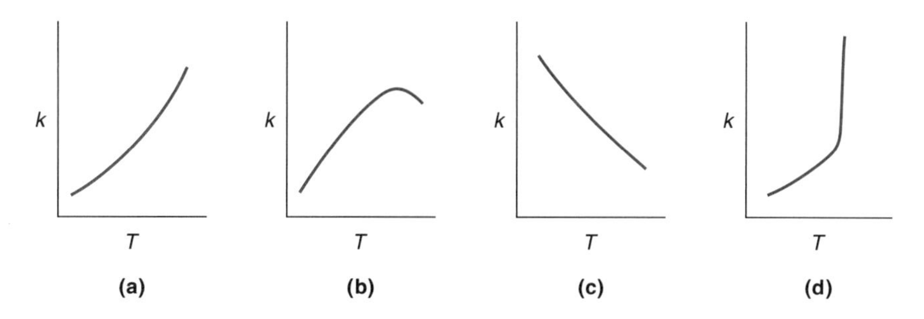

Figure 15.8
Four types of temperature dependence for rate constants. See text.

for the reaction. Both quantities should increase with increasing temperature. The complex nature of the reaction mechanism accounts for this ostensibly unusual behavior. For example, in an enzyme-catalyzed reaction, the enzyme molecule must be in a specific conformation to react with the substrate molecule. When the enzyme is in the native state, the reaction rate does increase with temperature. At higher temperatures, the molecule may undergo denaturation, thereby losing its effectiveness as a catalyst. Consequently, the rate will decrease with increasing temperature.

The behavior shown in Figure 15.8c is known for only a few systems. Consider the formation of nitrogen dioxide from nitric oxide and oxygen:

$$2NO(g) + O_2(g) \rightleftharpoons 2NO_2(g)$$

The rate law is

$$\text{rate} = k[NO]^2[O_2]$$

The mechanism is believed to involve two bimolecular steps:

This is an example of pre-equilibrium, discussed on p. 690.

$$\text{Rapid:}\ 2NO \rightleftharpoons (NO)_2 \qquad K = \frac{[(NO)_2]}{[NO]^2}$$

$$\text{Slow, rate determining:}\ (NO)_2 + O_2 \xrightarrow{k'} 2NO_2$$

Thus, the overall rate is

$$\begin{aligned}\text{rate} &= k'[(NO)_2][O_2] = k'K[NO]^2[O_2] \\ &= k[NO]^2[O_2]\end{aligned}$$

where $k = k'K$. Furthermore, the equilibrium between 2NO and $(NO)_2$ is exothermic from left to right. Because the decrease in K with temperature outweighs the increase in k' with temperature, the overall rate decreases with increasing temperature over a certain range of temperature.

Finally, the behavior shown in Figure 15.8d corresponds to a chain reaction. At first, the rate rises gradually with temperature. At a particular temperature, the chain propagation reactions become significant, and the reaction is literally explosive.

The Arrhenius Equation

In 1889, Arrhenius discovered that the temperature dependence of many reactions could be described by the equation,

$$k = Ae^{-E_a/RT} \tag{15.36}$$

where k is the rate constant, A is called the frequency factor or pre-exponential factor, E_a is the activation energy (kJ mol^{-1}), R is the gas constant, and T is the absolute temperature. The *activation energy* is the minimum amount of energy required to initiate a chemical reaction. The frequency factor, A, represents the frequency of collisions between reactant molecules. The factor $\exp(-E_a/RT)$ resembles the Boltzmann distribution law (see Equation 2.33); it represents the fraction of molecular collisions that

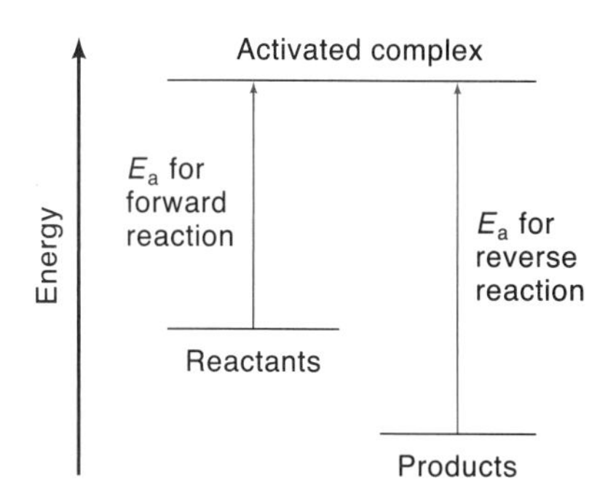

Figure 15.9
Schematic diagram of activation energy for an exothermic reaction.

have energy equal to or greater than the activation energy, E_a (Figure 15.9). Because the exponential term is a number, the units of A are the same as the units of the rate constant (s^{-1} for first-order rate constants; M^{-1} s^{-1} for second-order rate constants, and so on).

As we shall see later, because the frequency factor A is related to molecular collisions, it is temperature dependent. For a limited temperature range ($\leq$ 50 K), however, the predominant temperature variation is contained in the exponential term. Taking the natural logarithm of Equation 15.36, we obtain

$$\ln k = \ln A - \frac{E_a}{RT} \tag{15.37}$$

Thus, a plot of ln k versus $1/T$ yields a straight line whose slope, which is negative, is equal to $-E_a/R$ (Figure 15.10). Note that in Equation 15.37 k and A are treated as dimensionless quantities.

Alternatively, if we know the rate constants k_1 and k_2 at T_1 and T_2, we have, from Equation 15.37

$$\ln k_1 = \ln A - \frac{E_a}{RT_1}$$

$$\ln k_2 = \ln A - \frac{E_a}{RT_2}$$

Taking the difference between these two equations, we obtain

$$\ln \frac{k_2}{k_1} = -\frac{E_a}{R}\left(\frac{1}{T_2} - \frac{1}{T_1}\right) \tag{15.38}$$

The formulation is valid for any order rate constant.

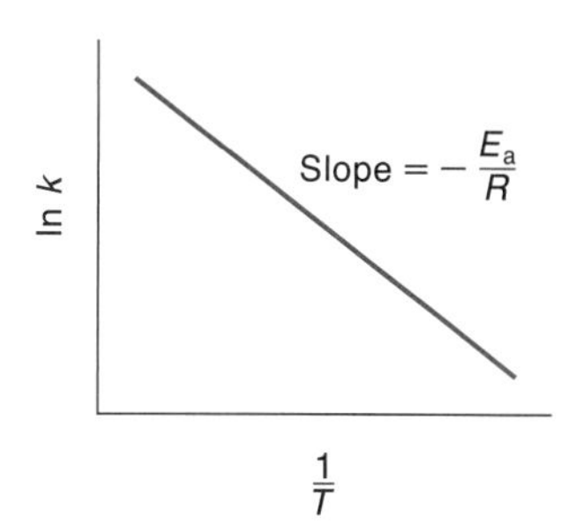

Figure 15.10
An Arrhenius plot. The slope of the straight line is equal to $-E_a/R$.

Equation 15.38 enables us to calculate the rate constant at a different temperature if E_a is known.

From the standpoint of Arrhenius's rate equation, a complete understanding of the factors determining the rate constant of a reaction requires that we be able to calculate the values of both A and E_a. Considerable effort has been devoted to this problem, as we shall see below.

15.6 Potential-Energy Surfaces

To discuss activation energy in more detail, we need to learn something about the energetics of a reaction. One of the simplest reactions is the combination of two atoms to form a diatomic molecule, such as $H + H \rightarrow H_2$. Basically, we would like to describe more complex reactions in terms of a potential-energy curve such as that shown in Figure 3.11. However, potential-energy diagrams are prohibitively complex for all but the simplest systems. One of the simplest and most studied systems is the exchange reaction between the hydrogen atom and the hydrogen molecule:

$$H + H_2 \rightarrow H_2 + H$$

Even for a three-atom system such as this, we need a four-dimensional plot, describing three bond lengths, or two bond lengths and a bond angle, versus energy. The problem is greatly simplified by assuming that the minimum energy configuration is linear so that only two bond lengths need to be specified. Consequently, only a

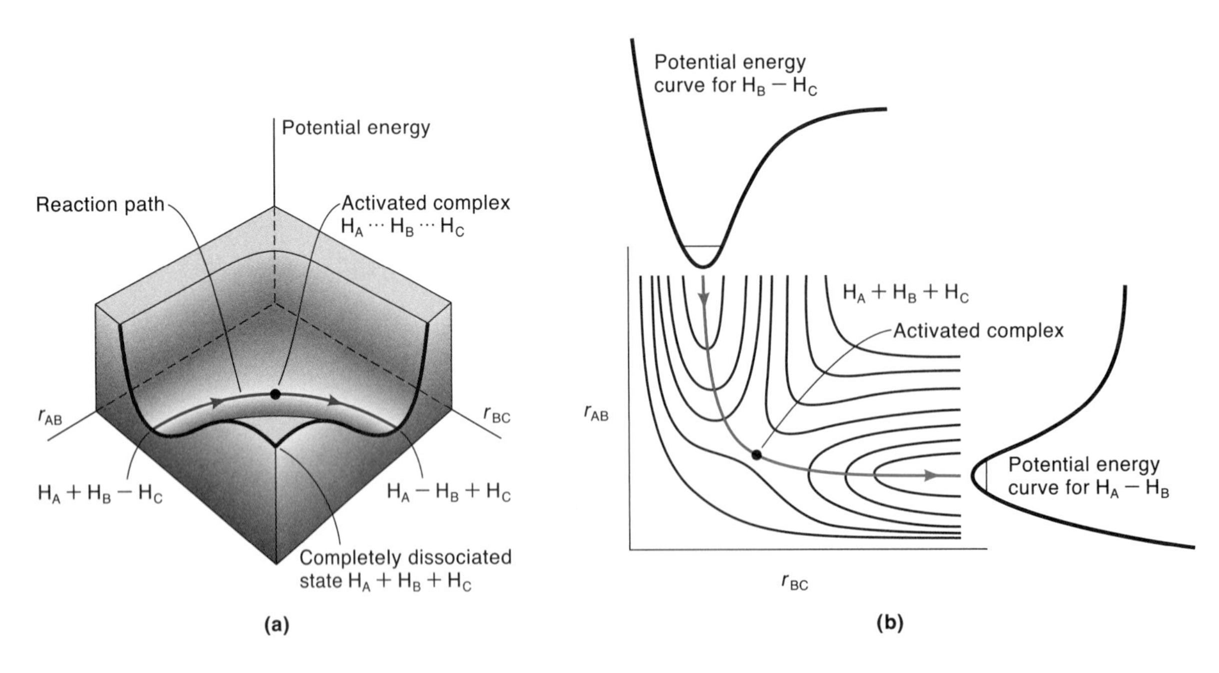

Figure 15.11
The $H + H_2 \rightarrow H_2 + H$ reaction. (a) Potential-energy surface. (b) Contour diagram of the potential-energy surface.

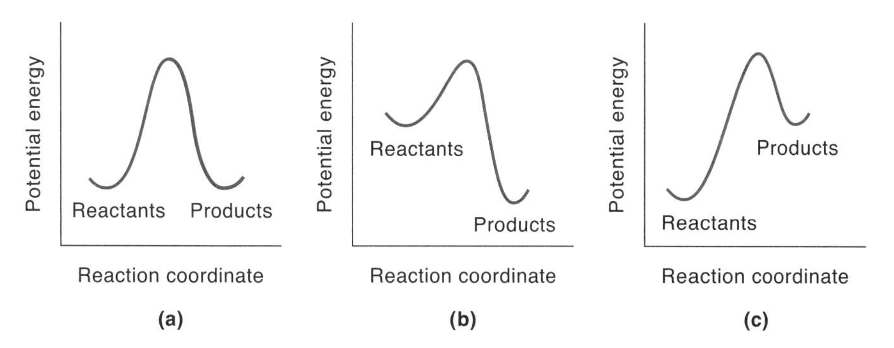

Figure 15.12
Potential-energy profile along the minimum energy path for (a) the $H + H_2$ reaction, (b) an exothermic reaction, and (c) an endothermic reaction.

three-dimensional plot is required (Figure 15.11). Labeling the atoms A, B, and C, we can represent the reaction as

$$H_A + H_B{-}H_C \rightarrow \underset{\text{activated complex}}{[H_A \cdots H_B \cdots H_C]} \rightarrow H_A{-}H_B + H_C$$

The plot, called the *potential-energy surface*, is a map of potential energies corresponding to different values of r_{AB} and r_{BC}, which are separations between the atoms. Although the reaction can proceed along any path, the one that requires the minimum amount of energy is shown by the red curve. The system travels along this path through the first valley and over the saddle point, which is the location of the activated complex, and then moves down the second valley. We represent this path in a plot of the potential energy versus the reaction coordinate, which describes the positions of the atoms during the course of a reaction. Figure 15.12a shows the plot for the $H + H_2$ reaction. The plots shown in Figures 15.12b and 15.12c are customarily employed for reactions in general, where the products differ from the reactants. You need to understand that these plots provide only a qualitative description for the reaction path because of the complexities involved for large molecules.

Much effort has gone into the calculation of the activation energy for the $H + H_2$ reaction. The correspondence between the calculated and measured values of E_a (36.8 kJ mol^{-1}) lends support to the validity of the model (i.e., linear activated complex).* An interesting side note is that if the reaction took the path involving the dissociation of the H_2 molecule followed by recombination, an activation energy of 432 kJ mol^{-1} would be required.

15.7 Theories of Reaction Rates

At this point, we are ready to consider two important theories of reaction rates: collision theory and transition-state theory. These theories provide us with greater insight into the energetic and mechanistic aspects of reactions.

* A theoretical analysis shows that the activation energy for this reaction is 40.2 kJ mol^{-1}. The calculation for this simple reaction required 80 days of computer time! See D. D. Diedrich and J. B. Anderson, *Science* **258**, 786 (1992).

Collision Theory

The collision theory of reaction rates is based on the kinetic theory of gases, discussed in Chapter 2. In its simplest form, it applies only to bimolecular reactions in the gas phase. Consider the bimolecular elementary reaction

$$\mathrm{A} + \mathrm{A} \rightarrow \text{product}$$

From Equation 2.17, the number of binary collisions per cubic meter per second between "hard-sphere" A molecules is given by

$$Z_{\mathrm{AA}} = \frac{\sqrt{2}}{2}\pi d^2 \bar{c}\left(\frac{N_{\mathrm{A}}}{V}\right)^2$$

According to Equation 2.14, the mean speed is

$$\bar{c} = \sqrt{\frac{8k_{\mathrm{B}}T}{\pi m}}$$

so that

$$Z_{\mathrm{AA}} = 2\left(\frac{N_{\mathrm{A}}}{V}\right)^2 d^2 \sqrt{\frac{\pi k_{\mathrm{B}}T}{m_{\mathrm{A}}}} \tag{15.39}$$

For a bimolecular reaction of the type

$$\mathrm{A} + \mathrm{B} \rightarrow \text{product}$$

the binary collision number is

$$Z_{\mathrm{AB}} = \left(\frac{N_{\mathrm{A}}}{V}\right)\left(\frac{N_{\mathrm{B}}}{V}\right) d_{\mathrm{AB}}^2 \sqrt{\frac{8\pi k_{\mathrm{B}}T}{\mu}} \tag{15.40}$$

where d_{AB} is the collision diameter between A and B, and μ, the *reduced mass*, is given by

$$\mu = \frac{m_{\mathrm{A}} m_{\mathrm{B}}}{m_{\mathrm{A}} + m_{\mathrm{B}}} \tag{15.41}$$

Now, if the collisions were 100% effective, that is, a product formed as a result of every binary collision, the rate of the reaction would be equal to either Z_{AA} or Z_{AB}. But this is not the case. In a gas at a pressure of 1 atm, the collision number is about $10^{31}\ \mathrm{L^{-1}\ s^{-1}}$ at 298 K. If every collision led to the formation of a product, all gas-phase reactions would be complete in about 10^{-9} s, which is contrary to our experience. The additional factor needed for Equations 15.39 and 15.40 is the term that contains the activation energy. For the A + B → product reaction, we write

$$\text{rate} = Z_{AB}e^{-E_a/RT}$$

$$= \left(\frac{N_A}{V}\right)\left(\frac{N_B}{V}\right)d_{AB}^2\sqrt{\frac{8\pi k_B T}{\mu}}e^{-E_a/RT} \tag{15.42}$$

Division of the rate by $(N_A/V)(N_B/V)$ gives a rate constant in molecular units (SI units: m^3 molecule^{-1} s^{-1}),*

$$k = \frac{\text{rate}}{(N_A/V)(N_B/V)} = z_{AB}e^{-E_a/RT} \tag{15.43}$$

where

$$z_{AB} = d_{AB}^2\sqrt{\frac{8\pi k_B T}{\mu}}$$

Comparing Equation 15.36 with Equation 15.43, we get

$$A = z_{AB} = d_{AB}^2\sqrt{\frac{8\pi k_B T}{\mu}}$$

Thus, the frequency factor, A, is temperature dependent. In practice, we usually treat it as a temperature-independent quantity in the calculation of E_a values. Doing so does not introduce any serious error, however, because the exponential term [i.e., $\exp(-E_a/RT)$] depends much more on temperature than the square-root term does on temperature.

Collision theory (Equation 15.43) predicts the value of the rate constant fairly accurately for reactions that involve atomic species or simple molecules if the activation energy is known. Significant deviations are found, however, for reactions involving complex molecules. For these reactions, the rate constants tend to be smaller than that predicted by Equation 15.43, sometimes by a factor of 10^6 or more. The reason is that the simple kinetic theory counts every sufficiently energetic collision as an effective one. In reality, the molecules may not approach each other in the right way for a reaction to occur, even if plenty of energy is available. To correct for the discrepancies observed, we modify Equation 15.43 as follows:

$$k = Pze^{-E_a/RT} \tag{15.44}$$

We use z to represent bimolecular reactions in general.

where P, called the *probability*, or *steric, factor*, takes into account the fact that in a collision complex, molecules must be properly oriented to undergo the reaction. This modification is an improvement, but the evaluation of P is rather difficult. The comparison of Equation 15.44 with Equation 15.36 shows that $A = Pz$.

* Equation 15.43 can be expressed in molar units (M^{-1} s^{-1}) by multiplication by the factor [6.022×10^{23} molecules mol^{-1}/(10^{-3} m^3 L^{-1})].

Transition-State Theory

Although collision theory is intuitively appealing and does not involve complicated mathematics, it does suffer from some serious drawbacks. Because it is based on the kinetic theory of gases, it assumes the reacting species are hard spheres and totally ignores the structures of molecules. For this reason, it cannot satisfactorily account for the probability factor at the molecular level. Furthermore, without quantum mechanics, we cannot use collision theory to calculate activation energy. A different approach, called transition-state theory (also known as activated complex theory), was developed by the American chemist Henry Eyring (1901–1981) and others in the 1930s, to provide greater insight into the details of a reaction on the molecular scale. It also enables us to calculate the rate constant with considerable accuracy.

The starting point of transition-state theory is similar to collision theory. In a bimolecular collision, an activated complex (also called a transition-state complex) of relatively high energy is formed. Consider the elementary reaction,

$$\mathrm{A} + \mathrm{B} \rightleftharpoons \mathrm{X}^{\ddagger} \xrightarrow{k} \mathrm{C} + \mathrm{D}$$

where A and B are reactants and $\mathrm{X}^{\ddagger}$ represents an activated complex. A fundamental assumption of transition-state theory (and one that differentiates it from collision theory) is that the reactants are always in equilibrium with $\mathrm{X}^{\ddagger}$. The activated complex should not be thought of as a stable, isolatable intermediate because it is assumed to be always in the process of decomposing. It is, in fact, neither stable nor isolatable.* Thus, the equilibrium between the reactants and the activated complex is not the conventional type. Nevertheless, we can write the equilibrium constant as

$$K^{\ddagger} = \frac{[\mathrm{X}^{\ddagger}]}{[\mathrm{A}][\mathrm{B}]} \tag{15.45}$$

The rate of the reaction is equal to the concentration of the activated complex at the top of the energy barrier, multiplied by the frequency, v, of crossing the barrier. Hence,

$$\begin{aligned}\text{rate} &= \text{number of activated complexes decomposing to form products per second}\\ &= v[\mathrm{X}^{\ddagger}]\\ &= v[\mathrm{A}][\mathrm{B}]K^{\ddagger}\end{aligned}$$

Because the rate can also be written as

$$\text{rate} = k[\mathrm{A}][\mathrm{B}]$$

where k is the rate constant, it follows that

$$k = vK^{\ddagger}$$

where v is the frequency of vibration of the activated complex in the degree of freedom leading to the formation of products; it has the unit s^{-1}. Calculating the value of k now

* This statement is not universally true. Using fast lasers, chemists in recent years have obtained spectroscopic evidence of an activated complex. See Section 14.4.

depends on our ability to evaluate both v and $K^{\ddagger}$. Using statistical thermodynamics (see Chapter 20), we can show that $v = k_B T/h$,* where h is Planck's constant, so that

$$k = \frac{k_B T}{h} K^{\ddagger}(M^{1-m}) \tag{15.46}$$

Note that to make the units equal on both sides of Equation 15.46, we added the term (M^{1-m}), where M is molarity and m is the molecularity of the reaction. For a unimolecular reaction, $m = 1$ and $(M^{1-1}) = 1$, so the first-order rate constant, k, has the same units as $k_B T/h$. (At 298 K, $k_B T/h = 6.21 \times 10^{12}\ s^{-1}$.) For bimolecular reactions, $m = 2$, and the units on the right side are $M^{-1}\ s^{-1}$, which are consistent with those of a second-order rate constant. The equilibrium constant $K^{\ddagger}$ can also be calculated from fundamental physical properties, such as bond lengths, atomic masses, and vibrational frequencies of the reactants and the activated complex. This approach has also been called the *absolute rate theory* because it enables us to calculate the value of k from absolute, or fundamental, molecular properties.

Thermodynamic Formulation of Transition-State Theory

The rate constant expressed in Equation 15.46, can be related to the thermodynamic properties of a reaction. We write

$$\Delta G^{\circ\ddagger} = -RT \ln K^{\ddagger}$$

Hence,

$$K^{\ddagger} = e^{-\Delta G^{\circ\ddagger}/RT} \tag{15.47}$$

where $\Delta G^{\circ\ddagger}$, the standard molar Gibbs energy of activation (Figure 15.13), given by

$$\Delta G^{\circ\ddagger} = G^{\circ}(\text{activated complex}) - G^{\circ}(\text{reactants})$$

* When the thermal energy $(k_B T)$ is comparable to the vibrational energy (hv), the activated complex dissociates into products. At 298 K, $k_B T = 208\ cm^{-1}$.

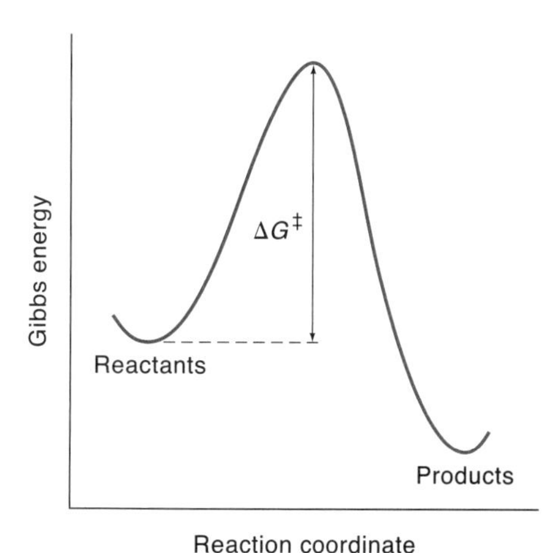

Figure 15.13
Definition of $\Delta G^{\ddagger}$ for a reaction.

The rate constant can be written as

$$k = \frac{k_B T}{h} e^{-\Delta G^{\circ\ddagger}/RT}(M^{1-m}) \tag{15.48}$$

Because $k_B T/h$ is independent of the nature of A and B, the rate of any reaction at a given temperature is determined by $\Delta G^{\circ\ddagger}$. Furthermore,

$$\Delta G^{\circ\ddagger} = \Delta H^{\circ\ddagger} - T\Delta S^{\circ\ddagger}$$

so Equation 15.48 becomes

$$k = \frac{k_B T}{h} e^{\Delta S^{\circ\ddagger}/R} e^{-\Delta H^{\circ\ddagger}/RT}(M^{1-m}) \tag{15.49}$$

where $\Delta S^{\circ\ddagger}$ and $\Delta H^{\circ\ddagger}$ are the standard molar entropy and standard molar enthalpy of activation, respectively. Equation 15.49 is the thermodynamic formulation of transition-state theory. A more rigorous approach includes a factor known as the transmission coefficient on the right side of Equation 15.49, but this factor is generally close to unity and may be ignored.

It is useful to compare the three expressions for rate constants discussed so far:

$$k = Ae^{-E_a/RT}$$

$$k = Pze^{-E_a/RT}$$

$$k = \frac{k_B T}{h} e^{\Delta S^{\circ\ddagger}/R} e^{-\Delta H^{\circ\ddagger}/RT}(M^{1-m})$$

The first equation (Equation 15.36) is an empirical one; both A and E_a must be determined experimentally. The second equation (Equation 15.44) is based in part on collision theory; the value of z can be calculated from the kinetic theory of gases. On the other hand, it is very difficult in general to estimate the magnitude of P accurately. The last equation (Equation 15.49) is based on transition-state theory. This equation provides us with the thermodynamic formulation of the reaction rate constant and is the most reliable of the three approaches. It is also the most difficult of the three equations to apply.

What is the significance of $\Delta S^{\circ\ddagger}$ and $\Delta H^{\circ\ddagger}$? Comparing Equation 15.49 with Equation 15.44 and assuming $\Delta H^{\circ\ddagger} = E_a$, we obtain

$$A = Pz = \frac{k_B T}{h} e^{\Delta S^{\circ\ddagger}/R} \tag{15.50}$$

This equation enables us to interpret the probability factor in terms of the standard molar entropy of activation. If reactants are atoms or simple molecules, then relatively little energy is redistributed among the various degrees of freedom in the activated complex. Consequently, $\Delta S^{\circ\ddagger}$ will be either a small positive or a small negative number, so that $\exp(\Delta S^{\circ\ddagger}/R)$—or P—is close to unity. But if complex molecules are involved in a reaction, $\Delta S^{\circ\ddagger}$ will be either a large positive or a large negative number. In the former case, the reaction will proceed much faster than predicted by collision theory; in the latter case, a much slower rate will be observed.

The standard molar enthalpy of activation, $\Delta H^{\circ\ddagger}$, is closely related to the ease of bond breaking and making in the generation of the activated complex. The lower the $\Delta H^{\circ\ddagger}$ value, the faster the rate. If we compare the coefficients of $1/T$ in Equations 15.49 and 15.44, we obtain $E_a = \Delta H^{\circ\ddagger}$. However, a more rigorous treatment (see Appendix 15.2 on p. 726) shows that

$$E_a = \Delta U^{\circ\ddagger} + RT \tag{15.51}$$

where $\Delta U^{\circ\ddagger}$ is the standard molar internal energy of activation. At constant pressure, we have

$$\Delta H^{\circ\ddagger} = \Delta U^{\circ\ddagger} + P\Delta V^{\circ\ddagger}$$

The quantity $\Delta V^{\circ\ddagger}$ is known as the standard molar volume of activation. Equation 15.51 can now be written as

$$E_a = \Delta H^{\circ\ddagger} - P\Delta V^{\circ\ddagger} + RT \tag{15.52}$$

For reactions occurring in solution, the term $P\Delta V^{\circ\ddagger}$ is quite small compared with $\Delta H^{\circ\ddagger}$ and can usually be neglected, so $\Delta H^{\circ\ddagger} \approx E_a - RT$ and Equation 15.49 can be written as

$$k = \frac{k_B T}{h} e^{\Delta S^{\circ\ddagger}/R} e^{-(E_a - RT)/RT} (M^{1-m})$$

$$= e\frac{k_B T}{h} e^{\Delta S^{\circ\ddagger}/R} e^{-E_a/RT} (M^{1-m}) \quad \text{in solution} \tag{15.53}$$

For reactions in the gas phase, we use the relationship $P\Delta V^{\circ\ddagger} = \Delta n^{\ddagger}RT$ in Equation 15.52 to get

$$E_a = \Delta H^{\circ\ddagger} - \Delta n^{\ddagger}RT + RT \tag{15.54}$$

For a unimolecular reaction, $\Delta n^{\ddagger} = 0$, and Equation 15.49 becomes

$$k = e\frac{k_B T}{h} e^{\Delta S^{\circ\ddagger}/R} e^{-E_a/RT} (M^{1-m}) \quad \text{unimolecular, gas phase} \tag{15.55}$$

which is the same as Equation 15.53. For a bimolecular reaction, $\Delta n^{\ddagger} = -1$, so $E_a = \Delta H^{\circ\ddagger} + 2RT$ and Equation 15.49 takes the form

$$k = e^2\frac{k_B T}{h} e^{\Delta S^{\circ\ddagger}/R} e^{-E_a/RT} (M^{1-m}) \quad \text{bimolecular, gas phase} \tag{15.56}$$

EXAMPLE 15.2

The pre-exponential factor and activation energy for the unimolecular reaction

$$CH_3NC(g) \rightarrow CH_3CN(g)$$

are 4.0×10^{13} s^{-1} and 272 kJ mol^{-1}, respectively. Calculate the values of $\Delta H^{\circ\ddagger}$, $\Delta S^{\circ\ddagger}$, and $\Delta G^{\circ\ddagger}$ at 300 K.

ANSWER

Equating the pre-exponential factor in Equation 15.55 to the experimental data, we have

$$e\frac{k_B T}{h} e^{\Delta S^{\circ\ddagger}/R} = 4.0 \times 10^{13}\ \text{s}^{-1}$$

so

$$e^{\Delta S^{\circ\ddagger}/R} = \frac{(4.0 \times 10^{13}\ \text{s}^{-1})h}{ek_B T}$$

$$= \frac{(4.0 \times 10^{13}\ \text{s}^{-1})(6.626 \times 10^{-34}\ \text{J s})}{(2.718)(1.381 \times 10^{-23}\ \text{J K}^{-1})(300\ \text{K})} = 2.354$$

$$\Delta S^{\circ\ddagger} = 7.1\ \text{J K}^{-1}\ \text{mol}^{-1}$$

From Equation 15.54, noting that $\Delta n^{\ddagger} = 0$, we write

$$\Delta H^{\circ\ddagger} = E_a - RT$$

$$= 272\ \text{kJ mol}^{-1} - \left[\frac{8.314}{1000}\ \text{kJ K}^{-1}\ \text{mol}^{-1}(300\ \text{K})\right]$$

$$= 270\ \text{kJ mol}^{-1}$$

Finally,

$$\Delta G^{\circ\ddagger} = \Delta H^{\circ\ddagger} - T\Delta S^{\circ\ddagger}$$

$$= 270\ \text{kJ mol}^{-1} - (300\ \text{K})\left(\frac{7.1}{1000}\ \text{kJ K}^{-1}\ \text{mol}^{-1}\right)$$

$$= 268\ \text{kJ mol}^{-1}$$

COMMENT

For unimolecular reactions, $\Delta S^{\circ\ddagger}$ is a small positive or negative quantity, so this is largely an enthalpy-driven process. ($\Delta S^{\circ\ddagger}$ will be a negative quantity for bimolecular gas-phase reactions because two molecules combine to form a single entity—the activated complex.) In general, regardless of the molecularity of the reactions, $\Delta H^{\circ\ddagger}$ is approximately equal to E_a.

15.8 Isotope Effects in Chemical Reactions

When an atom in a reactant molecule is replaced by one of its isotopes, both the equilibrium constant of the reaction and the rate constants may change. The term *equilibrium isotope effects* refers to changes in equilibrium constants that result from isotope substitution. Rate variations caused by this exchange are known as *kinetic isotope effects*. The study of isotope effects provides information on reaction mechanisms that has applications in many branches of chemistry. The underlying theory is complex, requiring both quantum mechanics and statistical mechanics; therefore, only a qualitative description will be given here.

Isotopic replacement in a molecule does not result in a change in the electronic structure of the molecule or in the potential-energy surface for any reaction the molecule might undergo, yet the rate of the reaction can be profoundly affected by the substitution. To see why, let us consider the H_2, HD, and D_2 molecules, whose zero-point energies, that is, the ground-state vibrational energies, are 26.5 kJ mol^{-1}, 21.6 kJ mol^{-1}, and 17.9 kJ mol^{-1}, respectively.* Because D_2 has the lowest zero-point energy (due to the fact that it has the largest reduced mass), more energy is required to dissociate this molecule than to break apart H_2 or HD (Figure 15.14a). Consequently, the reaction rate for $D_2 \rightarrow 2D$ will be the slowest compared with the other two corresponding dissociations. As a rough estimate, we can calculate the ratio of the rate constants for the dissociation of H_2 and D_2, k_H/k_D, as follows. According to Figure 15.14b, the activation energies for these two processes, E_H and E_D, are given by

$$E_H = E_{stretch} - E_H^0$$

$$E_D = E_{stretch} - E_D^0$$

* The zero-point energy is given by $E_{vib} = \frac{1}{2}h\nu$, where ν is the fundamental frequency of vibration. This frequency is given by $\nu = (1/2\pi)\sqrt{k/\mu}$, where k is the force constant of the bond and μ the reduced mass, given by $m_1m_2/(m_1 + m_2)$. Because D_2 has the largest value of μ, it possesses the smallest frequency of vibration and hence of E_{vib}. The reverse holds true for H_2.

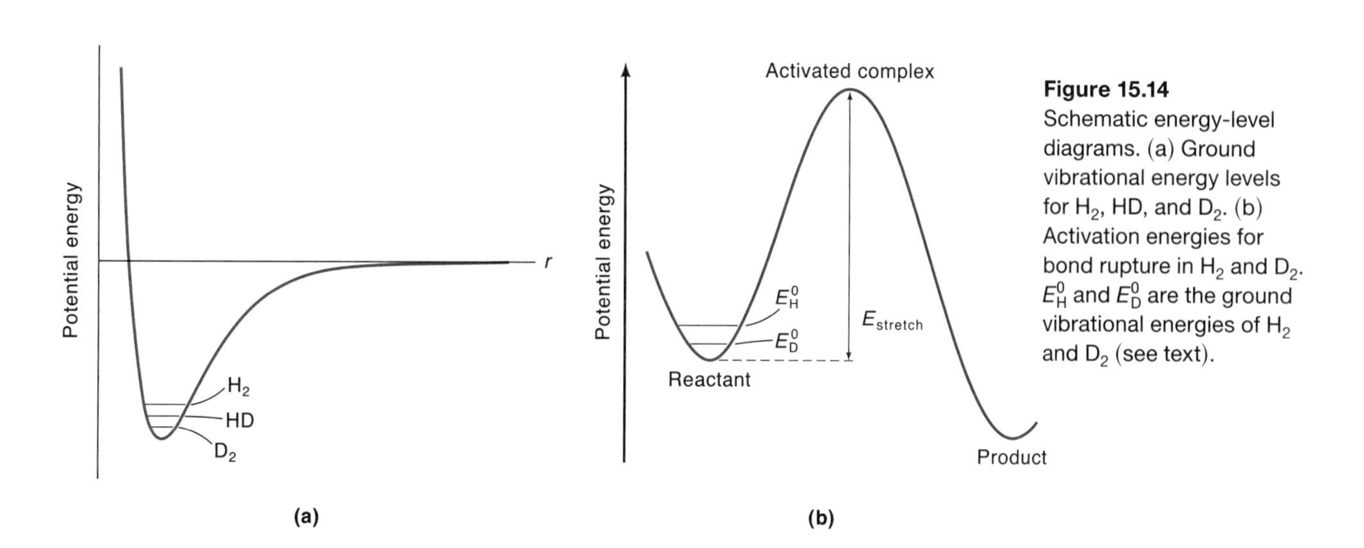

Figure 15.14 Schematic energy-level diagrams. (a) Ground vibrational energy levels for H_2, HD, and D_2. (b) Activation energies for bond rupture in H_2 and D_2. E_H^0 and E_D^0 are the ground vibrational energies of H_2 and D_2 (see text).

where E_H^0 and E_D^0 are the zero-point energies and $E_{stretch}$ is the difference between the lowest possible potential energy and the potential energy of the activated complex. Using the Arrhenius expression (Equation 15.36), we write

$$\frac{k_H}{k_D} = \frac{Ae^{-(E_{stretch} - E_H^0)/RT}}{Ae^{-(E_{stretch} - E_D^0)/RT}}$$

$$= e^{(E_H^0 - E_D^0)/RT}$$

$$= e^{(26.5 - 17.9) \times 1000 \text{ J mol}^{-1}/(8.314 \text{ J K}^{-1} \text{ mol}^{-1})(300 \text{ K})}$$

$$\approx 31$$

which is quite a large number.

The example above dramatizes the difference between the rate constants of D_2 and H_2. In practice, however, we are concerned more with the breaking of a bond between hydrogen and some other atom, such as carbon. Consider, for example, the reactions,

$$\geq\text{C—H} + \text{B} \longrightarrow \geq\text{C}\cdot + \text{H—B}$$

$$\geq\text{C—D} + \text{B} \longrightarrow \geq\text{C}\cdot + \text{D—B}$$

where B is some group that can take up a hydrogen atom. Again, we would predict a kinetic isotope effect, because the fundamental frequencies of vibration are different for the C—H and C—D bonds. But the difference is not as great as that between H_2 and D_2 because the reduced masses are closer to each other in this case (see Problem 15.39). Still, the ratio k_{C-H}/k_{C-D} might be appreciable, on the order of 5 or so.

We assume that this bond-breaking process is also the rate-determining step.

Isotope effects that reflect isotope substitution for an atom involved in a bond-breaking process are called primary kinetic isotope effects. Such effects are most pronounced for light elements such as H, D, and T. Reactions involving isotopes of mercury (^{199}Hg and ^{201}Hg), for example, would show hardly any detectable difference in rates. A secondary kinetic isotope effect occurs when the isotope is not directly involved in the bond rupture. We would expect a small change in the reaction rate in this case. Indeed, this prediction has been confirmed experimentally.

How does the isotope effect change an equilibrium process? Although the forward and reverse directions of an equilibrium process must trace the same reaction pathway, the isotope effect on the two rate constants need not be the same. Consequently, there can be an isotope effect on the equilibrium constant. As a simple example, let us consider the dissociation of a monoprotic acid, such as acetic acid, in H_2O and D_2O. The dissociations are

$$CH_3COOH \rightleftharpoons CH_3COO^- + H^+ \quad K_H = \frac{[H^+][CH_3COO^-]}{[CH_3COOH]}$$

$$CH_3COOD \rightleftharpoons CH_3COO^- + D^+ \quad K_D = \frac{[D^+][CH_3COO^-]}{[CH_3COOD]}$$

(In D_2O, all the ionizable protons are replaced by deuterons.) Experimentally, we find that $K_H/K_D = 3.3$. The greater acid strength of CH_3COOH over CH_3COOD can be explained by noting that the undeuterated molecule has a higher zero-point vibrational level (for the O−H bond), and less energy is required to dissociate the hydrogen than to dissociate the deuterium in CH_3COOD. A useful general rule with regard to the isotope effect on equilibria is that substitution with a heavier isotope will favor the formation of a stronger bond. As for acetic acid, when we replace H with D, the O−D bond becomes stronger, and the resulting molecule has less of a tendency to dissociate.

15.9 Reactions in Solution

The major difference between gas-phase reactions and reactions in solution lies in the role of the solvent. In many cases, the solvent plays a minor role, and rates do not differ much in the two phases. In terms of simple kinetic theory, the frequency of collisions between reacting molecules depends only on the concentrations of the reactants; it is not affected by solvent molecules. There is a difference, however, in the outcome of an encounter between reactant molecules in solution compared with the collision of molecules in the gas phase. If two molecules collide in the gas phase and do not react, they will normally move away from each other. There is very little likelihood that this same pair will collide again. In contrast, when two solute molecules diffuse together in a solution, they cannot move apart again quickly after the initial encounter because they are surrounded closely by solvent molecules. In this case, the reactants are temporarily trapped in a "cage" of solvent (Figure 15.15). To be sure, the cage is not rigid, as the solvent molecules are constantly in motion and changing positions. Nevertheless, the cage effect causes the reactant molecules to remain together for a longer time than they would in the gas phase, and they may collide with each other hundreds of times before they drift apart.* For reactions that have relatively low activation energies, the cage effect virtually ensures reaction during each encounter; the steric factor no longer plays an important role, because the reacting molecules would sooner or later become properly oriented for reaction during the collisions. Under

* We speak of molecular collisions in the gas phase and molecular encounters in solution. The difference is that after each encounter in solution, the molecules might collide many times before they move away from each other.

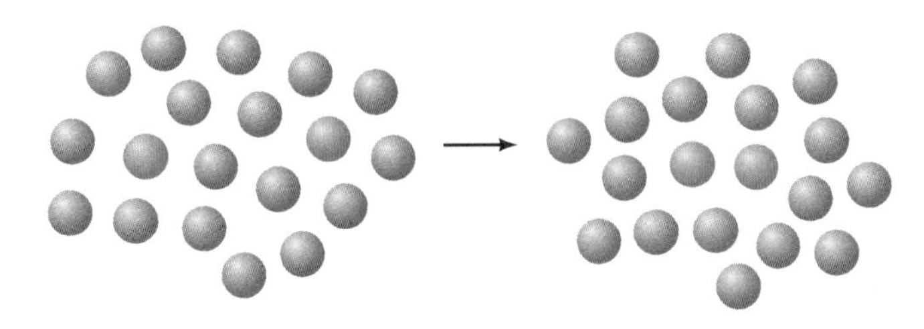

Figure 15.15
Solute molecules (red spheres) diffuse into a solvent "cage" and encounter each other. Hundreds of collisions between solute molecules occur before the cage is destroyed.

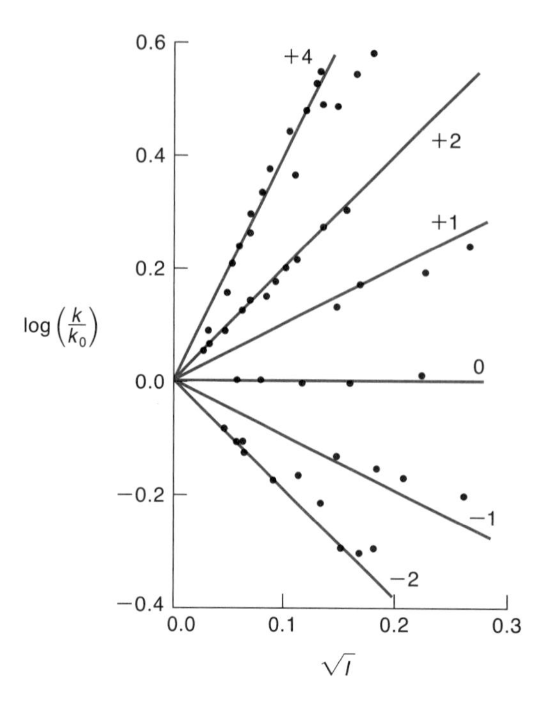

Figure 15.16
Effect of ionic strength on the rate of reaction between two ions. The reactions are:

+4: $Co(NH_3)_5Br^{2+} + Hg^{2+}$;
+2: $S_2O_8^{2-} + I^-$;
+1: $[NO_2NCO_2C_2H_5]^- + OH^-$;
0: $CH_3CO_2C_2H_5 + OH^-$;
−1: $H_2O_2 + H^+ + Br^-$;
−2: $Co(NH_3)_5Br^{2+} + OH^-$.

The slopes are determined by z_Az_B. [From V. K. LaMer, *Chem. Rev.* **10**, 179 (1932). Used with permission of Williams and Wilkins, Baltimore.]

these conditions, the rate of the reaction is limited only by how fast the reactants can diffuse together. We shall return to this type of reaction in the next section.

The situation is quite different if the reactants are charged species. The solvation of ions can be an appreciable factor in determining the sign and magnitude of $\Delta S^\ddagger$. In cases involving charged species, the value of $\Delta S^\ddagger$ depends on the relative net charge of the activated complex. If the activated complex has a greater charge than the reactants, we would expect $\Delta S^\ddagger$ to be negative because of the increase in solvation around the complex:

$$(C_2H_5)_3N + C_2H_5I \rightarrow (C_2H_5)_4N^+I^- \qquad \Delta S^\ddagger = -172 \text{ J K}^{-1} \text{ mol}^{-1}$$

On the other hand, if an activated complex carries a lower net charge than the reactants, we would predict a positive $\Delta S^\ddagger$ value:

$$Co(NH_3)_5Br^{2+} + OH^- \rightarrow Co(NH_3)_4Br(OH)^+ + NH_3 \qquad \Delta S^\ddagger = 83.7 \text{ J K}^{-1} \text{ mol}^{-1}$$

Steric and other factors also contribute to the large entropy changes.

As we would expect, the rate of a reaction involving ions depends strongly on the ionic strength of the solution. This dependence is known as the *kinetic salt effect*. The effect of ionic strength on the rate constant is given by*

$$\log \frac{k}{k_0} = z_A z_B B \sqrt{I} \qquad (15.57)$$

where B is a constant that depends only on the temperature and the nature of the solvent; k and k_0 are the rate constants at ionic strength I (of an inert salt) and at infinitely dilute

* Equation 15.57 is derived from the Debye-Hückel limiting law (see Section 7.5) and therefore applies only to dilute solutions.

concentration ($I = 0$), respectively; and z_A and z_B are the ionic charges of the reactants A and B. Equation 15.57 predicts that (1) if A and B have the same charges, $z_A z_B$ is positive and the rate constant k increases with $\sqrt{I}$; (2) if A and B have opposite signs, $z_A z_B$ is negative and k decreases with $\sqrt{I}$; and (3) if either A or B is uncharged, $z_A z_B = 0$ and k is independent of the ionic strength of the solution. Figure 15.16 confirms these predictions.

15.10 Fast Reactions in Solution

Roughly speaking, first- and second-order reactions that have rate constants between 10 and 10^9 can be described as fast reactions. Examples of fast reactions include recombination of reactive species in the gas phase and in solution, acid-base neutralization, and electron- and proton-exchange reactions. These reactions have engendered considerable interest because of their importance to chemistry and biology and because of a widespread desire to design experiments to measure processes whose half-lives are seconds or shorter.

How fast can a reaction occur in solution? The limit is set by the rate of approach by reacting molecules, which in turn is governed by the rate of diffusion. Thus, the fastest reaction is a *diffusion-controlled reaction* in which a reaction occurs with every encounter of reactant molecules. Suppose that we have a solution of two uncharged reactant molecules, B and C, with radii r_B and r_C The Polish physicist Marion Smoluchowski (1872–1917) showed that the rate constant k_D of the elementary diffusion-controlled reaction B + C → product is given by

$$k_D = 4\pi N_A(r_B + r_C)(D_B + D_C)$$

The equation for diffusion coefficient is derived in Section 19.4.

where N_A is Avogardro's constant and D_B and D_C are the diffusion coefficients.

If we assume that $D_B = D_C = D$, $r_B = r_C = r$, and use the expression $D = k_B T/6\pi\eta r$, where η is the viscosity of the solution, then

$$\begin{aligned} k_D &= 4\pi N_A(2D)(2r) \\ &= \frac{16\pi N_A k_B T r}{6\pi\eta r} = \frac{8}{3}\frac{RT}{\eta} \end{aligned} \qquad (15.58)$$

A truly diffusion-controlled reaction has two characteristics. First, such a reaction has zero activation energy [note the absence of the $\exp(-E_a/RT)$ term in Equation 15.58]. Second, the rate is inversely proportional to the viscosity of the medium. The dependence on viscosity is interesting in that η itself depends on temperature as follows:

$$\eta = Be^{E_a/RT}$$

where E_a is the "activation energy" of viscosity (note that η decreases with increasing temperature) and B is a constant characteristic of the solvent. Thus, Equation 15.58 can now be written as

$$k_D = \frac{8RT}{3B}\, e^{-E_a/RT} \qquad (15.59)$$

Equation 15.59 has the form of an Arrhenius equation.

EXAMPLE 15.3

Estimate the rate constant for a diffusion-controlled reaction in water at 298 K, given that the viscosity of water is 8.9×10^{-4} N s m^{-2}.

ANSWER

Because 1 J = 1 N m, the units of viscosity can also be expressed as J s m^{-3}. From Equation 15.58,

$$k_D = \frac{8(8.314 \text{ J K}^{-1} \text{ mol}^{-1})(298 \text{ K})}{3(8.9 \times 10^{-4} \text{ J s m}^{-3})}$$

$$= 7.4 \times 10^6 \text{ m}^3 \text{ mol}^{-1} \text{ s}^{-1}$$

$$= 7.4 \times 10^9 \, M^{-1} \text{ s}^{-1}$$

COMMENT

According to Table 15.1, the half-life for a diffusion-controlled process, assuming that the starting reactants are identical and their concentrations equal 1 *M*, is

$$t_{1/2} = \frac{1}{1 \, M \times 7.4 \times 10^9 \, M^{-1} \text{ s}^{-1}} = 1.4 \times 10^{-10} \text{ s} = 0.14 \text{ ps}$$

which is a very small number.

Many ingenious methods have been devised to study fast reactions. Two examples will be briefly discussed.

The Flow Method

There are two types of flow apparatus. In the continuous-flow apparatus, two reactant solutions are first brought together in a mixing chamber, and the mixed solution is then passed along an observation tube. By monitoring the concentration of either the reactant or product at different points along the tube spectrophotometrically (i.e., by measuring the absorption of light by either the reactant or the product), we can plot the extent of reaction versus time (Figure 15.17). The limiting factor here is the time required for mixing, which can be as short as 0.001 s. This technique uses large quantities of solutions for every run, a major disadvantage.

Figure 15.18 shows a stopped-flow apparatus. The advantage of the stopped-flow technique is that only small samples of reactants are needed; therefore, it is particularly suited for biochemical processes such as enzyme-catalyzed reactions.

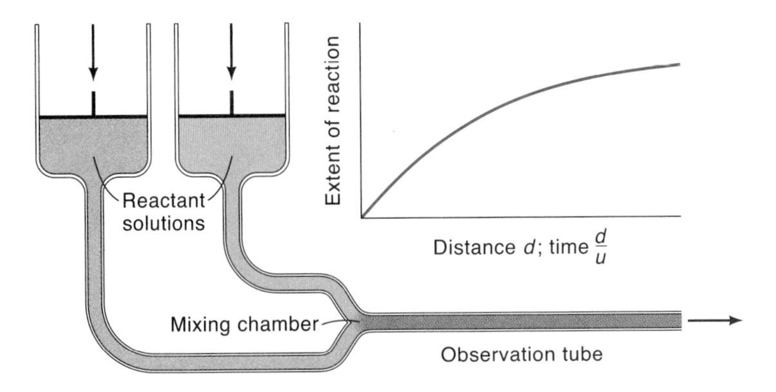

Figure 15.17
Schematic diagram for a continuous-flow experiment. The velocity of the mixed solution is u. Measurements are carried out along the length of the tube. (From E. F. Caldin, *Fast Reactions in Solution*, John Wiley & Sons, New York, 1964. Used by permission of Blackwell Scientific Publications, Oxford, England.)

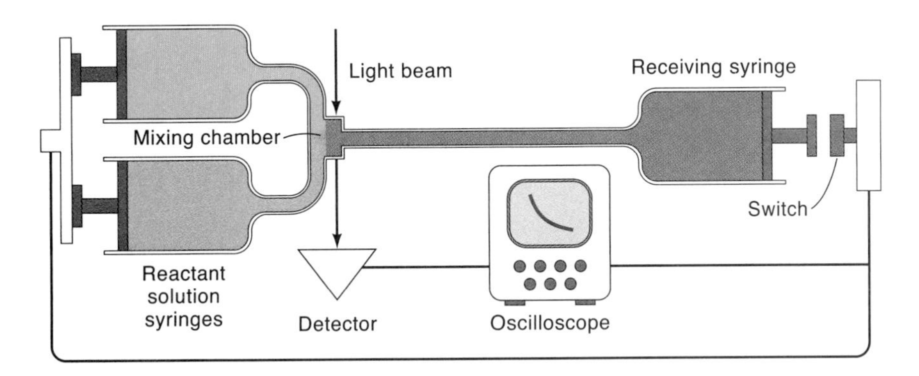

Figure 15.18
Schematic diagram for a stopped-flow experiment. As in the continuous-flow apparatus, two solutions containing different reactants are injected into the mixing chamber, usually with mechanically coupled hypodermic syringes. Another syringe on the right receives the effluent solution and is arranged so that when its plunger strikes a barrier, the flow is abruptly stopped. At the same instant, an oscilloscope is triggered to establish the time origin. The time scale is provided by the sweep frequency of the oscilloscope, which displays a plot of the transmitted light intensity against time. In this arrangement, the distance between the mixing region and observation point (i.e., the point at which absorption of light is monitored) is small.

The Relaxation Method

A system initially at equilibrium is subjected to an external perturbation, such as a temperature or pressure change. If the change is applied suddenly, there will be a time lag while the system approaches (or relaxes toward) a new equilibrium. This time lag, called the *relaxation time*, can be related to the forward and reverse rate constant. Depending on the systems, reactions with half-lives between 1 s and 10^{-10} s can be studied with relaxation techniques.

Any property (X) that varies linearly with the extent of the reaction (e.g., electrical conductance or spectroscopic absorption) is measured as a function of time following the disruption,

$$X_t = X_0 e^{-t/\tau}$$

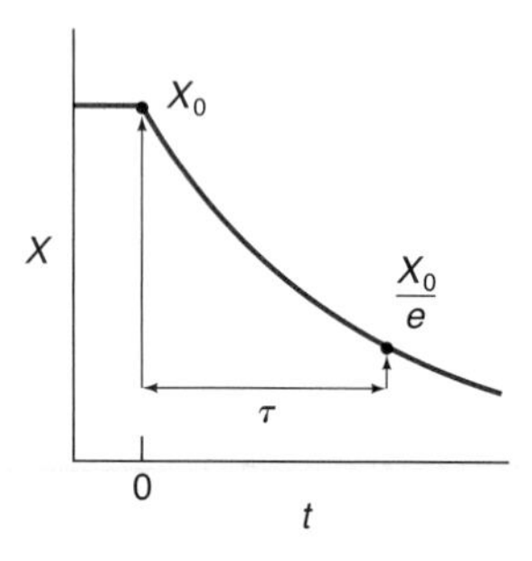

Figure 15.19
Definition of relaxation time, τ.

where X_t and X_0 are the values of the property at time $t = t$ and $t = 0$, respectively, and τ is the relaxation time. When $\tau = t$,

$$X_t = \frac{X_0}{e} = \frac{X_0}{2.718}$$

Thus, by measuring the time it takes for X_0 to decrease to $X_0/2.718$, we can determine the relaxation time (Figure 15.19).

We can derive a relationship between τ and the rate constant using an equilibrium chemical system in solution at some temperature:

$$\mathrm{A + B} \underset{k_\mathrm{r}'}{\overset{k_\mathrm{f}'}{\rightleftharpoons}} \mathrm{C}$$

At equilibrium, we have

$$\frac{d[\mathrm{C}]}{dt} = k_\mathrm{f}'[\mathrm{A}][\mathrm{B}] - k_\mathrm{r}'[\mathrm{C}] = 0$$

where k_f' and k_r' are the rate constants for the forward and reverse steps. In a temperature-jump experiment, the temperature of the solution can be increased by 5 K or so in a time as short as 10^{-6} s, either by discharging a capacitor through the solution or by applying a short, powerful laser pulse. After the temperature jump has occurred, the concentrations of A, B, and C will change as the system "relaxes" toward the new equilibrium at the elevated temperature, with altered rate constants k_f and k_r. Let x be the time-dependent reaction progress variable expressed as a concentration. According to the reaction stoichiometry, at any time t following the temperature jump we have

$$[\mathrm{A}] = [\mathrm{A}]_\mathrm{eq} + x$$

$$[\mathrm{B}] = [\mathrm{B}]_\mathrm{eq} + x$$

$$[\mathrm{C}] = [\mathrm{C}]_\mathrm{eq} - x$$

where the subscript eq denotes the new equilibrium concentration. Note that x could be positive or negative, depending on the direction in which the equilibrium is shifted. The rate of change of [C] is now given by

$$\frac{d[\mathrm{C}]}{dt} = \frac{d([\mathrm{C}]_\mathrm{eq} - x)}{dt} = -\frac{dx}{dt} = k_\mathrm{f}([\mathrm{A}]_\mathrm{eq} + x)([\mathrm{B}]_\mathrm{eq} + x) - k_\mathrm{r}([\mathrm{C}]_\mathrm{eq} - x)$$

$$= \underbrace{k_\mathrm{f}[\mathrm{A}]_\mathrm{eq}[\mathrm{B}]_\mathrm{eq} - k_\mathrm{r}[\mathrm{C}]_\mathrm{eq}}_{\text{first term}} + \underbrace{(k_\mathrm{f}[\mathrm{A}]_\mathrm{eq} + k_\mathrm{f}[\mathrm{B}]_\mathrm{eq} + k_\mathrm{r})x}_{\text{second term}} + \underbrace{k_\mathrm{f}x^2}_{\text{third term}}$$

The first term is equal to zero because at equilibrium the forward and reverse rates are equal; the third term can be neglected because x is a small number (due to a small rise in temperature), so $k_f x^2 \ll k_f[A]_{eq}x$ (or $k_f[B]_{eq}x$). Therefore,

$$\frac{dx}{dt} = -(k_f[A]_{eq} + k_f[B]_{eq} + k_r)x = -\frac{x}{\tau}$$

where τ is given by

$$\tau = \frac{1}{k_f([A]_{eq} + [B]_{eq}) + k_r} \tag{15.60}$$

Because $[A]_{eq}$ and $[B]_{eq}$ can be determined in separate experiments, τ can be measured by monitoring the change in concentrations during the time subsequent to the temperature jump (see Figure 15.19). Coupled with the equilibrium constant ($K = k_f/k_r$) of the reaction at the elevated temperature, these values enable us to determine the values of both k_f and k_r (two equations for two unknowns).

EXAMPLE 15.4

A sample of pure water was subjected to a temperature jump. The relaxation time for the system (water) to reach the new equilibrium at 25°C was found to be 36 μs (36×10^{-6} s). Calculate the values of k_f and k_r for the following reaction:

$$H^+ + OH^- \underset{k_r}{\overset{k_f}{\rightleftharpoons}} H_2O$$

ANSWER

From Equation 15.60,

$$\tau = \frac{1}{k_f([H^+]_{eq} + [OH^-]_{eq}) + k_r}$$

The equilibrium condition is

$$k_f[H^+]_{eq}[OH^-]_{eq} = k_r[H_2O]_{eq}$$

Substituting $k_r = k_f[H^+]_{eq}[OH^-]_{eq}/[H_2O]_{eq}$ and using $[H^+]_{eq} = [OH^-]_{eq}$, we get

$$\tau = \frac{1}{k_f(2[H^+]_{eq}) + k_f[H^+]^2_{eq}/[H_2O]_{eq}}$$

$$= \frac{1}{k_f(2[H^+]_{eq} + [H^+]^2_{eq}/[H_2O]_{eq})}$$

Using $[H^+]_{eq} = 1.0 \times 10^{-7}\ M$ and $[H_2O]_{eq} = 55.5\ M$, we write

$$36 \times 10^{-6}\ s = \frac{1}{k_f[2(1.0 \times 10^{-7}\ M) + (1.0 \times 10^{-7}\ M)^2/55.5\ M]}$$

or

$$k_f = 1.4 \times 10^{11}\ M^{-1}\ s^{-1}$$

To calculate the value of k_r, we need to first evaluate the equilibrium constant K, given by

$$K = \frac{k_r}{k_f} = \frac{[H^+]_{eq}[OH^-]_{eq}}{[H_2O]_{eq}} = \frac{(1.0 \times 10^{-7}\ M)(1.0 \times 10^{-7}\ M)}{(55.5\ M)} = 1.8 \times 10^{-16}\ M$$

Therefore,

$$k_r = Kk_f = (1.8 \times 10^{-16}\ M)(1.4 \times 10^{11}\ M^{-1}\ s^{-1})$$

$$k_r = 2.5 \times 10^{-5}\ s^{-1}$$

COMMENTS

(1) Note that the equilibrium constant, K, is related to the ion-product (K_w) by $K = K_w/[H_2O]_{eq}$. Watch out for the units: k_f is a second-order rate constant ($M^{-1}\ s^{-1}$), whereas k_r is a first-order rate constant (s^{-1}). The very large value of k_f shows that the combination of H^+ and OH^- ions in solution is diffusion controlled. (2) K normally is treated as a dimensionless quantity, but for calculating rate constants, it is assigned the unit M in this case. See K. J. Laidler, *J. Chem. Educ.* **67**, 88 (1990).

Reactions are seldom as simple as the one described above. Often, a reaction may have several relaxation times and the analysis can be very complex. Nevertheless, relaxation methods are among the most useful and versatile techniques in the study of fast chemical and biochemical processes.

15.11 Oscillating Reactions

Chemical reactions normally proceed until the reactants have been exhausted or until an equilibrium state is reached. For certain complicated reactions, however, the concentrations of intermediates can oscillate. Although such *oscillating reactions* have been known since the late 19th century, for a long time they were dismissed by most chemists as being nonreproducible phenomena or artifacts due to impurities. According to the second law of thermodynamics, at constant temperature and pressure in a closed system, the Gibbs energy, G, of a reacting mixture must continually

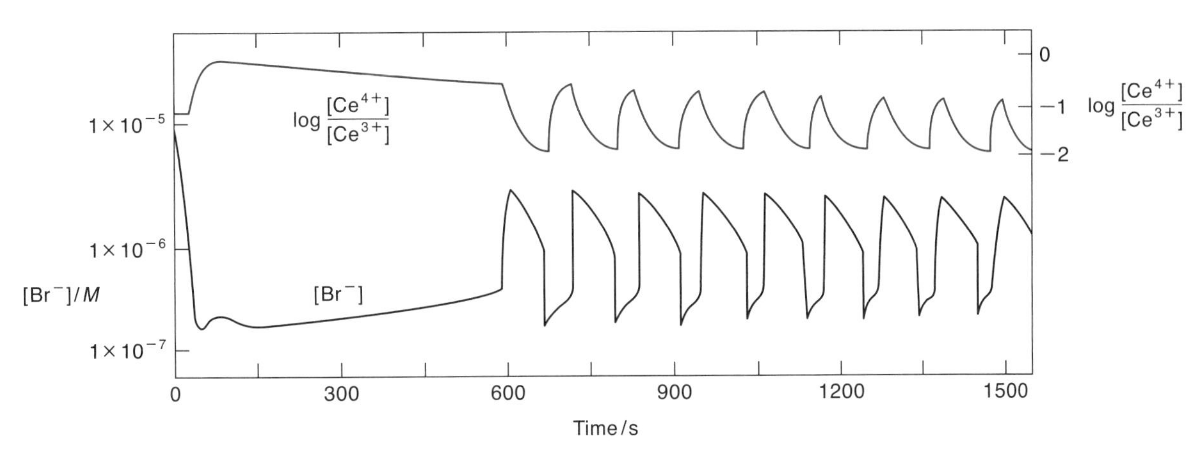

Figure 15.20
Oscillation in log [Br^-] and log ([Ce^{4+}]/[Ce^{3+}]) in the BZ reaction. Note that Br^-, Ce^{4+}, and Ce^{3+} are not reactants or products in the overall reaction. [Reprinted with permission from R. J. Field, E. Krönos, and R. M. Noyes, *J. Am. Chem. Soc.* **94**, 8649 (1972). Copyright 1972 American Chemical Society.]

decrease as the reaction approaches equilibrium. An oscillating reaction would seem to violate the second law.

The oscillating reaction that finally convinced the skeptics was discovered by the Russian chemist B. P. Belousov in 1958 and later studied in detail by the Russian chemist A. M. Zhabotinsky. The Belousov-Zhabotinsky reaction, or the BZ reaction as it is now commonly called, occurs when malonic acid [$CH_2(COOH)_2$] and sulfuric acid are dissolved in water with potassium bromate ($KBrO_3$) and a cerium salt (containing the ceric ion, Ce^{4+}). The overall reaction is represented by

$$2H^+ + 2BrO_3^- + 3CH_2(COOH)_2 \rightarrow 2BrCH(COOH)_2 + 3CO_2 + 4H_2O$$

The mechanism of this reaction has been investigated extensively over the past 30 years and is believed to involve 18 elementary steps and 20 different chemical species! During the reaction, the color of the solution changes periodically from pale yellow (Ce^{4+}) to colorless (Ce^{3+}). Figure 15.20 shows the oscillations in [Br^-] and [Ce^{4+}]/[Ce^{3+}].

The thermodynamic explanation of oscillating reactions comes from the Belgian chemist Ilya Prigogine (1917–2003). Reactions in a closed system cannot oscillate about their equilibrium state because such behavior is prohibited by the principle of microscopic reversibility (see p. 687). According to Prigogine, however, if a system is far from equilibrium, periodic oscillations in the concentrations of intermediate species in a chemical reaction can occur. Eventually, these oscillations die off as the system nears its equilibrium state. The initial reactants and final products cannot participate in oscillation because they are not intermediates. However, in open systems, where exchanges of both energy and mass with the surroundings are allowed, a steady state exists rather than an equilibrium state, so oscillations may occur and last indefinitely.

The study of oscillating reactions has added a new dimension to chemical kinetics and is one of the most rapidly growing branches of chemistry. It has yielded useful insight into chemical dynamics and reaction mechanisms. Such reactions may have considerable importance in biological systems. The periodic beating of the heart is one example. Oscillatory behavior has also been detected in glycolysis.* The atmosphere is another open system that shows periodic oscillations in the concentrations of various gas constituents (see Chapter 16).

15.12 Enzyme Kinetics

A catalyst is a substance that increases the rate of a reaction without itself being consumed by the process. A reaction in which a catalyst is involved is called a catalyzed reaction, and the process is called catalysis. In studying catalysis, keep in mind the following characteristics:

1. A catalyst lowers the Gibbs energy of activation by providing a different mechanism for the reaction (Figure 15.21). This mechanism enhances the rate and it applies to *both* the forward and the reverse directions of the reaction.
2. A catalyst forms an intermediate with the reactant(s) in the initial step of the mechanism and is released in the product-forming step. The catalyst does not appear in the overall reaction.
3. Regardless of the mechanism and the energetics of a reaction, a catalyst cannot affect the enthalpies or Gibbs energies of the reactants and products. Thus, catalysts increase the rate of approach to equilibrium, but cannot alter the thermodynamic equilibrium constant.

There are two types of catalysis: heterogeneous and homogeneous. In a heterogeneously catalyzed reaction, the reactants and the catalyst are in different phases

* A. Gosh and B. Chance, *Biochem. Biophys. Res. Commun.* **16**, 174 (1964).

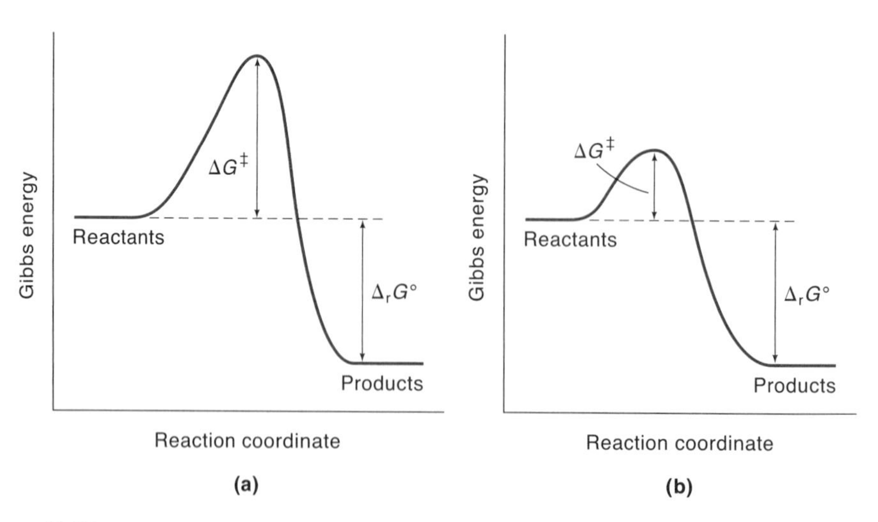

Figure 15.21
Gibbs energy change for (a) an uncatalyzed reaction and (b) a catalyzed reaction. A catalyzed reaction must involve the formation of at least one intermediate (between the reactant and the catalyst). The value of $\Delta_r G°$ is the same in both cases.

(usually gas/solid or liquid/solid). Well-known examples are the Haber synthesis of ammonia and the Ostwald manufacture of nitric acid. The bromination of acetone, catalyzed by acids,

$$CH_3COCH_3 + Br_2 \xrightarrow{H^+} CH_2BrCOCH_3 + HBr$$

is an example homogeneous catalysis because the reactants and the catalyst (H^+) are all present in the aqueous medium. Enzyme catalysis is usually homogeneous in nature.

Some enzymes are imbedded in cell membranes. In these cases, the reactions should be characterized as heterogeneous rather than homogeneous.

Enzyme Catalysis

Since 1926, when the American biochemist James Sumner (1887–1955) crystallized urease (an enzyme that catalyzes the cleavage of urea to ammonia and carbon dioxide), it has come to be known that most enzymes are proteins.* An enzyme usually contains one or more *active sites*, where reactions with substrates take place. An active site may comprise only a few amino acid residues; the rest of the protein is required for maintaining the three-dimensional integrity of the network. The specificity of enzymes for substrates varies from molecule to molecule. Many enzymes exhibit stereochemical specificity, that is, they catalyze the reactions of one configuration but not the other. For example, proteolytic enzymes catalyze only the hydrolysis of peptides made up of L-amino acids. Some enzymes are catalytically inactive in the absence of certain metal ions.

In the 1890s, the German chemist Emil Fischer (1852–1919) proposed a lock-and-key theory of enzyme specificity. According to Fischer, the active site can be assumed to have a rigid structure, similar to a lock. A substrate molecule then has a complementary structure and functions as a key. Although appealing in some respects, this theory has been modified to take into account the flexibility of proteins in solution and to explain the phenomenon of cooperativity. *Cooperativity* means that the binding of a substrate to an enzyme with multiple binding sites can alter the substrate's affinity for enzyme binding at its other sites.

Enzymes, like other catalysts, increase the rate of a reaction. An understanding of the efficiency of enzymes can be gained by examining Equation 15.49:

$$\begin{aligned} k &= \frac{k_B T}{h} e^{-\Delta G^{\circ\ddagger}/RT}(M^{1-m}) \\ &= \frac{k_B T}{h} e^{\Delta S^{\circ\ddagger}/R} e^{-\Delta H^{\circ\ddagger}/RT}(M^{1-m}) \end{aligned}$$

There are two contributions to the rate constant: $\Delta H^{\circ\ddagger}$ and $\Delta S^{\circ\ddagger}$. The enthalpy of activation is approximately equal to the energy of activation (E_a) in the Arrhenius equation (see Equation 15.36). Certainly a reduction in E_a by the action of a catalyst would enhance the rate constant. In fact, this is usually the explanation of how a catalyst works, but it is not always true for enzyme catalysis. Entropy of activation, $\Delta S^{\circ\ddagger}$, may also be an important factor in determining the efficiency of enzyme catalysis.

* In the early 1980s, chemists discovered that certain RNA molecules, called ribozymes, also possess catalytic properties.

Consider the bimolecular reaction,

$$A + B \rightarrow AB^{\ddagger} \rightarrow \text{product}$$

where A and B are both nonlinear molecules. Before the formation of the activated complex, each A or B molecule has three translational, three rotational, and three vibrational degrees of freedom. These motions all contribute to the entropy of the molecule.* At 25°C, the greatest contribution comes from translational motion (about 120 J K^{-1} mol^{-1}), followed by rotational motion (about 80 J K^{-1} mol^{-1}). Vibrational motion makes the smallest contribution (about 15 J K^{-1} mol^{-1}). The translational and rotational entropies of the activated complex are only slightly larger than those of an individual A or B molecule (these entropies increase slowly with size); therefore, there is a net loss in entropy of about 200 J K^{-1} mol^{-1} when the activated complex is formed. This loss in entropy is compensated for to a small extent by new modes of internal rotation and vibration in the activated complex. For unimolecular reactions, such as the cis–trans isomerization of an alkene, however, there is very little entropy change because the activated complex is formed from a single molecular species. A theoretical comparison of a unimolecular reaction with a bimolecular one shows a difference of as much as 3×10^{10} in the $e^{\Delta S^{\circ\ddagger}/R}$ term, favoring the unimolecular reaction.

Consider a simple enzyme-catalyzed reaction in which one substrate (S) is transformed into one product (P). The reaction proceeds as follows:

$$E + S \rightleftharpoons ES \rightleftharpoons ES^{\ddagger} \rightleftharpoons EP \rightleftharpoons E + P$$

In this scheme, the enzyme and the substrate must first encounter each other in solution to form the enzyme-substrate intermediate, ES. This is a reversible reaction, but when [S] is high, the formation of ES is favored. When the substrate is bound, forces within the active site can align the substrate and enzyme reactive groups into proper orientation, leading to the activated complex. The reaction takes place in the single entity enzyme–substrate intermediate to form the enzyme–substrate activated complex ($ES^{\ddagger}$), as in a unimolecular reaction, so the loss in entropy will be much less. In other words, the loss of the translational and rotational entropies occurred during the formation of ES, and not during the ES $\rightarrow$ $ES^{\ddagger}$ step. (This loss of entropy is largely compensated for by the substrate binding energy.) Once formed, $ES^{\ddagger}$ proceeds energetically downhill to the enzyme-product intermediate and finally to the product with the regeneration of the enzyme.

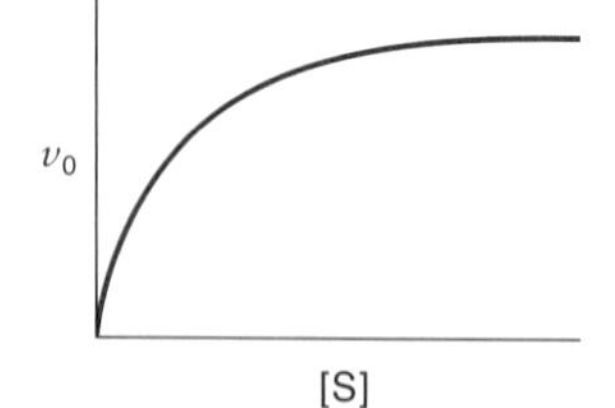

Figure 15.22
Plot of the initial rate (v_0) of an enzyme-catalyzed reaction versus substrate concentration [S].

The Equations of Enzyme Kinetics

In enzyme kinetics, it is customary to measure the *initial rate* (v_0) of a reaction to minimize reversible reactions and the inhibition of enzymes by products. Furthermore, the initial rate corresponds to a known fixed substrate concentration. As time proceeds, the substrate concentration will drop.

Figure 15.22 shows the variation of the initial rate (v_0) of an enzyme–catalyzed reaction with substrate concentration [S], where the subscript zero denotes the initial

* The relationship between molecular motion and thermodynamic functions will be discussed in Chapter 20.

value. The rate increases rapidly and linearly with [S] at low substrate concentrations, but it gradually levels off toward a limiting value at high concentrations of the substrate. In this region, all the enzyme molecules are bound to the substrate molecules, and the rate becomes zero order in substrate concentration. Mathematical analysis shows that the relationship between v_0 and [S] can be represented by an equation of a rectangular hyperbola,

$$v_0 = \frac{a[\mathrm{S}]}{b + [\mathrm{S}]} \tag{15.61}$$

where a and b are constants. Our next step is to develop the necessary equations to account for the experimental data.

Michaelis–Menten Kinetics

In 1913, the German biochemist Leonor Michaelis (1875–1949) and the Canadian biochemist Maud L. Menten (1879–1960), building on the work of the French chemist Victor Henri (1872–1940), proposed a mechanism to explain the dependence of the initial rate of enzyme-catalyzed reactions on concentration. They considered the following scheme, in which ES is the enzyme–substrate complex:

$$\mathrm{E} + \mathrm{S} \underset{k_{-1}}{\overset{k_1}{\rightleftharpoons}} \mathrm{ES} \overset{k_2}{\longrightarrow} \mathrm{P} + \mathrm{E}$$

The initial rate of product formation, v_0, is given by

$$v_0 = \left(\frac{d[\mathrm{P}]}{dt}\right)_0 = k_2[\mathrm{ES}] \tag{15.62}$$

To derive an expression for the rate in terms of the more easily measurable substrate concentration, Michaelis and Menten assumed that $k_{-1} \gg k_2$ so that the first step (formation of ES) can be treated as a rapid equilibrium process. The dissociation constant, K_S, is given by

$$K_\mathrm{S} = \frac{k_{-1}}{k_1} = \frac{[\mathrm{E}][\mathrm{S}]}{[\mathrm{ES}]}$$

The total concentration of the enzyme at a time shortly after the start of the reaction is

$$[\mathrm{E}]_0 = [\mathrm{E}] + [\mathrm{ES}]$$

so that

$$K_\mathrm{S} = \frac{([\mathrm{E}]_0 - [\mathrm{ES}])[\mathrm{S}]}{[\mathrm{ES}]} \tag{15.63}$$

Solving for [ES], we obtain

$$[\mathrm{ES}] = \frac{[\mathrm{E}]_0[\mathrm{S}]}{K_\mathrm{S} + [\mathrm{S}]} \tag{15.64}$$

Substituting Equation 15.64 into Equation 15.62 yields

$$v_0 = \left(\frac{d[\mathrm{P}]}{dt}\right)_0 = \frac{k_2[\mathrm{E}]_0[\mathrm{S}]}{K_\mathrm{S} + [\mathrm{S}]} \tag{15.65}$$

Thus, the rate is always proportional to the total concentration of the enzyme.

Equation 15.65 has the same form as Equation 15.61, where $a = k_2[\mathrm{E}]_0$ and $b = K_\mathrm{S}$. At low substrate concentrations, $[\mathrm{S}] \ll K_\mathrm{S}$, so Equation 15.65 becomes $v_0 = (k_2/K_\mathrm{S})[\mathrm{E}]_0[\mathrm{S}]$; that is, it is a second-order reaction (first order in $[\mathrm{E}]_0$ and first order in $[\mathrm{S}]$). This rate law corresponds to the initial linear portion of the plot in Figure 15.22. At high substrate concentrations, $[\mathrm{S}] \gg K_\mathrm{S}$, so Equation 15.65 can be written as

$$v_0 = \left(\frac{d[\mathrm{P}]}{dt}\right)_0 = k_2[\mathrm{E}]_0$$

Under these conditions, all the enzyme molecules are in the enzyme–substrate complex form; that is, the reacting system is saturated with S. Consequently, the initial rate is zero order in $[\mathrm{S}]$. This rate law corresponds to the horizontal portion of the plot. The curved portion in Figure 15.22 represents the transition from low to high substrate concentrations.

When all the enzyme molecules are complexed with the substrate as ES, the measured initial rate must be at its maximum value (V_max), so that

$$V_\mathrm{max} = k_2[\mathrm{E}]_0 \tag{15.66}$$

where V_max is called the *maximum rate*. Now consider what happens when $[\mathrm{S}] = K_\mathrm{S}$. From Equation 15.65 we find that this condition gives $v_0 = V_\mathrm{max}/2$, so K_S equals the concentration of S when the initial rate is half its maximum value.

Steady-State Kinetics

The British biologists George Briggs (1893–1978) and John Haldane (1892–1964) showed in 1925 that it is unnecessary to assume that enzyme and substrate are in thermodynamic equilibrium with the enzyme–substrate complex to derive Equation 15.65. They postulated that soon after enzyme and substrate are mixed, the concentration of the enzyme–substrate complex will reach a constant value so that we can apply the steady-state approximation as follows (Figure 15.23):*

$$\begin{aligned}\frac{d[\mathrm{ES}]}{dt} &= 0 = k_1[\mathrm{E}][\mathrm{S}] - k_{-1}[\mathrm{ES}] - k_2[\mathrm{ES}] \\ &= k_1([\mathrm{E}]_0 - [\mathrm{ES}])[\mathrm{S}] - (k_{-1} + k_2)[\mathrm{ES}]\end{aligned}$$

* Chemists are also interested in *pre-steady-state kinetics*, that is, the period before steady state is reached. Pre-steady-state kinetics is more difficult to study but provides useful information regarding the mechanism of enzyme catalysis. But steady-state kinetics is more important for the understanding of metabolism, because it measures the rates of enzyme-catalyzed reactions in the steady-state conditions that exist in the cell.

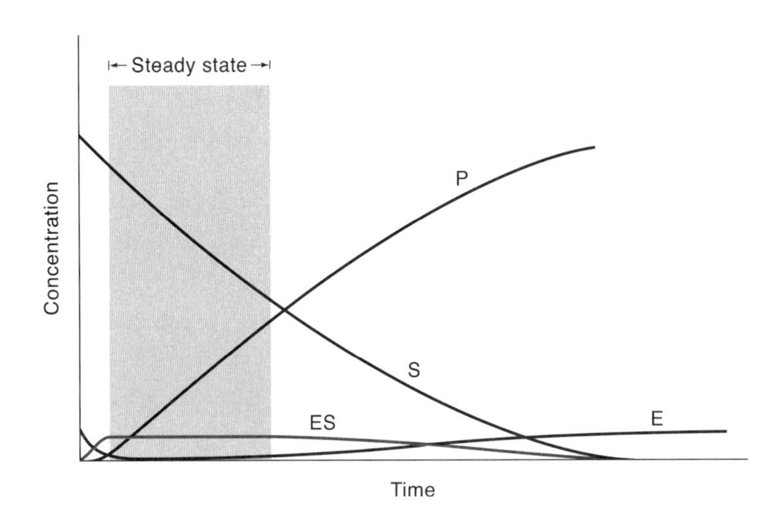

Figure 15.23
Plot of the concentrations of the various species in an enzyme-catalysed reaction $E + S \rightleftharpoons ES \rightarrow E + P$ versus time. We assume that the initial substrate concentration is much larger than the enzyme concentration and that the rate constants k_1, k_{-1}, and k_2 (see text) are of comparable magnitudes.

Solving for [ES], we get

$$[ES] = \frac{k_1[E]_0[S]}{k_1[S] + k_{-1} + k_2} \tag{15.67}$$

Substituting Equation 15.67 into 15.62 gives

$$\begin{aligned} v_0 = \left(\frac{d[P]}{dt}\right)_0 = k_2[ES] &= \frac{k_1 k_2[E]_0[S]}{k_1[S] + k_{-1} + k_2} \\ &= \frac{k_2[E]_0[S]}{[(k_{-1} + k_2)/k_1] + [S]} \\ &= \frac{k_2[E]_0[S]}{K_M + [S]} \end{aligned} \tag{15.68}$$

where K_M, the *Michaelis constant*, is defined as

$$K_M = \frac{k_{-1} + k_2}{k_1} \tag{15.69}$$

Comparing Equation 15.68 with Equation 15.65, we see that they have a similar dependence on substrate concentration; however, $K_M \neq K_S$ in general unless $k_{-1} \gg k_2$.

The Briggs–Haldane treatment defines the maximum rate exactly as Equation 15.66 does. Because $[E]_0 = V_{max}/k_2$, Equation 15.68 can also be written as

$$v_0 = \frac{V_{max}[S]}{K_M + [S]} \tag{15.70}$$

Equation 15.70 is a fundamental equation of enzyme kinetics. When the initial rate is equal to half the maximum rate, Equation 15.70 becomes

$$\frac{V_{max}}{2} = \frac{V_{max}[S]}{K_M + [S]}$$

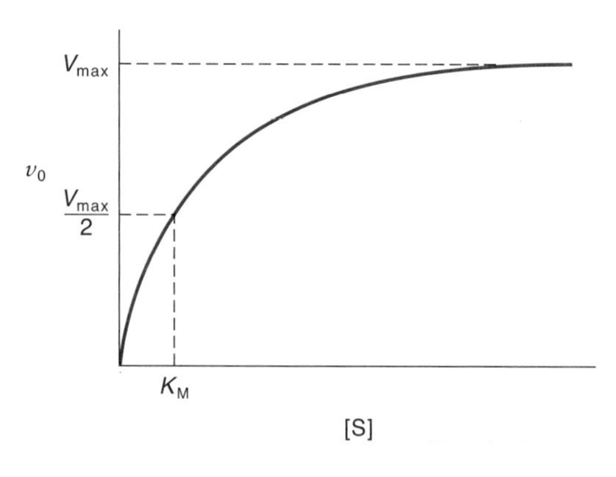

Figure 15.24
Graphical determination of V_{max} and K_M.

or

$$K_M = [S]$$

Thus, both V_{max} and K_M can be determined, at least in principle, from a plot such as the one in Figure 15.24. In practice, however, we find that the plot of v_0 versus $[S]$ is not very useful in determining the value of V_{max} because locating the asymptotic value V_{max} at very high substrate concentrations is often difficult. A more satisfactory approach, suggested by the American chemists Hans Lineweaver (1907–2009) and Dean Burk (1904–1988) is to employ the double-reciprocal plot of $1/v_0$ versus $1/[S]$. From Equation 15.70, we write

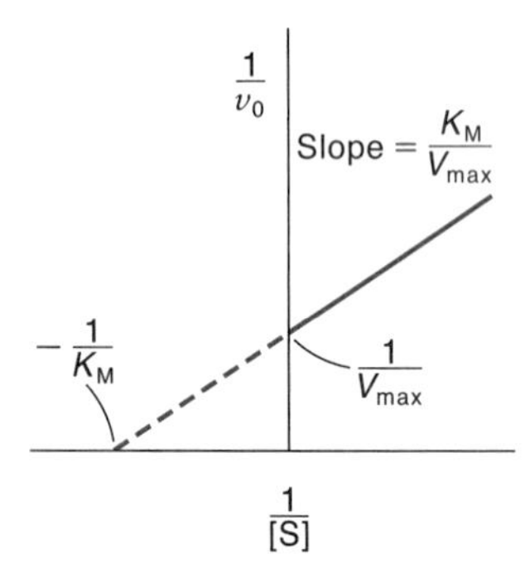

Figure 15.25
Lineweaver–Burk plot for an enzyme-catalyzed reaction obeying Michaelis–Menten kinetics.

$$\frac{1}{v_0} = \frac{K_M}{V_{max}[S]} + \frac{1}{V_{max}} \tag{15.71}$$

As Figure 15.25 shows, both K_M and V_{max} can be obtained from the slope and intercepts of the straight line.

Although useful and widely employed in enzyme kinetic studies, the Lineweaver–Burk plot has the disadvantage of compressing the data points at high substrate concentrations into a small region and emphasizing the points at lower substrate concentrations, which are often the least accurate. Of the several other ways of plotting the kinetic data, we shall mention the Eadie–Hofstee plot. Multiplying both sides of Equation 15.71 by $v_0 V_{max}$, we obtain

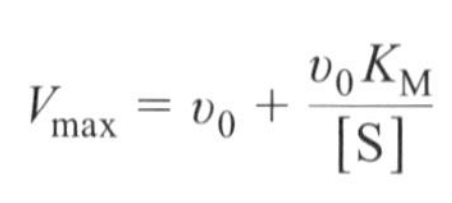

$$V_{max} = v_0 + \frac{v_0 K_M}{[S]}$$

Rearrangement gives

$$v_0 = V_{max} - \frac{v_0 K_M}{[S]} \tag{15.72}$$

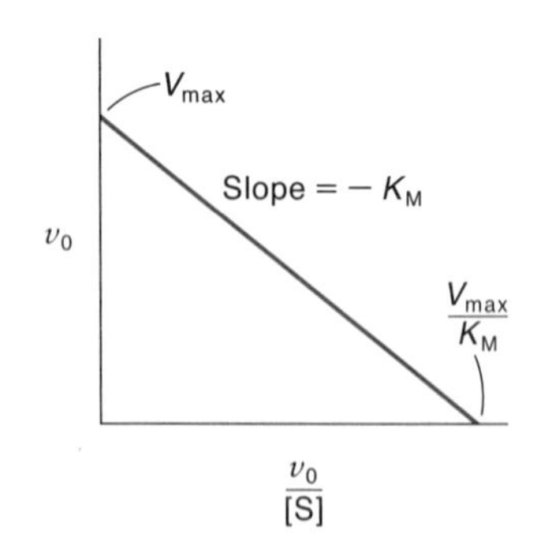

Figure 15.26
Eadie–Hofstee plot for the reaction graphed in Figure 15.24.

This equation shows that a plot of v_0 versus $v_0/[S]$, the so-called Eadie–Hofstee plot, gives a straight line with slope equal to $-K_M$ and intercepts V_{max} on the v_0 axis and V_{max}/K_M on the $v_0/[S]$ axis (Figure 15.26).

The Significance of K_M and V_{max}

The Michaelis constant, K_M varies considerably from one enzyme to another, and also with different substrates for the same enzyme. By definition, it is equal to the substrate concentration at half the maximum rate. Put another way, K_M represents the substrate concentration at which half the enzyme active sites are filled by substrate molecules. The value of K_M is sometimes equated with the dissociation constant of the enzyme–substrate complex, ES (the larger the K_M, the weaker the binding). As can be seen from Equation 15.69, however, this is true only when $k_2 \ll k_{-1}$, so that $K_M = k_{-1}/k_1$. In general, K_M must be expressed in terms of three rate constants. Nevertheless, K_M (in units of molarity) is customarily reported together with other kinetic parameters for enzyme-catalyzed reactions. To begin with, it is a quantity that can be measured easily and directly. Furthermore, K_M depends on temperature, the nature of the substrate, pH, ionic strength, and other reaction conditions; therefore, its value serves to characterize a particular enzyme–substrate system under specific conditions. Any variation in K_M (for the same enzyme and substrate) is often an indication of the presence of an inhibitor or activator. Useful information about evolution can also be obtained by comparing the K_M values of a similar enzyme from different species. For the majority of enzymes, K_M lies between 10^{-1} M and 10^{-7} M.

The maximum rate, V_{max}, has a well-defined meaning, both theoretically and empirically. It represents the maximum rate attainable; that is, it is the rate at which the total enzyme concentration is present as the enzyme–substrate complex. According to Equation 15.66, if $[E]_0$ is known, then the value of k_2 can be determined from the value of V_{max} measured by one of the plots mentioned earlier. Note that k_2 is a first-order rate constant and has the dimension of per unit time (s^{-1} or min^{-1}). It is called the *turnover number* (also referred to as k_{cat}, the *catalytic constant*). The turnover number of an enzyme is the number of substrate molecules (or moles of substrate) that are converted to product per unit time, when the enzyme is fully saturated with the substrate. For most enzymes, the turnover number varies between 1 and 10^6 s^{-1} under physiological conditions. Carbonic anhydrase, an enzyme that catalyzes the hydration of carbon dioxide and the dehydration of carbonic acid,

$$CO_2 + H_2O \rightleftharpoons H_2CO_3$$

has one of the largest turnover numbers known ($k_2 = 1 \times 10^6$ s^{-1}) at 25°C. Thus, a 1×10^{-6} M solution of the enzyme can catalyze the formation of 1 M H_2CO_3 from CO_2 (produced by metabolism) and H_2O per second; that is,

$$V_{max} = (1 \times 10^6 \text{ s}^{-1})(1 \times 10^{-6}\ M)$$
$$= 1\ M \text{ s}^{-1}$$

Without the enzyme, the pseudo first-order rate constant is only about 0.03 s^{-1}. Note that if the purity of the enzyme or the number of active sites per molecule is unknown, we cannot calculate the turnover number. In that case, the activity of the enzyme may be given as *units of activity per milligram of protein* (called the *specific activity*). One *international unit* is the amount of enzyme that produces one micromole (1 μmol) of product per minute.

Table 15.3
Values of K_M, k_{cat}, and k_{cat}/K_M for Some Enzymes and Substrates

Enzyme	Substrate	K_M/M	k_{cat}/s^{-1}	$(k_{cat}/K_M)/M^{-1}\ s^{-1}$
Acetylcholinesterase	Acetylcholine	9.5×10^{-5}	1.4×10^4	1.5×10^8
Catalase	H_2O_2	2.5×10^{-2}	1.0×10^7	4.0×10^8
Carbonic anhydrase	CO_2	0.012	1.0×10^6	8.3×10^7
Chymotrypsin	*N*-Acetylglycine ethyl ester	0.44	5.1×10^{-2}	0.12
Fumarase	Fumarate	5.0×10^{-6}	8.0×10^2	1.6×10^8
Urease	Urea	2.5×10^{-2}	1.0×10^4	4.0×10^5

As stated, we can determine the turnover number by measuring the rate under saturating substrate conditions; that is, when $[S] \gg K_M$ (see Equation 15.68). Under physiological conditions, the ratio $[S]/K_M$ is seldom greater than one; in fact, it is frequently much smaller than one. When $[S] \ll K_M$, Equation 15.68 becomes

$$\begin{aligned} v_0 &= \frac{k_2}{K_M}[E]_0[S] \\ &= \frac{k_{cat}}{K_M}[E]_0[S] \end{aligned} \tag{15.73}$$

Note that Equation 15.73 expresses the rate law of a second-order reaction. It is interesting that the ratio k_{cat}/K_M (which has the units $M^{-1}\ s^{-1}$) is a measure of the catalytic efficiency of an enzyme. A large ratio favors the formation of product. The reverse holds true for a small ratio.

Finally, what is the upper limit of the catalytic efficiency of an enzyme? From Equation 15.69, we find

$$\frac{k_{cat}}{K_M} = \frac{k_2}{K_M} = \frac{k_1 k_2}{k_{-1} + k_2}$$

This ratio is a maximum when $k_2 \gg k_{-1}$; that is, k_1 is rate determining and the enzyme turns over a product as soon as an ES complex is formed. However, k_1 can be no greater than the frequency of encounter between the enzyme and the substrate molecule, which is controlled by the rate of diffusion in solution.* The rate constant of a diffusion-controlled reaction is on the order of $10^8\ M^{-1}\ s^{-1}$. Therefore, enzymes with such k_{cat}/K_M values must catalyze a reaction almost every time they collide with a substrate molecule. Table 15.3 shows that acetylcholinesterase, catalase, fumarase, and perhaps carbonic anhydrase, have achieved this state of catalytic perfection.

* The rates of some enzyme-catalyzed reactions actually exceed the diffusion-controlled limit. When enzymes are associated with organized assemblies (e.g., in cellular membranes), the product of one enzyme is channeled to the next enzyme, much as in an assembly line. In such cases, the rate of catalysis is not limited by the rate of diffusion in solution.

Key Equations

$[\mathrm{A}] = [\mathrm{A}]_0 - kt$	(Integrated rate law for zero-order reaction)	(15.4)
$\ln \frac{[\mathrm{A}]}{[\mathrm{A}]_0} = -kt$	(Integrated rate law for first-order reaction)	(15.6)
$[\mathrm{A}] = [\mathrm{A}]_0 e^{-kt}$	(Integrated rate law for first-order reaction)	(15.7)
$t_{1/2} = \frac{\ln 2}{k}$	(Half-life of first-order reaction)	(15.8)
$t_{1/2} \propto \frac{1}{[\mathrm{A}]_0^{n-1}}$	(General expression for half-life)	(15.9)
$\frac{1}{[\mathrm{A}]} = kt + \frac{1}{[\mathrm{A}]_0}$	(Integrated rate law for second-order reaction)	(15.11)
$k = Ae^{-E_\mathrm{a}/RT}$	(Arrhenius equation)	(15.36)
$\ln k = \ln A - \frac{E_\mathrm{a}}{RT}$	(Arrhenius equation)	(15.37)
$\ln \frac{k_2}{k_1} = -\frac{E_\mathrm{a}}{R}\left(\frac{1}{T_2} - \frac{1}{T_1}\right)$	(Arrhenius equation)	(15.38)
$k = Pze^{-E_\mathrm{a}/RT}$	(Modified Arrhenius equation)	(15.44)
$k = \frac{k_\mathrm{B}T}{h} e^{\Delta S^{\circ\ddagger}/R} e^{-\Delta H^{\circ\ddagger}/RT} (M^{1-m})$	(Thermodynamic formulation of reaction rate)	(15.49)
$\log \frac{k}{k_0} = z_\mathrm{A} z_\mathrm{B} B \sqrt{I}$	(Kinetic salt effect)	(15.57)
$k_\mathrm{D} = \frac{8}{3}\frac{RT}{\eta}$	(Diffusion-controlled rate constant)	(15.58)
$V_\mathrm{max} = k_2[\mathrm{E}]_0$	(Maximum rate)	(15.66)
$K_\mathrm{M} = \frac{k_{-1} + k_2}{k_1}$	(Michaelis constant)	(15.69)
$\frac{1}{v_0} = \frac{K_\mathrm{M}}{V_\mathrm{max}} \frac{1}{[\mathrm{S}]} + \frac{1}{V_\mathrm{max}}$	(Lineweaver–Burk plot)	(15.71)
$v_0 = V_\mathrm{max} - \frac{v_0 K_\mathrm{M}}{[\mathrm{S}]}$	(Eadie–Hofstee plot)	(15.72)

APPENDIX 15.1

Derivation of Equation 15.9

Consider the reaction

$$\mathrm{A} \longrightarrow \text{product}$$

The rate is given by

$$-\frac{d[\mathrm{A}]}{dt} = k[\mathrm{A}]^n \tag{1}$$

where n is the order of the reaction. For now, we assume that $n \neq 1$ so that we can integrate Equation 1 as follows,

$$-\int \frac{d[\mathrm{A}]}{[\mathrm{A}]^n} = \int k dt$$

$$-\frac{[\mathrm{A}]^{1-n}}{1-n} = kt + C \tag{2}$$

where C is the constant of integration. At $t = 0$, $[\mathrm{A}] = [\mathrm{A}]_0$ so that

$$C = -\frac{[\mathrm{A}]_0^{1-n}}{1-n} \tag{3}$$

Equation 2 can now be expressed as

$$\begin{aligned} kt &= -\frac{[\mathrm{A}]^{1-n}}{1-n} - C \\ &= -\frac{[\mathrm{A}]^{1-n}}{1-n} + \frac{[\mathrm{A}]_0^{1-n}}{1-n} \end{aligned}$$

At $t = t_{1/2}$, $[\mathrm{A}] = [\mathrm{A}]_0/2$ and we have

$$\begin{aligned} kt_{1/2} &= -\frac{([\mathrm{A}]_0/2\,)^{1-n}}{1-n} + \frac{[\mathrm{A}]_0^{1-n}}{1-n} \\ &= -\frac{2^{n-1}[\mathrm{A}]_0^{1-n}}{1-n} + \frac{[\mathrm{A}]_0^{1-n}}{1-n} \\ &= \frac{1-2^{n-1}}{1-n}[\mathrm{A}]_0^{1-n} \end{aligned}$$

or

$$t_{1/2} = \frac{2^{n-1} - 1}{k(n-1)} \times \frac{1}{[A]_0^{n-1}} \qquad (4)$$

Because the first term on the right side of Equation 4 is a constant, it follows that

$$t_{1/2} \propto \frac{1}{[A]_0^{n-1}}$$

Equation 4 was derived assuming $n \neq 1$. We have previously shown, however, that for $n = 1$ (see Equation 15.8),

$$t_{1/2} = \frac{\ln 2}{k}$$

and there is no dependence on $[A]_0$. Thus, for any order, including first order, we have

$$t_{1/2} \propto \frac{1}{[A]_0^{n-1}} \qquad (5)$$

which is Equation 15.9.

APPENDIX 15.2

Derivation of Equation 15.51

We start by taking the natural logarithm of Equation 15.49:

$$\ln k = \ln\frac{k_B T}{h} + \frac{\Delta S^{\circ\ddagger}}{R} - \frac{\Delta H^{\circ\ddagger}}{RT} \tag{1}$$

Assuming ideal-gas behavior, the enthalpy of activation at constant pressure is given by

$$\begin{aligned}\Delta H^{\circ\ddagger} &= \Delta U^{\circ\ddagger} + P\Delta V^{\circ\ddagger} \\ &= \Delta U^{\circ\ddagger} + RT\Delta n^{\ddagger}\end{aligned} \tag{2}$$

As for any gaseous reaction,

$$\begin{aligned}\Delta n^{\ddagger} &= \text{number of moles of activated complex} - \text{number of moles of reactants} \\ &= 1 - m\end{aligned} \tag{3}$$

where m is the molecularity of the reaction. For example, for a unimolecular reaction, $m = 1$ and therefore $\Delta n^{\ddagger} = 1 - 1 = 0$, and so on. Thus, Equation 2 becomes

$$\Delta H^{\circ\ddagger} = \Delta U^{\circ\ddagger} + (1 - m)RT \tag{4}$$

We now differentiate Equation 1 with respect to T:

$$\frac{d\ln k}{dT} = \frac{1}{T} + \frac{\Delta H^{\circ\ddagger}}{RT^2} - \frac{1}{RT}\frac{d\Delta H^{\circ\ddagger}}{dT} \tag{5}$$

Substituting Equation 4 in Equation 5, we obtain

$$\begin{aligned}\frac{d\ln k}{dT} &= \frac{1}{T} + \frac{\Delta H^{\circ\ddagger}}{RT^2} - \frac{1}{RT}\frac{d[\Delta U^{\circ\ddagger} + (1 - m)RT]}{dT} \\ &= \frac{1}{T} + \frac{\Delta H^{\circ\ddagger}}{RT^2} - \frac{1}{RT}(1 - m)R \\ &= \frac{\Delta H^{\circ\ddagger} + mRT}{RT^2}\end{aligned} \tag{6}$$

Note that in deriving Equation 6, we assumed that both $\Delta U^{\circ\ddagger}$ and $\Delta S^{\circ\ddagger}$ are independent of temperature (often a fairly good approximation).

Next we differentiate Equation 15.37 with respect to T (assuming A to be temperature independent):

$$\frac{d \ln k}{dT} = \frac{E_a}{RT^2} \tag{7}$$

Comparison of the coefficients of the $(1/RT^2)$ term in Equations 6 and 7 gives

$$E_a = \Delta H^{\circ\ddagger} + mRT \tag{8}$$

Substituting Equation 4 in Equation 8 gives

$$\begin{aligned} E_a &= \Delta U^{\circ\ddagger} + (1 - m)RT + mRT \\ &= \Delta U^{\circ\ddagger} + RT \end{aligned} \tag{9}$$

which is Equation 15.51.

Suggestions for Further Reading

BOOKS

Brouard, M., *Reaction Dynamics*, Oxford University Press, New York, 1998.
Espenson, J. H., *Chemical Kinetics and Reaction Mechanism*, 2nd edition, WCB/McGraw-Hill, New York, 1995.
Fersht, A., *Enzyme Structure and Mechanism*, 2nd ed., W. H. Freeman, San Francisco, 1985.
Hague, D. N., *Fast Reactions*, Wiley-Interscience, New York, 1971.
Hammes, G. G., *Principles of Chemical Kinetics*, Academic Press, New York, 1978.
Houston, P. L., *Chemical Kinetics and Reaction Dynamics*, Dover, New York, 2006.
Laidler, K. J., *Chemical Kinetics*, 3rd ed., Harper & Row, New York, 1987.
Marangoni, A. G., *Enzyme Kinetics: A Modern Approach*, Wiley-Interscience, New York, 2002.
Nicholas, J., *Chemical Kinetics*, John Wiley & Sons, New York, 1976.
Pilling, M. J., and P. W. Seakins, *Reaction Kinetics*, Oxford University Press, New York, 1995.
Scott, S. K., *Oscillations, Waves, and Chaos in Chemical Kinetics*, Oxford University Press, New York, 1994.
Smith, I. W. M., *Kinetics and Dynamics of Elementary Gas Reactions*, Butterworths, London, 1990.
Steinfeld, J. I., J. S. Francisco, and W. L. Hase, *Chemical Kinetics and Dynamics*, 2nd ed., Prentice Hall, Englewood Cliffs, NJ, 1999.
Wright, M. R., *An Introduction to Chemical Kinetics*, John Wiley & Sons, New York, 2004.

ARTICLES

General

"Method for Determining Order of a Reaction," H. K. Zimmerman, *J. Chem. Educ.* **40**, 356 (1963).
"Concepts of Time in Chemistry," O. T. Benfey, *J. Chem. Educ.* **40**, 574 (1963).
"Unimolecular Gas Reactions at Low Pressures," B. Perlmutter-Hayman, *J. Chem. Educ.* **44**, 605 (1967).
"Tables of Conversion Factors for Reaction Rate Constants," D. D. Drysdale and A. C. Lloyd, *J. Chem. Educ.* **46**, 54 (1969).
"Mechanistic Ambiguities of Rate Laws," J. P. Birk, *J. Chem. Educ.* **47**, 805 (1970).
"Along the Reaction Coordinate," W. F. Sheehan, *J. Chem. Educ.* **47**, 853 (1970).
"Unconventional Applications of Arrhenius's Law," K. J. Laidler, *J. Chem. Educ.* **49**, 343 (1972).
"Drinking Too Fast Can Cause Sudden Death," C. D. Eskelson, *J. Chem. Educ.* **50**, 365 (1973).
"The Influence of Solvents on Chemical Reactivity," M. R. J. Dack, *J. Chem. Educ.* **51**, 231 (1974).
"Steady State and Equilibrium Approximations in Reaction Kinetics," L. Volk, W. Richardson, K. H. Lau, M. Hall, and S. H. Lin, *J. Chem. Educ.* **54**, 95 (1977).
"Some Common Oversimplifications in Teaching Chemical Kinetics," R. K. Boyd, *J. Chem. Educ.* **55**, 84 (1978).
"Chemical Reactions Without Solvation," R. T. McIver, Jr., *Sci. Am.* November 1980.
"What is the Rate-Determining Step of a Multistep Reaction?" J. R. Murdoch, *J. Chem. Educ.* **58**, 32 (1981).
"The Origin and Status of the Arrhenius Equation," S. R. Logan, *J. Chem. Educ.* **59**, 279 (1982).
"The Steady State and Equilibrium Assumptions in Chemical Kinetics," D. C. Tardy and and D. C. Cater, *J. Chem. Educ.* **60**, 109 (1983).
"The Extent of Reaction and Chemical Kinetics," H. Maskill, *Educ. Chem.* **21**, 122 (1984).
"Quantum Chemical Reactions in the Deep Cold," V. I. Goldanskii, *Sci. Am.* February 1986.
"The Meaning and Significance of 'the Activation Energy' of a Chemical Reaction," R. Logan, *Educ. Chem.* **23**, 148 (1986).
"A Time Scale for Fast Events," D. Onwood, *J. Chem. Educ.* **63**, 680 (1986).
"The Transition State," Anonymous, *J. Chem. Educ.* **64**, 208 (1987).
"Rate-Controlling Step: A Necessary and Useful Concept?" K. J. Laidler, *J. Chem. Educ.* **65**, 250 (1988).
"Just What Is a Transition State?" K. J. Laidler, *J. Chem. Educ.* **65**, 540 (1988).

"An Intuitive Approach to Steady-State Kinetics," R. T. Raines and D. E. Hansen, *J. Chem. Educ.* **65**, 757 (1988).
"The Birth of Molecules," A. H. Zewail, *Sci. Am.* December 1990.
"The Arrhenius Equation," H. Maskill, *Educ. Chem.* **27**, 111 (1990).
"Some Provocative Opinions on the Terminology of Chemical Kinetics," C. Reeve, *J. Chem. Educ.* **68**, 728 (1991).
"Applying the Principles of Chemical Kinetics to Population Growth Problems," G. F. Swiegers, *J. Chem. Educ.* **70**, 364 (1993).
"A Simplified Integration Technique for Reaction Rate Laws of Integral Order in Several Substances," G. Eberhardt and E. Levin, *J. Chem. Educ.* **72**, 193 (1995).
"Reaction Dynamics in Organic Chemistry," B. Carpenter, *Am. Sci.* **85**, 138 (1997).
"Anatomy of Elementary Chemical Reactions," A. J. Alexander and R. N. Zare, *J. Chem. Educ.* **75**, 1105 (1998).
"The Relationship between Stoichiometry and Kinetics Revisited," J. Y. Lee, *J. Chem. Educ.* **78**, 1283 (2001).
"How Accurate Is the Steady State Approximation?," L. O. Haustedt and J. M. Goodman, *J. Chem. Educ.* **80**, 839 (2003).
"Principle of Detailed Balance in Kinetics," R. A. Alberty, *J. Chem. Educ.* **81**, 1206 (2004).
"Ambiguities in Chemical Kinetics Rates and Rate Constants," K. T. Quisenberry and J. Tellinghuisen, *J. Chem. Educ.* **83**, 510 (2006).
"Gas Pressure Sensor Monitored Iodide-Catalyzed Decomposition Kinetics of Hydrogen Peroxide: An Initial Rate Approach," F. W. Nyasulu and R. Barlag, *Chem. Educator* [Online] **13**, 227 (2008) DOI 10.1333/s00897082150a.
"Determining Reaction Orders by Measuring Half-Life: A Simple Introduction to Experimental Kinetics," A. K. Grafton, *Chem. Educator* [Online] **14**, 19 (2009) DOI 10.1333/s00897092187a.
"Blue Bottle Light Experiments for Demonstrating Basic Kinetics Features of Homogeneous and Heterogeneous Photoinduced Redox Reactions," A. Mills, D. MacPhee, and K. Lawrie, *Chem. Educator* [Online] **15**, 150 (2010) DOI 10.1007/s00897102267a.
"Chemical Kinetics at the Single-Molecule Level," M. Levitus, *J. Chem. Educ.* **88**, 162 (2011).

Kinetic Isotope Effect

"The Exposition of Isotope Effects on Rates and Equilibria," M. M. Kreevoy, *J. Chem. Educ.* **41**, 636 (1964).
"Application of Isotope Effects," V. Gold, *Chem. Brit.* **6**, 292 (1970).
"The World of Isotope Effects," A. M. Rouhi, *Chem. Eng. News* December 22, 1997.
"Primary Kinetic Isotope Effect—A Lecture Demonstration," R. Chang, *Chem. Educator* [Online] **2**, 3 (1997) DOI 10.1333/s00897970121a.

Relaxation Kinetics

"Relaxation Kinetics," J. H. Swinehart, *J. Chem. Educ.* **44**, 524 (1967).
"The Temperature-Jump Method for the Study of Fast Reactions," J. E. Finholt, *J. Chem. Educ.* **45**, 394 (1968).
"Relaxation Methods in Chemistry," L. Faller, *Sci. Am.* May 1969.
"Temperature-Jump Techniques," E. Caldin, *Chem. Brit.* **11**, 4 (1975).

Oscillating Reactions

"Rotating Chemical Reactions," A. T. Winfree, *Sci. Am.* June 1974.
"Oscillating Chemical Reactions," I. R. Epstein, K. Kustin, P. De Kepper, and M. Orbán, *Sci. Am.* March 1983.
"Some Models of Chemical Oscillators," R. M. Noyes, *J. Chem. Educ.* **66**, 190 (1989).
"Oscillating Chemical Reactions and Nonlinear Dynamics," R. J. Field and F. W. Schneider, *J. Chem. Educ.* **66**, 195 (1989).
"Recipes for Belousov-Zhabotinsky Reagents," W. Jahnke and A. T. Winfree, *J. Chem. Educ.* **68**, 320 (1991).
"The Kinetics of Oscillating Reactions," R. F. Melka, G. Olsen, L. Beavers, and J. A. Draeger, *J. Chem. Educ.* **69**, 596 (1992).
"An Oscillating Reaction as a Demonstration of Principles Applied in Chemistry and Chemical Engineering," J. J. Weimer, *J. Chem. Educ.* **71**, 325 (1994).

"The BZ Reaction: Experimental and Model Studies in the Physical Chemistry Laboratory," O. Benini, R. Cervellat, and P. Fetto, *J. Chem. Educ.* **73**, 865 (1996).
"Introduction to Chemical Oscillations Using a Modified Lotka Model," E. Peacock-López, *Chem. Educator* [Online] **5**, 216 (2000) DOI 10.1333/s00897000413a.
"Chemical Oscillations: The Templator Model," E. Peacock-López, *Chem. Educator* [Online] **6**, 202 (2001) DOI 10.1333/s00897010483a.

Enzyme Kinetics
"Enzyme Catalysis and Transition-State Theory," G. E. Linehard, *Science* **180,** 149 (1973).
"An Introduction to Enzyme Kinetics," A. Ault, *J. Chem. Educ.* **51**, 381 (1974).
"What Limits the Rate of an Enzyme-Catalyzed Reaction?" W. W. Cleland, *Acc. Chem. Res.* **8**, 145 (1975).
"The Study of Enzymes," G. K. Radda and R. J. P. Williams, *Chem. Brit.* **12**, 124 (1976).
"Free Energy Diagrams and Concentration Profiles for Enzyme-Catalyzed Reactions," I. M. Klotz, *J. Chem. Educ.* **53**, 159 (1976).
"A Kinetic Investigation of an Enzyme-Catalyzed Reaction," W. G. Nigh, *J. Chem. Educ.* **53**, 668 (1976).
"Entropy, Binding Energy, and Enzyme Catalysis," M. I. Page, *Angew. Chem. Int. Ed.* **16**, 449 (1977).
"K_M as an Apparent Dissociation Constant," J. A. Cohlberg, *J. Chem. Educ.* **56**, 512 (1979).
"Enzyme Kinetics," O. Moe and R. Cornelius, *J. Chem. Educ.* **65**, 137 (1988).
"Homogeneous, Heterogeneous, and Enzymatic Catalysis," S. T. Oyama and G. A. Somorjai, *J. Chem. Educ.* **65**, 765 (1988).
"How Do Enzymes Work?," J. Krant, *Science* **242**, 533 (1988).
"Catalysis: New Reaction Pathways, Not Just a Lowering of Activation Energy," A. Haim, *J. Chem. Educ.* **66**, 935 (1989).
"Chemical Oscillations in Enzyme Kinetics," K. L. Queeney, E. P. Marin, C. M. Campbell, and E. Peacock-López, *Chem. Educator* [Online] **1**, S1430-4171 (1996) DOI 10.1333/s00897960035a.
"On the Meaning of K_M and V/K in Enzyme Kinetics," D. B. Northrop, *J. Chem. Educ.* **75**, 1153 (1998).
"How Do Approximations Affect the Solutions to Kinetic Equations?" J. M. Goodman, *J. Chem. Educ.* **76**, 275 (1999).
"A Simple Method for Demonstrating Enzyme Kinetics Using Catalase from Beef Liver Extract," A. K. Johnson, *J. Chem. Educ.* **77**, 1451 (2000).
"Kinetics of Alcohol Dehydrogenase-Catalyzed Oxidation of Ethanol Followed by Visible Spectroscopy," K. Bendinskas, C. DiJiacomo. A. Krill, and E. Vitz, *J. Chem. Educ.* **82**, 1068 (2005).
"Appreciating Formal Similarities in the Kinetics of Homogeneous, Heterogeneous, and Enzyme Catalysis," M. T. Ashby, *J. Chem. Educ.* **84**, 1515 (2007).
"Rapid-Equilibrium Enzyme Kinetics," R. A. Alberty, *J. Chem. Educ.* **85**, 1136 (2008).
"The Effect of Temperature on the Enzyme-Catalyzed Reaction: Insight From Thermodynamics," J. C. Aledo, S. Jiménez-Riveres, and M. Tena, *J. Chem. Educ.* **87**, 296 (2010).
"An Introduction to Enzyme Kinetics, Part Deux," A. Ault, *J. Chem. Educ.* **88**, 63 (2011).
"A Comprehensive Enzyme Kinetic Exercise for Biochemistry," J. S. Barton, *J. Chem. Educ.* **88**, 1336 (2011).

Problems

Reaction Order, Rate Law

15.1 Write the rates for the following reactions in terms of the disappearance of reactants and appearance of products:

(**a**) $3O_2 \rightarrow 2O_3$

(**b**) $C_2H_6 \rightarrow C_2H_4 + H_2$

(**c**) $ClO^- + Br^- \rightarrow BrO^- + Cl^-$

(**d**) $(CH_3)_3CCl + H_2O \rightarrow (CH_3)_3COH + H^+ + Cl^-$

(**e**) $2AsH_3 \rightarrow 2As + 3H_2$

15.2 The rate law for the reaction

$$NH_4^+(aq) + NO_2^-(aq) \rightarrow N_2(g) + 2H_2O(l)$$

is given by rate $= k[NH_4^+][NO_2^-]$. At 25°C, the rate constant is $3.0 \times 10^{-4}\ M^{-1}\ s^{-1}$. Calculate the rate of the reaction at this temperature if $[NH_4^+] = 0.26\ M$ and $[NO_2^-] = 0.080\ M$.

15.3 What are the units of the rate constant for a third-order reaction?

15.4 The following reaction is found to be first order in A:

$$A \rightarrow B + C$$

If half of the starting quantity of A is used up after 56 s, calculate the fraction that will be used up after 6.0 min.

15.5 A certain first-order reaction is 34.5% complete in 49 min at 298 K. What is its rate constant?

15.6 (**a**) The half-life of the first-order decay of radioactive ^{14}C is about 5720 years. Calculate the rate constant for the reaction. (**b**) The natural abundance of ^{14}C isotope is 1.1×10^{-13} mol % in living matter. Radiochemical analysis of an object obtained in an archeological excavation shows that the ^{14}C isotope content is 0.89×10^{-14} mol %. Calculate the age of the object. State any assumptions.

15.7 The first-order rate constant for the gas-phase decomposition of dimethyl ether,

$$(CH_3)_2O \rightarrow CH_4 + H_2 + CO$$

is $3.2 \times 10^{-4}\ s^{-1}$ at 450°C. The reaction is carried out in a constant-volume container. Initially, only dimethyl ether is present, and the pressure is 0.350 atm. What is the pressure of the system after 8.0 min? Assume ideal-gas behavior.

15.8 When the concentration of A in the reaction $A \rightarrow B$ was changed from 1.20 M to 0.60 M, the half-life increased from 2.0 min to 4.0 min at 25°C. Calculate the order of the reaction and the rate constant.

15.9 The progress of a reaction in the aqueous phase was monitored by the absorbance of a reactant at various times:

Time/s	0	54	171	390	720	1010	1190
Absorbance	1.67	1.51	1.24	0.847	0.478	0.301	0.216

Determine the order of the reaction and the rate constant.

15.10 Cyclobutane decomposes to ethylene according to the equation

$$C_4H_8(g) \rightarrow 2C_2H_4(g)$$

Determine the order of the reaction and the rate constant based on the following pressures, which were recorded when the reaction was carried out at 430°C in a constant-volume vessel:

Time/s	$P_{C_4H_8}$/mmHg
0	400
2000	316
4000	248
6000	196
8000	155
10,000	122

15.11 What is the half-life of a compound if 75% of a given sample of the compound decomposes in 60 min? Assume first-order kinetics.

15.12 The rate constant for the second-order reaction

$$2NO_2(g) \rightarrow 2NO(g) + O_2(g)$$

is 0.54 M^{-1} s^{-1} at 300°C. How long (in seconds) would it take for the concentration of NO_2 to decrease from 0.62 *M* to 0.28 *M*?

15.13 The decomposition of N_2O to N_2 and O_2 is a first-order reaction. At 730°C, the half-life of the reaction is 3.58×10^3 min. If the initial pressure of N_2O is 2.10 atm at 730°C, calculate the total gas pressure after one half-life. Assume that the volume remains constant.

15.14 The integrated rate law for the zero-order reaction A → B is $[A] = [A]_0 - kt$. **(a)** Sketch the following plots: **(i)** rate versus [A] and **(ii)** [A] versus *t*. **(b)** Derive an expression for the half-life of the reaction. **(c)** Calculate the time in half-lives when the integrated rate law is no longer valid, that is, when $[A] = 0$.

15.15 In the nuclear industry, workers use a rule of thumb that the radioactivity from any sample will be relatively harmless after 10 half-lives. Calculate the fraction of a radioactive sample that remains after this time period. (*Hint*: Radioactive decays obey first-order kinetics.)

15.16 Many reactions involving heterogeneous catalysis are zero order; that is, rate = *k*. An example is the decomposition of phosphine (PH_3) over tungsten (W):

$$4PH_3(g) \rightarrow P_4(g) + 6H_2(g)$$

The rate for this reaction is independent of $[PH_3]$ as long as phosphine's pressure is sufficiently high ($\geq$ 1 atm). Explain.

15.17 If the first half-life of a zero-order reaction is 200 s, what will be the duration of the next half-life?

15.18 Consider the following nuclear decay

$$^{64}\text{Cu} \rightarrow {}^{64}\text{Zn} + {}^{0}_{-1}\beta \qquad t_{1/2} = 12.8 \text{ h}$$

Starting with one mole of ^{64}Cu, calculate the number of grams of ^{64}Zn formed after 25.6 hours.

Reaction Mechanisms

15.19 The reaction

$$S_2O_8^{2-} + 2I^- \rightarrow 2SO_4^{2-} + I_2$$

proceeds slowly in aqueous solution, but it can be catalyzed by the Fe^{3+} ion. Given that Fe^{3+} can oxidize I^- and Fe^{2+} can reduce $S_2O_8^{2-}$, write a plausible two-step mechanism for this reaction. Explain why the uncatalyzed reaction is slow.

15.20 Derive Equation 15.35 using the steady-state approximation for both the H and Br atoms.

15.21 An excited ozone molecule, O_3^*, in the atmosphere can undergo one of the following reactions:

$$O_3^* \xrightarrow{k_1} O_3 \qquad \text{(1) fluorescence}$$

$$O_3^* \xrightarrow{k_2} O + O_2 \qquad \text{(2) decomposition}$$

$$O_3^* + M \xrightarrow{k_3} O_3 + M \qquad \text{(3) deactivation}$$

where M is an inert molecule. Calculate the fraction of ozone molecules undergoing decomposition in terms of the rate constants.

15.22 The following data were collected for the reaction between hydrogen and nitric oxide at 700°C:

$$2H_2(g) + 2NO(g) \rightarrow 2H_2O(g) + N_2(g)$$

Experiment	$[H_2]/M$	$[NO]/M$	Initial rate/M s^{-1}
1	0.010	0.025	2.4×10^{-6}
2	0.0050	0.025	1.2×10^{-6}
3	0.010	0.0125	0.60×10^{-6}

(a) What is the rate law for the reaction? **(b)** Calculate the rate constant for the reaction. **(c)** Suggest a plausible reaction mechanism that is consistent with the rate law. (*Hint*: Assume that the oxygen atom is the intermediate.) **(d)** More careful studies of the reaction show that the rate law over a wide range of concentrations of reactants should be

$$\text{rate} = \frac{k_1[NO]^2[H_2]}{1 + k_2[H_2]}$$

What happens to the rate law at very high and very low hydrogen concentrations?

15.23 The rate law for the decomposition of ozone to molecular oxygen

$$2O_3(g) \rightarrow 3O_2(g)$$

is

$$\text{rate} = k\frac{[O_3]^2}{[O_2]}$$

The mechanism proposed for this process is

$$O_3 \underset{k_{-1}}{\overset{k_1}{\rightleftharpoons}} O + O_2$$

$$O + O_3 \xrightarrow{k_2} 2O_2$$

Derive the rate law from these elementary steps. Clearly state the assumptions you use in the derivation. Explain why the rate decreases with increasing O_2 concentration.

15.24 The gas-phase reaction between H_2 and I_2 to form HI involves a two-step mechanism:

$$I_2 \underset{k_{-1}}{\overset{k_1}{\rightleftharpoons}} 2I$$

$$H_2 + 2I \xrightarrow{k_2} 2HI$$

(a) Assume the first step is a rapid equilibrium, and derive the rate law for the reaction.
(b) The rate of formation of HI increases with the intensity of visible light. How does this fact support the two-step mechanism given?

15.25 In recent years, ozone in the stratosphere has been depleted at an alarmingly fast rate by chlorofluorocarbons (CFCs). A CFC molecule such as $CFCl_3$ is first decomposed by UV radiation:

$$CFCl_3 \rightarrow CFCl_2 + Cl$$

The chlorine radical then reacts with ozone as follows:

$$Cl + O_3 \rightarrow ClO + O_2$$

$$ClO + O \rightarrow Cl + O_2$$

(a) Write the overall reaction for the last two steps. **(b)** What are the roles of Cl and ClO? **(c)** Why is the fluorine radical not important in this mechanism? **(d)** One suggestion for reducing the concentration of chlorine radicals is to add hydrocarbons such as ethane (C_2H_6) to the stratosphere. How will this approach work? **(e)** Draw potential energy versus reaction progress diagrams for the uncatalyzed and catalyzed (by Cl) destruction of ozone: $O_3 + O \rightarrow 2O_2$. Use the thermodynamic data in Appendix B to determine whether the reaction is exothermic or endothermic.

Activation Energy

15.26 Use Equation 15.36 to calculate the rate constant at 300 K for $E_a = 0$, 2, and 50 kJ mol^{-1}. Assume that $A = 10^{11}$ s^{-1} in each case.

15.27 Many reactions double their rates with every ten degrees rise in temperature. Assume that such a reaction takes place at 305 K and 315 K. What must its activation energy be for this statement to hold?

15.28 Over a range of about ±3°C from normal body temperature the metabolic rate, M_T, is given by $M_T = M_{37}(1.1)^{\Delta T}$, where M_{37} is the normal rate and ΔT is the change in T. Discuss this equation in terms of a possible molecular interpretation. [Source: "Eco-Chem," J. A. Campbell, *J. Chem. Educ.* **52**, 327 (1975).]

15.29 The rate of bacterial hydrolysis of fish muscle is twice as great at 2.2°C as at −1.1°C. Estimate a E_a value for this reaction. Is there any relation to the problem of storing fish for food? [Source: "Eco-Chem," J. A. Campbell, *J. Chem. Educ.* **52**, 390 (1975)].

15.30 The rate constants for the first-order decomposition of an organic compound in solution are measured at several temperatures:

k/s^{-1}	4.92×10^{-3}	0.0216	0.0950	0.326	1.15
$t/°\text{C}$	5.0	15	25	35	45

Determine graphically the pre-exponential factor and the energy of activation for the reaction.

15.31 The energy of activation for the reaction $2HI \rightarrow H_2 + I_2$ is 180 kJ mol^{-1} at 556 K. Calculate the rate constant using Equation 15.36. The collision diameter for HI is 3.5×10^{-8} cm. Assume that the pressure is 1 atm.

15.32 The rate constant of a first-order reaction is 4.60×10^{-4} s^{-1} at 350°C. If the activation energy is 104 kJ mol^{-1}, calculate the temperature at which its rate constant is 8.80×10^{-4} s^{-1}.

15.33 The rate at which tree crickets chirp is 2.0×10^2 per minute at 27°C but only 39.6 per minute at 5°C. From these data, calculate the "activation energy" for the chirping process. (*Hint*: The ratio of rates is equal to the ratio of rate constants.) Find the chirping rate at 15°C.

15.34 Consider the following parallel reactions

$$A \xrightarrow{k_1} B \qquad A \xrightarrow{k_2} C$$

The activation energies are 45.3 kJ mol^{-1} for k_1 and 69.8 kJ mol^{-1} for k_2. If the rate constants are equal at 320 K, at what temperature will $k_1/k_2 = 2.00$?

Thermodynamic Formulation of Transition-State Theory

15.35 The thermal isomerization of cyclopropane to propene in the gas phase has a rate constant of 5.95×10^{-4} s^{-1} at 500°C. Calculate the value of $\Delta G^{\circ\ddagger}$ for the reaction.

15.36 The rate of the electron-exchange reaction between naphthalene ($C_{10}H_8$) and its anion radical ($C_{10}H_8^-$) in an organic solvent is diffusion controlled:

$$C_{10}H_8^- + C_{10}H_8 \rightleftharpoons C_{10}H_8 + C_{10}H_8^-$$

The reaction is bimolecular and second order. The rate constants are

T/K	307	299	289	273
$k/10^9\ M^{-1}\ \text{s}^{-1}$	2.71	2.40	1.96	1.43

Calculate the values of E_a, $\Delta H^{\circ\ddagger}$, $\Delta S^{\circ\ddagger}$ and $\Delta G^{\circ\ddagger}$ at 307 K for the reaction. [*Hint*: Rearrange Equation 15.49 and plot $\ln(k/T)$ versus $1/T$.]

15.37 (**a**) The pre-exponential factor and activation energy for the hydrolysis of *t*-butyl chloride are 2.1×10^{16} s^{-1} and 102 kJ mol^{-1}, respectively. Calculate the values of $\Delta S^{\circ\ddagger}$ and $\Delta H^{\circ\ddagger}$ at 286 K for the reaction. (**b**) The pre-exponential factor and activation energy for the gas-phase cycloaddition of maleic anhydride and cyclopentadiene are 5.9×10^7 M^{-1} s^{-1} and 51 kJ mol^{-1}, respectively. Calculate the values of $\Delta S^{\circ\ddagger}$ and $\Delta H^{\circ\ddagger}$ at 293 K for the reaction.

Kinetic Isotope Effect

15.38 A person may die after drinking D_2O instead of H_2O for a prolonged period (on the order of days). Explain. Because D_2O has practically the same properties as H_2O, how would you test the presence of large quantities of the former in a victim's body?

15.39 The rate-determining step of the bromination of acetone involves breaking a carbon–hydrogen bond. Estimate the ratio of the rate constants $k_{\text{C-H}}/k_{\text{C-D}}$ for the reaction at 300 K. The wavenumbers of vibration for the particular bonds are $\tilde{\nu}_{\text{C-H}} \approx 3000$ cm^{-1} and $\tilde{\nu}_{\text{C-D}} \approx 2100$ cm^{-1}. [The wavenumber ($\tilde{\nu}$) is given by ν/c, where ν is the frequency and c is the velocity of light.]

15.40 Lubricating oils for watches or other mechanical objects are made of long-chain hydrocarbons. Over long periods of time they undergo auto-oxidation to form solid polymers. The initial step in this process involves hydrogen abstraction. Suggest a chemical means for prolonging the life of these oils.

Enzyme Kinetics

15.41 Explain why a catalyst must affect the rate of a reaction in both directions.

15.42 Measurements of a certain enzyme-catalyzed reaction give $k_1 = 8 \times 10^6$ M^{-1} s^{-1}, $k_{-1} = 7 \times 10^4$ s^{-1}, and $k_2 = 3 \times 10^3$ s^{-1}. Does the enzyme–substrate binding follow the equilibrium or steady-state scheme?

15.43 The hydrolysis of acetylcholine is catalyzed by the enzyme acetylcholinesterase, which has a turnover rate of 25,000 s^{-1}. Calculate how long it takes for the enzyme to cleave one acetylcholine molecule.

15.44 Derive the following equation from Equation 15.70,

$$\frac{v_0}{[\text{S}]} = \frac{V_{\text{max}}}{K_{\text{M}}} - \frac{v_0}{K_{\text{M}}}$$

and show how you would obtain values of K_{M} and V_{max} graphically from this equation.

15.45 An enzyme that has a K_{M} value of 3.9×10^{-5} M is studied at an initial substrate concentration of 0.035 M. After 1 min, it is found that 6.2 μM of product has been produced. Calculate the value of V_{max} and the amount of product formed after 4.5 min.

15.46 The hydrolysis of *N*-glutaryl-L-phenylalanine-*p*-nitroanilide (GPNA) to *p*-nitroaniline and *N*-glutaryl-L-phenylalanine is catalyzed by α-chymotrypsin. The following data are obtained:

$[\text{S}]/10^{-4}$ M	2.5	5.0	10.0	15.0
$v_0/10^{-6}$ M min^{-1}	2.2	3.8	5.9	7.1

where [S] = GPNA. Assuming Michaelis–Menten kinetics, calculate the values of V_{max}, K_{M}, and k_2 using the Lineweaver–Burk plot. Another way to treat the data is to plot v_0 versus

$v_0/[S]$, which is the Eadie–Hofstee plot. Calculate the values of V_{max}, K_M, and k_2 from the Eadie–Hofstee treatment, given that $[E]_0 = 4.0 \times 10^{-6}$ *M*. [*Source*: J. A. Hurlbut, T. N. Ball, H. C. Pound, and J. L. Graves, *J. Chem. Educ.* **50**, 149 (1973).]

15.47 The K_M value of lysozyme is 6.0×10^{-6} *M* with hexa-*N*-acetylglucosamine as a substrate. It is assayed at the following substrate concentrations: **(a)** 1.5×10^{-7} *M*, **(b)** 6.8×10^{-5} *M*, **(c)** 2.4×10^{-4} *M*, **(d)** 1.9×10^{-3} *M*, and **(e)** 0.061 *M*. The initial rate measured at 0.061 *M* was 3.2 μ*M* min^{-1}. Calculate the initial rates at the other substrate concentrations.

15.48 The hydrolysis of urea,

$$(NH_2)_2CO + H_2O \rightarrow 2NH_3 + CO_2$$

has been studied by many researchers. At 100°C, the (pseudo) first-order rate constant is 4.2×10^{-5} s^{-1}. The reaction is catalyzed by the enzyme urease, which at 21°C has a rate constant of 3×10^4 s^{-1}. If the enthalpies of activation for the uncatalyzed and catalyzed reactions are 134 kJ mol^{-1} and 43.9 kJ mol^{-1}, respectively, **(a)** calculate the temperature at which the nonenzymatic hydrolysis of urea would proceed at the same rate as the enzymatic hydrolysis at 21°C; **(b)** calculate the lowering of $\Delta G^\ddagger$ due to urease; and **(c)** comment on the sign of $\Delta S^\ddagger$. Assume that $\Delta H^\ddagger = E_a$ and that $\Delta H^\ddagger$ and $\Delta S^\ddagger$ are independent of temperature.

15.49 The initial rates at various substrate concentrations for an enzyme-catalysed reaction are as follows:

$[S]/M$	$v_0/10^{-6}\ M\ \text{min}^{-1}$
2.5×10^{-5}	38.0
4.00×10^{-5}	53.4
6.00×10^{-5}	68.6
8.00×10^{-5}	80.0
16.0×10^{-5}	106.8
20.0×10^{-5}	114.0

(a) Does this reaction follow Michaelis–Menten kinetics? **(b)** Calculate the value of V_{max} of the reaction. **(c)** Calculate the K_M value of the reaction. **(d)** Calculate the initial rates at $[S] = 5.00 \times 10^{-5}$ *M* and $[S] = 3.00 \times 10^{-1}$ *M*. **(e)** What is the total amount of product formed during the first 3 min at $[S] = 7.2 \times 10^{-5}$ *M*? **(f)** How would an increase in the enzyme concentration by a factor of 2 affect each of the following quantities: K_M, V_{max}, and v_0 (at $[S] = 5.00 \times 10^{-5}$ *M*)?

Additional Problems

15.50 A flask contains a mixture of compounds A and B. Both compounds decompose by first-order kinetics. The half-lives are 50.0 min for A and 18.0 min for B. If the concentrations of A and B are equal initially, how long will it take for the concentration of A to be four times that of B?

15.51 The term *reversible* is used in both thermodynamics (see Chapter 3) and in this chapter. Does it convey the same meaning in these two instances?

15.52 The recombination of iodine atoms in an organic solvent, such as carbon tetrachloride, is a diffusion-controlled process:

$$I + I \rightarrow I_2$$

Given that the viscosity of CCl_4 is 9.69×10^{-4} N s m^{-2} at 20°C, calculate the rate constant of recombination at this temperature.

15.53 The equilibrium between dissolved CO_2 and carbonic acid can be represented by

$$\begin{array}{ccc} H^+ + HCO_3^- & \underset{k_{21}}{\overset{k_{12}}{\rightleftharpoons}} & H_2CO_3 \\ k_{13} \downarrow\uparrow k_{31} & & k_{23} \downarrow\uparrow k_{32} \\ CO_2 & + & H_2O \end{array}$$

Show that

$$-\frac{d[CO_2]}{dt} = (k_{31} + k_{32})[CO_2] - \left(k_{13} + \frac{k_{23}}{K}\right)[H^+][HCO_3^-]$$

where $K = [H^+][HCO_3^-]/[H_2CO_3]$.

15.54 Polyethylene is used in many items, including water pipes, bottles, electrical insulation, toys, and mailing envelopes. It is a *polymer*, a molecule with a very high molar mass made by joining many ethylene molecules (the basic unit is called a *monomer*) together. The initiation step is

$$R_2 \xrightarrow{k_i} 2R\cdot \quad \text{initiation}$$

The R· species (called a radical) reacts with an ethylene molecule (M) to generate another radical

$$R\cdot + M \rightarrow M_1\cdot$$

Reaction of $M_1\cdot$ with another monomer leads to the growth or propagation of the polymer chain:

$$M_1\cdot + M \xrightarrow{k_p} M_2\cdot \quad \text{propagation}$$

This step can be repeated with hundreds of monomer units. The propagation terminates when two radicals combine

$$M'\cdot + M''\cdot \xrightarrow{k_t} M'\text{—}M'' \quad \text{termination}$$

The initiator in the polymerization of ethylene commonly is benzoyl peroxide $[(C_6H_5COO)_2]$:

$$[(C_6H_5COO)_2] \rightarrow 2C_6H_5COO\cdot$$

This is a first-order reaction. The half-life of benzoyl peroxide at 100°C is 19.8 min. **(a)** Calculate the rate constant (in min^{-1}) of the reaction. **(b)** If the half-life of benzoyl peroxide is 7.30 h, or 438 min, at 70°C, what is the activation energy (in kJ/mol) for the decomposition of benzoyl peroxide? **(c)** Write the rate laws for the elementary steps in the above polymerization process and identify the reactant, product, and intermediates. **(d)** What condition would favor the growth of long high-molar-mass polyethylenes?

15.55 In a certain industrial process involving a heterogeneous catalyst, the volume of the catalyst (in the shape of a sphere) is 10.0 cm^3. **(a)** Calculate the surface area of the catalyst. **(b)** If the sphere is broken down into eight spheres, each of which has a volume of 1.25 cm^3, what is the total surface area of the spheres? **(c)** Which of the two geometric configurations is the more effective catalyst? (*Hint*: The surface area of a sphere is $4\pi r^2$, where r is the radius of the sphere.)

15.56 Explain why grain dust in grain elevators can be explosive.

15.57 At a certain elevated temperature, ammonia decomposes on the surface of tungsten metal as follows:

$$NH_3 \rightarrow \tfrac{1}{2}N_2 + \tfrac{3}{2}H_2$$

The kinetic data are expressed as the variation of the half-life with the initial pressure of NH_3:

P/torr	264	130	59	16
$t_{1/2}$/s	456	228	102	60

(a) Determine the order of the reaction. **(b)** How does the order depend on the initial pressure? **(c)** How does the mechanism of the reaction vary with pressure?

15.58 The *activity* of a radioactive sample is the number of nuclear disintegrations per second, which is equal to the first-order rate constant times the number of radioactive nuclei present. The fundamental unit of radioactivity is the *curie* (Ci), where 1 Ci corresponds to exactly 3.70×10^{10} disintegrations per second. This decay rate is equivalent to that of 1 g of radium-226. Calculate the rate constant and half-life for the radium decay. Starting with 1.0 g of the radium sample, what is the activity after 500 years? The molar mass of Ra-226 is 226.03 g mol^{-1}.

15.59 The reaction X $\rightarrow$ Y has a reaction enthalpy of -64 kJ mol^{-1} and an activation energy of 22 kJ mol^{-1}. What is the activation energy for the Y $\rightarrow$ X reaction?

15.60 Consider the following parallel first-order reactions:

$$A \xrightarrow{k_1} B \qquad A \xrightarrow{k_2} C$$

(a) Write the expression for $d[B]/dt$ at time t, given that $[A]_0$ is the concentration of A at $t = 0$. **(b)** What is the ratio of $[B]/[C]$ upon completion of the reactions?

15.61 As a result of being exposed to the radiation released during the Chernobyl nuclear accident, a person had a level of iodine-131 in his body equal to 7.4 mCi (1 mCi $= 1 \times 10^{-3}$ Ci). Calculate the number of atoms of I-131 to which this radioactivity corresponds. Why were people who lived close to the nuclear reactor site urged to take large amounts of potassium iodide after the accident?

15.62 A certain protein molecule, P, of molar mass $\mathscr{M}$ dimerizes when it is allowed to stand in solution at room temperature. A plausible mechanism is that the protein molecule is first denatured before it dimerizes:

$$P \xrightarrow{k_1} P^* \text{ (denatured)} \qquad \text{slow}$$

$$2P^* \xrightarrow{k_2} P_2 \qquad \text{fast}$$

The progress of this reaction can be followed by making viscosity measurements of the average molar mass, $\bar{\mathcal{M}}$. Derive an expression for $\bar{\mathcal{M}}$ in terms of the initial concentration, $[P]_0$, and the concentration at time t, $[P]$, and $\mathcal{M}$. Write a rate equation consistent with this scheme.

15.63 The bromination of acetone is acid catalyzed:

$$CH_3COCH_3 + Br_2 \xrightarrow{H^+} CH_3COCH_2Br + H^+ + Br^-$$

The rate of disappearance of bromine was measured for several different concentrations of acetone, bromine, and H^+ ions at a certain temperature:

	$[CH_3COCH_3]/M$	$[Br_2]/M$	$[H^+]/M$	Rate of disappearance of $Br_2/M\ s^{-1}$
(1)	0.30	0.050	0.050	5.7×10^{-5}
(2)	0.30	0.10	0.050	5.7×10^{-5}
(3)	0.30	0.050	0.10	1.2×10^{-4}
(4)	0.40	0.050	0.20	3.1×10^{-4}
(5)	0.40	0.050	0.050	7.6×10^{-5}

(a) What is the rate law for the reaction? **(b)** Determine the rate constant. **(c)** The following mechanism has been proposed for the reaction:

$$CH_3-\overset{\overset{O}{\|}}{C}-CH_3 + H_3O^+ \rightleftharpoons CH_3-\overset{\overset{^+OH}{\|}}{C}-CH_3 + H_2O \text{ (fast equilibrium)}$$

$$CH_3-\overset{\overset{^+OH}{\|}}{C}-CH_3 + H_2O \longrightarrow CH_3-\overset{\overset{OH}{|}}{C}=CH_2 + H_3O^+ \text{ (slow)}$$

$$CH_3-\overset{\overset{OH}{|}}{C}=CH_2 + Br_2 \longrightarrow CH_3-\overset{\overset{O}{\|}}{C}-CH_2Br + HBr \text{ (fast)}$$

Show that the rate law deduced from the mechanism is consistent with that shown in **(a)**.

15.64 The rate law for the reaction $2NO_2(g) \longrightarrow N_2O_4(g)$ is rate $= k[NO_2]^2$. Which of the following changes will alter the value of k? **(a)** The pressure of NO_2 is doubled. **(b)** The reaction is run in an organic solvent. **(c)** The volume of the container is doubled. **(d)** The temperature is decreased. **(e)** A catalyst is added to the container.

15.65 For the cyclic reactions shown on p. 687, show that $k_1k_2k_3 = k_{-1}k_{-2}k_{-3}$.

15.66 Oxygen for metabolism is taken up by hemoglobin (Hb) to form oxyhemoglobin (HbO_2) according to the simplified equation

$$Hb(aq) + O_2(aq) \xrightarrow{k} HbO_2(aq)$$

where the second-order rate constant is $2.1 \times 10^6\ M^{-1}\ s^{-1}$ at 37°C. For an average adult, the concentrations of Hb and O_2 in the blood in the lungs are $8.0 \times 10^{-6}\ M$ and $1.5 \times 10^{-6}\ M$, respectively. **(a)** Calculate the rate of formation of HbO_2. **(b)** Calculate the rate of consumption of O_2. **(c)** The rate of formation of HbO_2 increases to $1.4 \times 10^{-4}\ M\ s^{-1}$ during exercise to meet the demand of an increased metabolic rate. Assuming the Hb concentration remains the same, what oxygen concentration is necessary to sustain this rate of HbO_2 formation?

15.67 Sucrose ($C_{12}H_{22}O_{11}$), commonly called table sugar, undergoes hydrolysis (reaction with water) to produce fructose ($C_6H_{12}O_6$) and glucose ($C_6H_{12}O_6$):

$$C_{12}H_{22}O_{11} + H_2O \rightarrow \underbrace{C_6H_{12}O_6}_{\text{fructose}} + \underbrace{C_6H_{12}O_6}_{\text{glucose}}$$

This reaction has particular significance in the candy industry. First, fructose is sweeter than sucrose. Second, a mixture of fructose and glucose, called *invert* sugar, does not crystallize, so candy made with this combination is chewier and not brittle as crystalline sucrose is. Sucrose is dextrorotatory (+), whereas the mixture of glucose and fructose resulting from inversion is levorotatory (−). Thus, a decrease in the concentration of sucrose will be accompanied by a proportional decrease in the optical rotation. **(a)** From the following kinetic data, show that the reaction is first order, and determine the rate constant:

time/min	0	7.20	18.0	27.0	∞
optical rotation (α)	+24.08°	+21.40°	+17.73°	+15.01°	−10.73°

(b) Explain why the rate law does not include $[H_2O]$ even though water is a reactant.

15.68 Thallium(I) is oxidized by cerium(IV) in solution as follows:

$$Tl^+ + 2Ce^{4+} \rightarrow Tl^{3+} + 2Ce^{3+}$$

The elementary steps, in the presence of Mn(II), are as follows:

$$Ce^{4+} + Mn^{2+} \rightarrow Ce^{3+} + Mn^{3+}$$

$$Ce^{4+} + Mn^{3+} \rightarrow Ce^{3+} + Mn^{4+}$$

$$Tl^+ + Mn^{4+} \rightarrow Tl^{3+} + Mn^{2+}$$

(a) Identify the catalyst, intermediates, and the rate-determining step if the rate law is rate = $k[Ce^{4+}][Mn^{2+}]$. **(b)** Explain why the reaction is slow without the catalyst. **(c)** Classify the type of catalysis (homogeneous or heterogeneous).

15.69 Derive expressions for the integrated rate law and half-life for the following third-order reaction

$$A \rightarrow \text{products}$$

15.70 The rate constants for the reaction

$$CH_2{=}CH{-}CH{=}CH_2 + CH_2{=}CH{-}CHO \longrightarrow \text{(cyclohexene-CHO)}$$

have been measured at several temperatures:

$10^3k/M^{-1}\,s^{-1}$	0.138	1.63	7.2	36.8	81
$t/°C$	155.3	208.3	246.5	295.8	330.8

Calculate the values of the pre-exponential factor, E_a, $\Delta S^{\circ\ddagger}$, and $\Delta H^{\circ\ddagger}$ for the reaction. Use 516 K as the mean temperature for your calculation. [Data taken from G. B. Kistiakowsky and J. R. Lacher, *J. Am. Chem. Soc.* **58**, 123 (1936).]

15.71 A gas mixture containing CH_3 fragments, C_2H_6 molecules, and He was prepared at 600 K with a total pressure of 5.42 atm. The elementary reaction

$$CH_3 + C_2H_6 \rightarrow CH_4 + C_2H_5$$

has a second-order rate constant of $3.0 \times 10^4\ M^{-1}\ s^{-1}$. Given that the mole fractions of CH_3 and C_2H_6 are 0.00093 and 0.00077, respectively, calculate the initial rate of the reaction.

15.72 To prevent brain damage, a drastic medical procedure is to lower the body temperature of someone who has suffered cardiac arrest. What is the physiochemical basis for this treatment?

15.73 The activation energy for the decomposition of hydrogen peroxide,

$$2H_2O_2(aq) \rightarrow 2H_2O(l) + O_2(g)$$

is 42 kJ mol^{-1}, whereas when the reaction is catalyzed by the enzyme catalase, it is 7.0 kJ mol^{-1}. Calculate the temperature that would cause the nonezymatic catalysis to proceed as rapidly as the enzyme-catalyzed decomposition at 20°C. Assume the pre-exponential factor to be the same in both cases.

15.74 The rate constant for the gaseous reaction

$$H_2(g) + I_2(g) \rightarrow 2HI(g)$$

is $2.4 \times 10^{-2}\ M^{-1}\ s^{-1}$ at 400°C. Initially an equimolar sample of H_2 and I_2 is placed in a vessel at 400°C and the total pressure is 1658 mmHg. **(a)** What is the initial rate (M min^{-1}) of the formation of HI? **(b)** What are the rate of formation of HI and the concentration of HI after 10.0 min?

15.75 When the concentration of A in the reaction A $\rightarrow$ B was changed from 1.20 M to 0.60 M, the half-life increased from 2.0 min to 4.0 min at 25°C. Calculate the order of the reaction and the rate constant.

15.76 The diameter of the methyl radical is 3.80 Å. Calculate the rate constant for the second-order gas phase reaction

$$2\ \cdot CH_3 \rightarrow C_2H_6$$

at 50°C. Is this the maximum possible rate constant? Explain.

CHAPTER 16

Photochemistry

Here comes the sun

—George Harrison*

A photochemist is interested in the fate of an electronically excited molecule. Depending on the system and conditions under which photoexcitation is carried out, such a molecule can undergo one of several processes. It can lose energy in collisions with other molecules, liberating heat. It can return to the ground state by emitting a photon; that is, it can fluoresce or phosphoresce. Alternatively, it can undergo a chemical reaction—isomerization, dissociation, or ionization, for example. Chapter 14 dealt with the phenomena of fluorescence and phosphorescence. In this chapter, we discuss several other types of photochemical processes.

16.1 Introduction

We begin our study of photochemical events with an introduction to some of the terminology used in this chapter.

Thermal Versus Photochemical Reactions

Chemical reactions can be categorized as thermal or photochemical. *Thermal reactions* (see Chapter 15) involve atoms and molecules in their electronic *ground* state. By definition, a *photochemical reaction* takes place in the presence of light, which usually means radiation from the visible and UV region or high-energy radiation such as X rays and γ rays.

If we take 4×10^{-19} J as the typical energy of an electronically excited state, then, using the Boltzmann distribution law (see Equation 2.33), we can show that at room temperature (25°C) $N_2/N_1 \approx 6 \times 10^{-43}$, and so only a negligible fraction of molecules are electronically excited. To achieve a mere 1% concentration of excited molecules would require a temperature of about 6000°C! At that temperature, practically all of the molecules would undergo rapid thermal decomposition in their ground electronic state, and it would be impossible to produce appreciable concentrations of electronically excited molecules.

On the other hand, if molecules absorb radiation at 500 nm, which roughly corresponds to the wavelength required for the electronic transition, then electronic excitation must occur. The concentration of the excited molecules depends on several factors,

* Title of a song written by George Harrison (1943–2001), a member of The Beatles.

including the intensity of irradiation and the rate at which the excited molecules return to the ground state. Further, if the electronic excitation energy can somehow be harnessed for bond breaking, then chemical change may occur. Thus, the energy of excitation for a photochemical reaction is analogous to the activation energy for a thermal reaction.

Primary Versus Secondary Processes

Photochemical reactions are subclassified as *primary* or *secondary processes.* Primary processes include vibrational relaxation, or loss of vibrational energy, by collision with other molecules; fluorescence; phosphorescence; isomerization; and dissociation. Dissociation of excited molecules may provide reactive intermediates that can undergo secondary processes of a thermal nature.

Let us illustrate the primary and secondary processes with the decomposition of hydrogen iodide in the gas phase. The overall reaction is

$$2HI \rightarrow H_2 + I_2$$

When light of the appropriate wavelength is applied, the reactions are

$$HI \xrightarrow{h\nu} H + I \quad \text{photochemical reaction (primary process)}$$

$$H + HI \rightarrow H_2 + I \quad \text{thermal reactions (secondary process)}$$

$$I + I \rightarrow I_2 \quad \text{thermal reactions (secondary process)}$$

$$\text{Overall:} \quad 2HI \rightarrow H_2 + I_2$$

where $h\nu$ represents the energy of the photon absorbed.

Quantum Yields

A useful ratio in the study of photochemical reactions is the *quantum yield* (Φ), which is the ratio of the number of molecules of product formed (or reactant molecules consumed) to the number of light quanta absorbed:

$$\Phi = \frac{\text{number of molecules produced}}{\text{number of photons absorbed}} \tag{16.1}$$

Equation 16.1 can be expressed in molar quantities as

$$\Phi = \frac{\text{number of moles of product formed}}{\text{number of einsteins absorbed}} \tag{16.2}$$

where an *einstein* is equal to 1 mole of photons.

The quantum yield of photochemical reactions varies greatly from one system to another, and the value of Φ often reveals the mechanism involved in the process. For the hydrogen iodide reaction discussed above, the quantum yield is 2 because the absorption of one photon leads to the removal of two reactant molecules (HI). When irradiated with UV light at about 280 nm, acetone forms a methyl and an acetyl radical with high yield:

$$(CH_3)_2CO \xrightarrow{h\nu} CH_3\cdot + CH_3CO\cdot$$

In the liquid phase, however, these radicals are likely to recombine because of the solvent cage effect (see p. 705). Therefore, the overall quantum yield for this reaction is below 0.1.

A mixture of gaseous hydrogen and chlorine is stable at room temperature. When exposed to visible or UV light ($\leq$ 400 nm), the gases react explosively to form hydrogen chloride. The mechanism is

$$Cl_2 \xrightarrow{h\nu} Cl + Cl$$

$$Cl + H_2 \rightarrow HCl + H \tag{a}$$

$$H + Cl_2 \rightarrow HCl + Cl \tag{b}$$

This is a chain reaction (p. 690) in which the propagation steps are (a) and (b). The quantum yield of this reaction is about 10^5! In general, a quantum yield greater than 2 is evidence of a chain mechanism.

Alternatively, a photochemical reaction can be analyzed in terms of rate constants. Consider the following situation,

$$A \xrightarrow{h\nu} A^*$$

$$A^* \xrightarrow{k_1} A$$

$$A^* \xrightarrow{k_2} \text{product}$$

where A is the reactant, and A^* is an electronically excited molecule. Assuming steady-state conditions, we write

$$\begin{aligned}\text{rate of formation of } A^* &= \text{rate of removal of } A^* \\ &= k_1[A^*] + k_2[A^*]\end{aligned}$$

The quantum yield of product formation is given by

$$\begin{aligned}\Phi_P &= \frac{\text{rate of product formation}}{\text{total rate of formation of } A^*} \\ &= \frac{k_2[A^*]}{k_1[A^*] + k_2[A^*]} = \frac{k_2}{k_1 + k_2}\end{aligned} \tag{16.3}$$

Note that photochemical efficiency as measured by Φ and photochemical reactivity as measured by rate constant are not fundamentally related. Two reactions may have very similar Φ values but differ greatly in their rate constants. Consider the following photochemical decompositions:

$$C_6H_5COCH_2CH_2CH_3 \xrightarrow{k} C_6H_5COCH_3 + CH_2{=}CH_2$$

$$\Phi = 0.40 \quad k = 3 \times 10^6 \text{ s}^{-1}$$

$$CH_3COCH_2CH_2CH_2CH_3 \xrightarrow{k} CH_3COCH_3 + CH_2{=}CHCH_3$$

$$\Phi = 0.38 \quad k = 1 \times 10^9 \text{ s}^{-1}$$

For an insight into the difference between Φ and rate, we need to look at the factors affecting the rate of a photochemical reaction, which can be expressed as

$$\text{rate} = IFf\Phi_P \tag{16.4}$$

where I is the intensity of incident light, F is the fraction of the total incident light that is absorbed, f is the fraction of absorbed light that produces the reactive state, and Φ_P is the quantum yield of product formation. Now we see why two reactions can have similar Φ values but very different rates if the reactants have different values of F and f.

Measurement of Light Intensity

Regardless of the mechanism involved, the rate of a photochemical reaction should be proportional to the rate of absorption of light. Thus, kinetic studies of photochemical reactions require accurate measurements of the intensity of light employed. Light intensity is measured with a chemical *actinometer*—a chemical system whose photochemical behavior is quantitatively understood. One of the most useful solution-phase actinometers is the potassium ferrioxalate system. When a sulfuric acid solution of $K_3Fe(C_2O_4)_3$ is irradiated with light in the range of 250 to 470 nm, the reduction of iron from Fe(III) to Fe(II) and oxidation of the oxalate ion to carbon dioxide occur simultaneously. The simplified equation for this process is

$$2Fe(C_2O_4)_3^{3-} \xrightarrow{h\nu} 2Fe^{2+} + 5C_2O_4^{2-} + 2CO_2$$

This reaction has been carefully studied, and its quantum yield is known as a function of wavelength. The amount of Fe^{2+} ions formed can be readily determined from the formation of the red 1,10-phenanthroline-Fe^{2+} complex ion whose molar absorptivity is known. In this way, the amount of photons absorbed in a given time period can be determined.

EXAMPLE 16.1

A 35-mL solution of $K_3Fe(C_2O_4)_3$ is irradiated with monochromatic light at 468 nm for 30 min. The solution is then titrated with 1,10-phenanthroline to form the red complex of 1,10-phenanthroline-Fe^{2+}. The absorbance of this complex ion measured in a 1-cm cell at 510 nm is 0.65 ($\varepsilon_{510} = 1.11 \times 10^4$ L mol^{-1} cm^{-1}). Assume that the quantum yield for the decomposition at this wavelength is 0.93, and calculate the number of einsteins absorbed per second and the total energy absorbed.

ANSWER

From Equation 11.20

$$c = \frac{A}{\varepsilon b} = \frac{0.65}{(1.11 \times 10^4 \text{ L mol}^{-1}\text{ cm}^{-1})(1 \text{ cm})}$$

$$= 5.86 \times 10^{-5}\, M$$

The number of einsteins absorbed is given by (see Equation 16.2)

$$\frac{\text{number of moles of Fe}^{2+}\text{ produced}}{\text{quantum yield}}$$

$$= \frac{(5.86 \times 10^{-5}\text{ mol/L})(1\text{ L/1000 mL})(35\text{ mL})}{0.93}$$

$$= 2.2 \times 10^{-6}\text{ mol}$$

$$= 2.2 \times 10^{-6}\text{ einstein}$$

The rate of absorption is given by

$$\frac{2.2 \times 10^{-6}\text{ einstein}}{30 \times 60\text{ s}} = 1.2 \times 10^{-9}\text{ einstein s}^{-1}$$

Finally,

$$\text{total energy absorbed} = \text{number of photons} \times h\nu$$

$$= (2.2 \times 10^{-6}\text{ mol})(6.022 \times 10^{23}\text{ mol}^{-1})$$

$$\times (6.626 \times 10^{-34}\text{ J s})\left(\frac{3.00 \times 10^{8}\text{ m s}^{-1}}{468 \times 10^{-9}\text{ m}}\right)$$

$$= 0.56\text{ J}$$

COMMENT

Light intensity is measured in photons $\text{cm}^{-2}\text{ s}^{-1}$ (or $\text{J cm}^{-2}\text{ s}^{-1}$). In photochemistry, we are more interested in the amount of light energy that is deposited in the sample, which is called absorbed intensity. Absorbed intensity is energy input into the reacting system per unit volume per unit time and has the units $\text{J cm}^{-3}\text{ s}^{-1}$. In our example, the absorbed intensity is

$$\frac{0.56\text{ J}}{(35\text{ cm}^3)(30 \times 60\text{ s})} = 8.9 \times 10^{-6}\text{ J cm}^{-3}\text{ s}^{-1}$$

Action Spectrum

Often, very useful information regarding the species responsible for photochemical and photobiological processes can be obtained if we measure the response or the effectiveness of the system as a function of the wavelength of the light employed. This procedure gives an *action spectrum*. In general, if a simple system contains only one type of molecule, then the action spectrum should and does resemble the absorption spectrum closely. In a complex biological system, there are usually several different compounds that strongly absorb the incident radiation over the range of wavelength of

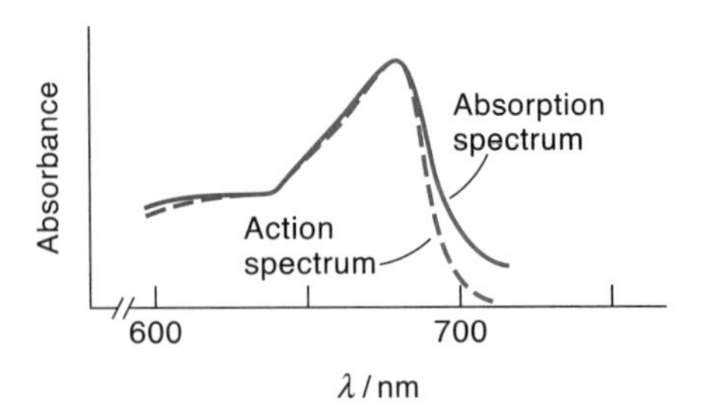

Figure 16.1
Comparison of the absorption spectrum with the action spectrum for the unicellular alga chlorella. The photosynthetic efficiency (measured by oxygen evolution) of light of different wavelengths (action spectrum) closely parallels the absorption spectrum of chlorophyll molecules. The discrepancy at about 700 nm is known as the "red drop." This comparison strongly suggests that chlorophyll plays a key role in photosynthesis. (From Clayton, R. K. *Light and Living Matter,* Vol. 2, Copyright 1971 by McGraw-Hill Book Company. Used with permission of McGraw-Hill, New York.)

interest. The molecules responsible for the photochemical reaction may be present in very low concentrations, so that their absorption spectra cannot always be detected. Their presence can be revealed, however, by recording the action spectrum instead of the usual absorption spectrum (Figure 16.1).

16.2 Earth's Atmosphere

The greenhouse effect, the formation of photochemical smog, and the depletion of ozone are the three photochemical processes that have major implications for the continuation of life on Earth. To a large extent, these important phenomena are byproducts of human activities, but ultimately they result from interactions between gases in our atmosphere and sunlight.

Composition of the Atmosphere

Earth is unique among the planets of our solar system in having an atmosphere that is chemically active and rich in oxygen. Mars, for example, has a much thinner atmosphere that is about 90% carbon dioxide. Jupiter, which has no solid surface, is made up of 90% hydrogen, 9% helium, and 1% other substances.

The total mass of Earth's atmosphere is about 5×10^{18} kg. Table 16.1 shows the composition of dry air at sea level. Water is excluded from this table because its concentration in air can vary drastically from location to location. Note that the concentrations of the gases that make up more than 99.9% of the atmosphere—nitrogen (N_2), oxygen (O_2), and the noble gases—have been nearly constant for much longer than human beings have been on Earth. The photochemical effects that we shall discuss are caused largely by changes, mainly increases, in the level of several minor constituents, or trace gases, including sulfur dioxide (SO_2), two nitrogen oxides known collectively as NO_x (NO and NO_2), and several chlorofluorocarbons (CFCs). Sulfur dioxide, the main source of acid rain, is present in the atmosphere at no more than 50 parts per billion by volume. The NO_x compounds are important in both forming

Table 16.1
Composition of Dry Air at Sea Level

Gas	Composition/% by volume
N_2	78.03
O_2	20.99
Ar	0.94
CO_2	0.040
Ne	0.0015
He	0.000524
Kr	0.00014
Xe	0.000006

acid rain and photochemical smog. The rising levels of CFCs, methane (CH_4), nitrous oxide (N_2O), and carbon dioxide (CO_2), which is by far the most abundant trace gas, contribute to the greenhouse effect.

Regions of the Atmosphere

Scientists divide the atmosphere into several layers according to temperature variation and composition (Figure 16.2). For visible events, the most active region is the *troposphere*, the layer that contains about 80% of the total mass of air and practically all of the atmosphere's water vapor. The troposphere is the thinnest layer of the atmosphere (10 km), but it is where all the dramatic events of weather—rain, lightning, hurricanes—occur. Temperature decreases almost linearly with increasing altitude in this region.

Above the troposphere is the *stratosphere*, which consists of nitrogen, oxygen, and ozone. In the stratosphere, the air temperature *rises* with altitude. This warming effect is the result of exothermic reactions triggered by UV radiation from the sun (see Section 16.5). One of the products of this reaction sequence is ozone (O_3), which, as we shall see shortly, prevents harmful UV rays from reaching Earth's surface.

The altitude of *tropopause*, the boundary between troposphere and stratosphere, varies with latitude and weather.

The concentration of ozone and other gases in the *mesosphere* above the stratosphere is low, and the temperature there decreases with increasing altitude. The *thermosphere*, or *ionosphere*, is the uppermost layer of the atmosphere. The rise in temperature in this region is the result of the bombardment of molecular oxygen and nitrogen and atomic species by energetic particles, such as electrons and protons, from the sun. Typical reactions are

$$N_2 \rightarrow 2N \qquad \Delta_r H° = 941.4\ \text{kJ mol}^{-1}$$

$$N \rightarrow N^+ + e^- \qquad \Delta_r H° = 1400\ \text{kJ mol}^{-1}$$

$$O_2 \rightarrow O_2^+ + e^- \qquad \Delta_r H° = 1176\ \text{kJ mol}^{-1}$$

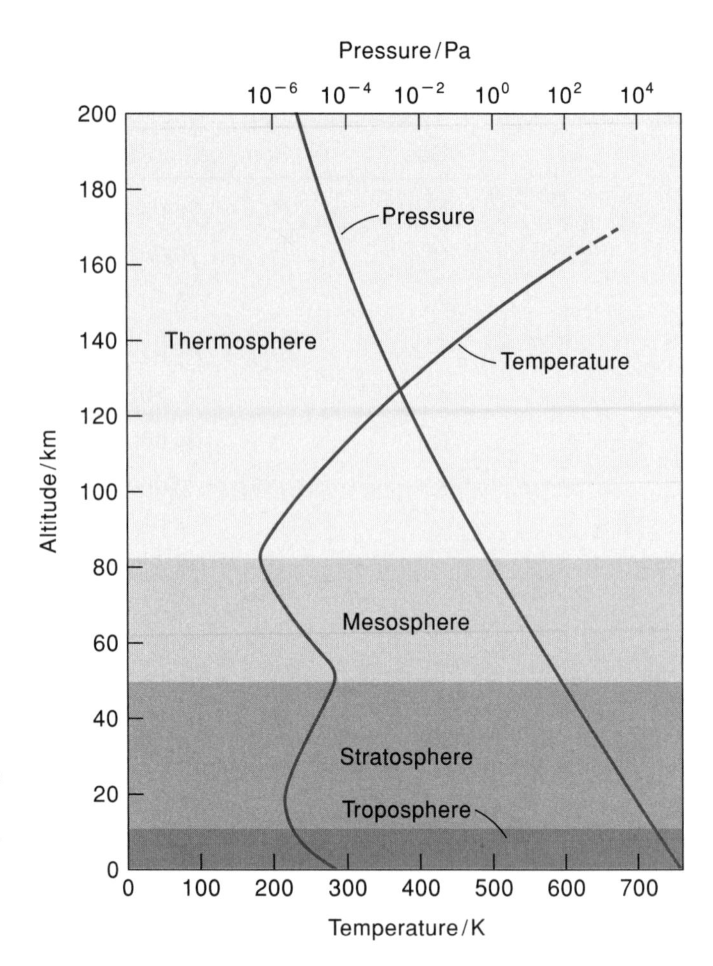

Figure 16.2
Different regions of the atmosphere. (Note that Mt. Everest has an altitude of about 8.8 km.) The pressure scale is logarithmic.

In reverse, these processes liberate the equivalent amount of energy, mostly as heat. Ionized particles are responsible for the reflection of radio waves back toward Earth.

Residence Time

Earth's atmosphere is a dynamic system; some components are constantly being produced within the atmosphere or released by sources at the surface of Earth. With a few exceptions, the overall composition of the atmosphere does not change very much because there are *sinks* that remove the gases from the atmosphere and balance the overall input and output. This steady-state situation is analogous to an overflowing bucket of water. If the bucket is full and the flow of water into it continues, the inflow will equal the outflow.

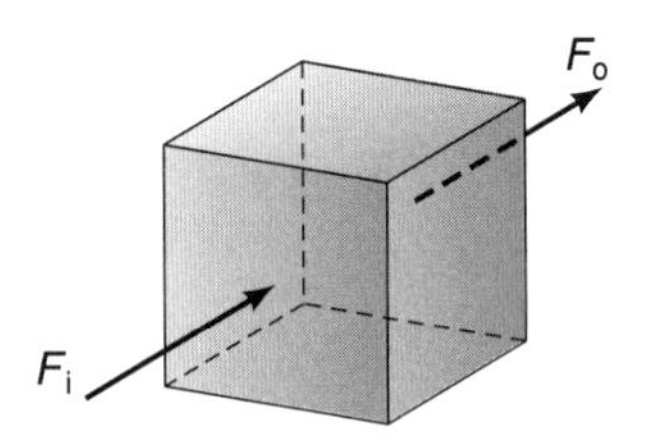

Figure 16.3
The mass flow rate of a substance into (F_i) and out of (F_o) a volume element of air.

Consider a volume element of air shown in Figure 16.3. F_i and F_o are the mass flow rates of a particular substance in and out of the element, respectively. In addition, we define P as the rate of introduction of the substance and R the rate of removal of the substance within this volume from activities on Earth. Let Q be the total mass of the substance in the volume. According to conservation of mass,

$$\frac{dQ}{dt} = (F_i - F_o) + (P - R)$$

Under steady-state conditions, $dQ/dt = 0$, so that

$$F_i + P = F_o + R$$

If the volume we are considering is the entire atmosphere, then $F_i = 0$, $F_o = 0$, and $P = R$. We define the *residence time* (τ) of a substance in the atmosphere as

$$\tau = \frac{Q}{P} = \frac{Q}{R} \tag{16.5}$$

A college has 2,000 students with 500 students admitted per year and 500 students graduating per year. Thus, the residence time is $2000/500\ \text{yr}^{-1} = 4$ years.

As an illustration of Equation 16.5, we note that the total sulfur-containing compounds in the atmosphere is about 4×10^{12} g (Q). The natural and human sources of sulfur result in a value of $2 \times 10^{14}\ \text{g yr}^{-1}$ for P; therefore, the residence time of sulfur compounds is given by

$$\tau = \frac{4 \times 10^{12}\ \text{g}}{2 \times 10^{14}\ \text{g yr}^{-1}} = 2 \times 10^{-2}\ \text{yr, or 1 week}$$

Due to their inertness, nitrogen and the noble gases have residence times exceeding millions of years. Oxygen is more reactive. It has a residence time of about 5000 years. The approximate residence times of other gases are: CO_2, 100 years; CH_4, 10 years; NO_x, days; N_2O, 200 years; SO_2, days to weeks; CFCs, 100 years.

16.3 The Greenhouse Effect

About 8 min after leaving the sun's surface, radiation traveling at $3 \times 10^8\ \text{m s}^{-1}$ reaches Earth (Figure 16.4). Estimates are that the incoming solar flux (also called solar irradiance), I_i, through a surface perpendicular to the beam is about $1.4 \times 10^3\ \text{J m}^{-2}\ \text{s}^{-1}$.

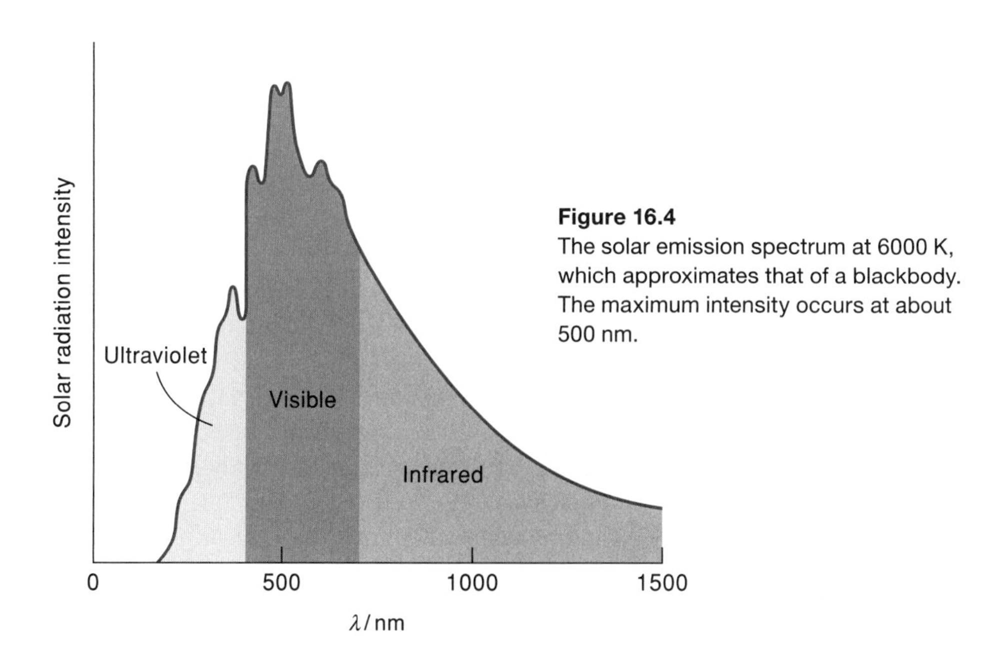

Figure 16.4
The solar emission spectrum at 6000 K, which approximates that of a blackbody. The maximum intensity occurs at about 500 nm.

Roughly one-third of solar radiation is reflected back by the surface and by the atmosphere (clouds, particles). The total energy received by Earth (R_i) per second is given by

$$R_i = (1 - 0.3)I_i(\pi r^2)$$
$$= 0.7I_i(\pi r^2) \tag{16.6}$$

where r is the radius of Earth. Note that the quantity πr^2 is not the total area of Earth; rather, it is the cross-sectional area exposed to the incoming radiation. Because Earth's temperature remains constant, the energy received must equal that radiated outward. We can estimate the effective temperature, T_e, of Earth by assuming that it is a blackbody radiator so that the radiation energy it emits per unit area per unit time (I_0) is given by the Stefan–Boltzmann law (after Boltzmann and the Austrian physicist Josef Stefan, 1835–1893),

See Section 10.2 for a discussion of blackbody radiation.

$$I_0 = \sigma T_e^4 \tag{16.7}$$

where σ is the Stefan–Boltzmann constant (5.67×10^{-8} J s^{-1} m^{-2} K^{-4}). The total surface area of Earth is $4\pi r^2$, which is also the total emitting area. Thus, the radiation energy emitted by Earth per second, R_0, is

$$R_0 = 4\pi r^2 \sigma T_e^4 \tag{16.8}$$

Energy balance requires that $R_i = R_0$, so

$$0.7I_i(\pi r^2) = 4\pi r^2 \sigma T_e^4$$

$$T_e^4 = \frac{0.7I_i(\pi r^2)}{4\pi r^2 \sigma} = \frac{0.7I_i}{4\sigma}$$

$$T_e = \left[\frac{0.7(1.4 \times 10^3 \text{ J m}^{-2}\text{ s}^{-1})}{4(5.67 \times 10^{-8}\text{ J s}^{-1}\text{ m}^{-2}\text{ K}^{-4})}\right]^{1/4}$$

$$= 256 \text{ K}$$

The average surface temperature of Earth, however, is 288 K, so there is a discrepancy of 32 K.

In the above calculation, we omitted the presence of gases that can trap some of the outgoing radiation. The temperature difference of 32 K is the result of the *greenhouse effect.* The glass roof of a greenhouse allows sunlight to stream in freely but blocks heat from escaping, mainly by preventing the warm air inside the greenhouse from mixing with outside air. Similarly, carbon dioxide and several other gases are relatively transparent to sunshine but trap heat by more efficiently absorbing the longer-wavelength IR radiation emitted by Earth.

Figure 16.5 shows the blackbody radiation curve of Earth's surface (at 288 K) and the IR absorption regions of the two most abundant greenhouse gases, water and carbon dioxide. As we saw in Chapter 11, all three vibrational modes of water and three

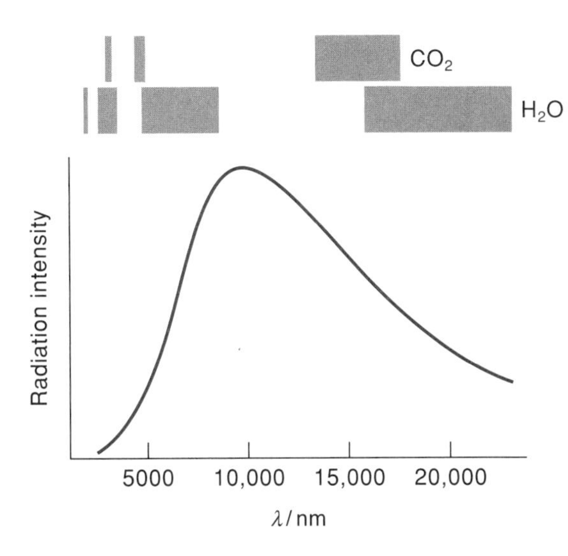

Figure 16.5
The blackbody radiation curve of Earth (at about 285 K) and the regions of IR absorption by carbon dioxide and water.

of the four vibrational modes of carbon dioxide are IR active. Upon receiving a photon in the IR region, these molecules are promoted to a higher vibrational energy level:

$$CO_2 \xrightarrow{h\nu} CO_2^*$$

$$H_2O \xrightarrow{h\nu} H_2O^*$$

where the asterisk denotes a vibrationally excited molecule. Such a molecule soon loses its excess energy either by the spontaneous emission of radiation or by collision with other molecules, resulting in an increase in the average translational energy. This trapped energy warms Earth's atmosphere and eventually its surface, by convection.

Although the total amount of water vapor in the atmosphere has not altered noticeably over the years, the concentration of CO_2 has been rising steadily since the turn of the last century due to the burning of fossil fuels (petroleum, natural gas, and coal). Figure 16.6 shows the rise in CO_2 level between 1960 and 2012. Looking at Figure 16.5, we see that terrestrial radiation has the best chance to escape to space at wavelengths between 2850 nm and 4000 nm and between 8300 nm and 12,500 nm. In addition to CO_2 and H_2O, however, CH_4, CFCs, N_2O (nitrous oxide), and other greenhouse gases contribute appreciably to global warming. Many have long residence times and they absorb IR radiation strongly in regions in which H_2O and CO_2 do not absorb. The overall effect is that only about 5% of the outward-directed radiation can escape into outer space. The remainder is absorbed by gases or by clouds, and more than 90% of that absorbed radiation is radiated back to Earth's surface.

Any polyatomic molecule has at least one vibrational mode that is IR active.

It is predicted that, should the buildup of greenhouse gases continue at the present pace. Earth's average temperature will increase by 2 to 4°C. This seemingly small increase in temperature will profoundly disturb the delicate thermal balance on Earth and could cause glaciers and icecaps to melt. Consequently, sea level would rise and coastal areas would be flooded. The most effective way to slow global warming is to lower emission of all greenhouse gases, particularly CO_2. More efficient use of fossil fuels, both in industry and in transportation, would accomplish this goal and would have other environmental benefits as well.

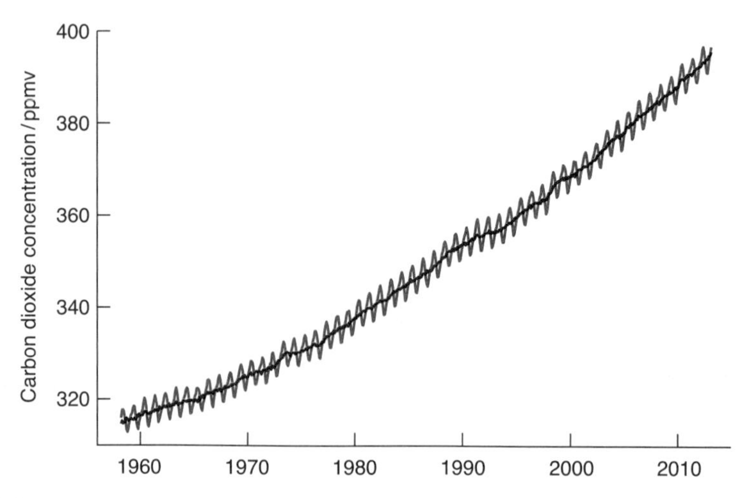

Figure 16.6
Yearly variation of carbon dioxide concentration at Mauna Loa, Hawaii. The general trend clearly points to an increase of carbon dioxide in the atmosphere. The mean concentration recorded in May 2013 was 400 ppm.

In response to the growing concern of global warming, an international agreement, called the Kyoto protocol, was adopted in 1997. The major feature of the Kyoto protocol is that it sets binding targets for industrialized countries for reducing greenhouse gas emissions. As of 2013, over 190 countries have signed and ratified the Kyoto protocol. The United States has signed but not ratified the protocol, meaning that it has not agreed to cap emissions in accordance with the protocol.

16.4 Photochemical Smog

Like the greenhouse effect, the formation of photochemical smog occurs in the troposphere. The word *smog* was originally coined to describe the combination of smoke and fog that blanketed London during the 1900s. The smog over London was caused by the presence in the atmosphere of sulfur dioxide produced mainly as a result of burning coal. Photochemical smog was discovered in the Los Angeles area in the 1950s. Although photochemical smog occurs in any city with heavy traffic, the city of Los Angeles appears to be uniquely suited for its formation. It has the combination of one of the world's greatest traffic densities and intense sunshine throughout most of the year. Moreover, the city is situated in a basin, encircled by mountains and the sea, a location in which air tends to stagnate, trapping pollutants. Consequently, much of the initial study of photochemical smog was based on data collected in this area. Today, the impact of photochemical smog on human health is a major issue worldwide.

Photochemical smog begins with primary pollutants, substances that are relatively unreactive by themselves. Secondary pollutants, formed photochemically from these primary pollutants, are responsible for the buildup of photochemical smog. Primary pollutants consist mainly of nitric oxide, carbon monoxide, and unburned aliphatic and aromatic compounds from automobile fuels, which are collectively called *volatile organic compounds* (VOCs).

Formation of Nitrogen Oxides

At ambient temperatures, nitrogen and oxygen gases have little tendency to form nitric oxide:

$$N_2(g) + O_2(g) \rightleftharpoons 2NO(g) \qquad \Delta_r G^\circ = 173.4 \text{ kJ mol}^{-1}$$

The large positive $\Delta_r G^\circ$ value corresponds to an equilibrium constant (K_P) of 4.0×10^{-31} at 25°C for the forward reaction. Lightning promotes the reaction, as does a running automobile engine, where the temperature may exceed 1000°C. Once emitted into the atmosphere, nitric oxide is oxidized to nitrogen dioxide (NO_2) in a complex series of gas-phase reactions involving species generated photochemically and the VOCs. For example, the VOC ethane (C_2H_6) converts NO to NO_2 via the steps summarized below:

$$CH_3CH_3 + \cdot OH \rightarrow CH_3CH_2\cdot + H_2O$$

$$CH_3CH_2\cdot + O_2 \rightarrow CH_3CH_2O_2\cdot$$

$$CH_3CH_2O_2\cdot + NO \rightarrow CH_3CH_2O\cdot + NO_2$$

$$CH_3CH_2O\cdot + O_2 \rightarrow HO_2\cdot + CH_3CHO$$

$$HO_2\cdot + NO \rightarrow \cdot OH + NO_2$$

where $CH_3CH_2O_2\cdot$ is an alkyl peroxy radical, $CH_3CH_2O\cdot$ is an alkoxide radical, $HO_2\cdot$ is a hydroperoxy radical, and $\cdot OH$ is a hydroxyl radical.

It is useful to compare the formation of NO_2 from NO in photochemical smog formation with the laboratory process. When a piece of copper wire is placed in a 30% nitric acid solution, the following reaction takes place:

$$3Cu(s) + 8HNO_3(aq) \rightarrow 3Cu(NO_3)_2(aq) + 4H_2O(l) + 2NO(g)$$

The colorless NO gas instantly turns brown, indicating the formation of NO_2:

$$2NO(g) + O_2(g) \rightarrow 2NO_2(g)$$

However, this reaction involves a rapid pre-equilibrium:

$$2NO(g) \rightleftharpoons N_2O_2(g)$$

$$N_2O_2(g) + O_2(g) \rightarrow 2NO_2(g)$$

Only at high NO concentrations is the NO dimer concentration sufficient to make the second step fast enough for NO_2 to be observed. In the atmosphere, the concentration of NO is so low that this reaction is insignificant. (The direct reaction $2NO + O_2$ is termolecular and is too slow to be of importance.) The other common oxide of nitrogen, nitrous oxide (N_2O), is a greenhouse gas but does not participate in photochemical smog formation.

Formation of O_3

Once NO_2 is formed, a variety of potential paths are available for the formation of other secondary pollutants such as ozone. Ozone is produced by the photodissociation of O_2 at wavelengths shorter than 242 nm (see Section 16.5). Because such radiation is present only in the stratosphere and above, no tropospheric ozone production is possible by this mechanism. Instead, in the presencc of light of wavelength $\lambda \leq 420$ nm, the following reactions take place:

$$NO_2 \xrightarrow{h\nu} NO + O^* \tag{a}$$

$$O^* + O_2 + M \rightarrow O_3 + M \tag{b}$$

where O^* is an electronically excited oxygen atom, and M is an inert molecule (N_2, for example) that serves to remove the excess energy from O_3 by collision to prevent it from dissociating back into O and O_2. Once formed, O_3 can readily oxidize NO to NO_2:

$$O_3 + NO \rightarrow O_2 + NO_2 \tag{c}$$

Reactions (a), (b), and (c) are cyclic and do not lead to a net production of O_3. The production of ozone is made possible, however, by the partial removal of its sink, that is, the conversion of NO to NO_2 by reaction with the VOCs shown earlier.

Formation of Hydroxyl Radical

The hydroxyl radical plays a central role in tropospheric chemistry because of its high reactivity with inorganic and organic compounds. It is formed when ozone is exposed to solar radiation at wavelengths shorter than about 320 nm:

$$O_3 \xrightarrow{h\nu} O^* + O_2$$

$$O^* + H_2O \rightarrow 2\ \cdot OH$$

Another source of $\cdot OH$ is the photolysis of nitrous acid (formed when NO_2 reacts with water vapor) at wavelengths below 400 nm:

$$HNO_2 \xrightarrow{h\nu} \cdot OH + NO$$

The hydroxyl radical has often been called "the atmosphere's detergent." It is a fragment of the stable water molecule, to which it can revert by abstracting a hydrogen atom from a molecule:

$$RH + \cdot OH \rightarrow R\cdot + H_2O$$

where RH can be an alkane such as C_2H_6 or C_3H_8. Once formed, the $R\cdot$ radical will undergo further reactions and eventually be removed from the atmosphere. In this way, the hydroxyl radical serves to purify the atmosphere by removing the various pollutants. It is remarkable that this cleansing action is accomplished by a very tiny concentration of hydroxyl radical, typically present at a ratio to air of about 2×10^{-14}. Without it, the composition of the trace gases in the atmosphere would be totally different and most likely hazardous to much of life on Earth.

The hydroxyl radical also oxidizes SO_2 to H_2SO_4 and NO_2 to HNO_3 the major components of acid rain. For example,

$$\cdot OH + SO_2 \rightarrow HOSO_2\cdot$$

The $HOSO_2\cdot$ radical is further oxidized to SO_3:

$$HOSO_2\cdot + O_2 \rightarrow HO_2\cdot + SO_3$$

The sulfur trioxide formed then rapidly reacts with water to form sulfuric acid:

$$SO_3 + H_2O \rightarrow H_2SO_4$$

Formation of Other Secondary Pollutants

In the oxidation of ethane leading to the conversion of NO to NO_2, acetaldehyde is formed during one step. This compound reacts with the hydroxyl radical as follows:

$$CH_3CHO + \cdot OH \rightarrow CH_3CO\cdot + H_2O$$

The acetyl radical then becomes oxidized by the following path:

$$CH_3CO\cdot + O_2 \rightarrow CH_3COO_2\cdot$$

$$CH_3COO_2\cdot + NO_2 \rightarrow CH_3COO_2NO_2\cdot$$

The final product, peroxyacetylnitrate (PAN), is one of the most harmful secondary pollutants.

Carbon monoxide is emitted in auto exhaust as a result of incomplete combustion, or it is formed in the atmosphere by reactions such as

$$HCHO \xrightarrow{h\nu} HCO\cdot + H\cdot$$

$$HCO\cdot + O_2 \rightarrow CO + HO_2\cdot$$

where the formaldehyde (HCHO) was generated by the oxidation of methane with the hydroxyl radical, in much the same way as ethane is oxidized.

The above discussion clearly shows that the reactions involved in photochemical smog formation are complex and interrelated. Furthermore, the rates and mechanisms are governed by sunlight and location. Nevertheless, the intense research effort over the past 40 years has yielded a much clearer picture of photochemical smog formation. Table 16.2 lists the trace constituents in photochemical smog, and Figure 16.7 shows the variation in pollutant concentrations with time on a smoggy day.

Harmful Effects and Prevention of Photochemical Smog

The secondary pollutants are harmful to both the biological and physical environment. The toxicity of ozone is well documented. A strong lung irritant, ozone causes pulmonary edema; it is also strongly irritating to the upper respiratory system. Ozone can cause leaves to turn brown and reduce the growth rate or physiological activity of plants. In addition, ozone attacks the C=C linkage in rubber:

$$R_2C{=}CR_2 + O_3 \longrightarrow \text{(ozonide: } R_2C\text{–O–}CR_2 \text{ with O–O bridge)} \xrightarrow{H_2O} R_2C{=}O + O{=}CR_2 + H_2O_2$$

In polluted urban areas, PAN is found at concentrations of several ppbv (parts per billion by volume).

Table 16.2
Concentrations of Trace Constituents in Photochemical Smog[a]

Constituent	Concentration/pphm[b]
Oxides of nitrogen	20
NH_3	2
H_2	50
H_2O	2×10^6
CO	4×10^3
CO_2	4×10^4
O_3	50
CH_4	250
Higher paraffins	25
C_2H_4	50
Higher olefins	25
C_2H_2	25
C_6H_6	10
Aldehydes	60
SO_2	20

[a] From R. D. Cadle, and E. R. Allen, *Science* **167**, 243–249 (1970). Copyright 1970 by the American Association for the Advancement of Science.

[b] Concentrations are measured in parts of constituents per hundred million parts of air by volume.

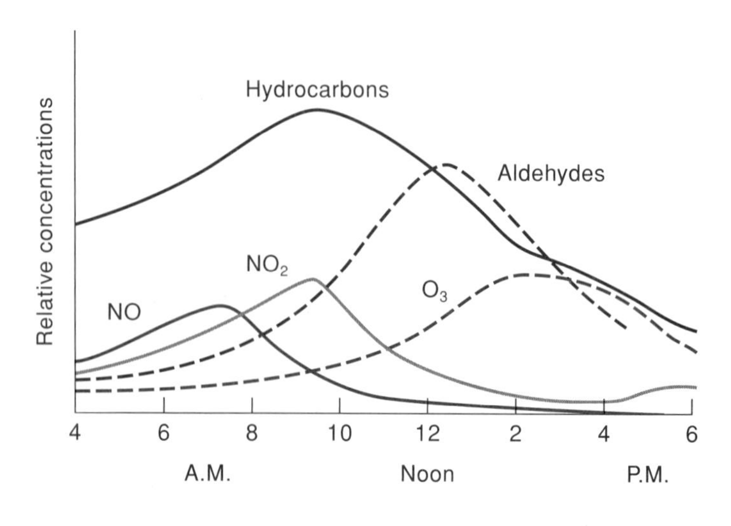

Figure 16.7
Average concentrations of pollutants during a smoggy day in a city with heavy traffic.

where R represents alkyl groups. In smog-ridden areas, this reaction can cause automobile tires to crack. Similar reactions are also damaging to lung tissues and other biological substances. PAN is a powerful lachrymator, or tear producer, and causes breathing difficulties. Because of its high affinity for hemoglobin, carbon monoxide can cause drowsiness and headache.

In principle, the solutions to photochemical smog are clear—devise more effective catalytic converters to minimize harmful auto exhausts, reduce traffic (e.g., by encouraging the use of public transportation), drive fuel-efficient cars, use less polluting fuels, and develop electric cars or cars powered by fuel cells for mass use. As in combating the greenhouse effect, many of these measures require appreciable changes in current lifestyle practices, which will be difficult at best to reach on a large scale in our society. Furthermore, developing countries must be persuaded not to repeat some of the costly environmental mistakes made by the industrialized nations.

16.5 Stratospheric Ozone

Ozone is Dr. Jekyll in the stratosphere and Mr. Hyde in the troposphere.

If all the ozone molecules in the atmosphere were compressed into a single layer at 1 bar and 25°C on Earth, the layer would be only about 3 mm thick! Yet the presence of ozone in the troposphere and stratosphere has profound consequences for the planet. We have already discussed the role of ozone as a secondary pollutant in the troposphere. In the stratosphere, however, ozone performs a beneficial function by absorbing the UV radiation in the sunlight that can cause skin cancer, genetic mutations, and other physiological problems. Here, we shall consider the chemistry of stratospheric ozone and the implications of its decreasing presence.

Formation of the Ozone Layer

Scientists generally believe that three to four billion years ago, Earth's atmosphere consisted mainly of ammonia, methane, and water. Little, if any, free oxygen was present. Ultraviolet (UV) radiation from the sun probably penetrated the atmosphere, rendering the surface of Earth sterile. The same UV radiation, however, may have triggered the chemical reactions (perhaps beneath the surface) that eventually led to the development of life to Earth.

Primitive organisms used energy from the sun to break down carbon dioxide (produced by volcanic activity) to obtain carbon, which they incorporated in their own cells. The major byproduct of this process, called *photosynthesis*, is oxygen. Another important source of oxygen is the *photodecomposition* of water vapor by UV light. Over time, the more reactive gases, such as ammonia and methane, largely disappeared, and today our atmosphere consists mainly of oxygen and nitrogen gases. Ozone was gradually produced by O_2 in the stratosphere by the absorption of solar radiation at $\lambda < 242$ nm:

$$O_2 \xrightarrow{h\nu} O + O$$

$$O + O_2 + M \rightarrow O_3 + M$$

Most of the solar radiation with wavelengths shorter than 100 nm is absorbed by N_2, O_2, N, and O at high altitudes (> 100 km). Absorption by O_2 limits radiation

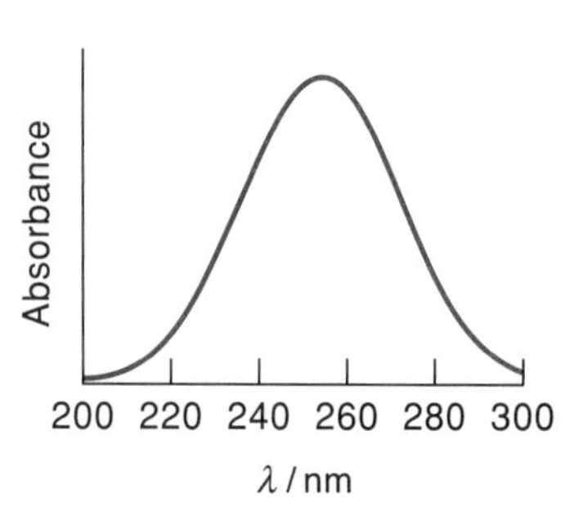

Figure 16.8
The absorption spectrum of ozone in the ultraviolet region.

of wavelengths shorter than 210 nm to a 50-km altitude and higher. Radiation with wavelengths longer than 210 nm is only weakly absorbed by O_2, so O_3 assumes the role of major absorber. As Figure 16.8 shows, O_3 absorbs effectively between 200 nm and 300 nm:

$$O_3 \xrightarrow{h\nu} O + O_2$$

Estimates are that at $\lambda = 250$ nm, for example, less than 1 part in 10^{30} of the incident solar radiation penetrates the "ozone layer." The recombination of O and O_2 to form O_3 is an exothermic reaction that warms the stratosphere.

Destruction of Ozone

The formation and destruction of ozone by natural processes is a delicate dynamic equilibrium that maintains a constant concentration of O_3 in the stratosphere. Scientists have long known that this equilibrium can be perturbed by a number of substances, among which are the nitrogen oxides, NO_x. The source of NO_x is N_2O, which is emitted from bacterial action in soils, especially those that contain a high concentration of fertilizers. N_2O is quite unreactive in the troposphere, so it gradually diffuses into the stratosphere, where it reacts with O atoms produced by the photodissociation of O_2 and O_3:

$$N_2O + O \rightarrow 2NO$$

Nitric oxide can then take part in a catalytic cycle that destroys O_3:

$$NO + O_3 \rightarrow NO_2 + O_2$$

$$O_3 \xrightarrow{h\nu} O + O_2$$

$$O + NO_2 \rightarrow NO + O_2$$

The overall reaction of these three steps is a net removal of O_3:

$$2O_3 \rightarrow 3O_2$$

Other sources of NO in the stratosphere include high-flying jets and rockets. High temperature combustion in air leads to formation of NO from N_2 and O_2.

In the 1970s, scientists became concerned about the harmful effects of certain chlorofluorocarbons (CFCs) on the ozone layer. CFCs, which are generally known by the trade name Freon, were first synthesized in the 1930s. Some of the common ones are $CFCl_3$ (Freon 11), CF_2Cl_2 (Freon 12), $C_2F_3Cl_3$ (Freon 113), and $C_2F_4Cl_2$ (Freon 114). Because these compounds are readily liquefied, relatively inert, nontoxic, noncombustible, and volatile, they have been used as coolants in refrigerators and air conditioners, in place of highly toxic sulfur dioxide and ammonia. Large quantities of CFCs were also used in the manufacture of disposable foam products such as cups and plates, as aerosol propellants in spray cans, and as a cleaning agent for newly soldered electronic circuit boards. In 1977, the peak year of production, nearly 1.5×10^6 tons of CFCs were produced in the United States. Most of the CFCs manufactured for commercial and industrial use are eventually discharged into the atmosphere.

Because of their relative inertness, the CFCs have a long residence time in the troposphere (about 100 years). They slowly diffuse to the upper stratosphere, where they encounter radiation of wavelengths between 175 nm and 220 nm and undergo dissociation:

$$CFCl_3 \xrightarrow{h\nu} CFCl_2 + Cl$$

$$CF_2Cl_2 \xrightarrow{h\nu} CF_2Cl + Cl$$

The reactive chlorine atoms then break ozone down into diatomic oxygen as follows:

$$Cl + O_3 \rightarrow ClO + O_2$$

$$ClO + O \rightarrow Cl + O_2$$

$$\text{Overall:}\quad O_3 + O \rightarrow 2O_2$$

where the O atom is supplied by the photochemical decomposition of O_2 or O_3 described earlier. The overall reaction results in a net removal of O_3 from the stratosphere. Note that Cl plays the role of a homogeneous catalyst, and ClO (chlorine monoxide) is an intermediate in these steps. Estimates are that, on average, one Cl atom can destroy 100,000 O_3 molecules before it is permanently removed by some other irreversible reaction.

The actual process is more involved because there are reactions that temporarily remove Cl and ClO. The major ones are as follows:

$$Cl + CH_4 \rightarrow HCl + CH_3$$

$$ClO + NO_2 + M \rightarrow ClONO_2 + M$$

$$Cl + HO_2 \rightarrow HOCl + O_2$$

where $ClONO_2$ is chlorine nitrate, and HOCl is hypochlorous acid. All three species HCl, $ClONO_2$, and HOCl serve as reservoir compounds of Cl. Under appropriate conditions, they release chlorine atoms as follows:

$$HCl + OH \rightarrow Cl + H_2O$$

$$ClONO_2 \xrightarrow{h\nu} Cl + NO_3$$

$$HOCl \xrightarrow{h\nu} Cl + OH$$

Bromine-containing compounds can also interfere with stratospheric ozone. The main source of bromine is methyl bromide, which is largely of natural origin (from the marine environment), although it is also used as a soil fumigant. Like the CFCs, CH_3Br diffuses into the stratosphere, where it is photolytically converted to Br and CH_3 and then to BrO. Chlorine monoxide and bromine monoxide then participate in a catalytic cycle leading to the destruction of ozone:

$$BrO + ClO \rightarrow Br + Cl + O_2$$

$$Br + O_3 \rightarrow BrO + O_2$$

$$Cl + O_3 \rightarrow ClO + O_2$$

$$\text{Overall:} \quad 2O_3 \rightarrow 3O_2$$

Br and BrO are not removed as easily as Cl and ClO, however, because the corresponding HBr and $BrONO_2$ molecules are very rapidly photolyzed. For this reason, bromine is a more effective catalyst for O_3 destruction than chlorine. Fortunately, bromine concentration in the atmosphere is still quite low, and so it does not play a major role in the destruction of ozone.

Polar Ozone Holes

In the mid-1980s, evidence began to accumulate that an "Antarctic ozone hole," which developed in late winter, had depleted the stratospheric ozone over Antarctica by as much as 50%. In the stratosphere, a stream of air known as the "polar vortex" circles Antarctica in winter. Air trapped within this vortex becomes extremely cold during the polar night. This condition leads to the formation of ice particles known as polar stratospheric clouds (PSCs). The significance of PSCs is that they act as heterogeneous catalysts, on whose surface some unusual reactions can take place. For example, the release of Cl atoms from the reservoir molecules is normally rather slow. But in the presence of ice particles, we have

$$ClONO_2 + HCl \rightarrow Cl_2 + HNO_3$$

The molecular chlorine is released as a gas, whereas the nitric acid remains in the ice particles. In the spring, when the sun returns, the following reactions ensue:

$$Cl_2 \xrightarrow{h\nu} Cl + Cl$$

$$Cl + O_3 \rightarrow ClO + O_2$$

The concentration of O atoms is too low (after a long absence of strong solar radiation), however, to make the following reaction significant:

$$ClO + O \rightarrow Cl + O_2$$

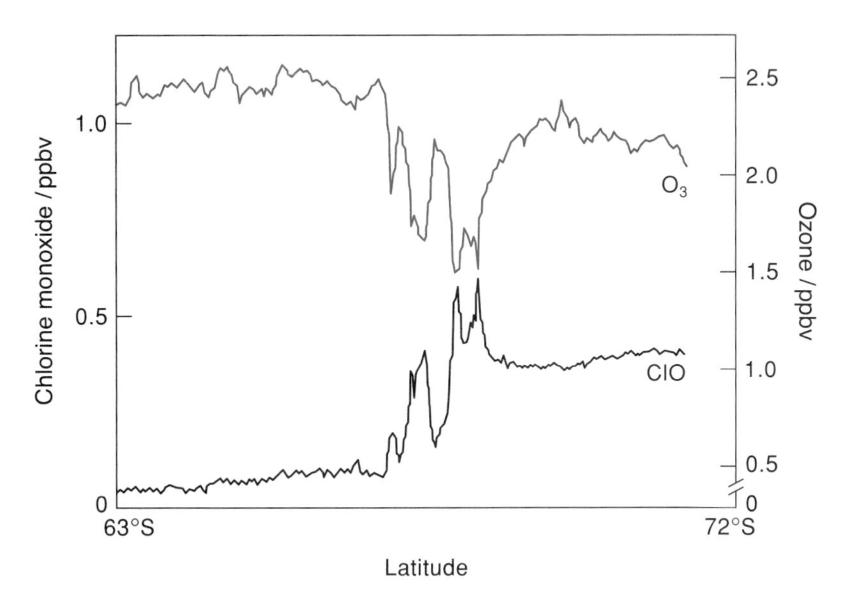

Figure 16.9
Variations in the concentrations of ClO and O_3 with latitude. (Courtesy of James G. Anderson.)

which would complete the catalytic cycle outlined earlier. Instead, a new catalytic cycle is believed to take place:

$$ClO + ClO + M \rightarrow (ClO)_2 + M$$

$$(ClO)_2 \xrightarrow{h\nu} Cl + ClOO$$

$$ClOO + M \rightarrow Cl + O_2 + M$$

$$2(Cl + O_3 \rightarrow ClO + O_2)$$

$$\text{Overall:} \qquad 2O_3 \rightarrow 3O_2$$

The key step here is the formation of the chlorine monoxide dimer, $(ClO)_2$, which is stable only at the low temperatures characteristic of the Antarctic polar air. Figure 16.9 shows a correlation between the concentrations of ClO and O_3 in the polar vortex.

The situation is not as severe in the warmer Arctic region, where the vortex does not persist quite as long. Recent studies show that similar processes also occur in that region, but to a lesser extent than over Antarctica.

Ways to Curb Ozone Depletion

Recognizing the serious implications of the loss of ozone in the stratosphere, nations throughout the world have acknowledged the need to curtail drastically or to stop totally the production of CFCs. An international treaty—the Montreal protocol—signed by most industrialized nations in 1987, set targets for cutbacks in CFC production. To make these reductions possible, an intense effort is under way to find CFC substitutes that are not harmful to the ozone layer. One group of compounds that has shown promise as replacements for CFCs are the hydrofluorocarbons (HFCs), including CF_3CFH_2, CF_3CF_2H, CF_3CH_3, and CF_2HCH_3. The presence of hydrogen atoms in these compounds makes them susceptible to oxidation in the troposphere by the hydroxyl radical; for example,

$$CF_3CFH_2 + \cdot OH \rightarrow CF_3CFH\cdot + H_2O$$

The $CF_3CFH\cdot$ fragment reacts with oxygen, eventually decomposing to CO_2, water, and hydrogen halides that are removed by rainwater. Because these compounds do not contain chlorine, HFCs will not promote the destruction of O_3, even if they diffuse into the stratosphere.

In spite of efforts to limit the amount of CFCs produced, chlorine levels in the atmosphere will continue to increase over the coming decades, because large quantities of CFCs currently in refrigerators, air conditioners, and foams will eventually be released into the atmosphere. The extent of the destruction of the ozone layer remains to be seen, but the consequences of not curbing CFCs are not in doubt.

16.6 Chemiluminescence and Bioluminescence

Instead of heat, energy released in a chemical reaction may appear as light. For example, a product molecule formed in an electronically excited state may return to the ground state with the emission of light, or it may transfer its excess energy to another molecule, which then becomes the light emitter. This type of luminescence differs from fluorescence and phosphorescence discussed in Chapter 14 in that light emission is a result of a chemical reaction, not from the direct irradiation of a molecule.

Chemiluminescence

The popular lightsticks are based on chemiluminescent reactions.

Perhaps the simplest chemiluminescent reaction, and one that has been studied extensively, is the reaction between nitric oxide and ozone:

$$NO + O_3 \rightarrow NO_2^* + O_2$$

$$NO_2^* \rightarrow NO_2 + h\nu$$

The electronically excited molecule (denoted by an asterisk) relaxes to the ground state with emission of orange light. A practical application of this reaction is in the monitoring of NO concentrations in the atmosphere from light intensity measurements.

Another well-studied chemiluminescent reaction is the oxidation of luminol (5-amino-2,3-dihydro-1,4-phthalazinedione). When luminol is treated with a base, hydrogen peroxide, and potassium ferricyanide, an intense blue light is produced. The steps are as follows:

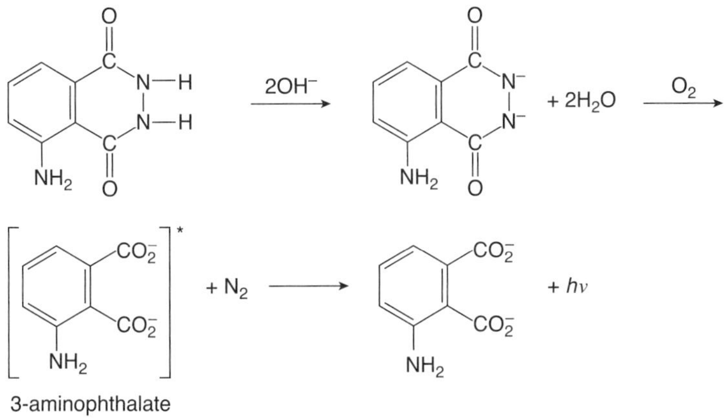

Spectroscopic studies show that the species responsible for light emission is the electronically excited 3-aminophthalate dianion, denoted by the asterisk. The transition is from the first excited singlet state to the ground state.

Bioluminescence

Bioluminescent reactions generally involve enzyme-catalyzed reactions with oxygen. A large number of organisms, such as bacteria, fungi, corals, clams, and insects, have developed the ability to emit light. Of these, the best known case is undoubtedly the firefly and its larvae, called glowworms. The chemistry of bioluminescence in fireflies has been fairly well clarified. The reaction involves luciferin, ATP, oxygen, and the enzyme luciferase (molar mass about 100,000 g). The first step is the formation of luciferyl adenylate:

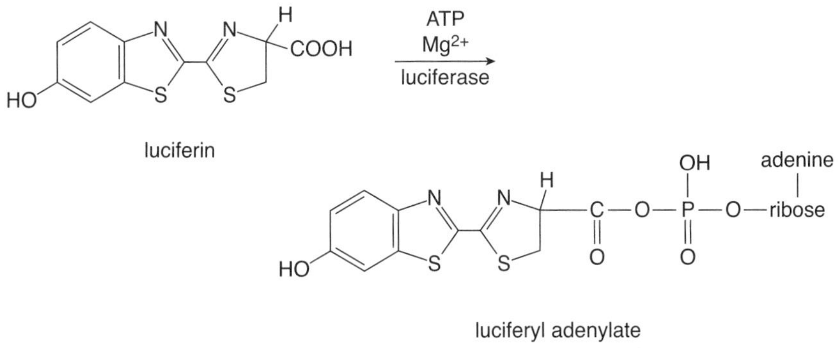

In the presence of molecular oxygen, luciferyl adenylate undergoes bioluminescence.* The firefly *Photinus pyralis* emits light at about 560 nm, which is in the yellow-green region, whereas other species of firefly emit at somewhat longer wavelengths. Since all the species have the same luciferin molecule, the variety of colors observed is probably caused by the difference in the enzyme luciferase, whose structure and/or

* The combination luciferin-luciferase can be used to assay ATP.

conformation may change from species to species. *In vitro* studies have shown that the actual wavelength of emission depends on the pH of the medium.

Bioluminescence has often been described as "cold light," meaning that the production of light from a biochemical reaction is complete and no heat is given off. This is an accurate description for fireflies, since their quantum efficiency is indeed close to 100%. On the other hand, a chemiluminescent reaction such as the oxidation of luminol has an overall efficiency of about 1%; that is, the ratio of photons emitted to the number of luminol molecules reacted is 0.01.

What is the significance of bioluminescence? A study of animal evolution provides a partial answer to this question. During the gradual appearance of oxygen in the Earth's atmosphere, the anaerobic organisms then in existence had to get rid of highly toxic O_2 molecules. One way to eliminate O_2 is to reduce it to water. The energy liberated in such a reaction is sufficient to excite certain molecules or intermediates, which can then emit light. The mechanism is obviously not necessary today, because these organisms have already made a transition from anaerobic to aerobic pathways. Yet this unnecessary byproduct, light, evolved a useful purpose, for the flashing light of fireflies now serves as a mating signal.*

16.7 Biological Effects of Radiation

Some of the detrimental effects of radiation have been mentioned elsewhere in this chapter. Radiation has also been used successfully to treat disease. In this section, we shall discuss both the harmful and beneficial effects of radiation.

Sunlight and Skin Cancer

In the United States, about 1 million new cases of skin cancer occur annually, rivaling the incidence of all other types of cancer combined. Of these, approximately 40,000 are malignant melanoma, which has an 18% fatality rate. In the vast majority of cases, skin cancer is attributable to solar radiation.

The harmful radiation from the sun is mainly in the UV range, which is divided into three regions called UV-C (200–280 nm), UV-B (280–320 nm), and UV-A (320–400 nm). The most harmful type is UV-C. Fortunately, most UV-C radiation is absorbed by the ozone layer in the stratosphere. UV-B reaches Earth's surface in small amounts and is responsible for the redness and blistered, peeling skin associated with sunburn. (The redness is due to increased flow of blood vessels beneath the skin, which widen in response to the radiation.) UV-B rays are blamed for skin cancer. The least energetic radiation, UV-A, causes what we call a "suntan."

When UV-A or UV-B strikes the pigment-producing melanocyte cells beneath the skin, they produce a UV-absorbing dark pigment called *melanin*. This substance screens out part of the radiation and helps to minimize damage to the underlying layers of skin. In addition, the melanocytes start dividing more rapidly than usual to replace damaged cells in the outer layer. Normally it takes a few weeks for the new cells to reach the surface, where they are shed as part of the skin's renewal cycle. Prolonged exposure to the sun speeds up this process, so that a large number of melanin-containing cells arrive at the surface in a few days, giving the skin a suntanned appearance.

---

* See "Synchronous Fireflies," J. Buck and E. Buck, *Sci. Am.* May 1976.

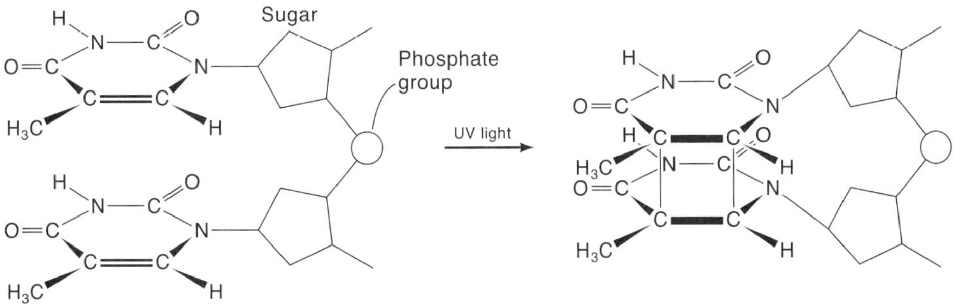

Figure 16.10
Dimerization of adjacent thymine bases on the same strand of a DNA molecule.

To understand sunlight-induced cancer, we must look at the effect of UV radiation on DNA. DNA molecules absorb radiation strongly between 200 nm and 300 nm, with a maximum at about 260 nm (see Figure 14.4). Unlike the pyrimidines, the purines (adenine and guanine) are much less sensitive to UV light. Experiments suggest that the dimerization of thymine is the most important photochemical reaction to occur in DNA molecules. Normally, a thymine solution is relatively insensitive to UV light, but when a frozen solution of thymine is irradiated with UV, thymine dimer is formed in high yield. The fact that thymine dimers are formed only in the frozen state suggests that the reaction requires not only that two thymine molecules be close to each other, but also that they be held in a certain orientation. Two adjacent thymine base pairs are both close and fixed in position on the *same* strand of a DNA molecule. We would then expect thymine dimers to form when DNA molecules are exposed to UV radiation, and this is indeed the case (Figure 16.10). This is probably the first step in the mutation of specific genes within skin cells. For example, a cell may reproduce excessively if the mutation turns a normal gene into a growth promoter (an oncogene). Alternatively, mutation may inactivate a gene that normally limits cell growth (a tumor suppresser gene).

The thymine dimer can be restored to its monomeric form through photoreactivation, a process in which light-absorbing enzymes called photoreactivating enzymes, or DNA photolyases, repair DNA by utilizing the energy of visible light to break the cyclobutane ring of the dimer. Photolyases are monomeric proteins with two flavin cofactors that act as chromophores. A photolyase enzyme binds the DNA substrate in a light-independent reaction. Then one chromophore of the bound enzyme absorbs a visible photon and, by dipole–dipole interaction, transfers energy to the second flavin, which, in turn, transfers an electron to the thymine dimer in DNA. Subsequently, the dimer breaks up. Back-electron transfer restores the functional form of the chromophore flavin, and the enzyme is ready for a new cycle of catalysis. There is no net redox change in the dimer splitting. Interestingly, photolyases are the only light-driven enzymes that are not involved in photosynthesis.

Photomedicine

Photomedicine is the application of the principles of photochemistry and photobiology to the diagnosis and therapy of disease. Interest in this subject dates back to the 19th century, when it was found that facial lesions resulting from tuberculosis could be cured by irradiation with UV light. It was reinforced by the discovery that UV light kills microorganisms and that sunlight is effective in the treatment and prevention of vitamin D deficiency (rickets). We shall briefly discuss two examples below.

Photodynamic Therapy. Photodynamic therapy utilizes light to generate a reactive species that can destroy cancerous cells. A patient is intravenously injected with a solution containing a compound called a *photosensitizer* (S). After a day or so, the solution has been distributed throughout the body. Specially designed fiber-optic probes are then inserted in the region containing affected cells, and the photosensitizer is irradiated with a laser. The following reactions take place:

Intersystem crossing is the radiationless transition of a molecule from one electronic state into another with a different spin multiplicity (see Section 14.2).

$S_0 \xrightarrow{h\nu} S_1$	singlet–singlet excitation
$S_1 \rightarrow S_0 + h\nu$	fluorescence
$S_1 \rightarrow T_1$	intersystem crossing
$T_1 + {}^3O_2 \rightarrow S_0 + {}^1O_2$	energy transfer to produce singlet oxygen

where S_0 and S_1 are the ground and first excited singlet state, and T_1 is the lowest triplet state of the photosensitizer, and ${}^3O_2$ and ${}^1O_2$ are the triplet state and singlet state of molecular oxygen.* Singlet oxygen is a highly reactive species, and it has the ability to destroy the neighboring tumor cells.

To be successful as an agent for photodynamic therapy, a photosensitizer must satisfy three requirements. First, it must be nontoxic and preferably water soluble. Second, it should absorb strongly in the red region of the visible spectrum or in the near-IR region. The reason is that after injection, the solution containing the photosensitizer is distributed throughout the body, including the skin. If the compound absorbs appreciably in the shorter wavelengths of the visible or UV light, the patient becomes sensitized to photodamage by sunlight, a side effect that is clearly undesirable. Third, to minimize damage to healthy tissues, the photosensitizer should be selectively retained by the cancerous cells. In this respect, the location of the compound can be monitored by studying its fluorescence.

The prospect of photodynamic therapy is promising. Besides treating cancer, photosensitizers also appear to be highly effective in killing bacteria. At present, much effort is being expended to the synthesis of photosensitizers (mostly compounds with complex structures containing the porphyrin ring system) with suitable photochemical and chemical properties for clinical applications.

Light-Activated Drugs

The ancient Egyptians recognized that a common plant called *Ammi majus* possesses medicinal properties that are elicited by light. *Ammi majus* is a weed that grows on the banks of the Nile. The physicians of the time found that people became unusually prone to sunburn after ingesting the plant. Consequently, the plant was used to treat certain skin disorders. Chemical analysis has shown that the active ingredients

---

* According to Hund's rule, the lowest or the ground electronic state of O_2 is a triplet, with two unpaired electrons.

in the plant belong to a class of compounds called psoralens, an example of which is 8-methoxypsoralen, or 8-MOP:

8-Methoxypsoralen (8-MOP)

Clinical studies have shown that 8-MOP is an effective anticancer drug that can be activated by light.

Cutaneous T-cell lymphoma (CTCL) is a malignancy of white blood cells; it has a poor prognosis. Treatment with 8-MOP and light, however, has yielded very promising results in CTCL. In a typical procedure, about 500 mL of blood (roughly the same volume as that given in a single blood donation) is drawn from a patient. Centrifugation separates the blood into three components: erythrocytes (red blood cells), leukocytes (white blood cells), and plasma (liquid portion of the blood). The leukocytes and plasma are combined with a saline solution to which 8-MOP is added. This solution is then irradiated with high-intensity UV-A light. After irradiation, the erythrocytes are recombined with the remainder of the original, drawn blood, which is then transfused into the patient. Without light, 8-MOP is inert and is totally harmless to the body.

Figure 16.11 shows that the size and shape of 8-MOP enable it to slide between base pairs of DNA molecules in the cell nucleus of a leukocyte. Upon irradiation, it forms chemical bonds with the bases on *both* strands. The strong chemical bonds now prevent the DNA from replicating, resulting in cell death. The treatment is nonspecific and damages both malignant cells and healthy ones. Interestingly, when the damaged malignant cells are returned to the patient's bloodstream, they somehow induce the immune system to destroy the malignant cells that have *not* been treated with 8-MOP and radiation. Although more work needs to be done to find drugs that have greater affinity for DNA and ways to activate them inside the body, there is little doubt that light-activated drugs will play an important role as therapeutic agents against cancer and other diseases.

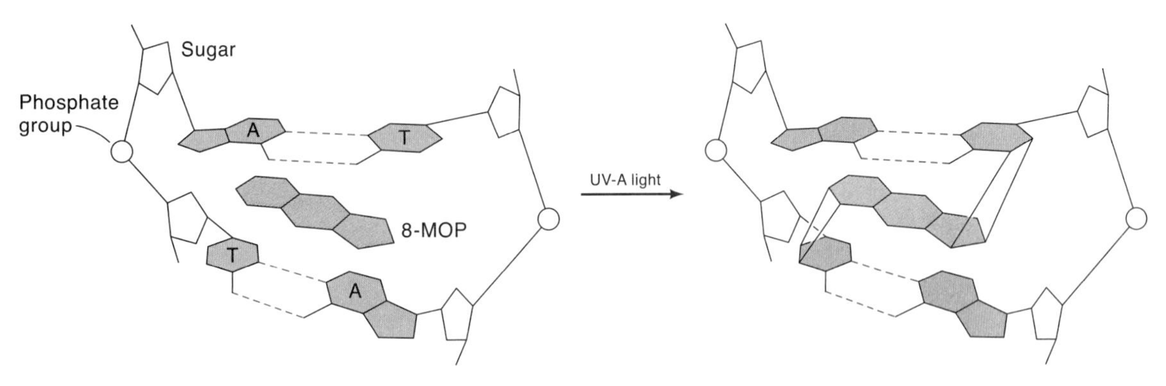

Figure 16.11
Schematic diagram showing chemical bond formation between 8-MOP and the thymine (T) molecules on different strands of a DNA molecule. This linkage keeps the strands from unwinding for replication.

Key Equations

$$\Phi = \frac{\text{number of molecules produced}}{\text{number of photons absorbed}} \quad \text{(Photochemical quantum yield)} \quad (16.1)$$

$$\Phi = \frac{\text{number of moles of product formed}}{\text{number of einsteins absorbed}} \quad \text{(Photochemical quantum yield)} \quad (16.2)$$

$$\Phi_P = \frac{\text{rate of product formation from A}^*}{\text{total rate of formation of A}^*} \quad \text{(Quantum yield of product formation)} \quad (16.3)$$

Suggestions for Further Reading

BOOKS

Birks, J. W., J. G. Calvert, and R. E. Sievers, Eds., *The Chemistry of the Atmosphere: Its Impact on Global Change*, American Chemical Society, Washington, DC, 1993.

Brasseur, G. P., J. J. Orlando, and G. S. Tyndall, *Atmospheric Chemistry and Global Change*, Oxford University Press, New York, 1999.

Brimblecombe, P., *Air Composition and Chemistry*, Cambridge University Press, New York, 1986.

Calvert, J. G., and J. N. Pitts, Jr., *Photochemistry*, John Wiley & Sons, New York, 1966.

Finlayson-Pitts, B. J., and J. N. Pitts, Jr., *Chemistry of the Upper and Lower Atmosphere*, Academic Press, New York, 1999.

Graedel, T. E., and P. J. Crutzen, *Atmospheric Change: An Earth System Perspective*, W. H. Freeman, New York, 1993.

Harm, W., *Biological Effects of Ultraviolet Radiation*, Cambridge University Press, New York, 1980.

Isidorov, V. A., *Organic Chemistry of the Earth's Atmosphere*, Springer-Verlag, New York, 1990.

Middlebrook, A. M., and M. A. Tolbert, *Stratospheric Ozone Depletion*, University Science Books, Sausalito, CA, 2000.

Seinfeld, J. H., and S. N. Pandis, *Atmospheric Chemistry and Physics: From Air Pollution to Climate Change*, John Wiley & Sons, New York, 1998.

Suppan, P., *Chemistry and Light*, Royal Society of Chemistry, London, 1994.

Turro, N. J., V. Ramamurthy, and J. C. Scaiano, *Principles of Molecular Photochemistry: An Introduction*, University Science Books, Sausalito, CA, 2009.

Wayne, R. P., *Chemistry of Atmospheres: An Introduction to the Atmospheres of Earth, the Planets, and Their Satellites*, 2nd ed., Clarendon Press, Oxford, 1991.

ARTICLES

General

"The Fates of Electronic Excitation Energy," H. H. Jaffé and A. L. Miller, *J. Chem. Educ.* **43**, 469 (1966).

"Photochemical Reactivity," N. J. Turro, *J. Chem. Educ.* **44**, 536 (1967).

"The Chemical Effects of Light," G. Oster, *Sci. Am.* September 1968.
"Photochemistry of Organic Compounds," J. S. Swenton, *J. Chem. Educ.* **46**, 7, 217 (1969).
"Photochemical Reactions of Natural Macromolecules," D. C. Neckers, *J. Chem. Educ.* **50**, 164 (1973).
"Photochemical Reactions of Tris(oxalato)Iron(III)," A. D. Baker, A. Casadavell, H. D. Gafney, and M. Gellender, *J. Chem. Educ.* **57**, 317 (1980).
"Photochemistry and Beer," A. Vogler and H. Kunkely, *J. Chem. Educ.* **59**, 25 (1982).
"Photochemistry in Organized Media," J. H. Fendler, *J. Chem. Educ.* **60**, 872 (1983).
"Atmospheric Physics," F. W. Taylor, *Encyclopedia of Applied Physics*, Trigg, G. L., Ed., VCH Publishers, New York (1994), Vol. 1, p. 489.
"Reactions Induced by Light," K. L. Stevenson and O. Horváth, *Encyclopedia of Applied Physics*, Trigg, G. L., Ed., VCH Publishers, New York (1996), Vol. 16, p. 117.
"Does a Photochemical Reaction Have a Kinetic Order?," S. Toby, *J. Chem. Educ.* **82**, 37 (2005).
"Photochemistry and Photophysics in the Laboratory. Showing the Role of Radiationless and Radiative Decay of Excited States," J. S. S. de Melo, C. Cabral, and H. D. Burrows, *Chem. Educator* [Online] **12**, 403 (2007) DOI 10.1333/s00897072096a.
"Resveratrol Photoisomerization: An Integrative Guided-Inquiry Experiment," E. Bernard, P. Britz-McKibbin, and N. Gernigon, *J. Chem. Educ.* **84**, 1159 (2007).
"Nanosecond Laser Induced Transient Absorption Flash Photolysis Experiment for Undergraduate Physical Chemistry," R. Oyola and R. Arce, *Chem. Educator* [Online] **15**, 365 (2010) DOI 10.1007/s00897102303a.

Chemistry in the Atmosphere
"The Carbon Cycle," B. Bolin, *Sci. Am.* September 1970.
"The Nitrogen Cycle," C. C. Delwiche, *Sci. Am.* September 1970.
"The Oxygen Cycle," P. Cloud and A. Gibor, *Sci. Am.* September 1970.
"Molecular Orbitals and Air Pollution," B. M. Fung, *J. Chem. Educ.* **49**, 26 (1972). See also p. 654 of the same volume.
"The Atmosphere," A. P. Ingersoll, *Sci. Am.* September 1983.
"Air Pollution: Components, Causes, and Cures," A. G. Russell, *Encyclopedia of Applied Physics* Trigg, G. L., Ed., VCH Publishers, New York (1991), Vol. 1, p. 489.
"Thermal Physics (and Some Chemistry) of the Atmosphere," S. K. Lower, *J. Chem. Educ.* **75**, 837 (1998).

The Greenhouse Effect
"The Carbon Dioxide Question," G. M. Woodwell, *Sci. Am.* January 1978.
"Carbon Dioxide and World Climate," R. Revelle, *Sci. Am.* August 1982.
"Climate Modeling," S. H. Schneider, *Sci. Am.* May 1987.
"How Climate Evolved on the Terrestrial Planets," J. F. Kasting, O. B. Toon, and J. B. Pollack, *Sci. Am.* February 1988.
"Modeling the Geochemical Carbon Cycle," R. A. Berner and A. C. Lasaga, *Sci. Am.* March 1989.
"Global Climate Change," R. A. Houghton and G. M. Woodwell, *Sci. Am.* April 1989.
"The Changing Climate," S. H. Schneider, *Sci. Am.* September 1989.
"Carbon Monoxide and the Burning Earth," R. E. Newell, H. G. Reichle, Jr., and W. Seiler, *Sci. Am.* October 1989.
"The Great Climate Debate," R. M. White, *Sci. Am.* July 1990.
"The Global Carbon Cycle," W. M. Post, T.-H.Peng, W. R. Emanuel, A. W. King, V. H. Dale, and D. L. DeAngelis, *Am. Sci.* **78**, 310 (1990).
"Global Warming Trends," P. D. Jones and T. M. L. Wigley, *Sci. Am.* August 1990.
"Methane, Plants, and Climate Change," F. Keppler, *Sci. Am.* February 2007.
"The Physical Science behind Climate Change," W. Collins, R. Colman, J. Haywood, M. R. Manning, and P. Mote, *Sci. Am.* August 2007.

Photochemical Smog

"The Control of Air Pollutants," A. J. Haagen-Smit, *Sci. Am.* January 1964.

"Computer Modeling of Photochemical Smog Formation," B. J. Huebert, *J. Chem. Educ.* **51**, 644 (1974).

"The Changing Atmosphere," T. E. Graedel and P. J. Crutzun, *Sci. Am.* September 1989.

"Tropospheric Air Pollution: Ozone, Airborne Toxics, Polycyclic Aromatic Hydrocarbons, and Particles," B. J. Finlayson-Pitts and J. N. Pitts, Jr., *Science* **276**, 1045 (1997).

Depletion of Ozone in the Stratosphere

"Chlorofluorocarbons and Stratospheric Ozone," S. Elliot and F. S. Rowland, *J. Chem. Educ.* **64**, 387 (1987).

"The Antarctic Ozone Hole," R. S. Stolarski, *Sci. Am.* January 1988.

"The Absorption of UV Light by Ozone," E. Koubek and J. O. Glanville, *J. Chem. Educ.* **66**, 338 (1989).

"Polar Stratospheric Clouds and Ozone Depletion," O. B. Toon and R. P. Turco, *Sci. Am.* January 1991.

"Polar Ozone Depletion," M. J. Molina, *Ang. Chem. Int. Ed.*, **35**, 1778 (1996).

Chemiluminescence and Bioluminescence

"Biological Luminescence," W. D. McElroy and H. H. Seliger, *Sci. Am.* December 1972.

"The Origin of Bioluminescence," H. H. Seliger, *Photochem. Photobiol.* **21**, 355 (1975).

"Biological Light: α-Peroxylates as Bioluminescent Intermediates," W. Adam, *J. Chem. Educ.* **52**, 97 (1975).

"Chemiluminescence," S. Gill and L. K. Brice, *J. Chem. Educ.* **61**, 713 (1984).

"Turning on the Light: Lessons from Luminescence," P. B. O'Hara, C. Engelson, and W. St. Peter, *J. Chem. Educ.* **82**, 49 (2005).

"Chemiluminescent Oscillating Demonstrations: The Chemical Buoy, the Lighting Wave, and the Ghostly Cylinder," H. E. Prypsztejn, *J. Chem. Educ.* **82**, 53 (2005).

"Luminescent Molecular Thermometer," S. Uchiyama, A. P. de Silva, and K. Iwai, *J. Chem. Educ.* **83**, 720 (2006).

Biological Effects of Radiation

"Ultraviolet Radiation and Nucleic Acid," R. A. Deering, *Sci. Am.* December 1962.

"The Repair of DNA," P. C. Hanawalt and R. H. Haynes, *Sci. Am.* February 1967.

"The Chemical Effects of Light," G. Oster, *Sci. Am.* September 1968.

"The Effects of Light on the Human Body," R. J. Wurtman, *Sci. Am.* July 1975.

"Biochemical Effects of Excited State Molecular Oxygen," J. Bland, *J. Chem. Educ.* **53**, 274 (1976).

"Inducible Repair of DNA," P. Howard-Flanders, *Sci. Am.* November 1981.

"Radiation Sensitization in Cancer Therapy," C. L. Greenstock, *J. Chem. Educ.* **58**, 156 (1981).

"The Biological Effects of Low-Level Ionizing Radiation," A. C. Upton, *Sci. Am.* February 1982.

"Phototherapy and the Treatment of Hyperbilirubinemia: A Demonstration of Intra- versus Intermolecular Hydrogen Bonding," A. C. Wilbraham, *J. Chem. Educ.* **61**, 540 (1984).

"Light Activated Drugs," R. L. Edelson, *Sci. Am.* August 1988.

"Effect of UV Irradiation on DNA as Studied by Its Thermal Denaturation," C. M. Lovett, Jr., T. N. Fitsgibbon, and R. Chang, *J. Chem. Educ.* **66**, 526 (1989).

"DNA, Sunlight, and Skin Cancer," J. S. Taylor, *J. Chem. Educ.* **67**, 835 (1990).

"A Simple UV Experiment of Environmental Significance," D. W. Daniel, *J. Chem. Educ.* **71**, 83 (1994).

"Structure and Function of DNA Photolyase," A. Sancar, *Biochemistry*, **33**, 2 (1994).

"Sunlight and Skin Cancer," D. J. Leffell and D. E. Brash, *Sci. Am.* July 1996.
"The Photochemistry of Sunscreens," D. R. Kimbrough, *J. Chem. Educ.* **74**, 51 (1997).
"The Spectrophotometric Analysis and Modeling of Sunscreens," C. Walters, A. Keeney, C. T. Wigal, C. R. Johnston, and R. D. Cornelius, *J. Chem. Educ.* **74**, 99 (1997).
"Let There Be Light and Let It Heal," A. M. Rouhi, *Chem. & Eng. News* **76**, 22 (1998).
"Photodynamic Therapy: The Sensitization of Cancer Cells to Light," J. Miller, *J. Chem. Educ.* **76**, 592 (1999).
"Photochemotherapy: Light-Dependent Therapies in Medicine," E. P. Zovinka and D. R. Sunseri, *J. Chem. Educ.* **79**, 1331 (2002).
"New Light on Medicine," N. Lane, *Sci. Am.* January, 2003.
"Thymine Dimerization in DNA is an Ultrafast Photoreaction," W. J. Schreier, T. E. Schrader, F. O. Koller, P. Gilch, C. E. Crespo-Hernández, V. N. Swaminathan, T. Carell, W. Zinth, and B. Kohler, *Science* **315**, 625 (2007).

Problems

General

16.1 In a photochemical reaction, 428.3 kJ mol^{-1} of energy input is required to break a chemical bond. What wavelength must be employed in the irradiation?

16.2 Convert 450 nm to kJ $einstein^{-1}$.

16.3 Design an experiment that would allow you to measure the rate of absorption of light by a solution.

16.4 An organic molecule absorbs light at 549.6 nm. If 0.031 mole of the molecule is excited by 1.43 einsteins of light, what is the quantum efficiency for this process? Also, calculate the total energy taken up in the process.

16.5 In the photochemical decomposition of a certain compound, light intensity of 5.4×10^{-6} einsteins s^{-1} was employed. Assuming the most favorable conditions, estimate the time needed to decompose 1 mole of the compound.

16.6 The first-order rate constants for the fluorescence and phosphorescence of naphthalene ($C_{10}H_8$) are 4.5×10^7 s^{-1} and 0.50 s^{-1}, respectively. Calculate how long it takes for 1.0% of fluorescence and phosphorescence to occur following termination of excitation.

Atmospheric Chemistry

16.7 A barometer that has a cross-sectional area of 1 cm^2 at sea level measures a pressure of 76.0 cm of mercury. The pressure exerted by this column of mercury is equal to the pressure exerted by all the air molecules on 1 cm^2 of Earth's surface. Given that the density of mercury is 13.6 g mL^{-1} and the average radius of Earth is 6371 km, calculate the total mass of Earth's atmosphere in kilograms. Is your result an upper or lower estimate of the mass? Explain. (*Hint:* The surface area of a sphere of radius r is $4\pi r^2$.)

16.8 Name the major source of heat that originates from Earth.

16.9 The highly reactive ·OH radical (a species with an unpaired electron) is believed to be involved in some atmospheric processes. Table 3.4 lists the bond enthalpy for the oxygen-to-hydrogen bond in OH as 460 kJ mol^{-1}. Determine the longest wavelength (in nm) of radiation that can bring about the following reaction:

$$\cdot\mathrm{OH}(g) \rightarrow \mathrm{O}(g) + \cdot\mathrm{H}(g)$$

16.10 The hydroxyl radical in the atmosphere is most effectively removed by hydrocarbons such as methane according to the second-order reaction

$$\cdot\mathrm{OH} + \mathrm{CH_4} \rightarrow \mathrm{H_2O} + \mathrm{CH_3}\cdot$$

Given that the second-order rate constant is 4.6×10^6 L mol^{-1} s^{-1}, calculate the lifetime of the radical at 25°C if the concentration of CH_4 is 1.7×10^3 ppb by volume. (*Hint:* The lifetime of the radical is given by $1/k[\mathrm{CH_4}]$.)

16.11 Describe three human activities that generate carbon dioxide. List two major mechanisms for the uptake of carbon dioxide.

16.12 Deforestation contributes to the greenhouse effect in two ways. What are they?

16.13 How does an increase in world population enhance the greenhouse effect?

16.14 Is ozone a greenhouse gas? Sketch three ways an ozone molecule can vibrate.

16.15 Suggest a gas other than carbon dioxide that scientists can study to substantiate the fact that CO_2 concentration is steadily increasing in the atmosphere.

16.16 Which of the following settings is the most suitable for photochemical smog formation? **(a)** Gobi Desert at noon in June, **(b)** New York City at 1 p.m. in July, or **(c)** Boston at noon in January. Explain your choice.

16.17 On a smoggy day in a certain city, the ozone concentration was 0.42 ppm by volume. Calculate the partial pressure of ozone (in atm) and the number of ozone molecules per liter of air if the temperature and pressure were 20.0°C and 748 mmHg, respectively.

16.18 The gas-phase decomposition of peroxyacetyl nitrate (PAN) obeys first-order kinetics:

$$CH_3COOONO_2 \rightarrow CH_3COOO + NO_2$$

with a rate constant of $4.9 \times 10^{-4}\ s^{-1}$. Calculate the rate of decomposition in $M\ s^{-1}$ if the concentration of PAN is 0.55 ppm by volume. Assume STP conditions.

16.19 Assume that the formation of nitrogen dioxide,

$$2NO(g) + O_2(g) \rightarrow 2NO_2(g)$$

is an elementary reaction. **(a)** Write the rate law for this reaction. **(b)** A sample of air at a certain temperature is contaminated with 2.0 ppm of NO by volume. Under these conditions, can the rate law be simplified? If so, write the simplified rate law. **(c)** Under the conditions described in **(b)**, the half-life of the reaction has been estimated to be 6.4×10^3 min. What would the half-life be if the initial concentration of NO were 10 ppm?

16.20 The safety limits of ozone and carbon monoxide are 120 ppb by volume and 9 ppm by volume, respectively. Why does ozone have a lower limit?

16.21 Ozone in the troposphere is formed by the following steps:

$$NO_2 \xrightarrow{h\nu} NO + O \qquad (1)$$

$$O + O_2 \rightarrow O_3 \qquad (2)$$

The first step is initiated by the absorption of visible light (NO_2 is a brown gas.) Calculate the longest wavelength required for step 1 at 25°C. (*Hint:* You need to first calculate the value of $\Delta_r H$ and hence the value of $\Delta_r U$ for step 1. Next, determine the wavelength for decomposing NO_2 from $\Delta_r U$.)

16.22 Given that the quantity of ozone in the stratosphere is equivalent to a 3.0-mm-thick layer of ozone on Earth at 1 atm and 25°C, calculate the number of O_3 molecules in the stratosphere and their mass in kilograms. See Problem 16.7 for other information.

16.23 Referring to the answer in Problem 16.22 and assuming that the level of ozone in the stratosphere has already fallen 6.0%, calculate the number of kilograms of ozone that must be manufactured on a daily basis so that we can restore the ozone to the original level in 100 years. If ozone is made according to the process $3O_2(g) \rightarrow 2O_3(g)$, how many kilojoules of energy input must be supplied to drive the reaction?

16.24 Why are CFCs not decomposed by UV radiation in the troposphere?

16.25 The average bond enthalpies of the C—Cl and C—F bonds are 340 kJ mol^{-1} and 485 kJ mol^{-1}, respectively. Based on this information, explain why the C—Cl bond in a CFC molecule is preferentially broken by solar radiation at 250 nm.

16.26 Like CFCs, certain bromine-containing compounds, such as CF_3Br, can participate in the destruction of ozone by a similar mechanism starting with the Br atom:

$$CF_3Br \xrightarrow{h\nu} CF_3 + Br$$

Given that the average C—Br bond enthalpy is 276 kJ mol^{-1}, estimate the longest wavelength required to break this bond. Will the decomposition of CF_3Br occur in the troposphere or in both the troposphere and stratosphere?

16.27 Draw Lewis structures for chlorine nitrate ($ClONO_2$) and chlorine monoxide (ClO).

16.28 Why are CFCs more effective greenhouse gases than methane and carbon dioxide?

16.29 One suggestion for slowing down the destruction of ozone in the stratosphere is to spray the region with hydrocarbons such as ethane and propane. How does this method work? What is the drawback of this procedure if used on a large scale for an extended period of time?

16.30 Calculate the standard enthalpy of formation ($\Delta_f \overline{H}^\circ$) of ClO from the following bond dissociation enthalpies: Cl_2: 242.7 kJ mol^{-1}; O_2: 498.8 kJ mol^{-1}; ClO: 206 kJ mol^{-1}.

Additional Problems

16.31 Why does one have to irradiate a sample for hours or even days to achieve acceptable yields in some photochemical reactions even though the lifetimes of excited electronic states are of the order of micro- or nanoseconds? Assume that the rate of light absorption is 2.0×10^{19} photons s^{-1}.

16.32 The transparency of a certain type of sunglasses to light depends on the intensity of light in the environment. The lenses are clear in dimly lit rooms but darken when the wearer goes outdoors. The material responsible for this change is the very tiny AgCl crystals incorporated in the glass. Suggest a photochemical mechanism that would account for this change.

16.33 Suppose that an excited singlet, S_1, can be deactivated by three different mechanisms whose rate constants are k_1, k_2, and k_3. The rate of decay is given by $-d[S_1]/dt = (k_1 + k_2 + k_3)[S_1]$. **(a)** If τ is the mean lifetime, that is, the time required for $[S_1]$ to decrease to $1/e$ or 0.368 of the original value, show that $(k_1 + k_2 + k_3)\tau = 1$. **(b)** The overall rate constant, k, is given by

$$\frac{1}{\tau} = k = k_1 + k_2 + k_3 = \frac{1}{\tau_1} + \frac{1}{\tau_2} + \frac{1}{\tau_3}$$

Show that the quantum yield Φ_i is given by

$$\Phi_i = \frac{k_i}{\sum_i k_i} = \frac{\tau}{\tau_i}$$

where i denotes the ith decay mechanism. **(c)** If $\tau_1 = 10^{-7}$ s, $\tau_2 = 5 \times 10^{-8}$ s, and $\tau_3 = 10^{-8}$ s, calculate the lifetime of the singlet state and the quantum yield for the path that has τ_2.

16.34 Consider the photochemical isomerization $A \rightleftharpoons B$. At 650 nm, the quantum yields for the forward and reverse reactions are 0.73 and 0.44, respectively. If the molar absorptivities at 650 nm of A and B are 1.3×10^3 L mol^{-1} cm^{-1} and 0.47×10^3 L mol^{-1} cm^{-1}, respectively, what is the ratio $[B]/[A]$ in the photostationary state?

16.35 The molar heat capacity of a diatomic molecule is 29.1 J K^{-1} mol^{-1}. Assuming the atmosphere contains only nitrogen gas and there is no heat loss, calculate the total heat intake (in kilojoules) if the atmosphere warms up by 3°C during the next 50 years. Given that there are 1.8×10^{20} moles of diatomic molecules present, how many kilograms of ice (at the North and South Poles) will this quantity of heat melt at 0°C? (The molar heat of fusion of ice is 6.01 kJ mol^{-1}.)

16.36 In 1991, it was discovered that nitrous oxide (N_2O) is produced in the synthesis of nylon. This compound, which is released into the atmosphere, contributes *both* to the depletion of ozone in the stratosphere and to the greenhouse effect. **(a)** Write equations representing the reactions between N_2O and oxygen atoms in the stratosphere to produce nitric oxide, which is then oxidized by ozone to form nitrogen dioxide. **(b)** Is N_2O a more effective greenhouse gas than carbon dioxide? Explain. **(c)** One of the intermediates in nylon manufacture is adipic acid [$HOOC(CH_2)_4COOH$], About 2.2×10^9 kg of adipic acid are consumed every year. Estimates are that for every mole of adipic acid produced, 1 mole of N_2O is generated. What is the maximum number of moles of O_3 that can be destroyed as a result of this process per year?

16.37 The hydroxyl radical is formed by the following reactions:

$$O_3 \xrightarrow{\lambda < 320\ \text{nm}} O^* + O_2$$

$$O^* + H_2O \rightarrow 2 \cdot OH$$

where O^* denotes an electronically excited atom. **(a)** Explain why the concentration of $\cdot OH$ is so small even though the concentrations of O_3 and H_2O are quite large in the troposphere. **(b)** What property makes $\cdot OH$ a strong oxidizing agent? **(c)** The reaction between $\cdot OH$ and NO_2 contributes to acid rain. Write an equation for this process. **(d)** The hydroxyl radical can oxidize SO_2 to H_2SO_4. The first step is the formation of a neutral HSO_3 species, followed by its reaction with O_2 and H_2O to form H_2SO_4 and the hydroperoxyl radical ($HO_2\cdot$). Write equations for these processes.

16.38 Given that the collision diameter of ozone is about 4.2 Å, calculate the mean free path of ozone at sea level (1 atm and 25°C) and in the stratosphere (3×10^{-3} atm and −23°C).

16.39 Comment on the comparison that the hydroxyl radical behaves in some ways like the white blood cells in our bodies.

16.40 Account for the oscillation in atmospheric CO_2 concentration shown in Figure 16.6.

16.41 Explain why phosphorescence of ethylene has never been observed.

16.42 A light source of power 2×10^{-16} W is sufficient to be detected by the human eye. Assuming the wavelength of light is at 550 nm, calculate the number of photons that must be absorbed by rhodopsin per second. (*Hint:* Vision persists for only 1/30 of a second.)

CHAPTER 17

Intermolecular Forces

United we stand.

In Chapter 13, we discussed the covalent bond and related the force holding atoms together to the overlapping of atomic orbitals. Interactions between molecules are best explained by several types of intermolecular forces, forces responsible for phenomena such as the liquefaction of gases and the stability of proteins. A special type of interaction, hydrogen bonding, plays an important role in determining the structure and properties of DNA and water.

17.1 Intermolecular Interactions

When two neutral molecules approach each other, various interactions between the electrons and nuclei of one molecule and the electrons and nuclei of the other molecule generate potential energy. At a very large distance of separation, where there is no intermolecular interaction, we can arbitrarily set the potential energy of the system at zero. As the molecules approach each other, electrostatic attractions outweigh electrostatic repulsions, so the molecules are pulled toward each other and the potential energy of interaction is negative. This trend continues until the potential energy reaches a minimum. Beyond this point (i.e., as the distance of separation decreases further), repulsive forces predominate and the potential energy rises (becoming more positive).

A simple demonstration of intermolecular forces is to ask why the far end of a walking-stick rises when you raise the handle.

For a discussion of molecular interaction, it is useful to distinguish between force and potential energy. In mechanics, work done is force times distance. The work done (dw) in moving two interacting molecules apart by an infinitesimal distance (dr) is given by

$$dw = -Fdr \tag{17.1}$$

The sign convention for work is that if the molecules attract each other (i.e., F is negative) and dr is positive (i.e., molecules are moved farther apart), then dw is positive. An expression for the potential energy (V) of the molecules separated by distance r can be obtained as follows. Imagine that one molecule is fixed in position, and we want to know how much work is done in bringing another molecule from infinite separation to a distance r from this molecule. This work is the potential energy acquired by the system, given by

$$V = \int_{\infty}^{r} dw$$

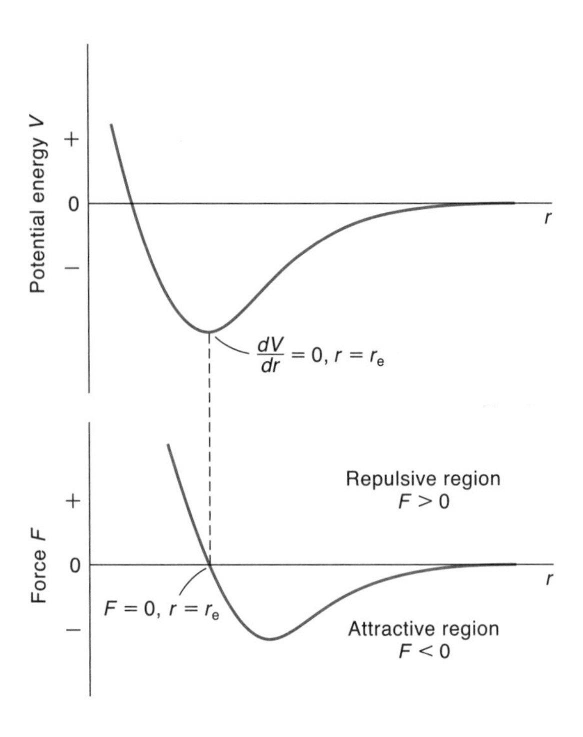

Figure 17.1
Relation between potential energy and force. Because $F = -dV/dr$, at the minimum of the $V(r)$ versus r curve, $F = 0$. Note that at $r < r_e$, where the potential energy is still negative, force becomes repulsive.

From Equation 17.1,

$$V = -\int_{\infty}^{r} F dr \tag{17.2}$$

Equation 17.2 relates the potential energy of interaction to the force between the two molecules. Differentiating Equation 17.2 with respect to r, we get

$$F = -\frac{dV}{dr} \tag{17.3}$$

Thus, force is the negative of the slope of the curve that describes the dependence of V on r. Figure 17.1 shows the relationship between potential energy and force.

17.2 The Ionic Bond

Before we study the interaction between neutral molecules, we should for comparison purposes examine the bond between a pair of ions. At elevated temperatures, ionic compounds such as NaCl vaporize to form ion pairs. The dipole moments of this and other similar alkali halide ion pairs are large, about 10 times that of hydrogen halides, indicating that the character of the bond is largely ionic. The potential energy due to the attraction between a pair of Na^+ and Cl^- ions is derived from Coulomb's law (see p. 272),

$$V = -\frac{q_{Na^+} q_{Cl^-}}{4\pi\varepsilon_0 \varepsilon r} \tag{17.4}$$

where r is the distance of separation and only the magnitude (not the signs) of the charges are shown. However, we must also include a term representing the repulsion between electrons and between nuclei on these ions. This term usually takes the form of be^{-ar} or b/r^n, where a and b are constants that are specific for a given ion pair and n is a number between 8 and 12. Using the latter term, we write the complete equation for potential energy as

$$V = -\frac{q_{\mathrm{Na}^+}q_{\mathrm{Cl}^-}}{4\pi\varepsilon_0 r} + \frac{b}{r^n} \tag{17.5}$$

We assume air is the medium, so the dielectric constant $\varepsilon = 1$ in all cases.

We can solve for b by realizing that, at the minimum of the potential-energy curve (see Figure 17.1), $dV/dr = 0$, so that

$$\frac{dV}{dr} = 0 = \frac{q_{\mathrm{Na}^+}q_{\mathrm{Cl}^-}}{4\pi\varepsilon_0 r_\mathrm{e}^2} - \frac{nb}{r_\mathrm{e}^{n+1}}$$

or

$$b = \frac{q_{\mathrm{Na}^+}q_{\mathrm{Cl}^-}}{4\pi\varepsilon_0 n} r_\mathrm{e}^{n-1} \tag{17.6}$$

where r_e is the equilibrium bond length of the ion pair. Substituting Equation 17.6 into 17.5, we write

$$\begin{aligned} V_0 &= -\frac{q_{\mathrm{Na}^+}q_{\mathrm{Cl}^-}}{4\pi\varepsilon_0 r_\mathrm{e}} + \frac{q_{\mathrm{Na}^+}q_{\mathrm{Cl}^-}}{4\pi\varepsilon_0 n r_\mathrm{e}} \\ &= -\frac{q_{\mathrm{Na}^+}q_{\mathrm{Cl}^-}}{4\pi\varepsilon_0 r_\mathrm{e}}\left(1 - \frac{1}{n}\right) \end{aligned} \tag{17.7}$$

Note that V_0 denotes the potential energy associated with the most stable separation (r_e). Using the NaCl(g) bond length of 2.36 Å (236 pm), 1.602×10^{-19} C for unit charge, and $n = 10$ for the repulsion, we write, for 1 mole of Na^+ and Cl^- ion pairs,

$$\begin{aligned} V_0 &= -\frac{(1.602 \times 10^{-19}\ \mathrm{C})^2(6.022 \times 10^{23}\ \mathrm{mol}^{-1})}{4\pi(8.854 \times 10^{-12}\ \mathrm{C}^2\ \mathrm{N}^{-1}\ \mathrm{m}^{-2})(236 \times 10^{-12}\ \mathrm{m})}\left(1 - \frac{1}{10}\right) \\ &= -5.297 \times 10^5\ \mathrm{N\ m\ mol}^{-1} \\ &= -529.7\ \mathrm{kJ\ mol}^{-1} \end{aligned}$$

This is the energy given off when 1 mole of NaCl ion pairs forms from Na^+ and Cl^- ions in the gaseous state:

$$\mathrm{Na}^+(g) + \mathrm{Cl}^-(g) \rightarrow \mathrm{NaCl}(g)$$

However, the ground state of the dissociated system consists of atoms rather than ions, that is,

$$\mathrm{NaCl}(g) \rightarrow \mathrm{Na}(g) + \mathrm{Cl}(g)$$

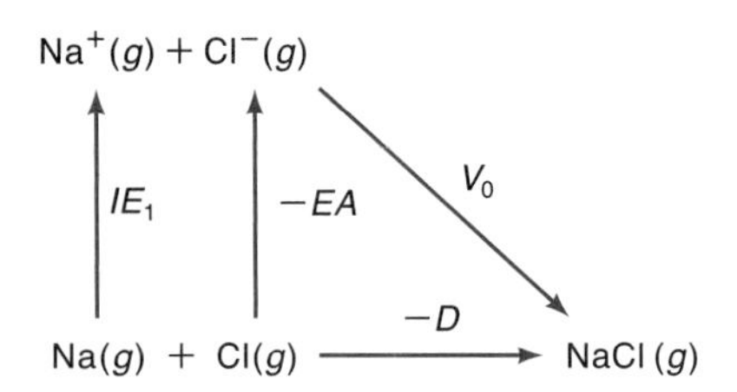

Figure 17.2
Born–Haber cycle for the formation of a gaseous NaCl ion pair.

To calculate the bond-dissociation enthalpy for this process, we apply the *Born–Haber cycle* (after Max Born and the German chemist Fritz Haber, 1868–1934) shown in Figure 17.2. Based on Hess's law, this procedure breaks the formation of NaCl(g) into separate steps involving the ionization energy of Na and electron affinity of Cl. Let D be the bond-dissociation enthalpy of NaCl(g) into atoms so that

$$-D = IE_1 - EA + V_0 \tag{17.8}$$

where IE_1 is the first ionization energy of Na, and EA is the electron affinity of Cl. Using data from Tables 12.7 and 12.8, we write

$$\begin{aligned} -D &= 495.9 \text{ kJ mol}^{-1} - 349 \text{ kJ mol}^{-1} - 529.7 \text{ kJ mol}^{-1} \\ &= -383 \text{ kJ mol}^{-1} \end{aligned}$$

Therefore, the bond-dissociation enthalpy for NaCl into Na and Cl atoms is 383 kJ mol^{-1}. This value differs somewhat from the experimentally measured value (414 kJ mol^{-1}) because the repulsion term is inexact, and the actual bond has some covalent character.

17.3 Types of Intermolecular Forces

We are now ready to survey the different types of intermolecular forces. For the sake of completeness, we shall discuss interactions between molecules as well as between ions and molecules.

Dipole–Dipole Interaction

An intermolecular interaction of the *dipole–dipole* type occurs between polar molecules, which possess permanent dipole moments. Consider the electrostatic interaction between the two dipoles μ_A and μ_B separated by distance r. In representative cases, these two dipoles can be aligned as shown in Figure 17.3. For the top example, the potential energy of interaction is given by

$$V = -\frac{2\mu_A\mu_B}{4\pi\varepsilon_0 r^3} \tag{17.9}$$

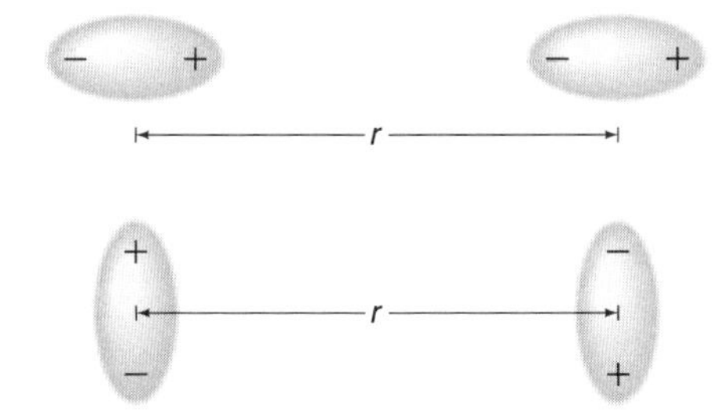

Figure 17.3
Schematic drawing showing the representative orientations of two permanent dipoles for attractive interaction.

and for the bottom pair we have

$$V = -\frac{\mu_A \mu_B}{4\pi\varepsilon_0 r^3} \tag{17.10}$$

where the negative sign indicates that the interaction is attractive; that is, energy is released when these two molecules interact. Reversing the charge signs of one of the dipoles makes V a positive quantity. Then the interaction between the two molecules is repulsive.

EXAMPLE 17.1

Two HCl molecules (μ = 1.08 D) are separated by 4.0 Å (400 pm) in air. Calculate the dipole-dipole interaction energy in kJ mol^{-1} if they are oriented end-to-end, that is, H−Cl···H−Cl.

ANSWER

We need Equation 17.9. The data are

$$\mu = 1.08 \text{ D} = 3.60 \times 10^{-30} \text{ C m (see Section 13.5)}$$

$$r = 4.0 \text{ Å} = 4.0 \times 10^{-10} \text{ m}$$

The potential energy due to this interaction is

$$\begin{aligned} V &= -\frac{2(3.60 \times 10^{-30} \text{ C m})(3.60 \times 10^{-30} \text{ C m})}{4\pi(8.854 \times 10^{-12} \text{ C}^2 \text{ N}^{-1} \text{ m}^{-2})(4.0 \times 10^{-10} \text{ m})^3} \\ &= -3.6 \times 10^{-21} \text{ N m} \\ &= -3.6 \times 10^{-21} \text{ J} \end{aligned}$$

To express the potential energy on a per mole basis, we write

$$\begin{aligned} V &= (-3.6 \times 10^{-21} \text{ J})(6.022 \times 10^{23} \text{ mol}^{-1}) \\ &= -2.2 \text{ kJ mol}^{-1} \end{aligned}$$

COMMENT

We have assumed that the dielectric constant of air is 1. In general, the dielectric constant (ε) of the medium through which the dipoles interact appears in the denominator of Equation 17.9 (see Section 7.2).

In a macroscopic system where all possible orientations of the dipoles are present, we might expect that the mean value of V would be zero, because there would be as many repulsions as attractions. But even under conditions of free rotation in a liquid or gaseous state, orientations giving rise to a lower potential energy are favored over those resulting in a higher potential energy, in accordance with the Boltzmann distribution law (see Chapter 2). A rather elaborate derivation shows that the average or net energy of interaction of permanent dipoles is given by

$$\langle V \rangle = -\frac{2}{3}\frac{\mu_A^2 \mu_B^2}{(4\pi\varepsilon_0)^2 r^6}\frac{1}{k_B T} \tag{17.11}$$

where k_B is Boltzmann's constant and T is absolute temperature. Note that V is inversely proportional to the sixth power of r, so that the energy of interaction falls off rapidly with distance. Also, V is inversely proportional to T, because at higher temperatures the average kinetic energy of the molecules is greater, a condition unfavorable to aligning dipoles for attractive interaction. In other words, the average dipole-dipole interaction will gradually approach zero with increasing temperature.

Ion–Dipole Interaction

The interaction between an ion and polar molecules was first discussed in Chapter 8 in relation to ionic hydration. The potential energy of interaction between an ion of charge q at a distance r from a dipole μ is given by

$$V = -\frac{q\mu}{4\pi\varepsilon_0 r^2} \tag{17.12}$$

Equation 17.12 holds only when the ion and the dipole lie along the same axis. This attractive interaction is mainly responsible for the dissolution of ionic compounds in polar solvents.

EXAMPLE 17.2

A sodium ion (Na^+) is situated in air at a distance of 4.0 Å (400 pm) from a HCl molecule with a dipole moment of 1.08 D. Use Equation 17.12 to calculate the potential energy of interaction in kJ mol^{-1}.

ANSWER

The data are

$$\mu = 1.08 \text{ D} = 3.60 \times 10^{-30} \text{ C m}$$

$$r = 4.0 \text{ Å} = 4.0 \times 10^{-10} \text{ m}$$

Recall that 1D = 3.336×10^{-30} C m

From Equation 17.12,

$$V = -\frac{(1.602 \times 10^{-19} \text{ C})(3.60 \times 10^{-30} \text{ C m})}{4\pi(8.854 \times 10^{-12} \text{ C}^2 \text{ N}^{-1} \text{ m}^{-2})(4.0 \times 10^{-10} \text{ m})^2}$$

$$= -3.2 \times 10^{-20} \text{ J}$$

$$= -19 \text{ kJ mol}^{-1}$$

Ion–Induced Dipole and Dipole–Induced Dipole Interactions

In a neutral nonpolar species such as the helium atom, the electron charge density is spherically symmetrical about the nucleus. If an electrically charged object, such as a positive ion, is brought near the helium atom, electrostatic interaction will cause a redistribution of the charge density (Figure 17.4). The atom will then acquire a dipole moment induced by the charged particle. The magnitude of the *induced dipole moment*, μ_{ind}, is directly proportional to the strength of the electric field, E,

$$\begin{aligned} \mu_{\text{ind}} &\propto E \\ &= \alpha' E \end{aligned} \tag{17.13}$$

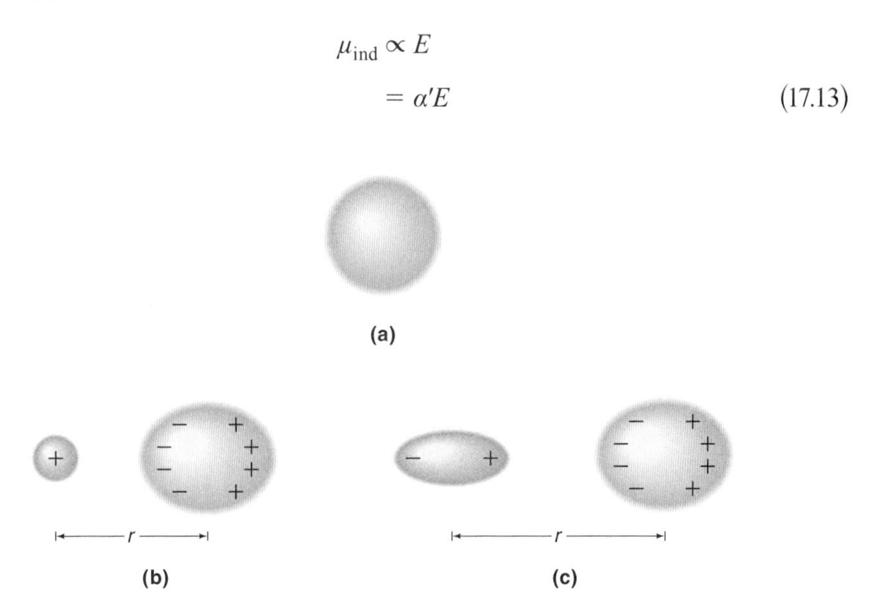

Figure 17.4
(a) An isolated helium atom has a spherically symmetric electron density. (b) Induced dipole moment in helium due to electrostatic interaction with a cation. (c) Induced dipole moment in helium due to electrostatic interaction with a permanent dipole. The plus and minus signs in helium represent partial charges due to shifts in electron density.

where α', the proportionality constant, is called the *polarizability*. The potential energy of interaction is given by the work done in bringing the helium atom from infinite distance ($E = 0$) to distance r ($E = E$); that is,

$$V = -\int_0^E \mu_{\text{ind}}\, dE$$

$$= -\int_0^E \alpha' E dE$$

$$= -\frac{1}{2}\alpha' E^2 \qquad (17.14)$$

The electric field exerted by the ion of charge q on the atom is (see Appendix 7.1)

$$E = \frac{q}{4\pi\varepsilon_0 r^2}$$

Substituting the expression for E in Equation 17.14, we get

$$V = -\frac{1}{2}\frac{\alpha' q^2}{(4\pi\varepsilon_0)^2 r^4} \qquad (17.15)$$

Qualitatively, polarizability measures how easily the electron density in an atom or a molecule can be distorted by an external electric field. Unsaturated bonds such as those in C=C, C=N, the nitro group ($-NO_2$), the phenyl group ($-C_6H_5$), the base pairs in DNA, and negative ions are highly polarizable groups. Generally, the larger the number of electrons and the more diffuse the electron charge cloud in the molecule, the greater its polarizability. As defined in Equation 17.13, however, α' has the rather awkward units of C m^2 V^{-1}. For this reason, using the polarizability α in units m^3 is more convenient, where

$$\alpha = \frac{\alpha'}{4\pi\varepsilon_0}$$

Equation 17.15 can now be expressed as

$$V = -\frac{1}{2}\frac{\alpha q^2}{4\pi\varepsilon_0 r^4} \qquad (17.16)$$

A permanent dipole can also induce a dipole moment in a nonpolar molecule (see Figure 17.4). The potential energy of interaction for the dipole–induced dipole interaction is given by

$$V = -\frac{\alpha' \mu^2}{(4\pi\varepsilon_0)^2 r^6} = -\frac{\alpha \mu^2}{4\pi\varepsilon_0 r^6} \qquad (17.17)$$

where α is the polarizability of the nonpolar molecule, and μ is the dipole moment of the polar molecule. Note that both Equations 17.16 and 17.17 are independent of temperature. This is so because the dipole moment can be induced instantaneously so that the value of V is unaffected by the thermal motion of the molecules. Table 17.1 lists the polarizability values of some atoms and simple molecules.

Table 17.1
Polarizabilities of Some Atoms and Molecules

Atom	$\alpha/10^{-30}$ m^3	Molecules	$\alpha/10^{-30}$ m^3
He	0.20	H_2	0.80
Ne	0.40	N_2	1.74
Ar	1.66	CO_2	2.91
Kr	2.54	NH_3	2.26
Xe	4.15	CH_4	2.61
I	4.96	C_6H_6	10.4
Cs	42.0	CCl_4	11.7

In general, both ion–induced dipole and dipole–induced dipole interactions are fairly weak compared with ion–dipole interactions. This is the reason that ionic compounds like NaCl and polar molecules like alcohols are not soluble in nonpolar solvents such as benzene or carbon tetrachloride.

EXAMPLE 17.3

A sodium ion (Na^+) is situated in air at a distance of 4.0 Å (400 pm) from a nitrogen molecule. Use Equation 17.16 to calculate the potential energy of ion–induced dipole interaction in kJ mol^{-1}.

ANSWER

The data are

$$\alpha(N_2) = 1.74 \times 10^{-30}\ \text{m}^3$$

$$r = 4.0\ \text{Å} = 4.0 \times 10^{-10}\ \text{m}$$

From Equation 17.16,

$$V = -\frac{1}{2}\frac{(1.74 \times 10^{-30}\ \text{m}^3)(1.602 \times 10^{-19}\ \text{C})^2}{4\pi(8.854 \times 10^{-2}\ \text{C}^2\ \text{N}^{-1}\ \text{m}^{-2})(4.0 \times 10^{-10}\ \text{m})^4}$$

$$= -7.8 \times 10^{-21}\ \text{J}$$

$$= -4.7\ \text{kJ mol}^{-1}$$

COMMENT

This value is about four times smaller than that obtained for the ion–dipole interaction in Example 17.2.

Dispersion, or London, Interactions

The cases considered thus far consist of at least one charged ion or one permanent dipole among the interacting species, and they can be satisfactorily treated by classical physics. We must now ask the following question: Because nonpolar gases such as helium and nitrogen can be condensed, what kind of attractive interaction exists between atoms and between nonpolar molecules?

When we speak of the spherical symmetry of the charge density in helium, we mean that averaged over a certain period of time (e.g., a time long enough for us to carry out a physical measurement on the system), the electron density at a fixed distance away from the nucleus is the same in every direction. If we could take snapshots of the instantaneous configuration of each individual helium atom, we would most likely find varying degrees of deviation from spherical symmetry, owing to interactions among the atoms. Nevertheless, the temporary dipole created at every instant can induce a dipole in its neighboring atom(s), so an attractive interaction will result. We expect this interaction to be a weak attraction; indeed, the low boiling point of helium (4 K) suggests that very weak forces hold the atoms together in the liquid state. For molecules with large polarizabilities, however, this interaction can be comparable to or even greater than dipole–dipole and dipole–induced dipole interactions. For example, carbon tetrachloride (CCl_4), a nonpolar molecule, has a large polarizability (see Table 17.1) and a considerably higher boiling point (76.5°C) than methyl fluoride (CH_3F), a polar molecule (−141.8°C).

A quantum mechanical treatment of the interactions between nonpolar molecules was given in 1930 by the German physicist Fritz London (1900–1954), who showed that the potential energy arising from the interaction of two identical atoms or nonpolar molecules is given by

$$V = -\frac{3}{4}\frac{\alpha'^2 I_1}{(4\pi\varepsilon_0)^2 r^6}$$

$$= -\frac{3}{4}\frac{\alpha^2 I_1}{r^6} \qquad (17.18)$$

where I_1 is the first ionization energy of the atom or molecule. For unlike atoms or molecules A and B, Equation 17.18 becomes

$$V = -\frac{3}{2}\frac{I_A I_B}{I_A + I_B}\frac{\alpha'_A \alpha'_B}{(4\pi\varepsilon_0)^2 r^6}$$

$$= -\frac{3}{2}\frac{I_A I_B}{I_A + I_B}\frac{\alpha_A \alpha_B}{r^6} \qquad (17.19)$$

The forces that arise from this kind of interaction are called *dispersion*, or *London*, *forces*.

EXAMPLE 17.4

Calculate the potential energy of interaction between two argon atoms separated by 4.0 Å in air.

ANSWER

The data are

$$\alpha = 1.66 \times 10^{-30}\ \text{m}^3 \qquad \text{(Table 17.1)}$$

$$I_1 = 1521\ \text{kJ mol}^{-1} \qquad \text{(Table 12.7)}$$

From Equation 17.18

$$V = -\frac{3}{4}\frac{(1.66 \times 10^{-30}\ \text{m}^3)^2(1521\ \text{kJ mol}^{-1})}{(4.0 \times 10^{-10}\ \text{m})^6}$$

$$= -0.77\ \text{kJ mol}^{-1}$$

Dipole–dipole, dipole–induced dipole, and dispersion forces are collectively called *van der Waals forces*. These forces are responsible for the deviation of gas behavior from ideality discussed in Chapter 1.

Repulsive and Total Interactions

In addition to the attractive forces discussed so far, atoms and molecules must repel one another; otherwise, they would eventually fuse. Fusion is prevented by strong repulsive forces between electron clouds and between nuclei.* The potential energy of repulsion is extremely short-range. It is proportional to $1/r^n$, where n is between 8 and 12. The British physicist Sir John Edward Lennard-Jones (1894–1954) proposed the following expression to represent the attractive and repulsive interactions in nonionic systems:

John Edward Jones married Kathleen Lennard in 1926, adding his wife's surname to his own to become Lennard-Jones.

$$V = -\frac{A}{r^6} + \frac{B}{r^{12}} \tag{17.20}$$

where A and B are constants for two interacting atoms or molecules. The first term on the right of Equation 17.20 represents attraction. (As we have seen, the dipole–dipole, dipole–induced dipole, and dispersion interactions all have $1/r^6$ dependence.) The second term, which is very short-range (depends on $1/r^{12}$), describes repulsion between molecules. A more common form of Equation 17.20, called the *Lennard-Jones 6–12 potential*, is given by

$$V = 4\varepsilon\left[\left(\frac{\sigma}{r}\right)^{12} - \left(\frac{\sigma}{r}\right)^{6}\right] \tag{17.21}$$

* The repulsion between atoms or molecules is a consequence of the Pauli exclusion principle (see Section 12.7), which prevents electrons from sharing the same region in space.

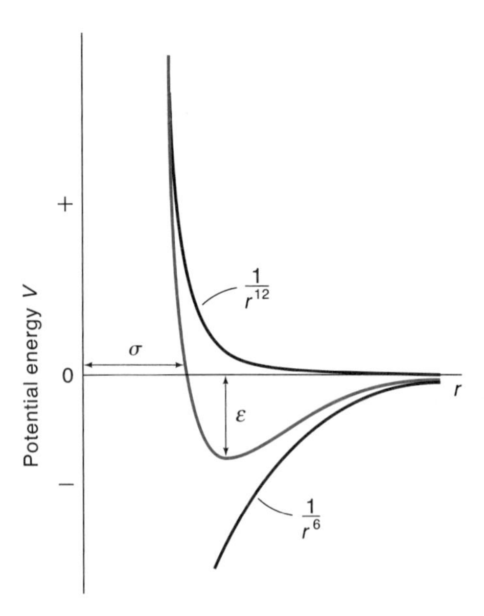

Figure 17.5
The potential energy curve (red) between two molecules or two nonbonded atoms is the sum of the $1/r^6$ (attraction) and $1/r^{12}$ (repulsion) terms. The depth of the well is given by ε and σ gives the distance between the centers of the molecules at $V = 0$.

Figure 17.5 shows the Lennard-Jones potential between two molecules. For a given pair of molecules, the quantity ε measures the depth of the potential well and σ is the separation at which $V = 0$. Table 17.2 lists ε and σ values for a few atoms and molecules. Although the Lennard-Jones potential has been used extensively in calculations, the $1/r^{12}$ term is a poor representation of the repulsive potential. For accurate work, a more satisfactory term is $e^{-ar/\sigma}$, where a is a constant.

Several points are worth noting about the total interaction potential. First, the second virial coefficient B' (see Section 1.7) can be related to the intermolecular potential. According to Equation 1.12, this coefficient can be measured experimentally

Table 17.2
Lennard-Jones Parameters for Atoms and Molecules

Particle	ε/kJ mol^{-1}	σ/Å
Ar	0.997	3.40
Xe	1.77	4.10
H_2	0.307	2.93
N_2	0.765	3.92
O_2	0.943	3.65
CO_2	1.65	4.33
CH_4	1.23	3.82
C_6H_6	2.02	8.60

from the slope of a plot of Z against P (if we ignore the third and higher virial coefficients). Therefore, knowledge of B' enables us to determine the intermolecular potential. Second, the value of σ gives a measure of how closely two nonbonded atoms can approach each other. Called the *van der Waals radius*, this measure is one-half the internuclear distance between two nonbonded atoms in a crystal. For example, in solid Cl_2, the average distance between two adjacent, nonbonded Cl atoms is 3.60 Å, so the van der Waals radius of Cl is taken to be 3.60 Å/2, or 1.80 Å. This value is considerably larger than the covalent radius of Cl, which is 1.01 Å. Table 17.3 lists the van der Waals radii of several atoms and the methyl group. Note that the size of atoms in space-filling models is based on their van der Waals radii. Third, we can make an interesting comparison of the relative magnitudes of intermolecular potential with intramolecular potential. Consider the interaction between a pair of H_2 molecules and the potential energy of a H_2 molecule. In the former case, the depth of the potential well is about 0.3 kJ mol^{-1}, and the "bond length" is 3.4 Å (340 pm). The corresponding quantities for H_2 are 432 kJ mol^{-1} and 0.74 Å (74 pm), respectively (see Figure 3.15). Thus, the stability of a normal chemical bond is two to three orders of magnitude greater than that of species held together by intermolecular forces, and bonded atoms are much closer in a molecule.

Table 17.3
van der Waals Radii of Atoms and the Methyl Group

Atom	Radius/Å
H	1.2
C	1.5
N	1.5
O	1.4
P	1.9
S	1.85
F	1.35
Cl	1.80
Br	1.95
I	2.2
$-CH_3$	2.0

17.4 Hydrogen Bonding

Table 17.4 summarizes different types of intermolecular interactions, including hydrogen bonding, which we shall discuss here. The hydrogen bond is a special type of interaction between molecules; it forms whenever a polar bond containing the hydrogen atom (e.g., O−H or N−H) interacts with an electronegative atom such as oxygen, nitrogen, or fluorine. This interaction is represented as A−H···B, where A and B are the electronegative atoms and the dotted line denotes the hydrogen bond.* Figure 17.6 shows several examples of hydrogen bonding. Although hydrogen bonds are relatively weak (about 40 kJ mol^{-1} or less), they play a central role in determining the properties of many compounds.

Early evidence of hydrogen bonding came from the study of the boiling points of compounds. Normally, the boiling points of a series of similar compounds containing elements in the same periodic group increase with increasing molar mass (and hence increasing polarizability). But, as Figure 17.7 shows, the binary hydrogen compounds of the elements in Groups 15, 16, and 17 do not follow this trend. In each of these series, the lightest compounds (NH_3, H_2O, HF) have the *highest* boiling points, because there is extensive hydrogen bonding between molecules in these compounds.

This type of bonding is unique to hydrogen primarily because the hydrogen atom has only one electron. When that electron is used to form a covalent bond with an electronegative atom, the hydrogen nucleus becomes partially unshielded. Consequently, its proton can interact directly with another electronegative atom on a different molecule. Depending on the strength of the interaction, such bonding can exist in the

* Detailed X-ray study of ice shows that the O···H hydrogen bond, and presumably other types of strong hydrogen bonds as well, has considerable covalent character. This conclusion is supported by quantum-mechanical calculations. See E. D. Isaacs, A. Shukla, P. M. Platzmann, et. al. *Phys. Rev. Lett.* **82**, 600 (1999). Also see *Science* **283**, 614 (1999).

Table 17.4
Interactions Between Molecules

Type of interaction	Distance dependence	Example	Magnitude of energy ($kJ\ mol^{-1}$)[a]
Covalent bond[b]	No simple expression	H–H	200–800
Ion–ion	$\dfrac{q_A q_B}{4\pi\varepsilon_0 r}$	Na^+Cl^-	40–400
Ion–dipole	$\dfrac{q\mu}{4\pi\varepsilon_0 r^2}$	$Na^+(H_2O)_n$	5–60
Dipole–dipole	$\dfrac{2}{3}\dfrac{\mu_A^2\mu_B^2}{(4\pi\varepsilon_0)^2 r^6}\dfrac{1}{k_B T}$	SO_2 SO_2	0.5–15
Ion–induced dipole	$\dfrac{1}{2}\dfrac{\alpha q^2}{4\pi\varepsilon_0 r^4}$	Na^+ C_6H_6	0.4–4
Dipole–induced dipole	$\dfrac{\alpha\mu^2}{4\pi\varepsilon_0 r^6}$	HCl C_6H_6	0.4–4
Dispersion	$\dfrac{3}{4}\dfrac{\alpha^2 I}{r^6}$	CH_4 CH_4	4–40
Hydrogen bond	No simple expression	$H_2O\cdots H_2O$	4–40

[a] The actual value depends on distance of separation, charge, dipole moment, polarizability, and the dielectric constant of the medium.

[b] This is listed for comparison purposes only.

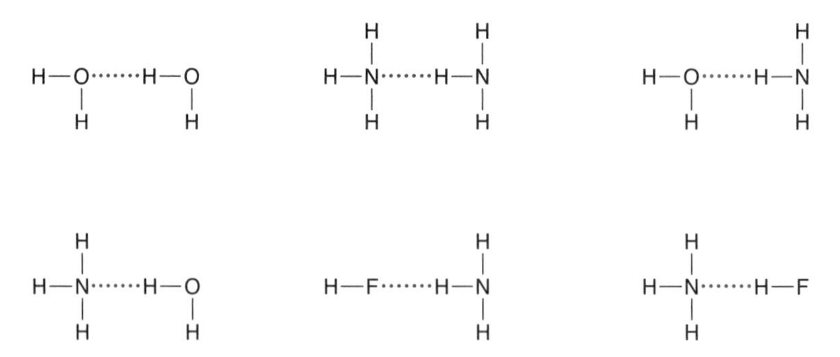

Figure 17.6
Some examples of hydrogen bonding. The dotted red lines represent hydrogen bonds.

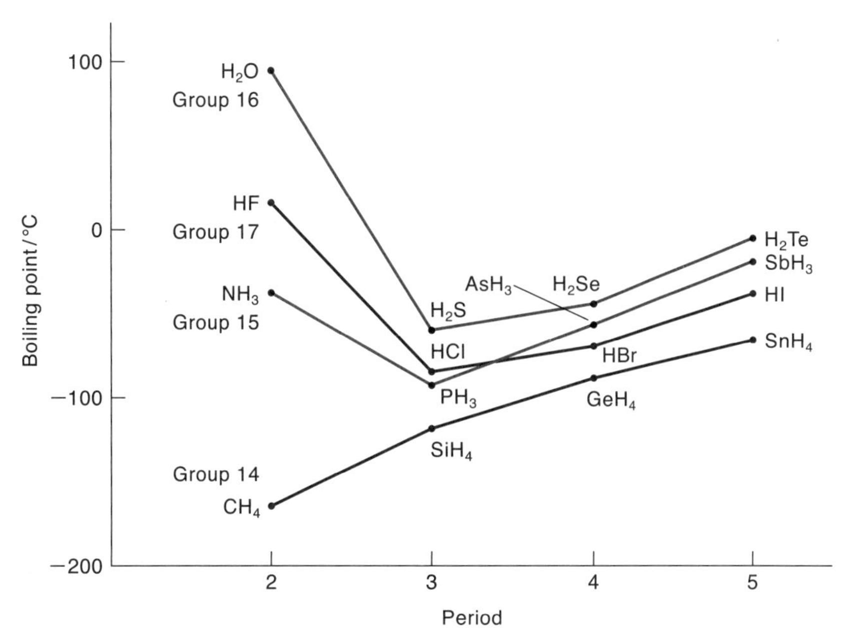

Figure 17.7
Boiling points of the hydrogen compounds of the Group 14, 15, 16, and 17 elements. Although normally we expect the boiling point to increase as we move down a group, we see that NH_3, H_2O, and HF behave differently, as a result of intermolecular hydrogen bonding.

gas phase as well as in the solid and liquid phases. In solids and liquids, HF forms a polymeric chain as follows:

$$\cdots F{-}H\cdots F{-}H\cdots F{-}H\cdots F{-}H\cdots F{-}$$

For maximum stability, the proton donor pair (AH) and the acceptor (B) are usually colinear (i.e., $\measuredangle AHB = 180°$), but deviations of up to 30° are known.

A molecule can also form *intramolecular* hydrogen bonds. Examples are fumaric and maleic acids, isomers for which the first and second acid dissociation constants, K_1 and K_2, are

Fumaric acid	Maleic acid
H(COOH)C=C(H)COOH (trans: HOOC and COOH on opposite sides)	H(COOH)C=C(H)COOH (cis: both COOH on same side)
$K_1 = 9.6 \times 10^{-4}$	$K_1 = 1.2 \times 10^{-2}$
$K_2 = 4.1 \times 10^{-5}$	$K_2 = 6.0 \times 10^{-7}$

The first dissociation constant for maleic acid is higher than that for fumaric acid because of steric interaction in the cis isomer, which facilitates the removal of a proton. Although the second dissociation constant of fumaric acid is only about 20 times lower than the first dissociation constant, K_2 is smaller than K_1 by a factor of

20,000 for maleic acid. This phenomenon can be explained by assuming the presence of a stable intramolecular hydrogen bond between the $-COOH$ and $-COO^-$ groups in maleic acid as follows:

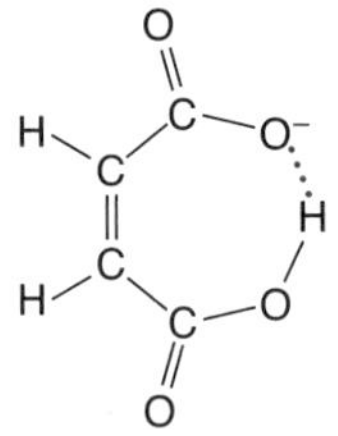

Several physical techniques are used to detect hydrogen bonds. For crystals, X-ray diffraction measurements usually provide the most direct evidence. A relatively strong hydrogen bond can shorten the expected distance between AH and B (the sum of their van der Waals radii) by two- to three-tenths of an angstrom. Infrared (IR) and nuclear magnetic resonance (NMR) spectroscopy facilitate the study of hydrogen bonding in liquids. For example, formation of a hydrogen bond shifts the stretching mode to lower IR frequencies and the bending modes to higher frequencies.

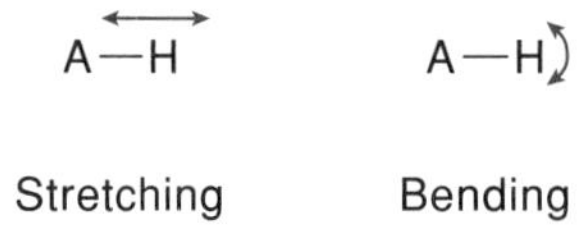

A—H (Stretching) A—H (Bending)

The width and intensity of the peak due to A–H stretching also may be enhanced by hydrogen-bond formation. In NMR studies, we predict and observe that the chemical shift of a proton can be appreciably altered by hydrogen bonding.

Hydrogen bonding is largely responsible for the stability of protein conformations. Intramolecular hydrogen bonds between the $\rangle C{=}O$ and $\rangle N{-}H$ groups of a polypeptide chain result in the α helix. On the other hand, intermolecular hydrogen bonds between two polypeptide chains account for β pleated-sheet structures. Here, let us consider the importance of hydrogen bonding in DNA.

DNA molecules are polymers that have molar masses in the millions to tens of billions of grams. They consist of three parts: phosphate groups, sugar groups (deoxyribose), and purine and pyrimidine bases (adenine, cytosine, guanine, and thymine). Figure 17.8 shows the Watson–Crick model of the DNA double helix [after the American biologist James Dewey Watson (1928–) and the British biologist Francis Harry Compton Crick (1916–2004)]. The molecule's backbone contains alternating sugar and phosphate residues. Each sugar residue is attached to a purine or pyrimidine base. Hydrogen bonds form between bases on two strands of DNA, generating the double-helical structure. The bases are roughly perpendicular to the axis of the helix; each one can form a strong hydrogen bond with only one of the four bases available. This specificity of base pairing gives the DNA structure the stability required for its function as the storage site for the genetic code.

Energetically, the most favorable pairings in DNA molecules are adenine (A) to thymine (T) and cytosine (C) to guanine (G), and as Figure 17.8b shows, the two strands are complementary. Although the amount of energy required to break a hydrogen bond is rather small (about 5 kJ mol^{-1}), the double-helical structure of DNA is

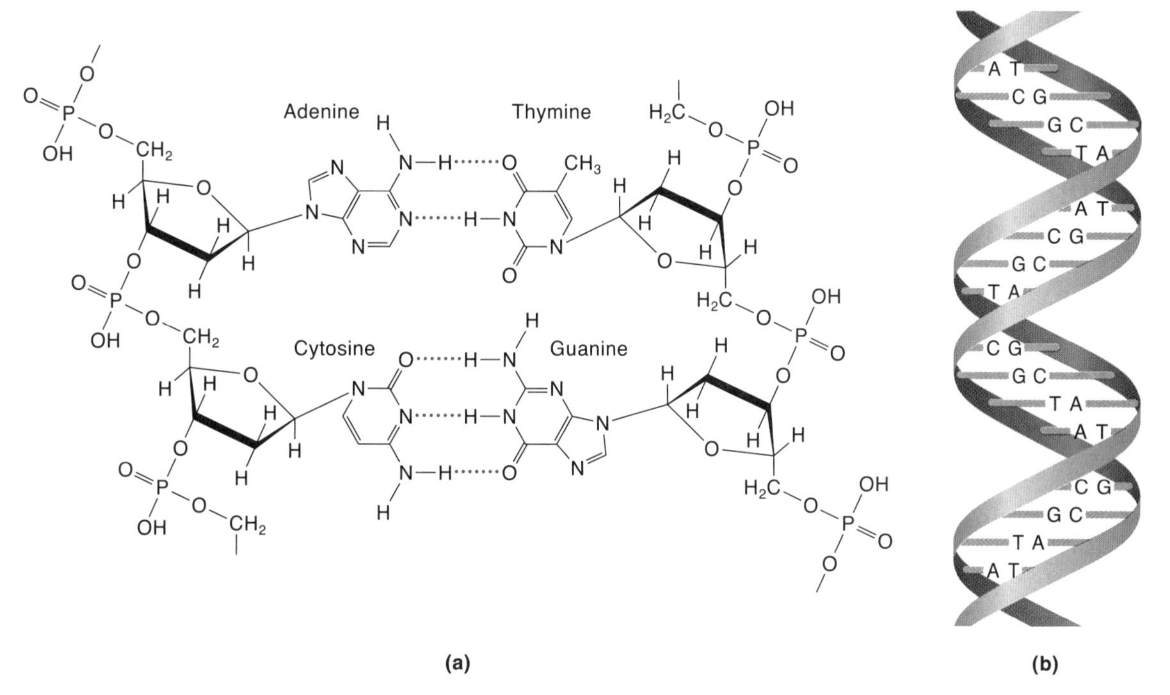

Figure 17.8
(a) Base-pair formation between adenine (A) and thymine (T) and between cytosine (C) and guanine (G). (b) The most common structure of DNA, which is a right-handed double helix. The two strands are held together by hydrogen bonds and other intermolecular forces.

stable under normal physiological conditions. The stability of the molecule rests on the cooperative nature of hydrogen-bond formation. Consider the pairing of two nucleotides, C and G, in solution at room temperature. The ratio of free bases to hydrogen-bonded base pair can be calculated from the Boltzmann distribution law:

The cytosine–guanine base pair has three hydrogen bonds.

$$\frac{(\text{C, G})_{\text{free bases}}}{(\text{C}-\text{G})_{\text{base pair}}} = e^{-\Delta E/RT} = \exp\left(\frac{-3 \times 5000\ \text{J mol}^{-1}}{8.314\ \text{J K}^{-1}\ \text{mol}^{-1} \times 300\ \text{K}}\right)$$

$$= 0.00244$$

Thus, there are 410 pairs of hydrogen-bonded bases to one pair of free bases. For dinucleotides in which the strands are made of two cytosines and guanines, respectively, the complexed form is favored over free bases by a factor of 410×410, or 1.68×10^5. Clearly, then, in a polynucleotide containing thousands of bases, equilibrium overwhelmingly favors the hydrogen-bonded structure.

The elucidation of the DNA structure immediately suggested how it could be replicated, or reproduced, each time a cell divides. During replication, the two strands part to form two identical DNA molecules. The primary reason that DNA is copied correctly during each replication lies in the specific hydrogen-bonding patterns between adenine and thymine and between cytosine and guanine.

So far, our discussion of hydrogen-bond formation has focused on the very electronegative atoms N, O, and F. Ample evidence suggests that hydrogen bonds also exist in compounds not containing these atoms. Shown below are examples of "weak"

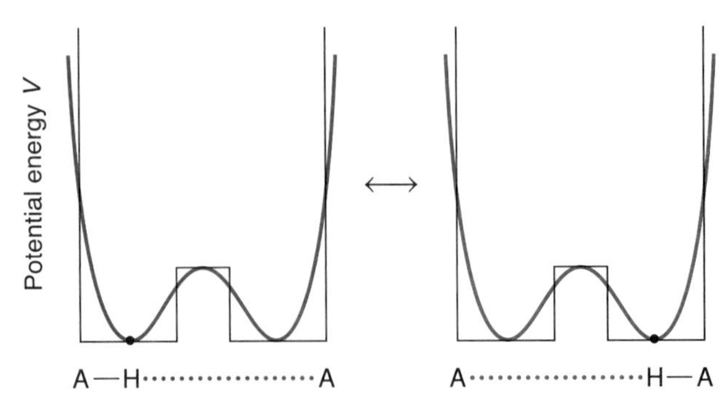

Figure 17.9
Potential energy curve for the A–H···A system as approximated by the model of the particle in a one-dimensional box. The particle can tunnel through the barrier to give the A···H–A configuration.

hydrogen bonds, so-called because of their diminished energy magnitude compared with hydrogen bonds in H_2O, NH_3, or HF.

```
                                                                          H
     Cl              CH3              Cl                                  |
     |              /                 |                                   C
Cl—C—H·······O=C               Cl—C—Cl·······H—F            F—H·······|||
     |              \                 |                                   C
     Cl              CH3              Cl                                  |
                                                                          H
```

The hydrogen bonds between chloroform and acetone are responsible for the azeotrope that attains a maximum boiling point (see Section 6.6). Interestingly, the electron-rich triple bond in acetylene can form a hydrogen bond with hydrogen fluoride. Similarly, the delocalized π electrons in benzene can form weak hydrogen bonds.

Finally we note that, because of the hydrogen atom's small mass, quantum-mechanical tunneling may occur in some hydrogen-bonded systems. In nonpolar solvents, carboxylic acids dimerize as follows:

```
       O—H·······O                      O·······H—O
      /           \\                   //           \
R—C                 C—R   ⟷   R—C                   C—R
      \\           /                   \            //
       O·······H—O                      O—H·······O
```

Figure 17.9 shows the double minimum potential well for this system. Proton tunneling should stabilize the hydrogen bonds because the phenomenon "lengthens" the box for the proton and therefore lowers the potential energy of interaction.

17.5 The Structure and Properties of Water

Water is so common a substance that we often overlook its unique properties. For example, given its molar mass, water should be a gas at room temperature, but due to hydrogen bonding, it has a boiling point of 373.15 K at 1 atm. In this section, we shall study the structure of ice and liquid water and consider some biologically significant aspects of water.

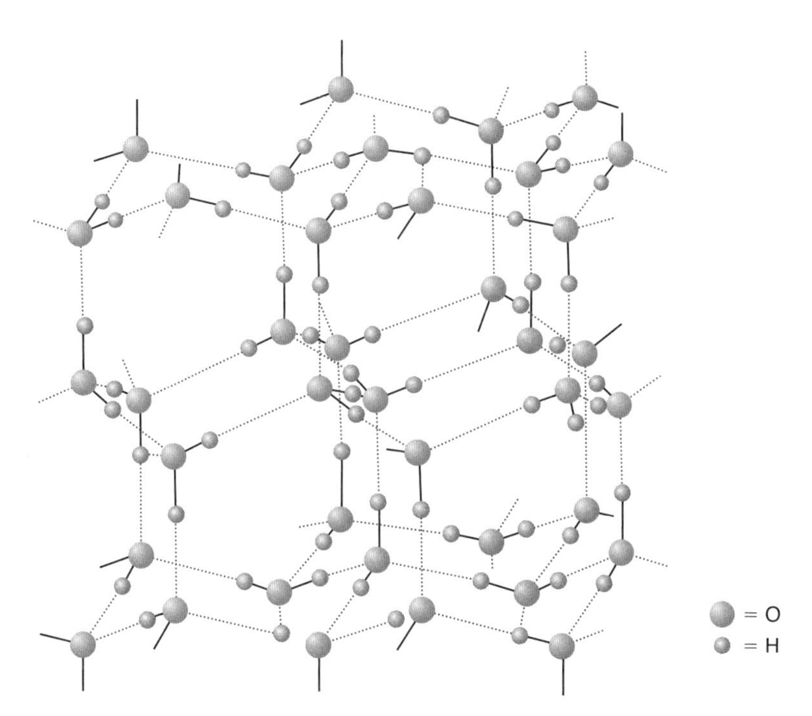

Figure 17.10
Structure of ice. The red dotted lines represent hydrogen bonds.

The Structure of Ice

To understand the behavior of water, we first investigate the structure of ice. There are more than 12 known crystalline forms of ice; most of them stable only at high pressures. Ice I, the familiar form, has been studied thoroughly. It has a density of 0.924 g mL^{-1} at 273 K and 1 atm pressure.

There is a significant difference between H_2O and other polar molecules, such as NH_3 and HF. The number of hydrogen atoms in a water molecule that can form the positive ends of hydrogen bonds is equal to the number of lone pairs on the oxygen atom that can form the negative ends:

O
H H
0.958 Å
104.45°

The result is an extensive three-dimensional network in which each oxygen atom is bonded tetrahedrally to four hydrogen atoms by means of two covalent bonds and two hydrogen bonds. This equality in number of protons and lone pairs is not characteristic of NH_3 and HF. Consequently, NH_3 and HF can form only rings or chains and not an extensive three-dimensional structure.

Figure 17.10 shows the structure of ice I. The distance between adjacent oxygen atoms is 2.76 Å. The O–H distance is between 0.96 Å and 1.02 Å, and the O···H distance is between 1.74 Å and 1.80 Å. Because of its open lattice, ice has a lower density than water, a fact that has profound ecological significance. Were it not for this unique type of hydrogen bonding, ice, like most other solid substances, would be denser than the corresponding liquid. On freezing, it would sink to the bottom of a lake or pond, causing all the water to freeze gradually and killing most live organisms within it. Fortunately, water reaches its maximum density at 277.15 K (4 K above freezing). Cooling below 277.15 K decreases the density of water and allows it to rise to thc surface, where freezing occurs. An ice layer formed on the surface does not sink; just as important, it acts as a thermal insulator to protect the biological environment beneath it.

The Structure of Water

Although using the word *structure* may seem strange when discussing liquids, most liquids possess short-range order. A convenient way to study the structure of liquids is to use the *radial distribution function*, $g(r)$. This function is defined so that $4\pi r^2 g(r)dr$ gives the probability that a molecule will be found in a spherical shell of width dr at distance r from the center of another molecule.* For a crystalline solid, a plot of $4\pi r^2 g(r)$ versus r gives a series of sharp lines because crystals have long-range order. In contrast, as Figure 17.11 shows, the radial distribution curve for liquid water at 4°C produces a major peak at 2.90 Å, with weaker peaks at 3.50 Å, 4.50 Å, and 7.00 Å. Beyond 7.00 Å, the function is essentially constant, meaning that the local order does not extend beyond this distance. X-ray diffraction studies of ice I show that the O−O distance is 2.76 Å. The strong peak at 2.90 Å suggests a very similar tetrahedral arrangement in the liquid. The peak at 3.50 Å does not correspond to any bond distance in ice I, which does, however, have interstitial sites at a distance of 3.50 Å from each O atom. Therefore, when ice melts, some of the water molecules break loose and become trapped in these interstitial sites, which are responsible for the peak at 3.50 Å. The peaks at 4.50 Å and 7.00 Å are also consistent with the tetrahedral arrangement.

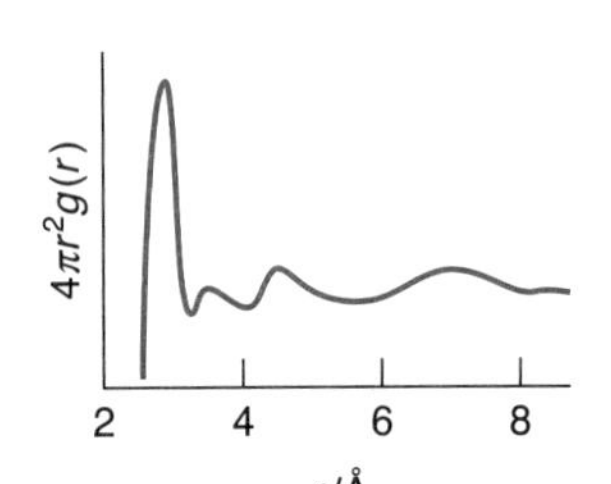

Figure 17.11 Experimental radial distribution curve for water at 4°C. The peaks become broader at higher temperatures.

The above discussion suggests that the extensive three-dimensional hydrogen-bonded structure that characterizes ice I is largely intact in water, although the bonds may become bent and distorted. On melting, monomeric water molecules occupy holes in the remaining "icelike" lattice, causing the density of water to be greater than that of ice. As temperature increases, more hydrogen bonds are broken, but at the same time the kinetic energy of molecules increases. Consequently, more water molecules are trapped, but the elevated kinetic energy decreases the density of water because the molecules occupy a greater volume. Initially, the trapping of monomeric water molecules outweighs the expansion in volume due to the increase in kinetic energy, so the density rises from 0°C to 4°C. Beyond this temperature, the expansion predominates, so the density decreases with increasing temperature (Figure 17.12).

Water Clusters. A powerful experimental approach to the study of water structure at the molecular level is to form solid and liquid water "one molecule at a time," that is, to generate *water clusters*. Eventually, the properties and behavior of the clusters converge with those of the bulk water. With this technique, water clusters are created and cooled to nearly absolute zero temperatures in supersonic molecular beams, and the vibrations of the hydrogen bonds between the water molecules comprising the clusters are studied by infrared spectroscopy. Figure 17.13 shows the structures of small water clusters.

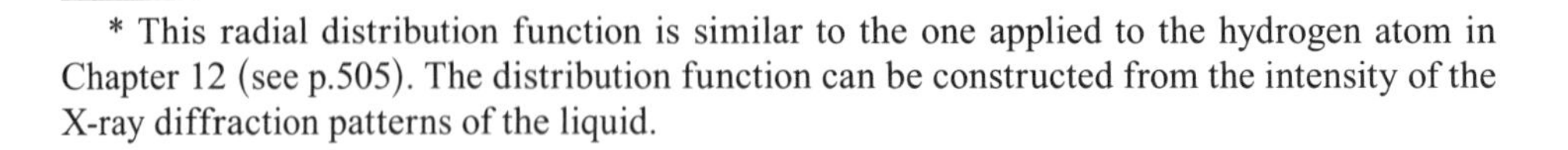

* This radial distribution function is similar to the one applied to the hydrogen atom in Chapter 12 (see p.505). The distribution function can be constructed from the intensity of the X-ray diffraction patterns of the liquid.

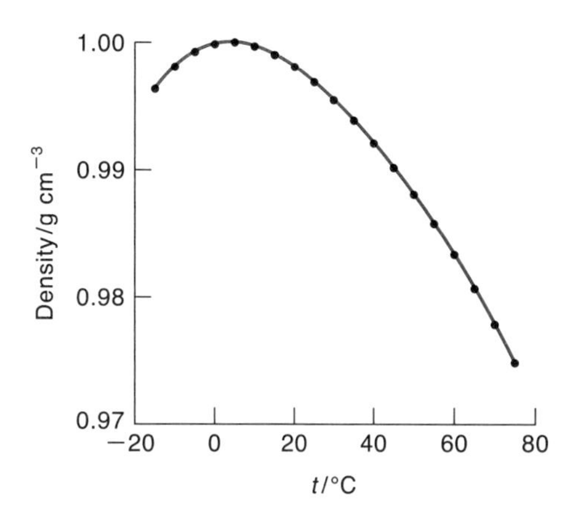

Figure 17.12
Plot of density versus temperature for liquid water. The maximum density of water is 0.99997 g cm^{-3} at 3.98°C. The density of ice at 0°C is 0.9167 g cm^{-3}.

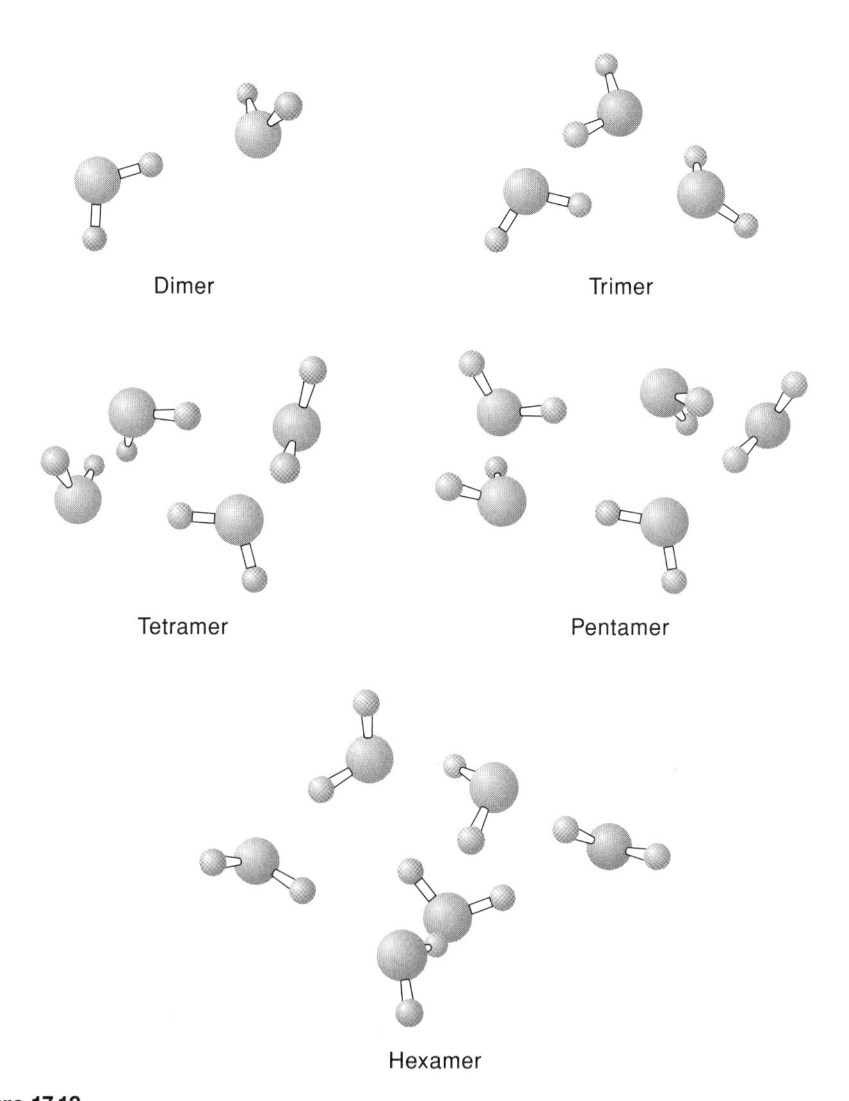

Figure 17.13
Structures of water clusters. The trimer, tetramer, and pentamer are cyclical water clusters, in which each water acts as both a donor and an acceptor of a hydrogen bond. The hexamer has a three-dimensional cage structure. (Graphics by Kun Liu, Mac Brown, and Jeff Cruzan of the University of California, Berkeley.)

Some Physiochemical Properties of Water

Table 17.5 lists some important physiochemical properties of water. The abnormally high values of several of its properties make water a unique solvent, particularly suited to the support of living systems. The reasons are briefly discussed below.

1. Water has one of the highest dielectric constants of all liquids (see Table 7.3). This property makes water an excellent solvent for ionic compounds. In addition, its ability to form hydrogen bonds enables it to dissolve carbohydrates, carboxylic acids, and amines.
2. Due to its extensive hydrogen bonds, water has a very high heat capacity. Because $\Delta H = C_P \Delta T$ (see Equation 3.18), $\Delta T = \Delta H / C_P$, which means that a large amount of heat is needed to raise the temperature of an aqueous solution by one kelvin. This property is important in regulating the temperature of a cell from the heat generated by metabolic processes. From an environmental point of view, the ability of water to absorb a lot of heat with little temperature rise greatly influences Earth's climate. Our lakes and oceans absorb solar radiation or give up large amounts of heat with only a small change in temperature. For this reason, the climate close to the ocean is more moderate than that inland.
3. Water has a high molar heat of vaporization (41 kJ mol^{-1}). Thus, sweating is an effective way to regulate body temperature. On average, a 60-kg person generates 1×10^7 J of heat daily from metabolism. If sweating were the only mechanism for cooling, then each of us would need to vaporize (1×10^7 J/41 $\times$ 1000 J mol^{-1}) = 244 mol, or nearly 4.4 liters of water, to maintain a constant temperature. Normally a person does not sweat this much (unless, for example, she is training for a marathon race in Houston, Texas,

Table 17.5
Some Physiochemical Properties of Water[a]

Melting point	0°C (273.15 K)
Boiling point	100°C (373.15 K)
Density of water	0.99987 g mL^{-1} at 0°C
Density of ice	0.9167 g mL^{-1} at 0°C
Molar heat capacity	75.3 J K^{-1} mol^{-1}
Molar heat of fusion	6.01 kJ mol^{-1}
Molar heat of vaporization	40.79 kJ mol^{-1} at 100°C
Dielectric constant	78.54 at 25°C
Dipole moment	1.82 D
Viscosity	0.001 N s m^{-2}
Surface tension	0.07275 N m^{-1} at 20°C
Diffusion coefficient	2.4×10^{-9} m^2 s^{-1} at 25°C

[a] Viscosity, surface tension, and diffusion coefficient are discussed in Chapter 19.

in mid-July). Part of the excess heat is radiated to the cooler surroundings. Although water's molar heat of fusion is not unusually high (6 kJ mol^{-1}), it is still sizable and helps protect the body against freezing.

4. The ecological significance of water's higher density than ice has been discussed.
5. Again, strong intermolecular interaction due to hydrogen bonding gives water a high surface tension (see Chapter 19). This property forces biological organisms to produce detergentlike compounds (called surfactants) to lower surface tension that would otherwise inhibit certain functions. For example, lung surfactants are needed to decrease the work required to open alveolar spaces and allow efficient respiration to take place (see p. 854).
6. The viscosity of water, unlike many of its properties, is comparable to that of many other liquids. Because the presence of macromolecules (proteins and nucleic acids, for example) appreciably increases the viscosity of a solution, the fact that water is not a viscous fluid facilitates blood flow and the diffusion of molecules and ions in the medium.

17.6 Hydrophobic Interaction

Experience tells us that oil and water do not mix. At first glance, the reason seems to be that the dipole–induced dipole and dispersion forces between water and nonpolar oil molecules are weak. From this observation, we might conclude that the enthalpy of mixing (ΔH) is positive, which causes ΔG to be positive ($\Delta G = \Delta H - T\Delta S$). Thus, the solubility of oil in water is very low. But this explanation is incorrect. The unfavorable interaction is primarily due to the *hydrophobic interaction* (also called the *hydrophobic effect*, or *hydrophobic bond*), which is the name given to influences that cause nonpolar substances to cluster together to minimize their contacts with water. This interaction forms the basis for many important chemical and biological phenomena, including the cleansing action of soaps and detergents, the formation of biological membranes, and the stabilization of protein structure.

Table 17.6 shows the thermodynamic quantities for the transfer of small nonpolar molecules from nonpolar solvents to water. The most striking feature of this table is

Table 17.6
Thermodynamic Quantities for the Transfer of Nonpolar Solutes from Organic Solvents to Water at 25°C

Process	ΔH/kJ mol^{-1}	ΔS/J K^{-1} mol^{-1}	ΔG/kJ mol^{-1}
$CH_4(CCl_4) \rightarrow CH_4(H_2O)$	−10.5	−75.8	12.1
$CH_4(C_6H_6) \rightarrow CH_4(H_2O)$	−11.7	−75.8	10.9
$C_2H_6(C_6H_6) \rightarrow C_2H_6(H_2O)$	−9.2	−83.6	15.7
$C_6H_{14}(C_6H_{14}) \rightarrow C_6H_{14}(H_2O)$	0.0	−95.3	28.4
$C_6H_6(C_6H_6) \rightarrow C_6H_6(H_2O)$	0.0	−57.7	17.2

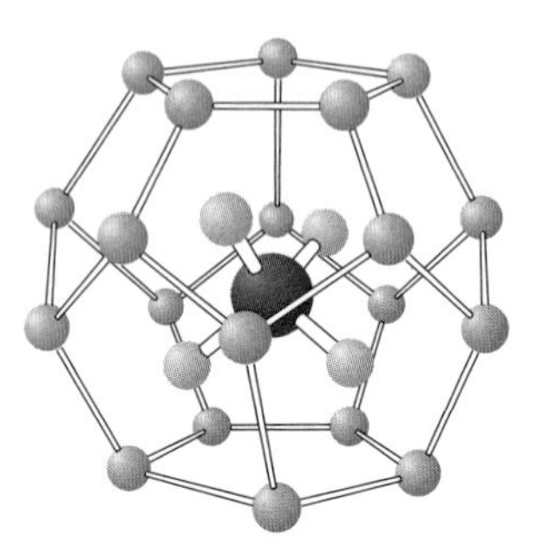

Figure 17.14
A structure for methane hydrate. The methane molecule is trapped in a cage of water molecules (red spheres) held together by hydrogen bonds.

that ΔS is negative for all of the compounds. When nonpolar molecules enter the aqueous medium, some hydrogen bonds must be broken to make room or create a cavity for the solutes. This part of interaction is endothermic because the broken hydrogen bonds are stronger than the dipole–induced dipole and dispersion interactions. Each solute molecule is now trapped in an icelike cage structure, referred to as the clathrate cage model, which consists of a specific number of water molecules held together by hydrogen bonds. The formation of the clathrate has two important consequences. First, the newly formed hydrogen bonds (an exothermic process) can partly or totally compensate for the hydrogen bonds that were broken initially to make the cavity. This explains why ΔH could be negative, zero, or positive for the overall process. Furthermore, because the cage structure is highly ordered, there is an appreciable decrease in entropy, which far outweighs the increase in entropy due to the mixing of solute and water molecules, so that ΔS is negative. Thus, the immiscibility of nonpolar molecules and water, or the hydrophobic interaction, is entropy driven rather than enthalpy driven.*

The formation of hydrocarbon clathrates has some practical consequences. Bacteria in the sediments on the ocean floor consume organic material and generate methane gas. At high-pressure and low-temperature conditions, methane's solubility in water increases (i.e., ΔG becomes more negative or less positive as T decreases). The methane clathrate, commonly called methane hydrate (Figure 17.14), looks like a gray ice cube, but if you put a lighted match to it, it will burn. Estimates are that the total reserve of the methane hydrate in the world's ocean floor contains about 10^{13} tons of carbon, about twice the amount of carbon in all the coal, oil, and natural gas on land. Harvesting the energy stored in methane hydrate, however, presents a tremendous engineering challenge. Interestingly, oil companies have known about methane hydrate since the 1930s, when they began using high-pressure pipelines to transport natural gas in cold climates. Unless water is carefully removed before the gas enters the pipeline, chunks of methane hydrate will impede the flow of gas.

Hydrophobic interaction has a profound effect on the structure of proteins. When the polypeptide chain of a protein folds into a three-dimensional structure in solution, the nonpolar amino acids (e.g., glycine, alanine, proline, phenylalanine, and valine) prefer the interior of the macromolecule with little or no contact with water, whereas the polar amino acid residues (such as aspartic acid, glutamic acid, arginine, and lysine) prefer the exterior. An insight into this entropy-driven process can be gained by considering just two nonpolar molecules in aqueous solution (Figure 17.15). The hydrophobic interaction causes the nonpolar molecules to come together into a single cavity to reduce the unfavorable interactions with water by decreasing the surface area. This response destroys part of the cage structure, resulting in an increase in ΔS and hence a decrease in ΔG. Moreover, enthalpy increases ($\Delta H > 0$) because some of the hydrogen bonds in the original cage structures are broken. Similarly, the folding of a protein is an example of this phenomenon because it minimizes the exposure of nonpolar surfaces to water.

---

* Comparing the solubility of nonpolar molecules with ionic compounds is interesting. In the latter, there is a large decrease in enthalpy ($\Delta H < 0$) due to the strong ion–dipole interaction, which outweighs the decrease in entropy ($\Delta S < 0$) when water becomes more organized around the charged ions, so that $\Delta G < 0$. Note that the structure of the hydration sphere surrounding ions is different from the clathrate cage structure.

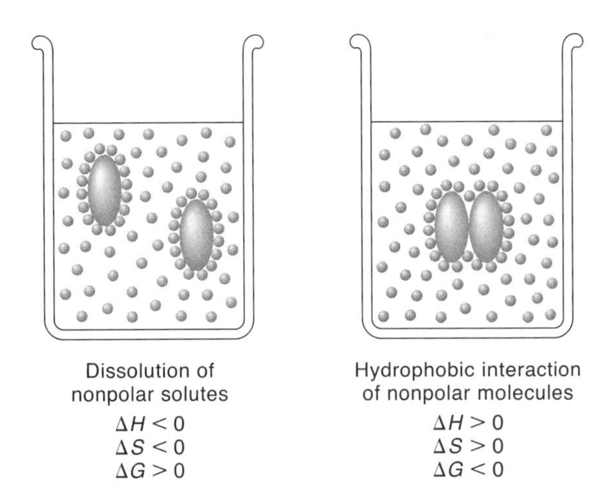

Figure 17.15
(a) The dissolution of nonpolar molecules in water is unfavorable because of the large decrease in entropy resulting from clathrate formation, even though the process is exothermic ($\Delta H < 0$). Consequently, $\Delta G > 0$. (b) As a result of hydrophobic interaction, the nonpolar molecules come together, releasing some of the ordered water molecules in the clathrate structure and thus increasing entropy. This is a thermodynamically favorable process ($\Delta G < 0$), even though it is endothermic ($\Delta H > 0$) because more hydrogen bonds are broken than formed.

Key Equations

$$V = -\frac{q_{\mathrm{Na^+}} q_{\mathrm{Cl^-}}}{4\pi\varepsilon_0 r} + \frac{b}{r^n}$$ (Potential energy of the ionic bond in a NaCl ion pair) (17.5)

$$\langle V \rangle = -\frac{2}{3}\frac{\mu_A^2 \mu_B^2}{(4\pi\varepsilon_0)^2 r^6}\frac{1}{k_B T}$$ (Average potential energy for dipole–dipole interaction in liquids) (17.11)

$$V = -\frac{q\mu}{4\pi\varepsilon_0 r^2}$$ (Potential energy for ion–dipole interaction) (17.12)

$$V = -\frac{3}{4}\frac{\alpha^2 I}{r^6}$$ (Potential energy for dispersion interaction between like molecules) (17.18)

$$V = -\frac{3}{2}\frac{I_A I_B}{I_A + I_B}\frac{\alpha_A \alpha_B}{r^6}$$ (Potential energy for dispersion interaction between unlike molecules) (17.19)

$$V = 4\varepsilon\left[\left(\frac{\sigma}{r}\right)^{12} - \left(\frac{\sigma}{r}\right)^{6}\right]$$ (Lennard-Jones 6–12 potential) (17.21)

Suggestions for Further Reading

BOOKS

Eisenberg, D., and W. Kauzmann, *The Structure and Properties of Water*, Oxford University Press, New York, 2005.

Franks, F., *Water: A Matrix of Life*, 2nd ed., The Royal Society of Chemistry, Cambridge, UK, 2000.

Israelachvili, J. N., *Intermolecular and Surface Forces*, 3rd ed., Elsevier, New York, 2011.

Jeffrey, G. A., *An Introduction to Hydrogen Bonding*, Oxford University Press, New York, 1997.

Jeffrey, G. A., and W. Saenger, *Hydrogen Bonding in Biological Structures*, Springer-Verlag, New York, 1994.

Kavanau, J. L., *Water and Water-Solute Interactions*, Holden-Day, San Francisco, 1964.

Pimentel, G. C., and A. L. McClellan, *The Hydrogen Bond*, W. H. Freeman, San Francisco, 1960.

Rigby, M., E. B. Smith, W. A. Wakeham, and G. C. Maitland, *The Forces Between Molecules*, Clarendon Press, Oxford, 1986.

Vinogrador, S. N., and R. H. Linnell, *Hydrogen Bonding*, Van Nostrand Reinhold, New York, 1971.

ARTICLES

General

"The Force between Molecules," B. V. Derjaguin, *Sci. Am.* July 1960.

"The Human Thermostat," T. H. Benzinger, *Sci. Am.* January 1961.

"A Molecular Theory of General Anesthesia," L. Pauling, *Science* **134**, 15 (1961).

"Inclusion Compounds," J. F. Brown, Jr., *Sci. Am.*, July 1962.

"Clathrates: Compounds in Cages," M. M. Hagan, *J. Chem. Educ.* **40**, 643 (1963).

"A Classical Electrostatic View of Chemical Forces," H. H. Jaffé, *J. Chem. Educ.* **40**, 649 (1963).

"Early Views on Forces between Atoms," L. Holliday, *Sci. Am.* May 1970.

"The Role of van der Waals Forces in Surface and Colloid Chemistry," P. C. Hiemenz, *J. Chem. Educ.* **49**, 164 (1972).

"Why Does a Stream of Water Deflect in an Electric Field?" G. K. Vemulapalli and S. G Kukolich, *J. Chem. Educ.* **73**, 887 (1996).

"Updated Principle of Corresponding States," D. Ben-Amotz, A. D. Gift, and R. D. Levine, *J. Chem. Educ.* **81**, 142 (2004).

"How Gecko Toes Stick," K. Autumn, *Am. Sci.* March-April 2006.

"Dancing Crystals: A Dramatic Illustration of Intermolecular Forces," D. W. Mundell, *J. Chem. Educ.* **84**, 1773 (2007).

"Using Molecular Dynamics Simulation to Reinforce Student Understanding of Intermolecular Forces," P. R. Burkholder, G. H. Purser, and R. S. Cole, *J. Chem. Educ.* **85**, 1071 (2008).

The Hydrogen Bond/Water Structure

"On Hydrogen Bonds," J. Donohue, *J. Chem. Educ.* **40**, 598 (1963).

"Ice," L. K. Runnels, *Sci. Am.* December 1966.

"The Significance of Hydrogen Bonds in Biological Structure," A. L. McClellan, *J. Chem. Educ.* **44**, 547 (1967).

"The Structure of Ordinary Water," H. S. Frank, *Science* **169**, 635 (1970).

"Hydrogen Bonding and Proton Transfer," M. D. Joesten, *J. Chem. Educ.* **59**, 362 (1982).
"Water Clusters," K. Liu, J. D. Cruzan, and R. J. Saykally, *Science* **271**, 929 (1996).
"Simulating Water and the Molecules of Life," M. Gerstein and M. Levitt, *Sci. Am.* November 1998.
"Hydrogen Bonds Involving Transition Metal Centers Acting As Proton Acceptors," A. Martin, *J. Chem. Educ.* **76**, 578 (1999).

Hydrophobic Interaction
"Hydrophobic Interactions," G. Némethy, *Angew. Chem. Intl. Ed.* **6**, 195 (1967).
"The Thermodynamic Parameters Involved in Hydrophobic Interaction," T. Aerts and J. Clauwaert, *J. Chem. Educ.* **63**, 993 (1986).
"The Hydrophobic Effect," E. M. Huque, *J. Chem. Educ.* **66**, 581 (1989).
"How Protein Chemists Learned About the Hydrophobic Effect," C. Tanford, *Protein Science*, **6**, 1358 (1997).
"The Real Reasons Why Oil and Water Don't Mix," T. P. Silverstein, *J. Chem. Educ.* **75**, 116 (1998).
"Flammable Ice," E. Suess, G. Bohrmann, J. Greinert, and E. Lavsch, *Sci. Am.* November 1999.
"Hydrophobic Solvation: Aqueous Methane Solutions," K. Oliver and L. Timm, *J. Chem. Educ.* **84**, 864 (2007).

Problems

Intermolecular Forces

17.1 List all the intermolecular interactions that take place in each of the following kinds of molecules: Xe, SO_2, C_6H_5F, and LiF.

17.2 Arrange the following species in order of decreasing melting points: Ne, KF, C_2H_6, MgO, H_2S.

17.3 The compounds Br_2 and ICl have the same number of electrons, yet Br_2 melts at −7.2°C, whereas ICl melts at 27.2°C. Explain.

17.4 If you lived in Alaska, which of the following natural gases would you keep in an outdoor storage tank in winter? Methane (CH_4), propane (C_3H_8), or butane (C_4H_{10}). Explain.

17.5 List the types of intermolecular forces that exist between molecules (or basic units) in each of the following species: **(a)** benzene (C_6H_6), **(b)** CH_3Cl, **(c)** PF_3, **(d)** NaCl, **(e)** CS_2.

17.6 The boiling points of the three different structural isomers of pentane (C_5H_{12}) are 9.5°C, 27.9°C, and 36.1°C. Draw their structures, and arrange them in order of decreasing boiling points. Justify your arrangement.

17.7 Two water molecules are separated by 2.76 Å in air. Use Equation 17.9 to calculate the dipole–dipole interaction. The dipole moment of water is 1.82 D.

17.8 Coulombic forces are usually referred to as long-range forces (they depend on $1/r^2$), whereas van der Waals forces are called short-range forces (they depend on $1/r^7$). **(a)** Assuming that the forces (F) depend only on distances, plot F as a function of r at $r = 1$ Å, 2 Å, 3 Å, 4 Å, and 5 Å. **(b)** Based on your results, explain the fact that although a 0.2 *M* nonelectrolyte solution usually behaves ideally, nonideal behavior is quite noticeable in a 0.02 *M* electrolyte solution.

17.9 Calculate the induced dipole moment of I_2 due to a Na^+ ion that is 5.0 Å away from the center of the I_2 molecule. The polarizability of I_2 is 12.5×10^{-30} m^3.

17.10 Differentiate Equation 17.21 with respect to r to obtain an expression for σ and ε. Express the equilibrium distance, r_e, in terms of σ and show that $V = -\varepsilon$.

17.11 Calculate the bond enthalpy of LiF using the Born–Haber cycle. The bond length of LiF is 1.51 Å. See Tables 12.7 and 12.8 for other information. Use $n = 10$ in Equation 17.7.

17.12 **(a)** From the data in Table 17.2, determine the van der Waals radius for argon. **(b)** Use this radius to determine the fraction of the volume occupied by 1 mole of argon at 25°C and 1 atm.

Hydrogen Bonding

17.13 Diethyl ether ($C_2H_5OC_2H_5$) has a boiling point of 34.5°C, whereas 1-butanol (C_4H_9OH) boils at 117°C. These two compounds have the same number and type of atoms. Explain the difference in their boiling points.

17.14 If water were a linear molecule, **(a)** would it still be polar and **(b)** would the water molecules still be able to form hydrogen bonds with one another?

17.15 Which of the following compounds is a stronger base: $(CH_3)_4NOH$ or $(CH_3)_3NHOH$? Explain.

17.16 Explain why ammonia is soluble in water but nitrogen trichloride is not.

17.17 Acetic acid is miscible with water, but it also dissolves in nonpolar solvents such as benzene or carbon tetrachloride. Explain.

17.18 Which of the following molecules has a higher melting point? Explain your answer.

(structures: o-nitrophenol — benzene ring with NO_2 and OH on adjacent carbons; p-nitrophenol — benzene ring with NO_2 and OH on opposite carbons)

17.19 What type of chemical analysis is needed to confirm the A—T and C—G pairing in DNA?

17.20 Assume the energy of hydrogen bonds per base pair to be 10 kJ mol^{-1}. Given two complementary strands of DNA containing 100 base pairs each, calculate the ratio of two separate strands to hydrogen-bonded double helix in solution at 300 K.

Additional Problems

17.21 The phrase "like dissolves like" has often been used to describe solubility. Explain what it means.

17.22 List all the intra- and intermolecular forces that could exist between hemoglobin molecules in water.

17.23 A small drop of oil in water usually assumes a spherical shape. Explain.

17.24 Which of the following properties indicates very strong intermolecular forces in a liquid? **(a)** A very low surface tension, **(b)** a very low critical temperature, **(c)** a very low boiling point, **(d)** a very low vapor pressure.

17.25 Figure 17.8 shows that the average distance between base pairs measured parallel to the axis of a DNA molecule is 3.4 Å. The average molar mass of a pair of nucleotides is 650 g mol^{-1}. Estimate the length in cm of a DNA molecule of molar mass 5.0×10^9 g mol^{-1}. Roughly how many base pairs are contained in this molecule?

17.26 Using values listed in Table 17.1 and a handbook of chemistry, plot the polarizabilities of the noble gases versus their boiling points. On the same graph, also plot their molar masses versus boiling points. Comment on the trends.

17.27 Given the following general properties of water and ammonia, comment on the problems that a biological system (as we know it) would have in developing in an ammonia medium.

	H_2O	NH_3
Boiling point	373.15 K	239.65 K
Melting point	273.15 K	195.3 K
Molar heat capacity	75.3 J K^{-1} mol^{-1}	8.53 J K^{-1} mol^{-1}
Molar heat of vaporization	40.79 kJ mol^{-1}	23.3 kJ mol^{-1}
Molar heat of fusion	6.0 kJ mol^{-1}	5.9 kJ mol^{-1}
Dielectric constant	78.54	16.9
Viscosity	0.001 N s m^{-2}	0.0254 N s m^{-2} (at 240 K)
Surface tension	0.07275 N m^{-1} (293 K)	0.0412 N m^{-1} (at 244 K)
Dipole moment	1.82 D	1.46 D
Phase at 300 K	Liquid	Gas

17.28 The HF_2^- ion exists as

$$\left[:\ddot{\underset{..}{F}}-H\cdots\cdots:\ddot{\underset{..}{F}}:\right]^-$$

The fact that both HF bonds are the same length suggests that proton tunneling occurs. **(a)** Draw resonance structures for the ion. **(b)** Give a molecular orbital description (with an energy-level diagram) of hydrogen bonding in the ion.

17.29 **(a)** Draw a potential-energy curve for two atoms based on a hard-sphere model. **(b)** A potential intermediate between the hard-sphere and the Lennard-Jones potentials is the square-well potential, defined by $V = \infty$ for $r < \sigma$, $V = -\varepsilon$ for $\sigma \leq r \leq a$, and $V = 0$ for $r > a$. Sketch this potential.

17.30 The potential energy of the helium dimer (He_2) is given by

$$V = \frac{B}{r^{13}} - \frac{C}{r^6}$$

where $B = 9.29 \times 10^4$ kJ Å^{13} (mol dimer)$^{-1}$ and $C = 97.7$ kJ Å^6 (mol dimer)$^{-1}$. **(a)** Calculate the equilibrium distance between the He atoms. **(b)** Calculate the binding energy of the dimer. **(c)** Would you expect the dimer to be stable at room temperature (300 K)?

17.31 The internuclear distance between two closest Ar atoms in solid argon is about 3.8 Å. The polarizability of argon is 1.66×10^{-30} m^3, and the first ionization energy is 1521 kJ mol^{-1}. Estimate the boiling point of argon. [*Hint:* Calculate the potential energy due to dispersion interaction for solid argon, and equate this quantity to the average kinetic energy of 1 mole of argon gas, which is $(3/2)RT$.]

17.32 The energy of a hydrogen bond between two water molecules is about 10 times that of van der Waals interaction between two xenon atoms. Yet water molecules dimerizes in air only about 30 percent more frequently than xenon atoms. Explain.

17.33 Calculate the work done in separating a Na^+ ion and a Cl^- ion from 5 Å to 10 Å in **(a)** vacuum and **(b)** water at 25°C.

CHAPTER 18

The Solid State

Since every solid substance contains parts that are crystalline, and since in many of them the whole is an aggregation of crystals, it will be readily understood that a knowledge of crystal structure often affords an explanation of the properties of the substance.

—Sir William Henry Bragg*

The gaseous state is characterized by complete randomness. At the opposite extreme, a crystalline solid has a highly ordered structure. There are four different types of crystals: metallic crystals, ionic crystals, covalent crystals, and molecular crystals. In this chapter, we discuss the determination of the three-dimensional structure of crystals by X-ray diffraction, which has played a central role in advancing our knowledge of the stability and function of proteins and nucleic acids. We shall also examine the structure of and bonding in these crystals.

18.1 Classification of Crystal Systems

What is a crystal? It is a substance in which the atoms or molecules are packed closely together in such a way that the total potential energy is at a minimum. These atoms and molecules make up a highly ordered structure, called a *crystal lattice*, in which they are arranged periodically in three dimensions. Let us start by considering a one-dimensional lattice, shown in Figure 18.1. The only geometric parameter is the repeating distance between the atoms, or lattice points,† which can be imagined to extend infinitely both to the left and right. In this case, each lattice point represents a *unit cell*, the basic repeating unit.

By definition, a two-dimensional lattice is a planar system; five different arrangements, or unit cells, can be constructed from the lattice points, also shown in Figure 18.1.

Figure 18.2 shows a unit cell and its extensions to form a three-dimensional lattice. A total of seven possible lattice types can be constructed based on the values of the lengths a, b, and c and angles α, β, and γ of the unit cell (Figure 18.3). Each of the unit cells is called a *primitive cell*, because all lattice points are at the corners of the unit cell. In 1850, the French physicist Auguste Bravais (1811–1863) showed that altogether there should be 14 different unit cells to account for lattice points at the center of the cell as well as at the centers of some of the faces. These 14 unit cells are now known as the *Bravais lattices*.

* *The Universe of Light*, Dover Publications, Sir W. H. Bragg, New York, 1940. Used by permission of G. Bell & Sons, London.

† Generally, a lattice point can be an atom, an ion, or a molecule.

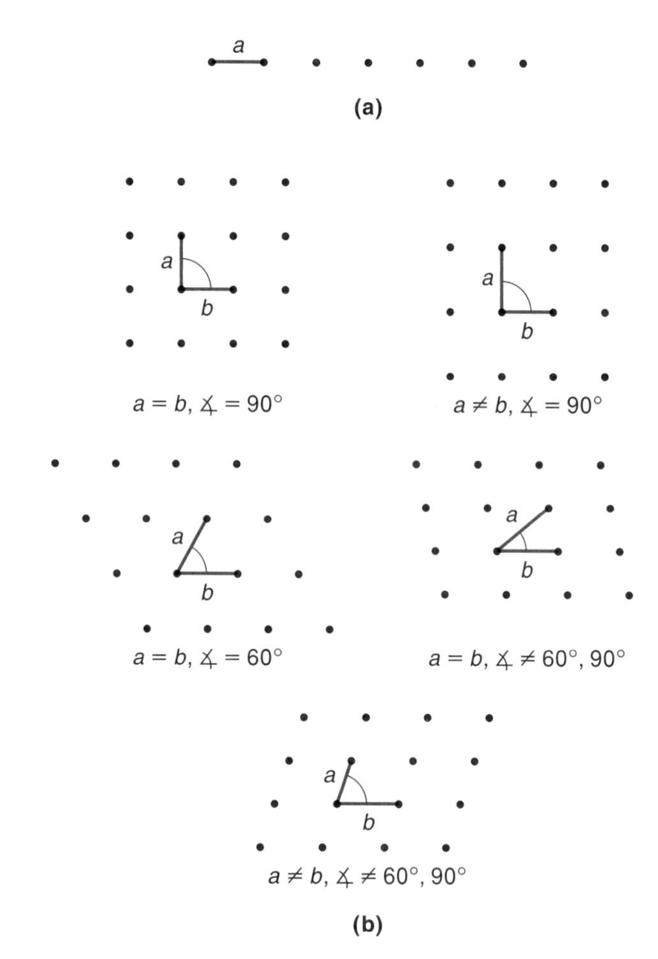

Figure 18.1
(a) One-dimensional lattice. (b) Five different arrangements of a two-dimensional lattice.

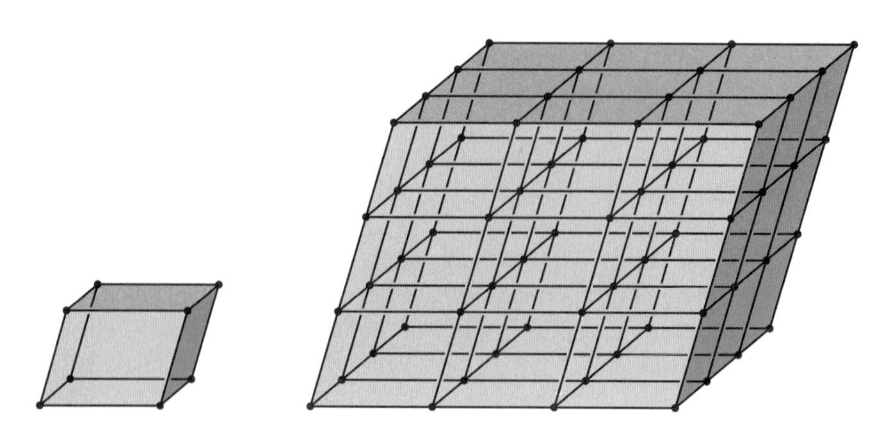

Figure 18.2
A unit cell and its extension in three dimensions.

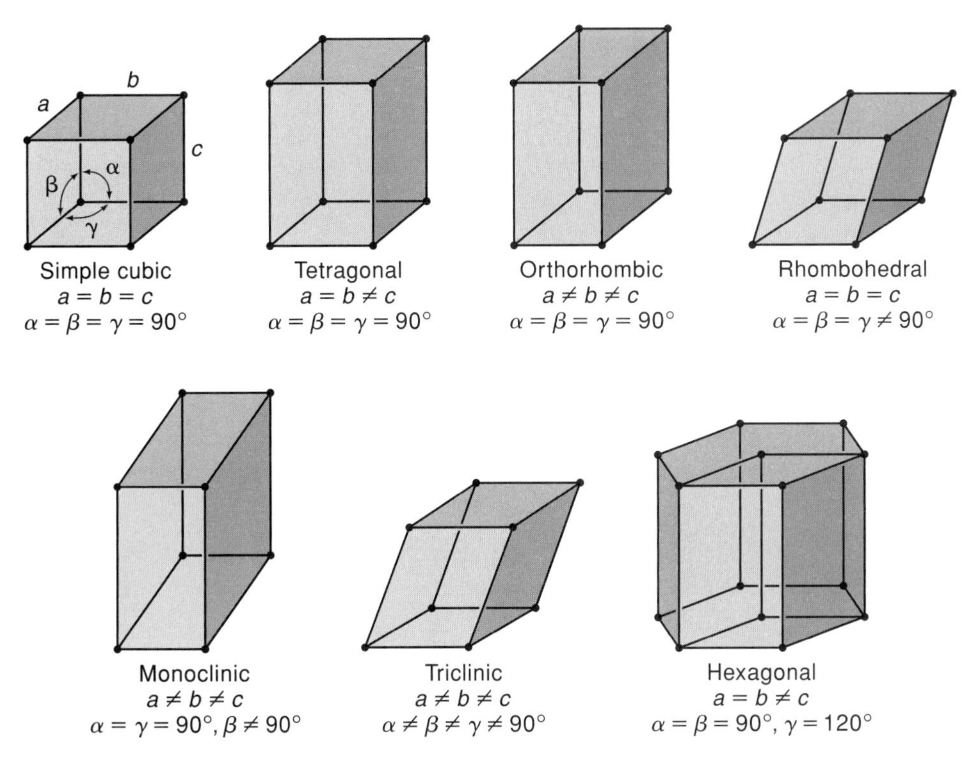

Figure 18.3
The seven possible lattice types having different cell lengths and angles.

The first task that confronts a crystallographer is measuring the size and shape of the unit cells. If a crystal is well formed, its external symmetry can be determined by measuring the interfacial angles, and hence, the crystal can be classified according to one of the seven types shown in Figure 18.3. The cell constants, or length and angles of the unit cell, must be determined by X-ray diffraction, as described below.

In X-ray crystallography, it is convenient to characterize a given crystal in terms of a set of planes. Consider a two-dimensional crystal lattice as shown in Figure 18.4. A number of planes with different orientations (AA', BB', CC', etc.) may be drawn so that each plane contains some of the lattice points. Parallel to any plane is a whole set of planes that can be generated from it by applying the unit lattice translation. The AA', BB', ... planes are labeled according to their intercepts on the a, b, and c axes, measured from an arbitrarily chosen origin. Thus, the CC' plane intercepts the axes at a, $4b$, and ∞ (since we are dealing with a two-dimensional crystal, this plane is parallel to the c axis and therefore the intercept is at infinity). Next, we take the reciprocals of the intercepts, which are $1/a$, $1/4b$, and $1/\infty$. We clear the fractions by multiplying each quantity by the largest common denominator, which is 4 in this case (we exclude infinity in this operation). This gives us three numbers (410), which are known as the *Miller indices* of the plane (after the British mineralogist William Miller, 1801–1880). Generally, the Miller indices (hkl) of any plane give the orientation of the plane in the crystal with reference to its three internal axes. Table 18.1 summarizes the steps for obtaining the Miller indices for the planes shown in Figure 18.4.

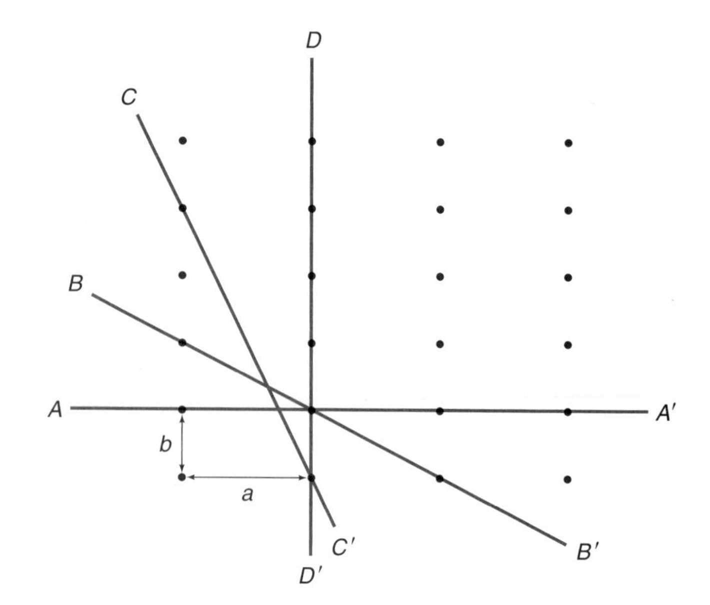

Figure 18.4
Characterization of a two-dimensional lattice in terms of a set of planes. The original is at the extreme lower left side.

Table 18.1
Miller Indices for a Two-Dimensional Lattice

Crystal face or plane	Intercepts	Reciprocals of multiples	Miller indices (*hkl*)
AA'	$\infty a, b, \infty c$	$\frac{1}{\infty}, \frac{1}{1}, \frac{1}{\infty}$	(010)
BB'	$2a, 2b, \infty c$	$\frac{1}{2}, \frac{1}{2}, \frac{1}{\infty}$	(110)
CC'	$a, 4b, \infty c$	$\frac{1}{1}, \frac{1}{4}, \frac{1}{\infty}$	(410)
DD'	$a, \infty b, \infty c$	$\frac{1}{1}, \frac{1}{\infty}, \frac{1}{\infty}$	(100)

The same procedure can be extended to a three-dimensional crystal. For the three types of cubic cells shown in Figure 18.5, for example, each set of Miller indices describes a series of parallel, equally spaced planes.

18.2 The Bragg Equation

Let us consider what happens when a monochromatic X-ray beam of wavelength λ strikes the face of a crystal (Figure 18.6). Because of the penetrating power of the X ray, the incident radiation can interact with atoms in many layers. On the first layer, the beam is reflected by the atoms in much the same manner as a mirror reflects visible light.

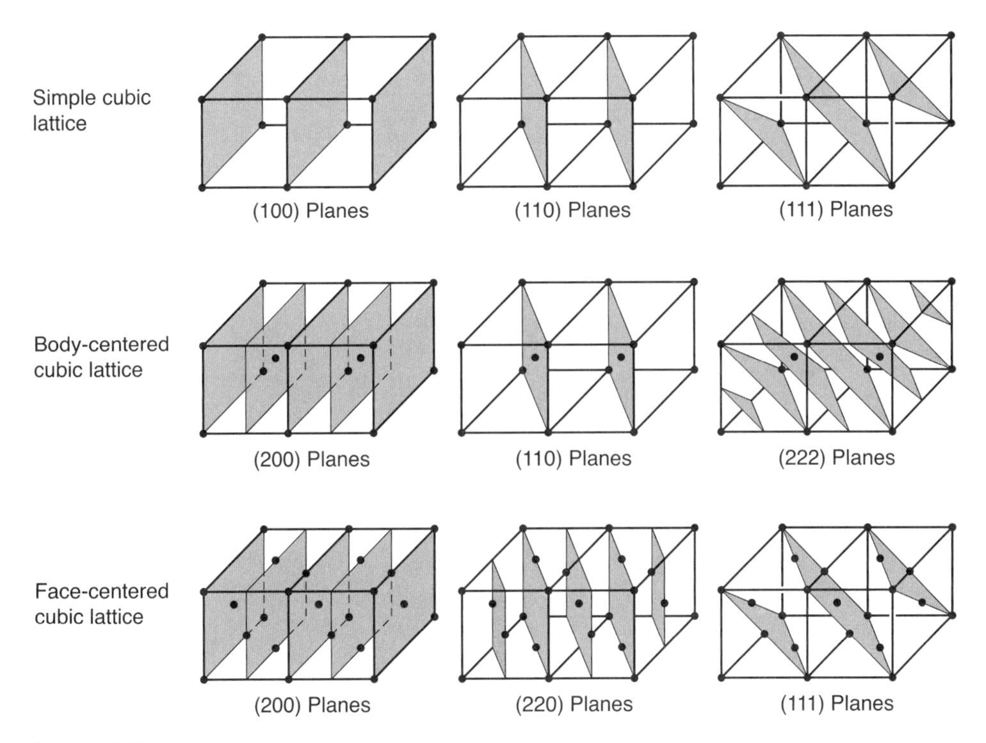

Figure 18.5
Miller indices for three types of cubic lattices.

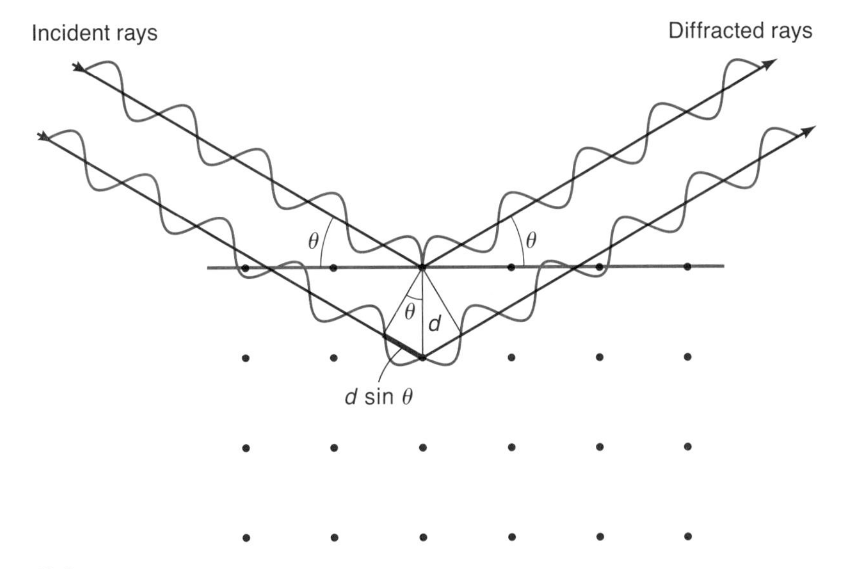

Figure 18.6
Reflection of X rays from different layers of atoms.

How does the reflected beam from the first plane interact with a reflected beam from a different plane separated by d from the first plane? For the scattered X-ray beams to interfere with one another constructively, that is, for the waves to be in phase with each other, the following condition must be met:

$$2d \sin \theta = n\lambda \qquad n = 1, 2, \ldots \tag{18.1}$$

where $2d \sin \theta$ is the difference in pathlength between two waves. A *diffraction pattern* of alternating intensities of reflected X-ray beams is obtained by placing a X ray detector at a certain angle. The number n represents the order of diffraction; $n = 1$ denotes first-order diffraction; $n = 2$, second-order diffraction; and so forth. From Equation 18.1, however, we see that the angle θ for a certain value of n and d is the same as the first-order diffraction for a set of planes with spacing d/n. For example, second-order diffraction from a (111) plane can be viewed as if it were a first-order diffraction from the (222) plane, even though this plane may not be present in a particular crystal. The Miller indices readily show that $d_{222} = d_{111}/2$ (see Figure 18.5) so that

$$\sin \theta = \frac{2\lambda}{d_{111}} = \frac{1\lambda}{d_{222}}$$

For this reason, we can always set $n = 1$ in Equation 18.1 and treat higher-order diffractions as if they were first-order diffractions from planes that are closer together. Thus, we have

$$2d_{hkl} \sin \theta = \lambda \tag{18.2}$$

which is known as the Bragg Equation, after the father-and-son British physicists Sir William Henry Bragg (1862–1942) and Sir William Lawrence Bragg (1890–1972).

18.3 Structural Determination by X-Ray Diffraction

The Bragg equation immediately provides us with a way of measuring cell dimensions. For a cubic lattice, the perpendicular distance, d_{hkl}, between adjacent members of the set of parallel planes represented by Miller indices (hkl) can be obtained as follows. From Figure 18.3, $\alpha = \beta = \gamma = 90°$ and $a = b = c$. The three-dimensional form of the Pythagorean theorem gives (see Appendix 18.1 on p. 836)

$$\begin{aligned}\frac{1}{d_{hkl}^2} &= \frac{h^2}{a^2} + \frac{k^2}{b^2} + \frac{l^2}{c^2} \\ &= \frac{h^2 + k^2 + l^2}{a^2}\end{aligned}$$

or

$$d_{hkl} = \frac{a}{\sqrt{h^2 + k^2 + l^2}} \tag{18.3}$$

From Equation 18.2, we have

$$\sin \theta_{hkl} = \frac{\lambda}{2d_{hkl}} = \frac{\lambda}{2a}\sqrt{h^2 + k^2 + l^2} \tag{18.4}$$

or

$$\sin^2 \theta_{hkl} = \frac{\lambda^2}{4a^2}(h^2 + k^2 + l^2) \tag{18.5}$$

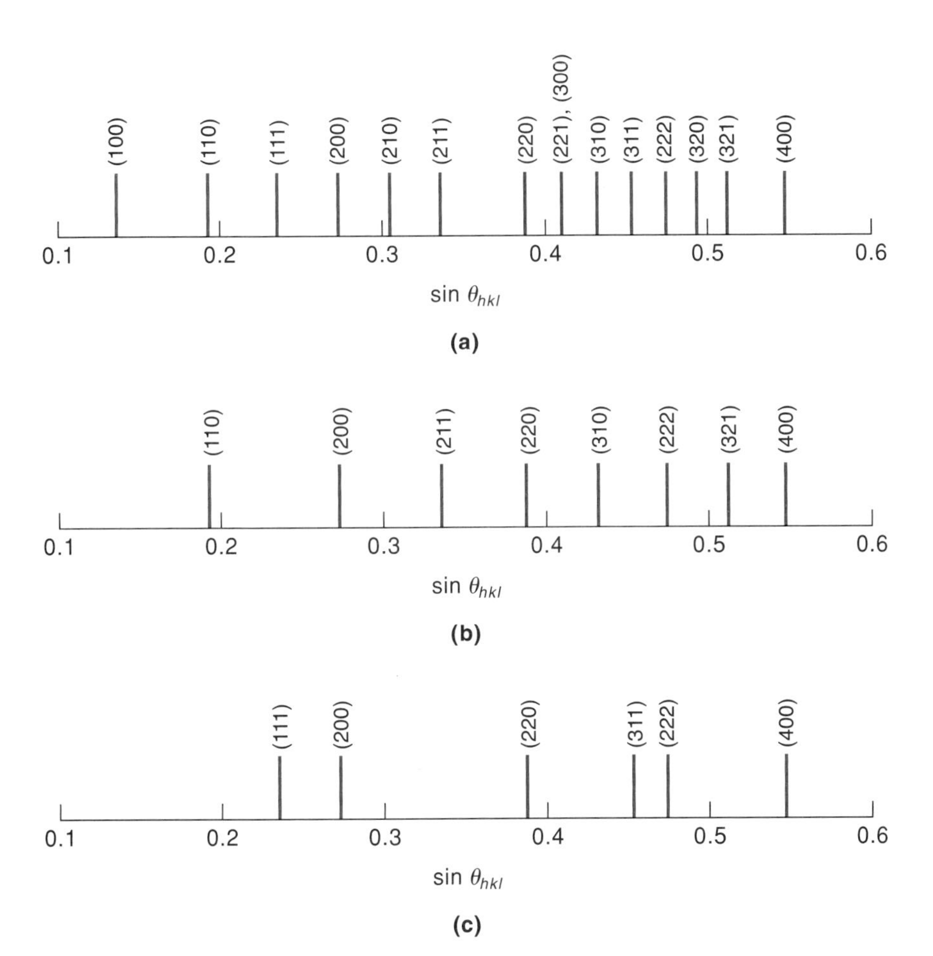

Figure 18.7
Theoretical plots of the X-ray diffraction pattern versus $\sin\theta_{hkl} = (\lambda/2a)\sqrt{h^2 + k^2 + l^2}$ on the horizontal axis for (a) simple-cubic, (b) body-centered cubic, and (c) face-centered cubic lattices. The (λ/a) term is taken to be 0.274. Each line represents a specific set of h, k, and l values. In reality, these lines have different intensities.

The values of $(h^2 + k^2 + l^2)$ are determined by the various planes:

(hkl)	(100)	(110)	(111)	(200)	(210)	(211)	(220)	(221)	···
$(h^2 + k^2 + l^2)$	1	2	3	4	5	6	8	9	···

If we know the angle θ_{hkl} for each plane, then we can construct a set of lines on the $\sin\theta_{hkl}$ scale (Figure 18.7a) for a given value of λ/a. Note that the seventh line is missing, because we cannot have $(h^2 + k^2 + l^2) = 7$. Similarly, the fifteenth line is also missing, as are others in the sequence.

Similar plots for body-centered and face-centered lattices are also shown in Figure 18.7b,c. Here, fewer lines are observed in comparison with the simple-cubic lattice. To see why some of the lines are missing, let us consider the body-centered cube. As Figure 18.5 shows, the (110) planes pass through all lattice points, and a strong diffraction pattern results. The situation is different for the (100) planes because they are

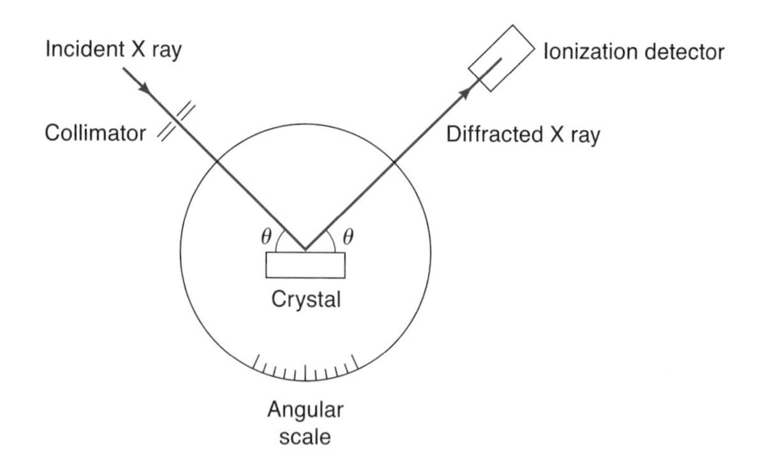

Figure 18.8
Practical arrangement for studying the diffraction of X rays by a crystal.

interleaved by another layer, the (200) plane, of atoms. The X rays diffracted by the (100) planes are in phase with one another but out of phase by half a wavelength with those diffracted by the (200) planes. Because a crystal contains many, many unit cells, there are essentially as many atoms in the (200) planes as there are in the (100) planes. Consequently, there is a total destructive interference, and the diffraction line from the (100) planes will be absent. On the other hand, diffraction from the (200) planes will be present because *all* atoms lie in these planes. The presence or absence of other lines can be similarly explained.

The Powder Method

The X-ray diffraction pattern of a crystal can be recorded in one of two ways. In the first method, a small single crystal is mounted with a particular axis perpendicular to the direction of the X-ray beam. The crystal is situated at the center of a table that is calibrated with an angular scale (Figure 18.8). An ionization detector is used to monitor the intensity of X rays. The procedure is to measure the intensity of the diffracted rays as a function of the angle θ; the intensity reaches a maximum whenever the angle satisfies the Bragg equation.* This method yields a number of lines, each one representing a particular plane, and measures the corresponding angles.

The crystal method was employed by the Bragg team early in the development of X-ray diffraction studies. Since then, several improvements have greatly aided data recording. The single-crystal method is essential for analyzing complex structures, but it requires carefully growing and mounting the crystal. An alternative approach, introduced by Debye and the Swiss physicist Paul Scherrer (1890–1969) and independently by the American physicist Albert Hull (1880–1966) enables structural determination from powdered samples rather than single crystals. In this technique, a beam of X rays is directed at a mass of finely ground powder of the substance under study. The powder sample is actually numerous small crystals, or *crystallites*. Because these crystallites are randomly oriented, the X-ray beam will meet the crystal planes at every possible value of θ for which the Bragg equation is satisfied. The diffracted

* The intensity depends on the number of atoms in the plane and the type of atoms present. X rays are almost entirely scattered by the electrons of atoms.

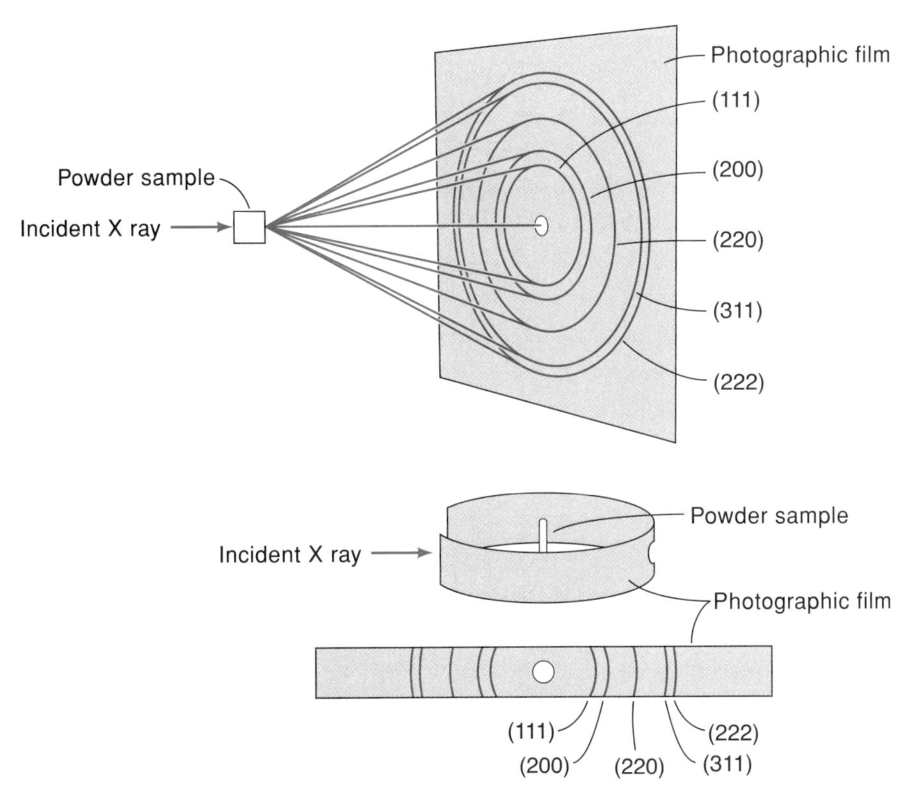

Figure 18.9
X-ray diffraction pattern produced by a powder sample containing face-centered crystallites.

beams from each plane actually form a cone, shown in Figure 18.9. One way to record the diffraction pattern is to use a cylindrical photographic film so arranged that its axis is perpendicular to the incident X-ray beam. From the distances between the lines and the dimensions of the film, the angle θ can be calculated for each line.

In practice, most of the X rays pass straight through the sample, undiffracted.

Determination of the Crystal Structure of NaCl

Let us now consider a specific example: the structural determination of NaCl. Table 18.2 gives the Bragg angles for some of the lines observed. These lines are positioned on the sin θ_{hkl} scale as shown in Figure 18.10. Comparing this pattern with that in Figure 18.7, we see that NaCl has a face-centered lattice. To determine the length of the cube, we need to find a common factor that divides all values of $\sin^2 \theta_{hkl}$. This factor turns out to be 0.0188, which is equal to $\lambda^2/4a^2$ in Equation 18.5. The experimental data in Table 18.2 are obtained by using X rays generated when a copper target is bombarded with high-energy electrons. The characteristic wavelength of X rays is 1.542 Å (0.1542 nm), so

$$0.0188 = \frac{\lambda^2}{4a^2} = \frac{(1.542\ \text{Å})^2}{4a^2}$$

or

$$a = 5.623\ \text{Å}$$

where a is the length of the cube.

Table 18.2
X-Ray Diffraction Data for NaCl

θ_{hkl}	$\sin^2\theta_{hkl}$	$\dfrac{\sin^2\theta_{hkl}}{0.0188}$	(hkl)
13.68°	0.0560	3	(111)
15.83°	0.0744	4	(200)
22.70°	0.1489	8	(220)
27.05°	0.2068	11	(311)
28.33°	0.2252	12	(222)
33.13°	0.2990	16	(400)

Also, from Equation 18.5, we can index each line as follows:

$$\frac{\sin^2\theta_{hkl}}{0.0188} = h^2 + k^2 + l^2$$

Because the ratio on the left side is known, each line can be indexed in a straightforward way as shown in Table 18.2.

To describe the NaCl structure completely, we must know how many atoms are in each unit cell. The density of NaCl is 2.16 g cm^{-3}, and the molar mass is 58.44 g; thus, the molar volume is given by 58.44 g mol^{-1}/2.16 g cm^{-3}, or 27.06 cm^3 mol^{-1}, and the volume of one NaCl formula unit is

$$\frac{27.06 \text{ cm}^3 \text{ mol}^{-1}}{6.022 \times 10^{23} \text{ mol}^{-1}} = 4.49 \times 10^{-23} \text{ cm}^3 = 44.9 \text{ Å}^3$$

where 1 cm^3 = 10^{24} Å^3. The volume of the unit cell is a^3 or 178 Å^3, and there must be 178 Å^3/44.9 Å^3, or 4 NaCl per unit cell. Figure 18.11 shows the crystal structure of sodium chloride. At first, there might appear to be more than four units of NaCl

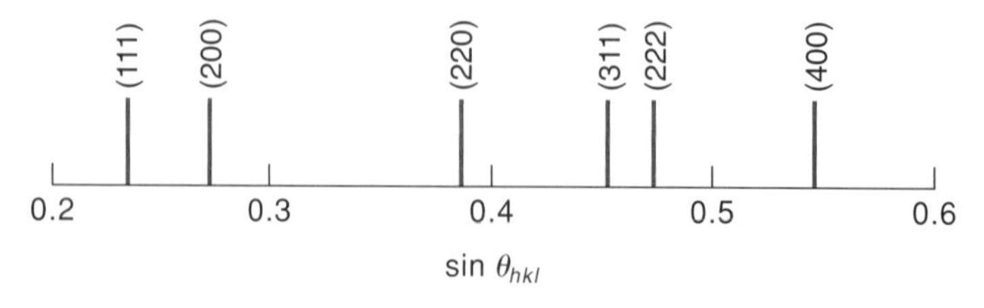

Figure 18.10
Bragg diffraction lines plotted on the sin θ_{hkl} scale for the NaCl powder sample, assuming a λ/a value of 0.274. In reality, these lines have different intensities.

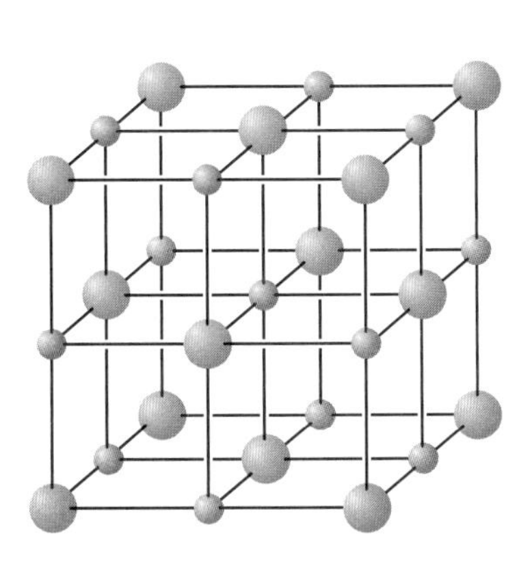
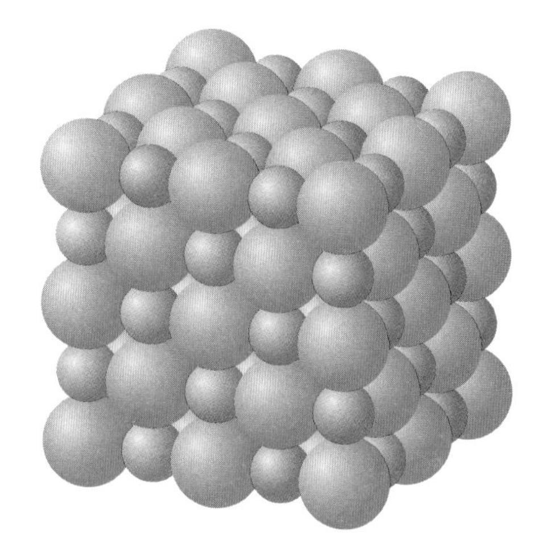

Figure 18.11
Two representations of the sodium chloride crystal lattice. The large spheres represent chloride ions and the smaller spheres represent sodium ions.

present in a unit cell. Except for the center sodium ion, however, every other ion is shared among adjacent cells. For example, each of the eight chloride ions at the corners is shared among eight unit cells, and each of the six chloride ions in the center of the outer faces is shared between two unit cells. Thus, the total number of chloride ions per unit cell is given by $(8 \times \frac{1}{8} + 6 \times \frac{1}{2}) = 4$. Similarly, we can show that the total number of sodium ions per unit cell is given by $(12 \times \frac{1}{4} + 1) = 4$, because each of the 12 sodium at the edges is shared among four cells, and there is a single unshared sodium ion in the center.

The powder method is most useful for crystals that have only one or two parameters to be determined—cubic, tetragonal, and rhombohedral crystals, for example. For other crystal systems, the task of indexing the lines becomes very difficult, if not impossible, by this technique.

EXAMPLE 18.1

For a simple cubic lattice with $a = 2.6$ Å (2.6×10^{-10} m), calculate the smallest diffraction angle for the (111) plane, given that $\lambda = 1.542$ Å (1.542×10^{-10} m).

ANSWER

From Equation 18.4,

$$\theta = \sin^{-1}\frac{\lambda}{2a}\sqrt{h^2 + k^2 + l^2}$$

$$= \sin^{-1}\frac{1.542 \times 10^{-10}\text{ m}}{2(2.6 \times 10^{-10}\text{ m})}\sqrt{3}$$

$$= 30.9°$$

The Structure Factor

For simple crystals, such as the alkali halide crystals, knowledge of the symmetry and unit-cell dimensions enables us to determine the exact structure. For most crystals, however, we must determine the array of atoms or ions within each unit cell. To do this, we need to measure and relate the observed intensity of X rays diffracted from a known set of planes (hkl) to the distribution of atoms within this set of planes. Theory shows that the measured intensity I_{hkl} is proportional to the square of the *structure factor*, F_{hkl}. Thus, by taking the square root of each observed intensity, we can obtain the observed structure factor, F_{hkl}^{obs}. The significance of F_{hkl} is that it can also be calculated if we know the positions and scattering power of the atoms within the unit cell. Consequently, if there are N atoms in the unit cell, we write

$$F_{hkl}^{\text{cal}} = \sum_{i=1}^{N} f_i \times Q(x_i, y_i, z_i) \tag{18.6}$$

where F_{hkl}^{cal} is the calculated structure factor for the set of planes (hkl) and $Q(x_i, y_i, z_i)$, known as the *geometric structure factor*, is a function of the coordinates x_i, y_i, z_i of the ith atom. The scattering factors, f_i, can be calculated from the wave functions of the electrons. We could try to guess the positions of all the atoms and then calculate the F_{hkl} values. Good agreement between our calculations and experimental data would confirm the correct structure. Unfortunately, this trial-and-error method is impractical because there are many possible positions for each atom in the unit cell, even though we could make some intelligent guesses. In practice, the reverse procedure is applied; that is, the atomic positions are determined from the measured intensities. This approach is aided by *Fourier synthesis*, a well-known mathematical technique named after the French mathematician Jean Fourier which maps out the electron density distribution within a unit cell. The positions of atoms can be determined by noting where the electron density rises to peak values.

In general, larger scattering factors f_i correspond to more electrons. Consequently, larger atoms or ions are more readily detected by X-ray diffraction.

A major obstacle in structural determination by Fourier synthesis is what is known as the *phase problem*. To construct the electron-density map, we must know both the signs and magnitudes of F_{hkl}, but only the magnitudes of F_{hkl} can be measured experimentally. The *isomorphous replacement technique*,* introduced by the Scottish chemist John Monteath Robertson (1900–1989) in his 1936 study of phthalocyanine, helps circumvent this difficulty. First, Robertson photographically measured the intensities, I_{hkl}, from various planes of phthalocyanine. Then, by substituting a heavy atom (e.g., mercury or gold) at the center of the phthalocyanine molecule, he redetermined the intensities. The signs of F_{hkl} could then be deduced as follows. If a spot on the photographic film became stronger (i.e., if I_{hkl} increased) after the heavy atom was introduced, its original F_{hkl} value with respect to the center of the molecule must have been positive; if the spot was weaker, the F_{hkl} value must have been negative. In this manner, Robertson determined all the signs of F_{hkl}. Figure 18.12 shows the Fourier electron-density map in the plane of phthalocyanine.

Phthalocyanine is a relatively small molecule, containing only 56 atoms (see Figure 18.12). With the X-ray analysis of vitamin B_{12} (181 atoms) completed by the British chemist Dorothy Hodgkin (1910–1994) in 1955, it seemed that we might

* This technique is based on the assumption that the presence of the heavy atom does not alter the crystal structure.

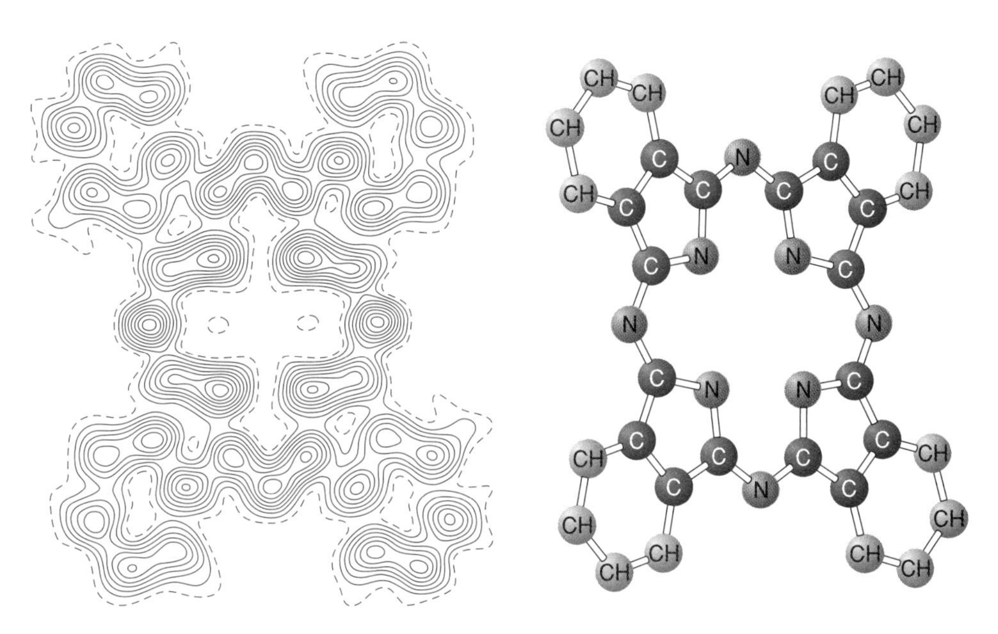

Figure 18.12
Electron density contour map (left) and molecular structure of phthalocyanine (right). [From J. M. Robertson, *J. Chem. Soc.* 1195 (1936). Used by permission.]

have reached the limit of complexity in structural determination by X rays. In 1953, however, the Austrian–British biochemist Max Perutz (1914–2002) realized that the isomorphous replacement technique was equally applicable to protein molecules containing thousands or even tens of thousands of atoms, such as hemoglobin. We usually speak of the *resolution* of the X-ray diffraction data. At 4.6 Å resolution, the electron-density map provides an overall shape of the protein molecule. At 3.5 Å resolution, it is often possible to discern the backbone, that is, the polypeptide chain. At 3.0 Å resolution, we begin to be able to identify the amino acid side chains and can therefore determine, in favorable cases, the primary sequence of the protein. At 2.5 Å resolution, the positions of atoms may be located with an accuracy of ±0.4 Å. Finally, at 1.5 Å resolution, the positions of atoms can be located to about ±0.1 Å.

The primary sequence of a protein molecule tells us how the amino acids are linked together.

Generally, the process of determining a protein structure consists of the following steps: (1) crystallization of the native protein (i.e., protein in its functional state) and collection of diffraction measurements; (2) collection of diffraction data from heavy-atom derivatives; (3) determination of the phases for the native protein data; (4) computer analysis and generation of an electron-density map from the data collected in steps 1, 2, and 3; and (5) construction of a structural model for comparison against the observed data. To determine the structure of such an enormously complex system, about 500,000 intensities must be accurately measured, and perhaps 1 million calculations must be performed! With the availability of desktop computers and improved instrumentation for recording and measuring intensities, however, the task of analyzing X-ray crystallographic data is no longer a major challenge to chemists.

Figure 18.13 shows the diffraction pattern of the enzyme lysozyme. As of 2013, over 80,000 macromolecular structures have been analyzed by X-ray diffraction. In fact, the present state of the art in X-ray crystallography is such that if we can grow suitable crystals of proteins for diffraction measurements—not an easy task by any

For X-ray diffraction study, single crystals that are at least 0.1 mm in each dimension are practical.

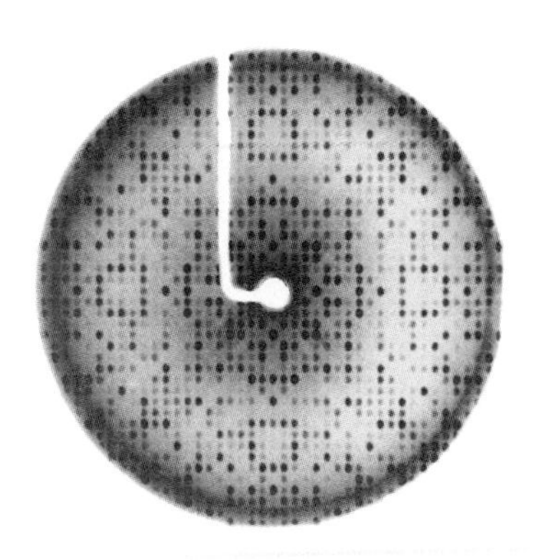

Figure 18.13
Photograph of an X-ray diffraction pattern of crystalline lysozyme. The white "L" is a shadow of the sample holder. (Courtesy of J. R. Knox.)

means—we can usually unravel their three-dimensional structures. Knowledge of the three-dimensional structure of proteins and nucleic acids has probably been the greatest contribution to our understanding of their stability and function.

Neutron Diffraction

As mentioned earlier, X rays are scattered primarily by electrons, and the intensity of scattered X rays increases strongly with increasing atomic number. For this reason, the X-ray diffraction technique is not useful for locating the positions of H atoms, which scatter X rays very weakly. In contrast, neutrons are not scattered by electrons; rather, they interact with the nuclei through the strong nuclear forces that are responsible for binding the nucleons (protons and neutrons) together. Thus, the neutron-diffraction technique supplements the X-ray method for molecules in which light elements (especially hydrogen) are present. In this respect, it is interesting to note that neutron diffraction intensities vary with isotope and that ^{1}H scatters significantly stronger than most common isotopes, including ^{2}H, ^{12}C, ^{14}N, and ^{16}O.

Neutrons generated from a nuclear reactor possess high kinetic energy. Through collisions with the moderator (a substance such as water that can reduce the kinetic energy of neutrons), neutrons can be slowed to *thermal velocities*, that is, velocities possessed by gaseous particles at room temperature. Using a velocity-selector device, we can obtain a monochromatic beam of neutrons suitable for diffraction study of crystals. We can calculate the wavelength of the thermal neutrons by using the de Broglie relation as follows. At temperature T, according to the equipartition of energy theorem (see Section 2.9),

$$\frac{1}{2}mu^2 = \frac{3}{2}k_BT$$

or

$$mu = \sqrt{3mk_BT} \tag{18.7}$$

From Equation 10.25

$$\lambda = \frac{h}{p} = \frac{h}{mu}$$

Therefore,

$$\lambda = \frac{h}{\sqrt{3mk_BT}} \tag{18.8}$$

Using $m = 1.675 \times 10^{-27}$ kg for a neutron, and $T = 298$ K, we write

$$\lambda = \frac{6.626 \times 10^{-34}\text{ J s}}{\sqrt{3(1.675 \times 10^{-27}\text{ kg})(1.381 \times 10^{-23}\text{ J K}^{-1})(298\text{ K})}}\left(\frac{\text{kg m}^2\text{ s}^{-2}}{\text{J}}\right)^{1/2}$$
$$= 1.46 \times 10^{-10}\text{ m} = 1.46\text{ Å}$$

This wavelength is comparable to chemical bond lengths, which is just the right magnitude needed for diffraction work.

In general, the neutron-diffraction technique is not as widely used as X-ray diffraction because the work must be done at a nuclear-reactor facility. In addition, a beam of neutrons is weak compared with a beam of X rays from an ordinary X-ray tube. Unlike X rays, neutrons cannot be measured easily and must be detected by counters. Nevertheless, in many ways the technique is complementary to X-ray diffraction, and its use in structural studies will continue to increase.

18.4 Types of Crystals

Having discussed the diffraction techniques used to study the structure of crystals, we shall now consider the four major types of crystals: metallic, ionic, covalent, and molecular. In particular, we shall concentrate on their structure, bonding, and stability.

Metallic Crystals

The crystal structure of metals is the simplest of the four types of crystals because all the atoms in a metal are the same size and they bear no charges. Metallic bonding is nondirectional; therefore, in most metals, the atoms are arranged in such a way as to achieve the most efficient packing (see below). We shall first systematically survey the ways that multiple identical spheres (Ping-Pong balls, for example), can be packed together.

Packing of Spheres. In the simplest case, a layer of spheres can be arranged as shown in Figure 18.14a. The three-dimensional structure can be generated by placing a layer

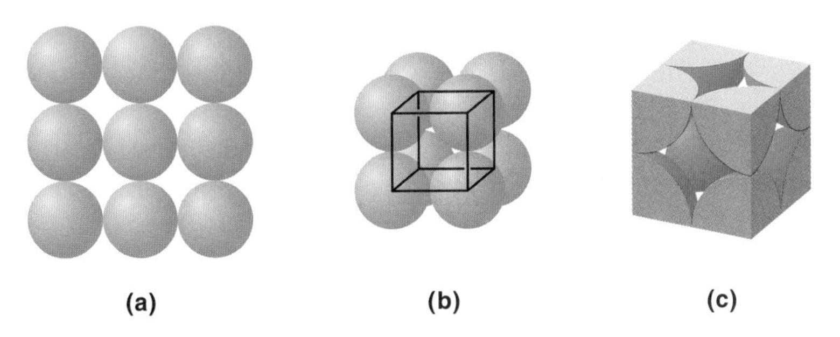

Figure 18.14
Arrangement of identical spheres in a simple cubic cell. (a) Top view of one layer of spheres. (b) Representation of a simple cubic cell. (c) Because each sphere is shared by eight unit cells and there are eight corners in a cube, there is the equivalent of one complete sphere inside a simple cubic unit cell.

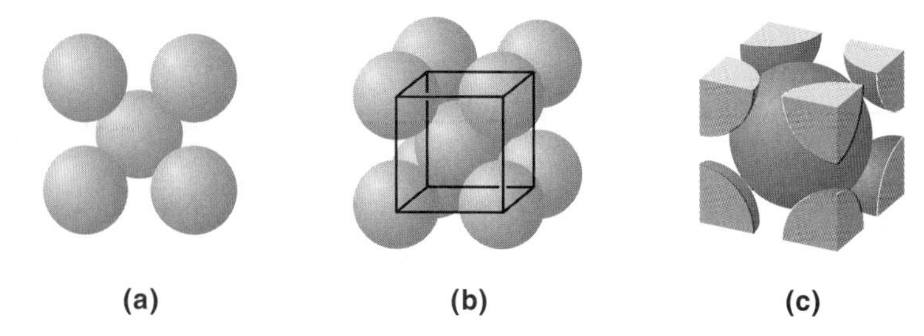

Figure 18.15
Arrangement of identical spheres in a body-centered cube. (a) Top view of a unit cell. (b) Representation of a body-centered cubic unit cell. (c) There is the equivalent of two spheres inside a body-centered cubic unit cell.

above and below this layer in such a way that spheres in one layer are directly over the spheres in the layer below it. This procedure can be extended to generate many, many layers, as in a crystal. Focusing on the sphere in the center, we see that it is in contact with four spheres in its own layer, one sphere in the layer above, and one sphere in the layer below. Each sphere is this arrangement is said to have a *coordination number* (CN) of 6 because it has six immediate neighbors. The coordination number is defined as the number of atoms (or ions) surrounding an atom (or ion) in a crystal lattice. Its value gives us a measure of how tightly the spheres are packed together—the larger the coordination number, the closer the spheres are to each other. The basic, repeating unit in the array of spheres described above is called a *simple cubic* (sc) cell (Figure 18.14b, c).

A more efficient way to pack spheres is that shown in Figure 18.15. The first layer is the same as that for a sc cell. The second-layer spheres fit into the depressions or notches of the first layer, and the third-layer spheres fit into the depressions of the second layer, and so on. This arrangement forms a *body-centered cubic* (bcc) cell. Each sphere has a CN of 8.

We can pack the spheres even closer (and hence increase the CN) by starting with a close-packed first-layer structure, which we call layer A, shown in Figure 18.16a. Focusing on the only enclosed sphere, we see that it has six immediate neighbors in that layer. In the second layer (which we call layer B), spheres are packed into the depressions between the spheres in the first layer so that all the spheres are as close together as possible (Figure 18.16b). There are two ways that a third-layer sphere may cover the second layer. The spheres may fit into the depressions so that each third-layer sphere is directly over a first-layer sphere (Figure 18.16c). Because there is no difference between the arrangement of the first and the third layers, we also call the third layer A. Alternatively, the third-layer spheres may fit into the depressions that lie directly over the depressions in the first layer (Figure 18.16d). In this case, we call the third layer C.

Figure 18.17 shows the "exploded views" and the structures resulting from the arrangements presented in Figure 18.16c and d. The ABA arrangement is known as the *hexagonal close-packed* (hcp) *structure*, and the ABC arrangement is the *cubic close-packed* (ccp) *structure*, which corresponds to a *face-centered cubic* (fcc) cell because there is a sphere at each of the six faces of the cube. Note that the spheres in every other layer of the hcp structure occupy the same vertical position (ABABAB ...), whereas in

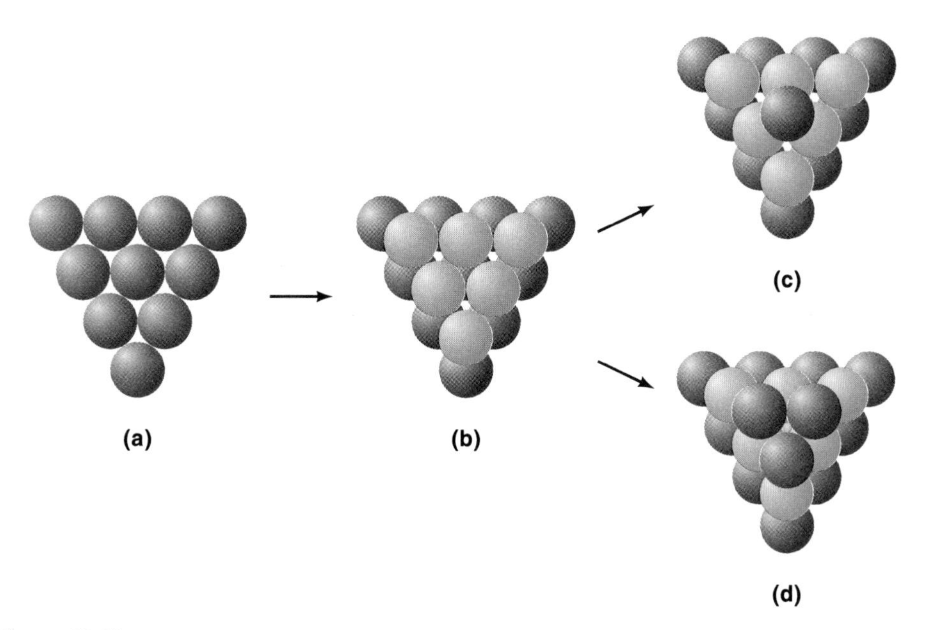

Figure 18.16
(a) In a close-packed layer, each sphere is in contact with six others. (b) Spheres in the second layer fit into the depressions between the first-layer spheres. (c) In the hexagonal close-packed structure, each third-layer sphere is directly over a first-layer sphere. (d) In the cubic close-packed structure, each third-layer sphere fits into a depression that is directly over a depression in the first layer.

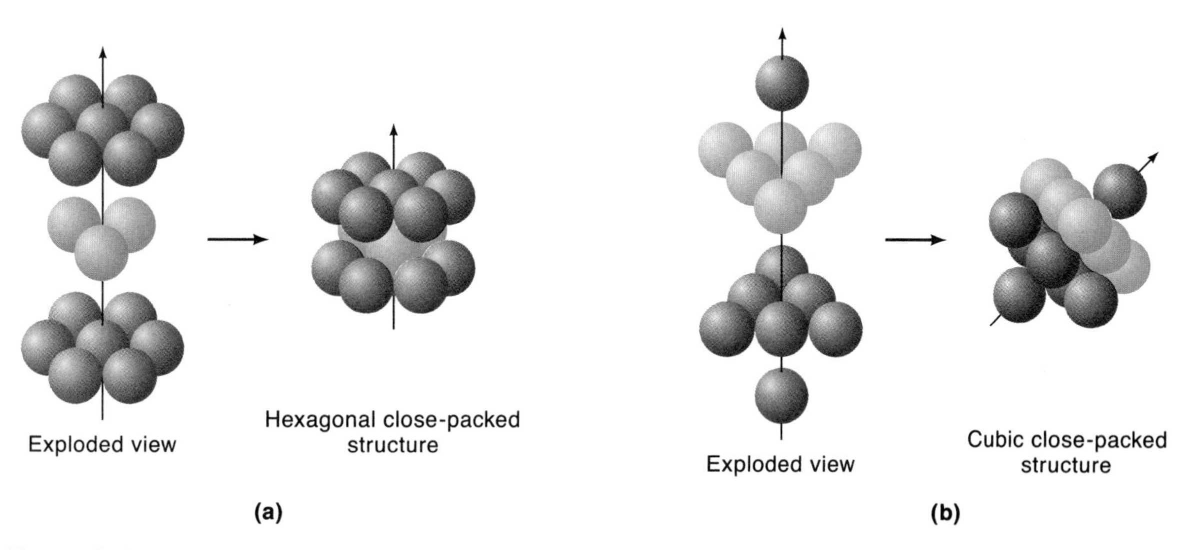

Figure 18.17
Exploded views of (a) a hexagonal close-packed structure and (b) a cubic close-packed structure. The arrow is tilted to show the cubic closed-packed structure more clearly. Note that this arrangement is the same as the face-centered cubic cell.

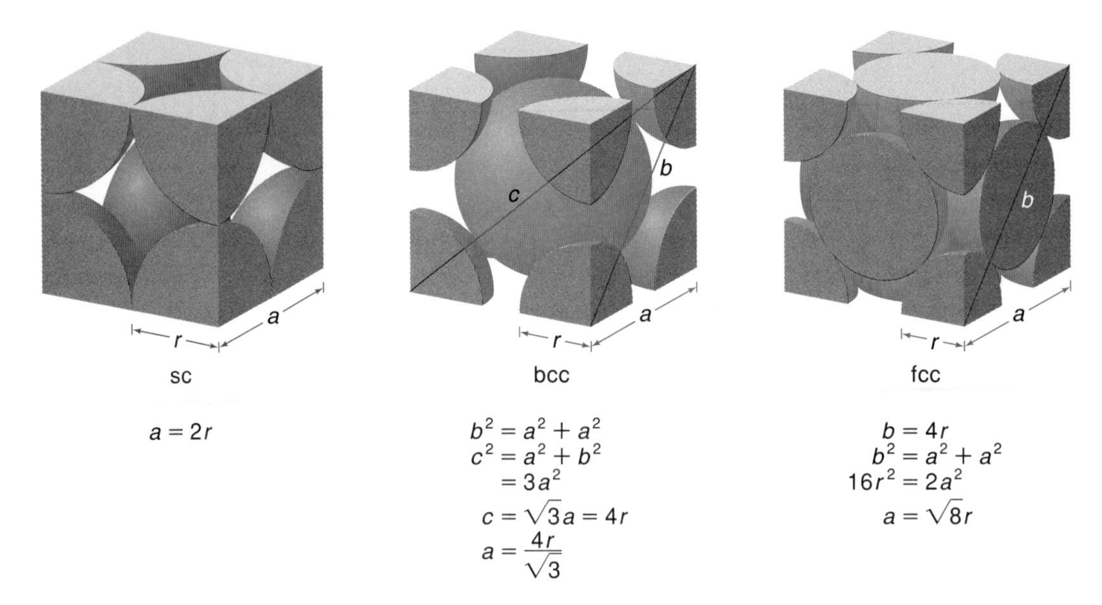

Figure 18.18
The relationship between the edge length (a) and radius (r) of atoms in the simple cubic (sc) cell, body-centered cubic (bcc) cell, and face-centered cubic (fcc) cell.

the ccp structure, the spheres in every fourth layer occupy the same vertical position (ABCABCA ...). In both structures, each sphere has a CN of 12 (each sphere is in contact with six spheres in its own layer, three spheres in the layer above, and three spheres in the layer below). Both the hcp and ccp structures represent the most efficient way of packing identical spheres in a unit cell, and there is no way to increase the coordination number beyond 12. For this reason, these two structures are called the *closest packing.*

Figure 18.18 shows the geometric relationship between the edge length (a) of the cube and the radius (r) of the spheres for the sc, bcc, and fcc cells. A useful quantity that tells us how efficiently the spheres are packed in a unit cell is the *packing efficiency* (*PE*), defined as

$$PE = \frac{\text{volume of spheres in the unit cell}}{\text{volume of unit cell}} \tag{18.9}$$

For the sc cell, there is the equivalent of one sphere within a unit cell, so

$$PE = \frac{(4/3)\pi r^3}{a^3} \tag{18.10}$$

We see from Figure 18.18 that $a = 2r$, so

$$PE = \frac{(4/3)\pi r^3}{(2r)^3} = 0.524, \text{ or } 52.4\%$$

Similarly, we can calculate the packing efficiencies of bcc and fcc structures (see Problem 18.5).

Table 18.3 lists the crystal structures of some metallic elements.

Table 18.3
Crystal Structures of Some Metals

Simple cubic cell	Po
Body-centered cubic cell	Li, Na, K, Rb, Cs, Ba, V, Nb, Cr, Mo, W, Fe
Face-centered cubic cell	Ca, Sr, Rh, Ir, Ni, Pd, Pt, Cu, Ag, Au, Al, Pb
Hexagonal-close-packed structure	Be, Mg, Sc, La, Ti, Zr, Ru, Os, Co, Zn, Cd, Tl

Bonding in Metals. Metals are ductile (can be drawn out into wires) and malleable (can be flattened), and they have the ability to conduct electricity. These properties derive from the special nature of the metallic bond. For a satisfactory explanation of metallic bonding, we turn to molecular orbital theory.

If we think of an entire block of metal as a giant molecule, all of the atomic orbitals of a particular type (1*s*, 2*s*, and so on) in the crystal interact to form a set of delocalized orbitals that extend throughout the system. Consider sodium metal. Figure 18.19 shows the successive overlaps of the 3*s* orbitals between neighboring Na atoms. The resulting molecular orbitals are so close together that they are appropriately called a *band*. The bands formed from Na molecular orbitals are shown in Figure 18.20. The 1*s*, 2*s*, and 2*p* bands are completely filled, but the 3*s* band is half-filled because each sodium atom has only one 3*s* electron. The presence of this partially-filled 3*s*

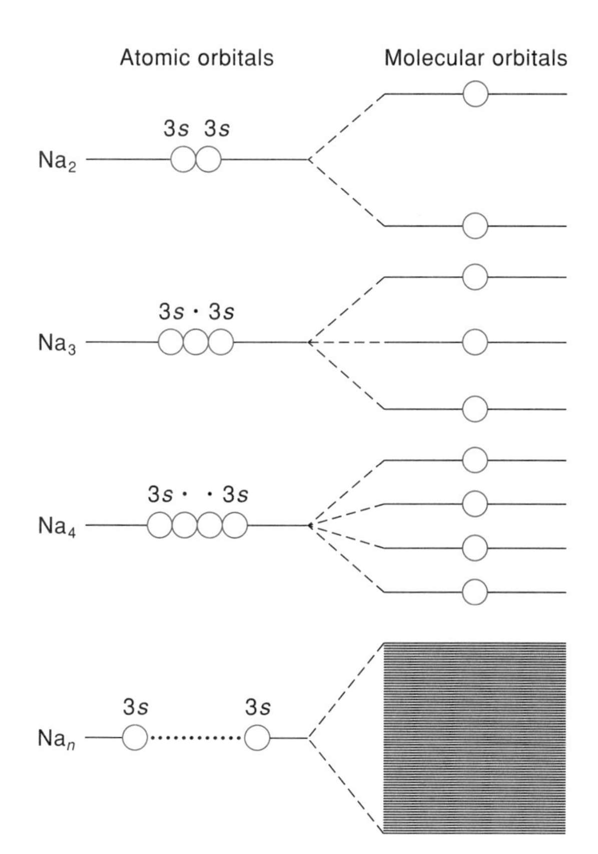

Figure 18.19
Formation of a delocalized molecular orbital band from the overlap of numerous 3*s* orbitals in sodium.

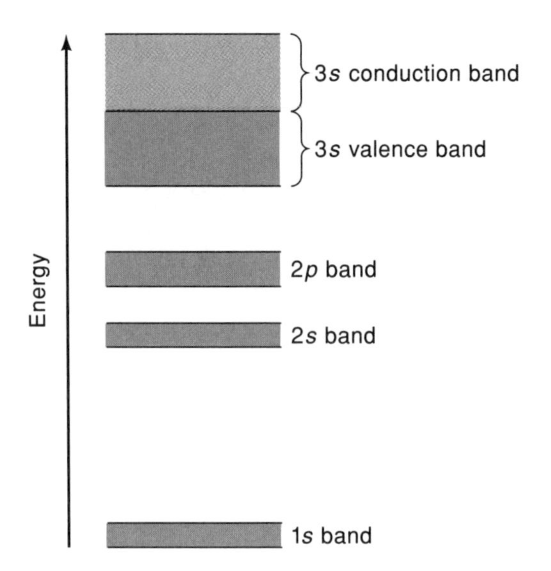

Figure 18.20
Delocalized molecular orbital bands in sodium. The 1s, 2s, and 2p bands are completely filled. The 3s band is half-filled. Only an infinitesimal amount of energy is required to excite an electron from the 3s valence band into the 3s conduction band, which extends over the entire metal. Thus, sodium is an electrical conductor.

band accounts for the stability of the metal. (There are more electrons in the bonding molecular orbitals than in the antibonding molecular orbitals.) Furthermore, only a minimal amount of energy is needed to promote an electron into the empty delocalized antibonding orbitals, where it is free to move through the entire metal, giving rise to the metal's electrical conductivity.

Figure 18.21 compares the energy gap between the *valence band* (the highest filled orbitals) and the *conduction band* (the lowest unfilled orbitals) of a metal, an insulator, and a semiconductor. In a metal, the energy gap is essentially nonexistent. In an insulator, the gap is large, so the promotion of an electron into the conduction band does not take place readily. A semiconductor has a smaller gap than an insulator does, so electrical conductance can be induced either by raising the temperature or by adding certain impurities that narrow the energy gap.

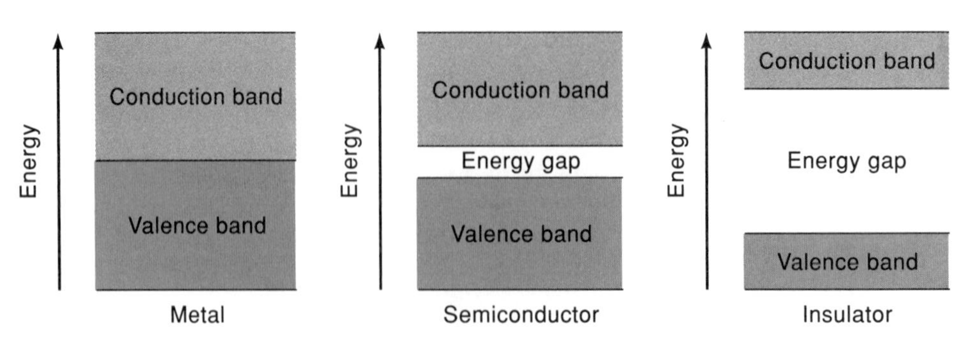

Figure 18.21
Comparison of the energy gaps between the valence band and the conduction band in a metal, a semiconductor, and an insulator.

Ionic Crystals

The packing of spheres serves as the basis for describing the structures of ionic compounds. Ionic crystals have two important characteristics: the anions and cations are quite different in size, and they are charged particles. Even in the rare cases in which the anions and cations are approximately the same size, we cannot have a closest-packed structure with a CN of 12 because electrical neutrality cannot be maintained with this arrangement. Consequently, ionic solids are generally less dense than metals.

A quantity of great interest in the study of ionic crystals is the ionic radius (also called the crystal radius). As we saw in Section 18.2, the length of the unit cell for NaCl is 5.623 Å. From Figure 18.22, we see that this length is equal to twice the sum of the ionic radii of Na^+ and Cl^-. There is no way to measure the radius of individual ions, but we can estimate the radius of a few ions and, by comparison, find others. For example, the radius of I^- in LiI is estimated to be 2.16 Å. Using this value, we can determine the radius of K^+ in KI, the radius of Cl^- in KCl, and so on. Table 18.4 lists the ionic radii of a number of ions. These are average values obtained from a large body of data. Thus, the sum of the cation and anion radii is not usually equal to the cell dimension in a particular ionic compound. For example, based on the values in Table 18.4, the length of the NaCl unit cell is $2(r_{Na^+} + r_{Cl^-}) = 2(0.98 + 1.81)$ Å = 5.58 Å, which is different from 5.623 Å. The reasons ions do not have unique radii in the solid state are that first, they are not hard spheres, so their electron densities are influenced by the type of counterions present, and second, the nature of bonding is never purely ionic, so the radii of ions are also affected by the percent of covalent character.

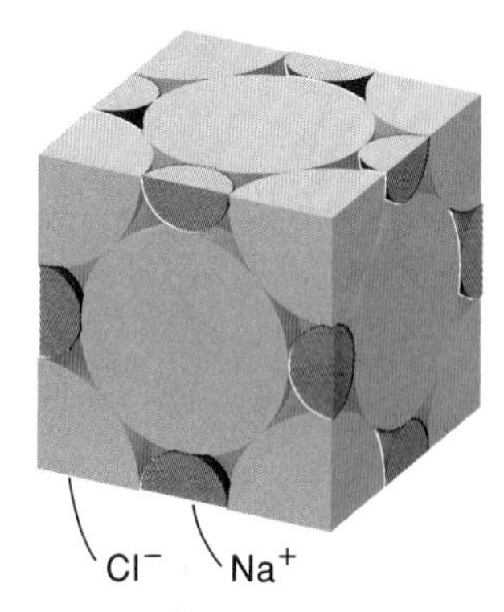

Figure 18.22
Portions of Na^+ and Cl^- ions within a face-centered cubic unit cell.

Radius Ratio Rule. The structure of many ionic crystals can be understood by examining the packing of the larger ions (usually the anions) in a tetrahedral, octahedral, or cubic arrangement, with the cations needed for charge balance occupying the "holes" formed by the larger spheres (Figure 18.23). These holes are not all the same size, and

Table 18.4
Radii of Some Ions/Å

						H^- 1.54
Li^+ 0.68	Be^{2+} 0.35	B^{3+} 0.23	C^{4+} 0.16	N^{3+} 1.71	O^{2-} 1.32	F^- 1.33
Na^+ 0.98	Mg^{2+} 0.66	Al^{3+} 0.51	Si^{4+} 0.42	P^{3-} 2.12	S^{2-} 1.84	Cl^- 1.81
K^+ 1.33	Ca^{2+} 0.99	Ga^{3+} 0.62	Ge^{4+} 0.53	As^{3-} 2.22	Se^{2-} 1.91	Br^- 1.96
Rb^+ 1.47	Sr^{2+} 1.12	In^{3+} 0.81	Sn^{4+} 0.71	Sb^{5+} 0.62	Te^{2-} 2.11	I^- 2.20
Cs^+ 1.67	Ba^{2+} 1.34	Tl^{3+} 0.95	Pb^{4+} 0.84	Bi^{5+} 0.74		

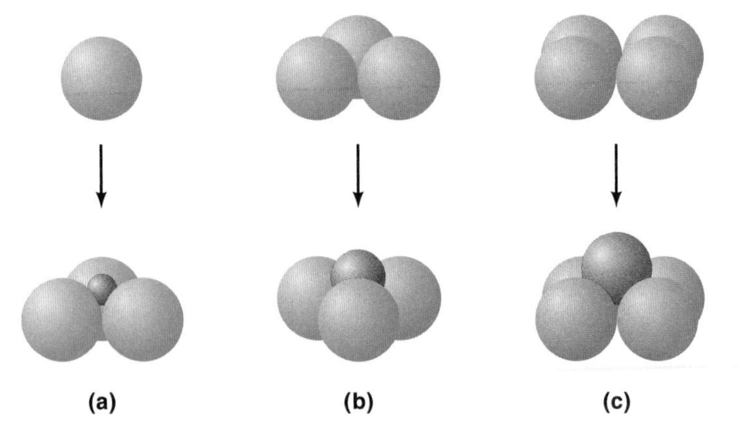

Figure 18.23
(a) A tetrahedral hole, (b) an octahedral hole, and (c) a cubic hole. In each case, the smaller ion is the cation.

the particular structure an ionic compound will assume can often be inferred by the magnitude of the *radius ratio rule*, defined by

$$\text{radius ratio} = \frac{r_{\text{smaller ion}}}{r_{\text{larger ion}}} \tag{18.11}$$

For example, the tetrahedral hole is quite small, so only small cations can fit into that site, achieving a CN of 4. As the size of the hole increases, so does the radius ratio and the CN (Table 18.5). Figure 18.24 shows the unit cells of ZnS, NaCl, and CsCl. These structures are characteristic of many ionic compounds.

The Stability of Ionic Crystals. In an ionic compound, there is no fixed, directed force of attraction. Although electrostatic forces are strong, as indicated by the high melting points of these compounds, ionic solids are quite brittle and cannot be easily bent or deformed.

The stability of an ionic crystal can be expressed in terms of the *lattice energy* (U_0), which is defined as the energy required to separate one mole of a crystal into its gaseous ions. In NaCl, for example, the lattice energy is equal to the enthalpy change for the reaction

$$\text{NaCl}(s) \rightarrow \text{Na}^+(g) + \text{Cl}^-(g)$$

Table 18.5
Radius Ratio Rules

Radius ratio	CN	Type of hole for cation	Example
0.225–0.414	4	tetrahedral	ZnS
0.414–0.732	6	octahedral	NaCl
0.732–1.000	8	cubic	CsCl

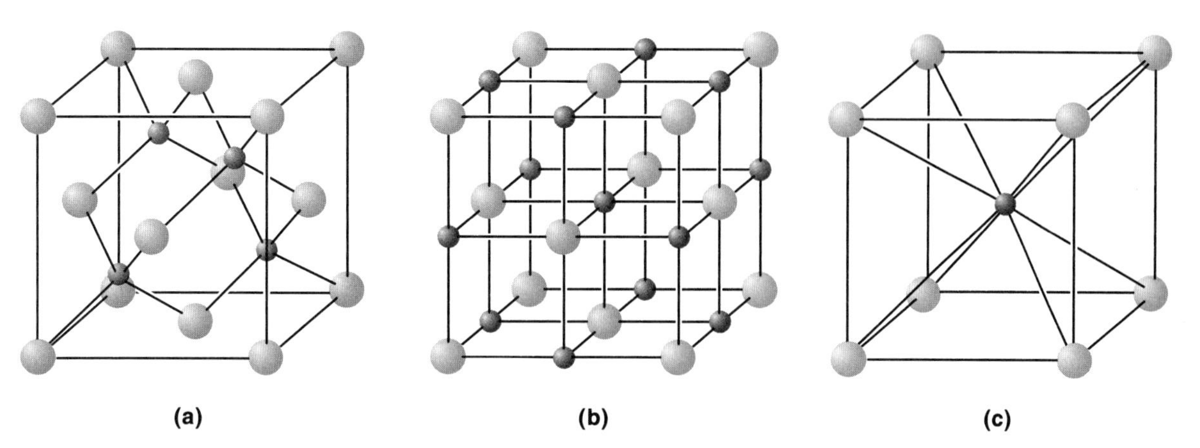

Figure 18.24
Crystal structures of (a) ZnS, (b) NaCl, and (c) CsCl. In each case, the smaller sphere is the cation.

Because this process is endothermic, U_0 is always a positive quantity. We can calculate the value of U_0 from the charges of the ions and the known cell dimensions by using an approach similar to that shown in Section 17.2. Recall, however, that Equation 17.7 applies to just a pair of ions. In a crystal of NaCl, each Na^+ ion interacts with *all* the ions present. Calculating the lattice energy of such a crystal may seem daunting, but it can be handled systematically as follows. Figure 18.25 shows the distances between one Na^+ ion and some of its immediate neighbors in the NaCl lattice. This Na^+ ion is surrounded by six Cl^- ions at distance r. The potential energy from this attractive interaction is

$$V_0 = -\frac{6e^2}{4\pi\varepsilon_0 r}\left(1 - \frac{1}{n}\right)$$

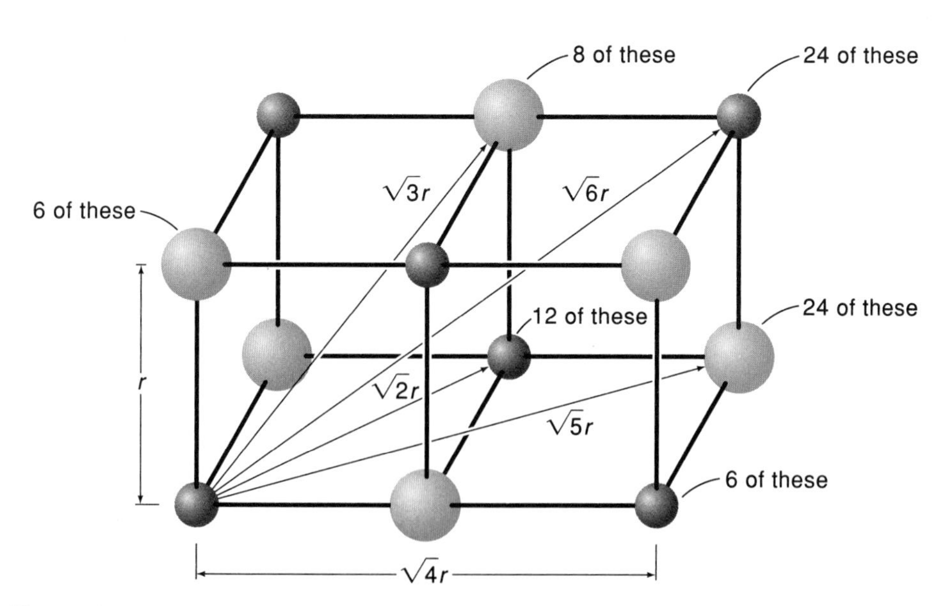

Figure 18.25
Distances between one Na^+ ion (lower left comer) and its neighbors in the NaCl lattice. The figure shows only one quarter of a unit cell. Color codes: gray is Na^+; red is Cl^-.

where n is a number between 8 and 12. Note that we have simplified Equation 17.7 by replacing $q_{Na^+}q_{Cl^-}$ with e^2, where e is the electronic charge. The next nearest neighbors to the Na^+ ion are 12 other Na^+ ions at a distance $\sqrt{2}r$, and the corresponding potential energy for this repulsive interaction is

$$V_0 = \frac{12e^2}{4\pi\varepsilon_0\sqrt{2}r}\left(1 - \frac{1}{n}\right)$$

Then come 8 Cl^- ions at a distance of $\sqrt{3}r$, 6 Na^+ ions at $\sqrt{4}r$, 24 Cl^- ions at $\sqrt{5}r$, 24 Na^+ ions at $\sqrt{6}r$, and so on. The potential energy (V) resulting from the interaction of the Na^+ ion of interest with all the ions in the lattice is given by the sum of the individual terms:

$$V = -\frac{e^2}{4\pi\varepsilon_0 r}\left(\frac{6}{1} - \frac{12}{\sqrt{2}} + \frac{8}{\sqrt{3}} - \frac{6}{\sqrt{4}} + \frac{24}{\sqrt{5}} - \frac{24}{\sqrt{6}} + \cdots\right)\left(1 - \frac{1}{n}\right) \quad (18.12)$$

Due to the long range of interionic forces, the series in the above equation converges very slowly and special techniques are needed to evaluate it. The sum of the series converges to the value $\mathcal{M} = 1.7476$, where $\mathcal{M}$ is called the *Madelung constant* [after the German physicist Erwin Madelung (1881–1972)] for the NaCl crystal lattice. For 1 mole of NaCl, Equation 18.12 becomes

$$\overline{V} = -\frac{N_A \mathcal{M} e^2}{4\pi\varepsilon_0 r}\left(1 - \frac{1}{n}\right) \quad (18.13)$$

where N_A is Avogadro's constant. From our definition of lattice energy it follows that $U_0 = -\overline{V}$. Comparing Equation 18.13 with Equation 17.7 for 1 mole of ion pairs, using the same value of n, shows that the lattice energy for the NaCl crystal is about 1.47 times greater than the energy holding one mole of NaCl ion pairs in the gas phase.

Madelung constants have been evaluated for other crystal structures besides NaCl. Thus, for CsCl, $\mathcal{M} = 1.76267$, and for ZnS, $\mathcal{M} = 1.63805$.

EXAMPLE 18.2

Calculate the lattice energy of NaCl, given that $n = 10$ for the repulsive term in the crystal and that the sum of Na^+ and Cl^- radii is 2.81 Å.

ANSWER

We use Equation 18.13 and the conversion factor 1 J = 1 N m:

$$\overline{V} = -\frac{(6.022 \times 10^{23}\ \text{mol}^{-1})(1.7476)(1.602 \times 10^{-19}\ \text{C})^2[1 - (1/10)]}{4\pi(8.854 \times 10^{-12}\ \text{C}^2\ \text{N}^{-1}\ \text{m}^{-2})(2.81 \times 10^{-10}\ \text{m})}$$

$$= -7.77 \times 10^5\ \text{J mol}^{-1}$$

$$= -777\ \text{kJ mol}^{-1}$$

Therefore, the lattice energy U_0 equals +777 kJ mol^{-1}.

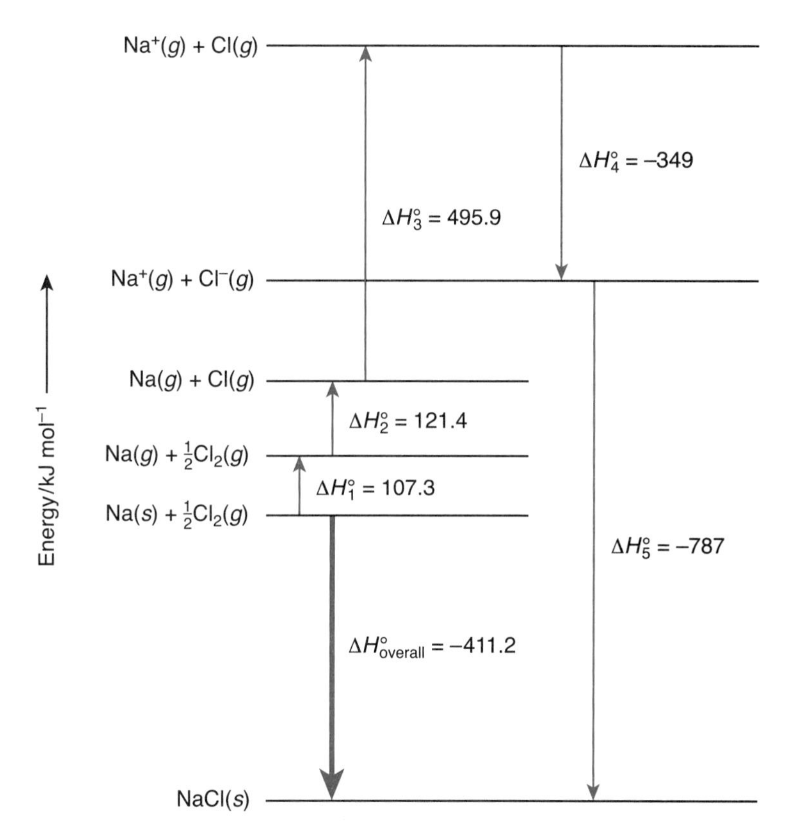

Figure 18.26
Born–Haber cycle for the formation of one mole of NaCl(s).

Lattice energy cannot be measured directly, but it can be evaluated by using the Born–Haber cycle. We start with the elements Na and Cl_2 in their standard states at 25°C and carry out the following steps shown in Figure 18.26:

1. Convert 1 mole of metallic sodium to sodium vapor. The enthalpy of sublimation, $\Delta H_1^\circ = 107.3$ kJ.
2. Dissociate $\frac{1}{2}$ mole of Cl_2 into Cl atoms. From the bond enthalpy of Cl_2 (see Table 3.4), we have $\Delta H_2^\circ = \frac{1}{2}$ (242.7 kJ) = 121.4 kJ.
3. Ionize 1 mole of Na atoms. From Table 12.7, $\Delta H_3^\circ = 495.9$ kJ.
4. Convert 1 mole of Cl atoms to Cl^- ions. From Table 12.8, $\Delta H_4^\circ = -349$ kJ.
5. Step 5 is

$$Na^+(g) + Cl^-(g) \rightarrow NaCl(s)$$

   which defines lattice energy; that is, $\Delta H_5^\circ = -U_0$.

In Appendix B, we find the standard enthalpy of formation for NaCl:

$$Na(s) + \tfrac{1}{2}Cl_2(g) \rightarrow NaCl(s) \qquad \Delta H^\circ_{\text{overall}} = -411.2 \text{ kJ mol}^{-1}$$

According to Hess's law, the enthalpy change for the overall process is equal to the sum of the individual steps so that

$$\Delta H^\circ_{\text{overall}} = \Delta H^\circ_1 + \Delta H^\circ_2 + \Delta H^\circ_3 + \Delta H^\circ_4 + \Delta H^\circ_5$$

or

$$\begin{aligned}\Delta H^\circ_5 &= -411.2 \text{ kJ mol}^{-1} - 107.3 \text{ kJ mol}^{-1} - 121.4 \text{ kJ mol}^{-1} \\ &\quad -495.9 \text{ kJ mol}^{-1} + 349 \text{ kJ mol}^{-1} \\ &= -787 \text{ kJ mol}^{-1}\end{aligned}$$

Therefore, the lattice energy of NaCl is given by $U_0 = 787$ kJ mol^{-1}.

The agreement between the calculated lattice energy and that obtained by the Born–Haber cycle is reasonably good. The discrepancy is mainly due to dispersion forces between the ions, the partial covalent character, and the zero-point energy of the crystal, all of which we did not take into account in Example 18.2.

Covalent Crystals

Covalent crystals are hard solids that possess very high melting points. They are usually poor conductors of electricity because they have no delocalized orbitals. In covalent crystals, atoms are held together by covalent bonds. Well-known examples are two of the allotropic forms of carbon, graphite and diamond (Figure 18.27). The structure of diamond is based on a fcc lattice. There are eight carbon atoms at the corners of the cube, six carbon atoms in the face centers, and four more carbon atoms within the unit cell. Each atom is tetrahedrally bonded to four other atoms. This tightly held lattice contributes to diamond's unusual hardness. The carbon-to-carbon distance is 1.54 Å, which is similar to that in ethane. In graphite, each carbon atom is bonded to three other carbon atoms. The carbon-to-carbon bond length is 1.12 Å, which is close to that in benzene. The layers are held together by rather weak dispersion forces. Consequently, graphite is easily deformed in directions parallel to the layers. Within a plane, graphite, like diamond, is a covalently bonded crystal. Another important example of a covalent crystal is quartz, or SiO_2, also shown in Figure 18.27.

The melting points of diamond and quartz are 3550°C and 1610°C, respectively. The sublimation point of graphite is 3652°C.

The 2010 Nobel Prize in Physics was awarded for work on the two-dimensional material *graphene*, which is essentially a single layer of graphite. Graphene is transparent, electrically conductive, dense, and surprisingly strong. It may be extracted from a graphite surface using ordinary adhesive tape!

Molecular Crystals

Molecular crystals are soft solids that possess low melting points. They are poor conductors of electricity. Example of substances that form molecular crystals include Ar, N_2, SO_2, I_2, and benzene. Generally, the molecules are packed together as closely as their size and shape allow. The attractive forces are primarily van der Waals interactions (dispersion forces and dipole–dipole forces). Hydrogen bonding is largely responsible for the crystal structure of ice (see Figure 17.10). The crystal structure of buckminsterfullerene, or buckyball (C_{60}), in which the C_{60} molecules are held together by dispersion forces only, has the fcc arrangement.

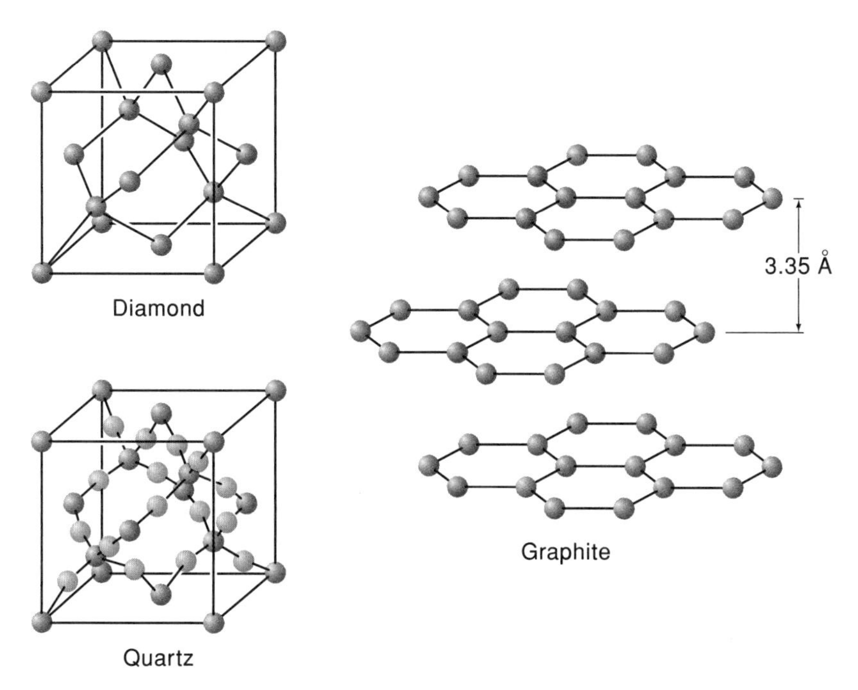

Figure 18.27
Structures of diamond, graphite, and quartz. The red spheres represent oxygen.

Finally, we should keep in mind that not all solids exist in crystalline forms. Solids that do not possess a regular structure are called *amorphous solids*; glass is a well-known example. The structure of such a substance is more difficult to study, because regular patterns are lacking.

Key Equations

$2d_{hkl} \sin \theta = \lambda$	(The Bragg equation)	(18.2)
$PE = \dfrac{\text{volume of spheres in the unit cell}}{\text{volume of unit cell}}$	(Packing efficiency)	(18.9)
$\text{radius ratio} = \dfrac{r_{\text{smaller ion}}}{r_{\text{larger ion}}}$	(Radius ratio)	(18.11)

APPENDIX 18.1

Derivation of Equation 18.3

To derive Equation 18.3, we consider *any* crystal system with orthogonal (perpendicular) axes. Consider, for example, the (210) planes shown in Figure 18.28a. The plane at the lower left goes through a lattice point. The next plane is displaced an amount a/h along the x, or a, axis and b/k along the y, or b, axis. An enlargement of the first two planes is shown in Figure 18.28b. For the two right-angled triangles, we write

$$\sin\alpha = \frac{d}{a/h} \tag{1}$$

$$\sin\alpha = \frac{b/k}{\sqrt{(a/h)^2 + (b/k)^2}} \tag{2}$$

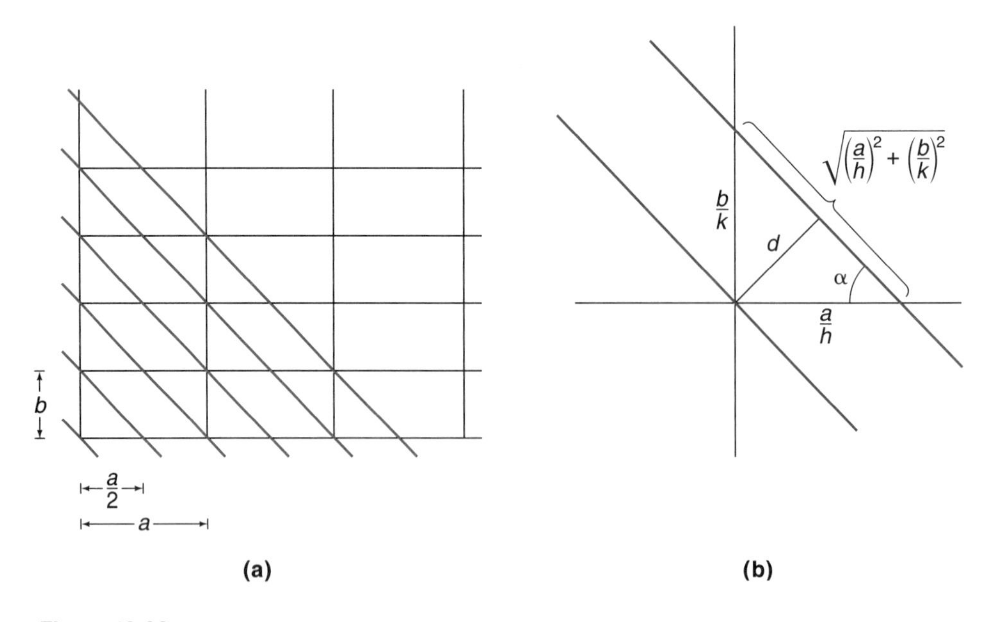

Figure 18.28
(a) The relation between (210) planes and the crystal lattice. (b) The relation between the (hkl) indices of the (210) planes and the spacing between these planes for systems with orthogonal axes.

Therefore,

$$\frac{d}{a/h} = \frac{b/k}{\sqrt{(a/h)^2 + (b/k)^2}}$$

$$d^2 = \frac{(a/h)^2(b/k)^2}{(a/h)^2 + (b/k)^2}$$

$$= \frac{1}{(h/a)^2 + (k/b)^2}$$

$$d = \frac{1}{\sqrt{(h/a)^2 + (k/b)^2}} \qquad (3)$$

For three dimensions and a cubic lattice where $a = b = c$,

$$d_{hkl} = \frac{1}{\sqrt{(h/a)^2 + (k/b)^2 + (l/c)^2}} \qquad (4)$$

$$= \frac{a}{\sqrt{h^2 + k^2 + l^2}}$$

which is Equation 18.3.

Suggestions for Further Reading

BOOKS

Blow, D., *Outline of Crystallography for Biologists*, Oxford University Press, Oxford, 2002.

Borchardt-Ott, W., *Crystallography: An Introduction*, 3rd ed., Springer-Verlag, New York, 2012.

Burdett, J. K., *Chemical Bonding in Solids*, Oxford University Press, New York, 1995.

Hammond, C., *The Basics of Crystallography and Diffraction*, 3rd ed., Oxford University Press, New York, 2009.

McPherson, A., *Introduction to Macromolecular Crystallography*, 2nd ed., Wiley-Blackwell, New York, 2009.

Rhodes, G., *Crystallography Made Crystal Clear: A Guide for Users of Macromolecular Models*, 3rd ed., Academic Press, New York, 2006.

Sands, D. E., *Introduction to Crystallography*, Dover, New York, 1994.

Smart, L., and E. Moore, *Solid State Chemistry: An Introduction*, 4th ed., CRC Press, Boca Raton, FL, 2012.

Tabor, D., *Gases, Liquids, and Solids*, 3rd ed., Cambridge University Press, New York, 1991.

Tilley, R., *Understanding Solids*, 2nd ed., John Wiley & Sons, New York, 2013.

Walton, A. J., *The Three Phases of Matter*, 2nd ed., Oxford University Press, New York, 1983.

Wormald, J., *Diffraction Methods*, Oxford University Press, New York, 1973.

ARTICLES

General

"The Solid State," Sir N. Mott, *Sci. Am.* September 1967.

"Calculation of Madelung Constants," W. B. Bridgman, *J. Chem. Educ.* **46**, 592 (1969).

"Crystal Lattice Energy and the Madelung Constant," D. Quane, *J. Chem. Educ.* **47**, 396 (1970).

"The Packing of Spheres," N. J. A. Sloane, *Sci. Am.* January 1984.

"Predictions of Crystal Structure Based on Radius Ratio," L. C. Nathan, *J. Chem. Educ.* **62**, 215 (1985).

"Quasicrystals," D. R. Nelson, *Sci. Am.* August 1986.

"Building Molecular Crystals," P. J. Fagan and M. D. Ward, *Sci. Am.* July 1992.

"Determination of ΔH for Reactions of the Born-Haber Cycle," R. S. Treptow, *J. Chem. Educ.* **74**, 919 (1997).

Report On "Integrating Materials Science into the Chemistry Curriculum," J. E. Bender, *Chem. Educator* [Online] **3**, S1430-4171 (1998) DOI 10.1333/s00897980166a.

"Space Subdivision and Voids Inside Body-Centered Cubic Lattices," C. Giomini and G. Marrow, *Chem. Educator* [Online] **16**, 232 (2011) DOI 10.1333/s00897112382a.

"An In-depth Look at the Madelung Constant for Cubic Crystal Systems," R. P. Grosso, Jr., J. T. Fermann, and W. J. Vining, *J. Chem. Educ.* **78**, 1198 (2001).

"Correspondence with Sir Lawrence Bragg Regarding Evidence for the Ionic Bond," N. C. Craig, *J. Chem. Educ.* **79**, 953 (2002).

"The Pythagorean Theorem and the Solid State," B. S. Kelly and A. G. Splittgerber, *J. Chem. Educ.* **82**, 756 (2005).

"Filling in the Hexagonal Close-Packed Unit Cell," R. C. Rittenhouse, L. M. Soper, and J. L. Rittenhouse, *J. Chem. Educ.* **83**, 175 (2006).

"Use of the Primitive Unit Cell in Understanding Subtle Features of the Cubic Close-Packed Structure," J. A. Hawkins and J. L. Rittenhouse, *J. Chem. Educ.* **85**, 90 (2008).

"The Origin of the Metallic Bond," W. B. Jensen, *J. Chem. Educ.* **86**, 278 (2009).
"Teaching Nanochemistry: Madelung Constants of Nanocrystals," M. D. Baker and A. D. Baker, *J. Chem. Educ.* **87**, 280 (2010).
"The Origin of the Ionic-Radius Ratio Rules," W. B. Jensen, *J. Chem. Educ.* **87**, 587 (2010).

X-ray Diffraction and Neutron Diffraction
"X-Ray Crystallography as a Tool for Structural Chemists," W. M. MacIntyre, *J. Chem. Educ.* **41**, 526 (1964).
"X-ray Crystallography Experiment," F. P. Baer and T. H. Jordan, *J. Chem. Educ.* **42**, 76 (1965).
"X-Ray Analysis of Crystal Structures," M. H. Harding, *Chem. Brit.* **4**, 548 (1968).
"X-Ray Crystallography," Sir L. Bragg, *Sci. Am.* July 1968.
"An Introduction to X-Ray Structure Determination," J. A. Kapecki, *J. Chem. Educ.* **49**, 231 (1972).
"Protein Molecular Weight by X-ray Diffraction," J. R. Knox, *J. Chem. Educ.* **49**, 476 (1972).
"Macromolecules, the X-ray Contribution," C. Bunn, *Chem. Brit.* **11**, 171 (1975).
"Protein Crystallography in a Molecular Biophysics Course," P. Argos, *Am. J. Phys.* **45**, 31 (1977).
"Teaching Crystallography to Noncrystallographers," J. P. Glusker, *J. Chem. Educ.* **65**, 474 (1988).
"Introducing Chemists to X-Ray Structure Determination," J. H. Enemark, *J. Chem. Educ.* **65**, 491 (1988).
"Teaching Biochemists and Pharmacologists How to Use Crystallographic Data," W. L. Daux, *J. Chem. Educ.* **65**, 502 (1988).
"Macromolecular Crystals," A. McPherson, *Sci. Am.* March 1989.
"Neutron Diffraction," T. Vogt, *Encyclopedia of Applied Physics*, Trigg, G. L., Ed., VCH Publishers, New York (1994), Vol. 11, p. 339.
"X-Ray Diffraction and the Bragg Equation," C. G. Pope, *J. Chem. Educ.* **74**, 129 (1997).
"The Incorporation of Single Crystal X-ray Diffraction into the Undergraduate Chemistry Curriculum Using Internet-Facilitated Remote Diffractometer Control," P. S. Szalay, A. Hunter, and M. Zeller, *J. Chem. Educ.* **82**, 1555 (2005).
"X-ray Diffraction and the Discovery of the Structure of DNA," D. T. Crouse, *J. Chem. Educ.* **84**, 803 (2007).

Problems

18.1 Construct a table that lists the h, k, l and $h^2 + k^2 + l^2$ values for the simple cubic, fcc, and bcc lattices. How would you use this table to deduce the nature of a crystal lattice from a series of experimentally determined *hkl* values?

18.2 When X rays with a wavelength of 0.85 Å are diffracted by a metallic crystal, the angle of first-order diffraction ($n = 1$) is measured to be 14.8°. What is the distance between the layers of atoms responsible for the diffraction?

18.3 When X rays of wavelength 0.090 nm are diffracted by a metallic crystal, the angle of first-order diffraction ($n = 1$) is measured to be 15.2°. What is the distance (in pm) between the layers of atoms responsible for the diffraction?

18.4 The distance between layers in a NaCl crystal is 282 pm. X rays are diffracted from these layers at an angle of 23.0°. Assuming that $n = 1$, calculate the wavelength of the X rays in nm.

18.5 Calculate the number of spheres in the simple cubic, body-centered cubic, and face-centered cubic cells. Also, calculate the packing efficiency of each type of cell.

18.6 Aluminum has a face-centered cubic lattice. The cell dimension is 4.05 Å. Calculate the closest interatomic distance and the density of the metal.

18.7 Silver crystallizes in a face-centered cubic lattice; the edge length of the unit cell is 4.08 Å, and the density of the metal is 10.5 g cm^{-3}. From these data, calculate the Avogadro constant.

18.8 Explain why diamond is harder than graphite. Why is graphite an electrical conductor but diamond is not?

18.9 Barium crystallizes in the body-centered cubic arrangement. Assuming a hard-sphere model, calculate the radius of a barium atom if the unit cell edge length is 5.015 Å.

18.10 Metallic iron crystallizes in a cubic lattice. The unit cell edge length is 287 pm. The density of iron is 7.87 g cm^{-3}. How many iron atoms are there within a unit cell?

18.11 Crystalline silicon has a cubic structure. The unit cell edge length is 543 pm. The density of the solid is 2.33 g cm^{-3}. Calculate the number of Si atoms in one unit cell.

18.12 Barium metal crystallizes in a body-centered cubic lattice (the Ba atoms are at the lattice points only). The unit cell edge length is 501.5 pm, and the density of the metal is 3.50 g cm^{-3}. Using this information, calculate the Avogadro constant.

18.13 Vanadium crystallizes in a body-centered cubic lattice (the V atoms occupy only the lattice points). How many V atoms are present in a unit cell?

18.14 Europium crystallizes in a body-centered cubic lattice (the Eu atoms occupy only the lattice points). The density of Eu is 5.26 g cm^{-3}. Calculate the unit cell edge length in pm.

18.15 Metallic iron can exist in the β form (bcc, cell dimension = 2.90 Å) and the γ form (fcc, cell dimension = 3.68 Å). The β form can be converted into the γ form by applying high pressures. Calculate the ratio of the densities of the β form to the γ form.

18.16 A face-centered cubic cell contains 8 X atoms at the corners of the cell and 6 Y atoms at the faces. What is the empirical formula of the solid?

18.17 Gold (Au) crystallizes in a cubic close-packed structure (face-centered cubic) and has a density of 19.3 g cm^{-3}. Calculate the atomic radius of gold.

18.18 Argon crystallizes in the face-centered cubic arrangement. Given that the atomic radius of argon is 191 pm, calculate the density of solid argon.

18.19 Given that the density of solid CsCl is 3.97 g cm^{-3}, calculate the distance between adjacent Cs^+ and Cl^- ions.

18.20 Use the Born–Haber cycle (see Section 17.2) to calculate the lattice energy of LiF. [The heat of sublimation of Li is 155.2 kJ mol^{-1} and $\Delta_f\overline{H}$(LiF) = −594.1 kJ mol^{-1}. Bond enthalpy for F_2 is 158.8 kJ mol^{-1}. Other data may be found in Tables 12.7 and 12.8.]

18.21 Calculate the lattice energy of calcium chloride, given that the heat of sublimation of Ca is 121 kJ mol^{-1} and $\Delta_f\overline{H}°(CaCl_2)$ = −795 kJ mol^{-1}. (See Tables 12.7 and 12.8 for other data.)

18.22 From the following data, explain why magnesium chloride in the solid state is $MgCl_2$ and not MgCl, whereas sodium chloride is NaCl and not $NaCl_2$.

	Mg	Na
First ionization energy	738 kJ mol^{-1}	496 kJ mol^{-1}
Second ionization energy	1450 kJ mol^{-1}	4560 kJ mol^{-1}

The lattice energy of $MgCl_2$ is 2527 kJ mol^{-1}.

18.23 Calculate the temperature at which the wavelength of a neutron is 1.00 Å.

18.24 Without referring to a handbook of chemistry, decide which of the following has a greater density: diamond or graphite.

18.25 Predict the influence of temperature on X-ray diffraction patterns of crystals.

18.26 Compare the temperature dependence of electrical conduction in an aqueous solution and in a metal.

18.27 Which of the following are molecular solids and which are covalent solids? Se_8, HBr, Si, CO_2, C, P_4O_6, B, SiH_4.

18.28 Classify the solid state of the following substances as ionic crystals, covalent crystals, molecular crystals, or metallic crystals: **(a)** SiO_2, **(b)** SiC, **(c)** S_8, **(d)** KBr, **(e)** Mg, **(f)** LiCl, **(g)** Cr.

18.29 Explain why most metals have a flickering appearance.

18.30 Zinc selenide (ZnSe) crystallizes in the zinc blende structure (see Figure 18.24) and has a density of 5.42 g cm^{-3}. **(a)** How many Zn^{2+} and Se^{2-} ions are in each unit cell? **(b)** What is the mass of a unit cell? **(c)** What is the edge length of a unit cell?

18.31 Copper crystallizes in a face-centered cubic lattice. The Bragg angles of the first two reflections in the powder pattern using X rays at 1.542 Å are 21.6° and 25.15°, respectively. Calculate the unit cell length and the radius of a copper atom.

18.32 The isolated O^{2-} ion is unstable, so it is not possible to measure the electron affinity of the O^- ion directly. Show how you can calculate the value by using the lattice energy of MgO (3890 kJ mol^{-1}) and the Born–Haber cycle. [Useful information: $Mg(s) \rightarrow Mg(g)$ $\Delta H°$ = 148 kJ mol^{-1}.]

18.33 For analysis of a newly-synthesized material, explain why neutron diffraction might be more valuable than X-ray diffraction.

18.34 Crystals of a pure compound and the same compound with deuterium atoms substituted for hydrogen atoms are separately grown. Compare and contrast how the neutron and X-ray diffraction patterns will look for these two isotopically-substituted crystals.

18.35 For each of the following compounds, predict whether the metal would occupy a tetrahedral, octahedral, or cubic hole. **(a)** KCl, **(b)** LiCl, **(c)** BaS, **(d)** InP.

CHAPTER 19

The Liquid State

No single thing abides; but all things flow, fragment to fragment clings—the things thus grow until we know and name them. By degrees they melt, and are no more the things we know.

—Titus Lucretius Carus*

The structure of liquids lies somewhere between the completely disordered gaseous state and the highly ordered crystalline state. This "in-between" quality makes accurate accounting for the intermolecular interactions difficult. In this chapter, we discuss the structure of liquids and consider three important topics: viscosity, surface tension, and diffusion. We shall also briefly study liquid crystals.

19.1 Structure of Liquids

The word *structure* applied to liquids may seem strange at first. A given amount of liquid has a fixed volume but assumes the shape of its container. At the molecular level, however, liquids do possess some degree of structure, or order, as evidenced by several physical measurements.

To see what we mean by structure of a liquid, let us imagine that we could take a series of snapshots of a liquid. Choosing an arbitrary point as the origin, we assign $\rho(r)$ as the mean or average number of atoms per unit volume at radius r from the origin, which can be viewed as a density function. Then, the total number of atoms whose centers lie within a spherical shell of radii r and $r + dr$ is given by†

$$\begin{aligned}\text{number of atoms in the shell} &= (\text{volume of shell}) \\ &\quad \times (\text{number of atoms per unit volume}) \\ &= 4\pi r^2 dr \rho(r) \qquad (19.1)\end{aligned}$$

The quantity $4\pi r^2 \rho(r)$ is called the *radial distribution function*, first introduced in Section 12.2.

* *No Single Thing Abides*, translated by W. H. Mallock, A & C Black Ltd., London, 1900. Used by permission.

† The volume of the spherical shell is given by $(4\pi/3)(r + dr)^3 - (4\pi/3)r^3 = 4\pi r^2 dr$ if we ignore the $(dr)^2$ and $(dr)^3$ terms.

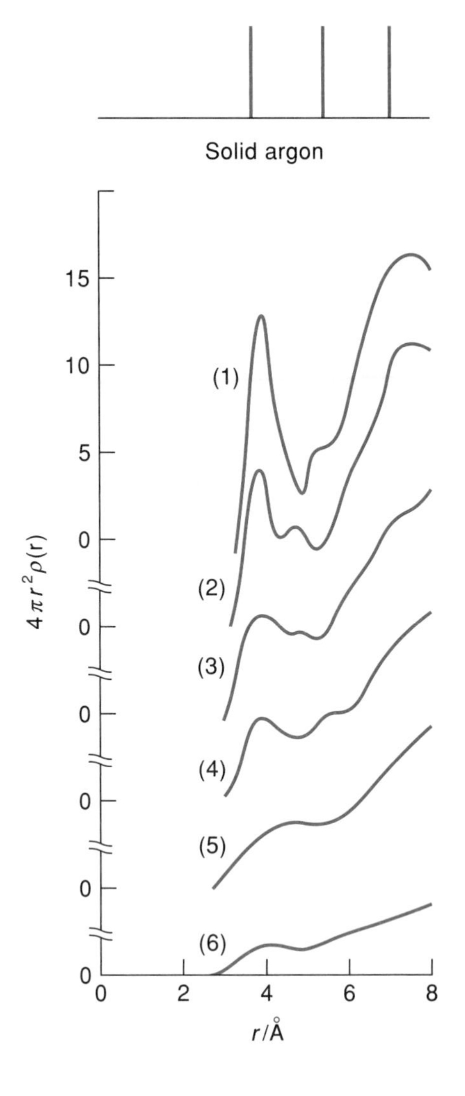

Figure 19.1
Radial distribution function of liquid argon at a series of temperatures and pressures. The conditions are (1) 84 K and 0.8 atm; (2) 91.6 K and 1.8 atm; and (3) 126.7 K and 18.3 atm; (4) 144.1 K and 37.7 atm; and (5) 149.3 K and 46.8 atm. Curve (6) is the radial distribution curve for gaseous argon at 149 K and 43.8 atm. The boiling point of argon is 87 K at 1 atm. The vertical lines on top represent the diffraction lines for solid argon. [Modified from A. Eisenstein and N. S. Gingrich, *Phys. Rev.* **62**, 261 (1940).]

Figure 19.1 shows radial distribution function plots for liquid argon at different temperatures and pressures. Crystalline argon is face-centered cubic; each argon atom has a coordination number of 12. In the crystalline state, the plot shows a series of sharp lines at different values of r that represent the distance of separation between nearest neighbors, next nearest neighbors, and so on. When argon melts, its volume increases by about 10%, and its coordination number decreases to about 10.* The maxima still have the same values as those for solid argon, however, although there is a rapid damping out of the amplitudes as distance increases from the center. As the temperature is raised further, these peaks become even broader and shift toward larger values of r. These observations are consistent with the fact that liquid argon, like all other liquids, possesses short-range order but lacks long-range order. Even its

* The first coordination number in the liquid state can be obtained by estimating the area under the first peak, given by

$$\int_{r_1}^{r_2} 4\pi r^2 \rho(r)\, dr$$

short-range order is disrupted by an increase in the kinetic energy of the atoms when the temperature is raised. Keep in mind that the word *order*, when applied to liquids, has a different meaning than when it is used to describe the solid state. In liquids, atoms are constantly in motion, so the X-ray diffraction pattern corresponds to their *time-averaged* positions.

We shall now examine three important properties of liquids: viscosity, surface tension, and diffusion.

19.2 Viscosity

The *viscosity* of a fluid—that is, a gas, a pure liquid, or a solution—is an index of its resistance to flow. In Chapter 2, we derived an expression for the viscosity of gases using the simple kinetic theory (Equation 2.27). This section considers the viscosity of liquids.

A viscometer is a device used to measure viscosity. Common viscometers monitor the ease with which fluids flow through capillary tubing. Let us derive an expression relating the viscosity of a liquid, η, to the experimental parameters. Consider a certain liquid flowing through a capillary tube of radius R and length L under constant pressure, P (Figure 19.2). The velocity of the liquid is zero at the wall and increases toward the center of the tube, reaching a maximum at the center. Imagine two concentric cylinders of radii r and $(r + dr)$. According to Equation 2.25, the frictional drag, F, between these two cylindrical layers is

$$F = -\eta(2\pi rL)\frac{dv}{dr} \tag{19.2}$$

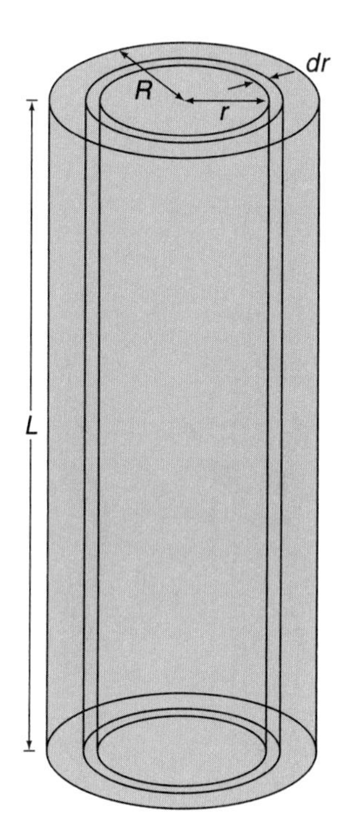

Figure 19.2
Flow of a liquid through a capillary tube of radius R.

where $2\pi rL$ is the surface area of the inner cylinder, and dv/dr is the velocity gradient. Because the velocity decreases as r increases, dv/dr is a negative quantity, and a negative sign has been included in Equation 19.2 to make F a positive number. For steady-state flow, the frictional drag must be exactly balanced by the downward force, which is given by the product of pressure, P, and area, πr^2. Thus,

$$P(\pi r^2) = -\eta(2\pi rL)\frac{dv}{dr}$$

$$dv = -\frac{P}{2\eta L}rdr$$

Integration between $v = 0$ (at $r = R$) and $v = v$ (at $r = r$) yields

$$\int_0^v dv = -\frac{P}{2\eta L}\int_R^r rdr$$

Hence,

$$v = \frac{P}{4\eta L}(R^2 - r^2) \tag{19.3}$$

$0 \le r \le R$

We see that the velocity of flow anywhere in the tube is a parabolic function of r. Keep in mind that Equation 19.3 holds only for *laminar flow*, which requires small diameters and low flow rates. Without these conditions, Equation 19.3 is not valid, and *turbulent flow* may result.* A useful distinction between turbulent flow and laminar flow is provided by the dimensionless *Reynolds number* (after the British physicist Osborne Reynolds, 1842–1912), defined as

$$\text{Reynolds number} = \frac{2Rv\rho}{\eta} \tag{19.4}$$

where ρ is the density of the liquid. A Reynolds number less than about 2000 indicates laminar flow; a number greater than 3500 indicates turbulent flow. Intermediate values (between 2000 and 3500) may accompany either type of flow, which must be determined experimentally.

Our next step is to calculate the total flow rate of the liquid through the capillary as a function of viscosity. The volume of liquid that flows through a cross-sectional element, $2\pi r dr$, per second is simply $(2\pi r dr)v$, and the total volume of the liquid flowing per second, Q, is given by

$$Q = \frac{V}{t} = \int_0^R v(2\pi r dr) = \frac{2\pi P}{4\eta L}\int_0^R (R^2 - r^2) r dr$$

$$= \frac{\pi P R^4}{8\eta L} \tag{19.5}$$

where V is the total volume, and t is the flow time. Equation 19.5, known as *Poiseuille's law* (after the French physician Jean Poiseuille, 1799–1869), applies to both liquids and gases.

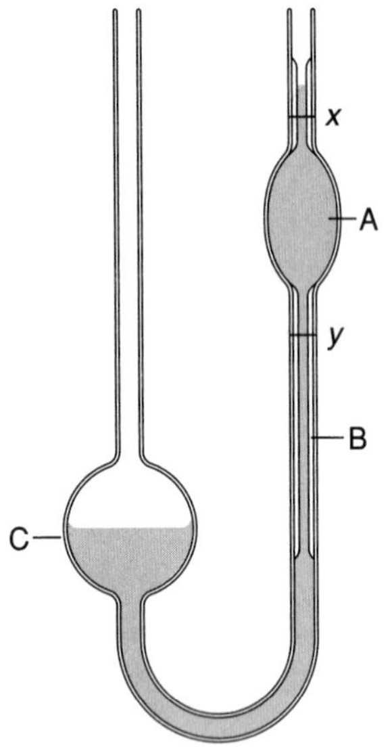

A relatively simple apparatus for measuring viscosity is the Ostwald viscometer (devised by the German chemist Wolfgang Ostwald, 1883–1943), shown in Figure 19.3. It consists of a bulb (A) with markings x and y, attached to a capillary tube (B) and a reservoir bulb (C). A definite volume of the liquid under study is introduced into C and drawn into A, and the time (t) the liquid takes to flow between x and y is recorded. Rearranging Equation 19.5 gives

$$\eta = \frac{\pi P R^4 t}{8VL} \tag{19.6}$$

The pressure, P, at any instant driving the liquid through B is equal to $h\rho g$, where h is the difference in height between the levels of the liquid in the two limbs, ρ is the density of the liquid, and g is the acceleration due to gravity. To be sure, this pressure varies during the experiment because h decreases. But because the initial and final values of h are the same in every case and g is a constant, the applied pressure is proportional to the density of the liquid.

Figure 19.3
An Ostwald viscometer. The time a liquid takes to flow between markings x and y is measured and compared with that of a reference liquid. A, bulb; B, capillary tube; C, reservoir bulb.

* For laminar flow, all particles of the liquid move parallel to the tube, and the velocity increases regularly from zero at the wall to a maximum at the center. These conditions are not satisfied if the flow is turbulent.

Table 19.1
Viscosity of Some Common Liquids at 293 K

Liquid	Viscosity/P[a] CGS units	Viscosity/N s m^{-2} SI units
Acetone	0.00316 (298 K)	0.000316
Benzene	0.00652	0.000652
Carbon tetrachloride	0.00969	0.000969
Ethanol	0.01200	0.001200
Diethyl ether	0.00233	0.000233
Glycerine	14.9	1.49
Mercury	0.01554	0.001554
Water	0.0101	0.00101
Blood plasma	0.015 (310 K)	0.0015
Whole blood	0.04 (310 K)	0.004

[a] 1.0 poise (P) = 0.1 N s m^{-2}.

In practice, we do not use Equation 19.6 to measure the value of η because of the uncertainties in determining the radius of the capillary tubing, R. (Note that the radius appears as R^4, so a small deviation in R can lead to a considerable error in η.) Instead, the viscosity of a liquid is most conveniently determined by comparison with a reference liquid of accurately known viscosity, as follows. The ratio of the viscosities of a sample and a reference liquid is given by

$$\frac{\eta_{\text{sample}}}{\eta_{\text{reference}}} = \frac{\pi R^4 (Pt)_{\text{sample}}}{8VL} \times \frac{8VL}{\pi R^4 (Pt)_{\text{reference}}}$$

Because V, L, and R values are the same if we use the same viscometer and P = constant $\times \rho$, the preceding equation reduces to

$$\frac{\eta_{\text{sample}}}{\eta_{\text{reference}}} = \frac{(\rho t)_{\text{sample}}}{(\rho t)_{\text{reference}}} \tag{19.7}$$

Thus, the viscosity of the sample can be obtained readily from the densities of the liquids and the times of flow if $\eta_{\text{reference}}$ is known. Table 19.1 lists the viscosity values of several common liquids.

Generally, the viscosity of a solution is greater than that of the pure solvent.* The presence of solute molecules disrupts the smooth flow pattern, or velocity gradient, of the fluid, resulting in an increase in viscosity. This viscosity change is particularly

* There are a number of cases in which the reverse is true. For example, the viscosities of many aqueous solutions containing alkali metal and ammonium ions and certain anions are *lower* than that of water (see Section 7.2).

true for solutions that contain macromolecules. As we would expect, the viscosities of such solutions also depend on the conformation of the macromolecules. The viscosity of a DNA solution, for example, can vary greatly depending on whether the solute molecules possess the native double-helical conformation or are arranged in random coils. Often, the kinetics of denaturation from helix to random coil can be conveniently measured in terms of changes in the solution's viscosity over a period of time.

Hot syrup pours more easily than cold syrup.

The viscosities of most liquids decrease with increasing temperature. A molecular interpretation is that liquids possess a number of holes or vacancies and molecules are continually moving into these vacancies. This process permits a liquid to flow but requires energy. To be able to move into a vacancy, a molecule must possess sufficient activation energy to overcome the repulsion by the molecules that surround the vacancy. At higher temperatures, more molecules possess the necessary activation energy, so the liquid flows more easily. In fact, there is an analogous Arrhenius equation for viscous flow, given by

$$\eta = \eta_0 e^{-E_v/k_B T} \tag{19.8}$$

where η_0 is a constant characteristic of the liquid and E_v is the "activation energy" for viscous flow. In contrast to liquids, the viscosity of a gas *increases* with temperature.* In the kinetic molecular treatment of gases, the origin of the viscous drag between two adjacent layers is the momentum transfer of the molecules from one layer to the other. The rate of transfer increases with temperature, and therefore the viscosity of the gas also increases.

Blood Flow in the Human Body

Equation 19.5 can be applied to the study of blood flow in our bodies. Figure 19.4 shows a schematic diagram of the various routes for blood circulation. The heart is, in effect, a single pump powering a double circuit. It contains four chambers—two atria and two ventricles—and four sets of values. Freshly oxygenated blood is conducted away from the *aorta*, the large artery, leading from the left chamber and into smaller arteries that carry it to the various parts of the body. These, in turn, branch into still smaller arteries, the smallest of which—the *arterioles*—break up into a complex network of capillaries. These minute structures thread their way to every part of the body to help the blood carry out its vital function: trading oxygen and other materials for carbon dioxide and waste with the cells. The capillaries join into tiny veins called *venules*, which in turn merge into increasingly larger veins that transport the deoxygenated blood back to the right atrium of the heart.

In the space of a single beat, the atria contract, forcing blood into the ventricles, and then the ventricles contract, forcing the blood out of the heart. Because of the pumping action of the heart, blood enters the arteries in spurts, or pulses. The maximum pressure at the peak of the pulse is called the *systolic pressure*; the lowest pressure between pulses is called *diastolic pressure*. In a healthy young adult, the systolic

* From Equation 2.27, we write

$$\eta = \frac{m\bar{c}}{3\sqrt{2}\pi d^2} = \frac{2}{3d^2}\sqrt{\frac{mk_B T}{\pi^3}}$$

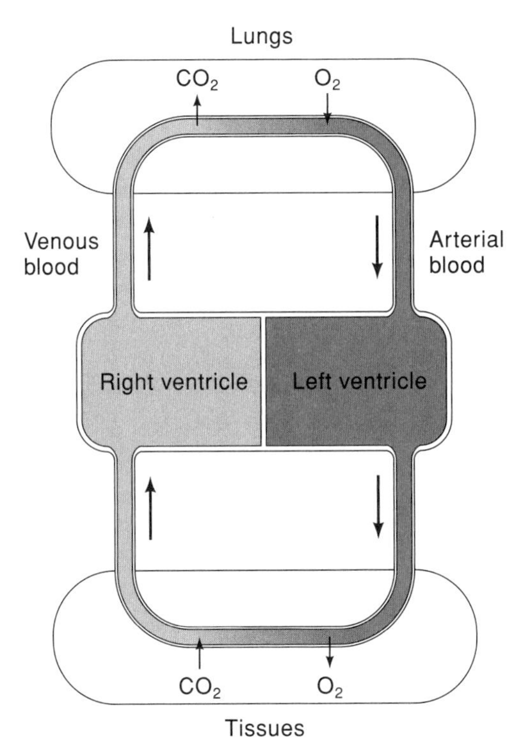

Figure 19.4
Schematic diagram of the human circulation system. Arterial blood, rich in oxyhemoglobin, is pumped through the left ventricle of the heart to the tissues, where oxygen is released and carbon dioxide is taken up. The venous blood, rich in dissolved carbon dioxide, is pumped by the right ventricle of the heart to the lungs, where the carbon dioxide is released and oxygen is taken up.

pressure is about 120 mmHg (120 torr) and the diastolic pressure is about 80 mmHg (80 torr).* These values represent the excess pressures over atmospheric pressure. Thus, the absolute systolic and diastolic pressures are 880 and 840 mmHg, respectively (assuming that atmospheric pressure is 760 mmHg), and the average value of the blood pressure is about 100 mmHg.

The radius of the aorta is sufficiently large (about 1 cm) that a small pressure difference is required to sustain normal blood flow through it. At rest, the rate of blood flow is roughly 0.08 L s^{-1}. Equation 19.5 can be rewritten as

$$Q = \frac{\pi \Delta P R^4}{8\eta L} \tag{19.9}$$

* To measure the blood pressure, a cuff is wrapped around a person's arm above the elbow, connected by a tube to the sphygmomanometer (the measuring device). The cuff is inflated until the main artery in the arm is squeezed tightly enough to shut off the flow of blood. The health professional listens with a stethoscope placed over the artery. Then the cuff is deflated slowly, reducing the pressure. When the first pulse sound is heard, the pressure reading is recorded. This is the person's systolic pressure. Another reading is taken when the pulse sound disappears. That is the diastolic pressure.

where ΔP is the difference in pressure at two points along the aorta, and L is the distance between the points. Setting $L = 0.01$ m and converting Q to 8×10^{-5} m^3 s^{-1}, we find that

$$\Delta P = \frac{8\eta L Q}{\pi R^4}$$

$$\Delta P = \frac{8(0.004 \text{ N s m}^{-2})(0.01 \text{ m})(8 \times 10^{-5} \text{ m}^3 \text{ s}^{-1})}{\pi(0.01 \text{ m})^4}$$

$$= 0.8 \text{ N m}^{-2}$$

$$= 6 \times 10^{-3} \text{ mmHg}$$

(The conversion factor is 1 N m^{-2} = 7.5×10^{-3} mmHg.) A drop in pressure equal to 6×10^{-3} mmHg per cm is negligibly small compared with the total blood pressure. The situation is different as the blood enters the other major arteries. Because these vessels have much smaller radii than the aorta, a pressure drop of about 20 mmHg is required to maintain the flow through them. Therefore, the pressure is only 80 mmHg when the blood enters the arterioles. Because these vessels have still smaller radii, there is another drop in pressure of about 50 mmHg. A further drop of 20 mmHg results when the blood flows through the capillaries. Note that although the capillaries have much smaller radii than arterioles, there are so many of them that the amount of blood passing through each one is very small. By the time the blood reaches the veins, its pressure has been reduced to about 10 mmHg. Fortunately, veins are fitted with cup-shaped valves to prevent backflow at this low pressure. The movement of blood in veins is promoted by the massaging effect of surrounding skeletal muscle or by the adjacent arteries. Finally, the blood returns to the right atrium, ready to be circulated again.

An interesting comparison can be made between Equation 19.9,

$$Q = \frac{\Delta P}{8\eta L/\pi R^4}$$

and Ohm's law,

$$\text{current} = \frac{\text{voltage}}{\text{resistance}}$$

By analogy, the resistance to flow is given by $8L\eta/\pi R^4$. Equation 19.9 can be applied to study the flow of blood in arteries, arterioles, and capillaries. Because resistance is inversely proportional to the fourth power of the radius, a decrease from 2×10^{-4} cm, a typical capillary radius, to 1.5×10^{-4} cm caused by cholesterol deposits, say, would result in an increase in the resistance by a factor of 3. Normal blood flow would then be sustained by higher blood pressure, causing a condition known as *hypertension*. On the other hand, if the resistance drops while the blood pressure is unchanged, the blood flow, Q, is increased. During vigorous exercise, both the blood pressure and the radii of blood vessels increase, a change known as *vasodilation*. These two changes facilitate a greater blood flow to meet the enhanced metabolic rate of the body.

19.3 Surface Tension

When the surface of a liquid expands, molecules that were originally in the interior region are brought out to the exterior. Work must be done to counteract the attractive forces among these molecules and their neighbors. This process is somewhat similar to vaporization of a liquid. In vaporization, however, the molecules are completely removed from the liquid, whereas molecules in a surface layer are still under the influence of strong intermolecular forces, except that the forces are away from the direction of the vapor phase (Figure 19.5). This unbalanced interaction experienced by surface-layer molecules results in a tendency for the liquid to minimize its surface area. For this reason, a small drop of liquid assumes a spherical shape.

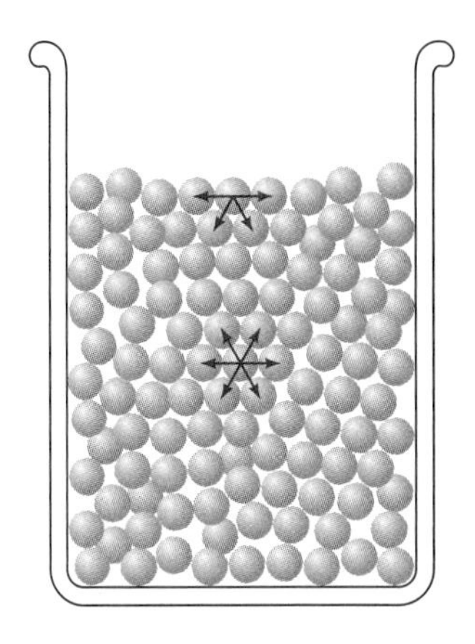

Figure 19.5
Intermolecular forces acting on a molecule in the surface layer and on a molecule in the interior region of a liquid.

Figure 19.6 shows a thin film, such as a soap film, stretched on a wire frame that has a movable side (called the piston) of length ℓ. The force (F) required to stretch the film is proportional to the length, ℓ. Because the film has two sides (i.e., two surfaces), the total dimension of the film is 2ℓ, so

$$\begin{aligned} F &\propto 2\ell \\ &= 2\gamma\ell \end{aligned} \tag{19.10}$$

where the proportionality constant γ is the *surface tension* of the liquid. Thus, surface tension can be viewed as the force exerted by a surface of unit length; it has the units N m^{-1}. Because N m^{-1} is equivalent to J m^{-2}, we can also interpret surface tension in terms of surface energy. The mechanical work done by moving the piston a distance dx is Fdx, and the change in surface area is $2\ell dx$. The ratio of work done over the increase in area is

$$\frac{Fdx}{2\ell dx} = \frac{2\gamma\ell dx}{2\ell dx} = \gamma \tag{19.11}$$

Surface tension can also be defined as the surface energy per unit area. The surface energy is of mechanical rather than thermal origin. The work done in stretching the film results in an increase in Gibbs energy, and the tendency of a surface to reduce its area is just another example of a system tending toward an arrangement of lower Gibbs energy (at constant temperature and pressure).

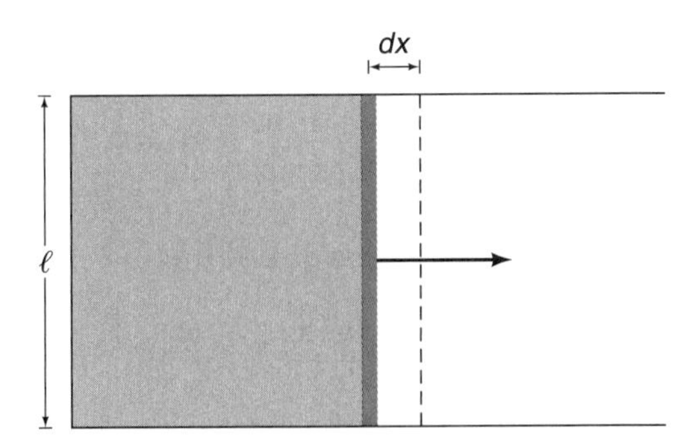

Figure 19.6
A wire frame supporting a thin liquid film. Work has to be done to expand the surface area of the film.

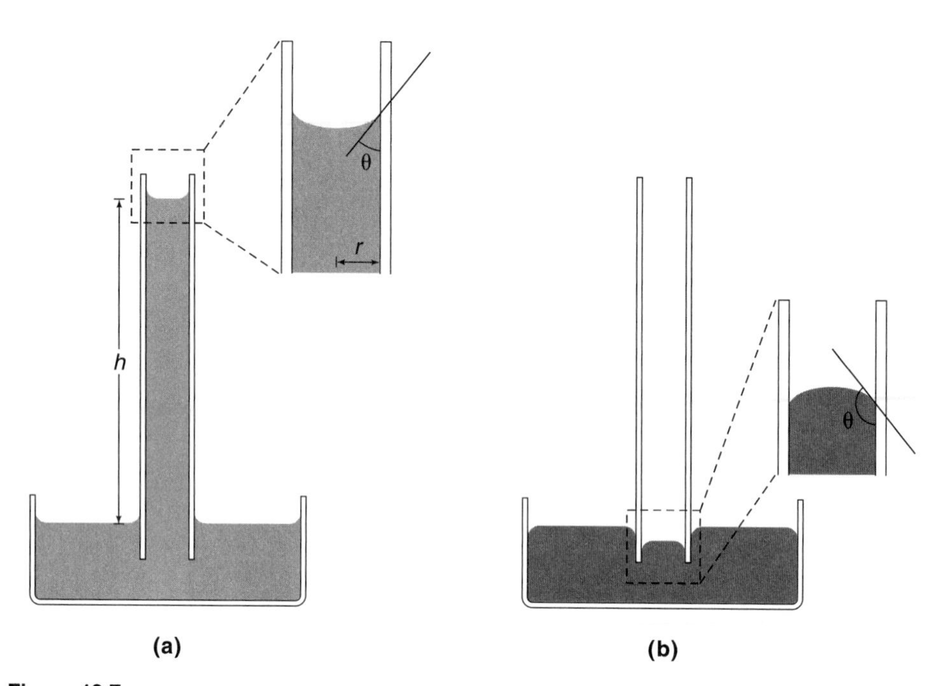

Figure 19.7
(a) Capillary-rise phenomenon for liquids in which adhesion is greater than cohesion. (b) When cohesion is greater than adhesion, the liquid in the capillary tube forms a depression.

The Capillary-Rise Method

The *capillary-rise method* provides a simple means for measuring the surface tension of liquids. In this arrangement, a capillary tube of radius r is dipped into the liquid under study (Figure 19.7a). The force acting downward is the gravitational pull on the liquid, given by $\pi r^2 h\rho g$, where $\pi r^2 h$ is the volume,* ρ is the density of the liquid, and g is the acceleration due to gravity. This weight must be balanced by an upward force caused by the liquid's surface tension. Acting along the periphery of the cylindrical bore, between the liquid and the glass wall, this force is given by $2\pi r\gamma \cos\theta$, where $2\pi r$ is the circumference of the bore, θ is the angle of contact between the liquid and the capillary tube in the meniscus, and $\cos\theta$ gives the vertical (upward) component of the force. Equating the upward and downward forces, we write

$$\pi r^2 h\rho g = 2\pi r\gamma \cos\theta$$

or

$$\gamma = \frac{rh\rho g}{2\cos\theta} \tag{19.12}$$

* We ignore here the small amount of liquid above the top of the meniscus. For accurate work, a correction term of $r/3$ can be added to h in the calculation.

Table 19.2
Surface Tension (γ) of Some Common Liquids at 293 K

Liquid	γ/N m^{-1}
Acetic acid	0.0276
Acetone	0.0237
Benzene	0.0289
Carbon tetrachloride	0.0266
Chloroform	0.0271
Ethanol	0.0223
Diethyl ether	0.0170
n-Hcxane	0.0184
Mercury	0.476 (298 K)
Water	0.07275

The surface tensions of several common liquids are listed in Table 19.2.

Although the rise of liquids up a capillary tube is commonly observed, it is by no means a universal phenomenon. For example, when a capillary tube is dipped into liquid mercury, the upper level of the liquid in the tube is actually lower than the surface of free liquid (Figure 19.7b). These two divergent behaviors can be understood by considering the intermolecular attraction between like molecules in the liquid, called *cohesion*, and the attraction between the liquid and the glass wall, called *adhesion*. If adhesion is stronger than cohesion, the walls become wettable, and the liquid will rise along the walls. Because the vapor-liquid interface resists being stretched, the liquid in the center of the column also rises. Conversely, if cohesion is greater than adhesion, the liquid in the capillary forms a depression.

Note that water has a rather large surface tension due to its strong hydrogen bonding.

EXAMPLE 19.1

The typical radius of a xylem vessel of a plant is about 0.020 cm. How high will water rise in such a vessel at 293 K?

ANSWER

From Equation 19.12,

$$h = \frac{2\gamma \cos \theta}{rg\rho}$$

Because the contact angle is usually quite small, we assume that $\theta = 0$, so $\cos\theta = 1$. The data are

$$\gamma = 0.07275 \text{ N m}^{-1}$$

$$r = 0.00020 \text{ m}$$

$$g = 9.81 \text{ m s}^{-1}$$

$$\rho = 1 \times 10^3 \text{ kg m}^{-3}$$

Hence

$$h = \frac{2(0.07275 \text{ N m}^{-1})}{(0.00020 \text{ m})(9.81 \text{ m s}^{-2})(1 \times 10^3 \text{ kg m}^{-3})}$$

$$= 0.074 \text{ N s}^2 \text{ kg}^{-1} \quad (1 \text{ N} = 1 \text{ kg m s}^{-2})$$

$$= 0.074 \text{ m}$$

COMMENT

The result shows that the capillary-rise phenomenon is partially responsible for the rise of water in plants and soils but cannot wholly account for it. The water is helped up by the major mechanism, osmosis, discussed in Chapter 6.

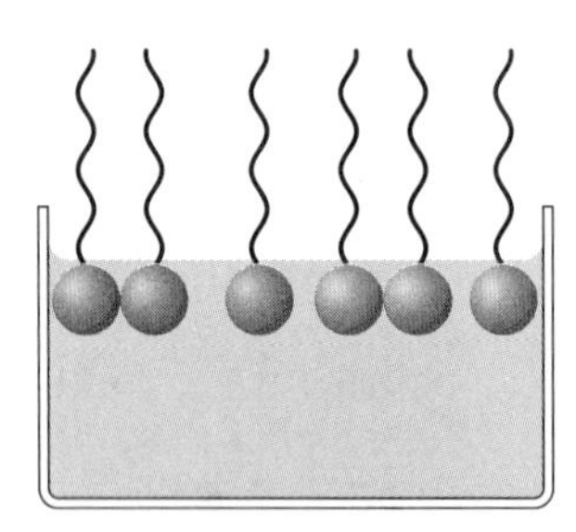

Figure 19.8
Schematic diagram of fatty acid molecules in water. The spheres represent polar groups and the zigzag lines represent the nonpolar hydrocarbon chains.

The surface tension of aqueous solutions is generally close to that of pure water if the solutes are salts, such as NaCl, or sucrose and other substances that do not preferentially collect at the air–water interface. On the other hand, surface tension can dramatically decrease if the dissolved substance is a fatty acid or a lipid. These molecules consist of two regions: a hydrophilic (water-loving) polar group such as −COOH at one end, and a long hydrocarbon chain that is nonpolar and therefore hydrophobic (water-fearing) at the other end. The nonpolar groups tend to line up together along the surface of water with the polar groups pointing toward the interior of the solution (Figure 19.8). Consequently, surface tension decreases. This effect depends on the nature of the solute molecule. Thus, although a 0.01 *M* solution of caproic acid $[CH_3{-}(CH_2)_4{-}COOH]$ lowers the surface tension by about 0.015 N m^{-1}, a decrease of about 0.025 N m^{-1} in surface tension is observed in a 0.0005 *M* solution of capric acid $[CH_3{-}(CH_2)_8{-}COOH]$. Any substance that causes a reduction in surface tension in this manner is called a *surfactant.* Among the most effective surfactants are soaps (salts of long-chain fatty acids) and denatured proteins.

Surface Tension in the Lungs

The action of surfactants also plays an important role in the breathing process. By far the most extensive surface of the human body in contact with the surroundings is the

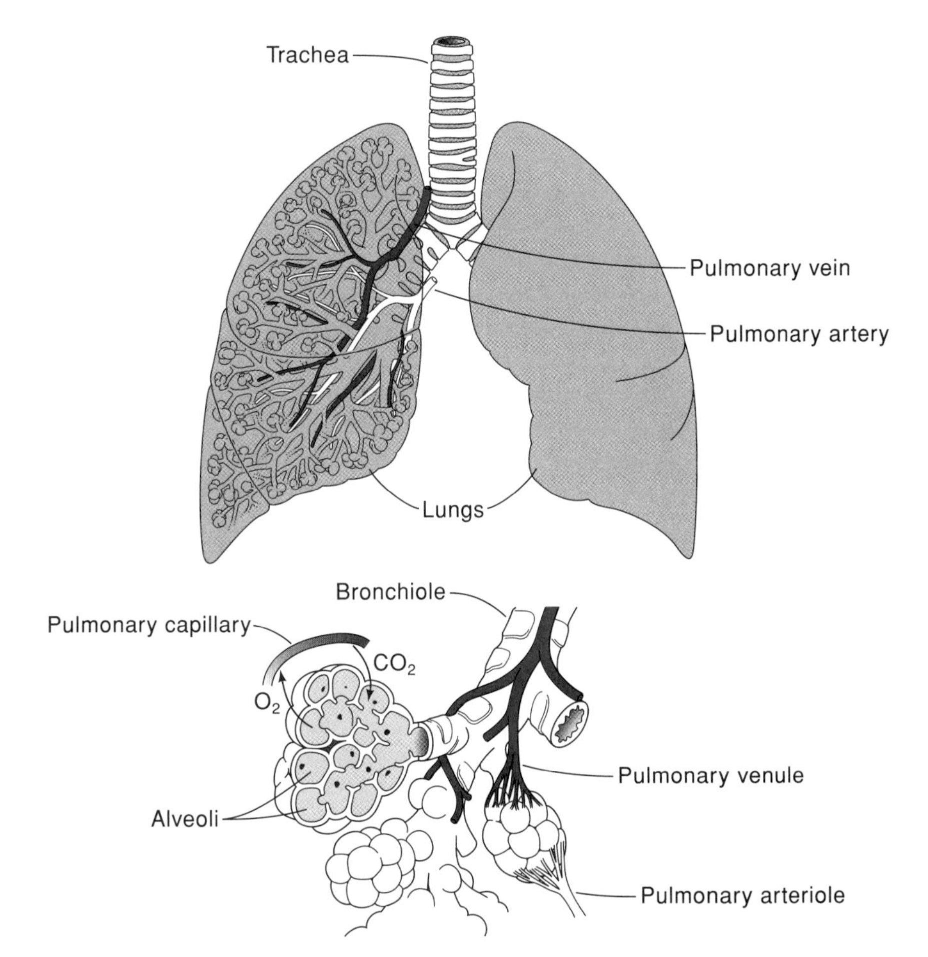

Figure 19.9
The relationships between respiratory airways and blood vessels. Alveoli are the air spaces in the lungs through which oxygen enters the blood and carbon dioxide leaves. An average alveolus expands and contracts about 15,000 times a day during breathing.

moist interior surface of the lungs. To carry on the active exchange of carbon dioxide and oxygen between the circulating blood and the atmosphere in an average adult requires a lung surface area roughly that of a tennis court. Such an area is encompassed in the relatively small volume of the chest by the compartmentalization of the lungs into hundreds of millions of tiny air spaces, or sacs, called *alveoli*. The alveoli have an average radius of about 50 μm; they are connected by confluent passages through the bronchial tree and the trachea to the atmosphere (Figure 19.9).

During normal inhalation, the pressure in the alveoli is about 3 mmHg below atmospheric pressure, and we say that they have a gauge pressure* of −3 mmHg, which enables air to flow into them through the bronchial tubes. The alveoli are lined with mucous tissue fluid that normally has a surface tension of 0.05 N m^{-1}. During

* Gauge pressure is the difference between the absolute pressure of a fluid (gas or liquid) and atmospheric pressure. For example, when we measure the pressure of a tire, the value corresponds not to the pressure of air inside the tire but its excess over the atmospheric pressure. The same applies to blood pressure, discussed earlier.

inhalation, the radius of the alveoli expands by about a factor of 2; the pressure difference required to inflate an alveolus is given by the equation (see Appendix 19.1 on p. 869 for derivation):

$$P_i - P_o = \frac{2\gamma}{r} \tag{19.13}$$

where P_i and P_o are the gauge pressures inside and outside of the alveoli, respectively, γ is the surface tension of mucous fluid, and r is the radius of the alveoli. For such an expansion to occur, the pressure difference must be at least

$$\begin{aligned} P_i - P_o &= \frac{2(0.05 \text{ N m}^{-1})}{5 \times 10^{-5} \text{ m}} \\ &= 2.0 \times 10^3 \text{ N m}^{-2} \\ &= 15 \text{ mmHg} \end{aligned}$$

The pressure in the space between the lungs and the pleural cavity that holds the lungs, P_o, is only -4 mmHg (or 756 mmHg in absolute value). Therefore, we have

$$\begin{aligned} P_i - P_o &= (-3 \text{ mmHg}) - (-4 \text{ mmHg}) \\ &= 1 \text{ mmHg} \end{aligned}$$

which is only 1/15 of the pressure required to expand an alveolus. To overcome this problem, the alveolar cells secrete a special type of surfactant (called dipalmitoyl lecithin), which effectively reduces the surface tension, so that the alveolus can expand without difficulty in the course of the 15,000 or so breaths that are drawn into the lungs of an adult each day. A striking example of what occurs when insufficient surfactant is present is the disorder known as respiratory-distress syndrome of the newborn, which frequently afflicts premature infants in whom the surfactant-synthesizing cells do not yet function adequately. Even in the lungs of a normal healthy baby, the alveoli are so collapsed at birth that a pressure difference as large as 25 to 30 mmHg is required to expand them the first time. Therefore, the first breath of life requires extraordinary effort to overcome the surface tension in the alveoli.

The surface activity discussed above also has a bearing on water conservation. Spreading a thin film of cetyl alcohol $[CH_3(CH_2)_{14}CH_2OH]$ over the surface of water can cut down the evaporation rate of water in reservoirs. A solid, cetyl alcohol is insoluble in water. It has surface solubility, however, in the sense that its molecules float on water and form a thin film (a monolayer) that spreads to cover the surface. If the film is disrupted by weather or other disturbances, it readily reforms. Only 30 g of the material is sufficient to cover about 10,000 m^2 (3 acres) of water surface.

19.4 Diffusion

Diffusion is the process by which concentration gradients in a solution spontaneously decrease until a uniform, homogenous distribution is obtained. The diffusion process is important to many chemical and biological systems. For example, it is the major mechanism by which carbon dioxide reaches the sites of photosynthesis in chloroplasts. Understanding the transportation of solute molecules across cell membranes

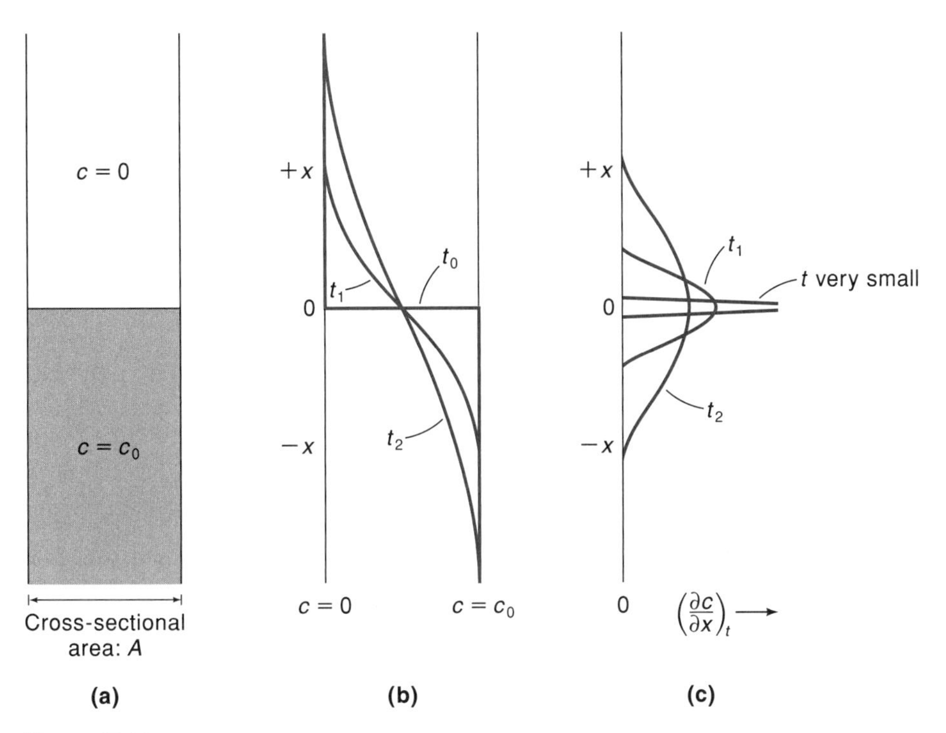

Figure 19.10
(a) Diffusion of a solute from a cell of uniform cross section into the pure solvent component. (b) Plots of concentration, c, versus x. At $t = 0$ (the t_0 curve), the boundary between the solution and pure solvent component is infinitely sharp. (c) Plots of concentration gradient $(\partial c/\partial x)_t$ versus x at various times t after diffusion has begun. At $t = 0$, the gradient is a vertical line of infinite height and no width centered at $x = 0$.

also requires a detailed knowledge of diffusion. In this section, we describe some characteristics of diffusion in solution.

Fick's Laws of Diffusion

Imagine a container with a solution on the bottom and the pure solvent on top, as shown in Figure 19.10a. Initially, there is a sharp boundary between the solution and the solvent. As time progresses, solute molecules gradually move upward by diffusion. This process continues until the entire system is homogeneous. In 1855, the German physiologist Adolf Eugen Fick (1829–1901) studied the diffusion phenomenon and found that the *flux* (J), that is, the net amount of solute that diffuses through a unit area per unit time, is proportional to the concentration gradient. Expressing this relationship mathematically in one dimension along the x axis, we write

The container is treated as an infinitely long tube.

$$J \propto -\left(\frac{\partial c}{\partial x}\right)_t$$

$$= -D\left(\frac{\partial c}{\partial x}\right)_t \tag{19.14}$$

Equation 19.14 is known as *Fick's first law of diffusion* in one dimension. The quantity $(\partial c/\partial x)_t$ is the concentration gradient of the diffusing substance at time t of diffusion,

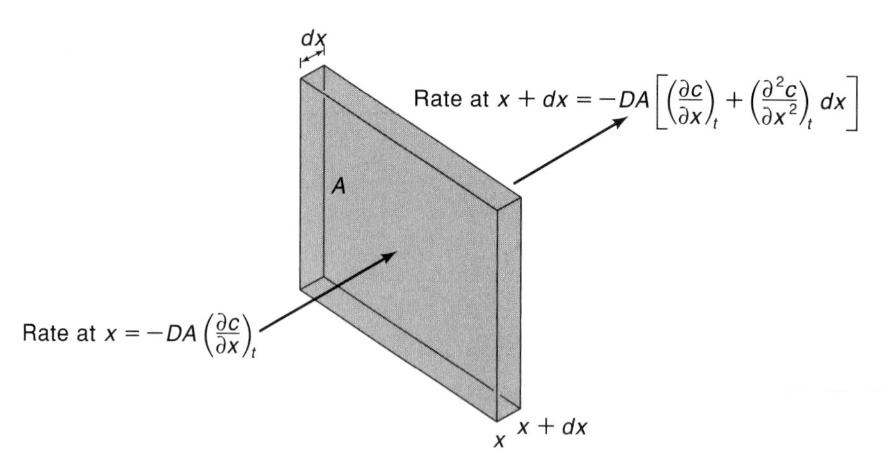

Figure 19.11
Rate of accumulation of solute in a volume element, Adx, during a diffusion process.

and D is the diffusion coefficient of the diffusing substance in the medium concerned. D has units $m^2\ s^{-1}$ or $cm^2\ s^{-1}$. The negative sign indicates that diffusion proceeds from higher to lower concentration, because the concentration gradient is negative in the direction of diffusion. Thus, the flux J is a positive quantity.

Let us investigate the diffusion process in a little more detail by asking the following question: What is the change of concentration with time at a given point along the x axis? Consider a volume element, Adx (where A is the cross-sectional area), shown in Figure 19.11. At distance x measured from the original boundary, the rate of solute molecules entering the volume element is $-DA(\partial c/\partial x)_t$. Because the rate at which the concentration gradient changes with x is given by

$$\frac{\partial}{\partial x}\left(\frac{\partial c}{\partial x}\right)_t = \left(\frac{\partial^2 c}{\partial x^2}\right)_t$$

the rate of solute molecules leaving the volume element, after having traveled distance dx, is

$$-DA\left(\frac{\partial c}{\partial x}\right)_t - DA\left(\frac{\partial^2 c}{\partial x^2}\right)_t dx = -DA\left[\left(\frac{\partial c}{\partial x}\right)_t + \left(\frac{\partial^2 c}{\partial x^2}\right)_t dx\right]$$

Thus, the rate of accumulation of solute in the volume element is the difference of the foregoing two quantities:

rate of accumulation of solute in the volume element	=	rate of solute entering the volume element	−	rate of solute leaving the volume element

$$= -DA\left(\frac{\partial c}{\partial x}\right)_t + DA\left[\left(\frac{\partial c}{\partial x}\right)_t + \left(\frac{\partial^2 c}{\partial x^2}\right)_t dx\right]$$

$$= DA\left(\frac{\partial^2 c}{\partial x^2}\right)_t dx \qquad (19.15)$$

Now, there is another way of arriving at an expression for the rate of accumulation. As time goes on, the concentration of solute in the volume element is steadily increasing as a result of diffusion. The rate of this increase is given by the product of the change of concentration with time and the volume element, that is, $(\partial c/\partial t)_x(Adx)$. Equating these two rates of solute accumulation, we obtain

$$\left(\frac{\partial c}{\partial t}\right)_x = D\left(\frac{\partial^2 c}{\partial x^2}\right)_t \tag{19.16}$$

Equation 19.16 is known as *Fick's second law of diffusion*. It says that the change of concentration with time at a certain distance, x, from the origin is equal to the product of the diffusion coefficient and the change of concentration gradient in the direction of x at time t.

Boundary conditions were first introduced for the particle-in-a-box discussion (p. 413).

Equation 19.16 is a fundamental diffusion equation; however, it must be integrated before we can apply it to practical systems. To obtain the value of D from suitable experimental measurements, the appropriate *boundary conditions* must be applied. If the columns of liquids shown in Figure 19.10 are effectively of infinite length so that the concentrations of solute at the top and bottom remain zero and c_0, respectively, throughout the experimental run, the following boundary conditions will hold:

$$\begin{array}{lll} \text{at } t = 0 & c = 0 & \text{for } x > 0 \\ & c = c_0 & \text{for } x < 0 \\ \text{at } t = t & c \to 0 & \text{as } x \to +\infty \\ & c \to c_0 & \text{as } x \to -\infty \end{array}$$

The solution of Equation 19.16 with the boundary conditions given above is*

$$c = \frac{c_0}{2}\left(1 - \frac{2}{\sqrt{\pi}}\int_0^{\beta} e^{-\beta^2} d\beta\right) \tag{19.17}$$

where

$$\beta = \sqrt{\frac{x^2}{4Dt}}$$

Using Equation 19.17, we can calculate the concentration of solute at a distance x from the origin after it has diffused for time t. Figure 19.10b shows a graphical representation of Equation 19.17 corresponding to various values of time t. We can also express Equation 19.17 in the differential form as

$$\left(\frac{\partial c}{\partial x}\right)_t = -\frac{c_0}{\sqrt{4\pi Dt}} e^{-x^2/4Dt} \tag{19.18}$$

Equation 19.18 enables us to plot the concentration gradient, $(\partial c/\partial x)_t$, versus x at different times, t (Figure 19.10c).

Determining the diffusion coefficient accurately is rather difficult. Optical methods, such as refractive index measurements, are normally employed to monitor

* For details, see Tanford, C. *Physical Chemistry of Macromolecules*, John Wiley & Sons, New York (1961), p. 354.

the concentration gradients at various distances from the origin after diffusion has started. One useful technique measures diffusion coefficients of biomacromolecules from laser light scattering.* Here, we shall describe a simple, although less accurate method for determining the value of D. From Equation 19.18, we see that the concentration gradient at the origin ($x = 0$) is given by

$$\left(\frac{\partial c}{\partial x}\right)_t = -\frac{c_0}{\sqrt{4\pi Dt}} \tag{19.19}$$

Additionally, we can rewrite Fick's first law (Equation 19.14) as

$$\frac{dn}{Adt} = -D\left(\frac{\partial c}{\partial x}\right)_t \tag{19.20}$$

where dn is the number of moles of solute that has diffused across the boundary (area A) in time dt. Thus,

$$\left(\frac{\partial c}{\partial x}\right)_t = -\frac{dn}{ADdt} \tag{19.21}$$

From Equations 19.19 and 19.21, we get

$$dn = \frac{ADc_0}{\sqrt{4\pi D}}\frac{dt}{\sqrt{t}} \tag{19.22}$$

Integrating between $t = 0$ and $t = t$ (and $n = 0$ and $n = n$) gives

$$n = \frac{ADc_0}{\sqrt{4\pi D}}\left(2\sqrt{t}\right) \tag{19.23}$$

Therefore,

$$D = \frac{n^2\pi}{A^2c_0^2t} \tag{19.24}$$

A specially constructed cell allows complete removal of the solvent column after a diffusion experiment so that it can be stirred to produce a homogeneous solution of concentration c. If the height of the solvent column is h, it follows that

$$n = cAh$$

Substituting this expression for n in Equation 19.24, we get

$$D = \frac{c^2h^2\pi}{c_0^2t} \tag{19.25}$$

Thus, by determining the concentration, c, after time t and knowing the original concentration, c_0, we can calculate the value of D. Two points are worth noting. First,

* See S. B. Dubin, J. H. Lunacek, and G. Benedek, *Proc. Natl. Acad. Sci. U.S.A.* **57**, 1164 (1967).

Table 19.3
Diffusion Coefficients (D) of Some Molecules in Water at 298 K

Molecule	$D/10^{-9}$ m^2 s^{-1}
Ethanol	1.10
Urea	1.18
Glucose	0.57
Sucrose	0.46
Myoglobin	0.113
Hemoglobin	0.069
DNA (calf thymus)	0.0013

our boundary conditions assume that the liquid columns are infinitely long, whereas in practice they are relatively short. If we keep t short, however, the concentrations of the solute at the extreme ends are still close to c_0 and zero at the end of the experiment. Second, strictly speaking, the diffusion coefficient is concentration dependent, so working with dilute solutions is preferable. Table 19.3 lists the diffusion coefficients of several molecules. We would expect that the larger the molecule, the slower its motion. The data in Table 19.3 qualitatively confirm this expectation.

In diffusion studies, it is important to determine the distance traveled by solute molecules from their place of origin in a given time, t. Although diffusion occurs in one definite direction, the movement of each individual molecule is completely random and unpredictable. Thus, the average, or net, distance, $\overline{x}$, traveled by the molecules is zero. For this reason, we need to consider the mean-square distance, $\overline{x^2}$, defined as (see Equation 19.18)

$$\overline{x^2} = \frac{\int_{-\infty}^{+\infty} x^2 \left(\frac{dc}{dx}\right) dx}{\int_{-\infty}^{+\infty} \left(\frac{dc}{dx}\right) dx}$$

These standard integrals are tabulated in the *Handbook of Chemistry and Physics*. The result is

$$\overline{x^2} = 2Dt$$

Hence, the root-mean-square distance, $\sqrt{\overline{x^2}}$, is given by

$$x_{\text{rms}} = \sqrt{\overline{x^2}} = \sqrt{2Dt} \tag{19.26}$$

Equation 19.26 gives us a simple, yet useful, relation for estimating mean diffusion distances.

For a liquid solution, we would expect the frictional force exerted by a solvent medium to affect the diffusion of a solute molecule. In 1905, Einstein proposed the following quantitative relationship (known as the Einstein-Smoluchowski relation):

$$D = \frac{k_B T}{f} \tag{19.27}$$

where k_B is Boltzmann's constant, and f is the frictional coefficient of the solute molecule. The units of f are N s m^{-1}. Thus, the product of f and the velocity of the solute molecule gives the frictional force of resistance (in newtons) exerted on the particle by the solvent. Stokes showed that for a spherical particle,

$$f = 6\pi\eta r \tag{19.28}$$

where η is the viscosity of the solvent, and r is the radius of the molecule. Equation 19.28 is known as Stokes' law.* Now Equation 19.27 becomes

$$D = \frac{k_B T}{6\pi\eta r} \tag{19.29}$$

which is called the Stokes–Einstein equation. Either Equation 19.27 or 19.29 provides a physical interpretation of the diffusion coefficient. The term $k_B T$ is a measure of the thermal or kinetic energy of the molecule, whereas f or η is a measure of the viscous resistance to diffusion. The ratio of these two opposing values determines how easily a solute molecule diffuses in solution.

Equation 19.29 suggests a way to measure the radius of the molecule if both D and η are known. We must realize, however, that Stokes' law is an idealized expression. Furthermore, even if a molecule is sufficiently symmetrical to be treated like a sphere, the radius measured does not necessarily correspond to the true radius because most solute molecules are solvated to a certain extent in solution. A measured radius, then, often tends to be greater than the true radius.

---

* Stokes' law also applies to nonspherical molecules. For a nonspherical molecule, the frictional coefficient is greater than that for a spherical molecule of the same volume. This is so because for the same volume, the sphere has the *smaller* surface area and experiences less frictional drag.

EXAMPLE 19.2

Calculate the root-mean-square distance traveled by a urea molecule in 1 hour by diffusion in water at 25°C.

ANSWER

The diffusion coefficient of urea in water is 1.18×10^{-9} m^2 s^{-1} (see Table 19.3). From Equation 19.26

$$\sqrt{\overline{x^2}} = \sqrt{2(1.18 \times 10^{-9}\ \text{m}^2\ \text{s}^{-1})(3600\ \text{s})}$$

$$= 2.9 \times 10^{-3}\ \text{m} = 2.9\ \text{mm}$$

COMMENT

In liquids, therefore, diffusion is an inefficient way to transport materials over large distances. Biological systems use diffusion for transport only over short distances, such as the dimension of a cell (about 10^{-2} mm in diameter).

EXAMPLE 19.3

Estimate the diffusion coefficient of a spherical molecule with a radius of 1.5 Å in water at 300 K.

ANSWER

We need Equation 19.29. The data are

$$k_B = 1.381 \times 10^{-23}\ \mathrm{J\ K^{-1}}$$

$$T = 300\ \mathrm{K}$$

$$\eta = 0.00101\ \mathrm{N\ s\ m^{-2}}$$

$$r = 1.5 \times 10^{-10}\ \mathrm{m}$$

Hence,

$$D = \frac{(1.381 \times 10^{-23}\ \mathrm{J\ K^{-1}})(300\ \mathrm{K})}{6\pi(0.00101\ \mathrm{N\ s\ m^{-2}})(1.5 \times 10^{-10}\ \mathrm{m})}$$

$$= 1.5 \times 10^{-9}\ \mathrm{J\ N^{-1}\ m\ s^{-1}}$$

$$= 1.5 \times 10^{-9}\ \mathrm{m^2\ s^{-1}}$$

19.5 Liquid Crystals

Ordinarily, there is a sharp distinction between the highly ordered state of a crystalline solid and the more random molecular arrangement of liquids. Crystalline ice and liquid water, for example, differ from each other in this respect. One class of substances, however, tends so greatly toward an ordered arrangement that a melting crystal first forms a milky liquid, called the *mesomorphic*, or *paracrystalline*, *state*, with characteristically crystalline properties. At higher temperatures, this milky fluid changes sharply into a clear liquid that behaves like an ordinary liquid. Such substances are known as *liquid crystals*.

Molecules that exhibit liquid crystallinity are usually long and rodlike. An example is *para*-azoxyanisol (PAA),

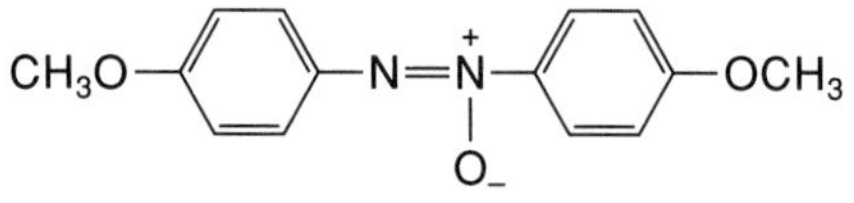

$$CH_3O-C_6H_4-N=\overset{+}{N}(O^-)-C_6H_4-OCH_3$$

which has the following "melting" or phase transition points:

$$\text{solid} \xrightarrow{118°C} \text{mesomorphic state} \xrightarrow{136°C} \text{liquid}$$

There are two known types of liquid crystals, called *thermotropic* and *lyotropic*. Thermotropic liquid crystals exhibit a liquid crystalline phase over a certain range of temperature, while lyotropic liquid crystals have two or more components and exhibit a liquid crystalline phase over a certain range of concentration.

Thermotropic Liquid Crystals

Thermotropic liquid crystals form when the solid is heated. They are subdivided into three common classes, called *smectic*, *nematic*, and *cholesteric*. Figure 19.12 shows schematic diagrams of the smectic and nematic structures. In smectic liquid crystals, the long axes of the molecules are perpendicular to the plane of the layers. The layers are free to slide over one another so that the substance has the structural properties of a two-dimensional solid. Optically, a smectic liquid crystal also behaves like a three-dimensional crystal such as quartz; that is, it possesses anisotropic properties (see discussion below). Nematic liquid crystals are less ordered. Although molecules in nematic liquid crystals are aligned with their long axes parallel to one another, they are not separated into layers.

Cholesterol itself ($R = -OH$) does not form liquid crystals.

Cholesteric liquid crystals resemble smectic liquid crystals in that molecules are arranged in layers, except that the long axes of the molecules are parallel to the layers (Figure 19.13). Cholesteric liquid crystals are formed by a variety of esters of cholesterol,

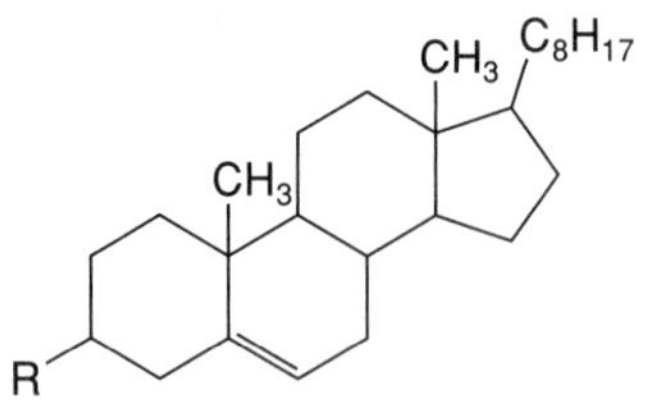

CH_3 C_8H_{17} CH_3 R

where R is an ester group. Within a layer, the molecules orient like those in the nematic phase. As we can see from Figure 19.13, the direction of molecular orientation rotates in a helical fashion. The pitch (the distance for a complete spiral) of the helix in cholesteric phase generally has the same order of magnitude as the wavelengths of visible light. Consequently, these phases diffract light in much the same way as crystals diffract X rays. This phenomenon gives rise to iridescent colors such as those observed in butterfly wings. Interestingly, a racemic mixture of cholesterol derivatives will not form a cholesteric phase, but a pure optical isomer will. Thus, the system possesses chirality.

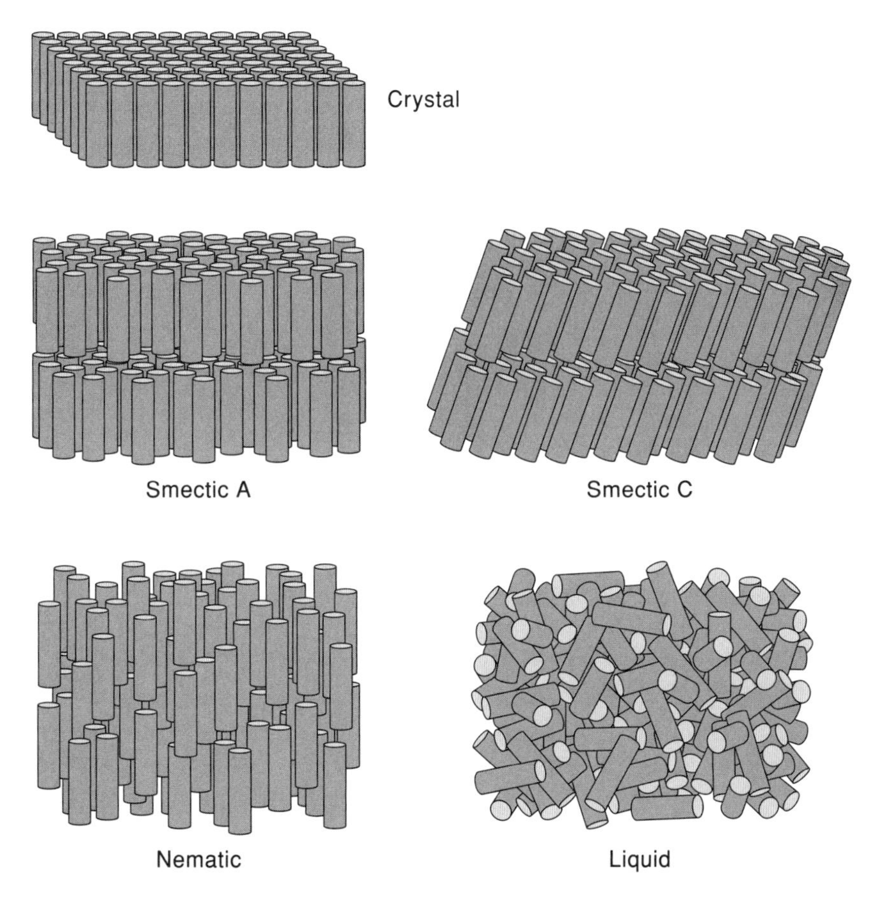

Figure 19.12
Two types of liquid crystal phases: smectic and nematic. The smectic phase has variations, including the smectic A and smectic C phases shown here. For simplicity, all the molecules in the liquid crystal phases are depicted as having the same tilts. In reality, there is a range of tilt angles within each phase. Smectic A and smectic C behave like a two-dimensional solid. Nematic liquid crystals behave like a one-dimensional solid. The perfectly-ordered crystalline solid state and the completely random liquid state are shown for comparison.

One property that distinguishes a liquid crystal of any type from an ordinary liquid is anisotropy. In liquid carbon tetrachloride, for example, the orientation of molecules is completely random, so all directions in space are equivalent. Consequently, any property measured along one direction of space is the same as that measured along any other direction. Thus, if we were to measure the speed of sound waves through such a liquid, we would find the same value no matter which direction the measurement was made. The property of obtaining the same result regardless of direction is called *isotropy*, and a medium that has this property is called an *isotropic phase*. All liquids and gases are isotropic. In contrast, liquid crystals (and crystalline solids) possess directional properties due to their ordered structure. Depending on the orientation of the molecules in liquid crystals, the speed of sound varies along the x, y, and z axes. This property is called *anisotropy* (meaning no isotropy), and the liquid crystal phase is called an *anisotropic phase*.

In Figure 19.12, the molecules in the liquid crystal phases are depicted as having identical orientations, or tilts. In reality, there are appreciable variations around a well-defined average tilt angle. For a particular molecule, the tilt angle is defined as

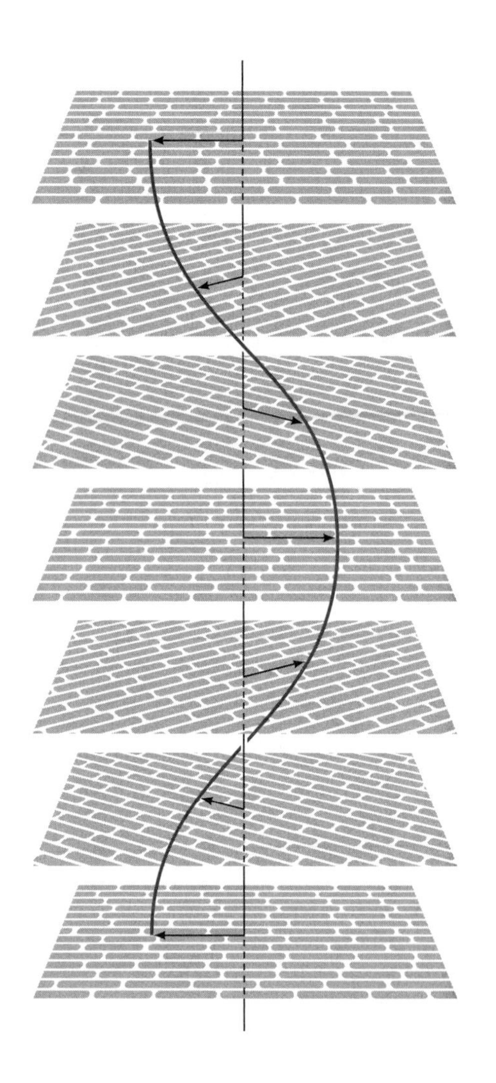

Figure 19.13
The cholesteric liquid crystal phase behaves like layers of two-dimensional solids. Note that the direction of molecular orientation rotates in a helical fashion. The orientation direction eventually points back in the original direction, then continues its spiral. The periodicity of the spiral (called the pitch) is hundreds of nanometers in length and is very sensitive to temperature.

the angle between the molecular axis and the director, which is the direction of the preferred orientation (which is also the optical axis):

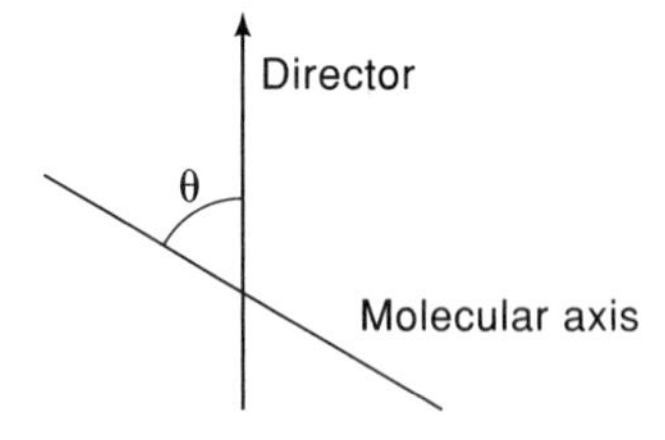

Instead of the angle θ, however, the function $(3\cos^2\theta - 1)/2$ is used to describe the degree of order. If the molecules are perfectly aligned, then $\theta = 0°$ and the function is 1, because $\cos 0° = 1$. For an isotropic liquid, this function is zero because the orientation is totally random. The *order parameter* (S) of a liquid crystal is defined as the average of this function, that is,

$$S = \left\langle \frac{3\cos^2\theta - 1}{2} \right\rangle \tag{19.30}$$

where the $\langle\ \rangle$ signs represent averaging. For most liquid crystal phases, S typically lies between 0.3 and 0.9; it decreases with increasing temperature.

Applications of Thermotropic Liquid Crystals. Thermotropic liquid crystals have many applications in science, technology, and medicine. The pitch length in the cholesteric phase is very sensitive to external parameters such as temperature and electric field. Consequently, the color of cholesteric liquid crystals, which depends on pitch length, changes over very small temperature ranges. For this reason, these liquid crystals are suitable for use as sensitive thermometers. In metallurgy, for example, they are used to detect metal stress, heat sources, and conduction paths. Medically, the temperature of the body at specific sites can be determined with the aid of liquid crystals. This technique has become an important diagnostic tool in treating infection and tumor growth (e.g., breast tumors). Because localized infections and tumors increase metabolic rate and hence temperature in the affected tissues, a thin film of liquid crystal can help a physician see whether an infection or tumor is present by responding to a temperature difference with a change in color.

The anisotropy of nematic liquid crystals causes light polarized along the director to propagate at a different speed than light polarized perpendicular to the director. Figure 19.14 shows how the familiar black-and-white displays in timepieces and calculators work. Transparent aligning agents made of tin oxide (SnO_2) doped with indium oxide (In_2O_3) applied to the lower and upper inside surfaces of the liquid crystal cell preferentially orient the molecules in the nematic phase by 90° relative to each other. In this way, the molecules become "twisted" through the liquid crystal phase (Figure 19.14a). When properly adjusted, this twist rotates the plane of polarization by 90° and allows

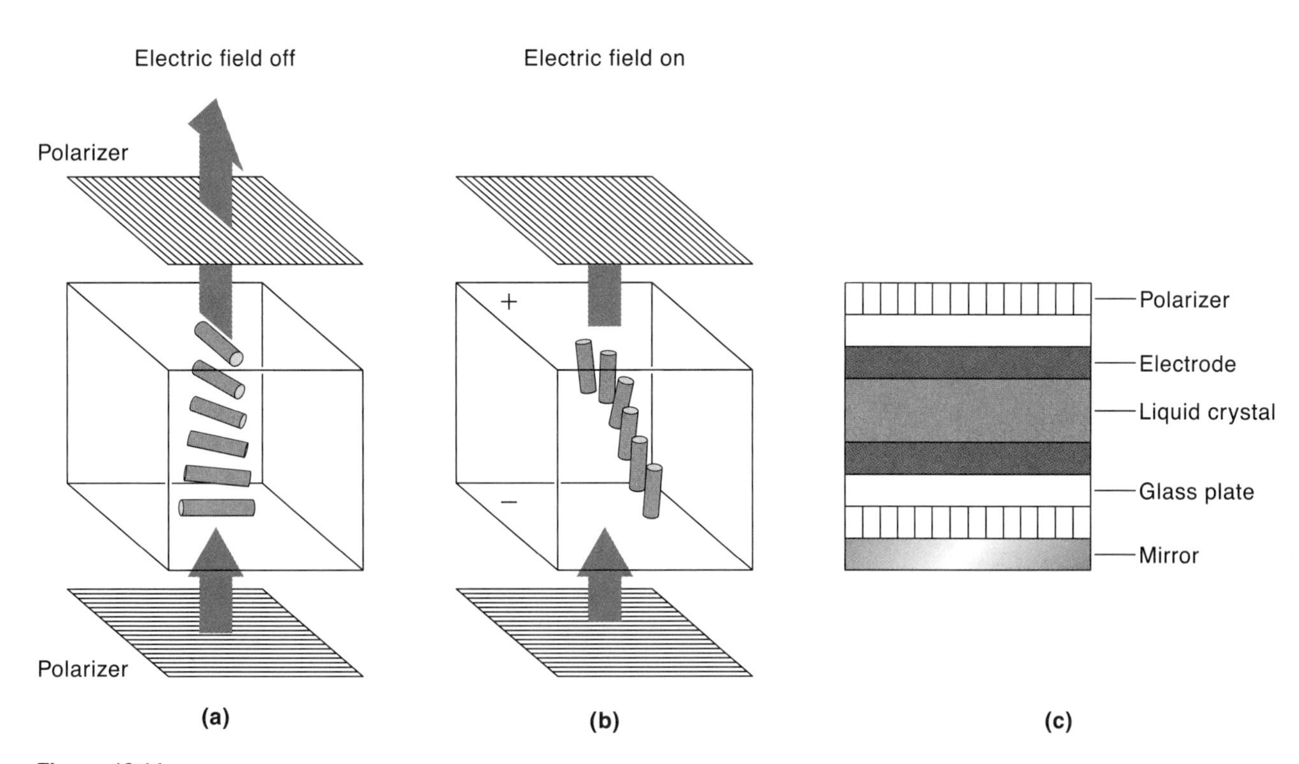

Figure 19.14
A black-and-white display using nematic liquid crystals. Molecules in contact with the bottom and top cell surfaces are aligned at right angles to one another. (a) The extent of twist in the molecular orientation between the surfaces is adjusted so as to rotate the plane of polarized light by 90°, allowing it to pass through the top polarizer. Consequently, the cell appears clear. (b) When the electric field is on, molecules orient along the direction of the field, the plane of polarized light can no longer pass through the top polarizer, and the cell becomes dark. (c) A cross section of a liquid crystal display.

the light to pass through the two polarizers (arranged at 90° to each other). In this mode, the display appears clear. When an electric field is applied (Figure 19.14b), the nematic molecules experience a torque (a torsion or rotation) that forces them to align along the direction of the field. Now the incident polarized light cannot pass through the top polarizer, and the cell appears dark. In watches and calculators, a reflector is placed under the bottom polarizer. In the absence of an electric field, the reflected light goes through both polarizers, and the cell looks clear from the top. When the electric field is turned on, the incident light from the top cannot pass through the bottom polarizer to reach the reflector, and the cell becomes dark. Typically, a few volts are applied across a nematic layer about 10 μm thick (1 μm = 10^{-6} m). The response time for molecules to align and relax when the electric field is turned on and off is in the millisecond range (1 ms = 10^{-3} s).

Contemporary flat panel displays for computers, cell phones, and televisions use TFT LCD (thin-film transistor liquid crystal display) technology to display a broad spectrum of colors.

Lyotropic Liquid Crystals

Lyotropic liquid crystals are mixtures of two or more compounds, one of which is usually a polar molecule such as water. Several synthetic polypeptides, including poly-γ-benzyl-L-glutamate and poly-β-benzyl-L-aspartate, form structures that resemble cholesteric liquid crystals when dissolved in water, dimethylformamide, or pyridine. The second type of lyotropic liquid crystals contains phospholipids and forms ordered membrane structures. Lyotropic liquid crystals are important not only in biological systems, but also as surfactants in cleaning products.

Key Equations

$\text{Reynolds number} = \dfrac{2Rv\rho}{\eta}$	(Laminar versus turbulent flow)	(19.4)
$Q = \dfrac{\pi P R^4}{8\eta L}$	(Poiseuille's law)	(19.5)
$\gamma = \dfrac{rh\rho g}{2\cos\theta}$	(Capillary-rise method)	(19.12)
$J = -D\left(\dfrac{\partial c}{\partial x}\right)_t$	(Fick's first law of diffusion)	(19.14)
$\left(\dfrac{\partial c}{\partial t}\right)_x = D\left(\dfrac{\partial^2 c}{\partial x^2}\right)_t$	(Fick's second law of diffusion)	(19.16)
$x_{\text{rms}} = \sqrt{\overline{x^2}} = \sqrt{2Dt}$	(Root-mean-square distance in diffusion)	(19.26)
$D = \dfrac{k_B T}{f}$	(Einstein-Smoluchowski relation)	(19.27)
$f = 6\pi\eta r$	(Stokes' law)	(19.28)
$D = \dfrac{k_B T}{6\pi\eta r}$	(Stokes–Einstein equation)	(19.29)

APPENDIX 19.1

Derivation of Equation 19.13

Consider a soap bubble of radius r. For the bubble to maintain its spherical shape, the internal force must be balanced by the external force. Figure 19.15 shows the bubble divided into two hemispheres. Looking at the upper hemisphere, we see that in addition to the outside force, there is a downward force, F, as a result of the surface tension. This surface force is given by the surface tension multiplied by the circumference. (Remember that surface tension has the units N m^{-1}.) The total downward force, then, is

$$F = P_o(\pi r^2) + 2(2\pi r\gamma) \quad (1)$$

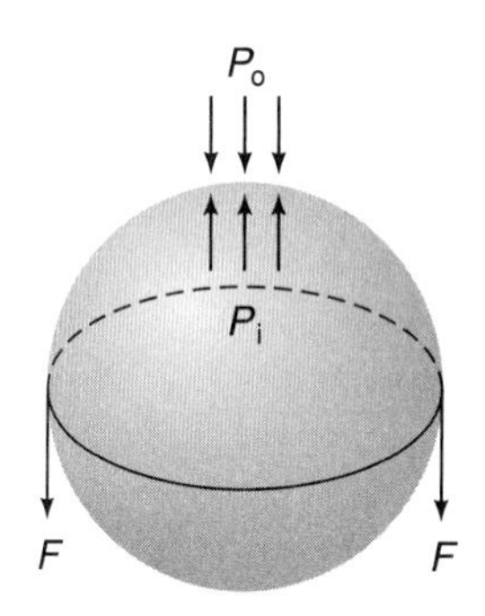

Figure 19.15
To maintain the shape of a soap bubble, the internal and external forces must balance.

where P_o is the outside pressure and πr^2 is the cross-sectional area. The extra factor 2 arises because the bubble has both an inner and an outer layer. The total upward force is $P_i(\pi r^2)$, where P_i is the inside pressure, so at equilibrium we have

$$P_i(\pi r^2) = P_o(\pi r^2) + 2(2\pi r\gamma) \quad (2)$$

or

$$P_i - P_o = \frac{4\gamma}{r} \quad (3)$$

Because an alveolus has only one layer, the equation above becomes

$$P_i - P_o = \frac{2\gamma}{r} \quad (4)$$

which is Equation 19.13. Note that the difference $(P_i - P_o)$ is large for small bubbles (when r is small) but decreases as r increases.

Suggestions for Further Reading

BOOKS

Chandrasekhar, S., *Liquid Crystals*, 2nd ed., Cambridge University Press, New York, 1992.
Collings, P. J., *Liquid Crystals*, Princeton University Press, Princeton, NJ, 1990.
Collings, P. J., and M. Hird, *Introduction to Liquid Crystals*, Taylor & Francis, Bristol, PA, 1997.
Tabor, D., *Gases, Liquids, and Solids*, 3rd ed., Cambridge University Press, New York, 1991.
Vogel, S., *Life in Moving Fluids*, Princeton University Press, Princeton, New Jersey, 1994.
Walton, A. J., *The Three Phases of Matter*, 2nd ed., Oxford University Press, New York, 1983.

ARTICLES

General

"The Structure of Liquids," J. D. Bernal, *Sci. Am.* August 1960.
"The Tensile Strength of Liquids," R. E. Apfel, *Sci. Am.* December 1972.
"The Fluid Phases of Matter," J. A. Barker and D. Henderson, *Sci. Am.* November 1981.
"A Two-Dimensional Model of a Liquid: The Pair-Correlation Function," M. Contreras and J. Valenzuela, *J. Chem. Educ.* **63**, 7 (1986).
"Reappearing Phases," J. Walker and C. A. Vanse, *Sci. Am.* May 1987.
"Why is Mercury a Liquid?" L. J. Norrby, *J. Chem. Educ.* **68**, 110 (1991).

Viscosity

"Negative Viscosity," V. P. Starr and N. E. Gaut, *Sci. Am.* July 1970.
"Life at Low Reynolds Number," E. M. Purcell, *Am. J. Phys.* **45**, 3 (1977).
"Aneurysms," K. Johansen, *Sci. Am.* July 1982.
"How Microorganisms Move Through Water," G. T. Yates, *Am. Sci.* **74**, 358 (1986).
"A Polymer Viscosity Experiment with No Right Answer," L. C. Rosenthal, *J. Chem. Educ.* **67**, 78 (1990).
"Do Cathedral Glasses Flow?" E. D. Zanotto, *Am. J. Phys.* **66**, 392 (1998).

Surface Tension

"Surface Tension in the Lungs," J. A. Clements, *Sci. Am.* December 1962.
"Convection," M. G. Verlade and C. Normand, *Sci. Am.* July 1980.
"Aqueous Foams," J. H. Aubert, A. M. Kraynik, and P. B. Rand, *Sci. Am.* May 1986.
"Walking on Water," R. B. Suter, *Am. Sci.* **87**, 154 (1999).
"A Demonstration of Surface Tension and Contact Angle," H. D. Gesser and P. Krause, *J. Chem. Educ.* **77**, 58 (2000).
"Using Surface Tension Measurements to Understand How Pollution Can Infuence Cloud Formation, Fog, and Precipitation," S. D. Brooks, M. Gonzales, and R. Farias, *J. Chem. Educ.* **86**, 838 (2009).

Diffusion

"Measurement and Interpretation of Diffusion Coefficients of Proteins," L. J. Gosting, *Advan. Protein Chem.* **11**, 429 (1956).
"A Quantitative Diffusion Experiment for Students," M. De Paz, *J. Chem. Educ.* **46**, 784 (1969). [See also *J. Chem. Educ.* **47**, A204 (1970).]
"Molecular Volumes and the Stokes-Einstein Equation," J. T. Edward, *J. Chem. Educ.* **47**, 261 (1970).

"The Diffusion Coefficient of Sucrose in Water," P. W. Linder, L. R. N'assimbeni, A. Polson, and A. L. Rodgers, *J. Chem. Educ.* **53**, 330 (1976).
"A Spectrophotometric Method for Measuring Diffusion Coefficients," J. Irina, *J. Chem. Educ.* **57**, 676 (1980).
"A Practical and Convenient Diffusion Apparatus: An Undergraduate Physical Chemistry Experiment," B. Clifford and E.-I. Ochiai, *J. Chem. Educ.* **57**, 678 (1980).
"Surface Diffusion," R. Gomer, *Sci. Am.* August 1982.
"Measurement of Diffusion Coefficients," J. E. Crooks, *J. Chem. Educ.* **66**, 614 (1989).
"A Laser Refraction Method for Measuring Liquid Diffusion Coefficients," M. E. King, R. W. Pitha, and S. F. Sontum, *J. Chem. Educ.* **66**, 787 (1989).
"Diffusion Confusion," L. C. Davis, *J. Chem. Educ.* **73**, 824 (1996).
"Learning Molecular Diffusion. A Laboratory Experiment," *Chem. Educator* [Online] **10**, 283 (2005) DOI 10.1333/s00897050935a.

Liquid Crystals
"Liquid Crystals," J. L. Ferguson, *Sci. Am.* August 1964.
"Liquid Crystals and Their Roles in Inanimate and Animate Systems," G. H. Brown, *Am. Sci.* **60**, 64 (1972).
"Liquid Crystals for Electro-optical Displays," G. Elliot, *Chem. Brit.* **9**, 213 (1973).
"Electronic Numbers," A. Sobel, *Sci. Am.* June 1973.
"Three Liquid-Crystal Teaching Experiments," J. R. Lalanne, and F. Hare, *J. Chem. Educ.* **53**, 793 (1976).
"Imposed Orientation of Dye Molecules by Liquid Crystals and an Electric Field," N. Sadlej-Sosnowska, *J. Chem. Educ.* **57**, 223 (1980).
"Liquid Crystals—The Chameleon Chemicals," G. H. Brown, *J. Chem. Educ.* **60**, 900 (1983).
"Liquid Crystals: A Colorful State of Matter," G. H. Brown and P. P. Crooker, *Chem. Eng. News* **61**(5), 24 (1983).
"Preparation and Properties of Cholesteric Liquid Crystals," G. Patch and G. A. Hope, *J. Chem. Educ.* **62**, 454 (1985).
"Liquid-Crystal Displays: Fabrication and Measurement of a Twisted Nematic Liquid-Crystal Cell," E. R. Waclawik, M. J. Ford, P. S. Hale, J. G. Shapter, and N. H. Voelcker, *J. Chem. Educ.* **81**, 854 (2004).
"Colors in Liquid Crystals," G. Lisensky and E. Boatman, *J. Chem. Educ.* **82**, 1360A (2005).
"Synthesis and Physical Properties of Liquid Crystals: An Interdisciplinary Experiment," G. R. Van Hecke, K. K. Karukstis, H. Li, H. C. Hendargo, A. J. Cosand, and M. M. Fox, *J. Chem. Educ.* **82**, 1349 (2005).
"Liquid Crystals Activity," D. L. Lewis and M. Warren, *J. Chem. Educ.* **83**, 1602 (2006).
"Liquid Crystals Activity Revisited," V. M. Petruševski, *J. Chem. Educ.* **84**, 1429 (2007).

Problems

Viscosity

19.1 The viscosity of a gas increases with increasing temperature (see Equation 2.22), yet the viscosity of a liquid decreases with increasing temperature. Explain.

19.2 At 293 K, the time of flow for water through an Ostwald viscometer is 342.5 s; for the same volume of an organic solvent, the time of flow is 271.4 s. Calculate the viscosity of the organic liquid relative to that of water. The density of the organic solvent is 0.984 g cm^{-3}.

19.3 For blood flowing in a capillary of radius 2.0×10^{-4} cm, estimate the maximum velocity for laminar flow at 37°C. (The density of whole blood is about 1.2 g cm^{-3}.)

19.4 An arteriole has a diameter of 2.4×10^{-5} m and a blood flow rate of 2.6×10^{-3} m s^{-1}. Calculate the pressure drop, ΔP, from one end to the other if the length of the arteriole is 5.0×10^{-3} m.

19.5 The viscosity of a liquid usually decreases with increasing temperature. An empirical equation is $\log \eta = A/T + B$. Determine the constants, A and B, for water from the following data:

T/K	273	293	310	373
η/P	0.01787	0.0101	0.00719	0.00283

19.6 Show that the Reynolds number (see Equation 19.4) is dimensionless.

19.7 From the definition of Reynolds number (Equation 19.4), calculate the maximum value of v for the laminar flow of water at 293 K along a tube with a radius of 0.60 cm.

19.8 The rate of flow of a liquid through a cylindrical tube that has an inner radius of 0.12 cm and a length of 26 cm is 364 cm^3 in 88 s. The pressure drop between the ends of the tube is 57 torr. Calculate the liquid's viscosity. Is the flow laminar? The density of the liquid is 0.98 g cm^{-3}.

Surface Tension

19.9 Water has an unusually large surface tension. Explain.

19.10 Give a molecular interpretation for the decrease in surface tension of a liquid with increasing temperature.

19.11 A glass capillary of diameter 0.10 cm is dipped into **(a)** water (contact angle 10°) at 293 K and **(b)** mercury (contact angle 170°) at 298 K. Calculate the level of the liquid in the capillary in each case.

19.12 Both ethanol and mercury are used in thermometers. Explain the difference between the menisci of the liquids in these two types of thermometers.

19.13 The surface tension of liquid naphthalene at 127°C is 0.0288 N m^{-1}, and its density at this temperature is 0.96 g cm^{-3}. What is the radius of the largest capillary that will permit the liquid to rise 3.0 cm? Assume that the angle of contact is zero.

19.14 The surface tension of quinoline is twice that of acetone at 20°C. If the capillary rise is 2.5 cm for quinoline, what is the rise for acetone in the same capillary? Assume that the angles of contact are zero. The densities of quinoline and acetone at 20°C are 1.09 g cm^{-3} and 0.79 g cm^{-3}, respectively.

19.15 A capillary tube that has an inner diameter of 0.40 mm is inserted vertically into a pool of mercury at 20°C (see Figure 19.7b). Calculate the depression in mercury given that the contact angle is 146°C. The density of mercury is 13.6 g cm^{-3}.

19.16 Two capillary tubes with inside diameters of 1.4 mm and 1.0 mm, respectively, are inserted into a liquid of density 0.95 g cm^{-3}. Calculate the surface tension of the liquid if the difference between the capillary rises in the tubes is 1.2 cm. Assume the contact angle is zero.

Diffusion

19.17 The diffusion coefficient of glucose is 5.7×10^{-10} m^2 s^{-1}. Calculate the time required for a glucose molecule to diffuse through **(a)** 10,000 Å and **(b)** 0.10 m.

19.18 The diffusion coefficient of sucrose in water at 298 K is 0.46×10^{-5} cm^2 s^{-1}, and the viscosity of water at the same temperature is 0.0010 N s m^{-2}. From these data, estimate the effective radius of a sucrose molecule.

19.19 From the diffusion coefficients listed in Table 19.3, estimate the radius and molecular volume of myoglobin and hemoglobin. What conclusion can you draw from the results?

19.20 Diffusion coefficients have been measured for many solid systems. If the diffusion coefficient of bismuth in lead is 1.1×10^{-16} cm^2 s^{-1} at 20°C, calculate how long it will take (in years) for a bismuth atom to travel 1.0 cm.

19.21 What is the diffusion coefficient of a membrane-bound protein of molar mass 80,000 g at 37°C if the viscosity of the membrane is 1 poise (0.10 N s m^{-2})? What is the average distance traveled by this protein in 1.0 s? Assume that this protein is an unhydrated, rigid sphere that has a density of 1.4 g cm^{-3}.

Additional Problems

19.22 Two soap bubbles of radii r_1 and r_2 ($r_2 > r_1$) are connected by a piece of tubing with a stopcock. Predict how the size of the bubbles will change when the stopcock is opened.

19.23 Swimming coaches sometimes suggest that a drop of alcohol (ethanol) placed in an ear plugged with water "draws out the water." Comment from a molecular point of view. [*Source:* "Eco-Chem," J. A. Campbell, *J. Chem. Educ.* **52**, 655 (1975).]

19.24 The carbon monoxide–hemoglobin complex has a diffusion coefficient of 0.062×10^{-9} m^2 s^{-1} in water at 298 K. In the more viscous cytoplasm, the diffusion coefficient is only 0.013×10^{-9} m^2 s^{-1}. How long would it take for such a complex to travel the 3.0-μm length of a bacterial cell?

19.25 Ozone (O_3) is a strong oxidizing agent that can oxidize all the common metals except gold and platinum. A convenient test for ozone is based on its action on mercury. When exposed to ozone, mercury becomes dull looking and sticks to glass tubing (instead of flowing freely through it). Write a balanced equation for the reaction. What properly of mercury is altered by its interaction with ozone?

19.26 A hypodermic syringe is filled with a solution of viscosity 1.6×10^{-3} N s m^{-2}. The plunger area of the syringe is 7.5×10^{-5} m^2, and the length of the needle is 0.026 m. The internal radius of the needle is 4.0×10^{-4} m. The gauge pressure in a vein is 1850 Pa (14 mmHg). Calculate the force in newtons that must be applied to the plunger so that 1.2×10^{-6} m^3 of the solution can be injected in 4.0 s.

19.27 A film of an organic liquid filled a rectangular wire frame similar to that shown in Figure 19.6. (**a**) Given that the wire frame is 9.0 cm wide and that a force of 7.2×10^{-3} N is needed to move the piston, calculate the surface tension of the liquid. (**b**) What is the work done in stretching the film to a distance of 0.14 cm?

19.28 How much work is required to break up 1 mole of water at 20°C into spherical droplets that have a radius of 4.16×10^{-3} m? [*Hint:* The volume of a sphere is $(4/3)\pi r^3$, and the surface area of a sphere is $4\pi r^2$, where r is the radius of the sphere.] The density of water is 1.0 g cm^{-3}.

19.29 A sphere of volume V falling through a fluid experiences a downward gravitational force mg, where m is the mass of the sphere, and g is the acceleration due to gravity. Simultaneously retarding the fall are the frictional force (see Equation 19.28) and an upward buoyant force given by $m_f g$, where m_f is the mass of the fluid of volume V. Calculate the terminal speed of fall of a steel ball of radius 1.2 mm and density 7.8 g cm^{-3} in water at 20°C. Based on your calculation, design an experiment that would enable you to measure the viscosity of a liquid.

19.30 The diffusion coefficient of oxygen in air is 0.20 cm^2 s^{-1}; the diffusion coefficient of the same gas in water is about 10^4 times smaller. (**a**) Explain the huge difference in the magnitude of these diffusion coefficients. (**b**) Most animal cells are bathed in fluids, so that a hemoglobin-like molecule and a circulatory system are necessary for the purpose of transporting O_2 to the cells and carrying CO_2 away. (The diffusion coefficients of CO_2 in air and in water are comparable in magnitude to those of oxygen.) Because plants do not have a circulatory system, explain how the O_2 and CO_2 gases are transported efficiently in these systems. (**c**) Insects do possess a circulating system but lack a hemoglobinlike molecule. Considering the diffusion coefficients of CO_2 and O_2 in water, do you think it likely that ants, bees, and cockroaches can grow to human size, as they sometimes do in horror movies?

19.31 For a liquid crystal, θ is the angle between the director and the molecular axis. At what average value of θ does the order parameter S equal 0.5?

19.32 A capillary tube of constant inner diameter d is inserted into an aqueous solution of concentration c at temperature T and pressure P. The solution rises within the tube to a height h. Describe at least five different ways to change this experiment to produce a larger rise h.

19.33 Due to cholesterol deposits, the radius of the anterior descending artery of a patient's heart has narrowed by 12.0%. Calculate the percent increase in the blood pressure drop across the artery necessary to maintain the normal blood flow through the artery.

19.34 Explain how a water strider is able to "walk" on water.

CHAPTER 20

Statistical Thermodynamics

Safety in numbers

Quantum mechanics has shown us how the energy levels of atoms and molecules can be calculated (in principle, at least) and measured spectroscopically. Thermodynamics, on the other hand, concerns macroscopic systems. How can knowledge of atomic and molecular energy levels be used to account for matter in bulk? The answer is provided by statistical thermodynamics, which gives us the link between the microscopic and the bulk properties of matter.

We begin with a derivation of the Boltzmann distribution law. The derivation leads to the concept of the partition function, which can be used to calculate all the thermodynamic quantities and the equilibrium constant.

20.1 The Boltzmann Distribution Law

Consider a system that contains N particles in which n_0 particles have energy ε_0, n_1 particles have energy ε_1, and so on. There are two restrictions (or constraints) on the permissible values of the numbers $n_0, n_1, \ldots$. The first constraint expresses the total number of particles as a sum of the number of particles occupying different energy levels:

$$n_0 + n_1 + n_2 + \cdots = \sum_i n_i = N \qquad (20.1)$$

Because the total number of particles cannot change, Equation 20.1 can be written in the differential form as

$$dN = \sum_i dn_i = 0 \qquad (20.2)$$

The other constraint concerns the energy of the system, E, given by

$$E = n_0\varepsilon_0 + n_1\varepsilon_1 + \cdots = \sum_i n_i\varepsilon_i \qquad (20.3)$$

Similarly, E is a constant, so

$$dE = \sum_i \varepsilon_i dn_i = 0 \qquad (20.4)$$

Note that because each ε_i is fixed, $d\varepsilon_i = 0$.

We saw in Section 4.7 that, at thermal equilibrium, there is a most probable distribution with W microstates. Because W is such an enormous number, however, looking for a maximum in ln W is mathematically more convenient than looking at W. Taking the natural logarithm of Equation 4.29 we get

$$\ln W = \ln N! - \ln \prod_i n_i!$$
$$= \ln N! - \sum_i \ln n_i! \qquad (20.5)$$

Note that "$\ln \prod_i n_i!$" becomes "$\sum_i \ln n_i$" in this step. The fact that N is a very large number means that all the n_i values are also large; therefore, we can apply *Stirling's approximation* (after the Scottish mathematician James Stirling, 1692–1770):

Many scientific calculators have the $x!$ function key. Test Equation 20.6 using $x = 5$ and $x = 50$.

$$\ln x! = x \ln x - x \qquad (20.6)$$

Equation 20.5 now becomes

$$\ln W = N \ln N - N - \sum_i (n_i \ln n_i - n_i) \qquad (20.7)$$

Because ln W is a maximum and N is a constant, it follows that

$$\left(\frac{\partial \ln W}{\partial n_i}\right) = 0 = -\sum_i \ln n_i$$

or

$$d \ln W = -\sum_i \ln n_i dn_i = 0 \qquad (20.8)$$

To solve Equation 20.8, we apply a mathematical technique called *Lagrange's method of undetermined multipliers* (after the French mathematician Joseph Louis Lagrange, 1736–1813). We multiply each constraint (Equation 20.2 or 20.4) by a constant and add them to the main equation (Equation 20.8). The variables (dn_i values) are then treated as though they are all independent, and the constants are evaluated at the end of the calculation. Applying this procedure, we multiply Equation 20.2 by α and Equation 20.4 by β, respectively, and then add them to Equation 20.8 to arrive at the final equation:

$$-\sum_i \ln n_i dn_i + \alpha \sum_i dn_i + \beta \sum_i \varepsilon_i dn_i = 0$$

or

$$\sum_i (-\ln n_i + \alpha + \beta \varepsilon_i) dn_i = 0 \qquad (20.9)$$

Now the dn_i values can all vary independently (by using proper values of α and β), so the only way of satisfying $d \ln W = 0$ is to require that, for each value of i,

$$-\ln n_i + \alpha + \beta \varepsilon_i = 0 \qquad (20.10)$$

Upon rearrangement,

$$\ln n_i = \alpha + \beta \varepsilon_i$$

or

$$n_i = e^{\alpha} e^{\beta \varepsilon_i} \qquad (20.11)$$

To evaluate α, we arbitrarily set the lowest energy level $\varepsilon_0 = 0$. Equation 20.11 now becomes

$$n_0 = e^{\alpha} \tag{20.12}$$

We see that e^{α} is just a number, so α is a dimensionless quantity.

To evaluate β, we start with the Boltzmann equation (Equation 4.30):

$$\begin{aligned} S &= k_B \ln W \\ &= k_B\Big[\ln N! - \sum_i \ln n_i!\Big] \\ &= k_B\Big[N \ln N - N - \sum_i (n_i \ln n_i - n_i)\Big] \end{aligned} \tag{20.13}$$

From Equation 20.11,

$$\ln n_i = \alpha + \beta\varepsilon_i \tag{20.14}$$

Substituting Equation 20.14 into Equation 20.13 gives

$$\begin{aligned} S &= k_B\Big[N \ln N - \sum_i n_i(\alpha + \beta\varepsilon_i)\Big] \\ &= k_B(N \ln N - \alpha N - \beta E) \end{aligned} \tag{20.15}$$

We can now identify energy, E, with the thermodynamic internal energy, U. Previously, we arbitrarily set $\varepsilon_0 = 0$. Therefore, to obtain the value of U from E, we must write $E = U - U_0$, where U_0 is the internal energy at absolute zero. According to Equation 5.9,

$$dU = TdS - PdV$$

so

$$\left(\frac{\partial U}{\partial S}\right)_V = T$$

or

$$\left(\frac{\partial S}{\partial U}\right)_V = \frac{1}{T} \tag{20.16}$$

From Equation 20.15, we write

$$\left(\frac{\partial S}{\partial E}\right)_V = -\beta k_B = \left(\frac{\partial S}{\partial U}\right)_V = \frac{1}{T}$$

Because U_0 is a constant, $dE = dU$.

Therefore,

$$\beta = -\frac{1}{k_B T} \tag{20.17}$$

Substitution of Equations 20.17 and 20.12 into 20.11 gives

$$\begin{aligned} n_i &= n_0 e^{\beta \varepsilon_i} \\ &= n_0 e^{-\varepsilon_i/k_B T} \end{aligned} \tag{20.18}$$

and the total number of particles, N, can be expressed as

$$N = \sum_i n_i = n_0 \sum_i e^{-\varepsilon_i/k_B T} \tag{20.19}$$

Dividing Equation 20.18 by 20.19, we obtain

$$\frac{n_i}{N} = \frac{e^{-\varepsilon_i/k_B T}}{\sum_i e^{-\varepsilon_i/k_B T}} \tag{20.20}$$

which is one form of the Boltzmann distribution law. The ratio of the number of particles (populations) in energy states 2 and 1, say, is given by

$$\begin{aligned} \frac{n_2}{n_1} &= \frac{e^{-\varepsilon_2/k_B T}}{e^{-\varepsilon_1/k_B T}} \\ &= e^{-(\varepsilon_2 - \varepsilon_1)/k_B T} = e^{-\Delta\varepsilon/k_B T} \end{aligned} \tag{20.21}$$

where $\Delta\varepsilon = \varepsilon_2 - \varepsilon_1$. Equation 20.21 is another form of the Boltzmann distribution law, introduced in Chapter 2 (see Equation 2.33).

Equation 20.21 was derived without considering degeneracy. In practice, it often happens that several states may have the same energy (see p. 422). For example, if g_i states have the same energy ε_i then that energy level is g_i-fold degenerate, and Equation 20.18 becomes

$$n_i = n_0 g_i e^{-\varepsilon_i/k_B T} \tag{20.22}$$

The Boltzmann distribution law (Equation 20.21) now takes the form

$$\begin{aligned} \frac{n_2}{n_1} &= \frac{g_2 e^{-\varepsilon_2/k_B T}}{g_1 e^{-\varepsilon_1/k_B T}} \\ &= \frac{g_2}{g_1} e^{-\Delta\varepsilon/k_B T} \end{aligned} \tag{20.23}$$

20.2 The Partition Function

The summation quantity in Equation 20.19 has great theoretical importance and is called the *partition function*, q, given by

$$q = \sum_i e^{-\varepsilon_i/k_B T} \tag{20.24}$$

where ε_i is the energy of state i, and the summation extends over all states. This definition shows that q is just a number; it has no units. As mentioned earlier, there may be several states corresponding to a given energy level, so we commonly write

$$q = \sum_i g_i e^{-\varepsilon_i/k_B T} \tag{20.25}$$

where i is now the label of an energy level, and g_i is its degeneracy.

The partition function plays a fundamental role in statistical thermodynamics. It tells us the number of states that are thermally accessible to a molecule at the temperature of interest, and it can be used to calculate the various thermodynamic properties. According to Equation 20.25, at absolute zero ($T = 0$), only the ground state is accessible and therefore $q = g_0$, or 1 if the ground state is nondegenerate. In the other extreme, as $T \to \infty$, q approaches the total number of states in the molecule, which is typically infinite.

The partition function is the link between thermodynamics and quantum mechanics.

EXAMPLE 20.1

A certain system has three energy levels at 0, 2.00×10^{-21} J, and 8.00×10^{-21} J and degeneracies 1, 3, and 5, respectively. Calculate the partition function of the system at 300 K.

ANSWER

First, we calculate the value of the term $k_B T$:

$$\begin{aligned} k_B T &= (1.381 \times 10^{-23}\ \text{J K}^{-1})(300\ \text{K}) \\ &= 4.14 \times 10^{-21}\ \text{J} \end{aligned}$$

The partition function of the system is given by Equation 20.25:

$$\begin{aligned} q &= g_0 e^{-\varepsilon_0/k_B T} + g_1 e^{-\varepsilon_1/k_B T} + g_2 e^{-\varepsilon_2/k_B T} \\ &= 1 + 3\exp(-2.00 \times 10^{-21}\ \text{J}/4.14 \times 10^{-21}\ \text{J}) \\ &\quad + 5\exp(-8.00 \times 10^{-21}\ \text{J}/4.14 \times 10^{-21}\ \text{J}) \\ &= 1 + 3 \times 0.617 + 5 \times 0.145 \\ &= 3.58 \end{aligned}$$

The significance of the partition function is that it can be used, in principle, to calculate various thermodynamic functions. To illustrate this point, let us derive an expression for the energy, E, of the system in terms of q. The average energy per molecule is given by

$$\frac{E}{N} = \frac{\sum_i n_i \varepsilon_i}{\sum_i n_i} \tag{20.26}$$

or

$$E = \frac{N\sum_i n_i\varepsilon_i}{\sum_i n_i}$$

Dividing the numerator and the denominator of the right side of the above equation by n_0 (a constant) and making use of Equation 20.18, we get

$$E = \frac{N\sum_i n_i\varepsilon_i/n_0}{\sum_i n_i/n_0} = \frac{N\sum_i \varepsilon_i e^{-\varepsilon_i/k_BT}}{\sum_i e^{-\varepsilon_i/k_BT}} = \frac{N\sum_i \varepsilon_i e^{-\varepsilon_i/k_BT}}{q} \tag{20.27}$$

Energy E is a function of volume.

From Equation 20.24, we take the partial derivative of q with respect to T at constant volume:

$$\left(\frac{\partial q}{\partial T}\right)_V = \left[\frac{\partial(\sum_i e^{-\varepsilon_i/k_BT})}{\partial T}\right]_V$$

$$= \frac{1}{k_BT^2}\sum_i \varepsilon_i e^{-\varepsilon_i/k_BT} \tag{20.28}$$

Upon rearrangement,

$$\sum_i \varepsilon_i e^{-\varepsilon_i/k_BT} = k_BT^2\left(\frac{\partial q}{\partial T}\right)_V \tag{20.29}$$

Substituting Equation 20.29 into Equation 20.27 gives

$$E = \frac{Nk_BT^2\left(\frac{\partial q}{\partial T}\right)_V}{q} \tag{20.30}$$

Because

$$\frac{\left(\frac{\partial q}{\partial T}\right)_V}{q} = \left(\frac{\partial \ln q}{\partial T}\right)_V$$

we can now write Equation 20.30 as

$$E = Nk_BT^2\left(\frac{\partial \ln q}{\partial T}\right)_V \tag{20.31}$$

For 1 mole of the gas, $k_BN_A = R$, so

$$\bar{E} = RT^2\left(\frac{\partial \ln q}{\partial T}\right)_V \tag{20.32}$$

Equation 20.32 gives us the molar energy of the system at some temperature T.

To understand the meaning of E, note that we chose the lowest energy to be zero (i.e., $\varepsilon_0 = 0$). At $T = 0$, q is a constant, and its derivative with respect to temperature is zero; that is, $E = 0$. We now add heat to the system at constant volume, leading to the population of upper levels, and the energy increases accordingly. But because no work is done (note that the volume is kept constant), this increase in energy is the internal energy, U. We can write E as

$$E = U - U_0 \tag{20.33}$$

where U_0 is the internal energy at $T = 0$ K. Equation 20.32 can now be expressed as

$$\bar{U} - \bar{U}_0 = RT^2\left(\frac{\partial \ln q}{\partial T}\right)_V \tag{20.34}$$

Equation 20.34 shows the connection between a thermodynamic quantity (U) and the partition function. It is one of many useful results in statistical thermodynamics.

20.3 Molecular Partition Function

To calculate any thermodynamic quantity, we first must evaluate the partition functions of the molecular system. Focusing on a single molecule, we see that its energy in the ith state, ε_i, is given by the sum of its various motions:

$$\varepsilon_i = (\varepsilon_i)_{\text{trans}} + (\varepsilon_i)_{\text{rot}} + (\varepsilon_i)_{\text{vib}} + (\varepsilon_i)_{\text{elec}} \tag{20.35}$$

where the subscripts denote translational, rotational, vibrational, and electronic motions, respectively. Except for translation, these motions are not completely independent of one another, so Equation 20.35 is only approximate, but it is satisfactory in most cases. Substituting Equation 20.35 into Equation 20.24 gives the molecular partition function (for simplicity, we omit the subscript i)

$$\begin{aligned} q &= \sum \exp[-(\varepsilon_{\text{trans}} + \varepsilon_{\text{rot}} + \varepsilon_{\text{vib}} + \varepsilon_{\text{elec}})/k_B T] \\ &= \sum \exp(-\varepsilon_{\text{trans}}/k_B T) \times \sum \exp(-\varepsilon_{\text{rot}}/k_B T) \\ &\quad \times \sum \exp(-\varepsilon_{\text{vib}}/k_B T) \times \sum \exp(-\varepsilon_{\text{elec}}/k_B T) \\ &= q_{\text{trans}} q_{\text{rot}} q_{\text{vib}} q_{\text{elec}} \end{aligned} \tag{20.36}$$

Thus, the molecular partition function is a *product* of the individual partition functions. Our next step is to derive expressions for q_{trans}, q_{rot}, q_{vib}, and q_{elec}.

Translational Partition Function

To calculate the translational energy of a molecule, we employ the particle-in-a-box model introduced in Chapter 10. For a one-dimensional system, Equation 10.50 gives the energy of the particle as

$$E_n = \frac{n^2 h^2}{8mL^2} \quad n = 1, 2, 3, \ldots$$

Therefore, the translational partition function can be expressed as

$$q_{\text{trans}} = \sum_{n=1}^{\infty} e^{-(n^2h^2/8mL^2k_BT)} \tag{20.37}$$

If L is of macroscopic dimension, then the translational energy levels are very close together (see Figure 2.14). Consequently, the above summation can be replaced by an integral,

$$q_{\text{trans}} = \int_0^{\infty} e^{-(n^2h^2/8mL^2k_BT)}\,dn \tag{20.38}$$

We make the simplifying substitution

$$x^2 = \frac{n^2h^2}{8mL^2k_BT}$$

or

$$x = \frac{nh}{(8mk_BT)^{1/2}L}$$

Differentiating x with respect to n, we obtain

$$\frac{(8mk_BT)^{1/2}L}{h}dx = dn$$

Equation 20.38 now becomes

$$q_{\text{trans}} = \frac{(8mk_BT)^{1/2}L}{h}\int_0^{\infty} e^{-x^2}dx \tag{20.39}$$

The definite integral in Equation 20.39 gives $\sqrt{\pi}/2$, so

$$q_{\text{trans}} = \frac{(2\pi mk_BT)^{1/2}L}{h} \tag{20.40}$$

In three dimensions, the translational partition function is

$$q^3_{\text{trans}} = \left[\frac{(2\pi mk_BT)^{1/2}L}{h}\right]^3 = \frac{(2\pi mk_BT)^{3/2}V}{h^3} \tag{20.41}$$

where the volume of the container $V = L^3$.

EXAMPLE 20.2

Calculate the translational partition function of a helium atom at 298 K in a container of volume 1.00 m^3.

ANSWER

We need Equation 20.41. The constants are

$$m = 4.003 \text{ amu} \times 1.661 \times 10^{-27} \text{ kg amu}^{-1} = 6.649 \times 10^{-27} \text{ kg}$$

$$k_B = 1.381 \times 10^{-23} \text{ J K}^{-1}$$

$$T = 298 \text{ K}$$

$$h = 6.626 \times 10^{-34} \text{ J s}$$

In three dimensions, the translational partition function is given by

$$q^3_{\text{trans}} = \frac{[2\pi(6.649 \times 10^{-27} \text{ kg})(1.381 \times 10^{-23} \text{ J K}^{-1})(298 \text{ K})]^{3/2}(1.00 \text{ m}^3)}{(6.626 \times 10^{-34} \text{ J s})^3}$$

$$= 7.75 \times 10^{30}$$

where we used the conversion factor: 1 J = 1 kg m^2 s^{-2}.

COMMENT

The very large partition function means that there are a very large number of thermally accessible translational energy states. This is because the translational energy levels are very closely spaced in a macroscopic container.

Rotational Partition Function

In Section 11.2, we derived an expression for the rotational energy of a diatomic molecule (Equation 11.43):

$$E_{\text{rot}} = \frac{J(J+1)h^2}{8\pi^2 I} \qquad J = 0, 1, 2, \ldots$$

As pointed out on p. 468, each rotational level has a degeneracy $(2J + 1)$, and so there are $(2J + 1)$ states at each level. The rotational partition function, including degeneracy, is given by

$$q_{\text{rot}} = \sum_{J=0}^{\infty}(2J+1)e^{-J(J+1)h^2/8\pi^2 I k_B T}$$

For the ground state, $J = 0$, so the first term of the summation reduces to 1. For the subsequent terms, with $J = 1, 2, \ldots$, we have

$$q_{\text{rot}} = 1 + 3e^{-2h^2/8\pi^2 Ik_{\text{B}}T} + 5e^{-6h^2/8\pi^2 Ik_{\text{B}}T} + \cdots \tag{20.42}$$

Unfortunately, this series does not sum to a closed analytical form. However, for molecules containing reasonably massive atoms (i.e., molecules with relatively large moments of inertia I) and at relatively high temperatures, the summation may be replaced by the following integral:*

$$q_{\text{rot}} = \int_0^\infty (2J + 1)e^{-J(J+1)h^2/8\pi^2 Ik_{\text{B}}T} dJ \tag{20.43}$$

The above integral can be evaluated by proper substitution and reference to standard integrals in the *Handbook of Chemistry and Physics*. The result is

$$q_{\text{rot}} = \frac{8\pi^2 Ik_{\text{B}}T}{h^2} \tag{20.44}$$

It turns out that symmetry consideration requires that Equation 20.44 be modified as follows:

$$q_{\text{rot}} = \frac{8\pi^2 Ik_{\text{B}}T}{\sigma h^2} \tag{20.45}$$

where σ is called the *symmetry number*, which is the number of indistinguishable orientations that can be produced by rotations less than or equal to 360° about any axis through the center of mass of the molecule. A homonuclear diatomic molecule, N≡N, say, has two indistinguishable orientations (rotations by 180° and 360°), so $\sigma = 2$. This reduces the number of different terms contributing to the partition function by the same factor. On the other hand, Br−Cl is distinguishable from Cl−Br (rotation by 360° only), so $\sigma = 1$.

Vibrational Partition Function

The harmonic oscillator treatment for a diatomic molecule gives the vibrational energy (see Equation 11.54) as

$$E_{\text{vib}} = (\upsilon + \tfrac{1}{2})h\nu \qquad \upsilon = 0, 1, 2, \ldots$$

and the zero-point energy is $\frac{1}{2}h\nu$. Because it is more convenient to set the lowest energy level to zero, we therefore write

$$E_{\text{vib}} = (\upsilon + \tfrac{1}{2})h\nu - \tfrac{1}{2}h\nu = \upsilon h\nu$$

* Exceptions are diatomic gases containing hydrogen, the lightest atom.

where E_{vib} is now measured from the $\upsilon = 0$ level. The vibrational partition function is given by

$$q_{\text{vib}} = \sum_{\upsilon=0}^{\infty} e^{-\upsilon h\nu/k_B T}$$
$$= 1 + e^{-h\nu/k_B T} + e^{-2h\nu/k_B T} + e^{-3h\nu/k_B T} + \cdots \quad (20.46)$$

Equation 20.46 is expressed as a geometric series (see Appendix A),

$$q_{\text{vib}} = 1 + x + x^2 + x^3 + \cdots$$

where $x = e^{-h\nu/k_B T}$. Unlike the translational and rotational cases, the vibrational energy levels have spaces between them that are normally large compared to $k_B T$ at room temperature; that is, $h\nu > k_B T$. We can apply the following formula for the geometric series:

$$1 + x + x^2 + x^3 + \cdots = \frac{1}{1 - x} \qquad |x| < 1$$

Now Equation 20.46 takes the form

$$q_{\text{vib}} = \frac{1}{1 - e^{-h\nu/k_B T}} \quad (20.47)$$

If there are several vibrational modes, as for a polyatomic molecule, we have to take the product of all the q_{vib} values, each with its own characteristic value of ν (see Problem 20.11). The vibrational partition function is usually close to unity at room temperature ($h\nu/k_B T > 1$), indicating that only the ground state is thermally accessible.

EXAMPLE 20.3

Evaluate q_{vib} for carbon monoxide at 300 K and 3000 K, given that the fundamental frequency of vibration for CO is $6.40 \times 10^{13}\ \text{s}^{-1}$.

ANSWER

First we calculate the quantity $h\nu/k_B T$ at 300 K:

$$\frac{h\nu}{k_B T} = \frac{(6.626 \times 10^{-34}\ \text{J s})(6.40 \times 10^{13}\ \text{s}^{-1})}{(1.381 \times 10^{-23}\ \text{J K}^{-1})(300\ \text{K})}$$
$$= 10.24$$

From Equation 20.47,

$$q_{\text{vib}} = \frac{1}{1 - e^{-10.24}}$$
$$= 1.00004$$

At 3000 K,

$$\frac{h\nu}{k_B T} = \frac{(6.626 \times 10^{-34}\ \text{J s})(6.40 \times 10^{13}\ \text{s}^{-1})}{(1.381 \times 10^{-23}\ \text{J K}^{-1})(3000\ \text{K})}$$
$$= 1.024$$

Here we have

$$q_{\text{vib}} = \frac{1}{1 - e^{-1.024}}$$
$$= 1.56$$

COMMENT

At 300 K, the only accessible vibrational state is the ground state because q_{vib} is very close to one. At the elevated temperature of 3000 K (assuming no dissociation of the molecule), a number of higher vibrational states become thermally accessible, and q_{vib} is appreciably greater than one.

Electronic Partition Function

The electronic partition function is given by

$$q_{\text{elec}} = \sum_i g_i e^{-\varepsilon_i/k_B T}$$
$$= g_0 e^{-\varepsilon_0/k_B T} + g_1 e^{-\varepsilon_1/k_B T} + g_2 e^{-\varepsilon_2/k_B T} + \cdots \quad (20.48)$$

As we saw earlier, the energy of the ground electronic level is taken as zero; that is, $\varepsilon_0 = 0$ so that

$$q_{\text{elec}} = g_0 + g_1 e^{-\varepsilon_1/k_B T} + g_2 e^{-\varepsilon_2/k_B T} + \cdots \quad (20.49)$$

Because the separations between the ground and excited electronic energy levels are typically very large, at room temperature, $\Delta\varepsilon \gg k_B T$ and all the terms after the first term in Equation 20.49 contribute negligibly to q_{elec} for temperatures up to about 5000 K. For most diatomic molecules, the ground electronic state is nondegenerate, so $g_0 = 1$. An important exception is O_2, which has a triplet ground state, so $g_0 = 3$.

The degeneracy of O_2 is given by $(2S + 1)$ or $(2 \times 1 + 1) = 3$.

20.4 Thermodynamic Quantities from Partition Functions

Having derived expressions for various partition functions, we are ready to calculate thermodynamic quantities of atomic and molecular systems. In this section, we focus on internal energy, heat capacity, and entropy. The next two sections will discuss chemical equilibrium and the transition-state theory of chemical kinetics.

Internal Energy and Heat Capacity

We consider two simple systems—a monatomic gas and a diatomic gas—and compare our results with those obtained in Section 2.9.

Monatomic Gas. Consider argon (Ar), a monatomic gas that has no rotational or vibrational motion. Therefore, we need consider only translational motion. Substituting the expression for q_{trans} (Equation 20.41) into Equation 20.34, we get, for 1 mole of the gas,

Electronic motion makes no contribution to heat capacity because argon has a singlet ($g_0 = 1$) ground state.

$$(\bar{U} - \bar{U}_0)_{\text{trans}} = RT^2\left[\partial\left(\frac{\ln(2\pi m k_B T)^{3/2}V}{h^3}\right)\Big/\partial T\right]_V$$

$$= RT^2\left(\frac{3}{2}\frac{1}{T}\right)$$

$$= \frac{3}{2}RT$$

The molar heat capacity, $\bar{C}_V$, is given by (see Equation 2.32)

$$\bar{C}_V = \left[\frac{\partial(\bar{U} - \bar{U}_0)}{\partial T}\right]_V = \frac{3}{2}R$$

This is the same result as that obtained from the equipartition of energy theorem.

Diatomic Gas. For a diatomic gas, such as nitrogen (N_2), we must examine the contributions to heat capacity by translational motion, rotational motion, and vibrational motion. The translational contribution is the same as that for an atomic system, that is,

$$(\bar{C}_V)_{\text{trans}} = \frac{3}{2}R$$

For rotation, we substitute Equation 20.45 into Equation 20.34:

$$(\bar{U} - \bar{U}_0)_{\text{rot}} = RT^2\left[\frac{\partial \ln(8\pi^2 I k_B T/\sigma h^2)}{\partial T}\right]_V$$

$$= RT^2\left(\frac{1}{T}\right)$$

$$= RT$$

Hence,

$$(\bar{C}_V)_{\text{rot}} = \left[\frac{\partial(\bar{U} - \bar{U}_0)}{\partial T}\right]_V = R$$

To evaluate the vibrational contribution to $\bar{C}_V$, we start with Equation 20.47,

$$q_{\text{vib}} = \frac{1}{1 - e^{-h\nu/k_B T}}$$

and consider two limiting cases. At low temperatures, $h\nu/k_BT \gg 1$. In the limit $T \to 0$, $e^{-h\nu/kT} \to 0$ so $q_{vib} = 1$. It follows, therefore, that $\overline{U} - \overline{U}_0 = 0$ (because q_{vib} is a constant) and $(\overline{C}_V)_{vib} = 0$. As expected, there is no contribution to heat capacity by molecular vibration at and below room temperature. In the other extreme, when T is large so that $h\nu/k_BT \ll 1$, we can expand the $e^{-h\nu/k_BT}$ term as follows (see Appendix A):

$$e^{-h\nu/k_BT} = 1 - \frac{h\nu}{k_BT} - \frac{1}{2}\left(\frac{h\nu}{k_BT}\right)^2 - \cdots$$

Ignoring $(h\nu/k_BT)^2$ and higher terms, we get

$$q_{vib} = \frac{1}{1 - (1 - h\nu/k_BT)} = \frac{k_BT}{h\nu}$$

Now

$$\begin{aligned}(\overline{U} - \overline{U}_0)_{vib} &= RT^2\left[\frac{\partial \ln(k_BT/h\nu)}{\partial T}\right]_V \\ &= RT\end{aligned}$$

and

$$(\overline{C}_V)_{vib} = \left[\frac{\partial(\overline{U} - \overline{U}_0)}{\partial T}\right]_V = R$$

which is in accord with the equipartition of energy theorem.

Entropy

To express the entropy of a system in terms of partition functions, we start with Equation 20.15 and recall that $\beta = -1/k_BT$, so

$$\begin{aligned}S &= k_B(N \ln N - \alpha N - \beta E) \\ &= k_B\left(N \ln N - \alpha N + \frac{E}{k_BT}\right)\end{aligned} \tag{20.50}$$

We need to find an expression for α. Taking the natural logarithm of Equation 20.12, we get

$$\alpha = \ln n_0 \tag{20.51}$$

From Equation 20.19,

$$n_0 = \frac{N}{\sum_i e^{-\varepsilon_i/k_BT}} = \frac{N}{q} \tag{20.52}$$

Substituting Equation 20.52 into 20.51 gives

$$\alpha = \ln N - \ln q \tag{20.53}$$

Using this expression for α in Equation 20.50, we write

$$S = k_B\left(N \ln N - N \ln N + N \ln q + \frac{E}{k_B T}\right)$$

$$= k_B \ln q^N + \frac{E}{T}$$

$$= k_B \ln Q + \frac{E}{T} \qquad (20.54)$$

where Q, the *canonical partition function*, is given by

Canonical means "according to a rule, or canon."

$$Q = q^N \qquad (20.55)$$

Thus, the canonical partition function of N particles is the product of the individual molecular partition functions. We can now use Equation 20.54 to calculate the molar entropies of monatomic and diatomic gases.

Monatomic Gas. As before, we need consider only the translational partition function. However, Equation 20.54 applies to N independent and *distinguishable* particles, as in a solid. In a gas, the molecules or atoms composing it are not localized and are therefore *indistinguishable*. For this reason, Equation 20.55 must be modified as follows (see Appendix 20.1 on p. 903 for justification):

$$Q = \frac{q^N}{N!} \qquad (20.56)$$

If we express Stirling's approximation as $\ln N! = N \ln N - N \ln e$, then $N! = (N/e)^N$ and Equation 20.56 becomes

$$Q = \left(\frac{qe}{N}\right)^N \qquad (20.57)$$

Using q^3_{trans} from Equation 20.41, we write

$$Q_{\text{trans}} = \left[\frac{(2\pi m k_B T)^{3/2} Ve}{Nh^3}\right]^N \qquad (20.58)$$

Substitution of Equation 20.58 into Equation 20.54 yields

$$S_{\text{trans}} = k_B N \ln\left[\frac{(2\pi m k_B T)^{3/2} Ve}{Nh^3}\right] + \frac{E_{\text{trans}}}{T} \qquad (20.59)$$

Equation 20.59 can be rearranged into a neater form by carrying out the following steps. First, for 1 mole of the gas, $N = N_A$ and $k_B N_A = R$. Second, as we saw in our discussion of heat capacity, $\bar{E}_{\text{trans}} = (\bar{U} - \bar{U}_0)_{\text{trans}} = (\frac{3}{2})RT$. Thus, $\bar{E}_{\text{trans}}/T = (\frac{3}{2})R$, which can be written as $R \ln e^{3/2}$. Third, assuming ideal gas behavior,

$$PV = nRT = RT = k_B N_A T \qquad (n = 1)$$

so

$$\frac{V}{N_A} = \frac{k_B T}{P} \tag{20.60}$$

Incorporating all these changes, we now write Equation 20.59 in molar form as

$$\overline{S}_{\text{trans}} = R \ln\left[\frac{(2\pi m k_B T)^{3/2}}{h^3} \frac{k_B T}{P} e^{5/2}\right] \tag{20.61}$$

Equation 20.61 is called the Sackur–Tetrode equation [after the German physical chemist Otto Sackur (1880–1914) and the Dutch physicist Hugo Martin Tetrode (1895–1931)]. As an illustration, let us calculate the molar entropy of argon at 1 bar and 298 K. The quantities we need are:

$$m = 39.95 \text{ amu} \times 1.661 \times 10^{-27} \text{ kg amu}^{-1} = 6.636 \times 10^{-26} \text{ kg}$$

$$k_B = 1.381 \times 10^{-23} \text{ J K}^{-1}$$

$$T = 298 \text{ K}$$

$$h = 6.626 \times 10^{-34} \text{ J s}$$

$$P = P^\circ = 1 \text{ bar} = 10^5 \text{ N m}^{-2}$$

For convenience, we calculate the separate terms in Equation 20.61 first.

$$\frac{(2\pi m k_B T)^{3/2}}{h^3} = \frac{[2\pi(6.636 \times 10^{-26} \text{ kg})(1.381 \times 10^{-23} \text{ J K}^{-1})(298 \text{ K})]^{3/2}}{(6.626 \times 10^{-34} \text{ J s})^3}$$

$$= 2.44 \times 10^{32} \text{ m}^{-3}$$

$$\frac{k_B T}{P} = \frac{(1.381 \times 10^{-23} \text{ J K}^{-1})(298 \text{ K})}{10^5 \text{ N m}^{-2}}$$

$$= 4.11 \times 10^{-26} \text{ m}^3$$

Finally, the molar entropy is given by

$$\overline{S}_{\text{trans}} = (8.314 \text{ J K}^{-1} \text{ mol}^{-1}) \ln[(2.44 \times 10^{32} \text{ m}^{-3})(4.11 \times 10^{-26} \text{ m}^3)\, e^{5/2}]$$

$$= 154.8 \text{ J K}^{-1} \text{ mol}^{-1}$$

which is in excellent agreement with the third-law entropy (154.8 J K^{-1} mol^{-1}).

Similarly, we find the molar entropy of neon at 1 bar and 298 K to be 146.3 J K^{-1} mol^{-1}. The difference in the entropy values for neon and argon can be explained as follows. According to the particle-in-a-box model, the spacing between energy levels is inversely proportional to the mass (see Equation 10.50). Being a heavier gas, argon has more closely spaced translational energy levels and consequently a larger q_{trans} value compared with neon at the same temperature.

Diatomic Gas. We choose nitrogen as our example. There are three contributions to nitrogen's entropy: translational, rotational, and vibrational. Using the Sackur–Tetrode equation for N_2, we find $\overline{S}_{trans} = 150.4$ J K^{-1} mol^{-1}.

For rotational contribution to entropy, we start with Equation 20.54:

$$S_{rot} = k_B \ln Q_{rot} + \frac{E_{rot}}{T}$$

Note that we do not need the correction factor $N!$ for Q_{rot} here as we did for translation because we are dealing with the internal motion of the molecule. A diatomic molecule has two rotational degrees of freedom, so for 1 mole of the gas,

$$\overline{S}_{rot} = R \ln q_{rot} + R \quad (Q_{rot} = q_{rot}^{N_A} \text{ and } \overline{E}_{rot} = RT)$$

From Equation 20.45,

$$q_{rot} = \frac{8\pi^2 I k_B T}{\sigma h^2}$$

Being linear, a diatomic molecule has only one moment of inertia I.

The bond length of N_2 is 1.09 Å, or 1.09×10^{-10} m. The mass of an N atom is 14.01 amu $\times$ 1.661×10^{-27} kg amu^{-1}, or 2.327×10^{-26} kg. From Equation 11.23, we find the reduced mass of N_2 to be

$$\mu = \frac{m_1 m_2}{m_1 + m_2} = \frac{m}{2} = \frac{2.327 \times 10^{-26}\text{ kg}}{2}$$
$$= 1.164 \times 10^{-26}\text{ kg}$$

and its moment of inertia to be

$$I = \mu r^2$$
$$= (1.164 \times 10^{-26}\text{ kg})(1.09 \times 10^{-10}\text{ m})^2$$
$$= 1.38 \times 10^{-46}\text{ kg m}^2$$

Therefore,

$$q_{rot} = \frac{8\pi^2(1.38 \times 10^{-46}\text{ kg m}^2)(1.381 \times 10^{-23}\text{ J K}^{-1})(298\text{ K})}{(2)(6.626 \times 10^{-34}\text{ J s})^2}$$
$$= 51.1$$

Note that we set $\sigma = 2$ because N_2 is a homonuclear diatomic molecule. Now we have

$$\overline{S}_{rot} = R \ln 51.1 + R$$
$$= 41.0\text{ J K}^{-1}\text{ mol}^{-1}$$

Finally, we come to S_{vib}. Again we need Equation 20.54. First, from Equation 20.31 we find that

$$E = Nk_BT^2\left(\frac{\partial \ln q}{\partial T}\right)_V$$

For 1 mole of N_2, Equation 20.54 becomes

$$\overline{S}_{\text{vib}} = R \ln q_{\text{vib}} + RT\left(\frac{\partial \ln q_{\text{vib}}}{\partial T}\right)_V \qquad \left(Q_{\text{vib}} = q_{\text{vib}}^{N_A}\right) \tag{20.62}$$

From Equation 20.47,

$$q_{\text{vib}} = \frac{1}{1 - e^{-h\nu/k_BT}}$$

so

$$\ln q_{\text{vib}} = -\ln(1 - e^{-h\nu/k_BT}) \tag{20.63}$$

and

$$\left(\frac{\partial \ln q_{\text{vib}}}{\partial T}\right)_V = \frac{h\nu}{k_BT^2}\frac{e^{-h\nu/k_BT}}{1 - e^{-h\nu/k_BT}} = \frac{h\nu}{k_BT^2}\frac{1}{e^{h\nu/k_BT} - 1} \tag{20.64}$$

Substituting Equations 20.63 and 20.64 into 20.62, we obtain

$$\overline{S}_{\text{vib}} = -R\ln(1 - e^{-h\nu/k_BT}) + R\frac{h\nu}{k_BT}\frac{1}{e^{h\nu/k_BT} - 1} \tag{20.65}$$

The wavenumber for the vibration of N_2 is 2360 cm^{-1}. Therefore, the frequency of vibration is

$$\begin{aligned}\nu = c\tilde{\nu} &= (3.00 \times 10^{10}\ \text{cm s}^{-1})(2360\ \text{cm}^{-1})\\ &= 7.08 \times 10^{13}\ \text{s}^{-1}\end{aligned}$$

Thus,

$$\frac{h\nu}{k_BT} = \frac{(6.626 \times 10^{-34}\ \text{J s})(7.08 \times 10^{13}\ \text{s}^{-1})}{(1.381 \times 10^{-23}\ \text{J K}^{-1})(298\ \text{K})} = 11.4$$

Using the value 11.4 for $h\nu/k_BT$ in Equation 20.65, we find that $(1 - e^{-h\nu/k_BT}) \approx 1$, so the first term on the right of Equation 20.65 is zero and the second term is approximately 1×10^{-3} J K^{-1} mol^{-1}. Thus, vibrational motion makes only a minute contribution to entropy at 298 K.

The total molar entropy of N_2 is given by the sum of all contributions, that is,

$$\begin{aligned}\overline{S} &= \overline{S}_{\text{trans}} + \overline{S}_{\text{rot}} + \overline{S}_{\text{vib}}\\ &= 150.4\ \text{J K}^{-1}\ \text{mol}^{-1} + 41.0\ \text{J K}^{-1}\ \text{mol}^{-1} + 1 \times 10^{-3}\ \text{J K}^{-1}\ \text{mol}^{-1}\\ &= 191.4\ \text{J K}^{-1}\ \text{mol}^{-1}\end{aligned}$$

which compares well with the third-law entropy (191.6 J K^{-1} mol^{-1}). The small discrepancy is due to rounding-off errors rather than to any inadequacy in the theory.

20.5 Chemical Equilibrium

In this section, we shall derive an expression for the thermodynamic equilibrium constant in terms of the partition functions and illustrate its use with a simple example.

Consider the equilibrium between A and B, whose energy levels are shown in Figure 20.1,

$$A \rightleftharpoons B$$

The number of molecules in the ith state of A or B is given by (see Equation 20.18)

$$n_i = n_0 e^{-\varepsilon_i/k_B T}$$

If only the lowest levels are accessible, then the equilibrium constant is given by

$$K = \frac{n_B}{n_A} = e^{-\Delta\varepsilon_0/k_B T} \tag{20.66}$$

where $\Delta\varepsilon_0$ is the energy difference between the lowest levels, as shown in Figure 20.1. If other (higher) energy levels are available, as is usually the case, the equilibrium constant should be expressed in terms of the partition functions:

$$K = \frac{q_B}{q_A} e^{-\Delta\varepsilon_0/k_B T} \tag{20.67}$$

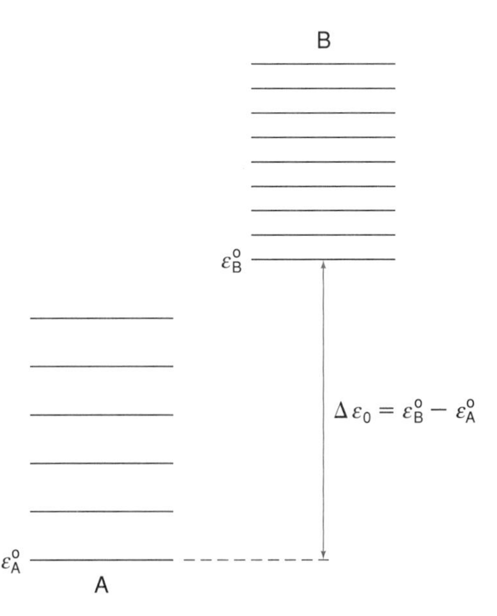

Figure 20.1
Energy levels of two species, A and B, at chemical equilibrium. The quantity $\Delta\varepsilon_0$ is the difference between the lowest energies of A and B.

An insight regarding factors affecting the equilibrium constant can be gained from Figure 20.1. As shown, the reaction A $\rightarrow$ B is endothermic. The energy levels are more closely spaced in B, however. From the equations, $\Delta_r G^\circ = -RT \ln K$ and $\Delta_r G^\circ = \Delta_r H^\circ - T\Delta_r S^\circ$, we can show that

$$K = e^{-\Delta_r H^\circ/RT} e^{\Delta_r S^\circ/R}$$

Although a positive $\Delta_r H^\circ$ (endothermic) value lowers the equilibrium constant, a large positive $\Delta_r S^\circ$ value due to the closely spaced energy levels favors B at equilibrium.

Let us now consider the following gaseous equilibrium system:

$$a\text{A}(g) \rightleftharpoons b\text{B}(g)$$

From Equations 8.7 and 8.8,

$$\Delta_r G^\circ = -RT \ln K_P$$

$$K_P = \frac{(P_\text{B}/1\ \text{bar})^b}{(P_\text{A}/1\ \text{bar})^a}$$

where

$$\Delta_r G^\circ = bG^\circ_\text{B} - aG^\circ_\text{A}$$

Our next goal is to derive an expression for $\Delta_r G^\circ$ (and hence K_P) in terms of partition functions. We start with Equation 20.54,

$$S = k_\text{B} \ln Q + \frac{E}{T}$$

$$= k_\text{B} \ln Q + \frac{U - U_0}{T} \tag{20.68}$$

From the definition of the Helmholtz energy and Equation 20.68,

$$A = U - TS$$

$$= U_0 - k_\text{B} T \ln Q$$

or

$$A - U_0 = -k_\text{B} T \ln Q \tag{20.69}$$

Equation 20.69 shows that at absolute zero, $A_0 = U_0$. From the definitions of Gibbs energy ($G = H - TS$) and enthalpy ($H = U + PV$) and Equation 20.69, we can readily show that, assuming ideal gas behavior,

$$G = A + PV$$

$$= A + Nk_\text{B}T \quad (nN_\text{A} = N \text{ and } Nk_\text{B} = nR)$$

$$= U_0 - k_\text{B} T \ln Q + Nk_\text{B}T \tag{20.70}$$

For indistinguishable molecules in the gas phase, Q is given by Equation 20.56:

$$Q = \frac{q^N}{N!}$$

Applying Stirling's approximation, we write Equation 20.70 as

$$\begin{aligned} G &= U_0 - k_BT[N \ln q - \ln N!] + Nk_BT \\ &= U_0 - k_BT[N \ln q - N \ln N + N] + Nk_BT \\ &= U_0 - Nk_BT \ln \frac{q}{N} \end{aligned} \tag{20.71}$$

For molar quantities ($N = N_A$) and under standard-state conditions,

$$\overline{G}^\circ = \overline{U}_0^\circ - RT \ln \frac{q}{N_A} \tag{20.72}$$

Note that in Equation 20.72 N_A is just 6.022×10^{23} (no units). The change in standard Gibbs energy can now be written as

Here N_A is used to represent both Avogadro's number and Avogadro's constant.

$$\begin{aligned} \Delta_r G^\circ &= b\overline{G}_B^\circ - a\overline{G}_A^\circ \\ &= b\overline{U}_{0,B}^\circ - a\overline{U}_{0,A}^\circ - RT \ln \frac{(q_B/N_A)^b}{(q_A/N_A)^a} \\ &= \Delta U_0^\circ - RT \ln \frac{(q_B/N_A)^b}{(q_A/N_A)^a} \\ &= -RT \ln\left(\frac{q_B^b}{q_A^a} N_A^{-\Delta n} e^{-\Delta U_0^\circ/RT}\right) \end{aligned} \tag{20.73}$$

where $\Delta U_0^\circ = b\overline{U}_{0,B}^\circ - a\overline{U}_{0,A}^\circ$ and $\Delta n = b - a$. Comparing Equation 20.73 with Equation 8.7 gives

$$K_P = \frac{q_B^b}{q_A^a} N_A^{-\Delta n} e^{-\Delta U_0^\circ/RT} \tag{20.74}$$

We now apply Equation 20.74 to the following equilibrium system at 1000 K:

$$Na_2(g) \rightleftharpoons 2Na(g)$$

To calculate the equilibrium constant, we first break Equation 20.74 into three parts for convenience.

Part I: q_B^b/q_A^a

The partition function of atomic sodium is a product of translational and electronic contributions:

$$q_{Na} = q_{trans}^3 q_{elec}$$

Assuming ideal gas behavior, the translational partition function is given by

$$q^3_{\text{trans}} = \frac{(2\pi m k_B T)^{3/2} V}{h^3}$$

$$= \frac{(2\pi m k_B T)^{3/2}(RT/P)}{h^3} \qquad (V = nRT/P = RT/P \text{ for 1 mole})$$

Using the following values,

$$m = 22.99 \text{ amu} \times 1.661 \times 10^{-27} \text{ kg amu}^{-1}$$

$$= 3.819 \times 10^{-26} \text{ kg}$$

$$k_B = 1.381 \times 10^{-23} \text{ J K}^{-1}$$

$$T = 1000 \text{ K}$$

$$R = 8.314 \text{ J K}^{-1} \text{ mol}^{-1}$$

$$h = 6.626 \times 10^{-34} \text{ J s}$$

$$P = P^\circ = 1 \text{ bar} = 10^5 \text{ N m}^{-2}$$

we find

$$q^3_{\text{trans}} = 5.452 \times 10^{31}$$

The sodium atom has one unpaired electron in the 3s orbital; therefore, it is a doublet ($S = \frac{1}{2}$), and so $q_{\text{elec}} = g_0 = (2S + 1) = 2$. The partition function of atomic sodium is given by

$$q_{\text{Na}} = (5.452 \times 10^{31}) \times 2$$

$$= 1.090 \times 10^{32}$$

The molecular partition of Na_2 is

$$q_{\text{Na}_2} = q^3_{\text{trans}} q_{\text{rot}} q_{\text{vib}} q_{\text{elec}}$$

To calculate the translational partition function, we use the constants above for Na and set $m = m_{\text{Na}_2} = 2m_{\text{Na}} = 2 \times 3.819 \times 10^{-26}$ kg $= 7.638 \times 10^{-26}$ kg, so that

$$q^3_{\text{trans}} = 1.542 \times 10^{32}$$

According to Equation 20.45, the rotational partition function is given by

$$q_{\text{rot}} = \frac{8\pi^2 I k_B T}{\sigma h^2}$$

The reduced mass of Na_2 is $(m_{Na}/2)$, which is 1.91×10^{-26} kg, and the bond distance is 3.078 Å, or 3.078×10^{-10} m, so the moment of inertia is 1.81×10^{-45} kg m^2 (see Equation 11.23). Therefore,

$$q_{rot} = \frac{8\pi^2(1.81 \times 10^{-45}\ \text{kg m}^2)(1.381 \times 10^{-23}\ \text{J K}^{-1})(1000\ \text{K})}{(2)(6.626 \times 10^{-34}\ \text{J s})^2} \times \frac{\text{J}}{\text{kg m}^2\ \text{s}^{-2}}$$
$$= 2248$$

Note that we set $\sigma = 2$ because Na_2 is a homonuclear diatomic molecule.

For the vibrational partition function, we note that the vibrational wavenumber of Na_2 is 159.1 cm^{-1}. First we calculate the quantity

$$\frac{h\nu}{k_B T} = \frac{hc\tilde{\nu}}{k_B T} = \frac{(6.626 \times 10^{-34}\ \text{J s})(3.00 \times 10^{10}\ \text{cm s}^{-1})(159.1\ \text{cm}^{-1})}{(1.381 \times 10^{-23}\ \text{J K}^{-1})(1000\ \text{K})}$$
$$= 0.229$$

From Equation 20.47,

$$q_{vib} = \frac{1}{1 - e^{-h\nu/k_B T}}$$
$$= \frac{1}{1 - e^{-0.229}}$$
$$= 4.89$$

The ground state of Na_2 is a singlet ($S = 0$), so $q_{elec} = g_0 = (2S + 1) = 1$.

Finally, we write

$$\frac{q_{Na}^2}{q_{Na_2}} = \frac{(1.090 \times 10^{32})^2}{(1.542 \times 10^{32})(2248)(4.89)(1)} = 7.01 \times 10^{27}$$

Part II: $N_A^{-\Delta n}$

Because

$$\Delta n = 2 - 1 = 1$$
$$N_A^{-\Delta n} = (6.022 \times 10^{23})^{-1}$$
$$= 1.661 \times 10^{-24}$$

Part III: $e^{-\Delta U_0^\circ/RT}$

The quantity ΔU_0°, which is the difference between the lowest energies of two Na atoms and the Na_2 molecule, is just the dissociation energy of Na_2 (Figure 20.2). Experimentally, this quantity is found to be 70.4 kJ mol^{-1}. Therefore,

$$e^{-\Delta U_0^\circ/RT} = \exp\left[-\frac{70.4 \times 1000\ \text{J mol}^{-1}}{(8.314\ \text{J K}^{-1}\ \text{mol}^{-1})(1000\ \text{K})}\right]$$
$$= 2.10 \times 10^{-4}$$

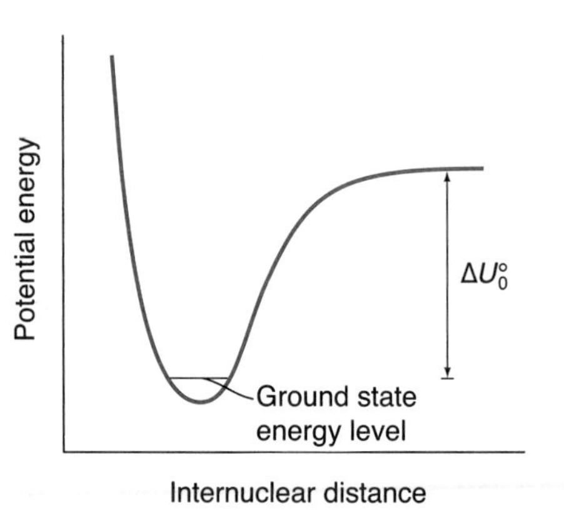

Figure 20.2
ΔU_0° is equal to the bond dissociation energy of Na_2.

Now we combine Parts I, II, and III to calculate the value of K_P at 1000 K using Equation 20.74:

$$K_P = (7.01 \times 10^{27})(1.661 \times 10^{-24})(2.10 \times 10^{-4})$$
$$= 2.45$$

which agrees well with the experimentally measured equilibrium constant. Note that K_P is a dimensionless quantity.

20.6 Transition-State Theory

In this final section, we shall apply the result from the previous section to derive an expression for the rate constant k. We start with the reaction first introduced in Section 15.7:

$$A + B \rightleftharpoons X^{\ddagger} \xrightarrow{k} C + D$$

We can write the equilibrium constant, which is the ratio of the concentration of the activated complex to that of A and B, according to Equation 20.67,

$$K^{\ddagger} = \frac{[X^{\ddagger}]}{[A][B]} = \frac{q^{\ddagger}}{q_A q_B} e^{-\Delta E_0/RT} \tag{20.75}$$

where ΔE_0 is the difference between the molar zero-point energy of the activated complex and that of the reactants. Upon rearrangement, Equation 20.75 becomes

$$[X^{\ddagger}] = [A][B]\frac{q^{\ddagger}}{q_A q_B} e^{-\Delta E_0/RT} \tag{20.76}$$

Note that $q^{\ddagger}$ is the partition function of the activated complex. If molecule A contains N_A atoms and molecule B contains N_B atoms, the activated complex must contain $(N_A + N_B)$ atoms. Assuming nonlinear geometry, the activated complex has three degrees of translational freedom, three degrees of rotational freedom, and

$[3(N_A + N_B) - 6]$ degrees of vibrational freedom. One of these vibrational modes has a different character because it corresponds to the breakup of the activated complex to form products. It is therefore a loose vibration such that $h\nu/k_BT \ll 1$. Following the procedure on p. 888, we can show that the partition function for this vibrational mode is given by $k_BT/h\nu$. The partition function for the activated complex can be written as

$$q^{\ddagger} = q_{\ddagger}\frac{k_BT}{h\nu} \tag{20.77}$$

where $q_{\ddagger}$ is the remaining product of the vibrational partition functions. Equation 20.76 now becomes

$$[X^{\ddagger}] = [A][B]\frac{k_BT}{h\nu}\frac{q_{\ddagger}}{q_Aq_B}e^{-\Delta E_0/RT} \tag{20.78}$$

The rate of the reaction is the change in $[X^{\ddagger}]$ with time, or $\nu[X^{\ddagger}]$ (see p. 698), so we can write

$$\begin{aligned}\text{rate} &= \nu[X^{\ddagger}]\\ &= [A][B]\frac{k_BT}{h}\frac{q_{\ddagger}}{q_Aq_B}e^{-\Delta E_0/RT}\end{aligned}$$

and the rate constant, defined by rate $= k[A][B]$, is given by

$$k = \frac{k_BT}{h}\frac{q_{\ddagger}}{q_Aq_B}e^{-\Delta E_0/RT} \tag{20.79}$$

Note that the factor

$$\frac{q_{\ddagger}}{q_Aq_B}e^{-\Delta E_0/RT}$$

resembles the one in Equation 20.75 for the equilibrium constant between the activated complex and the reactants. The difference is that in $q_{\ddagger}$ the contribution to the breakup of the activated complex has been omitted. The equilibrium constant in Equation 20.75 involves the complete partition functions of the activated complex. We can define a modified equilibrium constant (i.e., using $q_{\ddagger}$ rather than $q^{\ddagger}$) as

$$K^{\ddagger} = \frac{q_{\ddagger}}{q_Aq_B}e^{-\Delta E_0/RT} \tag{20.80}$$

so that Equation 20.79 becomes

$$k = \frac{k_BT}{h}K^{\ddagger} \tag{20.81}$$

From this equation, we can arrive at the thermodynamic formulation of transition-state theory as outlined on p. 699.

Comparison Between Collision Theory and Transition-State Theory

We mentioned in Chapter 15 that collision theory predicts reaction rates fairly well if the reactants are atoms, but significant deviations are observed if molecules are involved. Application of partition functions, in terms of transition-state theory, helps us understand this discrepancy. Two cases are considered below.

Case 1. Reactions Between Atoms. Consider the reaction between atoms A and B:

$$\mathrm{A} + \mathrm{B} \rightleftharpoons \mathrm{X}^{\ddagger} \rightarrow \text{product}$$

Each atom has three degrees of freedom. The activated complex has three translational degrees of freedom and two rotational degrees of freedom. A normal diatomic molecule would have one vibrational degree of freedom, but, because it is an activated complex, this mode of vibration corresponds to decomposition, and the corresponding partition function is therefore omitted. Thus, the partition function of the activated complex is

$$q_{\ddagger} = q_{\text{trans}}^3 q_{\text{rot}} = \frac{[2\pi(m_{\text{A}} + m_{\text{B}})k_{\text{B}}T]^{3/2}}{h^3}\left(\frac{8\pi^2 I k_{\text{B}}T}{h^2}\right)$$

where m_{A} and m_{B} are the masses of atoms A and B, and I is the moment of inertia of the activated complex. The moment of inertia is given by μd_{AB}^2, where μ is the reduced mass $[m_{\text{A}}m_B/(m_{\text{A}} + m_{\text{B}})]$, and d_{AB} is the bond distance. The translational partition functions of A and B are

All translational partition functions are given in per unit volume (m^{-3}).

$$q_{\text{A}} = \frac{(2\pi m_{\text{A}} k_{\text{B}}T)^{3/2}}{h^3} \quad \text{and} \quad q_{\text{B}} = \frac{(2\pi m_{\text{B}} k_{\text{B}}T)^{3/2}}{h^3}$$

According to Equation 20.79 the rate constant is given by

$$k_{\text{atoms}} = \frac{k_{\text{B}}T}{h}\frac{q_{\ddagger}}{q_{\text{A}}q_{\text{B}}}e^{-\Delta E_0/RT} = \frac{k_{\text{B}}T}{h}\left\{\frac{\dfrac{[2\pi(m_{\text{A}} + m_{\text{B}})k_{\text{B}}T]^{3/2}}{h^3}\left(\dfrac{8\pi^2\mu d_{\text{AB}}^2 k_{\text{B}}T}{h^2}\right)}{\dfrac{(2\pi m_{\text{A}}k_{\text{B}}T)^{3/2}}{h^3}\dfrac{(2\pi m_{\text{B}}k_{\text{B}}T)^{3/2}}{h^3}}\right\}e^{-\Delta E_0/RT}$$

where k_{atoms} is the rate constant for atoms. After cancellations and rearrangements of terms, we can show that

$$k_{\text{atoms}} = d_{\text{AB}}^2\sqrt{\frac{8\pi k_{\text{B}}T}{\mu}}\,e^{-\Delta E_0/RT} \tag{20.82}$$

Equation 20.82 is the same as Equation 15.43 (where $\Delta E_0 = E_{\text{a}}$), which was derived by collision theory. Therefore, both theories predict the same result for atoms.

Case 2. Reactions Between Molecules. We now consider the more complex situation, in which the reactants A and B are nonlinear molecules that contain N_A and N_B atoms, respectively:

$$\text{A} + \text{B} \rightleftharpoons \text{X}^{\ddagger} \rightarrow \text{product}$$

Each molecule has three translational degrees of freedom, three rotational degrees of freedom, and $(3N_A - 6)$ or $(3N_B - 6)$ vibrational degrees of freedom. The activated complex contains $(N_A + N_B)$ atoms. A normal molecule would have $[3(N_A + N_B) - 6]$ vibrational degrees of freedom, but one of these vibrational modes corresponds to the decomposition of the complex, so only $[3(N_A + N_B) - 7]$ vibrational degrees of freedom will be included in the partition function:

$$q_{\ddagger} = q_{\text{trans}}^3 q_{\text{rot}}^3 q_{\text{vib}}^{3(N_A+N_B)-7}$$

A nonlinear polyatomic molecule has three moments of inertia.

From Equation 20.79, we write the rate constant as

$$k_{\text{molecules}} = \frac{k_B T}{h} \frac{q_{\text{trans}}^3 q_{\text{rot}}^3 q_{\text{vib}}^{3(N_A+N_B)-7}}{q_{\text{trans}}^3 q_{\text{rot}}^3 q_{\text{vib}}^{3N_A-6} \, q_{\text{trans}}^3 q_{\text{rot}}^3 q_{\text{vib}}^{3N_B-6}} e^{-\Delta E_0/RT} \tag{20.83}$$

where $k_{\text{molecules}}$ is the rate constant for molecules. To avoid complicated calculations, we assume that all translational, rotational, and vibrational degrees of freedom for reactants and activated complex are equal, so Equation 20.83 reduces to

$$k_{\text{molecules}} \cong \frac{k_B T}{h} \frac{q_{\text{vib}}^5}{q_{\text{trans}}^3 q_{\text{rot}}^3} e^{-\Delta E_0/RT} \tag{20.84}$$

We can obtain a similar expression for Case 1, in which the reactants are atoms. From the partition functions,

$$q_A = q_{\text{trans}}^3 \qquad q_B = q_{\text{trans}}^3 \qquad q_{\ddagger} = q_{\text{trans}}^3 q_{\text{rot}}$$

and from Equation 20.79, the rate constant is given by

$$k_{\text{atoms}} \cong \frac{k_B T}{h} \frac{q_{\text{trans}}^3 q_{\text{rot}}}{q_{\text{trans}}^3 q_{\text{trans}}^3} e^{-\Delta E_0/RT} \tag{20.85}$$

Again, assuming that the translational degrees of freedom of the activated complex and atoms are equal, we write

$$k_{\text{atoms}} \cong \frac{k_B T}{h} \frac{q_{\text{rot}}}{q_{\text{trans}}^3} e^{-\Delta E_0/RT} \tag{20.86}$$

To compare the rate constants in Cases 1 and 2, we divide Equation 20.84 by Equation 20.86,

We also assume that ΔE_0 is of comparable magnitude for both atoms and molecules.

$$\frac{k_{\text{molecules}}}{k_{\text{atoms}}} \cong \frac{q_{\text{vib}}^5/q_{\text{trans}}^3 q_{\text{rot}}^3}{q_{\text{rot}}/q_{\text{trans}}^3} = \frac{q_{\text{vib}}^5}{q_{\text{rot}}^4} \tag{20.87}$$

As an estimate near 300 K, we see that q_{vib} is usually close to unity, whereas q_{rot} ranges between 10 and 100. Thus, Equation 20.87 predicts that the rate constant for molecules can be smaller by as much as 10^{-4} to 10^{-8} than the rate constant for atoms. This difference is in accord with experimentally observed values. On the other hand, because collision theory treats both atoms and molecules as hard spheres, it erroneously predicts the rate constants for molecules to be comparable to those for atoms (see p. 697). We see that statistical thermodynamics accounts for the difference between molecular and atomic rate constants without resorting to the use of a steric factor.

Key Equations

$$\ln x! = x \ln x - x$$ (Stirling's approximation) (20.6)

$$\frac{n_2}{n_1} = \frac{g_2}{g_1} e^{-\Delta\varepsilon/k_B T}$$ (Boltzmann distribution law) (20.23)

$$q = \sum_i g_i e^{-\varepsilon_i/k_B T}$$ (Partition function) (20.25)

$$\bar{U} - \bar{U}_0 = RT^2\left(\frac{\partial \ln q}{\partial T}\right)_V$$ (Molar internal energy) (20.34)

$$q_{trans}^3 = \frac{(2\pi m k_B T)^{3/2} V}{h^3}$$ (Translational partition function in three dimensions) (20.41)

$$q_{rot} = \frac{8\pi^2 I k_B T}{\sigma h^2}$$ (Rotational partition function) (20.45)

$$q_{vib} = \frac{1}{1 - e^{-h\nu/k_B T}}$$ (Vibrational partition function) (20.47)

$$q_{elec} = g_0 + g_1 e^{-\varepsilon_1/k_B T} + g_2 e^{-\varepsilon_2/k_B T} + \cdots$$ (Electronic partition function) (20.49)

$$Q = q^N$$ (Canonical partition function; distinguishable particles) (20.55)

$$Q = \frac{q^N}{N!}$$ (Canonical partition function; indistinguishable particles) (20.56)

$$\bar{S}_{trans} = R \ln\left[\frac{(2\pi m k_B T)^{3/2}}{h^3} \frac{k_B T}{P} e^{5/2}\right]$$ (Sackur–Tetrode equation) (20.61)

$$K_P = \frac{q_B^b}{q_A^a} N_A^{-\Delta n} e^{-\Delta U_0^\circ/RT}$$ (Equilibrium constant) (20.74)

$$k = \frac{k_B T}{h} \frac{q_\ddagger}{q_A q_B} e^{-\Delta E_0/RT}$$ (Rate constant) (20.79)

APPENDIX 20.1

Justification of $Q = q^N/N!$ for Indistinguishable Molecules

Consider first a situation in which there are three identical molecules with energies a, b, and c in a container. We assume that the locations of these molecules can be distinguished from one another, as indicated by the three boxes labeled 1, 2, and 3 (Figure 20.3). These boxes can be thought of as the lattice points in a solid, and each arrangement is a distribution. There are 6, or 3!, ways to distribute the molecules among these three boxes. For four molecules, the number of possible distributions becomes 24, or 4!, and for N molecules there are $N!$ possible distributions.

The situation is different in the gas phase. Because the molecules are not localized, all positions are possible for any molecule. Therefore, there is only one distribution for N particles, and the canonical partition function (Q) must be divided by $N!$ to avoid overestimating the number of distributions. We summarize the results as follows:

$$Q = q^N \text{ (distinguishable molecules)}$$

$$Q = \frac{q^N}{N!} \text{ (indistinguishable molecules)}$$

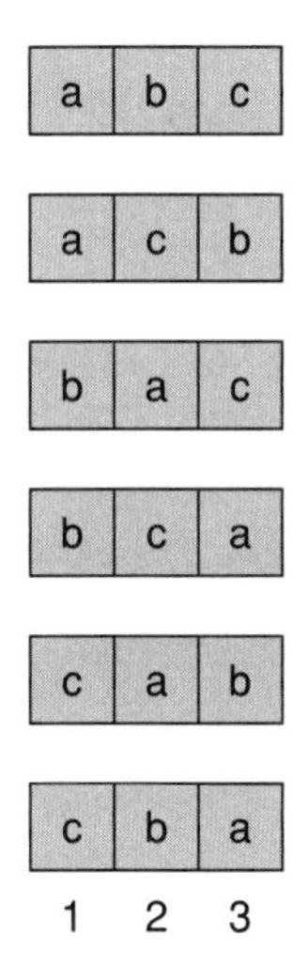

Figure 20.3
The six ways of distributing three molecules with energies a, b, and c among three boxes.

Suggestions for Further Reading

BOOKS

Guggenheim, E. A., *Boltzmann's Distribution Law*, Interscience, New York, 1959.
Maczek, A., *Statistical Thermodynamics*, Oxford University Press, New York, 1998.
Nash, L. K., *Elements of Statistical Thermodynamics*, Addison-Wesley, Inc., Reading MA, 1968.
Widom, B., *Statistical Mechanics: A Concise Introduction for Chemists*, Cambridge University Press, New York, 2002.

ARTICLES

"States, Indistinguishability, and the Formula $S = k \ln W$ in Thermodynamics," J. Braunstein, *J. Chem. Educ.* **46**, 719 (1969).
"Applications of Statistical Mechanics in Molecular Biology," V. A. Bloomfield, *J. Chem. Educ.* **49**, 462 (1969).
"Boltzmann Distribution and Boltzmann Hypotheses," G. C. Lie, *J. Chem. Educ.* **58**, 603 (1981).
"On the Boltzmann Distribution Law," L. K. Nash, *J. Chem. Educ.* **59**, 824 (1982).
"The Stabilization of Atomic Hydrogen," I. F. Silvera and J. Walraven, *Sci. Am.* January 1982.
"The Crystal and Gas Partition Functions and the Indistinguishability of Molecules," E. J. O'Reilly, *J. Chem. Educ.* **60**, 216 (1983).
"Derivation of the Second Law of Thermodynamics from Boltzmann's Distribution Law," P. G. Nelson, *J. Chem. Educ.* **65**, 390 (1988).
"Statistical Mechanical Interpretation of Entropy," P. G. Nelson, *J. Chem. Educ.* **71**, 103 (1994).
"A Statistical Mechanical Analysis of Energy and Entropy," K. Gardner, E. Croker, and S. Basu-Dutt, *Chem. Educator* [Online] **8**, 70 (2003) DOI 10.1333/s00897030648a.
"Introduction of Entropy via the Boltzmann Distribution in Undergraduate Physical Chemistry: A Molecular Approach," E. I. Kozliak, *J. Chem. Educ.* **81**, 1595 (2004).
"An Introduction to Statistical Mechanics," M. M. Francl, *J. Chem. Educ.* **82**, 867 (2005).
"Consistent Application of the Boltzmann Distribution to Residual Entropy in Crystals," E. I. Kozliak, *J. Chem. Educ.* **84**, 493 (2007).
"Deriving the Boltzmann Energy Distribution: An Alternate Approach," L. Eno, *Chem. Educator* [Online] **12**, 215 (2007) DOI 10.1333/s0089707204a.
"An Experimental Approach to Teaching and Learning Elementary Statistical Mechanics," D. C. Ellis and F. B. Ellis, *J. Chem. Educ.* **85**, 78 (2008).
"Overcoming Misconceptions about Configurational Entropy in Condensed Phases," E. I. Kozliak, *J. Chem. Educ.* **86**, 1063 (2009).
"Energy Distributions in Small Populations: Pascal versus Boltzmann," R. W. Kugel and P. A. Weiner, *J. Chem. Educ.* **87**, 1200 (2010).
"The Statistical Interpretation of Classical Thermodynamic Heating and Expansion Processes," S. F. Cartier, *J. Chem. Educ.* **88**, 1531 (2011).
"Quirks of Stirling's Approximation," R. M. Macrae and B. M. Allgeier, *J. Chem. Educ.* **90**, 731 (2013).

Problems

20.1 The population ratio between two energy levels separated by 1.5×10^{-22} J is 0.74. What is the temperature of the system?

20.2 What is the high temperature limit (i.e., as $T \to \infty$) of Equation 20.23?

20.3 Given that the bond length is 1.128 Å, calculate the ratio of $J = 1$ to $J = 0$ populations for carbon monoxide at **(a)** 300 K and **(b)** 600 K. **(c)** What is the limiting value of this ratio as $T \to \infty$? (*Hint:* See Example 11.1 on p. 466.)

20.4 The fundamental vibrational wavenumber for N_2 is 2360 cm^{-1}. For 1 mole of the molecules, calculate the number of N_2 molecules in the $\upsilon = 0$ and $\upsilon = 1$ levels at **(a)** 298 K and **(b)** 1000 K.

20.5 A system consists of three energy levels: a ground level ($\varepsilon_0 = 0, g_0 = 4$); a first excited level ($\varepsilon_1 = k_B T, g_1 = 2$); and a second excited level ($\varepsilon_2 = 4k_B T, g_2 = 2$). Calculate the partition function of the system. What is the probability for the second energy level?

20.6 Explain why q_{trans} increases with **(a)** m and **(b)** T.

20.7 Starting with the relation $P = -(\partial A/\partial V)_T$ (see Equation 12 on p. 201), show that $P = k_B T(\partial \ln Q/\partial V)_T$. Using argon as an example of an ideal monatomic gas, derive the ideal-gas equation ($PV = nRT$).

20.8 Calculate the entropy of HCl at 298 K and 1 bar, given that the bond length is 1.275 Å and the masses of ^{1}H and ^{35}Cl are 1.008 amu and 34.97 amu, respectively. The vibrational wavenumber is 2886 cm^{-1}.

20.9 Calculate the temperature at which $q_{vib} = 5.0$ for carbon monoxide. The vibrational wavenumber is $\tilde{\nu} = 2135$ cm^{-1}.

20.10 Calculate the translational partition function of helium at 1 bar in a 1.00-m^3 container. The large value of q_{trans} means that this motion can be treated classically. When $q_{trans} \leq 10$, however, the motion must be treated quantum mechanically. Calculate the temperature at which this change occurs.

20.11 Calculate the value of q_{vib} for the water molecule at 298 K. (*Hint:* See Figure 11.20.)

20.12 List the symmetry number (σ) for each of the following molecules: Cl_2, N_2O (NNO), H_2O, HDO, BF_3, CH_4, CH_3Cl. (*Hint:* For CH_4, note that each of the four C—H bonds represents a three-fold symmetry axis about which three successive 120° indistinguishable rotations are possible.)

20.13 Calculate the equilibrium constant for the following reaction at 1274 K:

$$I_2(g) \rightleftharpoons 2I(g)$$

The bond length of I_2 is 2.67 Å, and the vibrational wavenumber is 213.7 cm^{-1}. The bond dissociation energy of I_2 is 149.0 kJ mol^{-1}. [*Hint:* To calculate the degeneracy of the iodine atom in its ground electronic state, note that there is an unpaired electron in the 5p orbital. The degeneracy is given by $(2J + 1)$, where J, the total angular momentum, is given by the sum of the orbital angular momentum and spin angular momentum (see p. 524).]

20.14 Calculate the approximate value of $\Delta_r S°$ for

$$^{16}O_2(g) + {}^{18}O_2(g) \rightarrow 2\ {}^{16}O^{18}O(g)$$

Assume that differences in molar masses, moments of inertia, and vibrational frequencies are negligible.

20.15 Calculate the molar entropy of helium at 1.00 bar and 298 K and compare your results with those given in the text on p. 890 for neon and argon.

20.16 Provide a simple physical interpretation for the partition function.

20.17 Of the molecules H_2, O_2, and NO_2, which has **(a)** the largest value of the electronic partition function at 298 K? **(b)** the largest value of the translational partition function at 298 K?

20.18 Normally we expect the molar entropy to increase from N_2 to F_2 across the second row of the periodic table. According to Appendix B, however, we find the molar entropy of O_2 is actually greater than that of F_2. Explain.

APPENDIX A

Review of Mathematics and Physics Useful in Physical Chemistry

This appendix will briefly review some of the basic equations and formulas that are useful in physical chemistry.

Mathematics

Exponents and Powers

Many numbers are more conveniently expressed as powers of 10. For example,

$$1 = 10^0$$

$$0.1 = 10^{-1}$$

$$0.00023 = 2.3 \times 10^{-4}$$

$$100 = 10^2$$

$$100{,}000 = 10^5$$

$$3.1623 = 10^{0.5}$$

In general, we write a^n, where a is called the *base* and n the *exponent*. This expression is read as "a to the power of n." The following relations are useful:

Operation	Example
$a^m \times a^n = a^{m+n}$	$10^{0.2} \times 10^3 = 10^{3.2}$
$(a^m)^n = a^{m \times n}$	$(10^4)^2 = 10^8$
$\dfrac{a^m}{a^n} = a^{m-n}$	$\dfrac{10^3}{10^7} = 10^{-4}$

Note that a^0 (a to the power of zero) is equal to unity for all values of a except for $a = 0$; that is, $0^n = 0$ (for all values of n). Further, we have $1^n = 1$ for all values of n.

Logarithm. The concept of logarithm is a natural extension of exponents. The logarithm to the base a of a number x is equal to the exponent y to which the base number a must be raised so that $x = a^y$. Thus, if

$$x = a^y$$

then

$$y = \log_a x$$

For example, because $3^4 = 81$, we have

$$4 = \log_3 81$$

Similarly, for logarithm to the base 10, we write

Logarithm	Exponent
$\log_{10} 1 = 0$	$10^0 = 1$
$\log_{10} 2 = 0.301$	$10^{0.301} = 2$
$\log_{10} 10 = 1$	$10^1 = 10$
$\log_{10} 100 = 2$	$10^2 = 100$
$\log_{10} 0.1 = -1$	$10^{-1} = 0.1$

The logarithm to the base 10 is called the *common logarithm*. By convention, we use the notation log a instead of $\log_{10} a$ to denote the common logarithm of a.

Because the logarithms of numbers are exponents, they have the same properties as exponents. For simplicity, we express the following relations in terms of common logarithms:

Logarithm	Exponent
$\log AB = \log A + \log B$	$10^A \times 10^B = 10^{A+B}$
$\log \frac{A}{B} = \log A - \log B$	$\frac{10^A}{10^B} = 10^{A-B}$
$\log A^n = n \log A$	$(10^A)^n = 10^{A\times n}$

Logarithms taken to the base e are known as *natural logarithms*. The quantity e (Euler's constant) is a number given by

$$\begin{aligned} e &= 1 + \frac{1}{1!} + \frac{1}{2!} + \frac{1}{3!} + \cdots \\ &= 2.7182818286\cdots \\ &\simeq 2.7183 \end{aligned}$$

In physical chemistry, the exponential function $y = e^x$ is of great importance. Taking the natural logarithm on both sides, we get

$$\ln y = x \ln e = x$$

where "ln" represents $\log_e$. The relation between natural logarithm and common logarithm is as follows. We start with the equation

$$y = e^x$$

Taking the common logarithm on both sides, we obtain

$$\begin{aligned} \log y &= x \log e \\ &= \ln y \log e \end{aligned}$$

because $x = \ln y$. Now $\log e = \log 2.7183 = 0.4343$; thus,

$$\log y = 0.4343 \ln y$$

or

$$\ln y = 2.303 \log y$$

Simple Equations

Linear Equation. A linear equation is represented by

$$y = mx + b$$

A plot of y versus x gives a straight line with slope m and an intercept (on the y axis, that is, at $x = 0$) b.

Quadratic Equation. A quadratic equation takes the form

$$y = ax^2 + bx + c$$

where a, b, and c are constants and $a \neq 0$. A plot of y versus x gives a parabola.

Let us consider a particular quadratic equation,

$$y = 3x^2 - 5x + 2$$

A plot of y versus x is shown in Figure 1. The curve intercepts the x axis ($y = 0$) twice at $x = 1$ and $x = 0.67$. Alternatively, we can solve the equation using the quadratic formula. By setting the equation to be zero (that is, $y = 0$), we get

$$3x^2 - 5x + 2 = 0$$

$$\begin{aligned} x &= \frac{-b \pm \sqrt{b^2 - 4ac}}{2a} \\ &= \frac{5 \pm \sqrt{25 - 4 \times 3 \times 2}}{2 \times 3} \\ &= 1.00, \text{ or } 0.67 \end{aligned}$$

Mean Values

If we repeat an experimental measurement, we often obtain a value that is different from the previous reading, and it is appropriate to represent the result as a mean of

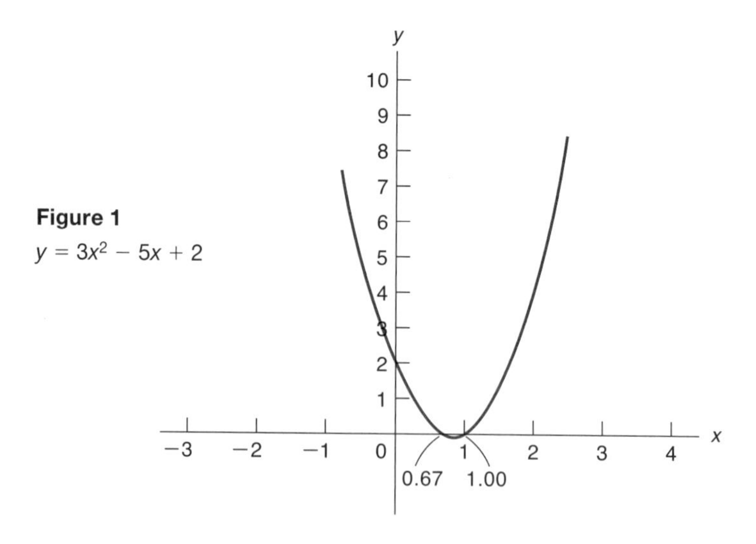

Figure 1
$y = 3x^2 - 5x + 2$

these two numbers. The most common mean value is the *arithmetic mean*. For two readings a and b, the arithmetic mean is given by $(a + b)/2$. There are occasions when the readings do not vary randomly. In such cases, we may try the *geometric mean*. The geometric mean of two numbers a and b is given by $\sqrt{ab}$. For n measurements of a quantity a, where a_i is the ith reading, we have

$$\text{arithmetic mean} = \frac{1}{n}\sum_{i=1}^{n} a_i$$

$$\text{geometric mean} = \sqrt[n]{a_1 a_2 \cdots a_n} = \left(\prod_{i=1}^{n} a_i\right)^{1/n}$$

Series and Expansions

Arithmetic Sequences

$$1, 2, 3, 4, \ldots$$

or

$$a, 2a, 3a, 4a, \ldots$$

or

$$a_{i+1} = a_i + \text{constant}$$

To find the nth term of an arithmetic sequence, we write

$$a_n = a_1 + (n - 1)d$$

where d is the common constant difference.

Arithmetic Series

The sum of terms of a sequence is called a series. It is given by

$$\frac{n}{2}(a_1 + a_n) = a_1 + a_2 + \cdots + a_n$$

Geometric Sequences

$$1, 2, 4, 8, \ldots$$

or

$$a, 2a, 4a, 8a, \ldots$$

or

$$\frac{a_{i+1}}{a_i} = \text{constant}$$

Geometric Series

$$\frac{1}{1-x} = 1 + x + x^2 + x^3 + \cdots \qquad |x| < 1$$

Binomial Expansion

$$(1 + x)^n = 1 + nx + \frac{n(n-1)}{2!}x^2 + \frac{n(n-1)(n-2)}{3!}x^3 + \cdots + x^n$$

Also see Table 14.7.

Exponential Expansions

$$e^x = 1 + \frac{x}{1!} + \frac{x^2}{2!} + \frac{x^3}{3!} + \cdots$$

$$e^{\pm ax} = 1 + \frac{(\pm ax)}{1!} + \frac{(\pm ax)^2}{2!} + \frac{(\pm ax)^3}{3!} + \cdots$$

Trigonometric Expansions

$$\sin x = x - \frac{x^3}{3!} + \frac{x^5}{5!} - \frac{x^7}{7!} + \cdots$$

$$\cos x = 1 - \frac{x^2}{2!} + \frac{x^4}{4!} - \frac{x^6}{6!} + \cdots$$

Logarithmic Expansion

$$\ln(1 + x) = x - \frac{x^2}{2} + \frac{x^3}{3} - \frac{x^4}{4} + \cdots \qquad |x| < 1$$

Angles and Radians

The common unit of angular measure is the *degree*, which is defined as $\frac{1}{360}$ of a complete circle. Often in physical chemistry, we find it more convenient to use another unit, called the *radian* (rad). The relation between angle and radian can be understood as follows. Consider a certain portion of the circumference of a circle of radius r. The length of the arc (s) is proportional to the angle θ and the radius r, so that

$$s = r\theta$$

where θ is measured in radians. Thus, 1 radian is defined to be the angle subtended when the arc length, s, is exactly equal to the radius.

If we consider the entire circle as the arc, then

$$s = 2\pi r = r\theta$$

or

$$2\pi = \theta$$

This means that $\theta = 2\pi$ radians corresponds to $\theta = 360°$. Thus,

$$1 \text{ rad} = \frac{360°}{2\pi} \simeq \frac{360°}{2 \times 3.1416} = 57.3°$$

On the other hand,

$$1° = \frac{2\pi}{360°} \simeq \frac{2 \times 3.1416}{360°} = 0.0175 \text{ rad}$$

Keep in mind that although the radian is a unit of angular measure, it does not have physical dimensions. For example, the circumference of a circle of radius 5 cm is given by $2\pi(\text{rad}) \times 5 \text{ cm} = 31.42 \text{ cm}$.

Areas and Volumes

Triangle. For a triangle with sides a, b, and c and height h (with side a as base), the *semiperimeter* s is given by

$$s = \frac{a + b + c}{2}$$

The area (A) of the triangle is

$$A = \tfrac{1}{2}ah = \sqrt{s(s-a)(s-b)(s-c)}$$
$$= \tfrac{1}{2}ab \sin \gamma$$

where angle γ is opposite side c. If a, b, and c are the sides of a right-angled triangle, c being the hypotenuse, then

$$c^2 = a^2 + b^2$$

which is the Pythagorean theorem.

Rectangle. The area of a rectangle of sides a and b is ab.

Parallelogram. The area of a parallelogram of sides a and b is ah, where h is the perpendicular distance between the two sides whose lengths are a.

Circle. The circumference of a circle is $2\pi r$, and the area of the circle is πr^2, where r is the radius.

Sphere. The area of the curved surface of a sphere of radius r is $4\pi r^2$, and the volume of the sphere is $\frac{4}{3}\pi r^3$.

Cylinder. The area of the curved surface of a cylinder of radius r and length h is $2\pi rh$, and the volume of the cylinder is $\pi r^2 h$.

Cone. The area of the curved surface of a cone is πrl, where r is the radius of the base and l is the slant height. The volume of the cone is $\frac{1}{3}r^2 h$, where h is the vertical height (from the apex to the base).

Operators

In Section 10.7, we mentioned the use of operators. An operator is a mathematical tool that tells us specifically what to do to a number or a function. Some examples of operators are as follows:

Operator	Function or Number	Final Form
log	24.1	$\log 24.1 = 1.382$
$\sqrt{\ }$	974.2	$\sqrt{974.2} = 31.21$
sin	61.9°	$\sin 61.9° = 0.882$
cos	x	$\cos x$
$\frac{d}{dx}$	e^{kx}	$\frac{de^{kx}}{dx} = ke^{kx}$

Differential and Integral Calculus

Functions of Single Variables. The following are derivatives of some common functions.

$y = f(x)$	dy/dx
x^n	nx^{n-1}
e^x	e^x
e^{kx}	ke^{kx}
$\sin x$	$\cos x$
$\sin(ax + b)$	$a\cos(ax + b)$
$\cos x$	$-\sin x$
$\cos(ax + b)$	$-a\sin(ax + b)$
$\ln x$	$1/x$
$\ln(ax + b)$	$\dfrac{a}{ax + b}$

Some Useful Integrals

$$\int x^n\,dx = \frac{1}{n+1}x^{n+1} + C \qquad \int \cos x\,dx = \sin x + C$$

$$\int \frac{dx}{x} = \ln x + C \qquad \int \ln x\,dx = x\ln x - x + C$$

$$\int \frac{dx}{ax + b} = \frac{1}{a}\ln(ax + b) + C \qquad \int e^x\,dx = e^x + C$$

$$\int \sin x\,dx = -\cos x + C \qquad \int e^{kx}\,dx = \frac{e^{kx}}{k} + C$$

Because all these integrals are indefinite integrals, a constant term C must be added to the results.

Physics

Mechanics

We now summarize the important physical quantities in mechanics.

Velocity. Velocity (v) is defined as the rate of change of position with time; that is,

$$v = \frac{dx}{dt}$$

It has the units cm s^{-1} or m s^{-1} (SI). The terms *velocity* and *speed* are often used interchangeably, although they are different quantities. Velocity is a *vector* quantity; it has both magnitude and direction. Speed is a *scalar* quantity; it has magnitude but no direction. The distinction between these two quantities is also discussed in Chapter 2.

Acceleration. Acceleration (a) is the rate of change of velocity with time; that is,

$$a = \frac{dv}{dt}$$

The units of a are cm s^{-2} or m s^{-2} (SI).

Linear Momentum. Linear momentum (p), or simply momentum of an object, is the product of its mass and its velocity:

$$p = mv$$

In SI units, p is in kg m s^{-1}.

Angular Velocity and Angular Momentum. Consider the motion of a particle of mass m about a circle of radius r shown in Figure 2. If the particle describes an angle θ in time t, then the angular velocity, ω, is given by

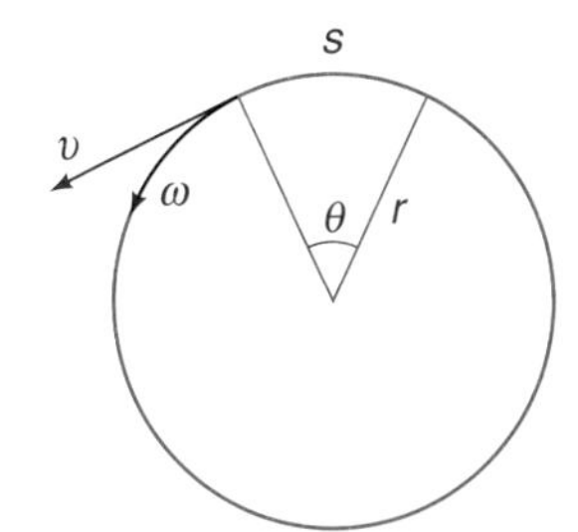

Figure 2

$$\omega = \frac{\theta}{t}$$

where θ is in radians and ω is in rad s^{-1}.

A relation between the angular velocity and the linear velocity can be derived as follows. The linear velocity is the instantaneous velocity of the particle; its direction is always tangential to the circle at every instant. From p. 912,

$$s = r\theta$$

Because distance = velocity × time, we have

$$s = vt$$

Thus,

$$r\theta = vt$$

or

$$\frac{r\theta}{t} = r\omega = v$$

The angular momentum of the particle is given by $m\omega r^2$ or mvr. Sometimes, it is expressed as $I\omega$, where I is the moment of inertia of the particle about the center of the circle (equal to mr^2).

In classical mechanics, the angular momentum of a system can vary continuously. Quantum mechanics imposes the restriction, however, that the angular momentum of an atomic or molecular system is quantized; that is, it can have only certain allowed values. This restriction is one of the fundamental postulates of Bohr's theory of the hydrogen atom, discussed in Chapter 10.

Force. The familiar definition of force (f) is Newton's second law of motion,

$$f = ma$$

This equation says that the force acting on an object is equal to the product of the mass of the object and its acceleration. The SI unit of force is the newton (N), where

$$1 \text{ N} = 1 \text{ kg m s}^{-2}$$

Alternatively, force can be defined as the rate of change of momentum. Thus,

$$f = \frac{dp}{dt}$$

Work. Work (w) is force times distance, that is,

$$w = fx$$

The units of work are N m or kg m^2 s^{-2}.

Newton's Law of Gravitational Attraction. According to Newton's law of gravitational attraction, the force between two masses, m_1 and m_2, separated by distance r is given by

$$f \propto -\frac{m_1 m_2}{r^2}$$

$$= -G\frac{m_1 m_2}{r^2}$$

where G is the universal gravitational constant, given by 6.673×10^{-11} N m^2 kg^{-2}. Note that the negative sign indicates that the force between the objects is *always* attractive.

Thermodynamic Data

Thermodynamic Data for Selected Elements and Inorganic Compounds at 1 bar and 298 K[a]

Substance	$\Delta_f\bar{H}°$/kJ mol^{-1}	$\Delta_f\bar{G}°$/kJ mol^{-1}	$\bar{S}°$/J K^{-1} mol^{-1}	$\bar{C}_p°$/J K^{-1} mol^{-1}
$Ag(s)$	0	0	42.7	25.4
$Ag^+(aq)$	105.9	77.1	72.68	—
$AgCl(s)$	−127.0	−109.8	96.2	56.8
$AgBr(s)$	−100.4	−96.9	107.1	52.38
$AgI(s)$	−61.8	−66.3	115.5	54.43
$AgNO_3(s)$	−124.4	−33.41	140.9	93.05
$Al(s)$	0	0	28.32	24.34
$Al^{3+}(aq)$	−524.7	−485	−313	—
$AlCl_3(s)$	−704.2	−628.8	110.67	91.84
$Al_2O_3(s)$	−1675.7	−1582.3	50.9	78.99
$Ar(g)$	0	0	154.8	20.79
$Ba(s)$	0	0	62.5	28.1
$Ba^{2+}(aq)$	−537.6	−560.8	10	—
$BaCl_2(s)$	−858.6	−810.9	123.7	75.31
$BaO(s)$	−548.0	−520.3	72.1	47.3
$BaSO_4(s)$	−1473.2	−1362.2	132.2	101.8
$BaCO_3(s)$	−1213.2	−1134.4	112.1	86.0
$Br_2(l)$	0	0	152.2	75.69
$Br_2(g)$	30.9	3.1	245.5	36.0

[a] Data are mostly from *The NBS Tables of Chemical Thermodynamic Properties*, D. O. Wagman et al., *J. Phys. Chem. Ref. Data* **11**, Supplement No. 2 (1982). The values for ions in aqueous solution (1 *M*), such as $Li^+(aq)$, are based on the convention that all the properties listed for $H^+(aq)$ are equal to zero.

Substance	$\Delta_f\overline{H}°$/kJ mol^{-1}	$\Delta_f\overline{G}°$/kJ mol^{-1}	$\overline{S}°$/J K^{-1} mol^{-1}	$\overline{C}_P°$/J K^{-1} mol^{-1}
$Br(g)$	111.7	82.4	175.0	20.8
$Br^-(aq)$	−121.6	−104.0	82.4	—
$HBr(g)$	−36.4	−53.4	198.7	29.12
C(graphite)	0	0	5.7	8.52
C(diamond)	1.90	2.87	2.4	6.11
$CO(g)$	−110.5	−137.3	197.9	29.14
$CO_2(g)$	−393.5	−394.4	213.6	37.1
$CO_2(aq)$	−413.8	−386.0	121	—
$CO_3^{2-}(aq)$	−677.1	−527.8	−56.9	—
$HCO_3^-(aq)$	−692.0	−586.8	91.2	—
$H_2CO_3(aq)$	−698.7	−623.1	191	—
$HCN(g)$	135.1	124.7	201.8	35.9
$CN^-(aq)$	151.6	172.4	94.1	—
$Ca(s)$	0	0	41.6	25.9
$Ca^{2+}(aq)$	−542.8	−553.6	−53.1	—
$CaO(s)$	−635.6	−604.2	39.8	42.8
$Ca(OH)_2(s)$	−986.6	−896.8	83.4	84.52
$CaCl_2(s)$	−795.8	−748.1	104.6	72.63
$CaCO_3$(calcite)	−1206.9	−1128.8	92.9	83.5
$Cl_2(g)$	0	0	223.0	33.93
$Cl(g)$	121.4	105.3	165.2	21.8
$Cl^-(aq)$	−167.2	−131.2	56.5	—
$HCl(g)$	−92.3	−95.3	186.5	29.12
$Cr(s)$	0	0	23.77	23.35
$Cr_2O_3(s)$	−1139.7	−1058.1	81.2	118.74
$CrO_4^{2-}(aq)$	−881.2	−727.8	50.2	—
$Cr_2O_7^{2-}(aq)$	−1490.3	−1301.1	261.9	—
$Cu(s)$	0	0	33.15	24.47
$Cu^+(aq)$	51.9	50.2	−26	—
$Cu^{2+}(aq)$	64.8	65.5	−99.6	—
$Cu_2O(s)$	−168.6	−146.0	93.14	63.6
$CuO(s)$	−157.3	−129.7	42.63	44.35

Substance	$\Delta_f\bar{H}°$/kJ mol^{-1}	$\Delta_f\bar{G}°$/kJ mol^{-1}	$\bar{S}°$/J K^{-1} mol^{-1}	$\bar{C}_p°$/J K^{-1} mol^{-1}
$CuS(s)$	−53.1	−53.6	66.53	47.82
$F_2(g)$	0	0	202.8	31.3
$F(g)$	79.4	62.3	158.8	22.7
$F^-(aq)$	−329.1	−276.5	−13.8	—
$HF(g)$	−273.3	−275.4	173.5	29.08
$Fe(s)$	0	0	27.2	25.23
$Fe^{2+}(aq)$	−89.1	−78.9	−137.7	—
$Fe^{3+}(aq)$	−48.5	−4.7	−315.9	—
$FeO(s)$	−272.0	−251.4	60.75	49.92
$Fe_2O_3(s)$	−824.2	−742.2	90.0	104.6
$H_2(g)$	0	0	130.6	28.8
$H(g)$	218.2	203.3	114.7	—
$H^+(aq)$	0	0	0	—
$H_3O^+(aq)$	−285.8	−237.2	69.9	—
$OH^-(aq)$	−229.6	−157.3	−10.75	—
$H_2O(g)$	−241.8	−228.6	188.7	33.6
$H_2O(l)$	−285.8	−237.2	69.9	75.3
$H_2O_2(l)$	−187.8	−120.4	109.6	89.1
$He(g)$	0	0	126.1	20.79
$Hg(l)$	0	0	75.9	27.98
$Hg(g)$	60.78	31.8	175.0	20.8
$Hg^{2+}(aq)$	171.1	164.4	−32.2	—
HgO(red)	−90.8	−58.5	70.29	44.06
$I_2(s)$	0	0	116.13	54.44
$I_2(g)$	62.4	19.3	260.7	36.9
$I^-(aq)$	55.19	51.57	111.3	—
$HI(g)$	26.48	1.7	206.3	29.16
$K(s)$	0	0	64.18	29.58
$K(g)$	89.2	60.7	160.2	—
$K^+(aq)$	−252.4	−283.3	102.5	—
$KOH(s)$	−424.8	−379.1	78.9	68.9
$KCl(s)$	−436.8	−409.1	82.6	51.3

Substance	$\Delta_f\overline{H}°$/kJ mol^{-1}	$\Delta_f\overline{G}°$/kJ mol^{-1}	$\overline{S}°$/J K^{-1} mol^{-1}	$\overline{C}_p^{\circ}$/J K^{-1} mol^{-1}
$KClO_3(s)$	−397.7	−296.3	143.1	100.3
$KNO_3(s)$	−494.6	−394.9	133.1	96.3
$Kr(g)$	0	0	164.1	20.79
$Li(s)$	0	0	28.03	23.64
$Li^+(aq)$	−278.5	−293.8	14.23	—
$LiOH(s)$	−487.2	−443.9	50.21	—
$Mg(s)$	0	0	32.68	24.9
$Mg^{2+}(aq)$	−466.9	−454.8	−138.1	—
$MgCO_3(s)$	−1095.8	−1012.1	65.7	75.5
$MgO(s)$	−601.8	−569.6	26.78	37.41
$MgCl_2(s)$	−641.3	−591.8	89.62	71.3
$N_2(g)$	0	0	191.6	29.12
$N(g)$	470.7	455.5	153.3	20.8
$NH_3(g)$	−46.3	−16.6	192.5	35.66
$NH_4^+(aq)$	−132.5	−79.3	113.4	—
$NH_4Cl(s)$	−314.4	−202.87	94.6	—
$N_2H_4(l)$	50.63	149.4	121.2	139.3
$NO(g)$	90.4	86.7	210.6	29.9
$NO_2(g)$	33.9	51.84	240.5	37.9
$N_2O_4(g)$	9.7	98.3	304.3	79.1
$N_2O(g)$	81.56	103.6	220.0	38.7
$HNO_3(l)$	−174.1	−80.7	155.6	109.9
$HNO_3(aq)$	−207.6	−111.3	146.4	—
$Na(s)$	0	0	51.21	28.41
$Na(g)$	107.5	77.0	153.7	20.8
$Na^+(aq)$	−239.7	−261.9	59.0	—
$NaF(s)$	−576.6	−546.3	51.1	46.9
$NaCl(s)$	−411.2	−384.1	72.13	50.5
$NaBr(s)$	−361.06	−348.98	86.82	51.4
$NaI(s)$	−287.8	−286.1	98.53	52.1
$Na_2CO_3(s)$	−1130.9	−1047.7	135.98	110.5
$NaHCO_3(s)$	−947.7	−851.9	102.1	87.6

Substance	$\Delta_f\overline{H}^\circ$/kJ mol^{-1}	$\Delta_f\overline{G}^\circ$/kJ mol^{-1}	$\overline{S}^\circ$/J K^{-1} mol^{-1}	$\overline{C}_p^\circ$/J K^{-1} mol^{-1}
$NaOH(s)$	−425.6	−379.5	64.46	59.54
$Ne(g)$	0	0	146.3	20.79
$O_2(g)$	0	0	205.0	29.4
$O_3(g)$	142.7	163.4	237.7	38.2
$O(g)$	249.4	231.7	161.0	21.9
P(white)	0	0	41.1	23.22
P(red)	−18.4	12.1	29.3	21.2
$PO_4^{3-}(aq)$	−1277.4	−1018.7	−221.8	—
$P_4O_{10}(s)$	−2984.0	−2697.0	228.86	211.7
$PCl_3(g)$	−287.0	−267.8	311.78	71.84
$PCl_5(g)$	−374.9	−305.0	364.6	112.8
$PH_3(g)$	5.4	13.5	210.2	37.1
S(rhombic)	0	0	31.88	22.59
S(monoclinic)	0.30	0.10	32.55	23.64
$SO_2(g)$	−296.1	−300.1	248.5	39.79
$SO_3(g)$	−395.2	−370.4	256.2	50.63
$SO_4^{2-}(aq)$	−909.3	−744.5	20.1	—
$H_2S(g)$	−20.63	−33.56	205.8	33.97
$H_2SO_4(l)$	−814.0	−690.0	156.9	—
$H_2SO_4(aq)$	−909.27	−744.53	20.1	—
$SF_6(g)$	−1209	−1105.3	291.8	97.3
$Si(s)$	0	0	18.83	19.87
$SiO_2(s)$	−910.9	−856.6	41.84	44.43
$Xe(g)$	0	0	169.6	20.79
$Zn(s)$	0	0	41.63	25.06
$Zn^{2+}(aq)$	−153.9	−147.1	−112.1	—
$ZnO(s)$	−348.3	−318.3	43.64	40.25
$ZnS(s)$	−202.9	−198.3	57.74	45.19
$ZnSO_4$	−978.6	−871.6	124.9	117.2

Thermodynamic Data for Selected Organic Compounds at 1 bar and 298 K[a]

Compound	State	$\Delta_f\bar{H}°$/kJ mol^{-1}	$\Delta_f\bar{G}°$/kJ mol^{-1}	$\bar{S}°$/J K^{-1} mol^{-1}
Acetic acid (CH_3COOH)	*l*	−484.2	−389.9	159.8
	aq	−485.8	−396.5	178.7
Acetaldehyde (CH_3CHO)	*l*	−192.3	−128.1	160.2
Acetone (CH_3COCH_3)	*l*	−248.1	−155.4	200.4
Acetylene (C_2H_2)	*g*	226.6	209.2	200.8
Benzene (C_6H_6)	*l*	49.04	124.5	172.8
	g	82.93	129.7	269.3
Benzoic acid (C_6H_5COOH)	*s*	−385.1	−245.3	167.6
Ethanol (C_2H_5OH)	*l*	−277.0	−174.2	161.0
Ethane (C_2H_6)	*g*	−84.7	−32.9	229.5
Ethylene (C_2H_4)	*g*	52.3	68.12	219.5
Formic acid (HCOOH)	*l*	−424.7	−361.4	129.0
Glucose ($C_6H_{12}O_6$)	*s*	−1274.5	−910.6	210.3
Methane (CH_4)	*g*	−74.85	−50.79	186.2
Methanol (CH_3OH)	*l*	−238.7	−166.3	126.8
Propane (C_3H_8)	*g*	−103.8	−23.5	269.9
2-Propanol (C_3H_7OH)	*l*	−317.9	−180.3	180.6
Sucrose ($Cl_{12}H_{22}O_{11}$)	*s*	−2221.7	−1544.3	360.2
Urea [$(NH_2)_2CO$]	*s*	−333.5	−197.3	104.6

[a] Data are mostly from *The NBS Tables of Chemical Thermodynamic Properties*, D. O. Wagman et al., *J. Phys. Chem. Ref. Data* **11**, Supplement No. 2 (1982). The concentration of all solutions is 1 *M*.

Glossary*

A

absolute zero Theoretically, the lowest attainable temperature. Zero kelvins. (1.5)

absolute zero temperature scale A temperature scale that uses the absolute zero of temperature as the lowest temperature. (1.5)

actinides Elements that have incompletely filled 5*f* subshells or readily give rise to cations that have incompletely filled 5*f* subshells. (12.8)

action potential A transient change in electric potential at the surface of a nerve or muscle cell occurring at the moment of excitation. (9.7)

action spectrum An absorption spectrum that displays the photochemical response or effectiveness of a system as a function of the wavelength of light employed. (16.1)

activated complex An energetically excited state that is intermediate between reactants and products in a chemical reaction. Also called the transition state. (15.7)

activation energy (E_a) The minimum energy required to initiate a chemical reaction. (15.5)

active site The site on an enzyme molecule where the substrate binds and the catalytic reaction is facilitated. (15.12)

activity (a) The activity is an effective thermodynamic concentration that takes into account deviation from ideal behavior. (6.5)

activity coefficient (γ) A characteristic of a quantity expressing the deviation of a solution from ideal behavior. It relates activity to concentration. (6.5)

adhesion Attraction between unlike molecules. (19.3)

adiabatic process A process in which there is no heat exchange between the system and its surroundings. (3.2)

anti-Stokes scattering In Raman spectroscopy, the anti-Stokes scattered light is of greater energy than the incident light, due to resonant energy transfer from excited vibrations and/or rotations of a molecule. (11.5)

antibonding molecular orbital A molecular orbital with an energy higher than that of its constituent atomic orbitals. (13.4)

Arrhenius equation An equation that relates the rate constant to the pre-exponential factor (A) and the activation energy (E_a); $k = A \exp(-E_a/RT)$. (15.5)

atomic force microscopy (AFM) A scanned-probe technique in which a very fine tip is used to spatially profile a surface. (10.12)

Aufbau principle The principle stating that as protons are added one at a time to the nucleus to build up the elements, electrons similarly are added to the atomic orbitals. (12.8)

Avogadro's law The law stating that at constant pressure and temperature, the volume of a gas is directly proportional to the number of moles of the gas present. (1.5)

Avogadro's constant (N_A) The number of particles in one mole. 6.022×10^{23} mol^{-1}. (1.4)

azeotrope A liquid mixture that boils to give a vapor of the same composition. (6.6)

B

Beer–Lambert law An equation that relates the absorbance (A) at a particular wavelength to the concentration of the solution (c) and pathlength of the cell (b); that is, $A = \varepsilon bc$, where ε is the molar absorptivity of the light-absorbing species at that wavelength. (11.1)

bimolecular reaction An elementary step that involves two molecules. (15.3)

Bohr radius (a_0) The radius of the smallest orbit of the hydrogen atom. It is equal to 0.529 Å. (10.4)

Boltzmann distribution law The law that expresses the population (the number of molecules) in a state of energy E_i in a system at thermal equilibrium at temperature T. It is often used to calculate the ratio of populations (N_2/N_1) in two states of energy, E_1 and E_2:

$$\frac{N_2}{N_1} = \exp[-(E_2 - E_1)/k_B T] \qquad (2.9)$$

bond dissociation enthalpy The enthalpy change that accompanies the breaking of a chemical bond in a diatomic molecule. (3.8)

bond enthalpy The enthalpy change that accompanies the breaking of a chemical bond in a polyatomic molecule. (3.8, 20.1)

bond moment The degree of polarity of a chemical bond. For a diatomic molecule, the bond moment is equal to the dipole moment and is given by the product of charge (magnitude only) and the distance between the charges. (13.5)

---

* The number in parentheses indicates the number of the chapter section or chapter appendix in which the term first appears.

bond order The difference between the number of electrons in bonding molecular orbitals and antibonding molecular orbitals, divided by two. (13.4)

bonding molecular orbital A molecular orbital with an energy lower than that of its constituent atomic orbitals. (13.4)

Born–Oppenheimer approximation Recognizing the large difference in mass and speed between electrons and nuclei, this approximation treats nuclei as stationary for the purposes of electronic structure calculations. (13.1)

Bose–Einstein condensate (BEC) A state of matter composed of integer-spin particles (bosons) occupying the same place and time, described by a single wave function. (12.7)

boson A particle with integer spin, such as a photon, a ^{4}He atom, or a Higgs boson. (12.7)

boundary-surface diagram A diagram of the region of an atomic orbital containing a substantial amount of electron density (about 90%). (12.3)

Boyle's law The law stating that the volume of a fixed amount of gas maintained at constant temperature is inversely proportional to the gas pressure. (1.5)

Bragg equation A relation between the angle of diffraction (θ) and the separation, d, of the lattice planes responsible for the diffraction when using X rays of wavelength λ:

$$n\lambda = 2d \sin\theta \qquad n = 1, 2, 3, \ldots \qquad (18.2)$$

Bravais lattice One of the 14 distinct crystal lattices that can exist in three dimensions. (18.1)

C

canonical partition function Product of all the molecular partition functions in a system. (20.3)

capacitance (C) The ratio of the charge on either of a pair of conductors of a capacitor to the potential difference between the conductors. (Chapter Appendix 7.1)

capillary action The action by which the surface of a liquid—where it contacts the inner walls of capillary tubing—is elevated or depressed, depending on whether adhesion is greater than or less than cohesion. (19.3)

Carnot cycle A hypothetical cycle that consists of four reversible processes in succession: an isothermal expansion with the absorption of heat, an adiabatic expansion, an isothermal compression with the release of heat, and an adiabatic compression. In such a cycle, some of the heat absorbed is converted to work. The Carnot cycle is used to show that entropy is a state function and to derive an expression for the thermodynamic efficiency. (4.3)

catalyst A substance that increases the rate of a reaction without itself being consumed. (15.12)

catalytic rate constant (k_{cat}) *See* turnover number.

centrifugal constant (D) In rotational spectroscopy, a parameter that describes the change in rotational energy levels due to stretching of bonds. (11.2)

chain reaction A reaction in which an intermediate generated in one step attacks another species to produce another intermediate, and so on. (15.4)

character table A mathematical collection of the representations of the motions of a molecule in a particular point group. (11.4)

Charles' law The law stating that the volume of a fixed amount of gas maintained at constant pressure is directly proportional to the absolute temperature of the gas. (1.5)

chemical potential (μ) Partial molar Gibbs energy. The chemical potential of the ith component in a mixture is defined by

$$\mu_i = (\partial G/\partial n_i)_{T,P,n_j}$$

Chemical potential is used to predict the direction of a spontaneous process in a mixture just as Gibbs energy is used for a pure component. (6.2)

chemical shift (δ) The difference between the NMR resonance frequency of the nucleus in question and that of a reference standard, divided by the resonance frequency of the standard. (14.6)

cholesteric phase A liquid crystal phase that resembles the smectic phase in that molecules are arranged in layers, except that the long axes of the molecules are parallel to the layers. (19.5)

chromophore A part of a molecule that absorbs light of a specific wavelength. (14.1)

Clapeyron equation A relation between the changes in pressure and temperature for two phases at equilibrium:

$$dP/dT = \Delta\overline{H}/T\Delta\overline{V}$$

It expresses the slope of a phase boundary line in a phase diagram. (5.5)

Clausius–Clapeyron equation An approximate form of the Clapeyron equation in which one phase is condensed and the other is a vapor that is treated as ideal. (5.5)

close-packed structure The packing of identical spheres into different layers to generate three-dimensional structures. In closest packing, each sphere attains a coordination number of 12. (18.4)

closed system A system that allows the exchange of energy (usually in the form of heat) but not mass with its surroundings. (1.2)

coherent A term used to describe electromagnetic waves that are in phase, such as those from a laser. (14.3)

cohesion Attraction between like molecules. (19.3)

colligative properties Properties of solutions that depend on the number of solute particles in solution and not on the nature of the solute particles. (6.7)

collision frequency (Z_1) Number of collisions made by a molecule per unit time. (2.5)

combination band In vibrational spectroscopy, a combination band is a transition in which two or more vibrational modes are excited. (11.3)

complex Mathematically, containing both real and imaginary parts. (10.7)

complex conjugate The complex conjugate of a complex number is obtained by replacing i (where $i = \sqrt{-1}$) with $-i$. (10.7)

component The minimum number of composition variables necessary to describe all possible variations in the composition of the system. (Chapter Appendix 5.2)

compressibility factor (Z) A quantity given by $(P\bar{V}/RT)$. Deviation of Z from 1 indicates nonideal behavior of the gas. (1.8)

conductance (C) The conductance of an electrolyte solution is a measure of the ability of its ions to transport an electric current. For a given medium of area A and length l, it is given by $C = \kappa A/l$, where κ is specific conductance. (7.1)

consecutive reactions Consecutive reactions are reactions of the type $A \rightarrow B \rightarrow C \ldots$. (15.4)

coordination number The number of atoms (or ions) surrounding an atom (or ion) in a crystal. (18.4)

correspondence principle The idea that in the limit of very large quantum numbers, quantum mechanical calculations agree with classical (Newtonian) results. (10.9)

Coulomb's law The mathematical relation that describes the force between charged particles.

$$F = \frac{q_1 q_2}{4\pi\varepsilon_0 r^2} \qquad (7.2)$$

coupled reaction A process in which an endergonic reaction is made to proceed by coupling it to an exergonic reaction. Biological coupled reactions are usually mediated with the aid of enzymes. (8.6)

covalent bond A chemical bond formed by the sharing of one or more pairs of electrons. (13.1)

covalent radius Half the distance between two identical nuclei joined by a covalent bond. (13.6)

critical temperature (T_c) The temperature above which a gas will not liquefy. (1.8)

crystal lattice A three-dimensional structure in which atoms, molecules, or ions are arranged in a highly ordered manner. (18.1)

D

Dalton's law of partial pressures The law stating that the total pressure of a mixture of gases is the sum of the pressures that each gas would exert if it were alone. (1.6)

de Broglie equation The wavelength of a particle with momentum p as given by the equation $\lambda = h/p$. (10.5)

Debye–Hückel limiting law A mathematical expression for calculating the mean ionic activity coefficient of an electrolyte solution in regions of low ionic strength. (7.5)

degeneracy (g) The number of states that have exactly the same energy. One energy level may have one or more states. (10.10)

degrees of freedom In the kinetic theory of gases and spectroscopy, it is the number of ways a molecule can execute its kinetic motion (translation, rotation, and vibration). (2.9) In Gibbs' phase rule, it is the number of intensive variables (pressure, temperature, and composition) that can be changed independently without disturbing the number of phases in equilibrium. (Chapter Appendix 6.1)

delocalized molecular orbital A molecular orbital that extends over more than two atoms. (13.7)

dialysis The process by which low-molar-mass solutes are added to or removed from a solution by means of diffusion across a semipermeable membrane. (8.5)

diamagnetic A diamagnetic substance contains only paired electrons and is slightly repelled by a magnet. (13.5)

dielectric constant (ε) The dielectric constant of a medium is the ratio of the capacitance (C) of a capacitor when the region between the plates is filled with the material to the capacitance (C_0) of the same capacitor when the region between the plates is a vacuum; that is,

$$\varepsilon = \frac{C}{C_0} \qquad (7.2)$$

diffusion The gradual mixing of molecules of one gas with the molecules of another by virtue of their kinetic properties. (2.8) Migration of particles down a concentration gradient. (19.4)

diffusion coefficient (D) A coefficient that indicates how quickly a substance will diffuse in a particular medium at a specific temperature and under the influence of a given concentration gradient. (19.4)

diffusion-controlled reaction A reaction whose rate-determining step is the rate of diffusion of reactant molecules in an encounter to form a product. (15.10)

dipole moment (μ) The vector sum of the bond moments in a molecule. It is a measure of the polarity of the molecule. (13.5)

dipole–dipole interaction The electrostatic interaction between the electric dipoles of two polar molecules. (17.3)

dipole–induced dipole interaction The electrostatic interaction between the electric dipole of a polar molecule and the induced electric dipole (by the polar molecule) of a nonpolar molecule. (17.3)

dispersion interaction Attractive interactions between molecules due to the fluctuating electron distributions in them. Also called instantaneous dipole–induced dipole or London forces. (17.3)

Donnan effect The unequal equilibrium distribution of small diffusible ions on the two sides of a membrane that is freely permeable to these ions but impermeable to macromolecular ions, in the presence of a macromolecular electrolyte on one side of the membrane. (7.6)

Doppler effect The change in the observed frequency of an acoustic or electromagnetic wave due to relative motion of source and observer. (11.1)

E

effusion The process by which a gas under pressure escapes from one compartment of a container to another by passing through a very small opening. (2.8)

eigenfunction A function that remains unchanged, except for a constant scalar multiplier, when processed by a mathematical operator. (10.7)

eigenvalue The constant scalar multiplier of an eigenfunction, after the eigenfunction has been acted on by a mathematical operator. (10.7)

einstein A unit for one mole of photons. (16.1)

elastic collision In an elastic collision, there is no transfer of energy from translational motion into internal modes of motion such as rotation and vibration. (2.1)

electromagnetic radiation Radiation that is emitted or absorbed in the form of electromagnetic waves. (10.1)

electromagnetic wave A wave that has an electric field component and a mutually perpendicular magnetic field component. (10.1)

electron affinity (*EA*) The negative of the energy change when an electron is accepted by an atom in the gaseous state to form an anion. (12.8)

electron configuration A description of the distribution of electrons among the various orbitals in an atom or molecule. (12.7)

electronegativity(χ) The ability of an atom to attract electrons toward itself in a chemical bond. (13.5)

elementary step A reaction that represents the progress at the molecular level. (15.3)

endergonic process A process that is accompanied by a positive change in Gibbs energy ($\Delta G > 0$) and is therefore not favored thermodynamically. (5.1)

endothermic reaction A reaction that absorbs heat from the surroundings. (3.6)

energy The capacity to do work or to produce change. (1.2)

energy level An energy level is an allowed energy of a quantized system. The level is said to be degenerate if several states possess the same energy. (10.4)

enthalpy (*H*) A thermodynamic quantity used to describe heat changes taking place at constant pressure. It is defined by the equation $H = U + PV$, where U is the internal energy, and P and V are the pressure and volume of the system. (3.3)

entropy (*S*) A measure of how dispersed the energy of a system is among the different ways that the system can contain energy. (4.2)

enzyme A biological catalyst that is either a protein or an RNA molecule. (15.12)

equation of state For fluids, an equation that provides the mathematical relationships among the properties that define the state of the system, such as n, P, T, and V. (1.5)

equilibrium vapor pressure The vapor pressure of a liquid in equilibrium with its vapor at a particular temperature. Frequently, it is referred to simply as vapor pressure. (1.8)

equipartition of energy theorem The theory that states that the energy of a molecule is equally divided among all types of motion (translational, rotational, and vibrational), or degrees of freedom. (2.9)

eutectic point The point in a two-component system at which the solid component melts at a single temperature and whose liquid component has the same composition as the solid component. (6.6)

exergonic process A process that is accompanied by a negative change in Gibbs energy ($\Delta G < 0$) and is therefore favored thermodynamically. (5.1)

exothermic reaction A reaction that gives off heat to the surroundings. (3.6)

expectation value The average value obtained for a series of measurements. (10.7)

extensive property A property that depends on how much matter is being considered. (1.2)

F

Faraday constant (*F*) The charge carried by one mole of electrons; it has the value 96,485 C mol^{-1}. (9.1)

fermion A particle with half-integer spin, such as an electron, a proton, or a deuterium atom. (12.7)

Fick's laws of diffusion Fick's first law of diffusion states that the flux of particles is proportional to the concentration gradient. Fick's second law of diffusion shows how the concentration changes in a volume element as a result of diffusion into and out of the volume element. (19.4)

first law of thermodynamics The law that states that energy can be converted from one form to another but cannot be created or destroyed. In chemistry, the first law is usually expressed by the equation $\Delta U = q + w$, where U is the internal energy of the system, q is the heat exchange between the system and its surroundings, and w is the work done on the system by the surroundings or by the system on the surroundings. (3.2)

first-order reaction A reaction whose rate depends on the reactant concentration raised to the first power. (15.2)

fluor A fluorescent molecule used in a scintillation detector that converts radioactive emission into UV–vis radiation. (14.2)

fluorescence The emission of electromagnetic radiation by a substance while the substance is illuminated. Fluorescence is characterized by a short lifetime (about 10^{-9} s) and by the fact that the emissive state and the ground state have the same spin multiplicity (usually a singlet state). (14.2)

force According to Newton's second law of motion, force is mass times acceleration. (1.2)

force constant (k) The force constant of a harmonic oscillator is the constant of proportionality between the restoring force and the displacement of a body (x) that obeys Hooke's law; that is, $F = -kx$. It is a measure of the stiffness of chemical bonds. (11.3)

Fourier transform (FT) A mathematical technique of converting a signal that is a function of time (or position) to/from a signal that is a function of frequency. Contemporary IR and NMR instruments employ FT. (Chapter Appendix 11.1 and 14.6)

fractional distillation A procedure for separating liquid components of a solution that is based on their different boiling points. (6.6)

Franck–Condon principle The principle stating that in any molecular system, the transition from one electronic state to another is so rapid that the nuclei of the atoms in the molecule can be considered stationary during the transition. Thus, the transitions are vertical and the probability of a transition is proportional to the overlap of the vibrational wave functions. (14.1)

free induction decay (FID) The decaying-in-time signal in FT-NMR, which is Fourier transformed to give a signal as a function of frequency. (14.6)

FRET (Förster resonance energy transfer) A method of dipole–dipole interaction that causes intermolecular energy transfer on a 1–10 Å scale. (14.2)

fugacity (f) The fugacity of a gas is its effective thermodynamic pressure. (8.1)

fugacity coefficient (γ) A quantity that relates the fugacity of a gas (f) to its pressure (P): $f = \gamma P$. (8.1)

G

gas constant (R) The universal constant that appears in the ideal gas equation. It has the value 0.08206 L atm K^{-1} mol^{-1}, or 8.314 J K^{-1} mol^{-1}. (1.5)

gerade (g) Gerade wave functions are symmetric with respect to inversion through the origin; inversion gives back the same wave function. (13.1)

Gibbs energy (G) A thermodynamic quantity defined by the equation $G = H - TS$, where H, T, and S are enthalpy, temperature, and entropy, respectively. The change in Gibbs energy of a system in a process at constant temperature and pressure is given by $\Delta G = \Delta H - T\Delta S \leq 0$, where the equal sign denotes equilibrium and the "less than" sign denotes a spontaneous process. (5.1)

Gibbs–Helmholtz equation An equation that expresses the temperature dependence of the Gibbs energy of a system in terms of its enthalpy. (5.4)

Gibbs phase rule A relationship among the number of degrees of freedom (f), the number of components (c), and the number of phases (p) of a system at equilibrium: $f = c - p + 2$. (5.5)

Graham's law of diffusion The law stating that the rate of diffusion of gas molecules is inversely proportional to the square root of the molar mass of the gas at constant temperature and pressure. (2.8)

Graham's law of effusion The law stating that the rate of effusion of gas molecules from a particular orifice is inversely proportional to the square root of the molar mass of the gas at constant temperature and pressure. (2.8)

greenhouse effect The warming of Earth's atmosphere (and surface) due to the transmission of visible sunlight and absorption of outgoing infrared radiation by certain gases (particularly CO_2) in the atmosphere. (16.3)

H

half-life ($t_{1/2}$) The time required for the concentration of a reactant to decrease to half of its initial concentration. (15.2)

Hamiltonian operator ($\hat{H}$) The quantum-mechanical operator that corresponds to the total energy of a system. The Hamiltonian operator appears in the Schrödinger equation $\hat{H}\psi = E\psi$. (10.7)

harmonic oscillator A body that obeys Hooke's law. In the study of their vibrational motions, molecules are treated (to a good approximation) as quantum-mechanical harmonic oscillators. (11.3)

hartree One atomic unit of energy. $E_h = 2R_H hc$. One hartree is approximately the potential energy of a ground-state hydrogen atom. (12.6)

Hartree–Fock method A self-consistent field method for solving the Schrödinger equation for multi-electron systems. Each electron is treated iteratively as being influenced by the average potential energy created by all the other electrons. (12.10)

heat (q) A process in which energy is transferred from one system to another as a result of a temperature difference between them. (3.1)

heat capacity (C) The amount of energy required to raise the temperature of a given quantity of the substance by one degree Celsius. (2.9)

heat capacity ratio (γ) The ratio given by C_P/C_V. (3.5)

Heisenberg uncertainty principle The principle stating that it is impossible to know simultaneously both the momentum and the position of a particle with certainty. The mathematical expression is $\Delta x \Delta p_x \geq h/4\pi$. (10.6)

Helmholtz energy (A) A thermodynamic quantity defined by the equation $A = U - TS$. The change in Helmholtz energy of a system in a process at constant temperature and volume is given by $\Delta A = \Delta U - T\Delta S \leq 0$, where the equal sign denotes equilibrium and the "less than" sign denotes a spontaneous process. (5.1)

Henry's law The law stating that the solubility of a gas in a liquid is proportional to the pressure of the gas over the solution. (6.4)

Hermitian operator In quantum mechanics, the eigenvalues of Hermitian operators are real numbers. (10.7)

Hess's law The law stating that when reactants are converted to products, the change in enthalpy is the same whether the reaction takes place in one step or a series of steps. (3.6)

hologram The image produced by the process of holography. (14.3)

holography A technique for producing a three-dimensional image using coherent laser light. (14.3)

HOMO Highest Occupied Molecular Orbital. (13.4)

Hooke's law The law stating that the restoring force (F) on a body is proportional to its displacement (x) from the equilibrium position: $F = -kx$, where k is the constant of proportionality, called the force constant. (11.3)

hot band In vibrational spectroscopy, a hot band is a transition which originates from an excited vibrational state. (11.3)

Hückel MO theory A semi-empirical molecular orbital method for calculating the π electron energies and wave functions for conjugated hydrocarbon systems. (13.8)

Hund's rule The rule that says the most stable arrangement of electrons in subshells is the one with the greatest number of parallel spins. (12.7)

hybrid orbitals Atomic orbitals obtained when two or more nonequivalent atomic orbitals on the same atom are combined. (13.6)

hybridization The process of mixing the atomic orbitals in an atom that have similar energies to generate a set of new atomic orbitals with different spatial distributions. (13.6)

hydration number The number of water molecules associated with a solute molecule or an ion in aqueous solution. (7.2)

hydrogen bond A special type of dipole–dipole and covalent interaction between the hydrogen atom bonded to an atom of an electronegative element and another atom of an electronegative element. (17.4)

hydrophobic interaction Influences that cause nonpolar substances to cluster together so as to minimize their contact with water. (17.4)

hypertonic solution A concentrated solution with a high osmotic pressure. (6.7)

hypochromism A reduction in the absorbance of UV light at 260 nm that accompanies the transition from denatured DNA strands to a double-stranded helix. This phenomenon is used to monitor the process of denaturation or renaturation of DNA molecules. (14.1)

hypotonic solution A dilute solution with a low osmotic pressure. (6.7)

I

ideal gas equation An equation expressing the relationships among pressure, volume, temperature, and amount of an ideal gas ($PV = nRT$, where R is the gas constant). (1.5)

ideal solution A solution in which both the solvent and the solute obey Raoult's law. (6.4)

ideal-dilute solution A solution in which the solvent obeys Raoult's law and the solute obeys Henry's law. (6.5)

imaginary Mathematically, including a factor of i ($i = \sqrt{-1}$). (10.7)

intensive properties A property that does not depend on how much matter is being considered. (1.2)

intermediate A species that appears in the mechanism of the reaction (that is, the elementary steps) but not in the overall balanced equation. (15.3)

internal energy (U) The internal energy of a system is the total energy of all its components. It consists of translational, rotational, vibrational, electronic, and nuclear energies, as well as energy resulting from intermolecular interactions. (3.2)

international system of units (SI) A particular choice of metric units that was adopted by the General Conference of Weights and Measures. (1.4)

intersystem crossing The radiationless transition of a molecule from one electronic state into another with a different spin multiplicity. (14.1)

ion–dipole interaction The electrostatic interaction between an ion and the electric dipole of a molecule. (17.3)

ion–induced dipole The electrostatic interaction between an ion and the induced electric dipole (by the ion) of a nonpolar molecule. (17.3)

ionic atmosphere A sphere of opposite charge surrounding each ion in an electrolyte solution. (7.5)

ionic mobility The ionic velocity per unit electric field (7.1)

ionic strength (I) A characteristic of an electrolyte solution, defined by

$$I = 1/2\sum_i m z_i^2 \qquad (7.5)$$

ionization energy (IE) The minimum energy required to remove an electron from an isolated atom (or ion) in its ground electronic state. (12.8)

isolated system A system that does not allow the transfer of either mass or energy to or from its surroundings. (1.2)

isomorphous replacement A technique in which a modification to the X-ray diffraction pattern is obtained by the substitution of an atom of one element with an atom of a heavier element. Chemists use this technique to determine the sign of the structure factor in X-ray analysis. (18.3)

isosbestic point The invariant point at a specific wavelength of an absorption spectrum where two chemical species possess the same molar absorptivity. The absorbance remains unchanged even though their concentrations may vary. The presence of one or more isosbestic points is an indication of chemical equilibrium process in solution. (14.1)

isotherm A plot of pressure versus volume of a gas at the same temperature. (1.5)

isothermal process A process that occurs at constant temperature. (3.5)

isotonic solution Solutions with the same concentration (of particles) and hence the same osmotic pressure. (6.7)

J

Jablonski diagram A schematic display of the relative energies of the electronic states of molecules and the vibrational levels associated with each state. The diagram also shows both the radiative and nonradiative transitions between electronic states. (14.1)

Joule (J) Unit of energy given by newton × meter. (1.4)

Joule–Thomson effect The temperature change as a result of gas expansion that occurs at constant enthalpy. (3.6)

K

kinetic isotope effect The change in the rate of a reaction by the replacement of an atom in a molecule by a different isotope. The effect is due to the change of zero-point energy. (15.8)

kinetic salt effect The effect of ionic strength on the rate of a reaction in solution. (15.9)

Kirchhoff's law In thermochemistry, the law stating that the difference between the enthalpies of a reaction at two different temperatures, T_1 and T_2 ($T_2 > T_1$), is just the difference in the enthalpies of heating the products and reactants from T_1 to T_2. (3.7)

Kohlrausch's law of independent migration The law stating that at infinite dilution, the equivalent conductance of an electrolyte is the sum of the equivalent ionic conductances of the anions and cations. (7.1)

Koopman's theorem The theorem states that the ionization energy of an electron is equal to the negative value of the energy of the orbital from which it originated. The theorem is only an approximation because it ignores the readjustment of the remaining electrons when one electron is removed. (14.5)

Krönecker delta (δ_{ij}) A symbol that takes the value zero when $i \neq j$ and unity when $i = j$. (10.7)

L

lanthanides Elements that have incompletely filled $4f$ subshells or that readily give rise to cations that have incompletely filled $4f$ subshells. Also called rare earth metals. (12.8)

Laplacian operator (∇^2) In Cartesian coordinates:

$$\nabla^2 = \frac{\partial^2}{\partial x^2} + \frac{\partial^2}{\partial y^2} + \frac{\partial^2}{\partial z^2} \qquad (10.8)$$

Larmor frequency (ω) The precession frequency of a magnetic moment about the axis of an applied magnetic field. (14.6).

laser An acronym for Light Amplification by Stimulated Emission of Radiation. The operation of a laser requires a population inversion; that is, a greater number of atoms (or molecules) in an upper energy level than in some lower level. (14.3)

laser-induced fluorescence (LIF) A spectroscopic technique that uses a laser to promote electronic excitation. (14.4)

lattice energy The enthalpy change when one mole of a solid is converted to a vapor. (7.3, 18.4)

lattice points The positions occupied by atoms, molecules, or ions that define the geometry of a unit cell. (18.1)

Le Châtelier's principle The principle stating that if an external stress is applied to a system at equilibrium, the system adjusts in such a way that the stress is partially offset as it tries to re-establish equilibrium. (8.4)

Lewis structure A representation of covalent bonding in which shared electrons are shown either as lines or as pairs of dots between two atoms, and lone pairs are shown as pairs of dots on individual atoms. (13.5)

line spectra Spectra produced when radiation is absorbed or emitted by substances only at certain wavelengths. (10.4)

liquid crystal A liquid that is not isotropic; it is a mesomorphic phase. (19.5)

liquid scintillation counting A method of assaying radioactive-tracer-labeled compounds by fluorescence. (14.2)

London interaction *See* dispersion interaction.

lone pairs Valence electrons that are not involved in covalent bond formation. (13.5)

LUMO Lowest Unoccupied Molecular Orbital (13.4)

lyotropic phase A liquid crystal phase prepared by mixing two or more components, one of which is polar in character, such as water. (19.5)

M

macrostate The state of a system as described by the macroscopic properties. (4.7)

Madelung constant A dimensionless constant that determines the lattice energy of an ionic crystal in terms of charges of the ions and distances between them. (18.4)

magic angle spinning (MAS) Physically rotating a solid sample at high speed (kHz) and at 54.74° with respect to the magnetic field to minimize dipolar broadening. (14.6)

magnetic resonance imaging (MRI) An NMR technique that provides a spatial image of an object. (14.6)

maximum rate (V_{max}) The rate of an enzyme-catalyzed reaction when all the enzymes are bound to substrate molecules. (15.12)

Maxwell distribution of speed A theoretical relationship that predicts the relative number of molecules at various speeds for a sample of gas at a particular temperature. (2.4)

Maxwell distribution of velocity A theoretical relationship that predicts the relative number of molecules at various velocities for a sample of gas at a particular temperature. (2.4)

mean ionic activity coefficient ($\gamma_{\pm}$) A quantity that describes the deviation from ideality in the behavior of ions in solution. (7.4)

mean free path (λ) Average distance traveled by a molecule between successive collisions. (3.5)

membrane potential A voltage difference that exists across a membrane due to differences in the concentrations of ions on either side of the membrane. (9.7)

mesomorphic phase A phase that is intermediate in character between a solid and a liquid. Also called paracrystalline phase. (19.5)

mesosphere The middle layer of Earth's atmosphere. (16.2)

Michaelis–Menton kinetics A mathematical treatment that assumes the initial step in enzyme catalysis is a pre-equilibrium between the substrate and the enzyme, followed by the conversion of the enzyme–substrate complex to product. (15.12)

microstate The state of a system as specified by the actual properties of each individual component (atoms or molecules). (4.7)

Miller indices (*hkl*) The Miller indices are used to label lattice planes in a crystal. (18.1)

molar absorptivity (ε) The constant of proportionality in the Beer–Lambert law. It measures the ability of a compound to absorb the electromagnetic radiation at a particular wavelength. (11.1)

molar conductance (Λ) The molar conductance of a solution of molar concentration c is given by $\Lambda = \kappa/c$, where κ is the specific conductance. (7.1)

molar mass ($\mathcal{M}$) The mass (in grams or kilograms) of one mole of atoms, molecules, or other particles. (1.4)

mole The amount of substance that contains as many elementary entities (atoms, molecules, or other particles) as there are atoms in exactly 12 grams (or 0.012 kilograms) of the carbon-12 isotope. (1.4)

mole fraction (x) The ratio of the number of moles of one component of a mixture to the total number of moles of all components in the mixture. (1.6)

molecular orbital (MO) A wave function describing an electron in a molecule. (13.4)

molecular orbital theory A theory of electronic structures of molecules. The electrons are considered to occupy the molecular orbitals that extend over the entire molecule. The molecular orbitals are generated by the linear combination of atomic orbitals of the atoms in the molecule. (13.4)

molecular partition function Product of individual partition functions due to different molecular motions. (20.3)

molecularity The number of molecules reacting in an elementary step. (15.3)

moment of inertia (I) The moment of inertia of a body composed of point masses m_i at a perpendicular distance r_i from a specified line (usually a line through its center of mass) is given by

$$I = \sum_i mr_i^2 \qquad (11.2)$$

monochromator A device which separates multi-frequency light into its component frequencies. (14.3)

most probable speed The speed possessed by the largest number of molecules in a gas at a given temperature. (2.4)

N

nematic phase A liquid crystal phase in which molecules are aligned parallel to one another but lack other spatial organization. (19.5)

Nernst equation An equation that expresses the cell potential (E) in terms of the standard cell potential ($E°$) and the reaction quotient for the cell reaction. (9.3)

neurotransmitter A molecule released at a nerve terminal that binds to and influences the function of other nerve or muscle cells. (9.7)

noble gas core The electron configuration of the noble gas element that most nearly precedes the element being considered. (12.8)

nodal theorem The idea that the energy of a wave function is proportional to the number of nodes, and that the ground-state wave function is nodeless. (10.9)

node A node is a point, a line, or a surface at which the wave function is zero. (10.9)

nonvolatile Does not have a measurable vapor pressure. (6.7)

normalization constant A term introduced to a function so that integrating the function over all space yields unity, and thus the function may be used as a probability density function. (10.7)

O

observable In quantum mechanics, a quantity that can be determined by a single measurement of a system. (10.7)

open system A system that can exchange mass and energy (usually in the form of heat) with its surroundings. (1.2)

operator A mathematical tool that converts a function to another function. Operators are indicated with a caret (^) above the symbol. (10.7)

orbital A one-electron wave function in an atom or molecule. (12.1)

orthonormal Two wave functions are said to be orthonormal when they are both orthogonal and normalized, that is, they satisfy the following mathematical relationship:

$$\int \psi_i^* \psi_j d\tau = \delta_{ij} \qquad (10.7)$$

oscillating reaction A reaction that shows a periodic variation in space or time of the concentration of one or more intermediates. (15.10)

osmosis The net movement of solvent molecules through a semipermeable membrane from a pure solvent or from a dilute solution to a more concentrated solution. (6.7)

osmotic pressure (Π) The pressure required to stop osmosis. (6.7)

overtone In vibrational spectroscopy, an overtone is a transition in which the vibrational quantum number changes by more than one unit; that is, the transition has $\Delta v > 1$. (11.3)

P

paracrystalline phase *See* mesomorphic phase.

paramagnetic A paramagnetic substance contains one or more unpaired electrons and is attracted by a magnet. (13.5)

partial molar volume ($\bar{V}_i$) It measures the change in the volume of a solution upon the addition of n_i moles of the component i at constant temperature, pressure, and moles of other components. (6.2)

particle in a box A model quantum-mechanical system in which a particle in contained by a potential-energy well in which the potential is zero inside the box (or plate, or line segment) and infinite everywhere else. (10.9)

particle-wave duality Particles may have wavelike properties and vice versa. (10.5)

partition function (q) A numerical quantity that indicates the number of states thermally accessible to the molecule. All the thermodynamic properties of a system can be derived from the partition functions of the molecules. (20.2)

Pauli exclusion principle The principle stating that no two electrons in an atom or molecule can have the same four quantum numbers. More rigorously, it states that the wave function of an atom (or molecule) is antisymmetric with respect to the exchange of a pair of electrons. (12.7)

perturbation method A mathematical approach to finding the energy and wave function for a complex quantum-mechanical system in terms of a simple system for which there is an analytical solution (such as a particle in a box, a rigid rotor, a harmonic oscillator, or a hydrogen atom). (12.11)

phase A homogeneous part of a system in contact with other parts of the system but separated from them by a well-defined boundary. (5.5) In wave theory, phase is an angle between 0° and 360° (0 and 2π) that describes the fraction of the wave cycle that has elapsed relative to the origin. (10.1)

phase diagram A diagram showing the conditions (temperature and pressure) at which a substance exists as a solid, liquid, or vapor. (5.5)

phosphorescence Emission of electromagnetic radiation by a substance after it has been illuminated. It is characterized by a long lifetime (on the order of seconds) and the fact that the emissive state and the ground state have different spin multiplicity (typically a triplet to singlet transition). (14.2)

photochemical reaction The reaction that occurs as a result of the excitation of reactant molecules by radiation (usually to a higher electronic state). (16.1)

photochemical smog The formation of smog by the reactions of automobile exhaust in the presence of sunlight. (16.4)

photoelectric effect A phenomenon in which electrons are ejected from the surface of a metal exposed to light of at least a minimum frequency. (10.3)

photomultiplier tube (PMT) A photon detector based on the photoelectric effect and a series of electron multipliers. (10.2)

photon A particle of light. Also known as a light quantum. (10.3)

pi (π) bond A covalent bond formed by sideways overlapping orbitals; its electron density is concentrated above and below the line joining the nuclei of the bonding atoms. (13.5)

Planck distribution law The equation that describes the distribution of radiation emitted per wavelength or frequency by a blackbody as a function of temperature. (10.1)

point group A mathematical description of the symmetry of a molecule, which contains all of the symmetry operations that apply to that molecule. (11.4)

polarizability (α) The ease with which the electron density in an atom (or molecule) can be distorted. Mathematically, it is the constant of proportionality between the strength of an applied electric field (E) and the induced electric dipole: $\mu_{\text{induced}} = \alpha E$. (17.3)

potential-energy curve A plot of the potential energy of a system (molecule) versus the atomic coordinates of the system. (3.7)

potential-energy surface A plot of the potential energy of a collection of atoms as their relative positions are allowed to range over all positions. (15.6)

powder method A method to study the symmetry and dimensions of the unit cell of a substance by causing a monochromatic beam of X rays to interact with the powder sample of the substance. (18.3)

pre-exponential factor (A) The factor that precedes the exponential term in the Arrhenius equation. (15.7)

pressure Force applied per unit area. (2.1)

principle of equal a *a priori* probability The principle stating that all states are equally likely to be occupied in a system at thermal equilibrium. (4.7)

principle of microscopic reversibility The principle stating that at equilibrium, the rates of the forward and reverse processes are equal for every elementary reaction. (15.4)

probability density function In quantum mechanics, a mathematical relation that describes the likelihood of finding a particle such as an electron in a certain location. (10.7)

Q

quantum mechanics The modern theory of matter, electromagnetic radiation, and the interaction between matter and radiation. It applies mainly to atomic and molecular systems. (10.7)

quantum number An integer or half-integer used to characterize an energy level in a certain state in an atom or molecule. (10.9)

quantum yield (Φ) The ratio of the number of molecules of product formed to the number of light quanta absorbed. (16.1)

quantum-mechanical tunneling The penetration of the wave function of a particle through a potential barrier into a classically forbidden region when the kinetic energy of the particle is less than the height of the potential barrier. (10.9)

R

radial distribution function The radial distribution, $4\pi r^2 R(r)^2$, that gives the probability of finding a particle in the range r to $r + dr$ regardless of direction, where r is the distance from the nucleus. (12.3, 19.1)

radius-ratio rule The radius of a cation to that of an anion. Such a ratio provides information regarding crystal lattice structure and coordination number. (18.4)

Raman spectroscopy A light scattering technique that probes molecular motions by the change in molecular polarizablility. (11.5)

Raoult's law The law stating that the partial pressure of a component of a solution is given by the product of the mole fraction of the component and the vapor pressure of the pure component. (6.4)

rare earth metals *See* lanthanides.

rate constant The constant of proportionality between the reaction rate and the concentrations of reactants. (15.2)

rate law An expression relating the rate of a reaction to the rate constant and the concentrations of the reactants. (15.2)

rate-determining step The slowest elementary step in the sequence of steps leading to the formation of products. (15.3)

reaction mechanism The sequence of elementary steps that leads to product formation. (15.3)

reaction order The sum of the powers to which all reactant concentrations appearing in the rate law are raised. (15.2)

reduced mass (μ) The reduced mass of two particles of masses m_1 and m_2 is defined by

$$\frac{1}{\mu} = \frac{1}{m_1} + \frac{1}{m_2} \qquad (11.2)$$

residence time The average time a species resides in the atmosphere. (16.2)

residual entropy The greater-than-zero entropy of a substance at absolute zero due to molecular disorder in a crystal. (4.8)

resonance The use of two or more Lewis structures to represent a particular molecule. (13.7)

resonance structure One of two or more Lewis structures for a single molecule that cannot be described fully with only one Lewis structure. (13.7)

reverse osmosis A process in which a solvent is forced by a pressure greater than the osmotic pressure to flow through a semipermeable membrane from a concentrated solution to a more dilute one. (6.7)

reversible process In a reversible process, a system is always infinitesimally close to equilibrium. Such a process is of theoretical interest but can never be realized in practice. (3.1) In chemical kinetics, a reversible process (reaction) is one in which the reaction can proceed in both directions. (15.4)

Reynolds number A dimensionless number whose value determines whether the flow of a fluid along a tube is laminar or turbulent. (19.2)

rigid rotor A model quantum-mechanical system with a mass rotating freely at a fixed distance about a point. (11.2)

root-mean-square velocity The square-root of the sum of all the squares of the speeds divided by the total number of molecules present. (2.4)

rotational constant (B) In rotational spectroscopy, the separation between adjacent transitions is equal to $2B$. (11.2)

rovibrational Refers to simultaneous rotational and vibrational transitions (or energy levels). (11.3)

rovibronic Refers to simultaneous rotational, vibrational, and electronic transitions (or energy levels). (11.3)

S

Sackur–Tetrode equation An equation for calculating the molar entropy of an ideal gas. (20.4)

salting-in effect The increase in solubility of an electrolyte at high ionic strengths. (7.5)

salting-out effect The decrease in solubility of an electrolyte at high ionic strengths. (7.5)

scanning tunneling microscopy (STM) A scanned-probe technique that relies on the tunneling of electrons from a very fine probe tip to a conducting surface. STM may achieve atomic-scale spatial resolution. (10.12)

Schrödinger wave equation ($\hat{H}\psi = E\psi$) The fundamental equation in quantum mechanics. It calculates the wave functions and energies of atoms and molecules. (10.8)

second law of thermodynamics The law that says the entropy of an isolated system increases in an irreversible process and remains unchanged in a reversible process; it can never decrease. The mathematical statement of the second law is

$$\Delta S_{\text{univ}} = \Delta S_{\text{sys}} + \Delta S_{\text{surr}} \geq 0 \qquad (4.4)$$

second-order reaction A reaction whose rate depends on the reactant concentration raised to the second power or on the concentrations of two different reactants, each raised to the first power. (15.2)

selection rule Describes whether a spectroscopic transition is allowed or forbidden. (11.1)

self-consistent field method An iterative procedure for obtaining wave functions of a many-electron atom or molecule. (12.10)

semipermeable membrane A membrane that allows solvent and certain solute molecules to pass through but blocks the movement of other solute molecules. (6.7)

SI units *See* international system of units. (1.2)

sigma (σ) bond A covalent bond formed by orbitals overlapping end to end; its electron density is concentrated between the nuclei of the bonding atoms along the internuclear axis. (13.5)

singlet state A state of an atom or molecule with zero electronic spin ($S = 0$). (11.1)

Slater determinant A matrix method of writing down a wave function that obeys the Pauli exclusion principle. (12.7)

smectic phase A form of liquid crystal in which molecules are arranged in layers that are free to glide over each other. (19.5)

spectral radiant energy density The amount of electromagnetic energy per unit volume per wavelength (or frequency) interval. (10.1)

spherical polar coordinates A coordinate system describing position as a function of distance from the origin, a longitude, and a latitude. (11.2)

spin multiplicity Spin multiplicity is given by $(2S + 1)$, where S is the total spin quantum number. (11.1)

spin–spin coupling A coupling between nuclear spins that gives rise to the fine structure of an NMR spectrum. (14.6)

spontaneous process A process that occurs on its own accord under a given set of conditions. (4.1)

standard enthalpy of reaction ($\Delta_r H°$) The enthalpy change at a certain temperature when reactants in their standard states are converted to products in their standard states. (3.7)

standard hydrogen electrode An electrode involving the reversible half-reaction

$$H^+(1\ M) + e^- \rightleftharpoons \tfrac{1}{2}H_2(g)$$

It is assigned a zero electrode potential when the gas is at 1 bar pressure and the concentration of the H^+ ion is at 1 M. (9.2)

standard molar enthalpy of formation ($\Delta_f \overline{H}°$) The enthalpy change when 1 mole of a compound is synthesized from its elements in their standard states of 1 bar at some temperature. (3.6)

standard molar Gibbs energy of formation ($\Delta_f \overline{G}°$) The Gibbs energy change when 1 mole of a compound is synthesized from its elements in their standard states of 1 bar at some temperature. (5.3)

standard reduction potential The electrode potential of a substance for the reduction half-reaction under standard state conditions. (9.2)

standard state A reference state with respect to which thermodynamic quantities are defined. The standard state of a solid or liquid is the most stable form of the solid or liquid at 1 bar and the specified temperature. The standard state of an ideal gas is the pure gas at 1 bar and the specified temperature. (3.6)

state The condition of a system that is specified as completely as possible by observations of a specific nature. An example is the thermodynamic state, which is described by thermodynamic properties such as temperature, pressure, and composition. (3.1)

state function A property that is determined by the state of a system. The change in any state function in a process is path independent. (3.1)

state of the system The values of all pertinent macroscopic variables (for example, composition, volume, pressure, and temperature) of a system. (1.2)

stationary state A state that does not evolve with time. (10.7)

statistical thermodynamics A theory of thermodynamic properties in terms of the average behavior of large assemblies of molecules. (20.1)

steady-state approximation Assumes that the concentration of a reaction intermediate remains constant during the main part of the reaction. (15.4)

Stirling's approximation For large values of N, where N is a positive integer, the approximation states that $\ln N! = N \ln N - N$. (20.1)

Stokes scattering In Raman spectroscopy, the Stokes scattered light is of lower energy than the incident light, due to resonant energy transfer to vibrations and/or rotations of a molecule. (11.5)

stratosphere The region of the atmosphere extending upward from the troposphere to about 50 km from Earth. (16.1)

supercritical fluid The state of a substance above its critical temperature. (1.8)

surface tension The amount of energy required to increase the surface area of a liquid by a unit area. (19.3)

surfactant Any substance that causes a reduction in surface tension. (19.3)

surroundings The rest of the universe outside the system. (1.2)

symmetry element A point, line, or plane with respect to which a symmetry operation can be performed. (11.4)

symmetry operation Moves a molecule to produce a new orientation indistinguishable from the original one. (11.4)

system Any specific part of the universe that is of interest to us. (1.2)

T

termolecular reaction An elementary step that involves three molecules. (15.3)

thermal motion The random, chaotic molecular motion. The more energetic the thermal motion, the higher the temperature. (2.3)

thermal reaction A reaction that occurs with the reactant molecules in their electronic ground states. The rate of the reaction is governed by the thermal motion of the reactant molecules. (16.1)

thermochemistry The study of heat changes in chemical reactions (3.6)

thermodynamic efficiency (η) The ratio of the work done by a heat engine to the heat absorbed by the engine. (4.3)

thermodynamic equation of state An equation that expresses the dependence of the internal energy of a system on volume at constant temperature. (Chapter Appendix 6.1)

thermodynamic equilibrium constant The equilibrium constant expression in which concentration terms are expressed either as activities (for solutes in solution) or fugacities (for gases). (8.1)

thermodynamics The scientific study of the interconversion of heat and other forms of energy. (3.1)

thermosphere The high-altitude region of the atmosphere in which the temperature increases continuously with altitude. (16.2)

third law of thermodynamics The law stating that every substance has a finite positive entropy, but at the absolute zero of temperature the entropy may become zero, and it does in the case of a pure, perfect crystalline substance. (4.6)

threshold frequency The minimum frequency of light required to eject an electron from a metal's surface. (10.3)

transformation of variables The mathematical process of converting from one coordinate system to another. (10.11)

transition dipole moment A measure of the dipolar character of the shift in electronic charge that occurs during a spectroscopic transition. A transition is allowed if the transition dipole moment is nonzero. (11.1)

transition metals Elements that have incompletely filled *d* subshells or readily give rise to cations that have incompletely filled *d* subshells. (12.8)

transition state *See* activated complex.

triplet state A state of an atom or molecule in which the total spin quantum number $S = 1$. Thus, the spin multiplicity of this state is $(2S + 1) = 3$. (14.2)

troposphere The lowest layer of the atmosphere. It contains about 80% of the total mass of air and practically all of the atmosphere's water vapor. (16.2)

Trouton's rule A rule stating that the molar entropy of vaporization of most liquids is approximately 88 J K^{-1} mol^{-1}. (4.5)

tunneling *See* quantum-mechanical tunneling.

turnover number The number of substrate molecules processed by an enzyme molecule per second when the enzyme is saturated with the substrate. Also referred to as k_{cat}, the catalytic rate constant. (15.12)

U

ultrafast process Occurring on the pico-, or femto-, or attosecond time scale. (14.4)

ungerade (u) Ungerade wave functions are antisymmetric with respect to inversion through the origin; inversion gives back negative one times the wave function. (13.1)

unimolecular reaction An elementary chemical reaction step that involves one molecule. (15.3)

UPS (ultraviolet photoelectron spectroscopy) An experimental technique that probes valence electron energies. Ultraviolet light ionizes a species, and the kinetic energy of the resulting photoelectron is measured. (14.5)

V

valence bond theory A theory of electronic structures of molecules. It describes each bond as being formed by spin-pairing of electrons in atomic orbitals. (13.3)

valence-shell electron-pair repulsion (VSEPR) theory A model that accounts for the geometrical arrangements of shared and unshared electron pairs around a central atom by the repulsion between electron pairs. (13.6)

valence electrons The outer electrons of an atom, which are those involved in chemical bonding. (12.8)

van der Waals equation An equation of state that accounts for the finite volume of molecules and intermolecular forces between them for a real gas. (1.7)

van der Waals forces The weak attractive forces: dipole–dipole, dipole–induced dipole, and dispersion forces. (17.3)

van't Hoff equation An equation that shows the temperature dependence of the equilibrium constant in terms of the enthalpy of reaction. (8.5)

van't Hoff factor The ratio of the actual number of ionic particles in solution after dissociation to the number of formula units initially dissolved in solution. (7.6)

variational method A method based on the variational principle for calculating the ground-state wave function. The variational parameter or parameters in a trial wave function is/are adjusted to minimize the expectation value of the energy. (12.9)

variational principle This principle is the basis of the variational method. When using the correct Hamiltonian operator in the Schrödinger equation with any trial wave function, the expectational value of the energy will always be greater than or equal to the true, correct energy of the system. (12.9)

virial equation of state An equation of state for real gases. The equation is expressed as an expansion in powers of the molar volume or pressure. (1.7)

viscosity A measure of a fluid's resistance to flow. (2.7)

volatile A term meaning that a substance has a measurable vapor pressure. (6.7)

W

wavenumber ($\tilde{\nu}$) The number of waves per unit length. Typical units are cm^{-1}. (11.1)

Wien effect An increase in the conductance of an electrolyte at very high potential gradients. (7.5)

work In mechanics, work is force times distance. In thermodynamics, the most common forms of work are gas expansion (or compression) and electrical work carried out in an electrochemical cell. (3.1)

workfunction (Φ) The amount of energy required to remove an electron from a metal surface. (10.3)

X

X-ray diffraction A crystal is illuminated with a beam of monochromatic X rays, and the repeating elements of the crystal structure scatter the X rays to form a diffraction pattern that gives information on the molecule's structure. (18.3)

XPS (X-ray photoelectron spectroscopy) An experimental technique that probes core electron energies. X-ray radiation ionizes a species, and the kinetic energy of the resulting photoelectron is measured. (14.5)

Y

YAG (yttrium aluminum garnet) A glass doped with Nd^{3+} ions to make the gain medium of a Nd:YAG laser. (14.3)

Z

zero-order reaction A reaction whose rate is independent of the concentrations of the reactants. (15.2)

zero-point energy The minimum energy that a system may possess. For a harmonic oscillator, the zero-point energy is $\frac{1}{2}h\nu$. (10.9)

zeroth law of thermodynamics The law stating that if system A is in thermal equilibrium with system B, and system B is in thermal equilibrium with system C, then system C is also in thermal equilibrium with system A. (1.3)

Answers to Even-Numbered Computational Problems

Chapter 1

1.6 32.0 g mol^{-1}
1.8 2.98 g L^{-1}
1.10 (**a**) 1.1×10^{-7} mol L^{-1}, (**b**) 18 ppm
1.12 (**a**) 0.85 L
1.14 (**a**) 4.9 L, (**b**) 6.0 atm, (**c**) 0.99 atm
1.16 N_2O
1.18 3.2×10^7 molecules; 2.5×10^{22} molecules
1.20 O_2: 28%; N_2: 72%
1.24 N_2: 88.9%; H_2: 11.1%
1.26 13 days
1.28 349 mmHg
1.30 0.45 g
1.32 4.8%
1.38 $P_T = 1.02$ atm, $P_{Ar} = 0.30$ atm, $P_{He} = 0.720$ atm, $x_{Ar} = 0.29$, $x_{He} = 0.71$
1.42 $P_c = 50.4$ atm, $T_c = 565$ K, $\overline{V}_c = 0.345$ L mol^{-1}
1.54 (**a**) 1.09×10^{44} molecules, (**b**) 1.18×10^{22} molecules, (**c**) 2.6×10^{30} molecules, (**d**) 2.4×10^{-14}; 3×10^8 molecules
1.56 (**b**) 0.54 atm
1.64 $x_{CH_4} = 0.789$, $x_{C_2H_6} = 0.211$
1.66 CH_4
1.68 46.6 m
1.70 45°C

Chapter 2

2.4 0.29 atm
2.6 6.07×10^{-21} J; 3.65×10^3 J mol^{-1}
2.8 460 K
2.10 42.6 K
2.14 N_2: 1.33×10^4 m; He: 9.31×10^4 m
2.16 $c_{rms} = 2.8$ m s^{-1}, $\bar{c} = 2.7$ m s^{-1}
2.20 $c_{rms} = 431$ m s^{-1}, $c_{mp} = 352$ m s^{-1}, $\bar{c} = 397$ m s^{-1}
2.26 3.53×10^{-8} m; 7.70×10^{31} collisions L^{-1} s^{-1}
2.28 12.0 K
2.30 At 1.0 atm: $Z_1 = 3.4 \times 10^9$ collisions s^{-1}, $Z_{11} = 4.0 \times 10^{34}$ collisions m^{-3} s^{-1}. At 0.10 atm: $Z_1 = 3.4 \times 10^8$ collisions s^{-1}, $Z_{11} = 4.0 \times 10^{32}$ collisions m^{-3} s^{-1}
2.32 466.2 m s^{-1}; 4.04 Å
2.36 4
2.38 0.43%
2.40 (**a**) 4.39×10^{21} molecules s^{-1}, (**b**) 2.70 min
2.44 7.25×10^{-21} J for all cases
2.46 H_2: 20.79 J K^{-1} mol^{-1}; CO_2: 20.79 J K^{-1} mol^{-1}; SO_2: 24.94 J K^{-1} mol^{-1}
2.48 Rotational: 0.89; vibrational: 5.4×10^{-6}; electronic: 0
2.54 (**a**) 61.3 m s^{-1}, (**b**) 4.57×10^{-4} s, (**c**) 328 m s^{-1}
2.56 Escape velocity = 1.1×10^4 m s^{-1}. He: 1.15×10^3 m s^{-1}; N_2: 435 m s^{-1}
2.58 16.3
2.66 CH_4

Chapter 3

3.4 (**a**) −112 J, (**b**) −230 J
3.6 -2.3×10^3 J
3.10 $\Delta U = 0$, $q = -20$ J
3.14 $\Delta U = 0$, $\Delta H = 0$ for both (**a**) and (**b**)
3.20 0.71 atm
3.22 50.8°C
3.26 Linear
3.34 (**a**) 207 K, (**b**) 226 K
3.36 24.8 kJ g^{-1}; 603 kJ mol^{-1}
3.38 25.0°C
3.40 (**a**) −2905.6 kJ mol^{-1}, (**b**) 1452.8 kJ mol^{-1}, (**c**) −1276.8 kJ mol^{-1}
3.42 (**a**) −167.2 kJ mol^{-1}, (**b**) −229.6 kJ mol^{-1}
3.44 −337 kJ mol^{-1}
3.46 −23.2 kJ mol^{-1}
3.48 500 J mol^{-1}
3.50 −197 kJ mol^{-1}
3.52 1.9 kJ mol^{-1}
3.54 −238.7 kJ mol^{-1}
3.56 0
3.58 (**b**) 79.4 kJ mol^{-1}
3.60 −2758 kJ mol^{-1}; −3119.4 kJ mol^{-1}
3.66 2.8×10^3 g
3.68 1.19×10^4 K
3.72 (**a**) −65.2 kJ mol^{-1}, (**b**) −9.4 kJ mol^{-1}
3.74 47.8 K; 4.1×10^3 g
3.76 7.60%
3.78 9.90×10^8 J; 305°C
3.80 4.10 L
3.86 (**a**) 561 J, (**b**) 810 J
3.88 0; −285.8 kJ mol^{-1}

Chapter 4

4.6 5.52×10^3 J
4.14 $\Delta U = 43.34$ kJ, $\Delta H = 46.44$ kJ, $\Delta S = 126.2$ J K^{-1}
4.16 4.5 J K^{-1}. Same
4.18 (**a**) 75.0°C, (**b**) $\Delta S_A = 22.7$ J K^{-1}, $\Delta S_B = -20.79$ J K^{-1}, $\Delta S_T = 1.9$ J K^{-1}
4.20 0.36 J K^{-1}
4.22 (**a**) $\Delta S_{sys}' = 5.8$ J K^{-1}, $\Delta S_{surr} = -5.8$ J K^{-1}, $\Delta S_{univ} = 0$. (**b**) $\Delta S_{sys} = 5.8$ J K^{-1}, $\Delta S_{surr} = -4.15$ J K^{-1}, $\Delta S_{univ} = 1.7$ J K^{-1}
4.24 0
4.26 (**a**) −543.8 J K^{-1} mol^{-1}, (**b**) −117.0 J K^{-1} mol^{-1}, (**c**) 284.4 J K^{-1} mol^{-1}, (**d**) 19.4 J K^{-1} mol^{-1}
4.28 (**a**) $\Delta S_{sys} = 5.8$ J K^{-1}, $\Delta S_{surr} = -5.8$ J K^{-1}, $\Delta S_{univ} = 0$. (**b**) $\Delta S_{sys} = 5.8$ J K^{-1}, $\Delta S_{surr} = -3.4$ J K^{-1}, $\Delta S_{univ} = 2.4$ J K^{-1}
4.30 $\Delta_r S° = 24.6$ J K^{-1} mol^{-1}, $\Delta S_{surr} = -607$ J K^{-1} mol^{-1}, $\Delta S_{univ} = -582$ J K^{-1} mol^{-1}
4.34 (**a**) 9.134 J K^{-1} mol^{-1}, (**b**) 11.53 J K^{-1} mol^{-1}, (**c**) 13.38 J K^{-1} mol^{-1}
4.40 0.20 J K^{-1}
4.42 $\Delta U = -1.25 \times 10^3$ J, $\Delta H = -2.08 \times 10^3$ J, $\Delta S = -15.1$ J K^{-1}
4.44 340°C
4.48 (**a**) 35 kJ, (**c**) 202 kJ
4.52 4×10^{-13} s
4.54 25.5 J K^{-1}
4.56 55 J K^{-1}
4.58 113400. 75600

Chapter 5

5.2 979.1 K
5.4 2.48 kJ mol^{-1}
5.6 -75.9 kJ mol^{-1}
5.8 **(a)** $\Delta_r H° = 1.90$ kJ mol^{-1}, $\Delta_r S° = -3.3$ J K^{-1} mol^{-1}, **(b)** 1.4×10^4 bar
5.10 -3198.4 kJ mol^{-1}
5.18 -2.20 K
5.26 88.9 torr
5.30 $\Delta H° = -11.5$ kJ mol^{-1}, $\Delta G° = 12.0$ kJ mol^{-1}
5.40 -6.24×10^3 J

Chapter 6

6.2 2.28 M
6.4 5.0×10^2 m; 18.3 M
6.8 10 m
6.10 **(a)** 11.53 J K^{-1}, **(b)** 50.45 J K^{-1}
6.14 2.6×10^{-4} mol kg^{-1}
6.20 0.85
6.22 $a = 0.9149$, $\gamma = 0.994$
6.26 **(b)** 67.2 mmHg, **(c)** $x^{\text{v}}_{\text{ethanol}} = 0.64$, $x^{\text{v}}_{\text{propanol}} = 0.36$
6.28 0.9261 M
6.36 $P^*_A = 1.9 \times 10^2$ mmHg, $P^*_B = 4.1 \times 10^2$ mmHg
6.38 **(a)** $x_A = 0.524$, $x_B = 0.476$, **(b)** $P_A = 50$ mmHg, $P_B = 20$ mmHg, **(c)** $x_A = 0.71$, $x_B = 0.29$, $P_A = 68$ mmHg, $P_B = 12$ mmHg
6.42 3.5 atm
6.54 **(b)** 1.5 L mol^{-1}
6.56 7.2×10^5 g mol^{-1}
6.58 -14 kJ
6.64 O_2: 8.7 mg (kg H_2O)$^{-1}$; N_2: 14 mg (kg H_2O)$^{-1}$

Chapter 7

7.2 1.5×10^2 Ω^{-1} mol^{-1} cm^2
7.4 390.71 Ω^{-1} mol^{-1} cm^2
7.6 2.8 Ω^{-1} mol^{-1} m^2
7.8 **(a)** 2.5×10^{-3} g L^{-1}, **(b)** 3.9×10^{-4} g L^{-1}
7.10 5.6×10^4 J mol^{-1}
7.12 **(a)** 5.1×10^{-5} M, **(b)** $[\text{Oxa}^{2-}] = 3.0 \times 10^{-7}$ M, $[Ca^{2+}] = 0.010$ M
7.14 **(a)** 0.10 m; 0.69, **(b)** 0.030 m; 0.67, **(c)** 1.0 m; 9.1×10^{-3}
7.16 $m_\pm = 0.32$ m, $a_\pm = 0.041$, $a = 7.0 \times 10^{-5}$
7.18 24.8 Å
7.20 4%
7.22 0.30 M; -0.55°C
7.24 **(a)** 0.150 atm, **(b)** 0.072 atm
7.26 1×10^{-14}
7.30 **(a)** 0.68, **(b)** $\gamma_+ = 0.88$, $\gamma_- = 0.32$

Chapter 8

8.2 **(a)** 0.49, **(b)** 0.23, **(c)** 0.036, **(d)** >0.036 mol
8.4 **(b)(i)** 1.4×10^5, **(ii)** CH_4: 2 atm; H_2O: 2 atm; CO: 13 atm; H_2: 38 atm
8.6 **(a)** $K_c = 1.07 \times 10^{-7}$, $K_P = 2.67 \times 10^{-6}$, **(b)** 22 mg m^{-3}
8.8 1.74×10^5 J mol^{-1}
8.10 2.59×10^4 J mol^{-1}
8.12 $\Delta_r G°$/kJ mol^{-1}: 23, 11, 0, -11, -23
8.14 $K_P = 0.116$, $\Delta_r G° = 5.33 \times 10^3$ J mol^{-1}
8.16 **(b)** 575 K
8.20 NO_2: 0.96 bar; N_2O_4: 0.21 bar
8.28 1.1×10^{-5}
8.32 **(b)** $\Delta_r G°' = 18.13$ kJ mol^{-1}. $\Delta_r G = -10.3$ kJ mol^{-1} for both conventions.
8.34 2.6×10^{-9}
8.38 $\Delta_r H° = -39$ kJ mol^{-1}, $\Delta_r S° = -1.3 \times 10^2$ J K^{-1} mol^{-1}
8.40 $[NH_3] = 0.042$ M, $[N_2] = 0.086$ M, $[H_2] = 0.26$ M
8.42 1.3 atm
8.44 4.0
8.46 **(a)** 0.075, **(b)** 5.6×10^{-3}, **(c)** 0.039. Not as effective.
8.48 157 mmHg

Chapter 9

9.2 1.125 V; 1.115 V
9.6 0.531 V
9.8 **(a)** $E° = 0.913$ V, $\Delta_r G° = -1.76 \times 10^5$ J mol^{-1}; $K = 7.34 \times 10^{30}$, **(b)** 0.824 V, -1.59×10^5 J mol^{-1}, 7.12×10^{27}, **(c)** 0.736 V, -3.55×10^5 J mol^{-1}; 1.60×10^{62}
9.10 0.50 bar
9.12 2.55; -2.32×10^3 J mol^{-1}
9.14 0.010 V
9.18 **(a)** -59.2 mV, **(b)** 5.92×10^{-8} C cm^{-2}, **(c)** 3.69×10^{11} K^+ ions cm^{-2}, **(d)** 6.022×10^{19} ions
9.20 **(b)** 4.2
9.24 Hg_2^{2+}
9.26 $\Delta_r G° = 1.19 \times 10^4$ J mol^{-1}, $K = 8.30 \times 10^{-3}$
9.28 **(b)** 2.92×10^{16}, **(c)** 0.428 V
9.32 Na: $E° = -2.71$ V; F_2: $E° = 2.87$ V
9.34 **(a)** 0.222 V, **(b)** 0.781
9.38 1.09 V
9.40 **(a)** 0.038 V, **(b)** Anode: 0.36 M, Cathode: 1.74 M

Chapter 10

10.2 5.66×10^{-19} J
10.4 2.34×10^{14} Hz, 1.28×10^3 nm
10.10 1×10^6 m s^{-1}
10.12 7.9×10^{-36} m s^{-1}
10.18 4.11×10^{23}
10.20 59.5094 cm. Radio frequency
10.28 **(a)** Orthogonal, **(b)** orthogonal, **(c)** orthogonal, **(d)** not, **(e)** not
10.30 **(a)** Normalized, **(b)** not, **(c)** normalized, **(d)** not, **(e)** normalized
10.36 $N = 6$: 354 nm; $N = 8$: 490 nm; $N = 10$: 626 nm
10.38 **(a)** 1×10^{-3}, **(b)** 0.002
10.42 3.16×10^{-34} J s. Vector points "down", perpendicular to the plane of the ring
10.44 $m = 0$, $u = 0$ m s^{-1}, 0%c; $m = 1$, $u = 1.2 \times 10^8$ m s^{-1}, 39%c; $m = 2$, $u = 2.3 \times 10^8$ m s^{-1}, 77%c
10.46 273 nm
10.48 419 nm
10.50 3.1×10^{19}
10.52 419 nm
10.56 2.75×10^{-11} m
10.58 2.8×10^6 K
10.62 1.1 mm. Microwave
10.64 **(a)** ψ_1: $\langle x \rangle = \frac{a}{2}$, $\langle x^2 \rangle = a^2\left(\frac{1}{3} - \frac{1}{2\pi^2}\right)$, $\langle p \rangle = 0$, $\langle p^2 \rangle = \frac{\hbar^2\pi^2}{a^2}$
(b) ψ_2: $\langle x \rangle = \frac{a}{2}$, $\langle x^2 \rangle = a^2\left(\frac{1}{3} - \frac{1}{8\pi^2}\right)$, $\langle p \rangle = 0$, $\langle p^2 \rangle = \frac{4\hbar^2\pi^2}{a^2}$
10.66 **(a)** B: $4 \rightarrow 2$; C: $5 \rightarrow 2$. **(b)** A: 41.1 nm; B: 30.4 nm. **(c)** 2.18×10^{-18} J
10.68 **(a)** $E_m = m^2 \times 17{,}640$ cm^{-1} **(b)** 5.29×10^4 cm^{-1}

10.72 $n_i = 5$ to $n_f = 3$
10.74 3×10^8 m s^{-1}
10.76 3.87×10^5 m s^{-1}

Chapter 11

11.2 2.2×10^4 cm^{-1}; 6.7×10^{14} s^{-1}
11.4 **(a)** 100%, **(b)** 76%, **(c)** 0.0025%
11.8 **(a)** 8×10^{11} s^{-1}, **(b)** 2×10^{10} s^{-1}
11.10 3.43×10^{-22} J; 17.3 cm^{-1}
11.12 **(a)** 2.7×10^{-12} s, **(b)** 0.16 s
11.14 63%
11.16 266 nm
11.20 0.88 cm^{-1}
11.22 **(a)** 1.64×10^{-46} kg m^2, **(b)** 6.78×10^{-23} J, **(c)** 1.02×10^{11} s^{-1}
11.26 $^{13}C^{16}O$: 1.10188×10^{11} s^{-1}; $^{12}C^{18}O$: 1.09768×10^{11} s^{-1}
11.28 **(a)** 3, **(b)** 7, **(c)** 9, **(d)** 30
11.30 3.5×10^2 N m^{-1}
11.36 H_2
11.38 2627 cm^{-1}
11.46 Rayleigh: 633 nm, very strong; Stokes: 675 nm, much weaker; anti-Stokes: 596 nm, even weaker still
11.50 **(c)** 0.109 Å, **(d)** 8.58%, **(e)** 0.0480 Å, 4.24%
11.62 **(a)** 1, **(b)** 1, **(c)** 2, **(d)** 1, **(e)** 3, **(f)** 1, **(g)** 1, **(h)** 3
11.64 Zero-point energy: $T_2 < DT < D_2 < HT < HD < H_2$. Dissociation energy: $H_2 < HD < HT < D_2 < DT < T_2$
11.70 $\ell = 2, m_\ell = -1, E = 3\hbar^2/I, |L| = \hbar\sqrt{6}, L_z = -\hbar$
11.72 ρ_λ: J m^{-4}; ρ_ν: J s m^{-3}

Chapter 12

12.4 Two, at $r = 230$ pm and $r = 510$ pm
12.10 1.058 Å
12.14 0.649
12.16 4.17×10^2 kJ mol^{-1}
12.18 343 nm. UV
12.24 $E_\phi = 0.209\dfrac{h^2}{mL^2}$, $E_{true} = 0.125\dfrac{h^2}{mL^2}$, E_ϕ is 67% larger than E_{true}
12.26 $\langle E\rangle_{variational} = \dfrac{k}{2c^2} = \sqrt{2}\dfrac{\hbar}{2}\sqrt{\dfrac{k}{m}} = \sqrt{2}\langle E\rangle_{real}$
12.30 $\dfrac{h^2}{8ma^2} + \dfrac{c}{2}$
12.32 $\dfrac{h}{2}\sqrt{\dfrac{k}{\mu}}$
12.36 $E_\phi = \dfrac{21h^2}{4\pi^2mL^2} = 0.532\dfrac{h^2}{mL^2}$, $E_{exact} = \dfrac{2^2h^2}{8mL^2} = 0.500\dfrac{h^2}{mL^2}$, and for an electron in a 0.8-nm box: $E_\phi = 241$ kJ mol^{-1}, $E_{exact} = 227$ kJ mol^{-1}
12.38 $\tilde{R}_H = 109737.3$ cm^{-1}, $\tilde{R}_\mu = 109677.5$ cm^{-1}, $\tilde{R}_D = 109707.3$ cm^{-1}
12.40 419 nm

Chapter 13

13.4 $1/\sqrt{2 + 2S}$
13.32 **(a)** sp^3d^2, **(b)** T-shaped
13.36 Two ethylene: $E = 4\alpha + 4\beta$; butadiene: $E = 4\alpha + 4.48\beta$; cyclobutadiene: $E = 4\alpha + 4\beta$. Butadiene has the lowest π energy.

Chapter 14

14.10 1.2 s. Phosphorescence
14.14 Quartet
14.18 1225 nm
14.22 355 nm, 266 nm, both are UV
14.24 40 mJ, 5 MW, 1.2×10^{17}
14.30 6.546 *eV*, 4.787 *eV*, 1.697 *eV*
14.32 533.7 *eV*, 402.7 *eV*, 285.4 *eV*, core
14.34 4.0 ppm
14.38 **(a)** One singlet, **(b)** one triplet and one quartet, **(c)** one singlet, **(d)** one doublet, **(e)** one singlet, one triplet, one quartet
14.40 **(a)** 7.05 T, **(b)** methyl protons: 700 Hz, aromatic protons: 2150 Hz, **(c)** 60 MHz: $\delta_{methyl} = 2.33$ ppm, $\delta_{aromatic} = 7.17$ ppm. 300 MHz: $\delta_{methyl} = 2.33$ ppm, $\delta_{aromatic} = 7.17$ ppm. The same.
14.46 310 nm
14.48 3.32×10^{-4} mol L^{-1}
14.50 **(a)** 8×10^{11} s^{-1}, **(b)** 2×10^{10} s^{-1}
14.56 **(a)** ^1H: 1.33×10^{-25} J; ^{13}C: 3.34×10^{-26} J, **(b)** ^1H: 200 MHz; ^{13}C: 50.3 MHz, **(c)** 11.7 T
14.58 **(b)** 2×10^6 s^{-1}, **(e)** 59 atm

Chapter 15

15.2 6.2×10^{-6} M s^{-1}
15.4 0.99
15.6 **(a)** 1.21×10^{-4} yr^{-1},**(b)** 2.1×10^4 yr
15.8 Second order. $k = 0.42$ M^{-1} min^{-1}
15.10 First order. 1.19×10^{-4} s^{-1}
15.12 3.6 s
15.18 47.95 g
15.22 **(a)** Rate $= k[NO]^2[H_2]$, **(b)** 0.38 M^{-2} s^{-1}
15.26 10^{11} s^{-1}; 4.5×10^{10} s^{-1}; 2.0×10^2 s^{-1}
15.30 $A = 3.38 \times 10^{16}$ s^{-1}, $E_a = 100$ kJ mol^{-1}
15.32 371°C
15.34 298 K
15.36 $E_a = 13.4$ kJ mol^{-1}, $\Delta H^{\circ\ddagger} = 10.8$ kJ mol^{-1}, $\Delta S^{\circ\ddagger} = -29.3$ J K^{-1} mol^{-1}, $\Delta G^{\circ\ddagger} = 19.8$ kJ mol^{-1}
15.46 $V_{max} = 1.3 \times 10^{-5}$ M min^{-1}, $K_M = 1.2 \times 10^{-3}$ M, $k_2 = 3.3$ min^{-1}. Both plots give the same values.
15.48 **(a)** 898 K, **(b)** 76.4 kJ mol^{-1}
15.50 56.2 min
15.52 6.71×10^9 M^{-1} s^{-1}
15.54 **(a)** 3.50×10^{-2} min^{-1}, **(b)** $E_a = 110$ kJ mol^{-1}
15.58 1.4×10^{-11} s^{-1}; 1.6×10^3 yr; 3.0×10^{10} s^{-1}
15.66 **(a)** 2.5×10^{-5} M s^{-1}, **(b)** 2.5×10^{-5} M s^{-1}, **(c)** 8.3×10^{-6} M
15.70 $A = 6.03 \times 10^5$ M^{-1} s^{-1}, $E_a = 79.0$ kJ mol^{-1}, $\Delta S^{\circ\ddagger} = -155$ J K^{-1} mol^{-1}, $\Delta H^{\circ\ddagger} = 70.4$ kJ mol^{-1}
15.74 **(a)** 1.13×10^{-3} M min^{-1}, **(b)** 6.83×10^{-4} M min^{-1}; 8.8×10^{-3} M
15.76 1.3×10^{11} M^{-1} s^{-1}

Chapter 16

16.2 266 kJ $einstein^{-1}$
16.4 0.022; 3.11×10^5 J
16.6 2.23×10^{-10} s, 2.01×10^{-2} s
16.10 3.1 s
16.18 1.2×10^{-11} $M\ s^{-1}$
16.22 3.8×10^{37} molecules; 3.0×10^{12} kg
16.26 434 nm
16.30 165 kJ mol^{-1}
16.34 4.6
16.36 **(c)** 3.01×10^{10} mol
16.38 5.2×10^{-8} m; 1.4×10^{-5} m
16.42 18

Chapter 17

17.12 **(a)** 1.70 Å, **(b)** 5.1×10^{-4}
17.20 0
17.30 **(a)** 2.98 Å, **(b)** -7.60×10^{-2} kJ (mol dimer)$^{-1}$, **(c)** Not stable at 300 K

Chapter 18

18.2 1.7 Å
18.4 0.220 mn
18.6 2.86 Å; 2.70 g cm^{-3}
18.10 2 atoms
18.12 6.22×10^{23} mol^{-1}
18.14 458 pm
18.16 XY_3
18.18 1.68 g cm^{-3}
18.20 1021 kJ mol^{-1}
18.30 **(a)** 4 each, **(b)** 9.59×10^{-22} g, **(c)** 5.61×10^{-8} cm
18.32 −844 kJ mol^{-1}

Chapter 19

19.2 7.88×10^{-4} N s m^{-2}
19.4 3×10^3 N m^{-2}
19.8 5.7×10^{-3} N s m^{-2}; yes
19.14 1.7 cm
19.16 0.20 N m^{-1}
19.18 4.7 Å
19.20 1.4×10^8 yr
19.24 0.35 s
19.26 0.23 N
19.28 7.1×10^{-4} J

Chapter 20

20.4 **(a)** $n_0 = 6.02 \times 10^{23}$; $n_1 = 6.75 \times 10^{18}$, **(b)** $n_0 = 5.82 \times 10^{23}$; $n_1 = 1.95 \times 10^{22}$
20.8 186.5 J K^{-1} mol^{-1}
20.10 7.75×10^{30}; 3.53×10^{-18} K
20.12 Cl_2: 2; N_2O: 1; H_2O: 2; HDO: 1; BF_3: 3; CH_4,: 12; CH_3Cl: 3
20.14 11.53 J K^{-1} mol^{-1}

Index

X

Y

Z

$134.95

Values of Some Fundamental Constants

Constant	Value
Atomic mass unit (amu)	$1.660\ 538\ 921 \times 10^{-27}$ kg
Avogadro's constant (N_A)	$6.022\ 141\ 29 \times 10^{23}$ mol^{-1}
Bohr radius (a_0)	$5.291\ 772\ 1092 \times 10^{-11}$ m
Boltzmann constant (k_B)	$1.380\ 6488 \times 10^{-23}$ J K^{-1}
Electron charge (e)	$1.602\ 176\ 565 \times 10^{-19}$ C
Electron mass (m_e)	$9.109\ 382\ 91 \times 10^{-31}$ kg
Faraday constant (F)	96485.3365 C mol^{-1}
Gas constant (R)	8.314 4621 J K^{-1} mol^{-1}
Neutron mass (m_n)	$1.674\ 927\ 351 \times 10^{-27}$ kg
Permittivity of vacuum (ε_0)	$8.854\ 187\ 817 \times 10^{-12}$ C^2 N^{-1} m^{-2}
Planck constant (h)	$6.626\ 069\ 57 \times 10^{-34}$ J s
Proton mass (m_p)	$1.672\ 621\ 777 \times 10^{-27}$ kg
Rydberg constant ($\tilde{R}_H$)	109737.315 685 39 cm^{-1}
Speed of light in vacuum (c)	299 792 458 m s^{-1} (exactly)

Pressure of Water Vapor at Various Temperatures

Temperature/°C	Water Vapor Pressure/mmHg
0	4.58
5	6.54
10	9.21
15	12.79
20	17.54
25	23.77
30	31.84
35	42.26
40	55.36
45	71.88
50	92.59
55	118.15
60	149.51
65	187.69
70	233.85
75	289.26
80	355.34
85	433.66
90	525.94
95	634.04
100	760.00